New Frontiers
in SCIENCES, ENGINEERING and the ARTS

Vol. III-B

The Chemistry of Initiation of Ringed, Ringed-Forming and Polymeric Monomers/Compounds

Sunny N.E. Omorodion

Professor of Chemical Engineering
University of Benin.

authorHOUSE®

AuthorHouse™
1663 Liberty Drive
Bloomington, IN 47403
www.authorhouse.com
Phone: 1 (800) 839-8640

Published by AuthorHouse   08/13/2018

ISBN: 978-1-5462-3490-6 (sc)
ISBN: 978-1-5462-3489-0 (e)

Print information available on the last page.

Any people depicted in stock imagery provided by Getty Images are models, and such images are being used for illustrative purposes only. Certain stock imagery © Getty Images.

This book is printed on acid-free paper.

Contents

Preface

Volume (III) titled "The chemistry of initiation of Ringed, Ring-forming and Polymeric Monomers/ Compounds" completes the initiation of compounds for chemical and homopolymeric reactions (Section D). The Volume which is a section contains six chapters, and is indeed a continuation of Volume (II). However, in view of the size of this Volume (Section D), it has been divided into two books – Volume (III)a and Volume (III)b. Volume (III)b which contains Part (II) and Part (III), the one contained herein, is a continuation of Volume (III)a which is Part (I).

Based on the unique characters of rings, far more numerous than non-ringed compounds, new concepts were introduced in this Volume. The characters of rings, the conditions favoring their openings, expansions and reduction in size and formations, the manners by which rings are opened or closed, etc. have been clearly explained. Having covered virtually every type of rings, all the different types of resonance stabilization and molecular rearrangement phenomena and other pheno-mena present in these systems would have been identified. For the first time for example, how meta, ortho-and para-substitutions take place in benzene rings, 2-, 5- positions or 3-, 4- positions substitu-tions take place in pyrrole, furan and thiophene rings etc. have been clearly explained based on the types of resonance stabilization phenomena favored by them. Interestingly enough, none of the substitution reactions is ionic in character as has been thought to be the case over the years. So also are most other reactions undergone by these members of fully unsaturated rings. For the first time, it will also be interesting to note that Silicon cannot carry ionic bonds or ionic charges and can therefore not undergo charged reactions. All the reactions favored by silicon are radical in character.

Chapter 10 which was the first chapter in the last Volume delt with cycloalkanes and only cyclo-propenes. These are rings without functional centers. For the first time, the chemistry of the compounds were clearly explained from a different point of view. Chapter 11 completed all types of only carbon-containing rings. Why some favor being used as monomers and others do not, have been clearly explained. Nucleophilic, electrophilic, radical-pushing and radical-pulling capacities of the compounds and their substituted groups (new) have been provided. Chapters 10 and 11 formed Part (I) of the last Volume which is Volume (III)a.

Chapters 12 and 13 which form Part (II) in Volume (III)b the second book herein contained, deal with cyclic ethers, acetals, esters, anhydrides, amides, oxaazocyclo-propane, pyrrole, carbon anhydrides and imino compounds. All the chemistry provided for these compounds are all new. These are compounds or rings with functional centers. Chapter 14 concludes with cyclosiloxanes, cyclic sulfides, cyclic disulfides, rhombic sulphur, and inorganic or semi-inorganic cyclic monomers etc. These are unique rings, some of which have functional centers which cannot be used. The routes favored by all the rings which favor opening of the rings, have been provided. Why and how some rings do not favor any existence or favor isolated existence, have been explained. All the concepts and chemistry provided here are complete departures from the past. Chapters 14 and 15 form Part (III) of the second book of this Volume (III), said to be Volume (III)b.

Chapter 15 deals with ring forming and polymeric monomers. All the cases above have largely been Inter-addition types of monomers, while the cases here include Intra-addition with Inter-addition. Different and extended new classifications for the groups of these monomers were introduced here, based on the mechanisms of ring formation. Some undergo Inter-Intra-additions, Di-inter-additions, or only Intra-additions. The routes favored for cyclization were also identified. This Volume has never been intended for copolymerization systems. However, in view of the need to provide convincing and unquestionable evidences for the new developments, some unique copolymerization reactions were considered beginning from Volume (II). As usual, at the end of every chapter, rules are proposal in order to provide convincing and clear picture of the entire concepts. Most of the new concepts, which were developed right from the beginning and in Volume (I), have been put into rules beginning from Volumes (II) into (III) and in particular in chapter 15 which is the last chapter in this volume.

In Volume (II), more than five hundred rules were proposed. In Volume (III), more than one thousand rules have been proposed bringing the total to one thousand seven hundred and twenty rules. Without the rules, no foundations can be established. With the rules already herein proposed, the chemical properties of any compound can readily be obtained without the further need for chemical analysis, once the structure of the compound can be ascertained. This completely eliminates the use of empirical rules, rule of thumb, trial and error methods, etc. methods on which the developments of most disciplines have largely depended on since the beginning of our time.

Before understanding this Volume, Volumes (I) and (II) must be read. This Volume will find most useful applications to the medical scientists, biochemists, chemical and other related disciplines, where little or nothing is indeed known about ringed compounds.

While all the works are original to the author, one will not continue to forget to thank many publishers whose works have helped to lay these new foundations for scientist, Engineers and the Arts. This Volume like the others still remains dedicated to humanity.

University of Benin

Sunny N. E. Omorodion.
(ETG)

About the Author

Sunny N. E. Omorodion has been a teacher mostly all his life. He started his teaching career at the age of nineteen in a High school teaching students many of whom were older than him. After graduating from the University of Ibadan with a BSc (Hons) in Chemistry at the age of twenty three, he left for Canada after teaching in two High schools again, to acquire another bachelor (B Eng.) in Chemical Engineering at the University of Alberta, since his dream career was Chemical Engineering since he was a child. In the same university, he acquired two Bachelors in one year in Mathematics and Physics, since when a student at Ibadan he only needed one year and two years to complete degrees in Mathematics and Physics respectively. It was the civil war in the country that made him to study Chemistry at a time when Chemical Engineering did not exist as a discipline in any of the Nigerian Universities, something which can be said to be a blessing in a different way, since in the process he was fully exposed to Mathematics and Physics at the tertiary level, while graduating with only Chemistry.

During the acquisition of four Bachelor degrees, he left for McMaster University in Ontario, Canada to acquire M Eng. and Ph. D degrees in an area which is a hybrid of Chemistry, Mathematics, Physics and Chemical Engineering-Polymer Engineering. He then worked in an industry (Polysar-Sania, Ontario, Canada) for about two years, before coming back to his teaching career at the University of Benin. After about twenty years of service as teacher/consultant, he left on sabbatical leave and leave of absence to teach at three universities- University of Regina, Saskatchewan, Canada, University of Windsor, Ontario, Canada, and University of Toledo, Ohio, USA. Introduction of three new courses at post-graduate level along with the teaching of other courses in Canada and USA made him one of the best professors in all the universities. Presently, he is now back to University of Benin to complete the cycle of one of the stages of life.

Sunny's research interests include Chemical and Polymer reaction Engineering very different from what exists in Present-day Science and Engineering, Environmental Science and Engineering with respect to Pollution Prevention, Waste Management, Enzymatic Chemistry and Engineering, Energy Sources and Conservation, Unit Separations and Process Control of Industrial Systems. Based on "The New Science", some research works which were thought not to be possible, have been made possible, such as oxidation of propane to propanol, polymerization of some monomers, which could not previously be polymerized to give useful products and so on. These are works which cannot be published without introducing "The New Science" in "The New Frontiers" universally.

Sunny is a member of American Institute of Chemical Engineers since 1972, Canadian Society of Chemical Engineers since 1974, Chemical Society of Canada since 1972, American Chemical Society since 2002, American Association for the Advancement of Science since 1996, African Academy of Science since 1990, Nigerian Society of Chemical Engineers since 1988, Polymer Institute of Nigeria since 1990, The Association of Professional Engineers and Geoscientists of Saskatchewan, since 2002. Because he has been so involved in writing, spending at least fourteen hours everyday since Jan 1st 1992, he has not been an active member of these bodies and even rejected serving the State Gov. as Commissioner and other appointments. Sunny is a Fellow of some Professional bodies such as Institute of Industrial Administration since 2008, American and Cambridge Biographical centers and Professional bodies since 1997, and Strategic Institute for Natural Resources and Human Development since 2012. Sunny has won more than twenty International awards with respect to Who is Who from ABI, IBC, The Marquis and more.

We are TENANTS not only in this WORLD, but also in our PHYSICAL BODY, for there is no DEATH of the BEING, BUT THE DEATH of the PHYSICAL BODY.

The Author

SECTION D
CONTS.

CHAPTER 12

TRANSFER OF TRANSFER SPECIES OF CYCLIC ETHERS, ACETALS, ESTHERS AND RELATED MONOMERS

12.0 Introduction

In the last chapter, non-benzenoid carbocyclic compounds were largely considered, for which some of the driving forces favoring the opening of their rings were provided. A monomer must possess the minimum required strained energy (MRSE) before the ring can be opened. With a monomer which does not have any door through which it can be opened like those just considered, the door can be introduced through the use of functional centers. The heterocyclic monomers are most advantaged by the presence of special types of functional centers (e.g. $\overset{\frown}{O}$, $\overset{O}{\underset{C}{\frown}}=O$) in the ring. If the MRSE cannot be attained through activation via the functional centers, then the ring remains intact.

While the manners by which strain energies in cycloalkanes have been determined are to some extent questionable, since there is no ring in which the strain energy can be zero, it has however been observed that:

(i)   Some five or six-membered rings have the smallest strain energies for majority of the families of cyclic monomers.

(ii) Three-membered rings have more than the MRSE.

(iii) Four-membered rings have very close to the MRSE for some families. Hence, some four-membered-rings will favor opening of the ring when a strong center such as that of a growing polymer chain is involved during co-polymerizations. Hence, most five- and six-membered heterocyclic rings are not known to favor opening of the rings.

(iv) All large-membered rings have small strain energy in them, otherwise they will not exist as rings.

For the families of monomers where there is no internally located double bond, only opening of the ring can favor their use as monomers for polymerizations. There are in general two common classes of cyclic ethers of interest - ***the epoxides and dioxanes.***

With the epoxides, there are 1,2-, 1,3-, 1,4-, 1,5- etc. epoxides of which the first members are shown below.

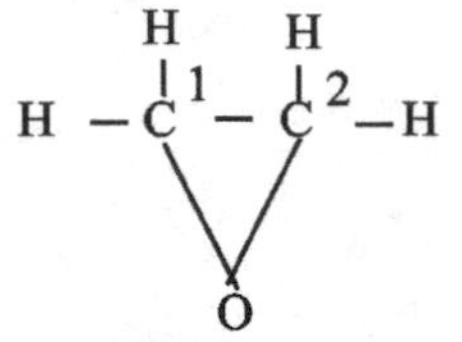

(I)

__1,2 - epoxides__

(oxiranes)

__Dimethylene oxide__

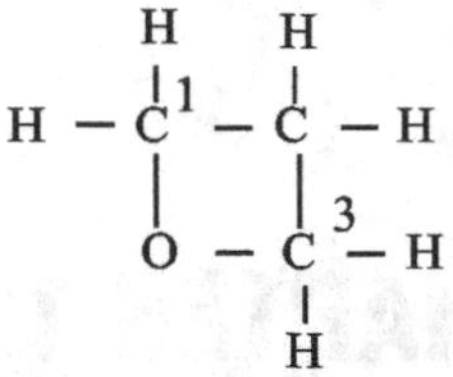

(II)

__1,3 - epoxide__

(oxetanes)

__Trimethylene oxide__

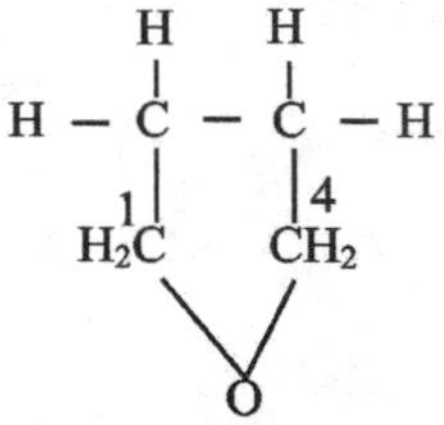

(III)

__1,4 - epoxide__

—

__Tetramethylene oxide__

[Tetrahydrofuran].

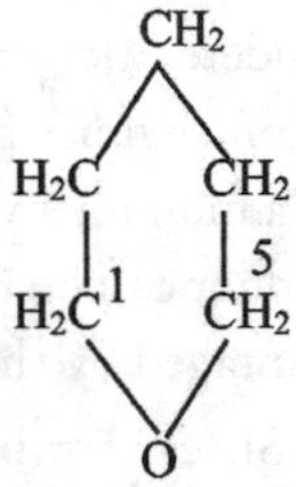

(IV)

__1,5 - epoxide__

—

__Pentamethylene oxide [Tetrahydropyran]__   12.1

As shown by the equation above, they go by varieties of names, which can be confusing. With the dioxanes, there are 1,3-, 1,4-, 1,5-, 1,6-, etc. dioxanes of which the first members are also shown below.

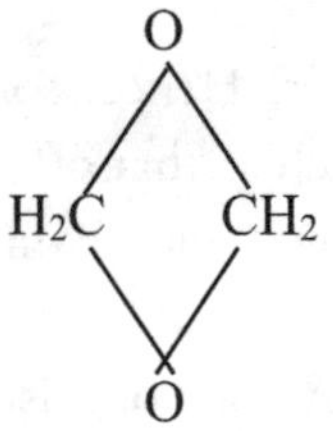

(I)

__1,3 - dioxane__

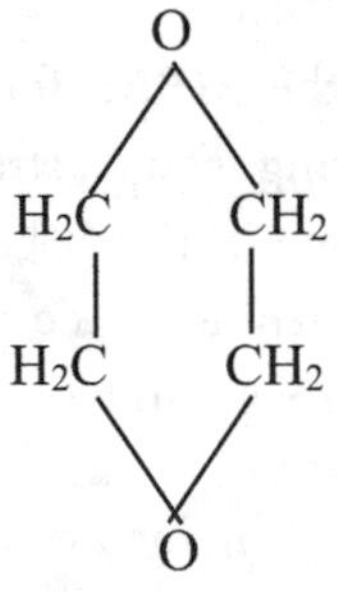

(II)

__1,4 - dioxane__

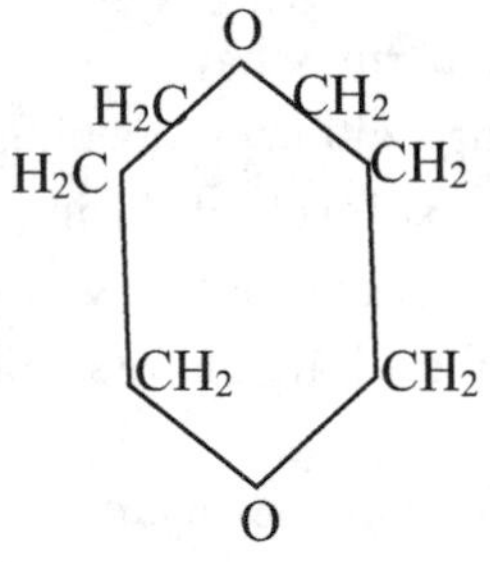

(III)

__1,5 - dioxane__   12.2

It is important to note the more symmetric placements in the dioxanes than in the epoxides. In the epoxides, there is only one functional center, while in the dioxanes, there are two functional centers of same type. 1,4- dioxane is like two 1,2 - epoxides put together and so on. Hence one should envisage the order of strain energy to be expected in the two families.

The ether linkage is known to be characteristically a strong one and is said to be basic in the Lewis

2

sense[1]. In general the ring opening polymerization of cyclic ethers is initiated only by cationic initiators, with 1,2 - epoxides being the only exception to this generalization during homo-polymerizations. The 1,2 - epoxides are the only ones that can be polymerized by both anionic and cationic initiators due to the existence of close to MRSE for three membered rings-strained rings, that is, their rings can be opened instantaneously. The oxetanes are also said to undergo the same types of chemical reactions as oxiranes but more slowly[2]. The five- and six- membered oxygen heterocyclic and still higher-membered rings are said to be "unsaturated" and therefore show the typical inertness of acyclic ethers[2]. The polymerization of simple cyclic ethers has therefore been said to be in practice generally limited to those 3, 4, 5- membered rings than six- membered rings with the five-membered ring being the most difficult to polymerize[1]. Substituted 5-membered cyclic ethers and acetals are said to be usually unreactive. For example 2-methyltetrahydrofuran does not homo-polymerize. The effect of substituents groups (type) in increasing or decreasing the stability and decreasing or increasing the reactivity of a cyclic compound seems to be well known.

Cyclic ethers such as tetrahydropyran and 1,4 - dioxane, both six-membered rings - (IV) and (II) of equations 12.1 and 12.2, - respectively, have been known to be completely unreactive to homopolymerizations under a wide range of reaction conditions. The fact that these rings have very little strain energy is reflected by the data on their enthalpies of polymerizations shown below in Table 12.1.

Table 12.1 Heat of Polymerizations for Cyclic Ethers[1,3-5]

	Monomer	Type	Ring size	- ΔH (kcal/mole)
1	Dimethylene oxide	1,2 - Epoxide	3	22.6
2	Oxacyclobutane	1,3 - Epoxide	4	16.1
3	3,3 - Bis(chloromethyl) Oxacyclobutane	1.3 - Epoxide	4	20.1
4	Tetrahydrofuran	1,4 - Epoxide	5	5.3
5	Tetrahydropyran	1,5 - Epoxide	6	0.4
6	1,4 – Dioxane	Dioxide	6	0.0

From the table, the followings seem to be obvious-

(i)   Presence of radical-pushing allylic groups of greater capacity than H in a ring tends to reduce the stability of hetero rings by increasing the strain energy in the ring of the same size (compare ② and ③).

(ii) The epoxides are more strained than the dioxanes for the same size of ring. In order words, tetrahydropyran is more strained than 1,4 - dioxane. This may not be the case with lower members. It must be realized that ΔH - heat of polymerization is not the major variable involved in determining the level of strain energy in a ring. However, it is a very strong measure of it.

The next Oxygen-heterocyclic monomers to be considered are ***acetals*** of the saturated types. There are two major types of cyclic acetals -

(a)   Trioxanes (cyclic trimmers of aldehydes and ketones).

(b)   Cyclic formals of the general structure shown below.

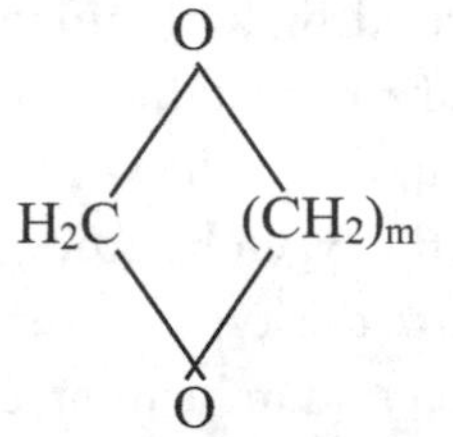

$$H_2C \quad (CH_2)_m \qquad m = 2, 3, 4, 5 \text{ etc.}$$

(cyclic formals)   12.3

Examples of the first members of Trioxanes and Cyclic formals are shown below.

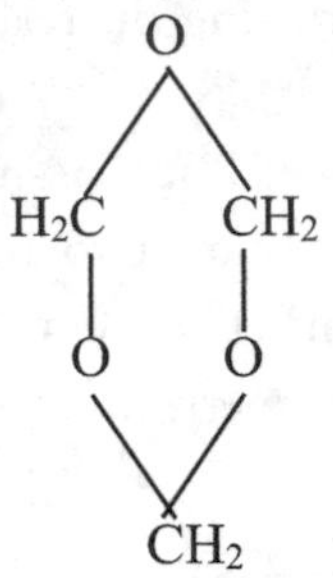

Trioxane   12.4

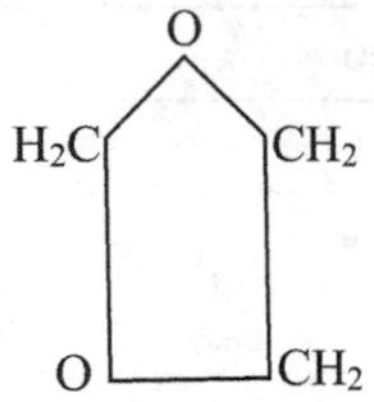

(I)
1,3 - dioxolane

(m = 2)

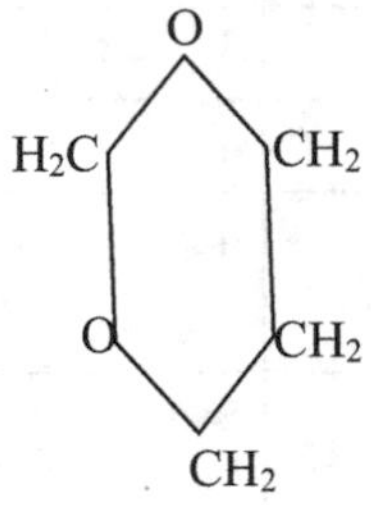

(II)

1,3 - dioxehane

(m = 3)

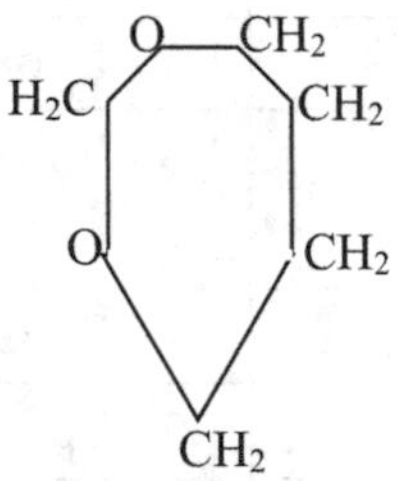

(III)

1,3 - dioxepane
(m = 4)

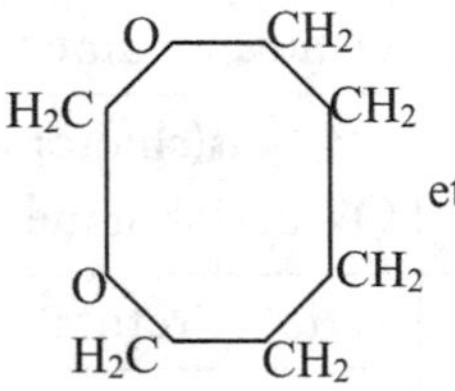

(IV)

1,3 - dioxocane

etc.

12.5

None of these **acetals** favor anionic routes, in the view of the fact that one cannot provide the minimum required strained energy to open the ring instantaneously. All the cases above can be homo-polymerized except the six-membered case. (II) of Equation 12.5 favors only the "cationic" i.e., the positively charged route. These are Nucleophiles and as such their only natural routes are the electro-free-radical or positively charged routes. Table 12.2 below also shows the heats of polymerization for some cyclic acetals compared with their linear counterparts.

Table 12.2 Heat of Polymerizations for Cyclic Acetals [1,3-5]

	Monomer	Type	Ring size	- ΔH (kcal/mole)
1	1,3 – Dioxolane	Cyclic formal	5	5.3
2	1,3 – Dioxepane	Cyclic formal	7	3.6

3	1.3 – Dioxocane	Cyclic formal	8	12.8
4	Formaldehyde	Linear	—	7.4
5	Chloral	Linear	—	9.0

Trioxane is so unique, since it is a cyclic trimmer of formaldehyde an unstable monomer. While trioxane is known to be used as a monomer, tetraoxane is not popularly known except as comonomer. Yet, some trioxane polymerizations are said to give rise to tetraoxane by a back-biting reaction [6].

Cyclic **esters**, unlike cyclic **ethers** and **acetals** undergo ring-opening polymeriza-tion to polyesters with the use of varieties of "cationic" and anionic initiators. There are two common classes of cyclic esters of interest-the lactides and lactones of which some of first members are shown below.

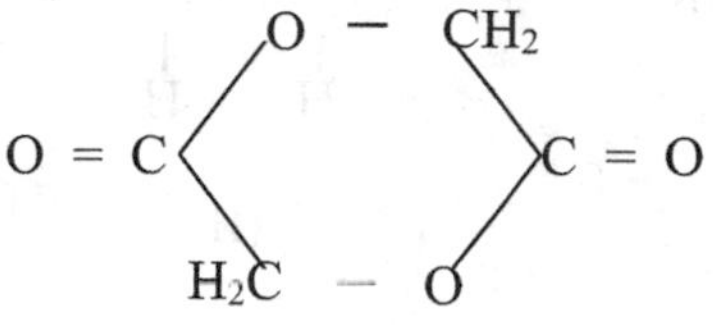

(I) (Glycolide) - a <u>Lactide</u>   12.6

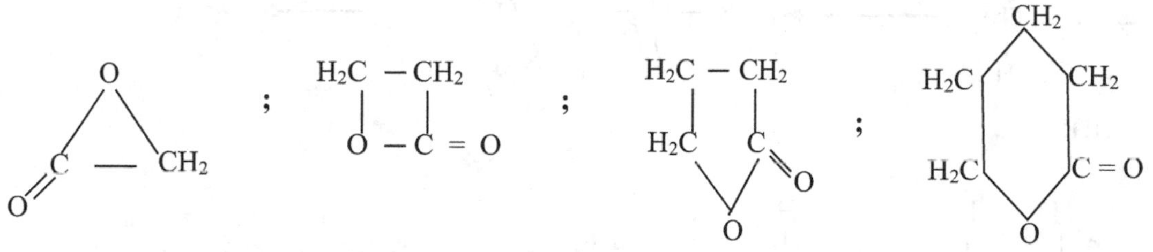

(II) (α - Lactone)   (III) (β - propiolactone)   (IV) (γ - butyro-lactone)   (V) (δ - valerolactone)

12.7

Of the cases above, only (IV) cannot be homo-polymerized, because it is said to be almost free of strain, that which is impossible. The six-membered counterpart δ - valerolactone are said to have more strain and therefore yield some polymers. However, the tendency to open the ring is reduced if substituent groups are present on the lactone- ring[7]. It is believed that substituents groups in question are of the radical-pushing types. While cyclic esters have two centers for activation $\overset{O}{C}=O$ and O,

cyclic ethers and acetals considered herein have only one center for activation O. It is in view of the

electrophilic character– $C=O$, the rings are difficult to be opened when radical-pushing groups

are placed on them. The O- center is nucleophilic (for positively charged and electro-free-radical routes), while the C = O center is electrophilic (for anionic and nucleo-non-free-radical routes) for ringed compounds. ***Thus, cyclic esters are Electro- philes (Males). While the C = O center is an activation center, the O- center is called a FUNCTIONAL center.***

The oxygen functional center can only be used with positively charged centers, since the paired unbonded radicals on the oxygen center can only attract positive charges and electro-free-radicals. However, the mode of opening of the ring is not of free-type, that is, a positive charged isolatedly placed such as a cation cannot only be involved, but must be paired.

In the considerations to follow, both charged and radical polymerizations will be considered, noting that some of these monomers can be made to undergo radical polymerizations, when suitable conditions exist.

12.1 Cyclic Ethers

12.1.1. Charged Character of Cyclic Ethers

Beginning with 1,2 - epoxides, consider the first member, dimethylene oxide.

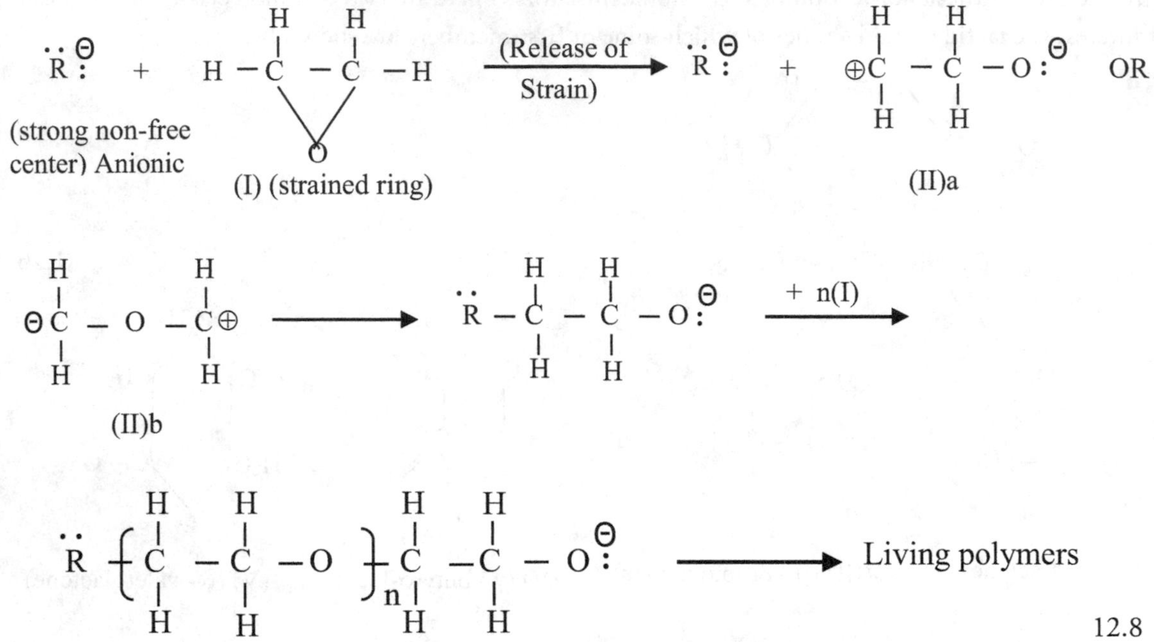

Since the ring is strained, that is, it has close to the minimum required strained energy, it is unzipped by electrostatic forces from the non-free anion. When unzipped, it is (II)a that is favored instead of (II) b, since the oxygen atom is more electronegative than carbon atom and since the larger radical-pushing potential difference exist on the bond.

A carbon center cannot carry a negative charge in the presence of electronegative groups. For its growing polymer chain, living polymers are produced in the absence of foreign agents. It should be noted that the anionic initiator chosen should be such that either -

(i)   The solvent is compatible with initiator in terms of polar/ionic classification
            Or
(ii)   Its counter-ion is shielded from the reaction zone via pairing. No free cation should be in the system.

With ion-paired initiators, the limitations above are not important when the solvents are adequately chosen.

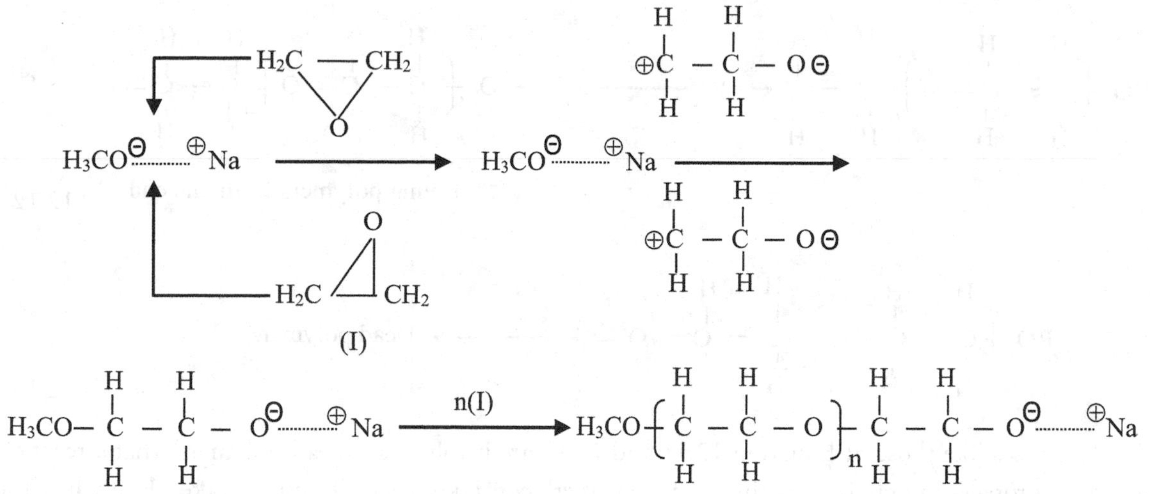

(I)

$$H_3CO-\overset{\overset{\displaystyle H}{|}}{\underset{\underset{\displaystyle H}{|}}{C}}-\overset{\overset{\displaystyle H}{|}}{\underset{\underset{\displaystyle H}{|}}{C}}-O^{\ominus}\cdots\cdots{}^{\oplus}Na \xrightarrow{\;n(I)\;} H_3CO\left[\overset{\overset{\displaystyle H}{|}}{\underset{\underset{\displaystyle H}{|}}{C}}-\overset{\overset{\displaystyle H}{|}}{\underset{\underset{\displaystyle H}{|}}{C}}-O\right]_n\overset{\overset{\displaystyle H}{|}}{\underset{\underset{\displaystyle H}{|}}{C}}-\overset{\overset{\displaystyle H}{|}}{\underset{\underset{\displaystyle H}{|}}{C}}-O^{\ominus}\cdots\cdots{}^{\oplus}Na$$

12.9

Anionic initiators which have been used with cyclic ethers include the followings as sources [8-11]-

(i)   Hydroxides (free ions)
(ii) Alkoxides e.g. $NaOCH_3$ (free ions)
(iii) Metal oxides (free ions)
(iv) Organometallic compounds (ion-paired)
(v)   Amides and other bases (free ions/ion-paired) etc.

Their abilities to favor the presence of full free-ions or full ion-paired initiators, largely depend on the type of solvent involved. If the solvent is more acidic than the monomer, then the following takes place.

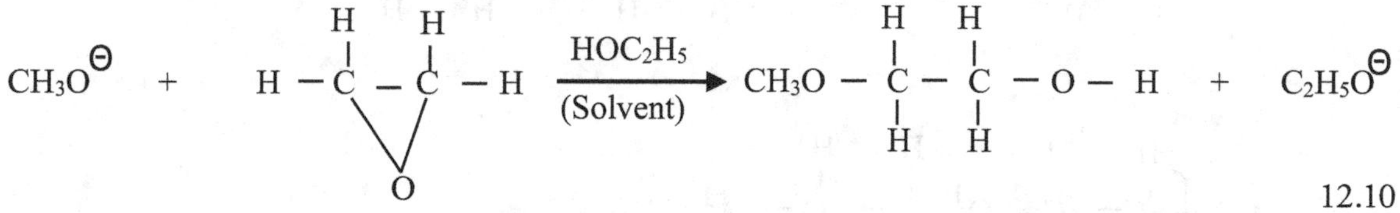

12.10

The reaction does not go beyond the initiation step, since C_2H_5OH has been said to be more acidic than monomer. In view of the non-free ionic character of the growing center, special terminating agents are required to kill the chain once polymerization is favored. With ion-paired growing polymer chain, an anion of greater capacity than that of the growing polymer chain will be required to displace the counter-ion center. Where $H^{\oplus}$ is involved in killing the growing chain, living polymers are still produced since the dead-polymer involved is very unstable from one end. If the initiators involved are hydroxides, then living polymers are produced from both ends.

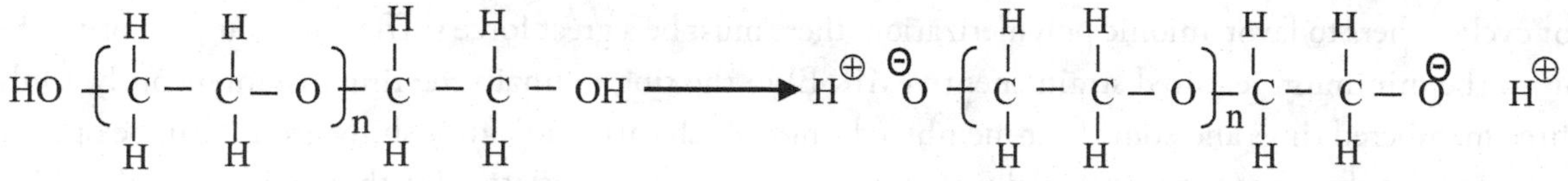

Living polymers from both ends

12.11

$$RO \left(\overset{\overset{\displaystyle H}{|}}{\underset{\underset{\displaystyle H}{|}}{C}} - \overset{\overset{\displaystyle H}{|}}{\underset{\underset{\displaystyle H}{|}}{C}} - O \right)_n \overset{\overset{\displaystyle H}{|}}{\underset{\underset{\displaystyle H}{|}}{C}} - \overset{\overset{\displaystyle H}{|}}{\underset{\underset{\displaystyle H}{|}}{C}} - OH \longrightarrow R - O \left(\overset{\overset{\displaystyle H}{|}}{\underset{\underset{\displaystyle H}{|}}{C}} - \overset{\overset{\displaystyle H}{|}}{\underset{\underset{\displaystyle H}{|}}{C}} - O \right)_n \overset{\overset{\displaystyle H}{|}}{\underset{\underset{\displaystyle H}{|}}{C}} - \overset{\overset{\displaystyle H}{|}}{\underset{\underset{\displaystyle H}{|}}{C}} - O^{\ominus} \quad H^{\oplus}$$

Living polymers from one end   12.12

$$RO \left(\overset{\overset{\displaystyle H}{|}}{\underset{\underset{\displaystyle H}{|}}{C}} - \overset{\overset{\displaystyle H}{|}}{\underset{\underset{\displaystyle H}{|}}{C}} - O \right)_n \overset{\overset{\displaystyle H}{|}}{\underset{\underset{\displaystyle H}{|}}{C}} - \overset{\overset{\displaystyle H}{|}}{\underset{\underset{\displaystyle H}{|}}{C}} - O - R \longrightarrow \text{Dead polymer.}$$

12.13

It is when cases like those of Equation 12.11 and 12.12 are involved as "dead polymers" that a reaction between a propagating chain and the "dead polymer" can take place. It cannot take place when the dead polymers are those of Equation 12.13. How these living polymers are terminated to produce more stable dead polymers will be considered at the appropriate time. Nevertheless, it is important to note that hydroxides such are KOH, metal oxides are not particularly suited for use as initiators and solvents during the polymerizations of these monomers anionically. When used as solvent, the disturbing factors are free characters of the species, the anionic route being not natural to it. Hence the strength of the growing polymer chains will keep decreasing for every monomer added; until a point is reached when the growing chain becomes less acidic than the alcohol at which point growth is terminated producing an unstable end and a low molecular weight polymer.

$$RO - \overset{\overset{\displaystyle H}{|}}{\underset{\underset{\displaystyle H}{|}}{C}} - \overset{\overset{\displaystyle H}{|}}{\underset{\underset{\displaystyle H}{|}}{C}} - O^{\ominus} \quad > \quad RO - \overset{\overset{\displaystyle H}{|}}{\underset{\underset{\displaystyle H}{|}}{C}} - \overset{\overset{\displaystyle H}{|}}{\underset{\underset{\displaystyle H}{|}}{C}} - O - \overset{\overset{\displaystyle H}{|}}{\underset{\underset{\displaystyle H}{|}}{C}} - \overset{\overset{\displaystyle H}{|}}{\underset{\underset{\displaystyle H}{|}}{C}} - O^{\ominus} \quad > >$$

$$RO \left(\overset{\overset{\displaystyle H}{|}}{\underset{\underset{\displaystyle H}{|}}{C}} - \overset{\overset{\displaystyle H}{|}}{\underset{\underset{\displaystyle H}{|}}{C}} - O \right)_n \overset{\overset{\displaystyle H}{|}}{\underset{\underset{\displaystyle H}{|}}{C}} - \overset{\overset{\displaystyle H}{|}}{\underset{\underset{\displaystyle H}{|}}{C}} - O^{\ominus}$$

<u>Decreasing order of strength of anionic active growing center</u>   12.14

A solvent which cannot split to produce charges or radicals is the best solvent to use, for example NR_3 with metallic alkoxide as initiators.

So far, it is important to note that the ⌒O⌒ functional center cannot be involved anionically. Hence for cyclic ethers to favor anionic polymerization, there must be a great force sufficient enough to bring the SE to the minimum required strain energy (MRSE) in the ring to unzip the ring instantaneously. Only three membered rings and some four-membered rings which carry radical-pushing groups can be opened instanta-neously. The four-membered rings will require stronger initiators for those with radical-pulling groups than for those with radical-pushing groups.

$$R \overset{\ominus}{\colon} \;+\; H-\underset{\underset{H}{|}}{\overset{\overset{H}{|}}{C}}-C(CH_2Cl)_2 \;\longrightarrow\; R\overset{\ominus}{\colon}\;+\; \oplus \underset{\underset{H}{|}}{\overset{\overset{H}{|}}{C}}-\underset{\underset{CH_2Cl}{|}}{\overset{\overset{CH_2Cl}{|}}{C}}-\underset{\underset{H}{|}}{\overset{\overset{H}{|}}{C}}-O^{\ominus}$$

(moderately
strong)

$$O-\underset{\underset{H}{|}}{\overset{}{C}}-H$$

(Nucleophile)

$$\longrightarrow \; R-\underset{\underset{H}{|}}{\overset{\overset{H}{|}}{C}}-\underset{\underset{CH_2Cl}{|}}{\overset{\overset{CH_2Cl}{|}}{C}}-\underset{\underset{H}{|}}{\overset{\overset{H}{|}}{C}}-O\colon^{\ominus} \qquad\qquad 12.15$$

$$R \overset{\ominus}{\colon} \;+\; H-\underset{\underset{H}{|}}{\overset{\overset{H}{|}}{C}}-\underset{\underset{H}{|}}{\overset{\overset{Cl}{|}}{C}}-Cl \;\longrightarrow\; \text{Not strong enough to release strain}$$

(moderately
strong)

$$O-\underset{\underset{H}{|}}{\overset{}{C}}-H$$

(Less nucleophilic in character) \qquad\qquad 12.16

$$R \overset{\ominus}{\colon} \;+\; H-\underset{\underset{H}{|}}{\overset{\overset{H}{|}}{C}}-\underset{\underset{}{}}{\overset{\overset{Cl}{|}}{C}}-Cl \;\longrightarrow\; R\overset{\ominus}{\colon}\;+\; \oplus \underset{\underset{H}{|}}{\overset{\overset{H}{|}}{C}}-\underset{\underset{Cl}{|}}{\overset{\overset{Cl}{|}}{C}}-\underset{\underset{H}{|}}{\overset{\overset{H}{|}}{C}}-O^{\ominus}$$

(very
strong)

$$O-\underset{\underset{H}{|}}{\overset{}{C}}-H$$

$$\longrightarrow \; R-\underset{\underset{H}{|}}{\overset{\overset{H}{|}}{C}}-\underset{\underset{Cl}{|}}{\overset{\overset{Cl}{|}}{C}}-\underset{\underset{H}{|}}{\overset{\overset{H}{|}}{C}}-O\colon^{\ominus} \qquad\qquad 12.17$$

It is important to note that the more nucleophilic the hetero- ring is, the easier it is for the ring to be unzipped by an anionic initiator or center, when the minimum required strain energy can be provided for the ring.

"Cationically" the situation is different. Most of the cyclic ethers can be involved due to the presence of the positively charged attracting functional center, through which the minimum required strain can be put into the ring. Consider using a free-cationic initiator such as protonic acids e.g. sulfuric acid. This can only be done radically.

$$H^{.e}\,^{nn.}OHSO_3 \;+\; H-\underset{\underset{}{}}{\overset{\overset{H}{|}}{C}}\!\!\diagdown\!\!\underset{\underset{}{}}{\overset{\overset{H}{|}}{C}}-H \;\xrightarrow[\text{Of strain)}]{\text{(Release}}\; H^{.e}\,^{nn.}OHSO_3 \;+\; {}^{.}e\,\underset{\underset{H}{|}}{\overset{\overset{H}{|}}{C}}-\underset{\underset{H}{|}}{\overset{\overset{H}{|}}{C}}-O.nn$$

$$O$$

$$\longrightarrow \; H-O-\underset{\underset{H}{|}}{\overset{\overset{H}{|}}{C}}-\underset{\underset{H}{|}}{\overset{\overset{H}{|}}{C}}\text{---}OHSO_3 \qquad OR \qquad H-O-\underset{\underset{H}{|}}{\overset{\overset{H}{|}}{C}}-\underset{\underset{H}{|}}{\overset{\overset{H}{|}}{C}}\,^{.e}\;^{nn.}OHSO_3$$

(I) \qquad\qquad\qquad\qquad (II) \qquad\qquad 12.18

In the reaction above, the instantaneous opening of the ring is via the anionic counter-ion which has the electrostatic forces. $H^{\oplus}$ or $H^{\cdot e}$ does not possess the electrostatic forces in view of the absence of paired unbonded radicals in the last shell. If the counter-center of $H\oplus$ or $H^{\cdot e}$ which is $^{\theta}OHSO_3$ or "nn•$OHSO_3$," is the only one present, then (I) above is favored, because the counter-center cannot diffuse. If monomers are present, then (II) is favored. With (II) polymerization is favored while with (I), a stable product is obtained.

For a larger sized ring where the MRSE cannot readily be provided, the followings are obtained.

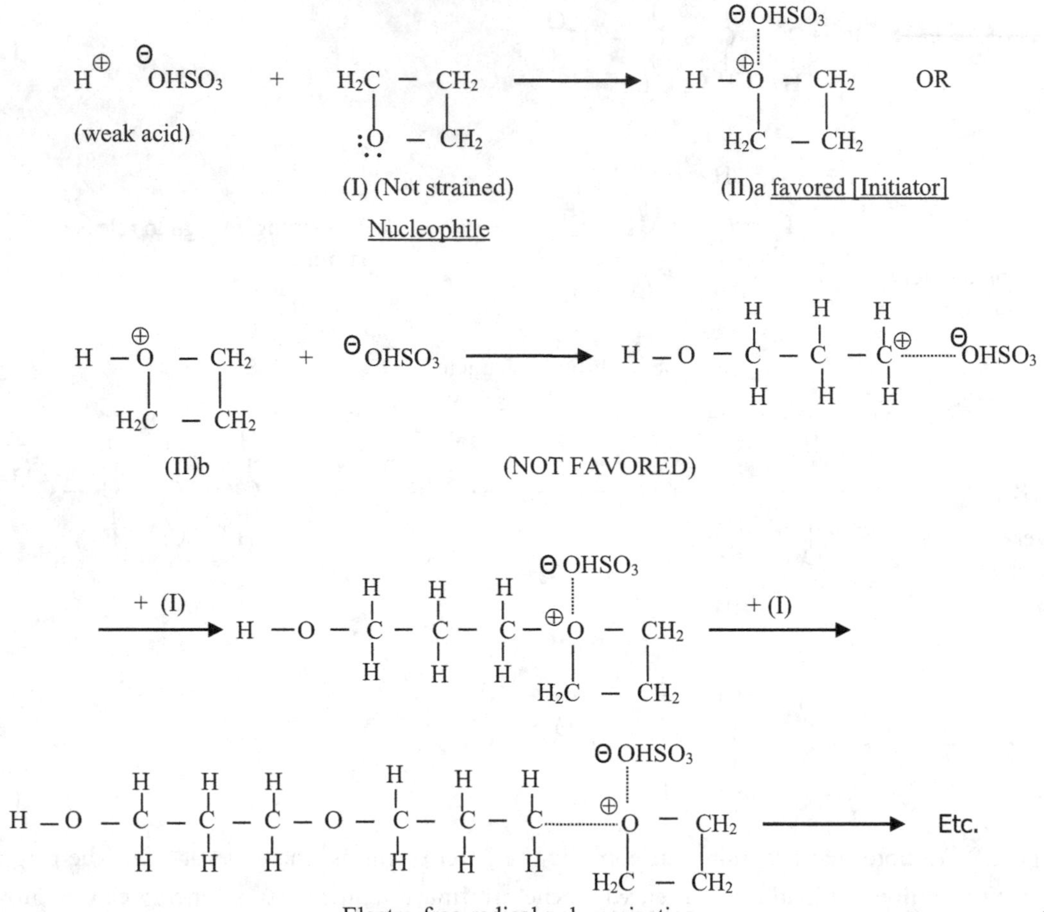

$$H^{\oplus} \quad ^{\ominus}OHSO_3 \quad + \quad \begin{matrix} H_2C - CH_2 \\ | \quad\quad | \\ :\overset{..}{O} - CH_2 \end{matrix} \longrightarrow \quad \begin{matrix} ^{\ominus}OHSO_3 \\ | \\ H - \overset{\oplus}{O} - CH_2 \\ | \quad\quad | \\ H_2C - CH_2 \end{matrix} \quad OR$$

(weak acid)   (I) (Not strained)   (II)a <u>favored [Initiator]</u>

                 <u>Nucleophile</u>

$$\begin{matrix} H - \overset{\oplus}{O} - CH_2 \\ | \quad\quad | \\ H_2C - CH_2 \end{matrix} \quad + \quad ^{\ominus}OHSO_3 \longrightarrow \quad H - O - \overset{\overset{H}{|}}{\underset{\underset{H}{|}}{C}} - \overset{\overset{H}{|}}{\underset{\underset{H}{|}}{C}} - \overset{\overset{H}{|}}{\underset{\underset{H}{|}}{\overset{\oplus}{C}}} \cdots ^{\ominus}OHSO_3$$

(II)b   (NOT FAVORED)

$$\xrightarrow{+ (I)} \quad H - O - \overset{\overset{H}{|}}{\underset{\underset{H}{|}}{C}} - \overset{\overset{H}{|}}{\underset{\underset{H}{|}}{C}} - \overset{\overset{H}{|}}{\underset{\underset{H}{|}}{C}} - \overset{\overset{\overset{\ominus OHSO_3}{|}}{\oplus}}{\underset{\underset{H_2C - CH_2}{|}}{O}} - CH_2 \quad \xrightarrow{+ (I)}$$

$$H - O - \overset{\overset{H}{|}}{\underset{\underset{H}{|}}{C}} - \overset{\overset{H}{|}}{\underset{\underset{H}{|}}{C}} - \overset{\overset{H}{|}}{\underset{\underset{H}{|}}{C}} - O - \overset{\overset{H}{|}}{\underset{\underset{H}{|}}{C}} - \overset{\overset{H}{|}}{\underset{\underset{H}{|}}{C}} - \overset{\overset{H}{|}}{\underset{\underset{H}{|}}{C}} \cdots \overset{\overset{\overset{\ominus OHSO_3}{|}}{\oplus}}{\underset{\underset{H_2C - CH_2}{|}}{O}} - CH_2 \longrightarrow Etc.$$

Electro-free-radical polymerization   12.19

Since the ring may be opened instantaneously the monomers being less nucleophilic than the type shown in Equation 12.15, all the steps indicated above are important. The free-cationic initiator is forced to assume the character of an ion-paired initiator, since the oxygen center cannot carry a positive ionic charged. Hence (II)b is not favored. Compare (II)a with an anionic ion-paired initiator of the electrostatic type shown below.

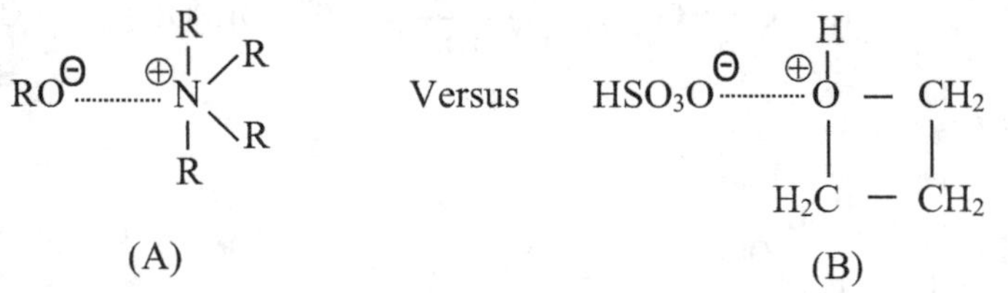

$$RO^{\ominus} \cdots \overset{\overset{R}{|}}{\underset{\underset{R}{|}}{\overset{\oplus}{N}}}\diagdown^{R}_{\diagdown R} \qquad\qquad Versus \qquad\qquad HSO_3O^{\ominus} \cdots \overset{\overset{H}{|}}{\underset{\underset{H_2C - CH_2}{|}}{\overset{\oplus}{O}}} - CH_2$$

(A)   (B)   12.20

Whereas the route is cationic, it seems that anionic ion-paired initiators are involved which indeed is not the case for (B), since $H^{\cdot e}$ is the first to attack or is the active center. Hence the route indicated above can be said to be electro-free-radical or Electrostatically anionically-paired initiator. With (B) types of initiators, both centers are active. For the case of (B) above, the center used depends on the character of the monomer (Male or Female). Males are Electrophiles while the Females are Nucleophiles as we already know. Very shortly, the mechanism above which is not Combination will be explained.

While the reaction of (II)a above may be possible with the four-membered 1,3 - epoxide, it may not readily be favored with five-membered 1,4- epoxide- tetrahydro-furan, partly because the minimum required strain energy cannot be provided, and also because the counter ion can be more nucleophilic than the monomers as shown below.

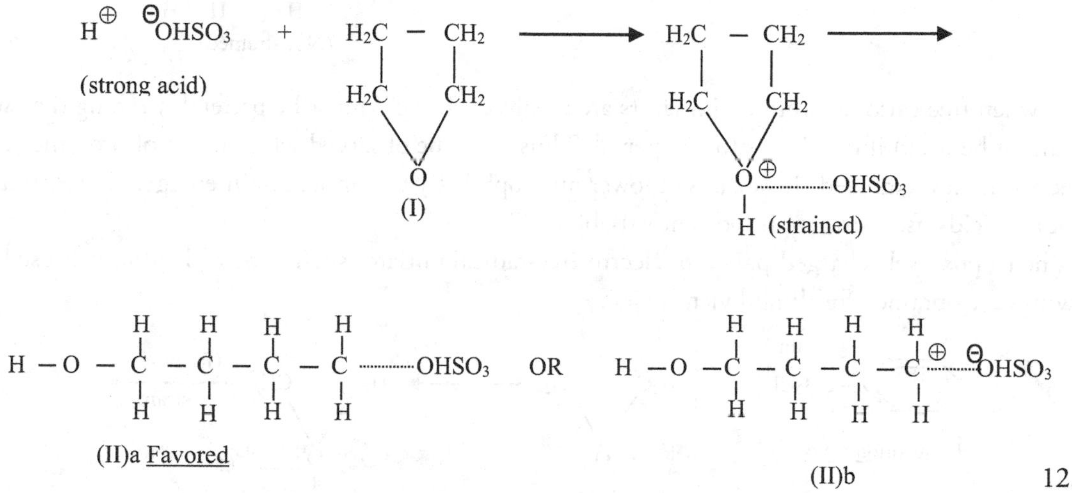

$$H^{\oplus} \quad {}^{\ominus}OHSO_3 \quad + \quad \underset{\substack{\big| \\ H_2C \quad CH_2 \\ \diagdown O \diagup \\ (I)}}{H_2C - CH_2} \longrightarrow \underset{\substack{\big| \\ H_2C \quad CH_2 \\ \diagdown O^{\oplus} \cdots OHSO_3 \\ \big| \\ H \;(strained)}}{H_2C - CH_2} \longrightarrow$$

(strong acid)

$$\underset{(II)a\ \underline{Favored}}{H - O - \overset{\overset{\displaystyle H}{|}}{\underset{\underset{\displaystyle H}{|}}{C}} - \overset{\overset{\displaystyle H}{|}}{\underset{\underset{\displaystyle H}{|}}{C}} - \overset{\overset{\displaystyle H}{|}}{\underset{\underset{\displaystyle H}{|}}{C}} - \overset{\overset{\displaystyle H}{|}}{\underset{\underset{\displaystyle H}{|}}{C}} \cdots OHSO_3} \quad OR \quad H - O - \overset{\overset{\displaystyle H}{|}}{\underset{\underset{\displaystyle H}{|}}{C}} - \overset{\overset{\displaystyle H}{|}}{\underset{\underset{\displaystyle H}{|}}{C}} - \overset{\overset{\displaystyle H}{|}}{\underset{\underset{\displaystyle H}{|}}{C}} - \overset{\overset{\displaystyle H}{|}}{\underset{\underset{\displaystyle II}{|}}{C}}{}^{\oplus} \; {}^{\ominus}OHSO_3$$

(II)b   12.21

It is (II)a that has more often been observed with tetrahydrofuran, from whence it can be observed that (I) is opened since it is strained, but cannot proceed beyond initiation because the concentration of the acid used is such that makes the counter-ion more nucleophilic than the monomer. It may be more nucleophilic at the beginning, but not during the course of propagation, since the positively charged route is natural to the monomer and with a cation under Equilibrium mechanism condition only a stable molecule can be produced.

Hence it is believed that, it is impossible for cyclic ethers to favor free cationic ring-opening polymerization via (II)a of Equation 12.19. However, as will shortly become obvious, the order of nucleophilicity of the monomer with all radical-pushing groups are as follows-

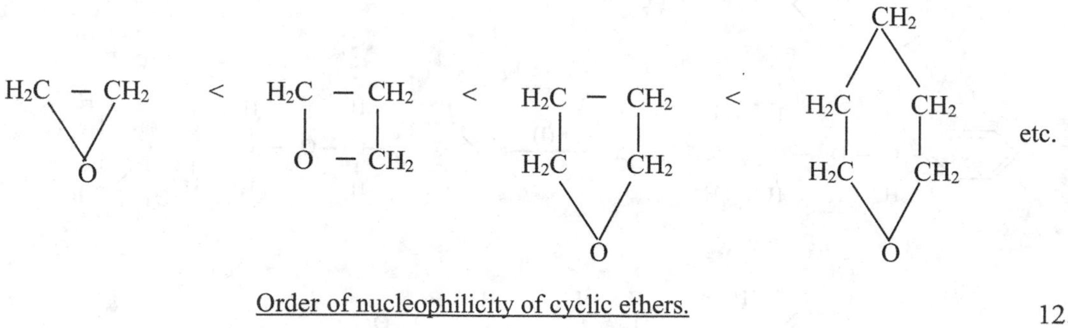

$$\underset{O}{\overset{H_2C - CH_2}{\diagdown \diagup}} \quad < \quad \underset{O - CH_2}{\overset{H_2C - CH_2}{|\qquad |}} \quad < \quad \underset{\underset{O}{H_2C \quad CH_2}}{\overset{H_2C - CH_2}{|\qquad |}} \quad < \quad \underset{\underset{O}{H_2C \quad CH_2}}{\overset{CH_2}{\underset{H_2C \quad CH_2}{\diagup \diagdown}}} \quad etc.$$

Order of nucleophilicity of cyclic ethers.   12.22

The moment one or more of the hydrogen atoms in (I) of Equation 12.21 is replaced with radical-pushing group, the monomer can now be homo-polymerized, since the ring becomes more strained. ***Thus, the***

more nucleophilic a ringed monomer is made to become, the more strained it is, and the less the energy required to attain the minimum required strain energy.

Unlike the four-membered ring, when a weak acid is involved in order to make the counter-ion less nucleophilic than the tetrahydrofuran, the following is obtained.

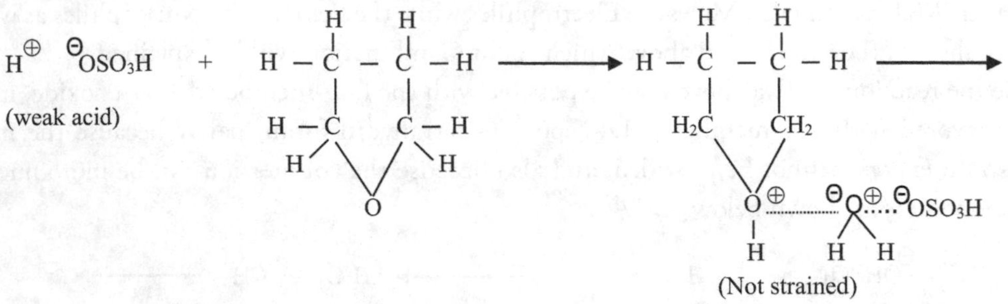

(Not strained)   12.23

Hence, when free cationic protonic initiators are involved, the acid must be preferably strong (i.e., weak acid cannot be used) if the ring is to be opened. Thus it can be observed why the use of protonic acid is limited to small membered rings-cases of lower nucleophilicity and more strain energy. Concentrations of protonic acids used are very important variables.

When a positively charged-paired or electro-free-radical initiator such as benzyl chloride is used, the followings are obtained for dimethylene oxide.

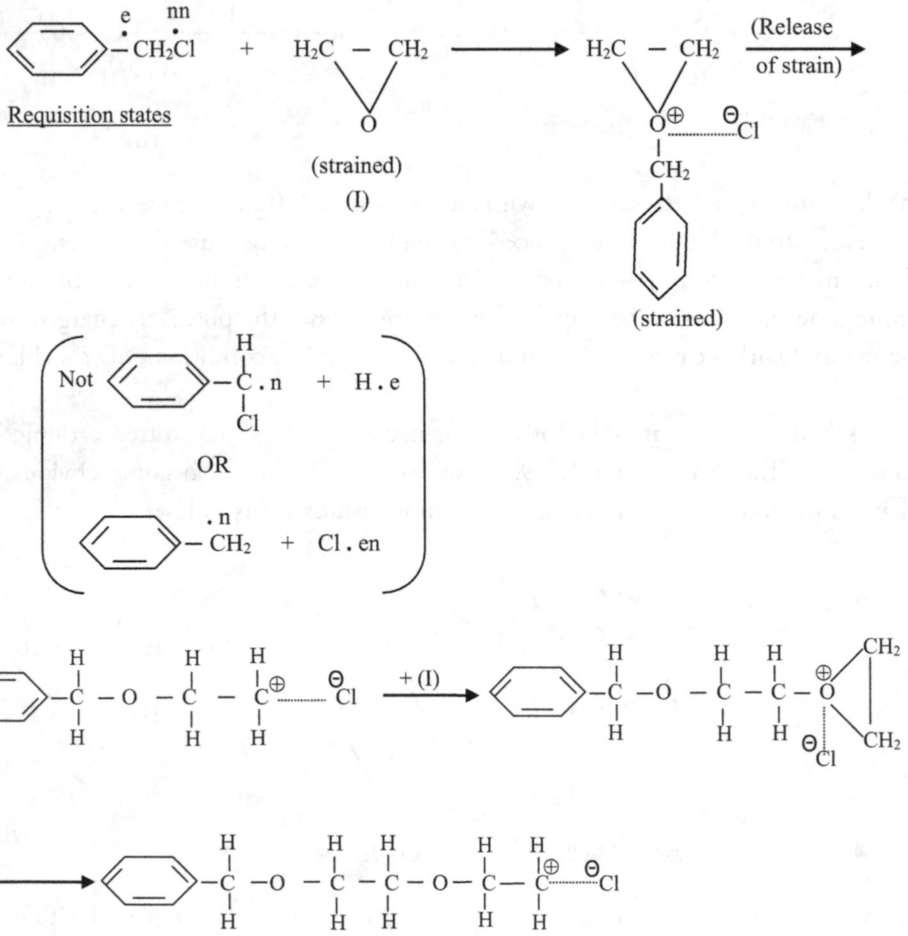

12.24

It is believed that when positively charged-paired or free-radical initiators are involved, the ring is never unzipped by the counter-ion center, but through the functional center. When opened, the propagation takes place as shown above, the route being natural to the monomer. The reaction above will be favored provided Cl .nn or $Cl^\ominus$ is less nucleophilic than the monomer and covalently placed. The use of benzyl group as initiator in place of H, makes the growing polymer chain more stable after the polymer has been produced, provided propagation takes place and an adequate terminating agent has been used.

When positively charged-paired initiators of the electrostatic types are involved, the followings are to be expected.

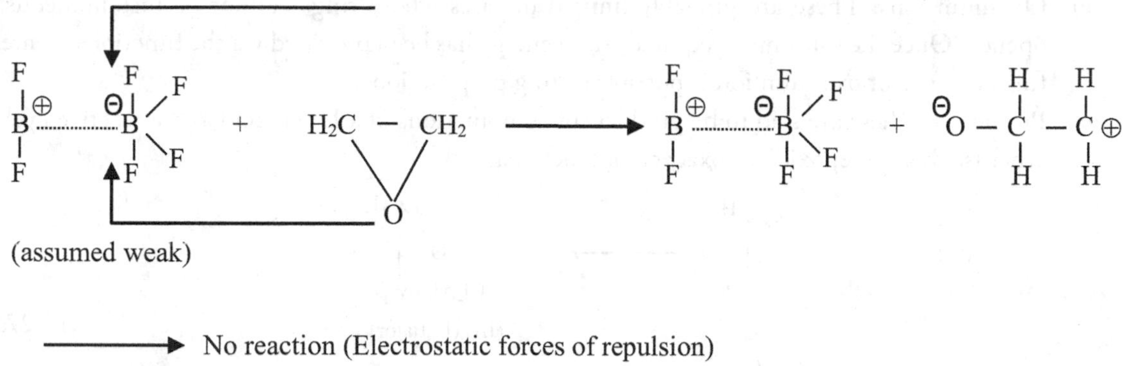

(assumed weak)

$$\longrightarrow \text{No reaction (Electrostatic forces of repulsion)}$$

12.25

(assumed strong)

(I)

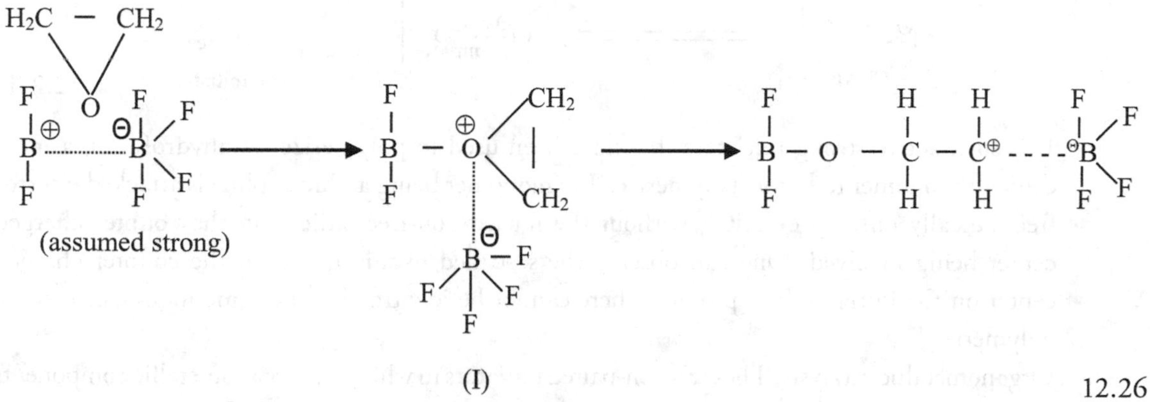

12.26

If the coordination is weak and very polar, then the initiation step may not be favored, due to electrostatic forces of repulsion between the paired unbonded radicals on the oxygen center and the paired unbonded radicals on the fluorine center. If however the coordination center is strong enough to unzip the ring, then the second equation above will be favored. This is the case wherein activation of the ring takes place after the positively charged center has made contact with the oxygen center. It should be noted that repulsion between two paired unbonded radicals is strong while that between two adequately located radical-pushing groups or between an electrostatic negative charge and paired unbonded radicals is zero. It should be observed that (I) in the last equation above is electrostatically bonded positively (i.e., of the positive type), while the initiator of Equation 12.24 is covalently bonded. There is only one active center in the initiator above. The initiator used above is a product of BF_3 serving the dual functions of catalyst and cocatalyst. From the reactions above, the polymerization of the three-membered rings is favored only when strong positively-charged coordination initiators are involved. For larger membered rings which do not favor opening of their rings instantaneously or via the functional center, very strong positively charged coordination initiators are more desired to add more energy to the ring.

"Cationic" initiators which have been reportedly used with cyclic ethers include the followings [10, 12,-16].

(i) Protonic acids such as dilute or concentrated sulfuric acids, trifluoroacetic acid or fluorosulfuric acids.

(ii) "Cationic ion-paired" catalyst/cocatalyst of Lewis acids type such as BF_3, $SnCl_4$, PF_5, $SbCl_5$ etc. preferably in the presence of added catalyst such as ether or water.

(iii) "Cationic ion-paired" catalyst and cocatalyst with carbonium centers. They are the same as when ether is used in (ii) above, such as ROR/BF_3.

(iv) Oxonium ions. These are probably limited to cases whose rings cannot be instantaneously opened. Once the minimum required strain energy has been provided via the functional center, full existence of oxonium ion is possible during propagation.

(v) Promoters. These are said to be products of combinations of a Lewis acid and a reactive cyclic ether (such as an epoxide or oxacyclobutane) - e.g.,

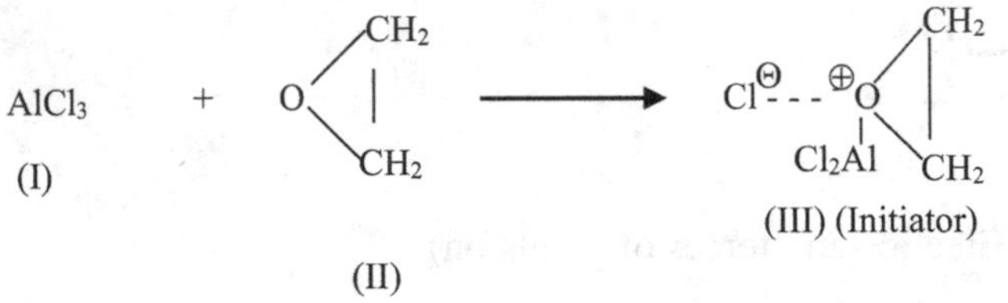

$$12.27a$$

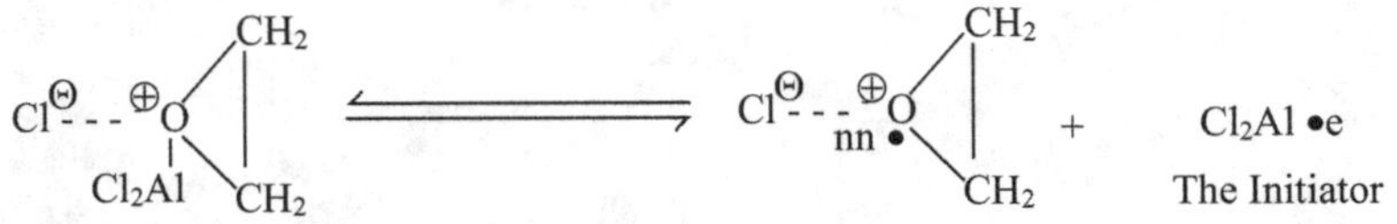

$$12.27b$$

It is such active strong initiators that have been used to polymerize tetrahydrofuran, a more difficult monomer to homo-polymerize. The monomer being a Nucleophile is attacked electro-free-radically only by $e \bullet AlCl_2$ without the nucleo-non-free-radical on the counter charged center being involved. One can observe the so-called oxonium ion on the counter charged center on the initiator. The promoter here cannot be identified if the same monomer is to be polymerized.

(vi) Organometallic catalysts. These are ion-paired initiators in which the organometallic components when used with other components such as water or alcohol, is the cocatalyst.

In the list above, no Ziegler-Natta initiators are involved. The initiators are largely charged-paired initiators either of the electrostatic or the covalent types. When free cationic initiators are involved, it been shown that no propagation is possible.

So far, the use of so-called free-cationic initiators has been considered. So also is the use of BF_3 as the only source. Considering the case of PF_5, when used alone, the ion-paired initiator generated is (I) shown below.

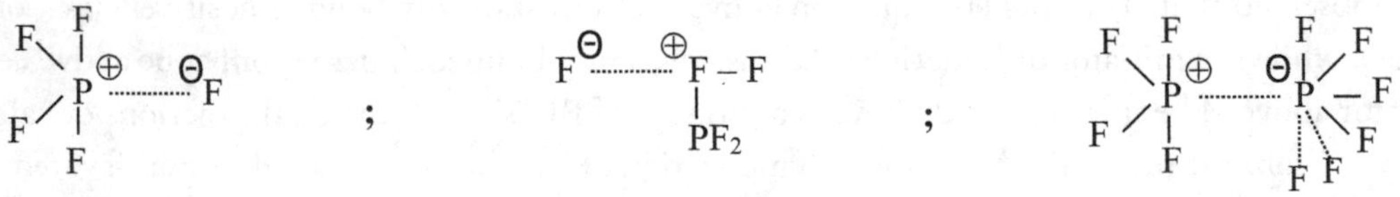

(I) Anionic ion-paired Initiator [FAVORED]

(II) Anionic ion-paired initiator [FAVORED]

(III) Cationic ion-paired initiator [NOT FAVORED]

Note: (II) is more favored than (I).

$$12.28$$

14

In (II) the positively charged center is electrostatically bonded, while in (III) both centers are electrostatically bonded, since the phosphorus atoms cannot carry more than eight radicals in the last shell. In the anionic center of (III), two fluorine atoms are electrostatically bonded to the phosphorus center while three fluorine atoms are covalently bonded to it. All these are impossible since the last shell in the P center has no vacant orbitals. When (I) above is involved, the followings are obtained only during initiation.

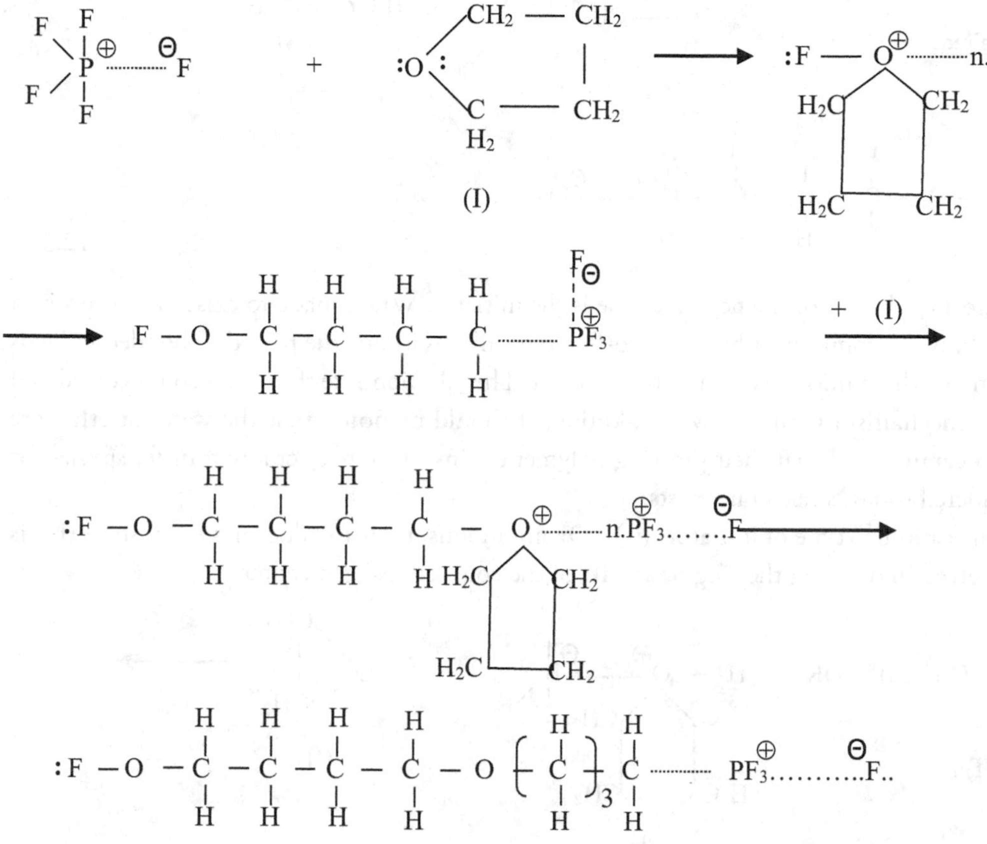

$$12.29$$

In the Initiation step, (I) first existed in Equilibrium state of existence, with fluorine being held as an electro-non-free-radical. This started the initiation step via Equilibrium mechanism and not Combination mechanism. After the initiation step, fluorine was next held in Equilibrium state of existence of the product from initiation to add another monomer to the chain. This addition continues backwardly as will be fully explained downstream. Of the three initiators in Equation 12.28, it is (I) and (II) that satisfy the conditions to be called electrostatically positively charged-paired initiator, while (III) cannot exist. The real structure of P_2F_{10} will be shown downstream, that wherein only the F element is also held. Despite the explorative use of (I) above, the real initiator is (II) above, in which the carrier of the chain is F_2P. The mechanism will remain the same as above. In (I), PF_3 was held in Stable state of existence, while in (II), F_2 is the component held in Stable state of existence, that which is to be expected, F_2 being a very stable molecule. The rings are opened with $F_2P.en$, the initiating species via Equilibrium mechanism.

When PF_5 is used in the presence of traces of water or ether, the followings are the initiator and reactions involved using water.

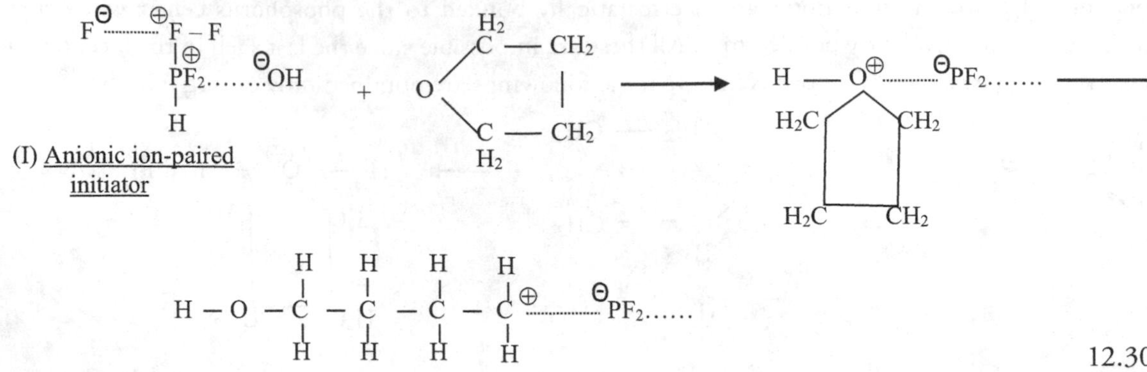

(I) Anionic ion-paired initiator

$$12.30$$

(I) with two electrostatic bonds of the negative type is the initiator. When made to exist in Equilibrium state of existence, H is the component held electro-free-radically. Without the paired unbonded radicals on the F and P centers, these initiators cannot be prepared. The addition like the others so far considered is via Equilibrium mechanism with backward addition. It should be noted that the water or ether are catalyst and not cocatalyst and that their growing polymer chains cannot reject any transfer species for all the cases considered so far, since none exists.

For the fourth "cationic" type of initiators (iv) - Oxonium ions, their use does not exist, since this is a part of steps involved in opening the ring or as part of the counter charged center.

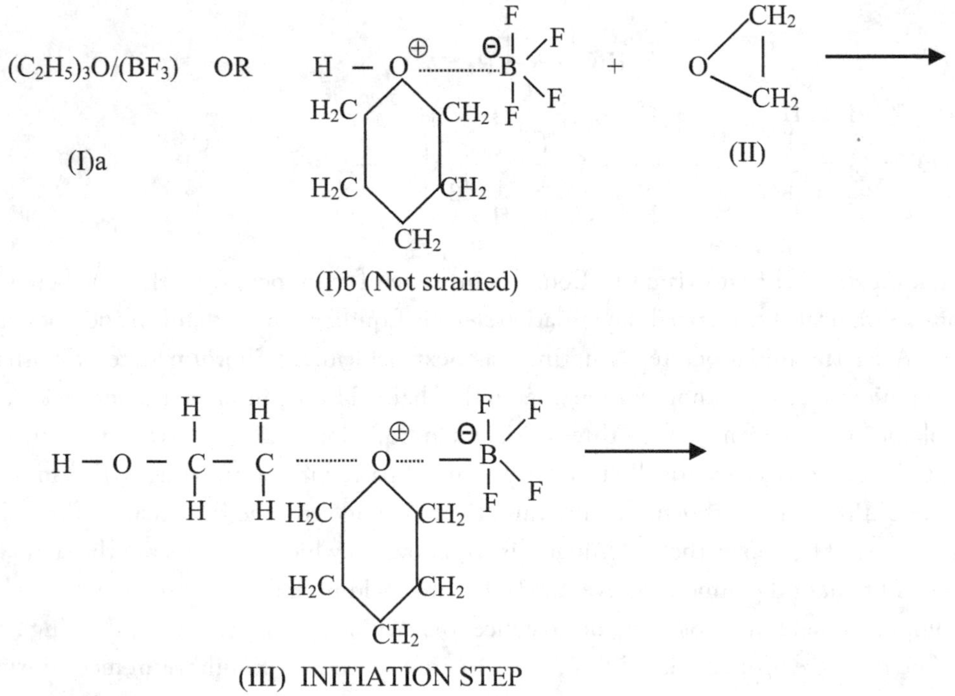

$$(C_2H_5)_3O/(BF_3)$$ OR

(I)a

(I)b (Not strained)

(II)

(III) INITIATION STEP

$$12.31$$

The oxonium ion (I)b shown above cannot be opened since the minimum required strain energy cannot be provided for this six-membered ring. It has therefore been used in forming a strong initiator since the oxonium center which is electrostatically bonded cannot be used. Addition of (II) above continues without the six-membered ring or ether being involved. From time to time, the ring may disturb the course of propagation.

When the fifth case of "cationic" initiator is involved, that is, promoters, the followings are obtained for tetrahydrofuran.

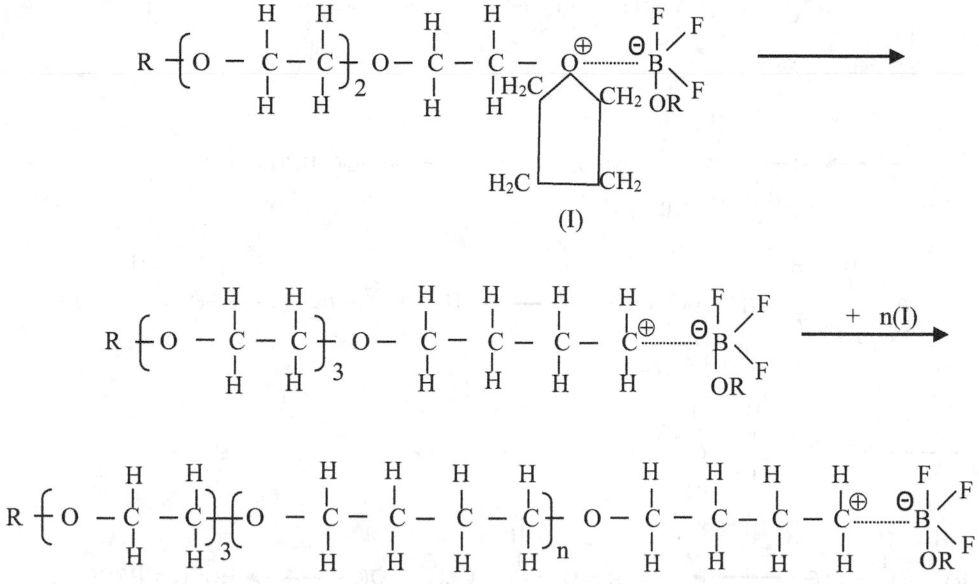

$$(I)$$

$$+ n(I)$$

$$12.32$$

The use of promoters is unique, since this is how some monomers which cannot readily be homopolymerized are copolymerized, noting that for every addition of the nucleophilic monomer, the stronger is the positively charged active center obtained. The promoter here is the three-membered ring. This is what cannot be attained with the use of Z|N initiators where the strength of their active centers remains the same if their use is made possible.

Other examples of so-called carbonium "ion" type of ion-paired initiators include those from the following combinations-

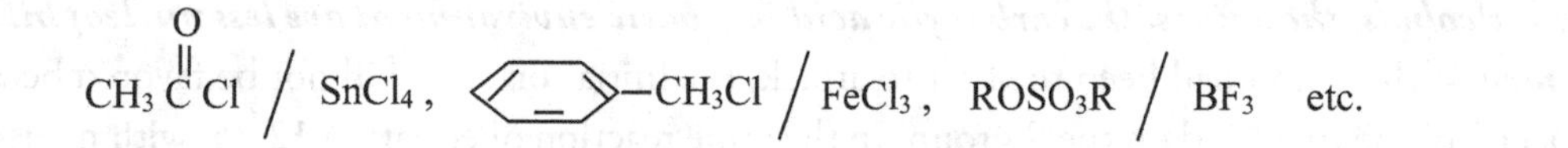

$$CH_3 \overset{O}{\overset{\|}{C}} Cl \;/\; SnCl_4, \quad \langle\!\!\!\rangle\!-\!CH_3Cl \;/\; FeCl_3, \quad ROSO_3R \;/\; BF_3 \quad etc. \qquad 12.33$$

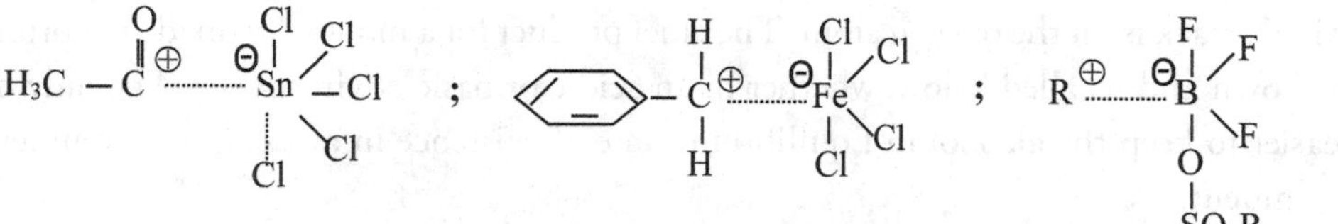

Electrostatically positively charged-paired initiators

$$12.34$$

These are cases which will favor the existence of a growing polymer chain.

Now consider replacing one of the hydrogen atoms with a radical-pushing alkylane groups in dimethylene oxide. There is need at this point in time to recall the chemistry of these monomers as already established. When a monoalkylsubstituted ethylene oxide is nucleophilically attacked through the release of strain energy to open the ring, it is said that the point of attacked is determined by the relative "electron" density and steric hindrance at the ring carbon atoms[2]; for which the following reactions have been observed -

$$R - \overset{\overset{\displaystyle H}{|}}{\underset{}{C}}{}^3 - \overset{\overset{\displaystyle H}{|}}{\underset{}{C}}{}^2 - H \;\; \underset{\displaystyle :\overset{}{O}:}{\diagdown\diagup} \;\; + \;\; HOR^1 \;\; \longrightarrow \;\; H^{\oplus} + {}^{\ominus}\underline{OR^1} \;\; + \;\; {}^{\oplus}\overset{\overset{\displaystyle H}{|}}{\underset{\underset{\displaystyle R}{|}}{C}}{}^3 - \overset{\overset{\displaystyle H}{|}}{\underset{\underset{\displaystyle H}{|}}{C}}{}^2 - O^{\ominus}$$

(alcohol)

$$\longrightarrow \;\; H - O - \overset{\overset{\displaystyle H}{|}}{\underset{\underset{\displaystyle H}{|}}{C}} - \overset{\overset{\displaystyle H}{|}}{\underset{\underset{\displaystyle R}{|}}{C}}{}^{\oplus}.......{}^{\ominus}OR^1 \;\; \longrightarrow \;\; HOCH_2CHROR^1 \qquad\qquad 12.35$$

$$R - \overset{\overset{\displaystyle H}{|}}{\underset{}{C}} - \overset{\overset{\displaystyle H}{|}}{\underset{}{C}} - H \;\; \underset{\displaystyle O}{\diagdown\diagup} \;\; + \;\; H_2NR^1 \;\; \longrightarrow \;\; H^{\oplus} + {}^{\ominus}\underline{NHR^1} \;\; + \;\; {}^{\oplus}\overset{\overset{\displaystyle H}{|}}{\underset{\underset{\displaystyle R}{|}}{C}} - \overset{\overset{\displaystyle H}{|}}{\underset{\underset{\displaystyle H}{|}}{C}} - O^{\ominus}$$

$$\longrightarrow H - O - \overset{\overset{\displaystyle H}{|}}{\underset{\underset{\displaystyle H}{|}}{C}} - \overset{\overset{\displaystyle H}{|}}{\underset{\underset{\displaystyle R}{|}}{C}}{}^{\oplus}..........{}^{\ominus}NHR^1 \;\; \longrightarrow \;\; HOCH_2CHRNHR^1 \qquad\qquad 12.36$$

$${}^{\oplus}\overset{\overset{\displaystyle H}{|}}{\underset{\underset{\displaystyle R}{|}}{C}} - \overset{\overset{\displaystyle H}{|}}{\underset{\underset{\displaystyle H}{|}}{C}} - O^{\ominus} \;\; \longrightarrow \;\; H - O - \overset{\overset{\displaystyle H}{|}}{\underset{\underset{\displaystyle H}{|}}{C}} - \overset{\overset{\displaystyle H}{|}}{\underset{\underset{\displaystyle R}{|}}{C}}{}^{\oplus}.......{}^{\ominus}OCOR^1 \longrightarrow HOCH_2\, CHROCOR^1$$

$$12.37$$

The rings are opened instantaneously by the underlined anions which cannot be the first to attack in the presence of a positive charge. As a matter of fact, based on the Old science, the monomers being nucleophiles should be attacked first by the "cations" when the source is acidic and by anions when the source is basic. Nevertheless, whether in the presence of basic or acidic environment, the point of scission is still determined by the bond with the highest radical-pushing potential difference and the natural route remains the only route. The point of scission remains the same. ***Molecular products are obtained above, because the alcohols, the oxides, the carboxylic acid in a basic environment are less nucleophilic than the monomer.*** If the anion had been the first to attack, the initiation step will not be favored because of transfer species of the first kind on the R group. In the same reaction of equation 12.35, with nucleophiles (anions) at a pH less than about 4, that is during "cationic" attacks, acidic catalysts is said to take over, and the initial attack is on the oxygen atom. The chief product for a monoalkyl oxide is a primary alcohol as already shown and recalled below, whether in an acidic or basic environment. The only difference is that it is easier to keep the alcohol in Equilibrium state of existence in an acidic environment than in a basic environment.

$$R - \underset{2}{\overset{\overset{\displaystyle H}{|}}{C}} - \underset{3}{\overset{\overset{\displaystyle H}{|}}{C}} - H \;\; \underset{\underset{\displaystyle 1}{\displaystyle O}}{\diagdown\diagup} \;\; \overset{+\; H^{\oplus}}{\underset{R^1O^{\ominus}}{\longrightarrow}} \;\; R - \overset{\overset{\displaystyle H}{|}}{\underset{}{C}} - \overset{\overset{\displaystyle H}{|}}{\underset{}{C}} - H \;\; \underset{\displaystyle :O^{\oplus}........{}^{\ominus}OR^1}{\diagdown\diagup} \;\; \longrightarrow \;\; H - O - \overset{\overset{\displaystyle H}{|}}{\underset{\underset{\displaystyle H}{|}}{C}} - \overset{\overset{\displaystyle R}{|}}{\underset{\underset{\displaystyle H}{|}}{C}}{}^{\oplus}.........{}^{\ominus}OR^1$$

$$pH \;\leq\; 4 \qquad\qquad\qquad \overset{\displaystyle |}{\underset{\displaystyle H}{}}$$

$$\longrightarrow \;\; H - O - \overset{\overset{\displaystyle H}{|}}{\underset{\underset{\displaystyle H}{|}}{C}} - \overset{\overset{\displaystyle R}{|}}{\underset{\underset{\displaystyle H}{|}}{C}} - OR^1 \;\; \longrightarrow \;\; RCH(OR^1)CH_2OH$$

(primary alcohol) $\qquad\qquad 12.38$

The positively charged center generated when the "cation" attacks the functional center is always on the carbon center carrying the most radical-pushing group under Equilibrium conditions. When the ring is opened anionically, this is done instanta-neously, and the positively charged carbon center generated still remains the same, since the point of scission is only determined by the bond with the largest radical-pushing potential difference. Hence anionically or with negatively charged initiator, initiation step can never be favored.

When radical-pushing groups such CH_2Cl are used in place of CH_3 group, the same reactions as above still remains favored, the point of scission remaining the same. Now consider replacing two hydrogen atoms located on two carbon centers in dimethylene oxide.

$$R:^{\ominus} \;+\; H-\overset{\overset{\displaystyle CH_3}{|}}{C}\underset{\diagdown \; \underset{O}{\diagup}}{-}\overset{\overset{\displaystyle CH_3}{|}}{C}-H \longrightarrow R:^{\ominus} \;+\; \oplus\overset{\overset{\displaystyle CH_3}{|}}{\underset{\underset{\displaystyle H}{|}}{C}}-\overset{\overset{\displaystyle H}{|}}{\underset{\underset{\displaystyle CH_3}{|}}{C}}-O^{\ominus} \longrightarrow$$

(Non-free anion)

$$RH \;+\; \overset{\overset{\displaystyle H}{|}}{\underset{\underset{\displaystyle H}{|}}{C}}=\overset{\overset{\displaystyle H}{|}}{\underset{\underset{\displaystyle H}{|}}{C}}-\overset{\overset{\displaystyle H}{|}}{\underset{\underset{\displaystyle CH_3}{|}}{C}}-O^{\ominus} \qquad OR \qquad R-\overset{\overset{\displaystyle CH_3}{|}}{\underset{\underset{\displaystyle H}{|}}{C}}-\overset{\overset{\displaystyle H}{|}}{\underset{\underset{\displaystyle CH_3}{|}}{C}}-O^{\ominus}$$

(I) FAVORED   (II) <u>Not favored</u>   12.39

$$R:^{\ominus} \;+\; H-\overset{\overset{\displaystyle C_2H_5}{|}}{C}\underset{\diagdown \; \underset{O}{\diagup}}{-}\overset{\overset{\displaystyle H}{|}}{C}-CH_3 \longrightarrow R:^{\ominus} \;+\; \oplus\overset{\overset{\displaystyle H}{|}}{\underset{\underset{\displaystyle C_2H_5}{|}}{C}}-\overset{\overset{\displaystyle CH_3}{|}}{\underset{\underset{\displaystyle H}{|}}{C}}-O^{\ominus} \longrightarrow$$

(Non-free anion)

$$RH \;+\; \overset{\overset{\displaystyle CH_3}{|}}{\underset{\underset{\displaystyle H}{|}}{C}}=\overset{\overset{\displaystyle H}{|}}{\underset{\underset{\displaystyle H}{|}}{C}}-\overset{\overset{\displaystyle H}{|}}{\underset{\underset{\displaystyle CH_3}{|}}{C}}-O^{\ominus} \qquad OR \qquad R-\overset{\overset{\displaystyle H}{|}}{\underset{\underset{\displaystyle C_2H_5}{|}}{C}}-\overset{\overset{\displaystyle CH_3}{|}}{\underset{\underset{\displaystyle H}{|}}{C}}-O^{\ominus}$$

(I) FAVORED   (II) <u>Not favored</u>   12.40

$$R^{\oplus}\cdots\cdots^{\ominus}\overset{\overset{\displaystyle F}{|}}{\underset{\underset{\displaystyle O}{|}}{\underset{\underset{\displaystyle R}{|}}{B}}}\overset{\diagup F}{\diagdown F} \quad H-\overset{\overset{\displaystyle CH_3}{|}}{C}\underset{\diagdown\underset{O}{\diagup}}{-}\overset{\overset{\displaystyle C_2H_5}{|}}{C}-CH_3 \longrightarrow R^{\oplus}\cdots\cdots^{\ominus}\overset{\overset{\displaystyle F}{|}}{\underset{\underset{\displaystyle O}{|}}{\underset{\underset{\displaystyle R}{|}}{B}}}\overset{\diagup F}{\diagdown F} \quad {}^{\ominus}O-\overset{\overset{\displaystyle H}{|}}{\underset{\underset{\displaystyle CH_3}{|}}{C}}-\overset{\overset{\displaystyle H}{|}}{\underset{\underset{\displaystyle C_2H_5}{|}}{C}}\oplus$$

(assumed strong)   (I)

$$R^{\oplus}\cdots\cdots^{\ominus}\overset{\overset{\displaystyle F}{|}}{\underset{\underset{\displaystyle O}{|}}{\underset{\underset{\displaystyle R}{|}}{B}}}\overset{\diagup F}{\diagdown F} \quad H-\overset{\overset{\displaystyle CH_3}{|}}{C}\underset{\diagdown\underset{O}{\diagup}}{-}\overset{\overset{\displaystyle C_2H_5}{|}}{C}-CH_3 \longrightarrow R^{\oplus}\cdots\cdots^{\ominus}\overset{\overset{\displaystyle F}{|}}{\underset{\underset{\displaystyle O}{|}}{\underset{\underset{\displaystyle R}{|}}{B}}}\overset{\diagup F}{\diagdown F} \quad {}^{\ominus}O-\overset{\overset{\displaystyle H}{|}}{\underset{\underset{\displaystyle CH_3}{|}}{C}}-\overset{\overset{\displaystyle H}{|}}{\underset{\underset{\displaystyle C_2H_5}{|}}{C}}\oplus$$

(assumed strong)   (I)

$$R - O - \overset{\overset{\displaystyle H}{|}}{\underset{\underset{\displaystyle CH_3}{|}}{C}} - \overset{\overset{\displaystyle H}{|}}{\underset{\underset{\displaystyle C_2H_5}{|}}{C}}^{\oplus} \quad \overset{\ominus}{B}\overset{\overset{\displaystyle F}{|}}{\underset{\underset{\displaystyle O}{|}}{}}\diagup^{F}_{\diagdown F} \qquad \xrightarrow{n(I)}$$

$$R \left\{ O - \overset{\overset{\displaystyle H}{|}}{\underset{\underset{\displaystyle CH_3}{|}}{C}} - \overset{\overset{\displaystyle H}{|}}{\underset{\underset{\displaystyle C_2H_5}{|}}{C}} \right\}_n O - \overset{\overset{\displaystyle CH_3}{|}}{\underset{\underset{\displaystyle H}{|}}{C}} - \overset{\overset{\displaystyle C_2H_5}{|}}{\underset{\underset{\displaystyle H}{|}}{C}}^{\oplus} \quad \overset{\ominus}{B}\overset{\overset{\displaystyle F}{|}}{\underset{\underset{\displaystyle O}{|}}{}}\diagup^{F}_{\diagdown F} \qquad \xrightarrow{}$$

$$R \left\{ O - \overset{\overset{\displaystyle H}{|}}{\underset{\underset{\displaystyle CH_3}{|}}{C}} - \overset{\overset{\displaystyle H}{|}}{\underset{\underset{\displaystyle C_2H_5}{|}}{C}} \right\}_n O - \overset{\overset{\displaystyle CH_3}{|}}{\underset{\underset{\displaystyle H}{|}}{C}} - \overset{\overset{\displaystyle H}{|}}{\underset{\underset{\displaystyle H}{|}}{C}} = \overset{\overset{\displaystyle H}{|}}{\underset{\underset{\displaystyle CH_3}{|}}{C}} + HOR + BF_3$$

(II)

Transfer of species of first kind of 1st type (Favored)   12.41

When (II) is activated, the positively charged route is favored by it. For the growing polymer chain, transfer species of the first kind of the first type is released to give a erminal double bond. The transfer species released is the same that prevented it being polymerized anionically. Radically, the situation is not different. Electro-free-radically, polymerization is favored. Nucleo-non-free-radically, polymerization is not possible.

$$N \bullet nn \quad + \quad H - \overset{\overset{\displaystyle CH_3}{|}}{\underset{\diagdown}{C}} - \overset{\overset{\displaystyle C_2H_5}{|}}{\underset{\diagup}{C}} - CH \longrightarrow nn \bullet O - \overset{\overset{\displaystyle H}{|}}{\underset{\underset{\displaystyle CH_3}{|}}{C}} - \overset{\overset{\displaystyle C_2H_5}{|}}{\underset{\underset{\displaystyle H}{|}}{C}} \bullet e \quad + \quad nn \bullet N$$

$$\underset{O}{}$$

(More nucleophilic)

$$\xrightarrow{} nn \bullet O - \overset{\overset{\displaystyle H}{|}}{\underset{\underset{\displaystyle CH_3}{|}}{C}} - \overset{\overset{\displaystyle H}{|}}{\underset{}{C}} = \overset{\overset{\displaystyle H}{|}}{\underset{\underset{\displaystyle CH_3}{|}}{C}} + NH \quad \begin{array}{l} \text{Transfer species of first kind} \\ \text{of first type} \end{array} \qquad 12.42$$

$$R^{\oplus} \overset{\ominus}{B}\overset{\overset{\displaystyle F}{|}}{\underset{\underset{\displaystyle R}{\underset{|}{O}}}{}}\diagup^{F}_{\diagdown F} \quad + \quad H - \overset{\overset{\displaystyle CH_2Cl}{|}}{\underset{\diagdown}{C}} - \overset{\overset{\displaystyle CH_3}{|}}{\underset{\diagup}{C}} - H \longrightarrow R - O^{\oplus} \overset{\ominus}{B}\overset{\overset{\displaystyle F}{|}}{\underset{\underset{\displaystyle R}{\underset{|}{O}}}{}}\diagup^{F}_{\diagdown F} \xrightarrow{}$$

$$\underset{O}{}$$

(Nucleophile)

$$HC - CH$$
$$\overset{\overset{\displaystyle }{}}{\underset{\underset{\displaystyle CH_2Cl \quad CH_3}{}}{}}$$

$$R - O - \overset{\overset{\displaystyle CH_2Cl}{|}}{\underset{\underset{\displaystyle H}{|}}{C}} - \overset{\overset{\displaystyle H}{|}}{\underset{\underset{\displaystyle CH_3}{|}}{C}}^{\oplus} \quad \overset{\ominus}{B}\overset{\overset{\displaystyle F}{|}}{\underset{\underset{\displaystyle R}{\underset{|}{O}}}{}}\diagup^{F}_{\diagdown F} \xrightarrow{} \begin{array}{l} \text{Transfer species of first kind} \\ \text{of first type} \end{array}$$

12.43

20

The point of scission has nothing to do with the type of initiators involved. Point of scission is determined by the type of substituted groups present in the ring. Nevertheless, the strong nucleophilic character of the last monomer can be observed. In view of the presence of a functional center, in the ring, unlike cycloalkanes, these monomers are therefore more nucleophilic in character. Consider reducing the nucleophilicity by replacing two differently located H atoms with radical-pushing groups of lower capacity than H.

$$R{:}^{\ominus} \quad + \quad \underset{\underset{\displaystyle O}{\diagdown\diagup}}{\overset{\overset{\displaystyle CF_3 \quad CF_3}{|\qquad|}}{HC - CH}} \quad\longrightarrow\quad R{:}^{\ominus} \quad + \quad {}^{\ominus}O - \overset{\overset{\displaystyle H}{|}}{\underset{\underset{\displaystyle CF_3}{|}}{C}} - \overset{\overset{\displaystyle H}{|}}{\underset{\underset{\displaystyle CF_3}{|}}{C}}{}^{\oplus} \quad\longrightarrow$$

(Non-free anion)

(I) (assumed strained)

$$R - \overset{\overset{\displaystyle H}{|}}{\underset{\underset{\displaystyle CF_3}{|}}{C}} - \overset{\overset{\displaystyle CF_3}{|}}{\underset{\underset{\displaystyle H}{|}}{C}} - O^{\ominus} \quad \xrightarrow{+\,n(I)} \quad R - \overset{\overset{\displaystyle H}{|}}{\underset{\underset{\displaystyle CF_3}{|}}{C}} - \overset{\overset{\displaystyle CF_3}{|}}{\underset{\underset{\displaystyle H}{|}}{C}} - O - \overset{\overset{\displaystyle H}{|}}{\underset{\underset{\displaystyle CF_3}{|}}{C}} - \overset{\overset{\displaystyle CF_3}{|}}{\underset{\underset{\displaystyle H}{|}}{C}} - O^{\ominus} \quad\longrightarrow$$

Living Polymers   12.44

$$R^{\oplus} \quad {}^{\ominus}\underset{\underset{\displaystyle R}{\underset{\displaystyle |}{\overset{\displaystyle |}{O}}}}{\overset{\diagup F}{\underset{\diagdown F}{B}}}\!\!\diagup\!\!F \quad + \quad {}^{\ominus}O - \overset{\overset{\displaystyle H}{|}}{\underset{\underset{\displaystyle CF_3}{|}}{C}} - \overset{\overset{\displaystyle CF_3}{|}}{\underset{\underset{\displaystyle H}{|}}{C}}{}^{\oplus} \quad\longrightarrow\quad R - O - \overset{\overset{\displaystyle H}{|}}{\underset{\underset{\displaystyle CF_3}{|}}{C}} - \overset{\overset{\displaystyle CF_3}{|}}{\underset{\underset{\displaystyle H}{|}}{C}}{}^{\oplus}\cdots$$

(I)

$$\xrightarrow{+\,n(I)} \quad R \left\{ O - \overset{\overset{\displaystyle H}{|}}{\underset{\underset{\displaystyle CF_3}{|}}{C}} - \overset{\overset{\displaystyle CF_3}{|}}{\underset{\underset{\displaystyle H}{|}}{C}} \right\}_n O - \overset{\overset{\displaystyle H}{|}}{\underset{\underset{\displaystyle CF_3}{|}}{C}} - \overset{\overset{\displaystyle CF_3}{|}}{\underset{\underset{\displaystyle H}{|}}{C}}\cdots{}^{\ominus}\underset{\underset{\displaystyle R}{\underset{\displaystyle |}{\overset{\displaystyle |}{O}}}}{\overset{\diagup F}{\underset{\diagdown F}{B}}}\!\!\diagup\!\!F \quad\longrightarrow\quad \text{Living polymer}$$

12.45

Nevertheless, in general, the following prevails, since C = O is more nucleophilic than C = C.

$$\text{Cyclic ethers} \quad > \quad \text{Cyclic alkanes}$$

<u>Order of nucleophilic</u>   12.46

With larger membered rings, functional oxygen center is always involved, since the SE in them are small.

However, assuming that the four-membered rings can be provided with the minimum required strain energy, then one should expect the followings.

$$R{:}^{\ominus} \quad + \quad \underset{\underset{\displaystyle O - CH_2}{|\qquad\quad|}}{\overset{\overset{\displaystyle CH_3 \quad CH_3}{|\qquad|}}{H - C - C - H}} \quad\longrightarrow\quad R{:}^{\ominus} \quad + \quad {}^{\ominus}O - \overset{\overset{\displaystyle H}{|}}{\underset{\underset{\displaystyle H}{|}}{C}} - \overset{\overset{\displaystyle H}{|}}{\underset{\underset{\displaystyle CH_3}{|}}{C}} - \overset{\overset{\displaystyle H}{|}}{\underset{\underset{\displaystyle CH_3}{|}}{C}}{}^{\oplus}$$

$$\xrightarrow{} \quad \overset{\overset{\displaystyle H}{|}}{\underset{\underset{\displaystyle H}{|}}{C}} = \overset{\overset{\displaystyle H}{|}}{\underset{\underset{\displaystyle H}{|}}{C}} - \overset{\overset{\displaystyle H}{|}}{\underset{\underset{\displaystyle CH_3}{|}}{C}} - \overset{\overset{\displaystyle H}{|}}{\underset{\underset{\displaystyle H}{|}}{C}} - O^{\ominus} \quad + \quad RH$$

12.47

21

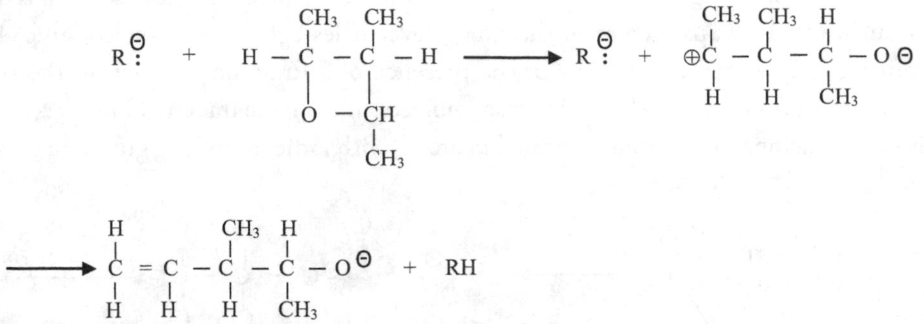

$$R\overset{\ominus}{:} \;+\; H-\underset{\underset{\underset{\underset{CH_3}{|}}{CH}}{\overset{\overset{CH_3}{|}}{C}}}{}-\overset{\overset{CH_3}{|}}{\underset{|}{C}}-H \longrightarrow R\overset{\ominus}{:} \;+\; \overset{\oplus}{C}\underset{\underset{H}{|}}{\overset{\overset{CH_3}{|}}{}}-\overset{\overset{CH_3}{|}}{\underset{\underset{H}{|}}{C}}-\overset{\overset{H}{|}}{\underset{\underset{CH_3}{|}}{C}}-O\ominus$$

$$\longrightarrow \quad \overset{\overset{H}{|}}{\underset{\underset{H}{|}}{C}}=\overset{}{\underset{\underset{H}{|}}{C}}-\overset{\overset{CH_3}{|}}{\underset{\underset{H}{|}}{C}}-\overset{\overset{H}{|}}{\underset{\underset{CH_3}{|}}{C}}-O\ominus \;+\; RH \qquad\qquad 12.48$$

When all the H atoms are replaced with radical-pushing substituent groups, the anionic route ceases to be favored, clear indication of the strong nucleophilic character of the monomer.

On the other hand, when the CH_3 groups are replaced with CF_3 groups, both routes are favored. Nevertheless, cyclic ethers are far more nucleophilic than alkenes. As an extension, the followings are valid.

$$\text{Cyclic ethers} \quad > \quad \text{Cyclic alkanes} \quad > \quad \text{Olefins}$$
$$\qquad\qquad 12.49$$

<u>Order of nucleophilicity</u>

$$\underset{\text{Olefins}}{\overset{\text{Electrophilic}}{}} \quad > \quad \underset{\text{Cyclic alkanes}}{\overset{\text{Electrophilic}}{}} \quad > \quad \underset{\text{Cyclic ethers}}{\overset{\text{Electrophilic}}{}}$$

<u>Order of electrophilicity</u> $\qquad\qquad 12.50$

This is the order when groups such as COOR, COR, CN, replaces one H atom. These general orders are worthy of note. That transfer species of the second first kind do not exist with these monomers are already obvious. Consider the case with etheric radical-pushing groups.

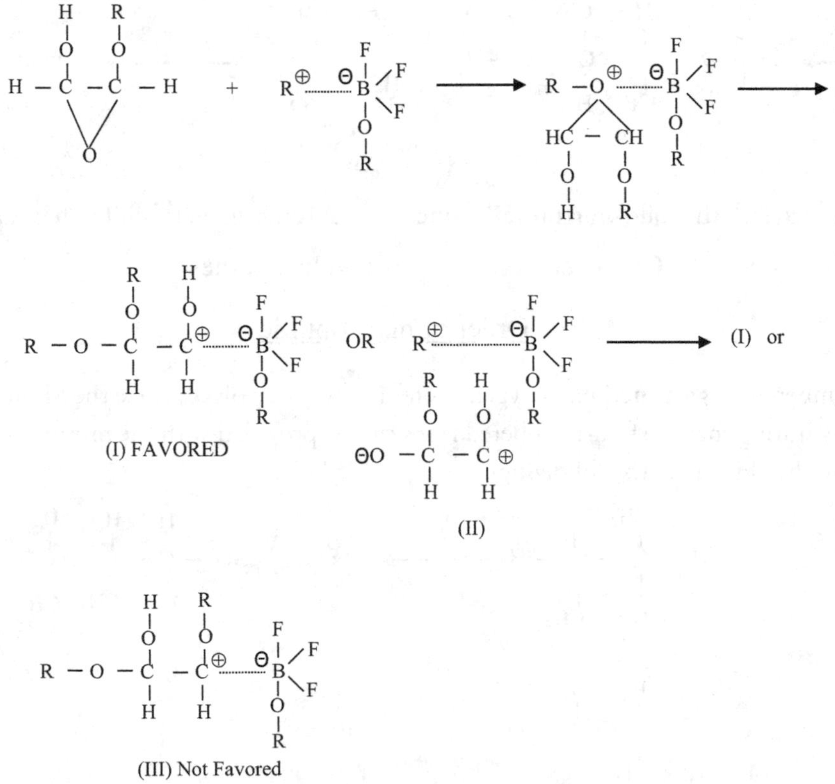

(I) FAVORED

(II)

(III) Not Favored $\qquad\qquad 12.51$

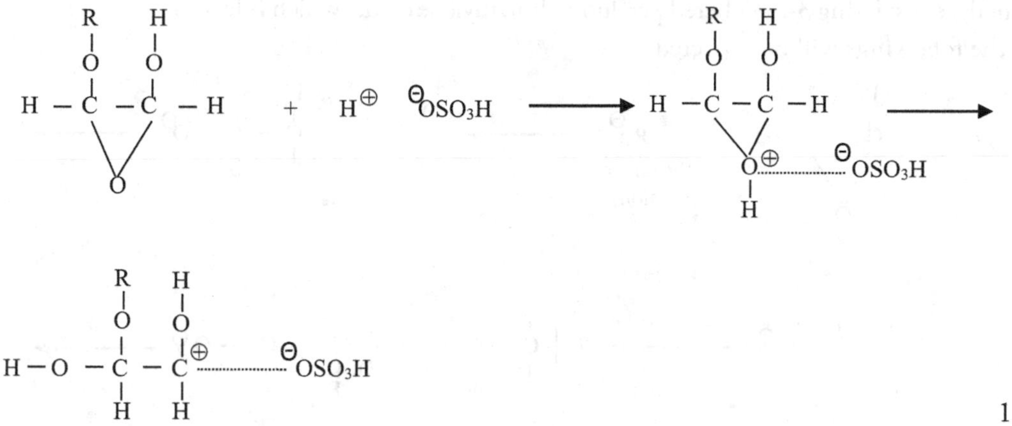

The structural formulas and reactions 12.52 are shown below:

$$
\underset{\text{(12.52)}}{
\begin{array}{c}
\text{H}-\overset{\overset{\text{R}}{|}\;\;\overset{\text{H}}{|}}{\underset{\underset{\text{O}}{\diagdown\;\;\diagup}}{\text{C}-\text{C}}}-\text{H}\;\;+\;\;\text{H}^{\oplus}\;{}^{\ominus}\text{OSO}_3\text{H}\;\longrightarrow
\end{array}
}
$$

The monomer being a strong nucleophile, will favor instantaneous opening of the ring when moderately strong cationic or anionic initiators are involved. Cationically, no polymerization will take place; but a stable product can be obtained. Since OH is more radical-pushing than OR, it is (I) or (II) that is favored and not (III), based on how the groups are placed.

With four membered ring for a similar type of nucleophilic monomer shown below, the ring can be opened instantaneously, because of the strong radical-pushing capacity of the groups carried.

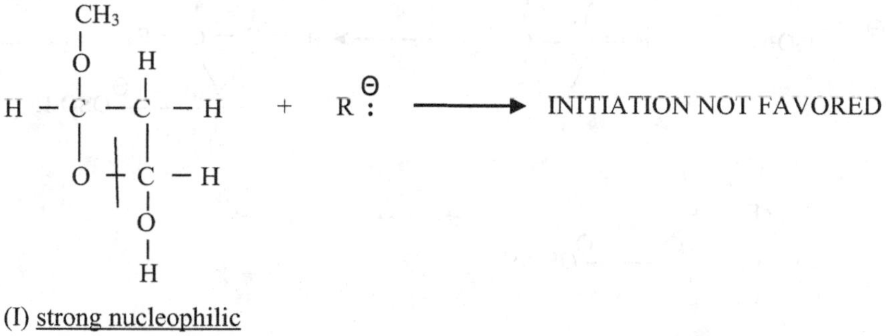

(I) <u>strong nucleophilic</u>

$$\text{(I)}\;+\;\text{H}^{\oplus}\;{}^{\ominus}\text{OSO}_3\text{H}\longrightarrow$$

12.53

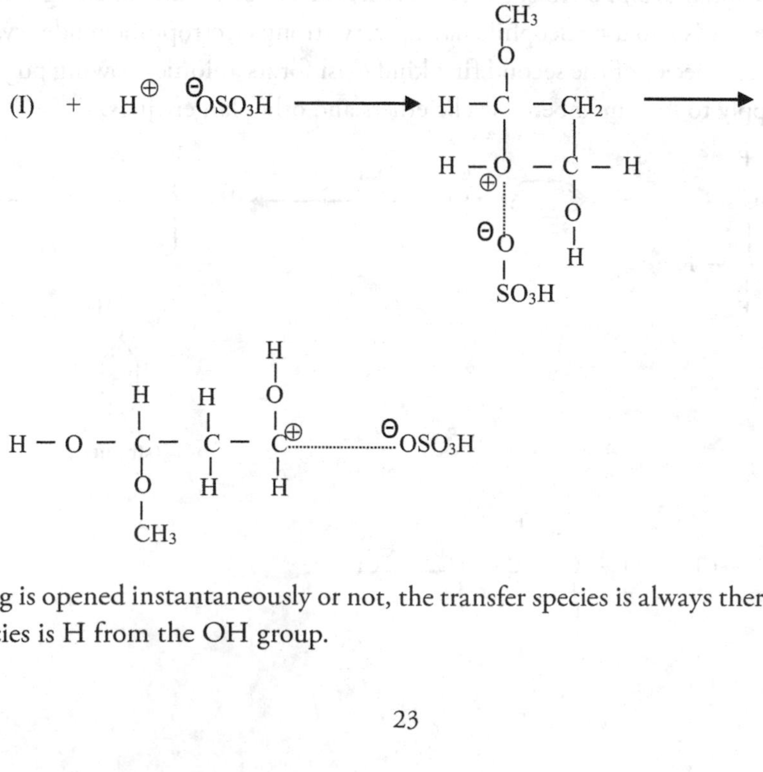

12.54

Whether the ring is opened instantaneously or not, the transfer species is always there. For the case above the transfer species is H from the OH group.

Finally, considering 3-membered perfluoro dimethylene oxide, which is less strained than dimethylane oxide, the followings will be expected.

$$F - \overset{\overset{\displaystyle F}{|}}{\underset{\diagdown}{C}} - \overset{\overset{\displaystyle F}{|}}{\underset{\diagup}{C}} - F \quad + \quad R: \overset{\ominus}{\underset{\text{(non-free anion)}}{}} \longrightarrow R: \overset{\ominus}{} \quad + \quad \oplus \overset{\overset{\displaystyle F}{|}}{\underset{\underset{\displaystyle F}{|}}{C}} - \overset{\overset{\displaystyle F}{|}}{\underset{\underset{\displaystyle F}{|}}{C}} - O \ominus \longrightarrow$$

(I)

$$R - \overset{\overset{\displaystyle F}{|}}{\underset{\underset{\displaystyle F}{|}}{C}} - \overset{\overset{\displaystyle F}{|}}{\underset{\underset{\displaystyle F}{|}}{C}} - O \ominus \xrightarrow{+ \; n(I)} R \left\{ \overset{\overset{\displaystyle F}{|}}{\underset{\underset{\displaystyle F}{|}}{C}} - \overset{\overset{\displaystyle F}{|}}{\underset{\underset{\displaystyle F}{|}}{C}} - O \right\}_n \overset{\overset{\displaystyle F}{|}}{\underset{\underset{\displaystyle F}{|}}{C}} - \overset{\overset{\displaystyle F}{|}}{\underset{\underset{\displaystyle F}{|}}{C}} - O \ominus \longrightarrow$$

$$R \left\{ \overset{\overset{\displaystyle F}{|}}{\underset{\underset{\displaystyle F}{|}}{C}} - \overset{\overset{\displaystyle F}{|}}{\underset{\underset{\displaystyle F}{|}}{C}} - O \right\}_n \overset{\overset{\displaystyle F}{|}}{\underset{\underset{\displaystyle F}{|}}{C}} - \overset{\overset{\displaystyle F}{|}}{C} = O \quad + \quad F: \ominus$$

<u>Transfer species of 2nd first kind</u>   12.55

$$H \overset{\oplus}{} \; \ominus OSO_3H \quad + \quad F - \overset{\overset{\displaystyle F}{|}}{\underset{\diagdown}{C}} - \overset{\overset{\displaystyle F}{|}}{\underset{\diagup}{C}} - F \longrightarrow F - \overset{\overset{\displaystyle F}{|}}{\underset{\underset{\overset{\displaystyle O\oplus}{|}}{}}{C}} - \overset{\overset{\displaystyle F}{|}}{C} - F \longrightarrow$$

$$O \overset{\oplus}{\cdots\cdots} \ominus OSO_3H$$
$$\underset{H}{}$$

$$H - O - \overset{\overset{\displaystyle F}{|}}{\underset{\underset{\displaystyle F}{|}}{C}} - \overset{\overset{\displaystyle F}{|}}{\underset{\underset{\displaystyle F}{|}}{C}} \overset{\oplus}{\cdots\cdots} \ominus OSO_3H$$

12.56

Unlike perfluoro-cycloalkanes and olefins, the cyclic ethers favor both routes of charged polymerizations, noting that the monomer is still a nucleophile having very strong electrophilic tendency. It is largely for that reason that transfer species of the second first kind exist for its anionic growing polymer chain. The same will similarly apply to four-membered cyclic ethers and other larger rings.

$$F - \overset{\overset{\displaystyle F}{|}}{\underset{\underset{\overset{\displaystyle O - \overset{\displaystyle F}{|}}{C} - F}{|}}{C}} - \overset{\overset{\displaystyle F}{|}}{\underset{}{C}} - F \quad + \quad \langle \bigcirc \rangle - \overset{e}{\underset{}{CH_2}} \; \overset{.nn}{Cl} \longrightarrow$$

(strained)

$$\langle \bigcirc \rangle - CH_2 - O - \overset{\overset{\displaystyle F}{|}}{\underset{\underset{\displaystyle F}{|}}{C}} - \overset{\overset{\displaystyle F}{|}}{\underset{\underset{\displaystyle F}{|}}{C}} - \overset{\overset{\displaystyle F}{|}}{\underset{\underset{\displaystyle F}{|}}{C}} \overset{\oplus}{\cdots\cdots} \ominus Cl$$

12.57

$$R \left\{ \begin{array}{c} F \\ | \\ C \\ | \\ F \end{array} - \begin{array}{c} F \\ | \\ C \\ | \\ F \end{array} - \begin{array}{c} F \\ | \\ C \\ | \\ F \end{array} - O \right\}_n \begin{array}{c} F \\ | \\ C \\ | \\ F \end{array} - \begin{array}{c} F \\ | \\ C \\ | \\ F \end{array} - \begin{array}{c} F \\ | \\ C \\ | \\ F \end{array} - O^{\ominus} \longrightarrow$$

$$R \left\{ \begin{array}{c} F \\ | \\ C \\ | \\ F \end{array} - \begin{array}{c} F \\ | \\ C \\ | \\ F \end{array} - \begin{array}{c} F \\ | \\ C \\ | \\ F \end{array} - O \right\}_n \begin{array}{c} F \\ | \\ C \\ | \\ F \end{array} - \begin{array}{c} F \\ | \\ C \\ | \\ F \end{array} - \begin{array}{c} F \\ | \\ C \\ | \\ F \end{array} = O \quad + \quad F^{\ominus}$$

<u>Transfer species of 2nd first kind</u>   12.58

With five membered-rings, the instantaneous opening of its ring may not be readily favored, for which after the fourth or fifth membered-rings, instantaneous opening may not exist, unless radical-pushing groups are placed on the ring.

Copolymerization between members in this family is limited mostly to positively charged or electro-free-radical routes. Only the three membered and some selected four-membered ringed ethers will favor anionic copolymerization via instantaneous opening of the ring. For example consider copolymerization between epichlorohydrin and 3,3- Bis(chloromethyl) oxetane anionically.

$$R \overset{\ominus}{:} \quad + \quad H - \underset{\underset{O}{\diagdown\diagup}}{\overset{H}{\underset{|}{C}}} - \overset{CH_2Cl}{\underset{|}{CH}} \quad + \quad H - \overset{H}{\underset{|}{C}} - \underset{\underset{O - CH_2}{|}}{\overset{CH_2Cl}{\underset{|}{C}}} - CH_2Cl \quad \longrightarrow \quad R \overset{\ominus}{:} \quad +$$

(I)   (II)

Note the nomenclature for the second epoxide above when compared to that of Equation 12.1. The numbering does not begin with O in Equation 12.1, like the others.

$$\overset{\oplus}{\underset{\underset{H}{|}}{\overset{\overset{H}{|}}{C}}} - \underset{\underset{H}{|}}{\overset{\overset{ClCH_2}{|}}{C}} - O^{\ominus} \quad + \quad (II) \quad \longrightarrow \quad \overset{\oplus}{\underset{\underset{H}{|}}{\overset{\overset{H}{|}}{C}}} - \underset{\underset{H}{|}}{\overset{\overset{ClCH_2}{|}}{C}} - O^{\ominus} \quad + \quad \overset{\oplus}{\underset{\underset{H}{|}}{\overset{\overset{H}{|}}{C}}} - \underset{\underset{CH_2Cl}{|}}{\overset{\overset{CH_2Cl}{|}}{C}} - \underset{\underset{H}{|}}{\overset{\overset{H}{|}}{C}} - O^{\ominus} \quad + \quad R \overset{\ominus}{:}$$

(A)   (A) Reactive   (B) Reactive

$$\overset{\oplus}{} \quad \overset{H}{\underset{\underset{H}{|}}{C}} - \underset{\underset{H}{|}}{\overset{\overset{CH_2Cl}{|}}{C}} - O - \overset{H}{\underset{\underset{H}{|}}{C}} - \underset{\underset{CH_2Cl}{|}}{\overset{\overset{CH_2Cl}{|}}{C}} - \overset{H}{\underset{\underset{H}{|}}{C}} - O^{\ominus} \quad + \quad R \overset{\ominus}{:} \quad \longrightarrow$$

(C)   A COUPLE

Couple can be initiated (No transfer species of the first kind)   12.59

Cationically or positively, both monomers are reactive. Therefore alternating placement can never be obtained. Anionically, both are also reactive. (II) being more nucleophilic than (I) diffuses to (I) to form a couple which is also reactive to the initiator. Hence alternating copolymers can never be obtained; but copolymerization will take place. Radically the situation is different, since only (II) can be homo-polymerized nucleo-non-free-radically and electro-free-radically the route natural to it. The activated state of (I) chargedly is worthy of note (A), since it is different from the radical state. If (I) had been more nucleophilic than (II), then radically some alternating placements would have been obtained along with random placement of (II) only. The two monomers above are nucleophiles, for which their natural route is positive or electro-free-radicals. Since the three-membered ring is nucleophilic and more strained, it is the first to be unzipped.

Though, it is not intended to start considering copolymerization reactions, the need here arises in order to confirm the observations which have been made over the years, but could not adequately be explained.

"Cationically", monomers which are stubborn to homo-polymerization due to inability to attain the minimum required strain energy for the ring, can be copolymerized. Such stubborn monomers include the 1,3-, 1,4- dioxanes, tetrahydropyran, etc. These can be copolymerized with for example 3,3-Bis(chloromethyl) oxetane using $BF_3/(C_2H_5)_2O$ at temperature as low as 0°C. A strong active positive center is first generated with the oxetane before the stubborn monomers are introduced or initiated into the chain.

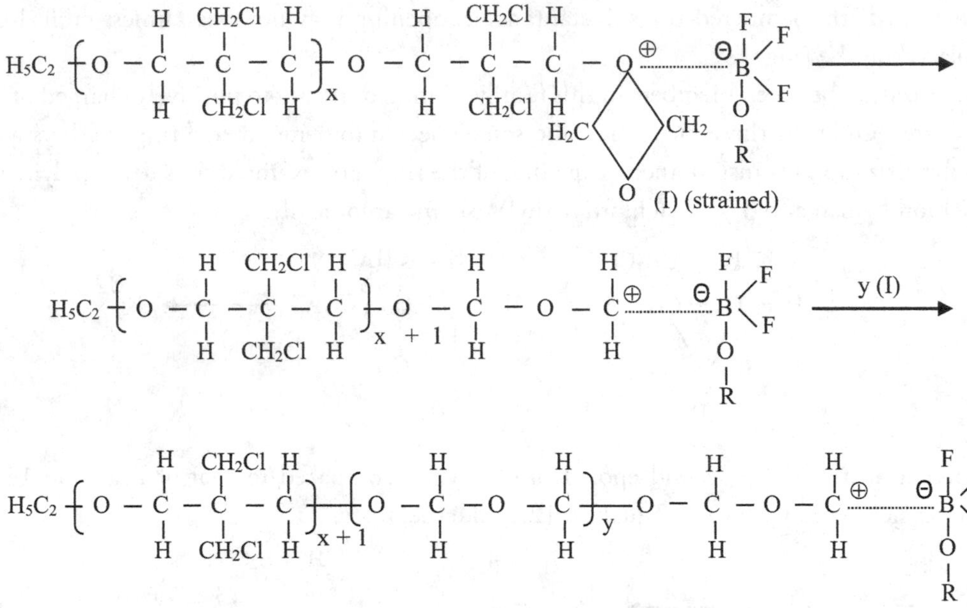

$$\text{Living Block Copolymers} \qquad 12.60$$

Without the strong active center, the 1,3 - dioxane cannot be activated. The growing polymer chain's active center is now able to provide the minimum required strain energy into the four-membered ring to open it. Block copolymers can be produced if 1,3 - dioxane is the only monomer later added to the system. Even then continued existence of 1,3 - dioxane along the chain is difficult as will be shown.

The growing polymer chain of oxetane which initiated the dioxane can only be said to be a promoter if the oxetane is used only in very small amounts relative to the cyclic dioxane, for which the product cannot be said to be a copolymer. Major aspects of the copolymerization of these monomers will be treated at the appropriate time. However, important aspects of propagation/depropagation phenomenon with respect to homo and copolymerization will be considered in this chapter.

12.1.2 Radical Characters of Cyclic Ethers

Little or nothing is known about the radical homopolymerizations of cyclic ethers. Based on the New Frontiers, these are monomers which can be radically polymerized when suitable conditions are chosen. Only non-free-radicals can activate some of the monomer instantaneously in view of the presence of paired unbonded radicals on the center. Therefore, considering the use of nucleo-non-free-radicals, the followings are to be expected only for three-membered cyclic ethers.

$$\ddot{N}.nn \quad + \quad H-\underset{\underset{\displaystyle O}{|}}{\overset{\overset{\displaystyle H}{|}}{C}}-\underset{}{\overset{\overset{\displaystyle CF_3}{|}}{C}}-H \longrightarrow \ddot{N}.nn \quad + \quad e.\underset{\underset{\displaystyle H}{|}}{\overset{\overset{\displaystyle H}{|}}{C}}-\underset{\underset{\displaystyle H}{|}}{\overset{\overset{\displaystyle CF_3}{|}}{C}}-O.nn$$

(I) (strained)   (II)

$$\longrightarrow \ddot{N}-\underset{\underset{\displaystyle H}{|}}{\overset{\overset{\displaystyle H}{|}}{C}}-\underset{\underset{\displaystyle H}{|}}{\overset{\overset{\displaystyle CF_3}{|}}{C}}-O.nn \quad \xrightarrow{+\ n(I)} \quad \ddot{N}\left(\underset{\underset{\displaystyle H}{|}}{\overset{\overset{\displaystyle H}{|}}{C}}-\underset{\underset{\displaystyle H}{|}}{\overset{\overset{\displaystyle CF_3}{|}}{C}}-O\right)_n \underset{\underset{\displaystyle H}{|}}{\overset{\overset{\displaystyle H}{|}}{C}}-\underset{\underset{\displaystyle H}{|}}{\overset{\overset{\displaystyle CF_3}{|}}{C}}-O.nn$$

No transfer species   12.61

Though the activated monomer, (II), carries an electro-free-radical center, the use of electro-free-radicals alone to unzip the ring is not favored since they do not possess the electrostatic forces. In general nucleo-non-free-radical centers are strong enough to open many of the three-membered rings. Strong nucleo-free-radical active centers can be obtained by using promoters, that is, from monomers which can be polymerized nucleo-non-free-radically as their natural route. Strong electro-free-radical centers can more readily be obtained using promoters. If CF_3 used above had been CH_2Cl, initiation will not be favored.

Consider copolymerizing the monomer above with CH_2Cl, with carbon monoxide and sulfur dioxide. The followings are to be expected.

$$\ddot{N}.nn \quad + \quad e.\underset{\underset{\displaystyle H}{|}}{\overset{\overset{\displaystyle ClCH_2}{|}}{C}}-\underset{\underset{\displaystyle H}{|}}{\overset{\overset{\displaystyle H}{|}}{C}}-O.nn \quad + \quad e.\overset{\overset{\displaystyle :O:}{\|}}{C}.n \longrightarrow \ddot{N}.nn \quad +$$

(I)   (II) (Unreactive)   (III) (Unreactive)

$$e.\overset{\overset{\displaystyle O}{\|}}{C}-\underset{\underset{\displaystyle H}{|}}{\overset{\overset{\displaystyle ClCH_2}{|}}{C}}-\underset{\underset{\displaystyle H}{|}}{\overset{\overset{\displaystyle H}{|}}{C}}-O.nn \qquad OR \qquad e\bullet\overset{\overset{\displaystyle ClCH_2}{|}}{\underset{\underset{\displaystyle H}{|}}{C}}-\overset{\overset{\displaystyle H}{|}}{\underset{\underset{\displaystyle H}{|}}{C}}-O-\overset{\overset{\displaystyle O}{\|}}{C}\bullet n \qquad \xrightarrow{\text{For (IV)}}$$

(IV)   (V)

$$\ddot{N}-\overset{\overset{\displaystyle O}{\|}}{C}-\underset{\underset{\displaystyle H}{|}}{\overset{\overset{\displaystyle ClCH_2}{|}}{C}}-\underset{\underset{\displaystyle H}{|}}{\overset{\overset{\displaystyle H}{|}}{C}}-O\bullet nn$$

(VI) Balanced   12.62

The ring is either opened by (I) or (III), both carrying paired unbonded radicals on their active centers or directly next to the active centers. (IV) is fully favored, that is, 1 to 1 or alternating placement only along the chain, only if (II) is more Nucleophilic than (III). If

$$\ddot{N}.nn \quad + \quad e.\underset{\underset{\displaystyle H}{|}}{\overset{\overset{\displaystyle ClCH_2}{|}}{C}}-\underset{\underset{\displaystyle H}{|}}{\overset{\overset{\displaystyle H}{|}}{C}}-O.nn \quad + \quad en.\overset{\overset{\displaystyle O\ominus}{|}}{\underset{}{S}}\overset{\oplus}{-}O.nn \longrightarrow \ddot{N}.nn \quad +$$

(I) (Unreactive)   (II) (Reactive)

otherwise, then (V) is obtained. (V) is unreactive to the initiator. Hence if copolymers can be obtained from the two monomers, this can only be done nucleo-non-free-radically, for which only alternating placements will be obtained, but less for the case shown below.

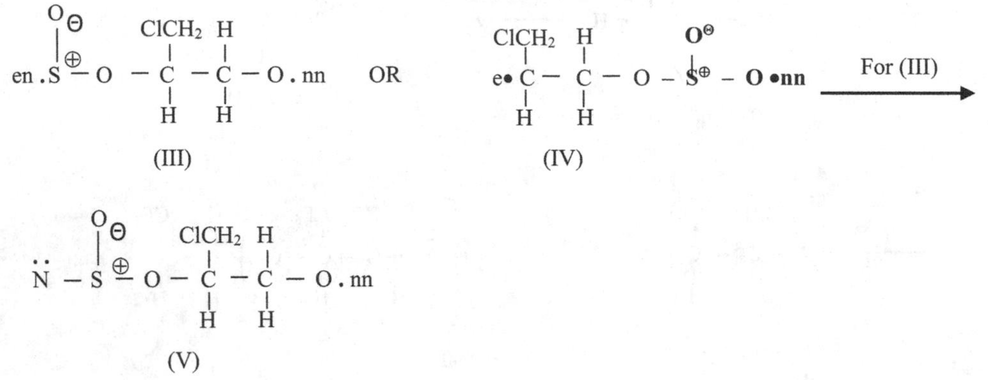

$$
\text{en} . \overset{\overset{\displaystyle O^{\ominus}}{|}}{\underset{}{S^{\oplus}}} = O - \overset{\overset{\displaystyle ClCH_2}{|}}{\underset{\underset{\displaystyle H}{|}}{C}} - \overset{\overset{\displaystyle H}{|}}{\underset{\underset{\displaystyle H}{|}}{C}} - O . nn \qquad \text{OR} \qquad e\bullet \overset{\overset{\displaystyle ClCH_2}{|}}{\underset{\underset{\displaystyle H}{|}}{C}} - \overset{\overset{\displaystyle H}{|}}{\underset{\underset{\displaystyle H}{|}}{C}} - O - \overset{\overset{\displaystyle O^{\ominus}}{|}}{\underset{}{S^{\oplus}}} - O \bullet nn \qquad \xrightarrow{\text{For (III)}}
$$

(III) (IV)

$$
\overset{\displaystyle \cdot\cdot}{N} - \overset{\overset{\displaystyle O^{\ominus}}{|}}{\underset{}{S^{\oplus}}} = O - \overset{\overset{\displaystyle ClCH_2}{|}}{\underset{\underset{\displaystyle H}{|}}{C}} - \overset{\overset{\displaystyle H}{|}}{\underset{\underset{\displaystyle H}{|}}{C}} - O . nn
$$

(V)   12.63

If (I) is more nucleophilic than (II), then the couple (III) is formed, and if otherwise, (IV) is formed. (IV) is unreactive to the initiator, due to presence of transfer species of the first kind. If copolymer can be obtained from them, then (III) is favored. The presence of (III) in alternating manner alone is not favored since (II) is reactive. Therefore, (II) will be randomly appear along the alternating chain.

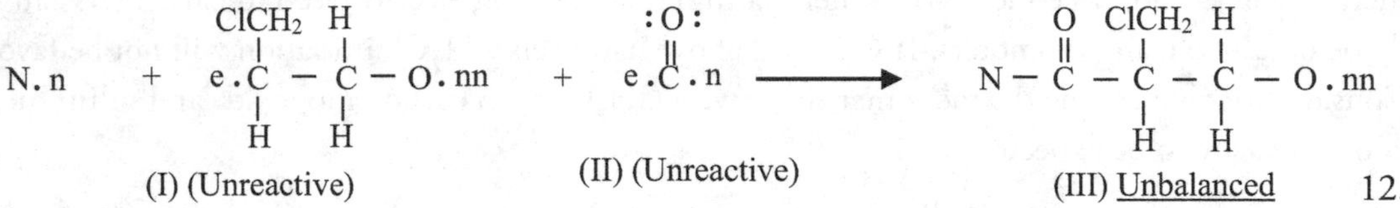

$$
N.n \quad + \quad e.\overset{\overset{\displaystyle ClCH_2}{|}}{\underset{\underset{\displaystyle H}{|}}{C}} - \overset{\overset{\displaystyle H}{|}}{\underset{\underset{\displaystyle H}{|}}{C}} - O.nn \quad + \quad e.\overset{\overset{\displaystyle :O:}{||}}{C}.n \quad \longrightarrow \quad N - \overset{\overset{\displaystyle O}{||}}{C} - \overset{\overset{\displaystyle ClCH_2}{|}}{\underset{\underset{\displaystyle H}{|}}{C}} - \overset{\overset{\displaystyle H}{|}}{\underset{\underset{\displaystyle H}{|}}{C}} - O.nn
$$

(I) (Unreactive)   (II) (Unreactive)   (III) <u>Unbalanced</u>   12.64

(II) cannot be homo-polymerized nucleo- or electro-free- radically due to electrostatic forces of repulsion. It can only add to (I) to form a couple which cannot be initiated nucleo-free-radically, because the equation will not be radically balanced. Hence, no polymerization can be favored nucleo-free-radically.

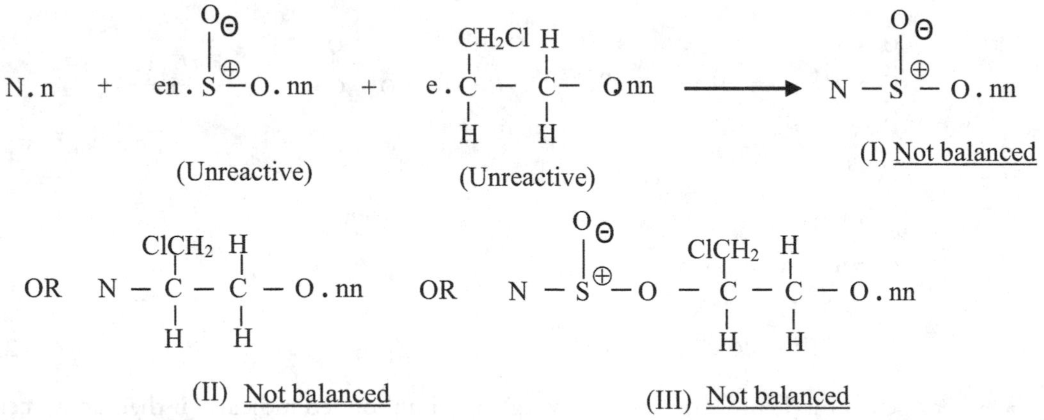

$$
N.n \quad + \quad \text{en}.\overset{\overset{\displaystyle O^{\ominus}}{|}}{\underset{}{S^{\oplus}}} = O.nn \quad + \quad e.\overset{\overset{\displaystyle CH_2Cl}{|}}{\underset{\underset{\displaystyle H}{|}}{C}} - \overset{\overset{\displaystyle H}{|}}{\underset{\underset{\displaystyle H}{|}}{C}} - O\,nn \quad \longrightarrow \quad N - \overset{\overset{\displaystyle O^{\ominus}}{|}}{\underset{}{S^{\oplus}}} = O.nn
$$

(Unreactive)   (Unreactive)   (I) <u>Not balanced</u>

$$
\text{OR} \quad N - \overset{\overset{\displaystyle ClCH_2}{|}}{\underset{\underset{\displaystyle H}{|}}{C}} - \overset{\overset{\displaystyle H}{|}}{\underset{\underset{\displaystyle H}{|}}{C}} - O.nn \quad \text{OR} \quad N - \overset{\overset{\displaystyle O^{\ominus}}{|}}{\underset{}{S^{\oplus}}} = O - \overset{\overset{\displaystyle ClCH_2}{|}}{\underset{\underset{\displaystyle H}{|}}{C}} - \overset{\overset{\displaystyle H}{|}}{\underset{\underset{\displaystyle H}{|}}{C}} - O.nn
$$

(II) <u>Not balanced</u>   (III) <u>Not balanced</u>   12.65

Looking at the radicals carried by the active centers, clearly indicates that nucleo-free-radicals cannot be used. Thus no homo- or copolymerization is favored nucleo-free-radically.

The copolymerization between propylene oxide and sulfur dioxide are said to be interesting in that they are ionic[17,18]. It has been stated that SO_2 cannot be activated ionically or chargedly, due to electrostatic forces of repulsion. As shown above in Equation 12.65, nucleo-free-radicals cannot be used. Only nucleo-non-free-radicals or invisible forces can be used to produce random copolymers as shown by Equation 12.63 and below.

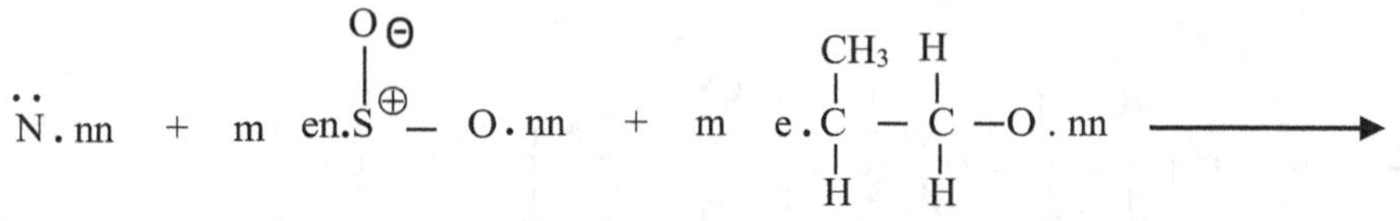

$$\ddot{N}.nn \quad + \quad m \ \text{en}.\overset{\overset{\displaystyle O}{|}\ominus}{\underset{}{S}}\overset{\oplus}{} - O.nn \quad + \quad m \ e.\overset{\overset{\displaystyle CH_3}{|}}{\underset{\overset{|}{H}}{C}} - \overset{\overset{\displaystyle H}{|}}{\underset{\overset{|}{H}}{C}} - O.nn \quad \longrightarrow$$

(I)

$$N\left\{\overset{\overset{\displaystyle O}{|}\ominus}{\underset{}{\overset{\oplus}{S}}} - O - \overset{\overset{\displaystyle CH_3}{|}}{\underset{\overset{|}{H}}{C}} - \overset{\overset{\displaystyle H}{|}}{\underset{\overset{|}{H}}{C}} - O\right\}_x\left\{\overset{\overset{\displaystyle O}{|}\ominus}{\underset{}{\overset{\oplus}{S}}} - O\right\}_y\overset{\overset{\displaystyle O}{|}\ominus}{\underset{}{\overset{\oplus}{S}}} - O - \overset{\overset{\displaystyle CH_3}{|}}{\underset{\overset{|}{H}}{C}} - \overset{\overset{\displaystyle H}{|}}{\underset{\overset{|}{H}}{C}} - O - \overset{\overset{\displaystyle O}{|}\ominus}{\underset{}{\overset{\oplus}{S}}} - O.nn \quad \text{etc.}$$

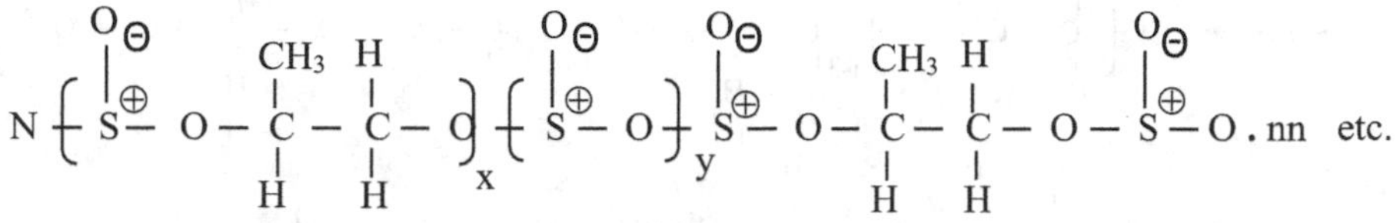

<u>Living Alternating/random copolymers</u>   12.66

Only SO_2 which is reactive, will appear randomly placed along the alternating chain.

12.1.3. Propagation /depropagation Phenomenon

Self-activating monomers readily undergo cyclization particularly of the trimeric types where in most cases there is less strain energy in the ring if the temperature is above the ceiling temperature of the monomer. When polymerized below their ceiling temperatures, they do not readily undergo depropagation during propagation, if the polymerization environment is adequately chosen. Some cyclic monomers are products of some self-activating monomers. Trioxane, 1,3 - dioxane, tetraoxane are products of formaldehyde. Of the three ringed monomers just mentioned, trioxane is the most unique in terms of the size. ***When some of the hydrogen atoms are replaced with alkylane groups, the strain energy is increased making it less stable. Hence acetaldehydes readily favor trimerization less than formaldehyde does.*** When cyclic products are formed, either the ring cannot be opened when strained or there is far below the minimum required strain energy in the ring.

Thus, when self-activating monomers and cyclic monomers in general are involved during polymerizations at conditions slightly above normal, they tend to depropagate to form rings, depending on the types of initiators involved. Their abilities to depropagate are due to the charged character of the linkages. Now, considering cyclic ethers, one will begin with dimethylene oxide anionically.

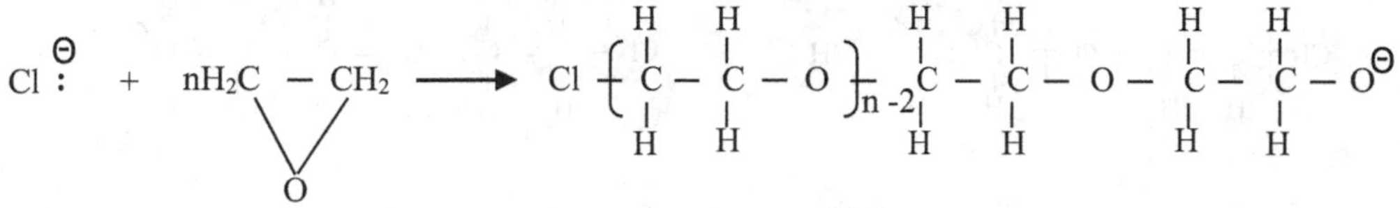

$$Cl\overset{\ominus}{:} \quad + \quad nH_2C\overset{\diagdown}{\underset{O}{\diagup}}CH_2 \quad \longrightarrow \quad Cl\left\{\overset{\overset{\displaystyle H}{|}}{\underset{\overset{|}{H}}{C}} - \overset{\overset{\displaystyle H}{|}}{\underset{\overset{|}{H}}{C}} - O\right\}_{n-2}\overset{\overset{\displaystyle H}{|}}{\underset{\overset{|}{H}}{C}} - \overset{\overset{\displaystyle H}{|}}{\underset{\overset{|}{H}}{C}} - O - \overset{\overset{\displaystyle H}{|}}{\underset{\overset{|}{H}}{C}} - \overset{\overset{\displaystyle H}{|}}{\underset{\overset{|}{H}}{C}} - O^{\ominus}$$

(I)   12.67

The anionic end of the growing polymer chain (I), cannot react with or be attracted to any part of its chain, since there are no driving forces. However, if polymerization environment is charged or has free-radical/non-free-radical, then the chain can be scissioned at specific points favoring the existence of a stable molecule or ring as shown below.

(I) + HCl $\longrightarrow$

$$Cl\left[\begin{array}{c}H\\|\\C\\|\\H\end{array}-\begin{array}{c}H\\|\\C\\|\\H\end{array}-O\right]_{n-2}\begin{array}{c}H\\|\\C\\|\\H\end{array}-\begin{array}{c}H\\|\\C\\|\\H\end{array}\overset{\oplus}{\underset{|}{\overset{\ominus Cl}{O}}}\begin{array}{c}H\\|\\C\\|\\H\end{array}-\begin{array}{c}H\\|\\C\\|\\H\end{array}-O-\begin{array}{c}H\\|\\C\\|\\H\end{array}-\begin{array}{c}H\\|\\C\\|\\H\end{array}-O^{\ominus}$$

$\longrightarrow Cl\left[-\begin{array}{c}H\\|\\C\\|\\H\end{array}-\begin{array}{c}H\\|\\C\\|\\H\end{array}-O-\right]_{n-3}\begin{array}{c}H\\|\\C\\|\\H\end{array}-\begin{array}{c}H\\|\\C\\|\\H\end{array}-O^{\ominus}$ + $\overset{\oplus}{C}\begin{array}{c}H\\|\\|\\H\end{array}-\begin{array}{c}H\\|\\C\\|\\H\end{array}-O-\begin{array}{c}H\\|\\C\\|\\H\end{array}-\begin{array}{c}H\\|\\C\\|\\H\end{array}-O^{\ominus}$ + HCl

$\longrightarrow Cl\left[\begin{array}{c}H\\|\\C\\|\\H\end{array}-\begin{array}{c}H\\|\\C\\|\\H\end{array}-O\right]_{n-3}\begin{array}{c}H\\|\\C\\|\\H\end{array}-\begin{array}{c}H\\|\\C\\|\\H\end{array}-O^{\ominus}$ + HCl + [1,4-dioxane ring]

(II) - 1, 4 - dioxane

Depropagation in a charged environment ionically   12.68

(II) is more stable than the original monomer in terms of the presence of lesser strain energy in the ring. On the hand, the ring formed cannot be opened anionically. This is a typical example of a strong depropagation reaction, since the part scissioned cannot be added to the chain. Without a strong charged or electro-free-radical/nucleo-non-free-radical environment, depropagation during anionic polymerization can never take place at temperature below the Ceiling temperature of the monomer.

Alternatively, in the reaction above, a new growing polymer chain could be gene-rated, if the conditions for the existence of the ring are not favored. With "ion-paired"

$$Cl\left[\begin{array}{c}H\\|\\C\\|\\H\end{array}-\begin{array}{c}H\\|\\C\\|\\H\end{array}-O\right]_{m}\begin{array}{c}H\\|\\C\\|\\H\end{array}-\begin{array}{c}H\\|\\C\\|\\H\end{array}\overset{\oplus}{\underset{**}{\overset{Cl^{\ominus}}{O}}}-\begin{array}{c}H\\|\\C\\|\\H\end{array}-\begin{array}{c}H\\|\\C\\|\\H\end{array}-O-\begin{array}{c}H\\|\\C\\|\\H\end{array}-\begin{array}{c}H\\|\\C\\|\\H\end{array}-O^{\ominus} \longrightarrow$$

$Cl\left[\begin{array}{c}H\\|\\C\\|\\H\end{array}-\begin{array}{c}H\\|\\C\\|\\H\end{array}-O\right]_{m}\begin{array}{c}H\\|\\C\\|\\H\end{array}-\begin{array}{c}H\\|\\C\\|\\H\end{array}-OH$ + $Cl-\begin{array}{c}H\\|\\C\\|\\H\end{array}-\begin{array}{c}H\\|\\C\\|\\H\end{array}-O-\begin{array}{c}H\\|\\C\\|\\H\end{array}-\begin{array}{c}H\\|\\C\\|\\H\end{array}-O^{\ominus}$

                                                                          12.69

$$Cl\left[\begin{array}{c}H\\|\\C\\|\\H\end{array}-\begin{array}{c}H\\|\\C\\|\\H\end{array}-O\right]_{m}\begin{array}{c}H\\|\\C\\|\\H\end{array}-\begin{array}{c}H\\|\\C\\|\\H\end{array}\overset{\oplus}{\overset{Cl^{\ominus}}{O}}-\begin{array}{c}H\\|\\C\\|\\H\end{array}-\begin{array}{c}H\\|\\C\\|\\H\end{array}-O-\begin{array}{c}H\\|\\C\\|\\H\end{array}-\begin{array}{c}H\\|\\C\\|\\H\end{array}-O^{\ominus}\cdots\cdots\overset{\oplus}{N}\begin{array}{c}R\\R\\R\end{array} \longrightarrow$$

$\longrightarrow Cl\left[\begin{array}{c}H\\|\\C\\|\\H\end{array}-\begin{array}{c}H\\|\\C\\|\\H\end{array}-O\right]_{m}\begin{array}{c}H\\|\\C\\|\\H\end{array}-\begin{array}{c}H\\|\\C\\|\\H\end{array}-O-H$ + NR_3 + [1,4-dioxane ring] + RCl

(II)   12.70

initiators, this may or may not be possible, noting however that a strong charged environment may not favor the existence of the anionic "ion-paired initiator".

With positive charges or electro-free-radicals, the mechanism for depropagation is quite different. Positively, the following are to be expected in a "non-ionic" environment with no paired center formed.

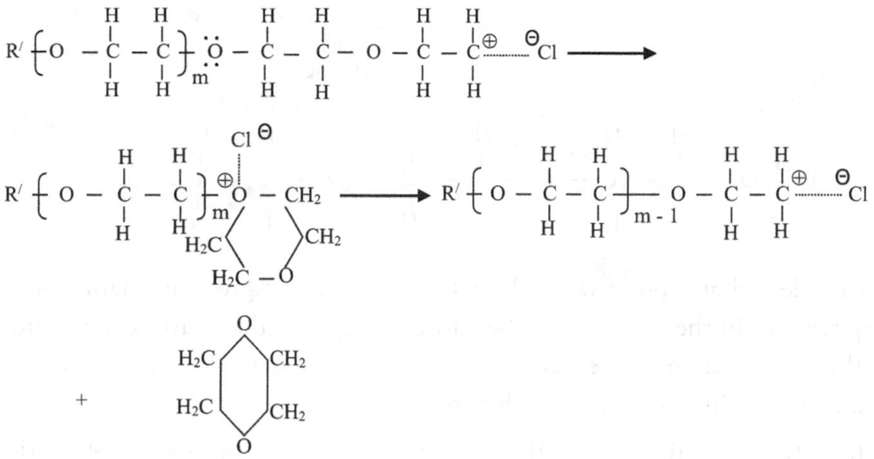

(II) 1,4- dioxane

<u>Depropagation in an uncharged environment positively</u>   12.71

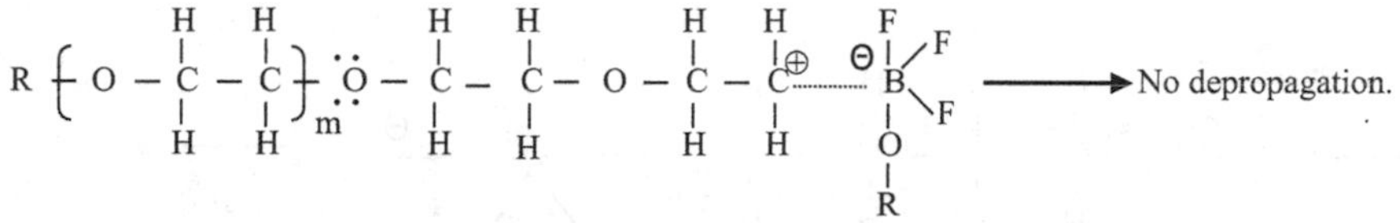

No depropagation.

12.72

Depending on the strength of the paired initiator or the distance between paired centers, which should decrease with every addition of monomer, the last reaction above unlike the positively charged covalent case before it, cannot favor depropagation. When there is depropagation when covalently positively charged-paired initiators are involved, then the variable involved is more of the polymerization environment, that is, more charged environment than necessary. It should be noted that with the covalent initiator of Equation 12.71, the 1,4 - dioxane obtained is less strained than if dimethylane oxide was obtained; though if the strength of the positive center is quite high, the dioxane can be activated and polymerized. This is the ideal propagation/depropagation/repropagation phenomenon or propagation/depropagation equilibria. The chain first propagates, then depropagates and then repropagates, depending on the strength of the active center on the growing chain, the size of the ring formed and the polymerization environment. While Equation 12.71 represents the propagation/depropagation reactions, the equation below represents the next step-the repropagation reaction, if the strength of the active center is strong enough to provide the MRSE.

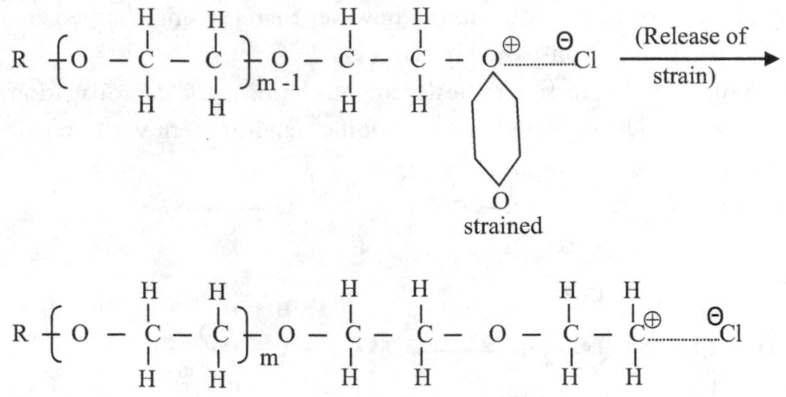

$$R \overbrace{\left[O - \underset{\underset{H}{|}}{\overset{\overset{H}{|}}{C}} - \underset{\underset{H}{|}}{\overset{\overset{H}{|}}{C}} \right]_{m-1}} O - \underset{\underset{H}{|}}{\overset{\overset{H}{|}}{C}} - \underset{\underset{H}{|}}{\overset{\overset{H}{|}}{C}} - O^{\oplus} \cdots \cdots {}^{\ominus}Cl \xrightarrow{\text{(Release of strain)}}$$

strained

$$R \overbrace{\left[O - \underset{\underset{H}{|}}{\overset{\overset{H}{|}}{C}} - \underset{\underset{H}{|}}{\overset{\overset{H}{|}}{C}} \right]_{m}} O - \underset{\underset{H}{|}}{\overset{\overset{H}{|}}{C}} - \underset{\underset{H}{|}}{\overset{\overset{H}{|}}{C}} - O - \underset{\underset{H}{|}}{\overset{\overset{H}{|}}{C}} - \overset{\overset{H}{|}}{C}{}^{\oplus} \cdots \cdots {}^{\ominus}Cl$$

12.73

When the ring is released at a point when the active center of the growing chain cannot provide the MRSE, the ring remains in the system until the critical strength of the active center is provided to open the ring. Note that also cases considered so far are polymeri-zation via Combination mechanism.

For larger membered rings, consider tetrahydrofuran.

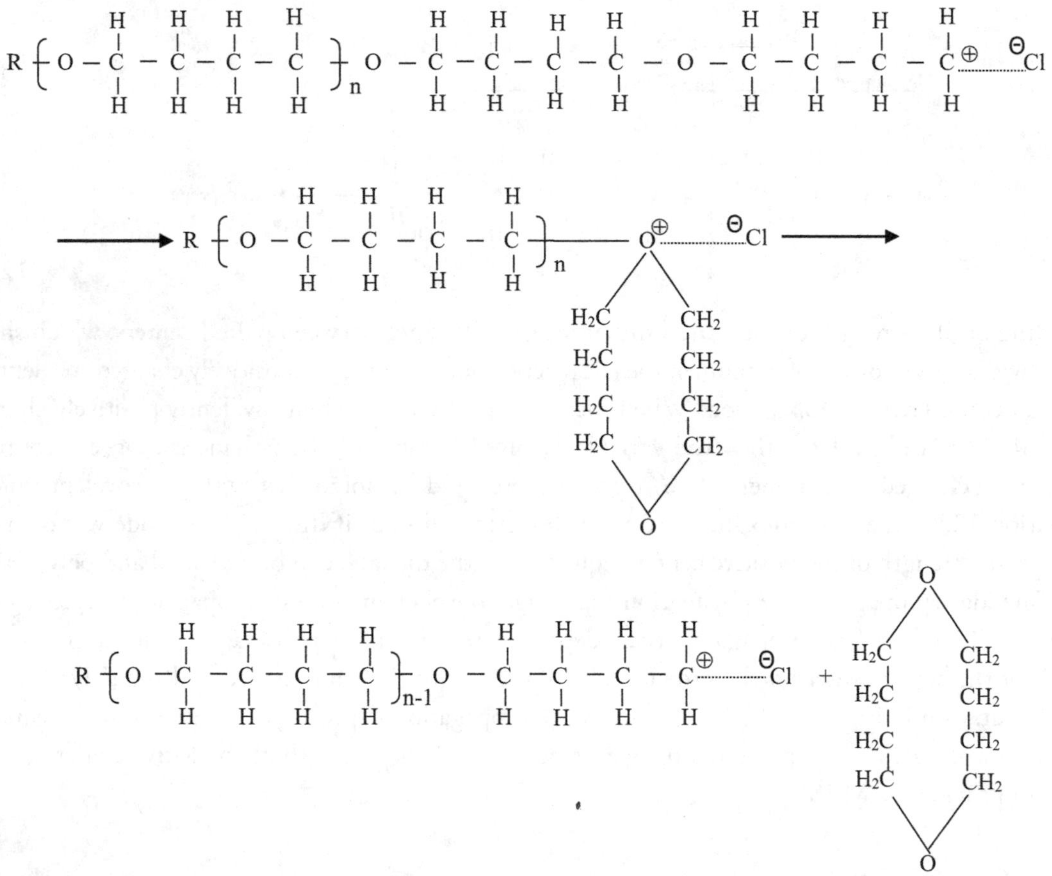

$$R \left[O - \underset{H}{C} - \underset{H}{C} - \underset{H}{C} - \underset{H}{C} \right]_{n} O - \underset{H}{C} - \underset{H}{C} - \underset{H}{C} - \underset{H}{C} - O - \underset{H}{C} - \underset{H}{C} - \underset{H}{C} - C^{\oplus} \cdots {}^{\ominus}Cl$$

$$\longrightarrow R \left[O - \underset{H}{C} - \underset{H}{C} - \underset{H}{C} - \underset{H}{C} \right]_{n} O^{\oplus} \cdots {}^{\ominus}Cl \longrightarrow$$

$$\begin{array}{cc} H_2C & CH_2 \\ H_2C & CH_2 \\ H_2C & CH_2 \\ H_2C & CH_2 \\ & O \end{array}$$

$$R \left[O - \underset{H}{C} - \underset{H}{C} - \underset{H}{C} - \underset{H}{C} \right]_{n-1} O - \underset{H}{C} - \underset{H}{C} - \underset{H}{C} - C^{\oplus} \cdots {}^{\ominus}Cl \quad + \quad \begin{array}{cc} & O \\ H_2C & CH_2 \\ H_2C & CH_2 \\ H_2C & CH_2 \\ H_2C & CH_2 \\ & O \end{array}$$

1,6 - dioxane   12.74

Either tetrahydrofuran or 1,6 - dioxane or even larger membered ring can be forced depending on the condition. For four-membered oxacyclobutane, eight-membered, sixteen-membered rings can be obtained, depending on the strain energy in the ring and

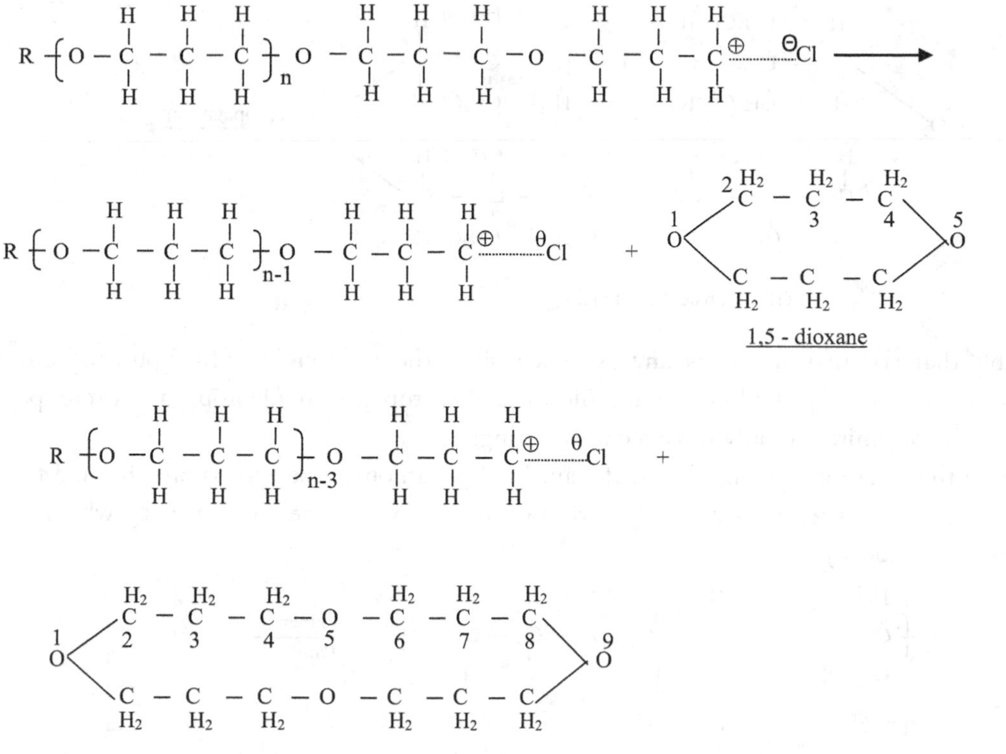

$$R \left[O-\overset{\overset{H}{|}}{\underset{\underset{H}{|}}{C}} - \overset{\overset{H}{|}}{\underset{\underset{H}{|}}{C}} - \overset{\overset{H}{|}}{\underset{\underset{H}{|}}{C}} \right]_n O - \overset{\overset{H}{|}}{\underset{\underset{H}{|}}{C}} - \overset{\overset{H}{|}}{\underset{\underset{H}{|}}{C}} - \overset{\overset{H}{|}}{\underset{\underset{H}{|}}{C}} - O - \overset{\overset{H}{|}}{\underset{\underset{H}{|}}{C}} - \overset{\overset{H}{|}}{\underset{\underset{H}{|}}{C}} - \overset{\overset{H}{|}}{\underset{\underset{H}{|}}{C}}{}^{\oplus} \quad {}^{\ominus}Cl \longrightarrow$$

$$R \left[O-\overset{\overset{H}{|}}{\underset{\underset{H}{|}}{C}} - \overset{\overset{H}{|}}{\underset{\underset{H}{|}}{C}} - \overset{\overset{H}{|}}{\underset{\underset{H}{|}}{C}} \right]_{n-1} O - \overset{\overset{H}{|}}{\underset{\underset{H}{|}}{C}} - \overset{\overset{H}{|}}{\underset{\underset{H}{|}}{C}} - \overset{\overset{H}{|}}{\underset{\underset{H}{|}}{C}}{}^{\oplus} \quad {}^{\ominus}Cl \quad + \quad$$

1,5 - dioxane structure (positions 1–5):

2 CH_2 — 3 CH_2 — 4 CH_2 ... O (1) ... O (5); CH_2 — CH_2 — CH_2

OR

$$R \left[O-\overset{\overset{H}{|}}{\underset{\underset{H}{|}}{C}} - \overset{\overset{H}{|}}{\underset{\underset{H}{|}}{C}} - \overset{\overset{H}{|}}{\underset{\underset{H}{|}}{C}} \right]_{n-3} O - \overset{\overset{H}{|}}{\underset{\underset{H}{|}}{C}} - \overset{\overset{H}{|}}{\underset{\underset{H}{|}}{C}} - \overset{\overset{H}{|}}{\underset{\underset{H}{|}}{C}}{}^{\oplus} \quad {}^{\ominus}Cl \quad +$$

1,5,9,13 - tetraoxane structure (positions 1–9):

1 O — CH_2(2) — CH_2(3) — CH_2(4) — O(5) — CH_2(6) — CH_2(7) — CH_2(8) — O(9); CH_2 — CH_2 — CH_2 — O — CH_2 — CH_2 — CH_2

1,5,9, 13 - tetraoxane

12.75

conditions of polymerization. The sixteen-membered ring is known to be incapable of further propagation. Substituted oxacyclobutanes, such as the 3,3 - Bis (chloromethyl) derivative are said not to undergo "this type of termination" to the bulky substituents. In the first case, most of the termination reactions in the literature and textbooks are no termination reactions as will be explained in subsequent volumes. Secondly, the reason proposed is not acceptable, since existence of its four-membered ring is favored. The four-membered ring may be more sterically hindered than a larger-membered ring. The propagation/depropagation reactions may take place with 3,3 - Bis(chloromethyl) oxetane, except that the four membered rings formed may not favor stable existence due to the less stable character of a ring with radical-pushing groups (CH_2Cl). There is no doubt that most of these larger membered rings can be provided with MRSE when the functional center is attacked. In order words 3,3-Bis(chloromethyl) oxetane may readily undergo propagation/depropagation phenomenon, when suitable conditions exist.

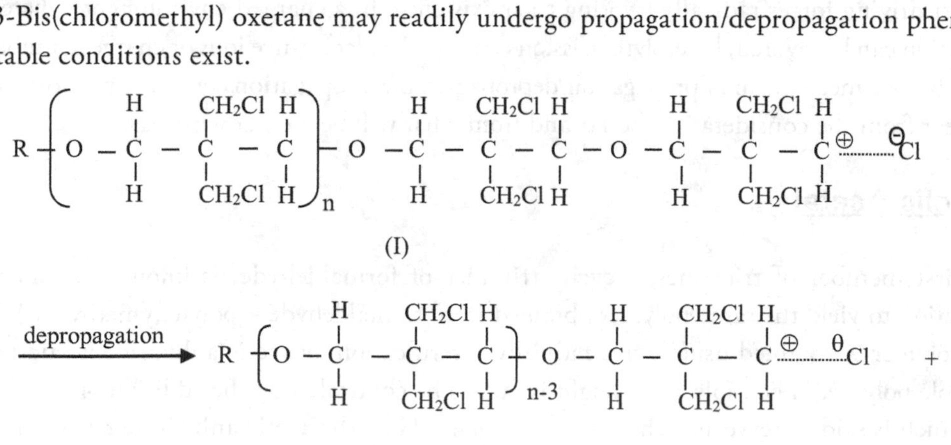

$$R \left[O-\overset{\overset{H}{|}}{\underset{\underset{H}{|}}{C}} - \overset{\overset{CH_2Cl}{|}}{\underset{\underset{CH_2Cl}{|}}{C}} - \overset{\overset{H}{|}}{\underset{\underset{H}{|}}{C}} \right]_n O - \overset{\overset{H}{|}}{\underset{\underset{H}{|}}{C}} - \overset{\overset{CH_2Cl}{|}}{\underset{\underset{CH_2Cl}{|}}{C}} - \overset{\overset{H}{|}}{\underset{\underset{H}{|}}{C}} - O - \overset{\overset{H}{|}}{\underset{\underset{H}{|}}{C}} - \overset{\overset{CH_2Cl}{|}}{\underset{\underset{CH_2Cl}{|}}{C}} - \overset{\overset{H}{|}}{\underset{\underset{H}{|}}{C}}{}^{\oplus} \quad {}^{\ominus}Cl$$

(I)

$$\xrightarrow{\text{depropagation}} R \left[O-\overset{\overset{H}{|}}{\underset{\underset{H}{|}}{C}} - \overset{\overset{CH_2Cl}{|}}{\underset{\underset{CH_2Cl}{|}}{C}} - \overset{\overset{H}{|}}{\underset{\underset{H}{|}}{C}} \right]_{n-3} O - \overset{\overset{H}{|}}{\underset{\underset{H}{|}}{C}} - \overset{\overset{CH_2Cl}{|}}{\underset{\underset{CH_2Cl}{|}}{C}} - \overset{\overset{H}{|}}{\underset{\underset{H}{|}}{C}}{}^{\oplus} \quad {}^{\ominus}Cl \quad +$$

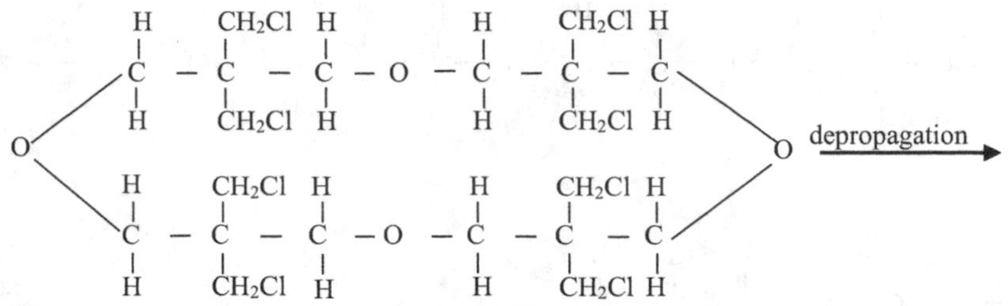

$$\text{(II) (Transient existence)}$$

(II) (Transient existence)   12.76

It is possible that (II) may not favor any existence, since there are eight radical-pushing groups in the sixteen-membered ring. Unlike the anionic case, the propagation /depropagation/repropagation phenomena are not limited to only three membered-rings.

Radically the situation is completely different. Like the anionic case, this is only limited to three-membered rings. Higher temperatures polymerization will be one of the few means by which the ring can be scissioned radically.

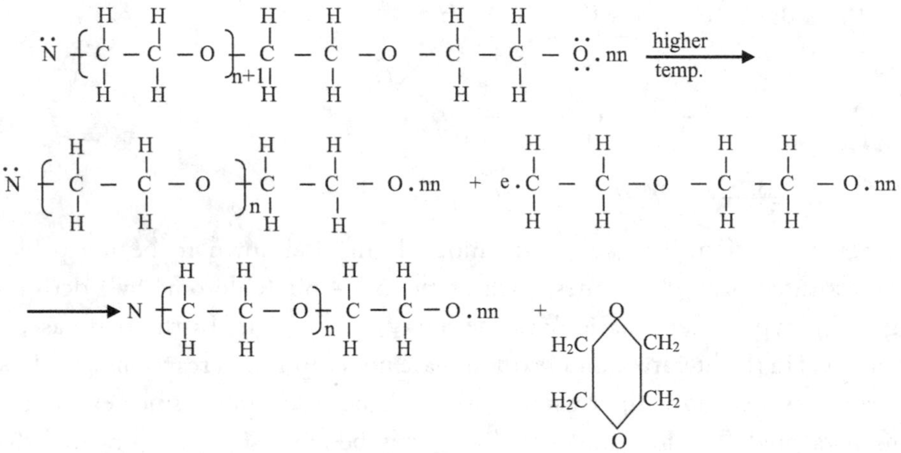

Depropagation radically   12.77

Since nucleo-non-free radical polymerization is limited to low temperature polymeri-zations, depropagation reactions radically will be far less common than even the positive cases at high temperatures, since there are no other driving forces radically favoring their existence. In a charged environment, where a radical polymerization can be favored, heterolytic scission cannot take place, since ions or charges cannot combine with radicals. The mechanism of propagation/depropagation/repropagation equilibria pheno-mena should become clear from the considerations so far and from what will be seen downstream.

12.2 Cyclic Acetal

The first member of trioxanes, a cyclic trimmer of formaldehyde, is known to undergo facile polymerization to yield the same polymer obtained from formaldehyde – poly(oxymethylene). When an unstable polymer is obtained usually free-radically, esterification by acetic anhydride is said to produce a more stable polymer. The molecular weight is said to be controlled by the addition of a small amount of water which is said to serve as a chain transfer agent[7]. How the acetic anhydride favors the existence of a more stable polymer during the ideal termination step, will be explained when the occasion arises.

Nevertheless, trioxane -a six-membered ring and the cyclic formals cannot readily be provided with the MRSE to favor the instantaneous opening of the ring chargedly or radically. It therefore will favor only cationic attack via the functional centers, of which many of the same type exist.

12.2.1 Charged feature of Cyclic Acetals

It has been observed without full identification that trioxanes favor "free cationic" route, as well as "cationic ion-paired" route. It is however during the "free-cationic" route that propagation/depropagation and repropagation phenomena have been observed. Unfortunately the "free-cationic" routes have largely involved the use of H^A on the active centers, that is, that involving the use of protonic acids which is not particularly ideal, since unstable molecules and polymers are said to be produced. Nevertheless, as usual, they will partly be used in order to illustrate some basic principles which indeed does not take place, but was thought to take place. Cationically for trioxane, the followings were thought to be expected.

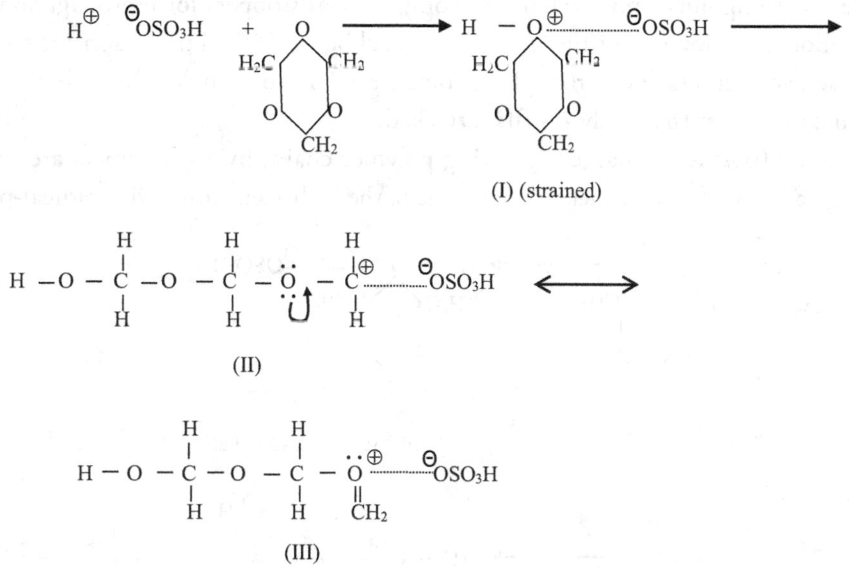

(I) (strained)

(II)

(III)

Impossible movement of paired unbonded radicals   12.78

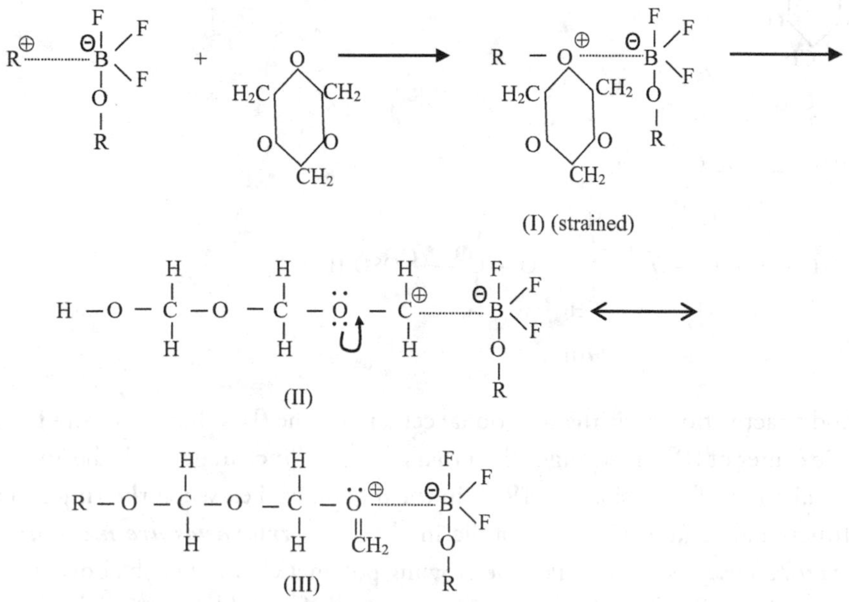

(I) (strained)

(II)

(III)

Impossible movement of paired unbonded radicals   12.79

35

It should be noted that, in this so-called type of resonance stabilization, paired unbonded radicals are the first to move. Whether "free-cationic" or "cationic ion" or positively charged -paired initiators are involved or not, electrostatic resonance stabilization phenomena cannot exist and the release of formaldehyde cationically or electro-free-radically is not possible.

Though the positively charged ends of the (II), are not electrostatic in character, it is latently present, since they are products of the (I), when the rings are opened. Aldehydes and ketones cannot however favor this type of resonance stabilization, based on the method of initiation, noting that addition takes place electro-free-radically backwardly. Cyclic ethers also do not allow for this type of resonance stabilization since the oxygen atom is not adjacently located to the positively charged active center. These observations are worthy of note, because while cyclic ethers cannot be copolymerized with alkenes, and carbonyl monomers favor very little or no copolymerization with alkenes, trioxanes, tetraoxane and cyclic formals favor copolymerization with alkenes. With carbonyl monomers, where little copolymerization exists, at best the alkenes of strong nucleophilic characters only serve as stoppers for repropagation phenomenon, with mostly carbonyl monomers on the main chain backbones [19, 20]. The reason for this is least to be expected because cyclic ethers and carbonyl monomers are far more nucleophilic than alkenes. It is the least nucleophilic monomer that is always first attacked.

For the cationic (positively charged) growing polymer chain, living polymers are produced in the absence of foreign agents. Now consider replacing one of the hydrogen atoms with radical-pushing groups.

$$12.80$$

Since the method of activation is via the functional center, it is the (II)s that are favored and not the (III)s, since the cationic center of (II)b is stronger than that of (III)b. Since in general, the anionic route cannot take place in the absence of provision of MRSE in the monomer (i.e., when the ring is not opened), and since oxygen functional centers are nucleophilic in character, **trioxanes are more nucleophilic than cyclic ethers as will shortly be shown.** For the growing polymer chain of (II)b above, like acetaldehydes, there is transfer species of the first kind here with no release of acetaldehyde.

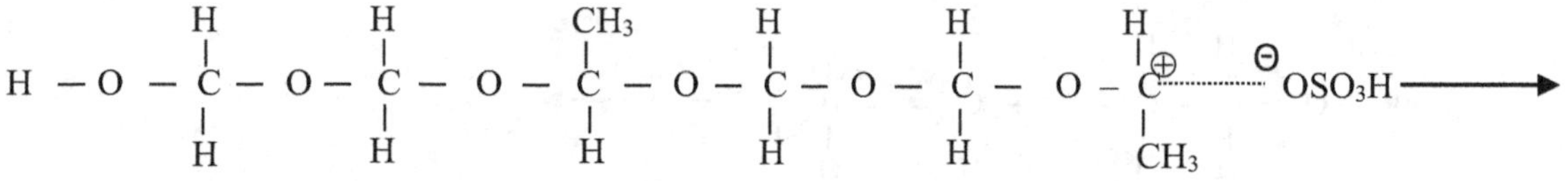

$$H - O - \overset{\overset{\displaystyle H}{|}}{\underset{\underset{\displaystyle H}{|}}{C}} - O - \overset{\overset{\displaystyle H}{|}}{\underset{\underset{\displaystyle H}{|}}{C}} - O - \overset{\overset{\displaystyle CH_3}{|}}{\underset{\underset{\displaystyle H}{|}}{C}} - O - \overset{\overset{\displaystyle H}{|}}{\underset{\underset{\displaystyle H}{|}}{C}} - O - \overset{\overset{\displaystyle H}{|}}{\underset{\underset{\displaystyle H}{|}}{C}} - O - \overset{\overset{\displaystyle H}{|}}{\underset{\underset{\displaystyle CH_3}{|}}{\overset{\oplus}{C}}} \cdots\cdots \overset{\ominus}{O}SO_3H \longrightarrow$$

Cannot reject acetaldehyde, but transfer species of the first kind   12.81

Note that the equations as have been shown so far are electro-free-radical in character and not as have been shown so far, because the real initiators are like the type shown in (I) of Equation 12.78 or (II)a or (III)a in Equation 12.80 where the rings are not opened. It is the H atom carried by the O center that is held in Equilibrium state of existence (H•e) that is the real initiator used for backward addition. Water can be used in place of the ring, such as in dilute H_2SO_4.

With one radical-pushing group, the followings are to be expected using electrostatically positively charged-paired initiators.

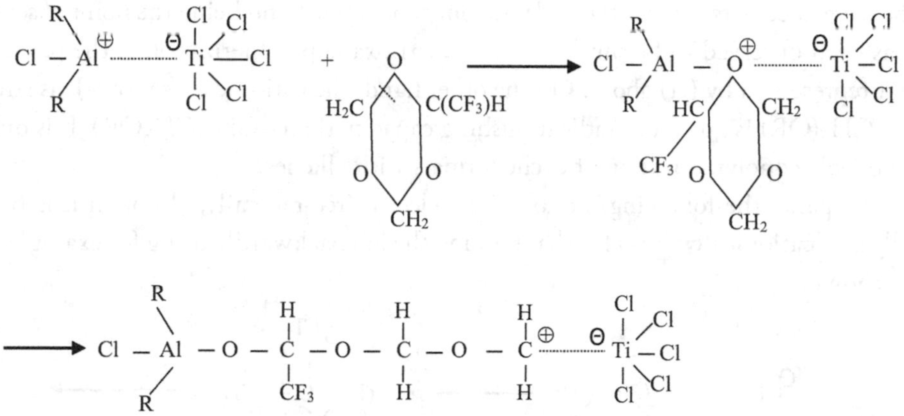

12.82

It is important to note the point of scission in these monomers. It is such that, the active center generated is the strongest positively. The type of positively charged initiator used above is also important to note- $AlR_3/TiCl_4$ combination with more of $TiCl_4$.

Now considering cyclic formals, one will begin with 1,3 - dioxolane using ROR/BF_3 combination.

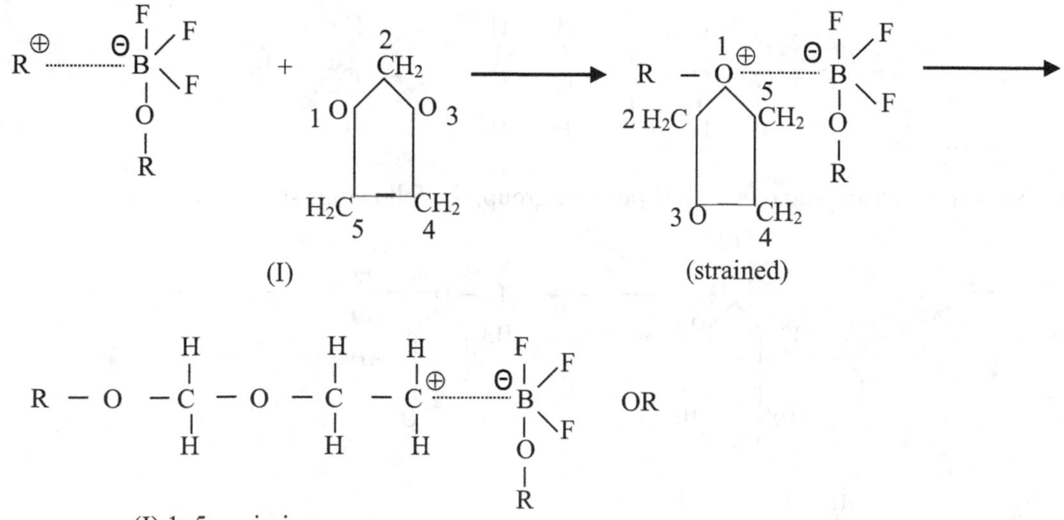

(I)   (strained)

(I) 1, 5 - scission

37

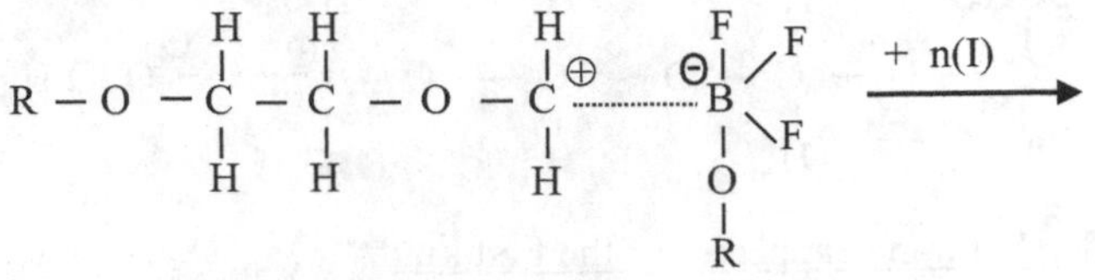

$$R - O - CH_2 - CH_2 - O - \overset{\oplus}{C}H \cdots\cdots \overset{\ominus}{B}\begin{smallmatrix} F \\ F \\ O-R \end{smallmatrix} \quad \xrightarrow{\; + \, n(I) \;}$$

(II) 1,2 - scission (favored)

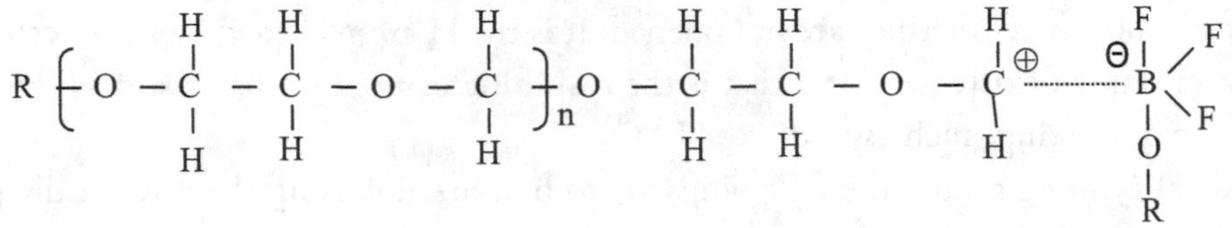

$$R \left[O - CH_2 - CH_2 - O - CH \right]_n O - CH_2 - CH_2 - O - \overset{\oplus}{C}H \cdots\cdots \overset{\ominus}{B}\begin{smallmatrix} F \\ F \\ O-R \end{smallmatrix}$$

12.83

Only one of the oxygen centers can be attacked with only one single bond being the point of scission. Since the reaction has been observed to be similar to that for trioxane polymerization[21], the point of scission therefore is that represented by (II) above. On the other hand, the cationic center of (II) is stronger than that of (I), i.e., $-CH_2(OR)$ is greater in radical-pushing capacity than $-CH_2(CH_2OR)$. It is only (II) that will favor the possible copolymerization of cyclic formals with alkenes.

For 1,3 - dioxepane, the followings are obtained electro-free-radically, although it is being shown below chargedly as if cationically. It is H•e that is doing the job backwardly using for example dilute HCl as will fully be shown.

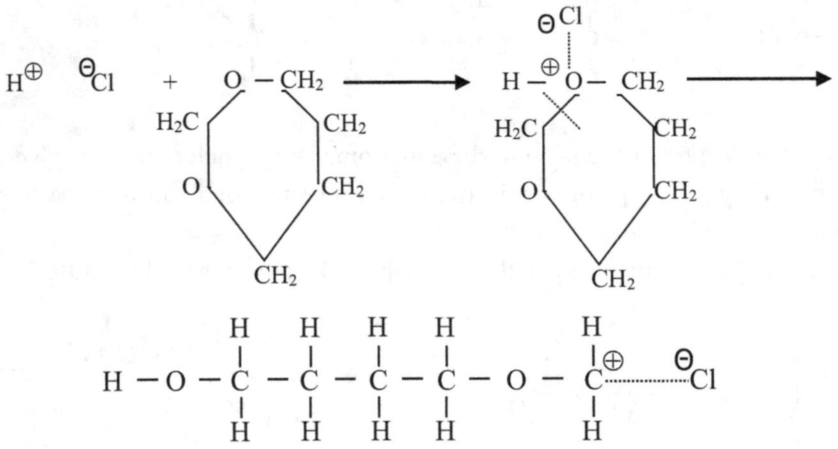

$$H - O - CH_2 - CH_2 - CH_2 - CH_2 - O - \overset{\oplus}{C}H \cdots\cdots \overset{\ominus}{C}l$$

12.84

With 1,3- dioxolane carrying one radical-pushing group, the followings should be expected.

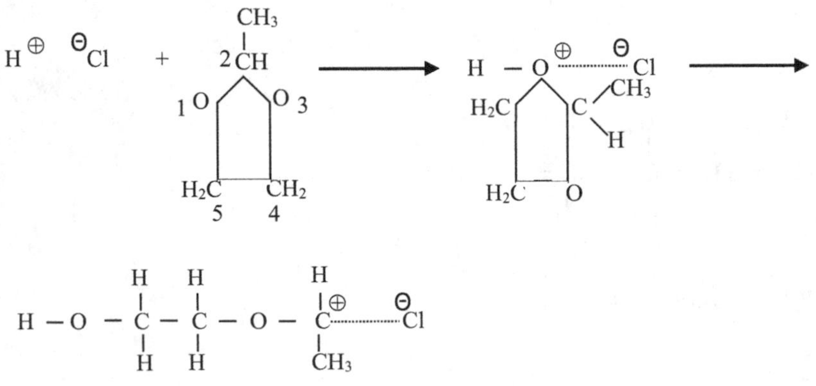

$$H - O - CH_2 - CH_2 - O - \overset{\oplus}{\underset{CH_3}{C}}H \cdots\cdots \overset{\ominus}{C}l$$

12.85

38

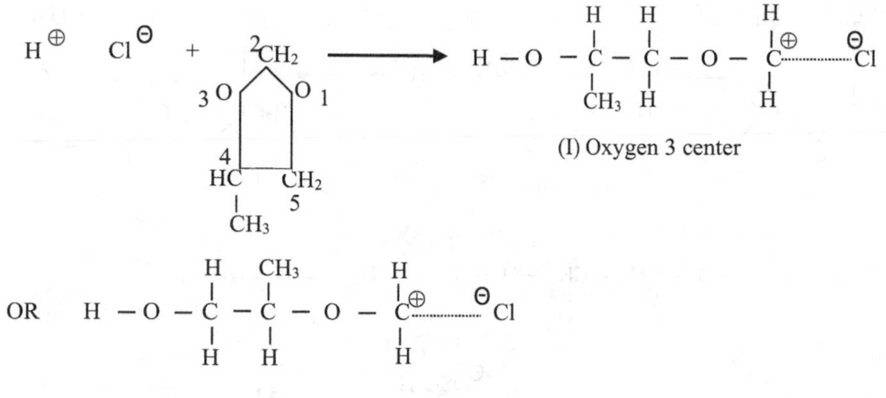

$$H^{\oplus} \quad Cl^{\ominus} \quad + \quad \underset{3O}{\overset{2CH_2}{\diamond}} \quad O1 \quad \longrightarrow \quad H-O-\overset{\overset{H}{|}}{\underset{\underset{CH_3}{|}}{C}}-\overset{\overset{H}{|}}{\underset{\underset{H}{|}}{C}}-O-\overset{\overset{H}{|}}{\underset{\underset{H}{|}}{C}}\cdots\cdots\overset{\ominus}{Cl}$$

(Structure with positions 4, 5, HC, CH$_2$, CH$_3$)

(I) Oxygen 3 center

$$OR \quad H-O-\overset{\overset{H}{|}}{\underset{\underset{H}{|}}{C}}-\overset{\overset{CH_3}{|}}{\underset{\underset{H}{|}}{C}}-O-\overset{\overset{H}{|}}{\underset{\underset{H}{|}}{C}}\cdots\cdots\overset{\ominus}{Cl}$$

(II) Oxygen 1 center (Favored)

12.86

It is the No 1 oxygen center that is attacked all the time, since the internal carbon center adjacently located to the oxygen center next to the active center, is positively stronger. If the CH_3 is replaced with a weaker radical-pushing group such as CF_3, then almost same type of (I) will be favored. Thus, it can be observed that the type and location of substituent groups in these monomers are important.

Ring opening coupled with "electron" rearrangement has been proposed to occur in the cationic polymerization of 1,4,6- trioxyspirononane (I) and 2- methyl - 4- methy-lene- 1, 3 - dioxolane (II) to low molecular weight products [22].

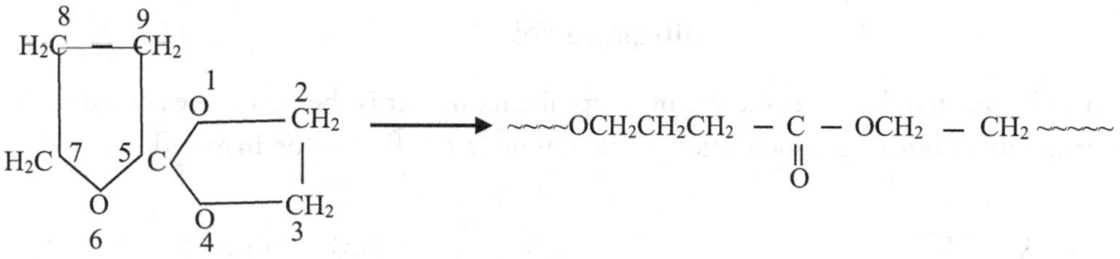

(Structure I) $\longrightarrow \sim\sim OCH_2CH_2CH_2 - \overset{\overset{}{C}}{\underset{\underset{O}{\parallel}}{}} - OCH_2 - CH_2 \sim\sim$

(I)

12.87

(Structure II) $H_2C = \overset{4}{C} - \overset{5}{CH_2} \longrightarrow \sim\sim\sim CH_2 - \overset{\overset{}{C}}{\underset{\underset{O}{\parallel}}{}} - CH_2 - OC\overset{\overset{H}{|}}{\underset{\underset{CH_3}{|}}{}} \sim\sim$

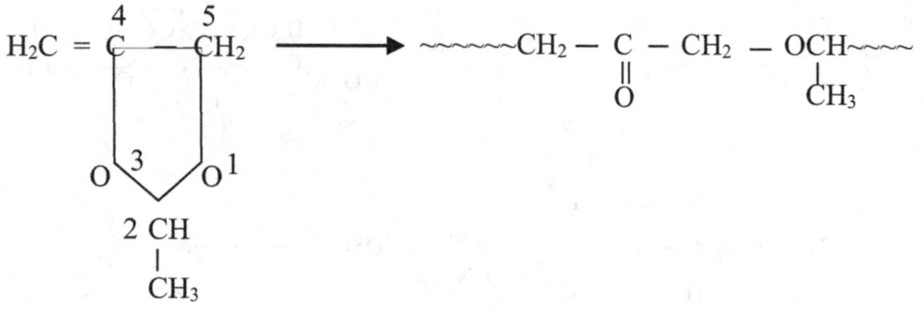

(II)

12.88

As is clearly obvious now, there is nothing like "electron rearrangement" in all these reactions. (I) above is not a fused ringed system. The rings are tetrahydrofuran (cyclic ether) and 1,3 - dioxane (cyclic formal) both five-membered rings, which cannot be opened instantaneously. The tetrahydrofuran ring seems to be less nucleophilic than the 1,3 - dioxolane ring which indeed is the case. If 1,3-dioxolane ring is more nucleophilic, it should not be the first to be attacked cationically, particularly with the presence of two functional centers (additive laws). When any of the two functional centers of the dioxolane is involved, the followings are obtained as shown below.

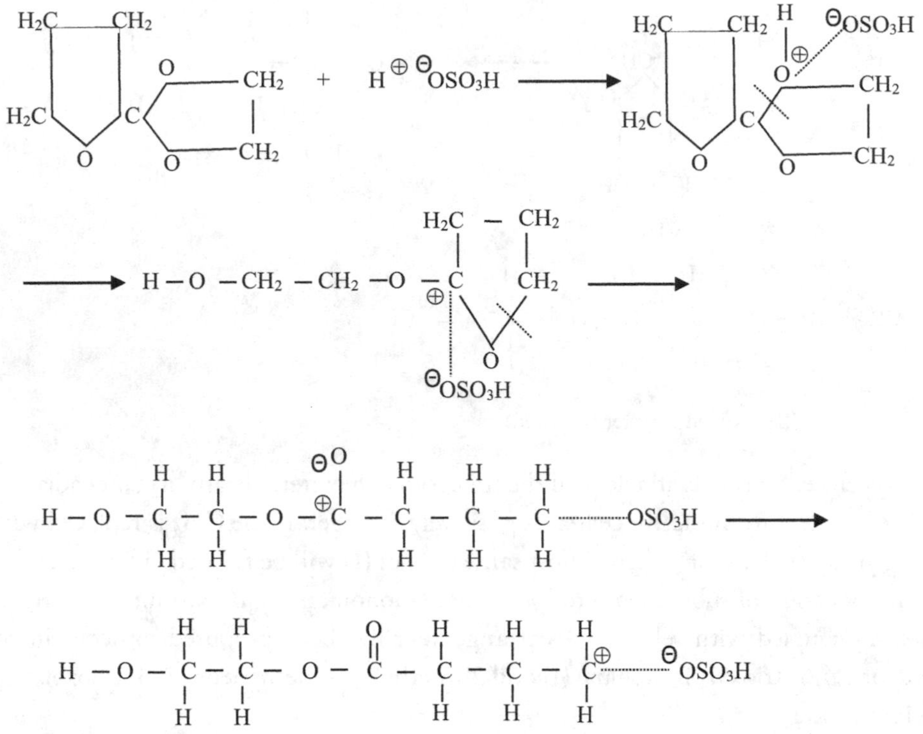

$$H - O - CH_2 - CH_2 - O - \overset{\oplus}{C} \quad CH_2 \longrightarrow$$

$$H - O - \overset{H\ H}{\underset{H\ H}{C - C}} - O - \overset{\ominus}{\underset{\oplus}{O}} \overset{H\ H\ H}{\underset{H\ H\ H}{C - C - C - C}} \cdots\cdots OSO_3H \longrightarrow$$

$$H - O - \overset{H\ H}{\underset{H\ H}{C - C}} - O - \overset{O}{\overset{\|}{C}} - \overset{H\ H\ H}{\underset{H\ H\ H}{C - C - C}} - \overset{\oplus}{C} \cdots\cdots \overset{\ominus}{O}SO_3H \longrightarrow$$

(III) <u>Not Favored</u>   12.89

It is said not to be favored, because, the less nucleophilic monomer is the first to be activated. Therefore, when the tetrahydrofuran ring which is less nucleophilic is the first to be involved, the followings are obtained.

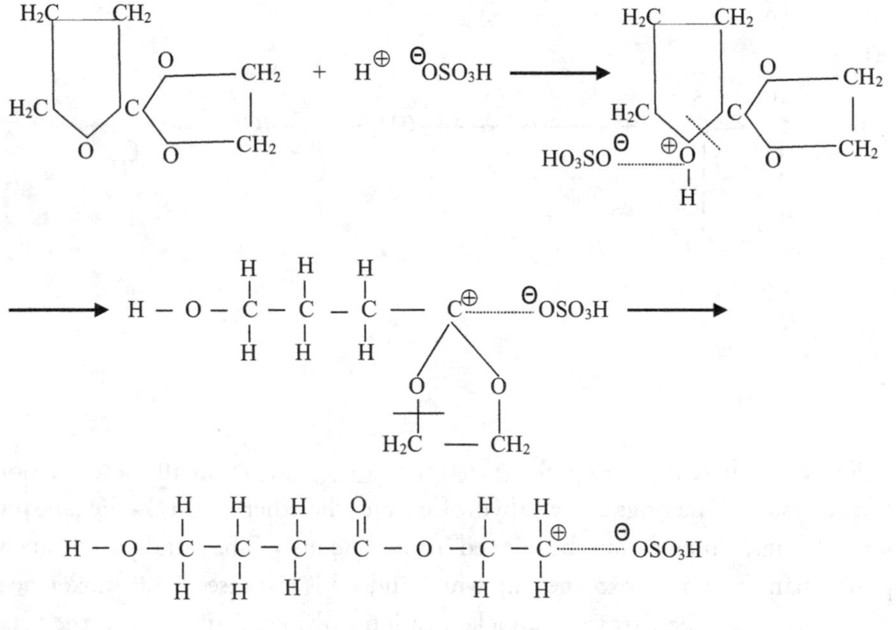

$$\longrightarrow H - O - \overset{H\ H\ H}{\underset{H\ H\ H}{C - C - C}} - \overset{\oplus}{C} \cdots\cdots \overset{\ominus}{O}SO_3H \longrightarrow$$

$$H - O - \overset{H\ H\ H\ O}{\underset{H\ H\ H}{C - C - C - C}} - O - \overset{H\ H}{\underset{H\ H}{C - \overset{\oplus}{C}}} \cdots\cdots \overset{\ominus}{O}SO_3H$$

(IV) <u>Favored</u>   12.90

In both reactions, there is nothing like "'electron' rearrangement" of any type. Nevertheless, it is (IV) above that is favored, since the positive active center of (IV) is far stronger than that of (III). This clearly indicates that tetrahydrofuran ring is less nucleophilic than the 1,3 - dioxolane ring. This is supported further by the fact that if the two rings were strained enough to favor their instantaneous opening, the followings are obtained.

For (V)(a) $\longrightarrow$

$$
\overset{\oplus}{\underset{H}{\overset{H}{C}}} - \underset{H}{\overset{H}{C}} - \underset{H}{\overset{H}{C}} - \overset{O}{\overset{\parallel}{C}} - O - \underset{H}{\overset{H}{C}} - \underset{H}{\overset{H}{C}} - O^{\ominus} \qquad \text{OR}
$$

(V)aa

For (V)(b) $\longrightarrow$

$$
{}^{\ominus}O - \underset{H}{\overset{H}{C}} - \underset{H}{\overset{H}{C}} - \underset{H}{\overset{H}{C}} - \overset{O}{\overset{\parallel}{C}} - O - \underset{H}{\overset{H}{C}} - \underset{H}{\overset{H}{C}}{}^{\oplus}
$$

(V) bb

$\longrightarrow$

$$
H - O - \underset{H}{\overset{H}{C}} - \underset{H}{\overset{H}{C}} - O - \overset{O}{\overset{\parallel}{C}} - \underset{H}{\overset{H}{C}} - \underset{H}{\overset{H}{C}} - \underset{H}{\overset{H}{C}}{}^{\oplus} \quad {}^{\ominus}OSO_3H \qquad \text{OR}
$$

(V) aaa

$$
H - O - \underset{H}{\overset{H}{C}} - \underset{H}{\overset{H}{C}} - \underset{H}{\overset{H}{C}} - \overset{O}{\overset{\parallel}{C}} - O - \underset{H}{\overset{H}{C}} - \underset{H}{\overset{H}{C}}{}^{\oplus} \quad {}^{\ominus}OSO_3H
$$

(V) bbb   12.91

It is (V)bbb that identifies with the case where the functional center was used. Hence, cyclic formals are less strained than cyclic ethers and it is also obvious that tetrahydrofuran ringed substituent group (i.e., 1,3-cyclic formalic group) is less radical-pushing than 1,3- dioxolane ringed substituent group (i.e., cyclic etheric group).

$$
\begin{array}{ccc}
H_2C - CH_2 & & H_2C - CH_2 \\
| \quad\quad | & & | \quad\quad | \\
O \quad\quad O & > & H_2C \quad C \\
\diagdown \;\; \diagup & & \diagdown \;\; \diagup \\
C & & O
\end{array}
$$

Order of radical-pushing capacity   12.92

Before one will begin delving into deeper frontiers with respect to (I) of Equation 12.87, one will complete the provision of the mechanism of (II) of Equation 12.88. With cycloalkanes, it was noted that $H_2C=$ groups can never be used as functional or activation centers when cumulatively placed to a ring. This however is not the case with hetero-ringed monomers, since it cannot be used for molecular rearrangements of the third kind.

If (II) of Equation 12.88 is instantaneously opened for example, the followings are obtained.

$$H_2C = C - CH_2 \longrightarrow \oplus C - \overset{H}{\underset{CH_2}{C}} - O - \overset{H}{\underset{CH_3}{C}} - O \ominus \longrightarrow$$

No transfer possible.

(12.93)

$$H_2C = C + CH_2 \longrightarrow \oplus C - O - \overset{H}{\underset{CH_3}{C}} - O - \overset{H}{\underset{H}{C}}\ominus \longrightarrow$$

No transfer possible.

Wrong point of Scission

(12.94)

The point of scission is that favored by the first reaction. Nevertheless, scission does not take place, since molecular rearrangement of the third kind is not possible chargedly or radically. The $H_2C = C\,O$ activation center being less nucleophilic than functional O center, the followings are obtained.

$$H^{\oplus}\ {}^{\ominus}OSO_3H \quad + \quad \left(\ominus\overset{H}{\underset{H}{C}} - \overset{\oplus}{C} - CH_2 \longleftrightarrow \ominus\overset{H}{\underset{H}{C}} - C - CH_2 \right) \longrightarrow$$

(I)

$$\underset{CH_3}{\underset{|}{CH}}$$

$$H - \overset{H}{\underset{H}{C}} - \overset{\oplus}{\underset{|}{C}} - CH_2 \longrightarrow H - \overset{H}{\underset{H}{C}} - \overset{O}{\underset{||}{C}} - \overset{H}{\underset{H}{C}} - O - \overset{H}{\underset{CH_3}{\overset{\oplus}{C}}} {}^{\ominus}OSO_3H \xrightarrow{+\ n(I)}$$

$$\underset{CH_3}{\underset{|}{CH}}$$

$$\longrightarrow H \left[\begin{array}{c} \\ \end{array} \right]_n - \overset{H}{\underset{H}{C}} - \overset{O}{\underset{||}{C}} - \overset{H}{\underset{H}{C}} - O - \overset{H}{\underset{H}{C}} = \overset{H}{\underset{H}{C}} \quad + \quad H_2SO_4$$

Transfer species of the first kind of first type

(12.95)

$$O \qquad > \qquad H_2C = C\,O$$

Order of nucleophilicity of Functional /Activation centers

(12.96)

Anionically, activation is not favored, in view of presence of transfer species of the first kind of the first type. In the growing polymer chain of Equation 12.95 which is the same as that reported in Equation 12.88, there is transfer species of the first kind of the first type, for these monomers chargedly or radically.

Now, coming back to (I) of Equation 12.87, one is going to use to advantage, such type of placement of same or different types of rings to determine the order of their nucleophilicities. It has already been shown that cyclic ethers are less nucleophilic than cyclic formals. To show similarly that cyclic ethers are less nucleophilic than trioxanes, the followings are obtained.

$$
\begin{array}{c}
\text{(I)} \quad + \quad H^{\oplus}\ {}^{\ominus}OSO_3H \quad \longrightarrow \quad \text{(II)}
\end{array}
$$

OR

$$
\text{(III)} \quad \longrightarrow \quad H-O-\overset{\overset{\displaystyle H}{|}}{\underset{\underset{\displaystyle H}{|}}{C}}-\overset{\overset{\displaystyle H}{|}}{\underset{\underset{\displaystyle H}{|}}{C}}-\overset{\overset{\displaystyle H}{|}}{\underset{\underset{\displaystyle H}{|}}{C}}- \quad \text{(II)(a)}
$$

OR

$$
H-O-\overset{\overset{\displaystyle H}{|}}{\underset{\underset{\displaystyle H}{|}}{C}}-O-\overset{\overset{\displaystyle H}{|}}{\underset{\underset{\displaystyle H}{|}}{C}}-O- \quad \text{(III)(a)} \quad \longrightarrow
$$

$$
H-O-\overset{\overset{\displaystyle H}{|}}{\underset{\underset{\displaystyle H}{|}}{C}}-\overset{\overset{\displaystyle H}{|}}{\underset{\underset{\displaystyle H}{|}}{C}}-\overset{\overset{\displaystyle H}{|}}{\underset{\underset{\displaystyle H}{|}}{C}}-\overset{\overset{\displaystyle O}{\|}}{C}-O-\overset{\overset{\displaystyle H}{|}}{\underset{\underset{\displaystyle H}{|}}{C}}-O-\overset{\oplus}{\underset{\underset{\displaystyle H}{|}}{C}}\cdots\cdots{}^{\ominus}OSO_3H \quad \text{OR}
$$

(II)(b)

$$
H-O-\overset{\overset{\displaystyle H}{|}}{\underset{\underset{\displaystyle H}{|}}{C}}-O-\overset{\overset{\displaystyle H}{|}}{\underset{\underset{\displaystyle H}{|}}{C}}-O-\overset{\overset{\displaystyle O}{\|}}{C}-\overset{\overset{\displaystyle H}{|}}{\underset{\underset{\displaystyle H}{|}}{C}}-\overset{\overset{\displaystyle H}{|}}{\underset{\underset{\displaystyle H}{|}}{C}}-\overset{\oplus}{\underset{\underset{\displaystyle H}{|}}{C}}\cdots\cdots{}^{\ominus}OSO_3H
$$

(III)(b)   12.97

In the (II), the cyclic ether is first attacked, while in the (III), the cyclic trioxane is first attacked. Since the positive end of (II)(b) is stronger than that of (III)(b), it is (II) that is favored; that is, **cyclic ether is less nucleophilic than trioxane.** Comparing (III)(b) above with (III) of Equation 12.89, it is obvious that, the former, (III)(b), is not favored. Also comparing (II)b with (V)bbb of Equation 12.91, it is obvious that **trioxane is more nucleophilic than cyclic formals.**

$$
\text{(IV)} \quad + \quad H^{\oplus}\ {}^{\ominus}OSO_3H \quad \longrightarrow \quad H-O-\overset{\overset{\displaystyle H}{|}}{\underset{\underset{\displaystyle H}{|}}{C}}-\overset{\overset{\displaystyle H}{|}}{\underset{\underset{\displaystyle H}{|}}{C}}-O-
$$

$$
H-O-\overset{\overset{\displaystyle H}{|}}{\underset{\underset{\displaystyle H}{|}}{C}}-\overset{\overset{\displaystyle H}{|}}{\underset{\underset{\displaystyle H}{|}}{C}}-O-\overset{\overset{\displaystyle O}{\|}}{C}-O-\overset{\overset{\displaystyle H}{|}}{\underset{\underset{\displaystyle H}{|}}{C}}-\overset{\oplus}{\underset{\underset{\displaystyle H}{|}}{C}}\cdots\cdots{}^{\ominus}OSO_3H
$$

12.98

$$\underset{(V)}{\overset{\displaystyle H_2C-O}{\underset{\displaystyle O}{H_2C}}\diagup C \diagdown \overset{\displaystyle O-CH_2}{\underset{\displaystyle O-CH_2}{}}O} \;+\; H^{\oplus}\,{}^{\ominus}OSO_3H \;\longrightarrow\; \underset{(V)(a)}{H-O-\overset{H}{\underset{H}{C}}-\overset{H}{\underset{H}{C}}-O-\overset{\oplus}{C}\diagup \overset{O-CH_2}{\underset{O-CH_2}{}}O \;\; {}^{\ominus}OSO_3H}$$

$$OR \quad \underset{(V)(b)}{H-O-\overset{H}{\underset{H}{C}}-O-\overset{H}{\underset{H}{C}}-O-\overset{\oplus}{C}\diagup \overset{O-CH_2}{\underset{O-CH_2}{}}\;\;{}^{\ominus}OSO_3H} \quad\longrightarrow$$

$$\underset{(V)aa\text{- FAVORED}}{H-O-\overset{H}{\underset{H}{C}}-\overset{H}{\underset{H}{C}}-O-\overset{O}{\overset{\|}{C}}-O-\overset{H}{\underset{H}{C}}-O-\overset{H}{\underset{H}{\overset{\oplus}{C}}}\cdots{}^{\ominus}OSO_3H} \qquad OR$$

$$\underset{(V)bb}{H-O-\overset{H}{\underset{H}{C}}-O-\overset{H}{\underset{H}{C}}-O-\overset{O}{\overset{\|}{C}}-O-\overset{H}{\underset{H}{C}}-\overset{H}{\underset{H}{\overset{\oplus}{C}}}\cdots{}^{\ominus}OSO_3H} \qquad\qquad 12.99$$

In (IV), the two rings are the same. Hence, any of the rings can be involved to favor the reactions in Equation 12.98. In (V), the two rings belonging almost to the same family are different. From the reactions above, (V)aa is the favored one than (V)bb, since the positive center of (V)aa is stronger than that of (V)bb. ***Hence cyclic formals are less nucleophilic than trioxanes***, and in general the following is obvious.

$$\text{Trioxanes} \quad > \quad \text{Cyclic formals} \quad > \quad \text{Cyclic ethers}$$

<u>Order of nucleophilicity</u> 12.100

Now, consider when two cyclic ether rings are placed side by side, without being fused or conjugatedly or cumulatively placed.

$$\underset{(I)}{\overset{H_2C}{\underset{O}{}}\diagup \underset{CH_2}{\overset{O}{}}\diagdown C} \xrightarrow{{}^{\ominus}:NH_2} \underset{}{{}^{\ominus}O-C-\overset{H}{\underset{O-CH_2}{\overset{\oplus}{C}}}} \longrightarrow \text{No further scission} \qquad 12.101$$

$$\underset{(I)}{\overset{H_2C}{\underset{O}{}}\diagup \underset{CH_2}{\overset{O}{}}\diagdown C} \xrightarrow[\substack{+\\ \text{(Non-ionic}\\ \text{metallic salt)}}]{\overset{nn}{\cdot}NH_2} nn\cdot O-\overset{H}{\underset{O-CH_2}{C}}-C\cdot e \longrightarrow nn\cdot O-\overset{H}{\underset{H}{C}}-\overset{O}{\overset{\|}{C}}-\overset{\mathbf{H}}{\underset{H}{C}}\cdot e \qquad 12.102$$

(II)

Chargedly, only one ring can be opened with possible polymerization in the absence of steric limitations. Radically, the two rings can be opened instantaneously one at a time to favor the existence of (II).

44

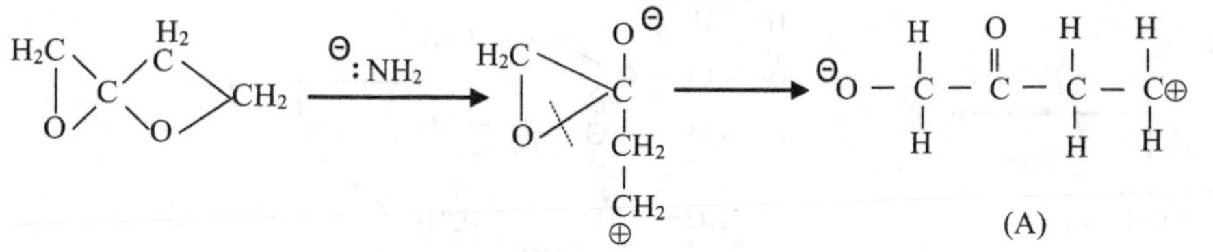

$$H_2C-C(H_2)-CH_2 \text{ (fused epoxide/oxetane)} \xrightarrow{^{\ominus}:NH_2} \cdots \longrightarrow {}^{\ominus}O-\overset{H}{\underset{H}{C}}-\overset{O}{\overset{\|}{C}}-\overset{H}{\underset{H}{C}}-\overset{H}{\underset{H}{C}}^{\oplus} \quad (A)$$

12.103

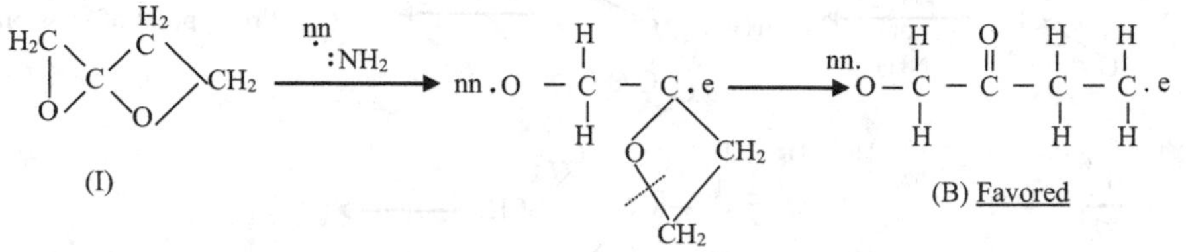

$$\text{(I)} \xrightarrow{nn \; :NH_2} nn \cdot O-\overset{H}{\underset{H}{C}}-C \cdot e \longrightarrow nn \cdot O-\overset{H}{\underset{H}{C}}-\overset{O}{\overset{\|}{C}}-\overset{H}{\underset{H}{C}}-\overset{H}{\underset{H}{C}} \cdot e \quad \text{(B) Favored}$$

12.104

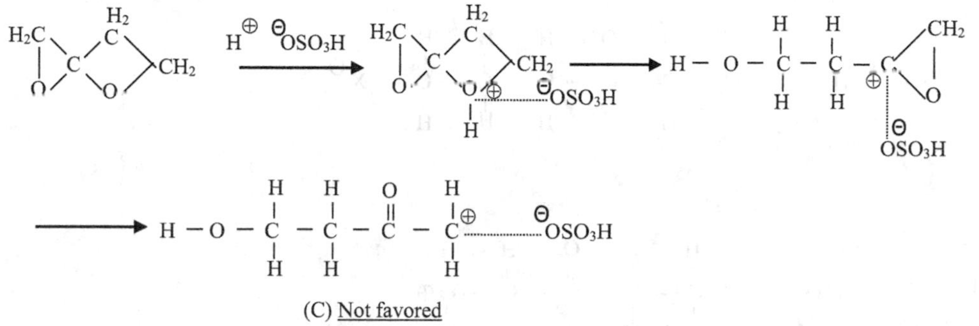

$$\xrightarrow{H^{\oplus} \; {}^{\ominus}OSO_3H} \cdots \longrightarrow H-O-\overset{H}{\underset{H}{C}}-\overset{H}{\underset{H}{C}}-C(\text{epoxide}) \quad {}^{\ominus}OSO_3H$$

$$\longrightarrow H-O-\overset{H}{\underset{H}{C}}-\overset{H}{\underset{H}{C}}-\overset{O}{\overset{\|}{C}}-\overset{H}{\underset{H}{C}}{}^{\oplus}\cdots\cdots {}^{\ominus}OSO_3H \quad \text{(C) Not favored}$$

12.105

$$\xrightarrow{H^{\oplus} \; {}^{\ominus}OSO_3H} \cdots \longrightarrow H-O-\overset{H}{\underset{H}{C}}-C(\text{oxetane}) \quad SO_3H$$

$$\longrightarrow H-O-\overset{H}{\underset{H}{C}}-\overset{O}{\overset{\|}{C}}-\overset{H}{\underset{H}{C}}-\overset{H}{\underset{H}{C}}{}^{\oplus}\cdots\cdots {}^{\ominus}OSO_3H \quad \text{(D) Favored}$$

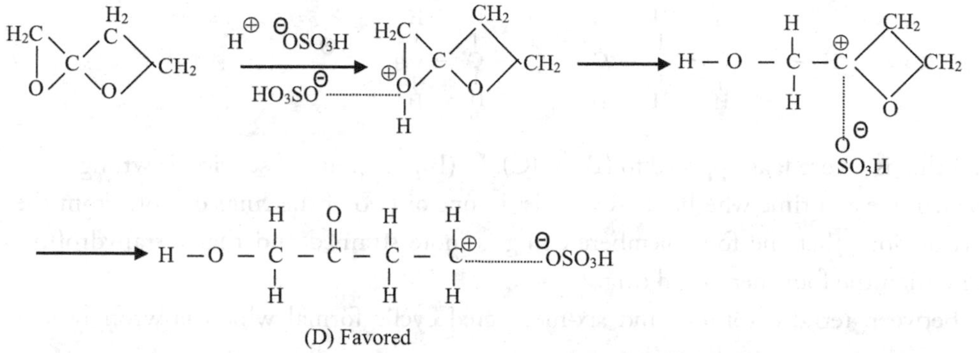

12.106

In the first equation above, it is the four-membered ring that is first opened instantaneously, which should not be the case. The three-membered ring should be the first to be opened instantaneously being less nucleophilic and more strained. Hence, as it seems for this case, both (A) and (B) are the same. Nevertheless, it is (B) that is favored radically or chargedly. In the absence of strong electrostatic forces, the four-membered ring being more nucleophilic, should be the last to be attacked. It is (D) that is therefore favored as opposed to (C). Hence, the following is valid.

Four-membered Cyclic ether　　>　　Three-membered Cyclic ether

Order of Nucleophilicity

12.107

$$+ \ H^{\oplus} \ X^{\ominus} \quad H-O-\overset{\overset{\textstyle H}{|}}{\underset{\underset{\textstyle H}{|}}{C}} -\overset{\overset{\textstyle H}{|}}{\underset{\underset{\textstyle H}{|}}{C}} -\overset{O-CH_2}{\underset{x^{\ominus}}{C^{\oplus}}} \overset{|}{\underset{CH_2-CH_2}{\quad}} \longrightarrow \text{(A)}$$

$$\xrightarrow[\underset{NH_2}{\ominus}]{:NR_3 \text{ or}} \quad \text{(B) Wrong point of scission}$$

$$+ \ H^{\oplus} \ \underset{X^{\ominus}}{} \quad H-O-\overset{\overset{\textstyle H}{|}}{\underset{\underset{\textstyle H}{|}}{C}} -\overset{\overset{\textstyle H}{|}}{\underset{\underset{\textstyle H}{|}}{C}} -\overset{\overset{\textstyle H}{|}}{\underset{\underset{\textstyle H}{|}}{C}} -\overset{CH_2}{\underset{x^{\ominus} \ O}{C^{\oplus}}} \overset{|}{\underset{}{CH_2}} \longrightarrow \text{(C)}$$

(II)

For (A)
$$H-O-\overset{\overset{\textstyle H}{|}}{\underset{\underset{\textstyle H}{|}}{C}}-\overset{\overset{\textstyle H}{|}}{\underset{\underset{\textstyle H}{|}}{C}}-\overset{\overset{\textstyle O}{\|}}{C}-\overset{\overset{\textstyle H}{|}}{\underset{\underset{\textstyle H}{|}}{C}}-\overset{\overset{\textstyle H}{|}}{\underset{\underset{\textstyle H}{|}}{C}}-\overset{\overset{\textstyle H}{|}}{\underset{\underset{\textstyle H}{|}}{C^{\oplus}}} \ X^{\ominus}$$

For (B)
$$^{\ominus}O-\overset{\overset{\textstyle H}{|}}{\underset{\underset{\textstyle H}{|}}{C}}-\overset{\overset{\textstyle H}{|}}{\underset{\underset{\textstyle H}{|}}{C}}-\overset{\overset{\textstyle H}{|}}{\underset{\underset{\textstyle H}{|}}{C}}-\overset{\overset{\textstyle O}{\|}}{C}-\overset{\overset{\textstyle H}{|}}{\underset{\underset{\textstyle H}{|}}{C}}-\overset{\overset{\textstyle H}{|}}{\underset{\underset{\textstyle H}{|}}{C^{\oplus}}}$$

For (C)
$$H-O-\overset{\overset{\textstyle H}{|}}{\underset{\underset{\textstyle H}{|}}{C}}-\overset{\overset{\textstyle H}{|}}{\underset{\underset{\textstyle H}{|}}{C}}-\overset{\overset{\textstyle H}{|}}{\underset{\underset{\textstyle H}{|}}{C}}-\overset{\overset{\textstyle O}{\|}}{C}-\overset{\overset{\textstyle H}{|}}{\underset{\underset{\textstyle H}{|}}{C}}-\overset{\overset{\textstyle H}{|}}{\underset{\underset{\textstyle H}{|}}{C^{\oplus}}} \ X^{\ominus}$$

12.108

It is the (A) that is favored, as opposed to (B) or (C). In (B), the point of scission is wrong. Only one ring can be opened one at a time whether MRSE exist in one or two of the rings or not. From the reactions above, it is obvious that the four membered ring is more strained and that tetrahydrofuran is more nucleophilic than the four-membered ring.

Now, between tetrahydrofuran and six-membered cyclic formal which is wrongly known to be strainless, the followings are to be expected.

$$\xrightarrow[\text{(MRSE in (a))}]{+ \ H^{\oplus} \ X^{\ominus} \ \text{OR}} \quad H-O-\overset{\overset{\textstyle H}{|}}{\underset{\underset{\textstyle H}{|}}{C}}-\overset{\overset{\textstyle H}{|}}{\underset{\underset{\textstyle H}{|}}{C}}-\overset{\overset{\textstyle H}{|}}{\underset{\underset{\textstyle H}{|}}{C}}-O-\overset{\overset{\textstyle O}{\|}}{C}-\overset{\overset{\textstyle H}{|}}{\underset{\underset{\textstyle H}{|}}{C}}-\overset{\overset{\textstyle H}{|}}{\underset{\underset{\textstyle H}{|}}{C}}-\overset{\overset{\textstyle H}{|}}{\underset{\underset{\textstyle H}{|}}{C^{\oplus}}} \ ^{\ominus}X$$

(A)

(MRSE) assumed present

$$^{\ominus}O-\overset{\overset{\textstyle H}{|}}{\underset{\underset{\textstyle H}{|}}{C}}-\overset{\overset{\textstyle H}{|}}{\underset{\underset{\textstyle H}{|}}{C}}-\overset{\overset{\textstyle H}{|}}{\underset{\underset{\textstyle H}{|}}{C}}-\overset{\overset{\textstyle O}{\|}}{C}-O-\overset{\overset{\textstyle H}{|}}{\underset{\underset{\textstyle H}{|}}{C}}-\overset{\overset{\textstyle H}{|}}{\underset{\underset{\textstyle H}{|}}{C}}-\overset{\overset{\textstyle H}{|}}{\underset{\underset{\textstyle H}{|}}{C^{\oplus}}}$$

(B)

12.109

46

By the reactions above in particular the first one, it is obvious that tetrahydrofuran or indeed cyclic ethers are less nucleophilic than cyclic formals. The fact that six-membered cyclic formal is more nucleophilic than five-membered cyclic formal is obvious when (B) above, is compared with (V)bb of Equation 12.91 or (III) of Equation 12.90.

$$
\begin{array}{ccccc}
\text{Seven-membered} & > & \text{Six-membered} & > & \text{Five-membered} \\
\text{cyclic formal} & & \text{cyclic formal} & & \text{cyclic formal}
\end{array}
$$

<u>Order of nucleophilicity</u>   12.110

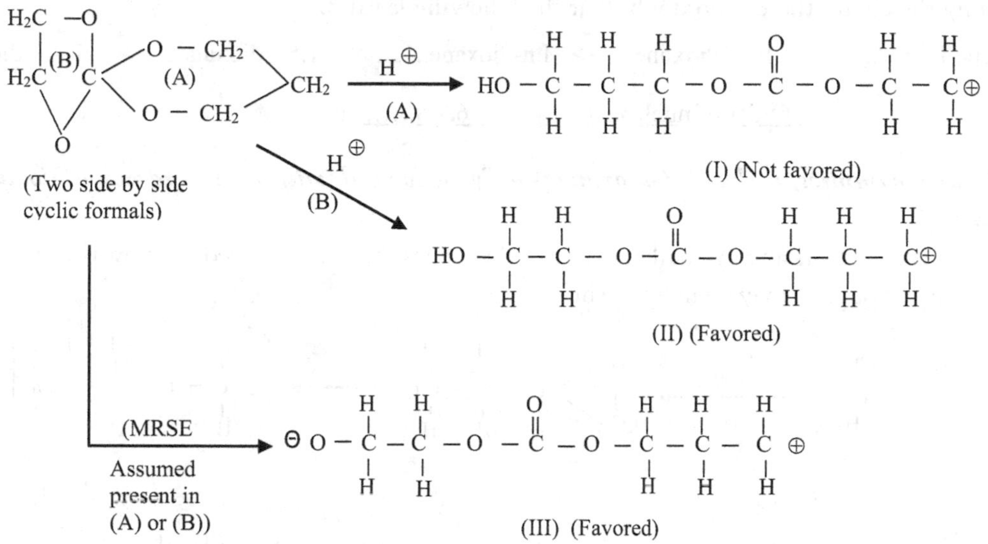

(Two side by side cyclic formals)

(I) (Not favored)

(II) (Favored)

(MRSE Assumed present in (A) or (B))

(III) (Favored)   12.111

It is the less nucleophilic ring that is first attacked, as indicated by (III) where artificial instantaneous opening of the ring has been done. The six-membered ring (A) is more nucleophilic than the five-membered ring, but less strained.

Even between tetrahydrofuran and 1,3- dioxane or 1,4-dioxane, two stubborn monomers, the followings are obtained.

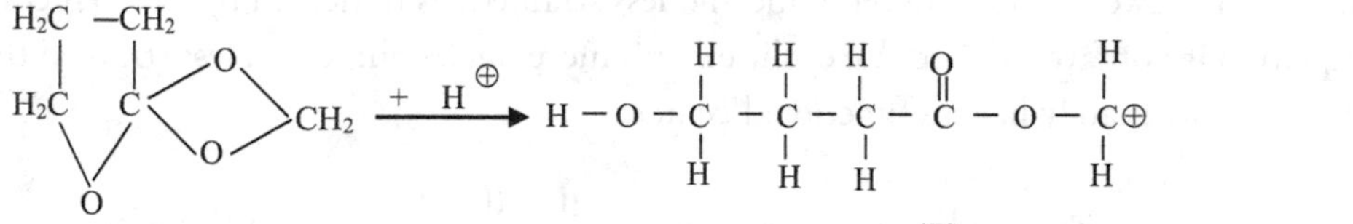

(B)   12.112

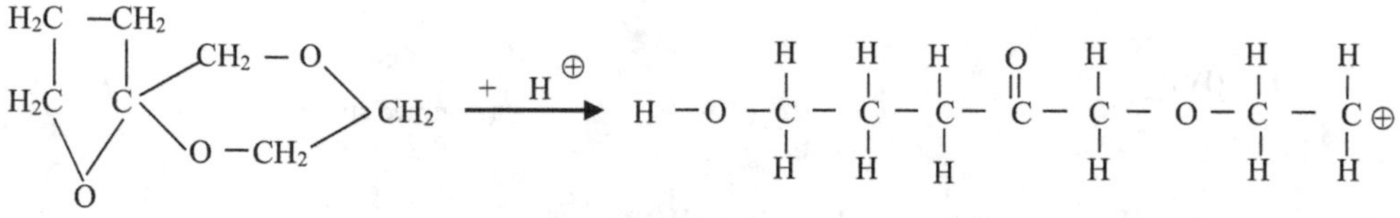

(C)   12.113

The fact that (B) from 1,3- dioxane is similar to (V)bb of Equation 12.91 for 1,3-dioxolane and (B) of Equation 12.109 for 1,3-dioxehane, all different from those obtained from 1,4-dioxane, 1,5-dioxane as shown by (C) of Equation 112.113 above and from the general considerations so far, the following is conclusively obvious.

$$--COOCH_3 \ (COOR) \quad > \quad -OCOCH_3 \ (OCOR)$$

<u>Order of radical- pulling capacity</u>   12.114

This order had long been observed; observed in the sense that, it was stated into rule that ***all free-radical-pulling groups are greater in capacity than all non-free-radical-pulling groups.***

From the product obtained in Equation 12.113, where in the (C), $-\overset{\overset{O}{\|}}{C}O$ OR $-\overset{\overset{O}{\|}}{C}-O-$ groups are absent along the chain, there is no doubt that the following is valid.

All Cyclic formals   >   1,3 - dioxanes   >   I,n-dioxane........ >   1,5 - dioxanes   >   1, 4 - dioxanes

<u>Order of nucleophilicity [n = 6,7,8,........]</u>   12.115

What the above implies, is that 1,3-dioxane is a Cyclic formal (with m=1) and not a Dioxane or of the Dioxane family.

    Between cyclic ethers and cycloalkanes and cycloalkenes etc., there is need to know the order of their nucleophilicity, though it may seem apparent.

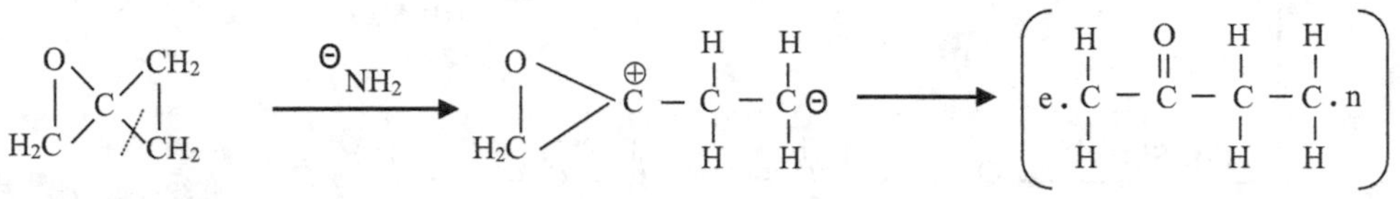

12.116

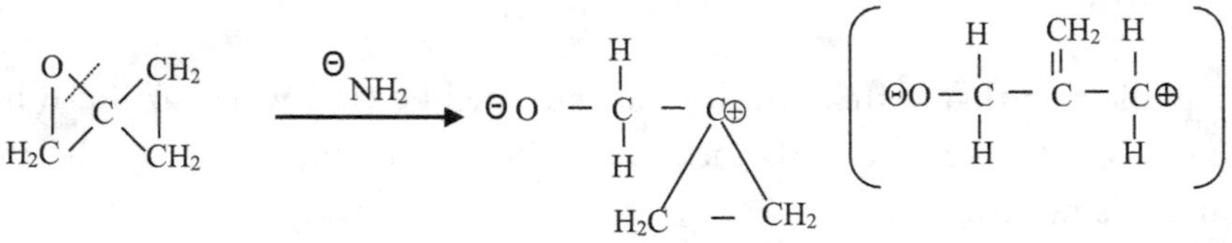

12.117

From the two equations above, it is obvious that monomers above can be opened chargedly or radically. The cyclic ether is however more nucleophilic and less strained, as indicated by the first equation above. The last equation is not favored, because cyclic ether is more nucleophilic and less strained than the cyclic alkane which unfortunately has no functional center.

(IV)   (assumed strained)

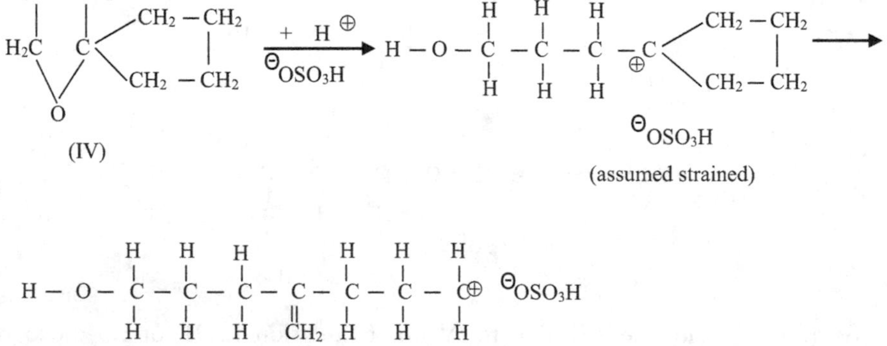

<u>Not favored</u>   12.118

The five-membered cycloalkane cannot be opened instantaneously. Secondly, it has no functional center. Hence, it is the cyclic ether ring that is first attacked, that which cannot take place. If MRSE can be provided, the reactions above cannot take place. Instead, similar reactions as in Equation 12.116 will take place, for which it is obvious that, while the cyclic ether is more nucleophilic, it is less strained than cycloalkane.

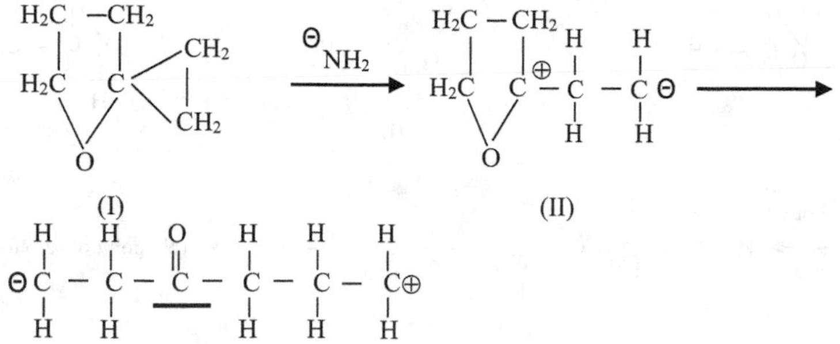

12.119

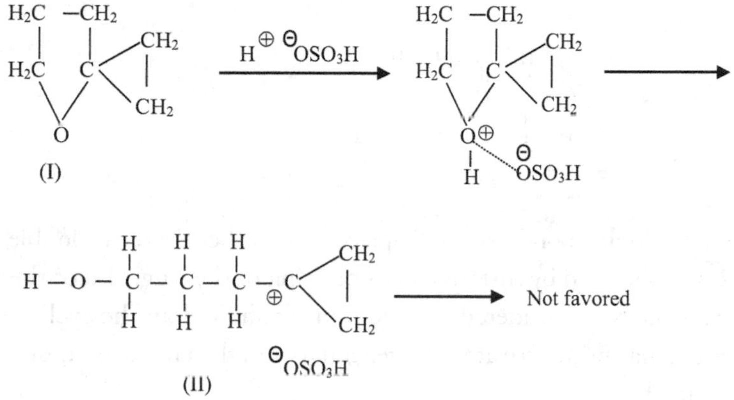

12.120

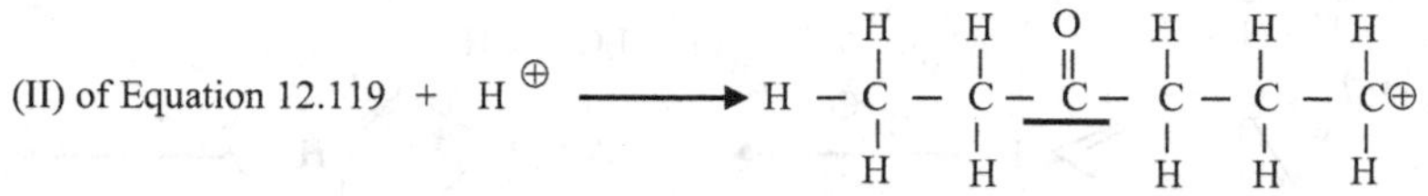

12.121

In the presence of strong electrostatic forces, the three-membered ring is opened instantaneously. In the presence of a positive charge, the oxygen center cannot be attacked though it is more nucleophilic. Hence, the reaction of equation 12.121 is favored if MRSE can be introduced into the tetrahydrofuran ring after attack by the cation on (II). In the presence of weak electrostatic forces, the more nucleophilic ring cannot be attacked as shown in Equation 12.120. Subsequently, there is no reaction. There is no doubt that cycloalkane rings are more strain than cyclic ether rings of the different sizes.

Replacing the cycloalkane with cycloalkene, the followings are obtained.

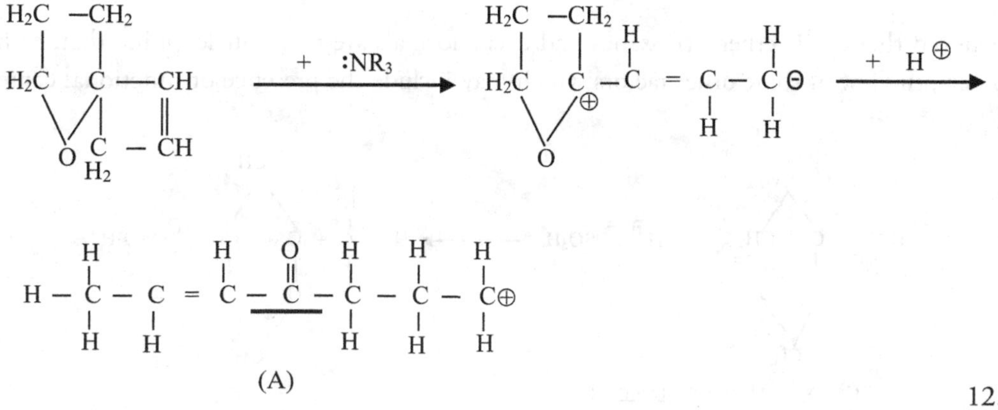

12.122

$$
\begin{array}{ccc}
\underset{\substack{\text{H}_2\text{C}-\text{CH}_2 \\ | \quad\quad | \\ \text{H}_2\text{C}\quad \text{C}-\text{CH} \\ \diagdown | \quad\quad \| \\ \text{O}\; \text{C}-\text{CH} \\ \text{H}_2}}{}
& \xrightarrow{+\;\text{H}^{\oplus}}
& \underset{\substack{\text{H}_2\text{C}-\text{CH}_2 \quad \text{H} \\ | \quad\quad | \quad | \\ \text{H}_2\text{C}\quad \text{C}-\text{C}-\text{H} \\ \diagdown | \quad\quad | \\ \text{O}\; \text{C}-\text{CH} \\ \text{H}_2 \;\oplus}}{(I)}
\quad \text{OR} \quad
\underset{\substack{\text{H}_2\text{C}-\text{CH}_2 \\ | \quad\quad | \\ \text{H}_2\text{C}\quad \text{C}-\text{CH} \\ | \quad\quad \| \\ \oplus\text{O}\; \text{C}-\text{CH} \\ \diagup\; \text{H}_2 \\ \text{X}^{\ominus}\diagup \text{H}}}{(II)}
\end{array}
$$

For (II)

$$
\xrightarrow{\quad} \text{H}-\text{O}-\underset{\text{H}}{\overset{\text{H}}{\underset{|}{\overset{|}{\text{C}}}}}-\underset{\text{H}}{\overset{\text{H}}{\underset{|}{\overset{|}{\text{C}}}}}-\underset{\text{H}}{\overset{\text{H}}{\underset{|}{\overset{|}{\text{C}}}}}-\overset{\oplus}{\text{C}}-\text{CH} \xrightarrow{\quad} \text{(No point of scission)}
$$
$$
\underset{\text{H}_2\text{C}-\text{CH}}{\|}
$$

(II)a

For (I)

$$
\xrightarrow{\quad} \underset{\text{H}}{\overset{\text{H}}{\underset{|}{\overset{|}{\text{C}}}}}=\underset{\text{H}}{\overset{\text{H}}{\underset{|}{\overset{|}{\text{C}}}}}-\underset{\text{H}}{\overset{\text{H}}{\underset{|}{\overset{|}{\text{C}}}}}-\underset{}{\overset{\text{O}}{\underset{\underline{\quad}}{\overset{\|}{\text{C}}}}}-\underset{\text{H}}{\overset{\text{H}}{\underset{|}{\overset{|}{\text{C}}}}}-\underset{\text{H}}{\overset{\text{H}}{\underset{|}{\overset{|}{\text{C}}}}}-\overset{\text{H}}{\underset{\text{H}}{\overset{|}{\underset{|}{\text{C}}}}}\oplus
$$

(I)a

12.123

Though, there seems to be a double bond externally placed to another ring, the double bond is in a ring. (A) of Equation 12.122 was obtained by instantaneous opening of the ring. (I)a of the last equation was obtained because cyclic ethene was considered to be less nucleophilic than the cyclic ether. It is (A) that identifies with (I)a after (I)a has been activated. In general, the cyclic ethers are more nucleophilic than cycloalkenes, but less strained.

$$
\begin{array}{ccc}
\underset{\substack{\text{H}_2\text{C}-\text{CH}_2 \\ | \quad\quad | \\ \text{H}_2\text{C}\quad \text{C}-\text{CH} \\ \diagdown | \quad\quad \diagdown \\ \text{O}\; \text{C}-\text{C}\;\;\text{CH} \\ \text{H}_2\;\text{H}_2}}{}
& \xrightarrow[\text{X}^{\ominus}]{\text{H}^{\oplus}}
& \underset{\substack{\text{H}_2\text{C}-\text{CH}_2 \\ | \quad\quad | \quad \text{CH} \\ \text{H}_2\text{C}\quad \text{C}\diagup \\ | \quad\quad \diagdown\text{CH} \\ \quad\quad \text{C}-\text{CH}_2 \\ \oplus\text{O} \\ \text{X}^{\ominus}\cdots\; | \\ \text{H}}}{}
& \xrightarrow{\quad\quad}
\end{array}
$$

$$
\text{H}-\text{O}-\left[\underset{\text{H}}{\overset{\text{H}}{\underset{|}{\overset{|}{\text{C}}}}}\right]_3\overset{\oplus}{\text{C}}-\text{CH} \xrightarrow{\quad} \text{H}-\text{O}-\left[\underset{\text{H}}{\overset{\text{H}}{\underset{|}{\overset{|}{\text{C}}}}}\right]_3\underset{\text{H}}{\overset{\text{CH}_2}{\underset{|}{\overset{\|}{\text{C}}}}}-\text{C}=\overset{\text{H}}{\underset{\text{H}}{\overset{|}{\underset{|}{\text{C}}}}}-\overset{\text{H}}{\underset{|}{\text{C}}}\oplus
$$
$$
\underset{\substack{\text{H}_2\text{C}\quad\quad\text{CH} \\ \diagdown\diagup \\ \text{CH}_2}}{\|}
\qquad\qquad \underline{\text{(Not favored)}}
$$

12.124

It is believed that cyclic ethers, trioxanes and cyclic formals are more nucleophilic than carbon-ringed monomers, in view of some other factors which may include the presence of functional centers in their rings.

$$
\underset{\substack{| \quad\quad | \\ \text{O}\quad\quad \text{O} \\ \diagdown\diagup \\ \text{CH}_2}}{\overset{\text{CH}_2}{\overset{\diagup\diagdown}{\text{H}_2\text{C}=\text{C}\quad\text{CH}_2}}} + \text{H}^{\ominus}\;{}^{\oplus}\text{OSO}_3\text{H} \xrightarrow{\quad} \text{H}-\underset{\text{H}}{\overset{\text{H}}{\underset{|}{\overset{|}{\text{C}}}}}-\overset{\oplus}{\underset{\substack{| \\ \text{O}}}{\text{C}}}\quad\underset{\substack{| \\ \text{O} \\ \diagdown\diagup \\ \text{CH}_2}}{\overset{\text{CH}_2}{\overset{\diagup\diagdown}{\text{CH}_2}}}\quad {}^{\ominus}\text{OSO}_3\text{H}
$$

(I) <u>Six-membered cyclic formal</u>

$$\longrightarrow \quad H - \overset{\overset{H}{|}}{\underset{\underset{H}{|}}{C}} - \underline{\overset{\overset{O}{\|}}{C}} - \overset{\overset{H}{|}}{\underset{\underset{H}{|}}{C}} - \overset{\overset{H}{|}}{\underset{\underset{H}{|}}{C}} - \underline{O} - \overset{\overset{H}{|}}{\underset{\underset{}{}}{C}}\oplus \quad {}^{\ominus}OSO_3H$$

(B)   12.125

$$H_2C = \overset{\overset{O}{\diagup\diagdown}}{\underset{\underset{CH_2}{\diagdown\diagup}}{C}} CH_2 \qquad \xrightarrow{H^{\oplus} \; {}^{\ominus}OSO_3H} \qquad H - \overset{\overset{H}{|}}{\underset{\underset{H}{|}}{C}} - \underline{\overset{\overset{O}{\|}}{C}} - O - \overset{\overset{H}{|}}{\underset{\underset{H}{|}}{C}} - O - \overset{\overset{H}{|}}{\underset{\underset{}{}}{C}}\oplus \; {}^{\ominus}OSO_3H$$

(C)

(II) <u>Six-membered Trioxane</u>   12.126

$$H_2C = \overset{\overset{CH_2}{\diagup\diagdown}}{\underset{\underset{O \quad CH_2}{\diagdown\diagup}}{C}} CH_2 \qquad \xrightarrow{H^{\oplus} \; {}^{\ominus}OSO_3H} \qquad H - \overset{\overset{H}{|}}{\underset{\underset{H}{|}}{C}} - \underline{\overset{\overset{O}{\|}}{C}} - \overset{\overset{H}{|}}{\underset{\underset{H}{|}}{C}} - \overset{\overset{H}{|}}{\underset{\underset{H}{|}}{C}} - \overset{\overset{H}{|}}{\underset{\underset{H}{|}}{C}} - \overset{\overset{H}{|}}{\underset{\underset{}{}}{C}}\oplus \; {}^{\ominus}OSO_3H$$

(D)

(III) <u>Six-membered Tetrahydropyran</u>   12.127

$$H_2C = \overset{\overset{CH_2}{\diagup\diagdown}}{\underset{\underset{H_2C \quad CH_2}{}}{C}} CH_2 \qquad \xrightarrow{H^{\oplus} \; {}^{\ominus}OSO_3H} \qquad \oplus C - \overset{\overset{H}{|}}{\underset{\underset{H}{|}}{C}} - \overset{\overset{H}{|}}{\underset{\underset{H}{|}}{C}} - \overset{\overset{H}{|}}{\underset{\underset{H}{|}}{C}} - \overset{\overset{H}{|}}{\underset{\underset{H}{|}}{C}} - C\ominus \quad + \quad H^{\oplus} \; {}^{\ominus}OSO_3H$$

(IV) <u>Six-membered cyclohexane</u>

$$\longrightarrow \quad H - \overset{\overset{H}{|}}{\underset{\underset{C_5H_{11}}{|}}{C}} = C\oplus \; {}^{\ominus}OSO_3H \qquad OR \qquad H - \overset{\overset{H}{|}}{\underset{\underset{H}{|}}{C}} - \overset{\overset{CH_2}{\|}}{\underset{\underset{H}{|}}{C}}\left\{ \overset{\overset{H}{|}}{\underset{\underset{H}{|}}{C}} \right\}_3 \overset{\overset{H}{|}}{\underset{\underset{H}{|}}{C}}\oplus \; {}^{\ominus}OSO_3H$$

(E)   (F)
                                              (Not favored)   12.128

If (IV) is instantaneously opened, (E) is obtained. Based on the activated state of (IV), existence of (F) is not possible, since the double bond in (IV) cannot be activated in the absence of a functional center in a ring.

$$\overset{\overset{O}{\diagup\diagdown}}{\underset{\underset{O \quad C = CH_2}{}}{H_2C \quad CH_2}} \qquad \xrightarrow{H^{\oplus} \; {}^{\ominus}OSO_3H} \qquad H - \overset{\overset{H}{|}}{\underset{\underset{H}{|}}{C}} - C\oplus \quad \overset{\overset{CH_2}{\diagup\diagdown}}{\underset{\underset{O}{}}{O \quad H_2C \quad CH_2}} \qquad \longrightarrow \text{(No point of scission).}$$

<u>Six-membered cyclic formal</u>   12.129

For the six-membered ring above, there will be no ring opening, whereas for the case of equation 12.125, the ring is opened, in view of the location of the $=CH_2$ group. This will however not show up with

51

five-membered cyclic formals. Nevertheless, from the products obtained in (B) of Equation 12.125, (C) of Equation 12.126 and (D) of Equation 12.127, the followings are valid.

> Six-membered cyclic formal   >   Five-membered cyclic formal   >

Order of nucleophilicity

12.130

Trioxane   >   Six-membered cyclic formal   >   Tetrahydropyran

Order of nucleophilicity

12.131

Non- Functional centers

12.132

One can thus observe, for the first time the unique features of these monomers. Since most of the rings of cyclic acetals do not favor instantaneous opening of their rings, radical characters of these family members will not be considered, since we know where the difference lies.

Based on side by side placements of two rings, corresponding equation to Equation 12.92 for order of radical-pushing capacities of groups can be obtained as follows. For cyclic ethers and acetals, the followings are obtained.

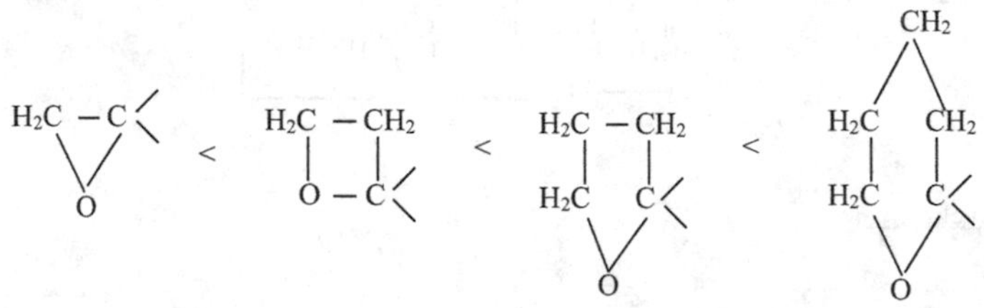

Order of radical-pushing capacity of cyclic ether groups

12.133

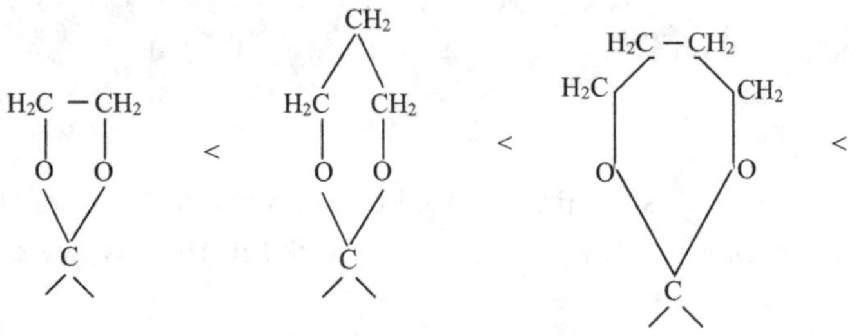

Order of radical-pushing capacity of cyclic formal groups

12.134

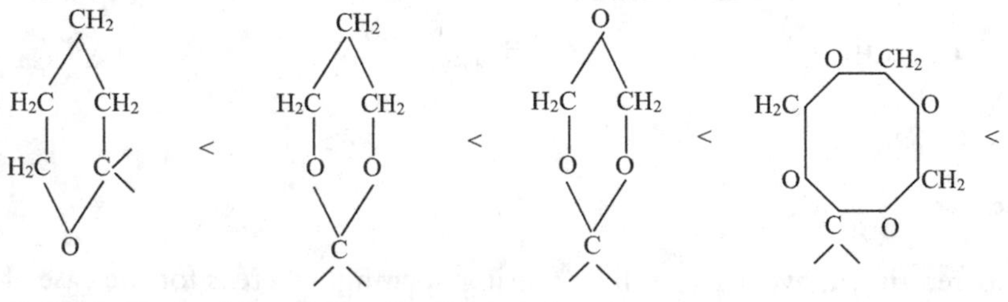

Order of radical-pushing capacity of cyclic ethers, formal, trioxane, and tetraoxane groups

12.135

Between cycloalkanes, cycloalkenes, etc. and cyclic oxygen hetero rings, the followings are obtained.

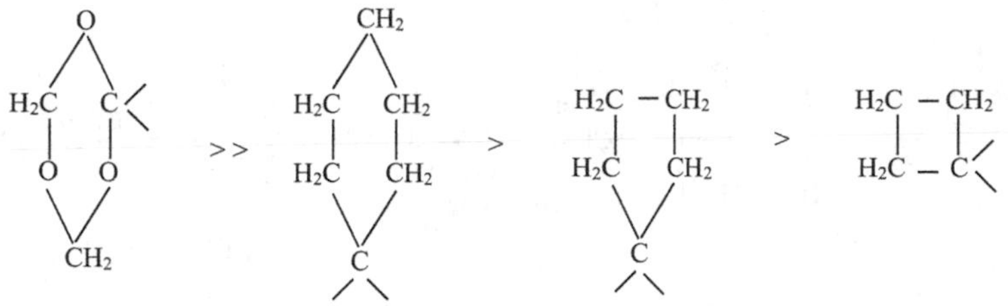

Order of radical-pushing capacity

12.136

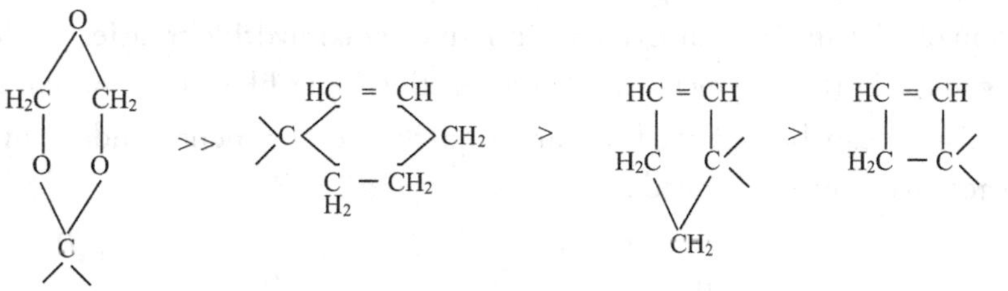

Order of radical-pushing capacity

12.137

These groups function when attached to other ring(s). As can be observed, the orders are measures of the nucleophilicity of the rings in the groups. It is important to note that the carbon center carrying the group externally located is not just any carbon center in the ring. It is the carbon center adjacently located to a hetero atom or an internally located double bond. Hence not all rings can possess much substituent groups. Cycloalkanes were considered, since they can be placed side by side to a ring. When placed side by side to a hetero-ring, depending on the size of the ring, the strain energy is considerably increased; and only instantaneous opening of the rings can favor the existence of a linear chain. The functional center on the second ring, when used can never open the cylcoalkane or cycloalkene rings or any carbon-carbon ringed monomer. Never can it be the first to be attacked.

12.2.2 Propagation/Depropagation Phenomena

1,3- dioxane is quite unreactive to homo-polymerization, while 1,3,5 - trioxane is reactive. The reason could partly be due to the fact that 1,3- dioxane readily undergoes instantaneous opening of its rings when activated to produce an etheric aldehyde radically. 1,3,5,7- tetraoxane is unreactive and less so than 1,3- dioxane. Despite this, existence of 1,3- dioxane during depropagation reactions of trioxane has never been reported. This is due to the type of ring which is more strained.

Similarly existence of even formaldehyde should be favored if polymerized in the route not natural to it. 1,3,5 - trioxane is a colorless highly refractive crystalline compound melting at 62°C and boiling without decomposition or "depolymerization" at 115°C. Strong acids are known to initiate "depolymerization" as will all compounds of this type and it is useful as a source of formaldehyde in reactions carried out under anhydrous conditions [23].

As has been indicated already, depropagation reactions take place largely when free centered initiators are involved. Free positively, the initiators known to be involved are acidic in character. When the initiator or acid is strong, then long induction periods exist in the system, because of equilibrium between the growing chain and formaldehyde as shown below, apart from the type of Electrostatic initiator involving the use of the monomer involved as will shortly be shown downstream. When this happens then the polymerization becomes Backward and electro-free-radical in character with more de-propagation. The existence of formaldehyde is however still favored from the type of growing chain shown below (I), its existence increasing with increasing acidity of the olymerization medium. [See Equations 12.28 and 12.29]

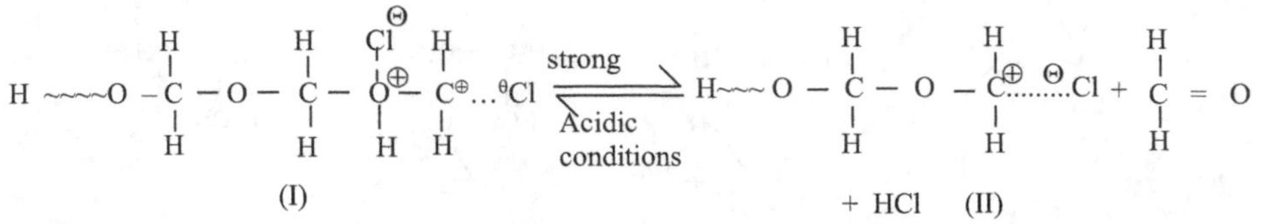

$$12.138$$

If a non-acid initiator such as benzyl chloride or a non-acid medium is provided, then the induction periods can be greatly reduced, if not completely eliminated. It is also said that an additional complication in the polymerization of trioxane is termination by intra-molecular hydride transfer resulting in **methoxyl and formate end groups** in the polymer[1]. While the reaction looks like a termination reaction, hydride ions $H^{\ominus}$ do not exist as can be observed so far in all the developments under any polymerization conditions. The reactions can be explained as follows.

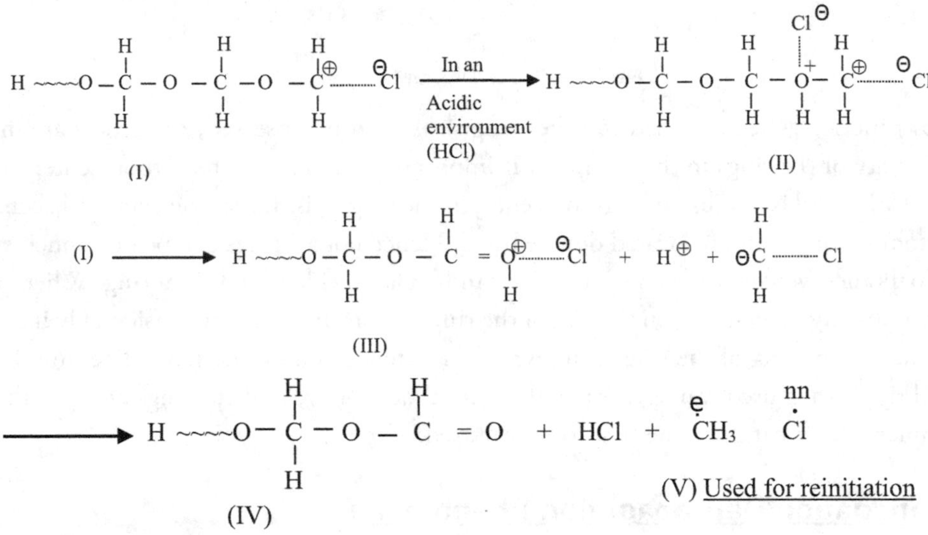

Transfer from growing polymer chain to monomer by Depropagation

$$12.139$$

It should be noted that the positive charge on the oxygen center in (I) and (III) is not ionic, but electrostatic in character, for which (IV) is favored after the release of a stable molecule (HCl). For dimethylene oxides, the followings are to be expected.

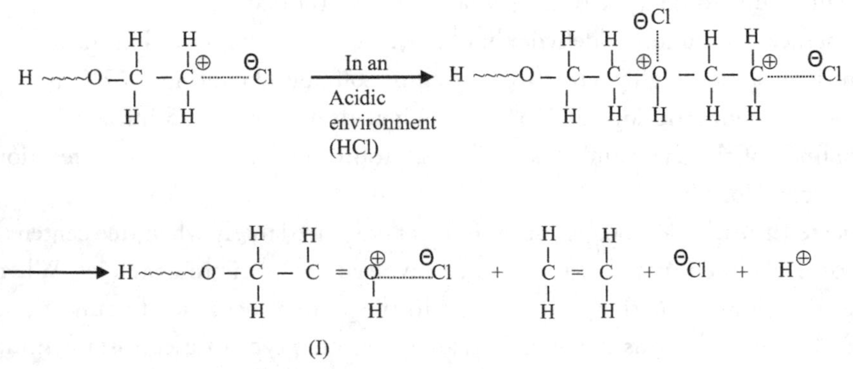

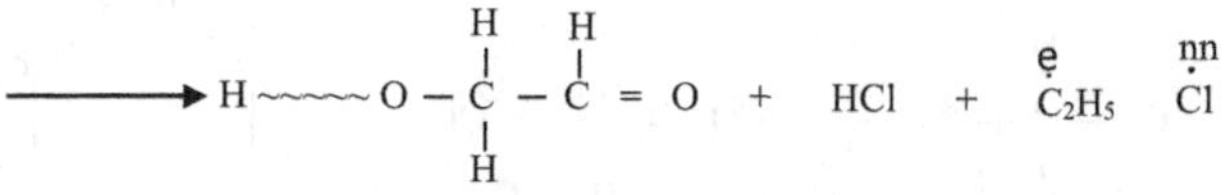

$$\longrightarrow H \sim\sim\sim O - \underset{\underset{H}{|}}{\overset{\overset{H}{|}}{C}} - \overset{\overset{H}{|}}{C} = O \quad + \quad HCl \quad + \quad \overset{e}{C_2H_5} \quad \overset{nn}{Cl}$$

(II)

<u>Transfer from growing chain to monomer by depropagation</u>   12.140

(II) above is not a formate end group like the case of Equation 12.139. In the reaction of Equation 12.139, the dead polymer obtained has a formate end group. That with a methyl end group can be obtained when the CH_3 groups of the equation initiates polymerization. The use of (I) of that equation or Equation 12.138 cannot favor their existence even radically as shown below via a different method of releasing H, because of electrostatic forces of repulsion.

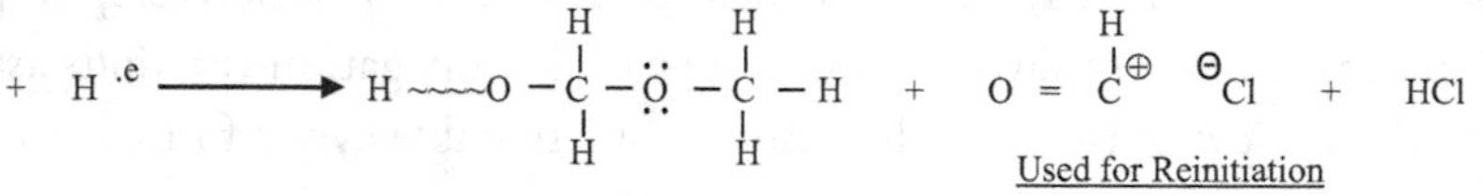

$$+ \quad H^{.e} \longrightarrow H\sim\sim\sim O - \underset{\underset{H}{|}}{\overset{\overset{H}{|}}{C}} - \overset{..}{\underset{..}{O}} - \underset{\underset{H}{|}}{\overset{\overset{H}{|}}{C}} - H \quad + \quad O = \overset{\overset{H}{|}}{C}{}^{\oplus} \quad {}^{\ominus}Cl \quad + \quad HCl$$

(IV) Not favored   <u>Used for Reinitiation</u>   12.141

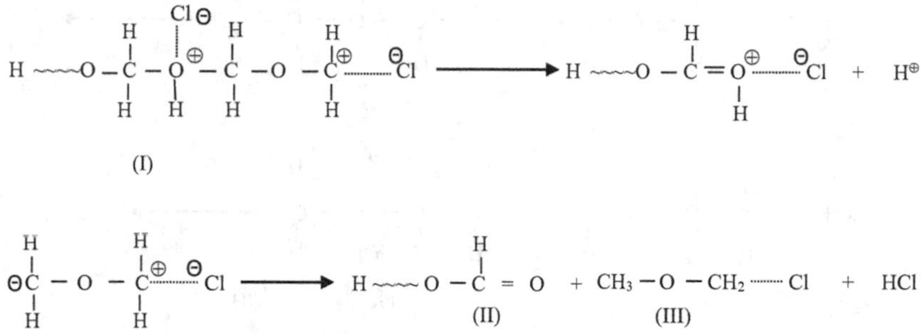

$$H\sim\sim\sim O - \underset{\underset{H}{|}}{\overset{\overset{H}{|}}{C}} - \overset{\overset{Cl^{\ominus}}{|}}{\underset{\underset{H}{|}}{O^{\oplus}}} - \underset{\underset{H}{|}}{\overset{\overset{H}{|}}{C}} - O - \underset{\underset{H}{|}}{C}{}^{\oplus}\cdots\cdots{}^{\ominus}Cl \longrightarrow H\sim\sim\sim O - \underset{\underset{H}{|}}{C} = O^{\oplus}\quad{}^{\ominus}Cl \quad + \quad H^{\oplus}$$

(I)

$$+ \quad {}^{\ominus}\overset{\overset{H}{|}}{\underset{\underset{H}{|}}{C}} - O - \underset{\underset{H}{|}}{C}{}^{\oplus}\quad{}^{\ominus}Cl \longrightarrow H\sim\sim\sim O - \overset{\overset{H}{|}}{C} = O \quad + \quad CH_3 - O - CH_2\cdots\cdots Cl \quad + \quad HCl$$

(II)   (III)

12.142

<u>Possible transfer from growing polymer chain to monomer by depropagation</u>

$$H\sim\sim\sim O - \underset{\underset{H}{|}}{\overset{\overset{H}{|}}{C}} - \overset{\overset{Cl^{\ominus}}{|}}{\underset{\underset{H}{|}}{O^{\oplus}}} - \underset{\underset{H}{|}}{\overset{\overset{H}{|}}{C}} - O - \underset{\underset{H}{|}}{C}{}^{\oplus}\quad{}^{\ominus}Cl \longrightarrow H\sim\sim\sim O - \underset{\underset{H}{|}}{\overset{\overset{H}{|}}{C}} - H \quad +$$

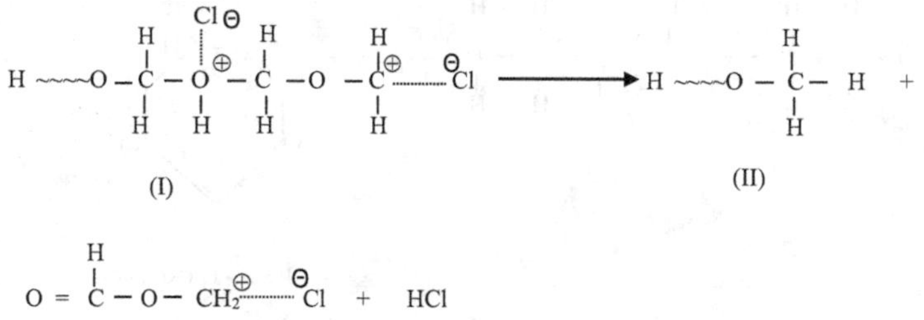

(I)   (II)

$$\overset{\overset{H}{|}}{O} = \underset{\underset{}{}}{C} - O - CH_2{}^{\oplus}\cdots\cdots{}^{\ominus}Cl \quad + \quad HCl$$

(III)   <u>NOT FAVORED</u>   12.143

(III) of Equation 12.142 generates chains with methyl and formate ends. (III) of Equation 12.143 though generates the same also, is not favored just like Equation 12.141. It can be observed that any of the oxygen centers can be attacked in an acidic environment to favor the existence of both types of end groups.

In a non-acidic environment, trioxane also undergoes depropagation reactions as shown below.

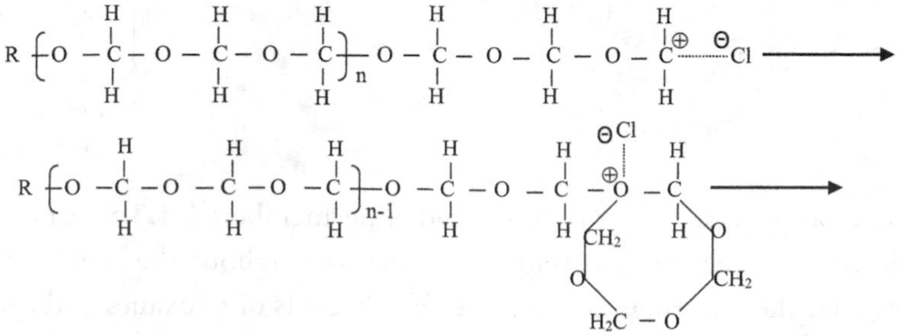

$$R \left(O - \underset{\underset{H}{|}}{\overset{\overset{H}{|}}{C}} - O - \underset{\underset{H}{|}}{\overset{\overset{H}{|}}{C}} - O - \underset{\underset{H}{|}}{\overset{\overset{H}{|}}{C}} \right)_n O - \underset{\underset{H}{|}}{\overset{\overset{H}{|}}{C}} - O - \underset{\underset{H}{|}}{\overset{\overset{H}{|}}{C}} - O - \underset{\underset{H}{|}}{C}{}^{\oplus}\quad{}^{\ominus}Cl \longrightarrow$$

$$R \left(O - \underset{\underset{H}{|}}{\overset{\overset{H}{|}}{C}} - O - \underset{\underset{H}{|}}{\overset{\overset{H}{|}}{C}} - O - \underset{\underset{H}{|}}{\overset{\overset{H}{|}}{C}} \right)_{n-1} O - \underset{\underset{H}{|}}{\overset{\overset{H}{|}}{C}} - O - \underset{\underset{H}{|}}{\overset{\overset{H}{|}}{C}} - \overset{\overset{{}^{\ominus}Cl}{|}}{O^{\oplus}} - \underset{\underset{H}{|}}{\overset{\overset{H}{|}}{C}} \longrightarrow$$

$$\begin{array}{c} CH_2 \quad\quad CH_2 \\ O \quad\quad\quad O \\ H_2C - O \end{array}$$

55

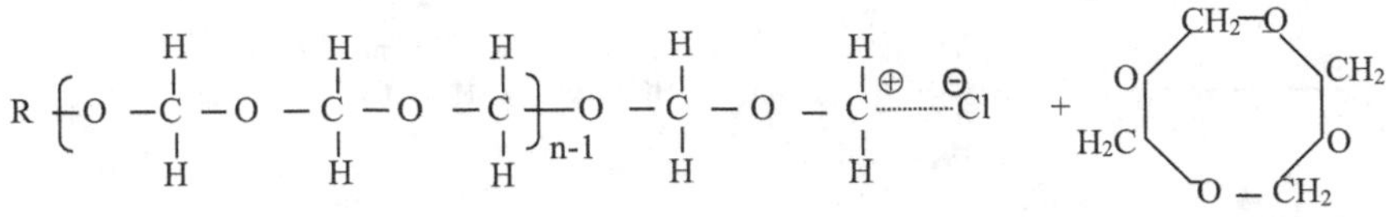

$$R \left[O - \underset{\underset{H}{|}}{\overset{\overset{H}{|}}{C}} - O - \underset{\underset{H}{|}}{\overset{\overset{H}{|}}{C}} - O - \underset{\underset{H}{|}}{\overset{\overset{H}{|}}{C}} \right]_{n-1} O - \underset{\underset{H}{|}}{\overset{\overset{H}{|}}{C}} - O - \underset{\underset{H}{|}}{\overset{\overset{H}{|}}{C}} {}^{\oplus} \overset{\ominus}{Cl} \quad + \quad \text{Tetraoxane}$$

$$\text{12.144}$$

Tetraoxane has been observed in the past during polymerization as one of the products[6]. Temporary existence of tetraoxane is favored only when the strength of the positive center of the growing chain is weak to provide the MRSE for the eight-membered ring. As propagation progresses some of the active centers become strong to consume the tetraoxane. When existence of larger membered rings which are very stable to polymer-ization is favored, than the mechanism is ***full depropagation.***

Cyclic formals undergo similar reactions of the type of transfer reactions step to produce methoxyl and formate end groups. They also favor the other type of depropagation reactions as clearly identified, but on a smaller scale in view of the possible existence of less stable cyclic formals than the oxanes. For example, consider the cases of 1,3 - dioxane and 1,3 - dioxepane.

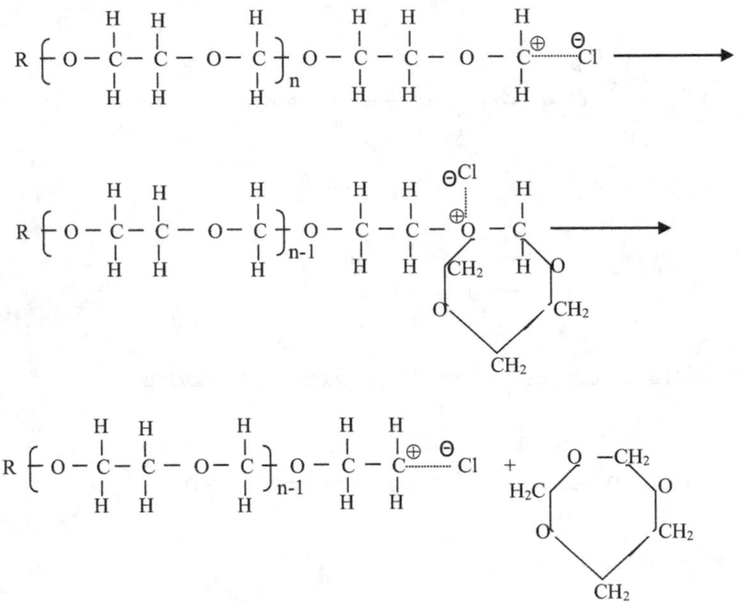

$$\text{12.145}$$

(I) 1,3,5 - trioxolane

$$\text{12.146}$$

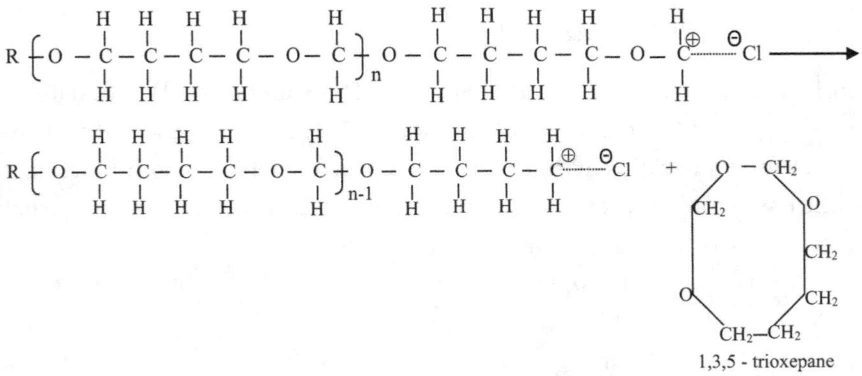

1,3,5 - trioxepane

If MRSE cannot be provided for the seven and nine membered 1,3,5 - trioxalane and 1,3, 5 - trioxepane respectively, then they exist as stable molecules throughout the course of polymerization. It is important to note that the two rings are cyclic acetal - hybrids of trioxanes and cyclic formals. Either the seven- or - nine-membered rings are favored or next ten- or fourteen-membered rings shown below are favored.

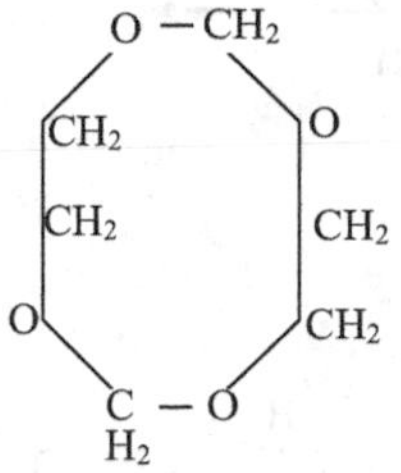

(I) <u>1,3,6,8 - tetraoxalane</u>

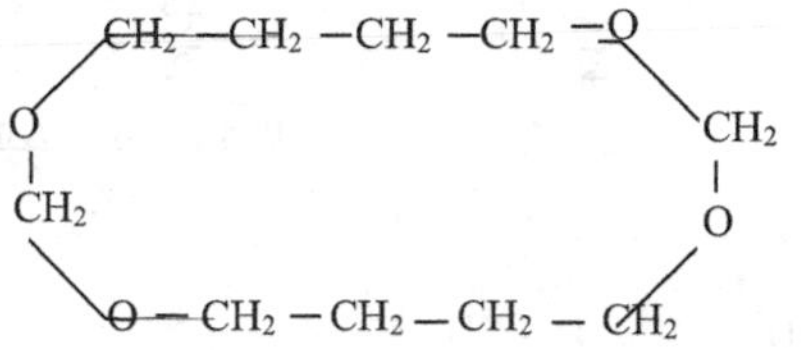

(I) <u>1,3,8,10 - tetraoxepane</u>

12.147

These are full cyclic formals, but no hybrids. It is of importance to note that presence of the di- or tetra- forms of oxanes cannot be favored during the depropagation reactions of cyclic formals. Only the tri - oxane forms can be obtained as a hybrid, based on the point of attack on the chain. The unique relationships between trioxanes and cyclic formals as full members of cyclic acetals can thus be observed.

The newly identified types of rings above, are worthy of note. Whether the growing polymer chains are of the electrostatic types or not, the depropagation reactions indicated above, particularly occur in the absence of formation of formaldehyde (see Equation 12.138). If presence of formaldehyde is favored, then there will be transfer from growing polymer chain to monomer sub-step, in addition to minor depropagation reaction all taking place simultaneously.

From all the considerations so far, the conditions favoring the existence of depropagation phenomena can largely be observed. The conditions seem to be determined most importantly by the type of monomer involved. While cyclic ethers can depropagate in both ionic and non-ionic environments, it is far less than those of cyclic acetals.

Why cyclic formals are more related to the trioxanes than dioxanes and tetraoxanes which are less reactive, have partly been explained. How these relationships were originally identified have largely been based on experimental data.

Based on the route favored for polymerization of cyclic acetals, only the positively charged routes will favor the copolymerization of these monomers with cyclic ethers and with themselves. Positively random copolymers will largely be obtained in the absence of influence of electrostatic forces from the counter-ion. The same will seem to apply to ion-paired initiators, except for probably the three-membered cyclic ether where instantaneous activation as opposed to attack on the functional oxygen center may predominate during propagation. These will largely depend on the strength of the activation center or coordination center or electro-free-radical initiator.

The importance of the influence of electrostatic forces cannot be under estimated, when the reaction below involving the use of a non-ionic amide as initiator is considered[21].

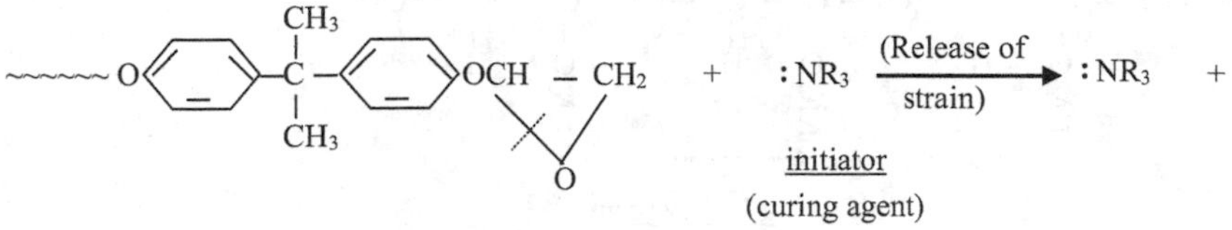

(I) <u>Linear condensation polymer</u>

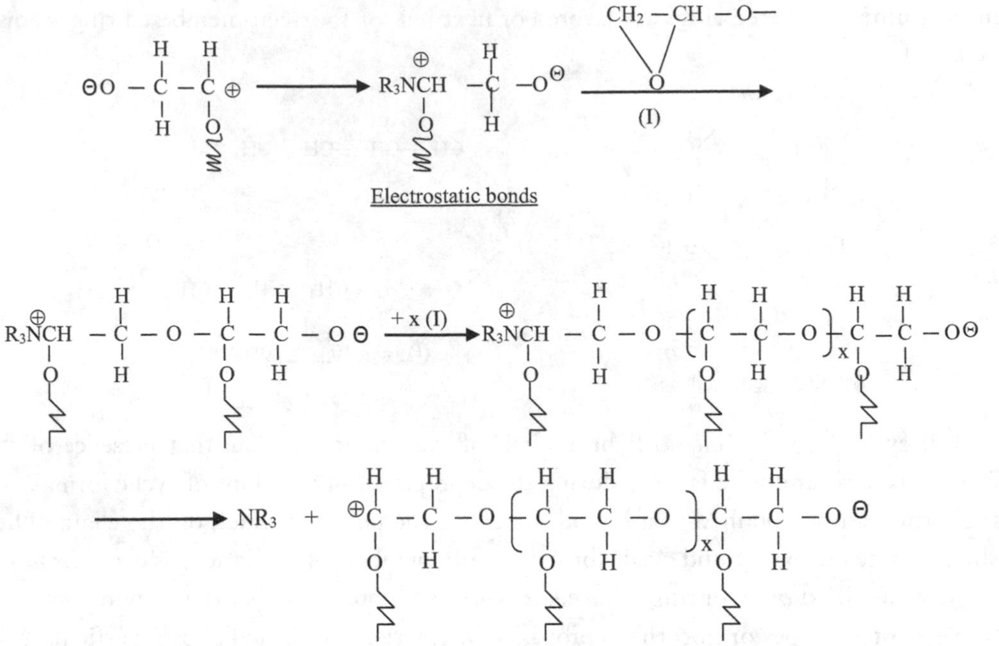

$$\Theta O - \overset{\overset{\displaystyle H}{|}}{\underset{\underset{\displaystyle H}{|}}{C}} - \overset{\overset{\displaystyle H}{|}}{\underset{\underset{\displaystyle O}{|}}{C}} \oplus \longrightarrow R_3NCH^{\oplus} - \overset{\overset{\displaystyle H}{|}}{\underset{\underset{\displaystyle H}{|}}{C}} - O^{\Theta} \quad \xrightarrow{\overset{\displaystyle CH_2 - CH - O -}{\diagdown \; O \; \diagup}}$$

Electrostatic bonds

$$R_3NCH^{\oplus} - \overset{H}{C} - O - \overset{H}{C} - \overset{H}{C} - O \ominus \quad \xrightarrow{+ \, x \, (I)} \quad R_3NCH^{\oplus} - \overset{H}{C} - O \left[\overset{H}{C} - \overset{H}{C} - O \right]_x \overset{H}{C} - \overset{H}{C} - O \ominus$$

$$\longrightarrow NR_3 \; + \; {}^{\oplus}\overset{H}{C} - \overset{H}{C} - O \left[\overset{H}{C} - \overset{H}{C} - O \right]_x \overset{H}{C} - \overset{H}{C} - O \ominus$$

Living Epoxy resin   12.148

The epoxy resins produced are said to have outstanding adhesive properties and are used for bonding to metals, glass, and ceramics. They are also used as casting resins in electrical assemblies. Their chief use has been in protective coatings because of their good adhesion, inertness, hardness, and unusual flexibility. The linear condensation polymer is a product of epichlorohydrin and bisphenol - A. The electrostatic forces for opening the terminal epoxy rings are the paired unbonded radicals on the nitrogen center of NR_3. It adds to the activated ring, forming an electrostatic bond with the anionic end of the monomer where addition takes place. The initiator is similar to anionic electrostatic charged-paired initiators with electrostatic bonds such as shown below

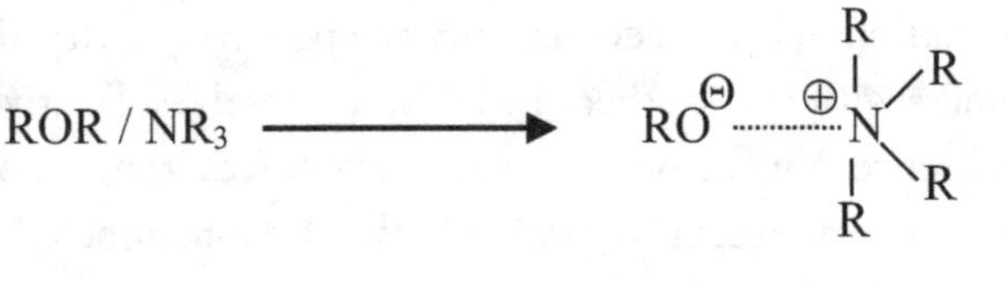

$$ROR \, / \, NR_3 \longrightarrow RO^{\Theta} \cdots \cdots \overset{\oplus}{\underset{\underset{\displaystyle R}{|}}{\overset{\overset{\displaystyle R}{|}}{N}}} \diagup \overset{R}{\diagdown R}$$

12.149

The epoxy resin produced can be observed to be a living one which has to be terminated at both ends to produce a dead polymer. Replacement of NR_3 with HOOCR (a "cationic" initiator) produces almost the same type of resin via electro-free-radical route, with however instantaneous opening of the ring.

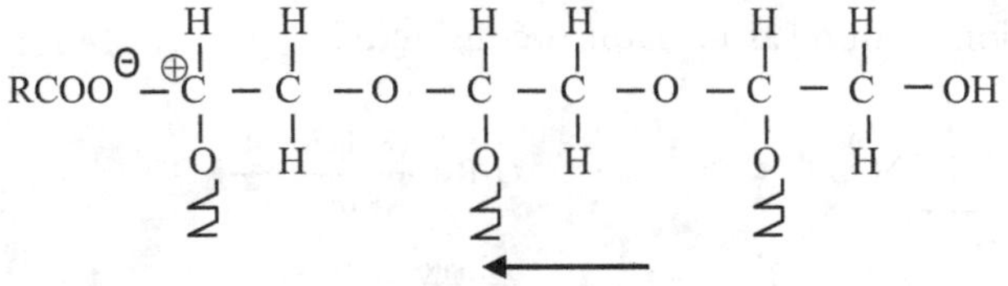

$$RCOO^{\Theta} \overset{\oplus}{-} \overset{H}{C} - \overset{H}{C} - O - \overset{H}{C} - \overset{H}{C} - O - \overset{H}{C} - \overset{H}{C} - OH$$

Half living Epoxy resin   12.150

The electro-free-radical route is favored here because that is the natural route. It can be observed why depropagation reaction will not occur for the first case above where NR_3 was used as initiator, since the paired centers are distantly placed.

Homopolymers of trioxane have long been known to be quite unstable due to presence of several types of depropagation phenomena. Another part of the reason is because; most of the homopolymers are living polymers. Polymerization in the presence of small amounts of ethylene oxide (that is dimethylene oxide) or dioxolane (a cyclic formal) are said to yield a stable copolymers [7]. The copolymers have been reported to be more stable toward thermal depolymerization than the end-capped homo-polymer of trioxane[24]. Esterification by acetic anhydride (end- capping) is also said to produce a more stable polymer[7]. Whichever the case, the end-capped product using the right terminating agent such as acetic anhydride is far more stable than the copolymers. The reduced presence of depropagation in the copolymerization reactions are due to -

(i) Less depropagation phenomena in cyclic ethers and cyclic formals than in trioxane.
(ii) Reduced presence of oxygen centers along the chain
(iii) The non-ionic or less ionic character of the polymerization media.

12.3 Cyclic esters

Two families have been selected, lactides and lactones. These are monomers which all favor both the anionic and "cationic" routes once the MRSE can be provided, via different mechanisms. The two different activation centers on the monomers are useful for providing the additional strain energy needed to attain the MRSE where possible. With the lactones where three -, four-membered rings exist, instantaneous opening of the rings can be favored anionically, nucleo-non-free-radically and with coordination centers, if there exists enough SE in their rings. In fact for most of the monomers, non-free-radical activation can be provided via the C = O activation center, for as long as the MRSE can be attained. Hence, both ionic and radical routes will be considered.

12.3.1 Charged Characters of Cyclic Esters

Beginning with lactides, consider glycolide using anionic initiator which are non-free in character. Free negatively charged initiators such as in $^QC_4H_9$ cannot be used, since the equation will not be chargedly balanced.

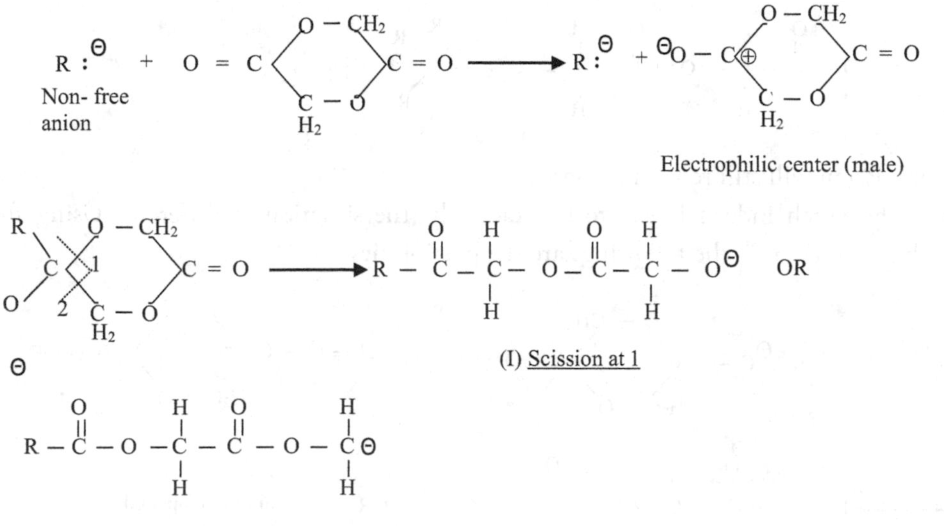

(I) Scission at 1

(II) Scission at 2 (NOT BALANCED)

12.151

It should be noted that the $\overset{\curvearrowleft}{C}$ center is a radical-pushing center. Since the oxygen center will favor carrying the anion in the presence of carbon, (I) is the favored reaction- that is *acyl oxygen cleavage, that is, cleavage of the bond between the carbonyl carbon and the non-carbonyl oxygen.* In the reaction above, it is important to note that the anion $R^{\ominus}$ did not abstract the H atom on the carbon atom next to the active carbon center as transfer species of the first kind due mainly to electrostatic forces of repulsion. Other reasons are also as follows-

(i)   The carbonyl center alone is not a functional center.

(ii) The active carbon center is carrying only one substituent group - the ring which is not shared with the other active center. When transfer species (H) is removed, a bond cannot be formed to balance the charge (i.e., the equation cannot be chargedly balanced). This is indeed is the main reason.

(iii) The carbonyl center cannot be used without opening the ring. If the center can be used without opening the ring, then transfer of species of the first kind will exist.

The fact that transfer species of the first kind cannot be abstracted from the ring, indicates the weak nucleophilic character of a ringed carbonyl center. Nevertheless as will be shown, lactides are far less nucleophilic than cyclic ethers in view of lactides being able to favor the anionic routes. In fact lactides are Electrophiles (Males)

Presence of radical-pushing or pulling groups in the ring, does not alter the route, except that the strain energies in the ring will differ. *With negatively charged-paired initiators (of the covalent types), the negative route cannot be favored, because the equation will not be chargedly balanced. Only the positive end can be used.* With electrostatically anionic charged-paired initiator, the followings are to be expected.

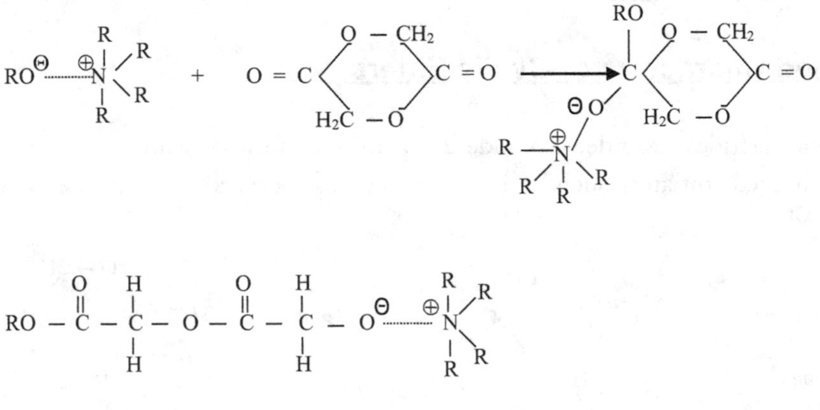

$$12.152$$

The point of scission will still remain the same.

Cationically which indeed is electro-free-radically, the situation is different. Using free cationic initiator such as dilute HCl, the followings are the possibilities.

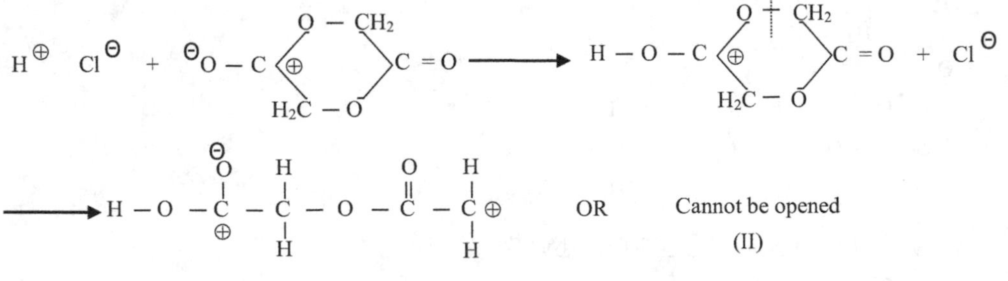

(I)   NOT FAVOURED

$$12.153$$

60

$$H^{\oplus} + Cl^{\ominus} + O = C \underset{H_2C - O}{\overset{O - CH_2}{<>}} C = O \longrightarrow O = C \underset{H_2C - O}{\overset{H - O \overset{Cl^{\ominus}}{\underset{\oplus}{|}} CH_2}{<\underset{1}{\overset{2}{}}>}} C = O \longrightarrow$$

$$H - O - \overset{O}{\overset{||}{C}} - \overset{H}{\underset{H}{\overset{|}{C}}} - O - \overset{O}{\overset{||}{C}} - \overset{H}{\underset{H}{\overset{|}{C}}}^{\oplus} \,{}^{\ominus}Cl \quad OR \quad H - O - \overset{H}{\underset{H}{\overset{|}{C}}} - \overset{O}{\overset{||}{C}} - O - \overset{H}{\underset{H}{\overset{|}{C}}} - \overset{O}{\overset{||}{C}}{}^{\oplus}\,{}^{\ominus}Cl$$

(I) <u>Scission at 2</u>   (II) <u>Scission at 1</u>

(Favored)   12.154

In the first equation where the carbonyl center is attacked, scission can only take place at one single bond to produce (I) in which the positive charge is carried by a weak carbon center, since $= O$ group is more radical pushing than alkylane groups. In fact it was a forbidden trip, because never will that center be activated by a cation or positive charge when the O center is there. In the second equation, oxygen center in the ring is the point of attack, for which only one point of scission exist in the ring. However, visibly one can identify two possible points of scission.

The favored point, based on the strength of the active center, is (II) of Equation 12.154. Hence, cationically due to the fact that the ⌐O center in a ring is less nucleophilic than the ⌐C$=O$ activation center, it is the former that is attacked all the time.

⌐:O: Less nucleophilic than   ⌐C$=O$

Center   Center   12.155

In the first equation (Equation 12.153) therefore, the ring cannot be opened and in fact there is no attack on that center. Depending on the type of substituent group carried by the carbon center in the ring, the activation center "anionically" and "cationically" may alter, once the MRSE can be provided. "Cationic ion-paired initiators" will favor the same mode of activation, noting that free cationically, the cationic center is electrostatic and electro-free-radical in character. Cyclic esters are in general more nucleophilic than alkenes via the oxygen ringed center ⌐O and more electrophilic than alkenes via the ⌐C$=O$ center, but less nucleophilic than cyclic ethers. Notice that the monomer units obtained cationically (Equation 12.154) and anionically (Equation 12.152) are the same.

Considering lactones, one will begin with a - lactones. Anionically, the followings are to be expected.

$$\underset{\underset{\text{"anion"}}{\underline{\text{Non-free}}}}{\overset{\ominus}{R:}} + \underset{O}{\overset{O}{\overset{\diagup\diagdown}{C}}} - CH_2 \xrightarrow[\text{of strain}]{\text{(Release}} R^{\ominus} + {}^{\oplus}\overset{O}{\overset{||}{C}} - \overset{H}{\underset{H}{\overset{|}{C}}} - O^{\ominus} \longrightarrow$$

$$R - \overset{O}{\overset{||}{C}} - \overset{H}{\underset{H}{\overset{|}{C}}} - O^{\ominus} \xrightarrow{+ n(I)} R \left\{ \overset{O}{\overset{||}{C}} - \overset{H}{\underset{H}{\overset{|}{C}}} - O \right\}_n \overset{O}{\overset{||}{C}} - \overset{H}{\underset{H}{\overset{|}{C}}} - O^{\ominus}$$

12.156

Only non-free negative charges, i.e. -anions or anionic-paired initiators can be used. It should be obvious that the three-membered ring with an externally located double bond adjacently placed to an oxygen center has more than or equal to the minimum required strain energy. If otherwise, the following is obtained.

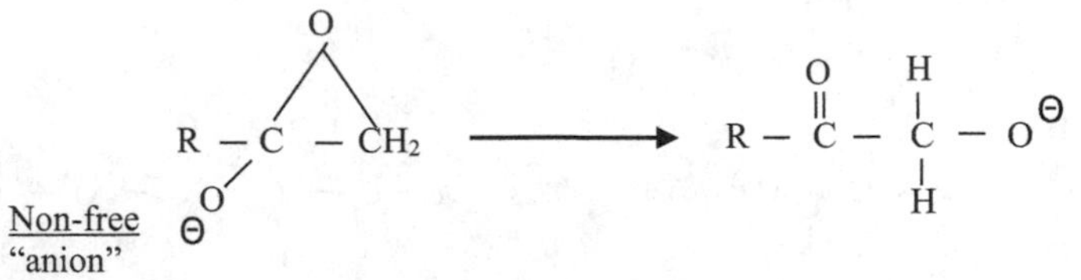

12.157

The initiated monomer still remains the same, though the steps are different. "Cationically", the followings are obtained.

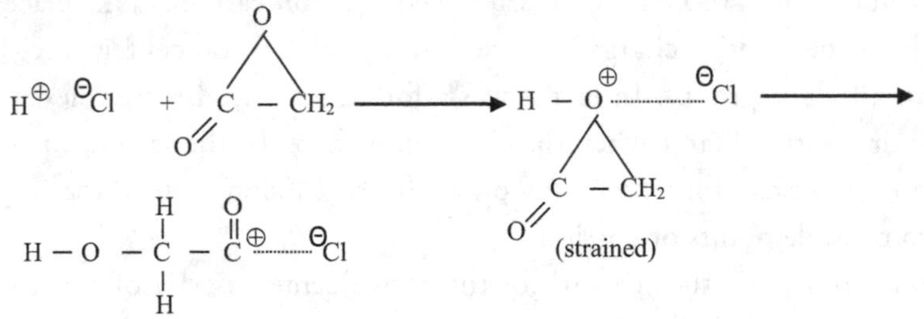

12.158

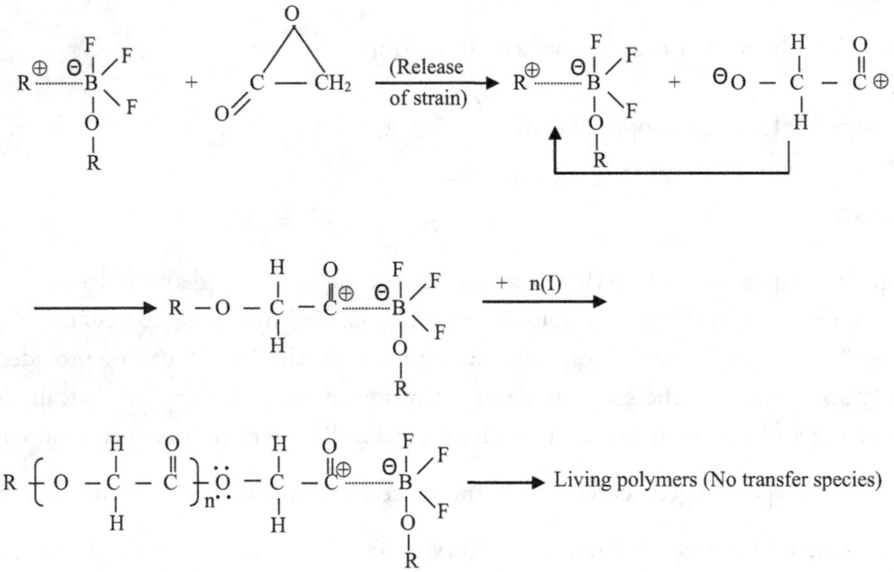

12.159

Apart from the three-membered and probably some four-membered rings, activation "anionically and cationically" are similar to those of lactides. For δ - valerolactone, the followings are to be expected.

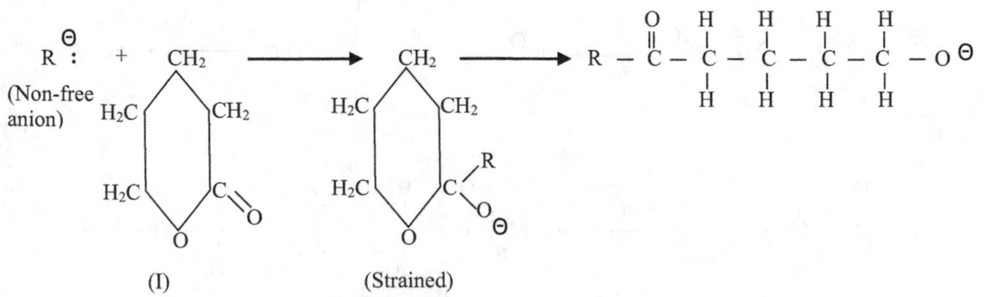

(I)   (Strained)

62

$$R : ^{\ominus} \ + \ \text{(CH}_2 \text{ ring I)} \longrightarrow \text{(Strained)} \longrightarrow R - \overset{\overset{O}{\|}}{C} - \overset{\overset{H}{|}}{\underset{\underset{H}{|}}{C}} - \overset{\overset{H}{|}}{\underset{\underset{H}{|}}{C}} - \overset{\overset{H}{|}}{\underset{\underset{H}{|}}{C}} - \overset{\overset{H}{|}}{\underset{\underset{H}{|}}{C}} - O^{\ominus}$$

(Non-free anion)   (I)   (Strained)

$$\xrightarrow{+ \ n(I)} \quad R - \left[\overset{\overset{O}{\|}}{C} - \overset{\overset{H}{|}}{\underset{\underset{H}{|}}{C}} - \overset{\overset{H}{|}}{\underset{\underset{H}{|}}{C}} - \overset{\overset{H}{|}}{\underset{\underset{H}{|}}{C}} - \overset{\overset{H}{|}}{\underset{\underset{H}{|}}{C}} - O \right]_n \overset{\overset{O}{\|}}{C} - \overset{\overset{H}{|}}{\underset{\underset{H}{|}}{C}} - \overset{\overset{H}{|}}{\underset{\underset{H}{|}}{C}} - \overset{\overset{H}{|}}{\underset{\underset{H}{|}}{C}} - \overset{\overset{H}{|}}{\underset{\underset{H}{|}}{C}} - O^{\ominus}$$

12.160

$$R^{\oplus} \ ^{\ominus}Cl \ + \ (I) \longrightarrow R - \overset{\oplus}{O} \cdots \cdots ^{\ominus}Cl \longrightarrow$$

(Strained)

$$R - O - \overset{\overset{H}{|}}{\underset{\underset{H}{|}}{C}} - \overset{\overset{H}{|}}{\underset{\underset{H}{|}}{C}} - \overset{\overset{H}{|}}{\underset{\underset{H}{|}}{C}} - \overset{\overset{H}{|}}{\underset{\underset{H}{|}}{C}} - \overset{\overset{O}{\|}}{\underset{}{C}}^{\oplus} \cdots \cdots ^{\ominus}Cl \xrightarrow{+ \ n(I)}$$

$$R - \left[O - \overset{\overset{H}{|}}{\underset{\underset{H}{|}}{C}} - \overset{\overset{H}{|}}{\underset{\underset{H}{|}}{C}} - \overset{\overset{H}{|}}{\underset{\underset{H}{|}}{C}} - \overset{\overset{H}{|}}{\underset{\underset{H}{|}}{C}} - \overset{\overset{O}{\|}}{C} \right]_n O - \overset{\overset{H}{|}}{\underset{\underset{H}{|}}{C}} - \overset{\overset{H}{|}}{\underset{\underset{H}{|}}{C}} - \overset{\overset{H}{|}}{\underset{\underset{H}{|}}{C}} - \overset{\overset{H}{|}}{\underset{\underset{H}{|}}{C}} - \overset{\overset{O}{\|}}{\underset{}{C}}^{\oplus} \cdots \cdots ^{\ominus}Cl$$

12.161

Though the positively charged route is natural to the monomer via the oxygen center, transfer species of the second kind can be rejected. The anionic route is also natural to the same monomer via the carbonyl center. The monomer cannot be resonance stabilized.

Positively the anionic oxygen center (C=O) cannot be used. If positive activation via $\overset{\frown}{C} = O$ the activation center had been favored, like acrolein, existence of copolymers from a single monomer would have been favored as indicated below for glycolide, β - propiolactone and acrolein.

$$R \sim\sim\sim O - \overset{\overset{O}{\|}}{C} - \overset{\overset{H}{|}}{\underset{\underset{H}{|}}{C}} - O - \overset{\overset{O}{\|}}{C} - \overset{\overset{H}{|}}{\underset{\underset{H}{|}}{C}} - O - \overset{\overset{H}{|}}{\underset{\underset{H}{|}}{C}} - \overset{\overset{O}{\|}}{C} - O - \overset{\overset{H}{|}}{\underset{\underset{H}{|}}{C}} - \overset{\overset{O}{\|}}{C}^{\oplus}$$

Via $\overset{\frown}{C} = O$ center   Via $\overset{\frown}{O}$ center

Copolymerization from glycolide (Not favored)

12.162

$$R \sim\sim\sim O - \overset{\overset{O}{\|}}{C} - \overset{\overset{H}{|}}{\underset{\underset{H}{|}}{C}} - \overset{\overset{H}{|}}{\underset{\underset{H}{|}}{C}} - O - \overset{\overset{H}{|}}{\underset{\underset{H}{|}}{C}} - \overset{\overset{H}{|}}{\underset{\underset{H}{|}}{C}} - \overset{\overset{O}{\|}}{C}\oplus$$

Via $C = O$ center   Via O center

<u>Copolymer from β - propiolactone (Not favored)</u>   12.163

$$R \sim\sim\sim \overset{\overset{H}{|}}{\underset{\underset{\underset{H}{|}}{C=O}}{C}} - \overset{\overset{H}{|}}{\underset{\underset{H}{}}{C}} - \overset{\overset{H}{|}}{\underset{\underset{\underset{H}{|}}{C=O}}{C}} - \overset{\overset{H}{|}}{\underset{\underset{H}{}}{C}} - O - \overset{\overset{H}{|}}{\underset{\underset{\underset{CH_2}{\|}}{CH}}{C}} - O - \overset{\overset{H}{|}}{\underset{\underset{\underset{CH_2}{\|}}{CH}}{C}}\oplus$$

Via $C = C$ center   Via $C = O$ center

<u>Copolymer from acrolein (favored)</u>   12.164

This clearly distinguishes $\overset{\overset{O}{\frown}}{C} = O$ activation center from $C = O$ activation center.

   Amongst the lactones, the fact that the first member, anionically can be activated in two different ways depending on the strength of the initiator, clearly indicates the less nucleophilic character than larger membered rings, where anionically the monomer can

$$\underset{\underset{O}{\diagdown\diagup}}{\overset{O\diagdown}{\underset{}{C}} - CH_2} \quad < \quad \overset{\overset{O}{\|}}{\underset{\underset{O - CH_2}{|}}{C} - CH_2} \quad < \quad \underset{\underset{H_2C - CH_2}{|}}{O\quad\ CH_2} \quad < \quad Etc.$$

<u>Order of nucleophilicity</u>   12.165

largely be activated in only one way. Hence the following order is to be expected for the members of lactone family, noting that these are electrophiles.

This is in fact based on the following order for their positive centers in the initiation step.

$$R - O - \overset{\overset{H}{|}}{\underset{\underset{H}{|}}{C}} - \overset{\overset{H}{|}}{\underset{\underset{H}{|}}{C}} - \overset{\overset{H}{|}}{\underset{\underset{H}{|}}{C}} - \overset{\overset{H}{|}}{\underset{\underset{H}{|}}{C}} - \overset{\overset{H}{|}}{\underset{\underset{H}{|}}{C}} - \overset{\overset{O}{\|}}{C}\oplus \quad > \quad R - O - \overset{\overset{H}{|}}{\underset{\underset{H}{|}}{C}} - \overset{\overset{H}{|}}{\underset{\underset{H}{|}}{C}} - \overset{\overset{H}{|}}{\underset{\underset{H}{|}}{C}} - \overset{\overset{H}{|}}{\underset{\underset{H}{|}}{C}} - \overset{\overset{O}{\|}}{C}\oplus \quad >$$

$$R - O - \overset{\overset{H}{|}}{\underset{\underset{H}{|}}{C}} - \overset{\overset{H}{|}}{\underset{\underset{H}{|}}{C}} - \overset{\overset{H}{|}}{\underset{\underset{H}{|}}{C}} - \overset{\overset{O}{\|}}{C}\oplus \quad > \quad R - O - \overset{\overset{H}{|}}{\underset{\underset{H}{|}}{C}} - \overset{\overset{H}{|}}{\underset{\underset{H}{|}}{C}} - \overset{\overset{O}{\|}}{C}\oplus \quad <$$

$$R - O - \overset{\overset{H}{|}}{\underset{\underset{H}{|}}{C}} - \overset{\overset{O}{\|}}{C}\oplus \quad > \quad R - O - \overset{\overset{O}{\|}}{C}\oplus$$

<u>Order of positive charge group capacity</u>   12.166

The positive charges on the centers are indeed electro-free-radicals, both of which are males. With lactones like others so far, it can be observed that no transfer species of the first kind can yet be identified. In all the mechanisms so far identified, it can be observed that-

(i) Presence of a cyclic oxonium electrostatic positive charge during addition i.e., transiently favored for all the polymerizable monomers cationically. It is not limited to cyclic ethers or cyclic esters alone as was thought to be the case[1].

(ii) Presence of carbonium electrostatic charge is favored for all cyclic ethers and acetal cationically; with cyclic acetals having in addition, presence of a non-cyclic oxonium electrostatic positive charge.

(iii) Presence of carbonium electrostatic charge is not favored for cyclic esters, but presence of acylium electrostatic charge cationically.

(iv) Presence of carboxylate anion is not favored anionically for cyclic esters, but for anhydrides as will be shown shortly.

(v) Presence of a carbonyl oxygen center $\left[\overset{..}{\underset{..}{C}} = O \right]$ is an electrophilic center for cyclic esters.

Anionically, the centers are fixed, independent of the method of activation, and the same applies cationically for all cyclic monomers.

12.3.2 Radical Characters of Cyclic Esters

Nucleo-non-free-radically for the first member of lactides, which is glycolide - a six-membered ring, the followings are to be expected.

(assumed stained)   12.167

Unlike the anionic case, the point of scission here is clearly defined, since the carbon centers cannot carry a nucleo-non-free-radical. Electro-free-radically, polymerization is not favored via that center, for the same reason as the "cationic" case. Thus, only nucleo-non-free-radical polymerization can be favored, since nucleo-free-radically, the equation cannot be radically balanced after the initiation step; noting that the activation center is a half-free-radical center. Nucleo-non-free-radically for its growing polymer chain, there is no transfer species. Electro-free-radically, the counter-center which becomes charged must be present. When both attack the oxygen center, the mechanism becomes positively-charged-paired polymerization with the positively charged end being electrostatic when attached to the ring.

For lactones, the situation as above is similarly obtained for larger membered rings except those of three and probably some four-membered rings. For three-membered lactone, the followings are to be expected nucleo-non-free-radically.

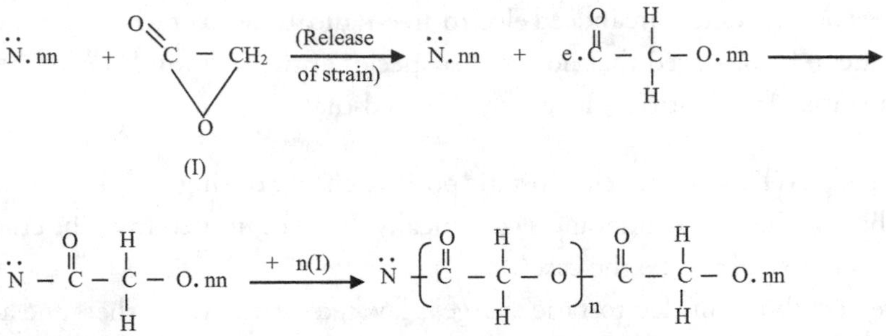

$$\ddot{N}.nn \quad + \quad \overset{O}{\underset{O}{\underset{\|}{C}}} - CH_2 \quad \xrightarrow{\text{(Release of strain)}} \quad \ddot{N}.nn \quad + \quad e.\overset{O}{\underset{H}{\underset{\|}{C}}} - \overset{H}{\underset{H}{\underset{|}{C}}} - O.nn \quad \longrightarrow$$

(I)

$$\ddot{N} - \overset{O}{\underset{H}{\underset{\|}{C}}} - \overset{H}{\underset{H}{\underset{|}{C}}} - O.nn \quad \xrightarrow{+ \ n(I)} \quad \ddot{N} \left\{ \overset{O}{\underset{H}{\underset{\|}{C}}} - \overset{H}{\underset{H}{\underset{|}{C}}} - O \right\}_n \overset{O}{\underset{H}{\underset{\|}{C}}} - \overset{H}{\underset{H}{\underset{|}{C}}} - O.nn$$

12.168

Electro-free-radically, such instantaneous opening of the ring in the absence of electrostatic forces, is impossible. There is also no transfer species for its growing polymer chain, though the route is not natural to the monomer, but to the center which incidentally is less nucleophilic than the male center, unlike non-ringed Electrophiles.

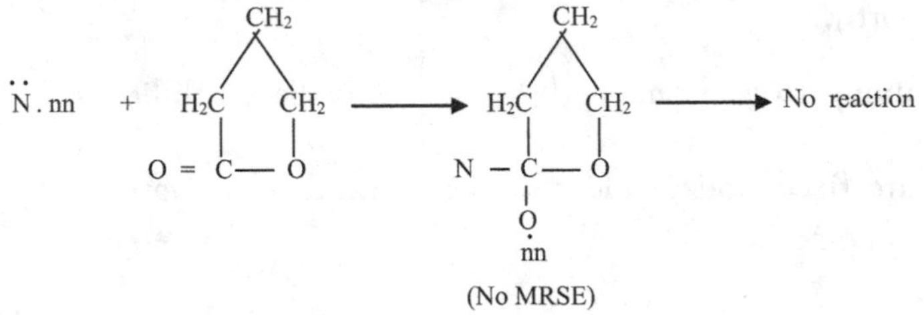

12.169

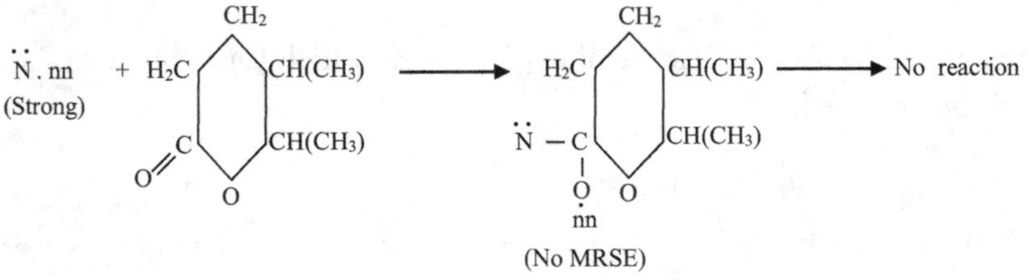

12.170

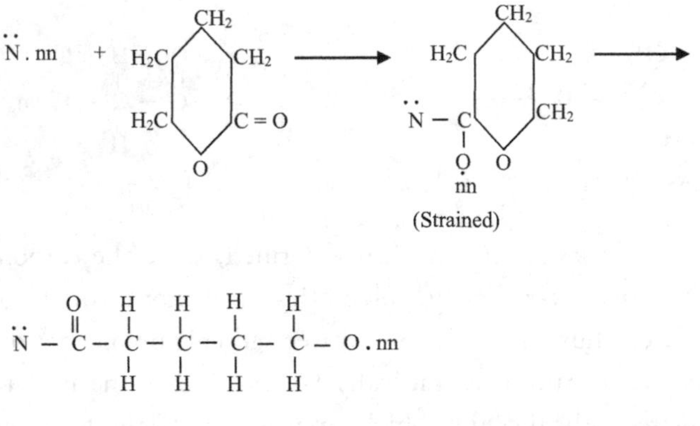

12.171

The five-membered and six-membered rings with radical-pushing substituent groups cannot in most cases be provided with the MRSE (that which is not questionable, being Electrophiles), except when very strong active centers are involved. Hence the reactions are as shown above. It can in general be observed that cyclic ethers, cyclic acetals and cyclic esters are half free radical monomers. Those of cyclic esters are of stronger capacity nucleophilically than non-ringed carbonyl centers, though while carbonyl centers can be made not to favor nucleo-non-free-radical route, cyclic esters can. If the six-membered ring can be opened as shown above, then what prevents the five-membered ring of Equation 12 169 not to be opened, since the five-membered ring is more strained than the six-membered ring?

12.3.3 Order of Nucleophilicity and Electrophilicity of Cyclic Esters

Between lactides and lactones, there is no doubt that the latters are less nucleophilic than the former. Consider placing the monomers side by side.

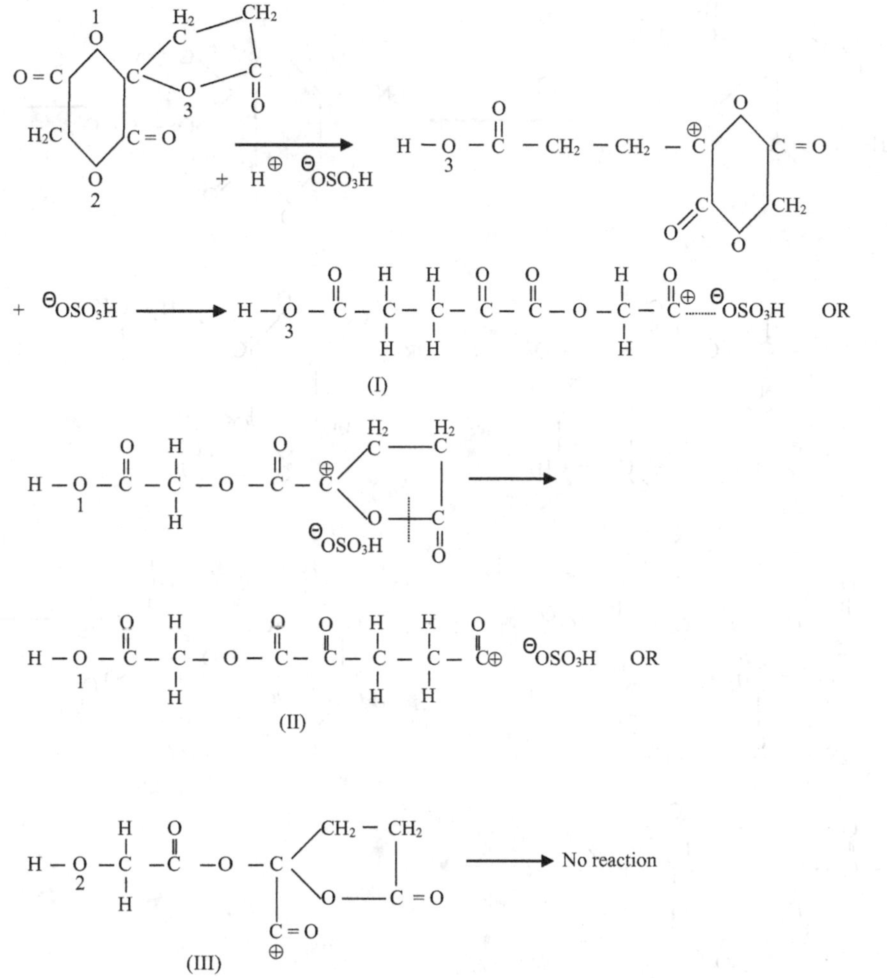

(I)

(II)

(III)   12.172

Though (I) and (II) above seem to be of almost the same electrophilicity, the positively charged center of (I) looks stronger than that of (II), as will shortly become obvious.

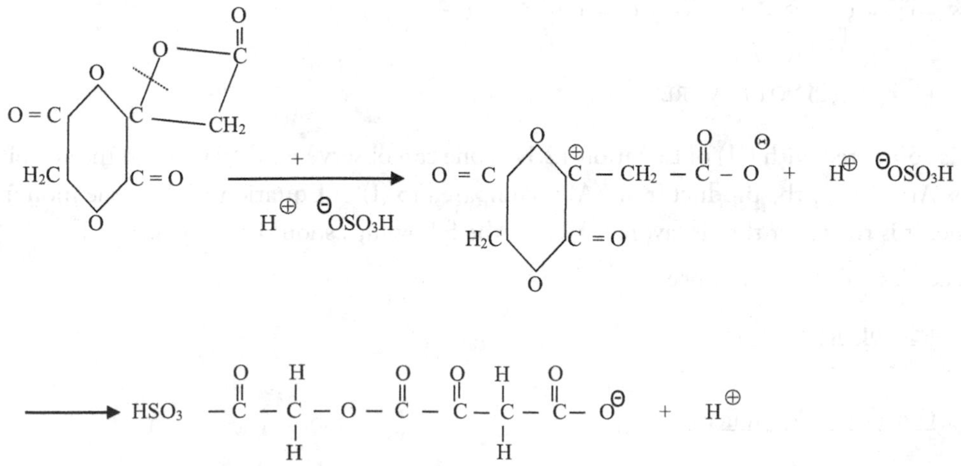

12.173

The four-membered ring is instantaneously opened anionically being more strained than the six-membered ring. This is then followed by the opening of the six-membered ring. Notice the change in the point of scission in the side by side rings. If the six-membered ring had been the first to be attacked anionically, only one ring -case (A) can be opened as shown below.

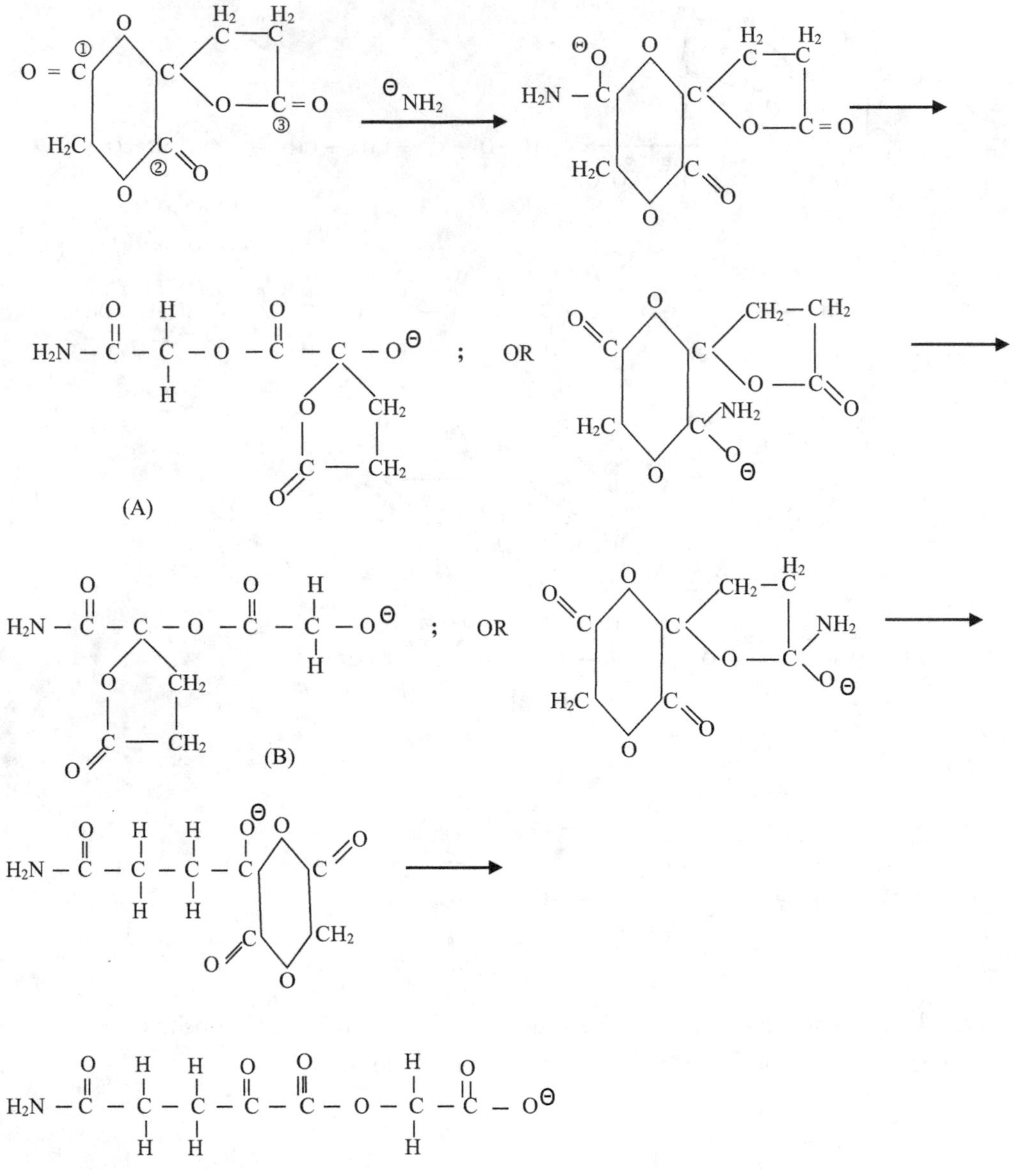

(A)

(B)

(C) NOT FAVORED

12.174

When (C) is compared with (II) of Equation 12.172, one can observe no difference in the activated states of the rings. Also, when the product from (A) is compared to (I) of Equation 12.172, the monomer units are the same. It is the latter that is favored. Hence, the followings should clearly be valid.

Lactides   >   Lactones

Order of electrophilicity   12.175

Lactides   >   Lactones

Order of nucleophilicity   12.176

This is supported by the presence of more functional centers in lactides.

When $H_2C=$ groups are introduced into the rings, the followings are obtained.

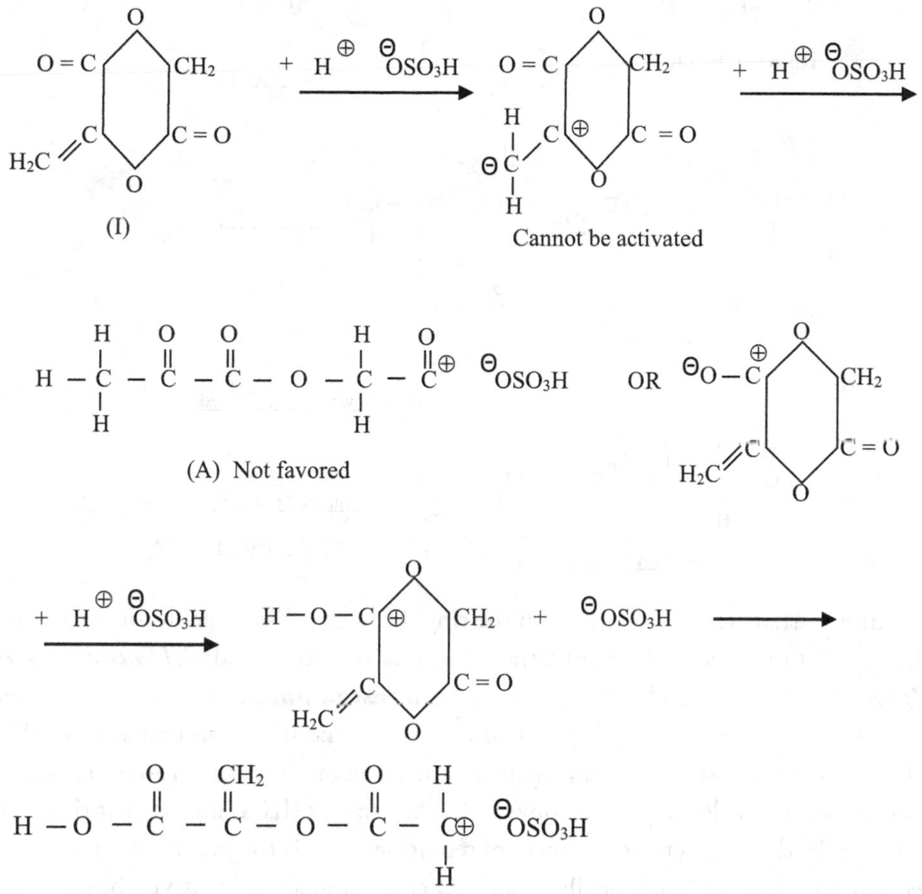

(I)   Cannot be activated

(A) Not favored

(B) <u>Not favored</u>   12.177

If the ring is not opened instantaneously, none of the above is favored for several reasons. $O = C$ ⌐ is electrophilic while $H_2C = C$ ⌐ is nucleophilic in character weaker than ⌐. The $H_2C=$ center the ring cannot be activated chargedly, due to electrostatic forces of repulsion. It has wrongly be activated above.

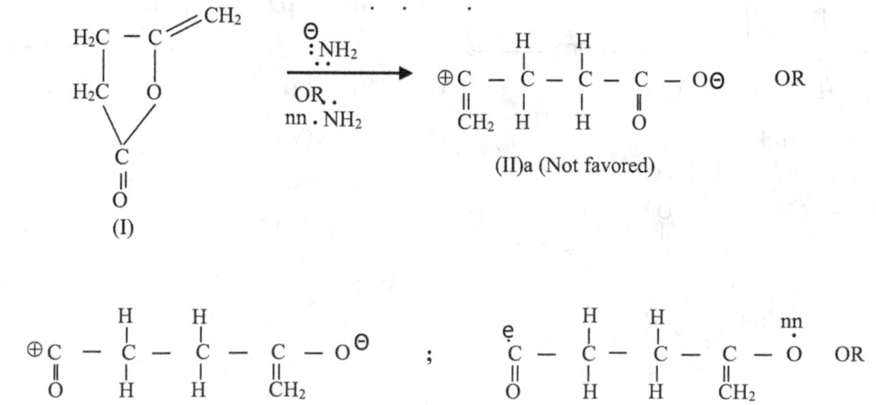

(II)a (Not favored)

(II)b (Favored)   (III)a (Favored)

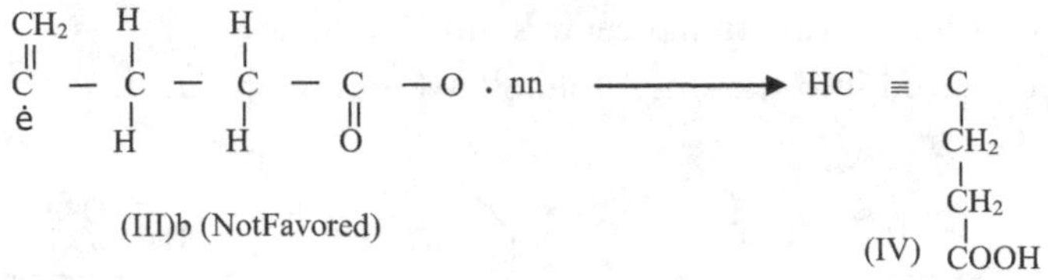

$$
\underset{\text{(III)b (NotFavored)}}{\overset{\displaystyle CH_2}{\underset{\displaystyle \dot{e}}{C}} = \overset{H}{\underset{H}{C}} - \overset{H}{\underset{H}{C}} - \overset{}{\underset{O}{C}} - O \cdot nn} \longrightarrow \underset{\text{(IV)}}{HC \equiv \overset{\displaystyle C}{\underset{\displaystyle \overset{\displaystyle CH_2}{\underset{\displaystyle \overset{\displaystyle CH_2}{\underset{\displaystyle COOH}{|}}}{|}}}{|}}
$$

12.178

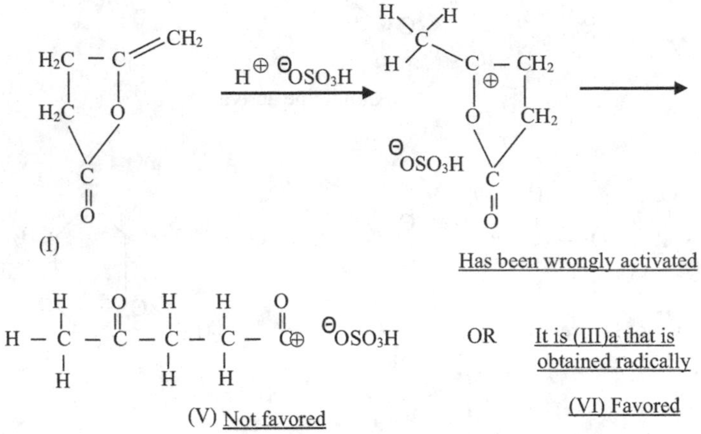

$$
\text{(I)} \quad \xrightarrow{H^{\oplus} \ {}^{\ominus}OSO_3H} \quad \text{Has been wrongly activated}
$$

12.179

$$
\underset{\text{(V) \underline{Not favored}}}{H - \overset{H}{\underset{H}{C}} - \overset{O}{\underset{}{C}} - \overset{H}{\underset{H}{C}} - \overset{H}{\underset{H}{C}} - \overset{O}{\underset{}{C}}{\oplus} \ {}^{\ominus}OSO_3H} \quad \text{OR} \quad \underset{\text{(VI) \underline{Favored}}}{\underline{\text{It is (III)a that is}}\ \underline{\text{obtained radically}}}
$$

It has been "assumed" that (I) has the MRSE in the ring. Hence (II)b-chargedly and (III)a- radically are obtained. (II)a will not undergo molecular rearrangement of the third kind. ***All four routes are favored by them [(II)b and (III)a], since the ring can be opened instantaneously,*** because of the large radical pushing capacities of $H_2C=$ and $O=$ groups. (II)b or (III)a is the true activated state of the monomer obtained if the ring can be instantaneously opened. The monomer unit electro-free-radically using the cation has been shown in the last equation above to be the same as (III)a, noting that that center cannot be activated chargedly, due to electrostatic forces of repulsion. This is the monomer unit that would have been obtained if it was attacked "cationically" via the O center and anionically via the carbonyl center on the ring. ***Hence, as shown by the (III)a for their monomer units, the Y (C = O) center and the X (O) center on the rings still remain as functional despite the presence of $H_2C=$ adjacently placed to the O center. Invariably, additional presence of $H_2C=$ to the ring, has not nullified the Electrophilic character of the monomer.*** Thus, when the ring is opened instantaneously radically (Something which is more readily possible based in the operating conditions), the (III)a (in Equation 12.178) obtained, could rearrange to give a ringed diketone as shown below.

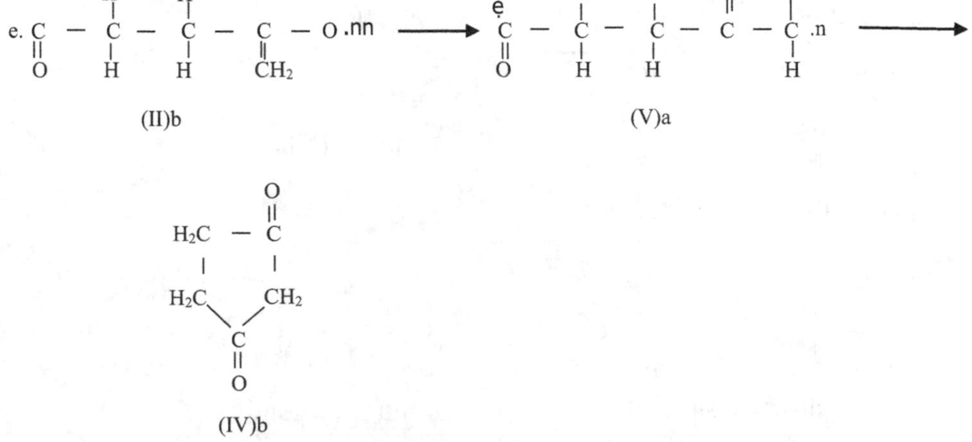

$$
\underset{\text{(II)b}}{e. \ \overset{}{\underset{O}{C}} = \overset{H}{\underset{H}{C}} - \overset{H}{\underset{H}{C}} - \overset{}{\underset{CH_2}{C}} - O \cdot nn} \longrightarrow \underset{\text{(V)a}}{\overset{e}{\underset{O}{C}} - \overset{H}{\underset{H}{C}} - \overset{H}{\underset{H}{C}} - \overset{O}{\underset{}{C}} - \overset{H}{\underset{H}{C}} \cdot n} \longrightarrow
$$

$$
\underset{\text{(IV)b}}{\overset{\displaystyle O}{\underset{\displaystyle H_2C - C}{\parallel}} \ \ \underset{\displaystyle \underset{\displaystyle \overset{C}{\underset{O}{\parallel}}}{H_2C \quad CH_2}}{|} }
$$

12.180

It is believed that during the instantaneous opening of the ring, the phenomenon above, takes place because only electro-free-radicals can move. Anionically and cationically, the followings are obtained for the monomer shown below if the ring is opened instant-aneously.

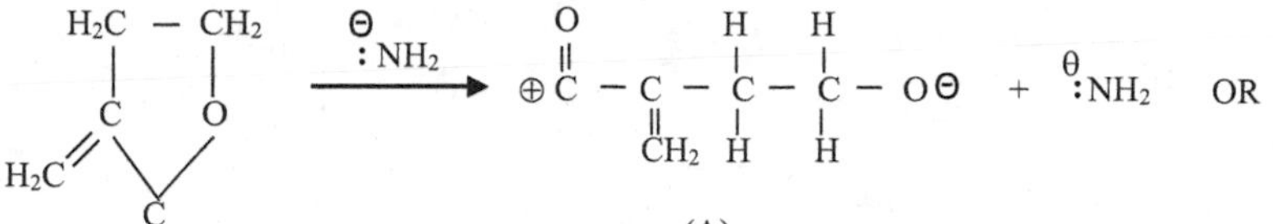

(A)

(VI)

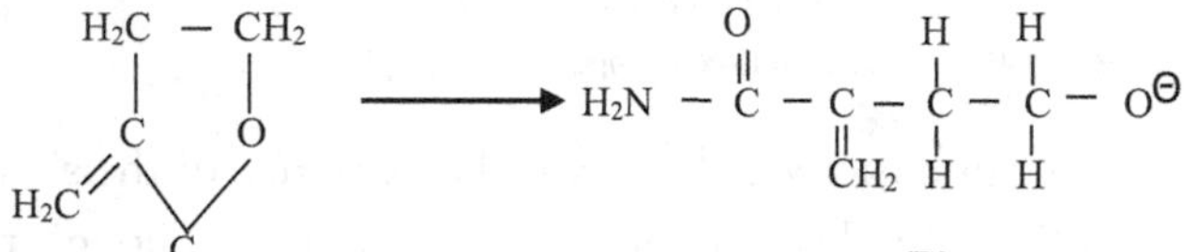

(B)

12.181

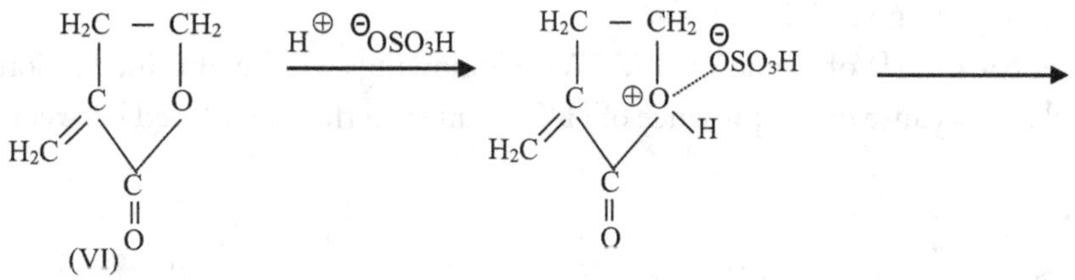

(VI)

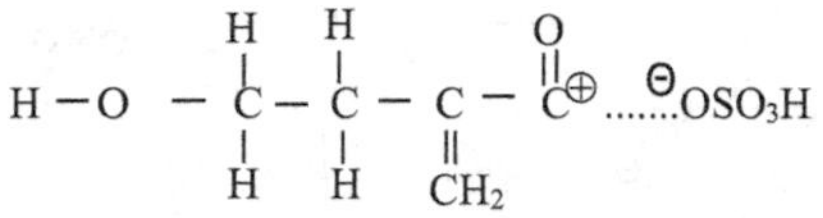

(V)

12.182a

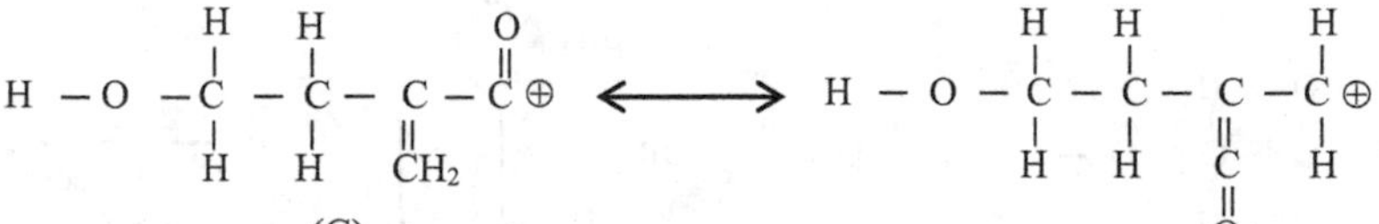

(C)

Impossible movement

12.182b

Since all the monomer units are the same above- i.e., instantaneously, anionically and positively, and since when the $H_2C=$ center is activated positively, the positively charged initiator will try to grab $O-CH_2-$ group as transfer species by opening the ring to give the same (C) above, the reactions above are favored. This clearly indicates that the Electrophilic character of the ring is not affected by the presence of $H_2C=$ group, in view of its location.

We already know that non-ringed centers such as shown below favor resonance stabilization of a different but very simple type– Radical resonance stabilization pheno-mena, taking place terminally along the chain nucleo-non-free-radically and anionically.

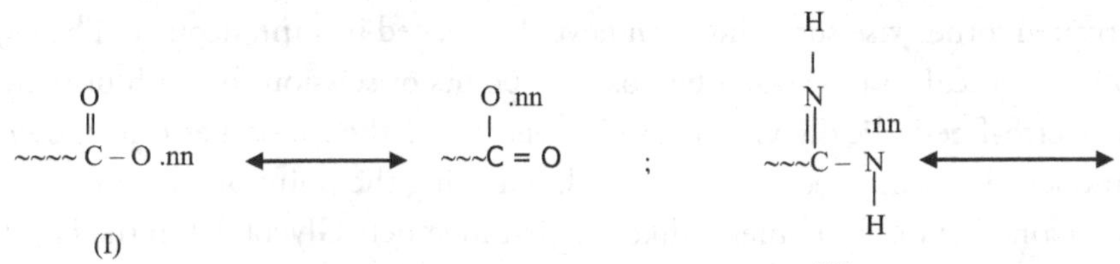

(I)

(II)

71

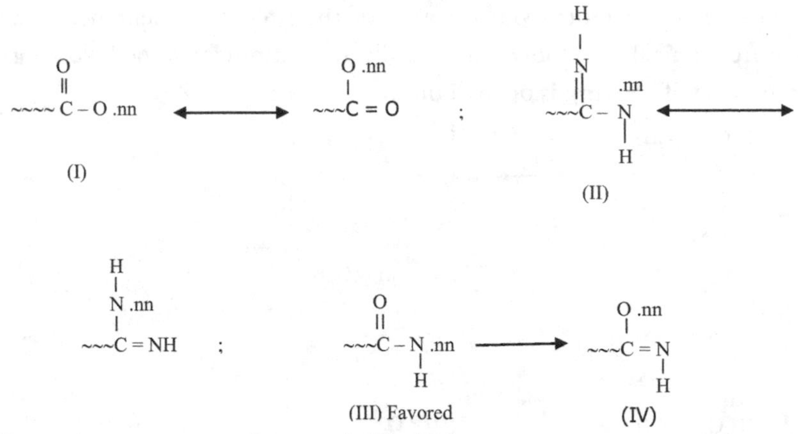

(III) Favored
Resonance Stabilized Components

(IV)

12.183

The activation centers in (I) and (II) can be activated to favor the existence of the same (I) or (II). With (III), the activation center when activated will favor the existence of (IV). Though the C=O center is more nucleophilic than C=N center, the oxygen center is more electronegative than the nitrogen center. Hence, (IV) cannot go to give (III).

Now, coming back to (I) of Equation 12.177, the followings will be obtained anionically, that which may never take place, because of the presence of $H_2C=$ center on the ring placed between two Electrophilic centers

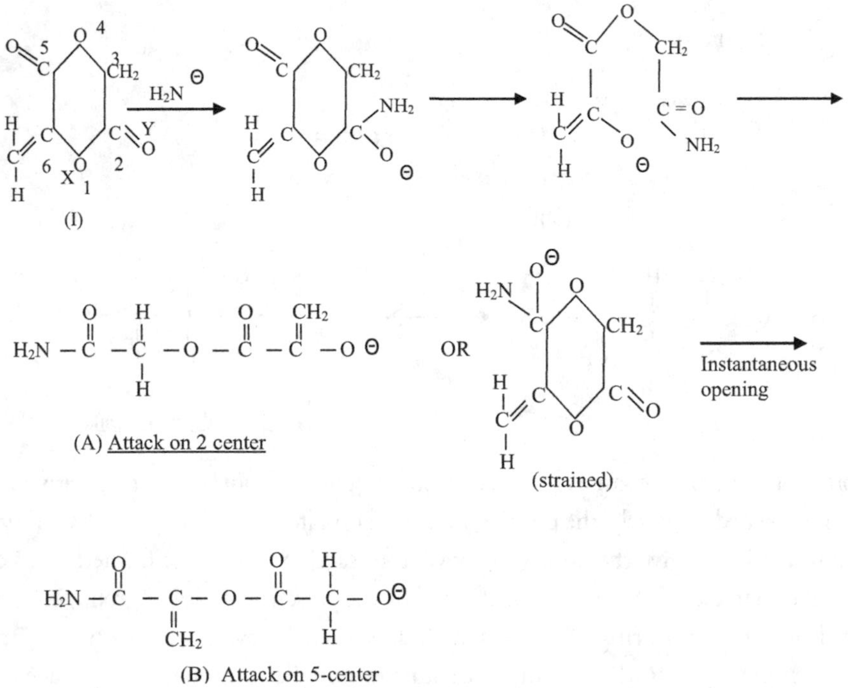

(A) Attack on 2 center

(strained)

Instantaneous opening

(B) Attack on 5-center

12.184

It can be observed that, when MRSE is present in such rings, before they are instantaneously opened, they must first be activated; otherwise such rings can never be opened instantaneously. The (A) above can be observed not to be identical to (B) above, because the points of scission for both of them are different. When attacked electro-free-radically via the $H_2C=$ center, it is the monomer unit shown in (A) that is obtained. When opened instantaneously, with 1,2- bond being the point of scission, the same monomer unit as in (A) is obtained. This monomer unlike the first member (Glycolide) in the family has only one Y and one X centers identified in (I) above.

There is also the need to confirm that cyclic ethers are more nucleophilic than cyclic esters, though the order is obvious.

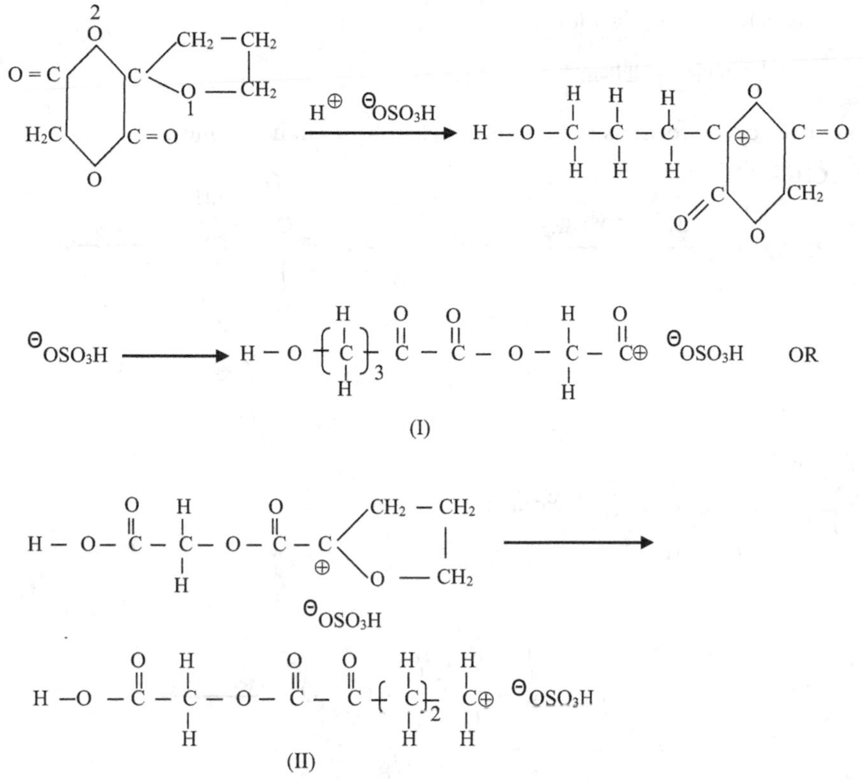

$$\text{(chemical structures)} \quad 12.185$$

(I)

(II)

When (II) is that favored, it is a confirmation of the more nucleophilic character of cyclic ether ring than the glycolide ring. But this is not the case when negatively charged initiators (Anionic), are used as shown below. Instead, it is (I) above that is obtained.

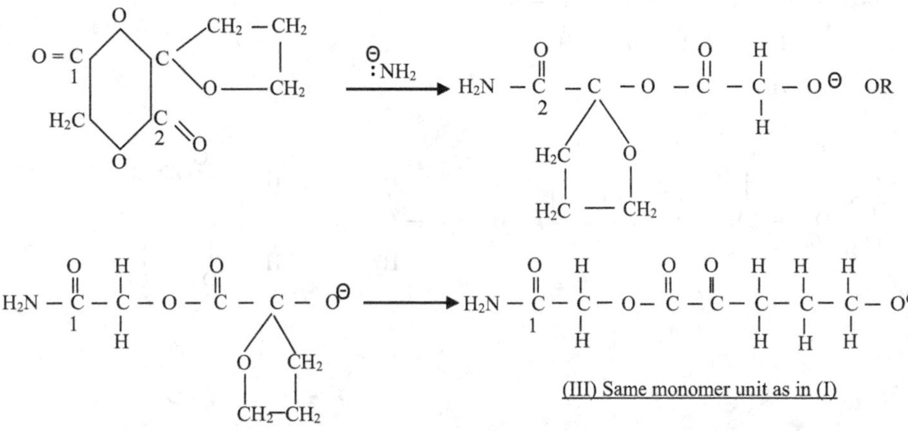

$$\text{(chemical structures)}$$

(III) Same monomer unit as in (I)

$$12.186$$

This is clear indication that when Males and Females are placed side by side, unlike with Females and Females or Males and Males placed side by side, the Female is attacked only in the route natural to it while the Male is attacked in the route natural to it. Hence the monomer unit of (III) above is the same as the monomer unit of (I) of Equation 12.185. Electrophilically, glycolides or lactones are stronger than cyclic ethers, with the C = O center being attacked, being the one farthest away from the C center carrying the other ring (See (I) of Equation 12.184. When Males and Males are placed side by side, it is the route natural to them that can used to test the order of electrophilicity.

73

$$\text{Glycolides} \quad > \quad \text{Cyclic ethers}$$

<u>Order of nucleophilicity</u>

$$\text{Glycolides} \quad > \quad \text{Cyclic ethers}$$

<u>Order of electrophilicity</u>

12.187

One should also expect cyclic formals to be more nucleophilic than lactones.

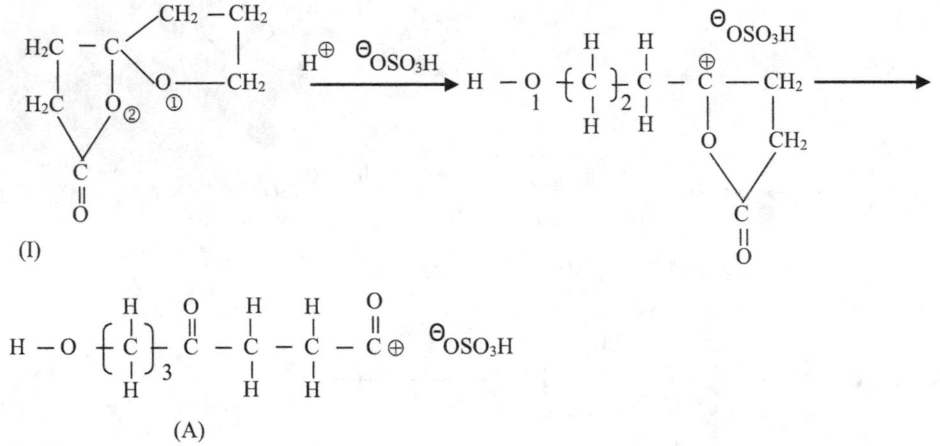

(I)

12.188

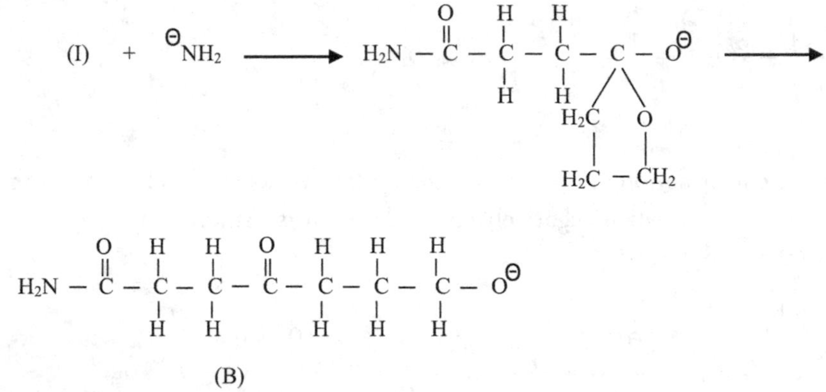

(A)

12.189

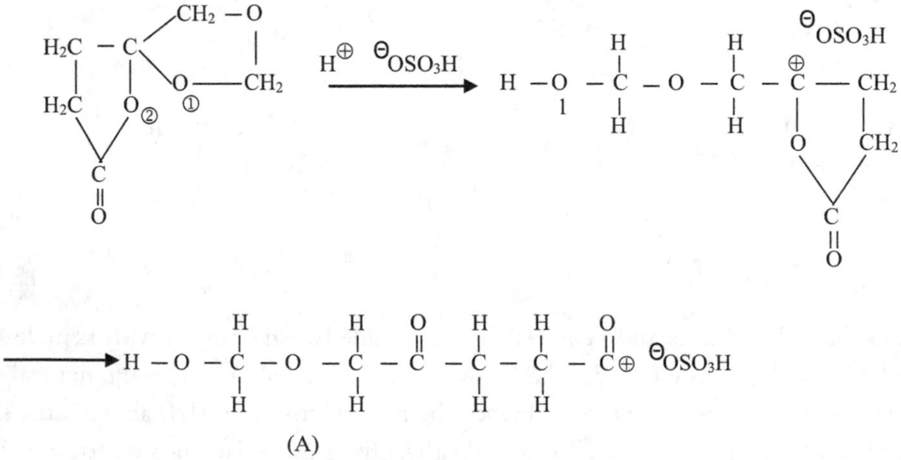

(B)

12.190

(A)

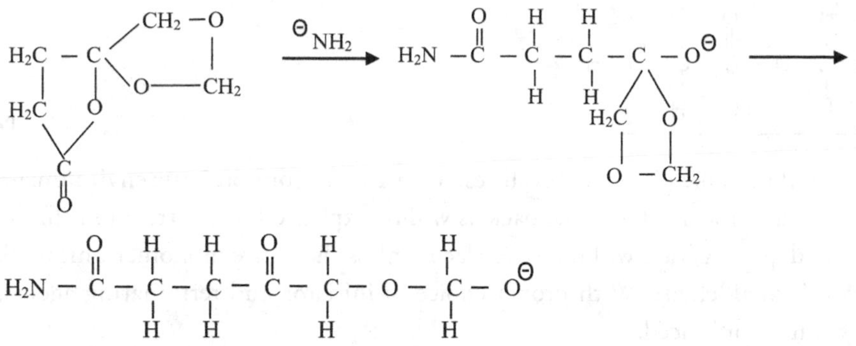

$$\text{(B)} \qquad\qquad 12.191$$

Thus, one can observe that the monomer unit must remain the same anionically and cationically. Hence only specific centers are involved. In general, the followings are valid.

Cyclic formals > Lactones
Order of nucleophilicity

Cyclic formals > Lactides
Order of nucleophilicity
$$12.192$$

Cyclic ethers > Lactones
Order of nucleophilicity

Lactones > cyclic formals
Order of electrophilicity
$$12.193$$

12.3.4 Propagation/Depropagation Phenomena

Because of the proximity of C = O center to O center along a linear chain, the oxygen center cannot be attacked by cations. The C = O center along the linear chain is now nucleophilic in character as shown below. These become sites for branching.

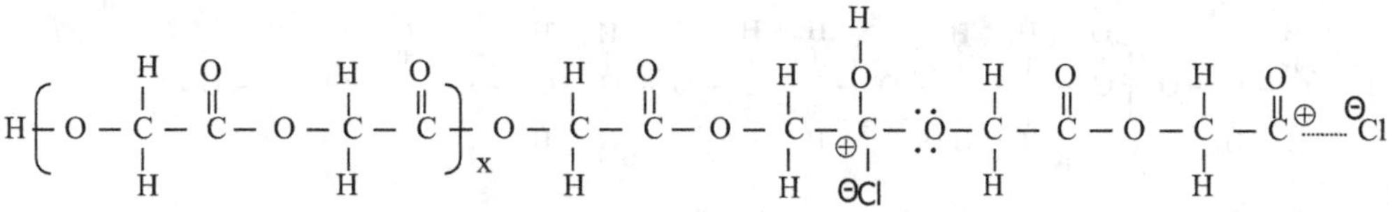

$$12.194$$

However, if dilute acids such as HCl, H_2SO_4 are used as initiators along with a cyclic ester, the chain also grows from behind electro-free-radically, that which generally is the case. The monomer being Electrophilic starts growing from the front in the absence of the cation wherein no depropagation can take place. When the optimum chain length is reached, growth begins from behind as shown below with no possibility of depropagation taking place.

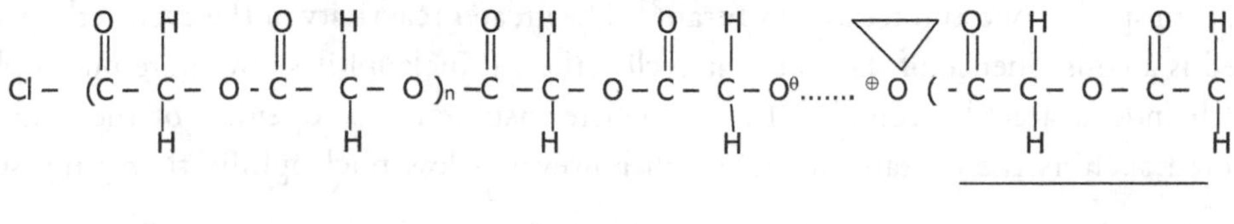

75

$$-O)_x - \overset{\overset{\displaystyle O}{\|}}{C} - \overset{\overset{\displaystyle H}{|}}{\underset{\underset{\displaystyle H}{|}}{C}} - O - \overset{\overset{\displaystyle O}{\|}}{C} - \overset{\overset{\displaystyle H}{|}}{\underset{\underset{\displaystyle H}{|}}{C}} - O - H$$

12.195

Depropagation reactions will not take place with cyclic esters in an ionic or non-ionic environment when it grows from the front, i.e., anionically. From the back as will be explained downstream in this Volume (with Backward addition), depropagation will not take place, unless the entire monomer unit is released. It is impossible to release formaldehyde. With proper choice of initiator and terminating agents, dead polymers with branches can be produced.

Lactones can be copolymerized with themselves once they favor homopoly-merization. For those that do not favor homopolymerization, strong active growing centers have to be first generated, before they can be activated. This can however only be done nucleo-non-free-radically or anionically, since lactones are more electrophilic in character. Anionically, lactones cannot be copolymerized with cyclic acetals and larger-membered cyclic ethers. They can only be copolymerized with three-membered cyclic ethers. For example, copolymerization between β-propiolactone and styrene oxide, β-propiolactone and epichlorohydrin have long been observed [25]. So also is that between δ-valerolactone and propylene oxide. While the tendency towards having blocks of the lactone monomers could not adequately be explained over the years, this can clearly be explained based on current developments. It has been shown that lactones are far more electrophilic (are Electrophiles) than the other cyclic monomers considered so far, that is, lactones are far more reactive than cyclic ethers to anionic attacks. It is largely for this reason that blocks of lactones are far more favored along the chain anionically or nucleo-non-free-radically.

$$R \overset{\ominus}{:} \quad + \quad \underset{\underset{\displaystyle O - C = O}{|\qquad|}}{H_2C - CH_2} \quad + \quad \underset{\underset{\displaystyle O}{\diagdown\diagup}}{H_2C - CH_2} \longrightarrow \underset{\underset{\displaystyle O - \underset{\underset{\displaystyle R}{|}}{C} - O^{\ominus}}{|\qquad|}}{H_2C - CH_2} \quad +$$

.(I) Electrophile
Reactive

(II) Nucleophile
Unreactive, but can be
opened instantaneously,

$$\ominus O - \overset{\overset{\displaystyle H}{|}}{\underset{\underset{\displaystyle H}{|}}{C}} - \overset{\overset{\displaystyle H}{|}}{\underset{\underset{\displaystyle H}{|}}{C}} \oplus \xrightarrow{\quad y[(I) + (II)] \quad}$$

$$R - \overset{\overset{\displaystyle O}{\|}}{C} - \overset{\overset{\displaystyle H}{|}}{\underset{\underset{\displaystyle H}{|}}{C}} - \overset{\overset{\displaystyle H}{|}}{\underset{\underset{\displaystyle H}{|}}{C}} - O \left(\overset{\overset{\displaystyle O}{\|}}{C} - \overset{\overset{\displaystyle H}{|}}{\underset{\underset{\displaystyle H}{|}}{C}} - \overset{\overset{\displaystyle H}{|}}{\underset{\underset{\displaystyle H}{|}}{C}} - O \right)_x \overset{\overset{\displaystyle H}{|}}{\underset{\underset{\displaystyle H}{|}}{C}} - \overset{\overset{\displaystyle H}{|}}{\underset{\underset{\displaystyle H}{|}}{C}} - O - \overset{\overset{\displaystyle H}{|}}{\underset{\underset{\displaystyle H}{|}}{C}} - \overset{\overset{\displaystyle H}{|}}{\underset{\underset{\displaystyle H}{|}}{C}} - O - \overset{\overset{\displaystyle O}{\|}}{C} - \overset{\overset{\displaystyle H}{|}}{\underset{\underset{\displaystyle H}{|}}{C}} - \overset{\overset{\displaystyle H}{|}}{\underset{\underset{\displaystyle H}{|}}{C}} - O - \sim\sim$$

12.196

It is possible that the four-membered b - propiolactone favors instantaneous opening of the ring like the epoxide. If that is the case, then the reaction with (I) will be faster than expected and this will be reflected in the relative reactivity ratios as will be explained in subsequent volume on copolymerizations. If in (II), one of the H atoms was CH_2Cl, (II) can never be found along the chain.

Cationically or indeed positively and electro-free-radically, lactones can be copolymerized with cyclic ether and acetal. BF_3 etherate initiates the copolymerizations of b-propiolactone and 3,3 - bis(chloromethyl) oxetane and, β-propiolactone and tetrahydrofuran [26]. The greater reactivity of the cyclic ethers which has been observed is a consequence of the fact that cyclic ethers (Nucleophiles) are more nucleophilic than lactones which indeed are Electrophiles. In fact, where instantaneous opening of the ring for cyclic ethers is favored such as the oxetane above which however is less nucleophilic than fully substituted oxetane itself, there is a large tendency of having long blocks of the oxetane in the copolymer positively or electro-free-radically.

Lactones as reported in general cannot be "cationically" copolymerized with alkenes. It cannot because Lactones are Electrophiles while alkenes are Nucleophiles. However, since Lactones can be polymerized positively in the route not natural to it, alkenes will largely exist along the chain, the route being natural to them. So also are cyclic ethers said not to favor copolymerization with alkenes, that which is not true, since both of them are Nucleophiles. Cyclic ethers being more Nucleophilic, will appear late along the chain. However, cyclic acetals which are more nucleophilic than cyclic ethers are said to favor copolymerization with alkenes. One observe a state of confusion. If cyclic acetals can copolymerize with alkenes, then why can cyclic ethers not copolymerize with alkenes? "Cationically" copolymerization is favored by them with the alkenes forming long blocks along the chain, while anionically, no copolymerization is possible.

β-propiolactone has been shown to favor weak copolymerization with γ-butyrolactone, a lactone which cannot be homo-polymerized using triethyl aluminum/ water as the source of initiator[27]. The initiator obtained from this combination as will be shown downstream is $(H_5C_2)_2Al^{\oplus}\ldots^{\theta}Al(C_2H_5)_2$ or $(H_5C_2)_2Al^{\cdot e}\ldots^{ne}Al(C_2H_5)_2$. It has a dual character, in the sense that both centers are fully active, one center for a Male and the other center for the Female. The negatively charged center and the nucleo-free-radical center in the initiator cannot be used, due to balancing of equations. Only the positively charged center or electro-free-radical center can be used for these two Male monomers which have strong abilities to favor both anionic/nucleo-non-free-radical and positively charged/electro-free-radical routes. The ability to initiate the γ- butyrolactone is due to the presence of a strong "cationic" active center of growing polymer chain of β-propiolactone first in the system followed by activation of the γ-butyrolactone. The use of so-called anionic - based initiators such as potassium hydroxide, aluminum isoprop-oxide to copolymerize the monomers, is very possible, "since the anionic center generated for a growing chain of β-propiolactone is strong enough to activate the γ-butyrolactone the monomers being Electrophiles." Thus, in general the great influence of characters of monomers and the routes favored by them and much more can be observed to be very important.

Imagine copolymerizing a lactone anionically with styrene a nucleophile. Only lactones can be homopolymerized. Lactones are strong electrophiles, while styrene is a "very weak electrophile" or indeed a nucleophile. Copolymerization can only take place positively and electro-free-radically. It is when all the styrenes in the system have been consumed in the route natural to it that lactones begin to appear along the chain, since the active center positively placed on styrene is strong enough to activate it leading to block copolymers.

During copolymerizations, involving lactones as comonomers, depropagation reactions will show up only electro-free-radically when more nucleophilic cyclic oxides of three and some four-membered rings are involved. When long blocks of these monomers exist along the chain, then if the conditions exist, depropagation reactions can take place.

Nucleo-non-free-radically, no depropagation can be favored during copolymeri-zation. Only, thermal scission will be required to achieve this! All nucleo-non-free-radical polymerizations are essentially low temperature polymerization reactions and these monomers do not have the features. Nucleo-non-free-radically, only three membered cyclic ethers will favor copolymerization with lactones, in almost the same manner as with the anionic route.

12.4 Related Monomers

The related monomers of interest are some anhydrides, where the rings can somehow be strained. Two cases will be considered - maleic anhydrides and phthalic anhydrides. Also to be considered is (III) shown below, that in which there is only $C = O$ in the ring and some others fairly unpopular or unknown monomers.

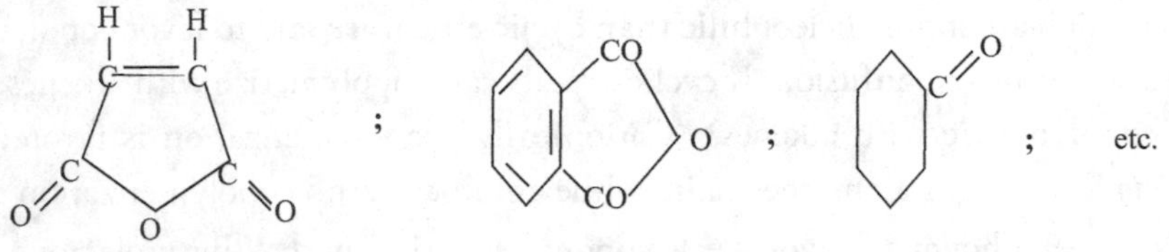

   (I) <u>Maleic anhydride</u>   (II) <u>Phthalic anhydride</u>   (III) <u>Cyclohexanone</u>   12.197

The five-membered rings are more strained. (II) is sterically hindered by the presence of benzene ring, otherwise there is little or no difference between (I) and (II), in terms of methods of activations. Both monomers have one functional center– O. There are three and five activation centers, $C = C$ and $C = O$. $C = O$ can be used anionically/positively and electro-free/nucleo-non-free-radically. Of all of them, only two can be used for opening the rings. These are $C = O$ and O centers. (III) has been used in the past.

Looking at the monomers themselves without the ring, they resemble the type of symmetric alkenes shown below.

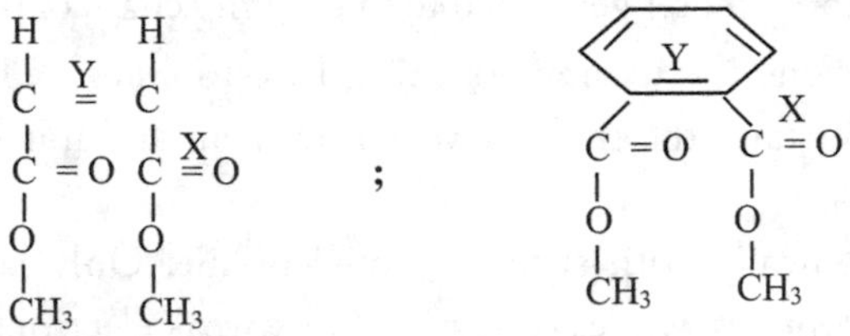

<u>Electrophiles</u>   12.198

In order words, the two monomers are more electrophilic in character via $C = C$ activation center. Hence for the monomers above, the order of "cationic" attack will be based on the following order of nucleophilicity for the center-

$$\overset{X}{C} = O \quad > \quad \overset{Y}{C} = C$$

Order of nucleophilicity for Non-ringed monomer

$$\overset{Y^O}{C} = O \quad > \quad \underset{}{\ddot{O}: X^O \quad > \quad C = C \; Y^C}$$

<u>Order of Nucleophilicity</u> (in Maleic anhydride)   12.199

With Maleic anhydride, there are two Ys, Y^O for ring opening and Y^C for use when the ring is not opened, i.e., when the ring is closed; for which the $C = O$ center becomes X. For phthalic anhydride, the benzene ring is the most nucleophilic center unlike the case of maleic anhydride. When the ring is about to be opened in phthalic anhydride, only the $C = O$ and O centers are first attacked anionically and positively respectively. In fact, with maleic anhydride, these are the only centers that can be used chargedly and radically while the other is largely used radically. It is also likely that the five-membered rings possess very close to the MRSE for their ring.

12.4.1 Charged Characters of the Anhydrides

Anionically for the monomers, the followings are to be expected, for which it is only the $\overset{O}{\underset{}{C}} = O$ center that can be involved.

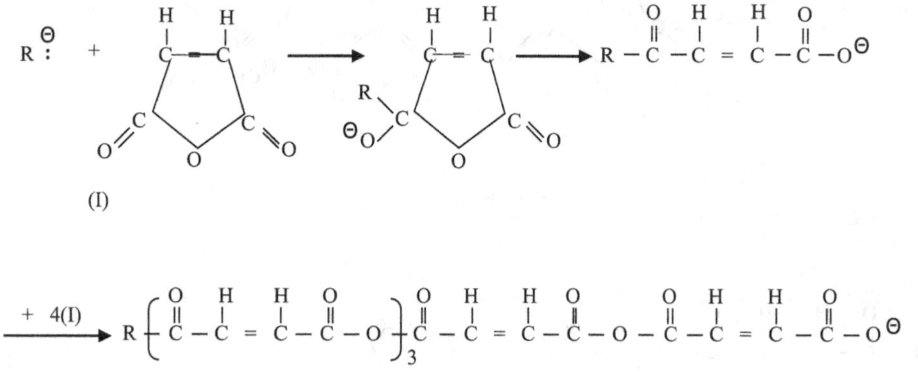

(I)

$$\xrightarrow{+ \ 4(I)} \quad R \left\{ \overset{O}{\overset{\|}{C}} - \overset{H}{\overset{|}{C}} = \overset{H}{\overset{|}{C}} - \overset{O}{\overset{\|}{C}} - O \right\}_3 \overset{O}{\overset{\|}{C}} - \overset{H}{\overset{|}{C}} = \overset{H}{\overset{|}{C}} - \overset{O}{\overset{\|}{C}} - O - \overset{O}{\overset{\|}{C}} - \overset{H}{\overset{|}{C}} = \overset{H}{\overset{|}{C}} - \overset{O}{\overset{\|}{C}} - O^{\ominus} \qquad 12.200$$

(I)

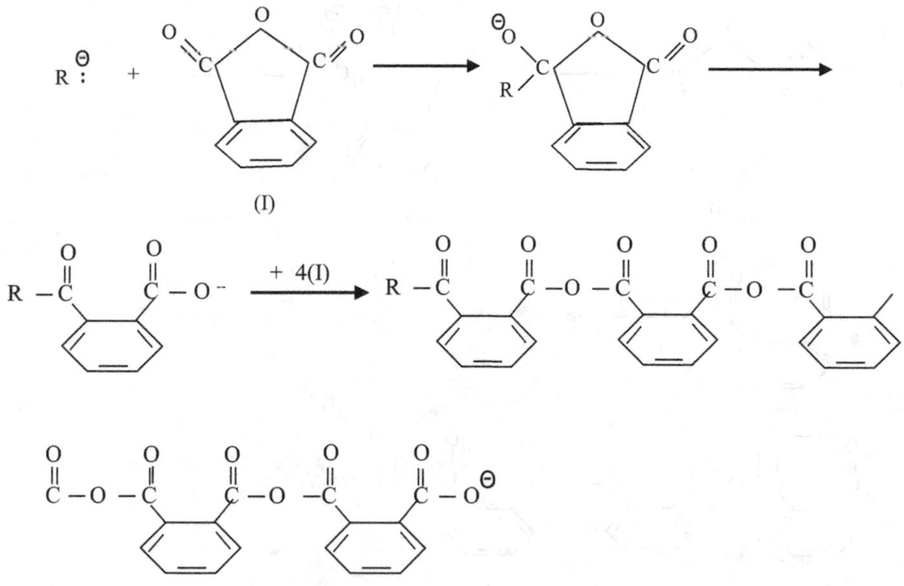

12.201

Phthalic anhydride favors being used more as a Step comonomer than an Addition monomer in view of its less steric limitation capacities in Step than in Addition polymerization. The presence of *carboxylate anionic growing chain* can be observed in these cases, unlike with lactones or cyclic ethers or cyclic formals. These are not however measures of the order of their nucleophilicity or electrophilicity, but something else. Based on what they are carrying, these anhydrides are more nucleophilic and electrophilic than the lactones.

Anhydrides   >   Lactones

Order of Electrophilicity and Nucleophilicity   12.202

Hence, when copolymerized anionically with three-membered cyclic ethers such as propylene oxide, mostly the anhydrides will be favored along the chain, being an Electrophile and the propylene oxide will only appear along the chain *alternatingly placed with the anhydride, since propylene oxide cannot be homopolymerized anionically.*

Cationically for the anhydrides, the followings are to be expected whether the C=C center is activated or not.

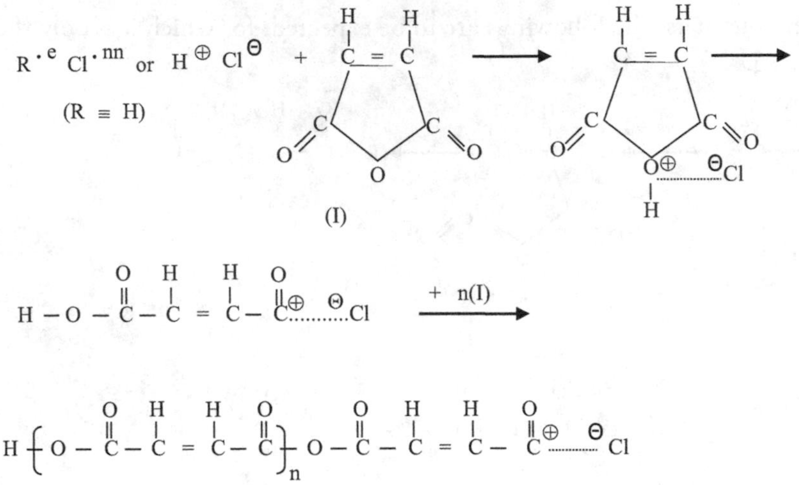

$$12.203$$

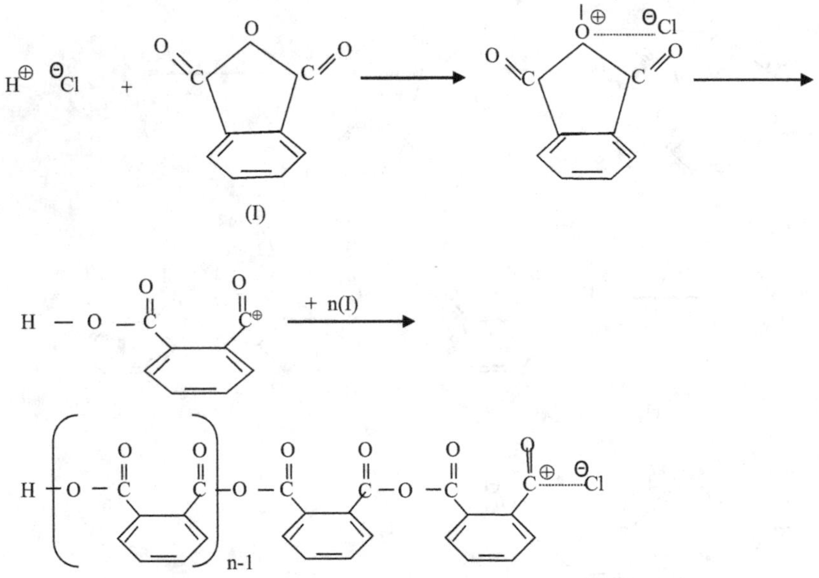

$$12.204$$

It is important to note that for anhydrides and lactones where both anionic and "cationic" routes are favored via different activation centers, the same repeating units must be maintained. Positively charged-paired initiators can also be used. Even then, they are all unstable because they are still living when Na, H and ionic metals are involved electro-free-radically. For initiators of transition metal types with two vacant orbitals, the benzene ring can be syndiotactically placed to reduce or eliminate any influence of steric hindrance if it exists. Shown below is a case where it is still unstable, though syndiotactic placement was obtained.

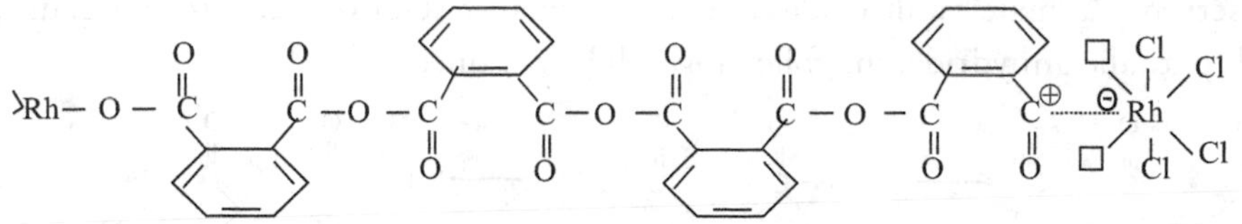

$$\text{>Rh}-\text{O}-\overset{\text{O}}{\underset{\|}{\text{C}}}\cdots\overset{\text{O}}{\underset{\|}{\text{C}}}-\text{O}-\text{C}\cdots\overset{\text{O}}{\underset{\|}{\text{C}}}\cdots\overset{\text{O}}{\underset{\|}{\text{C}}}-\text{O}-\overset{\text{O}}{\underset{\|}{\text{C}}}\cdots\text{C}^{\oplus}\cdots{}^{\ominus}\text{Rh}$$

12.205

When maleic anhydride is copolymerized with propylene oxide positively, unlike with lactones where blocks of propylene oxide will be more favored, here random copolymers are obtained, ***because the maleic anhydride is more reactive than the lactones.*** There will however be more monomer units of propylene oxide than the maleic anhydride, being less nucleophilic. While the reactivity of both of them is greatly increased by the presence of electrostatic bond created, the reactivity of anhydrides is decreased by the fact that it is an Electrophile. Terminal resonance stabilization of the electro-free-radical for electro-free-radical growing polymer chain of maleic anhydride and lactones are absent as shown below. The initiator that should be used should indeed be dilute acids growing from behind from the oxonium center and not that shown below.

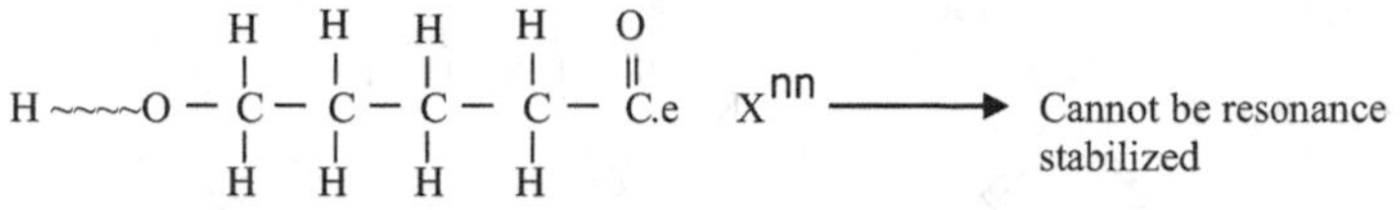

$$\text{H}\sim\sim\text{O}-\overset{\text{H}}{\underset{\text{H}}{\text{C}}}-\overset{\text{H}}{\underset{\text{H}}{\text{C}}}-\overset{\text{H}}{\underset{\text{H}}{\text{C}}}-\overset{\text{H}}{\underset{\text{H}}{\text{C}}}-\overset{\text{O}}{\underset{\|}{\text{C}}}\text{.e} \quad X^{nn} \longrightarrow \quad \text{Cannot be resonance stabilized}$$

<u>Growing chain of δ - valerolactone</u>

12.206

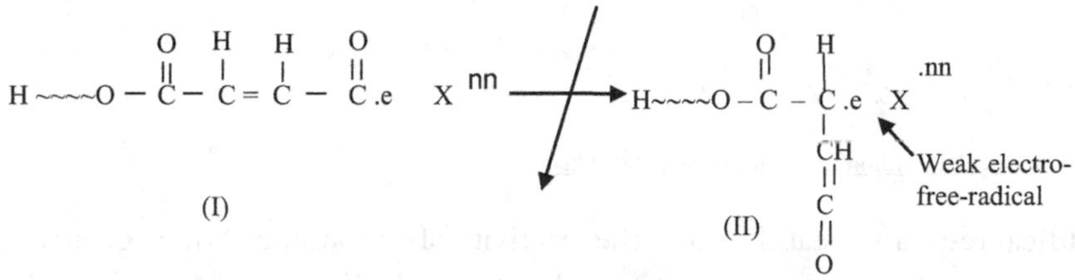

$$\text{H}\sim\sim\text{O}-\overset{\text{O}}{\underset{\|}{\text{C}}}-\overset{\text{H}}{\underset{\|}{\text{C}}}=\overset{\text{H}}{\underset{\|}{\text{C}}}-\overset{\text{O}}{\underset{\|}{\text{C}}}\text{.e} \quad X^{nn} \longrightarrow\!\!\!\!/ \quad \text{H}\sim\sim\text{O}-\overset{\text{O}}{\underset{\|}{\text{C}}}-\overset{\text{H}}{\underset{\underset{\text{O}}{\|}}{\text{C}}}\text{.e} \quad X^{.nn}$$

(I) (II) ← Weak electro-free-radical

12.207

Hence, when the lactone is copolymerized with maleic anhydride nucleo-non-free-radically, it is far less reactive than when copolymerized with propylene oxides, lactone and maleic anhydride belonging to the same family, while propylene oxide is unreactive nucleo-free- or non-free-radically, its presence as already said appearing alternatingly place to lactone. Electro-free-radically, lactone is far less reactive with maleic anhydride than lactone is with propylene oxide, the route being natural to propylene oxide, a Nucleophile. In fact, between propylene oxide and lactone, there are presence of long blocks of propylene oxide. Maleic anhydride because of the double bond is more reactive with propylene oxide than lactone is electro-free-radically.

$$\text{R}^{\oplus}\text{Cl}^{\ominus} + \underset{(I)}{\overset{\text{H} \quad \text{H}}{\underset{}{\text{C}=\text{C}}}} + \underset{(II)}{\overset{\text{CH}_3 \quad \text{H}}{\text{HC}-\text{CH}}} \longrightarrow$$

(R ≡ H)

$$\text{R}-\text{O}-\overset{\text{H}}{\underset{\text{H}}{\text{C}}}-\overset{\text{CH}_3}{\underset{\text{H}}{\text{C}}}\text{.e} + (I) \longrightarrow \text{R}-\text{O}-\overset{\text{H}}{\underset{\text{H}}{\text{C}}}-\overset{\text{CH}_3}{\underset{\text{H}}{\text{C}}}-\text{O}-\overset{\text{O}}{\underset{\|}{\text{C}}}-\overset{\text{H}}{\underset{\underset{\text{O}}{\|}}{\text{C}}}=\text{C}-\text{C}.\text{e} \longrightarrow$$

$$\text{R}\sim\text{O}-\overset{\text{O}}{\underset{\|}{\text{C}}}-\overset{\text{H}}{\underset{\text{H}}{\text{C}}}=\overset{\text{H}}{\underset{}{\text{C}}}-\overset{\text{O}}{\underset{\|}{\text{C}}}-\text{O}-\overset{\text{H}}{\underset{\text{H}}{\text{C}}}-\overset{\text{CH}_3}{\underset{\text{H}}{\text{C}}}(\text{O}-\overset{\text{O}}{\underset{\|}{\text{C}}}-\overset{\text{H}}{\underset{\text{H}}{\text{C}}}=\overset{\text{H}}{\text{C}}-\overset{\text{O}}{\underset{\|}{\text{C}}})_r\text{O}-\overset{\text{H}}{\underset{\text{H}}{\text{C}}}-\overset{\text{CH}_3}{\underset{\text{H}}{\text{C}}}-\text{O}-\overset{\text{H}}{\underset{\text{H}}{\text{C}}}-\overset{\text{CH}_3}{\underset{\text{H}}{\text{C}}}\sim$$

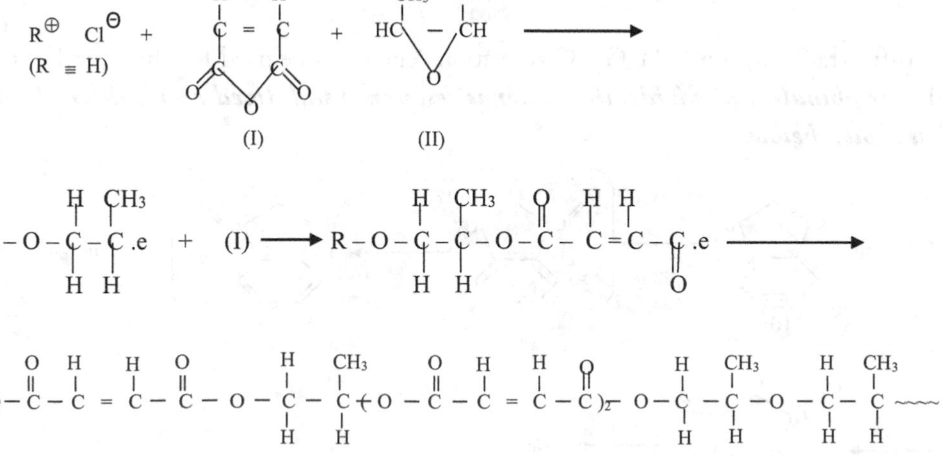

<u>Random copolymers</u>

12.208

81

Thus, it can be observed that maleic anhydride cannot be resonance stabilized electro-free-radically from the terminal. With phthalic anhydride, the following will be obtained.

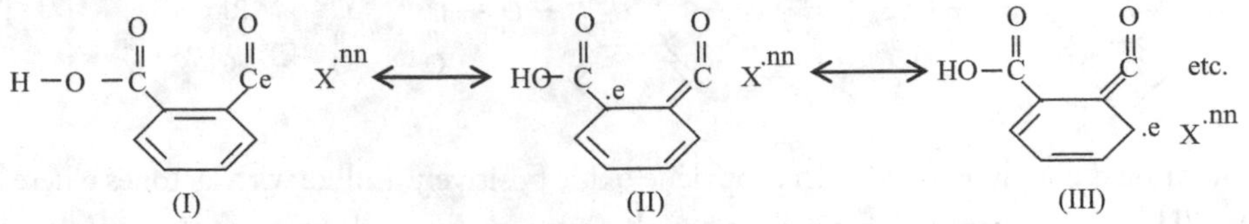

Resonance stabilization   12.209

Like styrene, only (I) which is also finally obtained above will favor being used during polymerization electro-free-radically. Chargedly, there is no such problem. An important difference between maleic anhydride and phthalic anhydride can be observed. Nucleo-non-free-radically, their growing chains are Equilibrium resonance stabilized as has already been explained. The C = O centers can however be activated to produce the same center

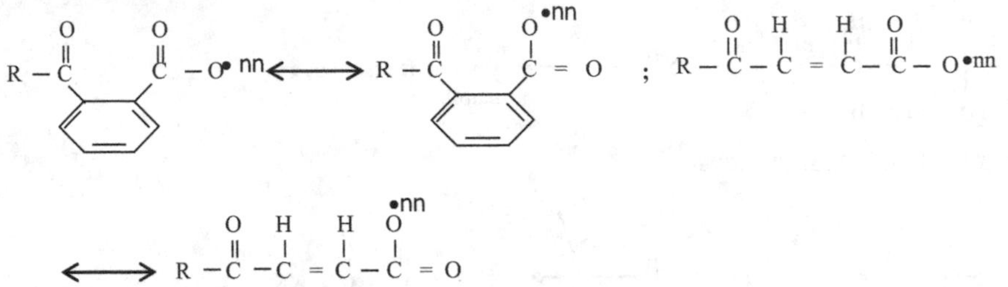

Identical Terminal resonance stabilization phenomena   12.210

When there is free radical resonance stabilization, the reactivity of a monomer can be greatly increased or decreased, depending on what is going on. It can largely be observed why the concept of copolymerization of monomers, little of which has only been considered, there is need to explain most of the observations which have been made in the past, but could not be explained in an orderly manner that which is NATURE.

There is need also to determine the order of nucleophilicity or electrophilicity of these cyclic monomers with respect to copolymerization. Fortunately, there are abun-dance of excellent data in the literature, very little of which has only been used to establish the new order.

12.4.2 Radical Characters of the Anhydrides

Nucleo-free-radically, only the C = C activation center can be used for the equation to be radically balanced. *For phthalic anhydride, the center is resonance stabilized and hidden. Hence it cannot be used as shown below.*

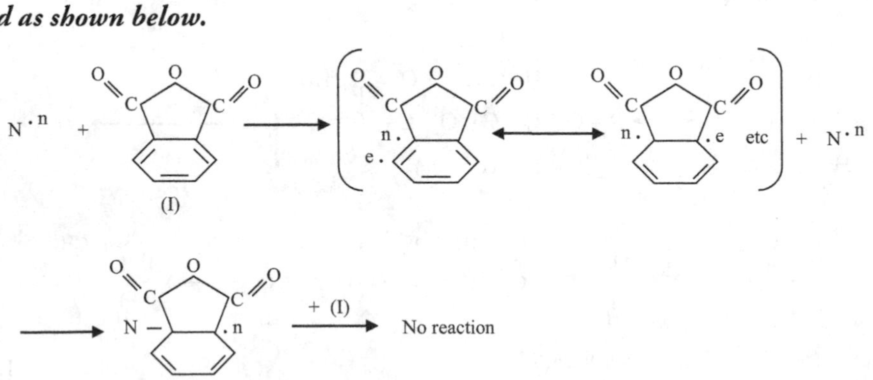

12.211

Indeed, since the activation centers are on the most nucleophilic section- the ring, never will the ring be activated as shown above. It was shown to highlight some fundamental principles. Electro-free-radically, the ring can be opened but not as fast as with maleic anhydride.

With maleic anhydride the followings are obtained.

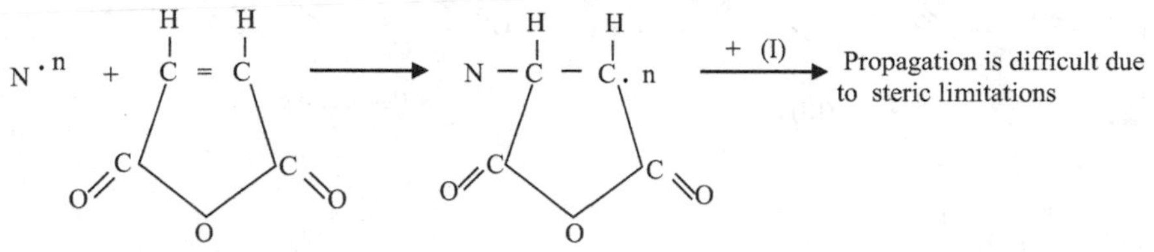

(I) Electrophile

12.212

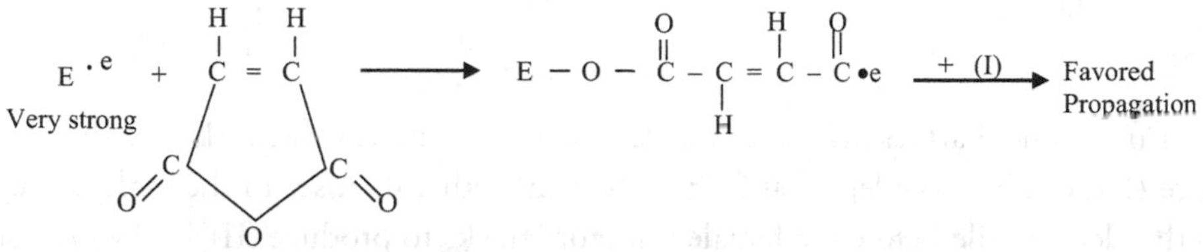

(I) Electrophile

12.213

There is limited propagation nucleo-free-radically after the initiation step because of electro-dynamic forces of repulsion and steric limitations, despite the fact that the route is natural to that center, but not to the monomer. Nevertheless, while maleic anhydride and diethyl fumarate do not homo-polymerize with the C = C center, they readily form very strongly alternating copolymers with monomers such as styrene and vinyl ethers[28]. The copolymerization of maleic anhydride and stilbene takes place, though neither monomer undergoes appreciable homopolymerization. Styrene and vinyl ethers are nucleophiles whose natural route is electro-free-radical route. Vinyl ethers favor more alternating placement than styrene, since vinyl ethers cannot favor the nucleo-free-radical route unlike styrene which does under harsh operating conditions. All the copolymerizations above were done using nucleo-free-radicals.

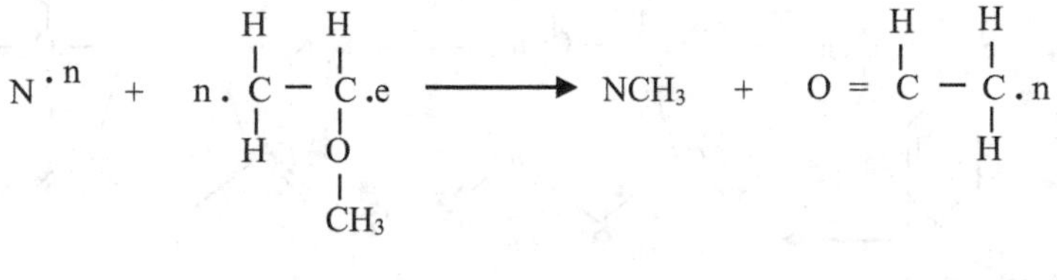

12.214

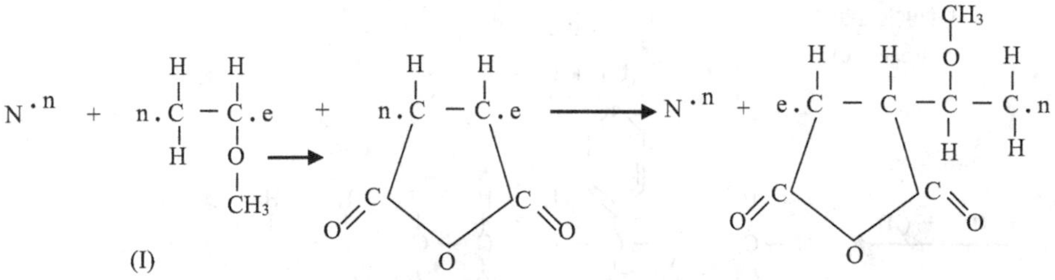

(I)

Nucleophile

(Unreactive)

(II)

Electrophile

(Unreactive)

(III)a

83

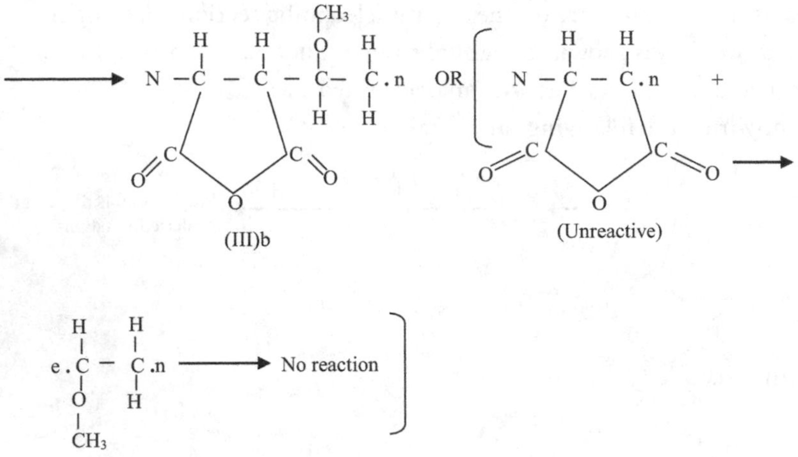

(III)b (Unreactive)

(IV)

12.215

Before the nucleo-free-radical attacks (II) which favors only the initiation step, (I) has already diffused to (II) to produce (III)a [The Couple]\, that is, it is the female that diffuses to the male or nucleophile that diffuses to the electrophile before the female initiator attacks to produce (III)b, (IV) is not favored because of presence of transfer species of the first kind. Hence when polymerizations exist, full alternating placements are obtained. With styrene which is resonance stabilized and can be homo-polymerized nucleo-free-radically, less alternating placement is favored.

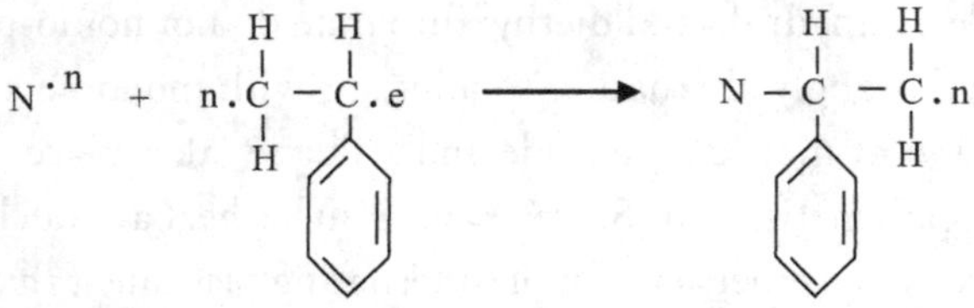

(Reactive)

12.216

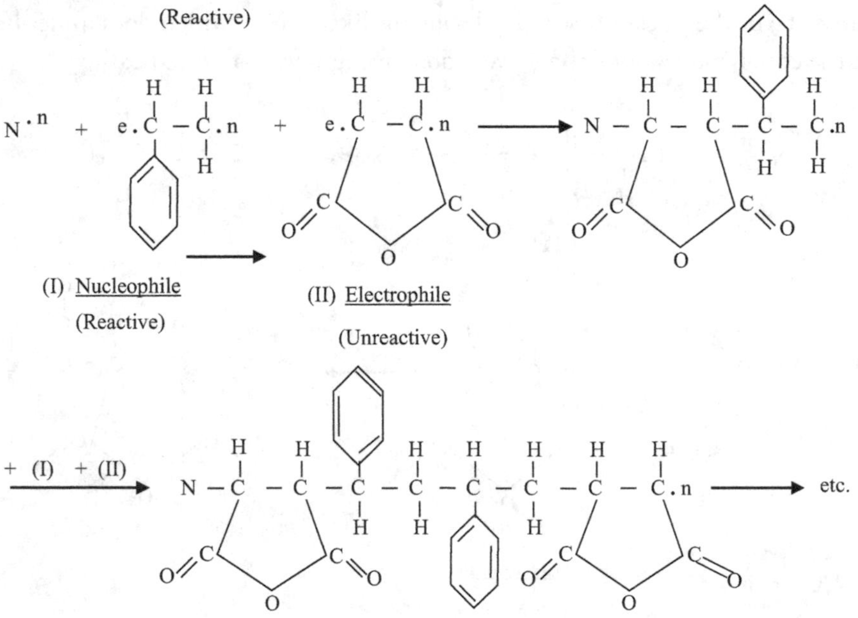

(I) Nucleophile (II) Electrophile
(Reactive) (Unreactive)

12.217

While two adjacently located maleic anhydrides cannot be possible, it is possible with styrene. Hence, full alternating placement cannot be formed. It is however strong because the nucleo-free-radical route is not natural to styrene. ***Thus while vinyl ethers are strong nucleophiles, styrene is however a***

weaker nucleophile being resonance stabilized. With stilbene which is unreactive homopolymerization-wise, due to electrostatic forces of repulsion and steric hindrance, full alternating placement is favored with maleic anhydride free-radically.

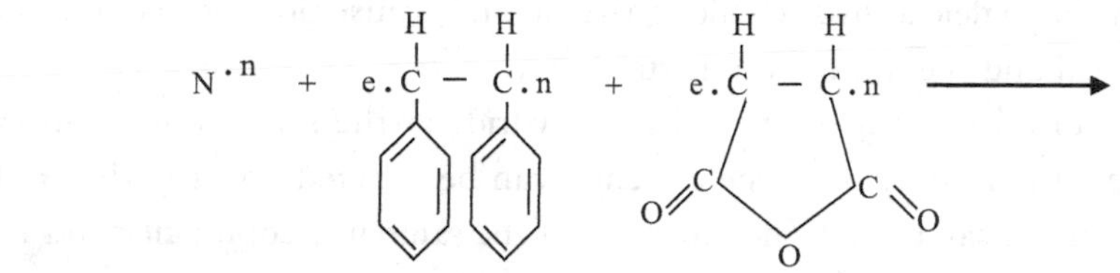

(I) <u>Nucleophile</u>
(Unreactive)

(II) <u>Electrophile</u>
(Unreactive)

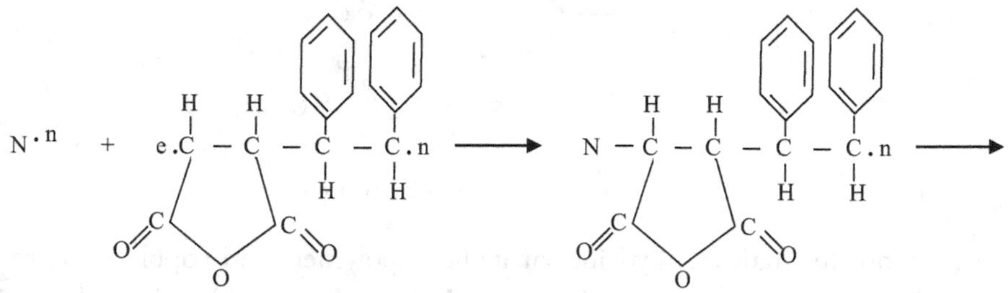

"THE COUPLE"

12.218

Electro-free-radically, same placement will be obtained here, while for styrene, block copolymers of styrene will largely be favored. Thus, in trying to explain some observations of the past, one has gone deep by already showing some of the driving forces favoring alternating placements in copolymerizations. This is as a result of an attempt to show that maleic anhydride undergoes free-radical polymerization. Nucleo-free-radically, it cannot be copolymerized with any of cyclic oxides or esters considered so far.

Non-free-radically, only the $\overset{O}{C} = O$ activation center can be used as follows.

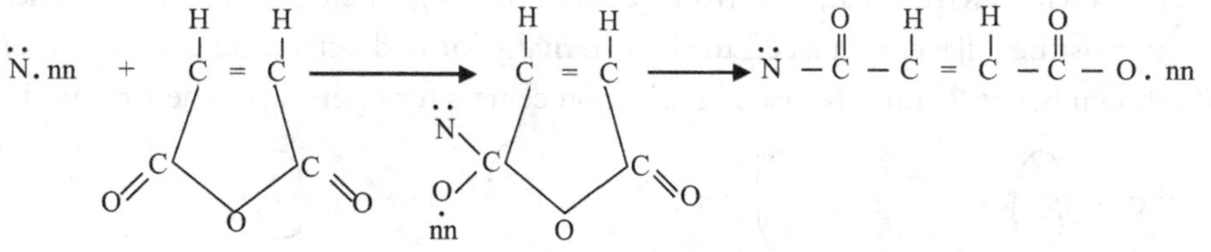

12.219

The $C = C$ activation center cannot be used, since the center is a full free-radical center. In addition, the following resonance stabilization shown below is not favored, since (I) is not chemically balanced.

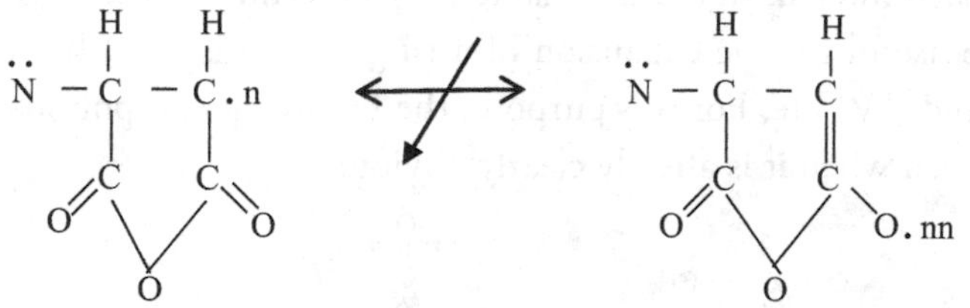

(I) (Not balanced)

12.220

That is, existence of (I) can never be favored. ***Like benzoquinone, maleic anhydride can be resonance stabilized from both ends with the O centers carrying the non-free-radicals. When this takes place, then the ring can never be opened and polymerization nucleo-non-free-radically will be favored.***

When the maleic anhydride is copolymerized with propylene oxide nucleo-non-free-radically, homopolymerization of the maleic anhydride or long blocks of maleic anhydride will largely be favored, since maleic anhydride is a strong Electrophile while propylene oxide is a strong Nucleophile of far more capacity than maleic anhydride. The propylene oxide will only appear alternatingly placed with maleic anhydride with more of the maleic anhydride along the chain, because propylene oxide is unreactive to the nucleo-non-free-radical and the route is not natural to it.

Charged activations of all the centers in maleic anhydride at the same time is impossible due to electrostatic forces of repulsion. Only one type of center can be ctivated one at a time. The two C=O centers can be activated at the same time since they are of the same nucleophilicity and if the activator is strong enough.

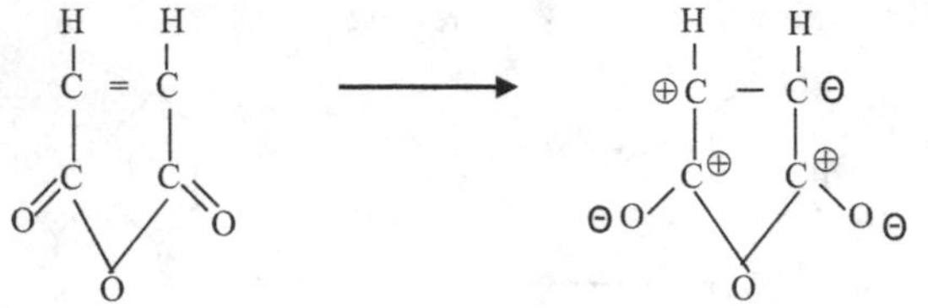

(Impossible existence)   12.221

Depropagation reactions for maleic anhydride for its homopolymers and copolymers are not possible. ***Thus, unlike cyclic esters which mistakenly are said not to be capable of copolymerization with alkenes, maleic anhydride can, only via the C = C activation.***

So far, the following functional/Activation centers have been identified.

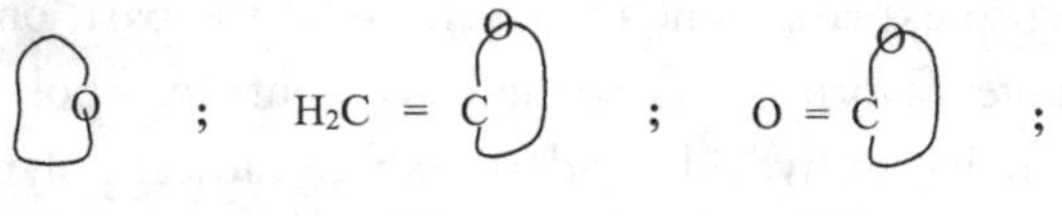

(I)   (II)   (III)

Functional center   Activation centers with Functional centers   12.222

These are centers via which MRSE can be introduced into the ring where possible. In (II) and (III) above the oxygen atom must be adjacently placed to the externally located activation centers. In order words, centers such as shown below, cannot be used as ctivation centers for opening of their rings via activation.

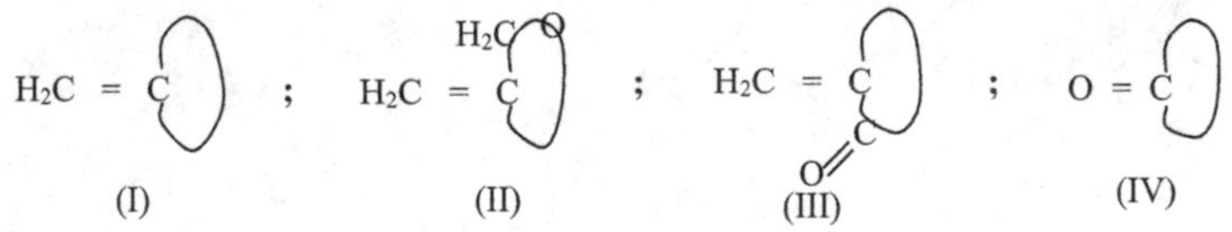

(I)   (II)   (III)   (IV)

Activation centers with no functional centers   12.223

(I) and (II) have already been fully encountered. In the last chapter, (IV) was introduced into the problem sections, where it was shown to be useful for the expansion of a ring. There is need however to fully identify what type of centers (III) and (IV) are. For this purpose, the following compounds and similar types will be considered for study, even when it is already clearly obvious.

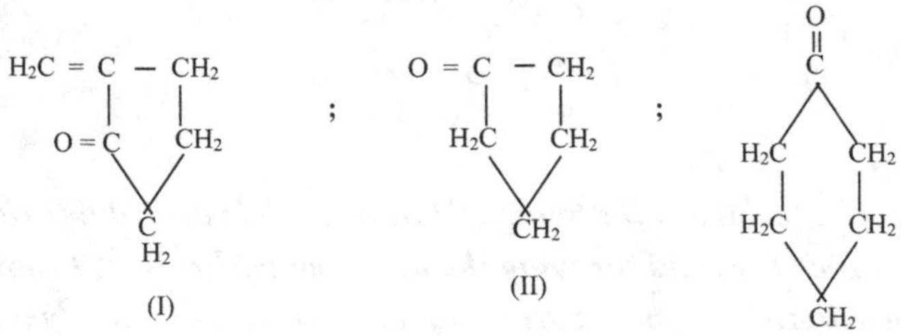

(I)   (II)   (III)

12.224

It is believed that (I) may not favor full existence, because the strain energy in the ring is more than the minimum required. Assuming its existence is favored, none of the activation centers can be used anionically or cationically to open the ring. Anionically, presence of transfer species of the tenth type of the first kind, does not allow for its initiation. In fact the C = O will never be activated. "Cationically", there seems to be no point of scission in the ring, after attack by the cation.

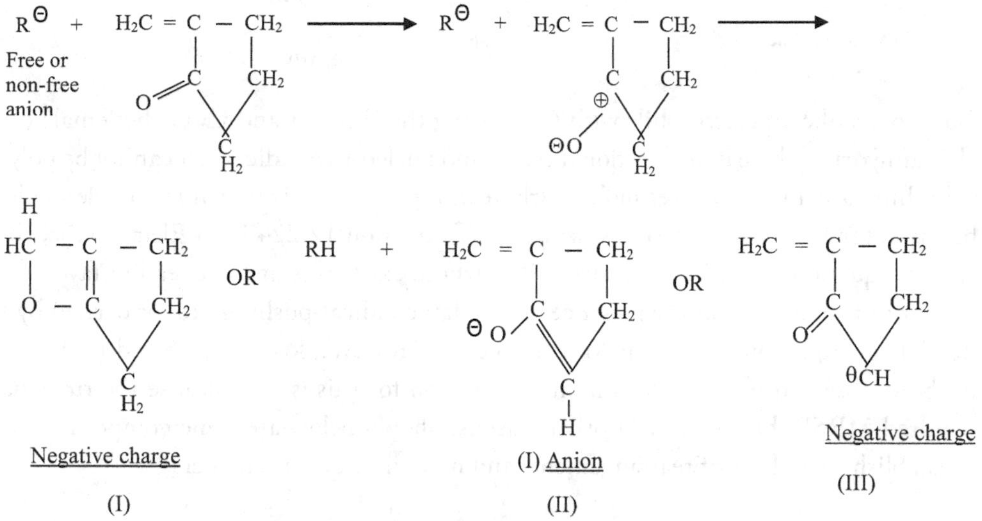

NON-FAVORED REACTIONS   12.225

In the reactions above anionically, none is favored, because $O = C$ center is more nucleophilic than $H_2C = C$ center. (I) of Equation 12.224, looks like a male (Acrolein) and indeed it is, since the C = C center is externally located cumulatively placed to the ring. Like acrolein, it has no transfer species. The C = C center is now the Y, while the C = O center is the X center.

Instantaneous opening of the ring may be favored, if the ring exists. However, when opened, a stable product cannot be obtained. Free-radically, the same will apply.

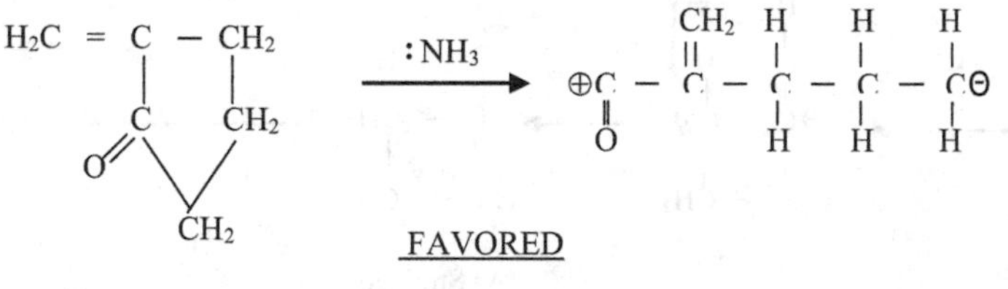

FAVORED   12.226

While larger-membered ringed compounds of (I) of Equation 12.224 may favor existence based on their size, some monomers such as shown below may not favor full existence.

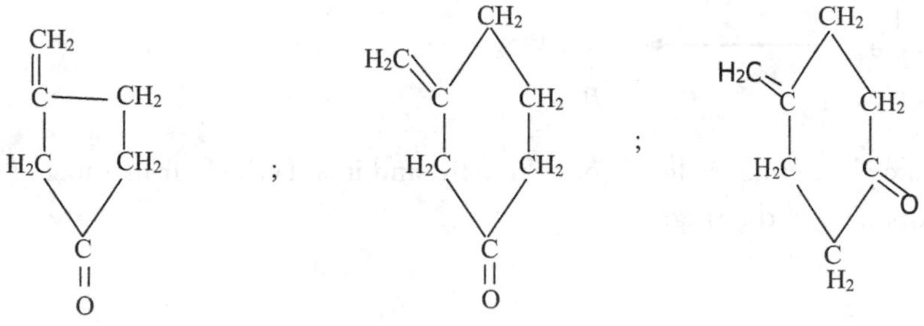

12.227

(I) for example may be a product of the following compound (A).

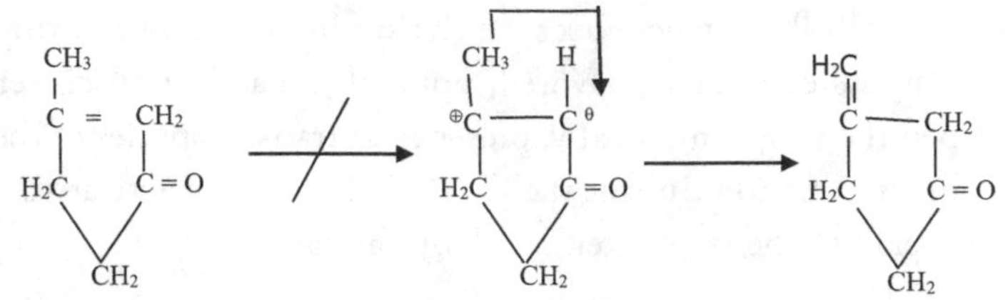

$$
\begin{array}{ccc}
\underset{\substack{| \\ \text{CH}_3 \\ | \\ \text{C} = \text{CH}_2 \\ | \\ \text{H}_2\text{C} \quad \text{C}=\text{O} \\ \diagdown \quad \diagup \\ \text{CH}_2}}{}
& \xrightarrow{\quad\diagup\quad} &
\underset{\substack{\text{CH}_3 \quad \text{H} \\ | \quad\quad | \\ {}^\oplus\text{C} - \text{C}\,{}^\theta \\ | \quad\quad | \\ \text{H}_2\text{C} \quad \text{C}=\text{O} \\ \diagdown \quad \diagup \\ \text{CH}_2}}{}
\xrightarrow{\quad\quad}
\underset{\substack{\text{H}_2\text{C} \\ \parallel \\ \text{C} \quad \text{CH}_2 \\ | \quad\quad | \\ \text{H}_2\text{C} \quad \text{C}=\text{O} \\ \diagdown \quad \diagup \\ \text{CH}_2}}{}
\end{array}
$$

.(A) Electrophile

(I)   12.228

The (A) above looks like an electrophile with C=C being the Y center and C=O the female center (X). Unlike maleic anhydride, there is no functional center and nucleo-free-radically, it cannot be polymerized. As an Electrophile, it cannot however molecularly rearrange to give (I) which is a nucleophile. (I) can probably be instantaneously opened. Thus, while (I) of Equation 12.224 is an Electrophile, the similar types shown in Equation 12.227 are Nucleophiles, whose existences are not readily favored either as intermediates or in full, in view of the presence of very large radical-pushing groups carried by them.

(II) and (III) of Equation 12.224 are known to exist. However, lower-membered rings of the same family, have been observed not to be common. The reason for this is not because the rings have more than the required MRSE, but because of other reasons. Shown below are some compounds which one has used to establish some laws of rearrangements and more in previous chapters.

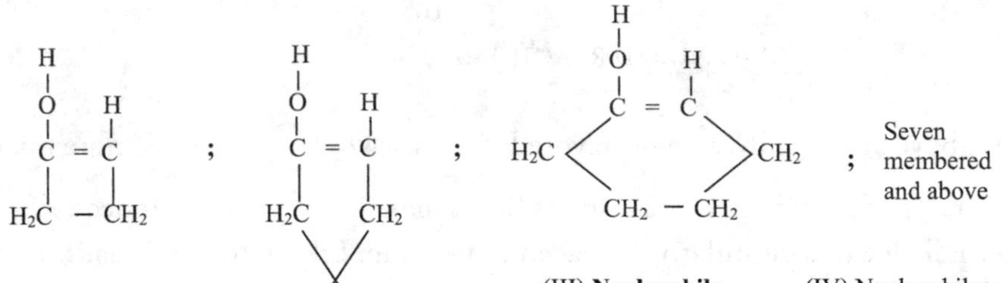

$$
\underset{\text{(I) \underline{Nucleophile}}}{\substack{\text{H} \\ | \\ \text{O} \quad \text{H} \\ | \quad\quad | \\ \text{C} = \text{C} \\ | \quad\quad | \\ \text{H}_2\text{C} - \text{CH}_2}}
\quad ; \quad
\underset{\substack{\text{(II) \underline{Nucleophile}}}}{\substack{\text{H} \\ | \\ \text{O} \quad \text{H} \\ | \quad\quad | \\ \text{C} = \text{C} \\ | \quad\quad | \\ \text{H}_2\text{C} \quad\quad \text{CH}_2 \\ \diagdown \quad \diagup \\ \text{CH}_2}}
\quad ; \quad
\underset{\text{(III) \textbf{Nucleophile}}}{\substack{\text{H} \\ | \\ \text{O} \quad\quad \text{H} \\ | \quad\quad | \\ \text{C} = \text{C} \\ \diagup \quad\quad \diagdown \\ \text{H}_2\text{C} \quad\quad\quad \text{CH}_2 \\ | \quad\quad\quad | \\ \text{CH}_2 - \text{CH}_2}}
\quad ; \quad
\underset{\text{(IV) \underline{Nucleophiles}}}{\substack{\text{Seven} \\ \text{membered} \\ \text{and above}}}
$$

12.229

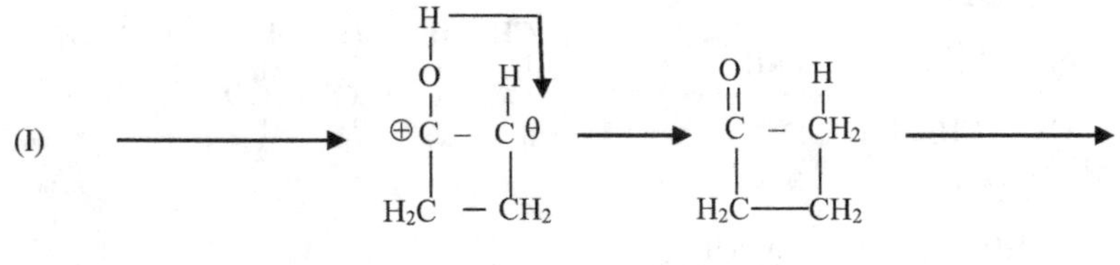

$$
\text{(I)} \quad \xrightarrow{\quad\quad} \quad
\underset{\text{(A) Strained}}{\substack{\text{H} - \\ | \\ \text{O} \quad \text{H} \\ | \quad\quad | \\ {}^\oplus\text{C} - \text{C}\,{}^\theta \\ | \quad\quad | \\ \text{H}_2\text{C} - \text{CH}_2}}
\xrightarrow{\quad\quad}
\underset{}{\substack{\text{O} \quad \text{H} \\ \parallel \quad\quad | \\ \text{C} - \text{CH}_2 \\ | \quad\quad | \\ \text{H}_2\text{C} - \text{CH}_2}}
\xrightarrow{\quad\quad}
$$

$$
\underset{\substack{\text{H} \quad \text{H} \quad \text{H} \\ | \quad | \quad | \\ {}^\oplus\text{C} - \text{C} - \text{C} - \text{C}\,{}^\theta \\ \parallel \quad | \quad | \quad | \\ \text{O} \quad \text{H} \quad \text{H} \quad \text{H}}}{}
\quad \xrightarrow{\quad\quad} \quad
\underset{}{\substack{\text{C}_2\text{H}_5 \\ | \\ \text{C} = \text{C} = \text{O} \\ | \\ \textbf{H}}}
$$

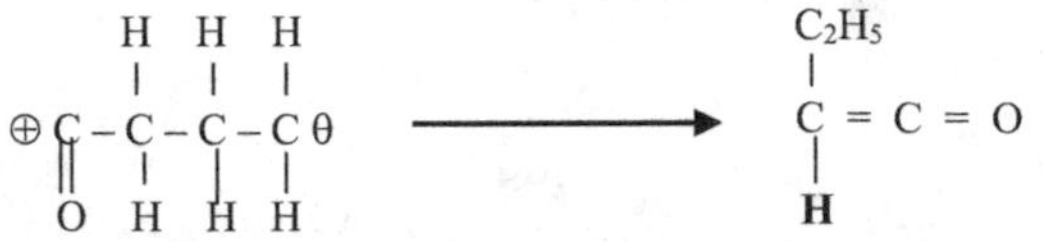

12.230

The reactions above can take place either radically or chargedly, and it is (I) the four membered ring that will preferable favor such opening of the ring.

88

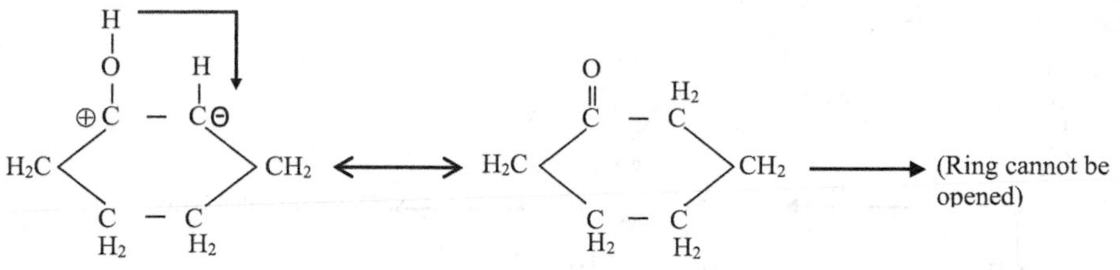

(B) (Unstrained)

<u>FAVORED</u>   12.231

Existence of (B) is favored after molecular rearrangement of the sixth type. When the smaller ring is strained molecular rearrangement of third kind takes place after instantaneous opening of the ring.

Therefore, considering the six-membered rings, that is (III) of Equation 12.224 or (B) of Equation 12.231, the followings are obtained.

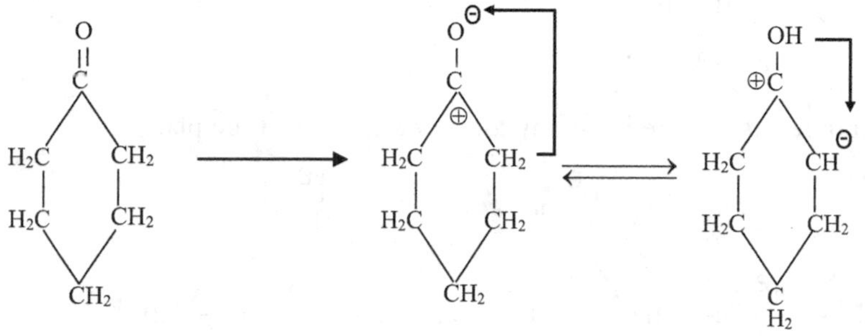

(I)   (II)   12.232

We have already seen lots of this in Chapters 7. This equation should be compared with that of the last equation. The fact that (II) is involved in ring expansion mechanism of (I), clearly indicates that (II) and (I) are equally stable, unlike the case of vinyl alcohol and acetaldehyde. Hence the equilibrium molecular rearrangement phenomenon of the first kind of the sixth type can be identified. In order words, (II) marks a point of transition whereby the capacity of OH group can be determined. Hence the following relationship is valid when OH group is carried.

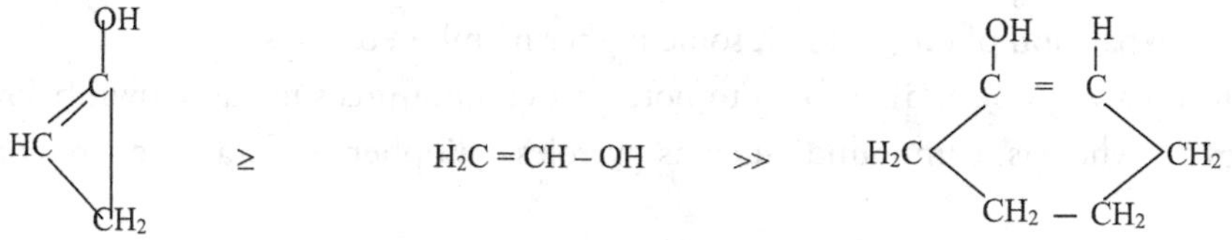

<u>Radical-pushing capacity of OH group</u>   12.233

At this point in the development, one seems to have found the capacity of OH groups.

Like the six-membered ring of cyclohexatrienes, this phenomenon does not take place with seven-membered rings and above. However, ring expansions from cycloheptanone to cyclooctanone and higher have now begun to be known as shown in Chapter 7.

It is believed that for rings lower than six in size, molecular rearrangements of the first kind takes place for the alcoholic monomers, that which cannot be reversed; while for the six-membered rings molecular rearrangement of the first kind takes place reversibly. For rings larger than six, the ketones are more stable than the alcohols, thus making the existence of alcohols possible only transiently. Four- and three-membered rings can be opened chargedly or radically, since they all have a point of scission (see Equation 12.230).

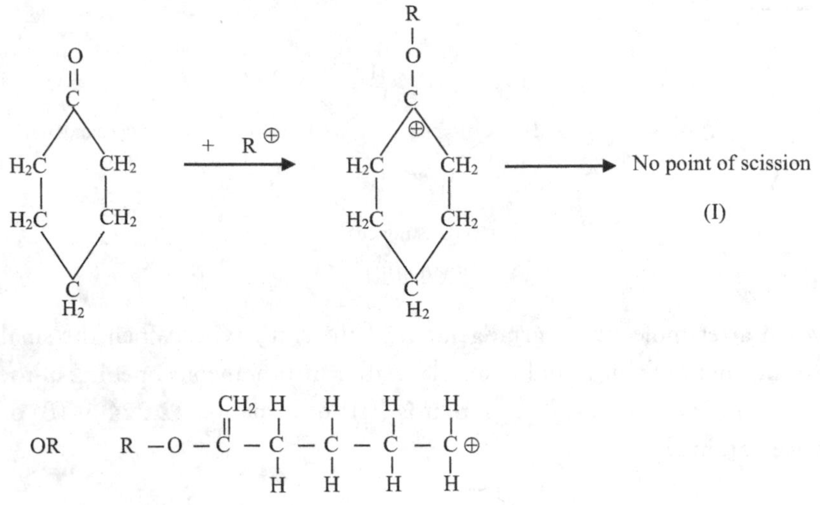

$$
\begin{array}{c}
O \\
\parallel \\
C \\
H_2C \diagup \diagdown CH_2 \\
H_2C \diagup \diagdown CH_2 \\
C \\
H_2
\end{array}
\quad + \; R^{\oplus} \quad \longrightarrow \quad
\begin{array}{c}
R \\
\mid \\
O \\
\mid \\
C^{\oplus} \\
H_2C \diagup \diagdown CH_2 \\
H_2C \diagup \diagdown CH_2 \\
C \\
H_2
\end{array}
\quad \longrightarrow \quad \text{No point of scission} \quad (I)
$$

$$
\text{OR} \qquad R-O-\overset{CH_2}{\underset{}{C}}-\overset{H}{\underset{H}{C}}-\overset{H}{\underset{H}{C}}-\overset{H}{\underset{H}{C}}-\overset{H}{\underset{H}{C}}^{\oplus}
$$

(II) <u>Not favored</u>

12.234

It must molecularly rearrange first before R^+ or any form of attack can take place.

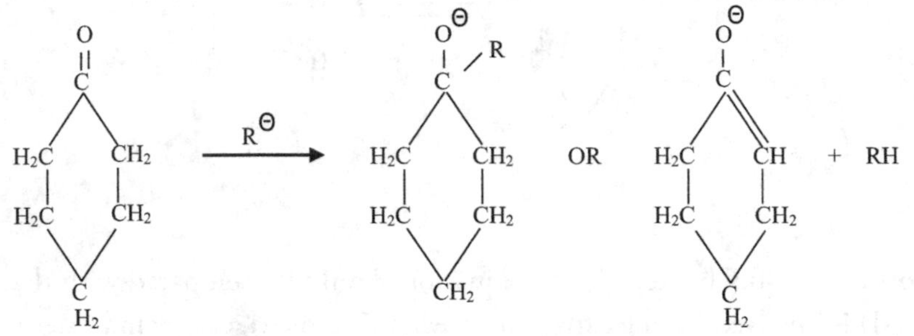

$$
\begin{array}{c}
O \\
\parallel \\
C \\
H_2C \diagup \diagdown CH_2 \\
H_2C \diagup \diagdown CH_2 \\
C \\
H_2
\end{array}
\;\xrightarrow{R^{\ominus}}\;
\begin{array}{c}
O^{\ominus} \\
\mid \diagup R \\
C \\
H_2C \diagup \diagdown CH_2 \\
H_2C \diagup \diagdown CH_2 \\
CH_2
\end{array}
\quad \text{OR} \quad
\begin{array}{c}
O^{\ominus} \\
\mid \\
C \\
H_2C \diagup \diagdown CH \\
H_2C \diagup \diagdown CH_2 \\
C \\
H_2
\end{array}
\; + \; RH
$$

(I)   (II)

12.235

It is (II) that is favored, with presence of transfer species of the first kind of the tenth type, the type involved in its molecular rearrangement when activated in the presence of a weak initiator. Thus, one can observe that while (IV) of Equation 12.223 cannot be used as a functional center, it can be used as an activation center for expansion of only six-and some higher membered rings.

Over the years, it has always been disturbing to note that compounds such as shown below are not common or do not exist, whereas, compounds such as pyrroles, thiophenes, furan etc. are common or exist.

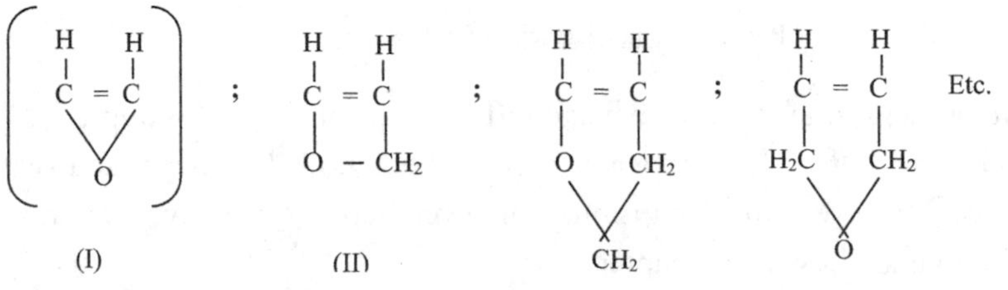

$$
\left(
\begin{array}{c}
H \quad H \\
\mid \quad\; \mid \\
C = C \\
\diagdown \;\; \diagup \\
O
\end{array}
\right)
\;;\;
\begin{array}{c}
H \quad H \\
\mid \quad\; \mid \\
C = C \\
\mid \quad\; \mid \\
O - CH_2
\end{array}
\;;\;
\begin{array}{c}
H \quad H \\
\mid \quad\; \mid \\
C = C \\
\mid \quad\; \mid \\
O \quad CH_2 \\
\diagdown \diagup \\
CH_2
\end{array}
\;;\;
\begin{array}{c}
H \quad H \\
\mid \quad\; \mid \\
C = C \\
\mid \quad\; \mid \\
H_2C \quad CH_2 \\
\diagdown \diagup \\
O
\end{array}
\quad \text{Etc.}
$$

(I)   (II)   (III)   (IV)

<u>Not commonly known</u>

12.236

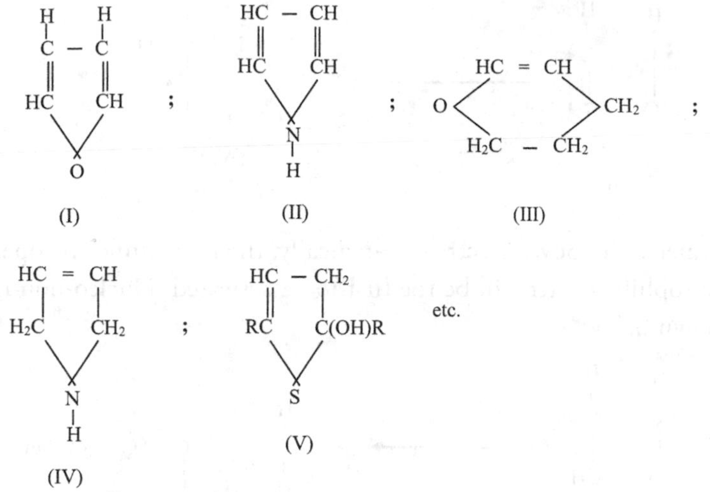

The structures labeled (I) through (V):

(I) — three-membered ring with O

(II) — ring with N–H

(III) — ring with O and CH_2, H_2C

(IV) — ring with N–H

(V) — ring with S, RC and C(OH)R

etc.

<u>Known to exist</u>

12.237

The non-existence of compounds of Equation 12.236 has nothing to do with presence of more than the MRSE in the rings, since some of those of Equation 12.237 are more strained.

It seems that under normal conditions, that is, standard temperature and pressure (STP) or above the ceiling temperature, the compounds of Equation 12.236 just like (I) of Equation 12.229 do not favor any stable existence, in view of the presence of more than the MRSE in the rings and most importantly, the presence of points of scission in the rings except for the three-membered rings. When the four- and five-membered rings are opened instantaneously, the followings are supposed to be obtained above the ceiling temperature.

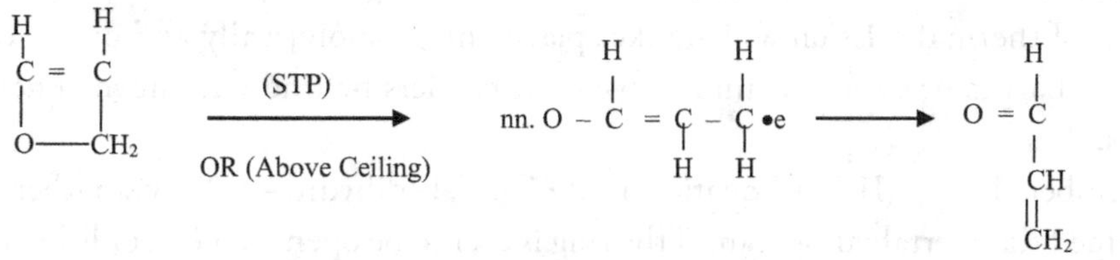

$$\text{(I)} \xrightarrow[\text{OR (Above Ceiling)}]{\text{(STP)}} \text{nn. } O-\overset{H}{\underset{H}{C}}=\overset{H}{\underset{H}{C}}-\overset{}{\underset{}{C}}\bullet e \longrightarrow O=\overset{H}{\underset{\overset{|}{CH}}{C}} \quad \text{(I)a Acrolein}$$

where in (I)a the bottom is $CH \overset{||}{} CH_2$

12.238a

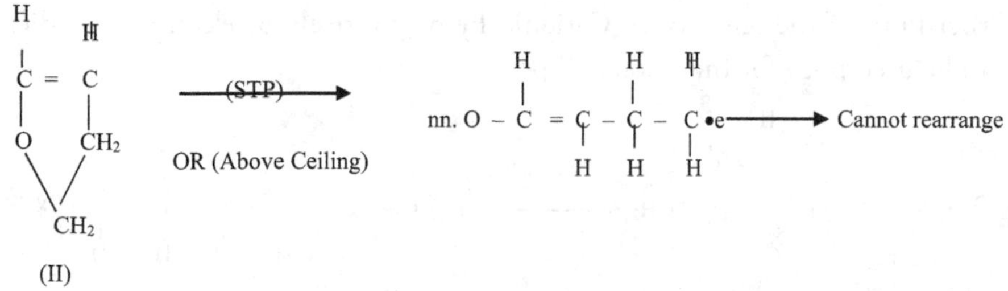

$$\text{(II)} \xrightarrow[\text{OR (Above Ceiling)}]{\text{(STP)}} \text{nn. } O-\overset{H}{\underset{H}{C}}=\overset{H}{\underset{H}{C}}-\overset{H}{\underset{H}{C}}-\overset{H}{C}\bullet e \longrightarrow \text{Cannot rearrange}$$

12.238b

As can be observed, (I)a is an electrophile. But (II) cannot molecularly rearrange or resonance stabilize. Where resonance stabilization is allowed to take place to form electrophiles, the reaction is favored only for four-membered ring. When (II) is activated nucleo-free-radically, its ring can be opened by scission at $O-CH_2$ bond as shown below for the five-membered ring. In the Initiation step, (A) is obtained.

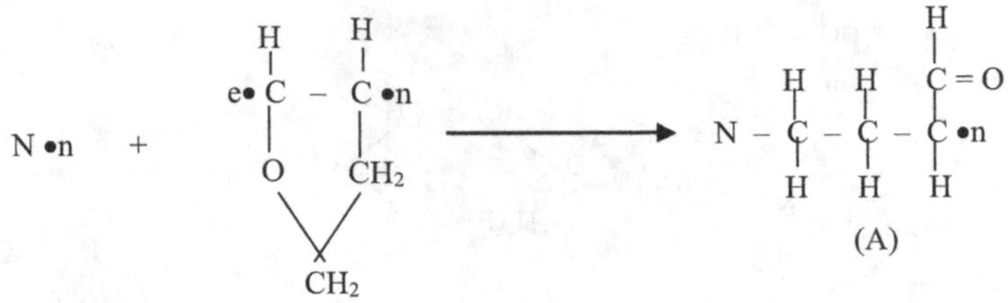

$$N \bullet n \quad + \quad \text{[epoxide monomer]} \longrightarrow N - \overset{H}{\underset{H}{C}} - \overset{H}{\underset{H}{C}} - \overset{\overset{H}{|}}{\underset{C=O}{C}} \bullet n \quad \text{(A)}$$

12.239a

Notice the type of monomer unit above. Electro-free-radically, the ring cannot be opened, because the double bond the less nucleophilic center will be the first to be activated. Nucleo-non-free-radically, the ring can be opened as shown below.

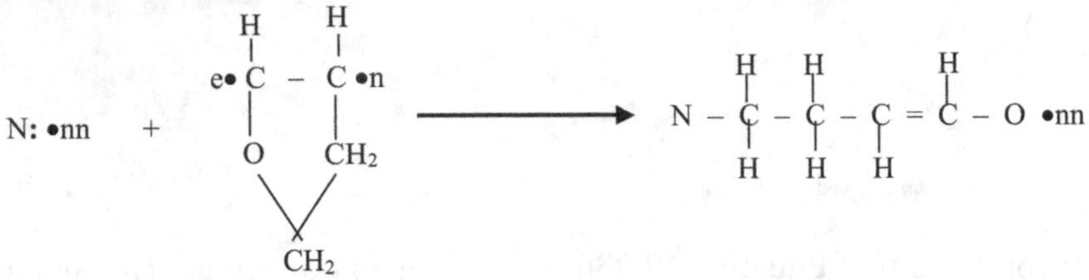

$$N: \bullet nn \quad + \quad \text{[epoxide monomer]} \longrightarrow N - \overset{H}{\underset{H}{C}} - \overset{H}{\underset{H}{C}} - \overset{H}{\underset{}{C}} = \overset{H}{\underset{}{C}} - O \bullet nn$$

12.239b

The ring has been opened as a result of transfer species of the first kind of the ***eleventh type.*** It was also observed electro-free-radically for opening of maleic anhydride. When it takes place, rings are opened.

While (IV) of equation 12.236 rearranges to a vinyl 1,2-epoxide when opened instantaneously, its six-membered ring will not favor instantaneous opening of its ring, since it is far less strained at STP or above the ceiling temperature of the monomer. Above STP conditions, when MRSE is large enough to allow for instantaneous opening of the ring, this can only be done radically as indicated above and this leads also to a vinyl 1,3-epoxide. These rearrangements are not possible chargedly, since this is a form of thermal scission which takes place only homolytically and only electro-radical can move. Positive charges or cations cannot leave their carriers behind and move. They move only with their carriers.

For the six-membered ring- (III) of Equation 12.237 or 3,4- dihydro - 1,2- pyran, there is no doubt that it is more strained than tetrahydropyran. If the ring is ever to be opened, it can only be done through the oxygen center as already shown in Equations 12.239a and 12.239b for five-membered case nucleo-free-radically and nucleo-non-free-radically. What is shown below can never take place via the O center, since the O- center is more nucleophilic than the double bond, but through the C = C center as transfer species of the first kind of the tenth type. Cationically or positively or electro-free-radically, only the double bond can be used both for Initiation and propagation.

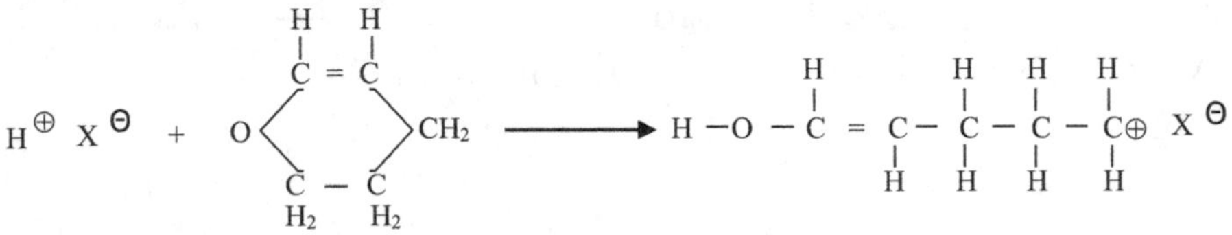

$$H^{\oplus} X^{\ominus} + \text{[dihydropyran ring]} \longrightarrow H - O - \overset{H}{\underset{}{C}} = C - \overset{H}{\underset{H}{C}} - \overset{H}{\underset{H}{C}} - \overset{H}{\underset{H}{C}}\oplus \ X^{\ominus}$$

Favored only via C=C activation

12.240

The reaction above is also favored if the ring can be opened instantaneously (that which will not take place). However in general, for the members of this family, the following is valid as has already been confirmed.

$$(I) \quad O \qquad > \qquad (II) \quad C = C$$

<u>Order of nucleophilicity</u>   12.241

If any of the members is to favor polymerization via the double bond without opening the ring, it will only be the three-membered ring - (I) of Equation 12.236, only free radically, since chargedly the oxygen center will be disturbing (Electrostatic forces of repulsion).

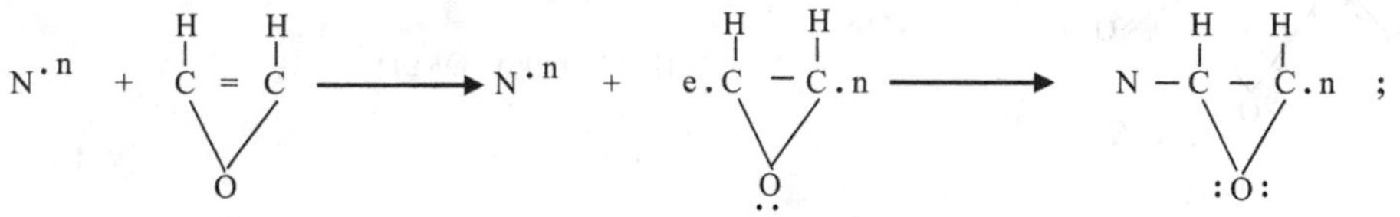

$$N \cdot^n + \underset{(I)}{C = C} \longrightarrow N \cdot^n + e.\underset{(II)}{C - C}.n \longrightarrow N - C - C.n \; ;$$

$$H \cdot^e + n(I) \longrightarrow H \left\{ C - C \right\}_{n-1} C - C.e$$

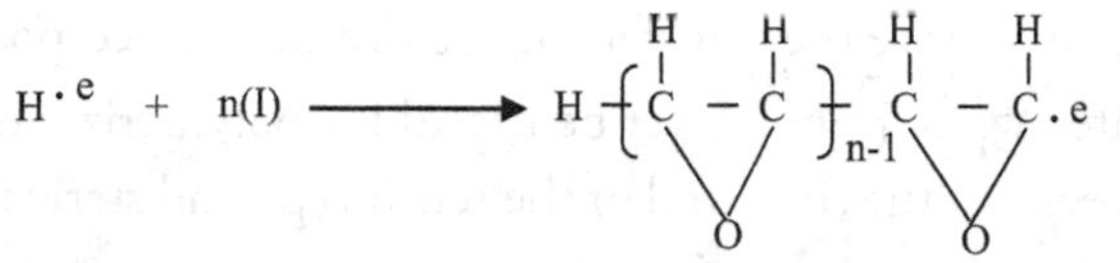

12.242

Maleic anhydride a strong electrophile can be both nucleophilic and electrophilic in character. This does not apply to the type shown below.

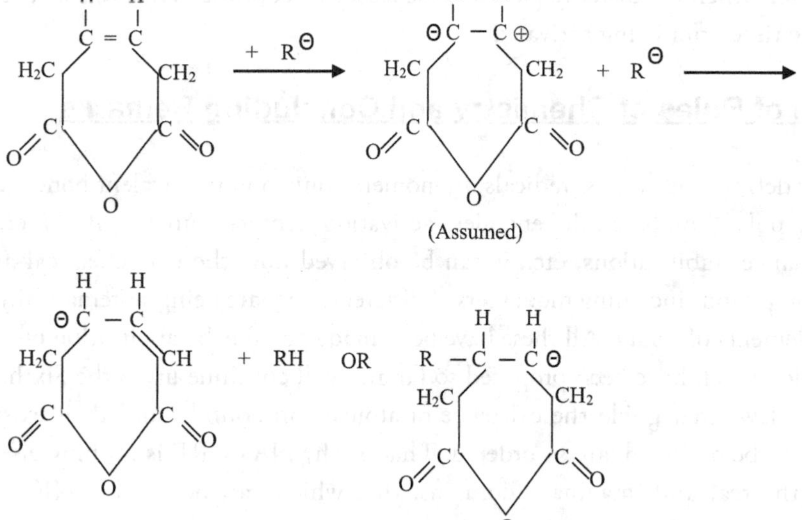

(Assumed)

(I) FAVORED   (II) NOT FAVORED   12.243

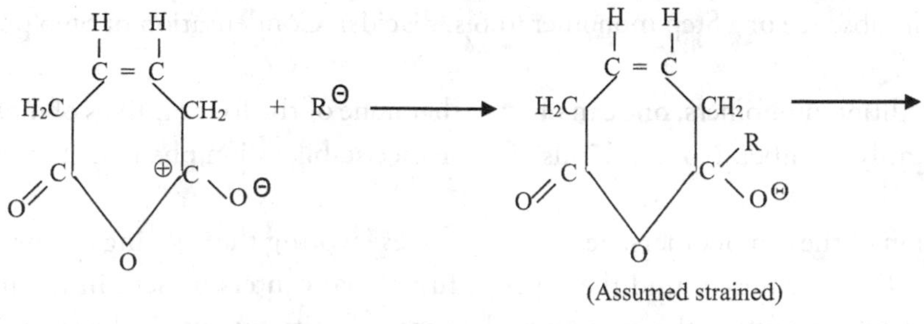

(Assumed strained)

$$R - \overset{\overset{O}{\|}}{C} - \overset{\overset{H}{|}}{\underset{\underset{H}{|}}{C}} - \overset{\overset{H}{|}}{C} = C - \overset{\overset{H}{|}}{\underset{\underset{H}{|}}{C}} - \overset{\overset{O}{\|}}{C} - O^{\ominus}$$

(III) NOT FAVORED

12.244

$$\longrightarrow H - O - \overset{\overset{O}{\|}}{C} - \overset{\overset{H}{|}}{\underset{\underset{H}{|}}{C}} - \overset{\overset{H}{|}}{C} = C - \overset{\overset{H}{|}}{\underset{\underset{H}{|}}{C}} - \overset{\overset{O}{\|}}{C} \oplus$$

(IV) NOT FAVORED

(Assumed strained)

12.245

In the first equation above, it is (I) that is favored since the transfer species can be rejected positively and abstracted anionically. Nevertheless the center $\overset{\frown}{C = C}$ will never be useful for polymerization radically and chargedly, due to presence of transfer species of the first kind of the tenth type and steric limitations.

On the other hand, the $\overset{\frown}{O}$ center is nucleophilic, while the $\overset{\frown}{C = O}$ center is electrophilic as indicated by the products obtained in the last two equations. However, those two last reactions will not take place because, the C=C center which is isolatedly placed is the least nucleophilic. That will be the center that will be attacked all the time after being activated.

12.5 Proposition of Rules of Chemistry and Concluding Remarks

Based on the new definitions of ions, radicals, monomers, ionic bonds, covalent bonds, electrostatic bonds, dative bonds, polar bonds, strain energies, activation centers, functional center, molecular rearrangements, resonance stabilizations, etc., it can be observed how the true chemical and physical characters of atoms, compounds including monomers of different types are being systematically elucidated, without bringing in elements of doubt. All these have been made possible by application of natural laws. Hence, the laws or rules which have been proposed so far and will continue up to the Sixth Volume are natural. These are the laws that guide the existence of atoms, compounds, and their existences with each other, which as can be observed, are so ordered. That is why NATURE is nothing else other than ORDERLINESS in the real and imaginary domains, that which can be SANTANIC or GODLY, based on..............

Cyclic ethers, acetals, esters, anhydrides and related monomers have been considered as Addition monomers or compounds for study. As Step monomers, only the anhydrides show these abilities which are yet to be shown. In the absence of a Step monomer (diols, diacids), Condensation or Step polymerization is impossible.

When used as Addition monomers, one can observe that none of the four families of monomers is the same with another family member. No new kinds of resonance stabilization phenomenon of importance have been identified.

With consideration of these monomers, new driving forces favoring the opening of ringed monomers have been identified. These center around the types of functional centers present in the monomer, the types of activation centers in and on the monomer, the types of initiators used, the types of initiations involved, the points of scissions on the monomers, the abilities to provide the MRSE in their rings and the character of the monomers. For years, there have been different proposals on the types of active centers carried by the growing polymer chains of these monomers. For the first time, all the type of active centers carried by them, have been fully ascertained. For the first time also, all the nucleophilic and electrophilic characters of the monomers and their orders with respect to members of the families, and to other families, have been fully identified. These orders are very important when considering not only their homopolymerization rates, but their copolymerization behaviors, where it will clearly become obvious that very little or nothing is known about these monomers, despite the important data which have been gathered over the years. The ability to analyze data is very important, which itself is a different subject development in Sciences and Engineering [29].

For the first time also, the mechanisms of propagation/depropagation and propagation/depropagation/repropagation phenomena have been provided. How and when they occur, have been identified. Why there exist **induction periods** in cyclic acetals polymerization, have partly been provided. The radical polymerizations of these monomers, little of which have been unknown, were fully considered. The four families of monomers considered, favor free and in particular non-free-radical, anionic, electro-free-radical (cationic) and positively charged homopolymerization reactions. Copolymer-ization reactions largely take place radically. While Forward Addition polymerization could be charged or radical additions, Backward Addition polymerization as will conclusively be shown is electro-free-radical in character. All the new revelations have been made possible based on the data which have been prov*ided over the years,* but could not clearly be explained.

Some of the ion-paired and electrostatically charged-paired initiators have been identified, of which for those of the charged types which are all electrostatic in character, three types have been used so far -

(i)   Those of non-transition metal types. They do not have vacant orbitals for stereospecific placement. How they are obtained have partly been identified.

(ii)   Those of transition metal types. Many of these have vacant orbitals for stereospecific placements. How they are obtained have been shown.

(iii)   Those of metallic types - e.g. PCl_5, $FeCl_3$, BF_3. These like the others are electrostatically bonded from only the negative center. In order words, these are all electrostatically positively charged-paired initiators. They do not have vacant orbitals for stereospecific placements. How they are obtained have partly been identified. Some like the so-called cationic types can only be used when they exist in Equilibrium state of existence.

More than one stable molecule are released or formed during the termination step. How these initiators can be used to favor full, half or no polymerization of the monomers have been identified. So also are how presence of full living, half living and dead polymers, favored by them, considered– that is, stable and unstable polymers.

Rule 983: This rule of Chemistry for **Ringed Monomers**, states that, the sixth driving forces favoring the opening of a ring, is existence of a functional center in the rings, through which the MRSE can be provided where possible to open the ring.
(Laws of Creations for Ringed compounds)

Rule 984: This rule of Chemistry for **CYCLIC ETHERS,** states that, first recognizable members of C/H/O- ringed compounds are the 1,2-, 1,3-, 1,4-, 1,5- etc. EPOXIDES of which the first members are shown below -

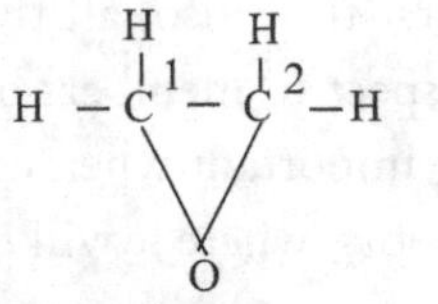

(I)

1,2 - epoxides

(oxiranes)

Dimethylene oxide

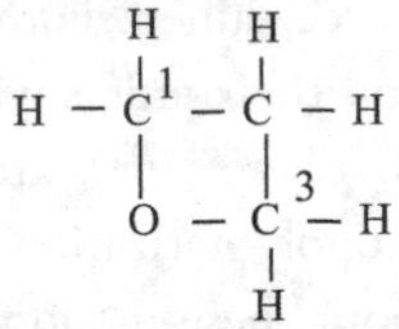

(II)

1,3 - epoxide

(oxetanes)

Trimethylene oxide

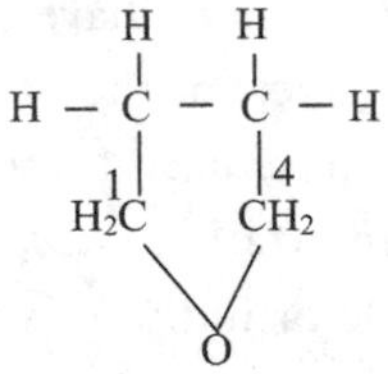

(III)

1,4 - epoxide

Tetramethylene oxide

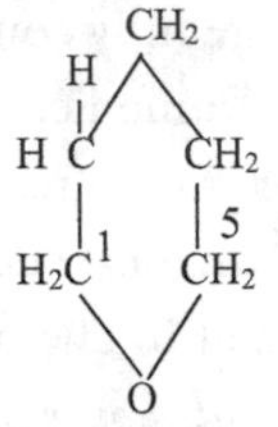

(IV)

1,5 - epoxide

Pentamethylene oxide

all of which carry one hetero atom [OXYGEN] –(Not numerically numbered) in an unsaturated ring, with decreasing strain energy in their rings as the size of the ring increases; noting that the only functional center on them is Nucleophilic in character.

(Laws of creations for Epoxides)

Rule 985: This rule of Chemistry for **CYCLIC ETHERS,** states that, the second recognizable members of C/H/O – ringed compounds are 1,3-, 1,4-, 1,5-, 1,6-, etc. unsaturated DIOXANES of which the first members are shown below-

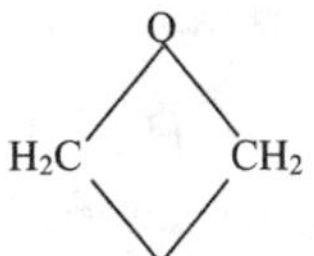

(I)

1,3 - dioxane

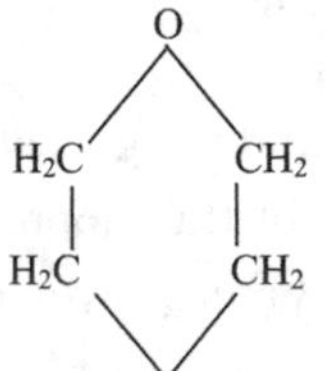

(II)

1,4 - dioxane

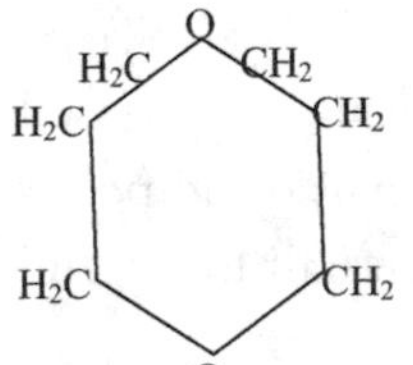

(III)

1,5 - dioxane

i)

all of which carry two OXYGEN hetero-atoms symmetrically placed as functional centers; the strain energy in the rings being small decreasing with increasing ring size, for which the following is valid-

$$\text{Dioxanes} \quad > \quad \text{Epoxides}$$
$$\text{Order of Nucleophilicity}$$

ii)

noting that the order of Strain energy in their rings is the reverse of the above.
(Laws of Creations for Dioxanes)

Rule 986: This rule of Chemistry for **OXYGEN containing ringed monomers or compounds,** states that, the fifth factor determining the ability of a ring to possess the MRSE, is the types of groups carried by the connecting centers of a ring; for which when the groups are all radical-pushing groups, the stability of the ring is decreased in terms of provision of MRSE than when at least one radical-pulling group is present.
(Laws of Creations for Cyclic O containing Ringed compounds)

Rule 987: This rule of Chemistry for **CYCLIC ACETALS all of which are unsatu-rated,** states that, the first recognizable members are cyclic trimmers (TRIOXANE), tetramers (TETRAOXANE) of aldehydes and ketones, of which the first members are as follows-

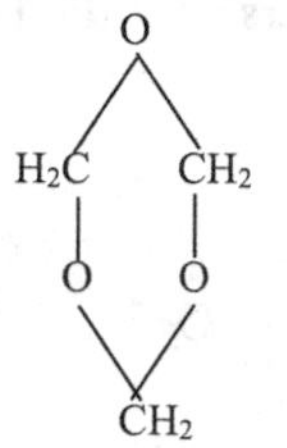

Trioxane 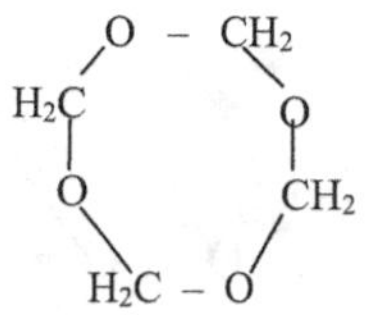   Tetraoxane

all of which carry three or more oxygen atoms conjugatedly placed in the rings; the order of nucleophilicity increasing with size of the ring with more ability to undergo propagation/depropagation reactions electro-free-radically.
(Laws of Creations for cyclic aldehydes and ketones)

Rule 988: This rule of Chemistry for **CYCLIC ACETALS,** states that, the second recognizable members are cyclic dimers of aldehydes and ketones conjugatedly placed, called CYCLIC FORMALS of the general structure shown below-

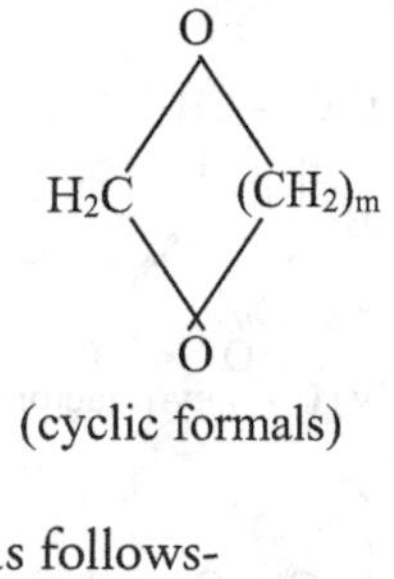

$m = 2, 3, 4, 5$ etc.

(cyclic formals)   i)

of which the first members are as follows-

(I)
1,3 - dioxolane
(m = 2)

(II)
1,3 - dioxehane
(m = 3)

(III)
1,3 - dioxepane
(m = 4)

(IV)
1,3 - dioxocane

etc.

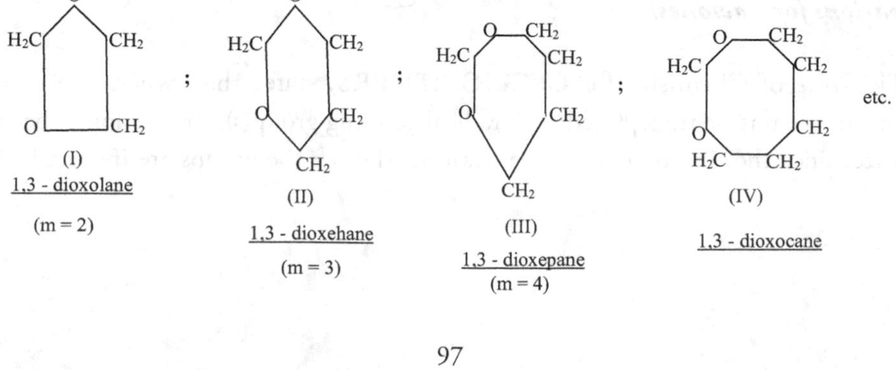

ii)

for which none of these acetals favor negatively charged or nucleo-non-free-radical routes, in the view of the small strain energy in their rings, and all the cases above favor only the positively charged or electro-free-radical route (with the exception of the six-membered case which is said not to favor homopolymerization, that which is not true).
(Laws of Creations for Cyclic Formals)

Rule 989: This rule of Chemistry for **OXYGEN containing ringed monomers or compounds,** states that, the oxygen center in the ring is a strong radical-pushing center whose capacity is equal to or greater than that of O = group, depending on the size of the ring.
(Laws of Creations for oxygen containing ringed compounds)

Rule 990: This rule of Chemistry for **CYCLIC ESTERS,** states that, unlike cyclic ethers and acetals, they undergo ring-opening polymerization to polyesters with the use of varieties of so-called "cationic" and anionic initiators; for which the first recognizable members are called -the LACTIDES of which the first member is shown below-

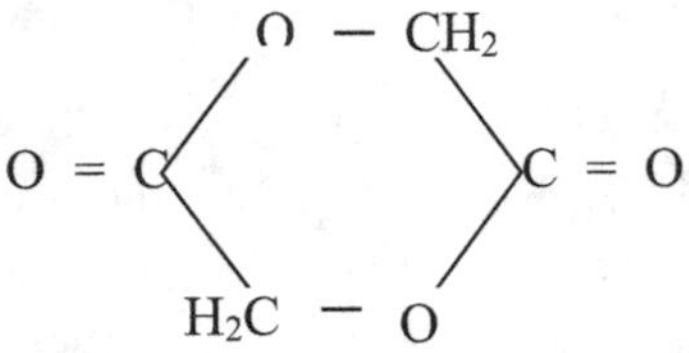

(I) (Glycolide) - a <u>Lactide</u>

all of which carry two male and two female centers conjugatedly placed.
(Laws of Creations for Lactides)

Rule 991: This rule of Chemistry for **CYCLIC ESTERS,** states that, unlike cyclic ethers and acetals, they undergo ring-opening polymerization to polyesters with the use of varieties of so-called "cationic" and anionic initiators; for which the second recognizable members are called -the LACTONES of which some of the first members are shown below-

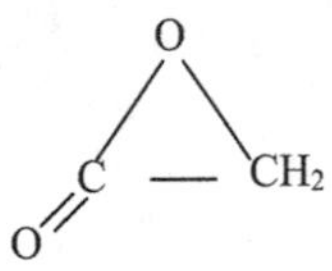

(II) (α - Lactone) 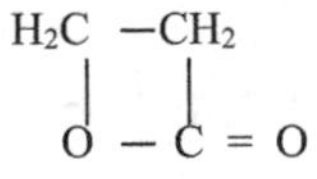 ; (III) (β - propiolactone) 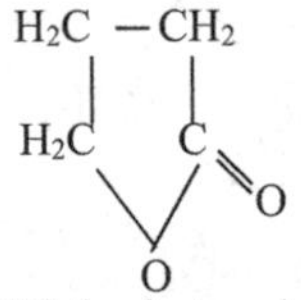 ; (IV) (γ - butyro-lactone) 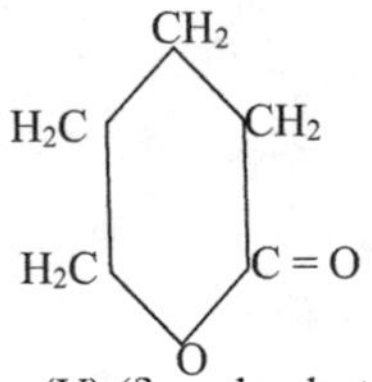 ; (V) (δ - valerolactone)

in which for all the cases above, there is only one male $C = O$ and one female O center.
(Laws of Creations for Lactones)

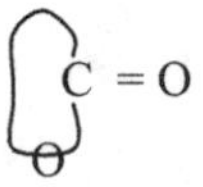

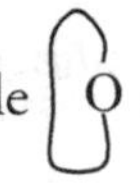

Rule 992: This rule of Chemistry for **CYCLIC ETHERS,** states that, when one or more of the hydrogen atoms in them is or are replaced with radical-pushing group(s), the monomer becomes easy to homopolymerize, since the ring becomes more strained whether the groups are internally or externally

located to the oxygen center; for which the more nucleophilic the ringed monomer becomes, the more strained it is, and the less the energy required to attain the minimum required strain energy.

$$
\underset{[External]}{\begin{array}{c} \text{CH}_3 \\ | \\ \text{CH}_3 - \text{C} - \text{CH}_2 \\ | \quad\quad | \\ \text{O} - \text{CH}_2 \end{array}} \;>\; \underset{[External]}{\begin{array}{c} \text{CH}_3 \\ | \\ \text{H} - \text{C} - \text{CH}_2 \\ | \quad\quad | \\ \text{O} - \text{CH}_2 \end{array}} \;>\; \underset{[Internal]}{\begin{array}{c} \\ \text{H}_2\text{C} - \text{CH(CH}_3) \\ | \quad\quad | \\ \text{O} - \text{CH}_2 \end{array}} \;>\; \underset{[None]}{\begin{array}{c} \\ \text{H}_2\text{C} - \text{CH}_2 \\ | \quad\quad | \\ \text{O} - \text{CH}_2 \end{array}}
$$

Order of Nucleophilicity of cyclic ethers

(Laws of Creations for different Cyclic Ethers)

Rule 993: This rule of Chemistry for **Ring opening monomers or compounds,** states that, the existence of a ring with no strain energy or a strainless ring, is impossible. [See Rule 583]
(Laws of Physics in Chemistry)

Rule 994: This rule of Chemistry for **CYCLIC ETHERS that do not readily favor instantaneous opening of the ring [from some four membered rings and above] during homo- or co-polymerization,** states that, when they do however, there is only one point of scission, and that point of scission is always the C – O single bond with the largest radical-pushing potential difference, when the functional center is attacked positively and electro-free-radically.
(Laws of Creations for Cyclic Ethers)

Rule 995: This rule of Chemistry for **CYCLIC ETHERS,** states that, only those which favor instantaneous opening of the ring can favor anionic and nucleo-non-free-radical homo- or co-polymerization in the absence of transfer species of the first kind and of the first type or nucleo-free-radical alternating copolymerization where possible, in addition to its natural routes- positively charged and electro-free-radical routes
(Laws of Creations for Cyclic Ethers)

Rule 996: This rule of Chemistry for **three or some four-membered CYCLIC ETHERS,** states that, when anionic (basic) initiators are involved, propagation after initiation is favored only if the initiator and/ or the solvent used is less acidic than the monomer; for if it is more acidic, stable molecules are produced.
(Laws of Propagations for Cyclic Ethers)

Rule 997: This rule of Chemistry for **Cyclic ETHERS/ACETALS,** states that, when one radical-pushing or pulling group is carried by the carbon center adjacently located to the O center, only one point of scission exists in the ring, and that point is such that *the active center generated is the strongest with positive charge or electro-free-radically, and the monomer unit generated must remain the same whether* instantaneously opened or not.

$$
\underset{O}{\overset{\text{CH}_3}{\underset{\diagdown\diagup}{\text{H}_2\text{C} - \text{CH}}}} \longrightarrow \text{e.}\underset{\text{H}}{\overset{\text{CH}_3}{\text{C}}} - \text{CH}_2 - \text{O .nn} \quad ; \quad \underset{O}{\overset{\text{CF}_3}{\underset{\diagdown\diagup}{\text{H}_2\text{C} - \text{CH}}}} \longrightarrow \text{e.}\underset{\text{H}}{\overset{\text{H}}{\text{C}}} - \underset{\text{H}}{\overset{\text{CF}_3}{\text{C}}} - \text{O .nn}
$$

(Laws of Creations for Points of Scission)

Rule 998: This rule of Chemistry for **CYCLIC ETHERS,** states that, when one radical-pushing or pulling groups is carried by the carbon center not adjacently located to the O center, there could be one or two points of scission depending in the size of the ring, and that point is that between the O center and adjacently located C center, and the monomer unit generated must remain the same whether instantaneously opened or not.

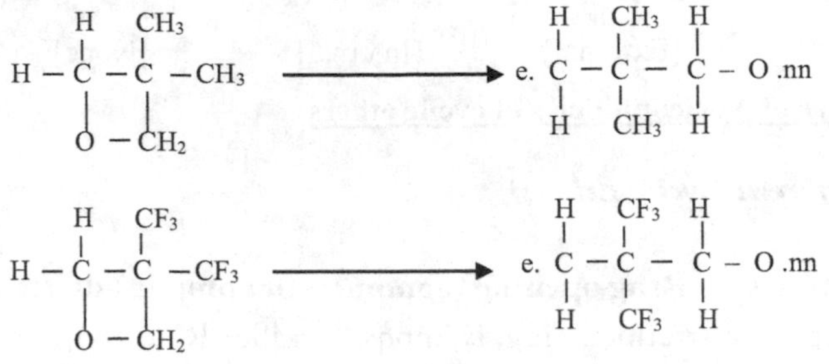

$$H - \underset{\underset{O - CH_2}{|}}{\overset{\overset{H}{|}}{C}} - \underset{\underset{CH_2}{|}}{\overset{\overset{CH_3}{|}}{C}} - CH_3 \longrightarrow e. \; \underset{\underset{H}{|}}{\overset{\overset{H}{|}}{C}} - \underset{\underset{CH_3}{|}}{\overset{\overset{CH_3}{|}}{C}} - \underset{\underset{H}{|}}{\overset{\overset{H}{|}}{C}} - O \; .nn$$

$$H - \underset{\underset{O - CH_2}{|}}{\overset{\overset{H}{|}}{C}} - \underset{\underset{CH_2}{|}}{\overset{\overset{CF_3}{|}}{C}} - CF_3 \longrightarrow e. \; \underset{\underset{H}{|}}{\overset{\overset{H}{|}}{C}} - \underset{\underset{CF_3}{|}}{\overset{\overset{CF_3}{|}}{C}} - \underset{\underset{H}{|}}{\overset{\overset{H}{|}}{C}} - O \; .nn$$

(Laws of Creations for Points of Scission)

Rule 999: This rule of Chemistry for **CYCLIC ACETALS,** states that, when a radical-pushing or pulling group is carried by the C center internally located, there is only one point of scission, and that point is still between the O center and the C center adjacently located to it as shown below, and that center remains the same whether the ring is instantaneously opened or not.

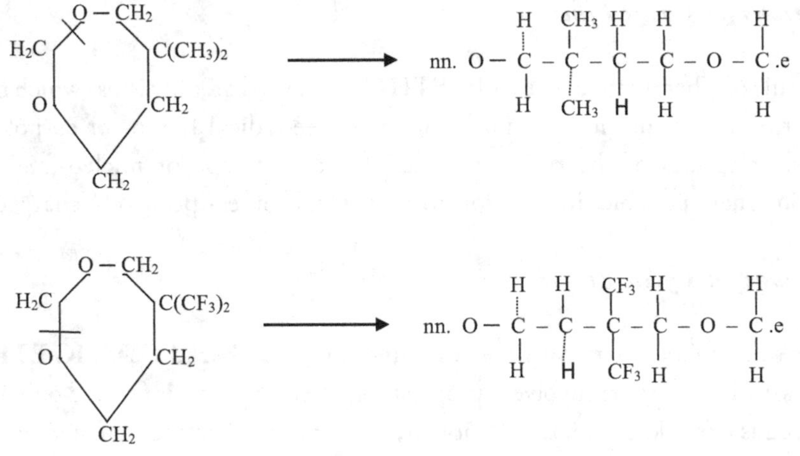

$$nn. \; O - \underset{\underset{H}{|}}{\overset{\overset{H}{|}}{C}} - \underset{\underset{CH_3}{|}}{\overset{\overset{CH_3}{|}}{C}} - \underset{\underset{H}{|}}{\overset{\overset{H}{|}}{C}} - \underset{\underset{H}{|}}{\overset{\overset{H}{|}}{C}} - O - \underset{\underset{H}{|}}{\overset{\overset{H}{|}}{C}} .e$$

$$nn. \; O - \underset{\underset{H}{|}}{\overset{\overset{H}{|}}{C}} - \underset{\underset{H}{|}}{\overset{\overset{H}{|}}{C}} - \underset{\underset{CF_3}{|}}{\overset{\overset{CF_3}{|}}{C}} - \underset{\underset{H}{|}}{\overset{\overset{H}{|}}{C}} - O - \underset{\underset{H}{|}}{\overset{\overset{H}{|}}{C}} .e$$

(Laws of Creations for Points of Scission)

Rule 1000: This rule of Chemistry for **CYCLIC ETHERS/CYCLIC ACETALS of the types shown below,** whether they exist or not, states that, there are three types of centers-

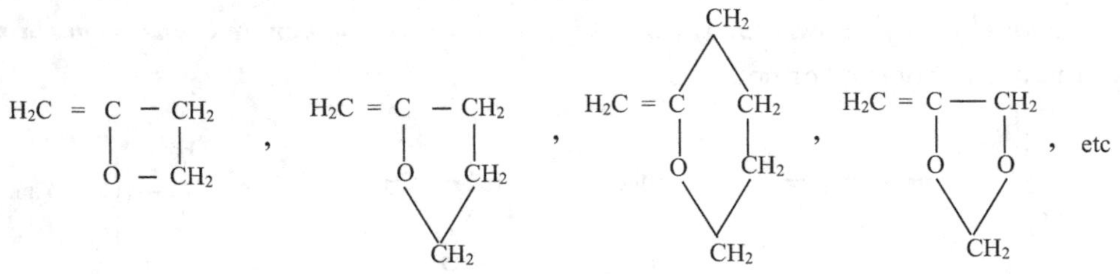

i)

functional center, Activation center and invisible p-bond carried by them, their nucleophilic capacities of which are shown below-

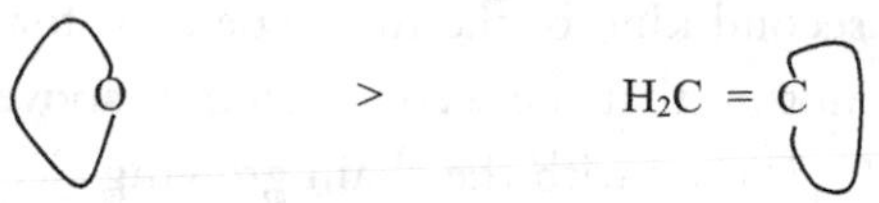

$$\text{(cyclic ether structure)} \quad > \quad H_2C = C \text{(structure)}$$

<u>Order of Nucleophilicity</u>   ii)

for which in general, the following prevails in nucleophilic character-

$$\text{Cyclic ethers} \quad > \quad \text{Cyclic alkanes}$$ iii)

noting that when one H atom on the C center adjacently located to the O center is replaced with a radical-pushing group, the anionic or nucleo-non-free-radical route ceases to be favored; while in general, the following is also valid.

$$\text{Cyclic ethers} \quad > \quad \text{Cyclic alkanes} \quad > \quad \text{Olefins}$$

<u>Order of Nucleophilicity</u>   iv)

$$\begin{array}{ccc} \text{Electrophilic} & & \text{Electrophilic} \\ \text{Olefins} & > & \text{Ringed monomers} \end{array}$$

<u>Order of Electrophilicity</u>   v)

(Laws of Creations for Methylene Cyclic Ethers and Cyclic Acetals)

<u>**Rule 1001:**</u> This rule of Chemistry for **CYCLIC ETHERS,** states that, considering 3-membered more electrophilic perfluoro dimethylene oxide, which are less strained compared to dimethylene oxide, the followings will be expected-

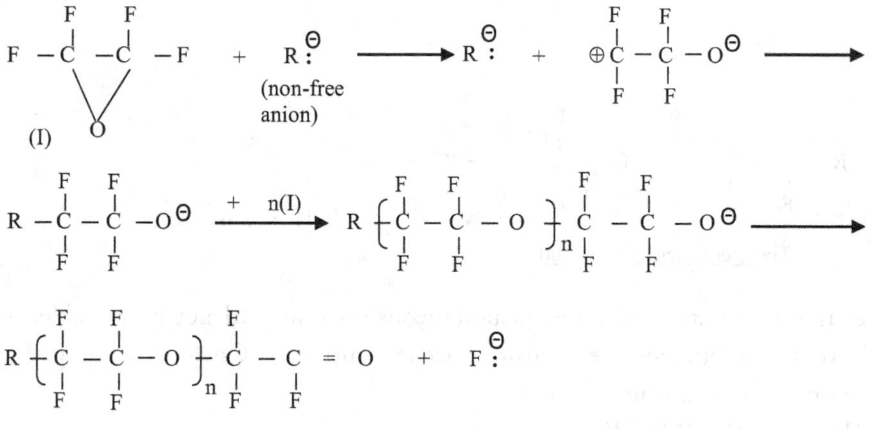

$$F-C(F)(F)-C(F)-F \text{ (I)} \quad + \quad R:^{\ominus} \text{ (non-free anion)} \longrightarrow R:^{\ominus} \quad + \quad {}^{\oplus}C(F)(F)-C(F)(F)-O^{\ominus} \longrightarrow$$

$$R-C(F)(F)-C(F)(F)-O^{\ominus} \quad \xrightarrow{+\ n(I)} \quad R\left[C(F)(F)-C(F)(F)-O\right]_n C(F)(F)-C(F)(F)-O^{\ominus} \longrightarrow$$

$$R\left[C(F)(F)-C(F)(F)-O\right]_n C(F)-C(F)(F)= O \quad + \quad F:^{\ominus}$$

<u>Transfer species of 2nd kind</u>   (a)

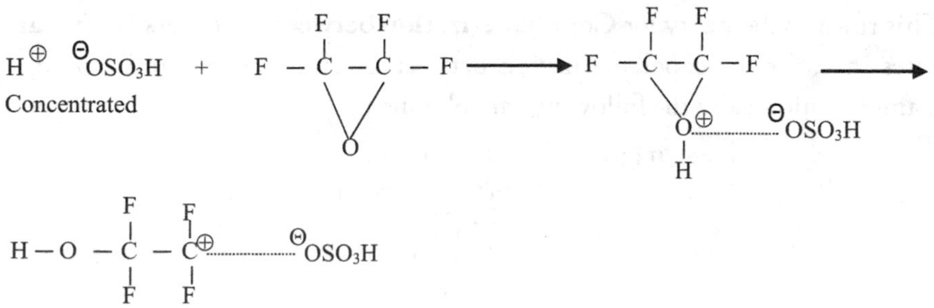

$$H^{\oplus}\ {}^{\ominus}OSO_3H \ \text{Concentrated} \quad + \quad F-C(F)(F)-C(F)-F \longrightarrow F-C(F)(F)-C(F)-F \ (O^{\oplus} H \cdots {}^{\ominus}OSO_3H) \longrightarrow$$

$$H-O-C(F)(F)-C^{\oplus}(F)(F) \cdots {}^{\ominus}OSO_3H$$

(b)

101

for which unlike perfluoro-cycloalkanes and olefins, the cyclic ethers favor charged polymerizations, noting that their monomers are still nucleophiles having very strong electrophilic tendency and it is largely for that reason that transfer species of the second kind of the first type exist for its non-free negatively charged (anionic) growing polymer chain; noting that the second reaction above is not favored, unless if the product is stable or when dilute H_2SO_4 is used with the chain growing electro-free-radically from behind.

(Laws of Creations for Perfluoro Cyclic Ethers)

Rule 1002: This rule of Chemistry for **Cyclic-ETHERS,** states that, just like 3-membered perfluoro dimethylene oxide, the same will similarly apply to four-membered cyclic ethers, if and only if the MRSE can be provided for the ring-

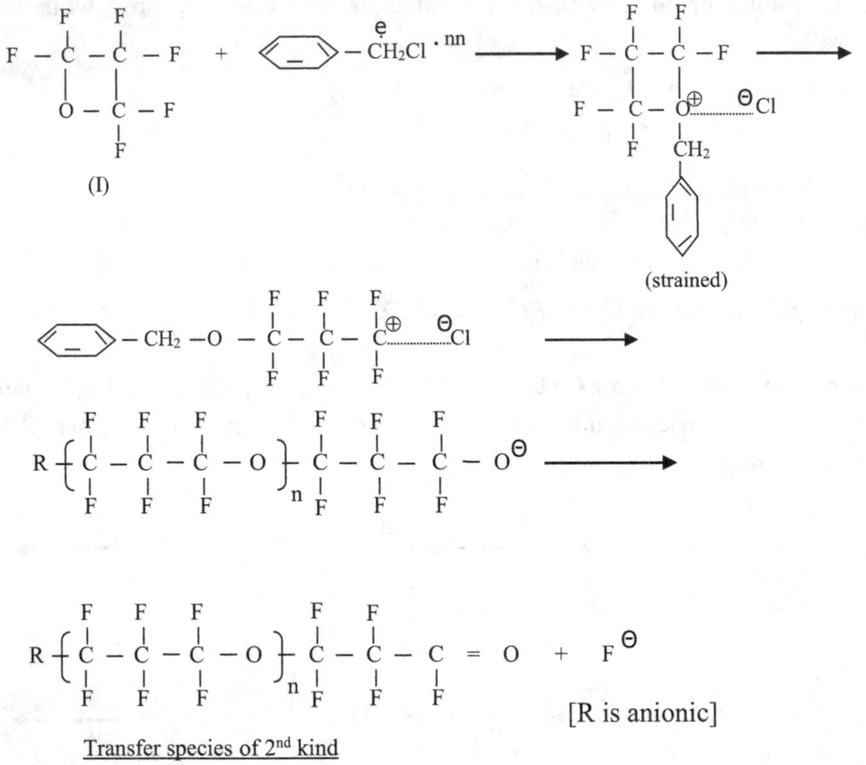

(a)

(b)

for which after the four-membered-rings, instantaneous opening will not be possible, noting that the mechanism above is Combination mechanism, since the initiator is Covalently charged-paired initiator for the first reaction, when pairing is favored.

(Laws of Perfluoro cyclic ETHERS)

Rule 1003: This rule of Chemistry for **Copolymerization between members in the families of cyclic ETHERS,** states that, for the copolymerization between epichlorohydrin and 3,3- Bis(chloromethyl) oxetane for example, anionically the followings are obtained-.

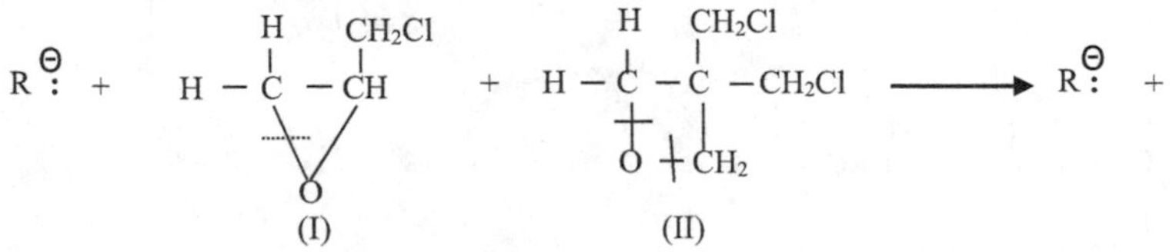

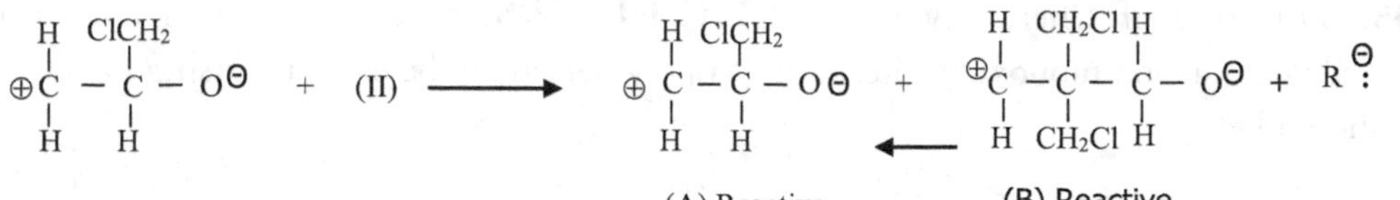

(A)   (A) Reactive   (B) Reactive

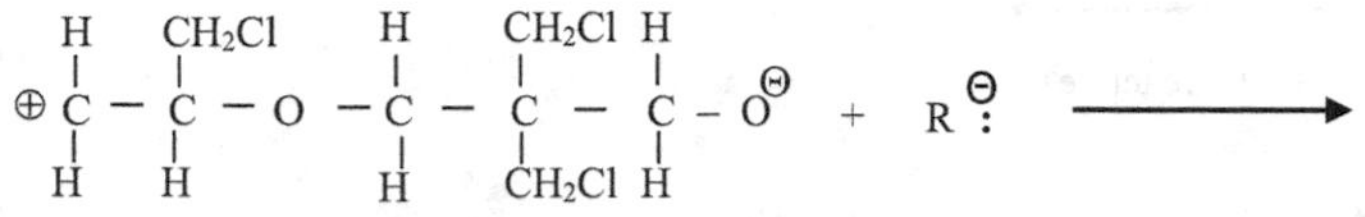

(C)   A COUPLE

Couple can be initiated (No transfer species of the first kind)

noting that (II) being more nucleophilic than (I) diffuses to (I) to form a couple (C) which like (I) and (II) are reactive to the initiators anionically and positively, clear indication that alternating placements can never be obtained, but blocks of copolymers with largely (I) positively or electro-free-radically along the chain; while radically the situation is different, since (I) can no longer favor the nucleo-radical routes, but only the electro-free-radical routes.

(Laws of Creations for Copolymerization of Cyclic Ethers)

Rule 1004: This rule of Chemistry for **Copolymerization between members in the families of cyclic ETHERS,** states that, though 3,3 –Bis(Chloromethyl) oxetane is limited mostly to positively charged or electro-free-radical routes, it can be copolymerized with monomers which are stubborn to homopolymerization such as the 1,3 -, 1,4 - dioxanes, tetrahydropyran, using $BF_3/(C_2H_5)_2O$ at temperature as low as 0°C to give a strong active cationic center first with the oxetane, after which the stubborn monomers are introduced or initiated into the chain.

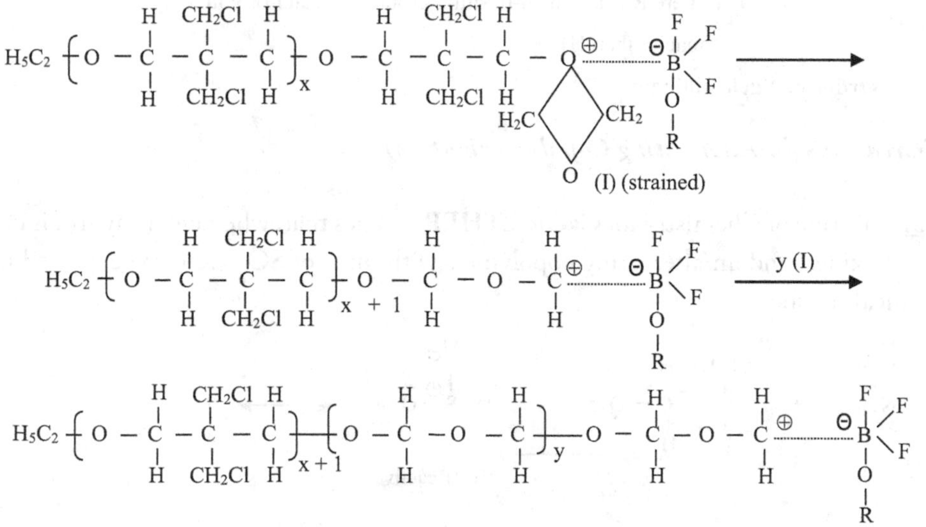

(I) (strained)

Living Block Copolymers

(Laws of Creations for Copolymerization between Cyclic Ethers and Dioxanes)

103

Rule 1005: This rule of Chemistry for **CYCLIC ETHERS,** states that, when chloro-hydrin is copolymerized with carbon monoxide, alternating copolymers are to be expected using nucleo-non-free-radicals as shown below-

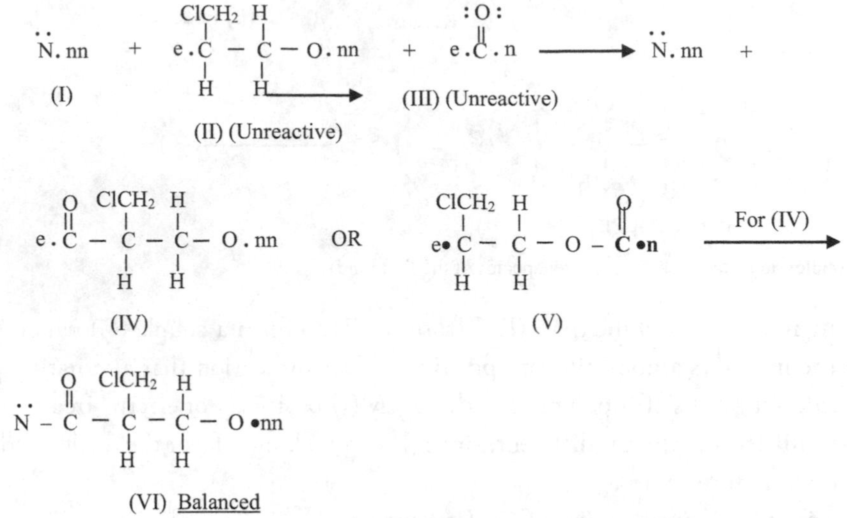

$$
\overset{..}{N}.nn \quad + \quad e.\overset{\overset{\displaystyle ClCH_2}{|}}{\underset{\underset{\displaystyle H}{|}}{C}} - \overset{\overset{\displaystyle H}{|}}{\underset{\underset{\displaystyle H}{|}}{C}} - O.nn \quad + \quad e.\overset{\overset{\displaystyle :O:}{\|}}{C}.n \longrightarrow \overset{..}{N}.nn \quad +
$$

(I)   (II) (Unreactive)   (III) (Unreactive)

$$
e.\overset{\overset{\displaystyle O}{\|}}{C} - \overset{\overset{\displaystyle ClCH_2}{|}}{\underset{\underset{\displaystyle H}{|}}{C}} - \overset{\overset{\displaystyle H}{|}}{\underset{\underset{\displaystyle H}{|}}{C}} - O.nn \qquad OR \qquad e\bullet\overset{\overset{\displaystyle ClCH_2}{|}}{\underset{\underset{\displaystyle H}{|}}{C}} - \overset{\overset{\displaystyle H}{|}}{\underset{\underset{\displaystyle H}{|}}{C}} - O - \overset{\overset{\displaystyle O}{\|}}{C}\bullet n \xrightarrow{\text{For (IV)}}
$$

(IV)   (V)

$$
\overset{..}{N} - \overset{\overset{\displaystyle O}{\|}}{C} - \overset{\overset{\displaystyle ClCH_2}{|}}{\underset{\underset{\displaystyle H}{|}}{C}} - \overset{\overset{\displaystyle H}{|}}{\underset{\underset{\displaystyle H}{|}}{C}} - O \bullet nn
$$

(VI) <u>Balanced</u>

ALTERNATING COPOLYMERS   i)

for which (IV) is fully favored, that is, 1 to 1 or alternating placement only along the chain, since (II) is more Nucleophilic than (III); for if otherwise, then (V) which is unreactive to the initiator is obtained; noting that the following is valid-

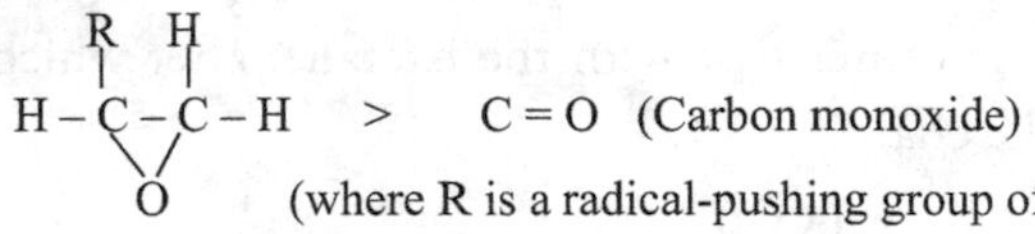

$$
H - \overset{\overset{\displaystyle R}{|}}{\underset{\underset{\displaystyle O}{\diagup \diagdown}}{C}} - \overset{\overset{\displaystyle H}{|}}{C} - H \quad > \quad C = O \quad \text{(Carbon monoxide)}
$$

(where R is a radical-pushing group of greater or equal capacity than H)

        Order of Nucleophilicity   ii)

(Laws of Creations for Alternating Copolymerization)

Rule 1006: This rule of Chemistry for **Cyclic ETHERS,** states that, when chlorohydrin is copolymerized with sulfur dioxide, random/alternating copolymers with more of SO_2 are to be expected using nucleo-non-free-radicals as shown below-

$$
\overset{..}{N}.nn \quad + \quad e.\overset{\overset{\displaystyle ClCH_2}{|}}{\underset{\underset{\displaystyle H}{|}}{C}} - \overset{\overset{\displaystyle H}{|}}{\underset{\underset{\displaystyle H}{|}}{C}} - O.nn \quad + \quad en.\overset{\overset{\displaystyle O\ominus}{|}}{\underset{}{S}}{}^{\oplus} - O.nn \longrightarrow \overset{..}{N}.nn \quad +
$$

                      (I) (Unreactive)   (II) (Reactive)

$$
en.\overset{\overset{\displaystyle O\ominus}{|}}{\underset{}{S}}{}^{\oplus} - O - \overset{\overset{\displaystyle ClCH_2}{|}}{\underset{\underset{\displaystyle H}{|}}{C}} - \overset{\overset{\displaystyle H}{|}}{\underset{\underset{\displaystyle H}{|}}{C}} - O.nn \qquad OR \qquad e\bullet\overset{\overset{\displaystyle ClCH_2}{|}}{\underset{\underset{\displaystyle H}{|}}{C}} - \overset{\overset{\displaystyle H}{|}}{\underset{\underset{\displaystyle H}{|}}{C}} - O - \overset{\overset{\displaystyle O\ominus}{|}}{S}{}^{\oplus} - O\bullet nn \xrightarrow{\text{For (III)}}
$$

(III)   (IV)

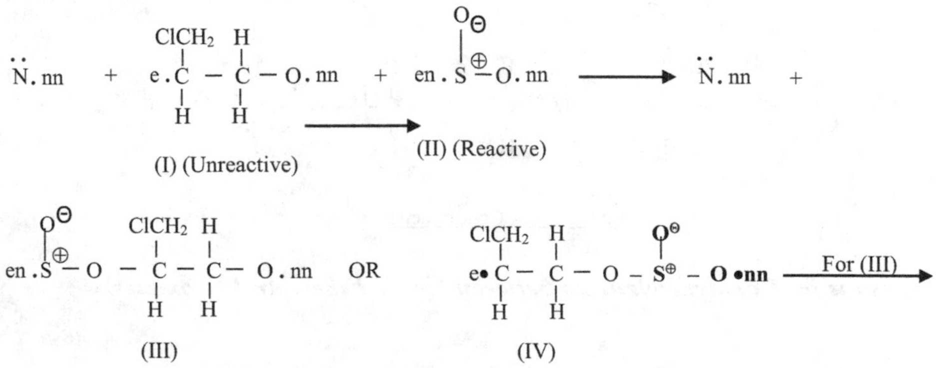

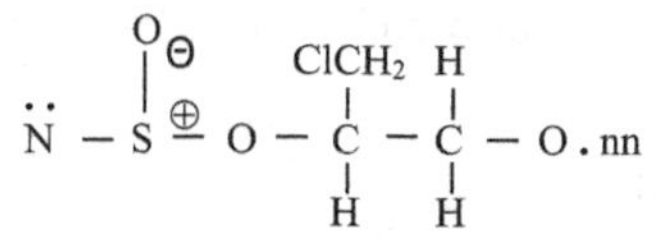

$$\ddot{N} - \overset{\overset{\displaystyle O}{\underset{\displaystyle \oplus}{|}}}{S} \oplus O - \overset{\overset{\displaystyle ClCH_2}{|}}{\underset{\underset{\displaystyle H}{|}}{C}} - \overset{\overset{\displaystyle H}{|}}{\underset{\underset{\displaystyle H}{|}}{C}} - O \cdot nn$$

(V)

i)

for if (I) is more nucleophilic than (II), then the couple (III) is formed, and if otherwise, (IV) which is unreactive to the initiator is formed, due to presence of transfer species of the first kind; noting however that (III) is the favored one present in an alternating manner with random appearance of (II) along the alternating chain since the route is not natural to it-.

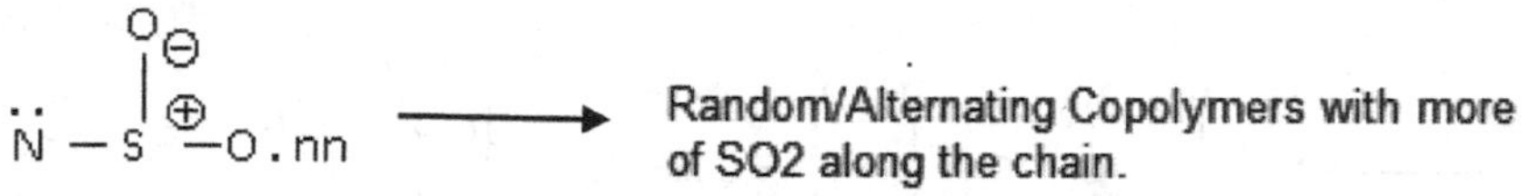

$$\ddot{N} - \overset{\overset{\displaystyle O}{\underset{\displaystyle \oplus}{|}}}{S} - O \cdot nn \longrightarrow$$ Random/Alternating Copolymers with more of SO2 along the chain.

ii)

noting also that the following is valid-

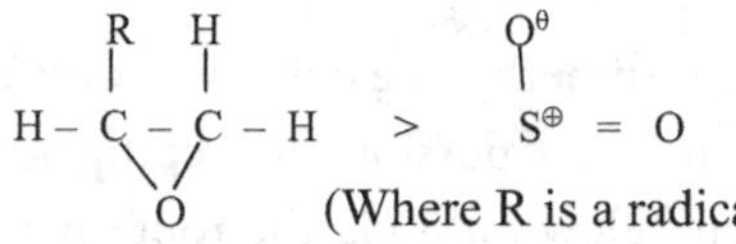

$$H - \overset{\overset{\displaystyle R}{|}}{\underset{}{C}} - \overset{\overset{\displaystyle H}{|}}{\underset{}{C}} - H \quad > \quad \overset{\overset{\displaystyle O^{\ominus}}{|}}{S^{\oplus}} = O$$

(Where R is a radical-pushing group of greater or equal capacity than H)

Order of Nucleophilicity

iii)

(Laws of Creations for Random/alternating Copolymerization)

Rule 1007: This rule of Chemistry for **Depropagation phenomenon in cyclic Ethers/ acetals,** states that, the first driving force favoring their existence, is the use of negatively charged paired (Anions) initiators such as (dilute acids-e.g. Ether/HCl or H_2O/HCl), wherein the growing chain grows electro-free-radically from behind as shown below-

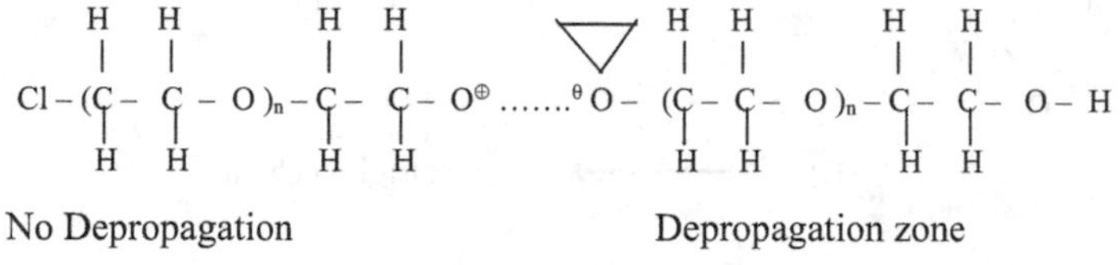

$$Cl - (\overset{\overset{\displaystyle H}{|}}{\underset{\underset{\displaystyle H}{|}}{C}} - \overset{\overset{\displaystyle H}{|}}{\underset{\underset{\displaystyle H}{|}}{C}} - O)_n - \overset{\overset{\displaystyle H}{|}}{\underset{\underset{\displaystyle H}{|}}{C}} - \overset{\overset{\displaystyle H}{|}}{\underset{\underset{\displaystyle H}{|}}{C}} - O^{\oplus} \ldots\ldots^{\theta}O - (\overset{\overset{\displaystyle H}{|}}{\underset{\underset{\displaystyle H}{|}}{C}} - \overset{\overset{\displaystyle H}{|}}{\underset{\underset{\displaystyle H}{|}}{C}} - O)_n - \overset{\overset{\displaystyle H}{|}}{\underset{\underset{\displaystyle H}{|}}{C}} - \overset{\overset{\displaystyle H}{|}}{\underset{\underset{\displaystyle H}{|}}{C}} - O - H$$

No Depropagation   Depropagation zone

(A) NAGTIVELY CHARGED PAIRED/FREE-MEDIA INITIATOR

i)

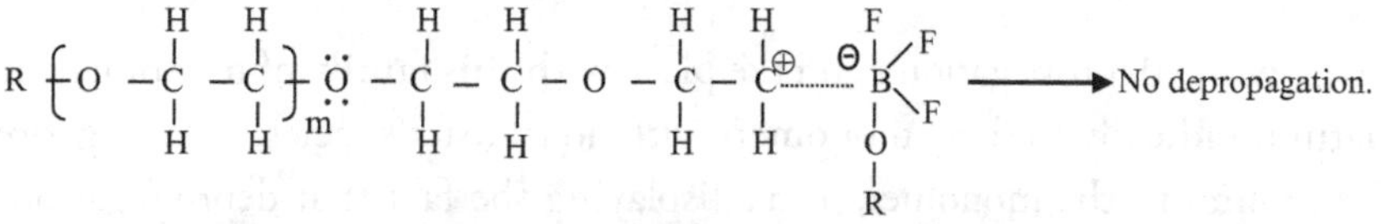

$$R \left(O - \overset{\overset{\displaystyle H}{|}}{\underset{\underset{\displaystyle H}{|}}{C}} - \overset{\overset{\displaystyle H}{|}}{\underset{\underset{\displaystyle H}{|}}{C}} \right)_m \ddot{O} - \overset{\overset{\displaystyle H}{|}}{\underset{\underset{\displaystyle H}{|}}{C}} - \overset{\overset{\displaystyle H}{|}}{\underset{\underset{\displaystyle H}{|}}{C}} - O - \overset{\overset{\displaystyle H}{|}}{\underset{\underset{\displaystyle H}{|}}{C}} - \overset{\overset{\displaystyle H}{|}}{\underset{\underset{\displaystyle O}{|}}{C}}^{\oplus} \overset{\theta}{\underset{\underset{\displaystyle R}{|}}{B}}\overset{\nearrow F}{\underset{\searrow F}{|}} \longrightarrow$$ No depropagation.

(A) POSITIVELY CHARGED PAIRED INITIATOR

ii)

for which as the H is held behind to activate another monomer, depropagation takes place with release of a cyclic ring- three- or six-membered rings.

(Laws of Creations for Depropagation via Backward Free-radical Addition Polymeri-zation)

Rule 1008: This rule of Chemistry for **Depropagation/Repropagation phenomenon in Cyclic Ethers/ Acetals,** states that, the second driving force favoring the existence of depropagation, is the use of higher polymerization temperatures far above the Ceiling temperature of the monomer beginning with the route which is not natural to the monomer as shown below nucleo-non-free-radically with three-membered cyclic ether-

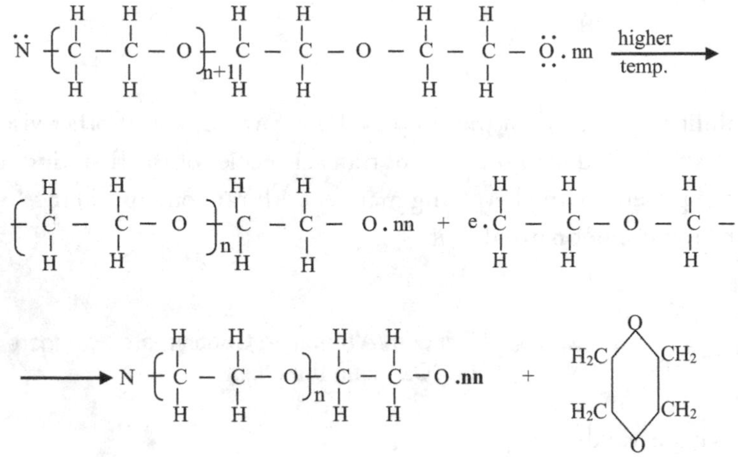

$$\ddot{N}\left\{\overset{H}{\underset{H}{C}} - \overset{H}{\underset{H}{C}} - O\right\}_{n+1}\overset{H}{\underset{H}{C}} - \overset{H}{\underset{H}{C}} - O - \overset{H}{\underset{H}{C}} - \overset{H}{\underset{H}{C}} - \ddot{\ddot{O}}.nn \xrightarrow[\text{temp.}]{\text{higher}}$$

$$\ddot{N}\left\{\overset{H}{\underset{H}{C}} - \overset{H}{\underset{H}{C}} - O\right\}_{n}\overset{H}{\underset{H}{C}} - \overset{H}{\underset{H}{C}} - O.nn + e.\overset{H}{\underset{H}{C}} - \overset{H}{\underset{H}{C}} - O - \overset{H}{\underset{H}{C}} - \overset{H}{\underset{H}{C}} - O.nn$$

$$\longrightarrow N\left\{\overset{H}{\underset{H}{C}} - \overset{H}{\underset{H}{C}} - O\right\}_{n}\overset{H}{\underset{H}{C}} - \overset{H}{\underset{H}{C}} - O.nn \quad + \quad \begin{array}{c} O \\ H_2C \diagup \diagdown CH_2 \\ H_2C \diagdown \diagup CH_2 \\ O \end{array}$$

(B)   Depropagation radically?   i)

noting that the 1,4 - dioxane obtained is less strained than if dimethylene oxide was obtained, for if the strength of the free-media active center is high, that which is not possible, no repropagation can take place, unless when Piared-media initiator is used as shown below only via the route natural to it (i.e., electrofree-radically) and this can be said to be propagation/depropagation/repropagation phenomenon-

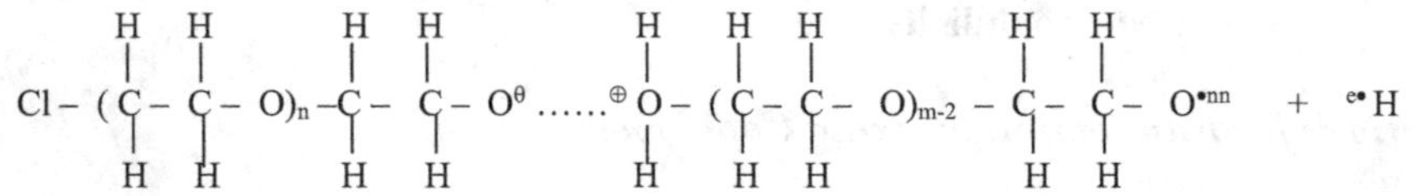

$$Cl - (\overset{H}{\underset{H}{C}} - \overset{H}{\underset{H}{C}} - O)_{n}\overset{H}{\underset{H}{C}} - \overset{H}{\underset{H}{C}} - O^{\ominus}......\overset{\oplus}{O} - (\overset{H}{\underset{H}{C}} - \overset{H}{\underset{H}{C}} - O)_{m-2}\overset{H}{\underset{H}{C}} - \overset{H}{\underset{H}{C}} - O^{\bullet nn} + {}^{e\bullet}H$$

(A)

$$H\bullet e \quad + \quad \begin{array}{c} O \\ H_2C \diagup \diagdown CH_2 \\ H_2C \diagdown \diagup CH_2 \\ O \end{array} \longrightarrow H - O - \overset{H}{\underset{H}{C}} - \overset{H}{\underset{H}{C}} - O - \overset{H}{\underset{H}{C}} - \overset{H}{\underset{H}{C}}\bullet e$$

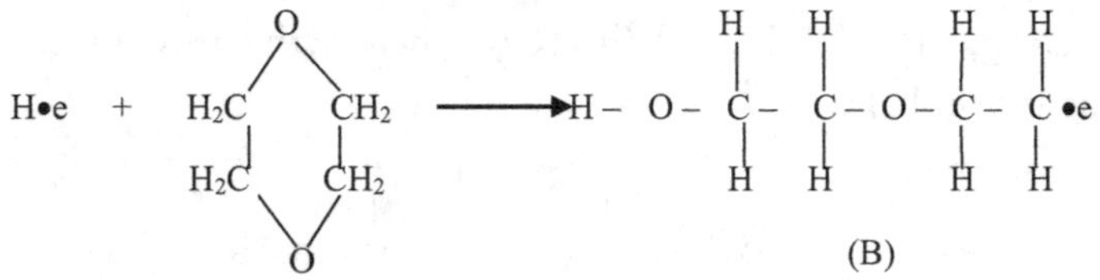

(B)   ii)

Back Part of (A)   +   (B) ⟶ Original chain

REPROPAGATION   iii)

clear indication that while depropagation can take place with this family of monomers in the route both natural and unnatural radically to these monomers with no transfer species, repropagation can only take place in the route natural to the monomer,; thus displaying the fact that depropagation as shown in i), the route unnatural to the monomer may not indeed take place radically.

(Laws of Creations for Propagation/Depropagation/Repropagation phenomenon)

Rule 1009: This rule of Chemistry for **Depropagation phenomenon in Cyclic Ethers/ Acetals,** states that, the third driving force favoring their existence, is an environment which is highly acidic chargedly or radically, for which a growing polymer chain can be scissioned at specific points along the chain by an acid favoring the existence of a stable molecule or ring as shown below-

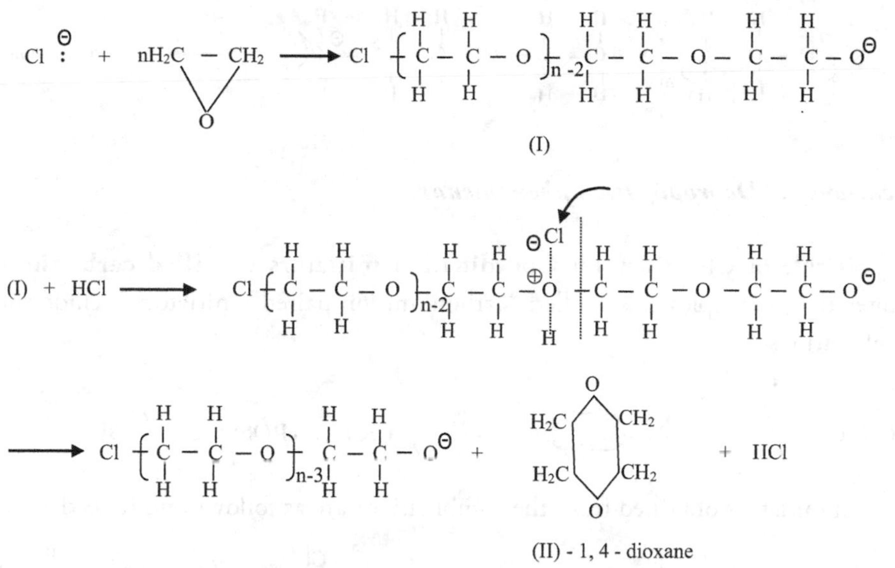

(I)

(II) - 1, 4 - dioxane

Depropagation in acidic environment

noting that the (II) removed is more stable than the original monomer in terms of the presence of lesser strain energy in the ring, and the ring formed cannot be opened anionically, typical example of a strong depropagation reaction, since the part scissioned cannot be added to the chain.

(Laws of Creations for Depropagation phenomenon)

Rule 1010: This rule of Chemistry for **Depropagation phenomena in Cyclic Ethers/ Acetals,** states that, the fourth driving force favoring their existence, is the use of free-/paired media initiators, via positively charged, or electro-free-radical routes, for which the followings are to be expected in a covalent type of paired-media environment where pairing is favored-

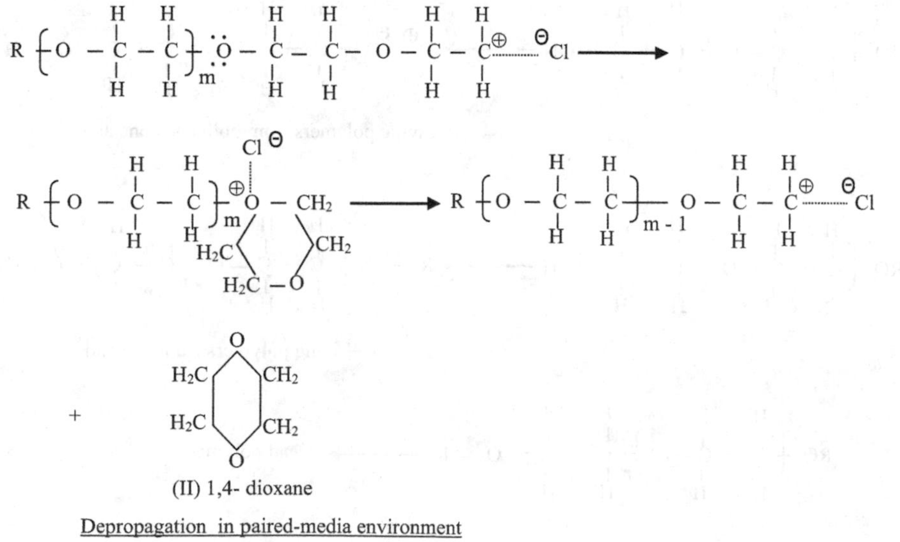

(II) 1,4- dioxane

Depropagation in paired-media environment

i)

107

noting that the ability to lose the 1,4-dioxane here is strong compared to the opposite route not natural to it; while in an electrostatic type of paired environment, the followings take place-

$$R\left[O-\overset{\overset{H}{|}}{\underset{\underset{H}{|}}{C}}-\overset{\overset{H}{|}}{\underset{\underset{H}{|}}{C}}\right]_m \overset{..}{\underset{..}{O}}-\overset{\overset{H}{|}}{\underset{\underset{H}{|}}{C}}-\overset{\overset{H}{|}}{\underset{\underset{H}{|}}{C}}-O-\overset{\overset{H}{|}}{\underset{\underset{H}{|}}{C}}-\overset{\overset{H}{|}}{\underset{\underset{H}{|}}{C}}{}^{\oplus}\quad {}^{\ominus}B\overset{F}{\underset{\underset{\overset{|}{R}}{O}}{\underset{|}{\overset{|}{}}}}{}^{F} \xrightarrow{\quad} \text{No depropagation.}$$

ii)

(Laws of Creations for Depropagation phenomena)

Rule 1011: This rule of Chemistry for **Coordination initiators so-called carbonium ion-paired initiators,** states that, examples of so-called "carbonium ion-paired" initiators include those from the following combinations-

$$\underset{CH_3\overset{\overset{O}{\|}}{C}Cl}{}\Big/ SnCl_4, \quad \langle\underline{\quad}\rangle{-}CH_3Cl \Big/ FeCl_3, \quad ROSO_3R \Big/ BF_3 \quad etc$$

i)

for which the real initiators obtained from the combinations are as follows and indeed

$$H_3C-\overset{\overset{O}{\|}\oplus}{C}{-}{-}{-}{-}{}^{\ominus}\overset{\overset{Cl}{|}}{\underset{\underset{Cl}{|}}{Sn}}\overset{Cl}{\underset{Cl}{\diagdown}} ; \quad \langle\underline{\quad}\rangle{-}\overset{\overset{H}{|}}{\underset{\underset{H}{|}}{C}}{}^{\oplus}{-}{-}{-}{}^{\ominus}\overset{\overset{Cl}{|}}{\underset{\underset{Cl}{|}}{Fe}}\overset{Cl}{\underset{Cl}{\diagdown}} ; \quad R{}^{\oplus}{-}{-}{}^{\ominus}\overset{F}{\underset{\underset{\overset{|}{SO_3R}}{O}}{\underset{|}{B}}}{}^{F}$$

<u>Positively-charged-paired initiators</u>

ii)

called *Electrostatically positively charged-paired initiators,* cases which will favor the existence of a dead polymer since the terminals of their growing polymer chain are electrostatically bonded.
(Laws of creations for Initiators)

Rule 1012: This rule of Chemistry for **Living and dead polymers for families of CYCLIC ETHERS/ ACETALS,** states that, these can clearly be identified as follows-

$$HO\left[\cdots\right]_n\cdots OH \longrightarrow H^{\oplus}\,{}^{\ominus}O\left[\cdots\right]_n\cdots\overset{\ominus}{O}\; H^{\oplus}$$

Living polymers from both ends one at a time (a)

$$RO\left[\cdots\right]_n\cdots OH \longrightarrow R-O\left[\cdots\right]_n\cdots\overset{\ominus}{O}\; H^{\oplus}$$

Living polymers from one end (b)

$$RO\left[\cdots\right]_n\cdots O-R \longrightarrow \text{Dead polymer.}$$

(c)

for which their ability of being dead or living is largely a function of their ability to exist in Equilibrium state of existence one at a time, for if none from both ends, the polymer is dead.
(Laws of Creations for Living and Dead polymers)

Rule 1013: This rule of Chemistry for **CYCLIC ETHERS,** states that, when polymerized anionically or nucleo-non-free-radically only after opening of the ring instantaneously, the following is valid for its growing polymer chain-

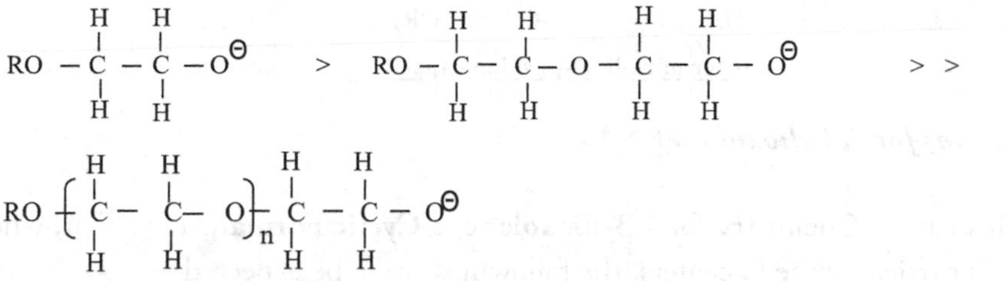

$$RO - \underset{\underset{H}{|}}{\overset{\overset{H}{|}}{C}} - \underset{\underset{H}{|}}{\overset{\overset{H}{|}}{C}} - O^{\ominus} \quad > \quad RO - \underset{\underset{H}{|}}{\overset{\overset{H}{|}}{C}} - \underset{\underset{H}{|}}{\overset{\overset{H}{|}}{C}} - O - \underset{\underset{H}{|}}{\overset{\overset{H}{|}}{C}} - \underset{\underset{H}{|}}{\overset{\overset{H}{|}}{C}} - O^{\ominus} \quad >>$$

$$RO \left\{ \underset{\underset{H}{|}}{\overset{\overset{H}{|}}{C}} - \underset{\underset{H}{|}}{\overset{\overset{H}{|}}{C}} - O \right\}_{n} \underset{\underset{H}{|}}{\overset{\overset{H}{|}}{C}} - \underset{\underset{H}{|}}{\overset{\overset{H}{|}}{C}} - O^{\ominus}$$

<u>Decreasing order of strength of anionic active growing center</u>

since the route is not natural to the monomer which has only one type of activation center that classifies it as a Nucleophile.

(Laws of Creations for Cyclic Ethers)

Rule 1014: This rule of Chemistry for **1,3-dioxolane a Cyclic formal,** states that, in view of the 1,3 placement of the O centers, when activated using ROR/BF$_3$ combination, the followings are to be expected-

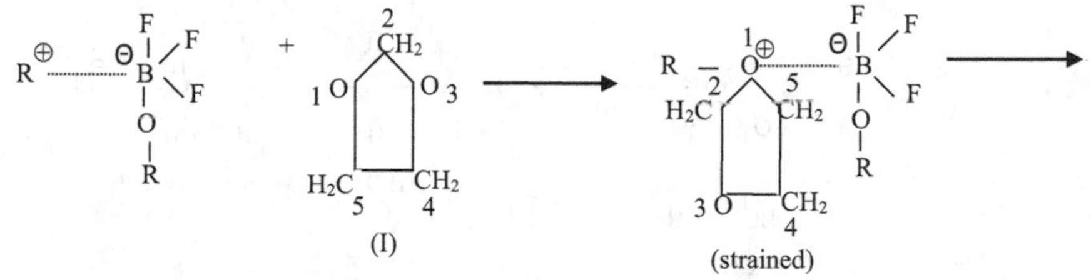

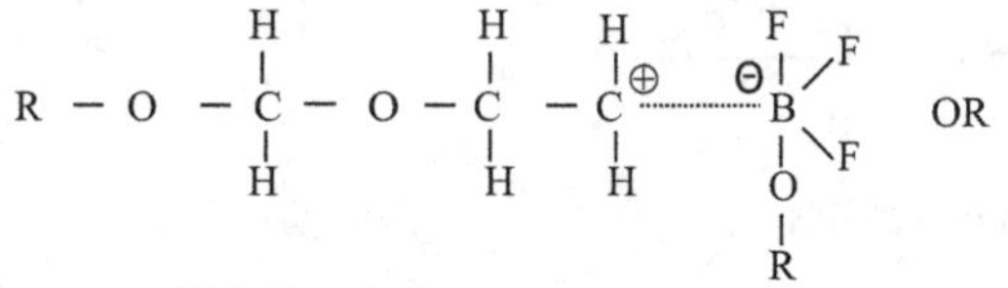

(I)   (strained)

(I) 1, 5 - scission

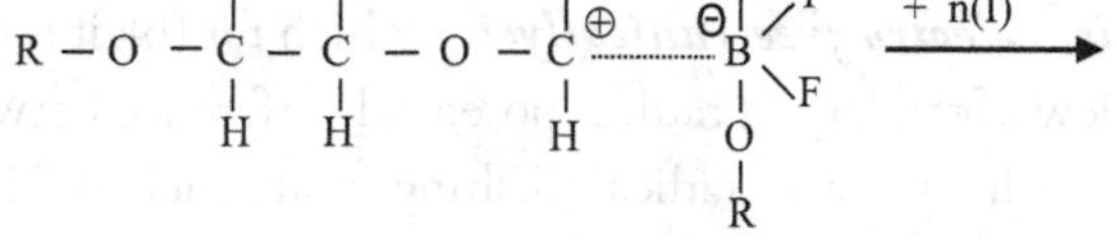

OR

(II) 1,2 - scission (favored)

+ n(I)

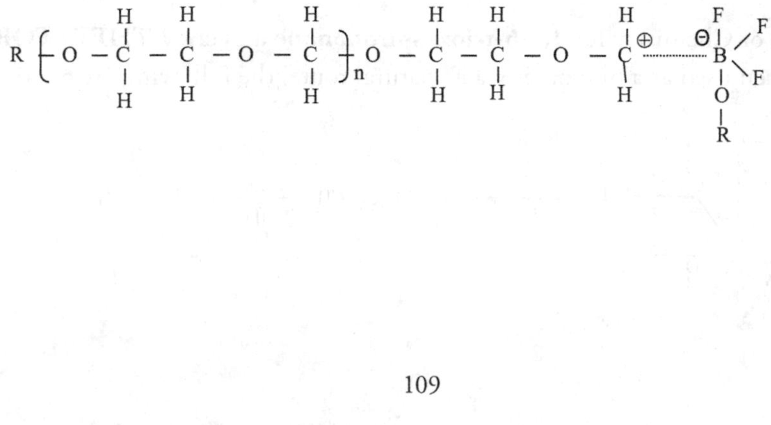

i)

for which only one of the oxygen centers can be attacked with only one single bond being the point of scission and that point of scission is that represented by (II) above, the positive center of which is stronger than that of (I), since the following is valid-

$$-CH_2(OR) \quad > \quad -CH_2(CH_2OR).$$

Order of radical-pushing capacity   ii)

(Laws of Creations for 1,3-dioxolane)

Rule 1015: This rule of Chemistry for **1,3-dioxolane, a Cyclic Formal**, states that, when one radical-pushing group is carried by the C centers, the followings are to be expected-

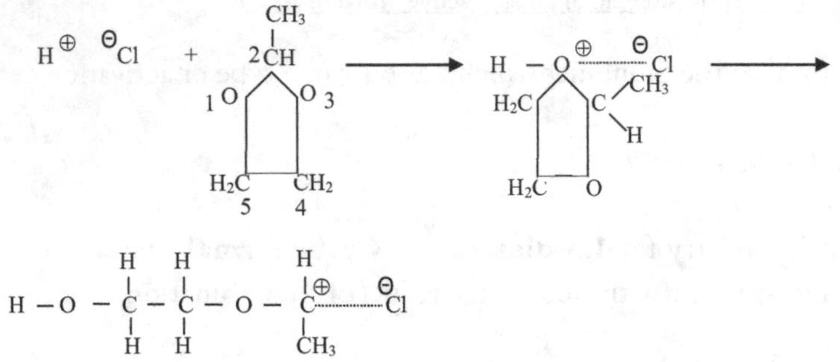

(a)

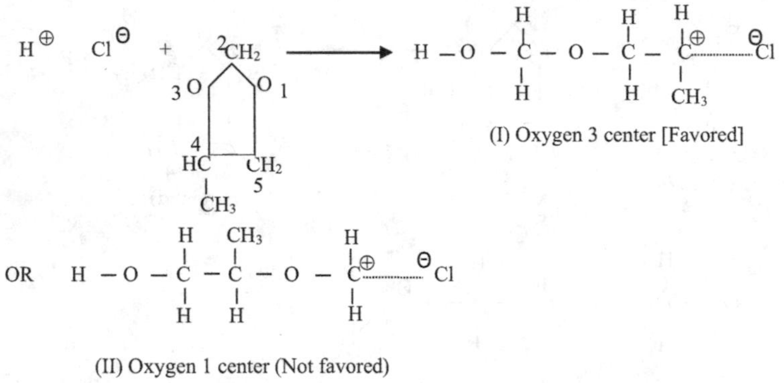

(I) Oxygen 3 center [Favored]

(II) Oxygen 1 center (Not favored)

(b)

noting first and foremost that the acid used is not concentrated as shown above, but dilute HCl, for if concentrated, stable molecules will be produced, *for which indeed the chain above is growing backwardly via Equilibrium mechanism electro-free-radically;* for which for (b), it is the No 3 oxygen center that is attacked all the time, in view of the largest radical potential difference between the 3 and 4 carbon centers; and if the CH_3 is replaced with a weaker radical-pushing group such as CF_3, the monomer unit will be different from all the above.

(Laws of Creations for 1,3-dioxolane)

Rule 1016: This rule of Chemistry for **1,4,6-trioxyspirononane a cyclic ETHER/ FORMAL shown below,** states that, when used as a monomer via its natural route, the following are obtained-

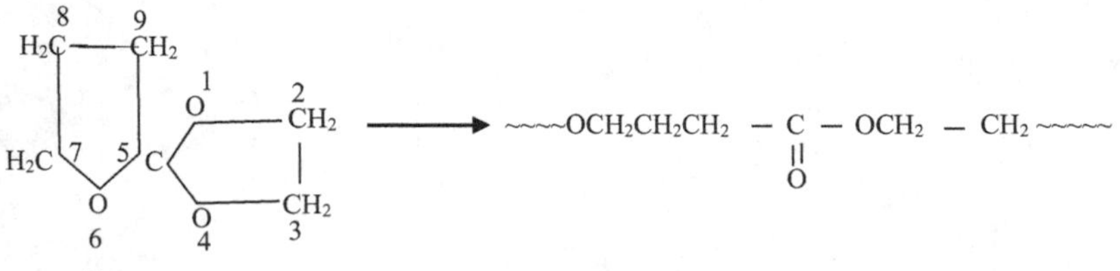

(I)

i)

110

the mechanism of which can be explained as follows using sulfuric acid as initiator-

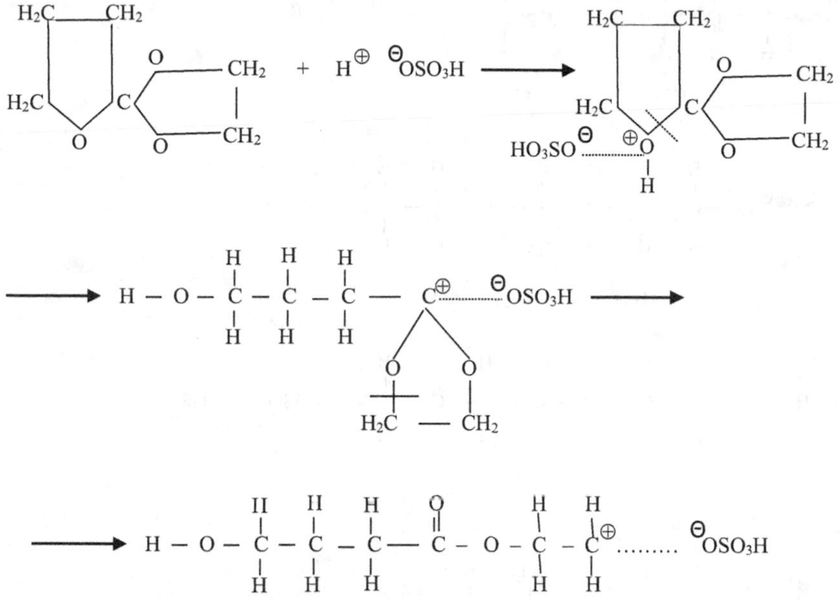

(IV) Favored   ii)

noting first and foremost that the acid used is not the concentrated one shown above, but dilute H_2SO_4, for which the chain above is growing backwardly via Equilibrium mechanism electro-free-radically; for which tetrahydrofuran is the first to be attacked being less nucleophilic and more strained than 1,3-dioxolane to give the polymer chain identified; noting that based on the nature of the growing chain, the possibility for depropagation to lose 1,4-epoxide does not exist, and the monomer unit is identical to that from g-butyro lactone/cyclo ethylene oxide.

(Laws of Creations for opening of coupled rings)

Rule 1017: This rule of Chemistry for **1,4,6-trioxyspirononane a cyclic ETHER/ FORMAL,** states that, though the two rings in 1,4,6- trioxyspirononane cannot be opened instantaneously under most operating conditions, when however made to take place, the followings are to be expected-

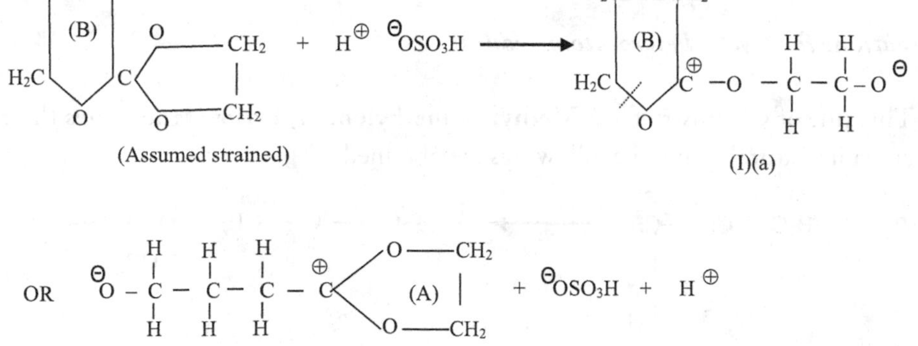

(Assumed strained)   (I)(a)

(I)(b)

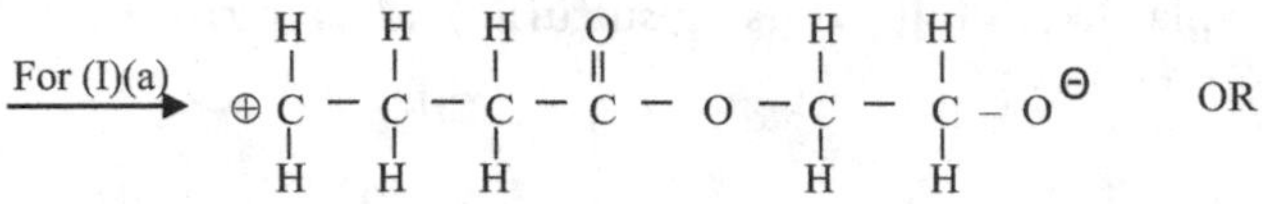

For (I)(a) →

$$\overset{\oplus}{C}H_2 - CH_2 - CH_2 - \overset{O}{\overset{\|}{C}} - O - CH_2 - CH_2 - O^{\ominus} \qquad OR$$

(I)aa

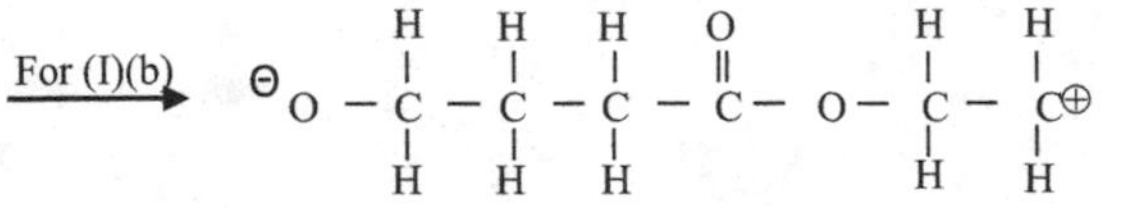

For (I)(b) →

$$^{\ominus}O - CH_2 - CH_2 - CH_2 - \overset{O}{\overset{\|}{C}} - O - CH_2 - \overset{\oplus}{C}H_2$$

(I) bb

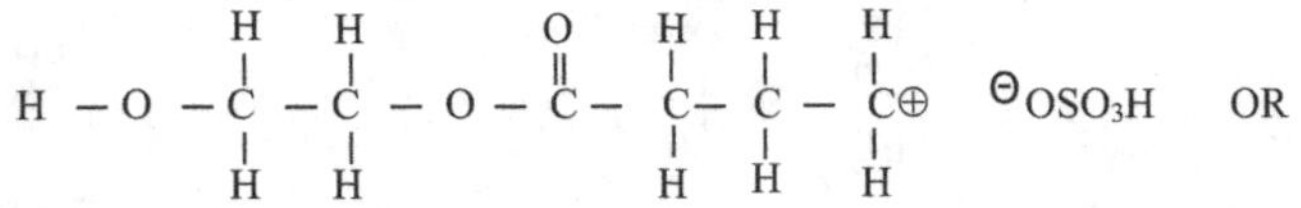

→

$$H - O - CH_2 - CH_2 - O - \overset{O}{\overset{\|}{C}} - CH_2 - CH_2 - \overset{\oplus}{C}H_2 \quad ^{\ominus}OSO_3H \qquad OR$$

(I) aaa

$$H - O - CH_2 - CH_2 - CH_2 - \overset{O}{\overset{\|}{C}} - O - CH_2 - \overset{\oplus}{C}H_2 \quad ^{\ominus}OSO_3H$$

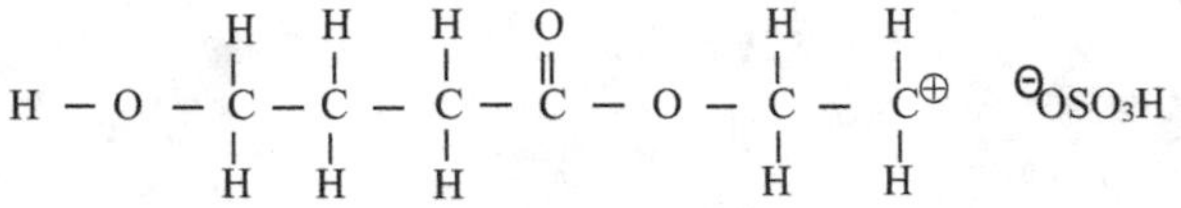

(I) bbb   i)

in which it is (I)bbb that identifies with the case where the functional center was used, clear indication of the fact that, ***the cyclic formal is less strained than the cyclic ether*** and it is also obvious that tetrahydrofuran ringed substituent group (i.e., 1,4-epoxide ringed substituent group) is less radical-pushing than 1,3- dioxolane ringed substituent group (i.e., cyclic etheric group).

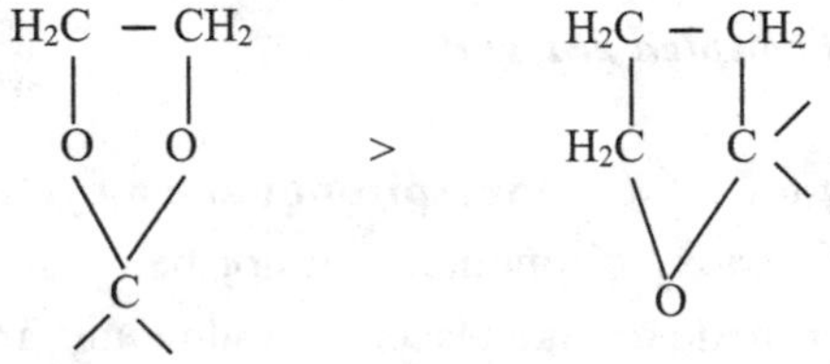

Order of radical-pushing capacity   ii)

(Laws of Creations for Cyclic hetero-atom groups)

Rule 1018: This rule of Chemistry for **2-Methyl–4-methylene–1,3-dioxolane,** states that, when used as a monomer via its natural route, the followings are obtained-

(II)   i)

the mechanism of which can be explained as follows-
not **VIA 1INSTANTANEOUS OPENING OF THE RING**

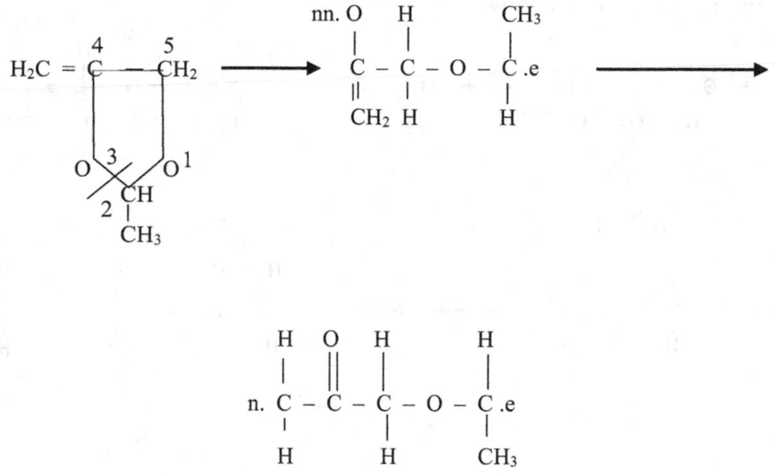

$$H_2C = \underset{\underset{\underset{\underset{CH_3}{|}}{\overset{CH}{2}}}{\overset{|}{O}\,\,\,\,\,O^1}}{\overset{4}{C}} - \overset{5}{CH_2} \longrightarrow \underset{\underset{CH_2\ \ H}{||\quad|}}{\overset{nn.\ O\quad H\qquad CH_3}{\underset{}{C - C - O - C}}.e} \longrightarrow$$

$$n.\ \underset{\underset{H}{|}}{\overset{\overset{H}{|}}{C}} - \underset{}{\overset{\overset{O}{||}}{C}} - \underset{\underset{H}{|}}{\overset{\overset{H}{|}}{C}} - O - \underset{\underset{CH_3}{|}}{\overset{\overset{H}{|}}{C}}.e$$

Electro-free-radical movement   ii)

since the point of scission above is the wrong bond,
but **VIA AN ACTIVATION CENTER**

for which the $H_2C = C$ activation center being less nucleophilic than functional O center, the followings are obtained.

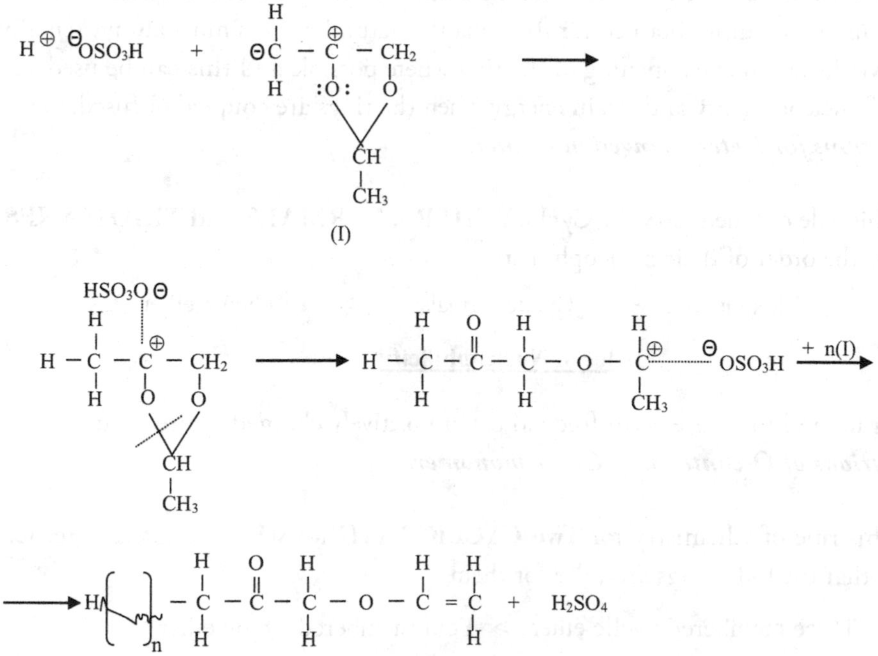

$$H^\oplus\ {}^\ominus OSO_3H\ +\ {}^\ominus\overset{\overset{H}{|}}{C} - \overset{\oplus}{C} - CH_2$$

(I)

$$\underset{\underset{CH_3}{|}}{\overset{HSO_3O^\ominus}{\underset{\underset{O}{|}}{\overset{\overset{H}{|}}{H - C}} - \overset{\oplus}{C} - CH_2}} \longrightarrow H - C - C - C - O - C {\cdots}^\ominus OSO_3H \xrightarrow{+\ n(I)}$$

$$\longrightarrow H \left[\right]_n - C - C - C - O - C = C\ +\ H_2SO_4$$

Transfer species of the first kind of **first** type   iii)

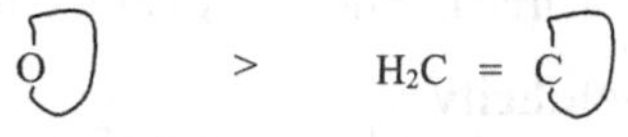

$$O\quad >\quad H_2C = C$$

Order of nucleophilicity of Functional /Activation centers   iv)

noting that when attacked via anionic or negatively charged or nucleo-free-radical or nucleo-non-free-radical route, the same transfer species above is abstracted with the ring opened via the $O - CHCH_3$ bond in the process of Initiation as shown below-

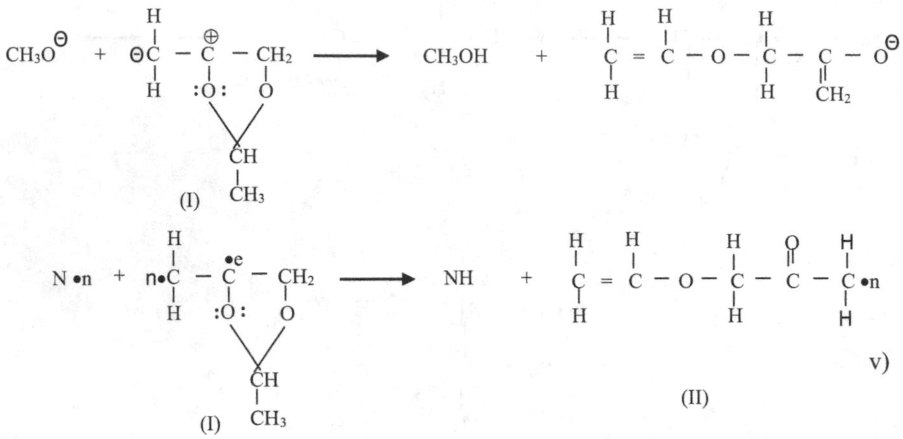

clear indication of the strong nucleophilic character of the monomer and the fact that it is (II) above that identifies with the monomer unit; noting that since a cation cannot be used via Combination mechanism, all the reactions above are taking place radically using H•e, i.e., electro-free-radical polymerization.
(Laws of Creations for 1,3-Dioxolane with methylene and methyl groups)

Rule 1019: This rule of Chemistry for **Ringed monomers, single or coupled or fused rings, all with at least one functional center or functional and cumulatively placed external activation centers,** states that, the monomer unit obtained for them via the natural centers must always remain the same as that obtained via instantaneous opening of the ring where possible and this can be used advantageously as a measure of nucleophilicity and strain energy when the rings are coupled or fused.
(Laws of Creations for Hetero-ringed monomers)

Rule 1020: This rule of Chemistry for **Cyclic ETHERS, FORMALS and TRIO-XANES,** states that, the following is the order of their nucleophilicities

Trioxanes > Cyclic formals > Cyclic ethers

Order of Nucleophilicity

for which their natural route is electro-free-radical or positively charged- paired route.
(Laws of Creations of O-containing Cyclic monomers)

Rule 1021: This rule of Chemistry for **Two CYCLIC ETHERS when coupled together,** states that, their use show that the followings are valid for them-

Three membered cyclic ether > Four membered cyclic ether

Order of Strain Energy   i)

Four membered cyclic ether > Three membered cyclic ether

Order of Nucleophilicity   ii)

(Laws of Creations for Cyclic Ethers)

Rule 1022: This rule of Chemistry for **Members of CYCLIC FORMALS,** states that, when coupled, the followings are valid-

Seven-membered cyclic formal	>	Six-membered cyclic formal	>	Five-membered cyclic formal

Order of Nucleophilicity

(Laws of Creations for Cyclic Formals)

Rule 1023: This rule of Chemistry for **CYCLIC FORMALS/ETHERS,** states that, based on the Coupling theory, the followings are valid for them, for rings of the same size- using tetrahydrofuran and 1,3-dioxolane, two five-membered rings-

1,3-Dioxolane   >   Tetrahydrofuran

Order of Nucleophilicity   i)   i)

Tetrahydrofuran   >   1,3-Dioxolane

Order of Strain Energy   ii)

(Laws of Creations for Cyclic Formals)

Rule 1024: This rule of Chemistry for **Tetrahydrofuran coupled with 1,3-dioxane, 1,3-dioxolane, 1,3-dioxelane, 1,3-dioxepane, and Cyclic formals,** states that, from the product obtained wherein -COO-, -OCO- groups are absent or present along the chain, there is no doubt that the followings are valid-

All Cyclic formals   >   1,3 - dioxanes ┊ > I,n-dioxane........ >   1,5 - dioxanes   >   1, 4 - dioxanes

Order of nucleophilicity [n = 6,7,8,........]

for which, it is important to note that 1,3-Dioxane is a Cyclic Formal (with m=1) and not a Dioxane or of the Dioxane family to which 1,4-Dioxane, 1,5-Dioxane, 1,6-Dioxane... belong to.
(Laws of creations for Dioxanes/Cyclic Formals)

Rule 1025: This rule of Chemistry for **CYCLIC ETHERS AND CYCLIC ALKANES coupled together,** states that, the followings are valid-

Cyclic Ethers   >   Cyclic Alkanes

Order of Nucleophilicity   i)

Cyclic Alkanes   >   Cyclic Ethers

Order of Strain energy   ii)

(Laws of Creation for Cyclic Ethers and Cyclic Alkanes)

Rule 1026: This rule of Chemistry for **CYCLIC ETHERS AND CYCLIC ALKENES coupled together,** states that, when molecular rearrangement takes place in the cyclic alkene when used as a monomer, the fact that the same products are obtained via instantaneous opening of the rings and via the activation center in the cyclic alkene, is a clear indication that the followings are valid-.

Cyclic Alkenes   >   Cyclic Ethers

Order of Strain Energy

Cyclic Ethers   >   Cyclic Alkenes

Order of Nucleophilicity

i)

ii)   ii)

(Laws of Creation for Cyclic Ethers and Cyclic Alkanes)

Rule 1027: This rule of Chemistry for **Cases where a double bond is cumulatively placed to a 1,3-dioxehane and some other rings as shown below,** the types of products obtained will depend on where the double is located in the ring-

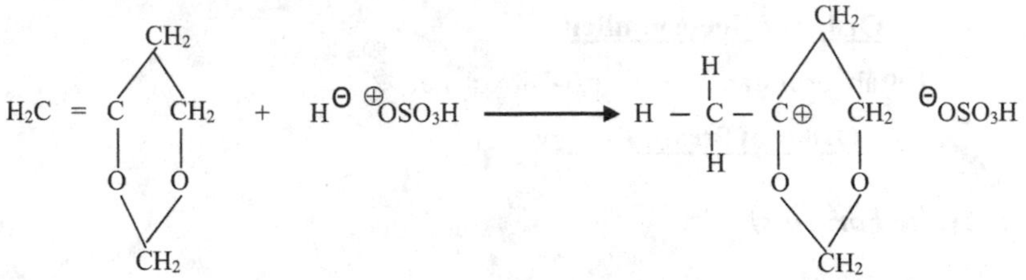

(I)   Six-membered cyclic formal (1.3-Dioxehane)

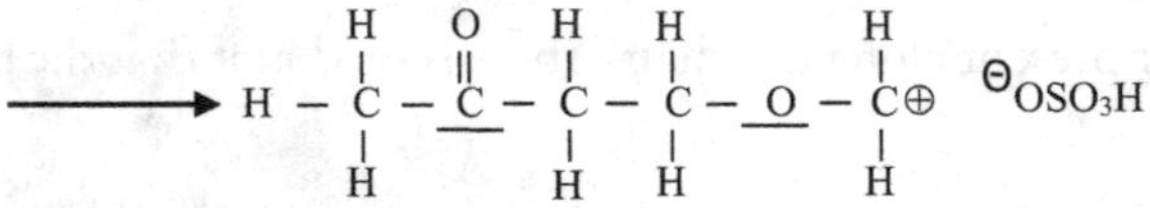

(B)

(a)

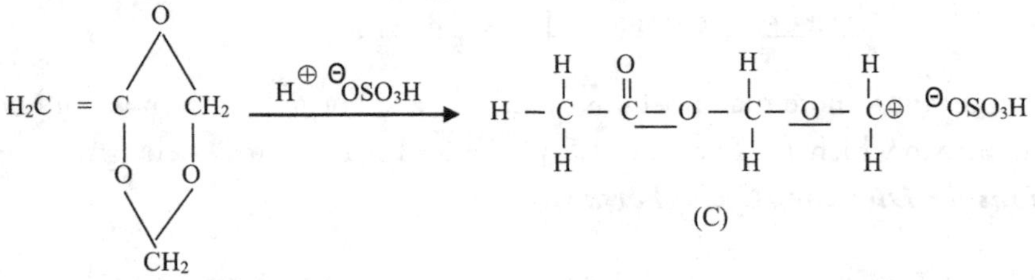

(II) Six-membered Trioxane

(C)

(b)

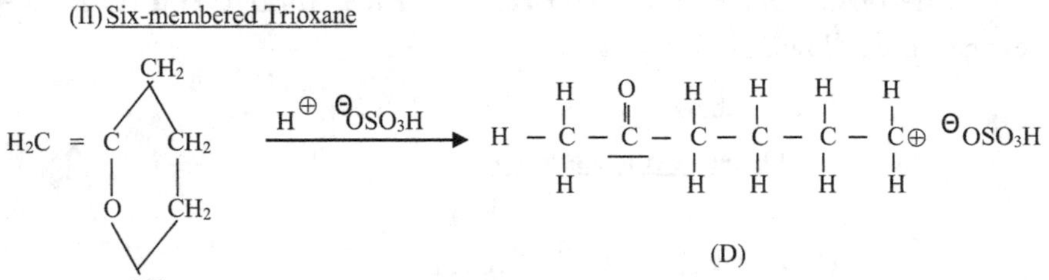

(III) Six-membered Tetrahydropyran

(D)

(c)

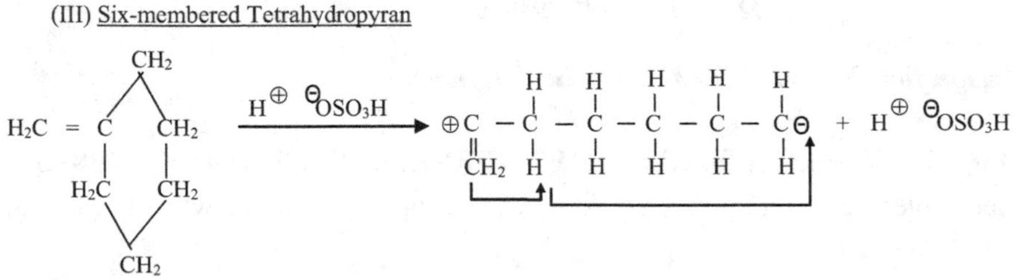

(IV) Six-membered cyclohexane

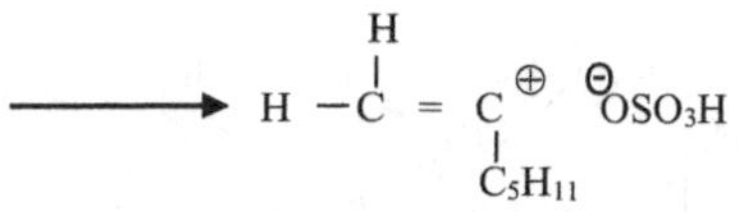

$$\longrightarrow \quad H - \overset{\overset{H}{|}}{C} = C^{\oplus} \; ^{\ominus}OSO_3H$$
$$\underset{C_5H_{11}}{|}$$

(E)

(d)

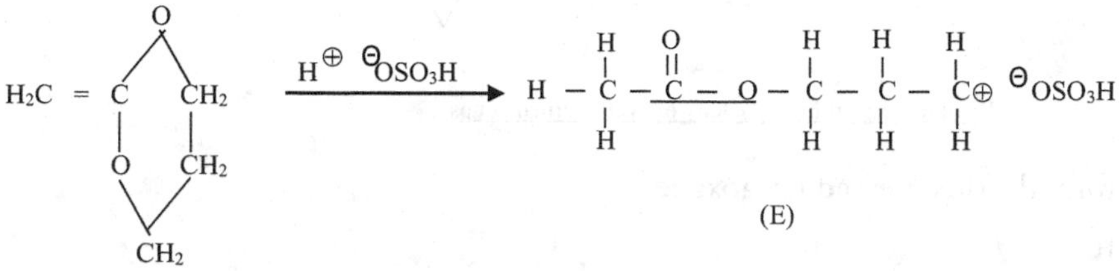

$$H_2C = \overset{O}{\underset{O}{\diamond}} \overset{CH_2}{\underset{CH_2}{}} \xrightarrow{H^{\oplus} \; ^{\ominus}OSO_3H} \quad H - \overset{H}{\underset{H}{C}} - \overset{O}{\underset{}{\overset{\|}{C}}} - O - \overset{H}{\underset{H}{C}} - \overset{H}{\underset{H}{C}} - \overset{H}{\underset{H}{C^{\oplus}}} \; ^{\ominus}OSO_3H$$

(E)

Six-membered cyclic formal [1,3-Dioxehane]

(e)

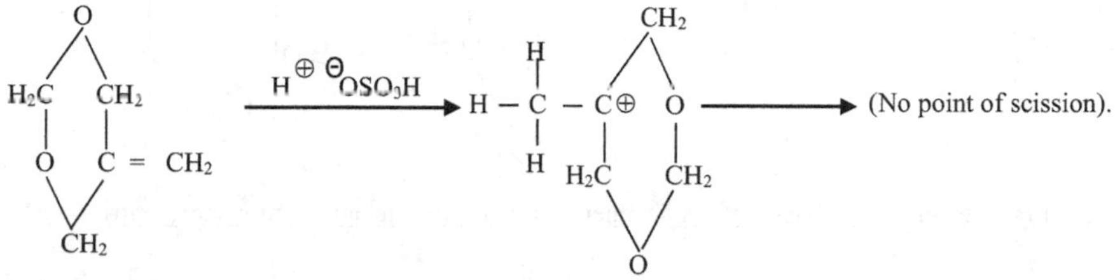

$$H_2C \overset{O}{\underset{O}{\diamond}} CH_2 \;\; \underset{C = CH_2}{} \xrightarrow{H^{\oplus} \; ^{\ominus}OSO_3H} \quad H - \overset{H}{\underset{H}{C}} - C^{\oplus} \;\; \overset{CH_2}{\underset{CH_2}{\diamond}} O \longrightarrow \text{(No point of scission).}$$

Six-membered cyclic formal [1,3-Dioxehane]

(f)

from which from the products obtained, tkhe followings are valid-

(i)   Trioxanes are carboxylic and etheric in character.
(ii)   Cyclic formals are carboxylic, ketonic and etheric in character.
(iii)   Cyclic ethers are ketonic in character.
(iv)   Cyclic alkanes (like the others) are unsaturated in character and the one wherein the methylene group cannot be activated in the absence of a functional center conjugatedly placed to it in the ring and the fact that the invisible p-bond is less nucleophilic than the visible one.

(Laws of Creations for O containing rings)

Rule 1028: This rule of Chemistry for **Ringed compounds,** states that, based on side by side placements of two rings (Coupling theory), corresponding equations for order of radical-pushing capacities of groups can be obtained as follows-
For cyclic ethers and acetals.

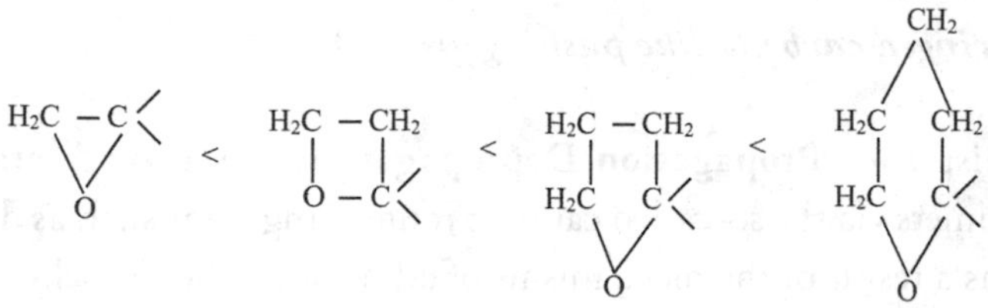

$$H_2C - C\overset{\diagup}{\diagdown} \atop \underset{O}{\diagdown\diagup} \;\; < \;\; \overset{H_2C - CH_2}{\underset{O - C\overset{\diagup}{\diagdown}}{|}} \;\; < \;\; \overset{H_2C - CH_2}{\underset{H_2C \quad C\overset{\diagup}{\diagdown}}{|}} \;\; < \;\; \overset{CH_2}{\underset{H_2C \quad C\overset{\diagup}{\diagdown}}{\overset{H_2C \quad CH_2}{}}}$$

Order of radical-pushing capacity of cylcic ether groups

(a)

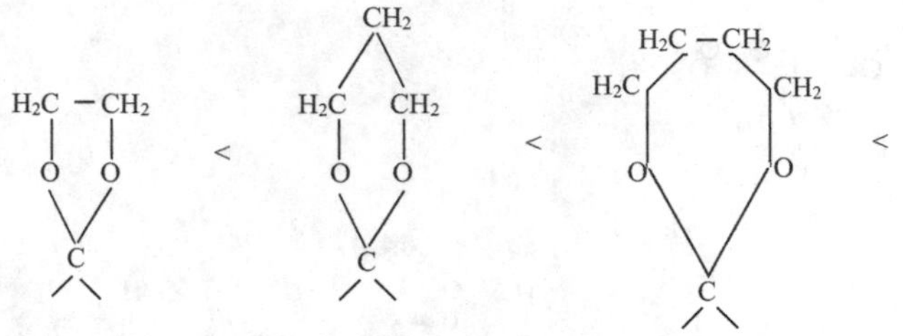

Order of radical-pushing capacity of cyclic formal goups (b)

For cyclic ethers, formal, trioxane and tetraoxane

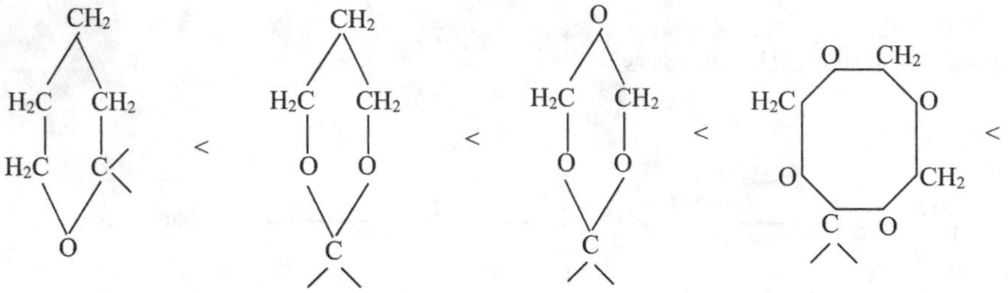

Order of radical-pushing capacity of cyclic ethers, formal, trioxane, and tetraoxane groups (c)

For trioxane, cyclohexane, cyclopentane, cyclobutane and their -enes,

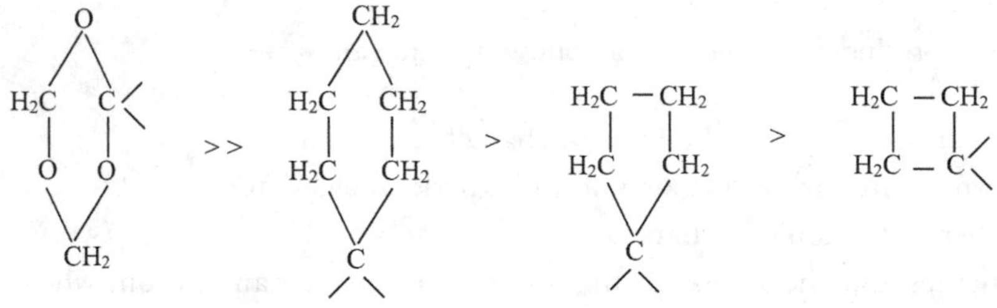

Order of radical-pushing capacity (d)

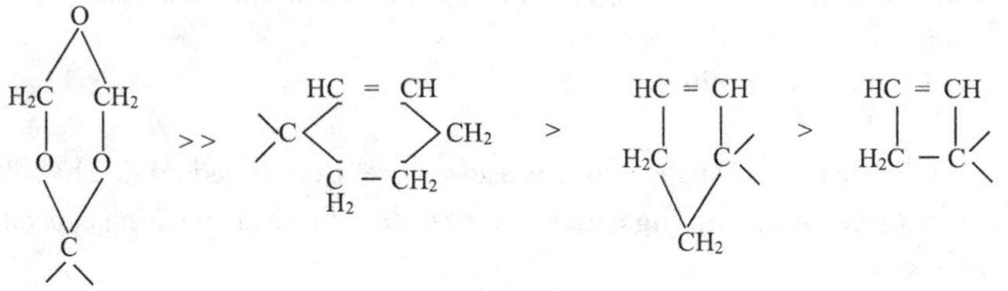

Order of radical-pushing capacity (e)

(Laws of Creation for Hetero-ringed carbene-like pushing groups)

Rule 1029: This rule of Chemistry for **Propagation/Depropagation of Cyclic Acetals,** states that, when Trioxanes are uses as monomers via the so-called cationic route using acids such as dilute HCl, long induction periods observed are as a result of the mechanisms of addition as shown below-

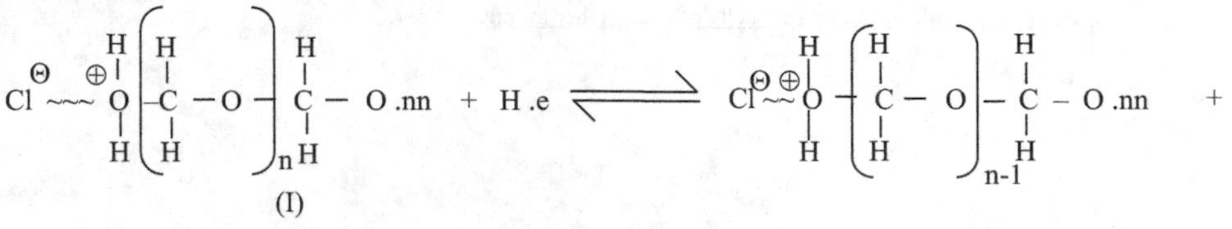

$$Cl^{\ominus} \sim\!\!\sim \overset{\oplus}{\underset{H}{\overset{H}{O}}} \overset{H}{\underset{H}{\overset{|}{C}}} - O \left[\overset{H}{\underset{H}{\overset{|}{C}}} - O \right]_n \overset{H}{\underset{H}{\overset{|}{C}}} - O .nn + H.e \rightleftharpoons Cl^{\ominus}\overset{\oplus}{O} \left[\overset{H}{\underset{H}{\overset{|}{C}}} - O \right]_{n-1} \overset{H}{\underset{H}{\overset{|}{C}}} - O .nn +$$

(I)

118

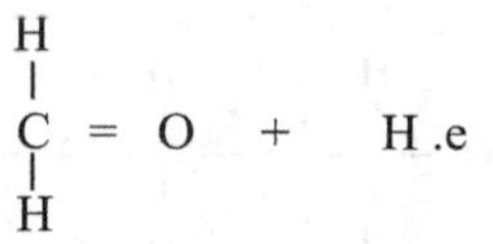

$$\begin{array}{c} H \\ | \\ C = O \ + \ \ H \cdot e \\ | \\ H \end{array}$$

that which is said to be Backward electro-free-radical polymerization, via Equilibrium mechanism (the source of induction), as opposed to Forward Combination mechanism for which as shown above, instead of adding another monomer unit, it is depropagating, due to higher operating conditions above the Ceiling temperature of the monomer; noting in general, that the same applies to 1,3-dioxane but not to 1,4-dioxane, 1,5-dioxane and higher, and all Cyclic formals wherein formaldehyde cannot be released and so also larger sized cyclic ethers. [See Rules 1007-1010]

(Laws of Creations for Propagation/Depropagation & Induction period in Cyclic Acetals)

Rule 1030: This rule of Chemistry for **Formation of Methoxyl and Formate end groups during the "cationic" polymerization of Trioxane,** states that, their presence can be explained as follows-

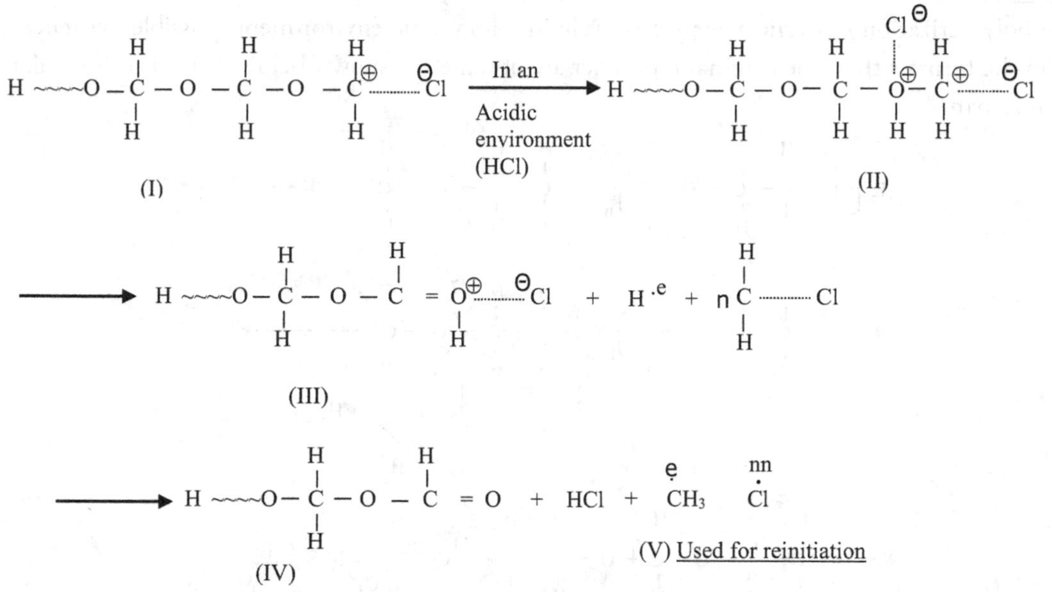

noting that the positive charge on the oxygen center in (II) and (III) is not ionic, but electrostatic in character, for which (IV) is favored after the release of a stable molecule (HCl) and the dead polymer obtained (IV), has a formate end group and that with a methyl end group can be obtained when the CH_3 groups of the equation initiates polymerization; this is that which cannot be obtained from cyclic ethers.

(Laws of Creations for presence of formate and methyl end groups in Trioxane polymerization)

Rule 1031: This rule of Chemistry for **Depropagation reactions of Cyclic Acetals,** states that, during polymerization of cyclic acetals positively in a non-ionic (non-acidic) environment, trioxane undergoes depropagation reactions as shown below-

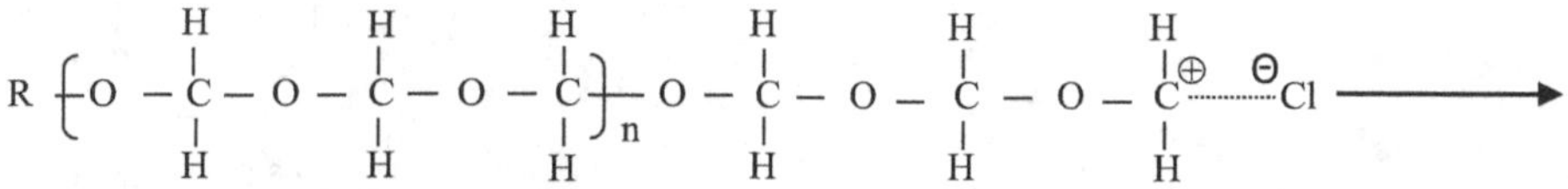

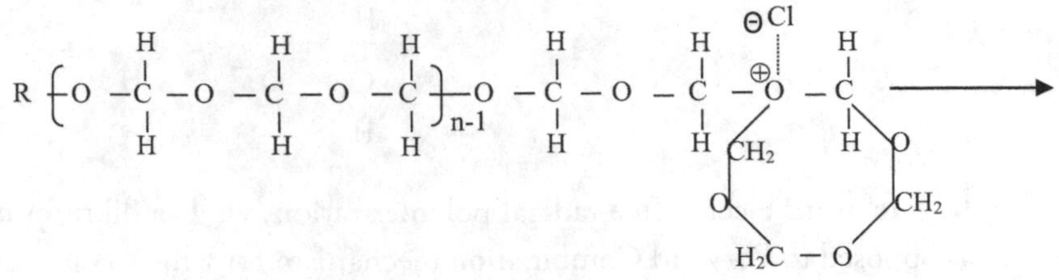

for which, since it is not electrostatically paired, temporary existence of tetraoxane is favored only when the strength of the positive center of the growing chain is weak to provide the MRSE for the eight-membered ring and as propagation progresses some of the active centers become strong to consume the tetraoxane; noting that when existence of larger membered rings which are very stable to polymerization is favored, then the *depropagation is said to be Full.*
(Laws of Creations for Depropagation of Cyclic Acetals)

Rule 1032: This rule of Chemistry for **Depropagation phenomena in Cyclic Acetals,** states that, during polymerization of cyclic acetals positively in a non-ionic environment, possible existence of less stable cyclic formals than the original monomer are obtained as shown below using 1,3 - dioxolane and 1,3 – dioxepane-.

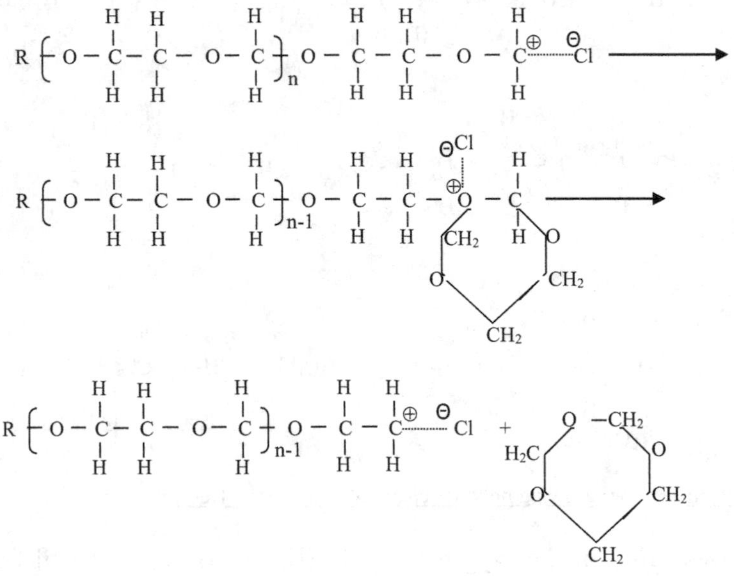

(I) 1,3,5 - trioxolane   (a)

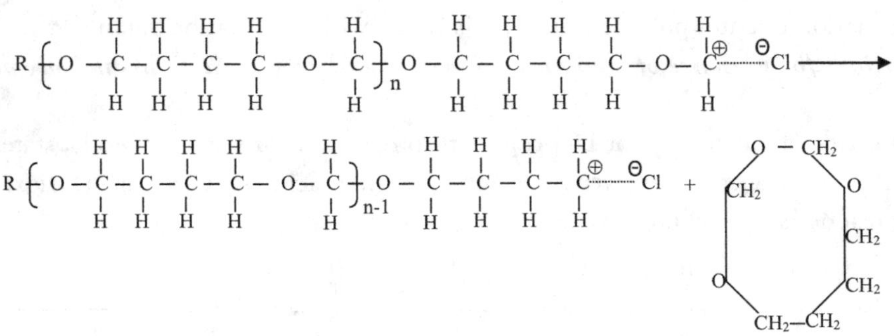

1,3,5 - trioxepane   (b)

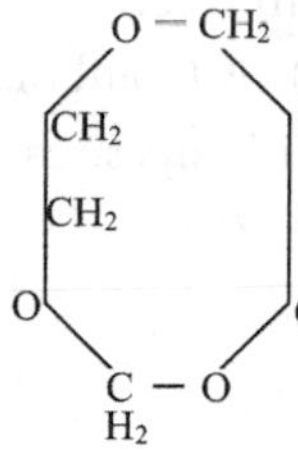

(I) 1,3,6,8 - tetraoxalane

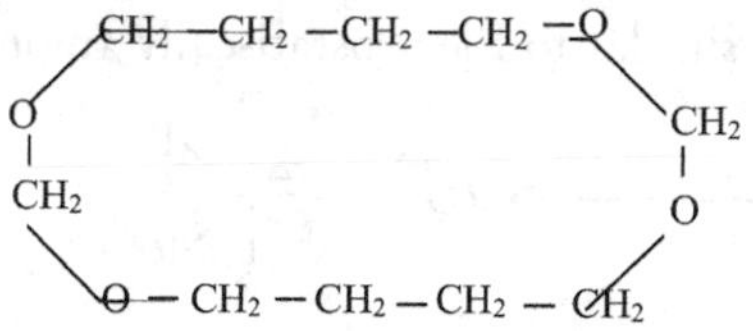

(I) 1,3,6,8, 10 - tetraoxepane

for which if MRSE cannot be provided for the seven and nine membered 1,3,5 - trioxalane and 1,3,5-trioxepane respectively, then they exist as stable molecules throughout the course of polymerization; noting that the two rings are cyclic acetal - hybrids of trioxanes and methylene or cycloalkanes, or in place of the rings above they could produce ten- or fourteen-membered rings which are full cyclic formals and no hybrids, noting that all these largely take place electro-free-radically.

(Laws of Creations for Depropagation of Cyclic formals)

Rule 1033: This rule of Chemistry for **Cyclic Ethers,** states that, when a Linear Condensation product of epichlorohydrin and bisphenol-A, is used as monomer involving the use of a non-ionic amine as initiator, the followings are to be expected-

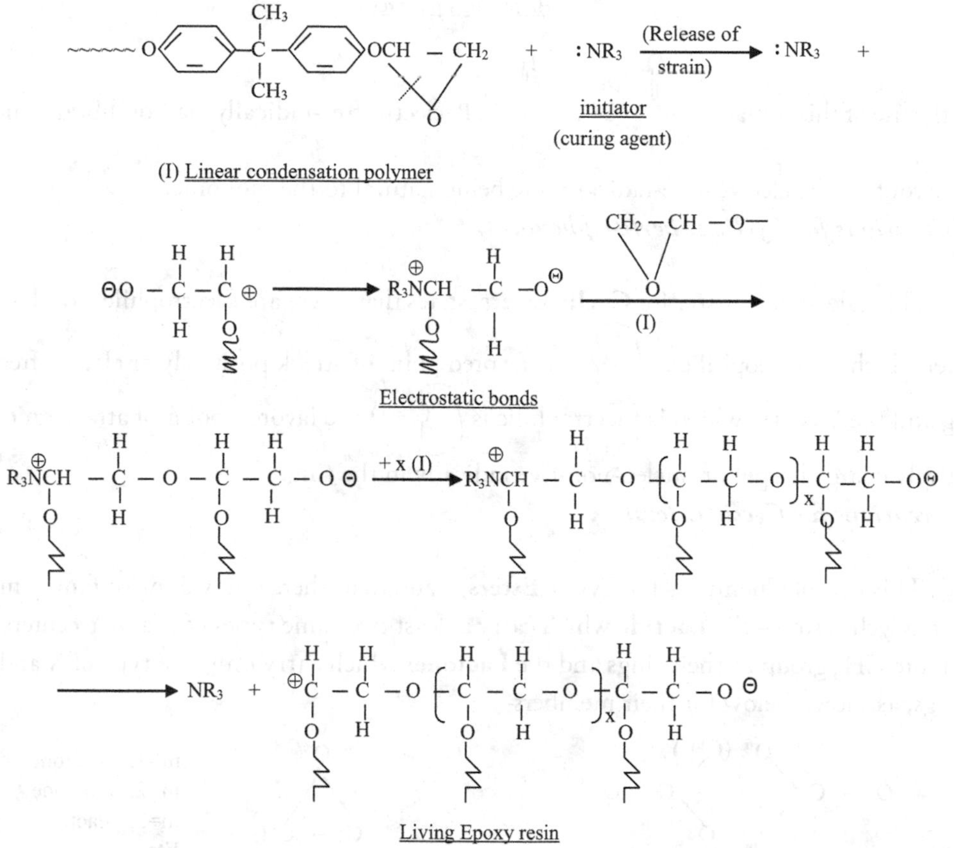

(I) Linear condensation polymer

Electrostatic bonds

Living Epoxy resin

i)

for which the epoxy resins (known to have outstanding adhesive properties and used for bonding to metals, glass, and ceramics and as casting resins in electrical assemblies and also as protective coatings in view of their good adhesion, inertness, hardness, and unusual flexibility), were produced via the use of electrostatic forces from the paired unbonded radicals on the nitrogen center of NR_3 for instantaneous opening of the

121

terminal epoxy rings, followed by addition to the activated ring to form an electrostatic bond with the anionic end of the monomer where addition takes place (i.e., the active center) via Combination mechanism, noting that the initiator is similar to Electrostati-cally anionically-paired initiators such as shown below-

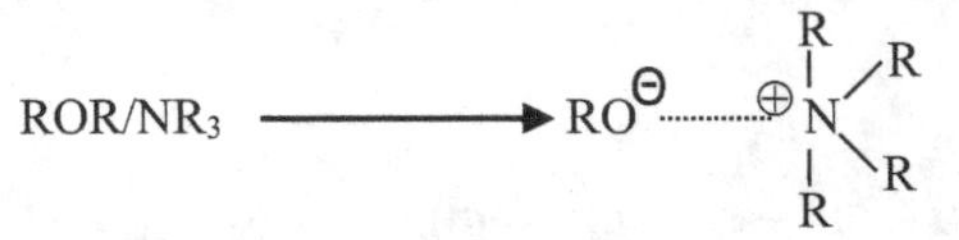

$$ROR/NR_3 \longrightarrow RO^{\ominus} \cdots {}^{\oplus}N \underset{R}{\overset{R}{\underset{|}{\cdots}}} \overset{R}{\underset{R}{\nwarrow}}$$

ii)

noting that, the epoxy resin is a living one which has to be terminated at both ends to produce a dead polymer.
(Laws of Creations for Cyclic Ether/Bisphenol-A)

Rule 1034: This rule of Chemistry for **Cyclic Ethers,** states that, when a Linear Condensation product of epichlorohydrin and bisphenol-A, is used as monomer involving the use of HOOCR as a "cationic" initiator, the same type of resin as produced anionically via instantaneous opening of the ring using the paired unbonded radicals on the initiator (NR_3) is produced, as shown below-

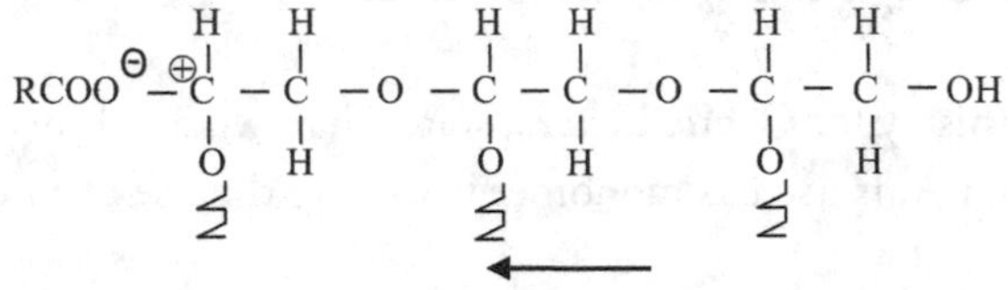

Half living Epoxy resin

or involve the use of this initiator– $RC - O^{\ominus} \cdots {}^{\oplus}O - CR$ electro-free-radically via Equilibrium mechanism.

wherein the route is the electro-free-radical route being natural to the monomer.
(Laws of Creations for Cyclic Ether/Bisphenol-A)

Rule 1035: This rule of Chemistry for **Cyclic Esters,** states that, these are Electrophiles (Males) for which the X center which is nucleophilic is $\left(\; :\overset{..}{O}: \right.$ the favored point of attack positively or electro-free-radically in the ring and the Y center which is Electrophilic is $\left(\; C = O \right.$ the favored point of attack anionically (not negatively charged type) or nucleo-non-free radically in the ring.
(Laws of Creations for Cyclic Esters)

Rule 1036: This rule of Chemistry for **Cyclic Esters,** states that, there are two major family members of the family of Cyclic esters– the **Lactide** which carry at least two same types of X and Y centers separated by one or more CH_2 group in their rings and the **Lactones** which carry only one type of X and Y centers in their rings, as shown below for their members-

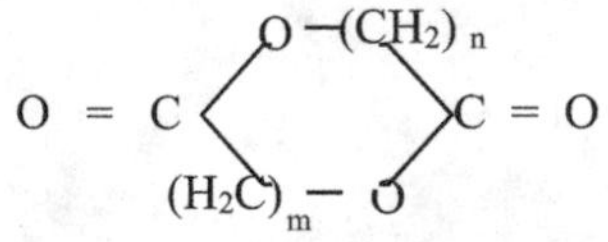

Lactides

When m = n = 1 ≡ Glycolide

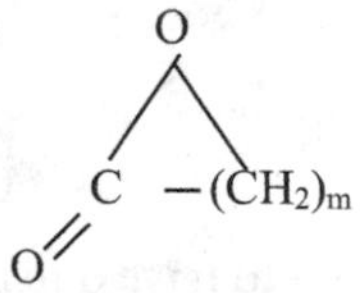

Lactones

m=1, α-lactone
m=2, β-lactone
m=3, γ-lactone
Etc.

(Laws of Creations for Cyclic Esters) [See Rule 990 and 991]

Rule 1037: This rule of Chemistry for **Cyclic Esters,** states that, while presence of radical-pushing or pulling groups in the ring does not alter the route, it does affect the SEs in the rings and also the point of scission in the ring may be altered depending on what the cyclic ester is carrying.
(Laws of Creations for Cyclic Esters)

Rule 1038: This rule of Chemistry for **Cyclic Esters,** states that, negatively charged-paired initiators (of the covalent types) and nucleo-free-radicals cannot be used for their initiations via the C = O center, because the equation will not be chargedly or radically balanced, noting that when the C = O center is activated, only the positive end carried by it when activated can be used anionically or nucleo-non-free-radically.
(Laws of Creations for Cyclic Esters)

Rule 1039: This rule of Chemistry for **Glycolide,** states that, when used as a monomer, the followings are to be expected anionically, nucleo-non-free-radically, positively, and electro-free-radically-

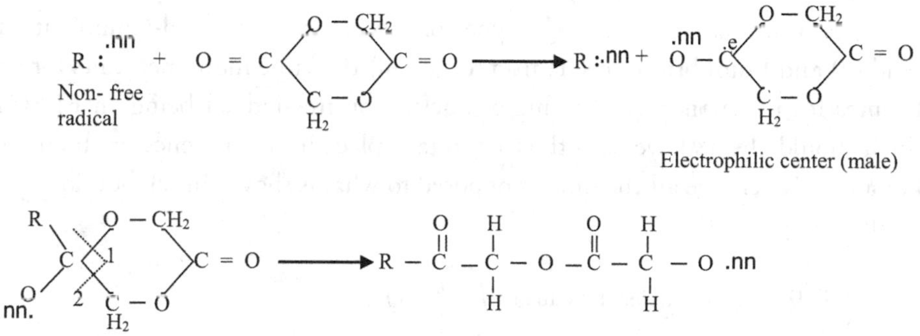

Electrophilic center (male)

Radically-paired initiator

(I) Scission at 1

(a)

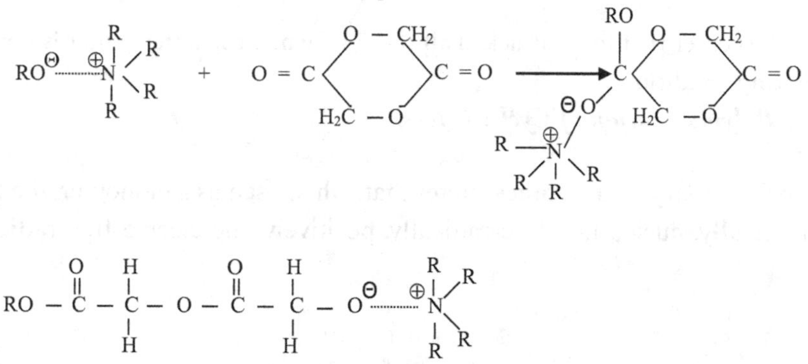

Electrostatically anionically charged-paired initiator

(b)

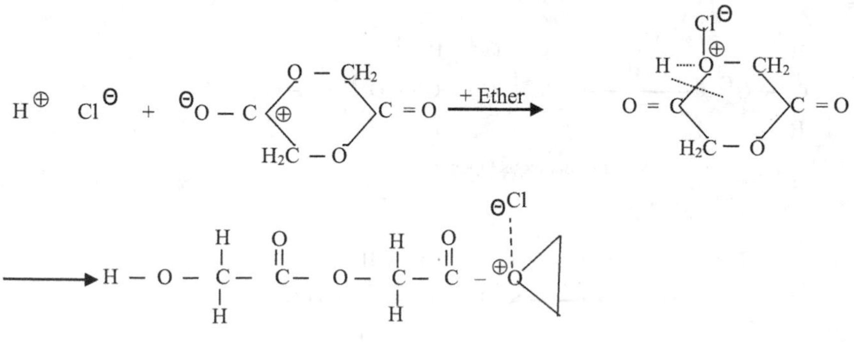

(C) Electro-free-radical Backward Addition

(c)

123

$$X^{\oplus} \cdots {}^{\ominus}Y \quad + \quad O = C \underset{H_2C - O}{\overset{O - CH_2}{\diagdown}} C = O \longrightarrow O = C \underset{H_2\overset{|}{C} - O}{\overset{\overset{\displaystyle X \diagdown \overset{\displaystyle Y^{\ominus}}{O} | CH_2}{}}{\diagup}} \overset{\oplus}{\underset{1}{C}} {}^2 \diagdown C = O \longrightarrow$$

$$X - O \; - \overset{\overset{\displaystyle O}{\|}}{\underset{\underset{\displaystyle H}{|}}{C}} - \overset{\displaystyle H}{\underset{\displaystyle H}{\overset{|}{C}}} - O - \overset{\overset{\displaystyle O}{\|}}{C} - \overset{\overset{\displaystyle H}{|}}{\underset{\underset{\displaystyle H}{|}}{C}} \overset{\oplus}{\underset{}{C}}{}^{\ominus}Y \qquad OR \qquad X - O - \overset{\overset{\displaystyle H}{|}}{\underset{\underset{\displaystyle H}{|}}{C}} - \overset{\overset{\displaystyle O}{\|}}{C} - O - \overset{\overset{\displaystyle H}{|}}{\underset{\underset{\displaystyle H}{|}}{C}} - \overset{\overset{\displaystyle O}{\|}}{C}\overset{\oplus}{}{}^{\ominus}Y$$

(I) <u>Scission at 2 [Not Favored]</u>   (II) <u>Scission at 1</u>

Where $X^{\oplus} \cdots {}^{\ominus}Y$ is $F_2\overset{\oplus}{B} \cdots {}^{\ominus}\overset{\overset{\displaystyle F}{|}}{\underset{\underset{\displaystyle F}{|}}{B}}\diagup^{\diagdown F}_{F}$

(d)

noting i) the types of initiators used, ii) the types of mechanisms involved-Combination mechanism for (a), (b), and (d) and Equilibrium mechanism for (c), iii) the same monomer units for all of them, iv) the natural routes for the monomer- Anionic or nucleo-non-free-radical being an ELECTROPHILE, the route which should always have been the first to take place in the presence or absence of an electro-free-radical or a positive charge all the time as opposed to what is shown in (c), but does not because the following is valid-

$:\overset{..}{O}:$ Less nucleophilic than $\overset{|}{\underset{|}{C}} = O$

Center   Center

(e)

for which the former is the center first to be attacked all the time when the latter center is not activated, and finally, v) the operating conditions.

(Laws of Creations for Polymerization of Cyclic Esters)

Rule 1040: This rule of Chemistry for **Lactones,** states that, when used as a monomer, the followings are to be expected anionically, nucleo-non-free-radically, positively and electro-free-radically using α - lactones

$$\underset{\substack{\text{Non-free}\\\text{"anion"}}}{\overset{\ominus}{R:}} \quad + \quad \overset{\overset{\displaystyle O}{\diagup\diagdown}}{\underset{O=}{C} - CH_2} \quad \xrightarrow[\text{of strain}]{\text{(Release}} \quad R^{\ominus} \quad + \quad \overset{\oplus}{\underset{}{C}} - \overset{\overset{\displaystyle O}{\|}}{\underset{\underset{\displaystyle H}{|}}{\overset{\overset{\displaystyle H}{|}}{C}}} - O^{\ominus} \quad \longrightarrow$$

$$R - \overset{\overset{\displaystyle O}{\|}}{C} - \overset{\overset{\displaystyle H}{|}}{\underset{\underset{\displaystyle H}{|}}{C}} - O^{\ominus} \quad \xrightarrow{+ \; n(I)} \quad R \left[\overset{\overset{\displaystyle O}{\|}}{C} - \overset{\overset{\displaystyle H}{|}}{\underset{\underset{\displaystyle H}{|}}{C}} - O \right]_n \overset{\overset{\displaystyle O}{\|}}{C} - \overset{\overset{\displaystyle H}{|}}{\underset{\underset{\displaystyle H}{|}}{C}} - O^{\ominus}$$

Via Instantaneous Opening

(a)

$$\underset{\substack{\text{Non-free}\\\text{"anion"}}}{R - \overset{\overset{\displaystyle O}{\diagup\diagdown}}{\underset{O^{\ominus}}{C} - CH_2}} \quad \longrightarrow \quad R - \overset{\overset{\displaystyle O}{\|}}{C} - \overset{\overset{\displaystyle H}{|}}{\underset{\underset{\displaystyle H}{|}}{C}} - O^{\ominus}$$

Via Activation of $C = O$

(b)

124

$$H^{\oplus} \; {}^{\ominus}Cl \quad + \quad \underset{O}{\overset{O}{C}} - CH_2 \quad \xrightarrow{+ \text{ Ether}} \quad H - \overset{\oplus}{O} \cdots {}^{\ominus}Cl \quad \xrightarrow[+ n(I)]{+ \text{ Ether}}$$

(I)

C — CH₂
(strained)

$$H \left[O - \overset{H}{\underset{H}{C}} - \overset{O}{\overset{\|}{C}} \right]_n O - \overset{H}{\underset{H}{C}} - \overset{O}{\overset{\|}{C}} \overset{\oplus}{\cdots} \underset{}{\triangle}$$

Electro-free-radical Backward Addition   (c)

$$R \cdots {}^{\ominus}\underset{\underset{R}{O}}{\overset{F}{\underset{}{B}}}{}^{F}_{F} \quad + \quad \underset{O}{\overset{O}{C}} — CH_2 \quad \xrightarrow[\text{of strain}]{(\text{Release})} \quad R \cdots {}^{\ominus}\underset{\underset{R}{O}}{\overset{F}{\underset{}{B}}}{}^{F}_{F} \quad + \quad {}^{\ominus}O - \overset{H}{\underset{H}{C}} - \overset{O}{\overset{\|}{C}}{}^{\oplus}$$

$$\xrightarrow{} \quad R - O - \overset{H}{\underset{H}{C}} - \overset{O}{\overset{\|}{C}}{}^{\oplus} \cdots {}^{\ominus}\underset{\underset{R}{O}}{\overset{F}{\underset{}{B}}}{}^{F}_{F} \quad \xrightarrow{+ \; n(I)}$$

$$R \left[O - \overset{H}{\underset{H}{C}} - \overset{O}{\overset{\|}{C}} \right]_n \overset{\cdot\cdot}{O} - \overset{H}{\underset{H}{C}} - \overset{O}{\overset{\|}{C}}{}^{\oplus} \cdots {}^{\ominus}\underset{\underset{R}{O}}{\overset{F}{\underset{}{B}}}{}^{F}_{F} \quad \xrightarrow{}$$

(No transfer species; but can be terminated to give a dead polymer by Starvation only radically)

(d)

noting i) the types of initiators used, ii) the types of mechanisms involved-Combination mechanism for (a), (b), and (d) and Equilibrium mechanism for (d), iii) the same monomer units for all of them, iv) the natural route for the monomer- Anionic or nucleo-non-free-radical being an ELECTROPHILE, the route which should always have been the first to take place in the presence or absence of an electro-free-radical or a positive charge all the time as opposed to what is shown in (c),but does not because the following is valid-

$$:\overset{\cdot\cdot}{O}: \quad \text{Less nucleophilic than} \quad \overset{}{C} = O$$

Center   Center   (e)

for which the former is the center first to be attacked all the time when the latter center is not activated, and finally, v) the operating conditions.
(Laws of Creations for Polymerization of Lactones)

Rule 1041: This rule of Chemistry for **Cyclic Esters,** states that, if cationic or electro-free-radical activation via the $\overset{}{C} = O$ activation center had been possible like with acrolein, existence of copolymers

125

from a single monomer would have been favored as indicated below for glycolide, β-propiolactone and acrolein.

$$R\sim\sim\sim O-\underset{H}{\overset{O}{\underset{|}{C}}}-\underset{H}{\overset{H}{\underset{|}{C}}}-O-\underset{H}{\overset{O}{\underset{|}{C}}}-\underset{H}{\overset{H}{\underset{|}{C}}}-O-\underset{H}{\overset{H}{\underset{|}{C}}}-\underset{H}{\overset{O}{\underset{|}{C}}}-O-\underset{H}{\overset{H}{\underset{|}{C}}}-\overset{O}{\underset{H}{\underset{|}{C}}}\oplus$$

Via $C = O$ center   Via O center

Copolymerization from glycolide (Not favored)   (a)

$$R\sim\sim\sim O-\underset{H}{\overset{O}{\underset{|}{C}}}-\underset{H}{\overset{H}{\underset{|}{C}}}-\underset{}{\overset{H}{\underset{|}{C}}}-O-\underset{H}{\overset{H}{\underset{|}{C}}}-\underset{H}{\overset{H}{\underset{|}{C}}}-\overset{O}{C}\oplus$$

Via $C = O$ center   Via O center

Copolymer from β - propiolactone (Not favored)   (b)

$$R\sim\sim\sim \underset{\underset{H}{\overset{\|}{C=O}}}{\overset{H}{\underset{|}{C}}}-\underset{H}{\overset{H}{\underset{|}{C}}}-\underset{\underset{H}{\overset{\|}{C=O}}}{\overset{H}{\underset{|}{C}}}-\underset{H}{\overset{H}{\underset{|}{C}}}-O-\underset{\underset{CH_2}{\overset{\|}{CH}}}{\overset{H}{\underset{|}{C}}}-O-\underset{\underset{CH_2}{\overset{\|}{CH}}}{\overset{H}{\underset{|}{C}}}\oplus$$

Via $C = C$ center   Via $C = O$ center

Copolymer from acrolein (favored)   (c)

for their not being favored gives a clear indication of the great distinction between ringed $C = O$ activation center and non-ringed $C = O$ activation center.
(Laws of Creations for Cyclic Esters and Acroleins)

Rule 1042: This rule of Chemistry for **Lactones,** states that, the fact that the first member, anionically can be activated in two different ways depending on the strength of the initiator, clearly indicates the less nucleophilic character than larger membered rings, where anionically the monomer can largely be activated in only one way for which the following order is to be expected for the members of lactone family, noting that these are electrophiles-

$$>\quad \underset{H_2C-CH_2}{\overset{\overset{\displaystyle O}{\|}}{\underset{O\quad CH_2}{\overset{C}{\triangle}}}}\quad >\quad \overset{\overset{\displaystyle O}{\|}}{\underset{O-CH_2}{\underset{|}{\overset{C-CH_2}{\underset{|}{}}}}}\quad >\quad \underset{O}{\overset{O}{\overset{\|}{\underset{}{\overset{C-CH_2}{\triangledown}}}}}$$

Order of Nucleophilicity

126

from which the following is valid-

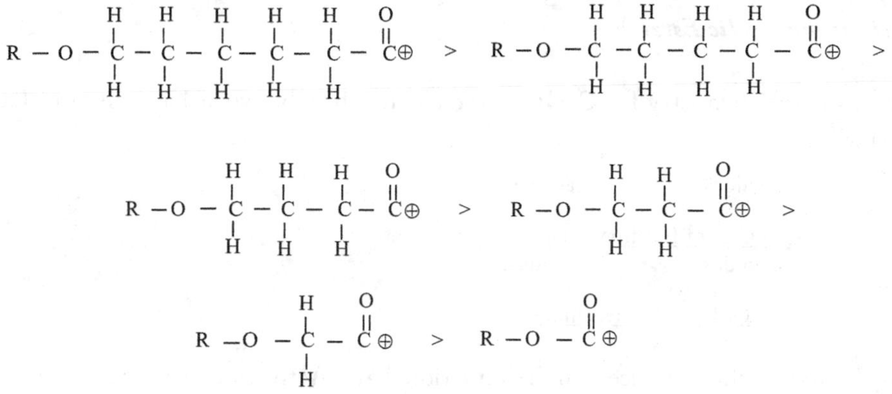

$$R-O-\overset{\overset{H}{|}}{\underset{\underset{H}{|}}{C}}-\overset{\overset{H}{|}}{\underset{\underset{H}{|}}{C}}-\overset{\overset{H}{|}}{\underset{\underset{H}{|}}{C}}-\overset{\overset{H}{|}}{\underset{\underset{H}{|}}{C}}-\overset{\overset{H}{|}}{\underset{\underset{H}{|}}{C}}-\overset{\overset{O}{\|}}{C}\oplus \quad > \quad R-O-\overset{\overset{H}{|}}{\underset{\underset{H}{|}}{C}}-\overset{\overset{H}{|}}{\underset{\underset{H}{|}}{C}}-\overset{\overset{H}{|}}{\underset{\underset{H}{|}}{C}}-\overset{\overset{H}{|}}{\underset{\underset{H}{|}}{C}}-\overset{\overset{O}{\|}}{C}\oplus \quad >$$

$$R-O-\overset{\overset{H}{|}}{\underset{\underset{H}{|}}{C}}-\overset{\overset{H}{|}}{\underset{\underset{H}{|}}{C}}-\overset{\overset{H}{|}}{\underset{\underset{H}{|}}{C}}-\overset{\overset{O}{\|}}{C}\oplus \quad > \quad R-O-\overset{\overset{H}{|}}{\underset{\underset{H}{|}}{C}}-\overset{\overset{H}{|}}{\underset{\underset{H}{|}}{C}}-\overset{\overset{O}{\|}}{C}\oplus \quad >$$

$$R-O-\overset{\overset{H}{|}}{\underset{\underset{H}{|}}{C}}-\overset{\overset{O}{\|}}{C}\oplus \quad > \quad R-O-\overset{\overset{O}{\|}}{C}\oplus$$

<u>Order of radical-pushing capacity and strength of growing center</u>

(Laws of Creation for Lactones)

<u>Rule 1043:</u> This rule of Chemistry for **Cyclic Esters,** states that, unlike Nucleophiles like Cyclic Ethers where when radical-pushing groups of greater capacity than H are put in place of H in their rings increases the SEs of their rings, for Electrophiles, the opposite is the case when the group is adjacently located to the O center as shown below for five- and six- membered lactones-

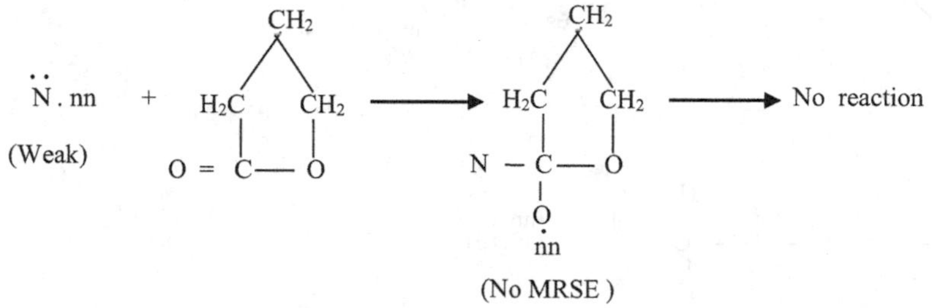

(No MRSE) (i)

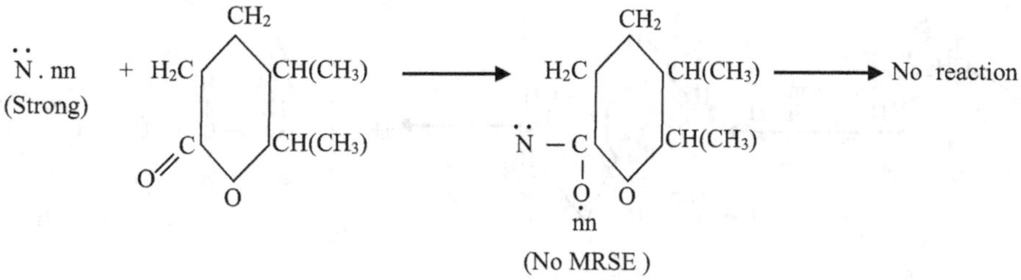

(No MRSE) (ii)

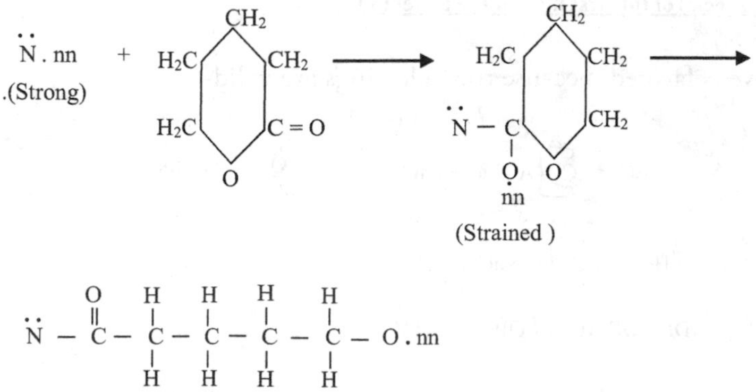

(Strained)

(iii)

127

while presence of radical-pulling groups on that center(s) adjacently located to O center will increase the SEs of their rings to favor the ease of opening their rings.
(Laws of Creations for Cyclic Esters)

Rule 1044: This rule of Chemistry for **Cyclic Esters,** states that, between Lactones and Lactides, the followings are valid-

$$Lactides \quad > \quad Lactones$$

<u>Order of Electrophilicity</u>
$$Lactides \quad > \quad Lactones$$

<u>Order of Nucleophilicity</u>

that which is supported by the presence of more Functional and Activation centers in lactides.
(Laws of Creations for Cyclic Esters)

Rule 1045: This rule of Chemistry for **Cyclic Esters,** states that, when $H_2C=$ group is introduced into the ring, the followings are obtained when the group is specially placed next to the O center –

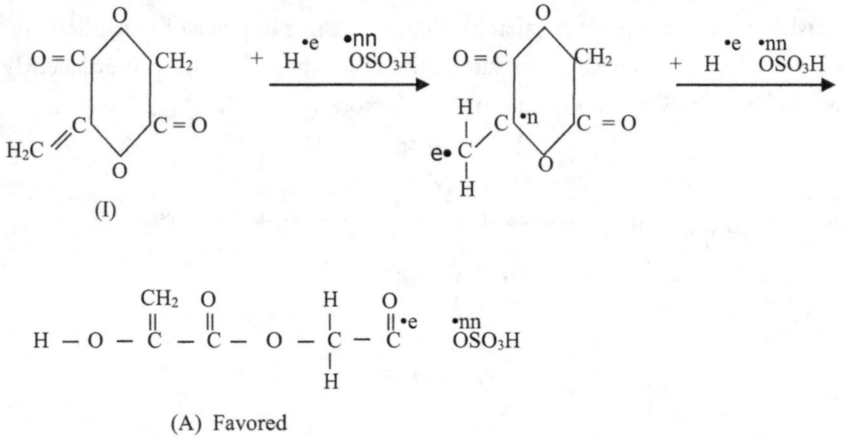

(I)

(a)

(A) Favored

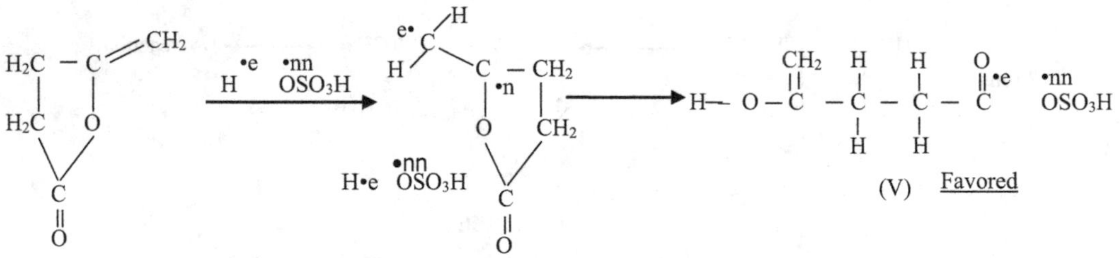

(I)   <u>**Transfer species of the first kind of the eleventh type**</u>

(V) <u>Favored</u>

(b)

for which the mechanism above is favored, because the followings are valid-

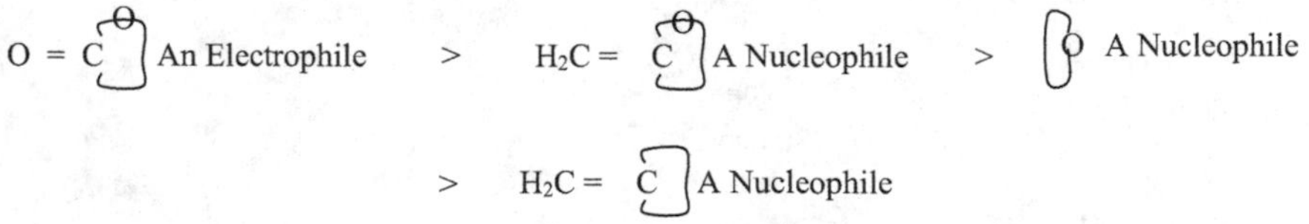

$$O = C \quad \text{An Electrophile} \quad > \quad H_2C = C \quad \text{A Nucleophile} \quad > \quad \text{A Nucleophile}$$

$$> \quad H_2C = C \quad \text{A Nucleophile}$$

ORDER OF NUCLEOPHILICITY

128

noting most importantly that, ***additional presence of H₂C= to the ring, has not nullified the Electrophilic character of these monomers, despite the fact that it was the H₂C= center that was used.; noting that the monomer cannot be activated chargedly due to electrostatic forces of repulsion, and can be activated nucleo-non-free-radically to give the same monomer unit.***
(Laws of Creation for Cyclic Esters)

Rule 1046: This rule of Chemistry for **Cyclic Esters,** states that, when H₂C= group is introduced into the ring, the followings are obtained when the group is specially placed next to the C=O center (whether the compound exists or not)-

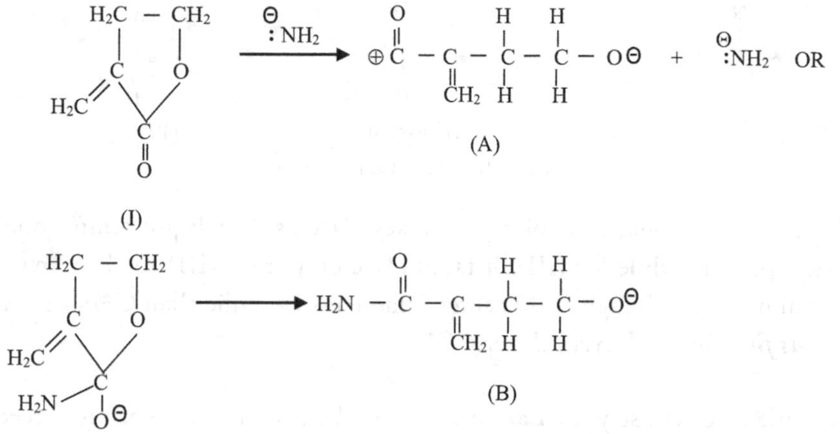

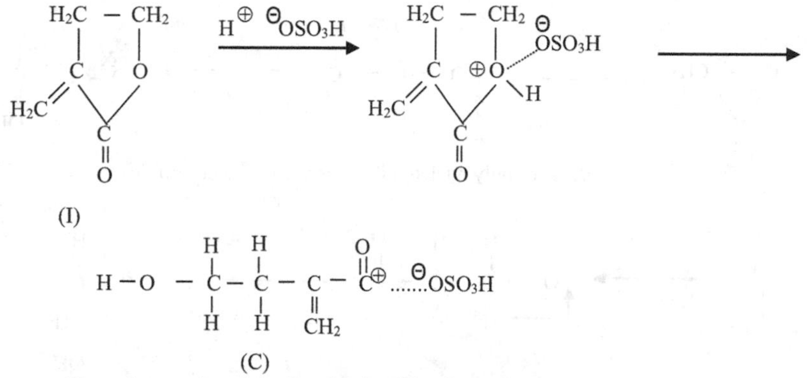

$$\text{Transfer species of the first kind of the eleventh type}$$

for which since the same monomer units are obtained instantaneously, anionically and positively, and since when the H₂C= center is activated chargedly, the positive charge attempts to grab the O-CH₂- group as transfer species of the first kind of the eleventh type by opening the ring to favor the existence of same (C) above, the reactions above are favored; clear indication that the Electrophilic character of the ring is still maintained, despite the presence of H₂C = as the least nucleophilic center.
(Laws of Creations for Cyclic Esters)

Rule 1047: This rule of Chemistry for **Cyclic Esters when their members are coupled and Cyclic Anhydride,** states that, during polymerization anionically or nucleo-non-free-radically, the terminals of the growing chain could be such that undergo movements of electro-free-radicals to and fro between two centers such as shown below–

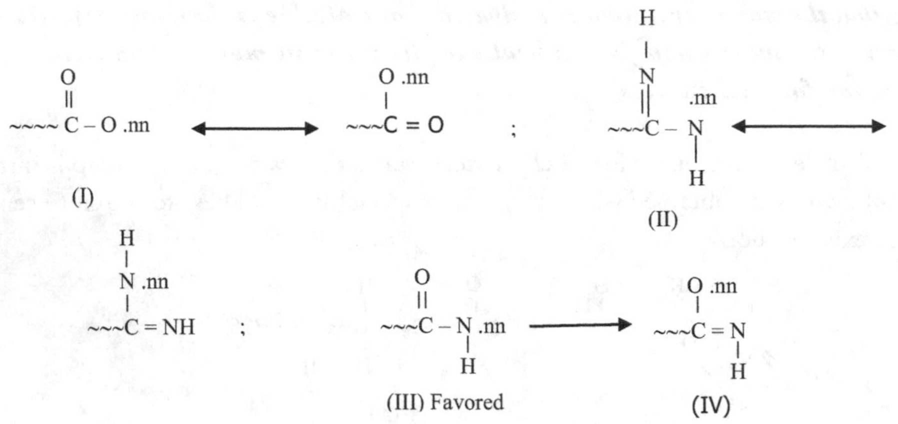

$$
\underset{\text{(I)}}{\sim\!\sim\!\sim \overset{\displaystyle O}{\overset{\|}{C}} - O \,.nn} \quad\longleftrightarrow\quad \sim\!\sim\!\sim \overset{\displaystyle O\,.nn}{\overset{\|}{C}} = O \quad ; \quad \sim\!\sim\!\sim \overset{\displaystyle \overset{H}{\underset{|}{N}}}{\overset{\|}{\underset{\underset{H}{|}}{C}}} - N \,.nn \quad\longleftrightarrow\quad
$$

(II)

$$
\underset{\sim\!\sim\!\sim C = NH}{\overset{\displaystyle \overset{H}{\underset{|}{N}}\,.nn}{}} \quad ; \quad \sim\!\sim\!\sim \overset{\displaystyle O}{\overset{\|}{C}} - \underset{\underset{H}{|}}{N}\,.nn \quad\longrightarrow\quad \sim\!\sim\!\sim \overset{\displaystyle O\,.nn}{\overset{|}{C}} = \underset{\underset{H}{|}}{N}
$$

(III) Favored   (IV)

Resonance Stabilized Components

for which for (I) and (II), resonance stabilization takes place as if in Equilibrium (one of the sources of minor induction period), while for (III), it takes place only from (III) to (IV), oxygen being more electronegative than nitrogen, though C=O center is more nucleophilic than C=N center.
(Laws of Creations for Coupled Cyclic Esters)

Rule 1048: This rule of Chemistry for **Lactones,** states that, when the rings are forced to be opened instantaneously, they undergo molecular rearrangement of the third kind to give Ketene as shown below-

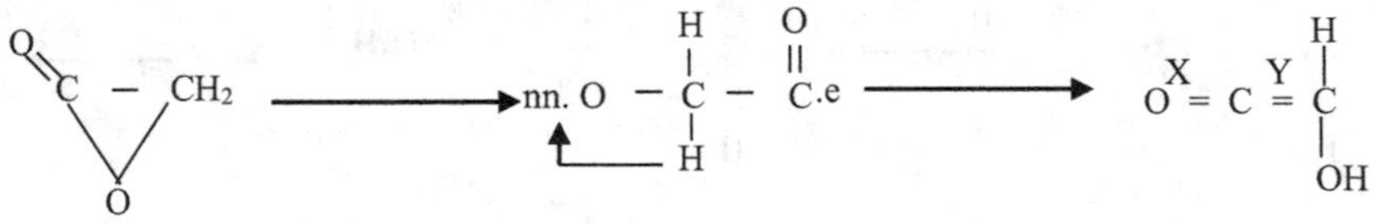

X is strongly nucleophilic;Y is weakly electrophilic

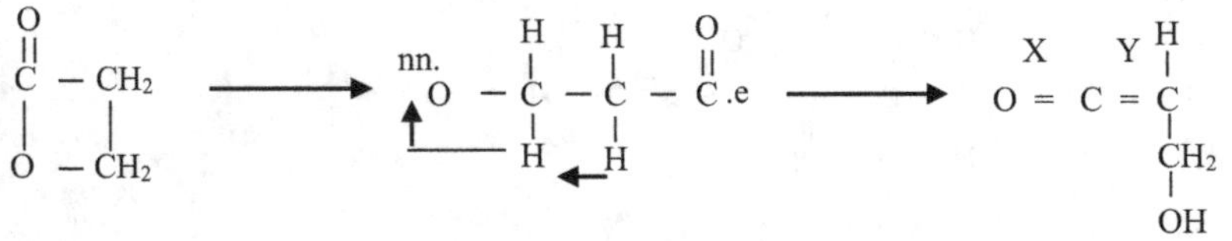

X is strongly nucleophilic; Y is strongly electrophilic

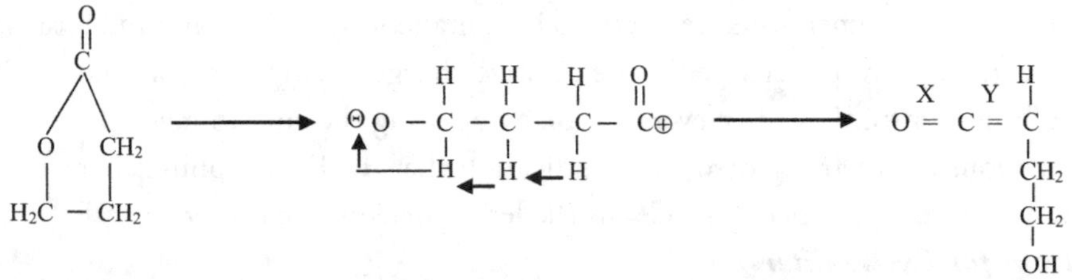

X is strongly nucleophilic; Y favors both routes

from which one can see that only the three- and four-membered rings are strongly electrophilic, while the higher membered rings are weakly electrophilic for which one can observe why five- and higher membered rings are unreactive as monomers; noting that this cannot fully be used as a complete measure of their order of electrophilicity, since not all the rings can be opened instantaneously.
(Laws of Creations for Lactones)

Rule 1049: This rule of Chemistry for **Lactides such as Glycolide,** states that, because of the presence of two or more Electrophilic centers in or on them, the followings can take place with them based on their sizes and operating conditions-

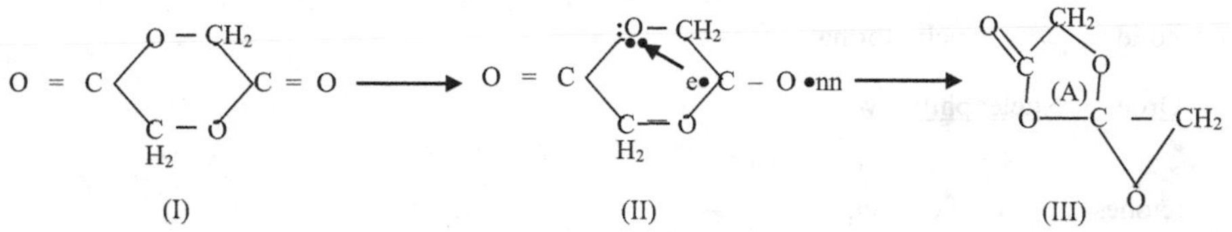

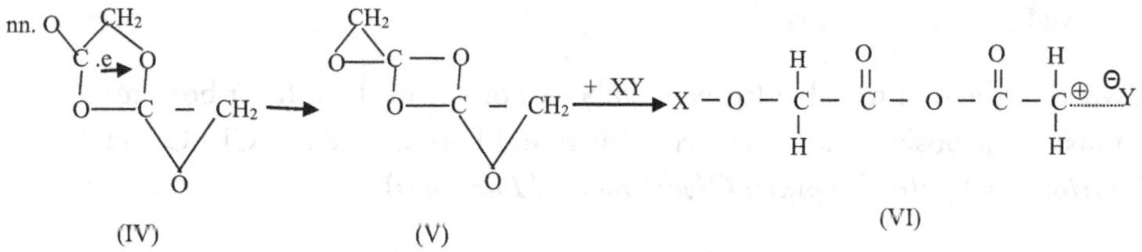

for which it is unique to notice **TRANS ANNULAR ELECTROSTATIC ADDITION,** between the C=O center of one Male center and the O center of another Male center to give in decreasing order stable (III) and (V), two coupled (not Fused) five- and three- membered rings and three coupled (not fused) three-, four- and three-membered rings respectively, which when any of them is used as a monomer gives a monomer unit [in (VI)] different from that obtained from (I); noting that in (III), the five membered ring is carrying both Male (X & Y) and Female (X) centers (similar to those carried by some Living organisms which have the ability to produce alone) and (V) obtained is a Complete Nucleophile, all the rings well strained to the point where their existence will be transient if they exist.
(Laws of Creations for Trans Annular Electrostatic Addition in Lactides)

Rule 1050: This rule of Chemistry for **Cyclic Esters,** states that, unlike cyclic ethers and alkanes, despite the presence of functional centers and activation center on the sides of their rings, as their rings becomes larger in size, it is almost impossible to open them no matter the harshness of the operating conditions, *because of exponentially deceasing SE with increasing size of ring caused by the increasing presence of radical pushing groups.*
(Laws of Creations for Cyclic Esters)

Rule 1051: This rule of Chemistry for **Glycolide and Cyclic Ethers,** states that, Cyclic esters are more electrophilic than Cyclic ethers (which are Nucleophiles), while nucleophilically the same is the case.

Glycolides   >   Cyclic ethers

Order of Nucleophilicity

Glycolides   >   Cyclic ethers

Order of Electrophilicity

(Laws of Creation for Cyclic Ethers and Glycolide)

Rule 1052: This rule of Chemistry for **Cyclic Formals, Glycolide and Lactones,** states that, the followings are valid for them-

> Lactones > Cyclic formals
>
> Glycolide > Cyclic formals
>
> Order of Nucleophilicity

> Lactones > cyclic ethers
>
> Lactones > cyclic formals
>
> Order of Electrophilicity

noting that, whether the first member (three-membered) or any of them favor both routes when opened instantaneously or not, all Cyclic Ethers and Formals are NUCLEO-PHILES.
(Laws of Creation for Cyclic Formals, Glycolide and Lactones)

Rule 1053: This rule of Chemistry for **Cyclic Esters,** states that, because of the proximity of C=O center to O center along a linear chain, the oxygen center cannot be attacked by positive charges since the C=O center along the linear chain is now nucleophilic in character as shown below, and these become sites for branching.

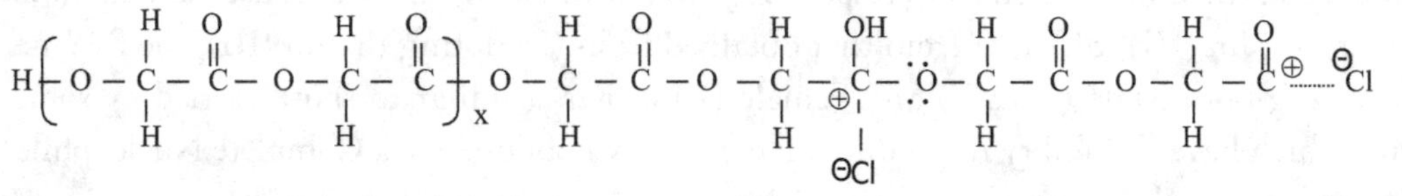

(Laws of Creations for Cyclic Esters)

Rule 1054: This rule of Chemistry for **Depropagation of Cyclic Esters,** states that, anionically or nucleo-non-free-radically, depropagation will not take place and electro-free-radically, depropagation will also not take place above or below the Ceiling temperature of the monomer via Backward Equilibrium Addition mechanism as shown below-

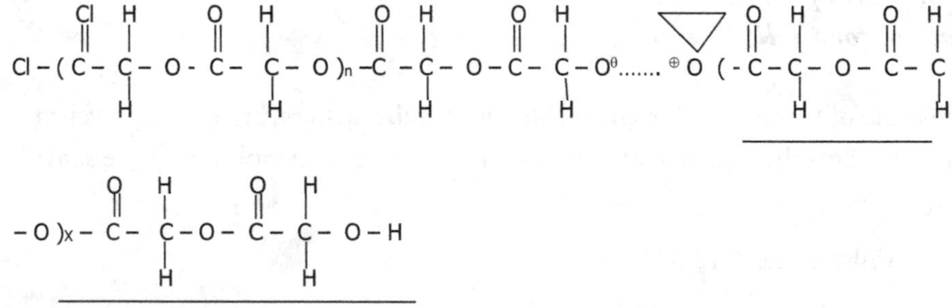

unless the entire monomer unit (containing formaldehyde which is impossible to be released, CO_2 and ketene both also impossible to be released) is released; noting the presence of full anionic center above, because the initiator was first made to exist in Stable state of existence and the C = O was activated during addition of the monomer, otherwise if the initiator was not kept in Equilibrium state of existence, only backward addition will take place.
(Laws of Creations for Cyclic Esters)

Rule 1055: This rule of Chemistry for **Lactones,** states that, they can be copolymerized with themselves once they favor homopolymerization and for those that do not favor homopolymerization, strong active growing centers have to be first generated, before they can be activated and this can be done nucleo-non-free-radically, anionically, positively, electro-free-radically, though lactones are electrophilic in character.
(Laws of Creations for Copolymerization of Lactones)

Rule 1056: This rule of Chemistry for **Lactones,** states that, though anionically lactones cannot be copolymerized with cyclic acetals and larger-membered cyclic ethers, they can only be copolymerized with three-membered cyclic ethers such as copolymerization between β-propiolactone and styrene oxide, β-propiolactone and epichlorohydrin, δ - valerolactone and propylene oxide, with tendency towards having blocks of the lactone monomers along the chain being due to the fact that lactones are Electrophiles while the three-membered cyclic ether which was opened instantaneously is a Nucleophile,

Long Block of Lactone

noting that if in (II), one of the H atoms was CH_3 (propylene oxide) or CH_2Cl (Epichlorohydrin - radically), (II) can never be found along the chain, unless when they (Propylene oxide or Epichlorohydrin) and the lactone are first instantaneously opened to favor the formation of a couple to give alternating placement along with random presence of the lactone along the chain and almost the same will apply to the case above if the lactone was instantaneously opened with both lactone and the ethylene oxide appearing randomly along with alternating placement, all these depending on the strength of the initiator; for with δ-valerolactone the situation as above will not fully apply, since the ring cannot readily be opened instantaneously.
(Laws of Creations for Copolymerization of Lactones with Cyclic Ethers/Acetals)

Rule 1057: This rule of Chemistry for **Lactones,** states that, positively or electro-free-radically, lactones can be copolymerized with cyclic ether and acetal when Electro-statically positively charged-paired initiator such as that from BF_3/Etherate combination (that in which the negatively charged center is not active) is used to initiate the copolymerizations of β-propiolactone and 3,3-bis(chloromethyl) oxetane, β-propio-lactone and tetrahydrofuran, for which the greater reactivity of the cyclic ether observed is a consequence of the fact that cyclic ethers are less nucleophilic than lactones and the fact that the route is natural to ethers and acetals, resulting to a large tendency of having long blocks of the oxetane in the copolymer.
(Laws of Creations for Copolymerization of Lactones)

Rule 1058: This rule of Chemistry for **Copolymerization of Cyclic formals, Trioxane, Tetraoxane, and Cyclic Ethers with Alkenes,** states that, positively copolymerization can take place, because the ringed ones are by far more nucleophilic than alkenes both of which are Nucleophiles, resulting in having long blocks of alkenes along the chain until all consumed followed by cyclic ones; while anionically, copolymerization is not favored either for the ringed ones except for those that can be opened instantaneously or the alkenes because of presence of transfer species of the first kind (unnatural route).
(Laws of Creations for Copolymerization of O-containing ringed Nucleophiles and Alkenes)

Rule 1059: This rule of Chemistry for **Copolymerization of Lactones with Alkenes,** states that, positively no copolymerization is possible, alkenes being far less nucleophilic than lactones which are Electrophiles, favoring only the homopolymerization of alkenes; while anionically only lactones will be homopolymerized, the route being natural to them.
(Laws of Creations for Copolymerization of Lactones)

Rule 1060: This rule of Chemistry for **Maleic anhydrides,** states that, these are unique Electrophiles with two different types of Electrophilic centers as shown below-

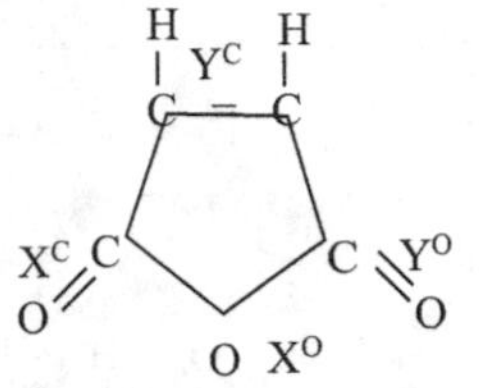

(I) Maleic anhydride 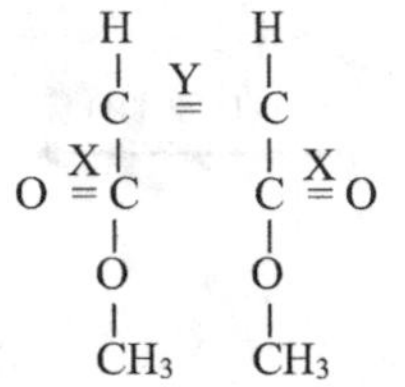   ;   (II) Vinyl-di methyl-acrylate

with one functional center- $\bigcirc$ O and three activation centers- one $C = C$ and two $C = O$, for which Y^C and X^C are used when the ring is closed, Y^O and X^O are used for opening the ring and looking at the monomer without the ring, it resembles the type of symmetric alkene shown above (II) and based on their chemical behaviors, the followings are valid for them-

$$\overset{X}{C = O} \quad > \quad \overset{Y}{C = C}$$

Order of Nucleophilicity in (II) Non-ringed monomer

Order of Nucleophilicity in Maleic anhydride (I)

noting that, when a strong electro-free-radical is involved, Y^C becomes X^O, that wherein the ring is opened with the Y bond still in place, ***via involvement of transfer species of the first kind of the eleventh type; and finally that both monomers can be resonance stabilized from Full-Free to Full Non-free.***
(Laws of Creations for Cyclic Anhydrides)

Rule 1061: This rule of Chemistry for **Phthalic Anhydride,** states that, this is a unique Electrophile with only one Electrophilic center as shown below-

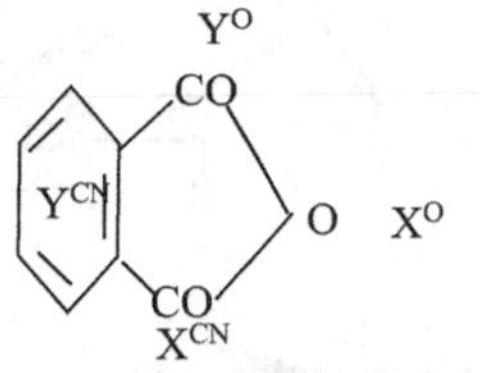

(I) Phthalic anhydride 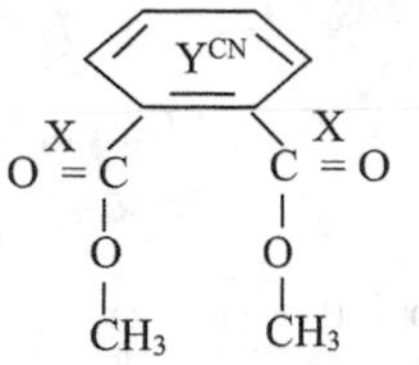   (II) Phenyl 1,2-Di-methyl acrylate

with one functional center $\left(\!\!\begin{array}{c}O\end{array}\!\!\right)$, two activation centers $\left(\!\!\begin{array}{c}C\end{array}\!\!\right)$ = O and one benzene ring fused to a five membered ring, for which Y^{CN} cannot be used without hydrogenating the ring or partially saturating the ring, Y^O and X^O are used for opening the five-membered ring and X^{CN} cannot be used and looking at the monomer without the ring, it resembles the type shown above (II) and based on their chemical behavior, the following are valid for them-

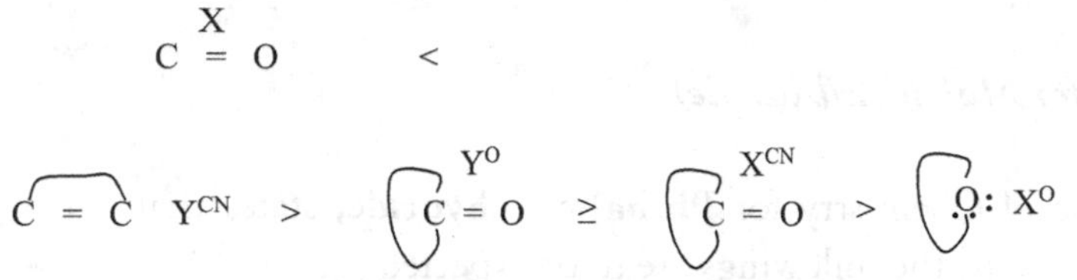

$$C \overset{X}{=} O \quad <$$

$$C = C \quad Y^{CN} \quad > \quad C \overset{Y^O}{=} O \quad \geq \quad C \overset{X^{CN}}{=} O \quad > \quad \underset{\cdot\cdot}{O} : X^O$$

Order of Nucleophilicity in Phthalic anhydride (I)

noting that since Y^{CN} is more nucleophilic than X^{CN}, electrophilicity can only be provided in (I) in the ring and not in (II); and finally they cannot be resonance stabilized.
(Laws of Creations for Phthalic Anhydride}

Rule 1062: This rule of Chemistry for **Maleic Anhydride,** states that, opening of the ring is favored only when the initiator of the anionic or nucleo-non-free-radical or electro-free-radical or cationic type involved is strong, while when the initiator is weak the ring cannot be opened and only the C = C center can be used nucleo-free-radically and electro-free-radically with limitations, noting that when used electro-free-radically it is done only when another activated nucleophilic monomer is adding to it (an Electro-phile) with no ability to open the ring since MRSE cannot be provided for it.
(Laws of Creations for Maleic Anhydride)

Rule 1063: This rule of Chemistry for **Maleic Anhydride,** states that, when used as a monomer anionically and positively, the followings are to be expected-

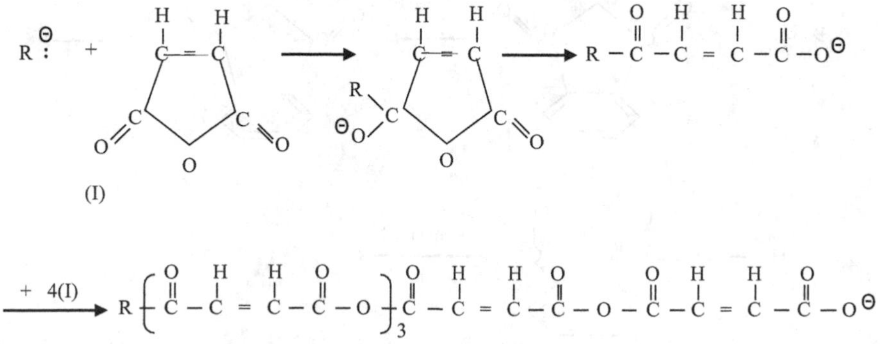

(a)

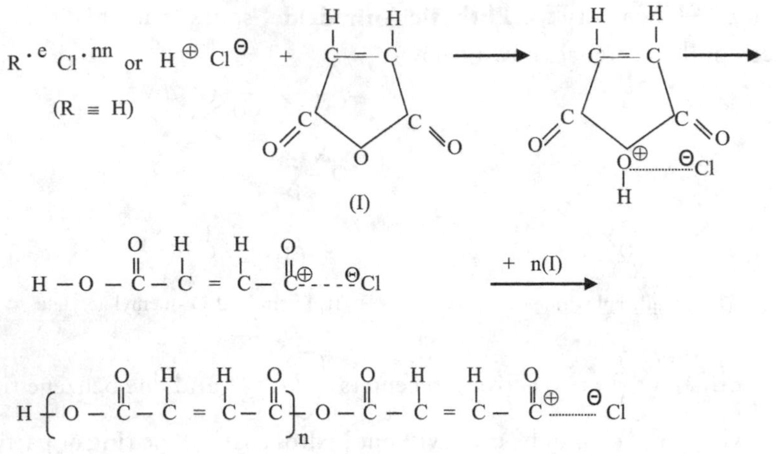

$$R \cdot^e \; Cl \cdot^{nn} \; \text{or} \; H^{\oplus} \; Cl^{\ominus} \; + \quad \text{(I)} \longrightarrow \longrightarrow$$

$$(R \equiv H)$$

$$H - O - \overset{O}{\overset{\|}{C}} - \overset{H}{\overset{|}{C}} = \overset{H}{\overset{|}{C}} - \overset{O}{\overset{\|}{C}}{}^{\oplus} \cdots {}^{\ominus}Cl \qquad \xrightarrow{+ \; n(I)}$$

$$H \left\{ O - \overset{O}{\overset{\|}{C}} - \overset{H}{\overset{|}{C}} = \overset{H}{\overset{|}{C}} - \overset{O}{\overset{\|}{C}} \right\}_n O - \overset{O}{\overset{\|}{C}} - \overset{H}{\overset{|}{C}} = \overset{H}{\overset{|}{C}} - \overset{O}{\overset{\|}{C}}{}^{\oplus} \cdots {}^{\ominus}Cl$$

(b)

for which the same monomer unit can be observed to be obtained, noting the anionic end of the growing chain wherein it is impossible to release CO_2, and the fact that positively, particularly when H•e is involved the mechanism of addition is electro-free-radically via Equilibrium mechanism, i.e., backward addition and not as shown above.

(Laws of Creations for Maleic Anhydride)

Rule 1064: This rule of Chemistry for **Phthalic Anhydride,** states that, when used as a monomer anionically and cationically, the followings are to be expected-

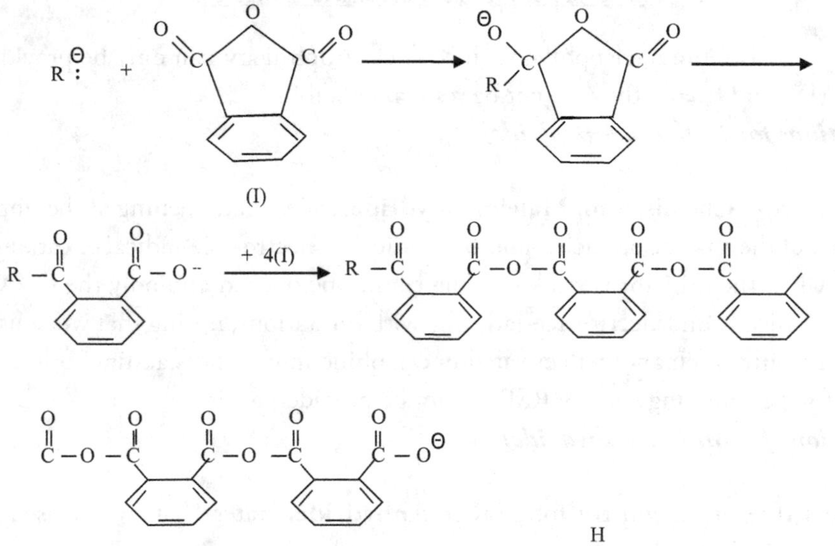

$$R :^{\ominus} \; + \quad \text{(I)} \longrightarrow \longrightarrow$$

$$R - \overset{O}{\overset{\|}{C}} \cdots \overset{O}{\overset{\|}{C}} - O^- \quad \xrightarrow{+ \; 4(I)} \quad R - \overset{O}{\overset{\|}{C}} \cdots \overset{O}{\overset{\|}{C}} - O - \overset{O}{\overset{\|}{C}} \cdots \overset{O}{\overset{\|}{C}} - O - \overset{O}{\overset{\|}{C}} \cdots$$

$$\overset{O}{\overset{\|}{C}} - O - \overset{O}{\overset{\|}{C}} \cdots \overset{O}{\overset{\|}{C}} - O - \overset{O}{\overset{\|}{C}} \cdots \overset{O}{\overset{\|}{C}} - O^{\ominus}$$

(a)

$$H^{\oplus} \; {}^{\ominus}Cl \; + \quad \text{(I)} \longrightarrow \longrightarrow$$

$$H - O - \overset{O}{\overset{\|}{C}} \cdots \overset{O}{\overset{\|}{C}}{}^{\oplus} \quad \xrightarrow{+ \; n(I)}$$

136

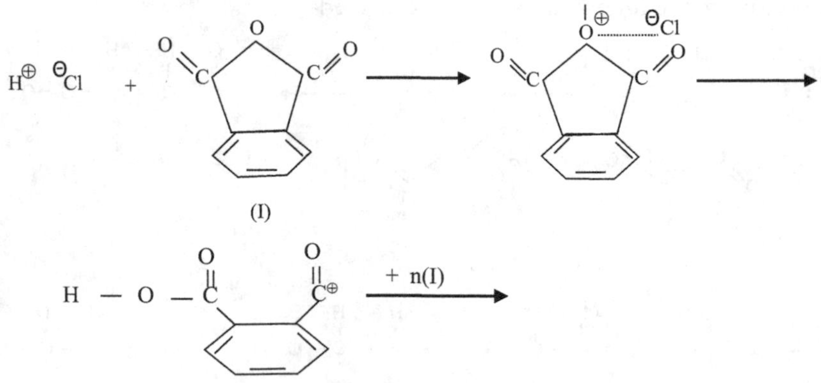

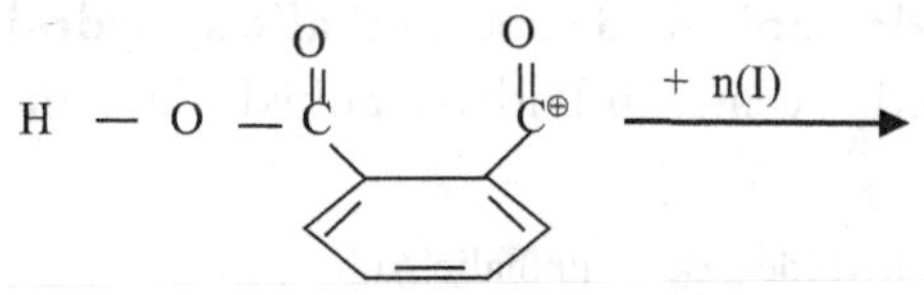

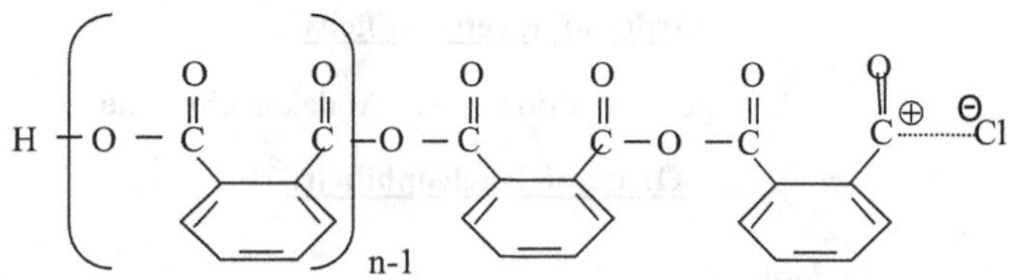

(b)

for which the same monomer unit can be observed to be obtained, noting the anionic end of the growing chain wherein it is impossible to release CO_2, and the fact that positively, particularly when H•e is involved the mechanism of addition is electro-free-radically via Equilibrium mechanism, i.e., backward addition and not as shown above.
(Laws of Creations for Phthalic Anhydrides)

Rule 1065: This rule of Chemistry for **Maleic and Phthalic Anhydrides,** states that, while Phthalic anhydride cannot be resonance stabilized, maleic anhydride like p- benzoquinone is resonance stabilized as shown below-

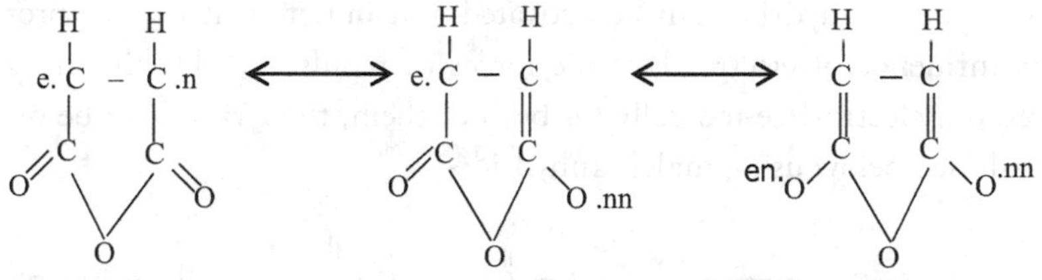

for which when it does, the ring cannot be opened and when used as a monomer under such conditions, only the nucleo-non-free-radical and electro-non-free-radical routes are favored.
(Laws of Creations for Resonance Stabilization in Cyclic Anhydrides)

Rule 1066: This rule of Chemistry for **Maleic Anhydride and Phthalic Anhydride,** states that, in view of the number and types of centers carried by them, when compared with Lactone, the following is valid for them-

Anhydrides   >   Lactones

Order of Electrophilicity and Nucleophilicity

for which when they are copolymerized anionically with lactones (both Electrophiles), mostly the anhydrides will be favored along the chain, being more electrophilic.
(Laws of Creations for Cyclic Anhydrides)

Rule 1067: This rule of Chemistry for **Maleic anhydride and Phthalic anhydride,** states that, in view of the more nucleophilic and less electrophilic center in Phthalic anhydride than in Maleic anhydride, the following is valid for them-

Maleic Anhydride   >   Phthalic Anhydride

Order of Electrophilicity

Phthalic Anhydride   >   Maleic Anhydride

Order of Nucleophilicity

(Laws of Creations for Cyclic Anhydrides)

Rule 1068: This rule of Chemistry for **Maleic Anhydride and Glycolide,** states that, in view of the number and types of centers carried by them, the following is valid for them-

Glycolide   >   Maleic Anhydride

Order of Electrophilicity and Nucleophilicity

(Laws of Creations for Maleic Anhydride and Glycolide)

Rule 1069: This rule of Chemistry for **Free-radical polymerization of Anhydrides,** states that, while nucleo-free-radically Maleic anhydride can be activated and initiated, it cannot propagate due steric limitations and less influence of electro-dynamic forces of repulsion, phthalic anhydride cannot be activated and reacted on; electro-free-radically for both of them, their rings can be opened, only if the initiator is strong as shown below using maleic anhydride-

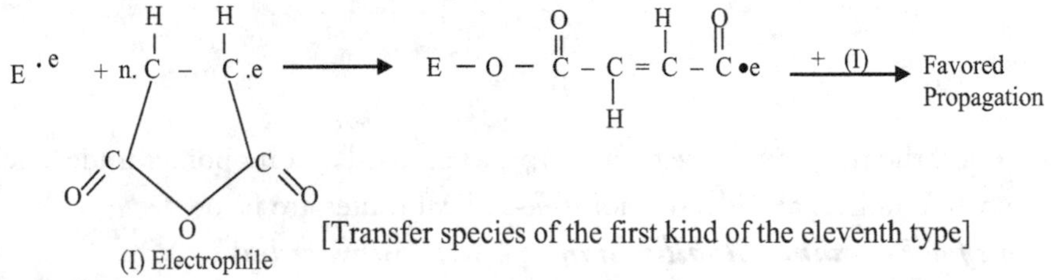

[Transfer species of the first kind of the eleventh type]   (a)

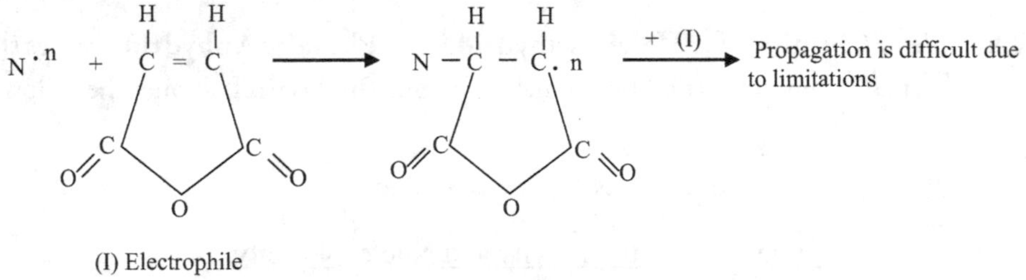

(b)

for which the ring can be opened in one way, via the O-center which when the C = C center is strongly activated, the electro-free-radical carrier grabs the O group as transfer species of ***the first kind of the eleventh type*** and thereby opens the ring [See Rule 1060].

(Laws of Creations for Maleic and Phthalic Anhydrides)

Rule 1070: This rule of Chemistry for **Maleic anhydride and Diethyl fumarate,** states that, while Maleic anhydride and Diethyl fumarate do not homopolymerize with the C = C center nucleo-free-radically, they readily form **very alternating copolymers** with a monomer such as **Vinyl ethers** as shown below-

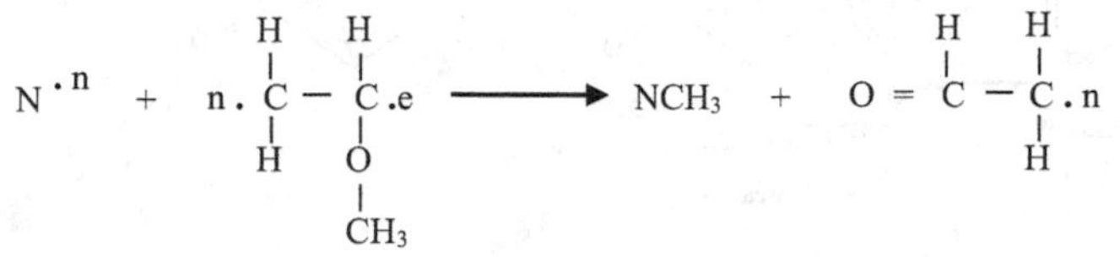

(a)

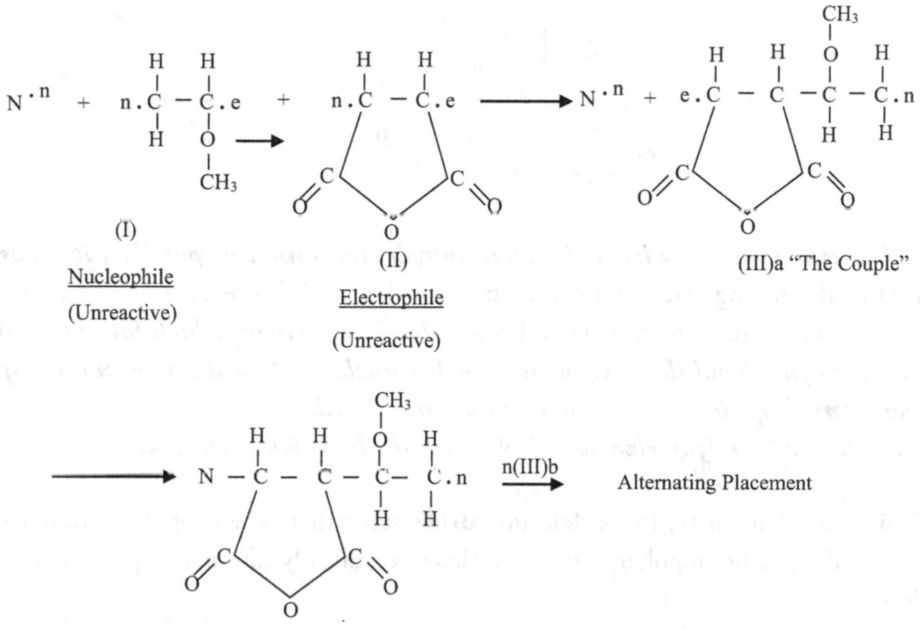

(b)

noting that before the nucleo-free-radical attacks (II) which favors only the initiation step, (I) has already diffused to (II) to produce (III)a [The Couple]\, that in which the female is the one diffusing using its electro-free-radical end to the male, after which the female initiator attacks to produce (III)b to complete the Initiation step and this is repeated for every couple added during the Propagation step; on the other hand, if the ring is opened when the couple is being formed, same as above will apply.

(Laws of Creations for Copolymerization of Maleic Anhydride with Vinyl Ether)

Rule 1071: This rule of Chemistry for **Maleic Anhydride,** states that, when copolymer-ized with **Styrene,** (which is resonance stabilized when activated and can be homopoly-merized nucleo-free-radically under higher operating conditions since the route is not natural to it) **less alternating placement** is favored as shown below-

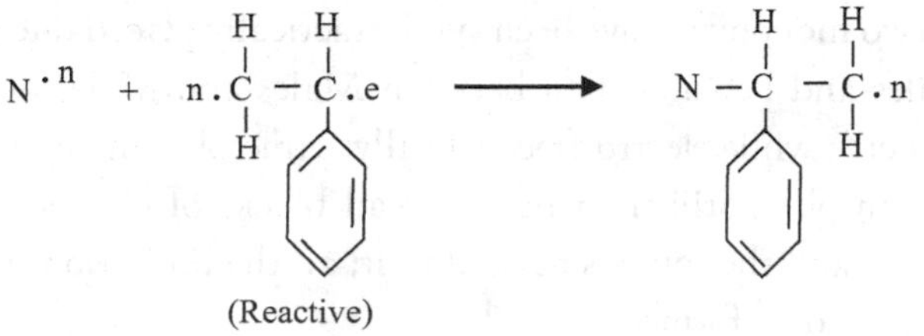

(a)

139

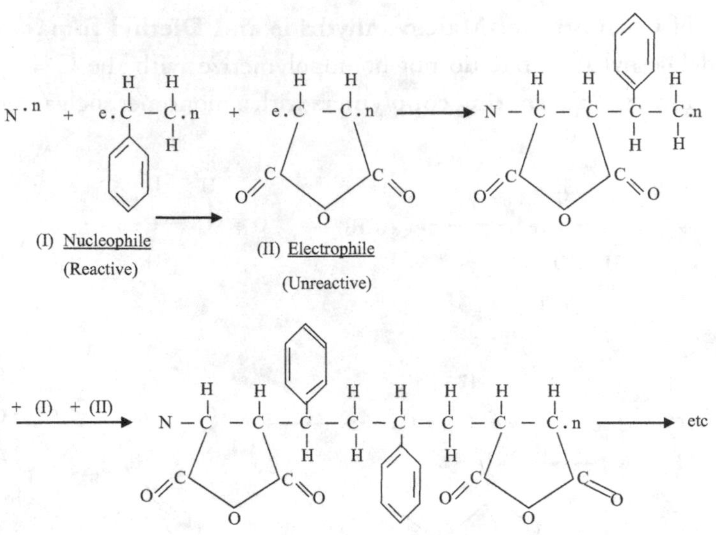

(b)

noting that while two adjacently located maleic anhydrides cannot be possible, it is possible with styrene, hence full alternating placement cannot be formed, yet it is however strong because the nucleo-free-radical route is not natural to styrene and *thus while vinyl ethers which favor full alternating placement are strong nucleophiles, styrene is a weaker nucleophile since it has no transfer species, but yet a stronger nucleophile being fully resonance stabilized.*
(Laws of Creations for Copolymerization of Maleic Anhydride with Styrene)

Rule 1072: This rule of Chemistry for **Maleic anhydride,** states that, when copolymerized with **Stilbene** neither of which undergoes homopolymerization nucleo-free-radically, alternating placement are obtained as shown below-

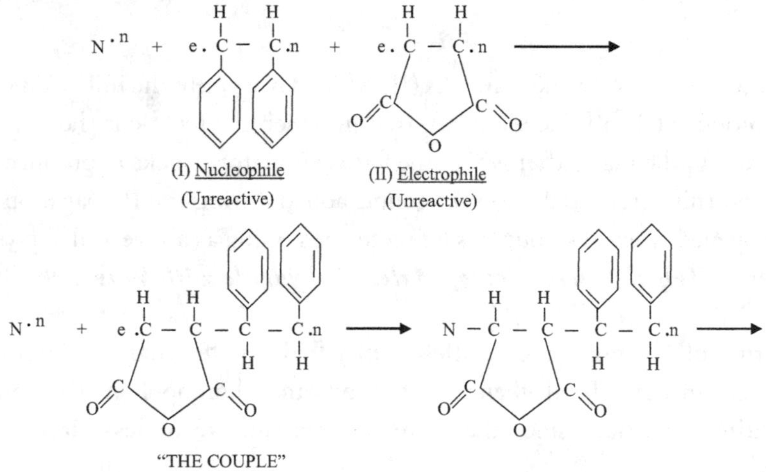

noting that in the couple formed, the two monomers have been syndiotactically placed due to the driving forces favoring addition between Males and Females than between Males and Males or Females and Females (for Non-halogenated monomers); while electro-free-radically, same placement as above will be obtained, but not when styrene is put in place Stilbene where instead blocks of styrene will largely be favored, since it is not sterically hindered and the route is natural to it; all the limitations above reduced if the ring is opened when the couple was to be formed.

(Laws of Creations for Copolymerization of Maleic Anhydride with Stilbene)

Rule 1073: This rule of Chemistry for **Cyclo-alkanols of the types shown below,** states that, while the three- to five-membered rings molecularly rearrange to give more nucleophilic cyclic ketones which are unstable, since they readily open instantaneously to give ketenes, the six- membered cyclic ketone rings rearrange to give the cyclo-alkanols as also shown below-

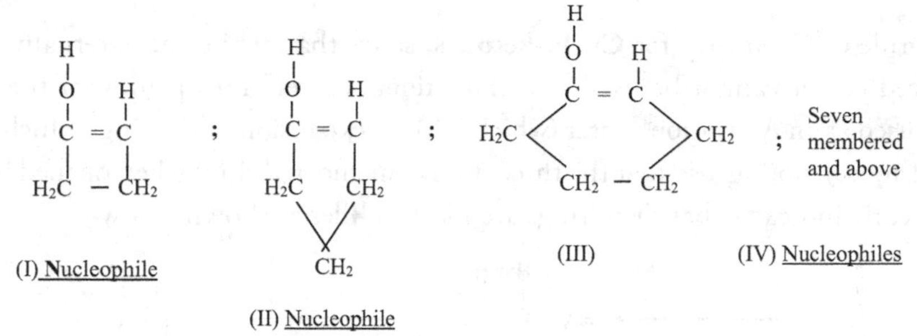

(I) Nucleophile ; (II) Nucleophile ; (III) ; (IV) Nucleophiles ; Seven membered and above

(a)

(I) ; (II) Activated Equilibrium state of existence ; (II)

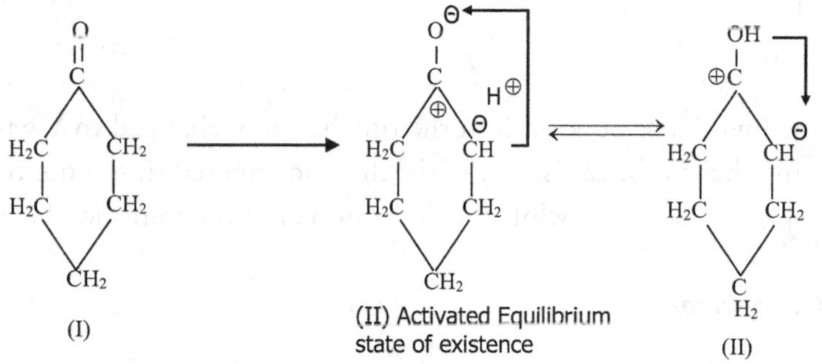

(b)

while the seventh and higher membered cyclic alkynols can also rearrange to give cyclic ketones which are very stable and cannot decompose to give ketenes; making the six-member ring being the point of transition for the cyclic alkynols; and from these observations, the followings can be seen to be valid-

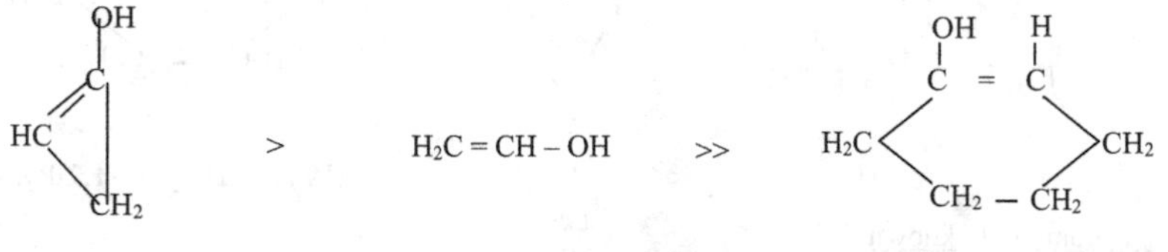

$H_2C = CH - OH$

Radical-pushing capacity of OH group

(c)

for which there is no doubt that cyclo-ketones from six-membered rings upwards favor more being in ACTIVATED/EQUILIBRIUM STATE of existence than any other state, for if not, the reaction shown below would have been impossible-

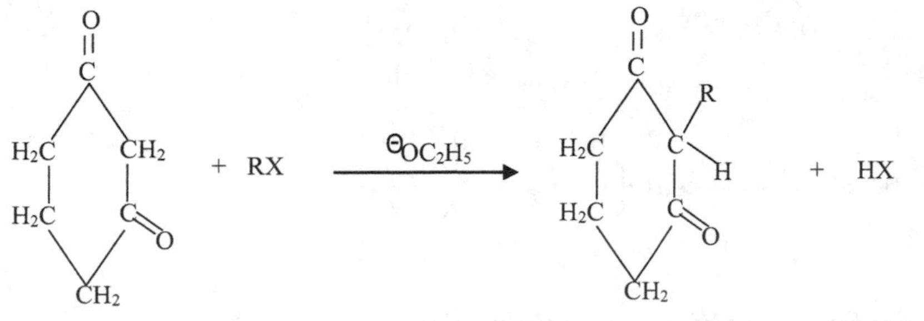

1,3-cyclohexanedione + RX $\xrightarrow{\ominus OC_2H_5}$ + HX

(d)

141

unless if NaCl and H_5C_2OH as opposed to HCl above are the side products if $NaOC_2H_5$ and RCl were used; noting that the capacity of OH group can be seen to be equivalent to that on the six-membered ring without OH group.

(*Laws of Creation for Cycloalkanols*)

Rule 1074: This rule of Chemistry for **Cyclo-ketones,** states that, while the externally located C = O cumulatively placed center cannot be used as a "Functional center" for opening of the ring for many reasons, it can be used as an Activation center (which it is) for expansion of the rings which are stable with the group (six and above); noting also that the three- to five-membered rings when opened instantaneously to give ketenes, clearly indicates that these rings are Electrophiles as shown below-

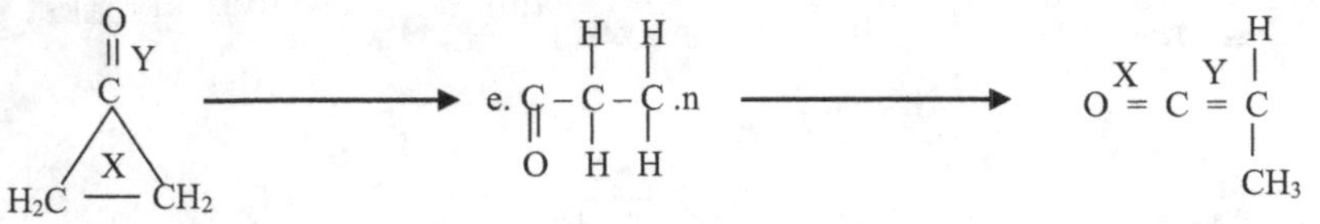

An Electrophile   Methyl ketene

where the X in the ring, the invisible π-bond (SE) in the ring has now changed to Y while the Y on the ring has changed to X; noting that the linear isomer of the three-membered ring is not the methyl ketene above, but $H_2C=CH(HC=O)$, i.e., acrolein, while the linear isomer of the four membered ring is methyl acrolein.

(*Laws of Creations for Cyclo-ketones*)

Rule 1075: This rule of Chemistry for **Cyclo O/C = C containing rings shown below,** states that, though (I) may not exist, it is the only one which has no point scission, and

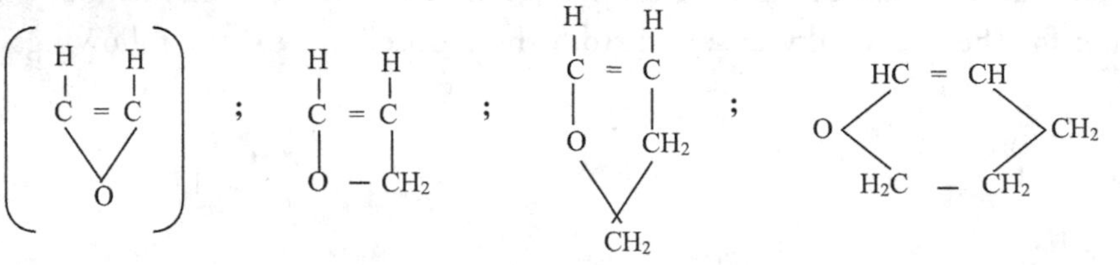

(I)   (II)   (III)   (IV) 3,4-Dihydro-1,2-Pyran

Not commonly known

the only one which can be polymerized via the double bond, only free radically both ways as shown below-

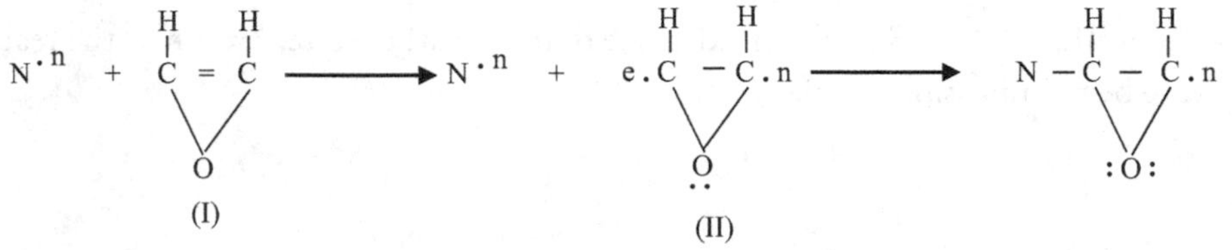

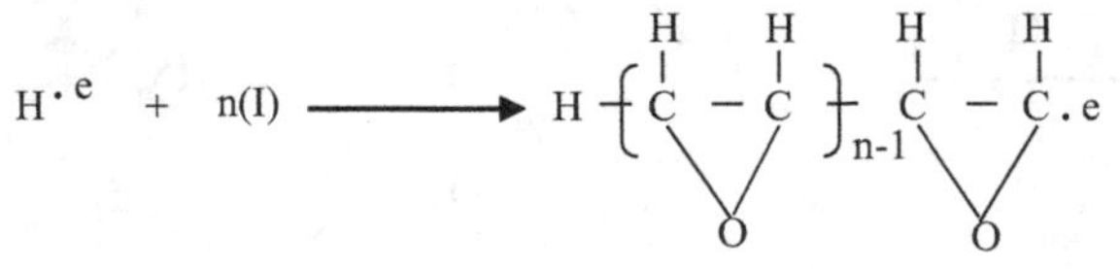

(I)   (II)

(*Laws of Creations for Cyclo O/C = C containing rings*)

Rule 1076: This rule of Chemistry for **Cyclic O/C = C containing rings shown below**, states that, of this family members, it is only (II) that may favor transient existence to

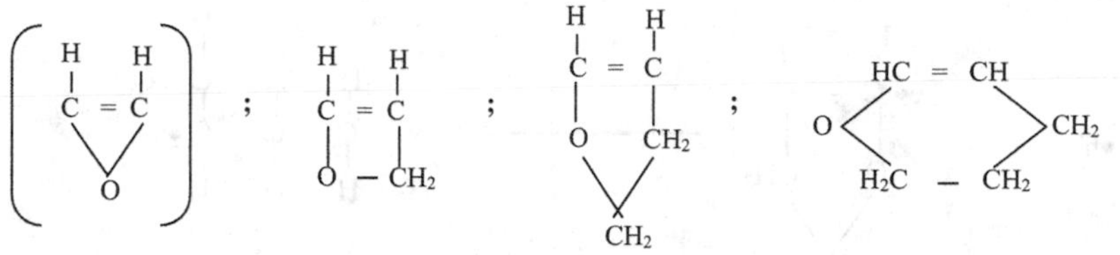

$$
\begin{array}{cccc}
\left[\begin{array}{c} \overset{H}{|} \;\; \overset{H}{|} \\ C = C \\ \diagdown\;\diagup \\ O \end{array}\right] &
\begin{array}{c} \overset{H}{|} \;\; \overset{H}{|} \\ C = C \\ | \quad\quad | \\ O - CH_2 \end{array} &
\begin{array}{c} \overset{H}{|} \;\; \overset{H}{|} \\ C = C \\ | \quad\quad | \\ O \quad\quad CH_2 \\ \diagdown\;\diagup \\ CH_2 \end{array} &
\begin{array}{c} HC = CH \\ O\diagup \quad\quad \diagdown CH_2 \\ H_2C - CH_2 \end{array} \\
(I) & (II) & (III) & (IV)\ 3,4\text{-Dihydro-1,2-Pyran} \\
\underline{\text{Not commonly known}} & & &
\end{array}
$$

give an Electrophile (Acrolein) when the ring is instantaneously opened at temperature above the Ceiling temperature as shown below-

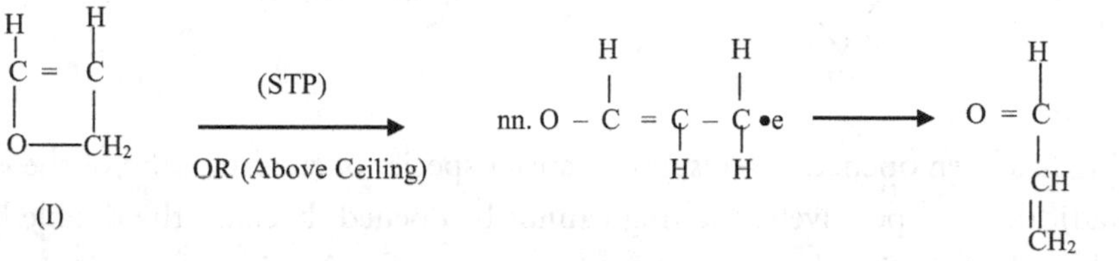

$$
\begin{array}{c} \overset{H}{|} \;\; \overset{H}{|} \\ C = C \\ | \quad\quad | \\ O - CH_2 \end{array}
\xrightarrow[\text{OR (Above Ceiling)}]{\text{(STP)}}
\quad nn.\ O - \overset{H}{\underset{H}{C}} = \overset{}{\underset{H}{C}} - \overset{}{\underset{H}{C}}\bullet e
\longrightarrow
\begin{array}{c} O = \overset{H}{C} \\ | \\ CH \\ \| \\ CH_2 \end{array}
$$

(I)   (I)a Acrolein

while when the same is done to (III), no rearrangement either via Molecular rearrangement of the third kind or resonance stabilization is possible as shown below, clear indication that this five-membered and other larger-membered rings cannot be opened instantaneously, unless a smaller membered ring is formed-

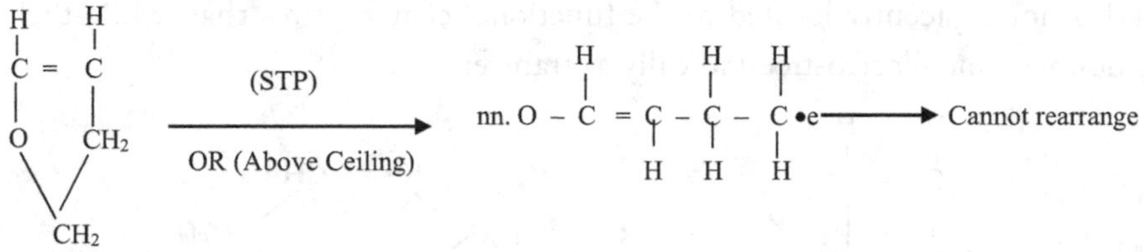

$$
\begin{array}{c} \overset{H}{|} \;\; \overset{H}{|} \\ C = C \\ | \quad\quad | \\ O \quad\quad CH_2 \\ \diagdown\;\diagup \\ CH_2 \end{array}
\xrightarrow[\text{OR (Above Ceiling)}]{\text{(STP)}}
\quad nn.\ O - \overset{H}{C} = \overset{H}{\underset{H}{C}} - \overset{H}{\underset{H}{C}} - \overset{H}{\underset{H}{C}}\bullet e \longrightarrow \text{Cannot rearrange}
$$

(II)

(Laws of Creations for Cyclo O-/C = C containing rings)

Rule 1077: This rule of Chemistry for **Transfer Species,** states that, Transfer species of the ***First kind of the eleventh type,*** is the type which exists in ringed monomers such as *3,4-dihydro-1,2-pyran* nucleo-free-radically and *Maleic anhydride* electro-free-radically, via activation of the C = C double bond, and leads to opening of the rings. [See Rules 1045,1046 and 1069]
(Laws of Creations for Transfer Species of the First hind of the Eleventh type)

Rule 1078: This rule of Chemistry for **Cyclic O/C = C containing rings shown below,** states that, of this family members, only (III) and (IV) can be opened by scission at

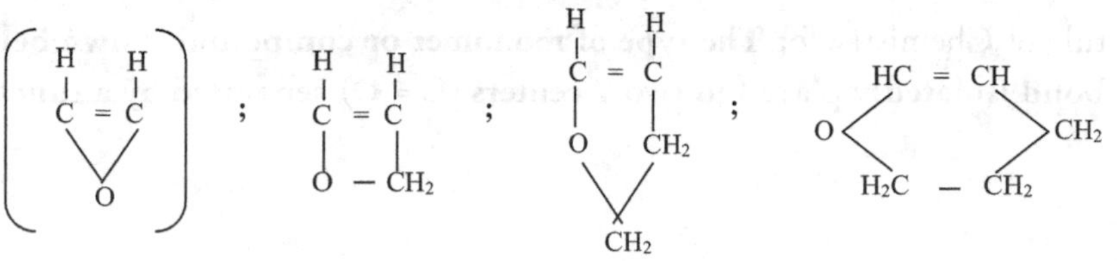

$$
\begin{array}{cccc}
\left[\begin{array}{c} \overset{H}{|} \;\; \overset{H}{|} \\ C = C \\ \diagdown\;\diagup \\ O \end{array}\right] &
\begin{array}{c} \overset{H}{|} \;\; \overset{H}{|} \\ C = C \\ | \quad\quad | \\ O - CH_2 \end{array} &
\begin{array}{c} \overset{H}{|} \;\; \overset{H}{|} \\ C = C \\ | \quad\quad | \\ O \quad\quad CH_2 \\ \diagdown\;\diagup \\ CH_2 \end{array} &
\begin{array}{c} HC = CH \\ O\diagup \quad\quad \diagdown CH_2 \\ H_2C - CH_2 \end{array} \\
(I) & (II) & (III) & (IV)\ 3,4\text{-Dihydro-1,2-Pyran} \\
\underline{\text{Not commonly known}} & & &
\end{array}
$$

O - CH_2 bond via the C = C double bond by activation, nucleo-free- and nucleo-non-free-radically, as shown below using the five-membered ring-

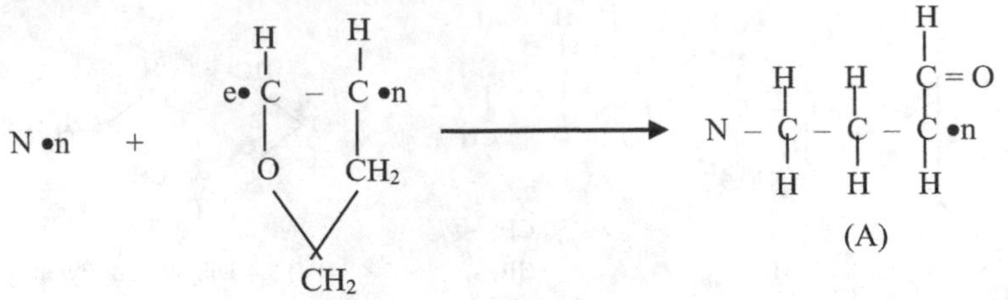

(a)

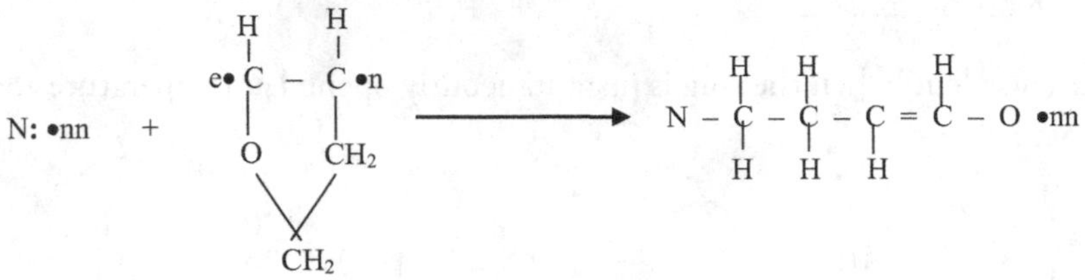

(b)

noting that the ring has been opened as a result of transfer species of the first kind of the ***eleventh type,*** and electro-free-radically and positively, the ring cannot be opened, because the double bond which is less nucleophilic than the functional center cannot favor its use for opening of the ring and if used after Initiation, propagation cannot take place due to steric limitations.
(Laws of Creations for Cyclo-O-/C = C containing rings)

Rule 1079: This rule of Chemistry for **The type of monomers or compounds shown below,** wherein the double bond is not adjacently located to the functional center, states that, while their rings can be opened via the double bond electro-free-radically as transfer

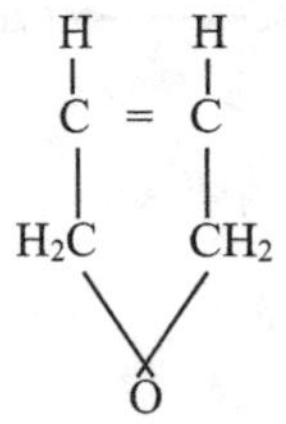

(I) 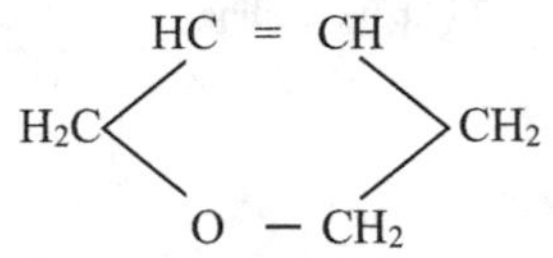 ; **(I I)**

species of the first kind of the eleventh type, and can be opened instantaneously and undergo molecular rearrangement of the third kind to give vinyl 1,2-epoxide for (I) and vinyl 1,3-epoxide for (II) free-radically via Decomposition mechanism, the functional center cannot be used, being more nucleophilic than the double bond in the ring.
(Laws of Creations for Cyclic monomers carrying a double bond and a functional center separated by CH₂ group)

Rule 1080: This rule of Chemistry for **The type of monomer or compound shown below, wherein there is a double bond isolatedly placed to two Y centers (C = O) separated by a functional center**

shared by them, states that, while Maleic anhydride a strong electrophile can be both nucleophilic and electrophilic in character, this does not apply to this type as shown below-

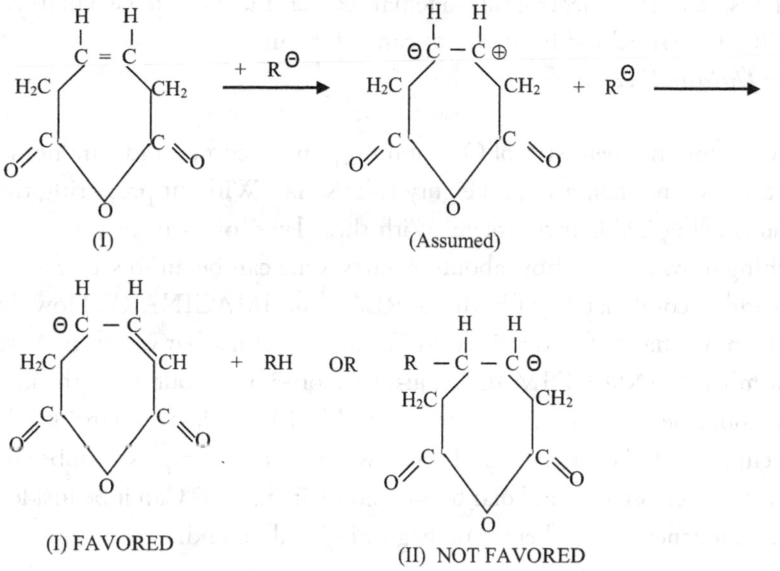

(I)   (Assumed)

(I) FAVORED   (II) NOT FAVORED   (a)

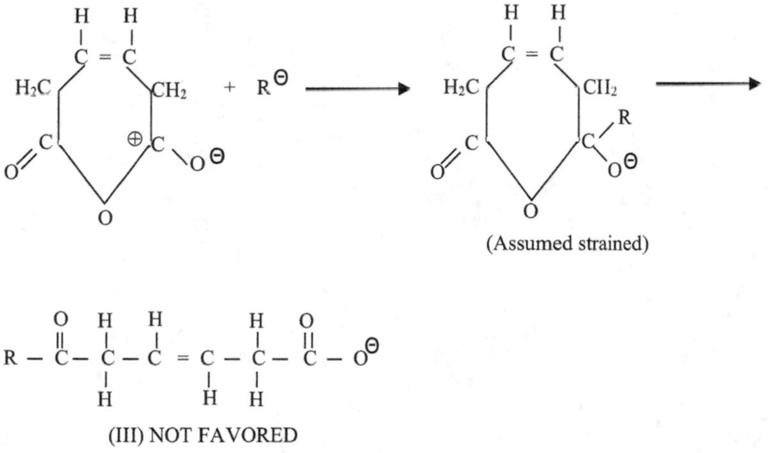

(Assumed strained)

$$R - \overset{O}{\underset{}{C}} - \overset{H}{\underset{H}{C}} - \overset{H}{\underset{}{C}} = C - \overset{H}{\underset{H}{C}} - \overset{O}{\underset{}{C}} - O^{\ominus}$$

(III) NOT FAVORED   (b)

(Assumed strained)

$$H - O - \overset{O}{\underset{}{C}} - \overset{H}{\underset{H}{C}} - \overset{H}{\underset{}{C}} = C - \overset{H}{\underset{H}{C}} - \overset{O}{\underset{}{C}} \oplus$$

(IV) NOT FAVORED

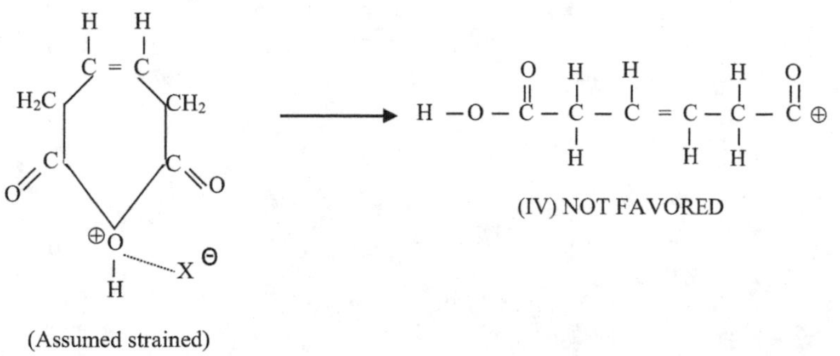

(c)

for which it can be observed that, the compound cannot be used as a monomer via all the centers, because the following is valid-

$$C = O \quad > \quad O \quad > \quad C = C$$

Order of Nucleophilicity

145

noting however, that its use as a monomer is only possible if the ring can be opened instantaneously to give a carboxylate anionic center and so-called carbonium center, thus making it display its Electrophilic character which indeed it is, "a unique Electrophile- a female center "married" to two male centers", when there is a female center (C=C) left behind unused, but can still be used!
(Laws of Creations for Unique Electrophile)

Ninety eight rules covering the chemistry of O containing ringed compounds /monomers have been proposed bringing the total to one thousand and eighty rules so far. Without proposing these rules, the job or task or foundations being laid is incomplete. With these laws, one can begin to clearly see how NATURE little or nothing of which we know about operates. One can begin to see very clearly that we live in a COMPLEX world, a combination of both the REAL and IMAGINARY. How do we identify this imaginary parts which one has so far identified in Chemistry with what we see in Mathematics (as the root of a negative number 1 ((INSERT IMAGE)), just as -2 or -3 or -4, but not 0 the most important of all numbers)? One is unique because all numbers are multiples of 1. For it is only inside CHEMISTRY, we see all disciplines including Mathematics and Physics. Where is this complex number in Chemistry? Where can it be found with Electrostatic and Polar bonds and their charges? Can it be inside the Nucleus? We have just only begun a journey where there is no beginning and no end.

References

1. G. Odian, "Principles of Polymer Systems", McGraw-Hill Book Company, (1970), pgs. 453 - 475.
2. C. R. Noller, "Textbook of Organic Chemistry", W. B. Saunders Company, (1966), pgs. 601 - 609.
3. J.Brandrup and E. H. Imnergut (eds.),"Polymer Handbook", pp. II - 363, Interscience Publishers, John Wiley & Sons, Inc. New York, (1966).
4. R. M. Joshi and B. J. Zwolinski, Heat of Polymerization and Their Structural and Mechanistic Implications, in G. E. Hams (eds.),"Vol. 1, part I, chap. 8, Marcel Dekker, Inc., New York, 1967.
5. P. H. Plesch and P. H. Westermann, Polymer, 10 : 105 (1969).
6. T. Miki, T. Higashimura, and S. Okamara, J. Polymer Sci., B5 : 583 (1967).
7. R. B. Seymour, "Introduction to Polymer Chemistry; McGraw-Hill Book Company, (1971), pgs. 121 - 123.
8. A. E. Gurgiolo, J. Macromol, Sci.Revs. Macromol Chem., 1(1) : 39 (1966).
9. E. C. Steiner, R. R. Pelletier, and R. O. Trucks, J. Am. Chem. Soc., 86; 4678 (1964).
10. J. Funikawa and T. Saegusa, "Polymerization of Aldehydes and Oxides", Chaps. III - VII, Interscience Publishers, John Wiley & Sons, Inc, New York, 1963.
11. G. Gee, W. C. E. Higginson, and G. T. Mewall, J. Chem. Soc., 1959; 1345.
12. R. A. Patsiga, J. Macromol. Sci. Revs. Macromol. Chem., C1(2) : 223 (1967).
13. R. C. Burrows, Polymer Preprints, 6(2) : 600 (1965).
14. H. Maerwein, D. Delf, and H. Morshel, Angew. Chem., 72 : 927 (1960).
15. D. Sims, MaKromol, Chem, 98 : 135, 245 (1966).
16. J. Kuntz, J. Polymers Sci., A - 1(5) : 193 (1967); Polymer Preprints, 9(1) : 508 (1968).
17. G. Odian, "Principles of Polymer Systems", McGraw-Hill Book Company, (1970), pgs. 507 - 513.
18. J. Schaefer, R. J. Kern, and R. J. Katrik, Macromolecules, 1 : 107 (1968).
19. G. Odian, "Principles of Polymers System", McGraw-Hill Book Company, (1970), pg. 443.
20. Y. P. Castille and V. Stannett, J. Polymer Sci., A - 1(4) : 2063 (1966).
21. Y. Yamashita, M. Okada, and K. Suyama, MaKromol. Chem., III : 277 (1968).
22. Murray Goodman, Akihiro Abe, "Couples vinyl and acetal ring-opening polymerization", Journal of Polymer Science Part A: General Papers, Volume 2, Issue 8, pages 3471-3490, (1964).
23. C. R. Noller, "Textbook of Organic Chemistry", W. B. Saunders Company, (1966), pg. 213.
24. M. B. Price and F. B. McAndrew, J. Macromol. Sci. (Chem.) A1(2) : 231 (1967).
25. V. H. Cherdron and H. Ohse, MaKromol. Chem., 92 : 213 (1966).
26. Y. Yamashita, T. Tsuda, M. Okada, and S. Iwatsuki, J. Polymer Sci. A - 1(4) : 2121 (1966).
27. K. Tada, Y. Numata, T. Saegusa, and J. Furukawa, MaKromol. Chem., 77 : 220 (1964).
28. M. G. Baldwin, J. Polymer Sci., A3 : 703 (1965).
29. S. N. E. Omorodion, A. E. Hamielec, Chromatographia Vol. 31, No 5/6 (1991).

Problems

12.1. Distinguish between the followings types of resonance stabilization phenomena where they exist. If it does not exist, give reason(s) why it does not exist.

    (a) Charged resonance stabilization.

    (b) Free-radical resonance stabilization.

    (c) Polar/Radical resonance stabilizations.

    (d) Electrostatic/polar resonance stabilizations.

12.2. Shown below are five membered cyclic ethers, cyclic acetal (formal), cyclic ester (lactone) and anhydride.

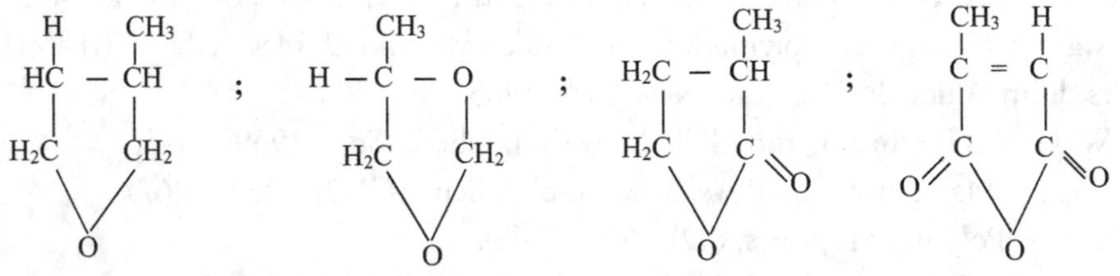

    (a) Show the order of their nucleophilicities, using side-by-side placement methods, where possible.

    (b) Show the monomers units during their positively charged polymeriza-tions.

    (c) Show the monomer units during anionic polymerization for those that favor the anionic route.

    (d) Which of the monomers will favor radical polymerizations? Identify the routes and the transfer species where they exit.

12.3. (a) Can (IV) of Q. 12.2 be copolymerized with stilbene? If it can, what types of copolymers are produced? Use all the routes where possible

    (b) Do the same as in (a) for (III) of Q.12.2.

    (c) What are the functions of the C = C double bond in maleic anhydride?

12.4. (a) Why is the form of resonance stabilization shown below for maleic anhydride favored? Compare with that of p-Benzoquinone.

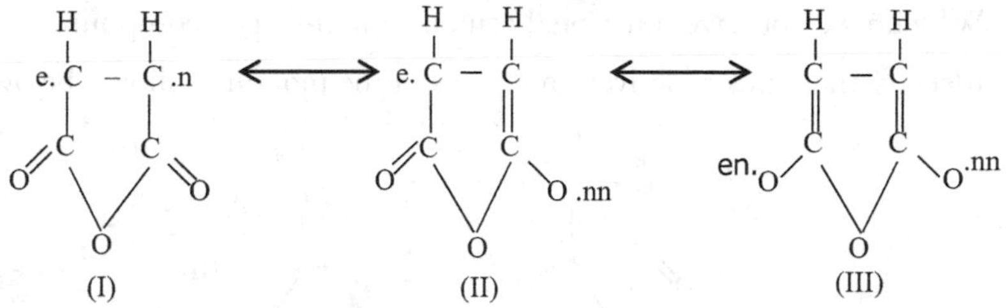

(I)   (II)   (III)

(b) What are the disadvantages of using hydroxides (OH) and protonic acids (H) as free ionic initiators in polymerization of cyclic oxygen containing monomers? Does it also apply to alkenes? What are the limitations imposed in using them as initiators? How are the limitations advanta-geously used for Step polymerization?

12.5. (a) Why can protonic acids not be used for homopolymerizations of tetrahy-drofuran, but can be used for homopolymerizations of 3- and 4-membered cyclic ethers as well as some larger-membered ones?

(b) Distinguish between propagation/depropagation and propagation/depro-pagation equilibria phenomena.

12.6. (a) Can PCl_3 be used to obtain a cationic ion-paired initiator? What type of initiator can be obtained from it?

(b) Using dimethylene oxide and tetrahydrofuran as monomers, show the initiation steps and subsequent addition of the monomers along the chain using (i) PCl_3 and (ii) PCl_3/$(C_2H_2)_2O$ as initiators.

12.7. Identify and distinguish between all the different types of anionic and "cationic" initiators and their sources, used for polymerization of cyclic oxygen containing monomers.

12.8. (a) Can cyclic ethers, esters and carbonyl monomers and maleic anhydride be copolymerized with alkene positively? Explain.

(b) Cyclic esters do not favor strong existence of random copolymers with three or four membered cyclic ethers positively, but maleic anhydride does. Explain why this is so?

(c) Only three membered and some four-membered cyclic ethers can be anionically copolymerized with lactones amongst members of cyclic ethers and none with cyclic acetals. Explain why this is so? What type of copolymers are produced?

(d) Can maleic anhydride be copolymerized with three membered cyclic ethers e.g. propylene oxide anionically? Show other ways by which they can copolymerized.

12.9. (a) What is a promoter?

(b) How can promoters be used for copolymerization reactions?

(c) Why do we not have functional centers in non-ringed compounds?

(d) Identify the types of activation centers in the monomer shown below.

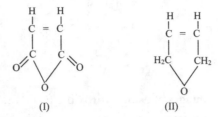

(I)   (II)

12.10. (a) Why is (II) of Q12.9 (d) not popularly known to exist even through hydrogenation of furan or dehydrogenation of tetrahydrofuran or the type of alkene shown below?

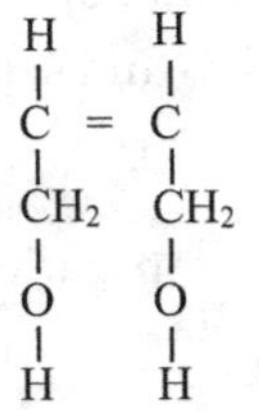

$$
\begin{array}{ccc}
H & & H \\
| & & | \\
C & = & C \\
| & & | \\
CH_2 & & CH_2 \\
| & & | \\
O & & O \\
| & & | \\
H & & H
\end{array}
$$

(b) Based on the activation of the activation centers in furan, how can the existence of (II) be guaranteed, if suitable conditions are used.

(c) Two moles of aniline condense with one of furfural with opening of the ring in the presence of acetic acid as shown below.

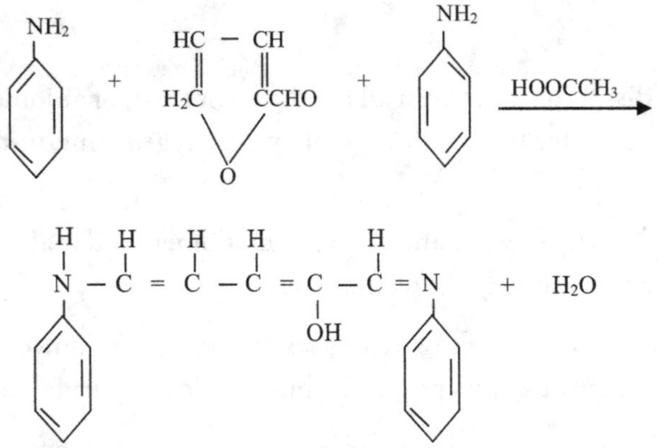

Explain the mechanism of the reactions. (see Chapter 14)

12.11.　(a)　Shown below is a reaction between maleic anhydride and hexachloro-cyclopentadiene.

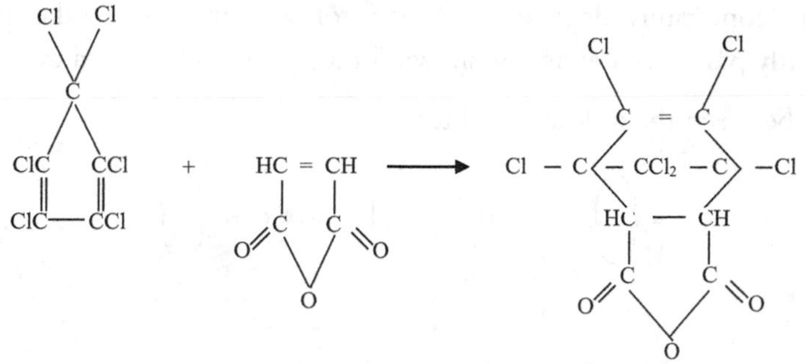

Explain the mechanism of the reaction.

(b)　Also shown below is polymerization of E - Caprolactone using a diol.

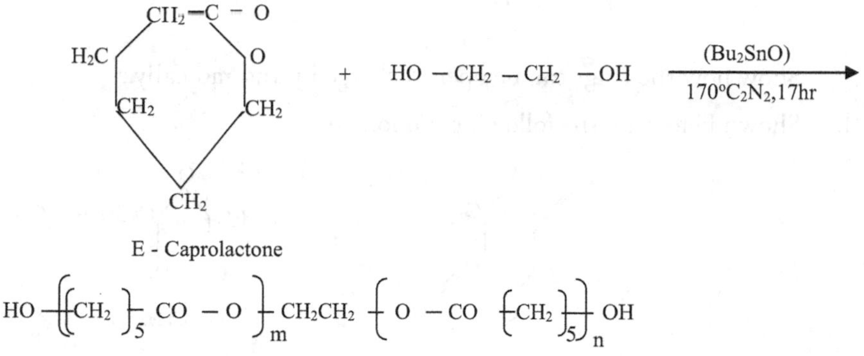

E - Caprolactone

$n + m = 10$

Explain the mechanism of the reactions. Is the polymer above stable?

(c)　Dihydropyran is made by the catalytic dehydration with ring enlargement of tetrahydrofurfuryl alcohol. Catalytic hydrogenation of dehydrogenation gives tetrahydropyran.

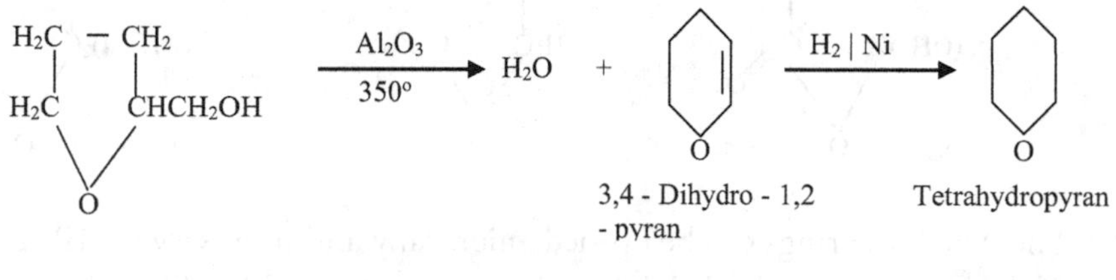

3,4 - Dihydro - 1,2
- pyran　　　　Tetrahydropyran

Explain the mechanism of the reactions.

12.12.　(a)　Why do acetaldehydes readily favor trimerization reaction more than formaldehyde does?

(b)　What are driving forces favoring the existence of propagation/ depropagation phenomenon in Cyclic oxides?

(c)　Distinguish between 1,3- dioxane, 1,3,5 - trioxane and 1,3,5,7 - tetraoxane as far as their abilities to homo-and co-polymerize are concerned.

12.13. (a) What is side-by-side placement of rings?

(b) Under what conditions does displacement of a component take place from an electrostatically paired center as was shown for a growing chain of a cyclic oxide?

(c) Distinguish between the following centers –

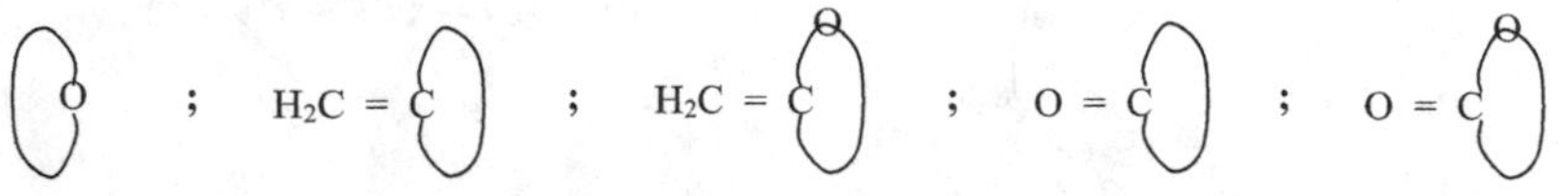

12.14. (a) Shown below are the following monomers

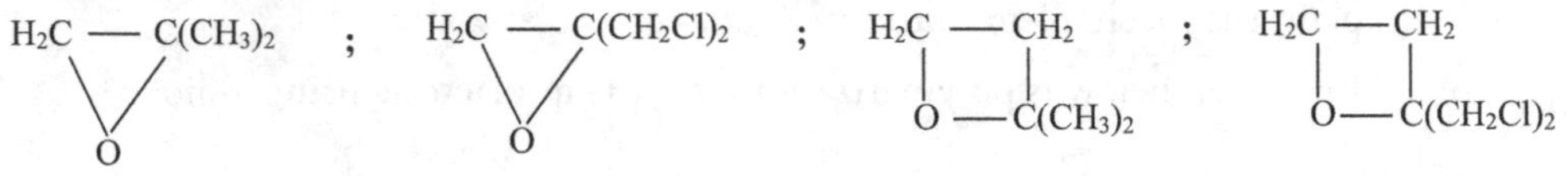

Show how the rings can be opened chargedly and radically.

(b) Shown below are the following monomers –

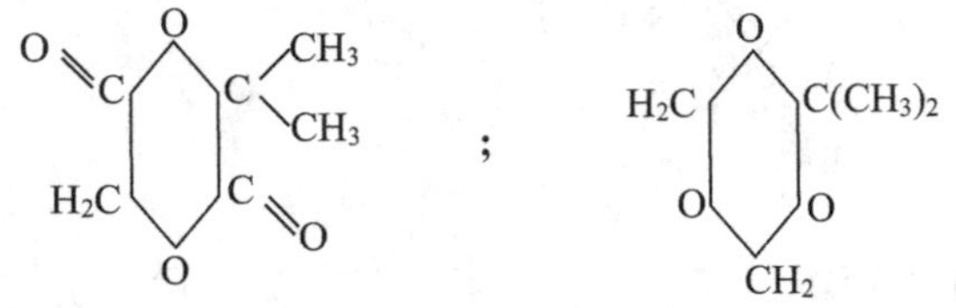

Show how the rings can be opened anionically and positively. Identify which is male and female.

12.15. (a) Shown below are the following monomers -

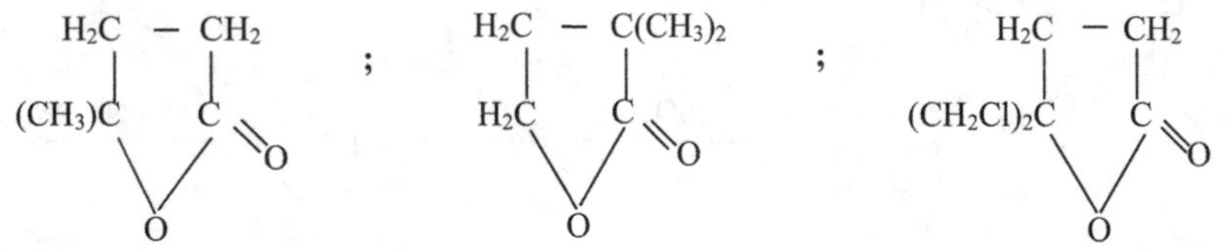

Show how the rings can be opened anionically and positively if MRSE can be provided. If they cannot be provided for some of them, explain why? What happens when the CH_3 groups are replaced with CF_3 groups.

(b) Why do these monomers not favor Terminal radical resonance stabilization phenomenon?

(c) Replace = O groups above with $H_2C=$ and show what happens to the monomers.

12.16. (a) What is equilibrium molecular rearrangement of Cyclic ketones?

(b) Show how it can be used to determine the radical-pushing capacity of OH group and the stability of members of this family.

12.17 (a) Distinguish between the three monomers below -

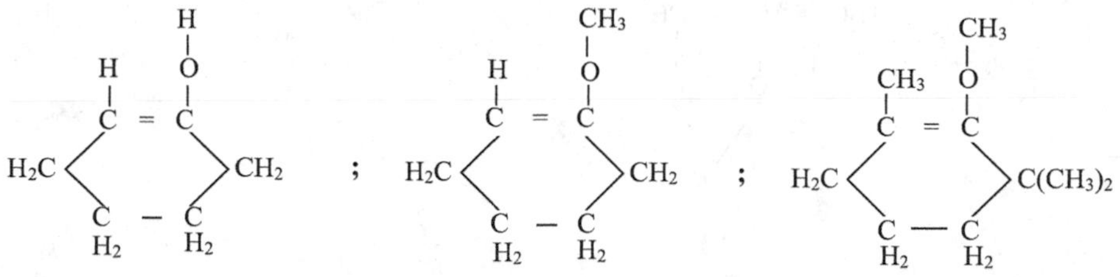

$$H_2C \underset{\underset{H_2}{C} - \underset{H_2}{C}}{\overset{\overset{H}{C} = \overset{\overset{H}{\overset{|}{O}}}{C}}{}} CH_2 \quad ; \quad H_2C \underset{\underset{H_2}{C} - \underset{H_2}{C}}{\overset{\overset{H}{C} = \overset{\overset{CH_3}{\overset{|}{O}}}{C}}{}} CH_2 \quad ; \quad H_2C \underset{\underset{H_2}{C} - \underset{H_2}{C}}{\overset{\overset{CH_3}{C} = \overset{\overset{CH_3}{\overset{|}{O}}}{C}}{}} C(CH_3)_2$$

in terms of their ability to favor equilibrium molecular rearrangements of Cyclic ketones.

(b) Why do some of their four and five-membered rings not favor any stable existence?

(c) Why do their cyclobutanone and cyclopentatone not favor any stable existence?

12.18 (a) Show which of (I) and (II) below is more nucleophilic.

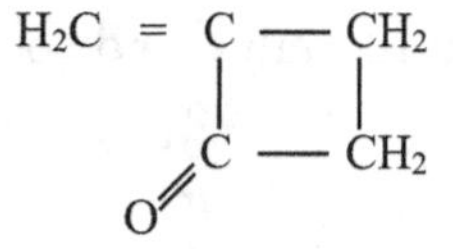

$$H_2C = \underset{\underset{O}{\overset{\|}{C}} - CH_2}{\overset{|}{C}} - CH_2 \qquad ; \qquad H_2C = \underset{\underset{O}{\overset{\|}{C}} \quad \underset{\underset{H_2}{C}}{CH_2}}{\overset{|}{C}} - CH_2$$

(I) 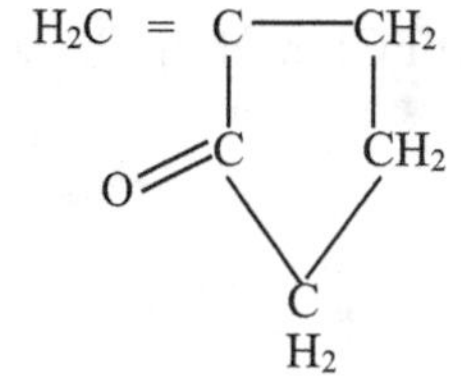   (II)

(b) Distinguish between the two monomers shown below in terms of means

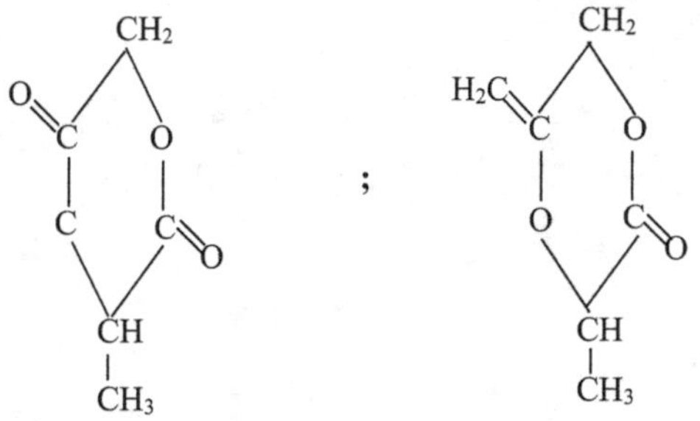

$$\underset{\underset{\underset{CH_3}{|}}{\overset{|}{CH}}}{\overset{CH_2}{O = \overset{}{C} \qquad O}} \underset{\overset{}{C} \overset{}{\underset{\|}{C}} O}{} \qquad ; \qquad \underset{\underset{\underset{CH_3}{|}}{\overset{|}{CH}}}{\overset{CH_2}{H_2C \overset{}{C} \qquad O}} \underset{O \qquad \overset{}{\underset{\|}{C}} O}{}$$

of opening the rings

12.19. Using side-by-side placement of rings, show that

(a) 1,4 - dioxanes are more nucleophilic than 1,3 – dioxanes.

(b) Lactides are more nucleophilic than lactones.

(c) Tetrahydropyran is more nucleophilic than 1,4 - dioxane.

12.20 (a) Shown below is 2-methyl - 4 - methylane - 1, 3 - dioxolane.

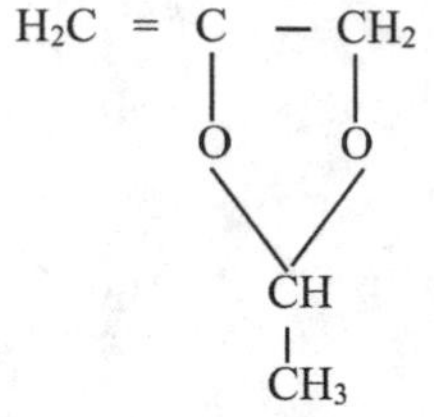

$$H_2C = C - CH_2$$

$$O \qquad O$$

$$CH$$

$$CH_3$$

(i) Why will the monomer not favor the anionic route?

(ii) Show how it can be polymerized to produce high molecular weight polymers.

(iii) How will the monomer be instantaneously opened when MRSE is assumed to exist in the ring?

(b) Can you identify the first driving force favoring the existence of electrostatic bonds?

(c) How many electrostatic bonds exist in the positively charged-paired initiators obtained from (i) PCl_3 and (ii) BF_3 as sources (that is catalyst and cocatalyst)?

CHAPTER THIRTEEN

TRANSFER OF TRANSFER SPECIES IN CYCLIC AMIDES, CYCLIC AMINES AND RELATED MONOMERS

13.0 Introduction

Unlike oxygen containing cyclic monomers, nitrogen containing cyclics are uniquely different, in view of their basic characters, their ionic characters, the trivalent character of the nitrogen center compared to the divalent character of oxygen center, and the presence of a carbonyl center adjacently located to the nitrogen and oxygen center for some members of these monomers. Unlike cycloalkanes and all carbon containing ringed monomers, and like oxygen containing cyclic monomers, these monomers have functional and activation centers $-\overset{\frown}{\underset{\ddot{N}}{\bigcirc}}$ and $\underset{}{\bigcirc}\text{C}=\text{O}$. To open the rings of these monomers, the minimum required strain energy (MRSE) must be present in the ring or be provided to the ring via the functional or activation center where possible.

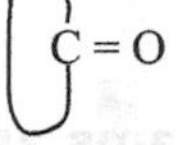

From the considerations of some ring-opening monomers so far, one can imagine the number of impossible reactions in the literatures of these monomers. The same also applies to the cases to be considered. Cyclic amides of interest are those formed by the intra-molecular amidation of amino acids. These are called lactams which begin with four-membered rings. The first members of some existing lactams are shown below.

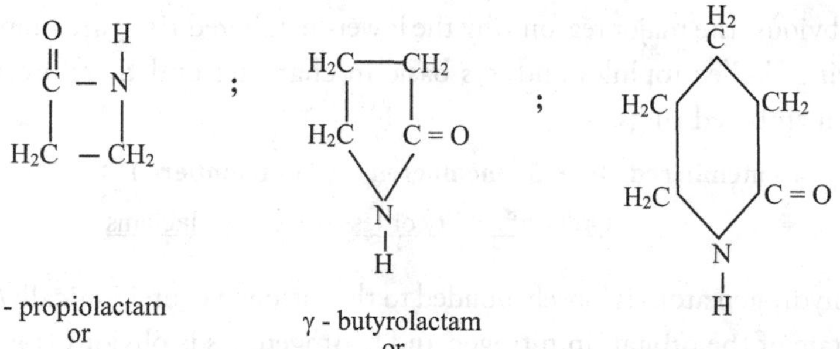

β - propiolactam or 2 - azetidinone 4 - membered ring	γ - butyrolactam or 2 - pyrrolidone 5 - membered ring	δ - valerolactam or 2 - piperidone 6 - membered ring

13.1

$$
\begin{array}{ccc}
\text{ε - caprolactam} & \text{Enantholactam} & \text{capryllactam} \\
\text{or} & \text{or} & \text{or} \\
\text{6 - hexanolactam} & \text{7 - heptanolactam} & \text{8 - octanolactam} \\
\underline{\text{7 - membered ring}} & \underline{\text{8 - membered ring}} & \underline{\text{9 - membered ring}}
\end{array}
$$

$$13.2$$

First it can be observed that existence of 3-membered rings is not favored, due to probable existence of maximum required strain energy. This may clearly indicate *that nitrogen containing ringed compounds are more strained than their corresponding oxygen containing ringed compound.* Secondly, only the existence of Nylon 4, Nylon 5, Nylon 6, Nylon 8, and Nylon 12 have been known to be commonly produced by ring-opening γ -butyrolactam, δ -valerolactam, ε - caprolactam, capryllactam and lauryllactam respectively. These are 5-, 6-, 7-, 9-, and 13-membered rings respectively. There could be more than just listed above, noting most importantly the unfavored use of β- propiolactam or unpopular presence of β- propiolactam as monomers for Nylon 3. In fact, the 5- and 6- membered rings are more difficult to polymerize than the larger membered rings, while the 4- membered ring does not favor any form of polymerization as shown in Table 13.1 below.

Table 13.1 Polymerization Reactivity of Lactams as a Function of Ring Size

	Types of initiation	Order of reactivity of different sized rings	Ref
1	Anionic (with acylating agent)	$7 \geq 5 > 6$	1
2	Protonic acid (HCl) (Cationic)	$8 > 7 > 11 >> 5, 6$	2
3	Water (Cationic)	$7 > 8 @ 9 >> 5, 6$	3

As will become obvious, the major reason why the lower-membered rings are more difficult to initiate is because they are ringed Electrophiles and less basic in character and therefore favor stronger ionic existence than larger-membered rings.

$$\textbf{4 - membered } >> \textbf{5 - membered } > \textbf{6 - membered } >$$
$$\underline{\text{Order of acidity or less basicity of lactams}} \qquad 13.3$$

For all ring sizes, the hydrogen atom is loosely bonded to the nitrogen center ionically/ radically in view of the non-hybridized state of the orbitals in nitrogen and hydrogen. As is obvious from the Table, lactams favor both anionic and "cationic" attacks via mechanisms different from most of those ever proposed in the past.

Table 13.2 below shows the exothermic heats of polymerization of some lactams, from which the general order with respect to size of the minimum required strain energy required for the rings, compared to other cyclic compounds is obviously clear. Hence, it can be observed from the values and the analysis to follow *that lactams, are more strained than cyclic oxygen containing corresponding counterparts.* The order of polymerization reactivity shown in Table13.1 seem to follow the order of heats of poly-

Table 13.2. Heats of Polymerizations For Lactams [4,5]

#	Monomer	Ring Size	- ΔH (kcal/mole)
1	γ - butyrolactam	5	1.3
2	δ - valerolactam	6	1.1
3	Caprolactam	7	3.0
4	Enantholactam	8	5.7
5	Capryllactam	9	9.6
6	12 – Dodecanolactam	10	1.5

merization anionically and "cationically", with minor variations, all of which will be explained. Substituted cyclic amides are said to show the expected lower reactivity relative to unsubstituted monomers[3]. The effect of a substituent group in lowering reactivity increases, however, as the ring size decreases if the substituent group is on the nitrogen[6]. Why these are so, will clearly be explained.

Unlike isocyanic acid, it is believed that four-membered lactams undergo equilibrium molecular rearrangement, radically and ionically as shown below under an Equilibrium/Activated state of existence in the absence of foreign agents, initiators or activators.

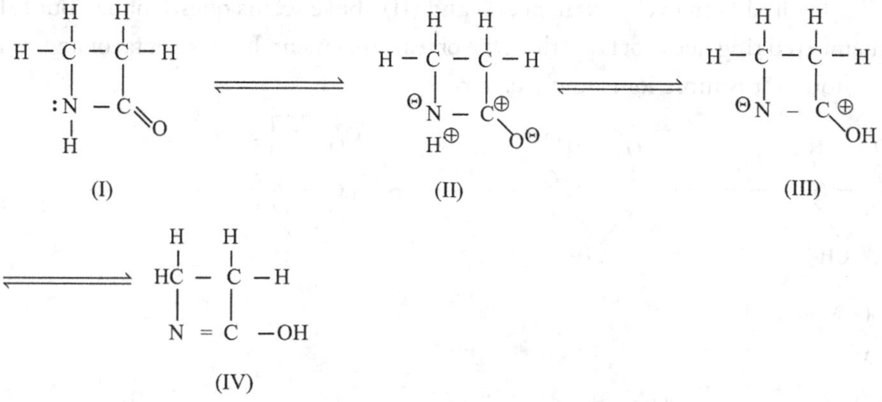

Ionic Equilibrium molecular rearrangement of monomer   13.4

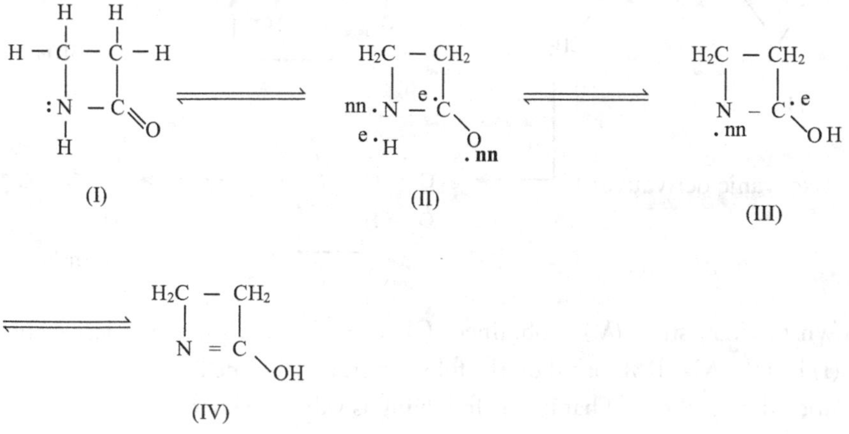

Radical Equilibrium molecular rearrangement of monomer   13.5

By the nature of this equilibrium rearrangement which could also be said to be Activated/Equilibrium states of existence, it is of the first kind of the sixth type; for which the followings are valid for the four-membered ring.

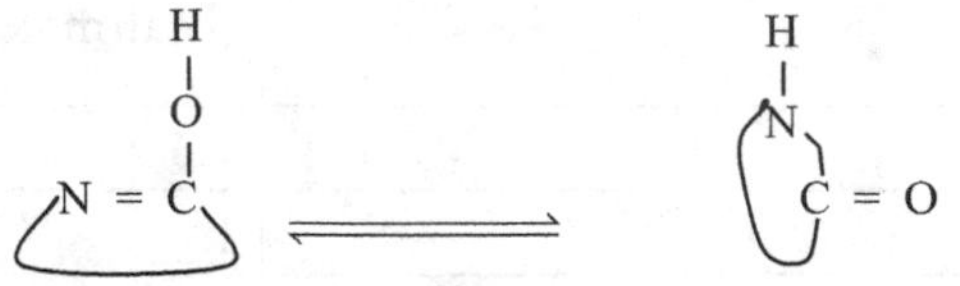

Order of Stability in the electrophile for four-membered ring 13.6

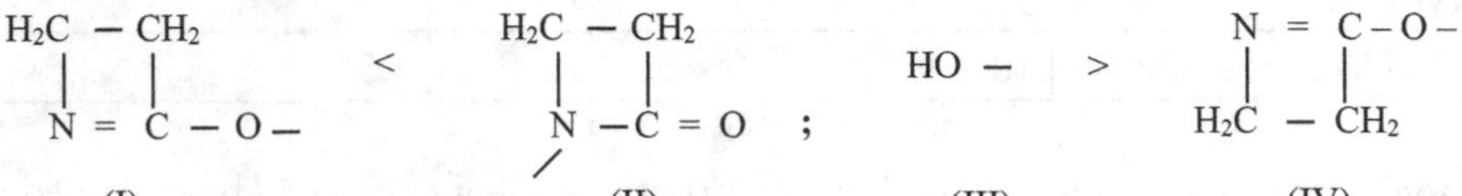

Order of radical-pushing capacity 13.7

(I) and (II) of Equation 13.6 are activation centers only for four-membered rings. (I) of Equation 13.7 is isocyanic in character, while (II) is amidic in character. Nevertheless, it is in view of the fact that b - propiolactam is always in Activated/Equilibrium state of existence, that makes the existence of Nylon 3 impossible, since (IV) of equations 13.4 and 13.5 if ever opened will not give Nylon 3. While (III) and (IV) of Equation 13.7 is favored as in cyclohexanone, (I) and (II) above seems questionable, but fully in place.

The three-membered ring does not favor this type of rearrangement. It does not favor any rearrangement as shown below, though it is more ionic in character.

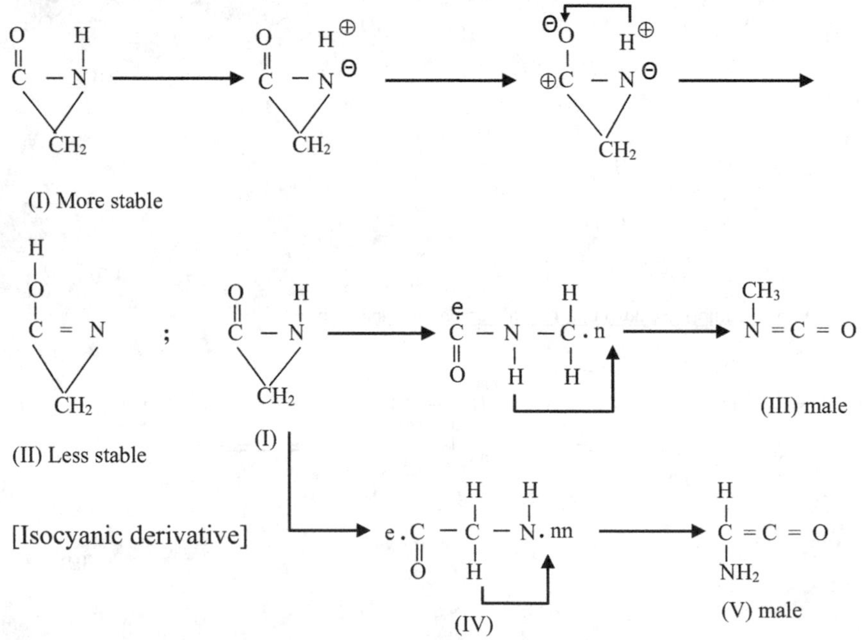

(I) is not known to exist, since (V) is obtained. (II) cannot exist, since it is less stable than (I). (V) is favored since (I) has the MaxRSE based on the followings. α- Lactone known to exist is less strained than the three-membered ring above. That is, the following is valid in general.

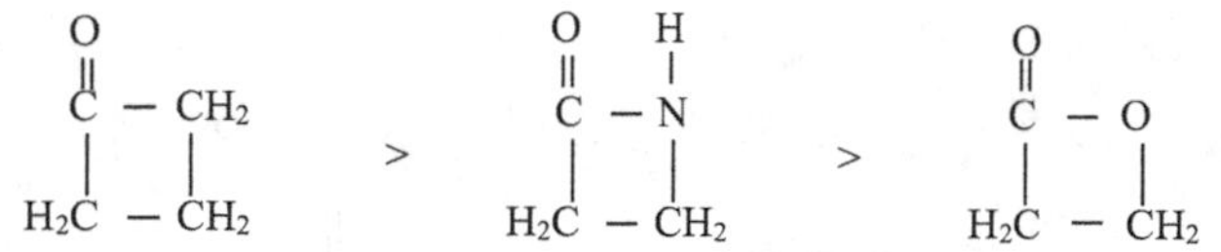

Order of strain energy

13.9

| Carbon - Carbon x - rings | > | Carbon - nitrogen x - rings | > | Carbon - Oxygen x - rings |

(where x is the size of the ring with CO in it)

Order of strain energy

13.10

The five-membered lactam rings and higher members do not undergo the equilibrium type of molecular rearrangement (i.e., they cannot exist in Activated/Equilibrium state of existence) since the lactams are more stable than their isocyanic derivatives.

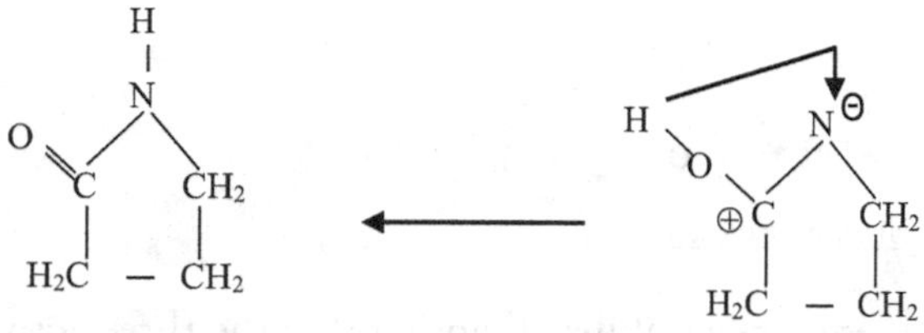

More stable / male

Isocyanic derivative

(Unstable existence / female)

13.11

However, the hydrogen atom on the nitrogen center is always loosely bonded radically or ionically in its unactivated state. On the other hand, it can be displaced by specific species.

For five-membered imine and higher and for lactams, like cyclic esters (lactones),

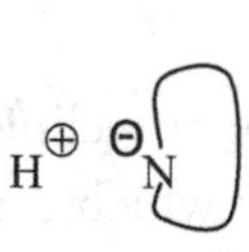

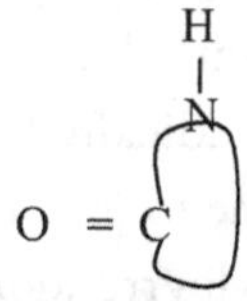

Imine   Lactam   13.12

there is one functional center on each of them, of which the former is more nucleophilic in character than the latter is. But unlike lactones, the methods of activations are uniquely different in view of the presence of the loosely bonded hydrogen atom to the nitrogen center ionically and radically to different levels for different sizes. If the H atom on the nitrogen center is replaced with substituent groups, then opening the ring will be favored, since the monomer can no longer exist in Equilibrium state of existence. Unlike five-membered lactams and above, **cyanuric acid** undergoes equilibrium molecular rearrangement of the first kind of the sixth type or readily exists in Activated/ Equilibrium state of existence as shown below.

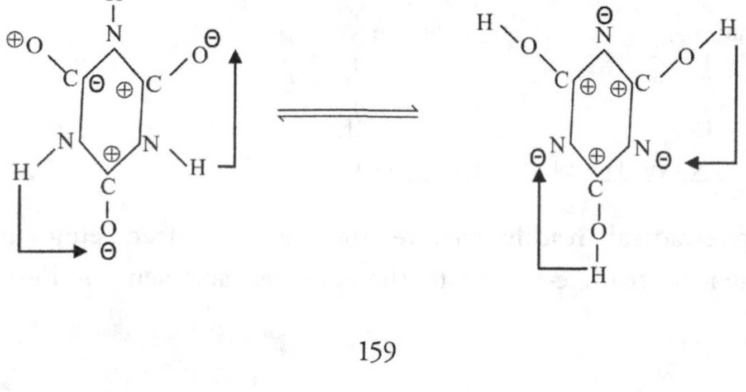

13.13

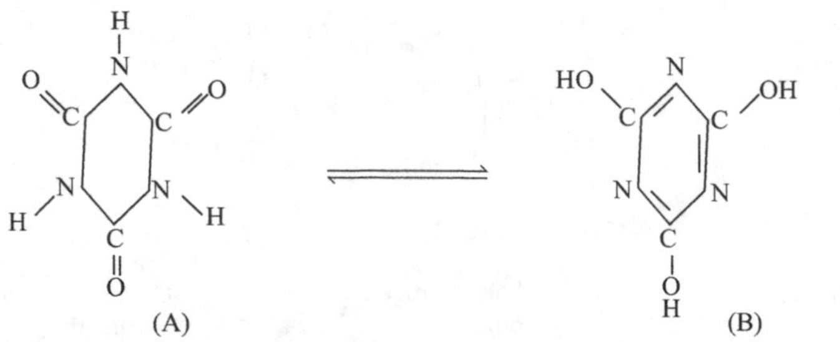

Equilibrium molecular rearrangement of sixth type for cyanuric acid   13.14

Hence the followings apply weakly for their substituent groups.

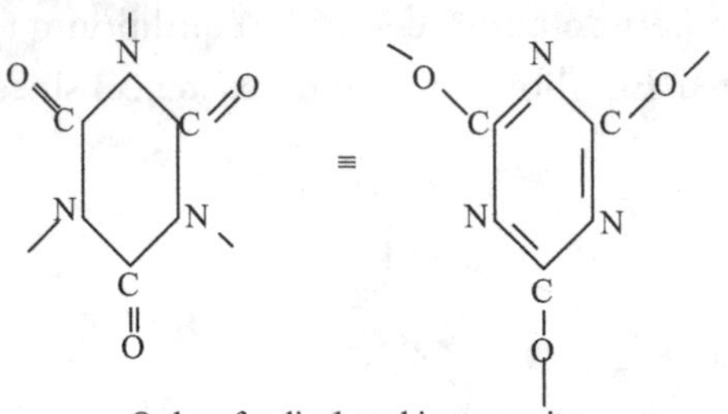

Order of radical-pushing capacity   13.15

The cyanuric acid itself cannot be resonance stabilized, noting that the three activation centers cannot be activated at the same time as was done stupidly above. It is only (B) that can be resonance stabilized. Both (A) and (B) of Equation 13.14 are of the different nucleophilicities. In view of the unique type of molecular rearrangement favored by cyanuric acid, in the absence of other phenomena, it cannot be used as a monomer to produce polymers. It is the isocyanic acid from which the cyanuric acid is obtained, that favors limited polymerization. The less the presence of CH_2 group in the ring, the more acidic is the ring, while the more the presence of CH_2 group in the rings, the more basic is the ring. This applies when C = O and N groups are adjacently located in the ring.

That the N - H bond is weakly radically bonded in character as well, is shown by the reaction of pyrrolidone for example with acetylene to give N-vinylpyrrolidone which can be positively and electro-free-radically polymerized to polyvinylpyrrolidone[7-8].

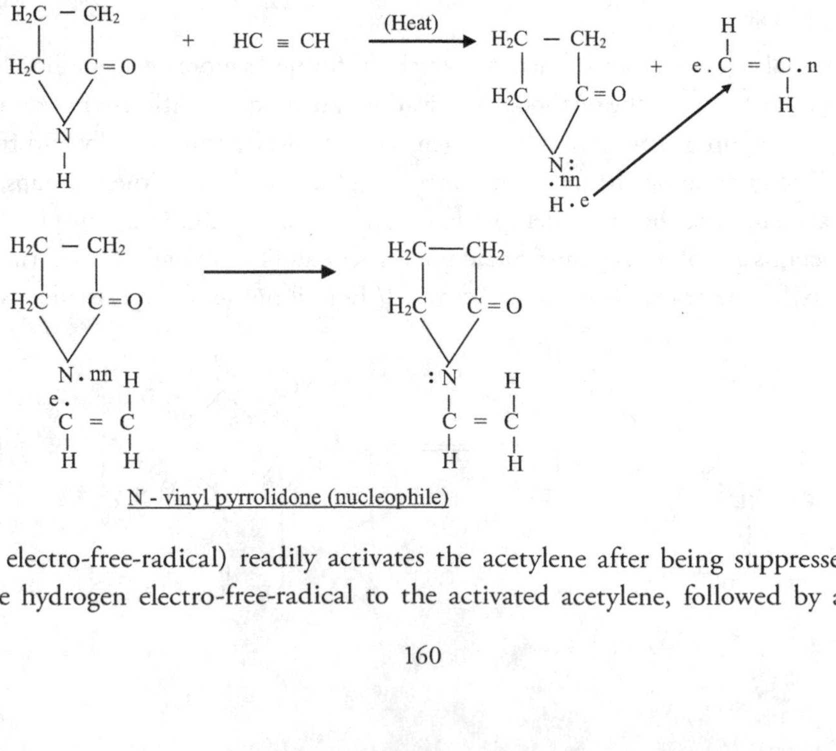

N - vinyl pyrrolidone (nucleophile)   13.16

H$^{\cdot e}$ (hydrogen electro-free-radical) readily activates the acetylene after being suppressed, followed by addition of the hydrogen electro-free-radical to the activated acetylene, followed by addition to the nucleo-non-free-radical. The N-vinyl pyrrolidone does favor positive attack via the C = C activation center because it is the least nucleophilic. The monomer with a nucleophilic $\left(\text{N} - \right.$ center in the ring, is more nucleophilic than the C = C activation center. The C = O center in the ring is the Y center. The N center is the X center. N-vinyl pyrrolidone favors electro-free-radical attacks via C = C activation center.

That the N - H bond is weakly free-radically bonded is further illustrated by the fact that alkali metals alone have been used to initiate the polymerization of some lactams by forming a lactam anion and hydrogen molecule[9]. That hydrogen molecules are produced, is clear indication of the radical character of the reaction, since $H^{\ominus}$ can never exist under any conditions in the absence of paired unbonded radicals in the last shell of H.

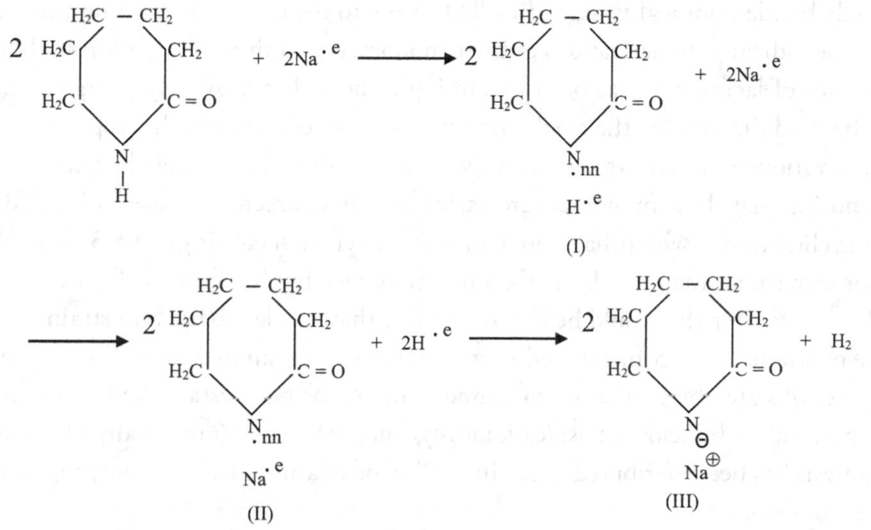

Via Equilibrium mechanism   13.17

For hydrogen to be displaced from the nitrogen center, it must be done preferably by an ionic metal, a metallic salt, alcohol or acid. Since $Na^{\cdot e}$ is more electropositive than $H^{\cdot e}$, $H^{\cdot e}$ is readily displaced to produce (II), followed by transfer of radicals, in view of the non-hybridized state of both centers to produce (III). (III) is an initiator which like $H_9C_4^{\ominus}........^{\oplus}Li$ (called Covalently charged-paired initiator) or $H_3CO^{\ominus}.....^{\oplus}Na$ (called Ionically charged-paired initiator) can be used negatively or ionically and electro-free-radically. The reaction above is a one Stage Equilibrium mechanism system. With ionic sodium solution, the following is to be expected.

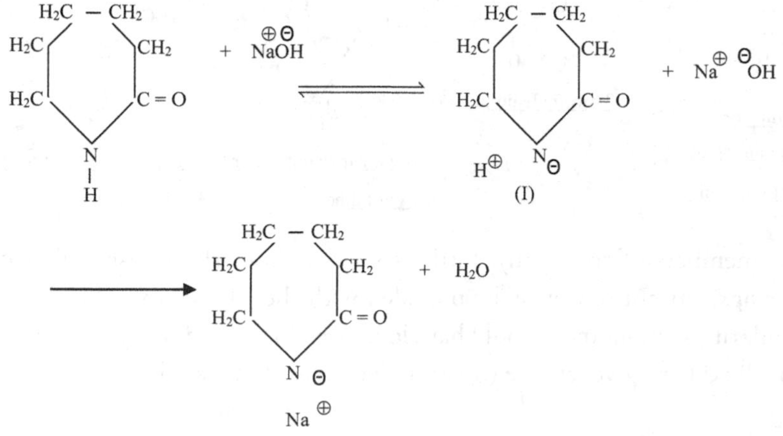

13.18

Equilibrium mechanism

It is also a one stage Equilibrium mechanism system. The H atom is more loosely bonded ionically than radically. Thus, the radical and ionic characters of the N - H bond can be observed to exist for these monomers, but to different levels. Unlike the isocyanates, the H on the nitrogen center does not have to be replaced by substituent groups for polymerization to be favored, in view of the unique differences in the states of existence of these monomers. Branched polymeric products have been obtained with these monomers, which broaden the molecular weight distributions of the polymeric products. They are said to occur in the later stages of reaction[10]. Why this is so, has never been clearly explained and this will be done when the need arises and in subsequent volumes.

Propagation/depropagation phenomena have never been observed with these monomers, in view of the presence of the adjacently located C = O center to the nitrogen center on the main chain backbones and presence of loosely bonded (ionically and radically) H atom to the nitrogen center on the chain. Nothing is known about the radical polymerizations of these monomers and this will be addressed in this chapter. With these families of lactams, it has been disturbing to note that positively charged-paired initiators cannot readily be used directly for their polymerizations. Secondly, very high temperatures are required for their polymerizations than the anionic case. Why this is so will definitely become clear by the end of the chapter, noting that these monomers are basic/acidic in character and are ELECTROPHILES.

Of all the cyclic amines which have no acidic character in their rings, the 3 - membered imines or aziridines (or ethyleneimine), has been the only monomer in this family of greatest attention and significance. The reason for this could be due to the fact that it is less basic and strained. It is said that the high degree of strain in the 3-membered ring is manifested by an extremely fast polymerization, for which violent reactions are obtained at room temperature. It has been established that cyclic amines are polymerized to polyamines by acid catalyst (cationically) but not by bases (anionically)[9,11-12]. Its unreactivity toward base catalysis has been attributed to the instability of an amine anion as the propagating species,[13] which indeed is not the case.

On the other hand, it is known that aziridines are somewhat weaker bases (pka = 8 to 9) than other similarly substituted aliphatic amines, but are stronger than pyridine. They are said to chemically undergo acid- and base- catalyzed nucleophilic dis-placement reactions with opening of the ring strictly analogous to those of the oxiranes[14]. As will become obvious however, they only undergo positively charged and electro-free radical polymerizations where possible.

The members of cyclic amines that will be considered for study, are shown below.

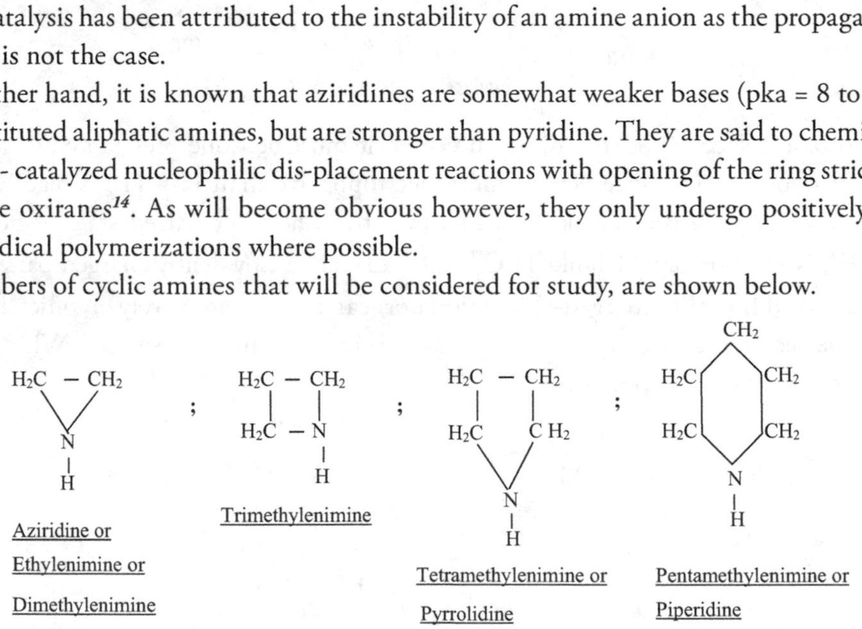

Aziridine or
Ethylenimine or
Dimethylenimine ; Trimethylenimine ; Tetramethylenimine or Pyrrolidine ; Pentamethylenimine or Piperidine

13.19

These are first four members of the family. Little is known about the four-membered ring. The five- and six-membered rings, cannot obviously be provided with the MRSE through the functional center, otherwise their popularity as monomers would have long been identified. Ethylenimine is known to react with aqueous sulfur dioxide to give taurine (2 - aminoethanesulfonic acid)[14].

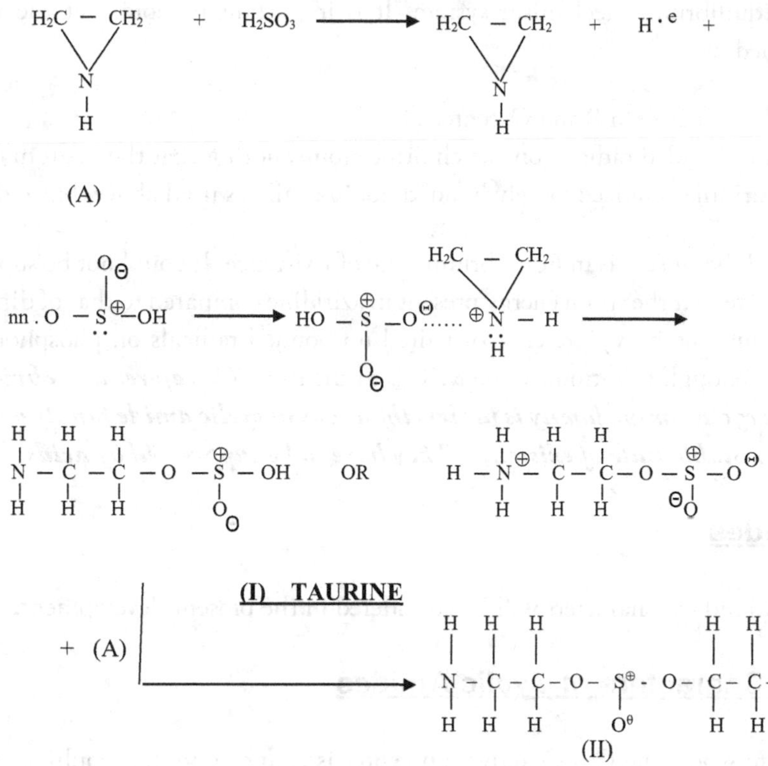

$$H_2C - CH_2 \quad + \quad H_2SO_3 \quad \longrightarrow \quad H_2C - CH_2 \quad + \quad H \cdot^e \quad +$$

(A)

(Reaction scheme showing sulfonate intermediate)

$$\underset{(I) \quad TAURINE}{N - C - C - O - S - OH} \quad OR \quad H - N - C - C - O - S - O$$

(I)   TAURINE

+ (A)

$$N - C - C - O - S \cdot O - C - C - N$$

(II)   13.20

The aqueous SO_2 is able to suppress the equilibrium state of existence of the imine, to give (I) which may continue further to give (II) in the presence of more aziridine.

Ethylenimine and polyethylenimine are said to react with hydroxyl groups of cellulose and are used to impact wet strength and abrasion resistance to paper and to decrease the tendency of rayon and cotton fiber to swell in water [14]. The reaction of the aziridine with the hydroxyl group is not anionic but cationic. Reaction of aziridine with phosphorus oxychloride gives triaziridyl - phosphine oxide (APO), $(C_2H_4N)_3PO$, which also reacts with the hydroxyl groups of cellulose and makes textiles fire resistant. This compound under the name *tepa* is used as a sterilant for insects.

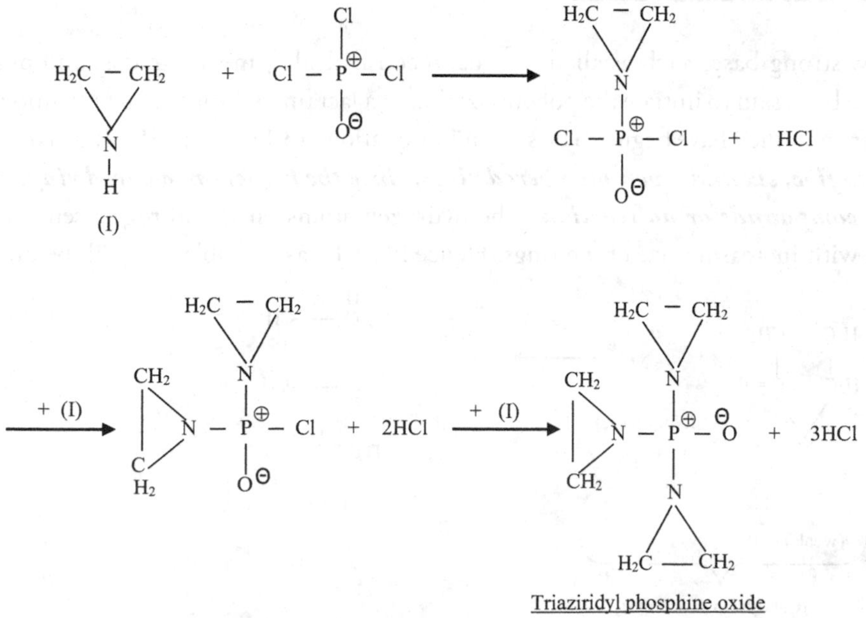

$$H_2C - CH_2 \quad + \quad Cl - P - Cl \quad \longrightarrow \quad \text{(aziridinyl-phosphorus intermediate)} \quad + \quad HCl$$

(I)

$$+ (I) \longrightarrow \text{(bis-aziridinyl intermediate)} + 2HCl \quad + (I) \longrightarrow \text{Triaziridyl phosphine oxide} + 3HCl$$

Triaziridyl phosphine oxide   13.21

163

Sunny N.E. Omorodion

It is a three stage Equilibrium mechanism system. It is important to note that the ring cannot be instantaneously opened by

(i)   The polar bond between P and O centers.
(ii) The paired unbonded radicals on the chlorine atoms, noting that the H atom on the nitrogen center of aziridine is indeed loosely bonded not ionically as used above, but radically.

It could not be opened, because it is in Equilibrium state of Existence. It could not be suppressed. (i) and (ii) above clearly indicate that the strain energy present in aziridine compared to that of dimethylene oxide (oxirane) is small. Thus, the heavy presence of paired un-bonded radicals on phosphorus oxychloride centers are not strong enough to introduce the MRSE to the ring. *Therefore, it is obvious that either the Strain energy in cyclic amine family is far less than that in cyclic amide family or aziridines are too strong to exist in Stable state of existence. They have to be suppressed by acids*

13.1 Cyclic Amides

Both charged and radical characters will be considered in the present development.

13.1.1 Charged Characters of Cyclic Amides

Unlike isocyanate where the C = O activation center is X and more nucleophilic than the N = C activation center which is Y, with cyclic amides, the $\left(\; N \; - H \right.$ functional center is X and nucleophilic, while the $\left(\; C \; = O \right.$ activation center is Y and more nucleophilic. Both anionic and "cationic" methods of polymerization will separately be considered.

13.1.1.1 Anionic Character

13.1.1.1(a) Use of Strong Bases

Anionically, strong bases such as alkali metals, metal hydrides, metal amides, and organometallic compounds have been said to initiate the polymerization of a lactam by forming lactam anion[6,9]. How the lactam anions are obtained have begun to be shown in Equations 13.17 and 13.18. *The reaction however seems limited to five, six and seven-membered rings, since the higher-membered rings readily form stable salts or compounds or no reaction.* The hydrogen atoms on the nitrogen centers are also less loosely bonded with increasing size of the rings. Hence in such cases, stable salts will be obtained.

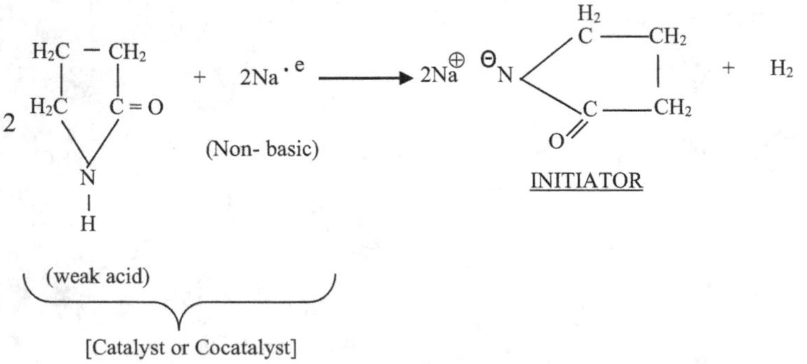

$$13.22$$

164

$$2H_2O + 2Na^{\cdot e} \rightleftharpoons 2NaOH + 2H^{\cdot e} \longrightarrow 2NaOH + H_2.$$

(Weak acid)

13.23

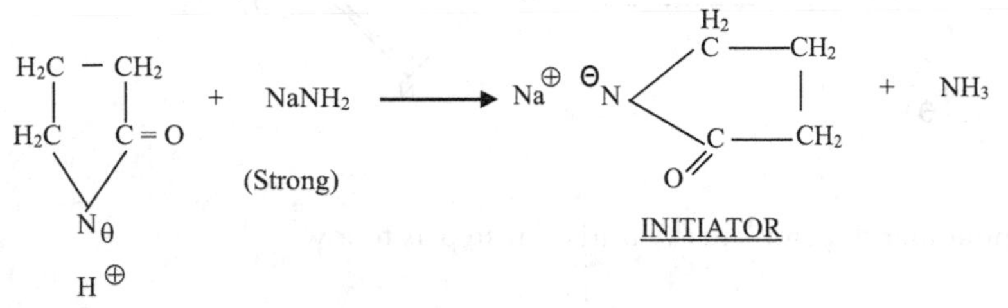

INITIATOR

13.24

(weak acid)

$$H_2O + NaNH_2 \longrightarrow NaOH + NH_3$$

(Strong base)

13.25

With metal hydrides for seven-membered ring, the following is obtained. The reaction has been considered, in order to show that metal hydrides can never be ionic in charac-

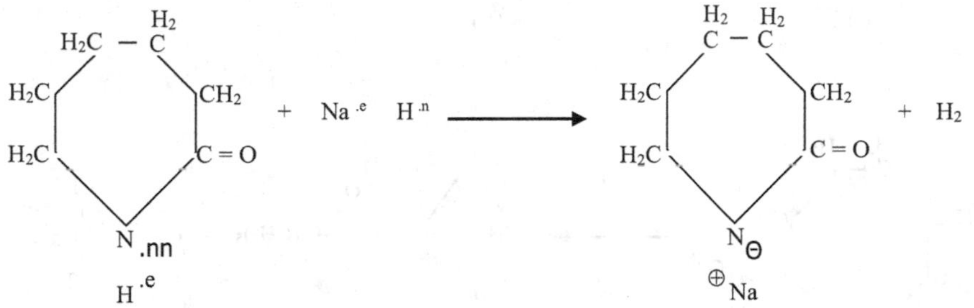

.(A) Ionically paired INITIATOR

13.26

ter, since the hydrogen center cannot carry an anion under any conditions; though one of the driving forces for anionic formation- last shell being full - is satisfied for "hydrogen anion", but however with no paired unbonded radicals to give it a polar character.

With organometallic initiators, the followings are obtained.

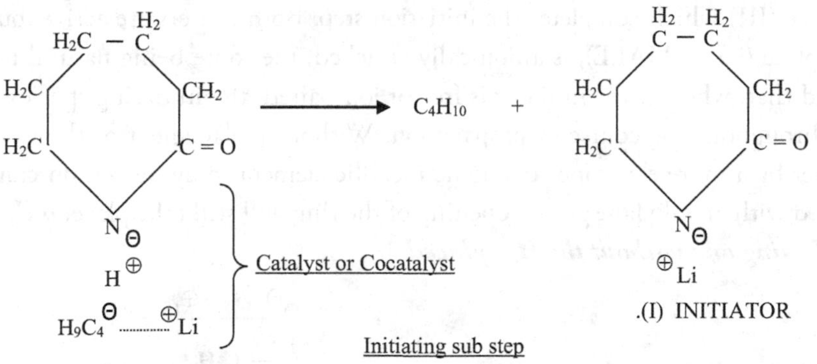

Catalyst or Cocatalyst

Initiating sub step

.(I) INITIATOR

13.27

(I) of Equation 13.27 is the true initiator involved in the polymerization of the monomer. (I) is only part of the initiation step. It does not undergo equilibrium molecular rearrangement phenomenon, i.e., it cannot exist in Activated/Equilibrium state of existence.

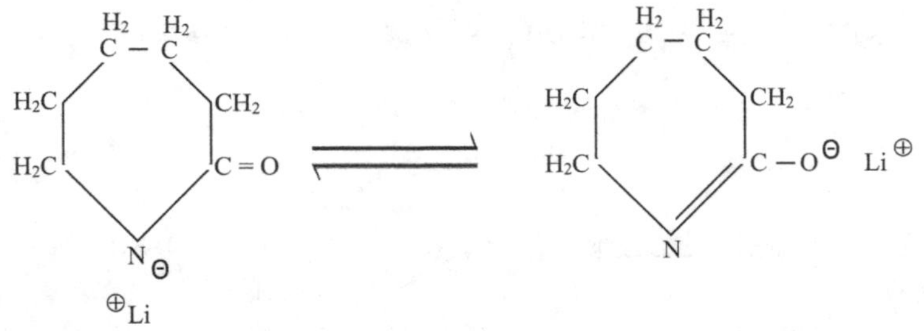

13.28

(I) now adds to a single monomer to complete the initiation step as follows.

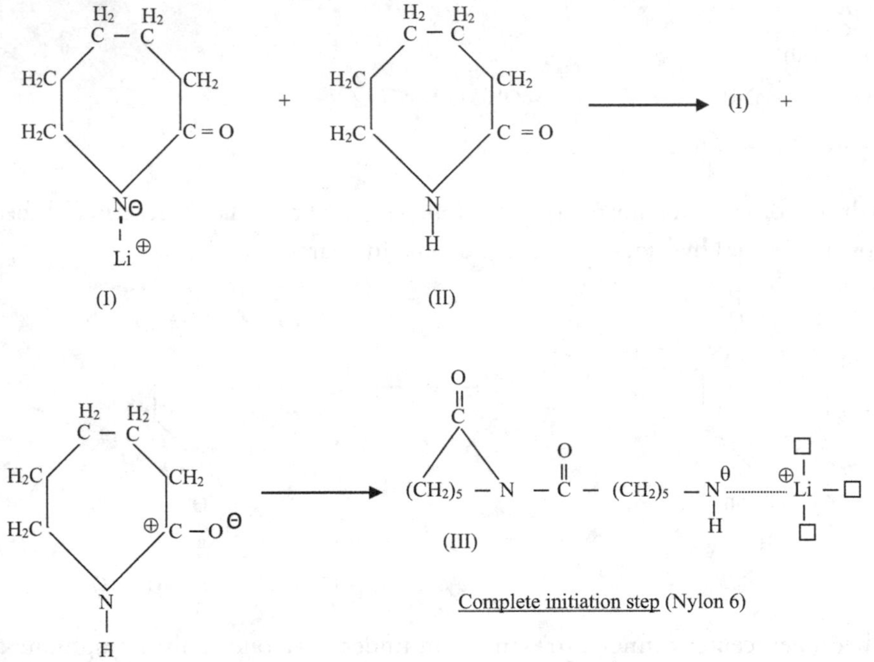

(I)   (II)

(III)

Complete initiation step (Nylon 6)

13.29

The lithium of (I) cannot be displaced by H. Therefore, (I) readily activates the $C = O$ center of (II) to add and produce (III) which completes the initiation step. Both centers are active, but the monomer being an Electrophile (i.e., a MALE), is anionically attacked, the route being natural to the monomer. It can be observed that, whether the initiator is free or ion-paired, the initiating species is ion-paired. It remains paired throughout the course of propagation. Without replacement of the hydrogen atom on the nitrogen center by a more electropositive ionic metallic element, polymerization cannot be favored. If the H is replaced with an alkylane group, opening of the ring will still take place, ***only if the initiator is first prepared using one without the H replaced.***

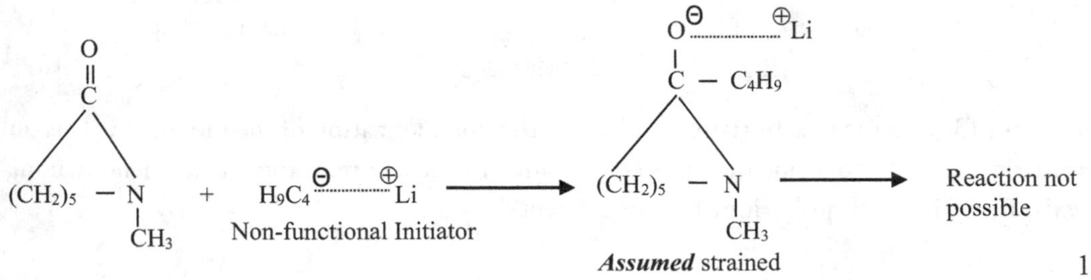

Non-functional Initiator   ***Assumed*** strained

13.30

The opening above is not favored, because the equation is not chargedly balanced. The negatively charged center is not anionic. As already said, the initiator must first be separately prepared with another monomer carrying H on the N center. ***The presence of substituent groups has been said to greatly lower the reactivity of the monomer when located on the nitrogen center more than on carbon centers, that which can be observed not to be the case as explained above.*** When the substituent group(s) of more radical-pushing capacity than H are placed on the C centers in the ring, then opening of the ring will become more difficult, since the ring will become less strained. For cyclic amides that show both acidic and basic characters, radical-pushing substituent groups seem to reduce the basicity and not reactivity of rings, when placed on the nitrogen centers because the nitrogen center can no longer be part of the initiator, anionically. If the metal is of the non-ionic type, e.g. Zn, Ti, Fe, the initiation step will not be favored, since covalently charged bonds will now exist between the N center and the metal and the first step of initiation which is ionic will not be favored.

(III) of Equation 13.29 now commences the propagation step as follows.

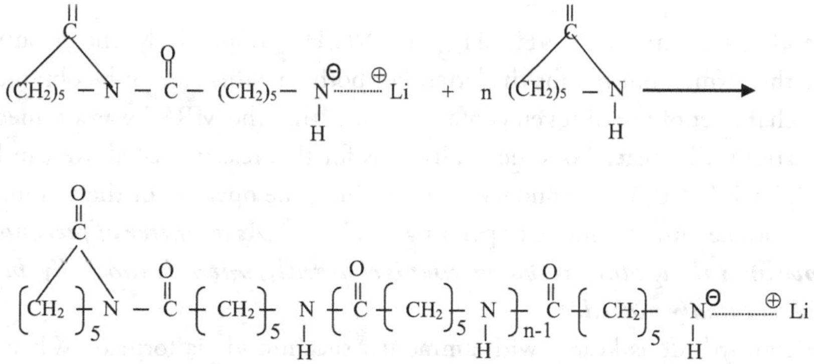

Nylon 6

13.31

The longer induction periods observed in this system is due to the less strained energy present in the monomer, ***the number of steps involved during initiation and in some cases the equilibrium state of the initial steps during initiation and most importantly the fact that a base is being used to attack a base.***

While the seven and larger membered rings favor the initiation and propagation steps, using strong bases, it is more difficult with the 5- and 6- membered rings. The reason is because the lower membered ones are more acidic and favor being more in Equilibrium state of existence. Apart from this fact, cyclic esters and cyclic amides are unique in the sense that, though g- butyrolactone cannot homopolymerize, the ring can be opened temporarily by ammonia as shown below, to produce g-butyrolactam.

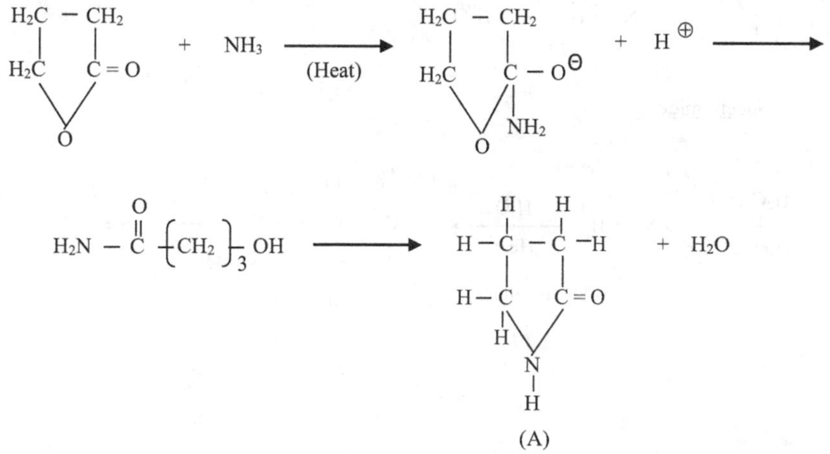

(A)

13.32

167

It is believed that it was the cation that opened the ring under Equilibrium mechanism conditions, and not as shown above since the anion cannot diffuse when not paired or alone.

With methylamine, N-methyl-g-butyrolactam (N-methyl-pyrrolidone) is obtained-(B)[8].

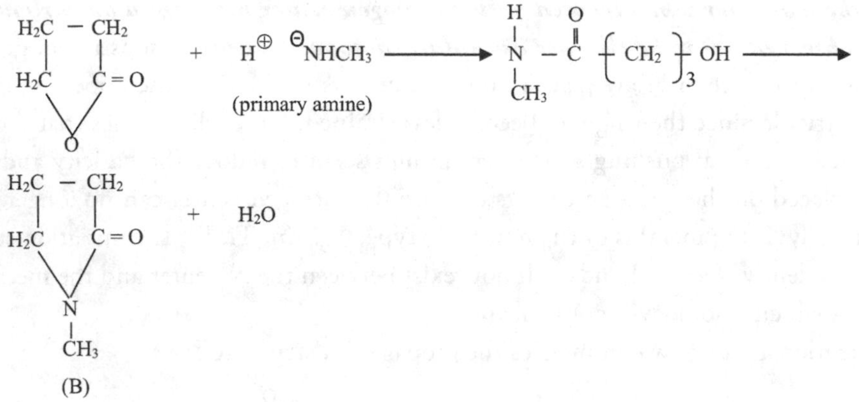

13.33

With secondary and tertiary amines - $NH(CH_3)_3$ and $N(CH_3)_3$ respectively, the γ- butyrolactone ring can be opened for the former, but not for the latter. For both, lactams cannot be obtained. It is in view of the more stable character of the nitrogen containing ring, that the MRSE was attained in the lactone to open the ring cationically instead of onionically. It is for this reason, (B) above can be likened to a tertiary acidic amine while (A), is a secondary acidic amine. The opening of the lactone ring is also be possible as already said, i.e., under different operating conditions. ***As a matter of fact, note that none of the reactions above are taking place as shown, that is chargedly, but only radically, because covalent charges cannot be isolatedly placed.***

When succinic anhydride is heated with ammonia, succinimide is formed. When succinimide is treated with bromine and alkali under the usual conditions of "Hofmann reaction", the product is b - alanine[15]. These reactions can be explained as follows.

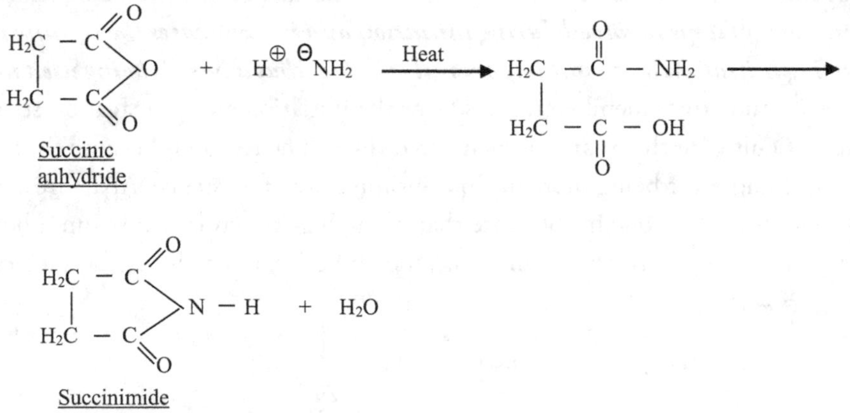

13.34

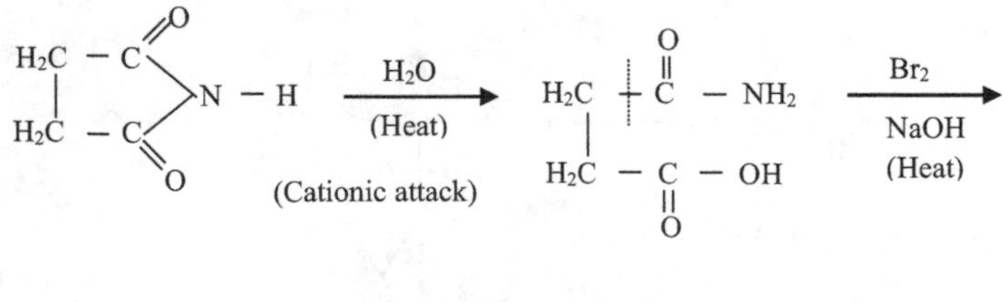

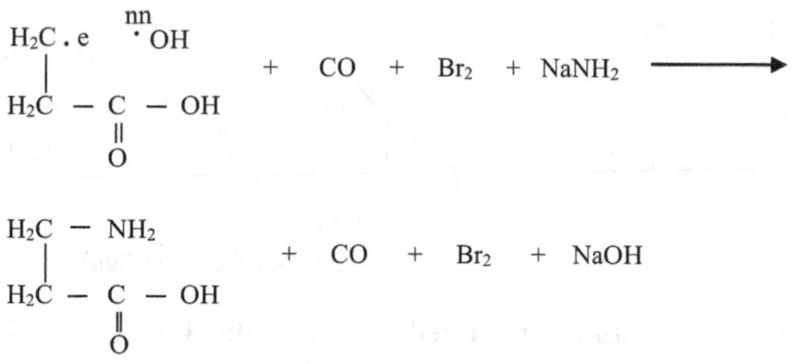

$$H_2C \cdot e \overset{nn}{\cdot} OH$$

$$H_2C - C - OH \quad + \quad CO \quad + \quad Br_2 \quad + \quad NaNH_2 \longrightarrow$$

$$\| \\ O$$

$$H_2C - NH_2$$

$$H_2C - C - OH \quad + \quad CO \quad + \quad Br_2 \quad + \quad NaOH$$

$$\| \\ O$$

$$\beta - Alanine \qquad\qquad\qquad\qquad\qquad\qquad\qquad 13.35$$

All these reactions are Equilibrium mechanism systems of one or more stages under different operating conditions including the manners by which components are added.

If however, bromine is added to an ice-cold alkaline solution of succinimide, N- bromosuccinimide precipitates in almost quantitative yield.

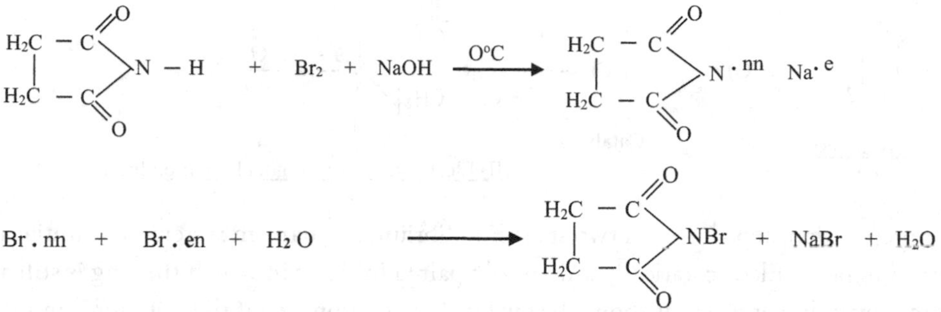

$$H_2C - C \overset{O}{\diagup} \\ \qquad\qquad N - H \quad + \quad Br_2 \quad + \quad NaOH \xrightarrow{0^\circ C} \quad H_2C - C \overset{O}{\diagup} \\ H_2C - C \diagdown O \qquad\qquad\qquad\qquad\qquad\qquad H_2C - C \diagdown \overset{}{N \cdot nn \quad Na \cdot e} \\ \qquad\qquad\qquad\qquad\qquad\qquad\qquad\qquad\qquad O$$

$$H_2C - C \overset{O}{\diagup} \\ Br \cdot nn \quad + \quad Br \cdot en \quad + \quad H_2O \longrightarrow \qquad \qquad NBr \quad + \quad NaBr \quad + \quad H_2O \\ \qquad\qquad\qquad\qquad\qquad\qquad\qquad\qquad\qquad H_2C - C \diagdown O$$

$$\qquad\qquad\qquad\qquad\qquad\qquad\qquad\qquad\qquad\qquad\qquad\qquad 13.36$$

Succinic anhydride is far less strained than maleic anhydride. Hence heat is added to achieve the MRSE to open the ring, which closes again to form another more nucleophilic ring. Radical steps can be observed to be largely involved. Nevertheless it can be observed that N has stronger tendency to form rings than O, an indication that nitrogen containing rings are less strained or more stable than oxygen containing rings. From the reactions above, presence of heat is a driving force for opening of some rings, which have functional centers. It is important to note that in Equation 13.35 above, the attack on the ring is cationic, since like five membered γ-butyro-lactam, anionic attack by $HO^{\ominus}$ from NaOH, H_2O and the like cannot take place when not paired or isolatedly placed. The radical route of the last reaction above is favored, in view of the fact that Br_2 molecule can never dissociate to produce ions under any conditions. When they dissociate they produce nucleo-non-free-and electro-non-free-radicals under Equilibrium State of Existence and during abstraction from Cl_2 molecule or two nucleo-non-free-radicals under Decomposition State of Existence. It is water that is first formed in the last reaction above and in most reactions where the possibility exists along with heat. This is then followed by formation of other products.

13.1.1.1(b) Use of Acylating Agents

To make the 5- and 6- membered rings favor anionic polymerization without any replacement of H on the nitrogen centers, and reduce the long induction periods to a bare minimum, the use of *acylating agents such as acid chlorides and anhydrides, isocyanates, inorganic anhydrides, and others have been reported to be very advanta-geous*[16-17]. In their use, it is said that, an imide is formed by their reactions with the monomers as shown below for an acylating agent.

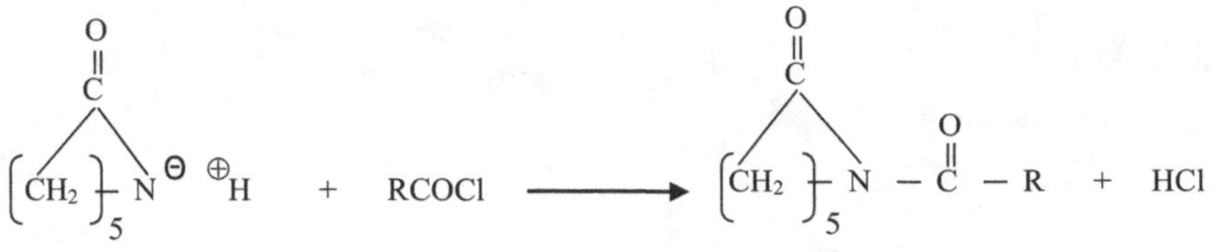

$$(I) \underline{\text{ the imide } (N \text{ - acyllactam})} \qquad 13.37$$

The N-acyllactam can be synthesized in situ or preformed and then added to the reaction system. The reaction system as has been observed by the mechanisms proposed in the past is one already containing the strong base which could not polymerize the 5 - and -6-membered rings. It could also be one involving the use of only the acylating agents alone. The preformed species which actually initiates polymerization is not (I) above or others proposed in literatures, but that shown below (II), that in which (I) is used.

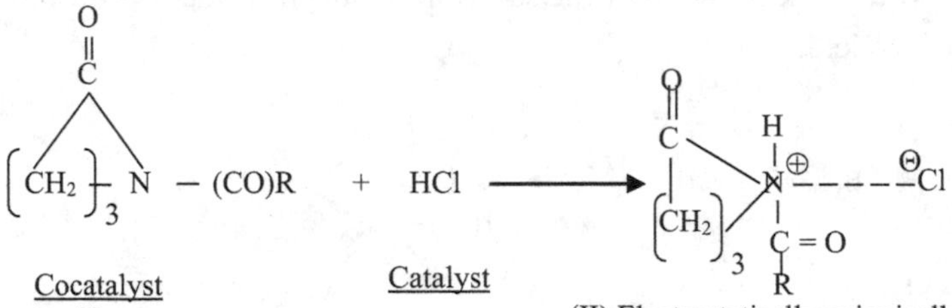

$$(II) \underline{\text{ Electrostatically anionically-paired Initiator}} \qquad 13.38$$

The Initiator preparation step above is a two stage Equilibrium mechanism system. The initiating species can be observed to be an Electrostatically anionically-paired initiator in which the ring is still a terminal part of the growing polymer chain as shown below for the initiation step if the initiator is in Equilibrium state of existence.

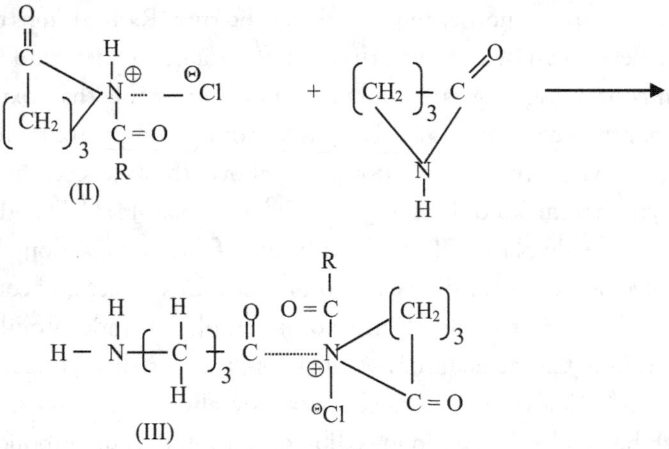

$$\underline{\text{Complete initiation step}} \text{ (Nylon 4)} \qquad 13.39$$

The formation of (III) marks the end of the initiations step. It should be noted that in the reaction above, the route is electro-free-radical the initiator being in Equilibrium state of existence, and it is the acylating agent alone for the five-membered ring that is involved in the presence or absence of strong bases. When 7- and larger membered rings are involved, in the presence of a strong base and an acylating agent, two modes of activations are simultaneously present-

(i)   That involving the use of the strong metallic (Li) base and
(ii)   That involving the use of the acylating agents.

Hence, 7-membered rings are noted to be more reactive than 5- and 6-membered rings as shown in Table 13.1, when acylating agents are involved in the presence of a strong base. In the absence of a strong base above, the unique natural order (of 5>6>7>8>9>) still remains. It is not surprising therefore, why it has been noted, that in a lactam polymerization involving the use of a strong base and an acylating agent, initiation may involve contributions from both the Electrostatically anionically-paired and Ionically-paired initiators and not from the imide dimer and the N-acyllatam[6].

As already anticipated by some schools of thought, but differently, ***the acylating agent is indeed the cocatalyst (but not a promoter), while the acid sitting on the monomer is the catalyst.*** When strong bases are involved, the bases are cocatalysts while the monomer is the catalyst. The bases is truly an activator for its role to form the initiating species or Initiator.

When anhydrides are involved, the followings are obtained.

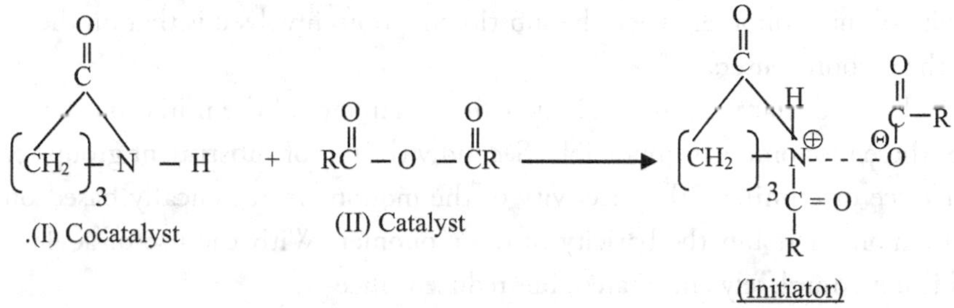

(Initiator)　　　　13.40

The Initiator preparation step above is a two stage Equilibrium mechanism system. In the first stage, (I) of Equation 13.37 is first prepared. In the second stage the acetic acid produced was then used to give the Initiator. One can see where the Induction period is coming from. After this Step which is Equilibrium mechanism (Slow), this is followed by the Initiation and Propagation steps all taking place via Combination mechanism (Very fast) when in stable state of existence or via Equilibrium mechanism when in Equilibrium state of existence (Very slow).

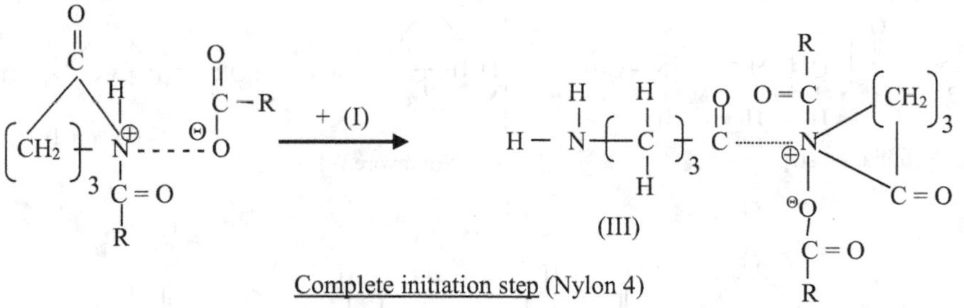

Complete initiation step (Nylon 4)　　　　13.41

When isocyanates are involved, the followings are also obtained.

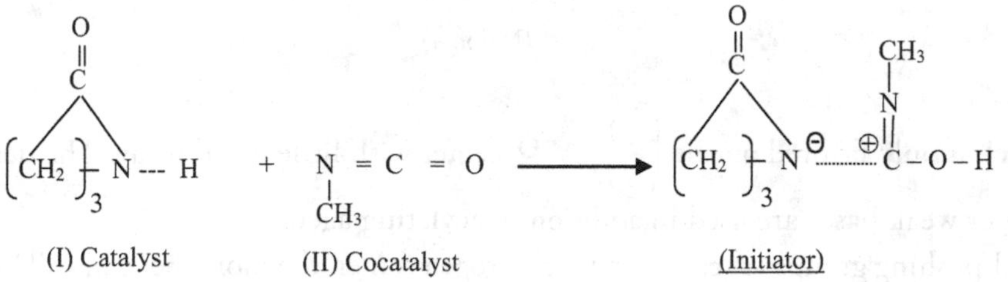

(I) Catalyst　　　(II) Cocatalyst　　　(Initiator)　　　　13.42

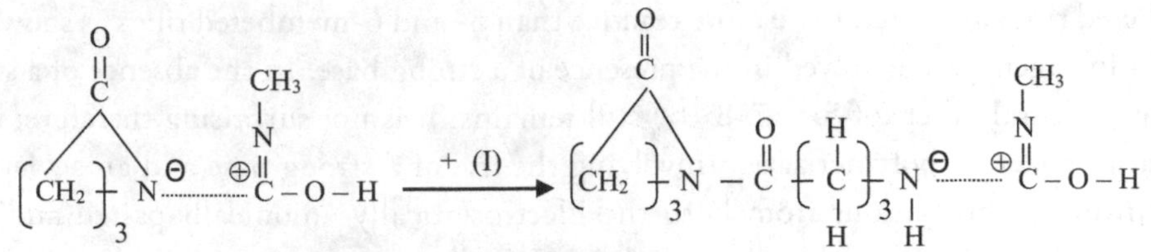

Initiation step (Nylon 4)   13.43

The isocyanate being Electrophilic in character, it is the C = O center that is activated to produce a ***covalently ionically charged-paired initiator, the first of its kind so far.*** It can be observed that if the monomer had the H atom already replaced with a substituent group, the use of these acylating agents to generate an Electrostatically paired initiator, will be impossible. This is partly what was thought to greatly reduce the reactivity of the monomer, when the substituent group involved is that on the nitrogen center, more so than on the carbon centers.

It can thus be observed that without the ionic-paired character of the initiation step anionically, the polymerization of these monomers is impossible. Secondly the use of substituent groups on the nitrogen center does not reduce or eliminate the reactivity of the monomers anionically based on the methods of initiation of the monomers but the basicity of the monomer. With the use of acylating agents, the induction period is not completely eliminated, but reduced since-

(i)   The monomer is always a part of the initiator, for which it take time to prepare (Initiator preparation step)

In general, no molecular rearrangement can take place and propagation follows as shown, that which is the real case as opposed to that of Equation 13.43.

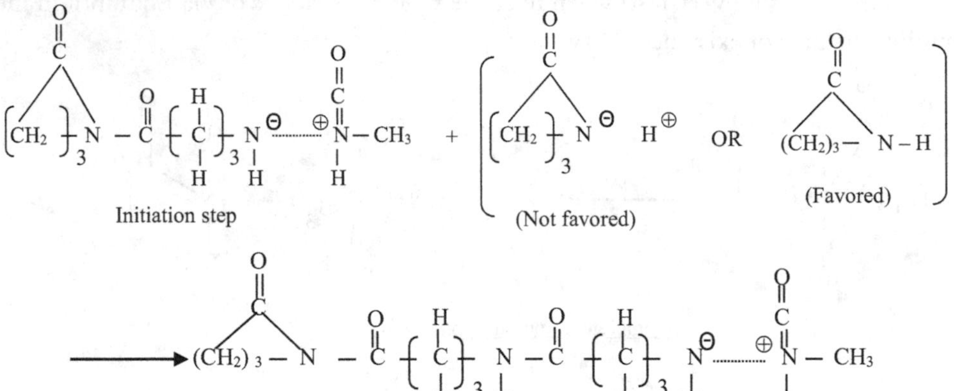

(Nylon 4)   13.44

Anionically, attack is only favored on the ⟨C = O center with little or no delay. The delay is greatly present, if strong or weak bases are used in addition to acylating agents.

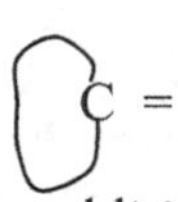

When radical-pushing groups are carried by the nitrogen center, the monomer can still be polymerized if MRSE can be introduced into the ring. To do this, part of the initiation step must be such that involves the use of the secondary acidic cyclic amide and a cocatalyst as follows.

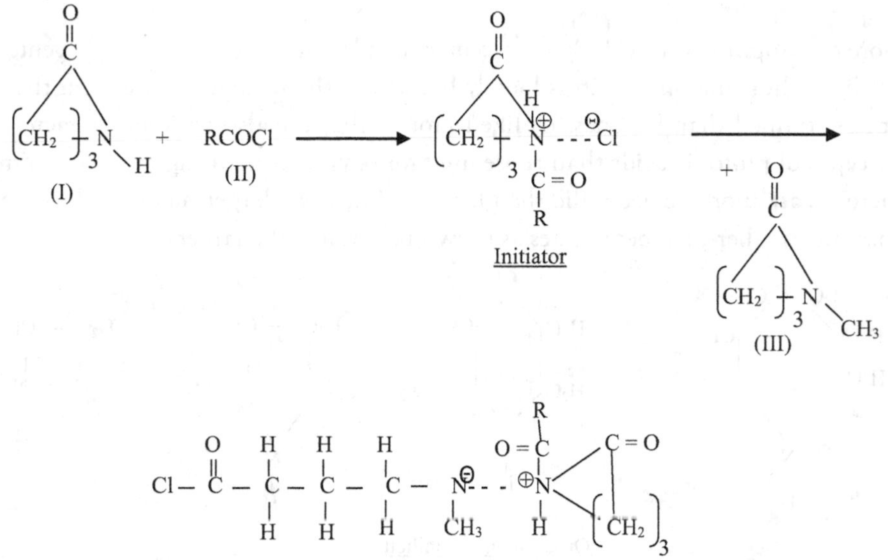

Initiation step 13.45

This is then followed by propagation of the monomer (III). If (II) was used alone with (III), no polymerization will be favored. Thus, it can be observed that the reactivity of (III) is not strongly disturbed by the presence of CH_3 group on the N center. All depend on the proper choice of initiators anionically and "cationically" also, and this is only possible when the mechanism of how NATURE OPERATES is understood.

When radical-pushing groups are internally located in the ring, the method of initiation will remain the same and polymerization will be favored if the MRSE can be provided for the ring under higher operating conditions than expected. Having considered the intricate aspects of the polymerizability of these monomers anionically, it is now the turn of the cationic possibilities, the mechanisms of which are different from what has been known to be the case in the past.

13.1.1.2. "Cationic" Characters

Unlike the anionic routes, there is no size limitation imposed by the use of these initiators ***cationically*** where possible. Some of the known initiators which have been used ***cationically*** for the polymerization of the monomers include-

(i)   Protonic acids such as phosphorus, hydrochloric, hydrobromic and carboxylic acids[20-22].
(ii)   Amines such as aniline, benzylamine[23].
(iii) Water[3,24].

"Cationically", it has been observed that, with these monomers, higher temperatures, of close to 250°C are required for their polymerization, unlike the anionic cases. These high temperatures are those that favor the opening of the ring by force due to less strain energy in their rings, and most importantly, the fact that the monomer is an Electrophile (i.e., a Male) whose natural route is the opposite (i.e., anionic and nucleo-non-free-radical).

In general however, lactams are less nucleophilic than lactones as will soon become obvious.

Lactones   >   lactams

Order of nucleophilicity 13.46

173

Lactams are more electrophilic via the $\overset{N}{\underset{}{C}} = O$ center than lactones via the $\overset{O}{\underset{}{C}} = O$ center, in view of the presence of the hydrogen atom which is loosely bonded to the nitrogen center. On the other hand, lactams are far less strained than lactones. Unlike lactones, they are also basic in character. Hence, the use of different types of protonic acids than those used for oxygen containing cyclic monomers, despite the fact that lactones are more nucleophilic than lactams. Since the larger membered lactams are more nucleophilic than the smaller-membered-ones as shown below, it is the larger-

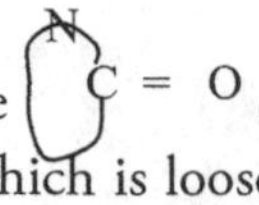

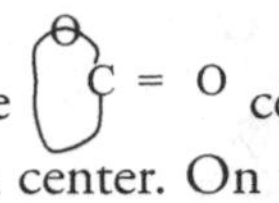

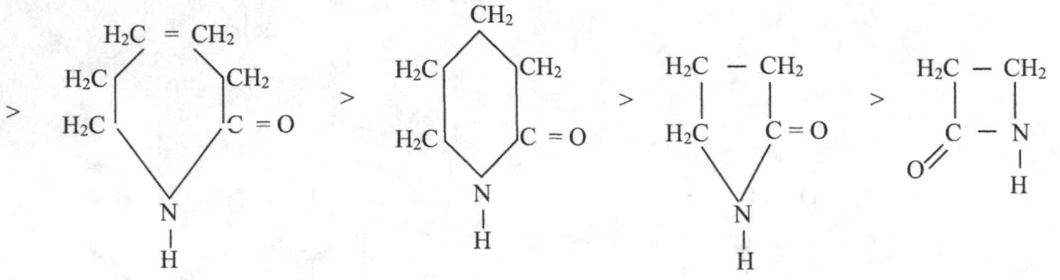

Order of nucleophilicity

13.47

membered rings that show stronger reactivity cationically than the five- and six-membered rings as already indicated in Table 13.1.

13.1.1.2(a). Use of protonic acid and positively charged-paired initiators

The efficiency of an acid in initiating polymerization is dependent on its acidity, for which the counter-ion of the acid must be less nucleophilic than the monomer. Though lactams are weak nucleophiles, only moderately strong acids can be used in view of their basic and acidic characters. In order words, the $H^{\oplus}$ of the acid must be of stronger capacity than the $H^{\oplus}$ of the monomer. In order words, the Equilibrium state of existence of the amide must be suppressed. When such acids are used in the absence of heat, the MRSE may never be provided for such less strained rings. Hence, high temperature polymerizations of these monomers are involved.

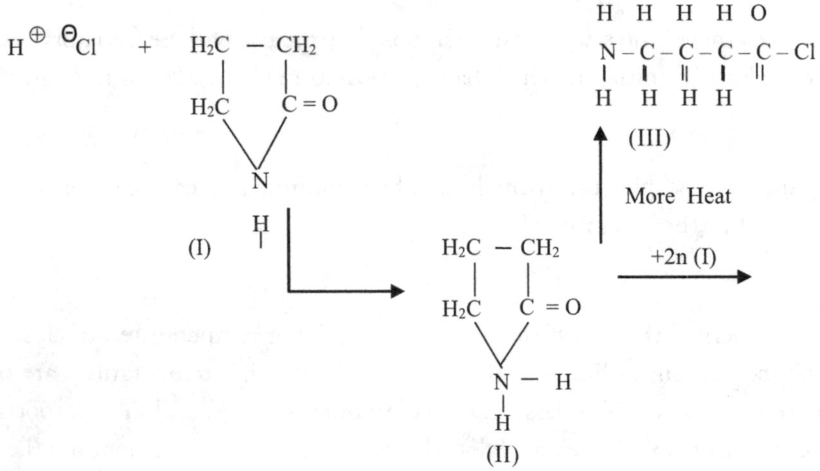

Equilibrium Mechanism

13.48a

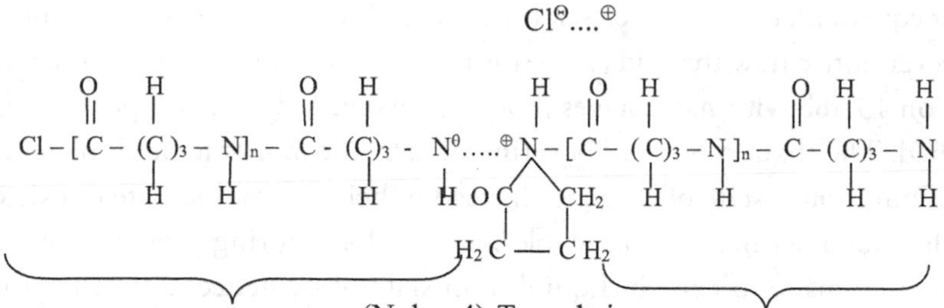

$$Cl^{\ominus}....^{\oplus}$$

$$Cl-[\overset{\overset{O}{\|}}{C}-(\overset{\overset{H}{|}}{C})_3-\overset{\overset{H}{|}}{N}]_n-\overset{\overset{O}{\|}}{C}-(\overset{\overset{H}{|}}{C})_3-\overset{\overset{}{}}{N}^{\ominus}.....^{\oplus}\overset{\overset{H}{|}}{N}-[\overset{\overset{O}{\|}}{C}-(\overset{\overset{H}{|}}{C})_3-\overset{\overset{H}{|}}{N}]_n-\overset{\overset{O}{\|}}{C}-(\overset{\overset{H}{|}}{C})_3-\overset{\overset{H}{|}}{\underset{H}{N}}$$

with bridging group:
$$H O \overset{}{C} \quad CH_2$$
$$H_2C - CH_2$$

(Nylon 4)-Two chains

Anionic Route	Electro-free-radical Route
Forward Addition	Backward Addition
Combination mechanism	Equilibrium mechanism

(IV)   13.48b

The $H^{\oplus}$ of the acid is the initiator which can only be used electro-free-radically. It cannot displace the $H^{\oplus}$ on the nitrogen, since the amide is in Stable state of existence in its presence. With (II) as the initiator when the ring is such that cannot yet be opened, then the route of propagation would be anionic at the beginning until optimum chain length has been reached, then followed by Electro-free-radical backward addition as shown in Equation 13.48b. *If the ring is opened in (II), (IV) cannot be obtained. Instead (III) is obtained. If the two H^e are of same capacity, then both will remain in equilibrium state of existence with no polymerization. If the H^e of the amide is greater in capacity than the H^e of acid, then there is no polymerization.* In the initiation step above, there is no induction period due to short time it took to prepare the initiator via Equilibrium mechanism in one stage. Induction period comes in when the mechanism of propagation is Equilibrium as against Combination mechanism which is fast.

If the ring is such that has enough energy to be opened as shown in (III) of Equation 13.48a, what take place as shown below is electro-free-radical propagation route of poly-merization. Electro-free-radical routes can only take place when there is no free negative charge or anion hanging around its vicinity in the system. This is only possible via Equilibrium mechanism.

$$\overset{\overset{H}{|}}{\underset{H}{N}}-(\overset{\overset{H}{|}}{\underset{H}{C}})_3-\overset{\overset{O}{\|}}{C}---Cl \quad = \quad [n+1](I)$$

(III) $\xrightarrow{\text{Electro-free-radical polymerization}}$
                         [Backward Addition]

$$\overset{\overset{H}{|}}{\underset{H}{N}}-(\overset{\overset{H}{|}}{\underset{H}{C}})_3-\overset{}{C}-\left[\overset{\overset{O}{\|}}{}\overset{\overset{H}{|}}{N}-(\overset{\overset{H}{|}}{\underset{H}{C}})_3-\overset{\overset{O}{\|}}{C}\right]_n\overset{\overset{H}{|}}{\underset{H}{N}}-(\overset{\overset{H}{|}}{\underset{H}{C}})_3-\overset{\overset{O}{\|}}{C}-Cl \quad \xrightarrow[\text{Termination}]{+(I)}$$

$$H-\overset{\overset{H}{|}}{\underset{H}{N}}\left\{\overset{\overset{H}{|}}{\underset{H}{C}}\right\}_3\overset{\overset{O}{\|}}{C}\left\{\overset{\overset{H}{|}}{\underset{H}{N}}\left\{\overset{\overset{H}{|}}{\underset{H}{C}}\right\}_3\overset{\overset{O}{\|}}{C}\right\}_n\overset{\overset{H}{|}}{\underset{H}{N}}\left\{\overset{\overset{H}{|}}{\underset{H}{C}}\right\}_3\overset{\overset{O}{\|}}{C}\cdots N\overset{CH_2-CH_2}{\underset{\underset{O}{\|}}{\underset{C}{}}-CH_2} \quad + \quad HCl$$

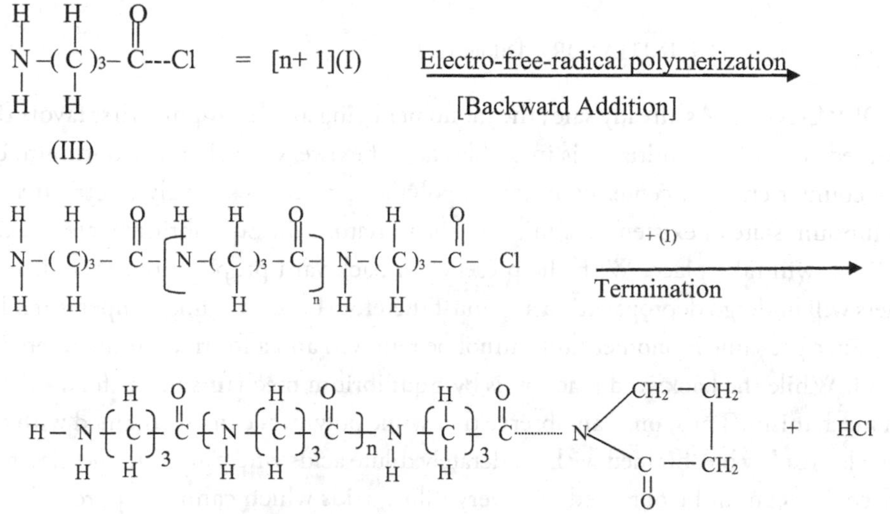

Termination step   13.49

Almost the same as above will apply to lactones or cyclic esters in general, though not shown therein, noting that unlike lactams, cyclic esters cannot exist in Equilibrium state of existence from the O center on the ring.

For the case above (the last equation), transfer species of the second kind cannot be rejected from such growing polymer chain. However notice how the acid catalyst is regenerated. In the same manner, it can be done for the case of Equation 13.48b with more stages probably involved. This takes place when the optimum chain length is reached. The chain could well be terminated by the monomer itself like the case above if it is made to exist in Equilibrium state of existence instead of being in Stable state of existence. Does this make sense when the monomer has been in Stable state all along during propagation? Yes, it does, because the monomer can be made to exist in Equilibrium state of existence, once the growing chain can no longer exist in Equilibrium state of existence. The rate of propagation will keep decreasing with backward addition but not with forward addition, since the route is not natural to the monomer. It is important to note that like some of the anionic cases, the presence of the ring as part of the polymeric product is favored here only at the beginning of the chain and under certain conditions.

As has already been seen with the case of aldehydes in Chapter 7 and from current developments so far, the (III) of Equation 13.48a is a stable salt, that obtained when concentrated acids such as HCl are used or when more heat is involved. With more heat and via this route, depropagation reaction cannot take place for this Electrophile.

When protonic acid is diluted with water, the situation is different as shown below.

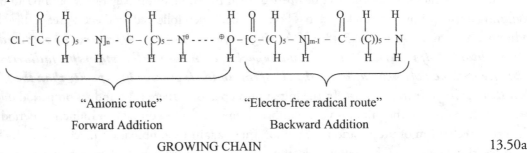

"Anionic route"   "Electro-free radical route"

Forward Addition   Backward Addition

GROWING CHAIN   13.50a

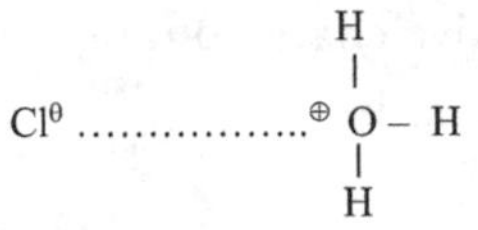

$$Cl^{\ominus} \ldots\ldots\ldots\ldots \overset{\oplus}{O} - H$$
with H above and H below the O

THE INITIATOR – Dilute HCl   13.50b

Nature is ORDERLINESS. As already said, the monomer being an electrophile first favors the use of the negatively charged center if the initiator is in Stable state of existence. When the optimum chain length is reached, the counter charged center commences polymerization backwardly electro-free- radically, if placed in Equilibrium state of existence. If initially, the initiator is in Equilibrium state of existence, only backward addition will take place. With the presence of backward propagation, there is no doubt that these monomers will undergo depropagation and must therefore have a Ceiling temperature. But this does not take place, since the same monomer unit cannot be removed and a fourteenth-membered ring cannot easily be formed. While the backward reaction is by Equilibrium mechanism, the forward reaction is by Combination mechanism. Thus, one can observe that while polymers can be obtained with concentrated acids, they can also readily be obtained with moderately dilute acids where anionic and electro-free radical routes take place, but cannot be obtained with very dilute acids which cannot suppress the Equilibrium state of existence of the amide. When the optimum chain length for the backward reaction is reached, both sides of the chain disengage to give two chains. HCl and water can now be recovered from the chains using the amide to give an amide group at the carbonyl end.

When so-called "ion-paired" initiators are involved, the situation is slightly different. In almost all the types, acids and a complex salt are produced. For example, consider the case of the use of $SnCl_4$ [25] as catalyst and cocatalyst i.e. $Cl_4Sn^{\oplus}.....^{\ominus}SnCl_5$.

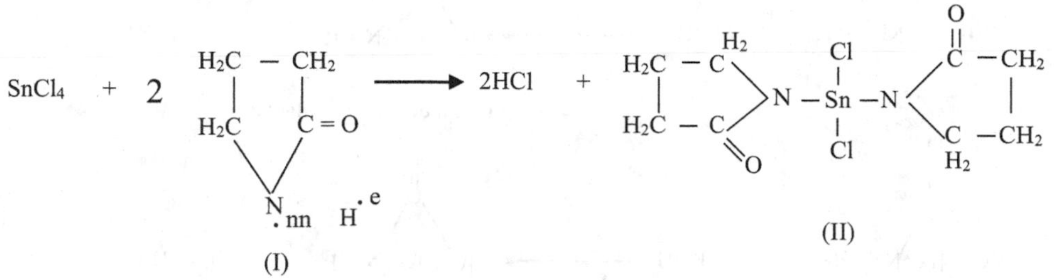

(I)   (II)   13.51

In view of the divalent and quadrivalent state of Tin (Sn) and the fact that it is a non-ionic metal, (II) above is produced if the presence of the first mole of HCl would not prevent further replacement. (II) above is the product which could not be identified by infrared analysis[25]. In the reaction above, before the presence of the cationic ion-paired (real name is Electrostatically positively charged paired) initiator obtained by transfer of chlorine atom on $SnCl_4$ to a Stable $SnCl_4$ molecule could be formed, HCl is formed, since the Equilibrium state of existence of $SnCl_4$ is suppressed by the presence of the amide, noting again that P and Sn, are non-ionic metals. If the presence of $\mathbf{H^{\oplus}.....^{\ominus}SnCl_5}$ (from $SnCl_4$/HCl combination), an electrostatically cationically charged-paired initiator is favored, then cationic polymerization cannot be made possible. The counter-negatively charged center cannot be used for backward addition nucleo-non-free radically, because the chlorine atom cannot readily be held in equilibrium from the carbonyl center. Otherwise, anionic/Electro-free radical polymerizations will take place if presence of (II) of Equation 13.48a is favored. In fact, the initiator underlined and highlighted above cannot be the final initiator used cationically by Combination mechanism, because it cannot be formed. $SnCl_4$ is insoluble and partially dissolves in HCl.

When BF_3 is used alone, similar reaction as above is obtained, with the formation of hydrofluoric acid.

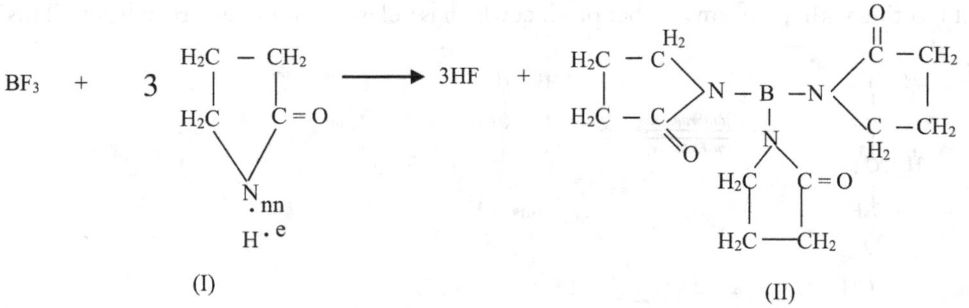

(I)   (II)   13.52

Note that the favored existence of the (II) where full replacement of F or Cl in $SnCl_4$ is made possible due to the presence of $SnCl_4$ and BF_3 in the system making it impossible to keep HCl or HF in Equilibrium state of existence, for which with the formation of the acids in the absence of $SnCl_4$ or BF_3, the followings are possible-

i)   Direct use of the lactam and the acid to produce an initiator or

ii)   No polymerization since an initiator cannot be obtained between HF and BF_3 or HCl and $SnCl_4$.

Hence long induction periods for initiator preparation step can be observed in their use.

With ROR/BF_3 combination, the reaction above and the case shown below are involved.

$$[\overset{\displaystyle O}{\overset{\|}{C}}]$$

$$[CH_2]_3 - \overset{\ominus}{N} \quad \overset{\oplus}{H} \quad + \quad ROR \quad \longrightarrow \quad [CH_2]_3 - N - R \quad + \quad ROH$$

.(I) Imide (N – alkyllactam)   13.53a

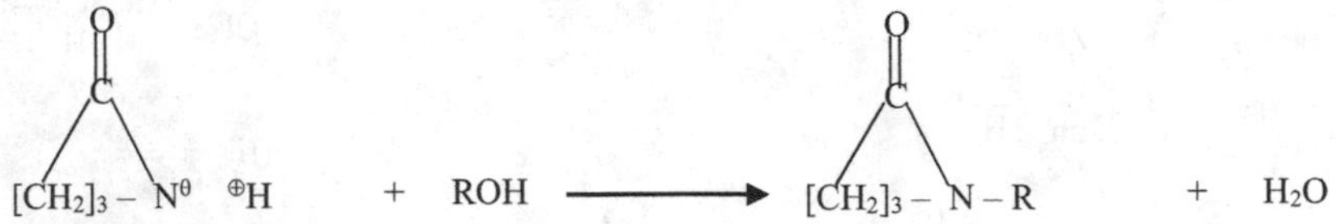

$$[CH_2]_3 - \overset{\ominus}{N} \quad \overset{\oplus}{H} \quad + \quad ROH \quad \longrightarrow \quad [CH_2]_3 - N - R \quad + \quad H_2O$$

.(I) Imide (N - alkyllactam   13.53b

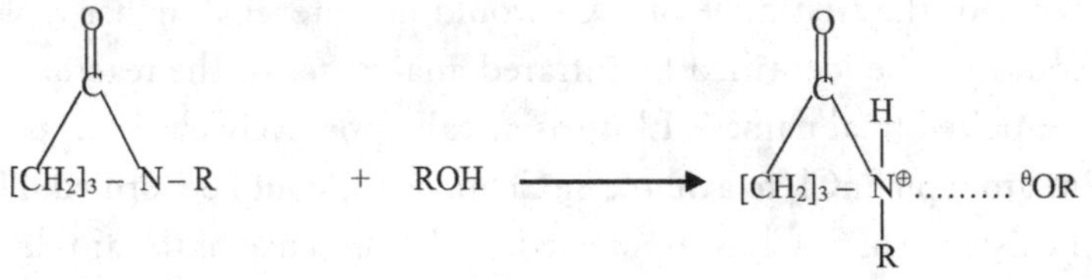

$$[CH_2]_3 - N - R \quad + \quad ROH \quad \longrightarrow \quad [CH_2]_3 - \overset{H}{\underset{R}{\overset{|}{\underset{|}{N^{\oplus}}}}} \cdots\cdots \overset{\ominus}{OR}$$

.(I) INITIATOR   13.53c

The presence of the last two equations will depend on the molar ratios of the components involved. For if there is more of the monomer, then (I) of Equation 13.53b will be obtained, and if not, (I) of Equation 13.53c will be obtained.

Thus, since **BF_3/HF** combination cannot be used, only H_2O or (I) can be used. Polyacrylamide is a water soluble polymer. Acrylamide and its polymer are polar/ionic. So also is water and organic alcohol such as CH_3OH. Yet when methanol a non-solvent is added continuously to an aqueous solution of polyacrylamide, another solution with increasing turbidity is obtained. The CH_3OH is precipitating the polyacrylamide from solution but reacting with it to form another product which is polar/non-ionic as shown below. This is similar

Stage 1:

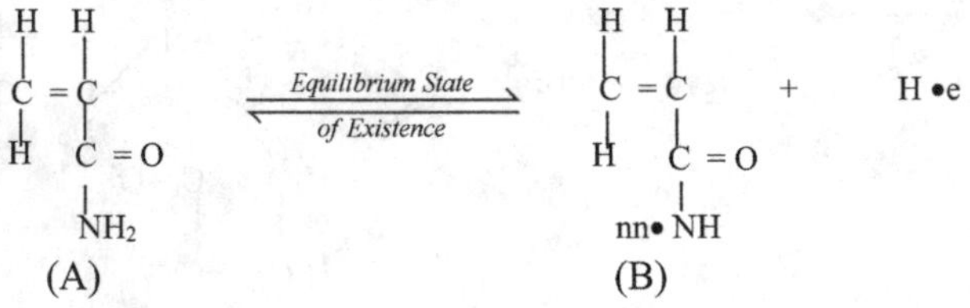

$$\underset{(A)}{\overset{\displaystyle H \quad H}{\underset{\displaystyle NH_2}{\overset{|\quad\ |}{C = C}}}} \quad \underset{of\ Existence}{\overset{Equilibrium\ State}{\rightleftharpoons}} \quad \underset{(B)}{\overset{\displaystyle H \quad H}{\underset{nn\bullet NH}{\overset{|\quad\ |}{C = C}}}} \quad + \quad H\bullet e$$

$$H\bullet e \quad + \quad CH_3OH \quad \overset{Abtraction}{\rightleftharpoons} \quad H_2O \quad + \quad e\bullet CH_3$$

$$H_3C\bullet e \quad + \quad (B) \quad \longrightarrow H_2C = CH - CONH(CH_3)$$

. (C)   13.54a

Stage 2:

$$(C) \quad \overset{Equilibrium\ State}{\underset{of\ Existence}{\rightleftharpoons}} \quad \underset{(D)}{H_2C = CH - \overset{CH_3}{\overset{|}{CON}} \bullet nn} \quad + \quad H\bullet e$$

$$H\bullet e \quad + \quad CH_3OH \quad \rightleftharpoons \quad H_2O \quad + \quad e\bullet CH_3$$

$$H_3C\bullet e \quad + \quad (D) \quad \longrightarrow \quad H_2C = CH - CON(CH_3)_2$$

(E)   13.54b

Overall Equation: (A) + $2CH_3OH \longrightarrow 2H_2O$ + (E)   13.54c

to Equations 13.53a and 13.53b. The formation of (E) for different sizes of the polymer chain, is the source of its turbidity.

The use of H_2O as the source of initiator is yet to be considered. So also is the use of amines. One can nevertheless observe so far why the direct use of "cationic ion-paired initiators" has never been clearly elucidated and understood.

13.1.1.2(b) Use of Amines.

Polymerization of lactams e.g. caprolactam by amines such as aniline or benzylamine is said to be "interesting"[6]. However, the mechanisms proposed are not interesting. Using aniline, the followings are obtained.

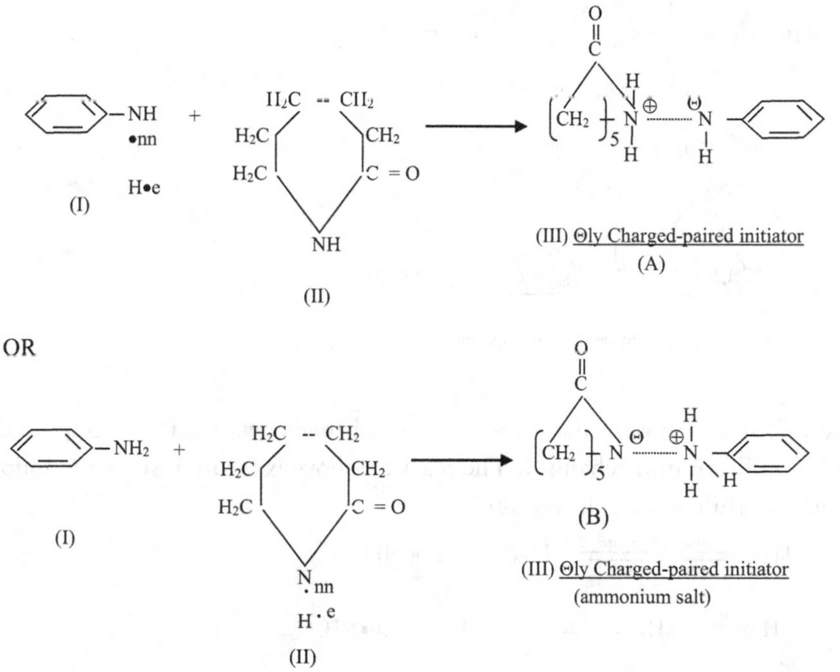

$$13.55$$

(A) is obtained when the Equilibrium state of existence of the amide is suppressed, while (B) is obtained when that of the aniline is suppressed. When both (A) and (B) are in Equilibrium or stable state of existence, polymerization can still take place anionically and electro-free-radically. As it seems, it is (A) that is favored, taking note that the two initiators above are not cationic, but Electrostatically anionically-paired initiators. On the other hand, however, as has already been shown and shortly be shown, electro-free radical backward addition can take place from the positively charged end, using H.e as initiator. Thus, when (A) is involved, the ring opens as shown below and the routes are both anionic and electro-free-radical, based on their state of existence.

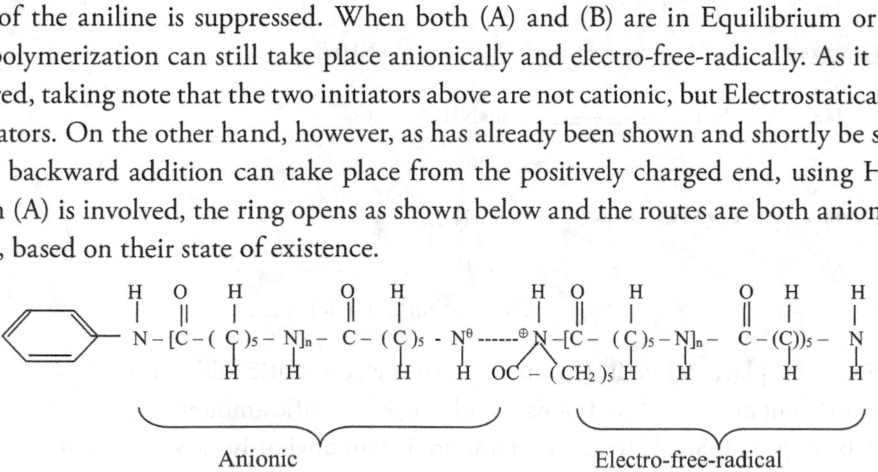

$$13.56$$

When aniline (I) was said to be used cationically, H_2 molecule was said to be released in the process radically from the nitrogen center[6,23]. Release of H_2 will take place only in the initiation step and not in propagation step as was thought to be the case. When H_2 is released, it will be under very different operating conditions and no initiator can immediately be obtained as shown below. This is the case where one of the components is suppressed, followed by hydride abstraction to give H_2 and (C) or (D) below. (D) is the weak covalently charged-paired initiator wherein N is made to carry a positive covalent charge which is non-free. This is impossible. Only anionic charge can be carried by a N center which is also impossible due to electrostatic forces of repulsion from the paired unbonded radicals on the adjacently placed N center. Therefore (D) cannot exist and be used as such.

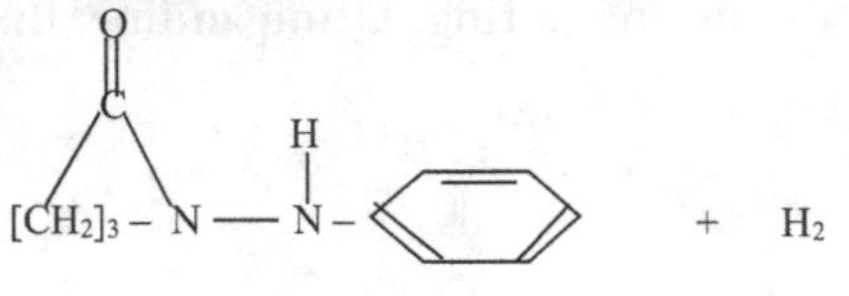

$$[CH_2]_3 - \overset{\overset{\displaystyle O}{\|}}{\underset{}{C}} \quad N - \overset{H}{\underset{}{N}} - \langle \rangle \quad + \quad H_2$$

.An Initiator Or Compound?
(C)

OR

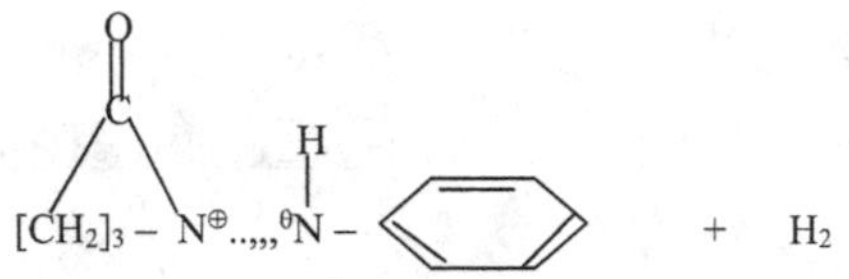

$$[CH_2]_3 - \overset{\overset{\displaystyle O}{\|}}{\underset{}{C}} \quad N^{\oplus}..,,^{\ominus}\overset{H}{\underset{}{N}} - \langle \rangle \quad + \quad H_2$$

.An Initiator Or Compound?
(D)   13.57

(C) is a stable compound with weak bond between two nitrogen centers. Hydrogen is indeed released in the first stage via Equilibrium mechanism. The reaction above is favored, since the followings can take place with ammonia. Aniline is a primary amine.

(A) **Stage 1:** $NH_3 \underset{\text{of Existence}}{\overset{\text{Equilibrium Sate}}{\rightleftarrows}} H\bullet e + nn\bullet NH_2$

$$H\bullet e + NH_3 \rightleftharpoons H_2 + en\bullet NH_2$$

$$en\bullet NH_2 + nn\bullet NH_2 \longrightarrow H_2N - NH_2 \qquad 13.58a$$

(B) **Stage 1:** $NH_3 \rightleftharpoons H\bullet e + nn\bullet NH_2$

$$H\bullet e + NH_3 \rightleftharpoons en\bullet NH_4$$

$$H_2N\bullet nn + en\bullet NH_4 \longrightarrow \overset{H}{\underset{H}{\overset{|}{N^{\ominus}}}}.....\overset{H \overset{H}{\diagup}}{\underset{H}{\overset{|}{\oplus N}}} - H$$

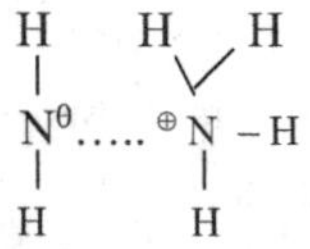

[Ammonium aminide]   13.58b

In (A), abstraction took place, while in (B) instantaneous electrostatic addition took place. They both take place under different operating conditions. While in (A), specific amount of heat will be required, in (B), no heat may be required. Stable fractions of ammonia will obviously vary under different operating conditions. After the release of H_2, the (C) of Equation 13.57 is used along with the monomer to produce an Electro-statically anionically-paired initiator similar to (A) of Equation 13.55, that in which part of (C) will be the carrier of the chain anionically.

With the secondary and tertiary types of aromatic amines, the followings are obtained if H_2 is to be formed.

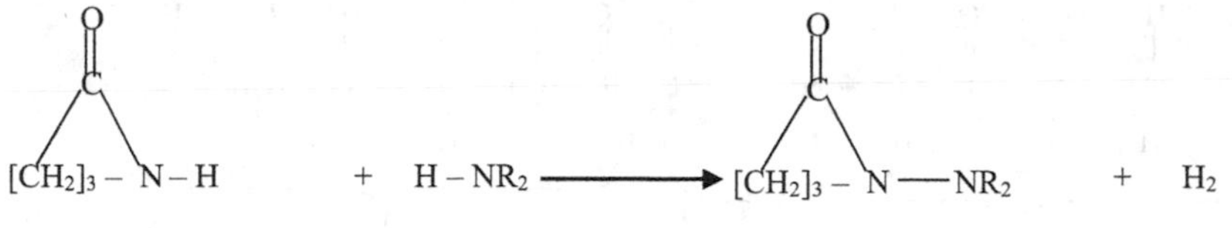

$$[CH_2]_3 - \overset{\overset{\displaystyle O}{\|}}{C} - N-H \quad + \quad H-NR_2 \longrightarrow [CH_2]_3 - \overset{\overset{\displaystyle O}{\|}}{C} - N-NR_2 \quad + \quad H_2$$

.(IV)

Note: R ≡ Phenyl or Benzyl group)   13.59

(IV) which is very stable can only be used as a counter charged center, if and only if the monomer can exist in Equilibrium state of existence or another secondary aromatic amine is present in the system. Otherwise secondary aromatic amine cannot be used if H_2 is to be released. It can only be used in a similar way shown for the primary aromatic amine (A) of Equation 13.55 When the tertiary amine (with no hydrogen atom on the nitrogen center) is used, there is no reaction, if the amide is still suppressed. The corresponding case of Equation 13.55 will also not be obtained. Hence, only primary and secondary amines have been observed to be effective as sources of charge-paired initiators. Thus, it can be observed that both acids and weak bases can suppress the Equilibrium state of existence of amides.

It is important to note that based on the mechanisms provided so far, induction periods are not present during addition of monomer to the growing chain. For each class of amine, the initiation efficiency has been noted to be directly dependent on the basicity of the amine[23,26]. The more moderately basic the amine, the greater the efficiency. The order of basicity for various amines has long be known ***to be ammonia < primary < secondary < tertiary.*** The actual order, however, can vary with the solvent and acid. If it is very basic such as NR_3, it is ineffective. If it is very weak such as ammonia, it is surprisingly ineffective. The use of ammonia is probably not favored for some reasons including its very weak basicity and the fact that it cannot suppress the Equilibrium State of Existence of the amides. Both are in Equilibrium State of existence. Thus, one has not shown here why primary aromatic amines have been noted to be more effective than secondary amines as cationic initiators, but something different. Their use has little to do with steric limitations more in the secondary than primary amines[6], since they are the carriers of the chain. Primary is more effective than secondary, because the Equilibrium state of existence of the primary is stronger than that of the secondary.

It has been observed that the so-called induction period observed during initiation step when amines are used, can be completely eliminated when amine hydrochloride is used, with a first order dependence of rate of polymerization (Rp) on the amine hydrochloride concentration[6]. This is to be expected since the initiator is not prepared here. The amine hyrochloride is the initiator as shown below.

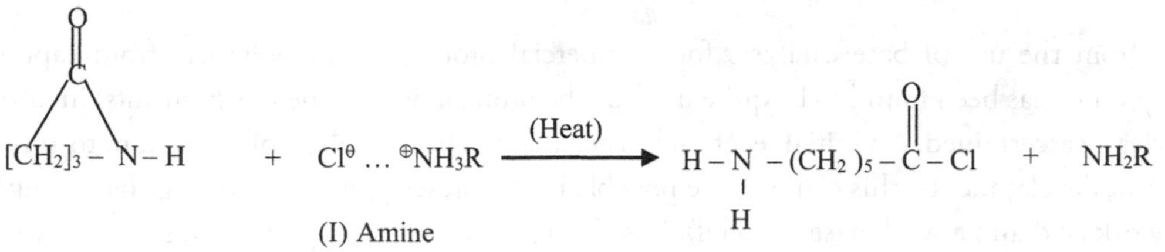

$$[CH_2]_3 - \overset{\overset{\displaystyle O}{\|}}{C} - N-H \quad + \quad Cl^{\ominus} \ldots {}^{\oplus}NH_3R \xrightarrow{\text{(Heat)}} H-\underset{H}{\overset{|}{N}} - (CH_2)_5 - \overset{\overset{\displaystyle O}{\|}}{C} - Cl \quad + \quad NH_2R$$

(I) Amine   13.60

If the amine hydrochloride could suppress the Equilibrium state of existence of the amide and itself could not exist in Equilibrium state of existence as it did above, one should indeed expect the followings-

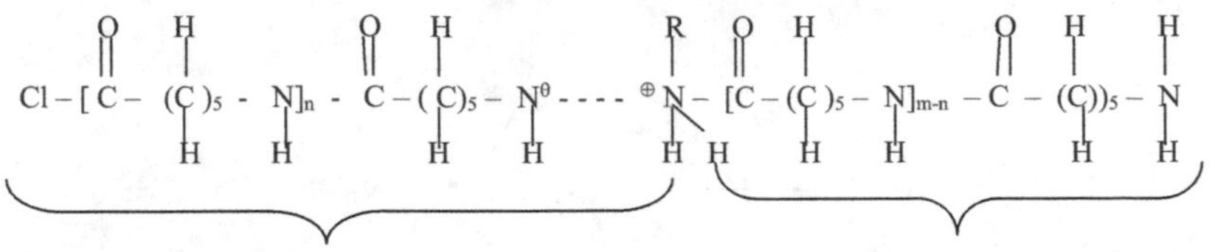

$$\text{Cl} - [\text{C} - (\text{C})_5 - \text{N}]_n - \text{C} - (\text{C})_5 - \text{N}^{\ominus} \cdots {}^{\oplus}\text{N} - [\text{C} - (\text{C})_5 - \text{N}]_{m-n} - \text{C} - (\text{C})_5 - \text{N}$$

"Anionic route"	"Cationic route"
Forward Addition	Backward Addition
First to take place	Second to take place

$$13.61$$

The amine hydrochloride is the catalyst and cocatalyst. There is no Equilibrium mechanism stage(s) involved in its preparation. Polymerization began anionically using $Cl^{\ominus}$ as the active center. After the optimum chain length is reached, this is then followed by Electro-free-radical polymerization. If the Equilibrium state of existence of the amine hydrochloride cannot be suppressed for anionic polymerization, then only electro-free-radical route will be favored and ***this seems to be the case with all these so-called "cationic" initiators considered so far.*** The Equilibrium state of existence of the nitiators are quite strong to be suppressed. It is important to note that in its use electro-free-radically (not cationically) under heat in Equation 13.60, a salt and primary amine are produced. With the backward addition, after propagation, the chain when it can no longer exist in Equilibrium state of existence, is killed using the monomer itself to give HCl, NH_2R and one or two living chains (because their nitrogen centers are still carrying two hydrogen atoms). HCl and NH_2R react to give the original initiator. Each chain carries the ring at its terminal. In the termination step, the amide was made to exist in Equilibrium state of existence in the absence of un-used initiators.

Tertiary amine hydrochloride shown below can be used electro-free-radically and anionically.

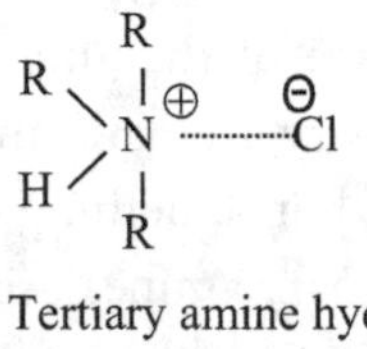

$$
\begin{array}{c}
\quad\quad R \\
\quad\quad | \\
R \diagdown \; {}^{\oplus} \quad\;\; {}^{\ominus} \\
\quad\quad N \cdots\cdots Cl \\
H \diagup \; | \\
\quad\quad R
\end{array}
$$

Tertiary amine hydrochloride

$$13.62$$

Thus, from all the consideration so far, one can observe the number of impossible mechanisms and reactions in the literature of the past and to the present. It is incomprehensible to imagine. Nevertheless without these universal data and observa-tions, the current developments would not have been possible

13.1.1.2(c) Use of Water/Amino Acids

Apart from the use of base catalysts for commercial production of polymers from caprolactam, catalysis by water has been found to be quite useful. The problem is that the mechanism(s) involved have not been clearly ascertained. An initial reaction has been said to be hydrolysis of the lactam to amino acid. From current developments, this can only be possible in the presence of heat, noting that though water is both a weak acid and a weak base (alcohol), it is able to suppress the Equilibrium state of Existence of amides (See Equation 13.35). As a base, it is too weak to favor any initiation anionically, since it is not paired as shown below. When paired and kept stable, the first route is anionic from the $HO^\ominus$ center. After this, the electro- free-radical route follows if kept in Equilibrium state of existence.

$$HO^\ominus \ldots\ldots\ldots \overset{\oplus}{O} \overset{\displaystyle H}{\underset{\displaystyle H}{|}} - H$$

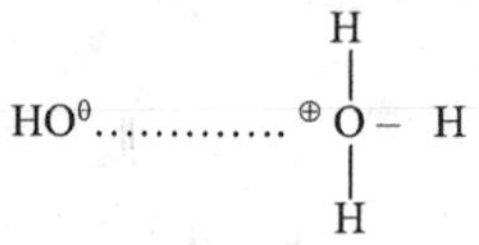

(A) Water as initiator

13.63a

When part of the water cannot be suppressed in the presence of the amide which is also already suppressed, the followings are to be expected.

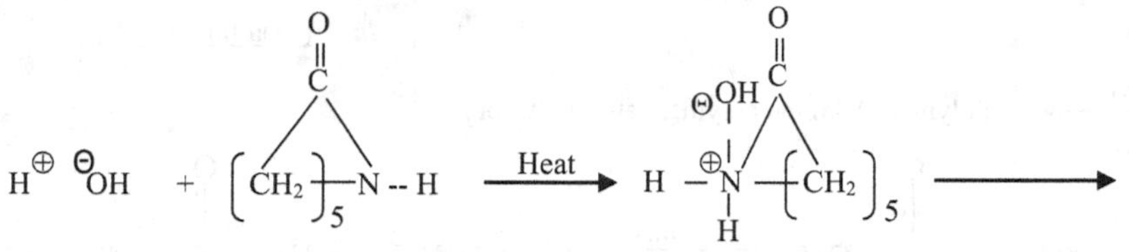

$$H^\oplus \ {}^\ominus OH \ + \ \left[CH_2 \right]_5 N \text{--} H \ \xrightarrow{\text{Heat}} \ H - \overset{\oplus}{\underset{H}{N}} \left(CH_2 \right)_5 \longrightarrow$$

(B) H₂O/Amide Initiator

$$H_2N - (CH_3)_5 - \overset{O}{\overset{||}{C}} - OH$$

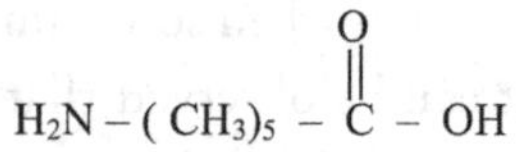

(I) An amino acid

13.63b

It is not easy to obtain (A) above in the absence of electrifying operating conditions. Water was able to suppress the Equilibrium state of existence of the amide and even open the ring to form a stable product- an amino acid. Existence of (I) will depend on the ionic intensity of the polymerization environment and amount of water used. If very small, then the counter-ion of $H^\oplus$, that is, $^\ominus OH$ will be less nucleophilic than the monomer which is in Equilibrium state of existence. If large amounts of water are used, then (I) is favored noting however that the monomer is more basic and acidic than water. If the amount is as large as that of the monomer, amino acids obtained is made to undergo <u>Step</u> rather than <u>Addition</u> as shown below.

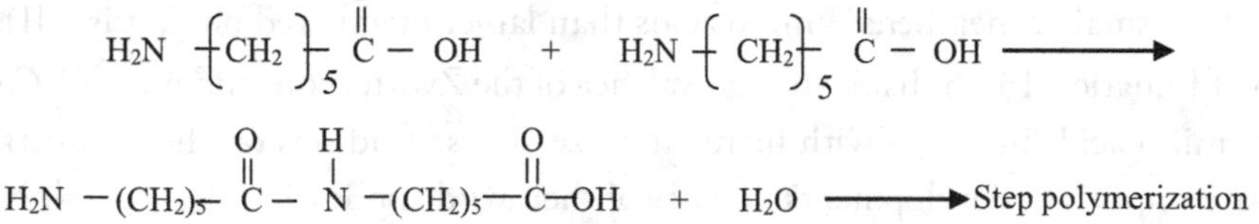

$$H_2N \left(CH_2 \right)_5 \overset{O}{\overset{||}{C}} - OH \ + \ H_2N \left(CH_2 \right)_5 \overset{O}{\overset{||}{C}} - OH \longrightarrow$$

$$H_2N - (CH_2)_5 - \overset{O}{\overset{||}{C}} - \overset{H}{\overset{|}{N}} - (CH_2)_5 - \overset{O}{\overset{||}{C}} - OH \ + \ H_2O \longrightarrow \text{Step polymerization}$$

13.64a

This takes place via Equilibrium mechanism in many stages in an acidic environment.

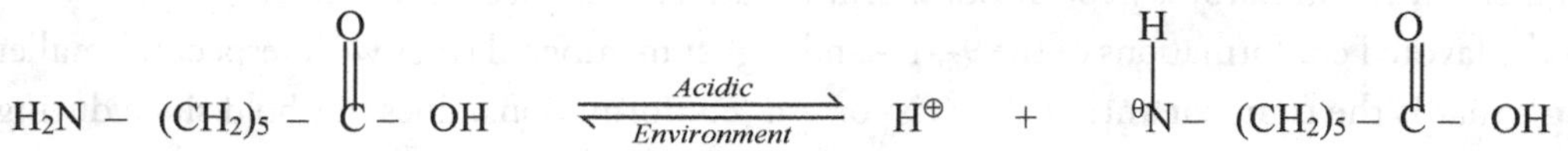

$$H_2N - (CH_2)_5 - \overset{O}{\overset{||}{C}} - OH \ \underset{\text{Environment}}{\overset{\text{Acidic}}{\rightleftharpoons}} \ H^\oplus \ + \ {}^\ominus N \overset{H}{\overset{|}{}} - (CH_2)_5 - \overset{O}{\overset{||}{C}} - OH \qquad 13.64b$$

183

When the amino acid is more acidic in an acidic environment (due to large concentration of water involved), then the amino-acid will add step-wisely. When the amino acid is in a basic environment, then the followings shown below take place.

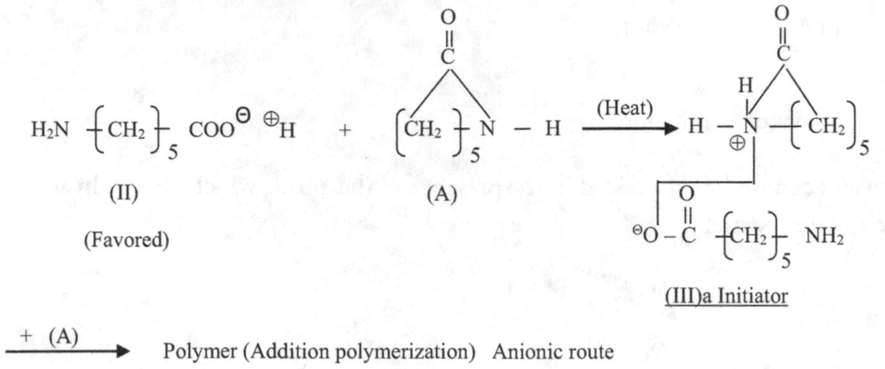

$$H_2N \left[CH_2\right]_5 COO^{\ominus} \ ^{\oplus}H \ + \ \left[CH_2\right]_5 N - H \xrightarrow{(Heat)} \ (III)a \ Initiator$$

$$\xrightarrow{+ \ (A)} \text{Polymer (Addition polymerization) Anionic route} \qquad 13.65a$$

$$H_2N - (CH_2)_5 - \overset{O}{\underset{\|}{C}} - OH \underset{Environment}{\overset{Basic}{\rightleftharpoons}} H_2N - (CH_2)_5 - \overset{O}{\underset{\|}{C}} - O^{\ominus} \qquad 13.65b$$

Thus, water alone cannot produce polymers, but amino acids can. The larger the size of the ring, the less acidic is the ring. The reaction above is a possible reaction when some amino acids alone are used as the source of initiator for opening of the ring. Thus, it can be observed that when water is involved as initiator, the followings routes are involved during initiation, depending on the concentration of water involved, the basic/acidic environment and the size of the ring-

(i) Existence of (I) of Equation 13.63b for use as a Step monomer in an acidic environment.

(ii) Existence of (III)a as initiator first used anionically followed by electro-free-radical route, if the environment is more basic than acidic.

(iii) Possible existence of (III)b shown below from (I) as initiator, if the environment is neutral or slightly more acidic

When (III)b is involved, there is no doubt that Step polymerization will be competing with Addition polymerization. It is in view of the above, that the unexpected reactivity orders, shown for lactams when water is involved as the source of initiators, in Table 13.1, were obtained. Though the existence of Zwitter-ion is[26-29] favored for smaller membered amino acids than larger membered ones, it is (III)a of Equation 13.65a or (III)b of Equation 13.66 that is the equivalence of the Zwitter-ion, and not $^{\ominus}O_2C(CH_2)_5NH_3^{\oplus}$. The basicity of amino acids increases with increasing size. These findings on the chemistry of this type of amino-acids are important developments, in view of their wild applications in general. Nevertheless, it can be observed that with the use of water as source of initiator, the route is largely anionic; with cationic or indeed electro-free-radical route following after the completion of the anionic route or only the latter (Electro-free-radical) if the initiator is kept in Equilibrium state of existence all the time.

The use of the favored conformations of the 9-, 11- and higher membered rings with respect to smaller membered-ring[6], that is the trans- or anti-, or the cis- or syn- conformations, does not hold since during charged paired polymerizations of the monomers, only specific conformations amongst the countless number of conformations favored by them, will be obtained, irrespective of the size, or even and odd numbered characters of their rings which have one functional and one activation centers.

From all the considerations so far, there is no doubt that, most of the doubts which have been raised over the years over the acceptability of the numerous mechanisms which have been proposed, have been completely laid to rest, based on current developments. Realization of the existence of female and male characters of compounds including all monomers are very important. While some monomers are part male-like and female-like in character, some females are part female-like and male-like in character. ***These observations can be found in nature with "so called" "living things". Indeed, all things known to exist (animate or inanimate) are living things, based on the chemistry of compounds and elements as currently developed. There are different types of living things, some with the senses of Life and others with little or no senses of Life. The senses of Life in question are the FIVE senses both real and imaginary- Sight, Sound, Smell, Taste and Feelings. These are all Forces of Nature, that which is Physics.***

In view of the electro-free-radical route of the polymerization of these monomers (wherein depropagation phenomenon will not manifest itself), one can begin to envisage what to expect when cyclic amides with radical-pushing groups are involved. When radical-pushing groups are internally located, the method of initiation will remain the same, except that more heat may be required to introduce more strain energy into the ring via the functional center or activation center. If the MRSE cannot be introduced then no polymerization will be favored.

13.1.2 Radical Characters of Cyclic Amides

Based on the functional centers present in cyclic amides, only $\overset{N}{\underset{C}{\bigcirc}} = O$ center can be considered for use nucleo-non-free-radically. Secondly, it has been observed that these monomers are weakly strained. Thirdly, it has been observed that the monomers have both electrophilic and nucleophilic tendencies, basic and acidic tendencies. Fourthly, the use of nucleo-non-free-radicals is favored since the anionic route is favored. Fifthly, at low temperatures, the ring cannot be opened, as reflected by the reactions of Equation 13.36.

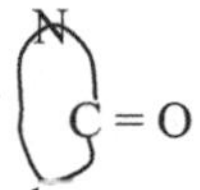

Nevertheless, the radical polymerization of these monomers will not be considered in view of the fact that all the reactions which take place anionically can also take place nucleo-non-free-radically and most of them which are said to take place cationically actually take place electro-free-radically. Indeed those that take place chargedly are radical in character. When ionic metallic centers are involved radically or ionically when initiators are prepared, anionic-paired initiators can be obtained.

When the radical-pushing group is located on the nitrogen center, by the method of initiation of these monomers wherein the monomer is the cocatalyst, polymerization is favored as along as the monomer and the initiator (water, protonic acid, amines) are the only ones present in the system. Hence, unlike what was done in the anionic case, where small amount of secondary acidic cyclic amides must be present in the system only during initiation, here this is not required as shown below. With the radical-pushing group placed on the nitrogen center, the monomer is no longer secondary, but tertiary and the monomer is therefore more basic than if the group had not been placed on the ring.

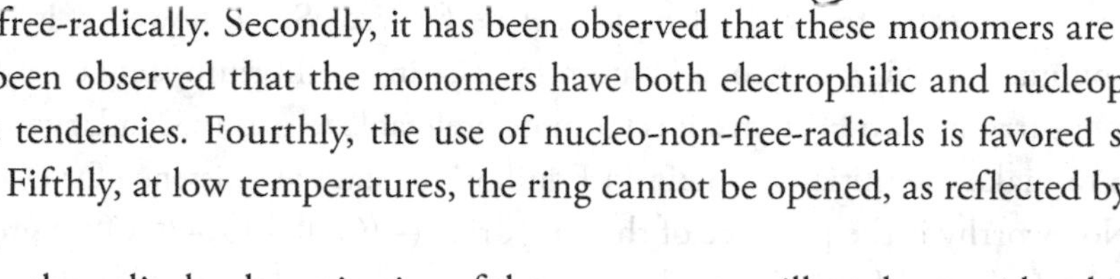

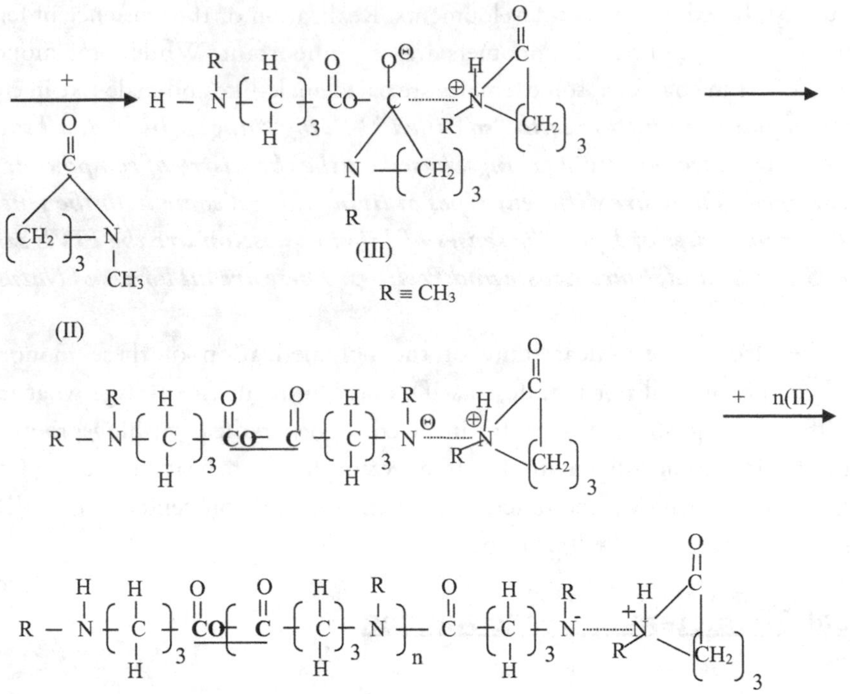

$$13.67$$

The route above is favored only if the initiator is in Stable state of existence. If in Equilibrium state of existence, then only the electro-free-radical route will be favored. Since however, it is more difficult for the tertiary monomer carrying only one H atom to exist in equilibrium state of existence than the secondary monomer carrying two H atoms, it is far more preferable to use a secondary monomer only for the initiation step to make the initiator remain in Equilibrium state of existence for electro-free-radical polymerization. Noteworthy is the presence of the anhydride (– (CO) – O – (CO) –) group (underlined above) at the terminal of one of the chains that grew anionically in an acidic environment. Less heat may be required to introduce MRSE into (III) anionically than electro-free-radically. Only the anionic route is natural to these monomers via the CO center, with the monomers being more electrophilic than nucleophilic in character. Hence, higher operating conditions required electro-free radically or cationically are to be expected, since the route is not natural to the monomer. Thus when the operating condition is normal, only the nucleo-non-free-radical or anionic route is allowed to take place, while higher operating conditions will be required for the electro-free-radical route.

13.1.3. Order of Nucleophilicity of Cyclic Amides

From the developments so far, one can observe the unique features of cyclic amides in which the ring is part of the initiator, in view of the weak strain in their rings and the unique qualities of a nitrogen center. Despite the weak strain in the rings compared to those of oxygen containing monomers and carbon-carbon-ringed monomers, they seem to be most nucleophilic amongst the three members. Side-by -side placement of rings have been found to be one of the methods by which the capacities of ring can be determined.

Considered placing two lactam rings of different sizes side by side.

Anionically, the followings are obtained.

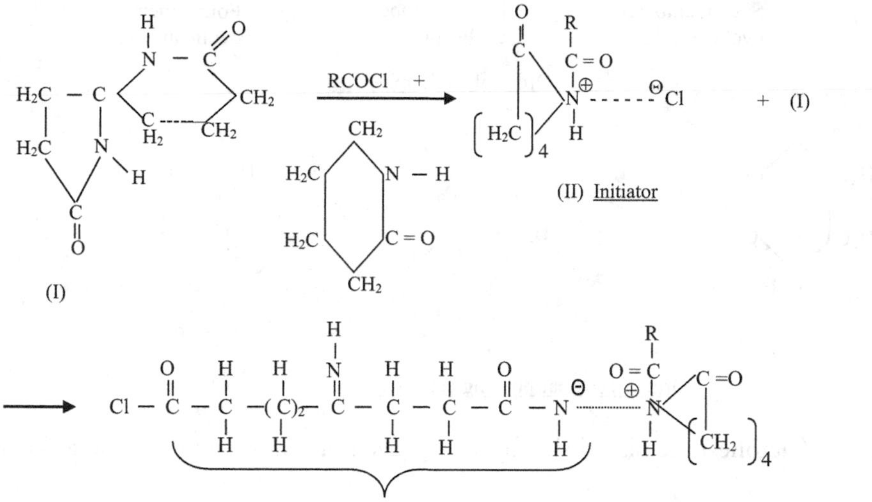

(III)Initiation Step   13.68a

In the reaction above, the initiator was first obtained using very small concentration of acid chloride and a six-membered lactam. (I) above is a five-membered lactam placed side-by-side to a six-membered lactam. In view of the size of (I), it could not be used as part of initiator. In the reaction above, it is the six-membered ring that was first attacked, since it is less electrophilic than the five-membered ring.

"Cationically", the followings are obtained using a protonic acid.

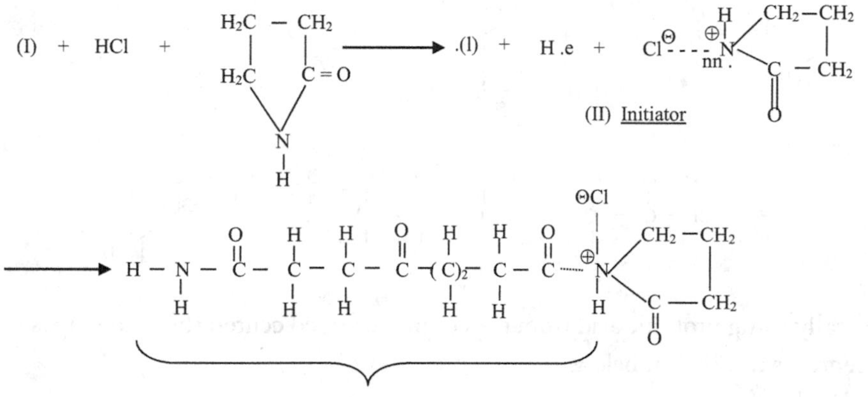

(III) INITIATION STEP   13.68b

(II) the initiator used above, must first be prepared before adding (I). As has already been shown when protonic acid is used, the route is anionic when the initiator is in Stable state of existence. So called cationic route which indeed is electro-free-radical route takes place only when the initiator is in Equilibrium state of existence. To obtain (III), like the case above, the five-membered ring was the first to be attacked, since it is less nucleophilic than the six-membered ring. When (III) above is compared to that obtained anionically, the same monomer unit shown below is obtained.

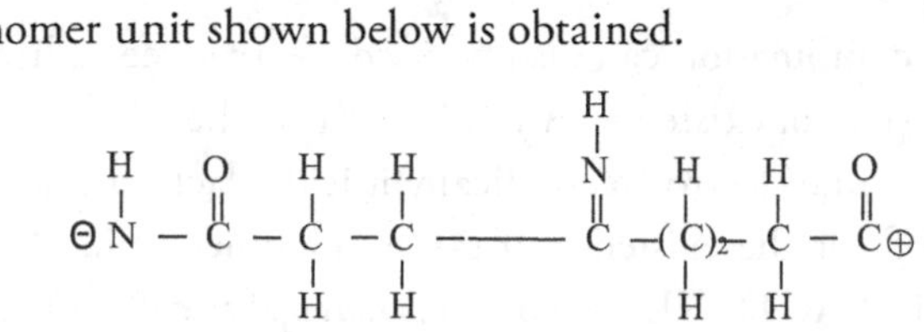

Monomer unit of (I)   13.69

This is to be expected, since the rings belong to the same family. It can be observed in general that-

$$> \quad \begin{array}{c} \text{Six - membered} \\ \text{cyclic amides} \end{array} \quad > \quad \begin{array}{c} \text{Five - membered} \\ \text{cyclic amides} \end{array} \quad > \quad \begin{array}{c} \text{Four - membered} \\ \text{cyclic amides} \end{array} \quad >$$

Order of nucleophilicity

13.70a

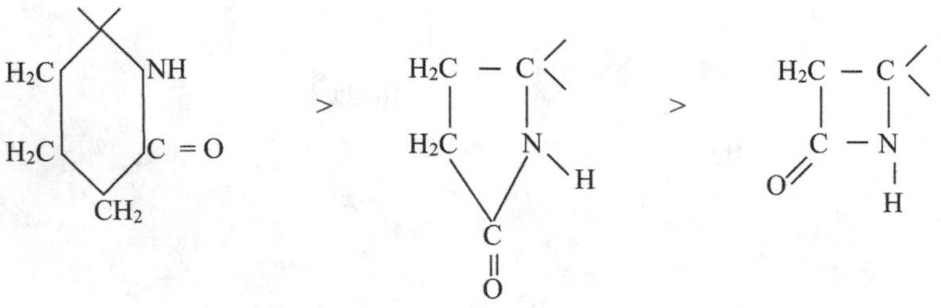

Order of radical-pushing capacity

13.70b

When cyclic esters (lactones) are placed side-by-side to cyclic amides (lactams), the followings are to be expected.

Anionically-

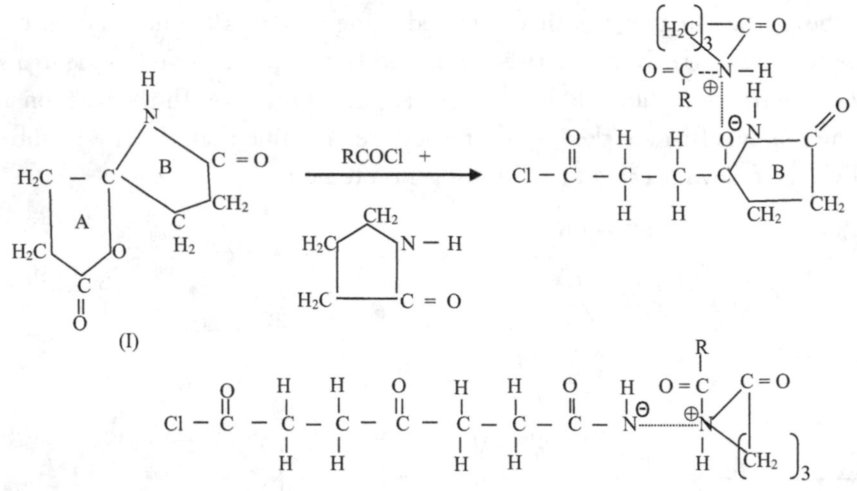

13.71a

Electro-free-radically using protonic acid from the counter charged center, the following is to be expected. The first used center is not shown below.

Electro-free-radically-

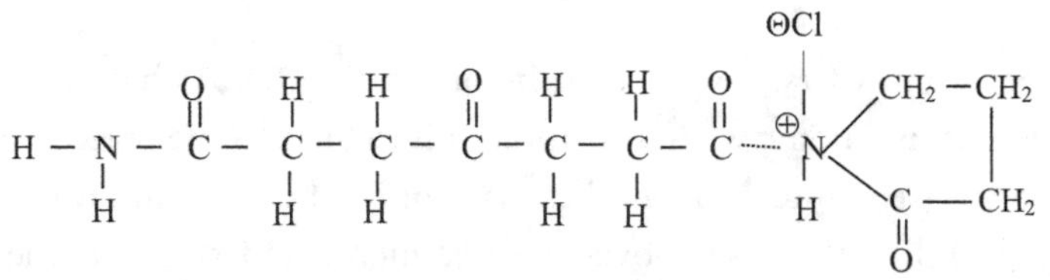

INITIATION STEP

13.71b

Note that the RCOCl/amide combination can also be used electro-free-radically if the initiator can be made to exist in Equilibrium state of existence. Anionically, it is the lactone that is first attacked being less electrophilic than lactam, while electro-free-radically it is the lactam that is first attacked being less nucleophilic than lactone, to favor the existence of same monomer units. Hence, the order shown in Equation 13.46 is valid, that is, LACTONES are *more Nucleophilic* than LACTAMS.

For the coupled ring shown below, the followings are obtained.

Anionically-

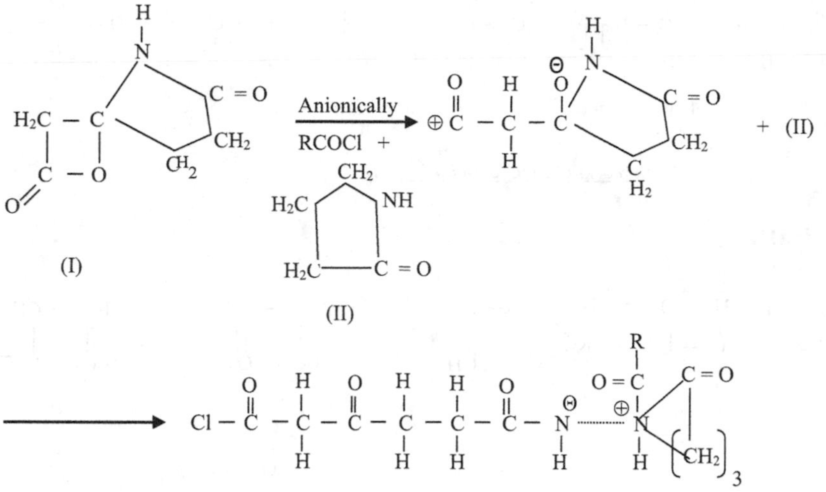

(I)

(II)

Lactams > Lactones [Order of Electrophilicity]   13.72

Electro-free-radically using protonic acid from the counter charged center, the following is to be expected. The first used center is not shown below.

Electro-free-radically-

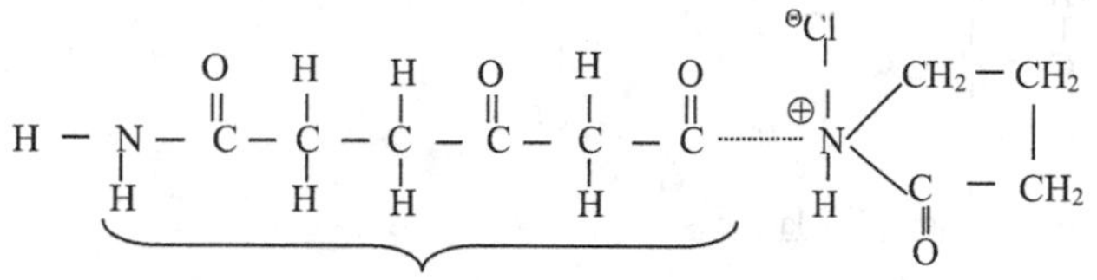

Lactones > Lactams   [Order of Nucleophilicity]   13.73

Electro-free-radically or so-called Cationically, it is important to note that the instanta-neous opening of the four membered ring by the initiator (II) above is still possible, though the four-membered ring is more nucleophilic than the five-membered ring and lactams are more strained than lactones. It is the less nucleophilic ring that is first attacked electro-free-radically. It was the one that was first opened instantaneously, despite the fact that lactams are more strained than lactones, because the size of the lactone above is smaller than the size of the lactam.

Based on the analysis above, it will not be surprising to note that lactam can be copolymerized with lactones anionically and cationically to produce long block of copolymers, provided the initiation is commenced using lactams and not lactones, in view of the unique charged-paired initiators involved with lactams throughout the entire course of polymerization.

Anionically-

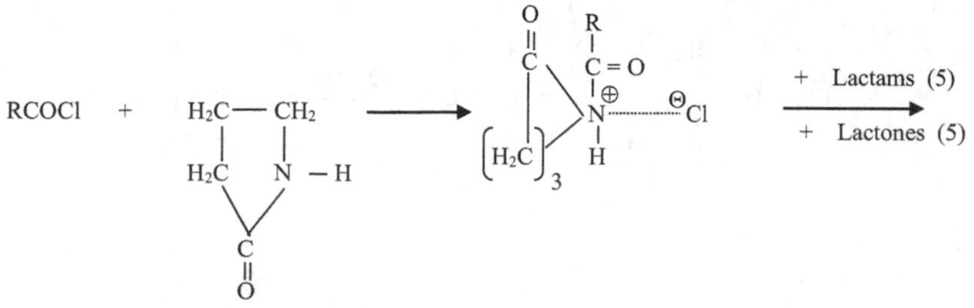

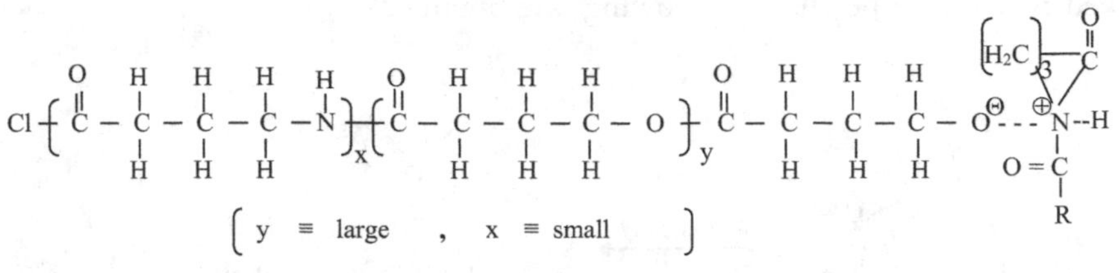

Long blocks of lactones (anionically)

$[y \equiv \text{large} \quad , \quad x \equiv \text{small}]$

13.74

Electro-free-radically-

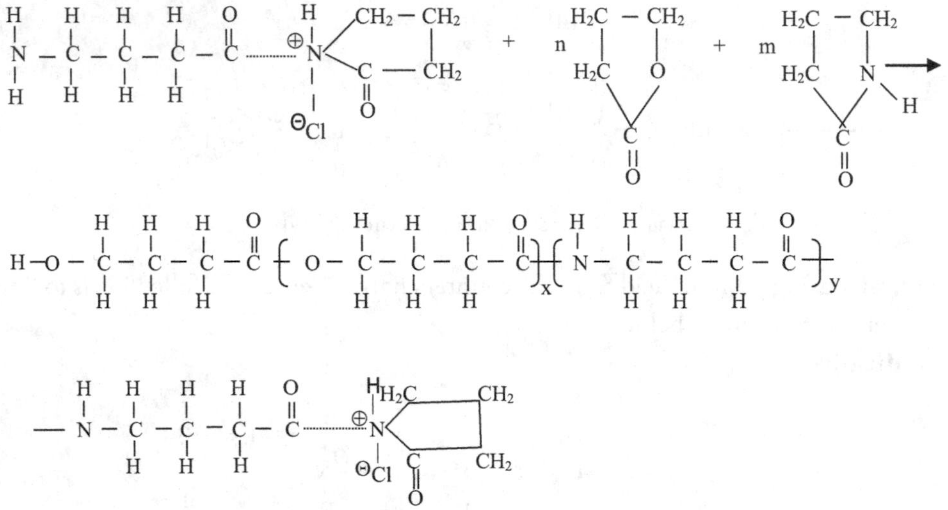

Long blocks of lactams than lactones

13.75

Note that the order of attack is independent of the sizes of the rings, but on the type of reactor, ratios of components involved and manners of addition of components.

When a cyclic ether ring is placed side by side to a cyclic amide, the followings are obtained.

Anionically-

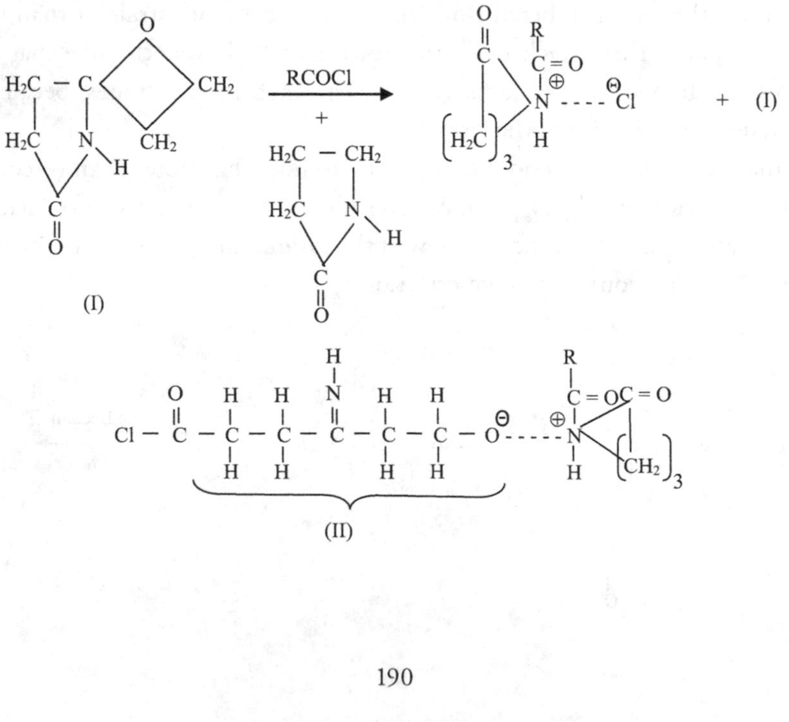

(I)

(II)

13.76

Electro-free-radically-

$$(I) \quad + \quad H\bullet e \quad + \quad nn \bullet \overset{H}{\underset{\underset{\Theta Cl}{\overset{\oplus}{|}}}{N}} \overset{CH_2-CH_2}{\underset{\underset{O}{\overset{||}{C}}-CH_2}{|}} \quad \longrightarrow$$

$$H-O-\overset{\overset{H}{|}}{\underset{\overset{|}{H}}{C}}-\overset{\overset{H}{|}}{\underset{\overset{|}{H}}{C}}-\overset{\overset{H}{\overset{|}{N}}}{\underset{}{C}}-\overset{\overset{H}{|}}{\underset{\overset{|}{H}}{C}}-\overset{\overset{H}{|}}{\underset{\overset{|}{H}}{C}}-\overset{\overset{O}{||}}{\underset{}{C}}\cdots\overset{H}{\underset{\underset{\Theta Cl}{\overset{\oplus}{|}}}{N}}\overset{CH_2-CH_2}{\underset{\underset{O}{\overset{||}{C}}-CH_2}{|}}$$

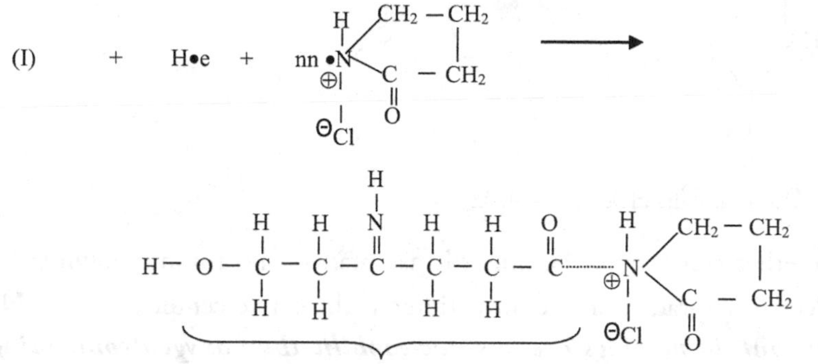

$$\underbrace{\qquad\qquad\qquad\qquad}_{(II)a}$$

13.77

Anionically, via instantaneous opening of the cyclic ether ring which is more stained in view of its size, we have the followings

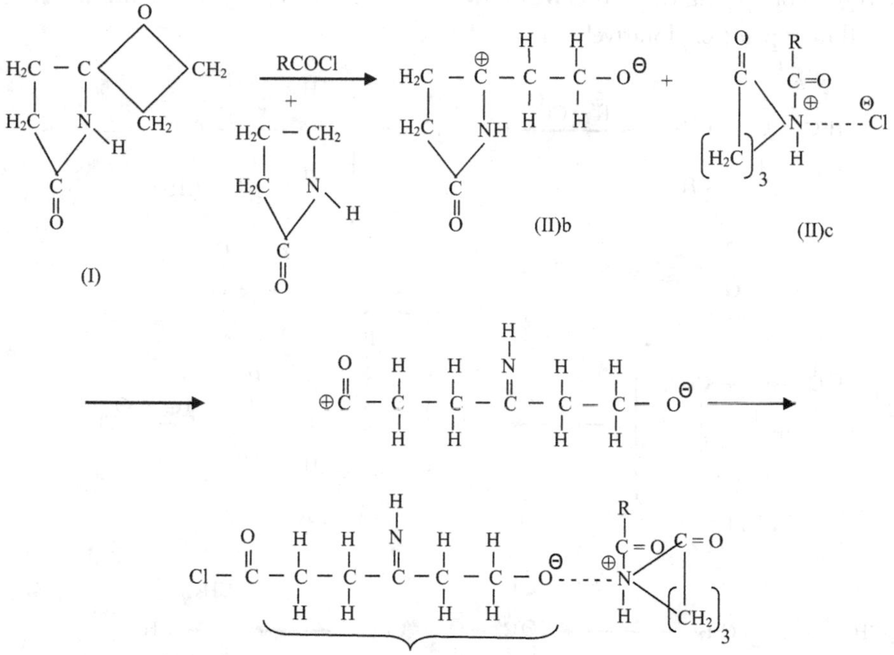

In (II), (II)a and (III), the monomer units are the same. Though the cyclic ether has no electrophilic activation center, the ring can be opened instantaneously as shown. Here, it is the cyclic ether ring, that is first attacked electro-free-radically. From the types of monomer units obtained, there is no doubt that the followings are valid.

Lactams > Cyclic ethers

Order of Nucleophilicity

Lactams > Cyclic ethers

Order of electrophilicity

13.79

191

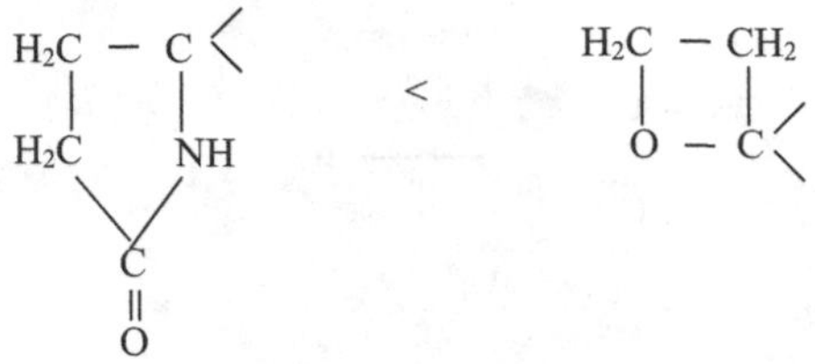

$$H_2C - C\langle \qquad\qquad H_2C - CH_2$$
$$| \qquad | \qquad\qquad\qquad | \qquad |$$
$$H_2C \quad NH \qquad < \qquad O - C\langle$$
$$\qquad |$$
$$\qquad C$$
$$\qquad \|$$
$$\qquad O$$

<u>Order of radical-pushing capacity</u>   13.80

When the cyclic ether ring cannot be opened instantaneously, anionic polymer-ization cannot be favored. Only the electro-free-radical route (or cationic if the active center is $H^\oplus$ or $Na^\oplus$,..) is favored. *In general nitrogen containing rings are less nucleophilic than oxygen containing rings whether in the absence or presence of adjacently located C = O group to the nitrogen center.* To be able to copolymerize cyclic amides with cyclic ethers, one must first begin with cyclic amide paired initiators before introducing mixtures of cyclic ethers and amides into the system.

Like the oxygen containing rings, one will now introduce the $H_2C =$ group into the ring, in order to identify what will take place exploratively.

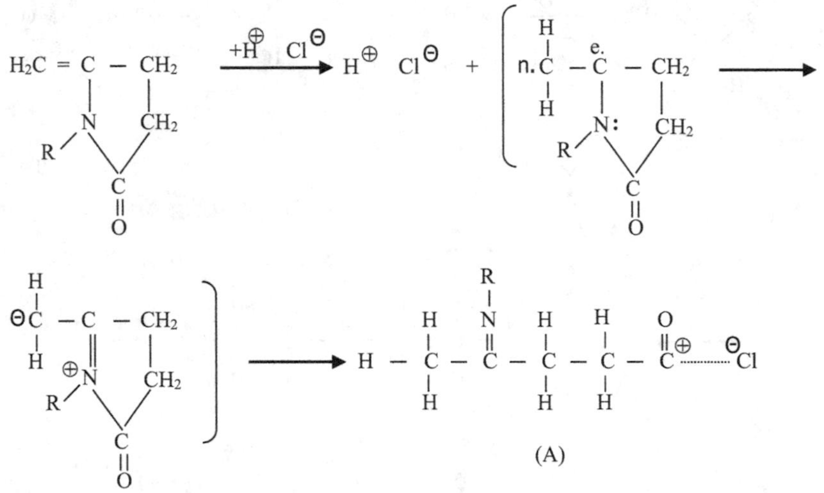

$$H_2C = C - CH_2 \xrightarrow{+H^\oplus \ Cl^\ominus} H^\oplus \ Cl^\ominus + \left(\begin{array}{c} H \\ | \\ n.C - \overset{e.}{C} - CH_2 \\ | \\ H \end{array} \right. \longrightarrow$$

(A)   13.81

$$H_2C = C - CH_2 \longrightarrow \ominus\overset{H}{C} - \overset{\oplus}{C} - CH_2 \longrightarrow \overset{CH_3}{C} - CH_2$$

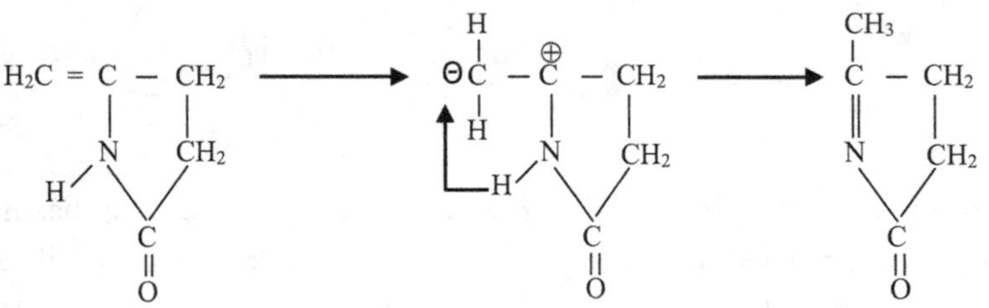

(I) <u>Electrophile</u>   (II) <u>Electrophile</u> (More stable)   13.82

When activated, it undergoes molecular rearrangement of the first type to produce (II) which is more nucleophilic in character than (I). (II) which looks resonance stabilized has two activation centers, of which only one can be used "cationically", since it is the only nucleophilic center, the second activation center being electrophilic. These are centers which are no functional centers that may be used for opening the ring. It is said to be expolatory, because it has been wrongly activated. The monomer cannot be activated chargedly due to electrostatic forces of repulsion. It can only be activated radically.

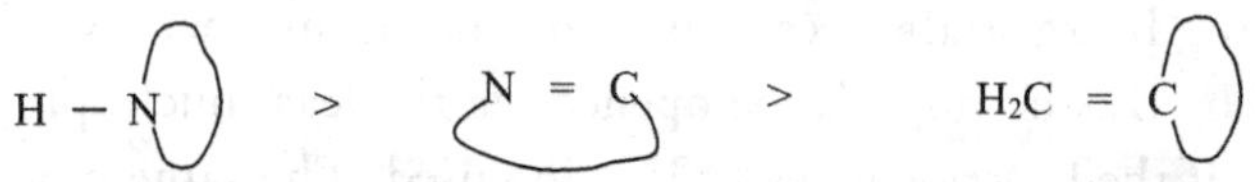

$$H - N \quad > \quad N = C \quad > \quad H_2C = C$$

13.83

When properly activated, the followings are obtained electro-free-radically.

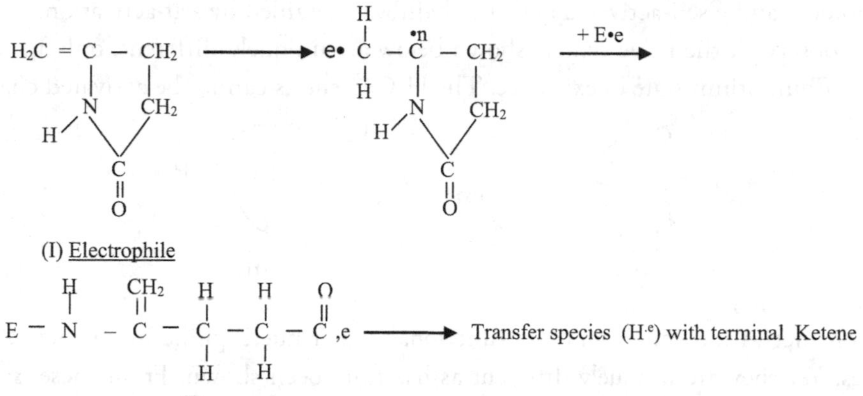

$$H_2C = C - CH_2 \longrightarrow e\bullet \overset{H}{\underset{H}{C}} - \overset{\bullet n}{C} - CH_2 \xrightarrow{+ E\bullet e}$$

(I) Electrophile

$$E - N - C - C - C - C, e \longrightarrow \text{Transfer species (H}^{\cdot e}\text{) with terminal Ketene}$$

(I)

13.84

It looks as if the functional center has been used. Nucleo-non-free-radically, the followings are obtained.

$$H_2C = C - CH_2 \longrightarrow e\bullet \overset{H}{\underset{H}{C}} - \overset{\bullet n}{C} - CH_2 \xrightarrow{+ N \bullet nn}$$

(I) Electrophile

$$N - C - C - C - C - N \cdot nn$$

(A) Favored

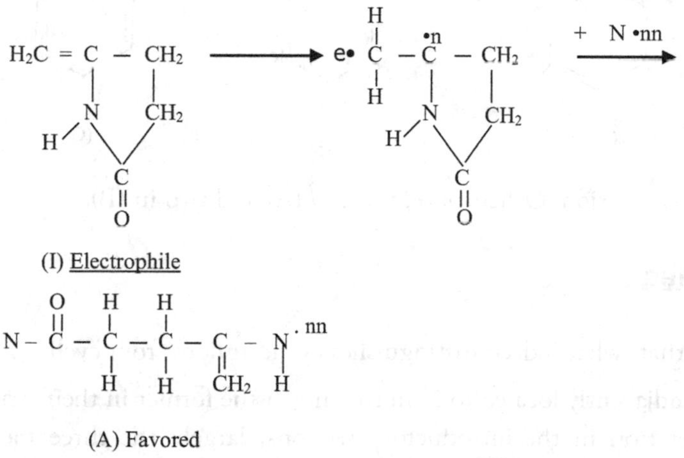

13.85

Chargedly the monomer cannot be activated due to electrostatic forces of repulsion. The electrophilic character of the monomer is still maintained. These observations are not different from those of oxygen containing rings with $H_2C=$ groups.

Consider changing the location of the $H_2C=$ group in the ring.

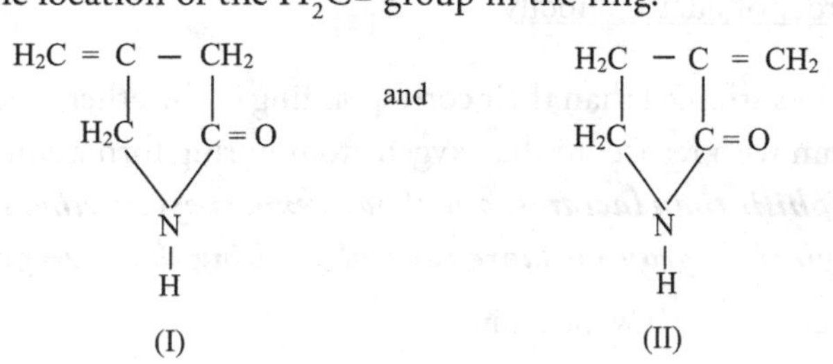

$$H_2C = C - CH_2 \qquad \text{and} \qquad H_2C - C = CH_2$$

(I)   (II)

13.86

For both cases above, there is one functional center like in other cyclic amides. However, it cannot be used in the same manner, since it is not the least nucleophilic center. The ring of (I) cannot be opened via the least nucleophilic center (the $H_2C=$ center on the ring) since there is no suitable point of scission on the ring. The ring being strained can favor instantaneous opening of the ring to reflect its electrophilic character. With (II), the ring can be opened via the least nucleophilic center electro-free-radically, making it look as if the functional center is being used. The same monomer unit is obtained nucleo-non-free-radically.

As has been said, propagation/depropagation phenomena do not take place during homopolymerization of cyclic amides electro-free-radically or anionically. With carbon–carbon chain monomers, it takes place only if the monomer can be self-activated and the chain was obtained by self-activation.

As has been observed, the three centers shown below are uniquely different, only in terms of their ability to exist in Equilibrium state of existence. The $H_2C=$ centers cannot be activated chargedly.

$$(I) \qquad\qquad (II) \qquad\qquad 13.87a$$

In (I), there is one activation center and one functional center nucleophilic in character, while in (II) the same applies. Yet they are uniquely different as has fully been shown. From these, so also are the compounds shown below in terms of point of scission.

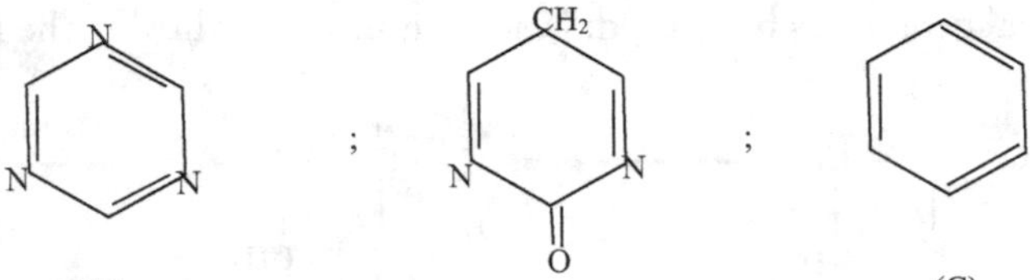

$$(A) \qquad\qquad (B) \qquad\qquad (C) \qquad\qquad 13.87b$$

Only (C) has no point of scission. Others have, three in (A) and two in (B).

13.2 Cyclic Amines

It has been noted that, what indeed distinguishes cyclic amides from cyclic amines, is the existence of a $C = O$ center adjacently located to N in the ring of the former in their structures.

From the consideration in the introductory sections, largely the three membered-ring will be considered. In view the absence of $C = O$ center, cyclic amides are more nucleophilic than cyclic amines, though cyclic amines are nucleophiles (Females).

$$\text{Cyclic amides} \quad > \quad \text{Cyclic amines}$$

$$\underline{\text{Order of nucleophilicity}} \qquad\qquad 13.88$$

They (Cyclic amines) look less strained than their corresponding cyclic ethers and cyclic acetals, in view of the role of the nitrogen atom with respect to the oxygen atom in ring formation. *As will be shown, since lactones are more nucleophilic than lactams, one should expect cyclic ethers to be more nucleophilic than cyclic amines, though HN= group is more radical-pushing than O= group.*

$$\text{Cyclic ethers} \quad > \quad \text{Cyclic amines}$$

$$\underline{\text{Order of nucleophilicity}} \qquad\qquad 13.89$$

The fact that three-membered ring favors copolymerization with carbon monoxide[30], does indicate that the monomer can be instantaneously opened using radical donating groups and free-radicals since large

SE exist in this unactivated ring. Hence both the charged and radical polymerization of cyclic amines will be considered in the present analysis.

13.2.1. Charged Characters of Cyclic Amines

The unreactivity of the monomer aziridine towards base catalysis was thought to be due to instability of an amine anionic growing polymer chain as propagating species. This is not fully truly the case. The reason is because the route is not natural to the monomer, for which as the chain continues to propagate, the intensity of the growing chain continues to decrease. Anionically only the three-membered ring can be considered, since instantaneous opening of the ring can be favored. Secondly and most importantly, the anionic route is not natural to cyclic amines, all being Females (i.e. Nucleophiles). The natural route is positively charged or Electro-free-radical route.

When the ring is opened in the presence of strong bases such as alkoxides, sodium, the initiation step is favored anionically as shown below. In view of the strong basic character of aziridines which are secondary amines, it will not be easy to instantaneously open the ring using weak bases. For some of them, salts some of which can be used as initiators are produced.

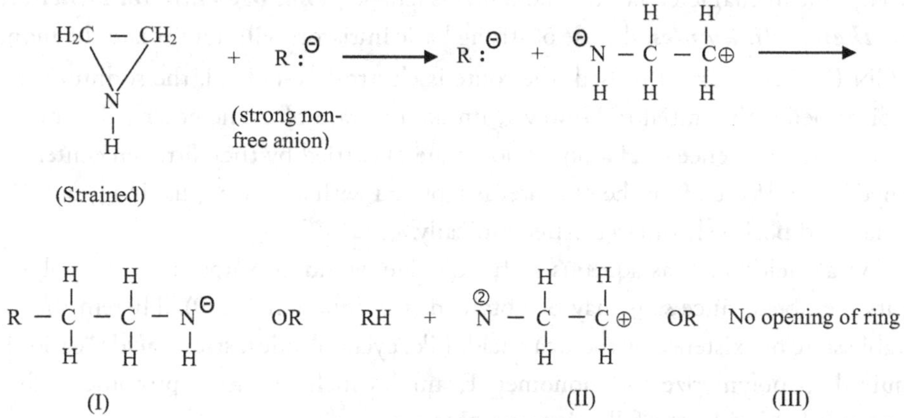

It is (I) that is most likely favored once the ring is opened instantaneously when stable. When the anionic initiator is weak, it cannot produce an energy sufficient enough to attain the MRSE for the ring, and the followings are obtained.

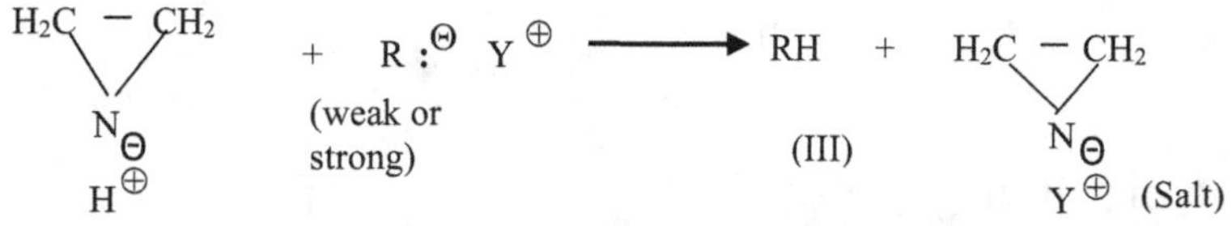

In (II) of Equation 13.90, there is no transfer species involved. Instead of what is shown in (II), it is the ring that is carrying the charge as shown below nucleo-non-free-radically. This is only favored when the ring is made to exist in Equilibrium state of existence.

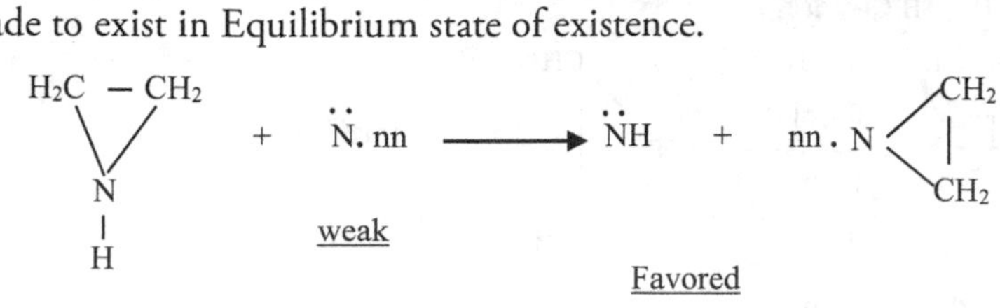

$$
\underset{\text{(I)}}{\underset{\overset{|}{\text{N}}}{\overset{\text{H}_2\text{C} - \text{CH}_2}{\diagdown \diagup}}} \quad + \quad \underset{\text{strong}}{\overset{..}{\text{N}}\cdot \text{nn}} \quad \xrightarrow{\begin{array}{c}\text{(Non-ionic}\\\text{environment)}\end{array}} \quad \overset{..}{\text{N}}\cdot \text{nn} \quad + \quad \underset{\text{(II)}}{e\cdot \overset{\overset{\text{H}}{|}}{\underset{\underset{\text{H}}{|}}{\text{C}}} - \overset{\overset{\text{H}}{|}}{\underset{\underset{\text{H}}{|}}{\text{C}}} - \overset{\text{H}}{\text{N}}\cdot \text{nn}} \longrightarrow
$$

$$
\text{NH} + \underset{\text{(III)}}{\underset{\underset{\text{nn}}{\overset{..}{\text{N}}}}{\overset{\overset{\text{nn}}{\bullet}}{\text{N}}} - \overset{\overset{\text{H}}{|}}{\underset{\underset{\text{H}}{|}}{\text{C}}} - \overset{\overset{\text{H}}{|}}{\underset{\underset{\text{H}}{|}}{\text{C}}}\cdot e} \quad \text{OR} \quad \underset{\text{(IV)}}{\overset{..}{\text{N}} - \overset{\overset{\text{H}}{|}}{\underset{\underset{\text{H}}{|}}{\text{C}}} - \overset{\overset{\text{H}}{|}}{\underset{\underset{\text{H}}{|}}{\text{C}}} - \overset{\text{H}}{\text{N}}\cdot \text{nn}} \quad \text{OR} \quad \overset{..}{\text{NH}} + \underset{\text{(V)}}{\overset{\text{nn}}{\underset{\overset{|}{\text{N}}}{\diagup}} \diagdown \underset{\text{CH}_2}{\overset{\text{CH}_2}{}}} \qquad 13.93
$$

It is (IV) that is favored above if the amine is in Stable state of existence. However, as it seems the monomer is more in Equilibrium state of existence, making it possible for (III) of Equation 13.91 and (V) of Equation 13.93 to be also favored. For instantaneous opening of the ring to be favored, this must be done by ***non-reactive electrostatic center such as in CO, NH₃. When gamma radiation is used, the initiator is nucleo-free-radical in character, for which initiation step will not be favored.***

"Cationically", the polymerization is favored due to the fact that, the only functional center present in the ring is nucleophilic in character, i.e., the monomer is female. ***It has been said that since cyclic amines are more basic than cyclic amides***, the use of strong basic initiators will not initiate the monomer. If Na (Na•e) or NaCN (Na$^{\bullet e}$.....$^{n\bullet}$C≡N) is used, the route is electro-free-radical, the sodium center carrying electro-free-radical being the initiator. However, unlike the case of cyclic ethers, the route will not be favored, because of the influence of H atom an ionic metal carried by the Nitrogen center. Na will keep replacing the hydrogen. If the H on the N center is replaced with a radical pushing group (R), the route will readily be favored positively or electro-free-radically.

The use of weak acids such as aqueous sulfur dioxide would also not allow for polymerization of aziridine and its members, since salts may be obtained (see Equation 13.20). Therein the aziridine was still kept in Stable state of existence by the weak acid. Like cyclic amides, strong acids "cationic" initiators would be required to polymerize the monomer. Examples include strong protonic acids and benzyl chloride. With benzyl chloride, the followings are obtained.

$$
\langle \text{Ph} \rangle - \text{CH}_2 - \text{Cl} \quad + \quad \underset{\text{(I)}}{\overset{..}{\underset{\text{H}}{\text{N}}} \overset{\text{nn}}{\underset{\diagdown}{\diagup}} \overset{\text{CH}_2}{\underset{\text{CH}_2}{|}}} \quad \longrightarrow \quad \langle \text{Ph} \rangle - \overset{\overset{\text{H}}{|}}{\underset{\underset{\text{H}}{|}}{\text{C}}} - \overset{\overset{\overset{\ominus}{\text{Cl}}}{|}}{\underset{\underset{\text{H}}{|}}{\overset{\oplus}{\text{N}}}} \overset{\diagup}{\underset{\diagdown}{}} \overset{\text{CH}_2}{\underset{\text{CH}_2}{|}}
$$

$$
\underset{\begin{array}{c}\text{(strained)}\\\underline{\text{Initiator}}\end{array}}{}
$$

$$
\text{+ (I)} \xrightarrow{} \overset{\overset{\text{H}\ \ \text{H}\ \ \text{H}}{|\ \ |\ \ |}}{\underset{\underset{\text{H}\ \ \text{H}\ \ \text{H}}{|\ \ |\ \ |}}{\text{N} - \text{C} - \text{C}}} - \overset{\overset{\overset{\ominus}{\text{Cl}}}{|}}{\underset{\underset{\text{H}_2\text{C} - \Phi}{|}}{\overset{\oplus}{\text{N}}}} \diagup\!\!\!\diagdown \xrightarrow{+(n-1)\,(\text{I})}
$$

$$
\text{H}\!\!-\!\!\underset{n}{\left[\underset{\underset{\text{H}\ \ \text{H}}{|\ \ |}}{\overset{\overset{\text{H}\ \ \text{H}\ \ \text{H}}{|\ \ |\ \ |}}{\text{N} - \text{C} - \text{C}}}\right]}\!\!-\! \overset{\overset{\text{H}\ \ \text{H}\ \ \text{H}}{|\ \ |\ \ |}}{\underset{\underset{\text{H}\ \ \text{H}}{|\ \ |}}{\text{N} - \text{C} - \text{C}}}\!\!-\!\! \text{N}\overset{\diagup}{\underset{\diagdown}{}}\overset{\text{CH}_2}{\underset{\text{CH}_2}{|}} \quad + \quad \text{ClH}_2\text{C}\text{-}\Phi
$$

$$
[\Phi \equiv \text{Phenyl group}] \qquad\qquad 13.94
$$

Note that the benzyl chloride was kept in Stable state of existence here by the monomer.

It should be noted that ion-paired initiator is first obtained. This is then followed by release of strain energy, just like with cyclic amides. This is backward electro-free-radical Addition polymerization via Equilibrium mechanism. The ***immonium center unlike the oxonium center*** favors less transient existence, because of the laws of the boundary limitations and types of groups carried by the N center. Once the MRSE exists in the monomer or activated monomer, it opens via the functional center. ***Cyclic amines and cyclic amides favor branch formations. But this is more present in cyclic amines than cyclic amides, for obvious reasons.*** While linear chain from cyclic amines will more readily favor existing in Equilibrium state of existence, that from cyclic amide may not but can be easily activated. Nevertheless, how these branches are formed will be considered from time to time and in subsequent volumes.

With cyclic amides, irrespective of the location of subsequent group on the carbon center or even the nitrogen center where possible, when instantaneous opening of ring is favored, the monomer unit remains the same. With aziridines, with the nitrogen center carrying a radical-pushing alkylane group, the followings are to be expected.

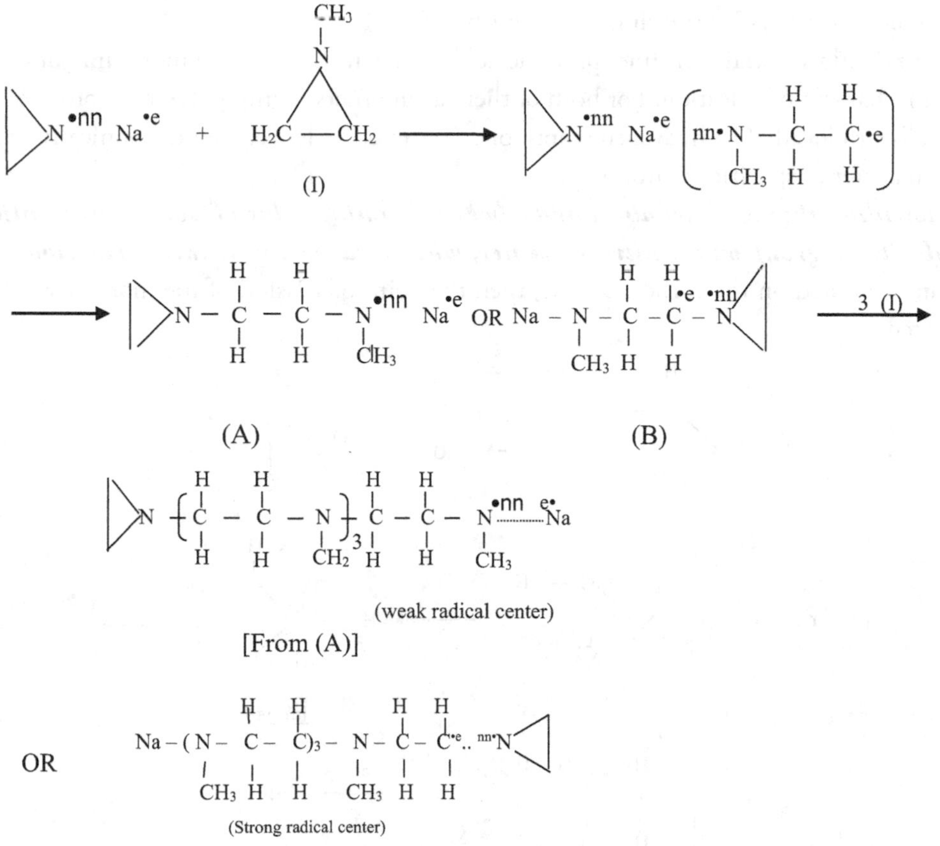

$$13.95$$

For (A), the route is not natural to the monomer and the strength of the nucleo-non-free-radical center of the growing chain keeps decreasing for every addition of monomer. A point is reached when the center can no longer activate the monomer instantaneously. Hence, low molecular weight polymers will be more favored anionically or nucleo-non-free-radically than electro-free-radically. Electro-free-radicaly it can grow only from the front and not from the back. As it seems, growth from the back is not possible since the initiator is covalent in character. In the monomer above, the CH_3 is strongly covalently bonded to the nitrogen center. (B) is the one indeed favored only electro-free-radically, the route being natural to the monomer, hoping that (B) will be ever growing as shown above from the front via Combination mechanism. In view of the basic character of the initiator and the monomer, activation will not indeed be favored anionically. However, note the type of initiator which has been used above as catalyst - the sodium salt of the monomer itself, obtained from the reaction of aziridine with Na. With the substituted monomer above, Na alone or $Na^{\bullet}...^{\ominus}C \equiv N$ cannot be used as initiators electro-free-radically. "Cationically" also, the followings are obtained using a strong acid.

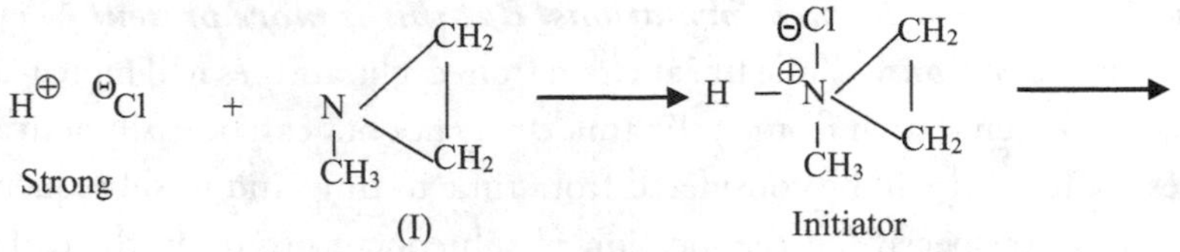

$$H^{\oplus} \quad ^{\ominus}Cl \quad + \quad \text{(I)} \quad \longrightarrow \quad \text{Initiator} \quad \longrightarrow \qquad\qquad 13.96$$

If the ring can be opened, propagation can continue electro-free-radically from the counter charged center backwardly to give a growing chain which can depropagate if the polymerization temperature is higher than the Ceiling temperature. In the reaction above, regardless the operating conditions, aziridine and CH_3Cl cannot be formed via Equilibrium mechanism.

Use of benzyl chloride and aziridine, protonic acids and aziridine require more time in view of the Equilibrium mechanism of Addition. For both of them as in others, living polymers from the initiating end are generally produced. The hydrogen atoms, one at a time, are loosely bonded ionically or radically to the nitrogen center in specific environments.

Unlike aziridine, the monomer above and other cyclic amines which favor polymerization when there is a substituent group on the nitrogen center, will not favor any branch formation. When an alkylane group is located on the carbon centers, then the point of scission of the monomer will depend on where located.

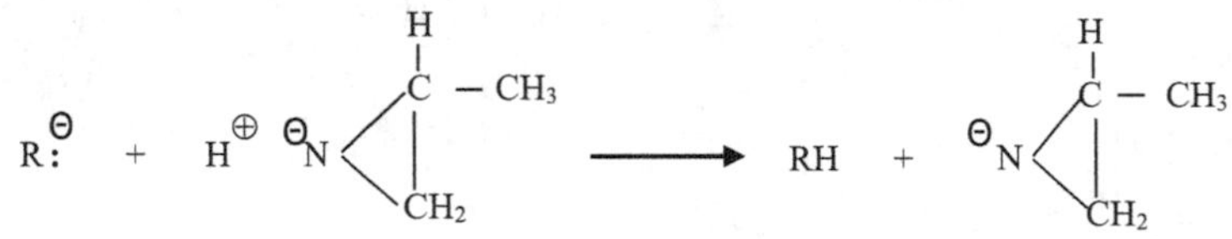

$$R:^{\ominus} \quad + \quad H^{\oplus} \quad ^{\ominus}N \quad \longrightarrow \quad RH \quad + \quad ^{\ominus}N \qquad\qquad 13.97$$

$$H^{\oplus} \quad ^{\theta}Cl \quad + \quad H\text{------}N \quad \longrightarrow \quad H\text{------}N \quad + \text{ (I)} \quad \longrightarrow$$

(I)   <u>Initiator</u>

<u>Initiation step</u>   + n(I) $\longrightarrow$

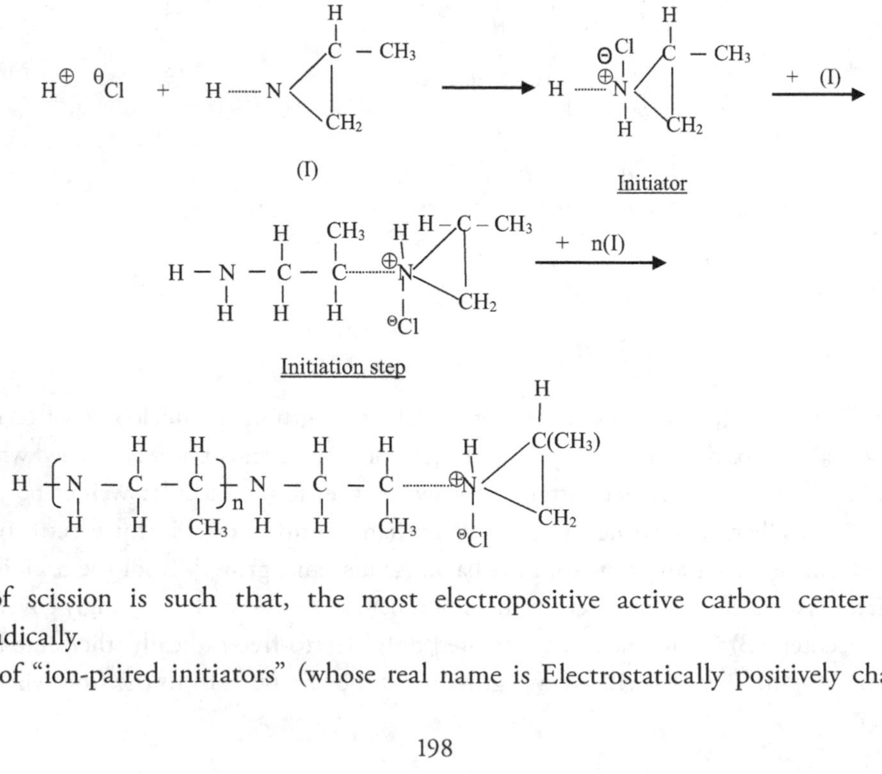

$$13.98$$

The point of scission is such that, the most electropositive active carbon center is obtained electro-free-radically.

The use of "ion-paired initiators" (whose real name is Electrostatically positively charged-paired initiators) for the polymerization of these monomers is not popularly known. Now, considering the use of BF_3 as the source of initiator, the followings are obtained.

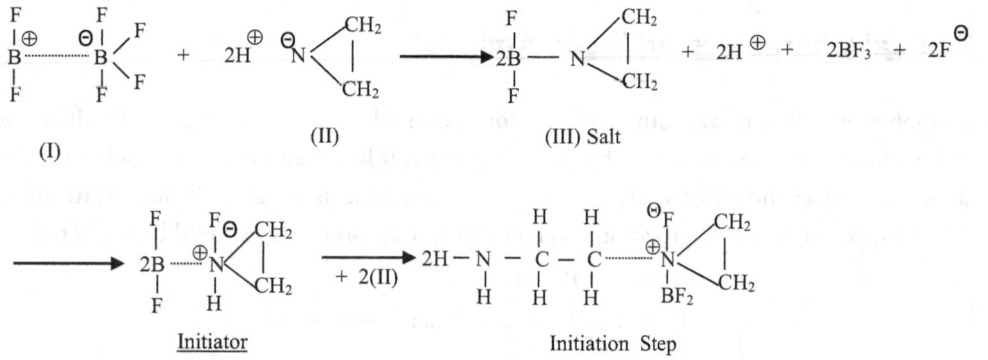

(I)   (II)   (III) Salt

Initiator   Initiation Step   13.99

Since hydrofluoric acid is strong enough (being an inorganic acid), subsequent polymeri-zation is favored by it, if the aziridine is suppressed after the initiator is formed. However, notice the control used in first preparing the initiator. BF_3 should not be present as a bye-product, otherwise the HF will not add to the monomer as wrongly shown above. Hence, to obtain the initiator above, the ratio of BF_3 to monomer should 1:1. Like the case of cyclic amides, use of the Electrostatically positively charged-paired initiator (I) may be favored if the right molar ratios are used, since an acid is produced in the process. This will depend on if the acid can exist in Equilibrium state of existence in the absence of the monomer. (See Equations 13.51 and 13.52 for amides). Note that never is there a time when (I) is allowed to exist in the system as shown above, because BF_3 is suppressed by the Equilibrium state of existence of the cyclic amine and in the process HF shown above is obtained. This adds to the salt formed to give the initiator above. The problems in their use is the long induction period during propagation via Equilibrium mechanism, and mostly our inability to know the molar ratios to use, the manner by which the components are to be added, and more. All these are to be expected, because we have since antiquity not understood the mechanisms of how chemical reactions take place or indeed how NATURE operates.

With ROR/BF_3 combination, the followings are obtained.

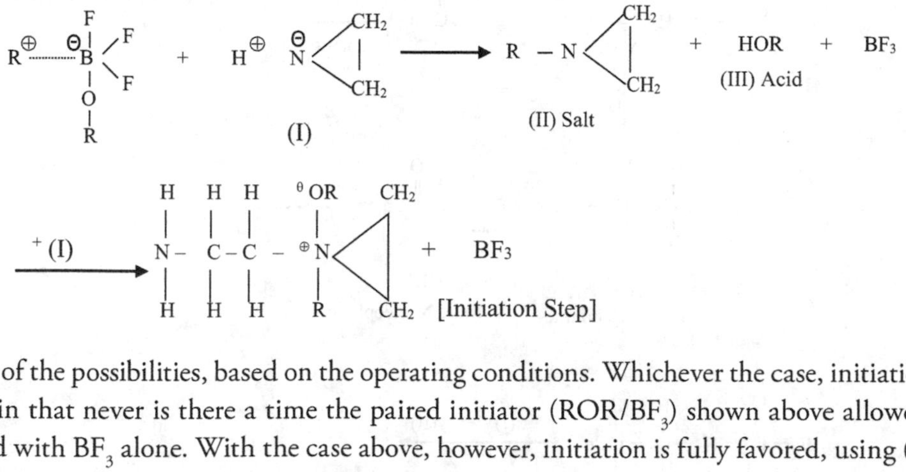

(I)   (II) Salt   (III) Acid

[Initiation Step]   13.100

This is one of the possibilities, based on the operating conditions. Whichever the case, initiation is favored, noting again that never is there a time the paired initiator (ROR/BF_3) shown above allowed to exist as already said with BF_3 alone. With the case above, however, initiation is fully favored, using (III) which is acidic in character along with (II) the salt. Then, (I) is added for initiation by backward addition followed by propagation, all via Equilibrium mechanism. If the Electrostatically positively charged-paired initiator had been favored from BF_3 alone or ROR/BF_3 combination, initiation would still have been favored. However, it is the acid that is formed. With the case above, it is hoped that the acid (ROH) obtained which is weak compared to HF, H_2SO_4, HCl, will favor the existence of the initiator. From the reactions above, it is not surprising why little is said about the use of these Electrostatically positively charged-paired initiators for the polymerization of aziridines.

13.2.2 Radical Characters of Cyclic Amines

In the copolymerization of aziridine with carbon monoxide, the route is not ionic, since the carbon center in carbon-monoxide cannot carry charges of any type. It has been said that unsubstituted aziridine will not favor the nucleo-non-free-radical route if the aziridine is in Equilibrium state of existence. However when suppressed, for its growing polymer chain, the followings should be expected.

$$\ddot{N} \left\{ \begin{matrix} H & H \\ | & | \\ C - C - N \\ | & | \\ H & H \end{matrix} \right\}_n \begin{matrix} H & H \\ | & | \\ C - C - N.nn \\ | & | \\ H & H \end{matrix} \longrightarrow H \cdot^n +$$

E.g. $\overset{\square}{N} \equiv H_6C_5 - (CO) - O \bullet nn$ from benzoyl peroxide

$$\ddot{N} \left\{ \begin{matrix} H & H \\ | & | \\ C - C - N \\ | & | \\ H & H \end{matrix} \right\}_n \begin{matrix} H \\ | \\ C - C = N \\ | \\ H \end{matrix}$$

<u>Transfer species of second kind of first type.</u>
[IMPOSSIBLE TRANSFER)

13.101

Since the route is not natural to the monomer, transfer species of second kind of the first types should be involved, but not here based on the operating conditions and since there is no receiving center for the H.n released from a nucleo-non-free-radical growing chain. When copolymerized with carbon monoxide, electro-free-radically, the followings are obtained. It should be noted here that the electro-free-radical catalyst involved is only that from an electro-free-radical generating initiator

$$E \cdot^e + H_2C - CH_2 + e.\overset{O}{\overset{||}{C}}.n \longrightarrow$$
(with N-H bridge)

$$E \cdot^e + e.\begin{matrix} H & H \\ | & | \\ C - C - N.nn \\ | & | \\ H & H \end{matrix} + e.\overset{O}{\overset{||}{C}}.n \longrightarrow$$

(I)   (II)
Less Nucleophilic   More Nucleophilic

$$E - \overset{O}{\overset{||}{C}} - N - \begin{matrix} H & H \\ | & | \\ C - C.e \\ | & | \\ H & H \end{matrix} \xrightarrow[y > x]{+ x(I) + y(II)}$$

$$E - \overset{O}{\overset{||}{C}} - N - \begin{matrix} H & H \\ | & | \\ C - C \\ | & | \\ H & H \end{matrix} - N - \begin{matrix} H & H \\ | & | \\ C - C \\ | & | \\ H & H \end{matrix} \left\{ \overset{O}{\overset{||}{C}} - N - \begin{matrix} H & H \\ | & | \\ C - C \\ | & | \\ H & H \end{matrix} \right\}_n N - \begin{matrix} H & H \\ | & | \\ C - C.e \\ | & | \\ H & H \end{matrix}$$

Weak alternating/Block copolymer

13.102

For the case above, CO has been assumed to be more nucleophilic than aziridine (Imine). If the reverse is the case, the followings are obtained.

$$
E \cdot^e \quad + \quad e \cdot \overset{\overset{H}{|}}{\underset{\underset{H}{|}}{C}} - \overset{\overset{H}{|}}{\underset{\underset{H}{|}}{C}} - \underset{\underset{H}{|}}{N}.nn \quad + \quad e \cdot \overset{\overset{O}{\|}}{C} . n \quad \longrightarrow
$$

(I)   (II)

More Nucleophilic   Less Nucleophilic

$$
E -(\underset{\underset{H}{|}}{N} - \overset{\overset{H}{|}}{\underset{\underset{H}{|}}{C}} - \overset{\overset{H}{|}}{\underset{\underset{H}{|}}{C}} - \overset{\overset{O}{\|}}{C})_n - \underset{\underset{H}{|}}{N} - \overset{\overset{H}{|}}{\underset{\underset{H}{|}}{C}} - \overset{\overset{H}{|}}{C} - \overset{\overset{O}{\|}}{C} . e
$$

Weak alternating copolymer   13.103

For the first case, it is the carbon monoxide that is activating the ring, while for the last case, it seems the ring was opened by the electro-free-radical. In both cases, there can never be more of carbon-monoxide than the ringed monomer along the chain, since CO cannot homopolymerize but aziridine can, the route being natural to them. However, the right initiator to use if alternating placement is desired, is that which cannot homopoly-merize any of the two monomers.

Nucleo-free-radically, if CO is more nucleophilic than the aziridine, then almost full alternating placements are obtained, with no transfer species. If CO is less nucleophilic than aziridine no copolymers or homopolymers can be obtained.

$$
N \cdot^n \quad + \quad e \cdot \overset{\overset{H}{|}}{\underset{\underset{H}{|}}{C}} - \overset{\overset{H}{|}}{\underset{\underset{H}{|}}{C}} - \underset{\underset{H}{|}}{N}.nn \quad + \quad e \cdot \overset{\overset{O}{\|}}{C} . n \quad \longrightarrow
$$

(I)   (II)

Less Nucleophilic   More Nucleophilic

$$
N - \overset{\overset{H}{|}}{\underset{\underset{H}{|}}{C}} - \overset{\overset{H}{|}}{\underset{\underset{H}{|}}{C}} - \underset{\underset{H}{|}}{N} - \overset{\overset{O}{\|}}{C} \bullet n \quad \xrightarrow{\quad + \ n\,(I)\ =\ x(II)\quad}
$$

$$
N - (\overset{\overset{H}{|}}{\underset{\underset{H}{|}}{C}} - \overset{\overset{H}{|}}{\underset{\underset{H}{|}}{C}} - \underset{\underset{H}{|}}{N} - \overset{\overset{O}{\|}}{C})_x - \overset{\overset{H}{|}}{\underset{\underset{H}{|}}{C}} - \overset{\overset{H}{|}}{\underset{\underset{H}{|}}{C}} - \underset{\underset{H}{|}}{N} - \overset{\overset{O}{\|}}{C} \bullet n
$$

Full Alternating Placement (Not Favored)   13.104

The monomer unit looks like Nylon 3, that which could not be obtained from four- membered lactam. The monomer unit from the lactam would have been $-NH-(CH_2)_2 - CO-$ slightly different from the case above. However, it said not to be favored because, this observation has never been reported from the literature and indeed CO is less nucleophilic than aziridine. Hence the couple formed will carry nucleo-non-free-radical, the route not natural to the monomer, but reactive with it. Hence, nucleo-non-free-radically, strong alternating placement is obtained, with the aziridine less randomly placed along the chain. Yet, this does not exist.

If the ringed monomer is a substituted aziridine with the N center carrying the substituent group, polymerization is still favored as shown below nucleo-free-radically, if instantaneous opening of the ring is favored and the CO is more nucleophilic than aziridine.

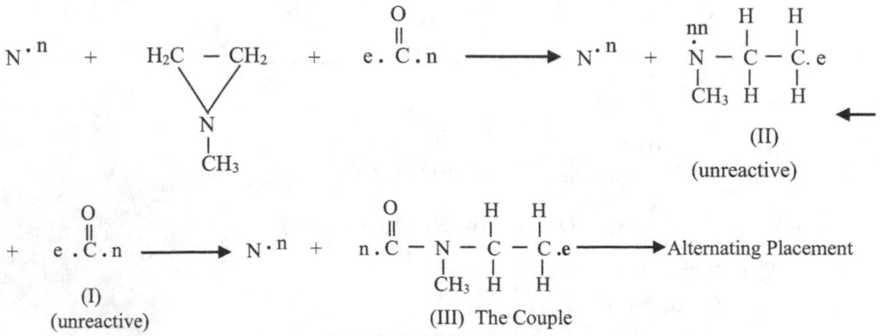

$$N\cdot^{n} \quad + \quad H_2C - CH_2 \quad + \quad e.\overset{O}{\overset{\|}{C}}.n \quad \longrightarrow \quad N\cdot^{n} \quad + \quad \overset{nn}{\underset{CH_3}{\overset{\cdot}{N}}} - \overset{H}{\underset{H}{\overset{|}{C}}} - \overset{H}{\underset{H}{\overset{|}{C}}}.e$$

(in the ring: N with CH₃)

(II)
(unreactive) ←

$$+ \quad e.\overset{O}{\overset{\|}{C}}.n \quad \longrightarrow \quad N\cdot^{n} \quad + \quad n.\overset{O}{\overset{\|}{C}} - \overset{H}{\underset{CH_3}{\overset{|}{N}}} - \overset{H}{\underset{H}{\overset{|}{C}}} - \overset{H}{\underset{H}{\overset{|}{C}}}.e \quad \longrightarrow \text{Alternating Placement}$$

(I)
(unreactive)

(III) The Couple
NOT FAVORED   13.105

Though we are yet to fully identify the order of nucleophilicity of the two monomers, there is no doubt that the following is valid.

Cyclic amines   >   CO

Order of Nucleophilicity   13.106

With the CH_3 group located on a carbon center, electro-free-radically, the following is obtained, if instantaneous opening of the ring is made to take place.

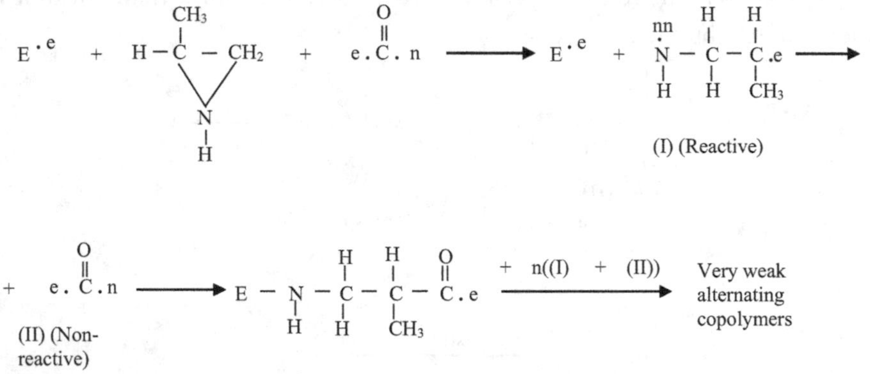

$$E\cdot^{e} \quad + \quad H - \overset{CH_3}{\underset{}{\overset{|}{C}}} - CH_2 \quad + \quad e.\overset{O}{\overset{\|}{C}}.n \quad \longrightarrow \quad E\cdot^{e} \quad + \quad \overset{nn}{\underset{H}{\overset{\cdot}{N}}} - \overset{H}{\underset{H}{\overset{|}{C}}} - \overset{H}{\underset{CH_3}{\overset{|}{C}}}.e \quad \longrightarrow$$

(in the ring: N with H)

(I) (Reactive)

$$+ \quad e.\overset{O}{\overset{\|}{C}}.n \quad \longrightarrow \quad E - \overset{H}{\underset{H}{\overset{|}{N}}} - \overset{H}{\underset{H}{\overset{|}{C}}} - \overset{H}{\underset{CH_3}{\overset{|}{C}}} - \overset{O}{\overset{\|}{C}}.e \quad \overset{+ \ n((I) \ + \ (II))}{\longrightarrow} \quad \begin{array}{l}\text{Very weak}\\ \text{alternating}\\ \text{copolymers}\end{array}$$

(II) (Non-
reactive)   13.107

The location of the CH_3 group is important to note. (I) can homopolymerize electro-free-radically and not nucleo-non-free-radically. ***Thus nucleo-non-free-radically, full alter-nating placement will be obtained as will fully be shown.*** Hence, if there is alternating placement electro-free-radically, it will be very weak. Nucleo-free-radically, there is no homopolymerization and copolymerization, because of the presence of transfer species of the first kind of the first type. With another CH_3 group introduced to the second carbon center, the monomer will still favor the electro-free-radical route.

Nucleo-non-free-radically for aziridine and CO, the followings are to be expected, if CO is assumed to be less nucleophilic than the monomer.

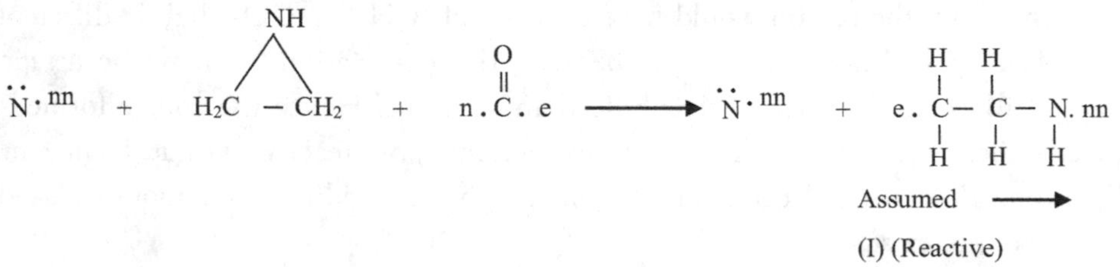

$$\overset{..}{N}\cdot^{nn} \quad + \quad H_2C - CH_2 \quad + \quad n.\overset{O}{\overset{\|}{C}}.e \quad \longrightarrow \quad \overset{..}{N}\cdot^{nn} \quad + \quad e.\overset{H}{\underset{H}{\overset{|}{C}}} - \overset{H}{\underset{H}{\overset{|}{C}}} - \overset{}{\underset{H}{\overset{|}{N}}}.nn$$

(in the ring: NH)

Assumed →

(I) (Reactive)

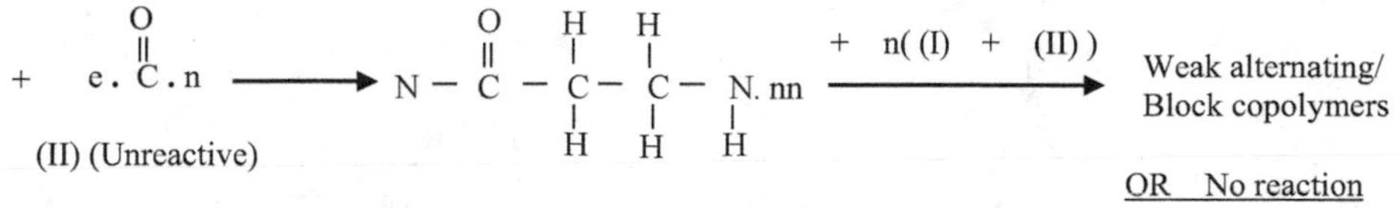

$$+ \quad e.\overset{\overset{O}{\|}}{C}.n \longrightarrow \quad \overset{\overset{O}{\|}}{N-C} \overset{H}{\underset{H}{\overset{|}{C}}} - \overset{H}{\underset{H}{\overset{|}{C}}} - N.nn \quad \overset{+ \ n(\ (I) \ + \ (II)\ )}{\longrightarrow} \quad \begin{array}{l}\text{Weak alternating/}\\ \text{Block copolymers}\end{array}$$

(II) (Unreactive)

OR   No reaction   13.108

The copolymer obtained will contain long block of the aziridine. If the CO is more nucleophilic than the aziridine, no copolymers can be obtained.

With substituted aziridine on the nitrogen center, the followings are obtained.

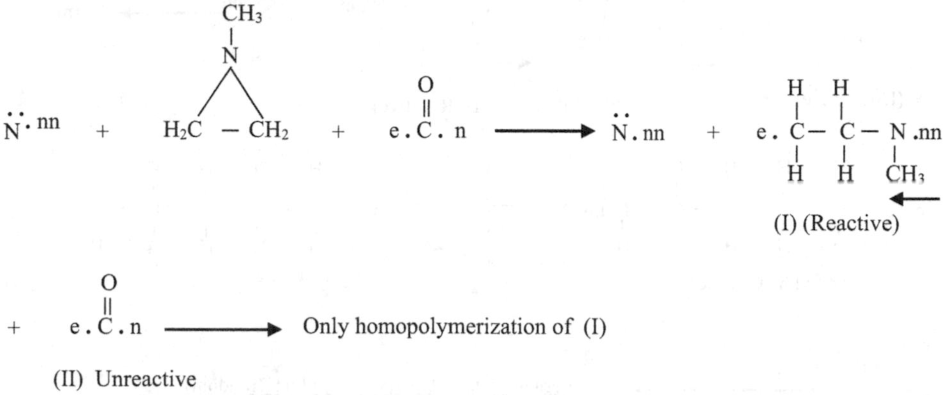

$$\ddot{N}.nn \ + \ \underset{CH_3}{\overset{CH_3}{\underset{\triangle}{N}}}(H_2C - CH_2) \ + \ e.\overset{\overset{O}{\|}}{C}.n \longrightarrow \ \ddot{N}.nn \ + \ e.\overset{H}{\underset{H}{\overset{|}{C}}} - \overset{H}{\underset{CH_3}{\overset{|}{C}}} - N.nn$$

(I) (Reactive)

$$+ \quad e.\overset{\overset{O}{\|}}{C}.n \longrightarrow \quad \text{Only homopolymerization of (I)}$$

(II) Unreactive   13.109

Nucleo-non-free-radically, only homopolymerization of aziridine will be favored if the ring can be opened instantaneously. With the monomer shown below, when instantaneous opening of the ring is favored, the followings are obtained.

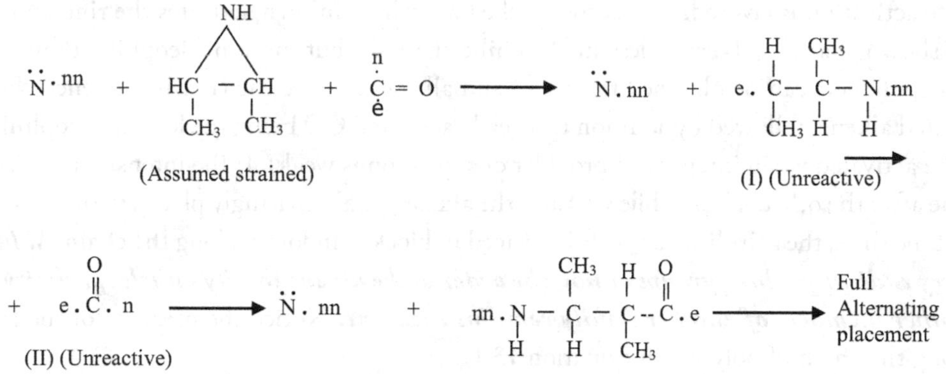

$$\ddot{N}.nn \ + \ \underset{CH_3 \ \ CH_3}{\overset{NH}{\overset{\triangle}{HC - CH}}} \ + \ \overset{n}{\underset{\overset{.}{e}}{C}} = O \longrightarrow \ \ddot{N}.nn \ + \ e.\overset{H}{\underset{CH_3}{\overset{|}{C}}} - \overset{CH_3}{\underset{H}{\overset{|}{C}}} - N.nn$$

(Assumed strained)   (I) (Unreactive)

$$+ \quad e.\overset{\overset{O}{\|}}{C}.n \longrightarrow \ \ddot{N}.nn \ + \ nn.\overset{CH_3}{\underset{H}{\overset{|}{N - C}}} - \overset{H}{\underset{CH_3}{\overset{|}{C}}} - \overset{O}{\overset{\|}{C}}.e \longrightarrow \begin{array}{l}\text{Full}\\ \text{Alternating}\\ \text{placement}\end{array}$$

(II) (Unreactive)   13.110

Nucleo-free-radically, no homo- or co-polymers can be produced.

Radical terpolymerization of ethylenimine, carbon monoxide and ethylene by azobisisobutyronitrile (AIBN) and gamma radiation has been reported [34]. AIBN cannot produce non-free-radicals when the scission is completed to release N_2 (see Volume 1). But however, nucleo-non-free-radicals can be produced via electro-radicalization as shown below.

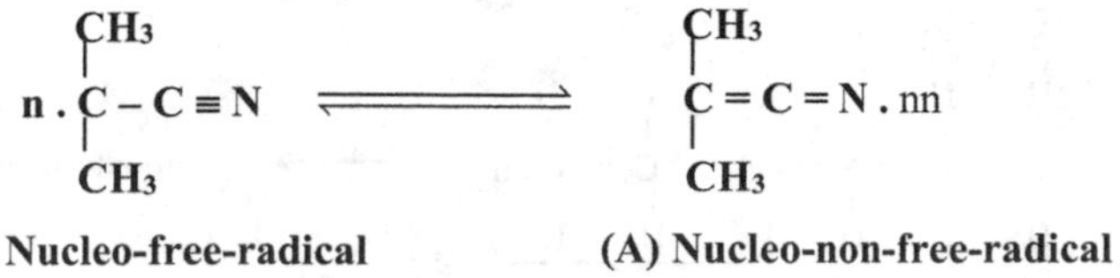

$$n.\underset{CH_3}{\overset{CH_3}{\overset{|}{C}}}{-C} \equiv N \rightleftharpoons \underset{CH_3}{\overset{CH_3}{\overset{|}{C}}} = C = N.nn$$

(A) Nucleo-free-radical   **(A) Nucleo-non-free-radical**

Nucleo-free-radical from AIBN   13.111

Therefore, nucleo-free-radical or nucleo-non-free-radical polymerization is involved as shown below.

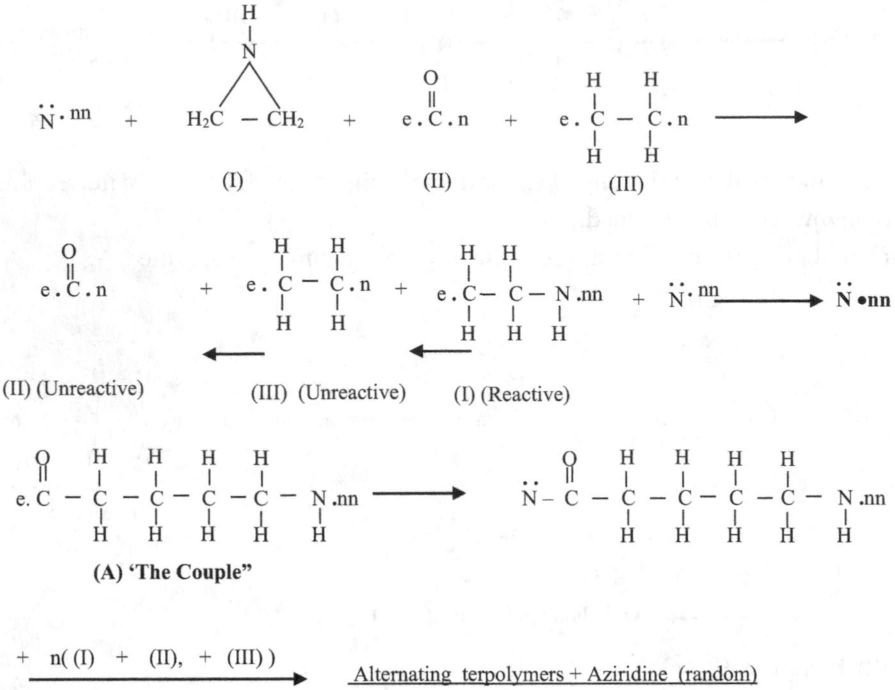

$$\ddot{N}\cdot^{nn} \quad + \quad H_2C - CH_2 \quad + \quad e.\overset{O}{\underset{||}{C}}.n \quad + \quad e.\overset{H}{\underset{|}{\underset{H}{C}}} - \overset{H}{\underset{|}{\underset{H}{C}}}.n \longrightarrow$$

(I)   (II)   (III)

with N–H on aziridine ring shown above (I).

$$e.\overset{O}{\underset{||}{C}}.n \quad + \quad e.\overset{H}{\underset{|}{\underset{H}{C}}} - \overset{H}{\underset{|}{\underset{H}{C}}}.n \quad + \quad e.\overset{H}{\underset{|}{\underset{H}{C}}} - \overset{H}{\underset{|}{\underset{H}{C}}} - \overset{H}{\underset{|}{N}}.nn \quad + \quad \ddot{N}\cdot^{nn} \longrightarrow \ddot{N}\bullet nn \quad +$$

(II) (Unreactive)   (III) (Unreactive)   (I) (Reactive)

$$e.\overset{O}{\underset{||}{C}} - \overset{H}{\underset{|}{\underset{H}{C}}} - \overset{H}{\underset{|}{\underset{H}{C}}} - \overset{H}{\underset{|}{\underset{H}{C}}} - \overset{H}{\underset{|}{\underset{H}{C}}} - N.nn \longrightarrow \ddot{N} - \overset{O}{\underset{||}{C}} - \overset{H}{\underset{|}{\underset{H}{C}}} - \overset{H}{\underset{|}{\underset{H}{C}}} - \overset{H}{\underset{|}{\underset{H}{C}}} - \overset{H}{\underset{|}{\underset{H}{C}}} - N.nn$$

(A) 'The Couple"

$$\xrightarrow{\quad + \quad n(\text{(I)} \; + \; \text{(II)}, \; + \; \text{(III)})\quad} \underline{\text{Alternating terpolymers + Aziridine (random)}} \qquad 13.112a$$

It is the initiator that decides what is to be done in the system. The initiator above is a nucleo-non-free-radical. (II) is unreactive to the initiator, but can be activated and activate. (I) cannot activate, but can homopolymerize nucleo-non-free-radically when ring is instantaneously opened. (III) is unreactive to the initiator once activation is favored. $N\cdot^{nn}$ activates the CO, which in turn activates the ring and ethylene. It has been shown that the alkene is less nucleophilic than (I), but more nucleophilic than CO, since alternating copolymers can be obtained from CO and alkenes. Hence the (I) when opened should first add to activated alkene, followed by addition to already activated CO being the least nucleophilic (always the first to be activated when intensity of provider of activation is weak). If the intensity is strong, all are activated one after the other. Thus, while CO and the alkene are alternatingly placed with aziridine along the chain all the time, the aziridine can be fully placed in blocks randomly along the chain. ***Whether the alkenes carry radical-pushing groups or not, the order of the alkene family's nucleophilicity between itself and other members of another family remains the same.*** Notice the presence of the features of Nylon 5 along the chain of polymer in Equation 13.112a.

With nucleo-free-radical as initiator, the followings are obtained.

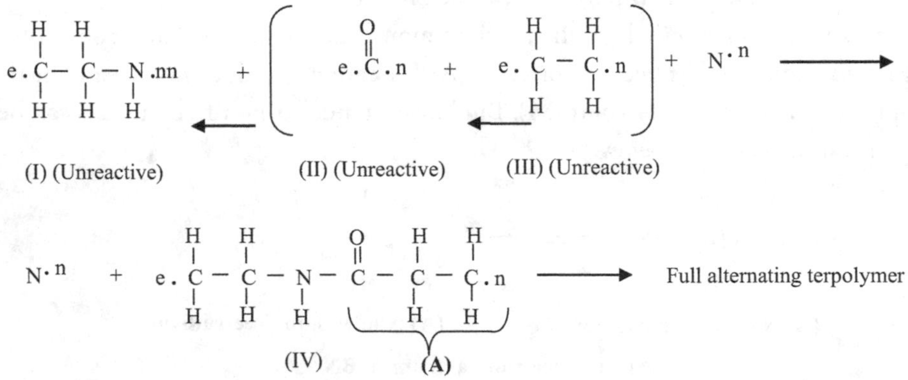

$$e.\overset{H}{\underset{|}{\underset{H}{C}}} - \overset{H}{\underset{|}{\underset{H}{C}}} - \overset{H}{\underset{|}{N}}.nn \quad + \quad \left(e.\overset{O}{\underset{||}{C}}.n \quad + \quad e.\overset{H}{\underset{|}{\underset{H}{C}}} - \overset{H}{\underset{|}{\underset{H}{C}}}.n \right) \quad + \quad N\cdot^{n} \longrightarrow$$

(I) (Unreactive)   (II) (Unreactive)   (III) (Unreactive)

$$N\cdot^{n} \quad + \quad e.\overset{H}{\underset{|}{\underset{H}{C}}} - \overset{H}{\underset{|}{\underset{H}{C}}} - \overset{H}{\underset{|}{N}} - \overset{O}{\underset{||}{C}} - \overset{H}{\underset{|}{\underset{H}{C}}} - \overset{H}{\underset{|}{\underset{H}{C}}}.n \longrightarrow \text{Full alternating terpolymer}$$

(IV)   (A)

$$13.112b$$

The initiator in control here is a nucleo-free-radical. Since NATURE abhors a vacuum and NATURE is orderliness, and since alternating placement can readily be obtained between CO and alkenes nucleo-free-radically, (III) first adds to (II) to give a three-membered Cyclic ketone which is too strained to exist as a ring. Since the cyclic ketone (A) is more nucleophilic than the aziridine which was formerly the most nucleophilic of the three of them, it diffuses to add to the aziridine after opening it to form (IV), the 'Couple'. Compare this couple (IV) with that of Equation 13.112a (A), when the nucleo-non-free-radical was in charge. With the formation of (IV), full alternating terpolymer is obtained. Presence of ethene randomly placed along the chain will be difficult, since the route is not natural to the nucleo-free-radical. Based on the routes involved for the two situations above, the reactivities and non-reactivities of the monomers have been indicated, noting however, that nucleo-non-free-radically, no copolymers can be produced as shown in Equation 13.112a, since the couple formed is (IV) of Equation 13.112b. One cannot have one couple as an alternative to another couple.

Electro-free-radically, the situation is completely different, since the route is natural to the monomers and ethene will not be partly but fully reactive. Hence, there will be no alternation along the chain and mostly ethene will exist along the chain. In all the cases considered so far, note that the growing chains do contain features of g-or b alanine or nylon 3 or 5 or higher as already shown.

From all the consideration so far, one has strongly begun to identify step by step, how different types of copolymers are produced, the order of the characters of monomers involved and the order of attack in the system, etc., all very important variables. Without identifying what is a monomer, an ion, a radical, an activation center, etc., the characters and the relative orders of the characters of monomers cannot be defined. This has been one of the most important countless fundamental differences between the past and the current developments in the Volumes.

13.2.3. Propagation/Depropagation Phenomena

Like cyclic ethers, aziridines readily undergo propagation/depropagation pheno-mena when polymerized electro-free-radically via backward addition. Most or all of the growing polymer chains cannot terminate itself in the absence of foreign terminating agents. Like cyclic ethers, the followings are obtained using aziridine.

Commencement of Depropagation

Mechanism of Depropagation

13.113

For this type of growing polymer chain, without the growing chain existing in Equilibrium state of existence (taking place from the end far from the initiating center), and the hydrogen atom held sitting on one of the nitrogen centers, depropagation to release a ring can never take place. The positive charge placed on the N center is now paired to the negative charge(anion) on the external N center. The same also applies to cyclic ethers when acid is not used. This happens when the growing chain has not reached its optimum chain length. For a forwardly growing polymer chain (taking place at the initiating center), depropagation can never take place. It can also take place if the chain is growing from both ends, that is, there is no initiator.

In view of the fact that the H atom on nitrogen center is loosely bonded, unreactive piperazine[11] is formed along with a living polymer. Piperazine was formed because the strain energy in its ring is smaller than that of aziridine. Only six- membered rings for three-membered ringed monomers are released in such reactions. These are rings that cannot be readily opened. 3 - membered ring can never be involved in such reactions.

The situation is not different for N substituted aziridines, since H is still loosely held to the nitrogen center during propagation, if polymerization is favored.

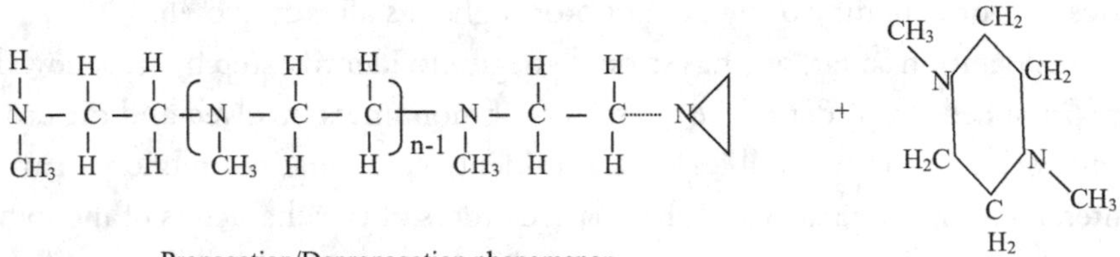

Propagation/Depropagation phenomenon   (II) Unreactive   13.114

Depropagation step is favored here, once the active growing center gets bonded to the nitrogen center followed by release of a stable molecule. By the presence of substituent group on the N center, the nitrogen center is now identical to the oxygen center in oxygen containing cyclic monomers. When ionic environment or species are strongly present in the system, then "depropagation reactions" similar to those of cyclic acetals are favored when aziridine is added. This can take place with both short and long chains.

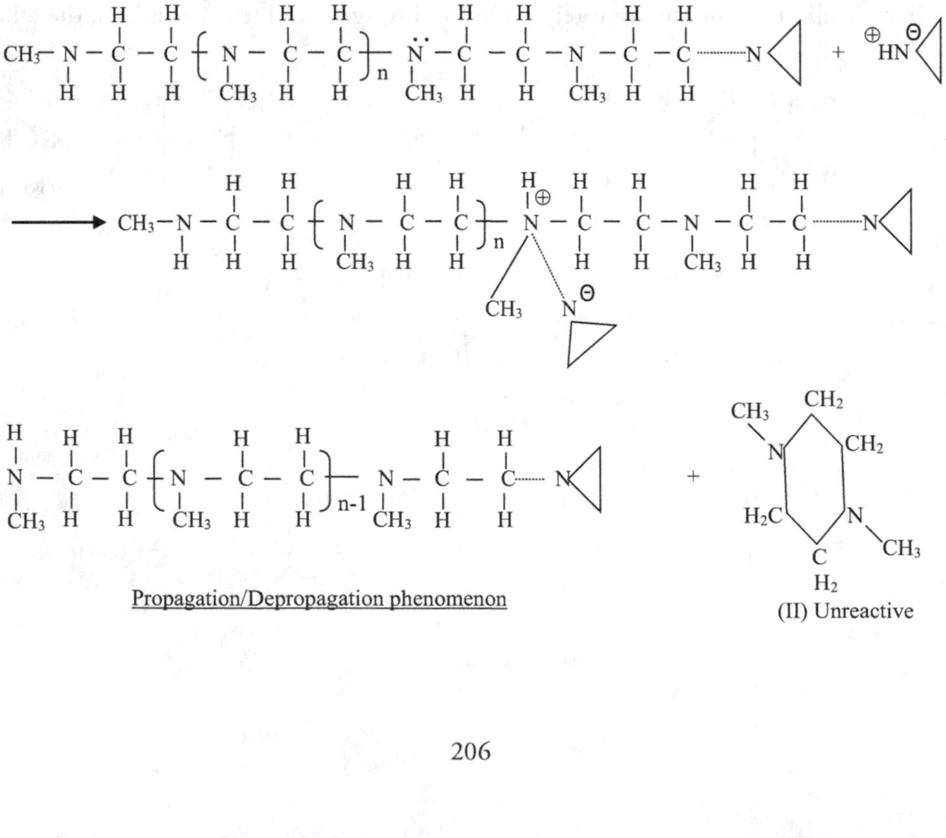

Propagation/Depropagation phenomenon   (II) Unreactive   13.115

This is not depropagation taking place during propagation, but degradation of a chain living or dead. If living, it does not have to be such that it can no longer grow linearly, that is, it has reached its optimum chain length. N-substituted piperazine or piperazine can only be obtained under such conditions. Thus, due to the electrostatic character of nitrogen and oxygen, the varieties of possible products obtained with these polymers in an ionic environment can be observed. Branches cannot be formed when the linear chain is living at the point where the optimum chain length has not been reached. When killed prematurely, then branches can be formed.

Nucleo-non-free-radically, no depropagation reactions will take place due to how polymerization if favored is taking place (Forward Addition). Free radically during copolymerization there may be some depropagation, if the temperature is higher than expected.

In the copolymerization of ethylenimine with b - propiolactone[32], the route can only beelectro-free-radical in character. Electro-free-radically and not "Cationically", since both monomers are reactive to the route, block copolymers will largely be obtained as shown below.

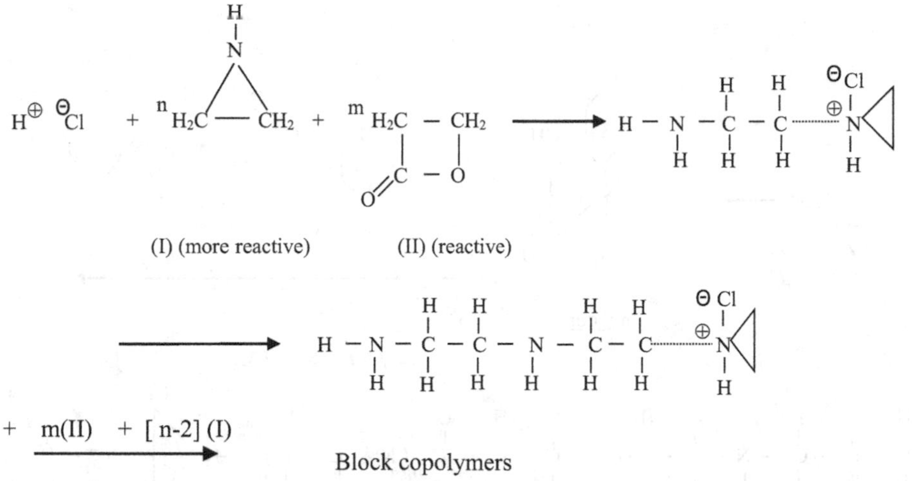

(I) (more reactive)   (II) (reactive)

Block copolymers   13.116

Only electro-free-radical initiator can conveniently be used for their copolymerizations. The above can take place when (I) is less nucleophilic than (II) in the presence of a protonic initiator. Thus, there will be more of (I) along the chain than (II). Depending on the polymerization conditions, there can be depropagation reactions. If (I) had been more nucleophilic than (II), the reverse is the case, that wherein the first block are from lactones. Nucleo-non-free-radically. if instantaneous activation of imine monomer is favored, (II) will be far more favored along the chain than the imines, (II) being more electrophilic than the imine, noting that both monomers are reactive if the nucleo-non-free-radical initiators are strong.

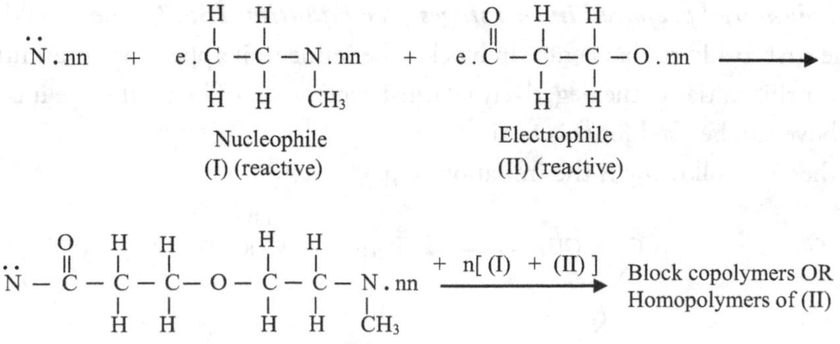

Nucleophile
(I) (reactive)

Electrophile
(II) (reactive)

+ n[(I) + (II)]   Block copolymers OR
                    Homopolymers of (II)

13.117

Of particular interest is the situation where it is believed that no catalyst are involved during the copolymerization of ethylenimine and b- propiolactone ***in various solvents such as toluene, ethyl*** *ether, ethyl acetate, ethylene dichloride, acetone, acetonitrile and dimethyl formamide*[32]. In the copolymerization, the solvents and the imine in some cases are the sources of the initiators. Of all the solvents above, only toluene is weakly non-polar. All are polar/non-ionic solvents. In order words, they can only produce radicals. Dimethyl formamide is the most polar of the solvents. The more polar the solvent, the easier it is for it to provide enough energy or electrostatic forces for the opening of a ring instantaneously. Solvents containing nitrogen centers such as in dimethyl formamide, acetonitrile are strong polar solvents of capacity equal to or greater than those with only nitrogen centers, in view of the more nucleophilic character of oxygen containing monomer. Hence more reactivity will be expected from such solvents, apart from other factors.

With toluene as solvent at temperatures as low as 0°C, the followings are obtained.

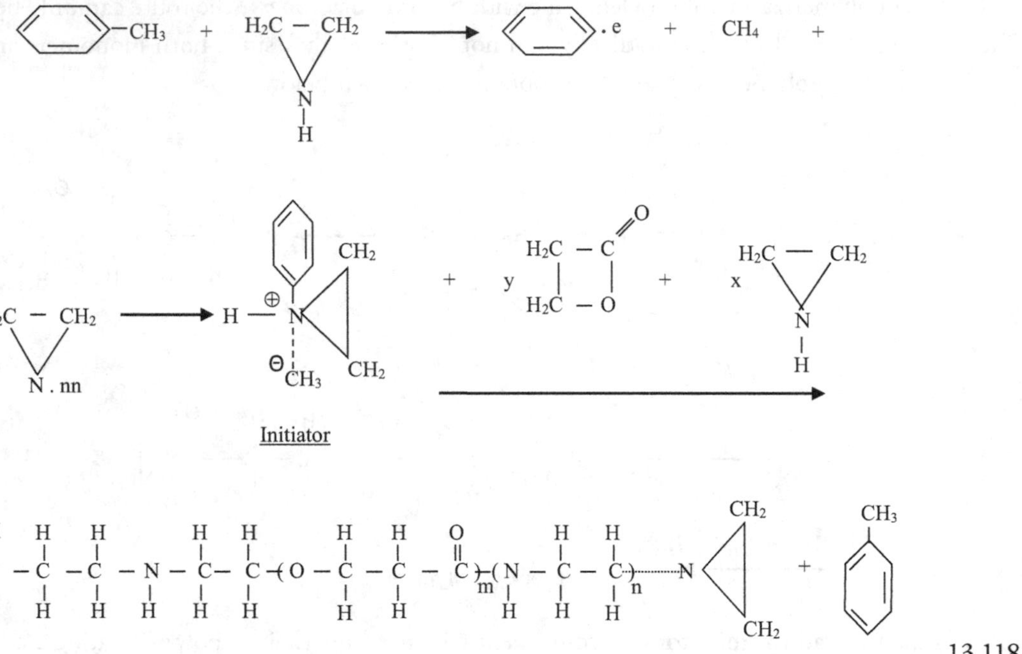

(Random Block Copolymers)

13.118

If the imine is less nucleophilic than the lactone, there will be more of the imine than the lactone along the polymer chain in the route natural to it. It is important to note that the imine is part of the initiator. Toluene alone cannot indeed initiate the polymerization of the lactone. It is after two molecules of imines have been used to complete the initiation step that the lactone begins to add intermittently. However, looking at the initiator *a negatively charged paired one of the electrostatic type, one of the first of its kind so far encountered prepared in two stages (See Equation 13.43),* one should expect that the lactone will be the first to add to that center in blocks, the imine unit appearing intermittently along the chain. However, for this initiator, the negatively charged center cannot be used since it is not ionic. Only the center used above can be used for this initiator.

With ethyl ether, the following is the initiation step.

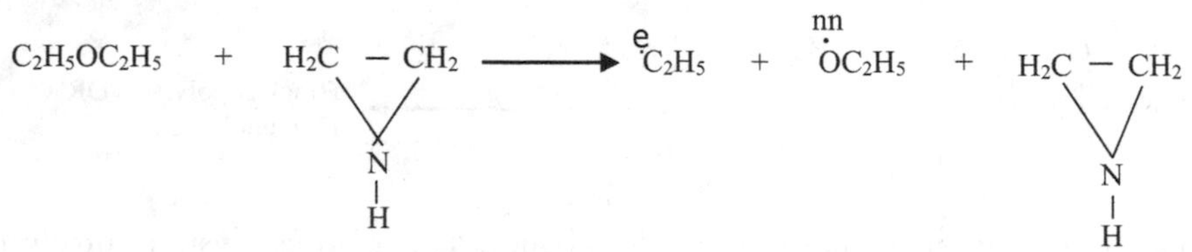

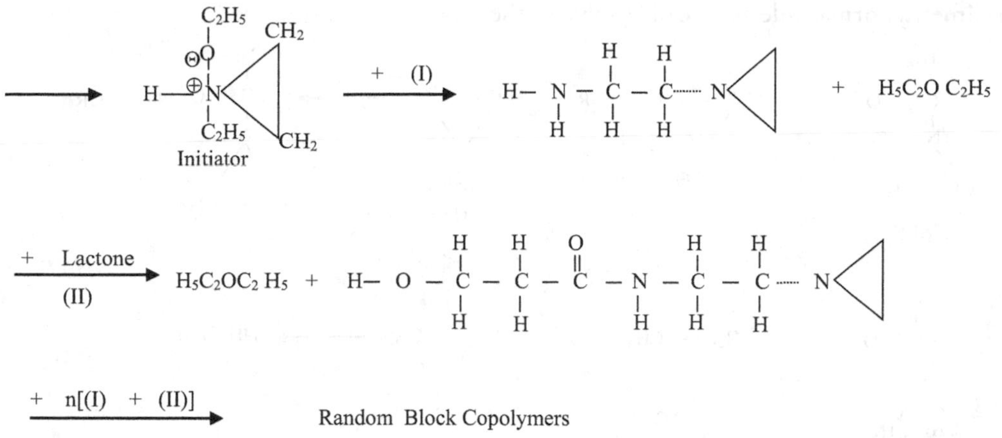

13.119

Unlike the case above (i.e., Equation 13.118), it is the negatively charged center (which is ionic in character) that is first and specifically used for the lactone which is an Electrophile, if the initiator is in Stable state of existence, that which is believed to be the case. This was not shown above because the initiator was in Equilibrium state of existence. Imine units will only show up intermittently along the chain. The ether can be observed to be the catalyst anionically, while the monomer is the cocatalyst, the center for electro-free-radical polymerization. That of ethyl acetate is similar to the case above, noting that the strength of the electro-free-radical $^{e\cdot}C_2H_5$ may not be the same as that from ethyl ether above. In ethyl acetate, CH_3COO is a radical-pulling type of group while in ethyl ether C_2H_5O is a radical-pushing type of group. Hence, the copolymer yield with ethyl acetate should be different from those with ethyl ether and different from those with toluene as initiators or so-called solvents.

With ethylene dichloride as so-called solvent, the followings are obtained, during initiation.

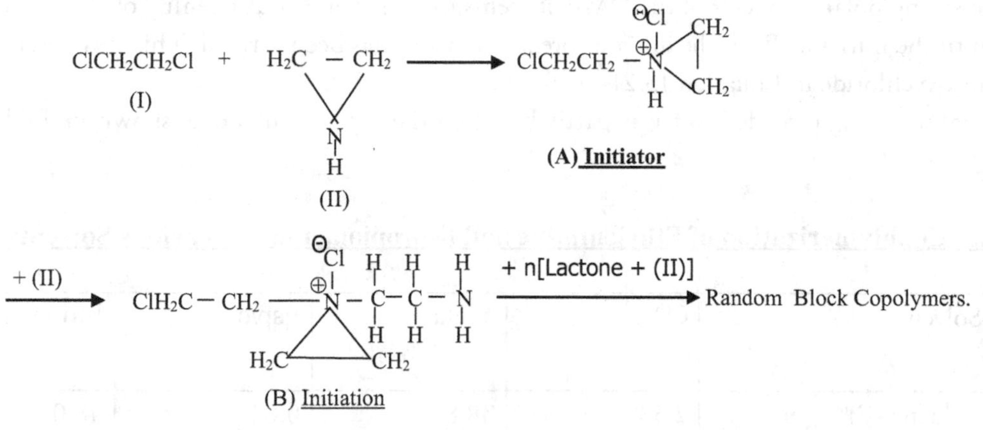

13.120

It should be noted again that it is the forward addition anionically that commences polymerization if the initiator is in Stable state of existence. Only the electro-free-radical route has all along been shown so far, because the initiators were "assumed" to be in Equilibrium state of existence. Like the cases considered so far, without the aziridine being involved during initiation the lactone cannot be copolymerized with it. (I) cannot be used chargedly. It is used free-radically to form the initiator (A) which like the others has dual centers of polymerization depending on their state of existence. Like other cases, both the initiator preparation and the initiation step are completed one after the other.

With dimethylformamide as so-called solvent, the followings are obtained.

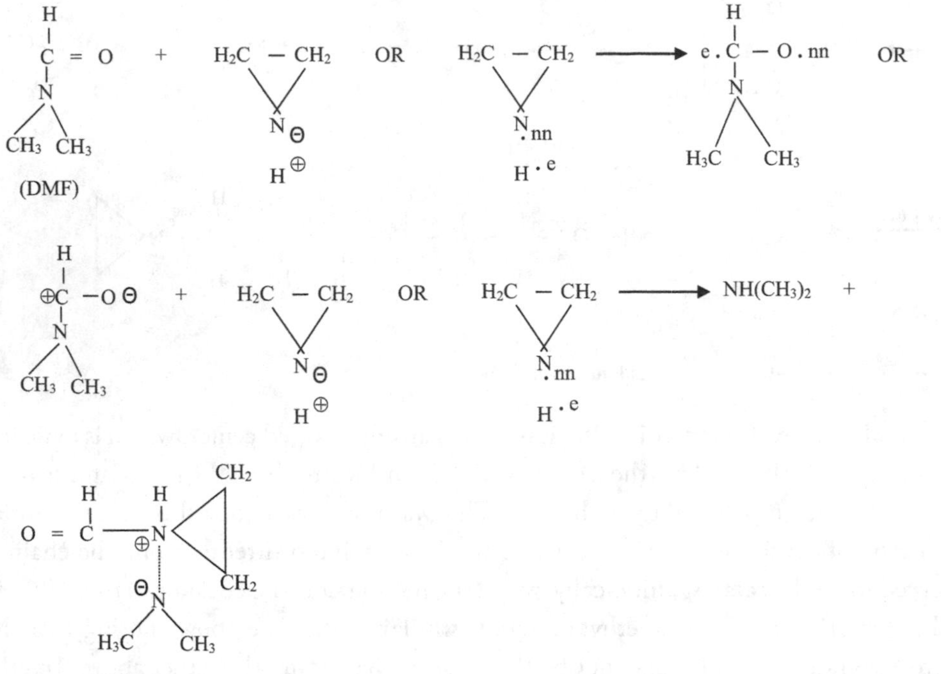

(DMF)

(B) Initiator

13.121

The DMF, like ethylene dichloride, acetone and acetonitrile (CH_3CN) cannot be used in full as the catalyst. Part of the monomer (imine) still remain as source of the cocatalyst. Without the comonomer aziridine, copolymerization in the absence of these so-called solvents would have been impossible. Thus, despite the strong polar character of the DMF, it seems that instantaneous opening of the rings are not favored, until the initiator (B) in the initiator preparation step has been formed. This was observed with phosphorus oxychloride in Equation 13.21.

The analysis being provided so far, is partly based on the experimental data shown in Table 13.3[32] below.

Table 13.3 Copolymerization of Ethylenimine and β-propiolactone in Various Solvents[a]

#	Solvent	ε[b]	Yield %	ηsp/c	Imine mole- %
1	Toluene (E)[c]	2.38	38.8	0.04	48.0
2	Ethyl ether (A&E)	4.34	37.6	0.03	50.3
3	Ethyl acetate (A&E)	6.02	22.2	0.09	53.7
4	Ethylene dichloride (A&E)	10.4	33.6	0.09	57.7
5	Acetone (E)	20.7	46.2	0.07	53.4
6	Acetonitrile (E)	37.5	54.5	0.07	55.3
7	Dimethyl–formamide (A&E)	37.6	83.8	0.06	55.6

(a) Monomers, 0.03 mole each; solvent 30ml; catalyst, none; polymerization temperature, 0°C.; polymerization time, 48hr.

(b) Dielectric constant of the solvent at room temperature.

(c) (A&E) is Anionic and/or Electro-free-radical route; (E) is electro-free-radical route only.

The non-ionic/non-polar, non-ionic/polar, ionic/polar characters of solvents, can be identified from the dielectric constants of the solvents in the second column. In fact, the dielectric measurements above are indeed direct measures of the polar and ionic characters. From the data indicated on the fifth column of the Table, it is obvious that imines are complete nucleophiles (carrying only X centers visible and the invisible ones) while lactones are electrophile (carrying both nucleophilic (with 2 Xs) and electrophilic (Y) centers). One cannot conclude from the data that cyclic ether is more nucleophilic than lactone, just because the lactone is an Electrophile. Secondly, it is obvious that, the solvents are part of the initiators playing the role of the catalyst, and therefore can be part of the products if the initiator was stable and anionic in character. Thirdly, it is obvious that, the routes favored are anionic particularly for the lactone only for those with anionic centers and electro-free-radical particularly for the amine. It is too early yet to address the data on the fourth column, except to say that the low viscosity is as a result of the large volume of solvents involved- too many small molecular weight chains in the system. Why the yields are what they are have been fully explained, noting that this is partly and largely dependent on the strength of the initiating center as shown below for some of them. All the initiators are ***Electrostatically negatively charged paired initiators with some as Electrostatically anionically-paired initiators***. Only the latter can be used more on the lactone than on the amine. It can only be used on the amine if the amine is opened instantaneously. The bonds carried by them are usually represented by dotted lines, because they are electrostatic (imaginary) in character.

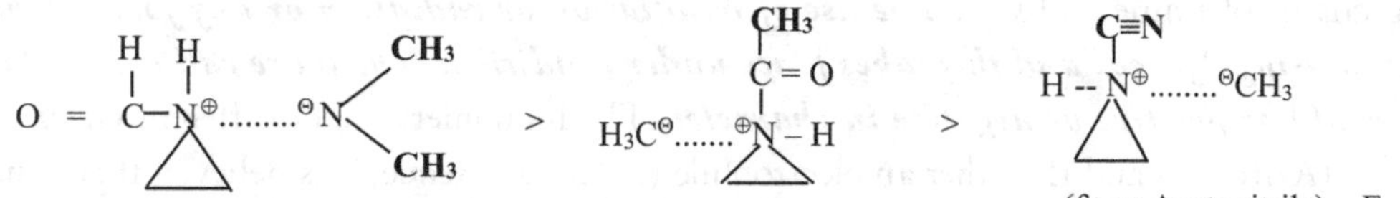

(from DMF) – E & A   (from acetone) –E   (from Acetonitrile) – E

Order of strength of active centers anionically and electro-free-radically   13.122

$$H_3C(CO)O^{\ominus} \quad > \quad Cl^{\ominus} \quad > \quad C_2H_5O^{\ominus} \quad > \quad (CH_3)_2N^{\ominus}$$

Order of strength of some Anionic Active centers   13.123

Notice that $^{\ominus}C \equiv N$ from the acetonitrile and $H_3C(OC)^{\ominus}$ from acetone could not exist as negatively charged centers due to electrostatic forces of repulsion. While the two cases are real, those carried by the counter charged centers are imaginary. Being real, hence they (unlike polar charges and the counter-charged center which do not repel or attract themselves or other charges) repel and attract. Hence they do not exist. Covalent and ionic charges which are real, undergo forces of repulsion and attraction. Radicals do not within themselves and not with charges, because they have identities invisibly present. Hence, it can be noticed that the charges and initiators obtained from them are as shown above, wherein the acetone, acetonitrile like the others were in stable state of existence. However, of the three initiators shown above, only the initiator from DMF can be used both anionically and electro-free-radically. The other two can only be used electro-free-radically. The routes favored by the use of initiators from the solvents have been shown in Table 13.3. Worthy of note is that when the initiators are prepared, some fractions are kept in Equilibrium state of existence while the remaining fractions are in Stable state of existence.

Those with negatively charged centers in stable state of existence cannot be used anionically and electro-free-radically. Those with anionic centers in stable state of existence can be used only anionically, while those in Equilibrium state of existence can be used only electro-free-radically. This is indeed what takes place. On the other hand, since the aziridine is part of the initiator in the presence of such large volume of solvent, then where is the remaining aziridine coming from for polymer formation in an ideal reactor? One should be very careful in choosing the molar ratios in these systems, the manners of addition, the operating conditions and more, because NATURE is orderliness.

From the analysis so far and the unique observations, the followings are valid for aziridines.

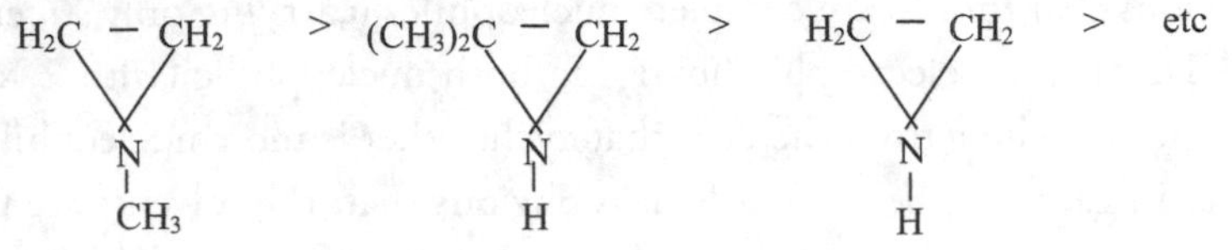

$$H_2C - CH_2 \quad > \quad (CH_3)_2C - CH_2 \quad > \quad H_2C - CH_2 \quad > \quad etc$$

$$\underset{\underset{CH_3}{|}}{N} \qquad\qquad \underset{\underset{H}{|}}{N} \qquad\qquad \underset{\underset{H}{|}}{N}$$

$$(I) \qquad\qquad\qquad (II) \qquad\qquad\qquad (III)$$

<u>Order of Nucleophilicity of aziridines</u>   13.124

As it seems, the solvents when used can be recovered after propagation is completed only when the route is electro-free-radical. In none of them are byproducts obtained. These simple but unique differences are worthy of note since they will help during analysis of polymers obtained and in reactor design consideration. *In the copolymer-ization of aziridines with carbon monoxide using gamma rays which instantaneously opened the aziridine rings and activated the carbon monoxide to produce living alternating copolymers clearly indicates that* **the gamma radiation is almost analogous to using nucleo-free-radicals, except that what is carrying the chain cannot be seen.** It is only with nucleo-free-radicals that full alternating placement is possible, while with electro-free-radicals, homopolymers of the imine must appear between some alternating chains of imine and CO. *The use of irradiation or radiation or any form of energy is analogous to using "force" and this takes place under conditions which are radical in character. The force could be positive or negative in character.* The monomers are of different characters, one a nucleophile (Aziridine) and the other an electrophile (Lactone). Hence, it is believed that, when non-chemical initiators are involved, the routes in general are specific depending on the type of force. That is addition of monomers will take place from a particular type of active center. In order words if a monomer such as propylene was to be initiated and polymerized using gamma rays, no polymerization will be favored, since there exists free-radical transfer species as shown below.

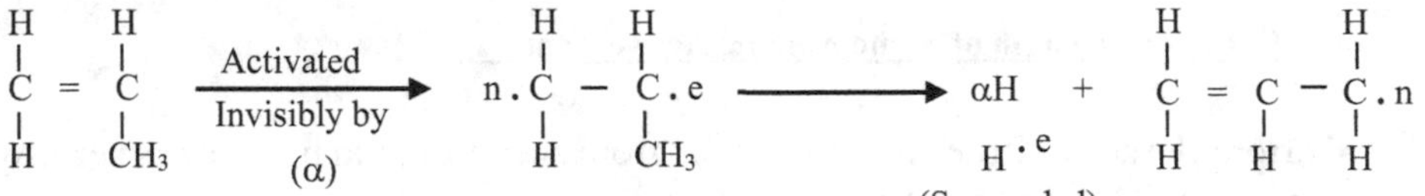

$$\underset{\underset{H}{|}}{\overset{\overset{H}{|}}{C}} = \underset{\underset{CH_3}{|}}{\overset{\overset{H}{|}}{C}} \quad \xrightarrow[\text{Invisibly by}]{\text{Activated}} \quad n.\underset{\underset{H}{|}}{\overset{\overset{H}{|}}{C}} - \underset{\underset{CH_3}{|}}{\overset{\overset{H}{|}}{C}}.e \quad \longrightarrow \quad \alpha H \quad + \quad \underset{\underset{H}{|}}{\overset{\overset{H}{|}}{C}} = \underset{\underset{H}{|}}{\overset{\overset{H}{|}}{C}} - \underset{\underset{H}{|}}{\overset{\overset{H}{|}}{C}}.n$$

$$(\alpha) \qquad\qquad\qquad (I) \qquad\qquad H \cdot e \quad (\text{Suspended}) \qquad\qquad\qquad\qquad 13.125$$

The α above is the "invisible force" which can be gamma radiation. These invisible forces can only activate a monomer or scission a compound radically that is homoly-tically and not heterolytically.

A monomer such as acrylamide can be polymerized using gamma irradiation since there is no free-radical transfer species via the C = C activation center as shown below.

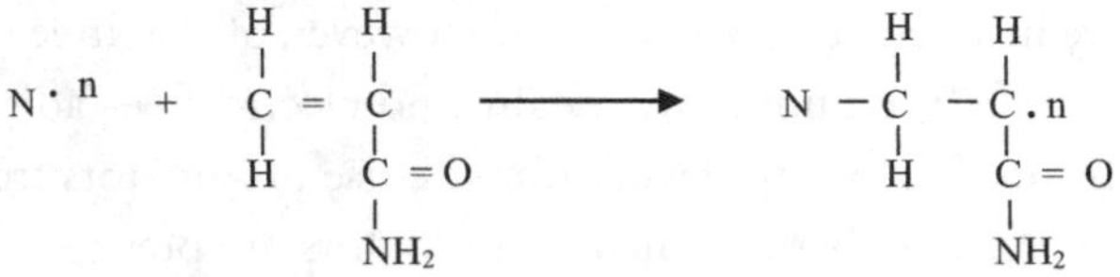

$$N \cdot n \quad + \quad \underset{\underset{H}{|}}{\overset{\overset{H}{|}}{C}} = \underset{\underset{\underset{NH_2}{|}}{\overset{C = O}{|}}}{\overset{\overset{H}{|}}{C}} \quad \longrightarrow \quad N - \underset{\underset{H}{|}}{\overset{\overset{H}{|}}{C}} - \underset{\underset{\underset{NH_2}{|}}{\overset{C = O}{|}}}{\overset{\overset{H}{|}}{C}}.n$$

$$13.126$$

$$E \cdot e \; + \; \begin{matrix} H & H \\ | & | \\ C = C \\ | & | \\ H & C=O \\ & | \\ & NH_2 \end{matrix} \longrightarrow \; E - \begin{matrix} H & H \\ | & | \\ C - C \cdot e \\ | & | \\ H & C=O \; H \\ & | \\ & NH_2 \end{matrix} \quad OR \quad ENH_2 \; + \; e \cdot \begin{matrix} CH_2 \\ || \\ CH \\ | \\ C = O \end{matrix} \qquad 13.127$$
$$\text{Favored}$$

$$\alpha \; + \; 3 \begin{matrix} H & H \\ | & | \\ C = C \\ | & | \\ H & C=O \\ & | \\ & NH_2 \end{matrix} \longrightarrow \; \alpha \; + \; 3 \; n \cdot \begin{matrix} H & H \\ | & | \\ C - C \cdot e \\ | & | \\ C=O & H \\ | \\ NH_2 \end{matrix} \longrightarrow$$

$$\alpha \; e \cdot \begin{matrix} H \\ | \\ C \\ | \\ H \end{matrix} - \begin{matrix} H \\ | \\ C \\ | \\ C=O \\ | \\ NH_2 \end{matrix} - \begin{matrix} H \\ | \\ C \\ | \\ H \end{matrix} - \begin{matrix} H \\ | \\ C \\ | \\ C=O \\ | \\ NH_2 \end{matrix} - \begin{matrix} H \\ | \\ C \\ | \\ H \end{matrix} - \begin{matrix} H \\ | \\ C \cdot n \\ | \\ C=O \\ | \\ NH_2 \end{matrix} \longleftarrow \quad \text{Active center}$$

$$13.128$$

By the reactions above, it is obvious that the "invisible initiator" is nucleophilic in character, since it is not the C = O activation center that is activated and is also free radical in character, since in for example ketene, it is the C = C center which is less nuc-leophilic that is activated instead of the C = O center.

13.2.4. Order of Nucleophilicity and Basicity of Cyclic Amines

So far it has clearly been established that –

$$\text{Cyclic amines} \quad > \quad \text{Alkenes} \quad > \quad \text{CO}$$

$$\underline{\text{Order of Nucleophilicity}} \qquad\qquad 13.129$$

Four, five and larger membered rings of cyclic amines are not popularly known to be useful as monomers because of the following reasons-

(i) They are far less strained than the three-membered rings and will therefore need very strong active initiators or centers and or heat to intro-duce the MRSE into the ring.

(ii) The larger the size of the ring, the less loosely bonded the hydrogen atom on the nitrogen center becomes; so that most of the time, the followings take place ionically or radically.

$$R^{\ominus} \; + \; \begin{matrix} H_2C - CH_2 \\ | \qquad | \\ H_2C - N \\ | \\ H \end{matrix} \longrightarrow \; \text{No reaction} \quad OR \quad \begin{matrix} H_2C - CH_2 \\ | \qquad | \\ H_2C - N^{\ominus} \end{matrix} \; + \; RH$$
$$\text{Impossible, unless with a metal} \qquad 13.130$$

$$R^{\oplus} \; + \; \begin{matrix} H_2C - CH_2 \\ | \qquad | \\ H_2C - N \\ | \\ H \end{matrix} \longrightarrow \; \text{No reaction} \quad OR \quad \begin{matrix} H_2C - CH_2 \\ | \qquad | \\ H_2C - N \\ | \\ R \end{matrix} \; + \; H^{\oplus}$$
$$\text{Impossible via abstraction} \qquad 13.131$$

$$H^{\oplus} \; + \; \begin{matrix} H_2C - CH_2 \\ | \qquad | \\ H_2C - N \\ | \\ H \end{matrix} \longrightarrow \; \text{No reaction}$$

$$13.132$$

When there is pairing positively, the rings cannot in most cases be opened, due to lack of MRSE. In most cases, the ring is very unreactive (No reaction). When the ring is of the tertiary type (more basic), the use of very strong acids will be required to favor their openings. In order words, these unsubstituted amines are analogous to secondary amines, for which the larger the size of the ring, the more basic it is.

$$6\text{ - membered} > 5\text{ - membered} > 4\text{ - membered} > 3\text{ - membered} >$$

<u>Order of basicity of cyclic amines</u>   13.133

These development for cyclic esters, amides and amines are analogous to the following relationships.

$$NR_3 > NR_2H > NRH_2 > NH_3$$

$$-NR_2 > -NRH > -NH_2$$

<u>Order of basicity (amines) and radical-pushing</u>
<u>capacity of their groups</u>   13.134a

$$HOH > CH_3OH > C_2H_5OH$$

$$HO- > CH_3O- > C_2H_5O-$$

<u>Order of acidity ("alcohols") and radical-pushing capacity</u>
<u>of their groups</u>   13.134b

$$C_2H_5OOC- > CH_3OOC- > HOOC-$$

<u>Order of radical-pulling capacity (esters)</u>   13.135

$$ROOC- > H_2NOC- > RHNOC- > R_2NOC-$$

<u>Order of radical-pulling capacity (amides)</u>   13.136

Thus, in general, when instantaneous opening of a ring is favored in the presence of a weak radical initiator, and molecular rearrangement of the third kind is applied, ***cyclic ethers cannot rearrange to aldehydes and ketones which are isomers. The same applies to cyclic amines. They cannot also rearrange to aldimines and ketimines***.

Side by side placement of rings of cyclic amines may favor polymerization, since the rings will be strongly nucleophilic in character and far more strained.

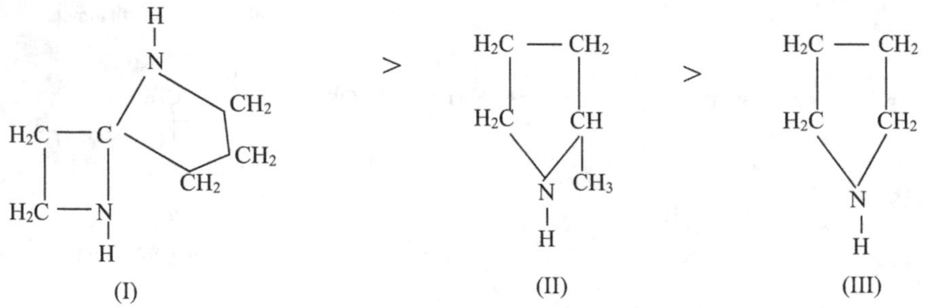

<u>Order of Nucleophilicity</u>   13.137

Hence, the four or five-membered rings of (I) will favor opening of its ring. However, only three-membered cyclic amines will be used in the remaining analysis. Between cyclic amines and cyclic ethers, the followings are obtained.

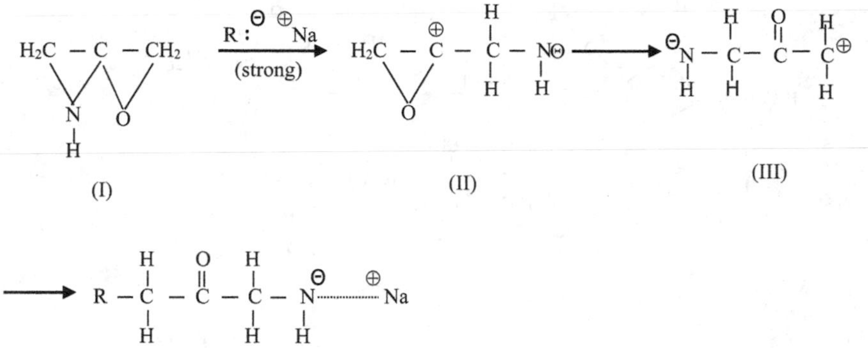

$$H_2C - C - CH_2 \xrightarrow[\text{(strong)}]{R:^{\ominus} \ ^{\oplus}Na} \quad H_2C - C - C - N\ominus \longrightarrow \ ^{\ominus}N - C - C - C^{\oplus}$$

(I)   (II)   (III)

$$\longrightarrow R - C - C - C - N^{\ominus} \text{------} \ ^{\oplus}Na$$

(IV) Anionically   13.138

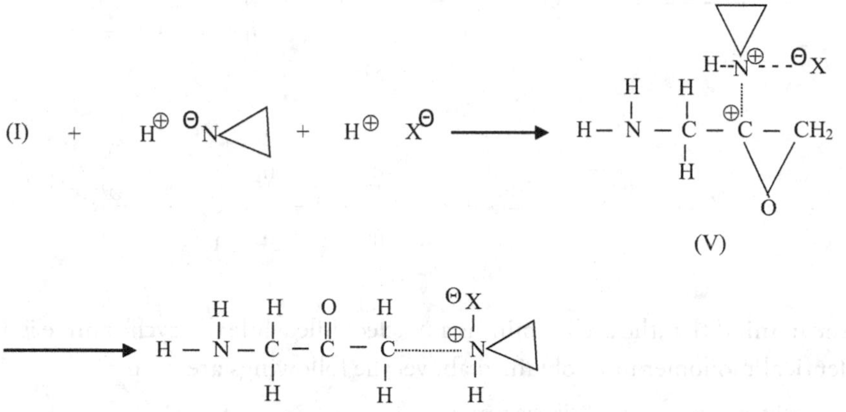

$$(I) \ + \ H^{\oplus} \ ^{\ominus}N \triangleleft \ + \ H^{\oplus} \ X^{\ominus} \longrightarrow H - N - C - C - CH_2$$

(V)

$$\longrightarrow H - N - C - C - C \text{------} \ ^{\oplus}N \triangleleft$$

(VI) Electro-free-radically   13.139

Based on the type of monomer units obtained for (IV) and (VI); there is no doubt that-

$$\text{Cyclic ethers} \quad > \quad \text{Cyclic amines}$$

Order of Nucleophilicity   13.140

Anionically, $R:^{\ominus}$ must be strong to favor the quick instantaneous opening of the nitrogen containing ring. Anionically or electro-free-radically, it is the cyclic amine ring that is first opened, since it is less nucleophilic and more strained than the cyclic ether ring. Electro-free-radically, no matter the strain energy in the ring, it is the less nucleophilic ring that is first attacked not instantaneously but via the functional center.

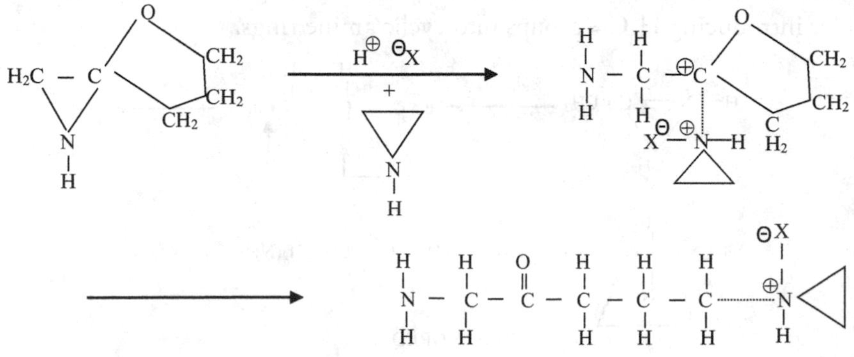

$$\longrightarrow H - N - C - C - C - C - C \text{------} \ ^{\oplus}N \triangleleft$$

13.141

215

Anionically, only instantaneous opening of the three membered cyclic amine, will favor any polymerization. It is important to note the type of initiator involved electro-free-radically.

Between cyclic amines and cyclic amides, the followings are obtained.

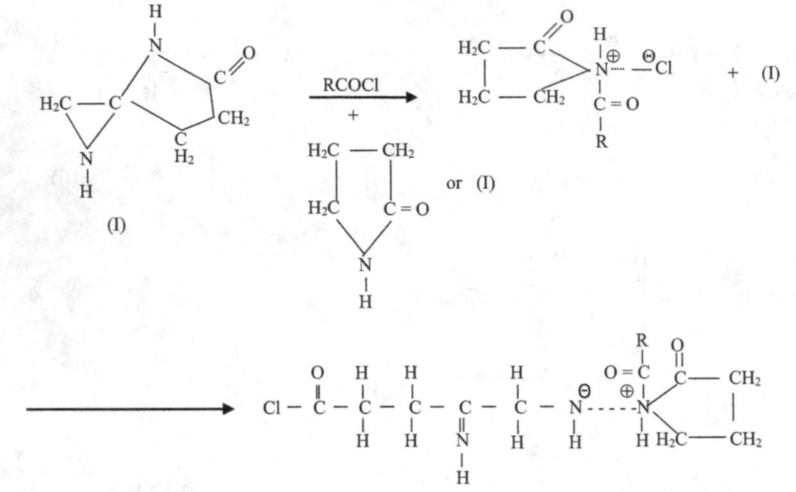

$$13.142$$

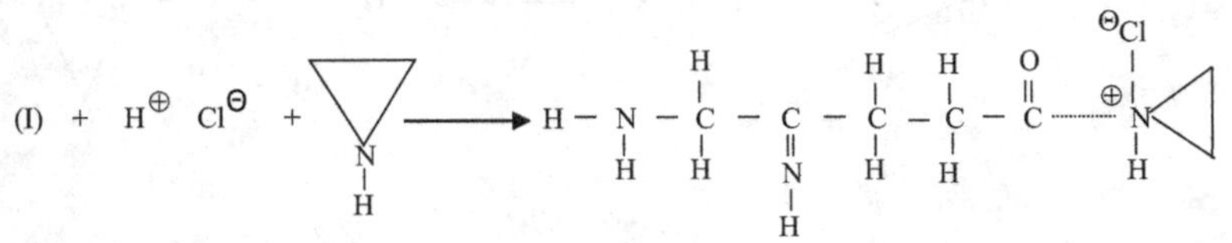

$$13.143$$

It should be borne in mind that the cyclic amine is a Nucleophile while the cyclic amide is an electrophile. Based on the identical monomer units obtained above, the followings are valid.

Cyclic amides > Cyclic amines

<u>Order of nucleophilicity</u>

$$13.144$$

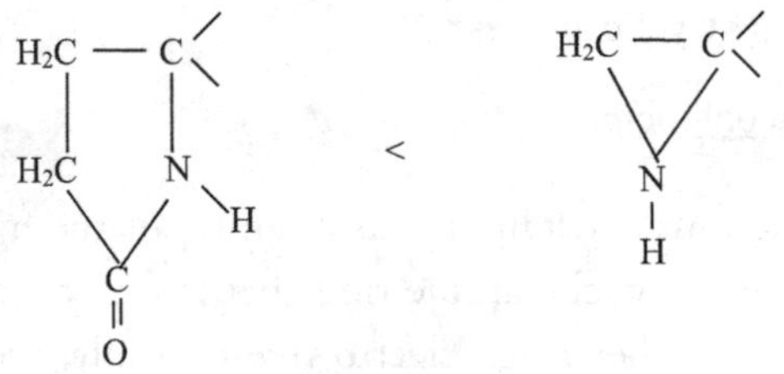

<u>Order of radical-pushing capacity</u>

$$13.145$$

Electro-free-radically and anionically, Electrostatically anionically-paired initiators are involved.

Now, consider introducing $H_2C =$ groups into cyclic amine rings.

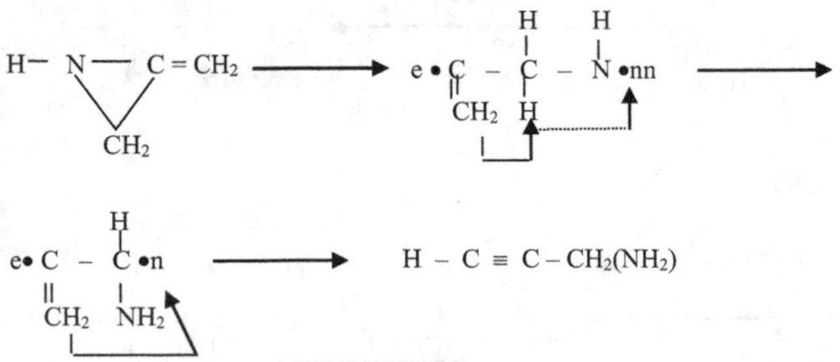

NOT FAVORED

$$13.146a$$

Above, the ring was instantaneously opened hoping that the ring has more than the MRSE. It rearranges to give an acetylenic amine if and only if $H_2C=$ group is greater than NH_2 group in radical-pushing capacity, that which is not the case. Hence, the rearrangement above is said not to be favored.

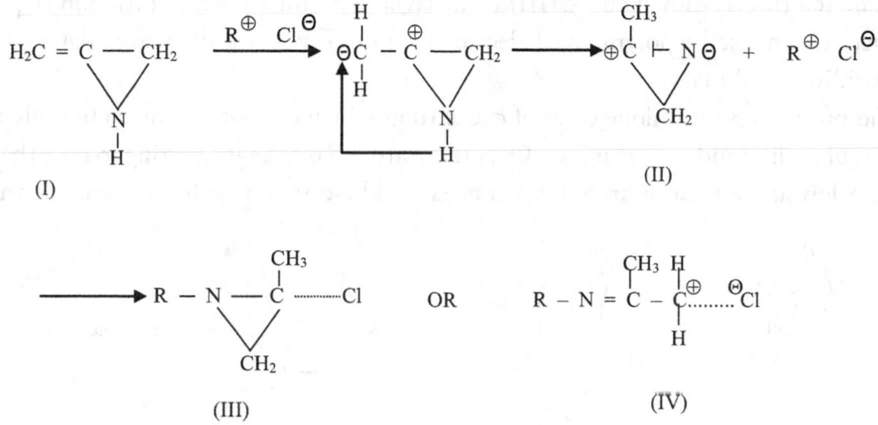

(13.146b)

When activated with a strong initiator, the ring instantaneously opens as already shown. When (I) is activated using weak anionic initiators, it favors molecular rearrangement of the sixth type to produce (II) which can only form a stable salt (III) and not (IV) whose existence is only favored by instantaneous opening of the ring and not as shown above. In (IV) the ring is opened at the only point of scission to give N=C bonds (branching sites) along the chain. Anionically, the following is obtained.

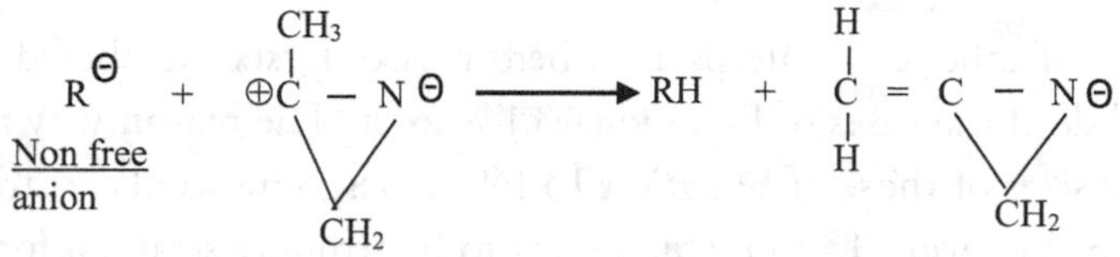

(13.147)

It can be observed that (I) is strongly nucleophilic and can be used a monomer.

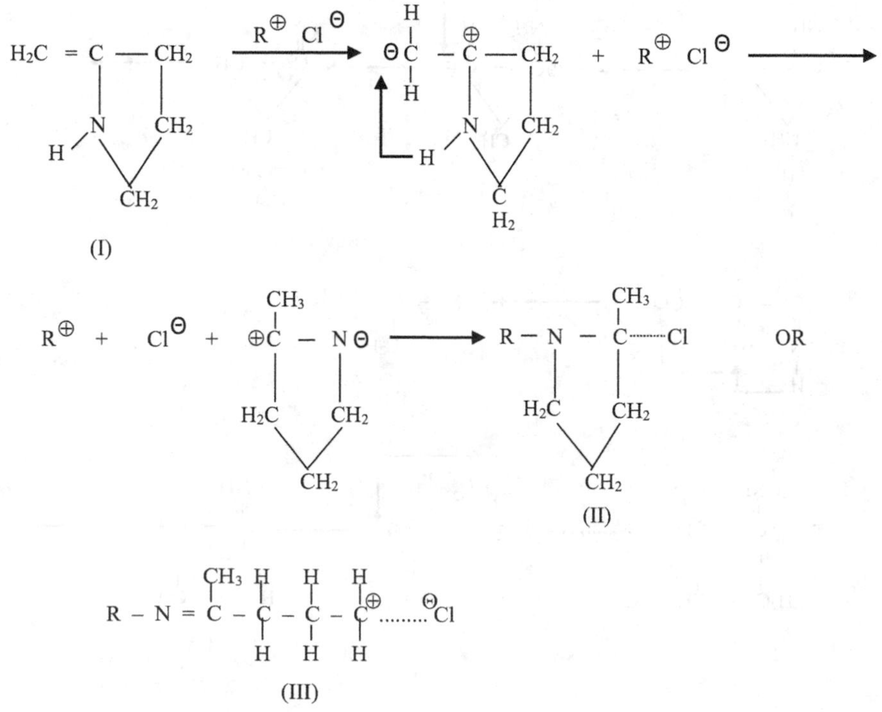

(13.148)

The presence of the $H_2C=$ group can be found to be more useful when placed on larger sized rings only if the ring can be opened, since when placed on three-membered ring, the ring may not exist as shown in Equation 13.146a. for which when instantaneously opened, it can only be used electro-free-radically. On a five-membered ring as shown above (III) is the favored product for the Initiation step if and only if it can be opened. It can also be polymerized electro-free-radically forwardly using NaCN or backwardly using HCl/Aziridine combination.

There is no need to revisit unique cases of cyclic rings which cannot be said to be fully related to the family members of cyclic amides and amines. Over the years, it has been interesting to note that compounds such as shown below are not common or known to exist. These are no cyclic amines, but amines.

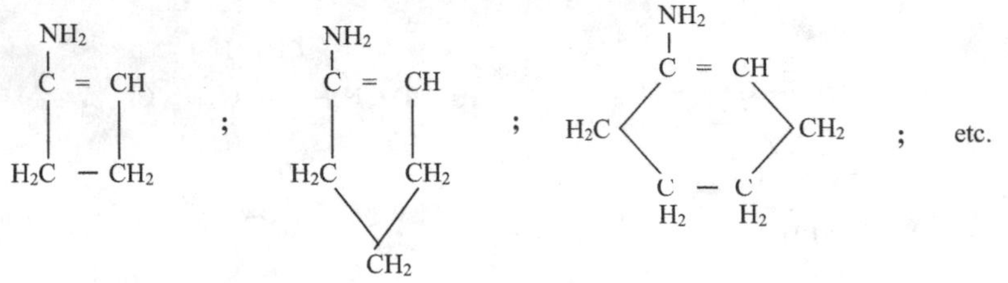

$$13.149$$

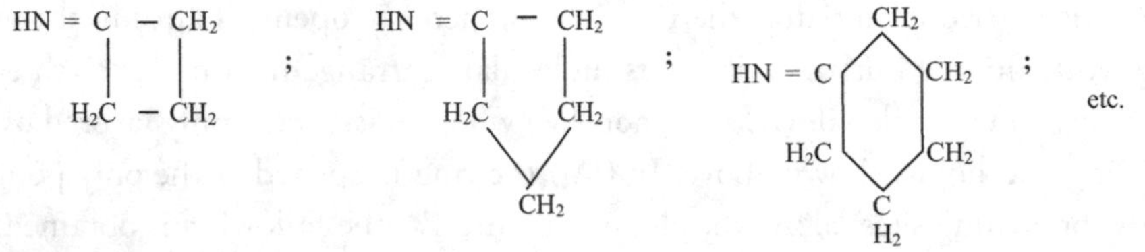

$$13.150$$

This is unlike the case of etheric counterparts, where it is only six-membered rings and above that rearrange, here they all do if the cases of Equation 13.149 exist. The reason why they are not known to exist, is because, all the sizes of those of Equation 13.149 undergo molecular rearrangement of the first kind to produce those of Equation 13.150 which seem to be strongly strained for the first ring, noting that each ring in the family like with other families all contain the same Strain energy.

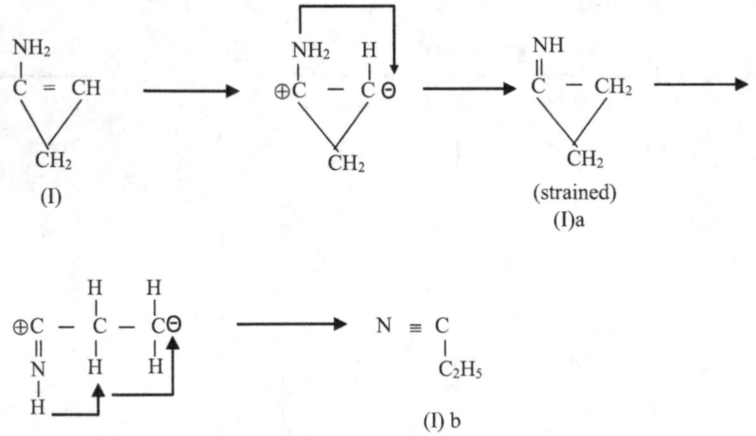

$$13.151$$

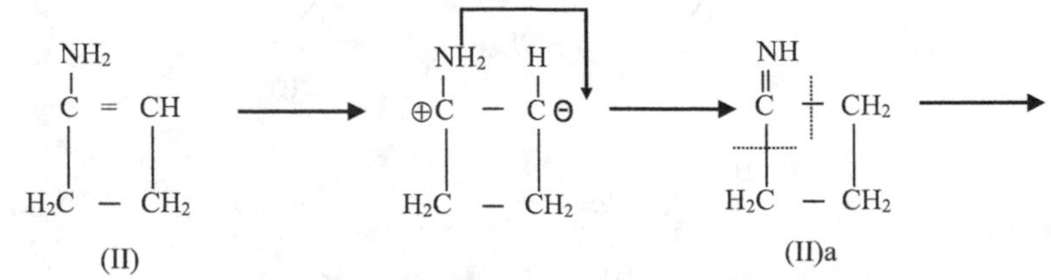

$$13.152$$

$$\underset{\substack{|\\N\\|\\H}}{\overset{\substack{H\\|}}{\oplus C}} - \underset{H}{\overset{H}{\underset{|}{C}}} - \underset{H}{\overset{H}{\underset{|}{C}}} - \underset{H}{\overset{H}{\underset{|}{C}\ominus}} \longrightarrow N \equiv \underset{C_3H_7}{\overset{}{C}}$$

(II) b

$$\underset{\substack{|\\CH_2\\|\\|}}{\overset{\substack{NH_2\\|}}{C}} = CH \qquad \longrightarrow \qquad \underset{H_2C \quad CH_2}{\overset{NH_2 \quad H}{\oplus C - C\ominus}} \qquad \longrightarrow \qquad \underset{H_2C \quad CH_2}{\overset{H}{\underset{||}{N}}} \underset{C}{\overset{}{C}} + CH_2 \longrightarrow$$

(III)   (III)a

$$\underset{\substack{|\\N\\|\\H}}{\overset{\substack{H\\|}}{\oplus C}} - \underset{H}{\overset{H}{\underset{|}{C}}} - \underset{H}{\overset{H}{\underset{|}{C}}} - \underset{H}{\overset{H}{\underset{|}{C}}} - \underset{H}{\overset{H}{\underset{|}{C}\ominus}} \longrightarrow N \equiv \underset{C_3H_7}{\overset{}{C}}$$

(II) b   13.153

Based on the types of nitriles obtained after molecular rearrangement of the third kind that is, (I)a, (II)a and (III)a from three, four and five-membered rings, it is obvious that the following is valid, **noting that these can only take place radically**.

$$> \quad \underset{\substack{C - C\\H_2 \quad H_2}}{\overset{\substack{NH_2 \quad H\\C = C}}{H_2C \diagdown \diagup CH_2}} \quad > \quad \underset{\substack{H_2C \quad CH_2\\ \diagdown \diagup \\ CH_2}}{\overset{\substack{NH_2 \quad H\\C = C}}{}} \quad > \quad \underset{H_2C - CH_2}{\overset{NH_2 \quad H}{C = C}} \quad > \quad \underset{CH_2}{\overset{\substack{NH_2 \quad H\\C = C}}{\diagdown \diagup}}$$

Order of Nucleophilicity   13.154

It must be recalled that HN = group is far more radical-pushing than H_2C = group. Hence, one should expect the rings with the HN = group to be more strained than the ring with NH_2 group, to the extent of making the three membered one favor instanta-neous opening of its ring at STP. At far low temperatures below STP, this may not be the case. Based on the analysis so far, it is obvious that the NH_2 group is greater than the OH group in capacity.

Thus, monomers or compounds of Equation 13.149 are self-activated compounds, since they molecularly rearrange to those of Equation 13.150 to produce nitriles, which are more nucleophilic than the cycloalkene rings carrying the groups. It is believed that the HN = C activation center, like the H_2C = C activation center but unlike O =C activation center, cannot be activated to favor molecular rearrangement of the sixth type, that is, to rearrange from (I)a and (II)a of Equations 13.151 and 13.152 respectively to (I) and (II).

Monomers such as shown below may look or resemble a cyclic amide, which indeed is not the case.

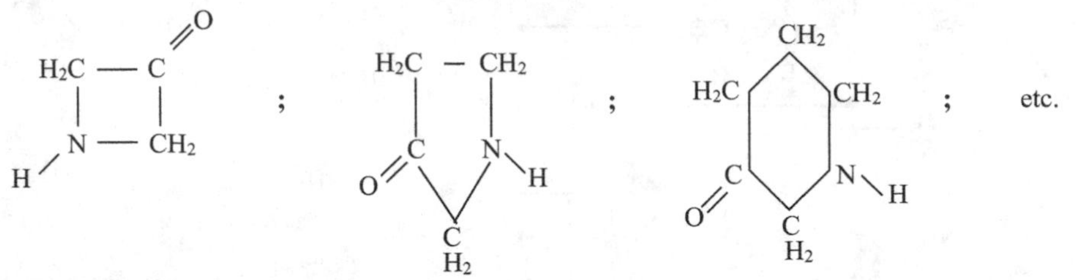

13.155

In the compounds above, there is only one functional center which can only be used in a manner similar to that of cyclic amines. But however, unlike larger membered rings of cyclic amines, the hydrogen atom on the nitrogen center is more loosely bonded than in conventional larger membered cyclic amines, because the compounds above are more nucleophilic and less basic (Secondary cyclic amine) in character. That is, the following is valid.

Order of Nucleophilicity   13.156

Unlike four, five and larger membered rings of cyclic amines, those of Equation 13.155 above, favor electro-free-radical polymerizations, since MRSE can readily be provided for the rings which are very strained. The four-membered ring may probably not exist, since it will rearrange to give a ketene when the ring is opened instantaneously.

When the four-membered ring of Equation 13.155 is placed side-by-side to an aziridine, the followings are to be expected.

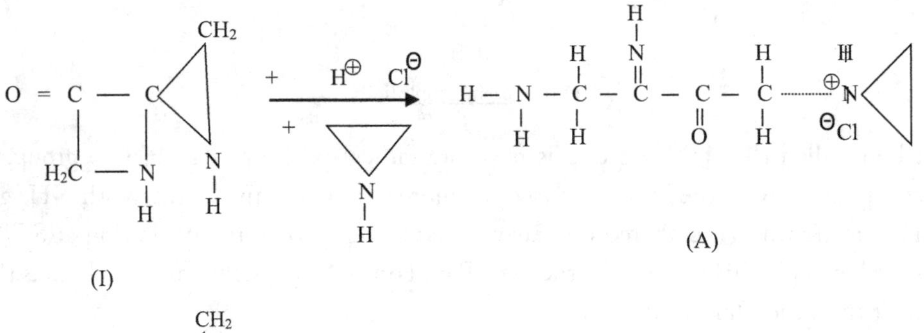

13.157

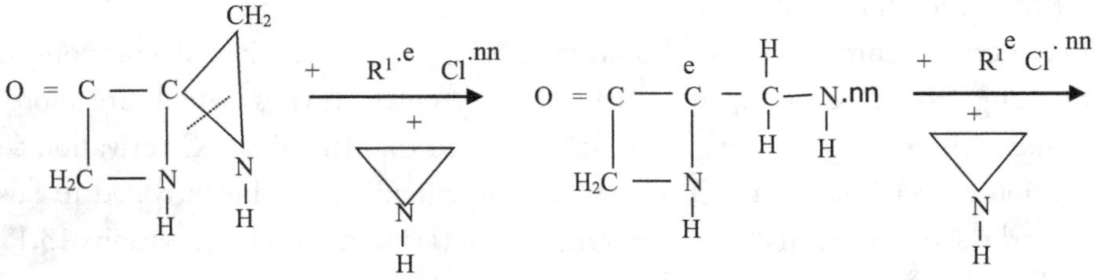

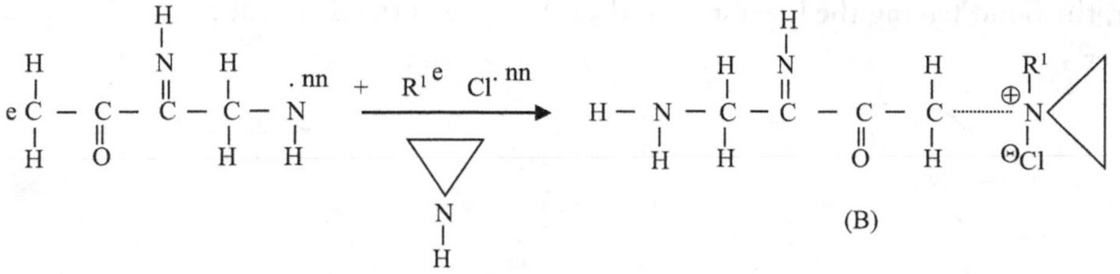

13.158a

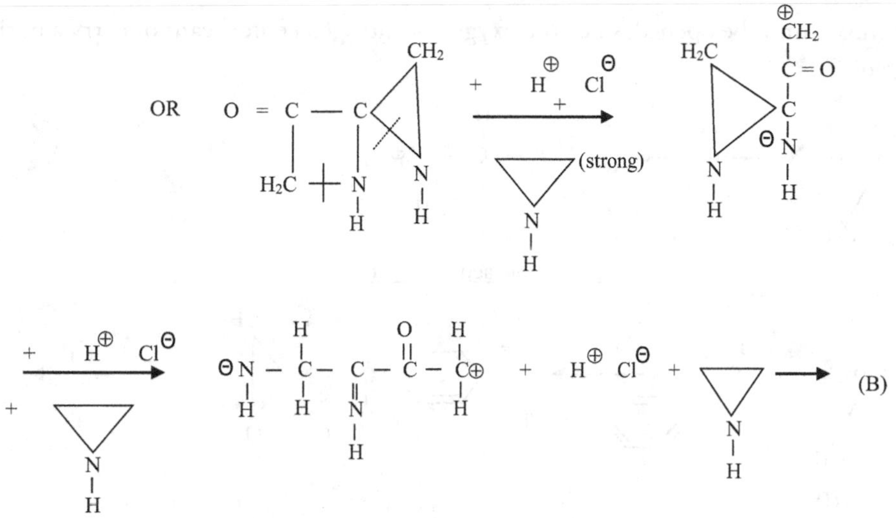

OR (strong)

(B)

13.158b

In the first equation, the ring was opened via the amine if less nucleophilic. In the second equation, the amine ring was instantaneously opened. In the third equation, the four membered ring was also instantaneously opened. The four membered ring is more nucleophilic than the three membered ring, and it looks as if it has equal strain energy with the three-membered ring. Imagine as an exercise what it will be when the N is replaced with O one at a time.

It can be observed how the reactions of so many years ago, some of which form the basic foundations for present day research, are being explained from different points of view, which are unquestionable. One has gone to the extent of explaining why some compounds do not favor any existence or favor transient existence based on the operating conditions.

13.3 Related Monomers

Related monomers of interest include

(i) Oxaazocyclopropane[33]
(ii) Pyrrole[33]
(iii) Carbaonhydrides and
(iv) Imino cyclic compounds

13.3.1 Oxaazacyclopropane

This is an example of a three-membered ring with two hetero atoms in the ring. Since the ring is small in size, it is supposed to have MRSE to favor its instantaneous opening. In the ring there are two nucleophilic functional centers adjacently located but cannot be used. As will be further explained in the next chapter, the bond having the highest radical-pushing potential difference is the N – O bond in the ring. However,

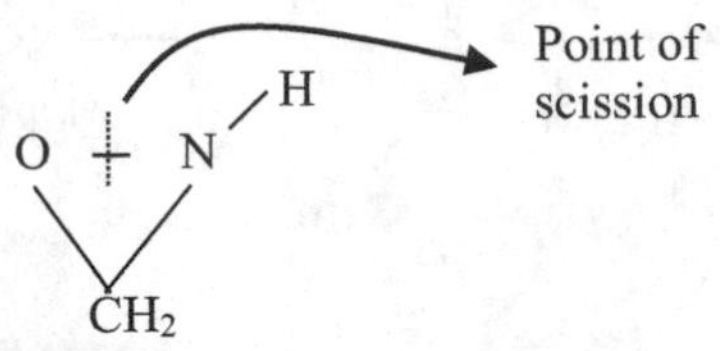

$$\text{O} \overset{|}{+} \text{N} \overset{H}{\diagdown} \qquad \substack{\text{Point of} \\ \text{scission}}$$
$$\text{CH}_2$$

13.159

chargedly, the ring cannot be opened, since the oxygen or nitrogen centers cannot carry a positive charge in the presence of C.

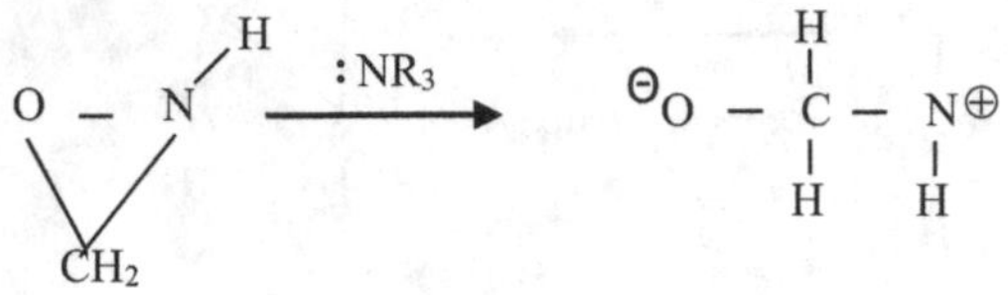

$$\text{O} - \text{N} \overset{H}{\diagdown} \quad \xrightarrow{\;: NR_3\;} \quad \overset{\ominus}{\text{O}} - \overset{H}{\underset{H}{\text{C}}} - \overset{\oplus}{\text{N}}$$
$$\text{CH}_2$$

Impossible activated state

13.160

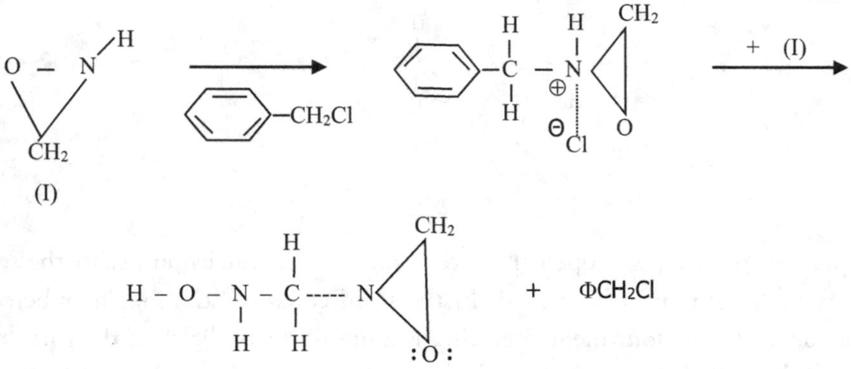

$$\text{O} - \text{N} \overset{H}{\diagdown} \quad \xrightarrow{\bigcirc-\text{CH}_2\text{Cl}} \quad \bigcirc-\overset{H}{\underset{H}{\text{C}}}-\overset{\oplus}{\text{N}}\diagup \overset{CH_2}{\diagdown_O} \quad \xrightarrow{\;+\;(I)\;}$$
$$\text{CH}_2 \qquad\qquad\qquad\qquad\qquad\qquad \overset{\ominus}{\text{Cl}}$$
(I)

$$\text{H} - \text{O} - \overset{H}{\underset{H}{\text{N}}} - \overset{H}{\underset{H}{\text{C}}} ----- \text{N} \diagup\overset{CH_2}{\diagdown} \qquad + \qquad \Phi\text{CH}_2\text{Cl}$$
$$: \text{O} :$$

13.161

Benzyl chloride has been used as the source of the initiator. The initiator a negatively charged-paired initiator, has also an electro-free-radical center on the positively charged counter center (Dual in character). The monomer is part of the initiator. Since the nega-tively charged (anionic) end is not natural to the monomer which is a Nucleophile, H•e from the center goes to the less nucleophilic center (N or O) to open the ring to complete the initiation step. This is then followed by propagation with all the centers still in place. Anionically, the following is obtained.

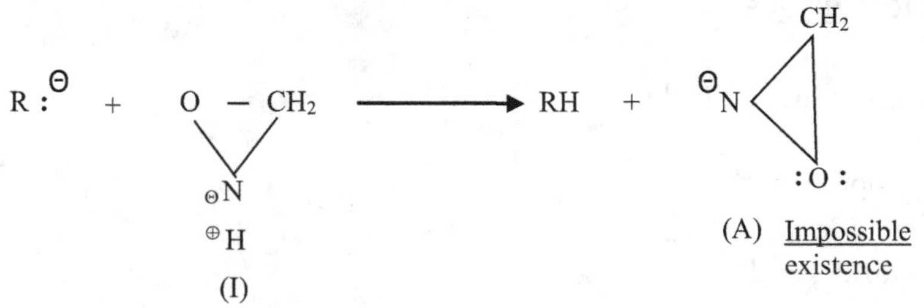

$$\text{R} : \overset{\ominus}{} \quad + \quad \text{O} - \text{CH}_2 \quad \longrightarrow \quad \text{RH} \quad + \quad \overset{\ominus}{\text{N}}\diagup\overset{CH_2}{\diagdown}$$
$$\underset{\oplus H}{\overset{\ominus N}{}} \qquad\qquad\qquad\qquad\qquad\qquad : \text{O} :$$
(I) \qquad\qquad\qquad\qquad\qquad\qquad (A) Impossible
existence

Cannot exist

13.162

Existence of (I) and (A) is not favored, due to electrostatic forces of repulsion.

When nucleo- or electro-non-free-radicals are involved, the following are to be expected.

$$\ddot{N}.nn \quad + \quad \underset{H_2\dot{C} - N\diagdown_H}{\overset{O}{\triangle}} \quad \longrightarrow \quad \ddot{N}.nn \quad + \quad e\bullet\underset{\underset{H}{|}}{\overset{\overset{H}{|}}{C}} - \underset{\underset{H}{|}}{N} - O.nn \quad \longrightarrow$$

$$\underset{(I)}{\ddot{N} - \underset{\underset{H}{|}}{\overset{\overset{H}{|}}{C}} - \underset{\underset{H}{|}}{N} - O.nn} \quad OR \quad \ddot{N}.nn \quad + \quad \underset{\underset{H\bullet e}{\underset{|}{nn}}}{\underset{\overset{\bullet}{N} - CH_2}{\overset{O}{\triangle}}} \quad \longrightarrow \quad NH \quad + \quad \underset{(II)}{\overset{nn}{\overset{\bullet}{N}}\underset{CH_2}{\overset{O}{\triangle}}}$$

$$\text{(I)} \qquad\qquad\qquad\qquad\qquad \text{(II)} \qquad\qquad\qquad\qquad 13.163$$

The half non-free character of the monomer can be observed which should remain the same whether the ring is opened instantaneously or via a functional center. It seems the nucleo-non-free-radical route as reflected above are not favored, since the monomer is a strong nucleophile and may be the monomer has not been properly scissioned at the right bond.

$$E.en \quad + \quad \underset{\underset{(I)}{H_2\dot{C} - N\diagdown_H}}{\overset{O}{\triangle}} \quad \longrightarrow \quad E.en \quad + \quad nn.O - \underset{\underset{H}{|}}{\overset{\overset{H}{|}}{N}} - \underset{}{\overset{\overset{H}{|}}{C}}.e \quad \longrightarrow$$

$$E - O - \underset{\underset{H}{|}}{\overset{\overset{H}{|}}{N}} - \underset{\underset{H}{|}}{\overset{\overset{H}{|}}{C}}.e \quad + \quad (I) \quad \longrightarrow \quad E - O - \underset{\underset{H}{|}}{\overset{\overset{H}{|}}{N}} - \underset{\underset{H}{|}}{\overset{\overset{H}{|}}{C}} - O - \underset{\underset{H}{|}}{\overset{\overset{H}{|}}{N}} - \underset{\underset{H}{|}}{\overset{\overset{H}{|}}{C}}.e$$

$$\longrightarrow \quad \text{No transfer species} \qquad\qquad\qquad 13.164$$

Thus, only the electro-free-radical route is favored by it being a strong Nucleophile, if and only if the point of scission is valid.

By the nature of the instantaneous opening of the ring, it seems that when N and O are adjacently placed in a ring, the following is valid.

$$\underset{\underline{\textbf{N--H}}}{\overset{O}{\bigcirc}} \qquad > \qquad \underset{\text{N--H}}{\overset{\underline{\textbf{O}}}{\bigcirc}}$$

Order of Nucleophilicity $\qquad\qquad\qquad\qquad 13.165$

This seems to look contrary to what has been observed so far, wherein the reverse is the case. *However, there are clear indications that cyclic amines are less nucleophilic than oxaaza-cyclopropane, since the number of nucleophilic or electrophilic centers adjacently located in a ring is additive and is a measure of their strong or weak characters.* When an aziridine ring is placed side by side to oxaazacyclopropane, polymerization may be favored radically, though aziridine is a half free ionic or radical monomer, while oxaazacyclopropane is a full non-free-radical monomer.

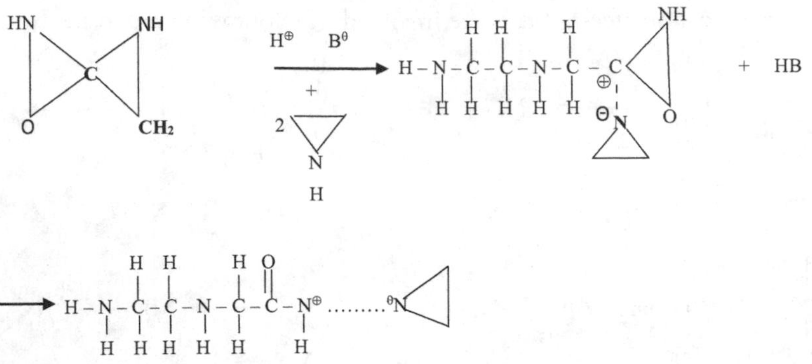

IMPOSSIBLE EXISTENCE   1   13.166

Ionically, the reaction above cannot take place. The cyclic amine ring is more strained than the oxaazacyclopropane ring because of the presence of a weak N – O bond in the latter. *It seems that, in general however, when weak bonds are absent in a ring, the less nucleophilic and the more strained it is.* The point of scission above can be observed to be an explosive case, for which this ring cannot be placed side by side to another ring. The point of scission as shown in Equations 13.163 and 13.164 are not valid. Based on Equation 13.160, when opened instantaneously only radically, a nitrogedizing agent

en.N.nn agent and formaldehyde may be formed as may be confirmed downstream. It could be methylene and HN=O. Just as there exist full free ionic or radical, full non-free-radical and half free ionic or radical non-ringed monomers, so also they strongly exist with ringed monomers. The only way they can change from one to the other where possible is via tautomerism.

13.3.2. Pyrrole

Pyrrole like furan and furfural cannot readily be opened. Hence it is not a ring opening monomer. However, it can still be mildly used to produce mostly resins without the need to open the ring. Nevertheless, it is being considered here, since the structure of pyrrole is strongly based on its formation by distillation of succinimide with zinc dust[33] and since very little is known about these compounds. Succinimide is said to undergo some form of enolization phenomena. These "phenomena" can differently be explained as follows.

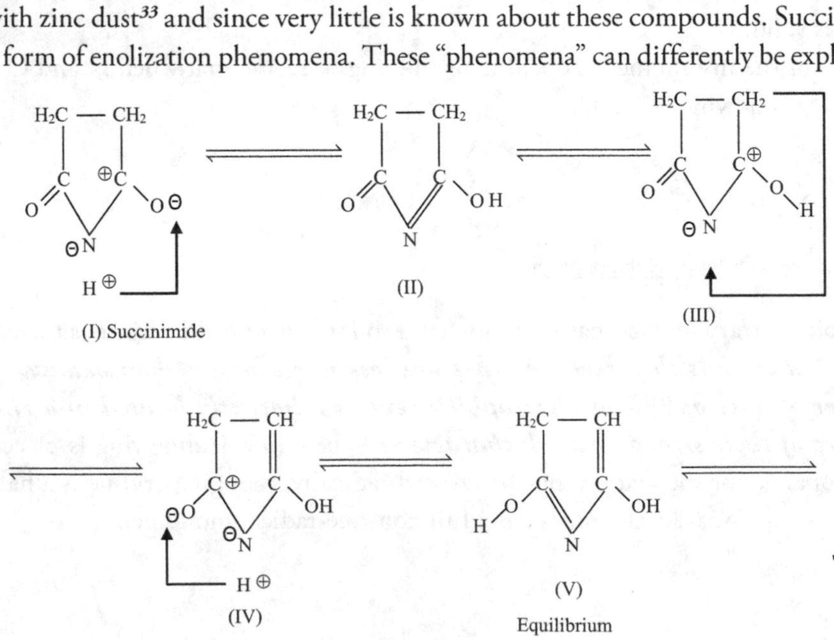

224

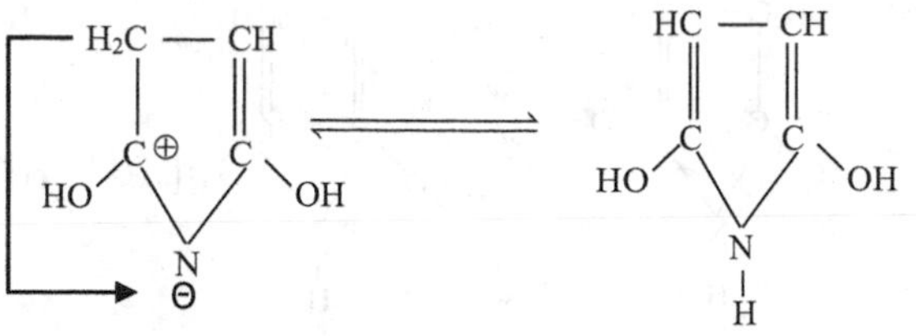

(VI)   (VII)
Enolized form   13.167

The number of molecular rearrangements involved in the so-called enolization reaction are important to note. It is indeed Enolization/molecular rearrangement phenomena, enolization taking place via Activated/Equilibrium state of existence of the C = O center and molecular rearrangement taking place via activation of the C = N center. It is for this reason that succinimide does not favor being used as a monomer, as will shortly be fully explained.

When (VII) or (I) is heated with Zn dust pyrrole is produced. The reaction is only radical in character, since Zn is a non-ionic metal.

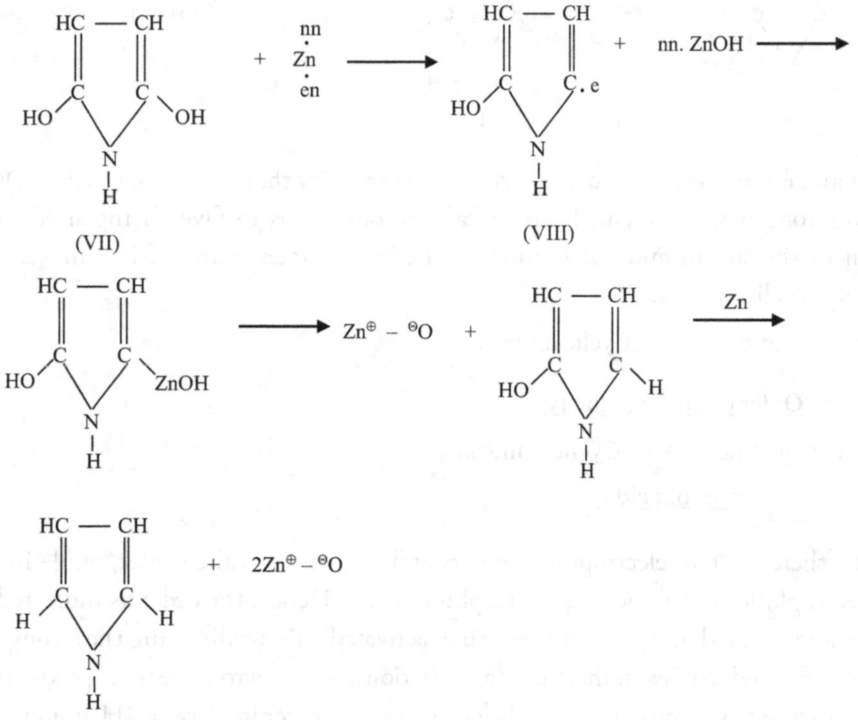

(VII)   (VIII)

Pyrrole   13.168

The reaction above is Equilibrium mechanism in four stages, with no form of distillation.

Based on the reaction of Equation 13.167, the followings are valid from the rearrangements.

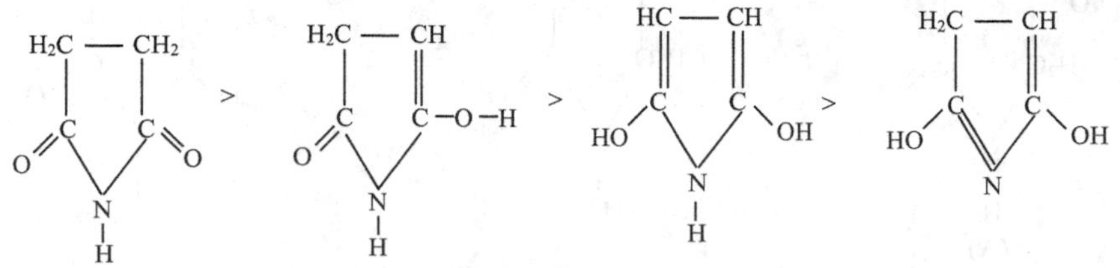

Order of Nucleophilicity   13.169

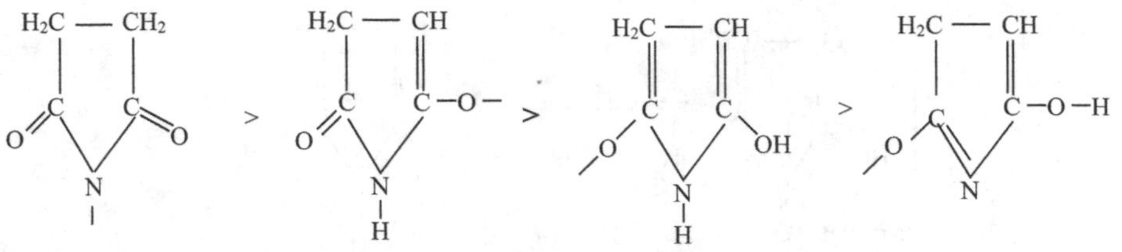

Order of Radical-pushing capacity

13.170

It is interesting to note that while **2-pyrrolidone** can be polymerized "cationically", succinimide cannot be polymerized "cationically". As shown by Equation 13.35, the ring of the succinimide can indeed be opened under heat. Yet, it cannot undergo further polymerization. The catalyst involved in that reaction is water, one of the types also used for 2-pyrrolidone. With 2-pyrrolidone, the amino acid obtained where possible is less acidic in character, hence the initiation step will be identical to the use of weak acids. When the water attacks the succinimide under heat, the following is obtained.

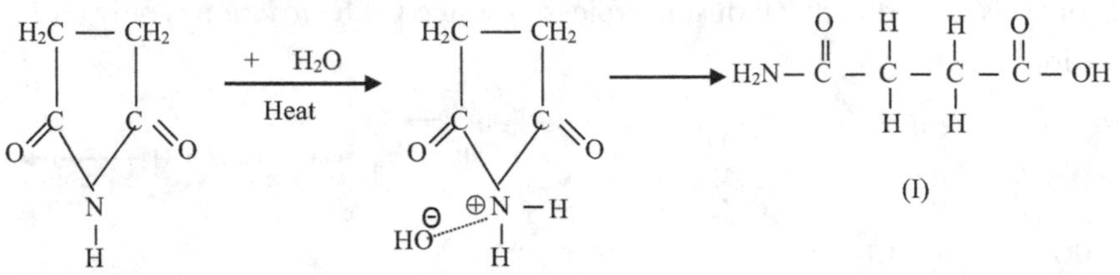

13.171

(I) above is not an amino acid, but a sort of diacid of less capacity than $HOOC\,CH_2CH_2COOH$. However, the acid is too strong when compared with weak protonic acids to favor being used for the cationic polymerization of the succinimide. It is more acidic in character than cyclic amides, and also more nucleophilic than cyclic amides.

Succinimide   >   Cyclic amides

Order of nucleophilicity 13.172

Succunimide   >   Cyclic amides

Order of acidity 13.173

In succinimide, there are two electrophilic centers and one nucleophilic center, while in cyclic amides, there is one electrophilic center and one nucleophilic center. Hence, the order as indicated above.

Since it is more acidic than cyclic amides, when activated anionically using the strongest of bases, no polymerization is favored. At best if there is or are reactions, only micro molecular products are obtained. In view of the presence of O = group adjacently located to a N center in place of 2H atom in 2-pyrrolidone, succinimide is far less strained than 2-pyrrolidone. If it was separately or distantly located as shown in (A), below, then it will be more strained.

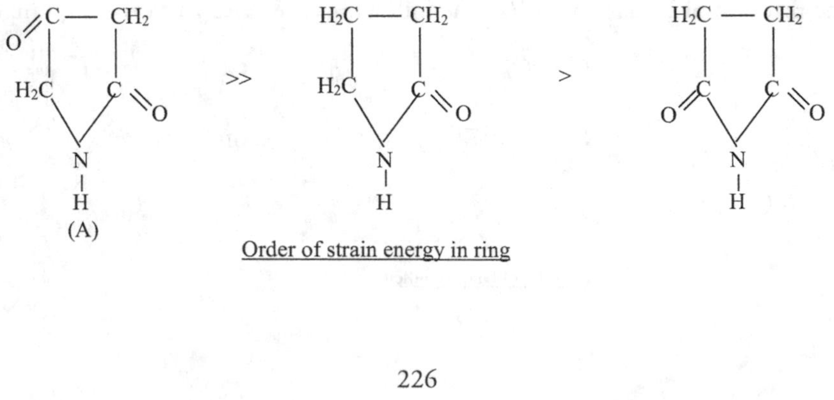

Order of strain energy in ring

13.174

Hence succinimide are not popularly known to undergo anionic polymerization, despite the presence of two electrophilic centers. Heat must be added to introduce more strain into the ring.

Now, coming back to pyrrole, it is said that the formation of red resinous polymers in the presence of mineral acids may be explained if it is assumed that the s complex with a proton is sufficiently stable to initiate a chain reaction[33]. Like other cases considered so far the H atom on the nitrogen center in pyrrole is both weakly ionically or radically covalently bonded to the nitrogen center.

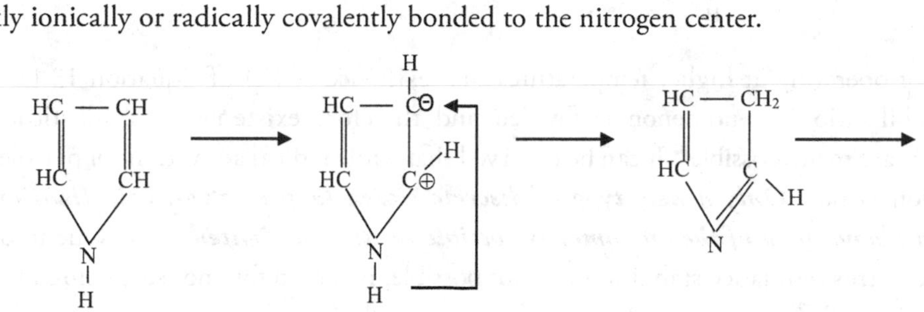

Molecular rearrangement phenomenon of 1st Kind (Chargedly)-Incomplete   13.175

Chargely the rearrangement above looks favored since the right center has been activated. However, in view of the fact that resonance stabilization cannot take place chargedly, it is said to be incomplete as shown below.

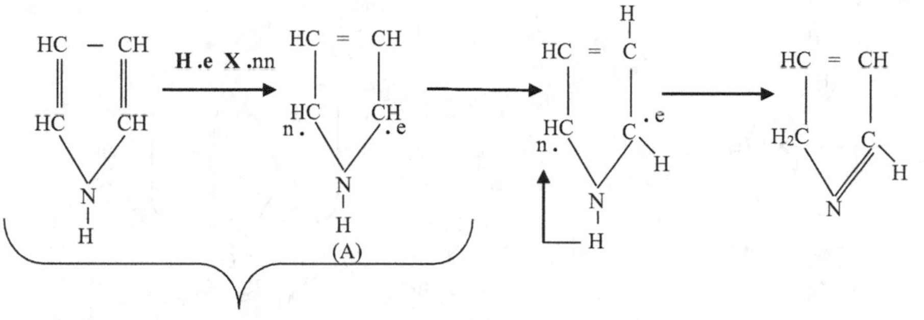

Discrete Radical molecular rearrangement/resonance stabilization phenomena   13.176
(Favored)

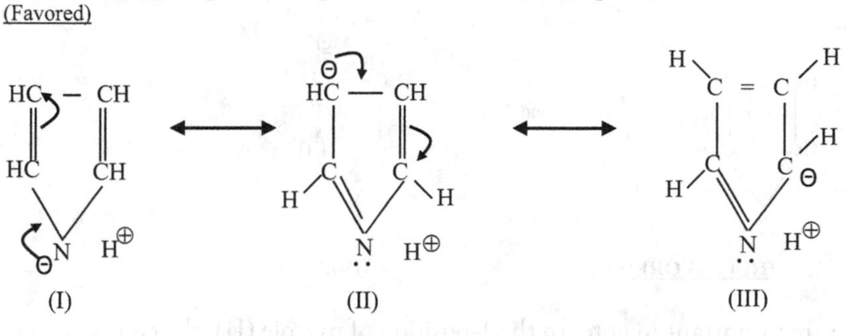

Ionic close-loop resonance stabilization phenomenon (Not favored)   13.177

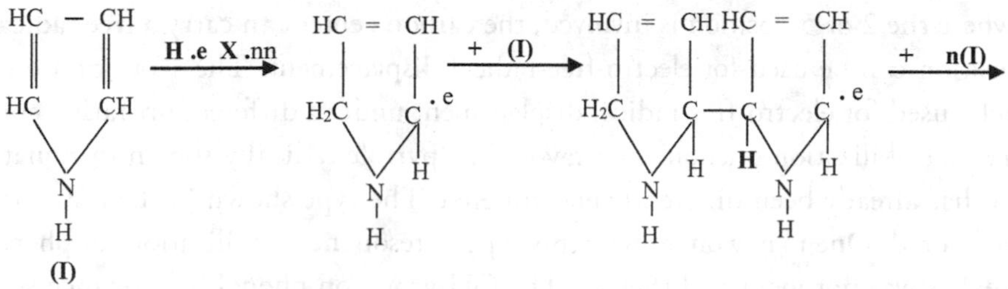

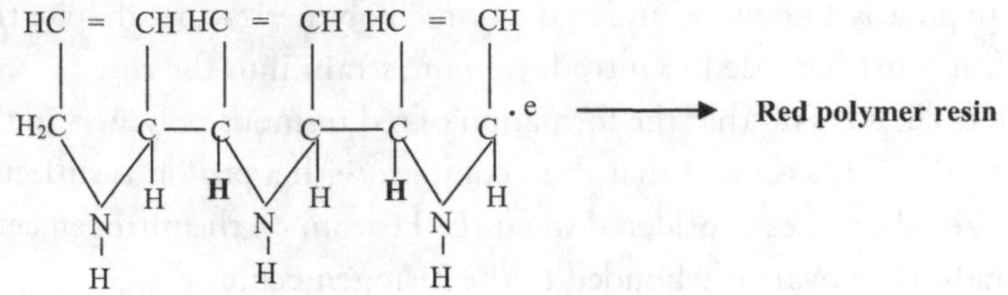

$$\text{(structure)} \longrightarrow \text{Red polymer resin}$$

13.178

It is in view of operating at higher temperatures that existence of (A) of Equation 13.176 via discrete resonance stabilization phenomenon is favored and therefore existence of 2-substituted alkyl- or acylpyrrole etc. are made possible[3]. It can be observed that while radical activations of pyrrole are favored, ionic activation is not. ***Only a new type of discrete closed loop electro-radicalization/resonance stabilization phenomena of the monomer is possible in its unactivated state.*** Due to other factors, the ionic type of this resonance stabilization is not possible. It is therefore not surprising to note why the most commonly used activator is methyl magnesium bromide[33], a non-ionic activator

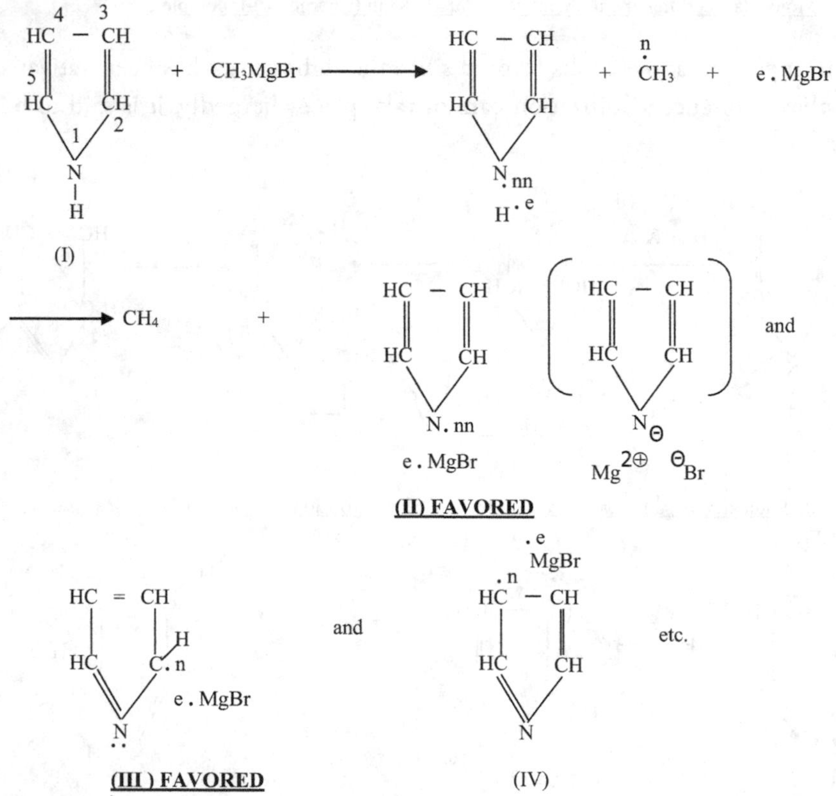

$$\text{(reaction scheme)}$$

(I)

(II) FAVORED

(III) FAVORED   (IV)

13.179

The reactions above are important to note. In the 1-position of pyrrole (II), the centers can only be radical and not ionic in character, since the magnesium centers are divalent. The ionic state does not exist here. Also in (III), where the 2 or 5 position is involved, the carbon center can carry a free radical. Therefore, the 2- or 5- position can be used for electro-free-radical displacement. The 1-position on the nitrogen center can also be used for electro-free-radical displacement under a different operating condition.

The resonance stabilization phenomenon favored by pyrrole radically shown in Equation 13.176 is not new since it has already been differently encountered. The type shown in the last part of Equation 179 is new and favored. One may wonder why this type of resonance stabilization for phenol in Chapter 11 which also exists was not identified therein. The OH group on phenol is resonance stabilized when

228

the monomer is activated. When the monomer is not activated, the radical carried by it when the H is held can be resonance stabilized radically as shown below.

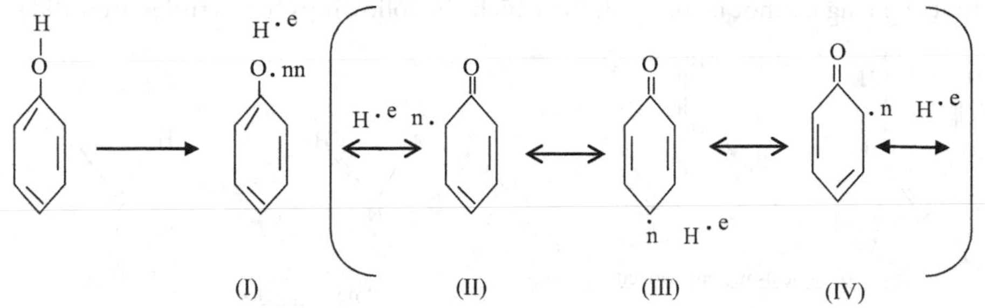

$$\text{(I)} \quad \text{(II)} \quad \text{(III)} \quad \text{(IV)}$$

Continuous closed-loop Electro-radicalization phenomena (favored)   13.180

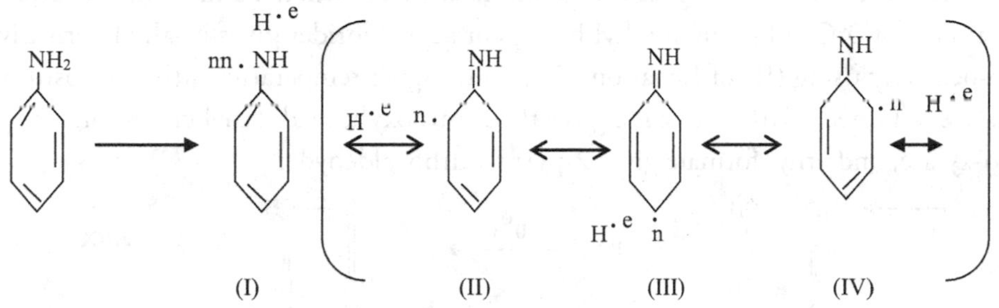

$$\text{(I)} \quad \text{(II)} \quad \text{(III)} \quad \text{(IV)}$$

Continuous closed-loop Electro-radicalization phenomena (favored)   13.181

In this type of resonance stabilization favored by pyrrole it is the electro-free-radical in the π - radicals that moves, after existing in Equilibrium state of existence. It is important to note that, if the benzene rings favors this type of resonance stabilization it would be continuous in character as shown in Equations 13.180 and 13.181 above. With pyrrole when activated as shown in Equations 13.176 (A) and 13.178, the transfer of electro-free-radical takes place but stops on the C center next to the nitrogen center present inside the ring. This is also discrete as opposed to continuous. The 2- or 5- placement can be obtained on a discrete basis in two ways. However, the following should also be expected for mostly 2- or 5- positions where higher operating conditions must prevail, that is, when pyrrole is kept in Equilibrium state of existence.

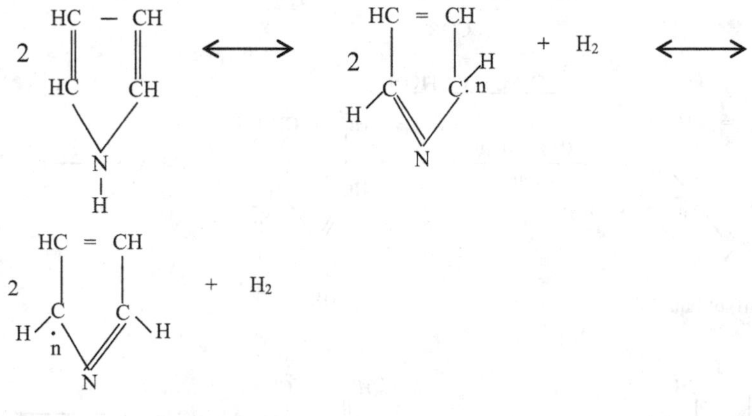

13.182

It should be noted also that since nucleo-free-radicals are obtained from nucleo-non-free-radicals after application of heat, the 2- or 5- position in pyrrole is highly favored in their displacement reactions, only when heat is involved. Without heat, the 1-position is largely involved.

    One can observe, why there is always need to look deeply into every reaction and the steps involved, step by step, without illusion. This is a complete departure from what has been known to be the case for very long, basically because of the new definitions for ions, radicals, monomers etc. It is in view of the presence of Electro-radicalization phenomenon a form of molecular rearrangement followed by resonance stabilization, that the ring cannot be opened; for which the followings for pyrroles are valid.

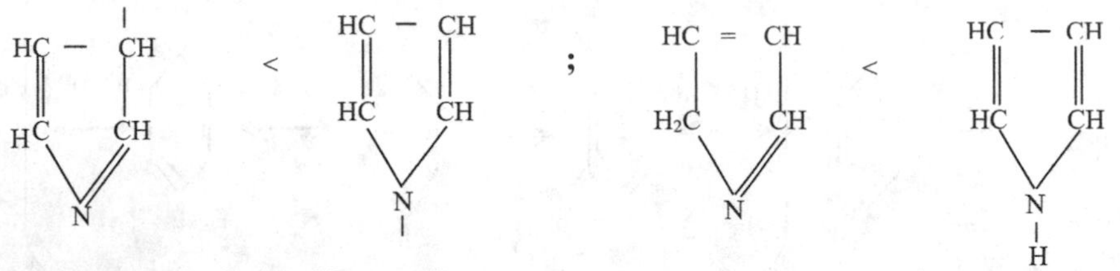

$$
\begin{array}{ccc}
\underset{\text{(radical-pushing)}}{\text{pyrrole structure A}} & < & \text{pyrrole structure B} \quad ; \quad \text{pyrrole structure C} & < & \text{pyrrole structure D}
\end{array}
$$

<u>Order of radical-pushing capacity</u>   <u>Order of nucleophilicity</u>   13.183

After reaction with methyl magnesium bromide with evolution of methane to give the bromo-magnesium amide, at 0°C, addition of alkyl halides or acyl chlorides gives N-alkyl derivative or N-acyl derivative respectively using (II) of Equation 13.179. At higher temperatures, the 2-substituted alkyl- or acyl pyrroles are obtained. Carbon dioxide gives the 2-carboxylic acid; ethyl chloroformate gives ethyl-2 pyrrolecarbo-xylate, and ethyl formate gives 2-pyrrolecarboxaldehyde.

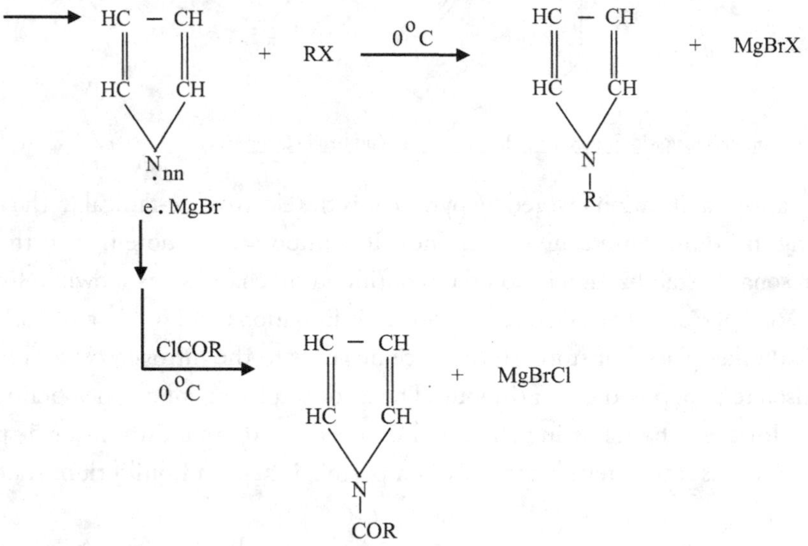

$$
\text{e} \cdot \text{MgBr (pyrrolyl)} \; + \; RX \xrightarrow{0^{\circ}C} \text{N-R pyrrole} \; + \; MgBrX
$$

$$
\xrightarrow[0^{\circ}C]{ClCOR} \text{N-COR pyrrole} \; + \; MgBrCl
$$

<u>NO ACTIVATION</u>   13.184

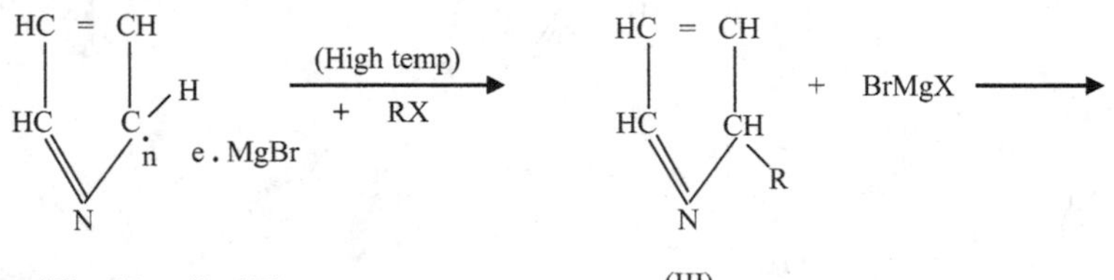

$$
\text{(III) of Equ. 13 . 179} \xrightarrow[\text{+ RX}]{\text{(High temp)}} \text{(III)} \; + \; BrMgX \longrightarrow
$$

$$
\text{(IV)} \; + \; BrMgX \longrightarrow \text{(V)} \; + \; BrMgX \longrightarrow
$$

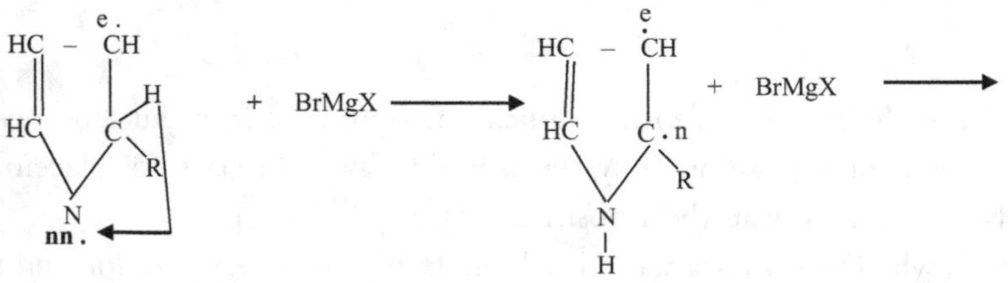

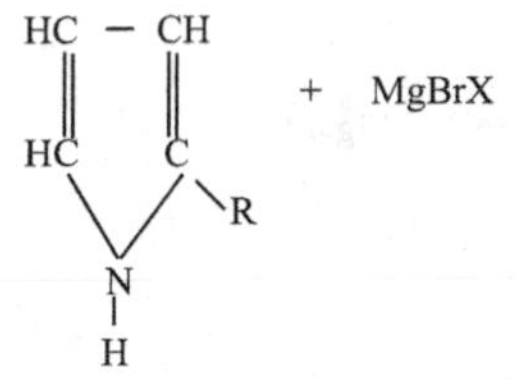

$$
\begin{array}{c}
HC - CH \\
\| \quad \| \\
HC \quad C \\
\diagdown \; \diagup \diagdown R \\
N \\
| \\
H
\end{array}
\quad + \quad MgBrX
$$

(VI)   <u>ACTIVATION PRESENT</u>   13.185

Without the molecular rearrangement of the sixth type on (III), presence of (VI) would be impossible. (III) when activated, resonance stabilized to give (IV). This molecularly rearranged to give (V). The $N = C$ center being less nucleophilic than the $C = C$ center, is the first to be activated. After (V) is formed, this was followed by deactivation to give (VI).

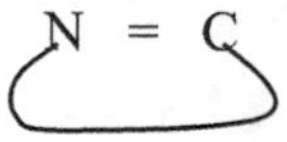

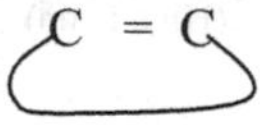

For ROCCl, the followings are obtained.

$$
\xrightarrow{\quad}
\begin{array}{c}
HC = CH \\
| \quad | \\
HC \quad C \diagup^{H}_{\cdot n} \\
\diagdown \diagup \\
N \\
e.MgBr
\end{array}
\xrightarrow[+ \; ROCCl]{(High\;temp)}
\begin{array}{c}
HC = CH \\
| \quad | \\
HC \quad C \diagup^{H}_{O} \\
\diagdown \diagup \quad \| \\
N \quad C - R
\end{array}
\quad + \quad ClMgBr
$$

$$
\xrightarrow[\text{Stabilization/Rearrangement}]{\text{Activation / Resonance}}
\begin{array}{c}
HC - CH \\
\| \quad \| \\
HC \quad C \\
\diagdown \diagup \diagdown \\
N \quad C = O \\
| \qquad | \\
H \qquad R
\end{array}
\quad + \quad MgBrCl
$$

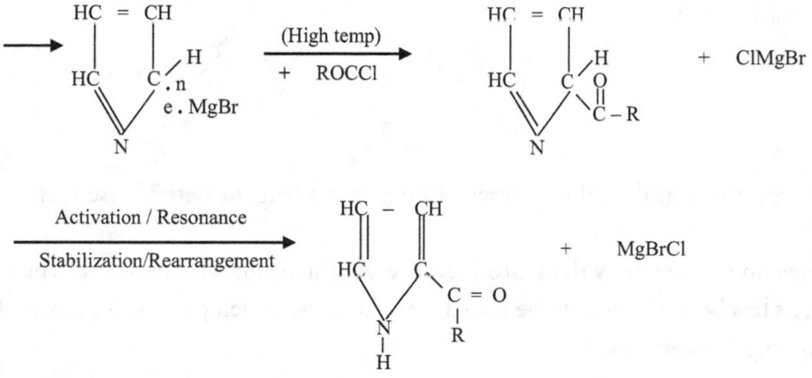

13.186

For CO_2 and water, the followings are to be expected.

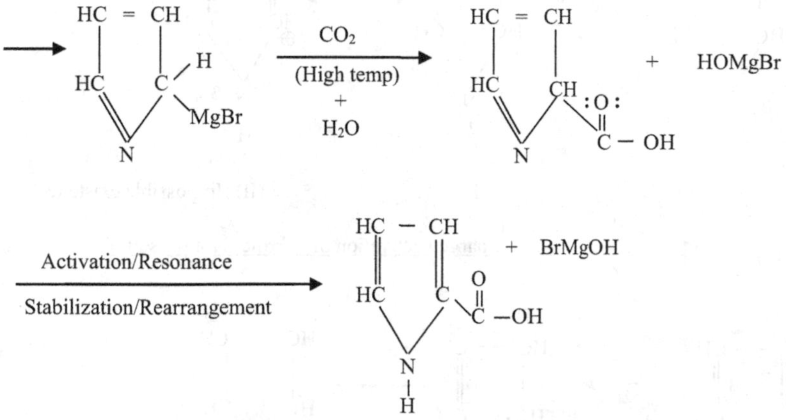

$$
\xrightarrow{\quad}
\begin{array}{c}
HC = CH \\
| \quad | \\
HC \quad C \diagup^{H} \\
\diagdown \diagup \diagdown \\
N \quad MgBr
\end{array}
\xrightarrow[\substack{(High\;temp) \\ + \\ H_2O}]{CO_2}
\begin{array}{c}
HC = CH \\
| \quad | \\
HC \quad CH \quad :O: \\
\diagdown \diagup \quad \| \\
N \qquad C - OH
\end{array}
\quad + \quad HOMgBr
$$

$$
\xrightarrow[\text{Stabilization/Rearrangement}]{\text{Activation/Resonance}}
\begin{array}{c}
HC - CH \\
\| \quad \| \\
HC \quad C \quad O \\
\diagdown \diagup \| \\
N \quad C - OH \\
| \\
H
\end{array}
\quad + \quad BrMgOH
$$

13.187

For $ClCOOC_2H_5$, the followings are obtained.

$$
\xrightarrow{\quad}
\begin{array}{c}
HC = CH \\
| \quad | \\
HC \quad CH \\
\diagdown \diagup \diagdown \\
N \quad MgBr
\end{array}
\xrightarrow[(High\;temp)]{Cl \; \overset{O}{\overset{\|}{C}} - O - C_2H_5}
\begin{array}{c}
HC = CH \\
| \quad | \\
HC \quad C \diagup^{H}_{O} \\
\diagdown \diagup \quad \| \\
N \quad C - O - C_2H_5
\end{array}
\quad + \quad ClMgBr
$$

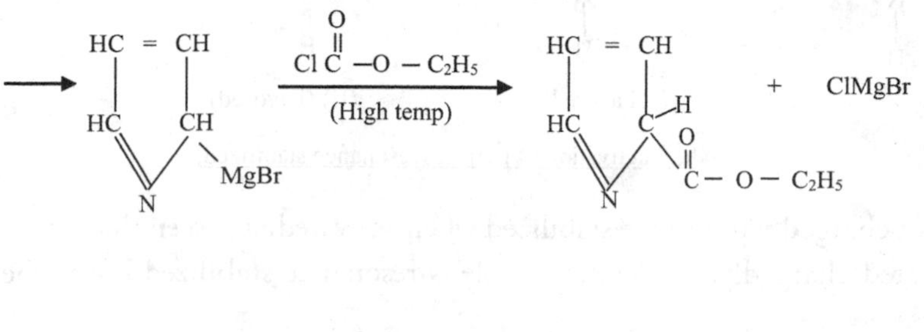

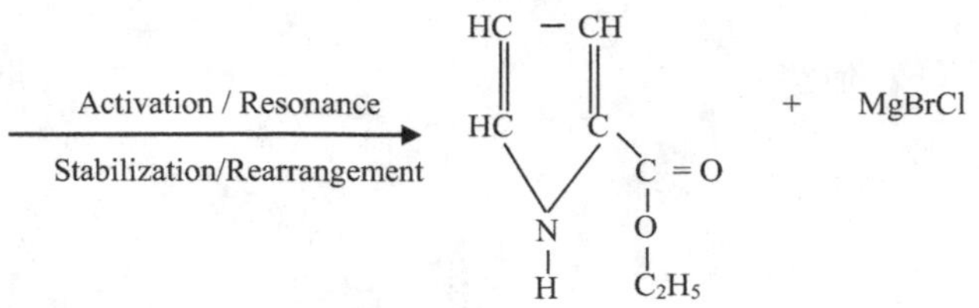

13.188

With $HCOOC_2H_5$, the followings are obtained.

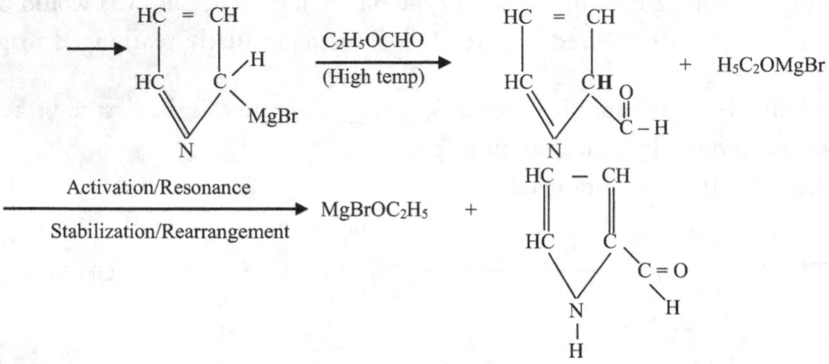

13.189

One can observe the unique and dual functions of Grignard's reagent here whose real structure is yet to be shown.

In general, one can observe how these products are obtained, for which none of the reactions can be ionic in character, as has been thought to be the case in the past. When pyrrole is activated chargedly and radically, the followings are obtained.

(I)   (II) (Impossible existence)

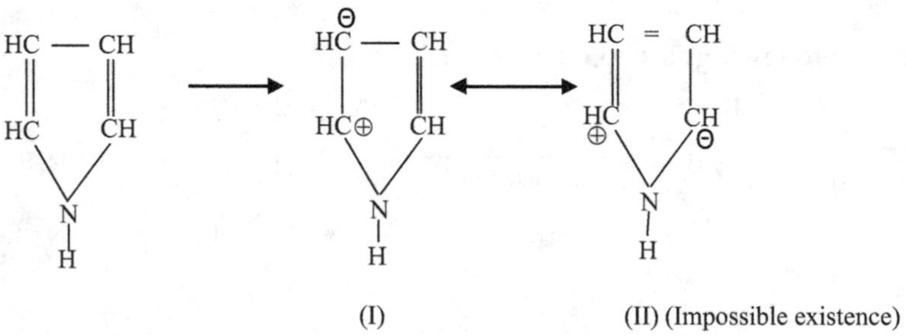

Charged activation of Pyrrole (Not possible)   13.190

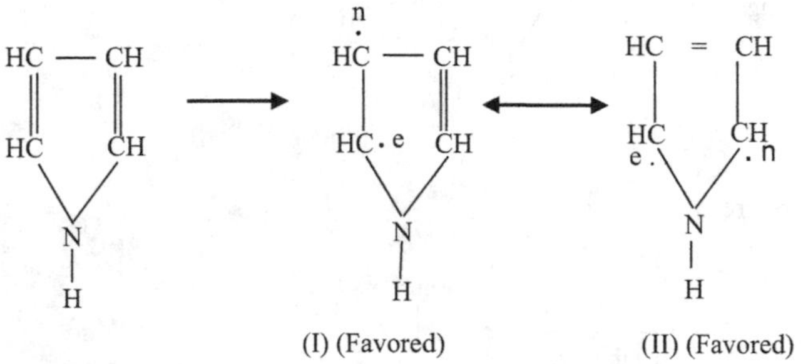

(I) (Favored)   (II) (Favored)

Radical activation of pyrrole (resonance stabilized)   13.191

Though pyrrole is not chargedly resonance stabilized when activated, it is seen that none of the activation centers can be activated chargedly. Radically, pyrrole is resonance stabilized both when activated and when in Equilibrium state of existence, a state which is easy for her to stay in at STP. That was what the Grignard reagent did, the Mg center being an ionic metal. It suppressed its Equilibrium state of existence.

In the presence of mineral acids, the followings are some of the possibilities obtained during the polymerization of pyrrole.

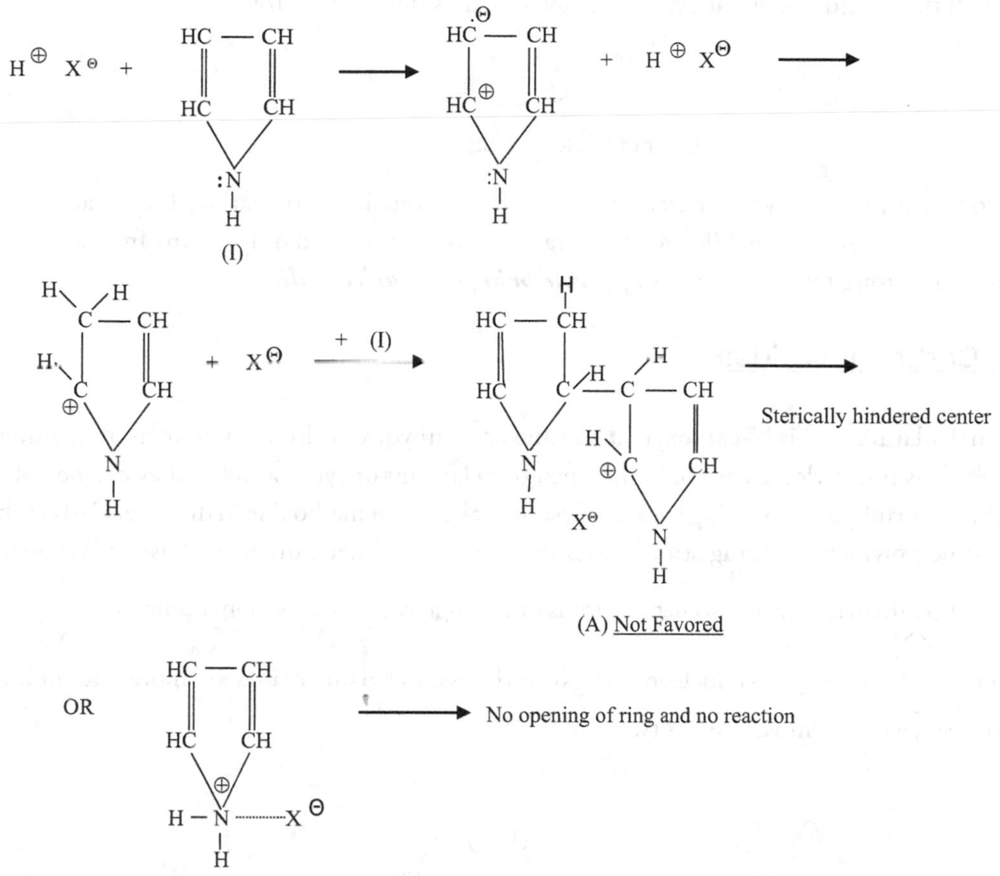

(I)

Sterically hindered center

(A) <u>Not Favored</u>

OR

No opening of ring and no reaction

(B) Initiator

13.192a

The reaction above which is cationic or positively charged, polymerization is not favored due to steric limitations. Electro-free-radically, the following as already shown is obtained.

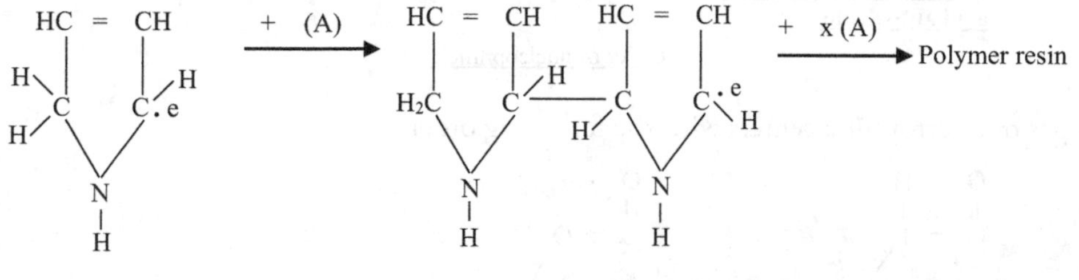

Polymer resin

13.192b

(A) above was obtained after activation of (I) of Equation 13.192a using the hydrogen electro-free-radical, which adds to it. Therefore, there is no molecular rearrangement here, unlike using RMgBr. Thus, in the use of the mineral acids, in addition to the need of having to operate at higher temperatures, it is only the electro-free-radical route that is favored, because of presence of transfer species nucleo-free-radically. Due to electro-dynamic forces of repulsion and steric limitations, the monomers are syndiotactically placed and not as indeed shown above, noting that the syndiotactic placement is not easy in the absence of coordination. Resins are obtained also partly because the H atom on the C center in the ring has a

1:1 ratio of C to H. The ring has no point of scission and therefore cannot be opened. In the reactions of Equation 13.192b, there is also no molecular rearrangement since the electro-free-radical carrier is strong, giving no time for transfer to take place. This is slightly different from the case of using furan as monomer to produce furan resins in the presence of acids[34].

From all the considerations above, the followings are valid with pyrroles-

$$\text{[N]} > \text{C=C} > \text{C=N}$$

Order of Nucleophilicity 13.193

Obviously, if pyrrole favors discrete type of radical resonance stabilization, furan can, but not the second type favored by pyrrole. Like pyrrole, furan can favor being used only electro-free-radically. Furan and pyrrole are strong nucleophiles with **pyrrole being less nucleophilic.**

13.3.3. Carboanhydrides

Of particular interest is N-carboxyl - α - amino acid anhydrides also referred to as oxazolidine - 2, 5-diones [19, 35, 36]. While Nylon 2 cannot be obtained from lactams or cyclic amides, they can be obtained by heating the carboanhydride or using basic catalyst such as sodium methoxide in dioxane. Carboanhydrides can indeed be polymerized using strong bases in the same manner similar to those of lactams. Unlike lactams, carboanhydrides have two hetero atoms or two nucleophilic functional centers-- (O and (N

of which the (N center is less nucleophilic. Nevertheless, carboanhydrides are more nucleophilic than the lactams since carboanhydride has two

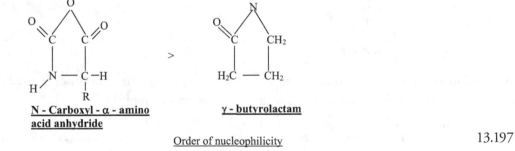

N - Carboxyl - α - amino acid anhydride   γ - butyrolactam

Order of nucleophilicity 13.197

different types of electrophilic centers with the following order.

Order of electrophilicity 13.198

Thus, it can be observed that like lactams, carboanhydrides will favor both anionic and cationic routes. Anionically, using "so called" strong bases such as Na, metal hydrides, metal amides, the followings are obtained

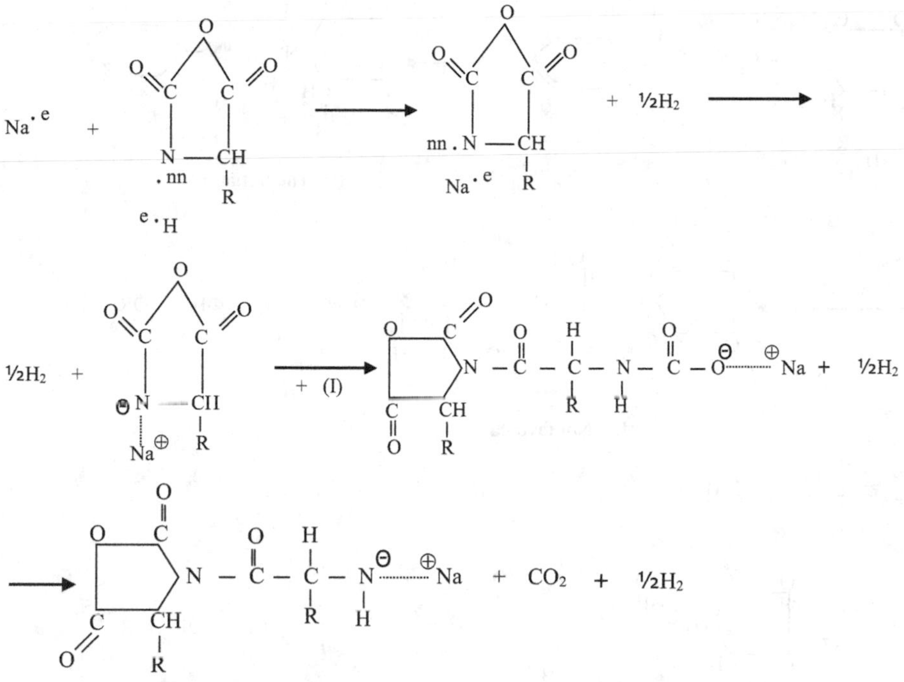

Initiation step   13.199

In view of the type of initiation step involved anionically, there is bound to be some induction period in the polymerization of the monomer. Unlike lactams where the possibility does not exist, carbon dioxide can be released here for one of the several reasons-

(i)   The nitrogen center on the monomer is less nucleophilic than the oxygen center on the monomer which should be the case, since oxygen is more electronegative than nitrogen.
(ii) Higher polymerization temperature.
(iii) Presence of some phenomena.

The release of CO_2 continues throughout the propagation step as shown below, to produce nylon 2 anionically with a terminal ring after termination using a foreign agent. That is, the monomer is part of the initiator.

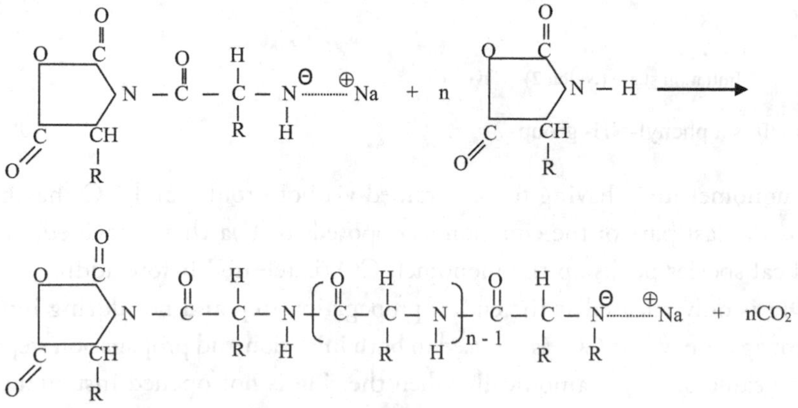

Nylon 2   13.200

235

When weak bases such as non-metallic amine are involved, the route is cationic for which higher polymerization temperatures will be desired for addition of more energy to the ring.

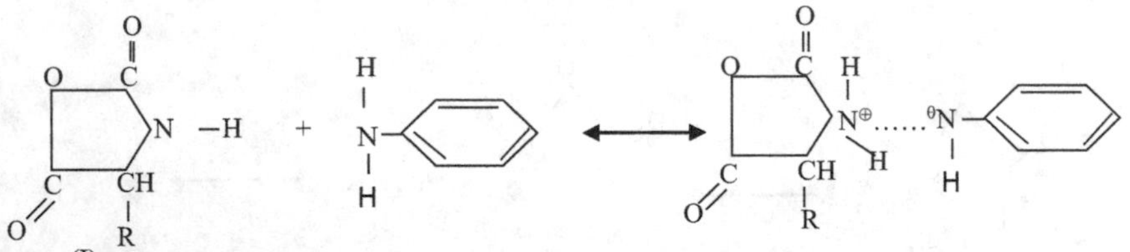

(I) + (II) The Initiator

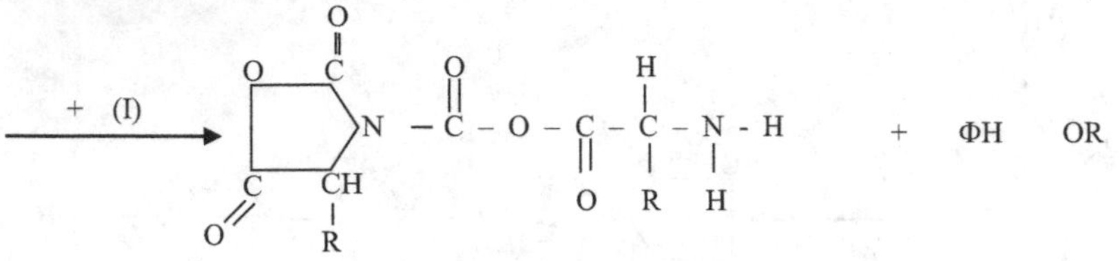

(II)a Not favored

(I)

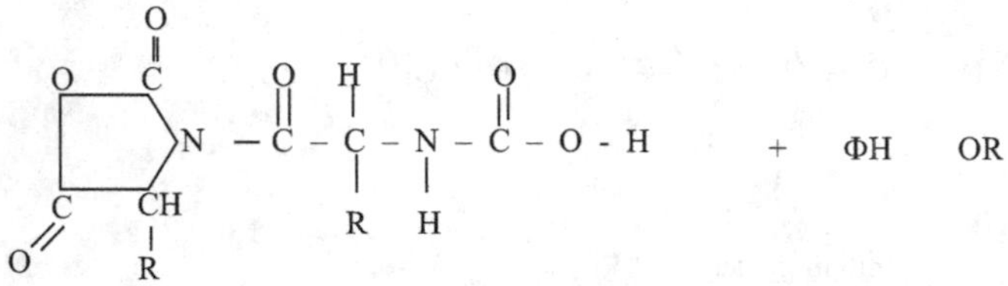

(II)b Not favored

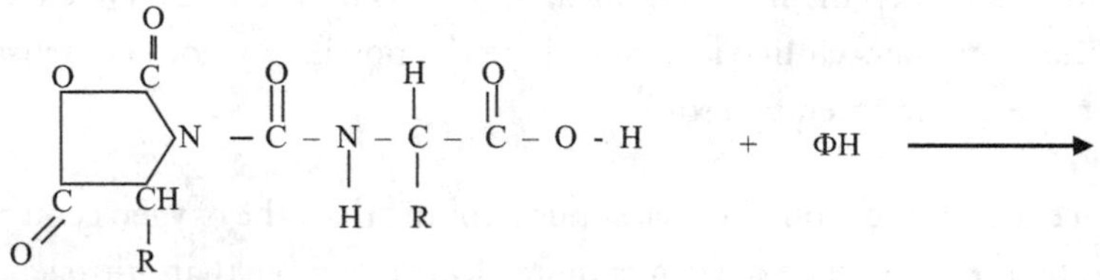

(II)c Not favored

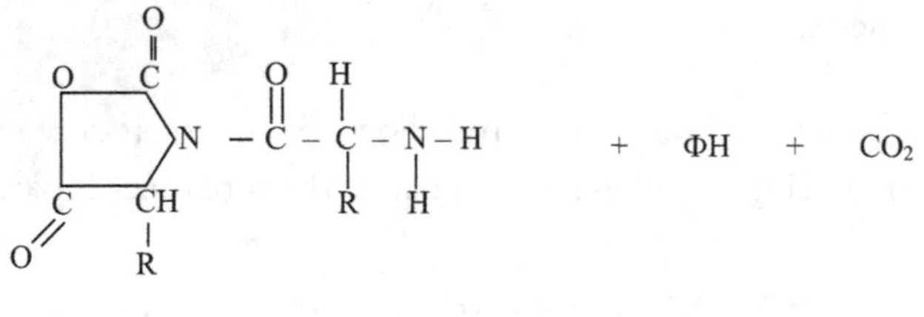

Initiation step (Nylon 2) (favored)

Where Φ is a phenyl-NH- group .   13.201

Due to the same monomer units having to be obtained via both routes and CO_2 having to be released in the process, it is the last part of the equation as opposed to (II)a that is favored, because when the H electro-free-radical species peaks up the monomer, CO_2 is released before adding to the backwardly growing center. ΦH is only released at the end of propagation step and not during initiation as shown above or during propagation, while CO_2 is released in both initiation and propagation steps. The monomer unit of above is the same obtained anionically when the ring is not opened instantaneously. It can be observed, that favored release of CO_2, cationically and anionically is largely due to (iii) that is one of the reasons provided above and not heat (ii) or (i), as will be explained. For the propagation step, the followings are obtained.

No transfer species

13.202

Only primary and secondary amines can be used as source of initiators electro-free-radically when amines are involved. This is only possible when the initiator when prepared remains in Equilibrium state of existence. If the initiator is in Stable state of existence, then only the anionic route will be involved. The anionic route is very fast compared to the electro-free-radical route, because two different mechanisms are involved. It is important to note that the initiator (II) of Equation 13.201 can also be used anionically when it can be made to exist in Stable state of existence.

In almost the same similar pattern with lactams, water can be used. CO_2 will be released for every addition both during initiation and propagation steps. Temperature of polymerization may be higher than expected in view of the fact that lactams are less nucleophilic than carboanhydrides; provided the MRSE can be attained. Since its amino acid is less basic than acidic, the mechanism will be similar to those of amides. With the use of weak protonic acids, if the counter-ion is less nucleophilic than the monomer, polymerization will be favored. If the acid is very weak, opening of the ring may not be favored. It is obviously not surprising to note why the use of weak protonic acids and water are not popularly known as source of cationic initiators for carboanhydrides.

The fact that, they are less basic and more electrophilic than lactams is indicated by the fact that, they can be polymerized anionically using sodium methoxide, noting that sodium methoxide is not mildly basic. It is more so than non-metallic alkoxides (ethers). Anionically, whether the lactam or the carboanhydride is a strong base or not,

(I)

13.203

the use of non-metallic free-ionic initiator is not favored. Depropagation phenomenon with carboanhydrides may take place unlike with lactams, in view of the existence of three membered cyclic lactam along the chain The $C = O$ center is too close to the nitrogen center for comfort. Six-membered rings can readily be removed electro-free-radically during propagation at high operating conditions.

In the reactions above, the $NaOCH_3$ could be the initiator, if it can be made not to exist in

Equilibrium state of existence and this can only be done radically since ionically in an ionic environment, ionic compounds are always in Equilibrium state of existence. However, it has been established that-

Carboanhydride   >   lactams (cyclic amides)

<u>Order of nucleophilicity</u>   13.204

In order to determine which is more nucleophilic between the two members, though this is already obvious after all the considerations so far, there is need to conclude the analysis with side-by-side placements of their rings.

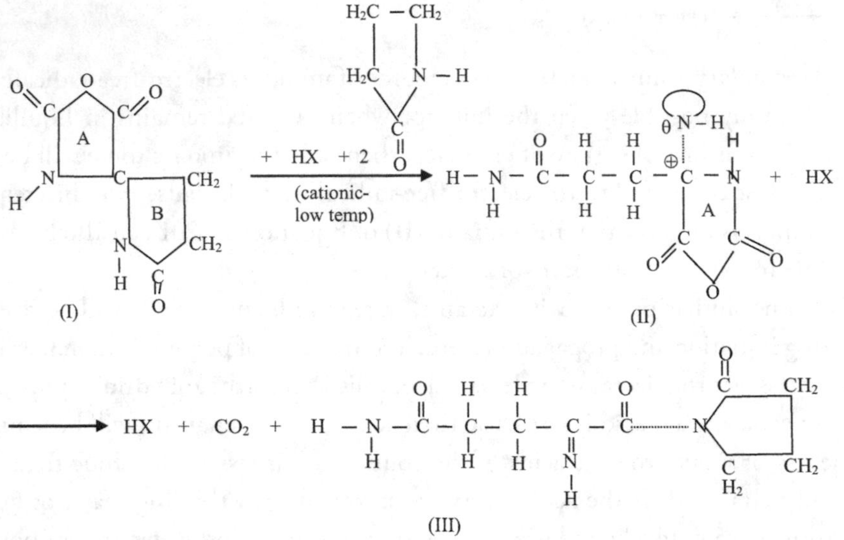

$$(I) \quad + \ HX \ + \ 2 \ \text{(cationic-low temp)} \quad \longrightarrow \quad (II) \quad + \ HX$$

$$\longrightarrow \quad HX \ + \ CO_2 \ + \ (III) \qquad\qquad\qquad 13.205a$$

$$(I) \quad + \ NaOCH_3 \quad \longrightarrow$$

$$(IV) \quad + \ CO_2 \ + \ CH_3OH \quad \longrightarrow$$

$$(V) \quad + \ CO_2 \ + \ CH_3OH \qquad\qquad\qquad 13.205b$$

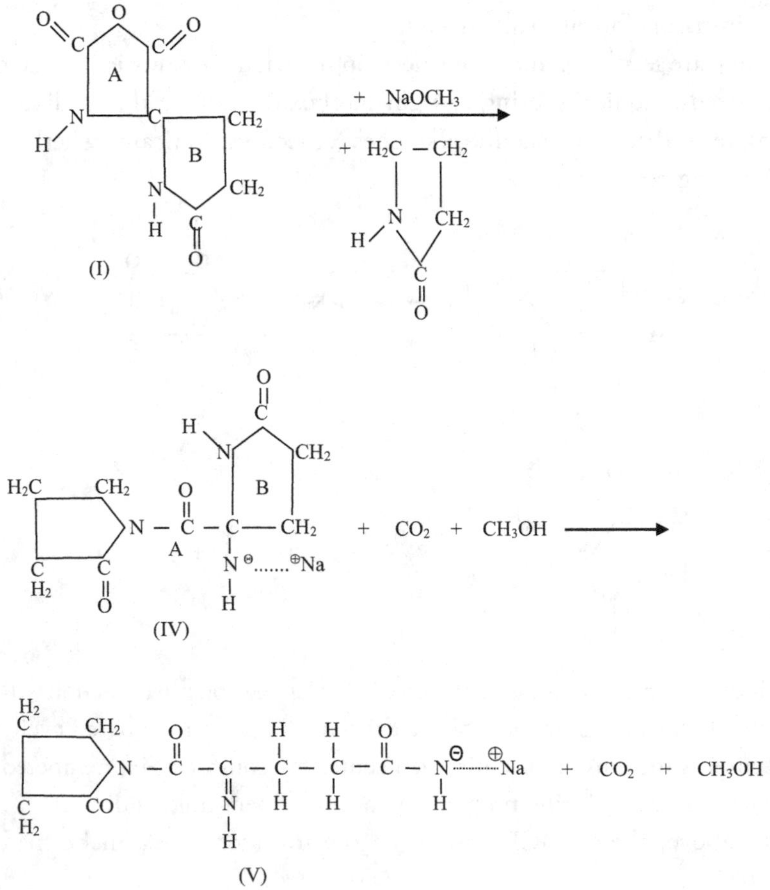

238

(V) and (III) above have same monomer units. Electro-free-radically, it is (B) in (I) that is first attacked, while anionically, the route natural to the monomer, it is (A) in (I) that is first attacked to first release CO_2. Hence the following is valid.

$$\text{Carboanhydride} \quad > \quad \text{Cyclic amides}$$

<u>Order of nucleophilicity</u>   13.206

Other possibilities include –

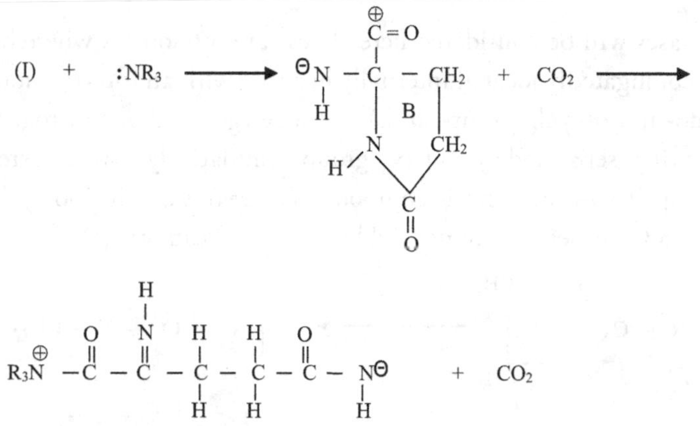

13.207

Identical monomer units are obtained, clear indication of the validity of Equations 13.206 and 13.207 and the fact that Carboanhydride is less strained than cyclic amides.

For the type of carboanhydride shown below, the followings are obtained.

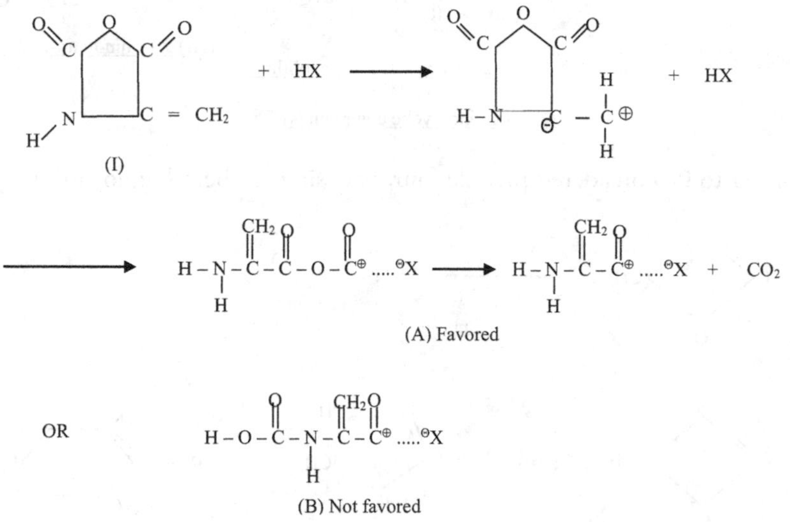

(A) Favored

(B) Not favored   13.208

It is (A) that is favored, since the least nucleophilic center has been activated and the right transfer species has been abstracted. With (B), the wrong transfer species was abstracted based on the laws of Nature. The ring was opened at the wrong bond. In (A), CO_2 was released as to be expected. Anionically, the same monomer unit can be obtained if the carbonyl center is attacked. With the presence of the $H_2C=$ group in the ring where placed, the Electrophilic character of the monomer has not been eliminated. The reason is because the following is valid.

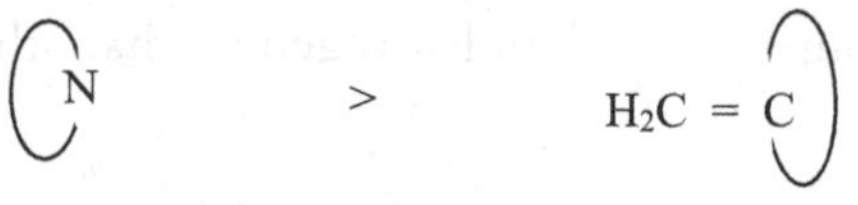

Order of Nucleophilicity   13.209

Radically, the behaviors of carboanhydrides are similar to those of lactams electro-free-radically and those of lactones anionically. Based on the New Frontiers, one knows what to expect radically in general. The use of nucleo-non-free-radicals and electro-free-radicals, rather than charged routes are far more advantageous. Most reactions have to be considered very closely using the already proposed rules, in order to understand what actually goes on in chemical and polymeric systems of some difficult monomers.

13.3.4 Imino Cyclic Compounds[37,38]

Only few known cases will be considered here. These are monomers where both a nitrogen and an oxygen atom are both conjugatedly located internally in a ring with an internal double bond, but without a carbonyl group (Endo-imino-cyclic compounds) or one oxygen and one nitrogen or two oxygen atoms alone are located in the ring, separated by an RN= group cumulatively bonded externally to the ring, also without a carbonyl group (Exo-imino-cyclic compounds). Various exo-imino cyclic compounds have in the past been reported to favor being polymerized by "cationic" catalysts [13,37].

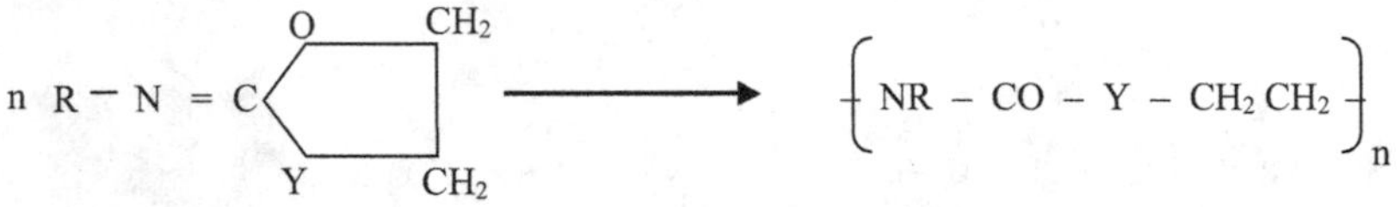

$$13.210a$$

where the Y is as shown below for three cases- $Y \equiv O$, or NR^1 or CH_2.

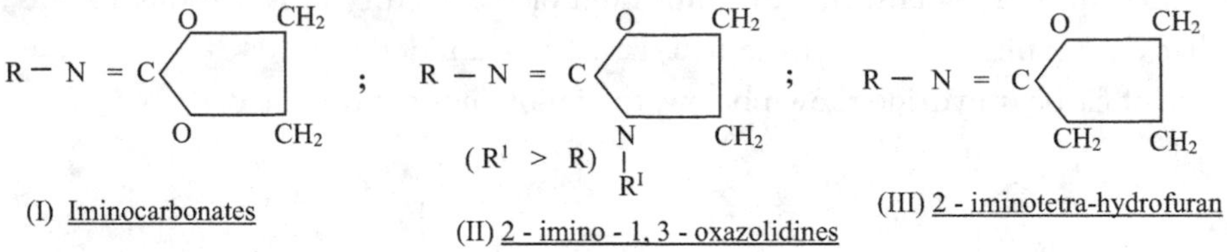

(I) Iminocarbonates ; (II) 2 - imino - 1, 3 - oxazolidines ($R^1 > R$) ; (III) 2 - iminotetra-hydrofuran

(Exo - imino-cyclic compounds)

$$13.210b$$

Examples of the former to be considered include four, five, six membered endo-imino cyclic ethers[13,38].

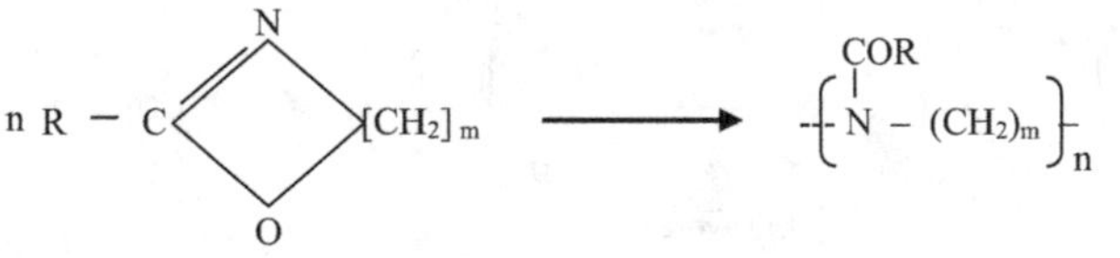

$$13.211a$$

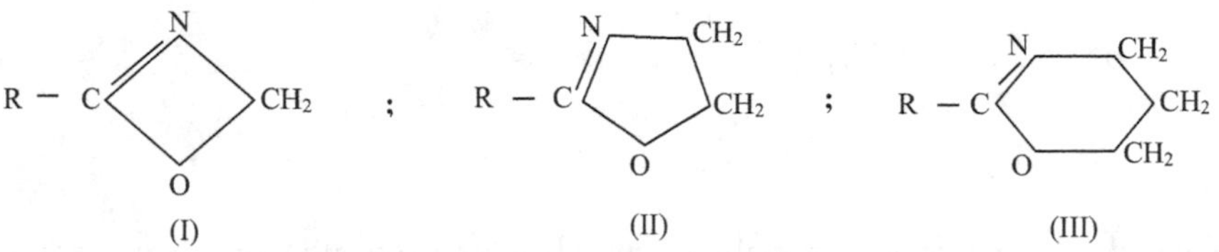

(I) ; (II) ; (III)

(Endo - imino cyclic ethers)

$$13.211b$$

Various cyclic compounds have been known to be polymerized via "cationic" routes[37]. Looking at the three exo-imino compounds to be considered for study, there are two, two and one functional centers for (I), (II) and (III) of Equation 13.210b respectively. It is also important to note that, the N centers are carrying substituent groups rather than hydrogen atoms; otherwise the situation would be different.

The activation/functional centers on the monomers- $R-N=C$, O and N are all nucleophilic in character with the following orders.

$$> \quad O \quad > \quad N \quad > \quad R-N=C$$

Exo -

Order of nucleophilicity (R is a radical-pushing group)   13.212

Thus, while $R-N=C$ center can be observed to be weakly electrophilic, the same does not apply to $C=O$ center which is more electrophilic. The reason why this is so will become obvious after all considerations.

For endo-imino cyclic ethers of Equation 13.211b, there is only one activation center on the monomers- $C=N$. This center is nucleophilic in character far less so than the $C=O$ center.

$$C=O \quad >> \quad C=N$$

Endo -

Order of nucleophilicity   13.213

The order being indicated still have to be confirmed based on literature data and application of new rules. Of particular concern is the exclusion of certain members in the family of monomers to be considered for study.

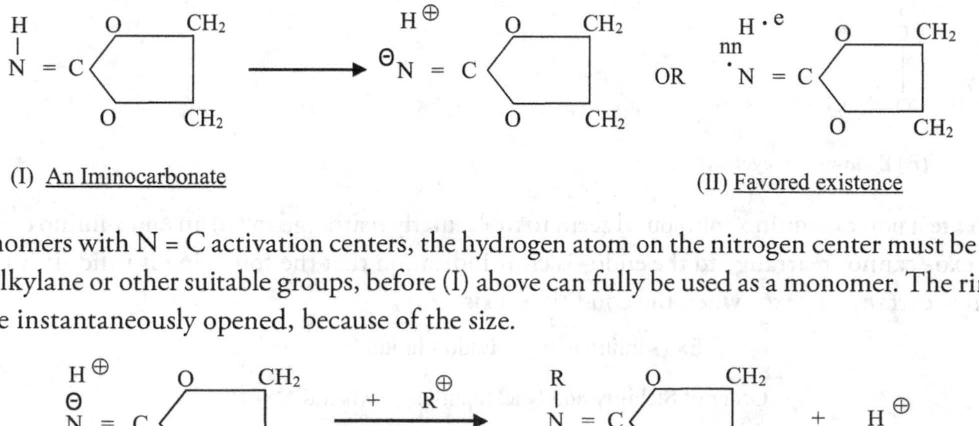

(I) An Iminocarbonate   (II) Favored existence   13.214

Like monomers with N = C activation centers, the hydrogen atom on the nitrogen center must be replaced with an alkylane or other suitable groups, before (I) above can fully be used as a monomer. The ring above cannot be instantaneously opened, because of the size.

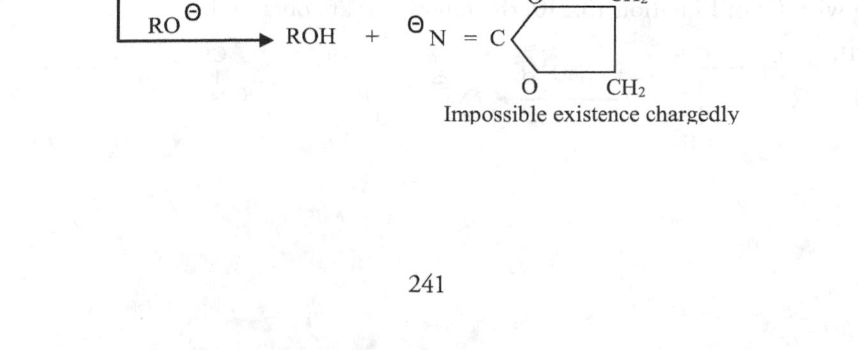

Impossible existence chargedly   13.215

For 2-imino-1,3-oxazolidines, consider the following cases.

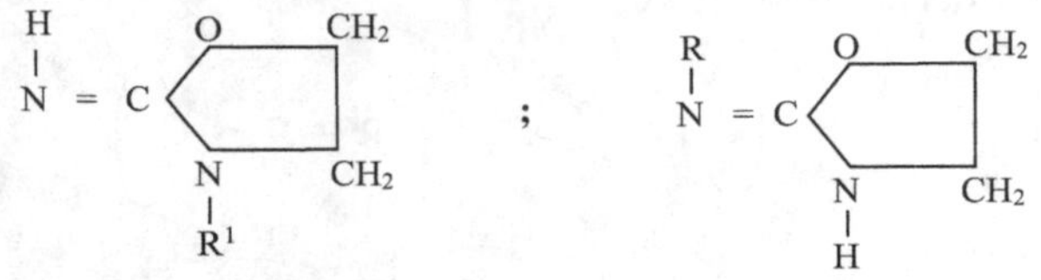

(I)   ;   (II)   13.216

The hydrogen atom in (I) above, like (I) of Equation 13.214 must be replaced before it can fully be used as a monomer. With (II) above, it seems the followings may take place when activated.

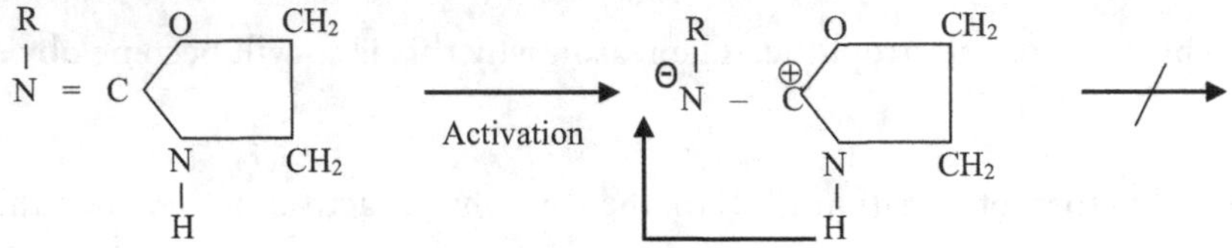

(II) <u>Exo - imino</u>

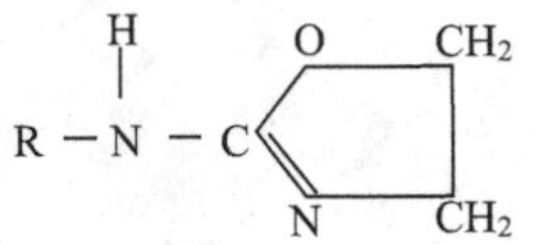

(III) <u>Endo - imino</u> 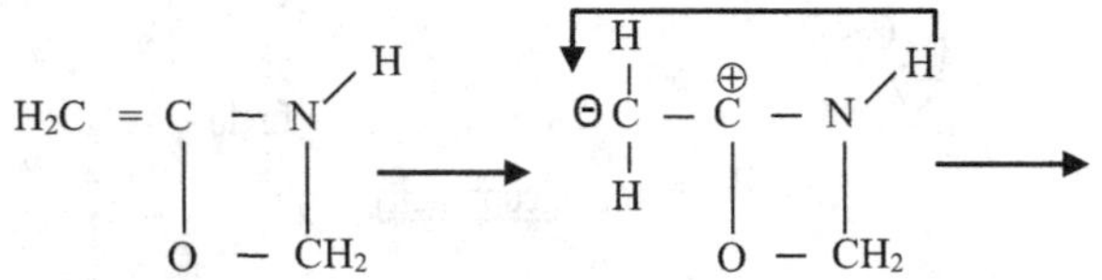   ;   (A) 2-methylene 1,3- oxazolidene

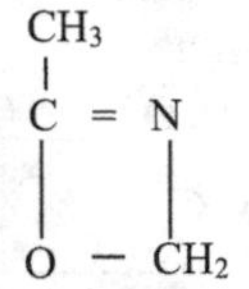

(B) Endo-imino cyclic ether

13.217

(A) which are a non exo-imino compound seem to molecularly rearrange to (B) an endo-imino compound. That the exo- cannot rearrange to the endo- is clear indication that the following is valid. It is the endo- that should rearrange to exo- when the conditions exist.

$$\text{Exo - imino} \quad > \quad \text{Endo - imino}$$

Order of Stability and Nucleophilicity (when $R > R^1$)   13.218

13.3.4.1. Exo-imino Cyclic Compounds

Beginning with (I) of Equation 13.210, the followings are obtained.

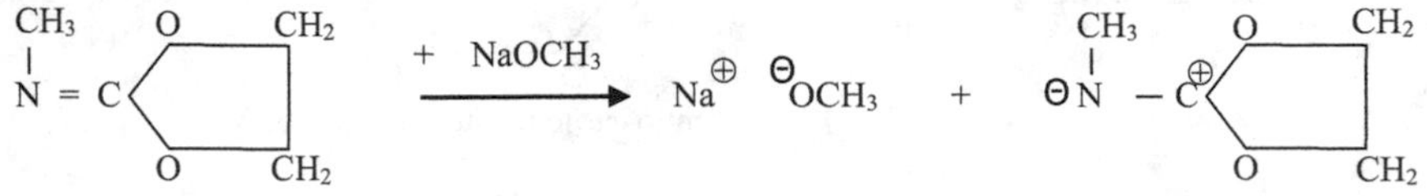

242

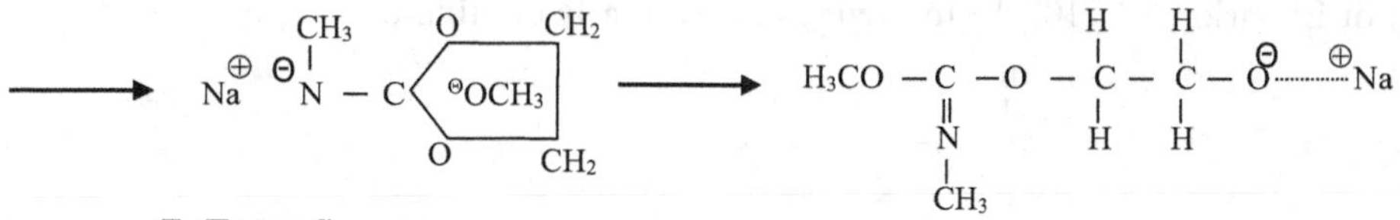

(I) (Favored)

(II) (Not favored)

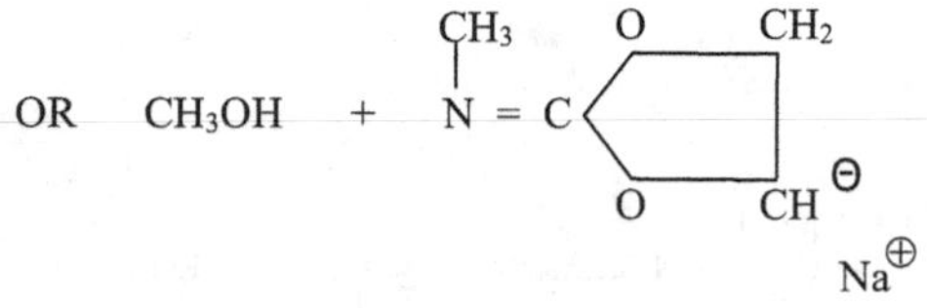

OR CH_3OH +

(III) (Not favored)

13.219

OR CH_3OH +

(III) (Not favored)

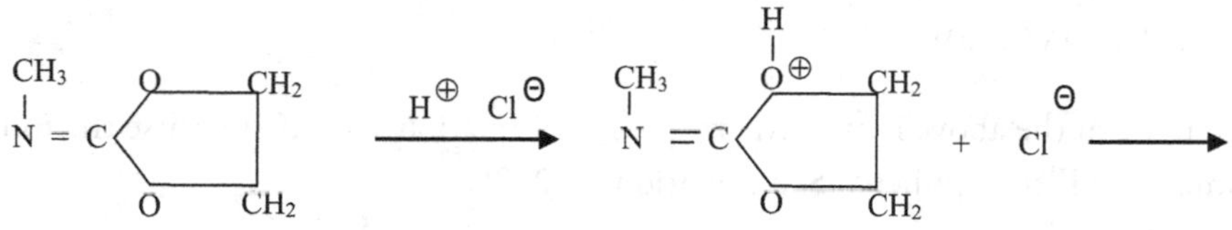

(I)

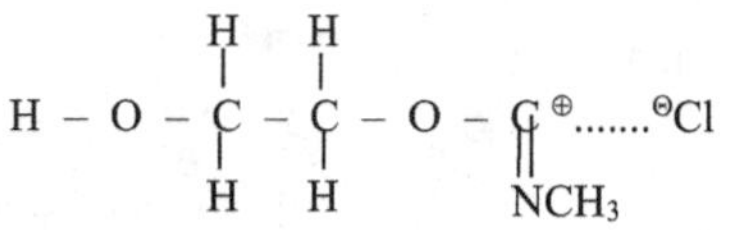

(II) Not favored

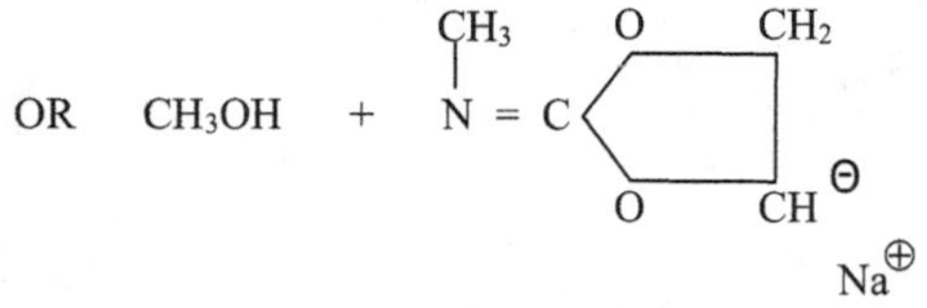

13.220

Unlike $H_2C = C$ center, $N = C$ center has been used as an electrophile. Though both the cationic and anionic routes look favored above, to give the same monomer unit, the reactions above are not favored except (I) of the first equation above as will shortly become obvious, because the monomer is an Electrophile of a different kind.

Radically, the situation is the same if the monomer is an Electrophile.

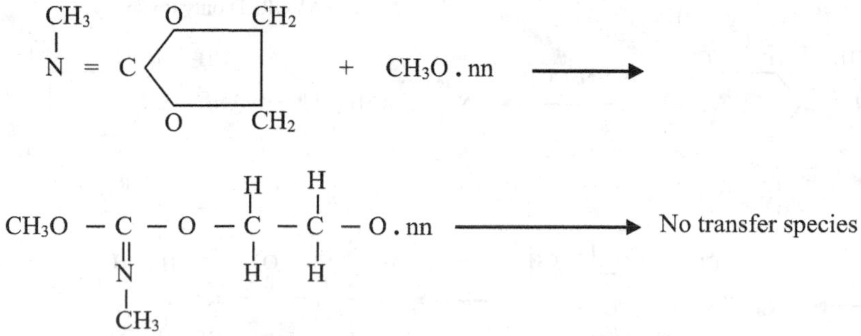

(Not Favored)

13.221

With (II) of Equation 13.210, the followings are similarly obtained.

$$H-N-\underset{\underset{H}{|}}{\overset{\overset{H}{|}}{C}}-\underset{\underset{H}{|}}{\overset{\overset{H}{|}}{C}}-O-\underset{\underset{CH_3}{\underset{|}{N}}}{\overset{\oplus}{C}}\ ....^{\ominus}Cl$$

(NOT FAVORED)

13.222

(II) $\xrightarrow{CH_3ONa}$ $CH_3O-\underset{\underset{CH_3}{\underset{|}{N}}}{\overset{\overset{O}{\|}}{C}}-O-\underset{\underset{H}{|}}{\overset{\overset{H}{|}}{C}}-\underset{\underset{R}{|}}{\overset{\overset{H}{|}}{C}}-N^{\ominus}\ ....^{\oplus}Na$ OR CH_3OR +

(NOT FAVORED)

$$\underset{nn.N-CH_2}{\overset{\overset{CH_3}{\underset{|}{N}}=\overset{O-CH_2}{\underset{|}{C}}}{}}$$

(NOT FAVORED)

13.223

As already said, none of the above is favored except (I) of Equation 13. 219, because the following is valid and the monomers are Electrophiles - [See Equation 13.212]

(I) > (II) > (III)

Order of Nucleophilicity

13.224

Coming back to the two cases above, the followings are truly obtained "cationically".

$$\underset{O\quad CH_2}{\overset{CH_3\quad O\quad CH_2}{\underset{|}{N}=C}} + NaOCH_3 \longrightarrow Na^{\oplus}\ ^{\ominus}OCH_3 + \underset{O\quad CH_2}{\overset{CH_3\quad O\quad CH_2}{\overset{\oplus}{\underset{|}{N}}-C}}$$

$$\longrightarrow Na^{\oplus}\ \underset{O\quad CH_2}{\overset{CH_3\quad O\quad CH_2}{\overset{\ominus}{\underset{|}{N}}-C\langle\ ^{\ominus}OCH_3}} \longrightarrow Na-\underset{H}{\overset{CH_3\ O}{\underset{|}{N}}}-\overset{\overset{O}{\|}}{C}-O-\underset{\underset{H}{|}}{\overset{\overset{H}{|}}{C}}-\underset{\underset{H}{|}}{\overset{\overset{H}{|}}{\overset{\oplus}{C}}}......^{\ominus}OCH_3$$

(I) (II) (FAVORED only Radically)

13.225

$$\underset{N\quad CH_2}{\overset{CH_3\quad O\quad CH_2}{\underset{|}{N}=C}} + NaOCH_3 \longrightarrow Na^{\oplus}\ ^{\ominus}OCH_3 + \underset{\underset{R^1}{\underset{|}{N}}\quad CH_2}{\overset{CH_3\ \oplus\ O\quad CH_2}{\underset{|}{N}-C}}$$

$$\longrightarrow Na^{\oplus}\ \underset{\underset{R^1}{\underset{|}{N}}\quad CH_2}{\overset{CH_3\quad O\quad CH_2}{\overset{\ominus}{\underset{|}{N}}-C\langle\ ^{\ominus}OCH_3}} \longrightarrow Na-\underset{R^1}{\overset{CH_3\ O}{\underset{|}{N}}}-\overset{\overset{O}{\|}}{C}-N-\underset{\underset{H}{|}}{\overset{\overset{H}{|}}{C}}-\underset{\underset{H}{|}}{\overset{\overset{H}{|}}{\overset{\oplus}{C}}}......^{\ominus}OCH_3$$

(I) R^1 (II) (FAVORED only Radically)

13.226

It is favored only electro-free-radically because with a cation the equation is not chargedly balanced under Combination mechanism. The initiator is radically paired, i.e., $H_3CO^{\bullet nn}\ldots\ldots e^{\bullet}Na$ and the active center should be OCH_3 and not Na, the monomer being an Electrophile.

Note the points of scission on the ring after the activation of that bond. Anionically, the first ring can be opened via transfer species abstraction, and the same applies to the second and third rings. Since the transfer species cannot be rejected from the electro-free-radical chain, then it is a different type of transfer species- of the eleventh type as will be shown when rules are stated. The route is favored being natural to the monomers. With the first case, there are two points of scission, while in the second and third cases there is only one point of scission. The R^1 above must be greater than CH_3.

For (III) of Equation 13.210, the followings are obtained.

$$
\begin{array}{l}
\text{(III)} \quad + \quad H^{\oplus} X^{\ominus} \longrightarrow \quad \ldots \quad + \quad X^{\ominus}
\end{array}
$$

(III)

$$
\longrightarrow \quad H - \overset{CH_3}{\underset{}{N}} - \overset{O}{\underset{}{C}} - \overset{H}{\underset{H}{C}} - \overset{H}{\underset{H}{C}} - \overset{H}{\underset{H}{C}}{\cdots}X \quad + \quad n\text{(III)} \longrightarrow
$$

$$
H - \left(\overset{CH_3}{\underset{}{N}} - \overset{O}{\underset{}{C}} - \overset{H}{\underset{H}{C}} - \overset{H}{\underset{H}{C}} - \overset{H}{\underset{H}{C}} \right)_n \overset{CH_3}{\underset{}{N}} - \overset{O}{\underset{}{C}} - \overset{H}{\underset{H}{C}} - \overset{H}{\underset{H}{C}} - \overset{H}{\underset{H}{C}}{\cdots}X
$$

<u>No transfer species</u>   13.227

In the monomer (III), there are two nucleophilic centers in which the one externally located is less nucleophilic. It can be observed that the three exo-imino-cyclic compounds are weak Electrophiles with no transfer species of the first kind electro-free-radically. These rings cannot be opened instantaneously, because the point of scission will be different and may give an electrophilic character to some of the monomers. The O functional center cannot be used, because of Equation 13.212 or 13.224.

Thus, between the three members, based on what is carried by them, the follows are valid.

Iminocarbonates > 2-imino-1,3-oxazolidenes > 2-iminotetrahydrofuran

(I)   **(II)**   **(III)**

Order of Nucleophilicity   13.228

Considering side by side placement of them, the followings are obtained for (I) and (III).

$$
\longrightarrow \quad H^{\oplus} X^{\ominus} \longrightarrow
$$

$$
\longrightarrow \quad H - \overset{CH_3}{\underset{}{N}} - \overset{O}{\underset{}{C}} - \overset{H}{\underset{H}{C}} - \overset{H}{\underset{H}{C}} - \overset{O}{\underset{}{C}} - \overset{H}{\underset{\underset{CH_3}{N}}{C}} - O - \overset{\oplus}{\underset{}{C}}{\cdots}X^{\ominus}
$$

13.229

The two rings placed side by side are (I) and (III), of which (III) is the first to be attacked being less nucleophilic than (I).

For (I) and (II) when placed side by side, the followings are obtained.

$$\text{(I) and (II) structures} \quad \xrightarrow{H^{\oplus} X^{\ominus}} \quad \text{product}$$

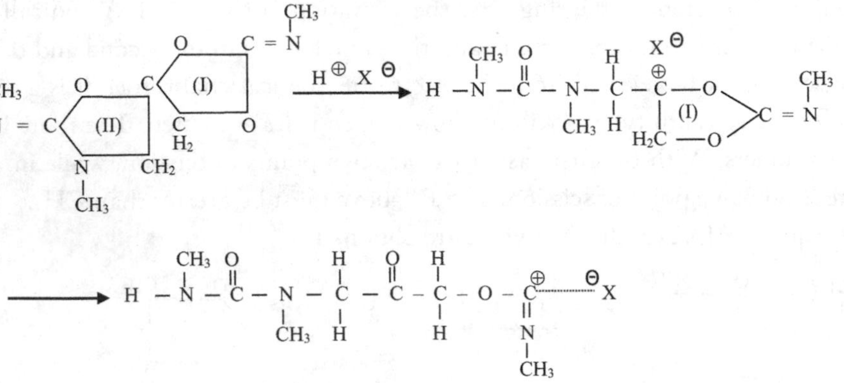

13.230

No functional center is "cationically" involved above, that which is unique, based on Equation 13.224. When used, the followings are to be expected for the last case above.

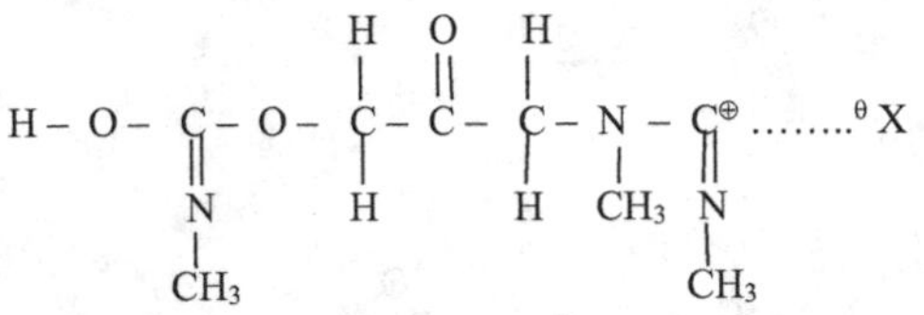

$$\text{H}-\text{O}-\underset{\underset{\text{CH}_3}{|}}{\overset{}{\underset{\text{N}}{\text{C}}}}-\text{O}-\underset{\underset{\text{H}}{|}}{\overset{\text{H}}{\text{C}}}-\overset{\text{O}}{\overset{||}{\text{C}}}-\underset{\underset{\text{CH}_3}{|}}{\overset{\text{H}}{\text{C}}}-\text{N}-\underset{\underset{\text{CH}_3}{|}}{\overset{}{\underset{\text{N}}{\text{C}^{\oplus}}}}\cdots\cdots^{\theta}\text{X}$$

(NOT FAVORED) From (I)

13.231

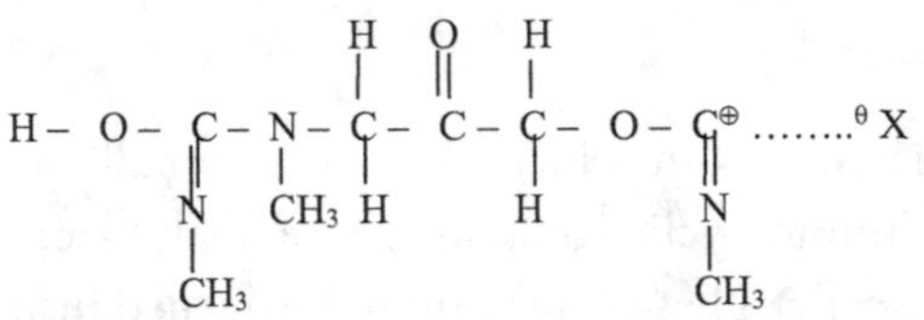

$$\text{H}-\text{O}-\underset{\underset{\text{CH}_3}{|}}{\overset{}{\underset{\text{N}}{\text{C}}}}-\text{N}-\underset{\underset{\text{H}}{|}}{\overset{\text{H}}{\underset{\text{CH}_3}{\text{C}}}}-\overset{\text{O}}{\overset{||}{\text{C}}}-\underset{\underset{\text{H}}{|}}{\overset{\text{H}}{\text{C}}}-\text{O}-\underset{\underset{\text{CH}_3}{|}}{\overset{}{\underset{\text{N}}{\text{C}^{\oplus}}}}\cdots\cdots^{\theta}\text{X}$$

(NOT FAVORED) From (II)

13.232

None is favored above, because based on current observations the N center should be point of attack "cationically".

When (I) and (II) are differently placed side by side, the followings are also to be expected.

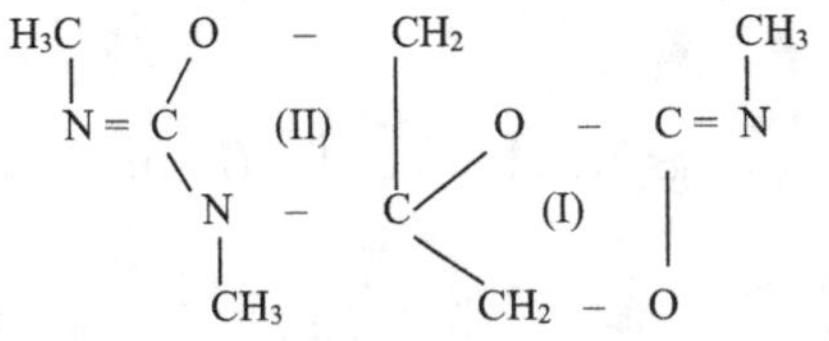

Couple between (I) and (II)

13.233a

"Cationically", the followings are obtained.

$$\text{H}-\text{N}-\underset{\underset{\text{N}}{\overset{}{\underset{\text{CH}_3}{|}}}}{\overset{\text{CH}_3}{\underset{}{\text{C}}}}-\text{O}-\underset{\underset{\text{H}}{|}}{\overset{\text{H}}{\text{C}}}-\overset{\text{O}}{\overset{||}{\text{C}}}-\underset{\underset{\text{H}}{|}}{\overset{\text{H}}{\text{C}}}-\text{O}-\underset{\underset{\text{CH}_3}{|}}{\overset{}{\underset{\text{N}}{\text{C}^{\oplus}}}}\cdots\cdots^{\theta}\text{X}$$

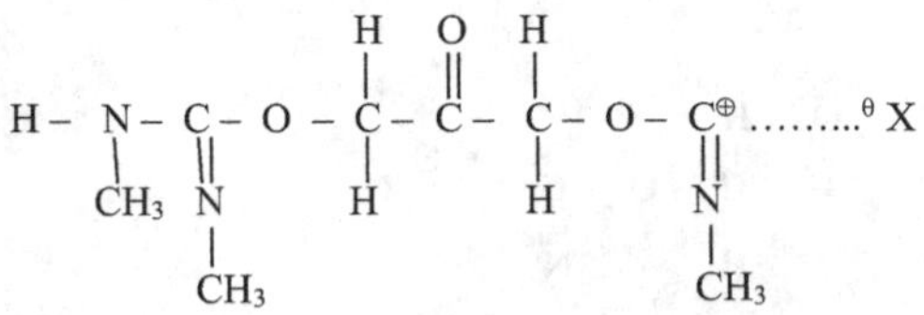

13.233b

If the couple was Electrophilic, then using $NaOCH_3$ would clearly show that it is electrophilic, because this initiator is dual in character, and this the case as shown below.

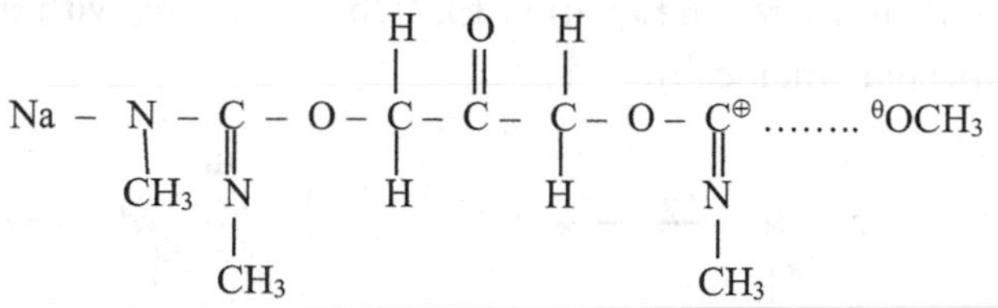

$$Na - \underset{\underset{CH_3}{|}}{\overset{\overset{H}{|}}{N}} - \underset{\underset{\underset{\underset{CH_3}{|}}{N}}{\|}}{C} - O - \underset{\underset{H}{|}}{\overset{\overset{H}{|}}{C}} - \underset{}{\overset{\overset{O}{\|}}{C}} - \underset{\underset{H}{|}}{\overset{\overset{H}{|}}{C}} - O - \underset{\underset{\underset{CH_3}{|}}{N}}{\overset{\oplus}{C}} \quad \ldots\ldots \quad {}^{\ominus}OCH_3$$

13.233c

"Cationically", both HCl and $NaOCH_3$ have been used as initiators. In fact, when HCl is used, the route is largely electro-free-radical with backward addition. However, one can observe that the manners by which two monomers are placed side by side is very important. Compare the monomer unit here with that of Equation 13 230. The orders of Nucleophilicity of the three exo-monomers as shown in Equation 13.228 have been clearly confirmed. These will become clear when the rules are stated.

Between exo-imino compounds and cyclic amines, the followings are obtained.

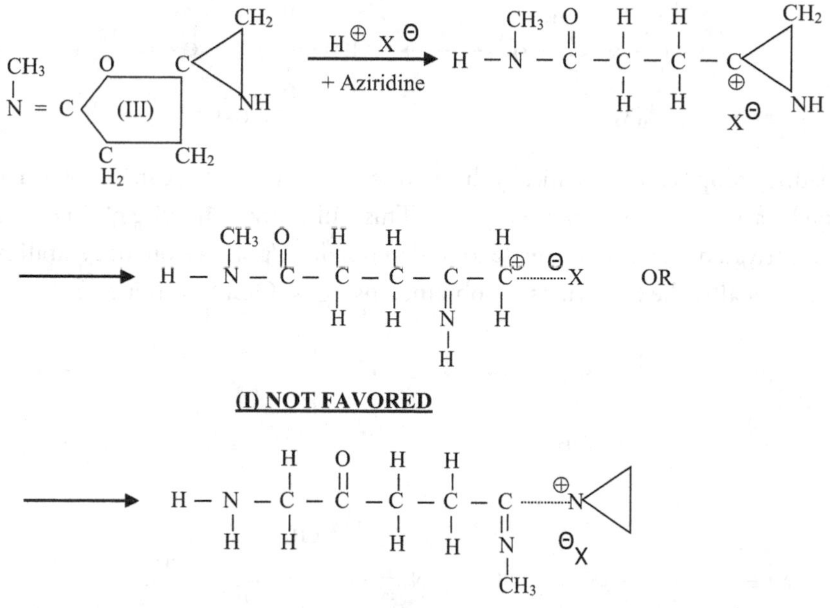

(I) NOT FAVORED

(II) _Favored_

13.234a

(II) is obtained when the aziridine ring is the first to be attacked using a different type of initiation. Based on the types of centers in the rings, it is obvious that (I) is not the favored reaction. Anionically, the same monomer unit will be obtained if the cyclic amine is instantaneously opened, being more strained than the exo-imino compound.

Exo - imino compounds   >   Cyclic amines

Order of nucleophilicity

13.234b

The fact that the exo-imino-compounds largely favor the electro-free-radical route, clearly indicates that-

Cyclic amines   >   Cyclic amides   >   Exo - imino compounds

Order of basicity

13.234c

13.3.4.2 Endo-imino Cyclic Ethers

These are the cases already indicated in Equation 13.211b. Beginning with the four-membered rings, the followings are to be expected anionically.

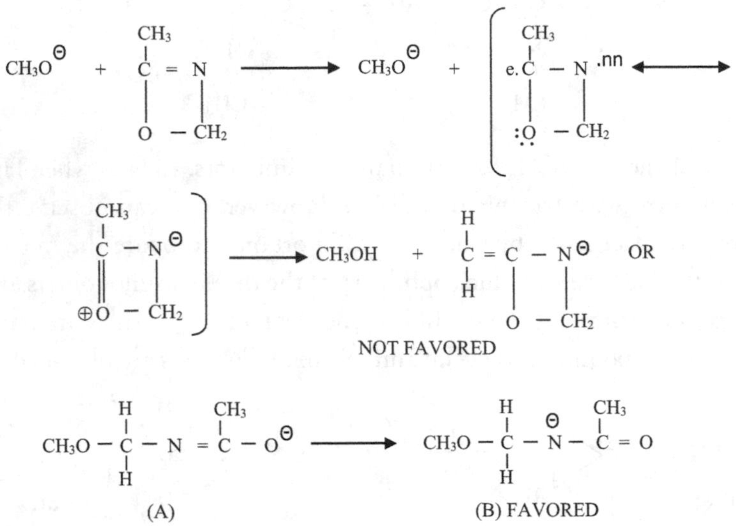

$$CH_3O^{\ominus} \;+\; \underset{\substack{| \\ O - CH_2}}{\overset{\substack{CH_3 \\ |}}{C}} = N \;\longrightarrow\; CH_3O^{\ominus} \;+\; \left[\underset{\substack{| \\ :O - CH_2}}{\overset{\substack{CH_3 \\ |}}{e.C}} - N^{.nn} \longleftrightarrow \right.$$

$$\left. \underset{\substack{|| \\ \oplus O - CH_2}}{\overset{\substack{CH_3 \\ |}}{C}} - N^{\ominus} \right\} \longrightarrow CH_3OH \;+\; \underset{\substack{| \\ H}}{\overset{\substack{H \\ |}}{C}} = \underset{\substack{| \\ O - CH_2}}{C} - N^{\ominus} \quad OR$$

$$\text{NOT FAVORED}$$

$$CH_3O - \underset{\substack{| \\ H}}{\overset{\substack{H \\ |}}{C}} - N = \underset{\substack{| \\ H}}{\overset{\substack{CH_3 \\ |}}{C}} - O^{\ominus} \longrightarrow CH_3O - \underset{\substack{| \\ H}}{\overset{\substack{H \\ |}}{C}} - \overset{\ominus}{N} - \overset{\substack{CH_3 \\ |}}{C} = O$$

$$\text{(A)} \qquad\qquad\qquad \text{(B) FAVORED} \qquad\qquad 13.235$$

With a radical-pushing group (CH_3), anionically the route is favored if the ring can be opened instantaneously or the ring is opened via abstraction of transfer species. This will apply to any size of ring. It can be observed that the ring when activated, can also be made to undergo radical/polar resonance stabilization as shown above. Electro-free-radically, the followings are obtained using $NaOCH_3$ as initiator.

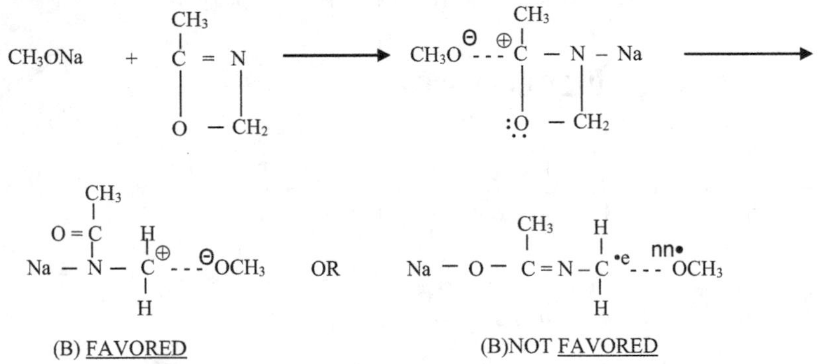

$$CH_3ONa \;+\; \underset{\substack{| \\ O - CH_2}}{\overset{\substack{CH_3 \\ |}}{C}} = N \;\longrightarrow\; CH_3O^{\ominus} - - - \overset{\oplus}{\underset{\substack{| \\ :O - CH_2}}{\overset{\substack{CH_3 \\ |}}{C}}} - N - Na \;\longrightarrow$$

$$\underset{\substack{| \\ H}}{Na - N - \overset{\substack{O = C \quad H \\ | \quad\quad |}}{C^{\oplus}}} - - - {}^{\ominus}OCH_3 \quad OR \quad Na - O - \underset{\substack{| \\ H}}{\overset{\substack{CH_3 \quad H \\ | \quad\quad |}}{C}} = N - C^{\bullet e} - - - {}^{nn\bullet}OCH_3$$

$$\text{(B) \underline{FAVORED}} \qquad\qquad\qquad \text{(B)NOT \underline{FAVORED}} \qquad 13.236$$

Note that the reaction above is electro-free-radical and not cationic as shown above.

When endo- and exo- imino rings are placed side by side, whichever way, only one ring can be opened if the laws of Nature are not fully applied.

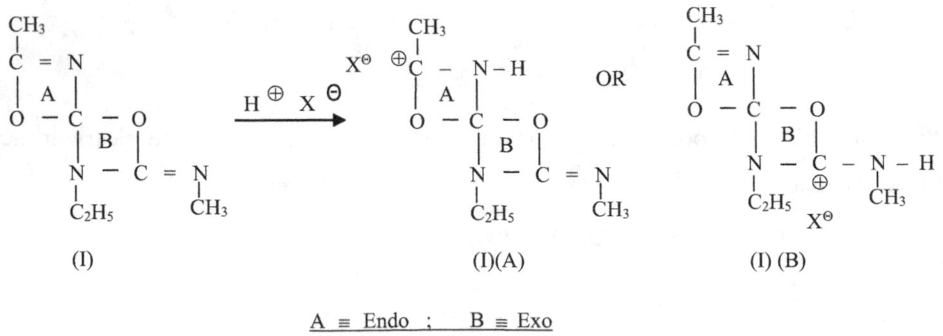

$$\begin{array}{ccc}
\text{(I)} & \text{(I)(A)} & \text{(I) (B)}
\end{array}$$

$$A \equiv Endo \;;\quad B \equiv Exo$$

$$
\begin{array}{ccc}
\underset{(I)}{
\begin{array}{c}
CH_3 \\
| \\
C = N \\
| \quad A \quad | \\
O - C - O \\
\quad | \; B \; | \\
\quad N - C = N \\
\quad | \qquad | \\
\quad C_2H_5 \quad CH_3
\end{array}
}
&
\xrightarrow{\; H^{\oplus} \; X^{\ominus} \;}
&
\begin{array}{cc}
\underset{(I)(A)}{
\begin{array}{c}
CH_3 \\
X^{\ominus} \;\; {}^{\oplus}| \\
C - N - H \\
| \quad A \quad | \\
O - C - O \\
\quad | \; B \; | \\
\quad N - C = N \\
\quad | \qquad | \\
\quad C_2H_5 \quad CH_3
\end{array}
}
&
\text{OR}
\qquad
\underset{(I)\,(B)}{
\begin{array}{c}
CH_3 \\
| \\
C = N \\
| \quad A \quad | \\
O - C - O \\
\quad | \; B \; | \\
\quad N - C - N - H \\
\quad | \quad {}^{\oplus} \quad | \\
\quad C_2H_5 \qquad CH_3 \\
\qquad X^{\ominus}
\end{array}
}
\end{array}
\end{array}
$$

$$\underline{A \equiv Endo \; ; \quad B \equiv Exo}$$

$$
\longrightarrow
\begin{array}{cc}
\underset{\text{FOR (I)(A)}}{
\begin{array}{c}
CH_3 \\
| \\
O = C \quad O \quad C_2H_5 \\
| \qquad || \quad | \\
H - N - C - N - C^{\oplus} \!\!-\!\!-\!\!-^{\ominus}\! X \\
\qquad\qquad\qquad || \\
\qquad\qquad\qquad N \\
\qquad\qquad\qquad | \\
\qquad\qquad\qquad CH_3
\end{array}
}
&
\text{OR}
\quad
\underset{\text{FOR (I)(B) NOT FAVORED}}{
\begin{array}{c}
CH_3 \quad\; CH_3 \\
| \qquad | \\
H - N - C - N - C - N = C^{\oplus} \!\!-\!\!-\!\!-^{\ominus}\! X \\
\qquad || \quad || \qquad | \\
\qquad O \quad O \quad CH_3
\end{array}
}
\end{array}
\qquad 13.237
$$

Since C = N center is less nucleophilic than the O center in (A), it is the former center that should be attacked cationically by $H^{\oplus}$. (B) ring being a weak Electrophile (whether $R^1 > R$ or not), carrying three centers, all of which are stronger than the two centers carried by (A), clearly indicates that the exo-imino is more nucleophilic than the endo- imino compound (See Equation 13.218).

$$Exo \text{-} imino \quad > \quad Endo \text{-} imino$$

$$\underline{Order\ of\ nucleophilicity} \qquad\qquad 13.238$$

When an exo- ring of the same size is placed side by side with an endo- ring, polymers can be produced. The conditions under which polymers are produced as seen so far are numerous to list. However, when two rings are placed side by side, it is important to note how the rings are placed.

$$
\begin{array}{c}
NR_2 \\
| \\
C = N \\
\diagup \qquad \diagdown \\
O \qquad C \qquad N {-} R \\
\diagdown \quad \diagup \quad | \\
CH_2 \qquad\qquad | \\
\qquad\qquad\qquad \\
H_2C \qquad C = N - R \\
\diagdown \quad \diagup \\
O
\end{array}
\quad
\xrightarrow{\; + \; H^{\oplus} \; X^{\ominus} \;}
\quad
\begin{array}{c}
X^{\ominus} \\
{}^{\oplus} \\
CH_2 \\
| \\
H - N - C - NR \\
\quad | \\
\quad C = O \\
\quad | \\
\quad NR_2 \\
\qquad H_2C \qquad C = NR \\
\qquad\; \diagdown \quad \diagup \\
\qquad\qquad O
\end{array}
$$

$$\text{Ring cannot be opened} \qquad\qquad 13.239$$

With five-membered rings placed side-by-side, it looks as if either only one of the rings can be opened or none at all. This is not indeed true as will shortly be shown. With four-membered rings placed side-by-side with the types of group carried above [NR_2 group in place of R and R^1 equal to R], the followings are obtained.

$$
\text{Endo}
\underset{(I)}{
\begin{array}{c}
NR_2 \\
| \\
C = N \\
| \quad A \quad | \\
O - C - O \\
\quad | \; B \; | \\
\quad N - C = N \\
\diagup \qquad\qquad | \\
R \qquad\qquad\quad R \\
\qquad \text{Exo}
\end{array}
}
\quad
\xrightarrow[\text{(Attack on A)}]{\; + \; H^{\oplus} \; X^{\ominus} \;}
\quad
\begin{array}{c}
\qquad O \\
\qquad || \\
H - N - C - N - C^{\oplus} \\
\quad | \qquad\quad | \qquad \ddot{} \\
\quad C = O \quad R \quad N: \\
\quad | \qquad\qquad\quad | \\
\quad NR_2 \qquad\qquad R
\end{array}
\quad
\xrightarrow{\; + \; n(I) \;}
$$

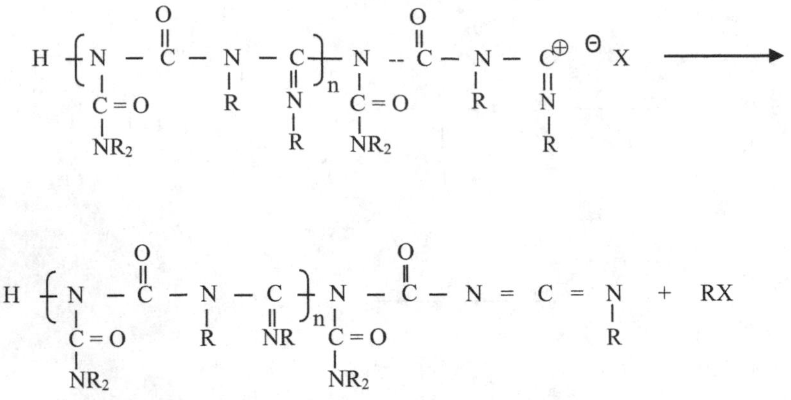

$$H \left\{ \begin{array}{c} N - \overset{\overset{\textstyle O}{\|}}{C} - N - C \\ | \\ C=O \\ | \\ NR_2 \end{array} \right\}_n N \cdots \overset{\overset{\textstyle O}{\|}}{C} - N - \overset{\oplus}{C} \; \overset{\ominus}{X} \longrightarrow$$

$$H \left\{ \begin{array}{c} N - \overset{\overset{\textstyle O}{\|}}{C} - N - C \\ | \\ C=O \\ | \\ NR_2 \end{array} \right\}_n N - \overset{\overset{\textstyle O}{\|}}{C} - N = C = N \quad + \quad RX$$

<u>Transfer species of 2nd kind [FAVORED only radically]</u> 13.240a

Just like the case of Equation 13.236, it is the N (Not the O center) center that is carried by H as shown above electro-free-radically.

Anionically, the followings would have been obtained.

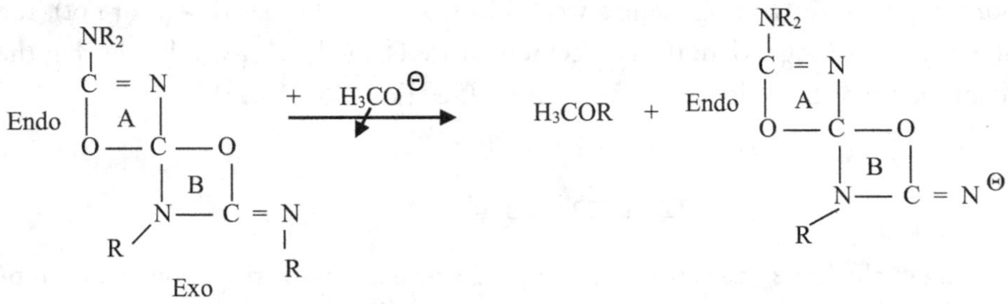

$$\text{Endo} \quad \begin{array}{c} NR_2 \\ | \\ C = N \\ | \quad A \quad | \\ O - C - O \\ | \; B \; | \\ N - C = N \\ R \qquad | \\ \qquad R \\ \text{Exo} \end{array} \quad + \quad H_3CO^{\ominus} \quad \longrightarrow \quad H_3COR \quad + \quad \text{Endo} \quad \begin{array}{c} NR_2 \\ | \\ C = N \\ | \quad A \quad | \\ O - C - O \\ | \; B \; | \\ N - C = N^{\ominus} \\ R \end{array}$$

(I) 13.240b

It is not favored, because the wrong transfer species has been abstracted. Anionically, initiation is favored with (B) being the first to be opened to give the same monomer unit. Indeed nucleo-non-free-radically, initiation is favored. If (B) was allowed to molecularly rearrange, that which is not possible, the followings would be obtained cationically only under Equilibrium conditions.

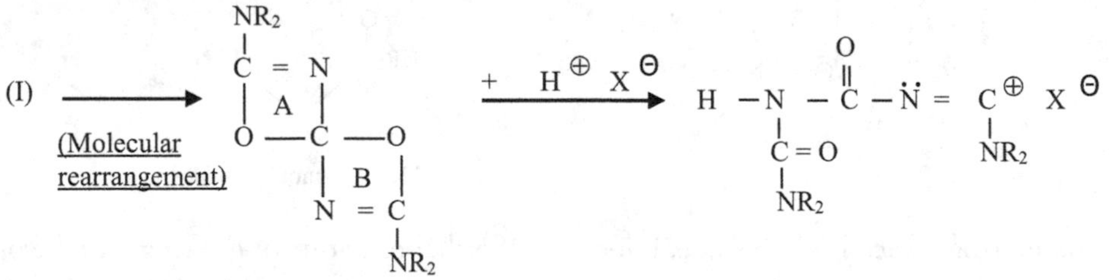

$$(I) \xrightarrow[\substack{\underline{\text{(Molecular}} \\ \underline{\text{rearrangement)}}}]{} \begin{array}{c} NR_2 \\ | \\ C = N \\ | \quad A \quad | \\ O - C - O \\ | \; B \; | \\ N = C \\ | \\ NR_2 \end{array} \xrightarrow{\; + \; H^{\oplus} X^{\ominus} \;} H - N - \overset{\overset{\textstyle O}{\|}}{C} - \overset{..}{N} = \overset{\oplus}{C} \; X^{\ominus} \\ \qquad\qquad\qquad \begin{array}{cc} | & | \\ C=O & NR_2 \\ | \\ NR_2 \end{array}$$

13.241

It is the exo- that should rearrange to the endo- as shown above. The monomer unit obtained will contain no aldehyde group placed along the chain as shown above. Based on the reactions above and the analysis so far, molecular rearrangement from exo- to endo- does take place when the conditions exist i.e., $R^1 \leq$ R. This is rearrangement from a less nucleophilic center to a more nucleophilic center.

The five-membered rings of Equation 13.239 were not properly placed. As shown below, when placed on the second C center close to O, the followings are obtained.

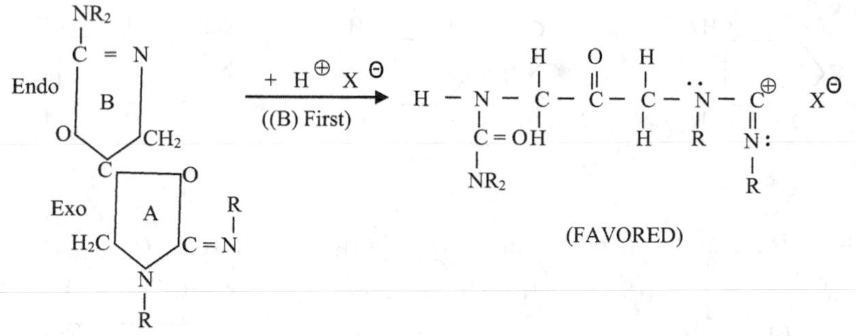

$$\text{(Endo / Exo cyclic structure B, A)} \quad \xrightarrow[\text{((B) First)}]{+\ H^{\oplus}\ X^{\ominus}} \quad H-N-\overset{H}{\underset{C=OH}{C}}-\overset{O}{\underset{H}{C}}-\overset{H}{\underset{R}{C}}-\overset{\oplus}{N}-\overset{\oplus}{\underset{N:}{C}}\ \ X^{\ominus}$$

(FAVORED)

13.242

Anionically, the followings are obtained.

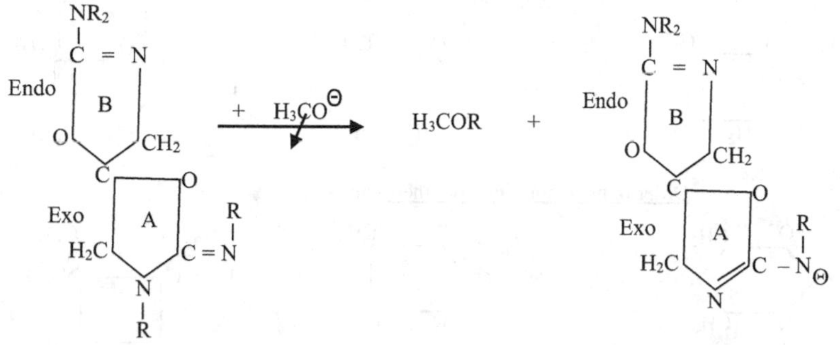

$$\text{(Endo / Exo structure)} \quad \xrightarrow{+\ H_3CO^{\ominus}} \quad H_3COR\ +\ \text{(Endo / Exo structure with } C-N^{\ominus})$$

13.243

When (A) molecularly rearranges to (B), the anionic route is not favored. But if it does not, then the anionic route is favored, since when transfer species is released electro-free-radically according to Equation 13.240a, it is the second kind and not the first kind, the route being unnatural to the exo-, but natural to the endo-.

There is need to know the corresponding order for exo- compounds when members of the same family carrying alkylane groups are involved.

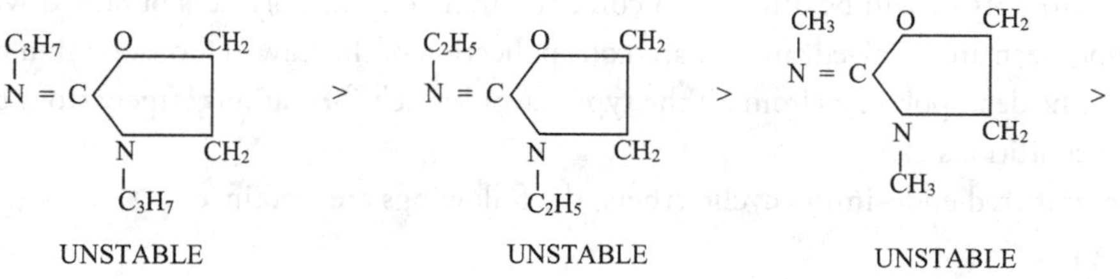

$$\underset{\text{UNSTABLE}}{[C_3H_7,\ C_3H_7]} \ > \ \underset{\text{UNSTABLE}}{[C_2H_5,\ C_2H_5]} \ > \ \underset{\text{UNSTABLE}}{[CH_3,\ CH_3]} \ >$$

13.244

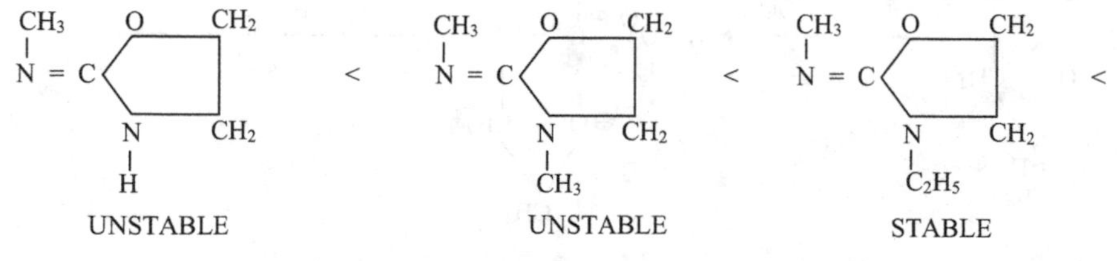

$$\underset{\text{UNSTABLE}}{[CH_3,\ H]} \ < \ \underset{\text{UNSTABLE}}{[CH_3,\ CH_3]} \ < \ \underset{\text{STABLE}}{[CH_3,\ C_2H_5]} \ <$$

Order of nucleophilicity of exo - imino compounds

13.245

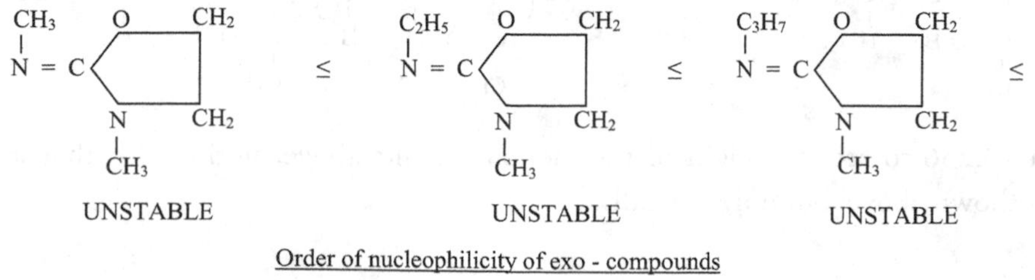

$$\underset{\text{UNSTABLE}}{[CH_3,\ CH_3]} \ \le \ \underset{\text{UNSTABLE}}{[C_2H_5,\ CH_3]} \ \le \ \underset{\text{UNSTABLE}}{[C_3H_7,\ CH_3]} \ \le$$

Order of nucleophilicity of exo - compounds

13.246

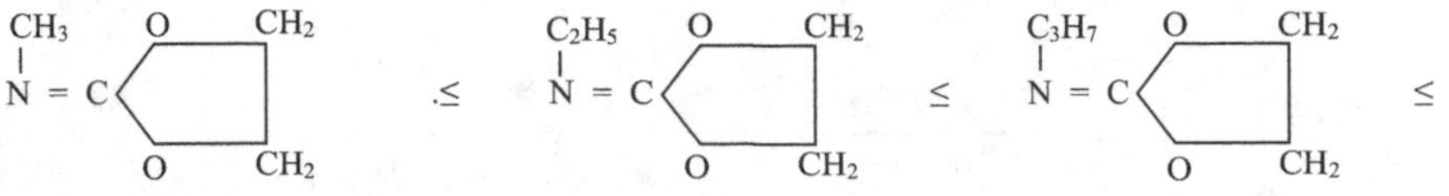

$$13.247$$

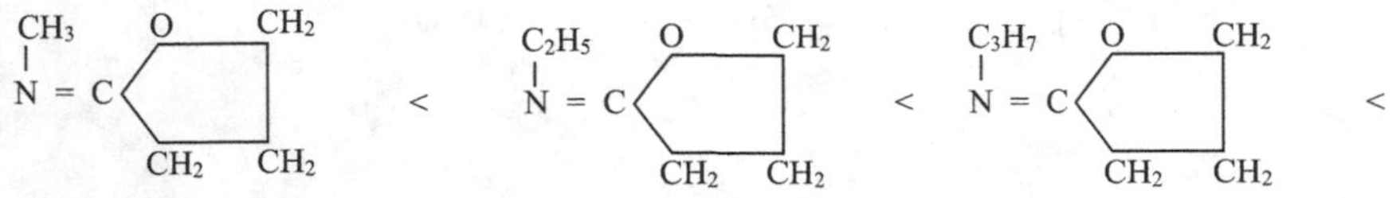

Order of nucleophilicity of exo - compounds

$$13.248$$

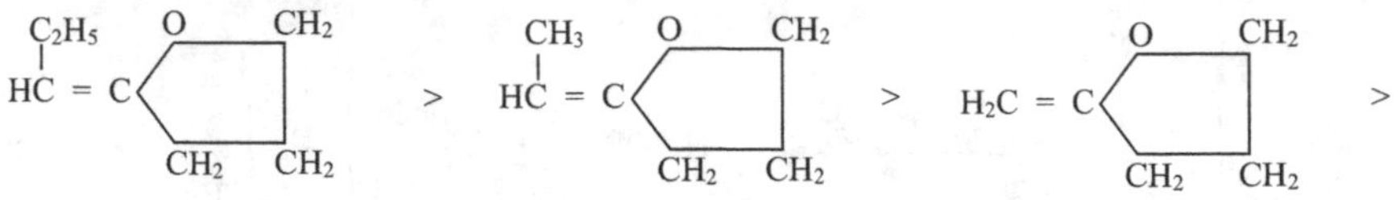

Order of nucleophilicity of ether- compounds

$$13.249$$

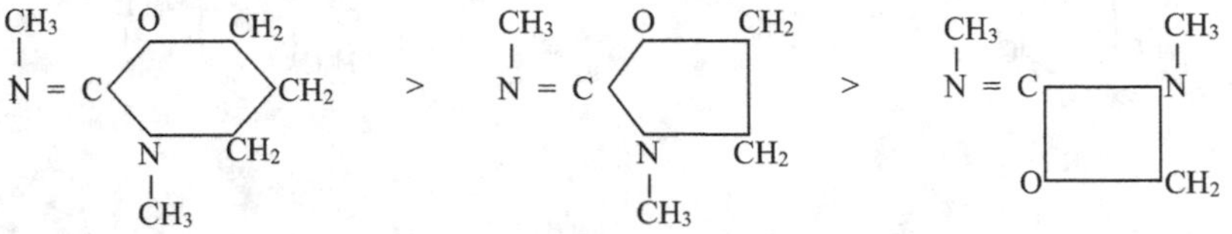

Order of nucleophilicity of exo - compounds

$$13.250$$

Note that all these but one (Equation 13.249) are all weak Electrophiles. What is unique about ringed monomers which favor being used as monomers and favor both positive and negative routes is that, the same monomer units are always obtained. This is one of the essences of what is called a cycle i.e., cyclic – Zero- with no beginning and no end. Without application of the rules which have previously been proposed in totality, it will be difficult to come so far. The dead polymers obtained when transfer species of the types seen are involved, are all based on application of the Law of Conservation of transfer of transfer species. The dead polymers from all the types favor branch formations largely from the carbonyl center when the conditions exist.

With five-membered endo-imino cyclic ethers, the followings are obtained.

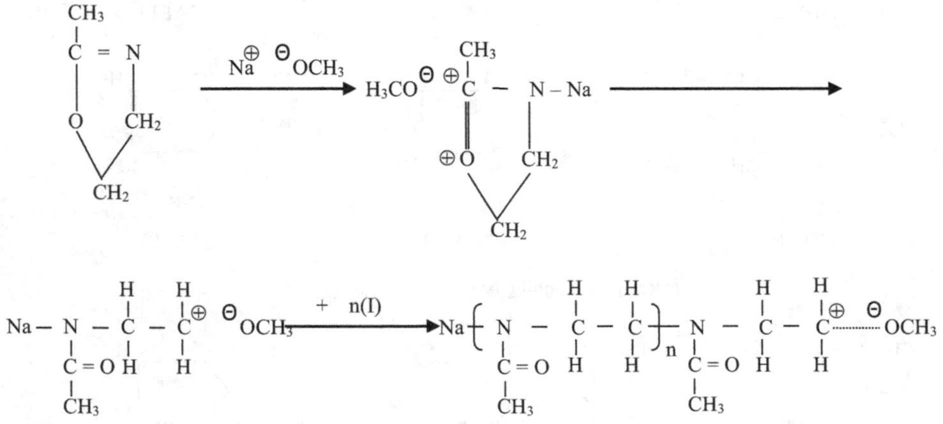

$$13.251$$

Use Equation 13.236 to see the origin of the monomer units above, noting also that it cannot take cationically as shown above, but only radically.

When radical-pulling group such as CF_3 is introduced into the rings in place of CH_3, the followings should be expected.

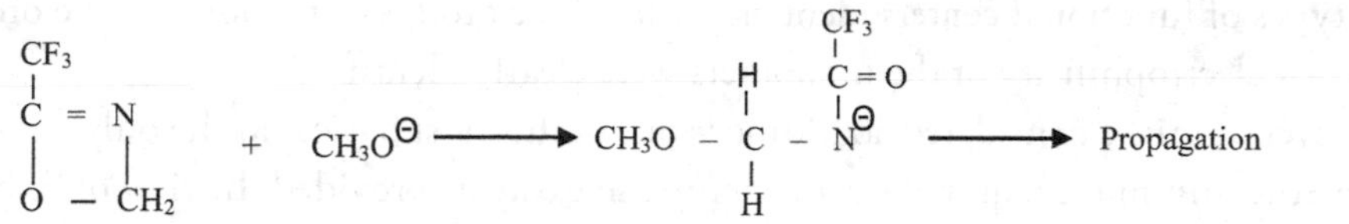

$$
\begin{array}{c}
CF_3 \\
| \\
C = N \\
| \quad\quad | \\
O - CH_2
\end{array}
\; + \; CH_3O^{\ominus} \longrightarrow CH_3O - \begin{array}{c} CF_3 \\ | \\ C \\ | \\ H \end{array} - \begin{array}{c} C = O \\ | \\ N^{\ominus} \end{array} \longrightarrow \text{Propagation}
\qquad 13.252
$$

$$
\begin{array}{c}
CF_3 \\
| \\
C - N^{\ominus} \\
\| \quad\quad | \\
{}^{\oplus}O - CH_2 \\
(I)
\end{array}
\; + \; H^{\oplus} \; X^{\ominus} \longrightarrow H - \begin{array}{c} H \\ | \\ N \\ | \\ C = O \end{array} - \begin{array}{c} C^{\oplus} \\ | \\ H \\ | \\ CF_3 \end{array} \; {}^{\ominus}X \xrightarrow{\; + \; n(I) \;}
$$

$$
H \left[\begin{array}{c} H \\ | \\ N \\ | \\ C=O \\ | \\ CF_3 \end{array} - \begin{array}{c} H \\ | \\ C \\ | \\ H \end{array} \right]_n \begin{array}{c} H \\ | \\ N \\ | \\ C=O \\ | \\ CF_3 \end{array} - \begin{array}{c} H \\ | \\ C^{\oplus} \cdots X \\ | \\ H \end{array} \; {}^{\ominus}X \longrightarrow \text{Living polymers}
$$

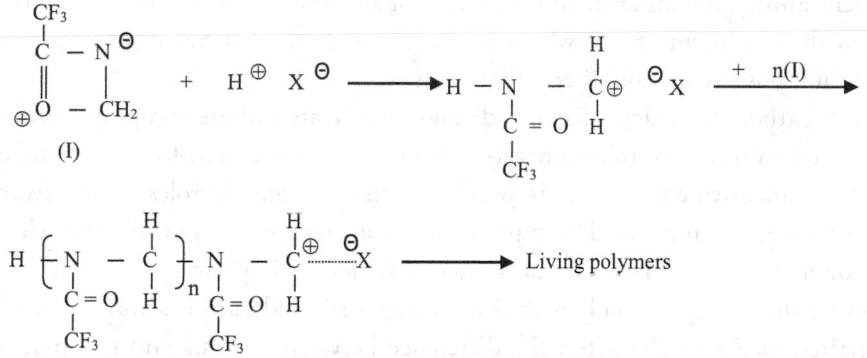

$$\qquad 13.253$$

As will become obvious when the rules are stated, use Equation 13.236 to show that the monomer units above are valid, that wherein nitrogen is the center carrying the negative charge. Note that, the reactions above can only take place radically and not chargedly as shown above, because charges cannot move in the real domain. Though there is no transfer species, the nucleophilic character of the monomer is still maintained. If one had used COOR in place of CF_3, polymerization will still take place, once MRSE can be provided for opening of the rings. With exo-imino cyclic compounds, one should know what to expect. On the whole, from considerations of these members of cyclic imino compounds, the followings are valid.

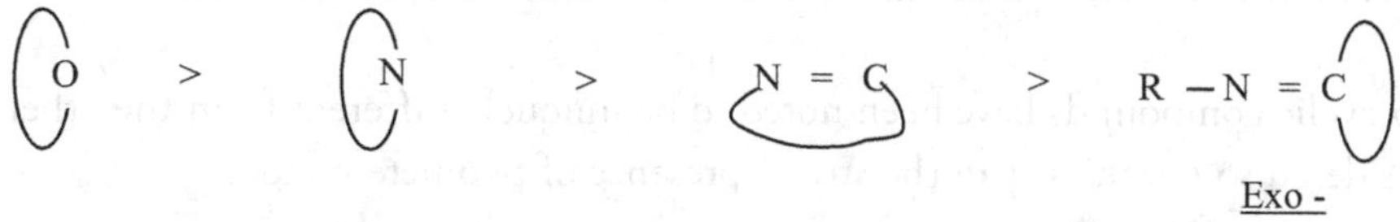

$$
\left(\!\! \begin{array}{c} \\ O \\ \\ \end{array} \!\!\right) > \left(\!\! \begin{array}{c} \\ N \\ \\ \end{array} \!\!\right) > \left(\!\! \begin{array}{c} N = C \end{array} \!\!\right) > \left(\!\! \begin{array}{c} R - N = C \end{array} \!\!\right)
$$
$$\qquad\qquad\qquad\qquad\qquad\qquad\qquad\qquad\qquad\qquad\text{Exo -}$$

<u>Order of nucleophilicity</u> (R is a radical-pushing group with limitations) $\qquad$ 13.254

From all the considerations so far, one can observe why there has been the strong need to consider most existing and non-existing monomers and compounds, based on the new definitions provided for many concepts which were thought to be known. Based on the manners by which one has been moving, all these orders can be fully confirmed quantitative and qualitatively using well designed experimental systems.

13.4 Proposition of Rules of Addition and Concluding Remarks.

Having critically examined how cyclic amides, amines and related monomers undergo different types of reactions to favor the opening of their rings, there was need in the process to go to the past and provide the one and only mechanism of some chemical reactions, such as the case of pyrrole, succinic anhydride, succinimide etc., in order to produce convincing and unquestionable evidence of the new foundations. In the consideration of these groups of monomers, more additional driving forces favoring the openings of rings, have been identified.

Based on the basic character of these groups of monomers, the more basic ones can be used as catalyst anionically and cocatalyst electro-free-radically. In view of the charged-paired electrostatic character of addition of monomer, the hydrogen atom which is loosely bonded to the nitrogen center does not need to be replaced for all of them, unlike the case with linear or non- ringed or non-cyclic monomers. In view of the several types of functional centers identified with these groups of monomers, the order of basicity, nucleophilicity or electrophilicity of the monomers were clearly identified,

Of all the cyclic amines considered, aziridine is one of the most basic and the only polymerizable one with that where the minimum required strain energy can easily be provided. In view of its basicity, higher members of cyclic amines do not favor anionic attack of any kind, even when the H on the N center has been replaced with substituent groups which can still provide the MRSE. Aziridines or cyclic amines, and the likes do not possess an anionic activation center.

The lactams, carboanhydrides, succinimide and pyrrole are unique compounds in different ways. Unlike phenol and aniline, pyrrole undergoes discrete radical closed-loop resonance stabilization phenomena in its unactivated state. It is partly for this reason, pyrroles favor electro-free-radical polymerization with mineral acids. The types of reactions favored by pyrrole were clearly identified. The anionic polymerization of lactams and carboanhydrides using strong bases, acylating agents are unique. So also are their "cationic" polymerizations using weak acids, amines and water. All the initiators involved are paired-media in character; the difference between lactams and carbo-anhydrides being the release of CO_2 during the latter's polymerization. While the five-membered carboanhydrides favor the existence of Nylon 2, the six-membered one should favor Nylon 3 and so on, once the MRSE can be provided for them during activation. In this way, all the different types of Addition nylons can be obtained – Nylon 2, 3, 4, 5, 6, etc. using either the lactams or carboanhydrides. This may have been practiced already with the commercially based industries, but unknown to the academia, a phenomenon which does not augur well for commercial industries, since the commercial industries know little or nothing about the chemistry of monomers or even how the rings are opened. The same also apply to the academia, but much less so, nothing however that without the academia, the commercial industries cannot exist.

The imino cyclic compounds have been noted to be uniquely different from the other hetero-cyclic monomers considered so far, in view of the strong presence of two hetero atoms.

<u>Rule 1081:</u> This rule of Chemistry for **Cyclic amides (formed by the intra-molecular amidation of amino acids),** states that, these are called **Lactams** beginning with four-membered ring followed by other higher membered ones as shown below-

β - propiolactam
or
2 - azetidinone
<u>4 - membered ring</u>

;

γ - butyrolactam
or
2 - pyrrolidone
<u>5 - membered ring</u>

;

δ - valerolactam
or
2 - piperidone
<u>6 - membered ring</u>

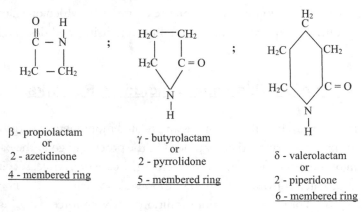

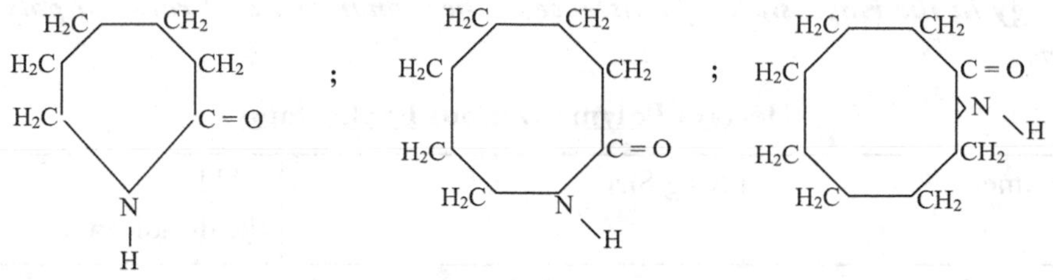

ε - caprolactam	Enantholactam	Capryllactam
or	or	or
6 - hexanolactam	7 - heptanolactam	8 - octanolactam
7 - membered ring	8 - membered ring	9 - membered ring

(Laws of Creations for Lactams)

Rule 1082: This rule of Chemistry for **Lactams,** states that, existence of 3-membered rings is not favored, due to existence of maximum required strain energy in the ring, clear indication *that nitrogen containing ringed compounds are <u>more strained</u> than their corresponding oxygen containing ringed compound.*
(Laws of Creations for Lactams)

Rule 1083: This rule of Chemistry for **Lactams,** states that, while the production of Nylon 4, Nylon 5, Nylon 6, Nylon 8, and Nylon 12 by ring-opening of g-butyrolactam, d-valerolactam, ε-caprolactam, capryllactam and lauryllactam- 5-, 6-, 7-, 9- and 13- membered lactams respectively are possible, the production of Nylon 2, and Nylon 3 from 3- and 4- membered lactams respectively are impossible.
(Laws of Creations for Lactams)

Rule 1084: This rule of Chemistry for **The polymerization reactivity of Lactams,** states that, based on the data shown below-

<u>**Polymerization Reactivity of Lactams as a Function of Ring Size.**</u>

	Types of initiation	Order of reactivity of different sized rings	Ref
1	Anionic (with acylating agent)	$7 \geq 5 > 6$	1
2	Protonic acid (HCl) (Cationic)	$8 > 7 > 11 >> 5, 6$	2
3	Water (Cationic)	$7 > 8 \cong 9 >> 5, 6$	3

one of the major reasons why the lower-membered rings are more difficult to initiate is because they are ringed Electrophiles which are more acidic (i.e., readily favor existing in Equilibrium state of existence) in character as shown below-

4 - membered > > 5 - membered > 6 - membered >

<u>Order of acidity of lactams</u>

(Laws of Creations for Lactams)

Rule 1085: This rule of Chemistry for **The exothermic heats of polymerization of some lactams,** states that, based on the data shown below, the general order of exothermic heats of polymerization *is a function* *of strain energy in the rings, states of existence of the monomers, and route of polymerization of the monomer.*

Heats of Polymerizations For Lactams

#	Monomer	Ring Size	$-\Delta H$ (kcal/mole)
1	γ - butyrolactam	5	1.3
2	δ - valerolactam	6	1.1
3	Caprolactam	7	3.0
4	Enantholactam	8	5.7
5	Capryllactam	9	9.6
6	12 – Dodecanolactam	13	1.5

(Laws of Creations for Lactams)

Rule 1086: This rule of Chemistry for **Only four-membered lactams,** states that, four-membered lactam the first member of the family cannot be used as a monomer because it is always under the state of Equilibrium Enolization/molecular rearrangement radically or ionically as a result of its existence in ACTIVATED/EQUILIBRIUM STATE OF EXISTENCE as shown below

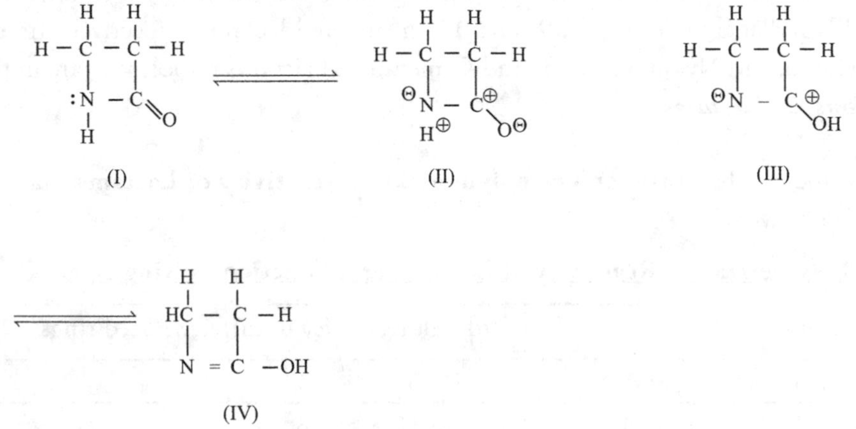

Ionic Equilibrium Enolization/molecular rearrangement of monomer

(a)

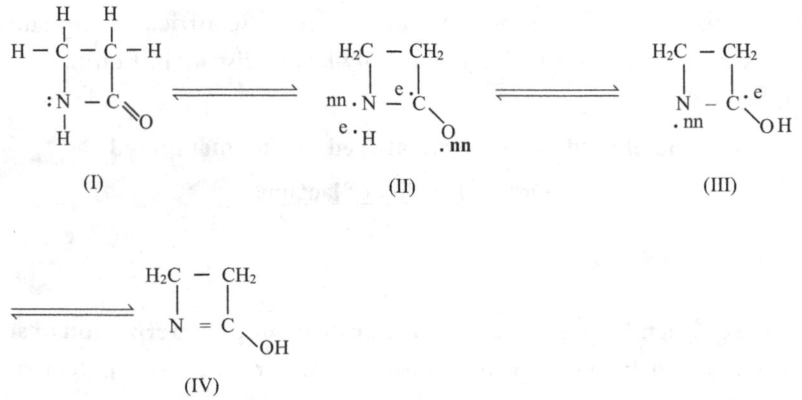

Radical Equilibrium Enolization/molecular rearrangement of monomer

(b)

for which by the nature of this equilibrium rearrangement, the followings are valid for the four-membered ring.

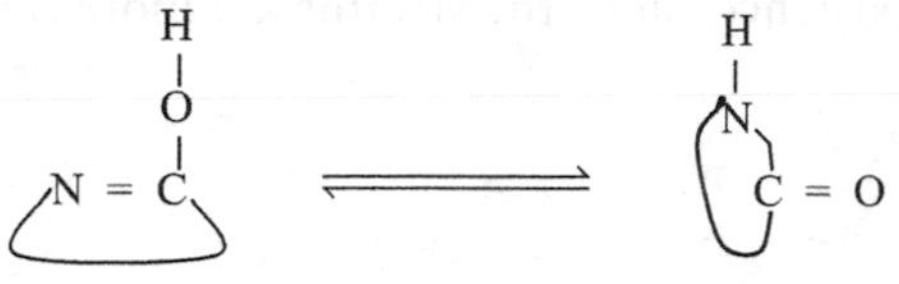

<u>Order of Stability for only four-membered- ring</u>

(Laws of Creations for Lactams-for 4-membered ring)

Rule 1087: This rule of Chemistry for **Three-membered lactams,** state that, this ring does not have any time to favor any type of molecular rearrangement, since it opens up instantaneously to give a different compound (V) as shown below; (I) is not known to

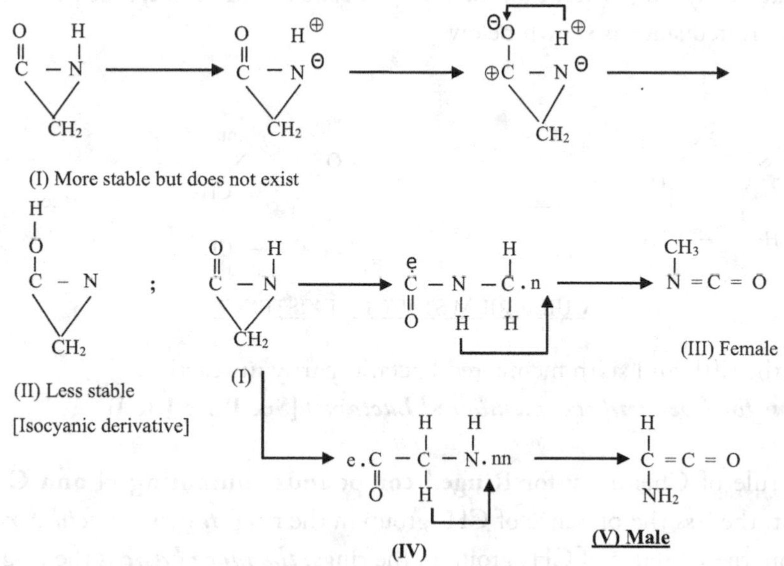

(I) More stable but does not exist

(II) Less stable
[Isocyanic derivative]

exists, for which (V) and not (III) is obtained and (II) cannot exist, since it is less stable than (I). [See Rule 1082]
(Laws of Creations for Lactams- Non Existence of 3-membered ring)

Rule 1088: This rule of Chemistry for **Lactams,** states that, just as α - lactone known to exist is less strained than three-membered lactam ring, the following is valid in general.

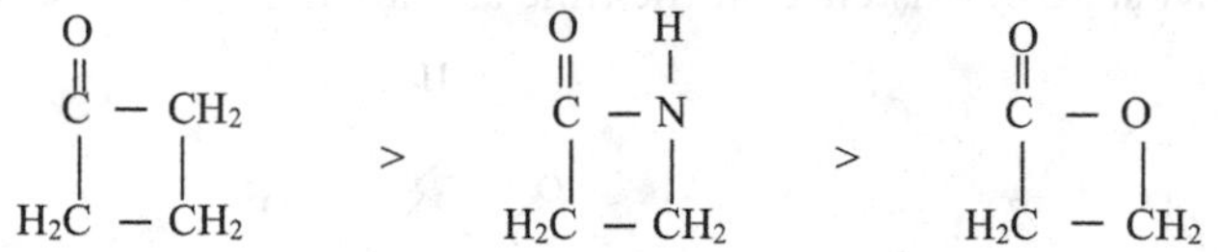

<u>Order of strain energy</u>   (a)

Carbon - Carbon x - rings	>	Carbon - nitrogen x - rings	>	Carbon - Oxygen x - rings

(where x is the size of the ring with CO in it)
<u>Order of strain energy</u>   (b)

(Laws of Creations for Lactams)

257

Rule 1089: This rule of Chemistry for **Five- and six-membered lactams,** states that, five-and six-membered lactam rings do not undergo the equilibrium type of molecular rearrangement (i.e., they cannot exist in Activated/Equilibrium state of existence) since the lactams are more stable than their isocyanic derivatives as shown below-

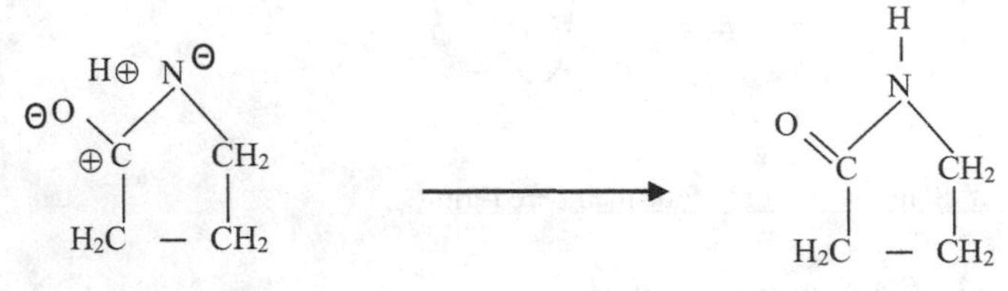

Isocyanic derivative

(Unstable existence/male)

More stable/male

ACTIVATED/EQUILIBRIUM STATE OF EXISTENCE-not possible.   (a)

noting however that the hydrogen atom on the nitrogen center is still always loosely bonded radically or ionically in its unactivated state as shown below-

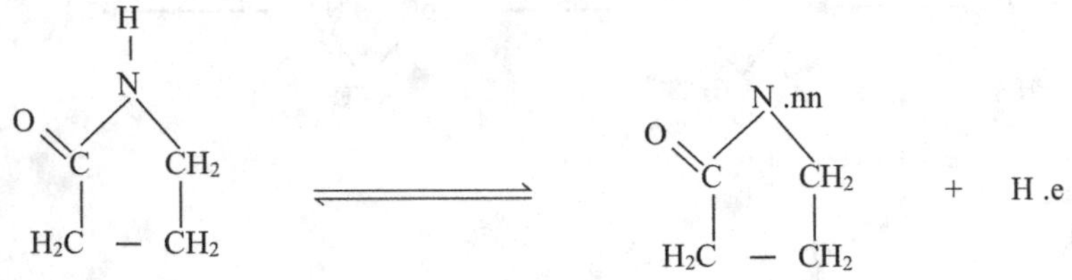

EQUILIBRIUM STATE OF EXISTENCE   (b)

that which makes the fifth and sixth membered Lactams partly unreactive.
(Laws of Creations for five- and six- membered Lactams) **[See Rule 1084]**

Rule 1090: This rule of Chemistry for **Ringed compounds containing N and C = O adjacently located,** states that, the less the presence of CH_2 group in the ring, *the more acidic or less basic* is the ring, while the more the presence of CH_2 group in the rings, *the more basic* is the ring.
(Laws of Creations for Lactams) **[See Rule 1084]**

Rule 1091: This rule of Chemistry for **Cyanuric acid a six-membered ring with three electrophilic centers adjacently located,** states that, this compound cannot be used as a monomer, because it readily undergoes Equilibrium Enolization/molecular rearrange-ment phenomena, since it largely exists in ACTIVATED/EQUILIBRIUM state of existence all the time as shown below-

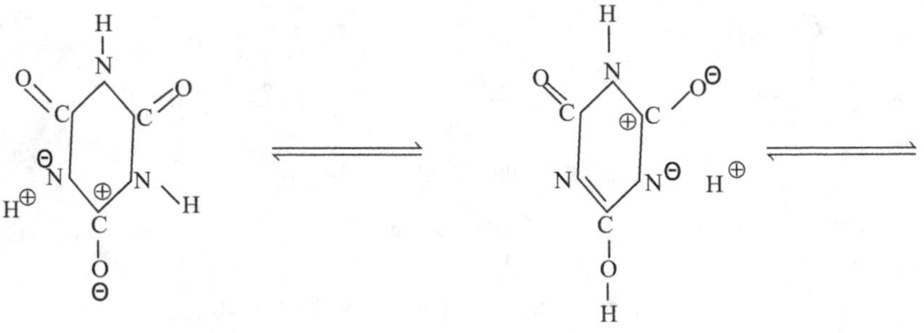

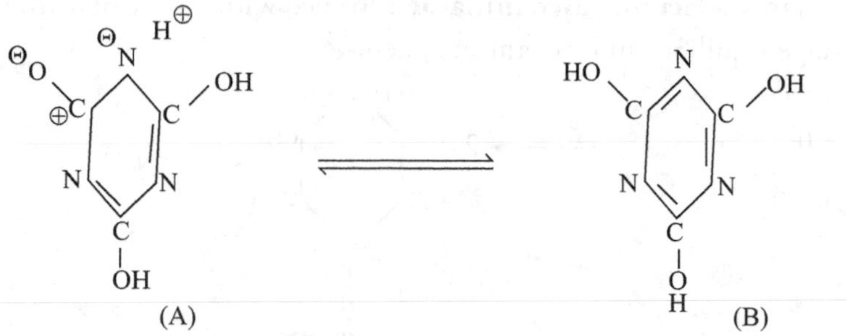

Equilibrium Enolization/molecular rearrangement phenomena of sixth type for cyanuric acid

(Laws of Creations for Cyanuric acid) [See Rule 1086]

Rule 1092: This rule of Chemistry for **Polymerizable lactams,** states that, that the N - H bond is weakly radically bonded in character, is shown *by the reaction of pyrrolidones for example with acetylene to give N - vinylpyrrolidone* which can be positively or electro-free-radically polymerized to polyvinylpyrrolidone.

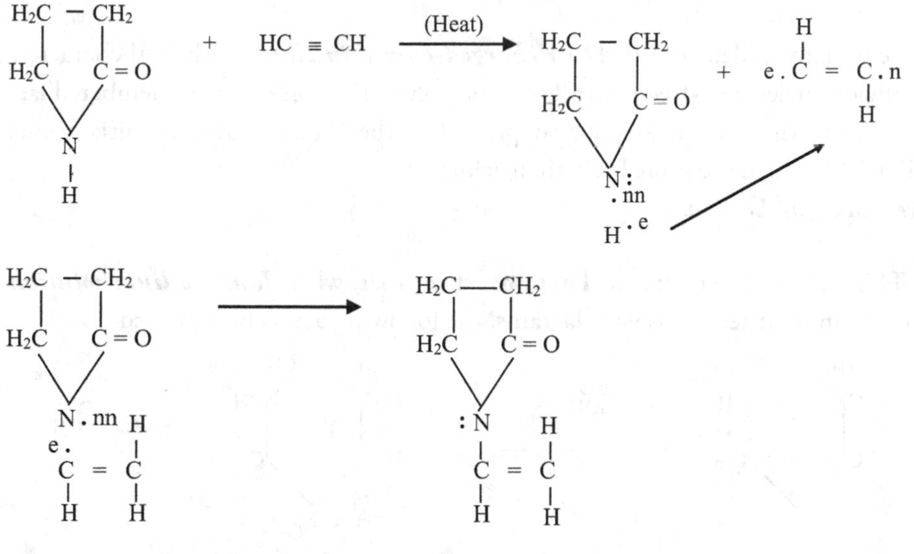

N - vinyl pyrrolidone (Nucleophile)

a one stage Equilibrium mechanism system, noting that the N-vinyl pyrrolidone is an electrophile from which the following is valid.

Order of Nucleophilicity

(Laws of Creations for Lactams)

Rule 1093: This rule Chemistry for **Lactams,** states that, that the N - H bond is not weakly free-radically bonded for larger sized lactams is illustrated by the fact that when *Alkali metals alone* are used to initiate the polymerization of some lactams, **a lactam based initiator** is formed with release of hydrogen molecule as shown below in a single stage Equilibrium mechanism system-

(I) Stabilized

(II)

(III) AN INITIATOR

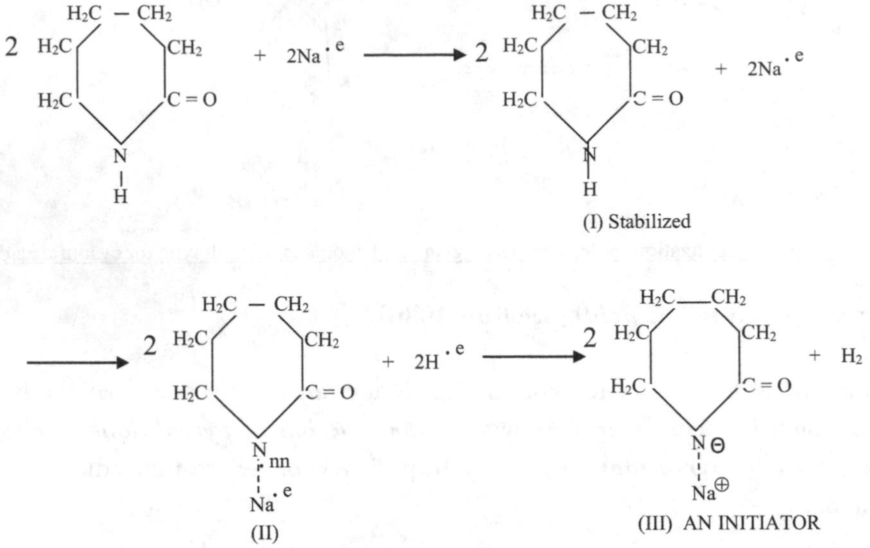

noting that the initiator called *IONICALLY charged-paired initiator* with dual character, can only be used anionically or nucleo-non-free-radically for the polymerization of seven-membered lactams (whose equilibrium state of existence can now be suppressed by the Na atom and its initiator), lactams being ELECTROPHILES which are more basic than acidic.
(Laws of Creations for Lactams)

Rule 1094: This rule of Chemistry for **Lactams,** states that, when *Ionic sodium solution* are used in place of Na metal in their reactions with lactams, the following are to be expected-

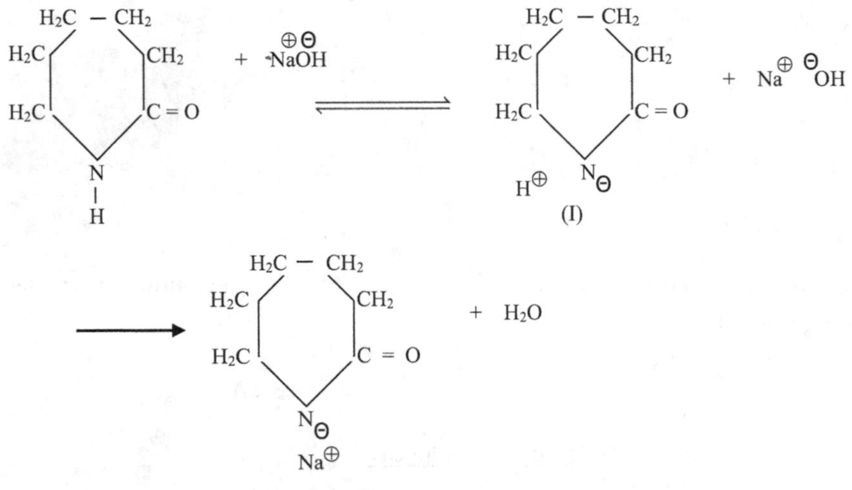

(I)

THE INITIATOR

for which the single stage Equilibrium mechanism system that in which the sodium hydroxide was suppressed by the lactam with release of heat, produces an *IONICALLY charged–paired initiator which can also be radically paired in character,* dual in character, but can only be used anionically or nucleo-non-free-radically for polymerization of lactams whose Equilibrium state of existence can now be suppressed, lactams being Electrophiles, more basic than acidic.
(Laws of Creations for Lactams)

Rule 1095: This rule of Chemistry of the **First member of Cyclic amines (aziridines),** states that, aziridines are weak bases inorganically, for which the following is valid organically-

$$NR_3 \text{ (pKa = 10-11)} \quad > \quad \text{Aziridines (pKa = 8-9)} \quad > \quad \text{Pyridine (pKa = 5.2)}$$

<u>Order of Basicity</u>

(Laws of Creations for Cyclic amines)

Rule 1096: This law of Chemistry for **The first four members of the family of Cyclic amines,** states that, of the four, the three-membered aziridine is the most unique, because the four, five- and six-membered and higher ones cannot easily be provided with the MRSE through their functional centers in view of their strong basic character.

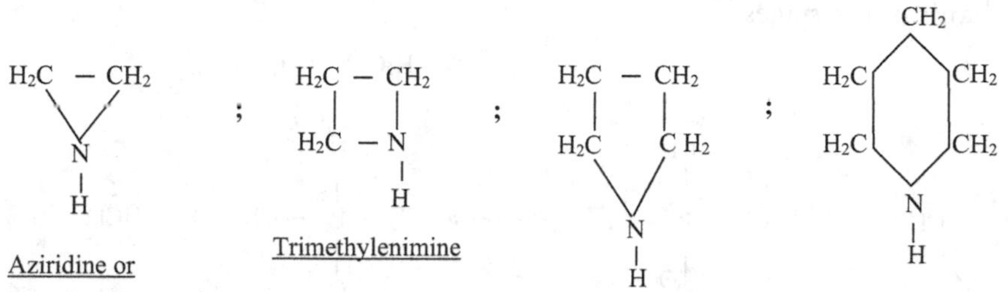

<u>Aziridine or Ethylenimine or Dimethylenimine</u> ; <u>Trimethylenimine</u> ; <u>Tetramethylenimine or Pyrrolidine</u> ; <u>Pentamethylenimine or Piperidine</u>

(Laws of Creations for Cyclic imine)

Rule 1097: This rule of Chemistry for **Ethylenimine,** states that, this compound is known to react with *Aqueous sulfur dioxide* to give taurine (2–aminoethanesulfonic acid), a stable compound as shown below,

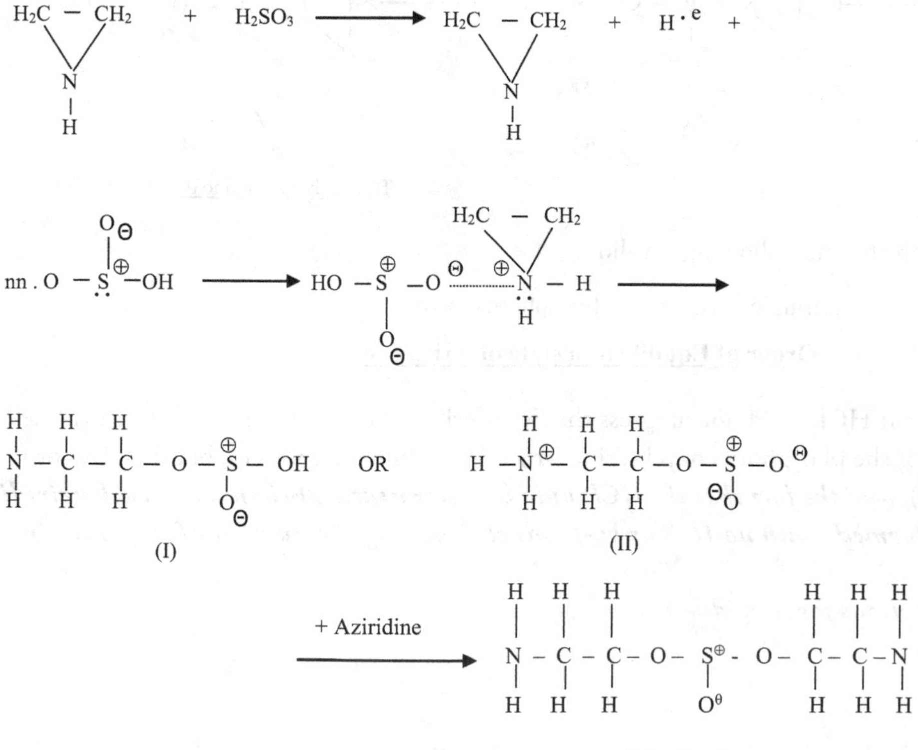

(I)   OR   (II)

+ Aziridine

(III) <u>Taurine</u>

261

wherein (I) is first formed when the Equilibrium state of existence of the imine is suppressed by the acid and if (I) is strong enough to exist in equilibrium state of existence suppressing that of imine, (III) is formed as the Taurine instead of (II), all via Equilibrium mechanism, noting that the following is valid.

$$\text{Sulphurous acid} \quad > \quad \text{Aziridine}$$

Order of Equilibrium state of Existence

(Laws of Creations for Aziridine)

Rule 1098: This rule of Chemistry for **Aziridine (Ethylenimine),** states that, when aziridine reacts with **Phosphorus oxychloride,** triaziridyl-phosphine oxide (APO), $(C_2H_4N)_3PO$ is formed as shown below via Equilibrium mechanism in 3 stages,

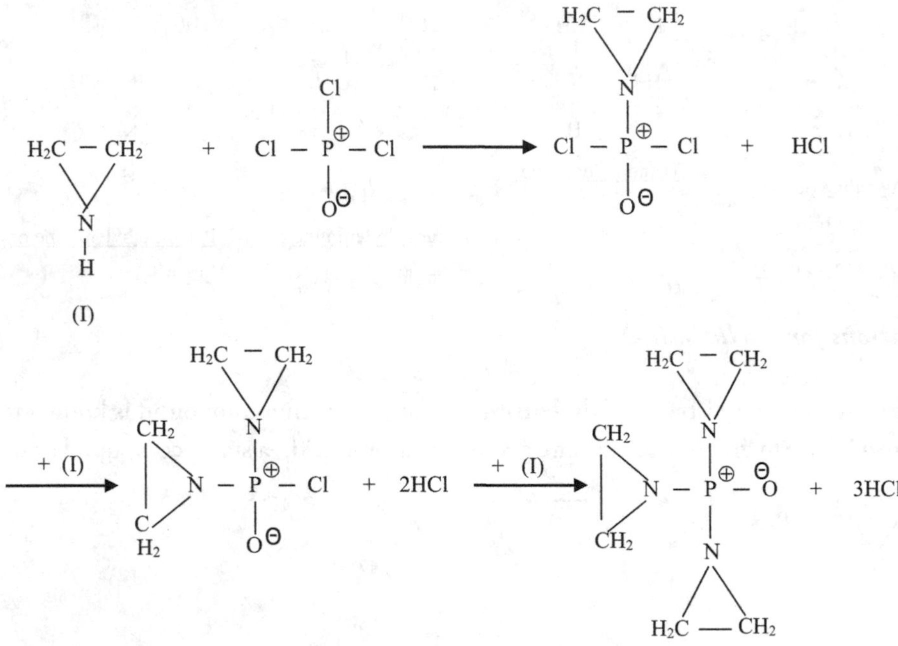

Triaziridylphosphine oxide

clear indication that the following is valid-

$$\text{Aziridine} \quad > \quad \text{Phosphorus oxychloride}$$

Order of Equilibrium state of Existence

and the fact that HCl could not suppress the Equilibrium state of existence of the aziridine because of the presence of the phosphorus oxychloride which is insoluble in the acid, but dissolves or miscibilizes in it (the acid), *and the fact that the HCl could be electrostatically bonded to each aziridine group after being formed, with no HCl as by-product, bringing the number of stages to six instead of three.*

(Laws of Creations for Aziridine)

Rule 1099: This rule of Chemistry for **Lactams,** states that, when *Sodium amide* is used in place of Na metal in their reactions with lactams, the following are to be expected-

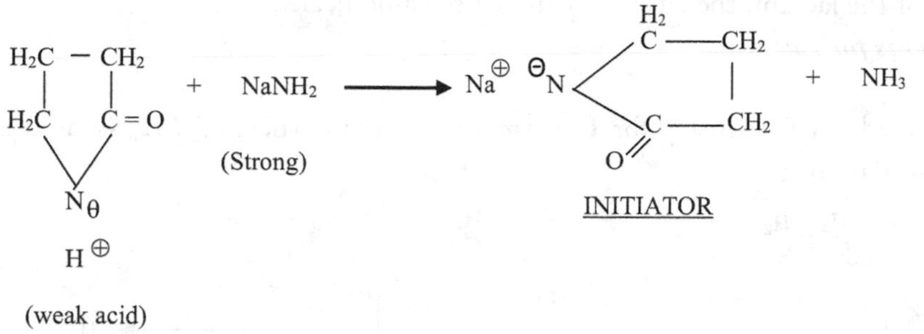

for which the single stage Equilibrium mechanism system, produces an ***IONICALLY charged–paired initiator,*** dual in character, but can only be used anionically or nucleo-non-free-radically for polymerization of lactams whose Equilibrium state of existence can be suppressed, lactams being Electrophiles, more basic than acidic in character.
(Laws of Creations for Lactams)

Rule 1100: This rule of Chemistry for **Lactams,** states that, when *Metal hydrides* are made to react with lactams such as that of the seven-membered ring, the following is obtained-

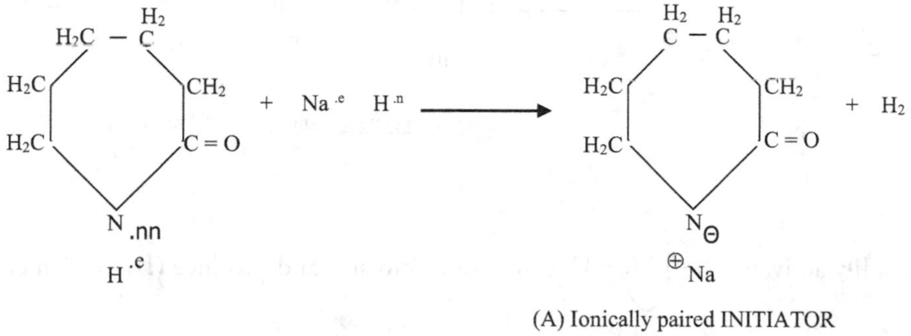

(A) Ionically paired INITIATOR

for which the single stage Equilibrium mechanism system, produces an **IONICALLY charged–paired initiator,** dual in character, but can only be used anionically or nucleo-non-free-radically for polymerization of lactams, lactams being Electrophiles, more basic than acidic in character.
(Laws of Creations for Lactams)

Rule 1101: This rule of Chemistry for **Lactams,** states that, when *Organometallic initiators* are used for their polymerizations, the followings are obtained using LiC_4H_9-

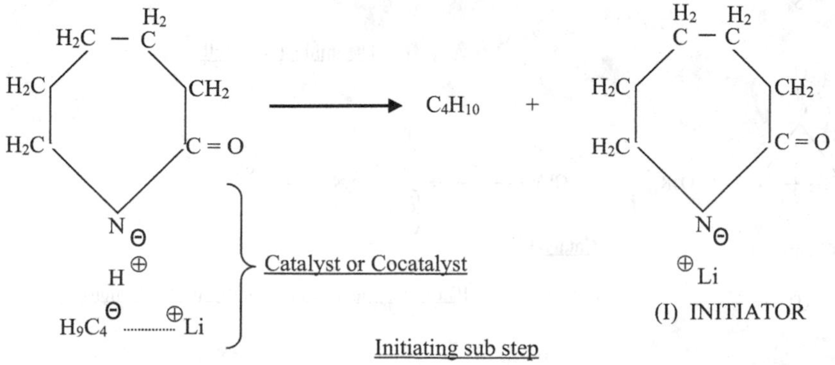

Catalyst or Cocatalyst

Initiating sub step

(I) INITIATOR

263

for which the ***Covalently charged-paired initiator*** (LiC_4H_9) has been converted to an ***Anionically charged-paired initiator,*** both of which are differently dual in character; noting that while the former cannot be used for the lactam, the latter (I) can be used anionically.
(Laws of Creations for Lactams)

Rule 1102: This rule of Chemistry for **Lactams,** states that, when LiC_4H_9 is used as initiator, the following is the initiation step-.

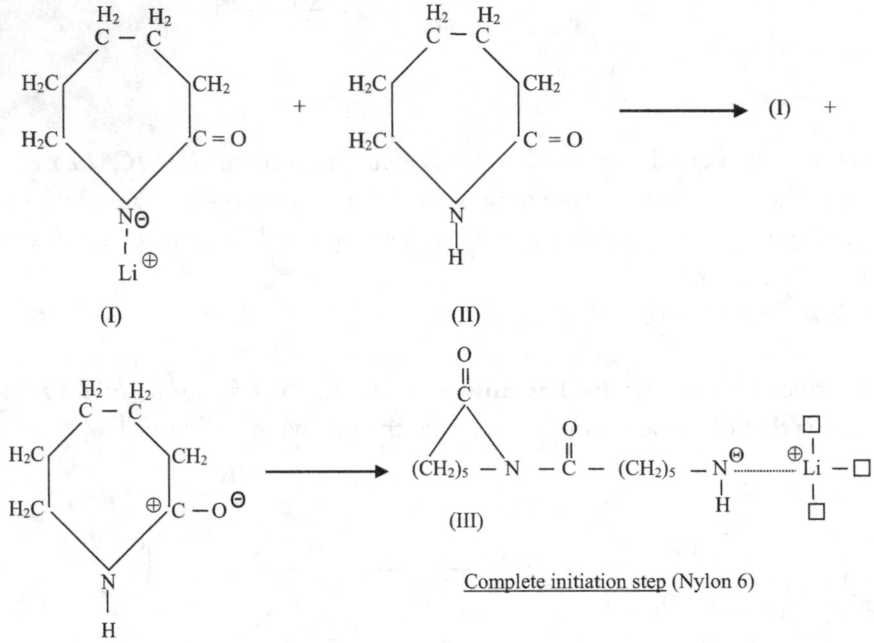

(I)   (II)

Complete initiation step (Nylon 6)

for which (I) readily activates the $C = O$ center of (II) to add and produce (III) which completes the initiations step.
(Laws of Creations for Lactams)

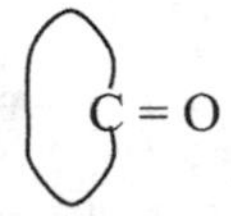

Rule 1103: This rule of Chemistry for **Lactams,** states that, when ***Acylating agents*** are used, the followings are obtained using ***Acid chloride*** as shown below-

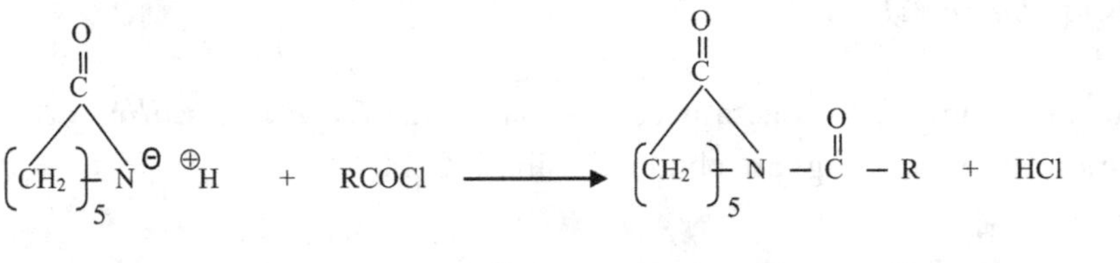

(I) The imide (N - acyllactam)   (a)

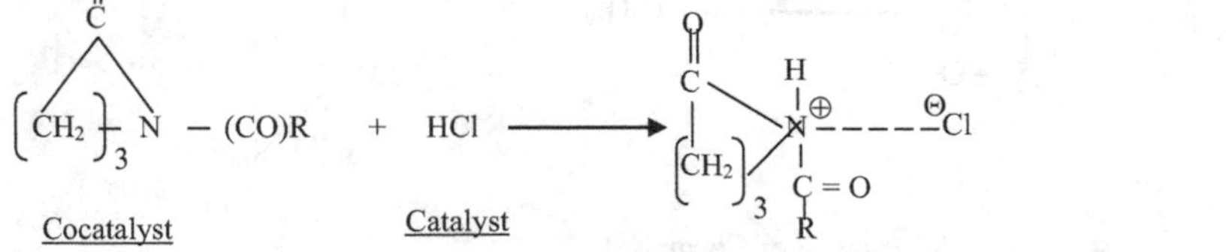

Cocatalyst   Catalyst

(I)   (II) Electrostatically anionically-paired Initiator   (b)

264

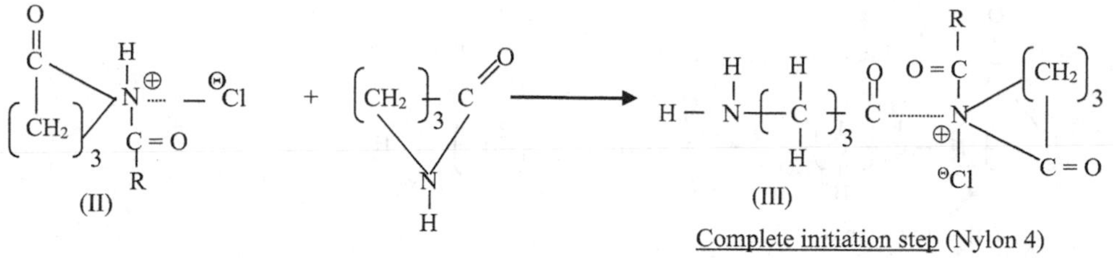

Complete initiation step (Nylon 4) (c)

with formation of (III) marking the end of Initiation step, noting that the route is electro-free-radical, since the initiator is the component in Equilibrium state of existence all the time making it impossible to use it anionically.
(Laws of Creations for Lactams)

Rule 1104: This rule of Chemistry for **Lactams,** states that, when ***Acylating agents*** are used, the followings are obtained using ***Anhydrides*** as shown below-

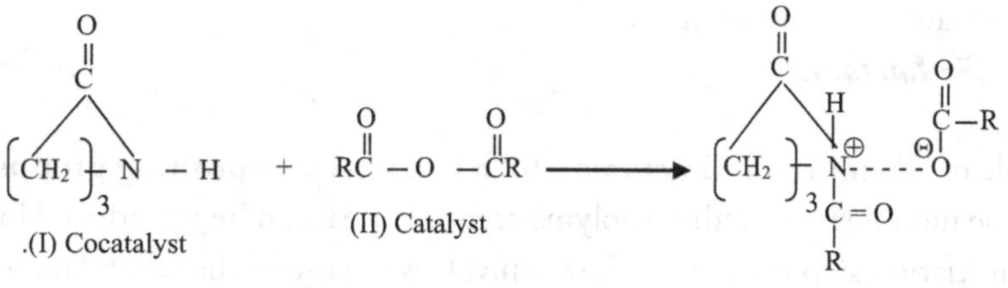

(Initiator) (a)

noting that the Initiator preparation step above is a two stage Equilibrium mechanism system (An induction period), in which in the first stage an imide- N-acyllactam is first prepared and in the second stage the acetic acid produced is then used to give the Initiator above followed by the Initiation step as shown below-

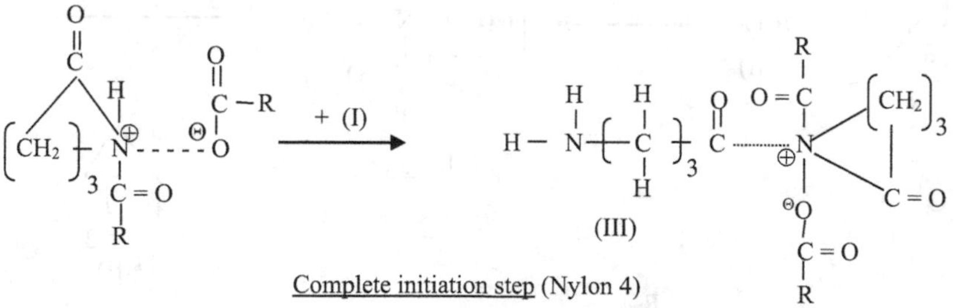

Complete initiation step (Nylon 4) (b)

noting that the route is electro-free-radical, since the initiator is the component in Equilibrium state of existence all the time making it impossible to use it anionically.
(Laws of Creations for Lactams)

Rule 1105: This rule of Chemistry for **Lactams,** states that, when ***Acylating agents*** are used, the followings are obtained using ***Isocyanates*** as shown below-

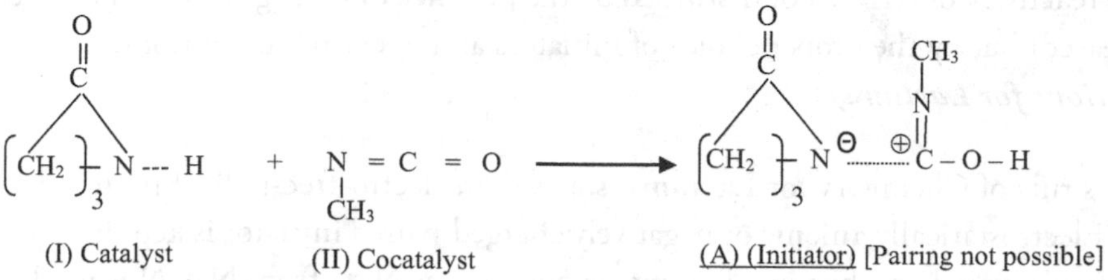

(A) (Initiator) [Pairing not possible]

OR i)

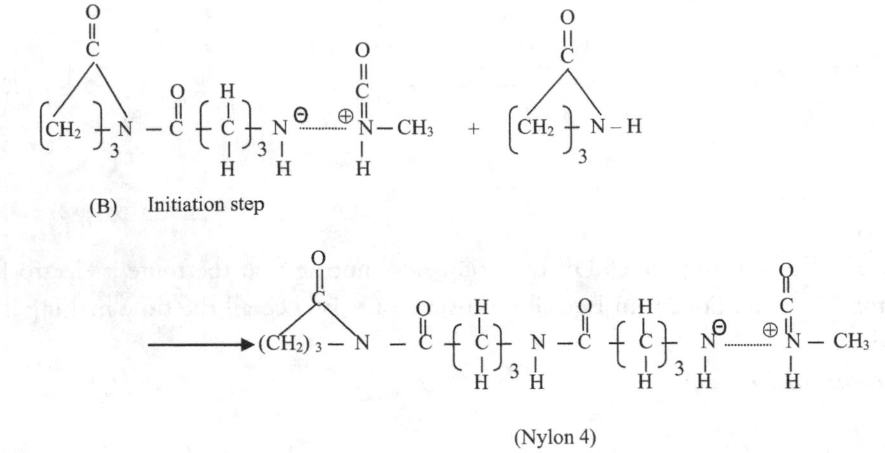

(B)   Initiation step

(Nylon 4)   ii)

for which it is (B) that is favored being an acylating agent with the mechanism of addition beng anionic via Combination mechanism as shown above.
(Laws of Creations for Lactams)

Rule 1106: This rule of Chemistry for **Lactams which carry radical-pushing groups on the nitrogen center,** states that, the monomer can still be polymerized if MRSE can be introduced into the ring and to do this, part of the initiation step must be such that involves the use of the secondary acidic cyclic amide as catalyst and a cocatalyst as follows.

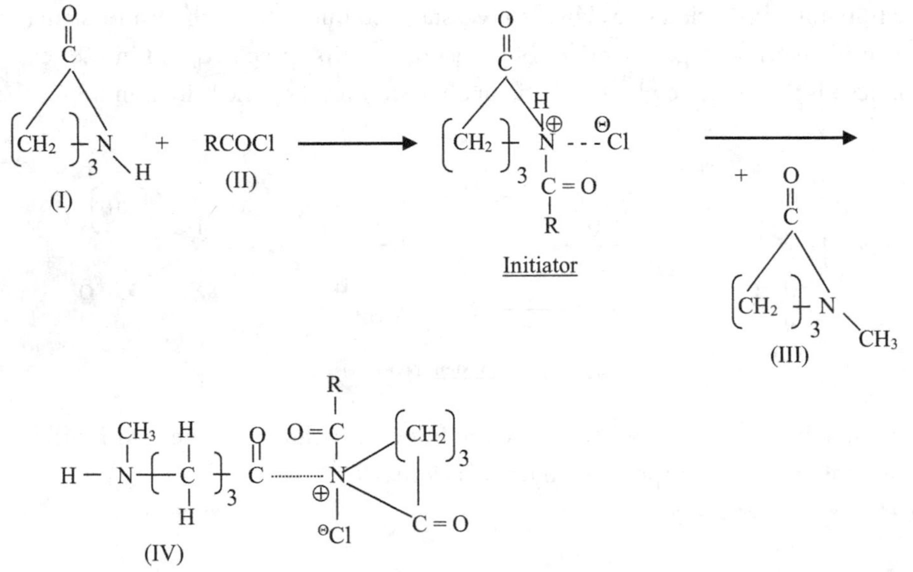

(I)   (II)   Initiator   (III)

(IV)

Complete initiation step (Nylon 4)

noting that the reactivity of (III) is not disturbed by the presence of CH_3 group on the N center (though basicity is increased) but by the proper choice of initiators and operating conditions.
(Laws of Creations for Lactams)

Rule 1107: This rule of Chemistry for **Lactams,** states that, electro-free-radical route can only be used, either when an Electrostatically anionic or negatively charged-paired initiator is kept in Equilibrium state of existence when prepared or when just free-media one such as Na•e from NaCN is used.
(Laws of Creations for Lactams)

266

Rule 1108: This rule of Chemistry for **Lactams and Lactones,** states that, in general lactams are less nucleophilic than lactones-.

Lactones   >   Lactams

Order of Nucleophilicity

and lactams are more electrophilic via the C = O center than lactones via the C = O center, in view of the presence of the hydrogen atom which is loosely bonded to the nitrogen center and on the other hand, *lactams are far more strained than lactones* and unlike lactones, lactams are also basic in character.

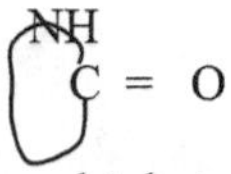

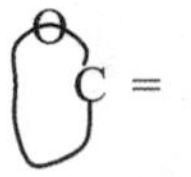

(Laws of Creations for Lactams and Lactones)

Rule 1109: This rule of Chemistry for **Lactams,** states that, between members of the family, the followings are valid-

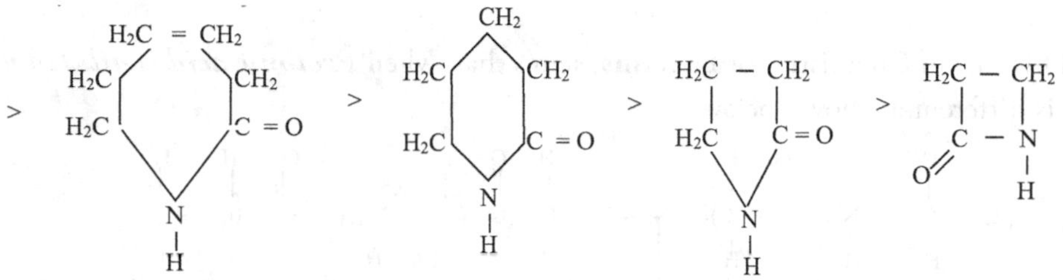

Order of Nucleophilicity

for which it not surprising to note that it is the larger membered rings that show stronger reactivity positively or electro-free-radically than the five- and six-membered rings.

(Laws of Creations for Lactams)

Rule 1110: This rule of Chemistry for **Lactams,** states that, when *Moderately strong inorganic acids such as HCl acids* are used in the absence of heat, the MRSE may never be provided for such less strained rings for which higher temperature polymerizations of these monomers are involved as shown below-

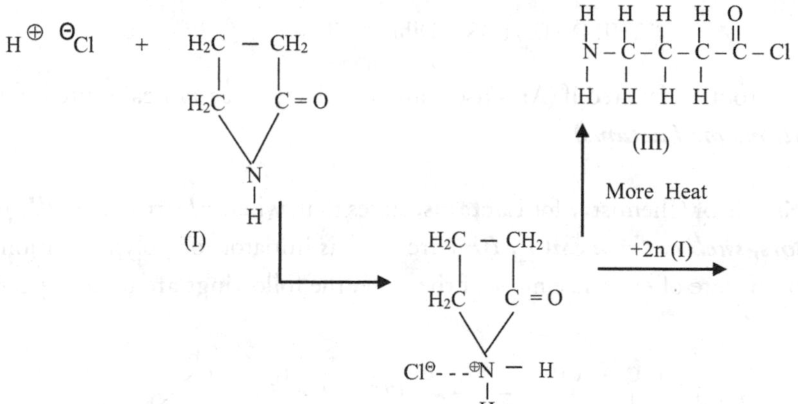

Equilibrium Mechanism

Nylon 4

Electro-free-radical Route
Backward Addition
Equilibrium Mechanism

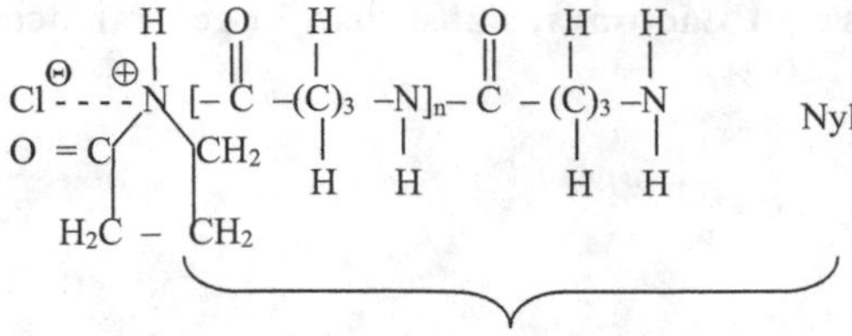

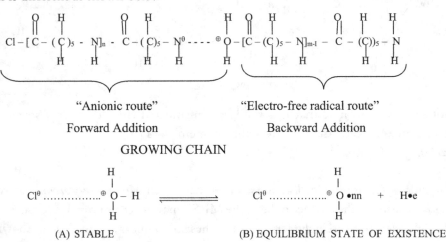

for if the ring is opened in (II), Nylon 4 cannot be obtained and If the two $H^{\cdot e}$ on the acid and monomer are of same capacity, then both will remain in equilibrium state of existence with no polymerization and If the $H^{\cdot e}$ of the amide is greater in capacity than the $H^{\cdot e}$ of acid, then there is no polymerization; noting that (II) when formed via Equilibrium mechanism, must be in Equilibrium state of existence for electro-free-radical polymerization to be favored.
(Laws of Creations for Lactams)

Rule 1111: This rule of Chemistry for **Lactams,** states that, when ***Protonic acid is diluted with water***, the situation is different as shown below-

"Anionic route" "Electro-free radical route"
Forward Addition Backward Addition

GROWING CHAIN

(A) STABLE (B) EQUILIBRIUM STATE OF EXISTENCE

THE INITIATOR – Dilute HCl

for which anionic route is favored if (A) is used, and the electro-free-radical route is favored if (B) is used.
(Laws of Creations for Lactams)

Rule 1112: This rule of Chemistry for **Lactams,** states that, when ***Electrostatically positively charged-paired initiators, such as from $SnCl_4$, BF_3***, are used as initiator for polymerization of Lactams which are in Equilibrium state of existence most of the time, the followings are to be expected-

(I) (II) (a)

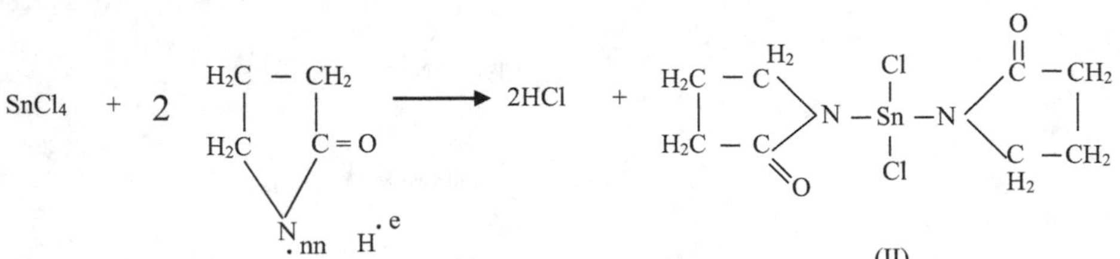

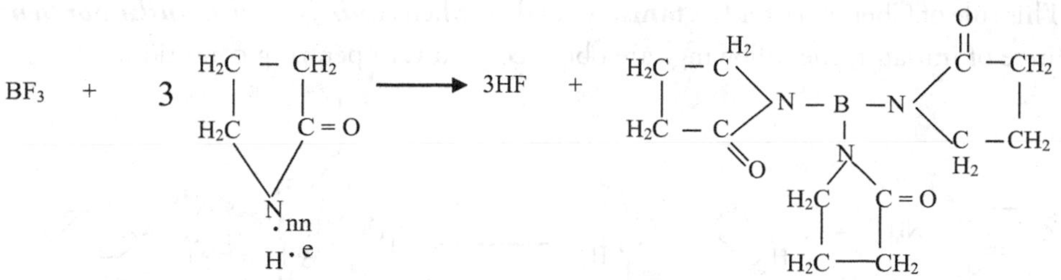

$$BF_3 \ + \ 3 \ \text{(I)} \longrightarrow 3HF \ + \ \text{(II)}$$

(b)

noting that the favored existence of the (II) where full replacement of Cl and F is made possible due to the presence of $SnCl_4$ and BF_3 in the system making it possible to keep HCl or HF in Equilibrium state of existence to form electrostatic bonds with the lactam groups (Not shown above), for which their use as initiators will be sterically hindered, otherwise with the formation of the acids in the absence of $SnCl_4$ or BF_3, the followings are possible-

i) Direct use with the lactam and the acid if still present to produce an initiator or
ii) No polymerization since an initiator cannot be obtained between HF and BF_3 or HCl and $SnCl_4$.

(Laws of Creations for Lactams)

Rule 1113: This rule of Chemistry for **Lactams,** states that, with *ROR/BF₃ combination as initiator,* some of the reactions shown below are involved-

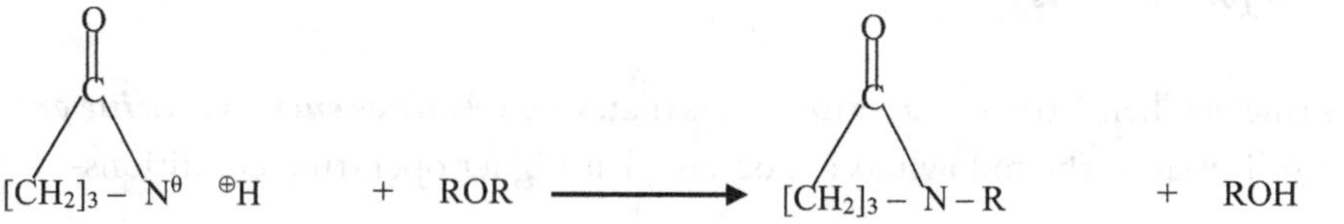

$$[CH_2]_3 - N^{\ominus} \ ^{\oplus}H \ + \ ROR \longrightarrow [CH_2]_3 - N - R \ + \ ROH$$

(I) Imide (N – alkyllactam)   (a)

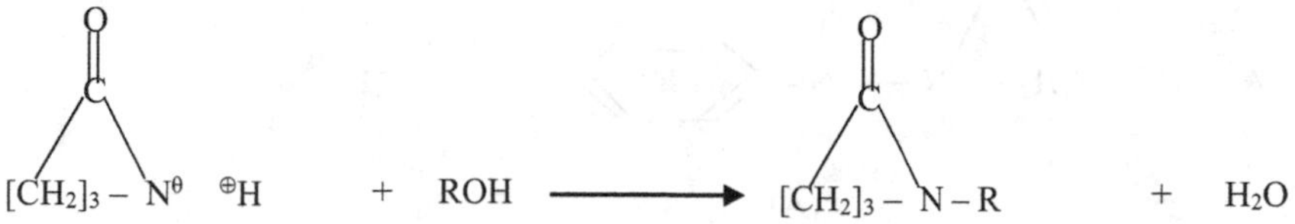

$$[CH_2]_3 - N^{\ominus} \ ^{\oplus}H \ + \ ROH \longrightarrow [CH_2]_3 - N - R \ + \ H_2O$$

(I) Imide (N – alkyllactam)   (b)

$$[CH_2]_3 - N - R \ + \ ROH \longrightarrow [CH_2]_3 - N^{\oplus} \cdots \ ^{\ominus}OR$$

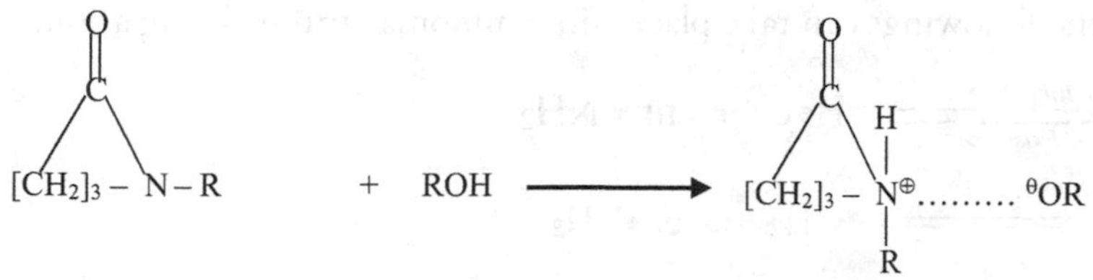

(I) INITIATOR   (c)

for which the presence of the last two equations will depend on the molar ratios of the components involved; if there is more of the monomer, then (I) of Equation (b) will be obtained and if not, (I) of Equation (c) will be obtained.

(Laws of Creations for Lactams)

Rule 1114: This rule of Chemistry for **Lactams,** states that, when *Amines such as aniline or benzylamine* is used as source of initiator, the followings are obtained at lower operating conditions-

(III) $\ominus$ly Charged-paired initiator
(A)

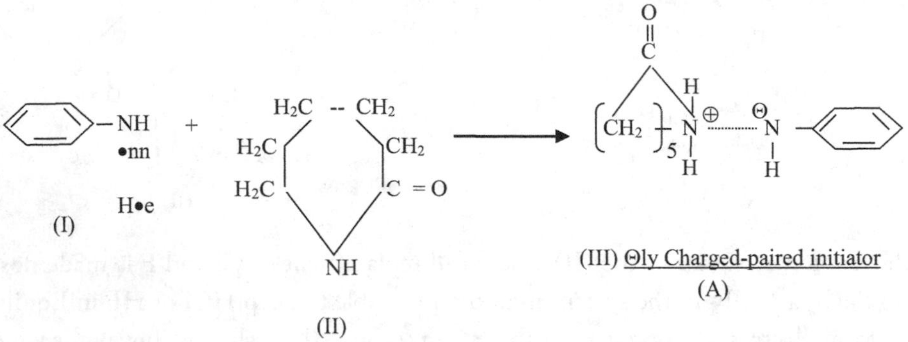

i)

and when (A) is involved, the ring opens as shown below and the routes can either be anionic or electro-free-radical depending on the state of existence of (A).

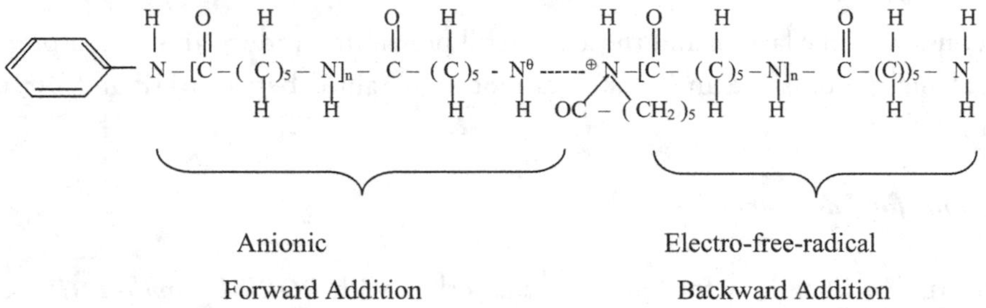

Anionic

Forward Addition

Electro-free-radical

Backward Addition

ii)

Laws of Creations for Lactams

Rule 1115: This rule of Chemistry for **Lactams,** states that, when *Amines such as aniline or benzylamine* is used as source of initiator, the followings are obtained at higher operating conditions-

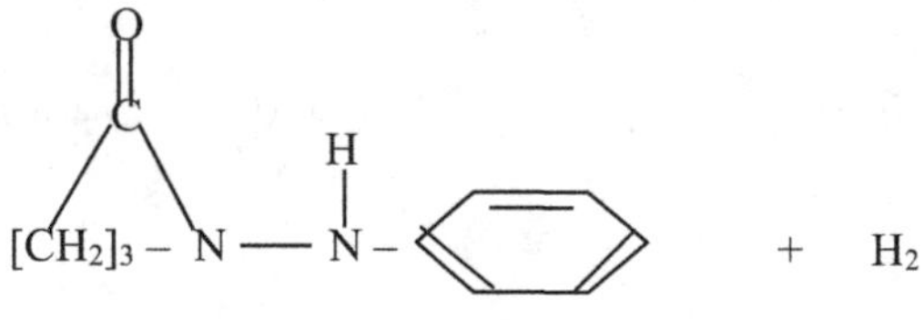

(A)

i)

for which H_2 is indeed released in the first stage via Equilibrium mechanism, noting that the reaction above is favored, since the followings can take place with ammonia, aniline being a primary amine,

Stage 1: $NH_3 \xrightleftharpoons[\text{of Existence}]{\text{Equilibrium Sate}} H \bullet e + nn \bullet NH_2$

$H \bullet e + NH_3 \rightleftharpoons H_2 + en \bullet NH_2$

$en \bullet NH_2 + nn \bullet NH_2 \longrightarrow H_2N - NH_2$

ii)

with (A) now playing the same role as aniline when it was used under mild operating conditions to produce the Initiator with the chain carrier being (A) anionically if in Stable state of existence or H electro-free-radically if in Equilibrium state of existence.

(Laws of Creations for Lactams).

Rule 1116: This rule of Chemistry for **Lactams,** states that, when *Secondary and Tertiary* types of aromatic amines are used as source of initiator, the followings are obtained if H_2 is to be formed-

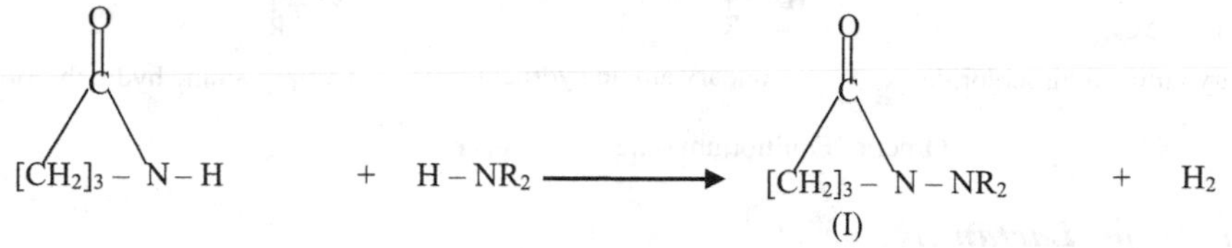

$$[\overset{\overset{\displaystyle O}{\parallel}}{\underset{\diagdown}{C}}CH_2]_3 - N-H \quad + \quad H-NR_2 \longrightarrow [\overset{\overset{\displaystyle O}{\parallel}}{\underset{\diagdown}{C}}CH_2]_3 - N-NR_2 \quad + \quad H_2$$

(I)

(Where R is Phenyl or Benzyl group)

for which (I) which is very stable can only be used as a counter charged center, if and only if the monomer can exist in Equilibrium state of existence or another secondary aromatic amine is present in the system; otherwise secondary aromatic amine cannot be used if H_2 is to be released, but can only be used in a similar way shown for the primary aromatic amine when H_2 is not released, while with tertiary amine (with no hydrogen atom on the nitrogen center), no initiator can be produced if the amide is still suppressed.

(Laws of Creations for Lactams)

Rule 1117: This rule of Chemistry for **Lactams,** states that, when *Amine hydrochloride* is used as source of initiator, there is no induction period since the initiator is used as it is, with a first order dependence of rate of polymerization (Rp) on the amine hydrochloride concentration as shown below-

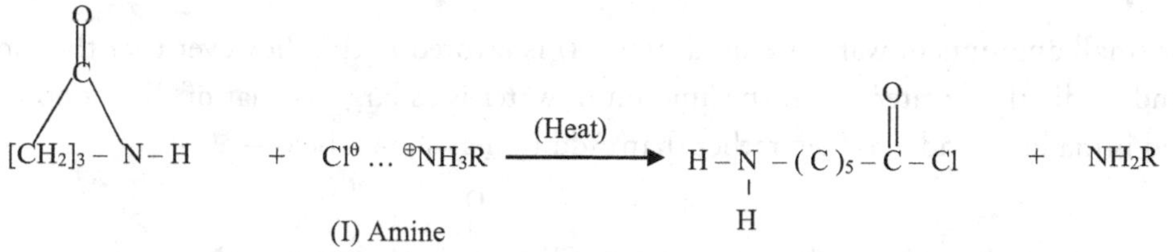

$$[\overset{\overset{\displaystyle O}{\parallel}}{\underset{\diagdown}{C}}CH_2]_3 - N-H \quad + \quad Cl^\ominus \ldots {}^\oplus NH_3R \xrightarrow{\text{(Heat)}} H-N-(C)_5-\overset{\overset{\displaystyle O}{\parallel}}{C}-Cl \quad + \quad NH_2R$$

(I) Amine

$\qquad\qquad\qquad\qquad\qquad\qquad\qquad\qquad\qquad\qquad\qquad$ i)

for which when mild operating condition re involved, one should expect the followings-

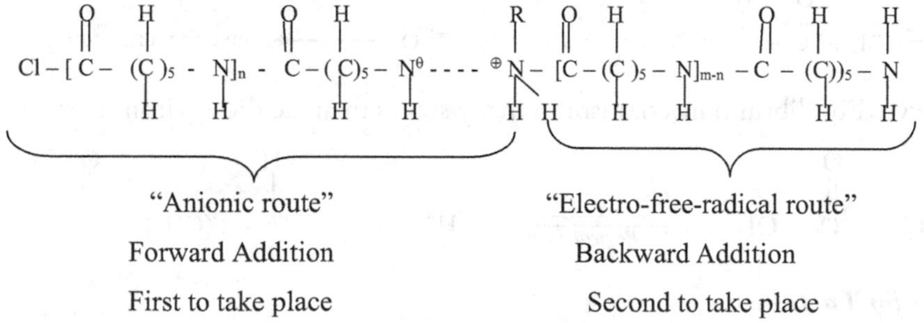

$$Cl-[\overset{\overset{\displaystyle O}{\parallel}}{C}-(\overset{\overset{\displaystyle H}{\mid}}{\underset{H}{C}})_5-\overset{\overset{\displaystyle H}{\mid}}{N}]_n - \overset{\overset{\displaystyle O}{\parallel}}{C}-(\overset{\overset{\displaystyle H}{\mid}}{C})_5-N^\ominus \cdots {}^\oplus \overset{\overset{\displaystyle R}{\mid}}{N}-[\overset{\overset{\displaystyle O}{\parallel}}{C}-(\overset{\overset{\displaystyle H}{\mid}}{C})_5-\overset{\overset{\displaystyle H}{\mid}}{N}]_{m-n} - \overset{\overset{\displaystyle O}{\parallel}}{C}-(\overset{\overset{\displaystyle H}{\mid}}{C})_5-\overset{\overset{\displaystyle H}{\mid}}{N}$$

"Anionic route"   "Electro-free-radical route"

Forward Addition   Backward Addition

First to take place   Second to take place   ii)

with backward addition being the only means of polymerization, since this initiator is most of the time in Equilibrium state of existence.

(Laws of Creations for Lactams)

Rule 1118: This rule of Chemistry for **Lactams,** states that, when *Primary, Secondary and Tertiary Amine hydrochloride* are used as initiators for lactams, they will all favor being used electro-free-radically in decreasing order from primary to secondary and to tertiary while anionically the reverse is the case, since the initiator must be in Stable state of existence.

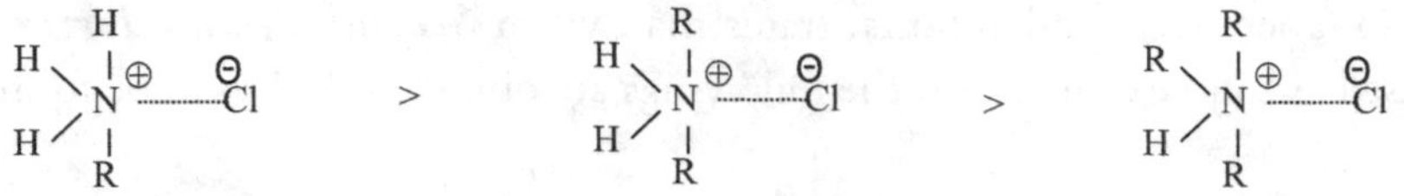

$$H \underset{H}{\overset{H}{\underset{\big|}{\overset{\big|}{N}}}} \overset{\oplus}{} \overset{\ominus}{} Cl \quad > \quad H \underset{R}{\overset{R}{\underset{\big|}{\overset{\big|}{N}}}} \overset{\oplus}{} \overset{\ominus}{} Cl \quad > \quad R \underset{R}{\overset{R}{\underset{\big|}{\overset{\big|}{N}}}} \overset{\oplus}{} \overset{\ominus}{} Cl$$

Primary amine hydrochloride   Secondary amine hydrochloride   Tertiary amine hydrochloride

Order of Equilibrium state of Existence

(Laws of Creations for Lactams)

Rule 1119: This rule of Chemistry for **Lactams,** states that, when *Water* is used as source of initiator for lactams, the followings are to be expected-

$$H^{\oplus} \quad {}^{\ominus}OH \quad + \quad \left[CH_2 \right]_5 N \text{--} H \quad \xrightarrow{\text{Heat}} \quad H - \overset{\oplus}{\underset{H}{N}} \left[CH_2 \right]_5 \longrightarrow$$

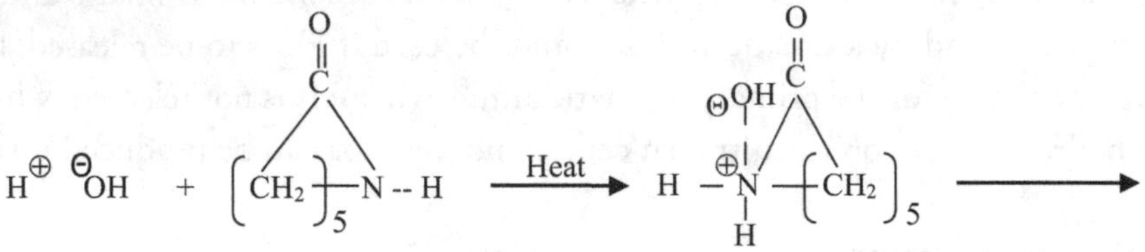

(A)   (B) H_2O/Amide Initiator

$$H_2N - (CH_3)_5 - \overset{O}{\overset{\|}{C}} - OH$$

(I) An amino acid   (a)

for if large or small amounts of water are used, then (I) is favored noting however that the monomer is more basic and acidic than water and if the amount of water is as large as that of the monomer, amino acids obtained is made to undergo <u>Step</u> rather than <u>Addition</u> as shown below-

$$H_2N \left[CH_2 \right]_5 \overset{O}{\overset{\|}{C}} - OH \quad + \quad H_2N \left[CH_2 \right]_5 \overset{O}{\overset{\|}{C}} - OH \longrightarrow$$

$$H_2N - (CH_2)_5 \overset{O}{\overset{\|}{C}} - \overset{H}{\underset{\|}{N}} - (CH_2)_5 - \overset{O}{\overset{\|}{C}} - OH \quad + \quad H_2O \longrightarrow \text{Step polymerization} \qquad (b)$$

and this takes place via Equilibrium mechanism in many stages in an acidic environment.

$$H_2N - (CH_2)_5 - \overset{O}{\overset{\|}{C}} - OH \quad \underset{Environment}{\overset{Acidic}{\rightleftarrows}} \quad H^{\oplus} \quad + \quad {}^{\ominus}\overset{H}{\underset{\|}{N}} - (CH_2)_5 - \overset{O}{\overset{\|}{C}} - OH \quad (c)$$

(Laws of Creations for Lactams)

Rule 1120: This rule of Chemistry for Lactams, states that, when *Water* is used as source of initiator for lactams, the followings are to be expected-

$$H^{\oplus} \quad {}^{\ominus}OH \quad + \quad \left[CH_2 \right]_5 N \text{--} H \quad \xrightarrow{\text{Heat}} \quad H - \overset{\oplus}{\underset{H}{N}} \left[CH_2 \right]_5 \longrightarrow$$

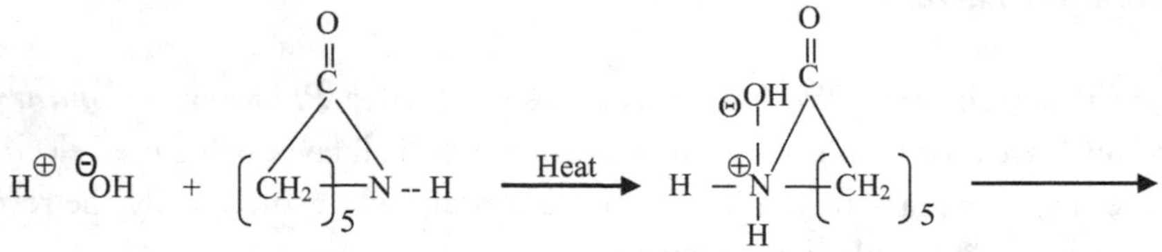

(A)   (B) H_2O/Amide Initiator

$$H_2N - (CH_3)_5 - \overset{\overset{\textstyle O}{\|}}{C} - OH$$

(I) An amino acid

(a)

for if small amounts of water are used, then (I) is favored and with the presence of large concentration of the monomer, that is, when the amino acid is in a basic environment, then the followings shown below take place-

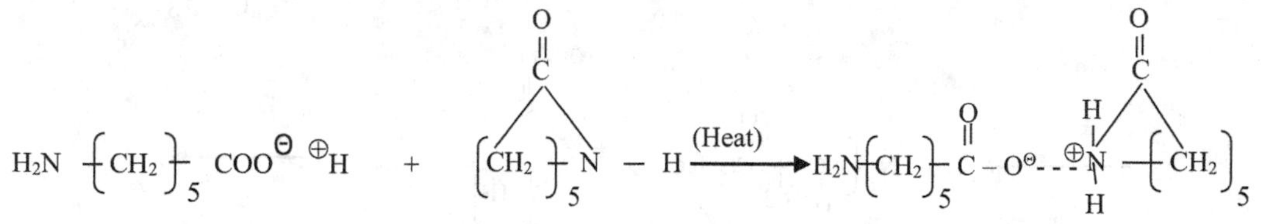

$$H_2N \left[CH_2 \right]_5 COO^{\ominus} {}^{\oplus}H \quad + \quad \left[CH_2 \right]_5 N - H \xrightarrow{\text{(Heat)}} H_2N \left[CH_2 \right]_5 \overset{\overset{\textstyle O}{\|}}{C} - O^{\ominus} - \cdot {}^{\oplus}\underset{H}{\overset{H}{N}} \left[CH_2 \right]_5$$

(II)
(Favored)

(A)

(III) Initiator

$$\xrightarrow{\quad + \ (A) \quad} \text{Polymer (Addition polymerization)} \quad \text{Anionic route}$$

(b)

$$H_2N - (CH_2)_5 - \overset{\overset{\textstyle O}{\|}}{C} - OH \underset{\textit{Environment}}{\overset{\textit{Basic}}{\rightleftharpoons}} H_2N - (CH_2)_5 - \overset{\overset{\textstyle O}{\|}}{C} - O^{\ominus} + H^{\oplus}$$

(c)

for which when (III) is in Equilibrium state of existence, only the electro-free-radical route is favored.
(Laws of Creations for Lactams)

Rule 1121: This rule of Chemistry for **Lactams,** states that, when ***Water*** is used in ***a neutral environment*** for lactams, after amino acid is formed, the followings are to be expected-

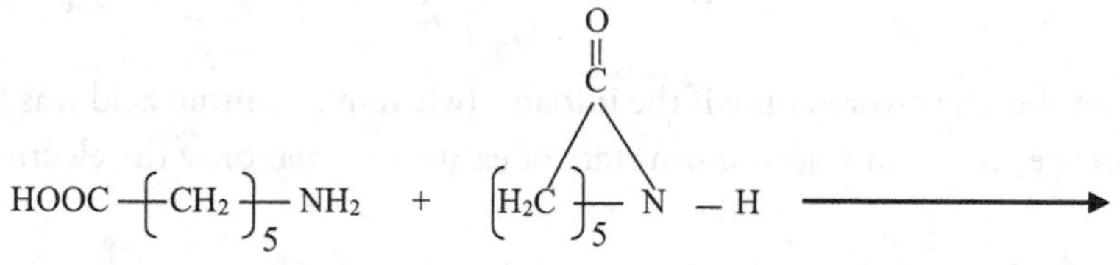

$$HOOC \left[CH_2 \right]_5 NH_2 \quad + \quad \left[H_2C \right]_5 N - H \longrightarrow$$

(I)

$$HOOC \left[CH_2 \right]_5 \underset{H}{\overset{}{N}}{}^{\ominus} - - - - - - - - - {}^{\oplus}\underset{H}{\overset{H}{N}} \left[H_2C \right]_5$$

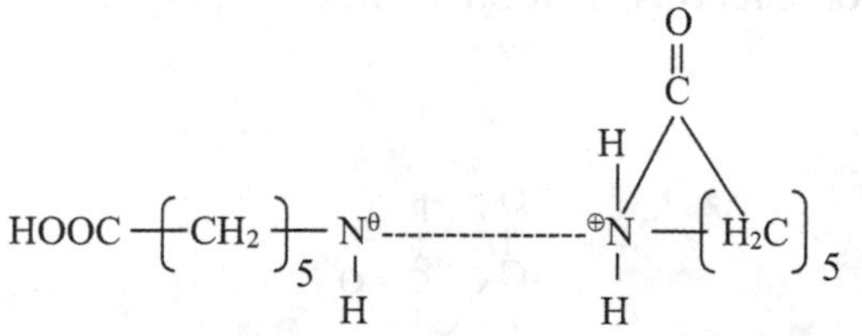

(II) Paired Initiator – Negatively Charged

for which the initiator (II) is formed, and when formed, there is no doubt that Step polymerization will be competing with Addition polymerization to give low molecular weight products.
(Laws of Creations for Lactams)

Rule 1122: This rule of Chemistry for **Lactams,** states that, when *Water* is used as source of initiator for Lactams wherein the N center is carrying a radical-pushing group, the followings are to be expected-

$$2 \left[CH_2\right]_3 - N - R + H_2O \longrightarrow H - N \left\{ C \right\}_3 CO^- \oplus N \left(CH_2\right)_3$$

$$\longrightarrow H - N \left\{ C \right\}_3 CO - C \cdots \oplus N \left(CH_2\right)_3$$

(II)

(III)

$$R \equiv CH_3$$

$$R - N \left\{ C \right\}_3 \underline{CO - C} \left\{ C \right\}_3 N^\ominus \oplus N \left(CH_2\right)_3 \xrightarrow{+ n(II)}$$

$$R - N \left\{ C \right\}_3 \underline{CO - C} \left\{ C \right\}_3 N \right\}_n C \left\{ C \right\}_3 N^\ominus \oplus N \left(CH_2\right)_3$$

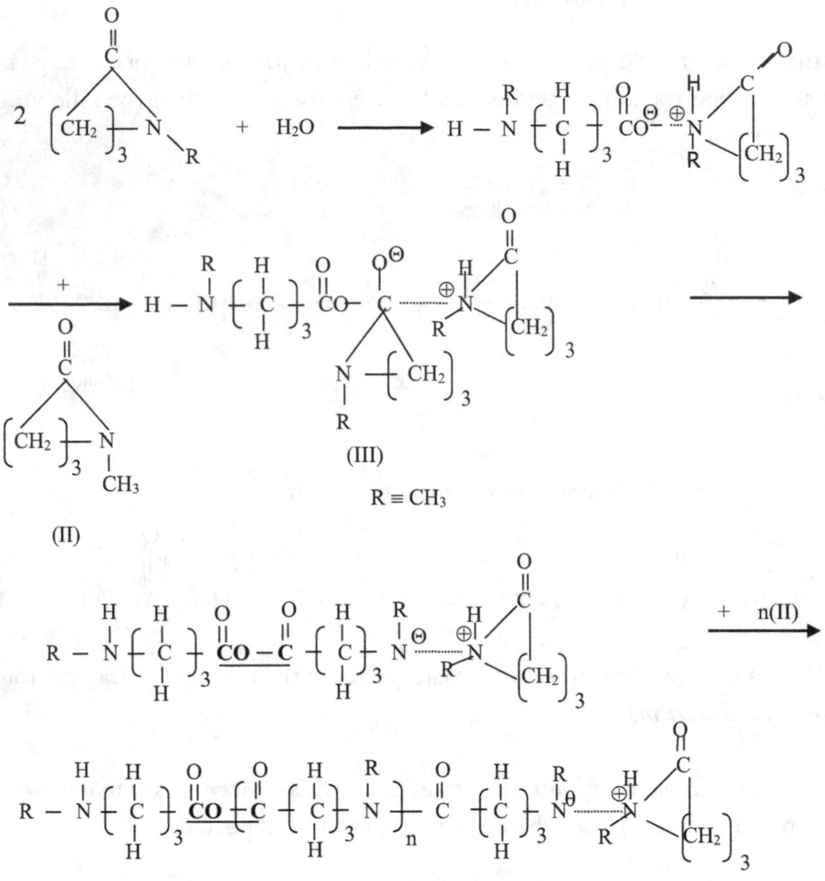

noting that the route above is favored only if the initiator (wherein an amino acid was first prepared) is in Stable state of existence, and if in Equilibrium state of existence, then only the electro-free-radical route will be favored.

(Laws of creations for Lactams)

Rule 1123: This rule of Chemistry for **Lactams,** states that, when copolymerized with **Lactones,** the followings are to be expected-

Anionically-

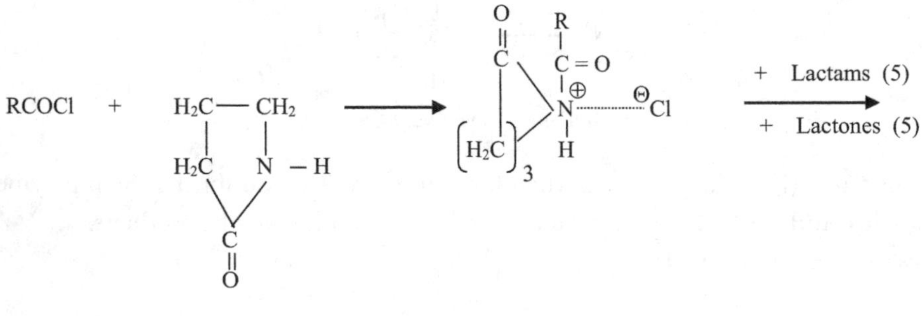

$$RCOCl + H_2C-CH_2 \big/ H_2C \quad N-H \longrightarrow \left[H_2C\right]_3 \oplus N \cdots \ominus Cl \quad \begin{array}{l} + \text{ Lactams (5)} \\ + \text{ Lactones (5)} \end{array} \longrightarrow$$

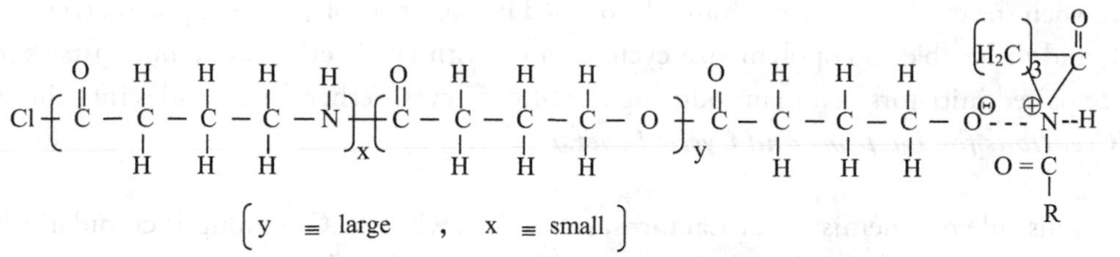

$$\left[\; y \;\equiv\; large \;,\; x \;\equiv\; small \;\right]$$

<u>Long blocks of lactones (anionically)</u> (a)

Electro-free-radically-

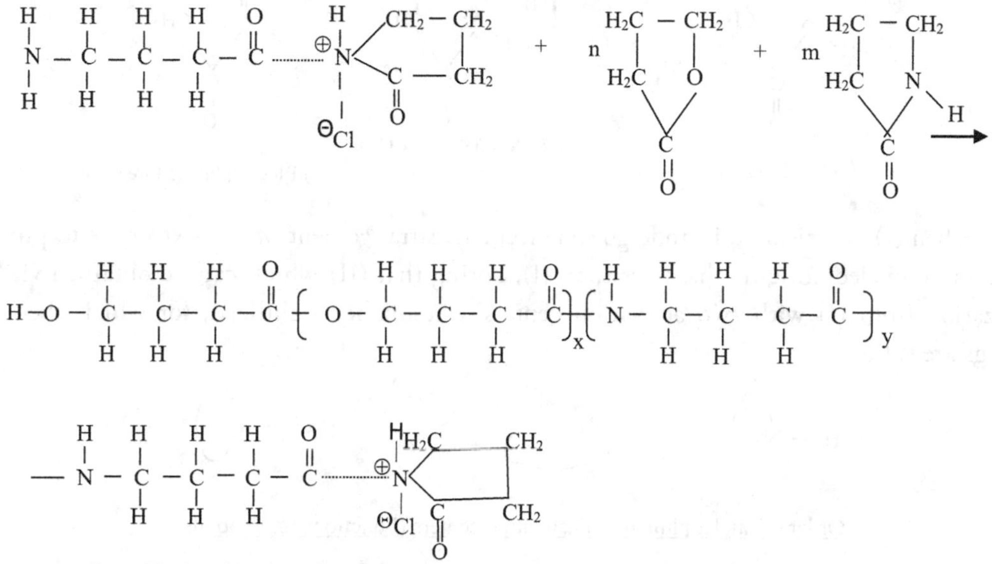

<u>Long blocks of lactams than lactones</u> (b)

for which long blocks of lactones and lactams are produced anionically and electro-free-radically respectively, noting that the order of attack is independent of the sizes of the rings, but on other factors, and the type of initiator used is that in which lactam is a part of it as shown above.
(Laws of Creations for Lactams and Lactones)

<u>Rule 1124:</u> This rule of Chemistry for **Lactams,** states that, between *Lactams and Cyclic ethers* based on side by side placement and what they are carrying, the followings are valid-

Lactams > Cyclic ethers

<u>Order of Nucleophilicity</u>

Lactams > Cyclic ethers

<u>Order of electrophilicity</u> (a)

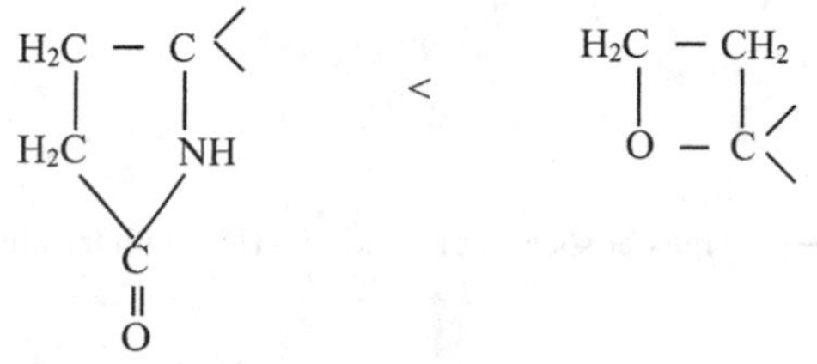

<u>Order of radical-pushing capacity</u> (b)

275

noting that when the cyclic ether ring cannot be opened instantaneously, anionic polymerization cannot be favored, and to be able to copolymerize cyclic amides with cyclic ethers, one must first begin with cyclic amide paired initiators before introducing mixtures of cyclic ethers and amides into the system. ***(Laws of Creations for Lactams and Cyclic Ethers)***

Rule 1125: This rule of Chemistry for **Lactams,** states that, when H_2C = group is cumulatively placed on the ring next to the N center, the followings are to be expected exploratively–

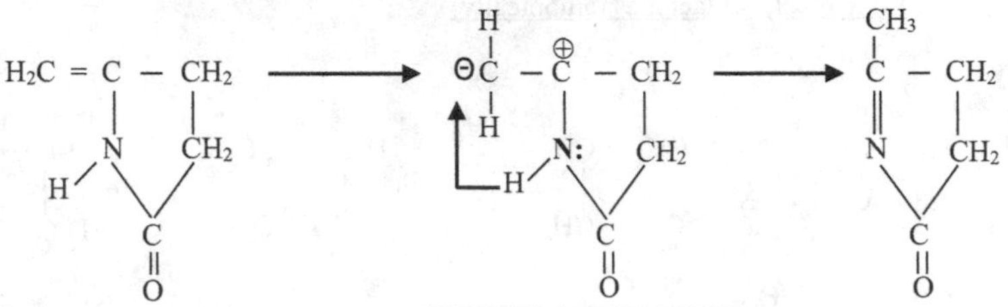

WRONG ACTIVATION

(I) Electrophile (II) Electrophile (More stable) (a)

in which when (I) is activated, it undergoes molecular rearrangement of the sixth type to produce (II) which is more nucleophilic in character than (I), noting that (II) which can be obtained via Electro-radicalization from (I) with two activation centers is resonance stabilized, for which therefore, the followings are valid-

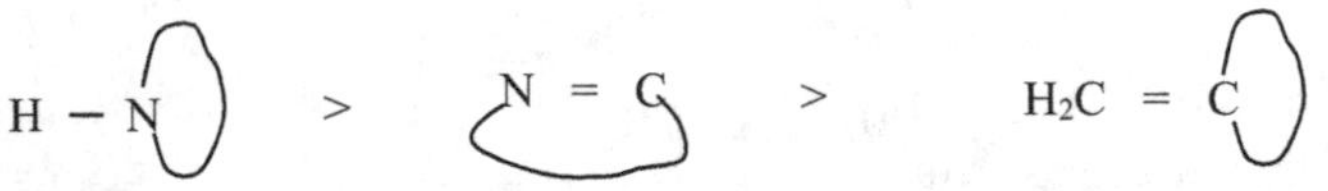

Order of nucleophilicity of activation centers (Functional & π-bonds) (b)

while in reality, the monomer cannot be activated chargedly due to electrodynamic forces of repulsion, since the charges are the reverse of what is shown above, for which electro-free-radically and nucleo-non-free-radically, the Electrophilic character of the monomer is still maintained; from which it can be seen that ***–NHCOR group is a radical-pulling group.***
(Laws of Creations for Lactams)

Rule 1126: This rule of Chemistry for **Lactams,** states that, when H_2C = group is cumulatively placed on the ring next to the N center, the monomer or compound still remains an Electrophile as shown below-

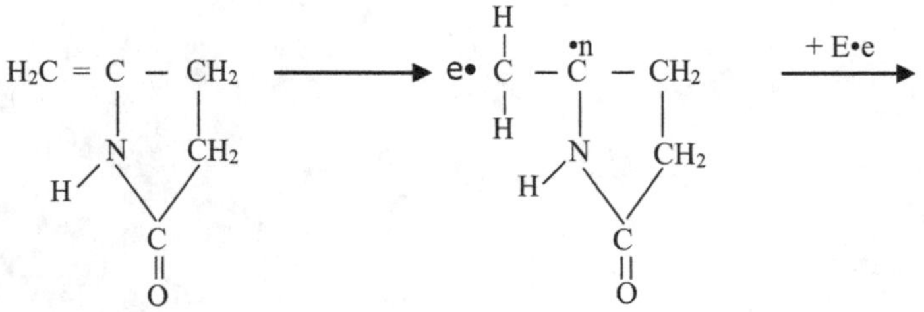

(I) Electrophile

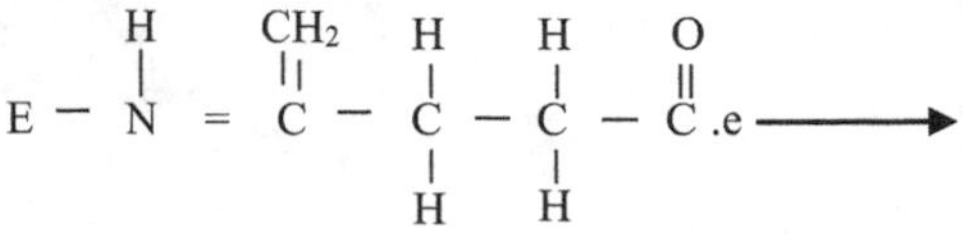

Transfer species of 2nd kind ($H^{\cdot e}$), with terminal Ketene

(I) (Transfer species of 1st kind of the 11th type) (a)

noting that if the ring had instantaneously been opened, the point of scission would have been the same as above-

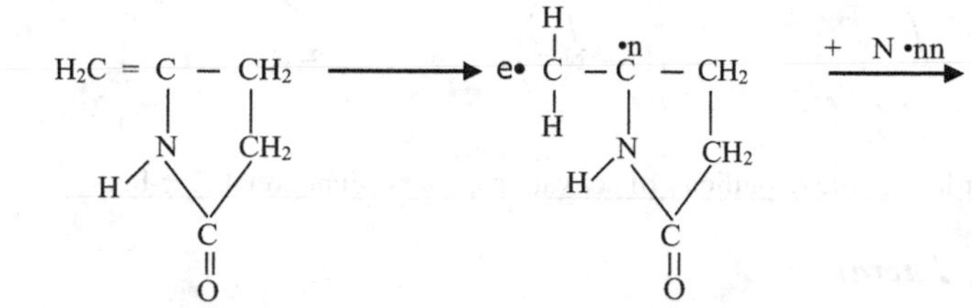

(I) Electrophile

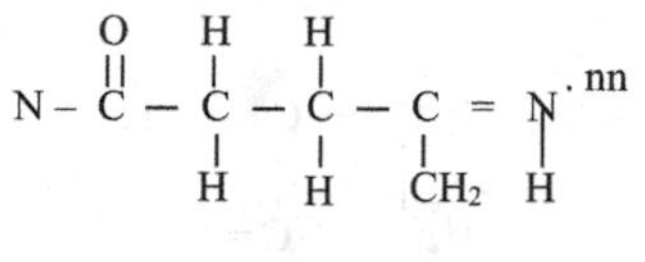

(B) Favored

(b)

for which (I) can only be activated radically, providing the same monomer unit electro-free-radically and nucleo-non-free-radically.

(Laws of Creations for Lactams)

Rule 1127: This rule of Chemistry for **Lactams,** states that, when H_2C = group is cumulatively placed on the ring next to the N center carrying a radical-pushing alkanyl group, the monomer or compound still remains an Electrophile as shown below-

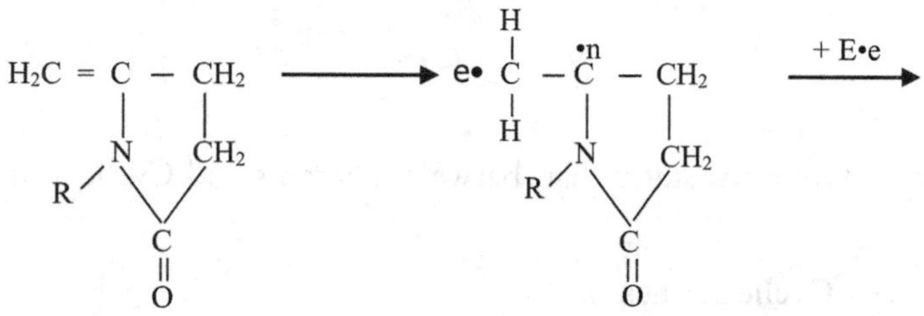

(I) Electrophile

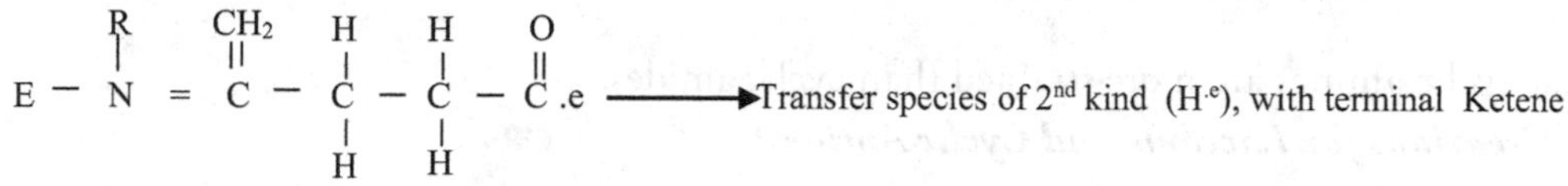

(I) (Transfer species of the 1st kind of the 11th type)

(a)

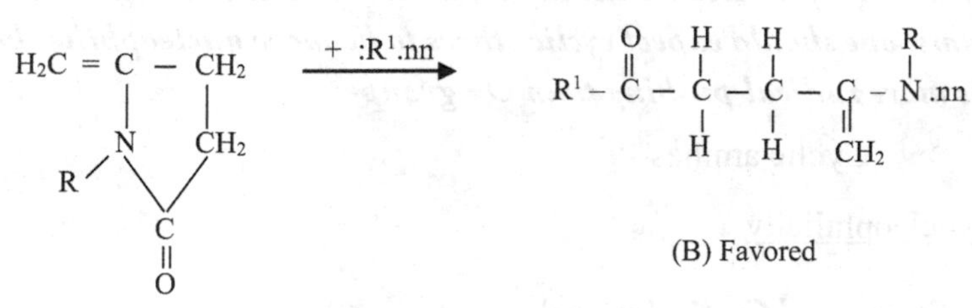

(B) Favored

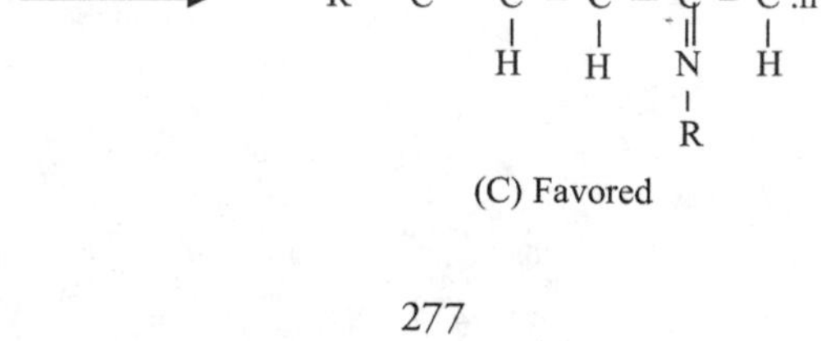

(C) Favored

(b)

noting that for the same monomer unit to be obtained electro-free-radically and nucleo-non-free-radically, the following must be valid-

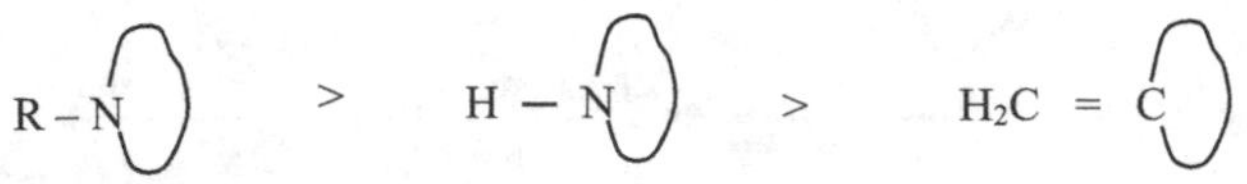

$$R - N \bigcirc \quad > \quad H - N \bigcirc \quad > \quad H_2C = C \bigcirc$$

<u>Order of Nucleophilicity of activation centers (Functional & π-bonds)</u>　　　　　(c)

(Laws of Creations for Lactams)

<u>**Rule 1128:**</u> This rule of Chemistry for **Lactams,** states that, when $H_2C=$ is cumulatively placed on the ring not next to a N center as shown below-.

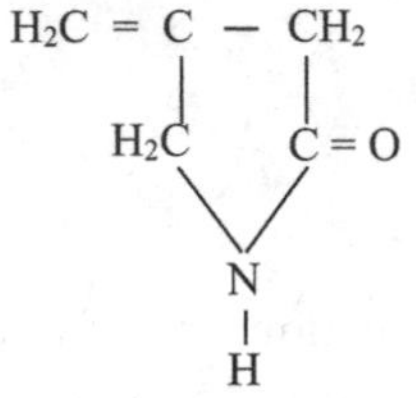

$$H_2C = C - CH_2$$
$$| \qquad |$$
$$H_2C \qquad C = O$$
$$\diagdown \quad \diagup$$
$$N$$
$$|$$
$$H$$
(I)

and

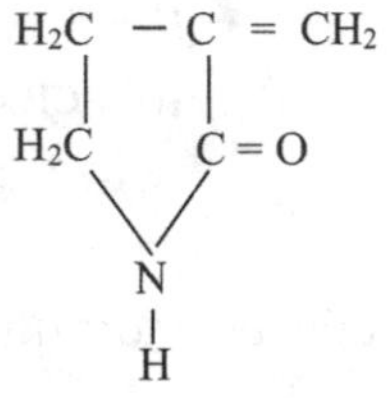

$$H_2C - C = CH_2$$
$$| \qquad |$$
$$H_2C \qquad C = O$$
$$\diagdown \quad \diagup$$
$$N$$
$$|$$
$$H$$
(II)

while (I) becomes a non-productive Electrophile, because the ring cannot be opened via the least nucleophilic center (the $H_2C=$ center on the ring) since there is no suitable point of scission on the ring, being strained, the ring can only be opened instantaneously to favor being used as an Electrophile; (II) can be opened via the $H_2C=$ center on the ring electro-free-radically, with the same monomer unit obtained nucleo-non-free-radically.
(Laws of Creations for Lactams)

<u>**Rule 1129:**</u> This rule of Chemistry for **Lactams,** states that, between Lactams and Cyclic amines, the following is valid-

　　　　　Cyclic amides　　　>　　　Cyclic amines

　　　　　　　<u>Order of Nucleophilicity</u>

noting that cyclic amines are more strained than cyclic amides.
(Laws of Creations for Lactams and Cyclic Amines)

<u>**Rule 1130:**</u> This rule of Chemistry for **Cyclic Ethers and Cyclic amines,** states that, *since Lactones are more nucleophilic than Lactams, one should expect cyclic ethers to be more nucleophilic than cyclic amines, though HN= group is more radical-pushing than O= group-*

　　　　　Cyclic ethers　　　>　　Cyclic amines

　　　　　　　<u>Order of Nucleophilicity</u>

(Laws of Creations for Cyclic Ethers and Cyclic Amines)

Rule 1131: This rule of Chemistry for **Aziridine,** states that, in the absence of charged-paired types of initiators or isolatedly placed radicals, it can never be used as a monomer as shown below-

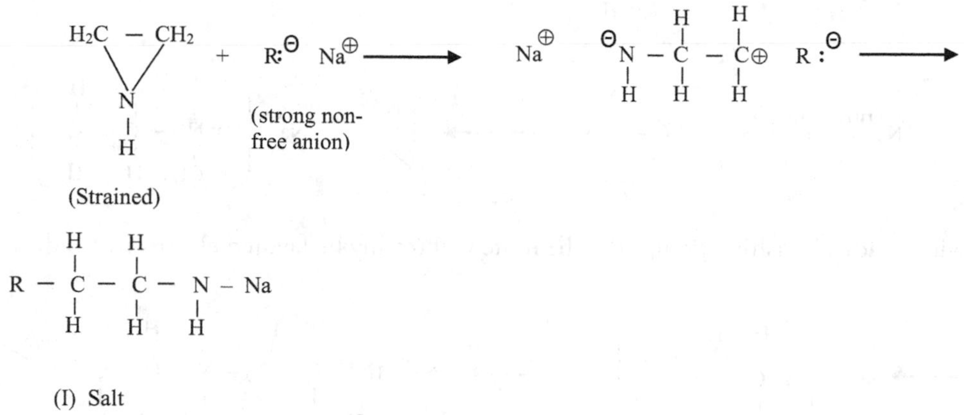

$$H_2C - CH_2 \overset{\displaystyle \diagdown N \diagup}{\underset{\displaystyle |}{}} \quad + \quad R\!:\!^{\ominus} \ Na^{\oplus} \longrightarrow \quad Na^{\oplus} \ \overset{\ominus}{N} - \overset{H}{\underset{H}{C}} - \overset{H}{\underset{H}{C}}\!\oplus \quad R\!:\!^{\ominus} \longrightarrow$$

(Strained)   (strong non-free anion)

$$R - \overset{H}{\underset{H}{C}} - \overset{H}{\underset{H}{C}} - \overset{H}{\underset{}{N}} - Na \qquad\qquad\qquad\qquad (a)$$

(I) Salt

$$H_2C - CH_2 \overset{\displaystyle \diagdown \underset{\ominus}{N} \diagup}{\underset{\displaystyle \overset{|}{H}\!\oplus}{}} + \ R\!:\!^{\ominus} \ Y^{\oplus} \longrightarrow RH \ + \ H_2C - CH_2 \overset{\displaystyle \diagdown N \diagup}{\underset{\displaystyle \overset{|}{Y}}{}}$$

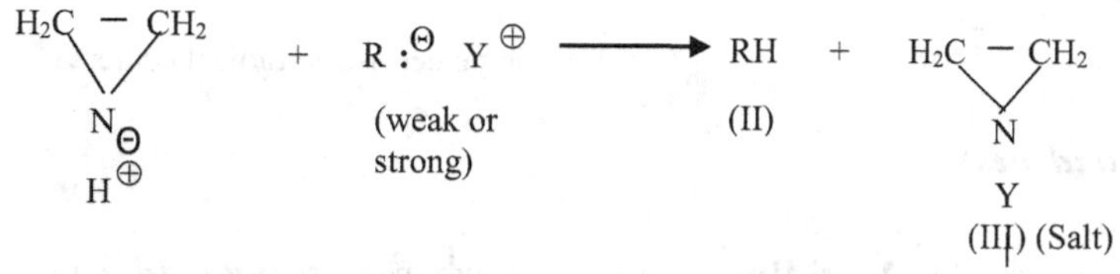

(weak or strong)   (II)   (III) (Salt)   (b)

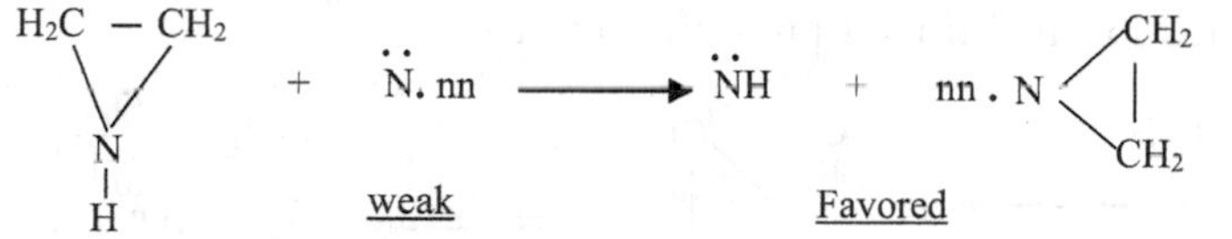

$$H_2C - CH_2 \overset{\displaystyle \diagdown N \diagup}{\underset{\displaystyle \overset{|}{H}}{}} + \ \overset{\cdot\cdot}{N}.\,nn \longrightarrow \overset{\cdot\cdot}{N}H \ + \ nn.N \overset{\diagup CH_2}{\underset{\diagdown CH_2}{\underset{|}{}}}$$

weak   Favored   (c)

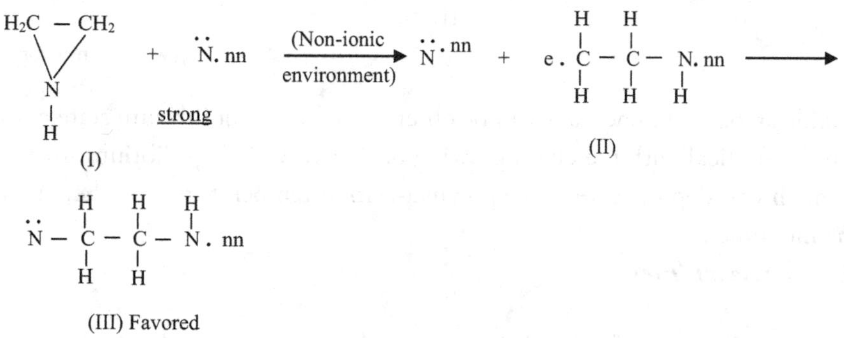

$$H_2C - CH_2 \overset{\displaystyle \diagdown N \diagup}{\underset{\displaystyle \overset{|}{H}}{}} + \ \overset{\cdot\cdot}{N}.\,nn \xrightarrow[\text{environment}]{\text{(Non-ionic}} \overset{\cdot\cdot}{N}\!\cdot nn \ + \ e.\overset{H}{\underset{H}{C}} - \overset{H}{\underset{H}{C}} - \overset{H}{\underset{}{N}}.\,nn \longrightarrow$$

(I)   strong   (II)

$$\overset{\cdot\cdot}{N} - \overset{H}{\underset{H}{C}} - \overset{H}{\underset{H}{C}} - \overset{H}{\underset{}{N}}.\,nn$$

(III) Favored   (d)

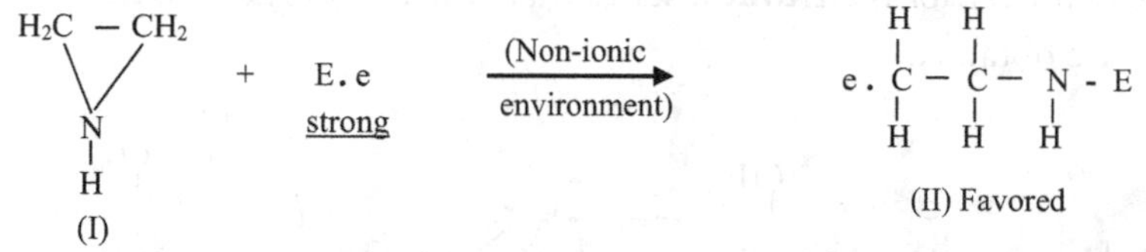

$$H_2C - CH_2 \overset{\displaystyle \diagdown N \diagup}{\underset{\displaystyle \overset{|}{H}}{}} + \ E.e \xrightarrow[\text{strong}]{\text{(Non-ionic environment}} e.\overset{H}{\underset{H}{C}} - \overset{H}{\underset{H}{C}} - \overset{H}{\underset{}{N}} - E$$

(I)   (II) Favored   (e)

for which for the use of radicals, the opposite counterpart must be absent in the system otherwise telomers or salts will be produced, and anionically and nucleo-non-free-radically, the ring must be opened instantaneously, since the routes are not natural to the monomer.
(Laws of Creations for Aziridine)

Rule 1132: This rule Chemistry for **Aziridine,** states that, when strong bases such as *alkoxides, sodium, sodium cyanide and the likes* are used as initiators, the initiation step is not favored because of the influence of ionic metals on H atom carried by the Nitrogen center with Na replacing the hydrogen all the time; for if the H on the N center is

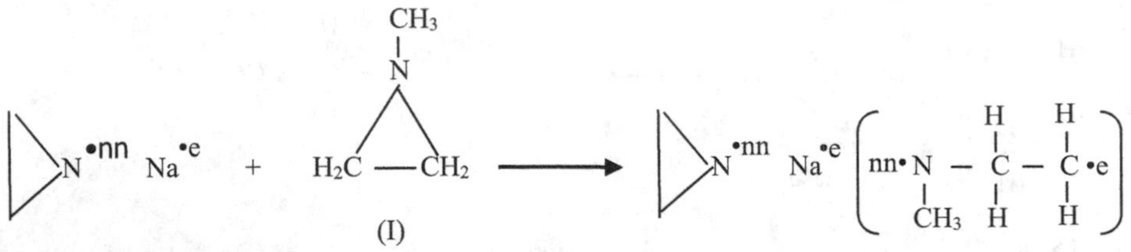

(I)

replaced with a radical pushing group (R), the route will readily be favored electro-free-radically but not chargedly.

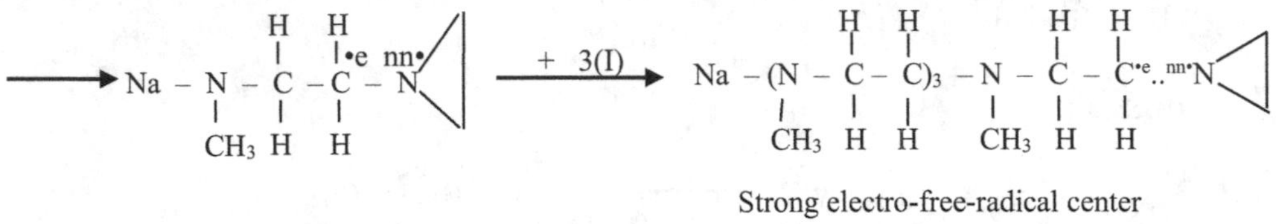

Strong electro-free-radical center

(Laws of Creations for Aziridines)

Rule 1133: This rule of Chemistry for **Aziridines,** states that, when *a strong acid* is used as source of initiator, the following is obtained for the initiator preparation step-

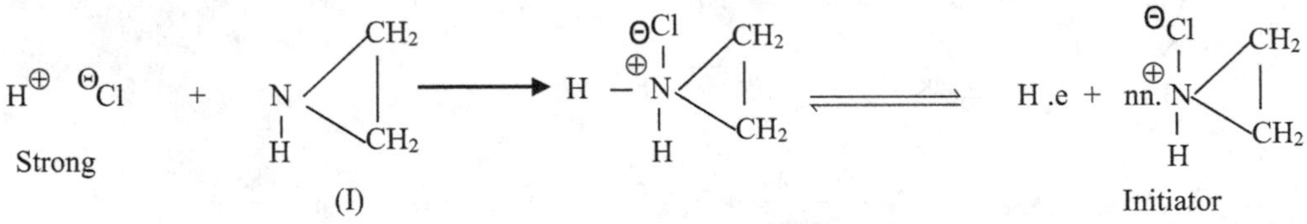

Strong   (I)   Initiator

for which the initiator prepared in one stage can be observed to be in Equilibrium state of existence, and the route is electro-free-radical with the chain growing backwardly via Equilibrium mechanism to give a growing chain which can depropagate if the polymer-ization temperature is higher than the Ceiling temperature of the monomer.

(Laws of Creations for Aziridine)

Rule 1134: This rule of Chemistry for **Aziridines**, states that, when *Benzyl chloride* is used as source of initiator, the followings are obtained-

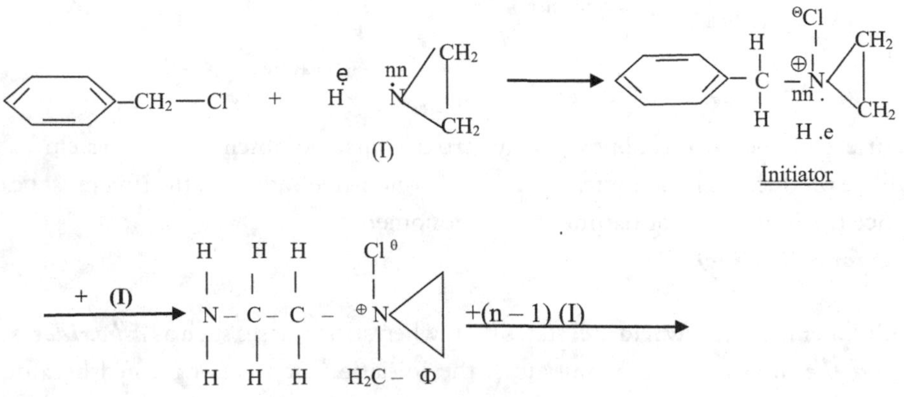

(I)   Initiator

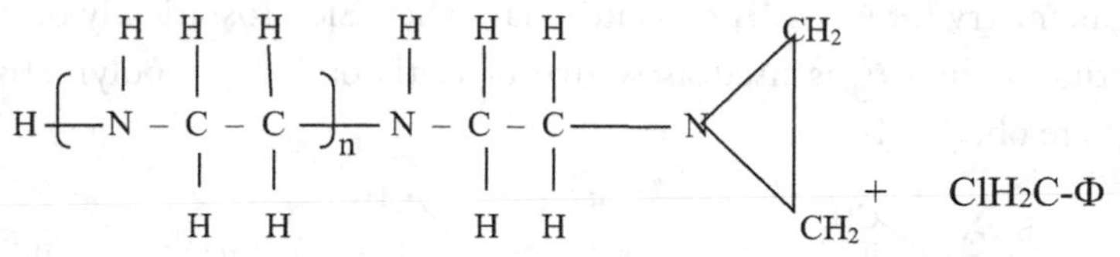

$$[\Phi \equiv \text{Phenyl group}]$$

for which the initiator prepared in two stages can be observed to be in Equilibrium state of existence, and the route is electro-free-radical with the chain growing backwardly via Equilibrium mechanism to give a growing chain which can depropagate if the polymer-ization temperature is higher than the Ceiling temperature of the monomer.

(Laws of Creations for Aziridines)

Rule 1135: This rule of Chemistry for **Aziridines,** states that, when an alkylane group is located on the carbon centers, then the point of scission of the monomer will depend on where located as shown below-

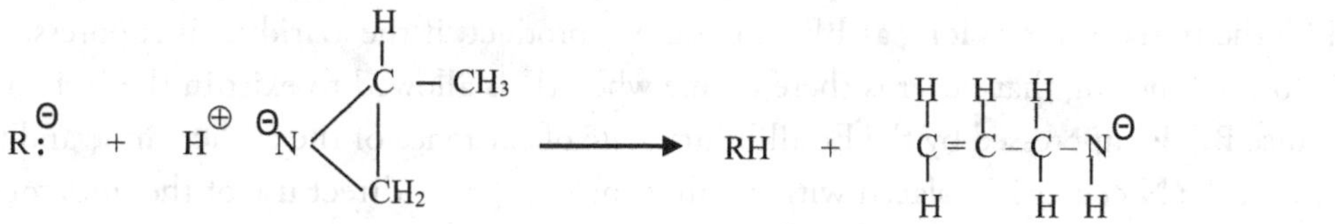

(a)

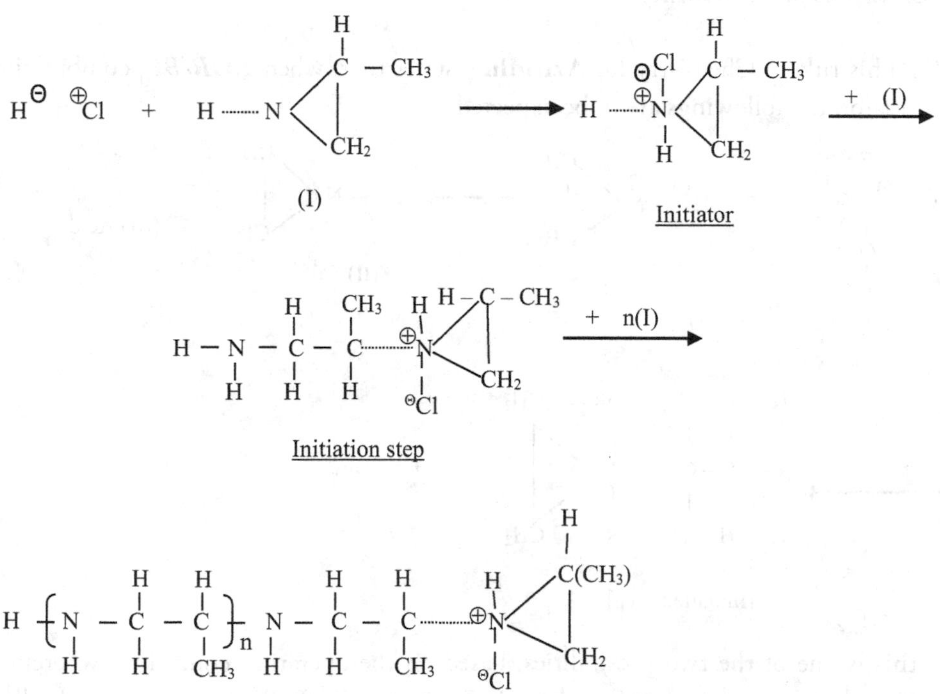

(I)   Initiator

Initiation step

(b)

in which the point of scission is such that, the most electropositive active carbon center is obtained electro-free-radically, noting that since the anionic route is not favored, clearly indicates that the monomer is a complete Nucleophile, more nucleophilic than aziridine.

(Laws of Creations for Aziridines)

Rule 1136: This rule of Chemistry for **Aziridines,** states that, when Electrostatically positively charged-paired initiators such as that from BF_3 is used as source of initiator for the polymerization of these monomers, the followings are obtained-

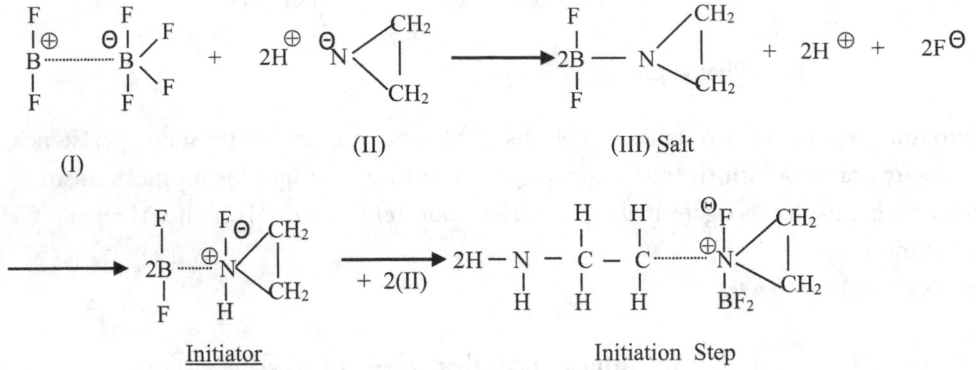

(I)   (II)   (III) Salt

Initiator   Initiation Step

for which since hydrofluoric acid is strong enough (being an inorganic acid), subsequent polymerization is favored by the initiator for as long as BF_3 is not a bye-product, if the aziridine is suppressed after the initiator is formed; noting that never is there a time when (I) is allowed to exist in the system as shown above, because BF_3 is suppressed by the Equilibrium state of existence of the cyclic amine and that when the H atom on the N center is replaced with an alkylane group, the direct use of the initiator from BF_3 becomes possible.

(Laws of Creations for Aziridine)

Rule 1137: This rule of Chemistry for **Aziridine,** states that, when ROR/BF_3 combination is used as source of initiator, the followings are to be expected-

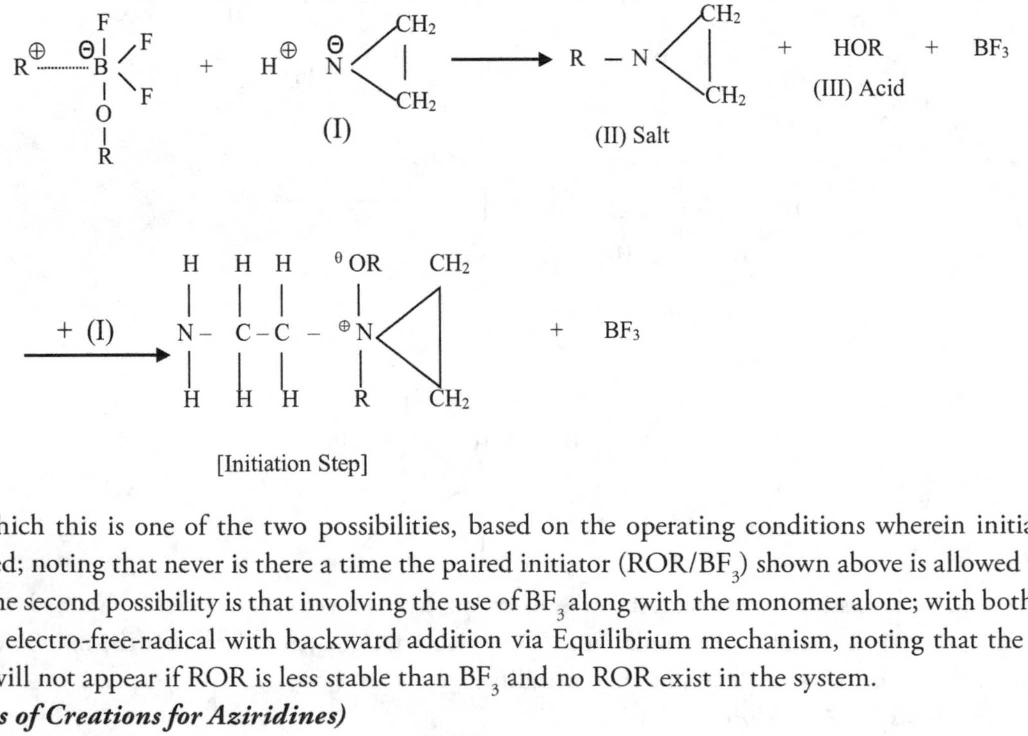

(I)   (II) Salt   (III) Acid

[Initiation Step]

for which this is one of the two possibilities, based on the operating conditions wherein initiation is favored; noting that never is there a time the paired initiator (ROR/BF_3) shown above is allowed to exist and the second possibility is that involving the use of BF_3 along with the monomer alone; with both routes being electro-free-radical with backward addition via Equilibrium mechanism, noting that the second case will not appear if ROR is less stable than BF_3 and no ROR exist in the system.

(Laws of Creations for Aziridines)

Rule 1138: This rule of Chemistry for **Aziridines or Cyclic amines and Carbon monoxide,** states that, between both of them, based on the nature of reactions favored by them, the following is valid-

Cyclic amine   Carbon monoxide

Order of Nucleophilicity

(Laws of Creations for Cyclic Amines and CO)

Rule 1139: This rule of Chemistry for **Aziridine,** states that, when copolymerized with *Carbon monoxide,* the followings are obtained-

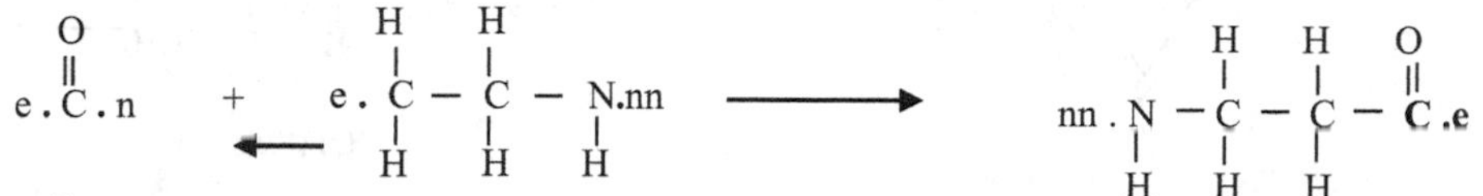

(I)   (II)

Less Nucleophilic   More Nucleophilic   "THE COUPLE"   (a)

Nucleo-non-free-radically:

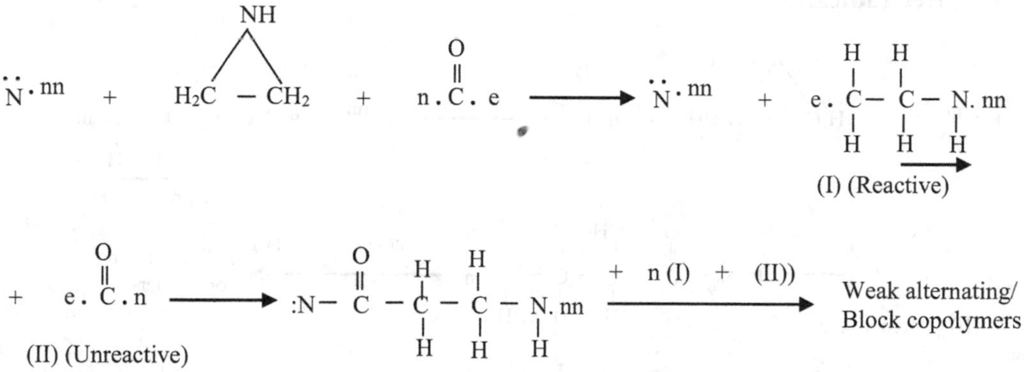

(I) (Reactive)

(II) (Unreactive)   Weak alternating/ Block copolymers

(b)

Electro-free-radically:

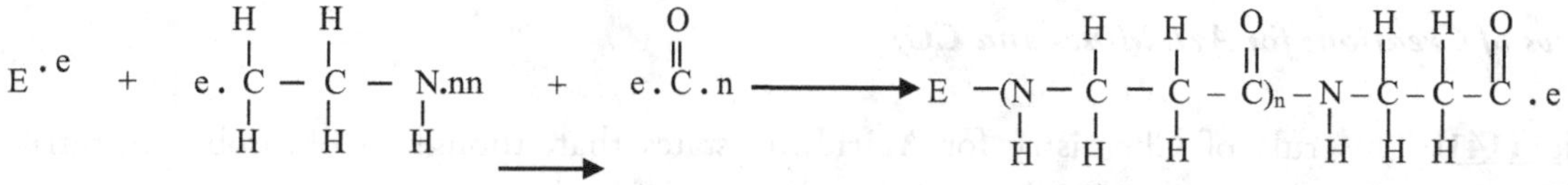

(I)   (II)

More Nucleophilic   Less Nucleophilic   **Weak alternating copolymer**   (c)

for which the possibility of having full alternating placement is impossible; noting that in (a) for the couple when formed, and based on what it is carrying, only electro-free-radicals and nucleo-non-free-radicals can be used, and aziridine is reactive to both of them, with their products favoring the existence of blocks of aziridine along the chain.

(Laws of Creations for Aziridine and CO)

Rule 1140: This rule of Chemistry for **Aziridines and Carbon monoxide**, states that, when CH_3 group is located on a carbon center of aziridine, and copolymerized with CO, the followings are to be expected-

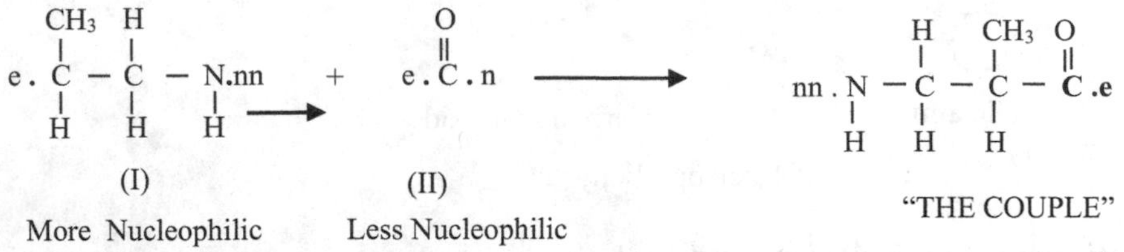

(I)   (II)
More Nucleophilic   Less Nucleophilic   "THE COUPLE"   (a)

Electro-free-radically

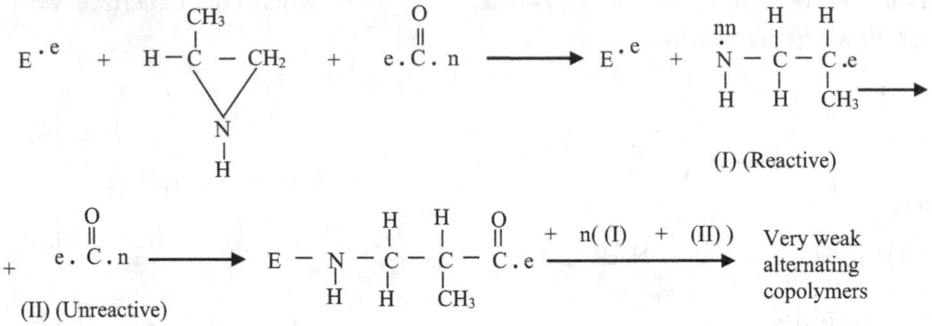

(I) (Reactive)

(II) (Unreactive)   Very weak alternating copolymers   (b)

Nucleo-non-free-radically:

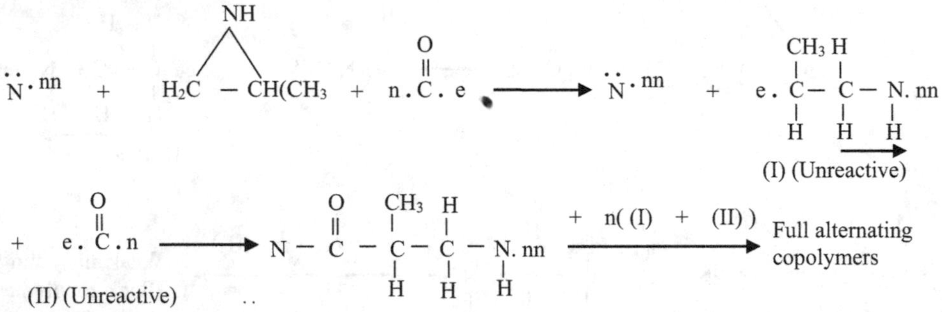

(I) (Unreactive)

(II) (Unreactive)   Full alternating copolymers   (c)

noting the conditions in which full alternating placement has been obtained, for which the location of the CH_3 group is important to note; (I) can homopolymerize electro-free-radically and not nucleo-non-free-radically.

(Laws of Creations for Aziridines and CO)

Rule 1141: This rule of Chemistry for **Aziridine,** states that, though Azobisisobutyro-nitrile or 2,2'Dicyano-2,2'-azopropane (AIBN) a well-known source of nucleo-free-radicals cannot produce non-free-radicals when the scission is completed to release N_2, it can however be made to produce nucleo-non-free-radicals via electro-free-radical movement as shown below.

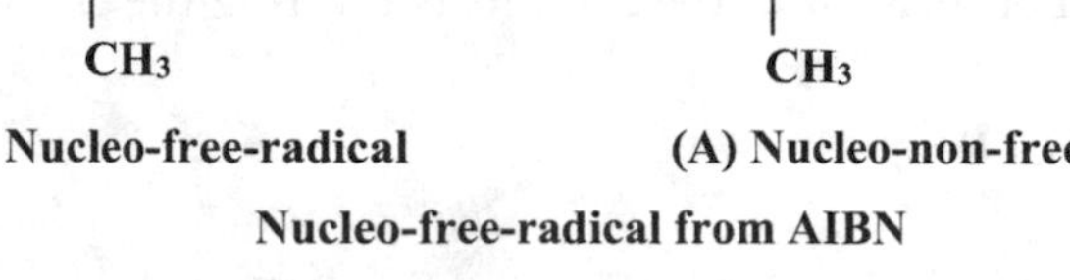

(A) Nucleo-free-radical   (A) Nucleo-non-free-radical
Nucleo-free-radical from AIBN

(Laws of Creations for AIBN)

Rule 1142: This rule of Chemistry for **Aziridine, Carbon monoxide and Ethene,** when they are all polymerized together, using AIBN as initiator, the followings are obtained for them nucleo-non-free-radically-

$$\ddot{N}\cdot nn \quad + \quad H_2C - CH_2 \quad + \quad e.\overset{\overset{O}{\|}}{C}.n \quad + \quad e.\underset{H}{\overset{H}{\underset{|}{C}}} - \underset{H}{\overset{H}{\underset{|}{C}}}.n \longrightarrow$$

(I)   (II)   (III)

$$e.\overset{\overset{O}{\|}}{C}.n \quad + \quad e.\underset{H}{\overset{H}{\underset{|}{C}}} - \underset{H}{\overset{H}{\underset{|}{C}}}.n \quad + \quad e.\underset{H}{\overset{H}{\underset{|}{C}}} - \underset{H}{\overset{H}{\underset{|}{C}}} - \underset{H}{\overset{}{N}}.nn \quad + \quad \ddot{N}\cdot nn \longrightarrow \ddot{N}\bullet nn \quad +$$

(II) (Unreactive)   (III) (Unreactive)   (I) (Reactive)

$$e.\overset{\overset{O}{\|}}{C} - \underset{H}{\overset{H}{\underset{|}{C}}} - \underset{H}{\overset{H}{\underset{|}{C}}} - \underset{H}{\overset{H}{\underset{|}{C}}} - \underset{H}{\overset{H}{\underset{|}{C}}} - \underset{}{\overset{}{N}}.nn \longrightarrow \ddot{N} - \overset{\overset{O}{\|}}{C} - \underset{H}{\overset{H}{\underset{|}{C}}} - \underset{H}{\overset{H}{\underset{|}{C}}} - \underset{H}{\overset{H}{\underset{|}{C}}} - \underset{H}{\overset{H}{\underset{|}{C}}} - \underset{H}{\overset{}{N}}.nn$$

(A) 'The Couple"

$$\xrightarrow{\quad + \quad n(\,(I) \; + \; (II), \; + \; (III)\,) \quad} \underline{\text{Alternating terpolymers + Aziridine (randomly}}$$
$$\underline{\text{placed)}}$$

                                                                  (a)

for which the order for addition has been done because the following is valid for them-

$$\overset{\frown}{\underset{\smile}{(\,N\,)}} \quad > \quad C = C \quad > \quad \overset{\overset{O}{\|}}{C:}$$

Cyclic amine   Alkenes   Carbon monoxide

Order of Nucleophilicity   (b)

(Laws of Creations for Aziridine, CO and Alkenes)

Rule 1143: This rule of Chemistry for **Aziridine, Carbon monoxide and Ethene,** when they are all polymerized together, using AIBN as initiator, the followings are obtained for them nucleo-free-radically-

$$e.\underset{H}{\overset{H}{\underset{|}{C}}} - \underset{H}{\overset{H}{\underset{|}{C}}} - \underset{H}{\overset{}{N}}.nn \quad + \quad \left(e.\overset{\overset{O}{\|}}{C}.n \quad + \quad e.\underset{H}{\overset{H}{\underset{|}{C}}} - \underset{H}{\overset{H}{\underset{|}{C}}}.n \right) \quad + \quad N\cdot n \longrightarrow$$

(I) (Unreactive)   (II) (Unreactive)   (III) (Unreactive)

$$N\cdot n \quad + \quad e.\overset{\overset{O}{\|}}{C} - \underset{H}{\overset{H}{\underset{|}{C}}} - \underset{H}{\overset{H}{\underset{|}{C}}} - \underset{H}{\overset{}{N}} - \overset{\overset{O}{\|}}{C} - \underset{H}{\overset{H}{\underset{|}{C}}} - \underset{H}{\overset{H}{\underset{|}{C}}}.n \longrightarrow \text{Full alternating terpolymer}$$

                                  (IV)   **(A)**

                                                                     (a)

for which the initiator in control here is a nucleo-free-radical and since NATURE abhors a vacuum and NATURE is orderliness, and since alternating placement can readily be obtained between CO and alkenes nucleo-free-radically, (III) first adds to (II) to give a three-membered Cyclic ketone which is too strained to exist as a ring, and since the cyclic ketone (A) is more nucleophilic than the aziridine which was formerly the most nucleophilic of the three of them, it diffuses to add to the aziridine to form (IV), the "Couple", and since the three are unreactive to the route, full alternating terpolymer is obtained; from which the following is valid-

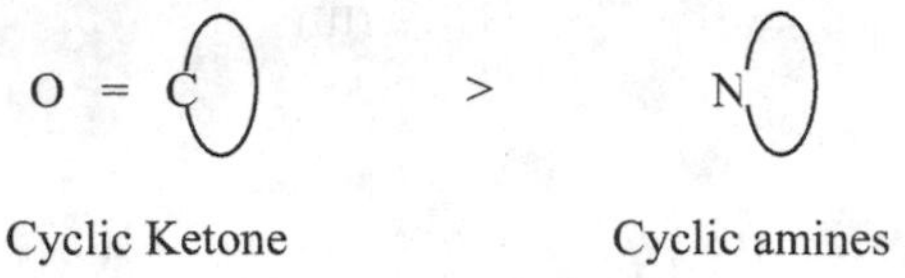

$$O = C \overbrace{\hspace{1cm}} \quad > \quad N \overbrace{\hspace{1cm}}$$

Cyclic Ketone   Cyclic amines

Order of Nucleophilicity   (b)

(Laws of Creations for Aziridine, CO and Alkenes)

<u>**Rule 1144:**</u> This rule of Chemistry for **Aziridines**, states that, like cyclic ethers, the followings are obtained for Propagation/depropagation phenomena-

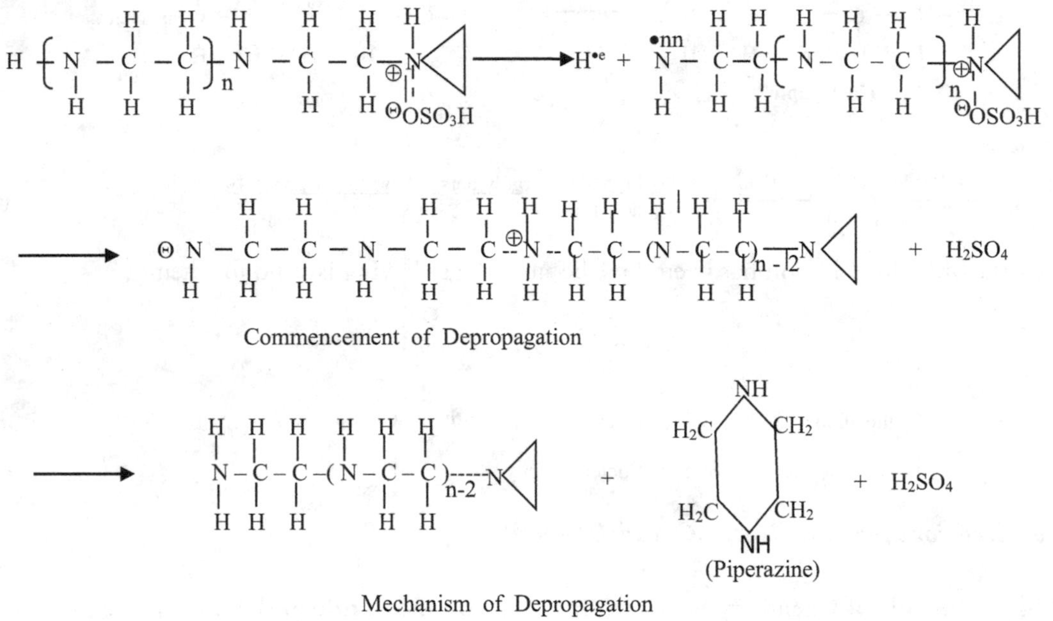

Commencement of Depropagation

(Piperazine)

Mechanism of Depropagation

noting that for this type of growing polymer chain which cannot terminate itself, without the growing chain existing in Equilibrium state of existence (taking place from the end far from the initiating center), and the hydrogen atom held sitting on one of the nitrogen centers, depropagation to release a ring can never take place; the positive charge if well placed on the N center is now paired to the negative charge (anion) on the external N center to release a ring.

(Laws of Creations for Propagation/Depropagation phenomenon in Aziridine)

Rule 1145: This rule of Chemistry for **Aziridines,** states that, unlike cyclic ethers, the followings can be obtained for *Degradation phenomenon-*

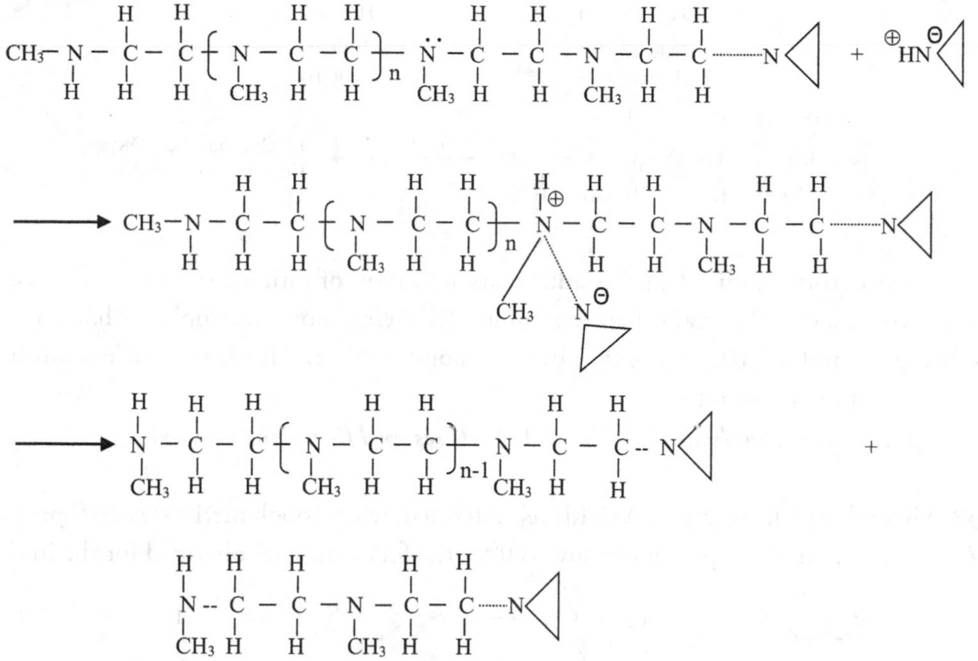

for which the presence of aziridine in the system during or after the end of polymeriza-tion can lead to shorter polymer chains and oligomers, noting that this is not depropagation taking place during propagation, but **degradation of a chain living or dead.**
(Laws of Creations for Degradation of Aziridines polymer chains)

Rule 1146: This rule of Chemistry for **Aziridines,** states that, when copolymerized with β-propiolactone, the route can either be electro-free-radical or nucleo-non-free-radical in character as shown below-

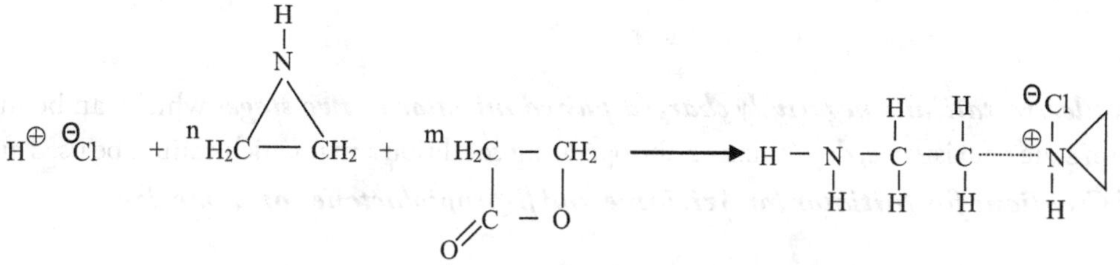

(I) (more reactive)   (II) (reactive)

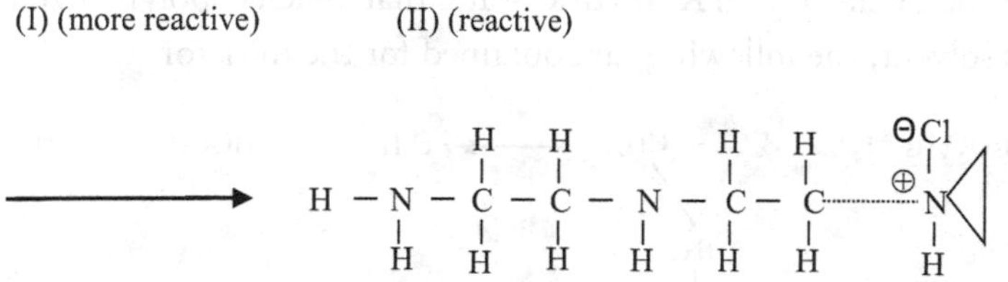

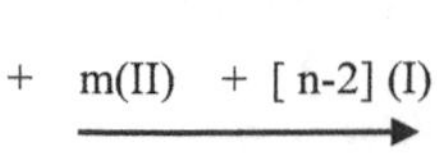

+   m(II)   + [n-2] (I)

Block copolymers   (a)

noting that (I) is less nucleophilic than (II) in character and that (I) is a Nucleophile while (II) is an Electrophile for which there will be more of (I) along the chain than (II) electro-free-radically;

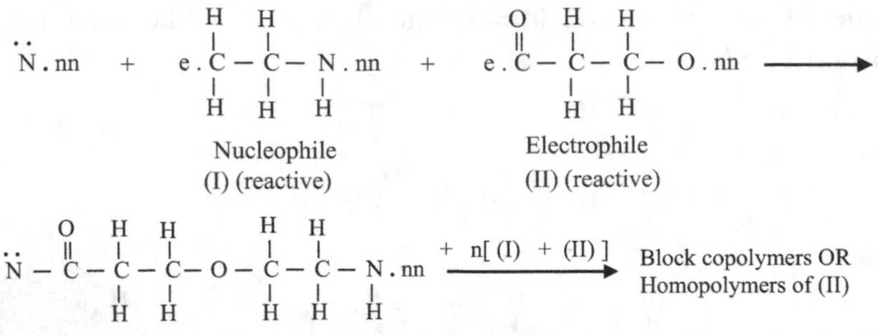

$$\ddot{N}.nn \quad + \quad e.\overset{\overset{H}{|}}{\underset{\underset{H}{|}}{C}} - \overset{\overset{H}{|}}{\underset{\underset{H}{|}}{C}} - N.nn \quad + \quad e.\overset{\overset{O}{\|}}{C} - \overset{\overset{H}{|}}{\underset{\underset{H}{|}}{C}} - \overset{\overset{H}{|}}{\underset{\underset{H}{|}}{C}} - O.nn \quad \longrightarrow$$

Nucleophile   Electrophile
(I) (reactive)   (II) (reactive)

$$\ddot{N} - \overset{\overset{O}{\|}}{C} - \overset{\overset{H}{|}}{\underset{\underset{H}{|}}{C}} - \overset{\overset{H}{|}}{\underset{\underset{H}{|}}{C}} - O - \overset{\overset{H}{|}}{\underset{\underset{H}{|}}{C}} - \overset{\overset{H}{|}}{\underset{\underset{H}{|}}{C}} - N.nn \xrightarrow{+\ n[\ (I)\ +\ (II)\ ]} \begin{array}{l}\text{Block copolymers OR}\\\text{Homopolymers of (II)}\end{array}$$

(b)

for which nucleo-non-free-radically, if instantaneous activation of imine monomer is favored, (II) will be far more favored along the chain than the imines, (II) being more electrophilic than the imine and the route being natural to (II), noting that both monomers are reactive if the nucleo-non-free-radical initiators are used and are strong.

(Laws of Creations for Copolymerization of Aziridines and β-propiolactone)

Rule 1147: This rule of Chemistry for **Aziridine,** states that, when copolymerize*d with β-propiolactone* using *toluene* as solvent at temperature as low as $0^{\circ}C$, the followings are obtained for the initiator-

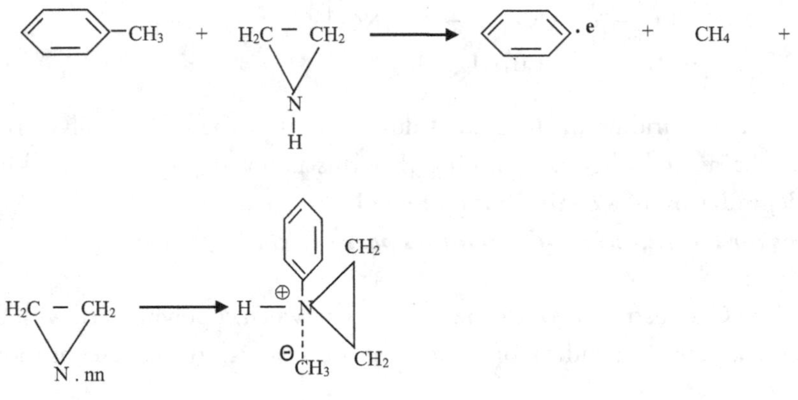

Initiator

to give an ***Electrostatically negatively charged-paired initiator in two stages*** which can be in Stable or Equilibrium state of existence, depending on the operating conditions; that which cannot be used anionically.

(Laws of Creations for Initiator for Aziridine and β-propiolactone copolymerization)

Rule 1148: This rule of Chemistry for **Aziridine**, states that, when copolymerized with ***β-propiolactone*** using *ethyl ether* as solvent, the followings are obtained for the initiator-

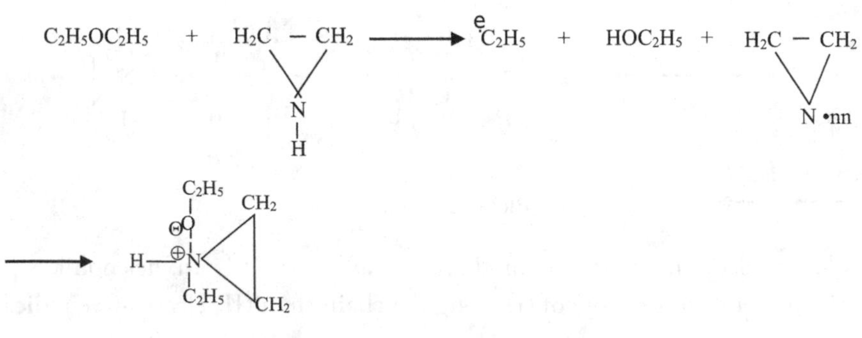

Initiator

288

to give ***Electrostatically anionically charged-paired initiator in two stages*** which can either be in Stable or Equilibrium state of existence, depending on the operating conditions.
(Laws of Creations for the iniator for Aziridine and β-propiolactone copolymerization)

Rule 1149: This rule of Chemistry for **Aziridine,** states that, when copolymerized with ***β-propiolactone*** using ***ethylene dichloride*** as solvent, the followings are obtained for the initiator-

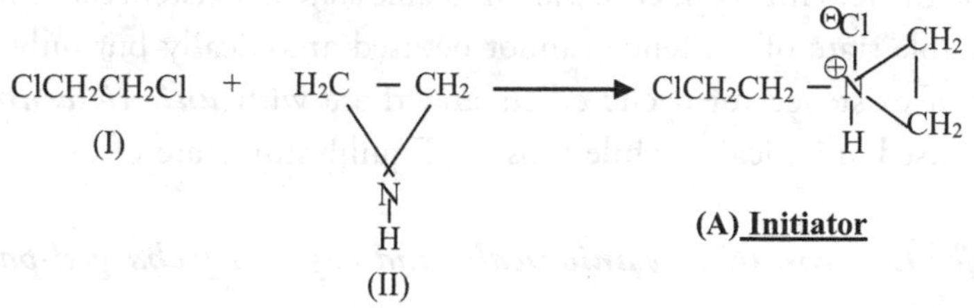

$$ClCH_2CH_2Cl \quad + \quad H_2C - CH_2 \longrightarrow ClCH_2CH_2 - N$$

(I)

(II)

(A) **Initiator**

to give an ***Electrostatically uniunically charged-paired initiator in two stages*** which can either be in Stable or Equilibrium state of existence, depending on the operating conditions.
(Laws of Creations for initiator for Aziridine and β-propiolactone copolymerization)

Rule 1150: This rule of Chemistry for **Aziridine,** states that, when copolymerized with ***β-propiolactone*** using ***dimethyl formamide*** as solvent, the following is the initiator involved-

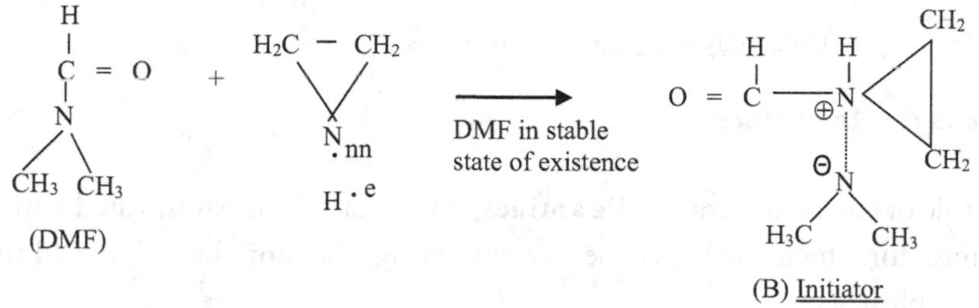

(DMF)

DMF in stable
state of existence

(B) Initiator

to give an ***Electrostatically anionically charged-paired initiator*** and for without the comonomer aziridine, copolymerization in the presence of these so-called solvents would have been impossible; noting that this like other cases where so-called solvents are used is a two stage Equilibrium mechanism system.
(Laws of Creations for Initiator for Aziridine and β-propiolactone copolymerization)

Rule 1151: This rule of Chemistry for **Aziridine,** states that, when initiators are prepared from it using the following solvents (DMF, Acetonitrile, Acetone), the followings are the orders of the strength of the active centers of the initiators-

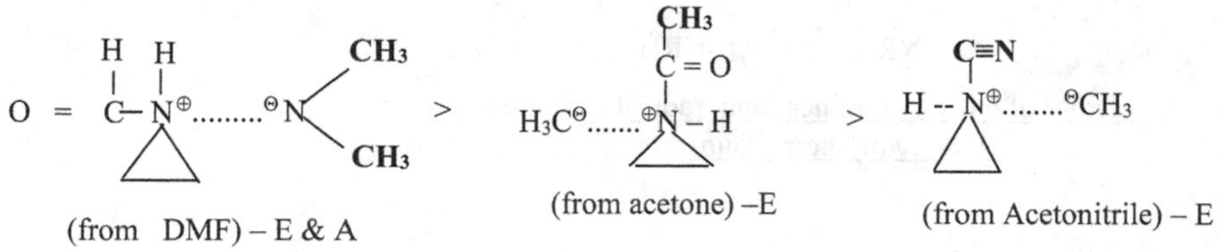

(from DMF) – E & A

(from acetone) –E

(from Acetonitrile) – E

<u>**Order of strength of active centers anionically (A) and electro-free-radically (E)**</u>

(a)

$$Cl^\ominus \quad > \quad H_3C(CO)O^\ominus \quad > \quad C_2H_5O^\ominus \quad > \quad (CH_3)_2N^\ominus$$

<u>**Order of strength of some Anionic Active centers**</u>

(b)

noting that $^{\ominus}C\equiv N$ from the acetonitrile and $H_3C(OC)^{\ominus}$ from acetone cannot exist as negatively charged centers due to electrostatic forces of repulsion and that the last two cannot be used anionically.
(Laws of Creations for Some Initiators for Aziridine polymerization)

Rule 1152: This rule of Chemistry for **Electrostatically anionically and negatively charged-paired initiators,** states that, when the initiators are prepared, all or some fractions are kept in Equilibrium state of existence while the remaining fractions are in Stable state of existence and *those with negatively charged centers* in stable of existence cannot be used anionically but only electro-free-radically in Equilibrium state of existence for cyclic ethers and those with *anionic centers* in stable state of existence can only be used anionically, while those in Equilibrium state of existence can only be used electro-free-radically.
(Laws of Creations for Electrostatically anionically and negatively charged-paired initiators)

Rule 1153: This rule of Chemistry for **Aziridines,** states that, for the members of the family, the following is valid-

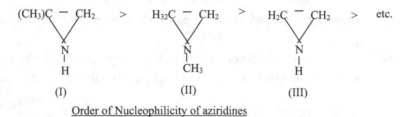

Order of Nucleophilicity of aziridines

(Laws of Creations for Aziridines).

Rule 1154: This rule of Chemistry for **Cyclic amines,** states that, the unsubstituted amines are analogous to secondary amines, for which the larger the size of the ring, the more basic it is and this is the same as their order of nucleophilicity

6 - membered   >   5 - membered   >   4 - membered   >   3 - membered   >

Order of Basicity and Nucleophilicity of cyclic amines

(Laws of Creations for Cyclic Amines)

Rule 1155: This rule of Chemistry for **Cyclic Esters, Amides and Amines,** states that, from their chemistry, the following relationships are valid-

$$NR_3 \quad > \quad NR_2H \quad > \quad NRH_2 \quad > \quad NH_3$$

$$-NR_2 \quad > \quad -NRH \quad > \quad -NH_2$$

Order of basicity (amines) and radical-pushing
capacity of their groups   (a)

$$HOH \quad > \quad CH_3OH \quad > \quad C_2H_5OH$$

$$HO- \quad > \quad CH_3O- \quad > \quad C_2H_5O-$$

Order of acidity ("alcohols") and radical-pushing capacity
of their groups   (b)

$$C_2H_5OOC - \quad > \quad CH_3OOC - \quad > \quad HOOC -$$

<u>Order of radical-pulling capacity (esters)</u> (c)

$$\underline{\mathbf{\textit{ROOC}} -} \quad > \quad H_2NOC - \quad > \quad RHNOC - \quad > \quad R_2NOC -$$

<u>Order of radical-pulling capacity (amides)</u> (d)

from which one can see the order of Electrophilicities of the acrylates and acrylamides, i.e.,

$$H_2C=CH(COOC_2H_5) \quad > \quad H_2C=CH(COOCH_3) \quad > \quad H_2C=CH(COOH)$$

$$H_2C=CH(CONH_2) \quad > \quad H_2C=CH(CONHR) \quad > \quad H_2C=CH(CONR_2)$$

<u>Order of Electrophilicity</u> (e)

(Laws of Creations for Radical pushing and pulling groups)

<u>Rule 1156:</u> The rule of Chemistry for **Cyclic Ethers and Amines,** states that, when instantaneous opening of a ring is favored in the presence of a weak radical initiator, and molecular rearrangement of the third kind is applied, *cyclic ethers cannot rearrange to aldehydes and ketones which are isomers and cyclic amines cannot also rearrange to aldimines and ketimines.*
(Laws of Creations for Cyclic Ethers and Amines)

<u>Rule 1157:</u> This rule of Chemistry for **Cyclic Amines,** states that, Side by side placement of rings of cyclic amines may favor their use as monomers for polymerization, since the rings will be strongly nucleophilic and more strained in character based on the order shown below-

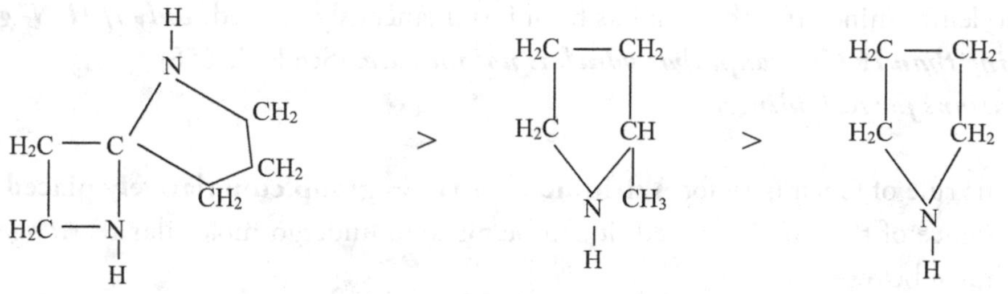

<u>Order of Nucleophilicity</u>

for which four or five-membered rings of (I) will now favor opening of their rings.
(Laws of Creations for Cyclic Amines)

<u>Rule 1158:</u> This rule of Chemistry for **Cyclic Amines and Cyclic Amides,** states that, though cyclic amines are Nucleophiles and cyclic amides are Electrophiles, based on Side–by-side placement of rings, the followings are valid.

$$\text{Cyclic amides} \quad > \quad \text{Cyclic amines}$$

<u>Order of Nucleophilicity</u> (a)

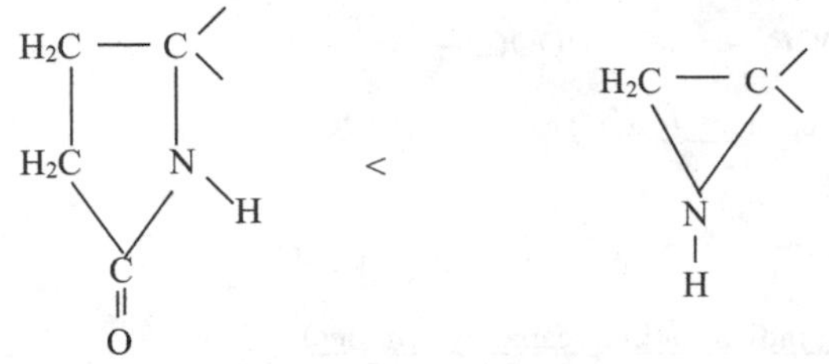

$$H_2C - C\diagdown$$

(structures depicting order of radical-pushing capacity)

$$H_2C - C\diagdown$$

Order of radical-pushing capacity

(b)

(Laws of Creations for Cyclic Amides and Amines)

Rule 1159: This rule of Chemistry for **Cyclic amines,** states that, when $H_2C=$ group is cumulatively placed on the ring next to the N center, the followings are obtained for aziridine-

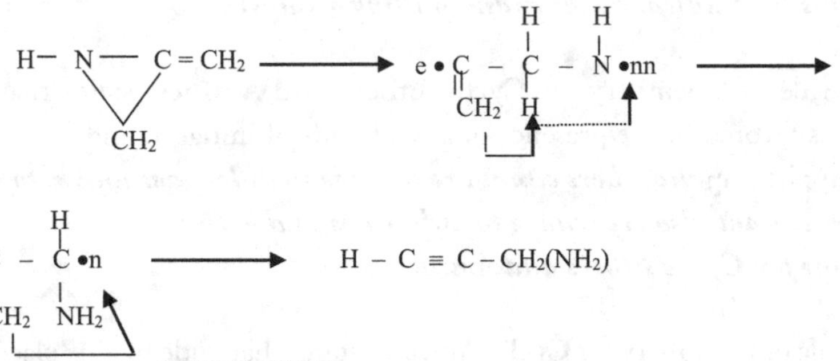

$$H- N - C = CH_2 \longrightarrow e\bullet C - C - N\bullet nn \longrightarrow$$

$$e\bullet C - C\bullet n \longrightarrow H - C \equiv C - CH_2(NH_2)$$

NOT FAVORED

for which its existence on aziridine is impossible, since it is too strained to exist, and therefore rearranges to give an acetylenic amine after the ring has been instantaneously opened, ***only if H_2N- group is less radical-pushing than $H_2C=$ group, that which is not the case***. [See Rule 658]
(Laws of Creations for Aziridines)

Rule 1160: This rule of Chemistry for **Aziridine with $H_2C=$ group cumulatively placed on it,** states that, if the existence of the ring is favored, it can be made to undergo molecular rearrangement of the sixth type as shown below-

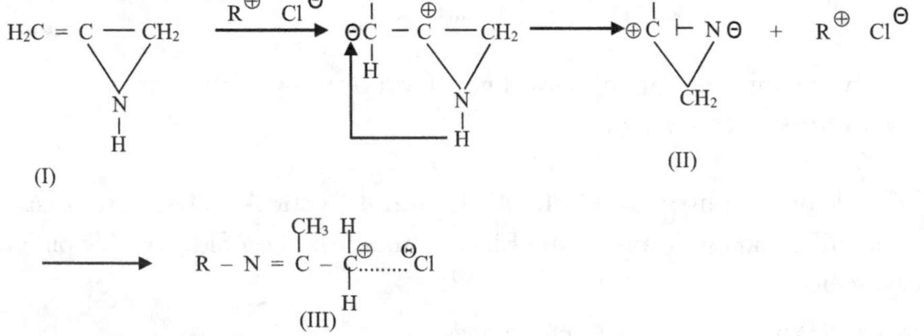

$$H_2C = C - CH_2 \xrightarrow{R^{\oplus} \; Cl^{\ominus}} {}^{\ominus}C - \overset{\oplus}{C} - CH_2 \longrightarrow {}^{\oplus}C \vdash N^{\ominus} + R^{\oplus} \; Cl^{\ominus}$$

(I)   (II)

$$\longrightarrow R - N = C - \overset{\oplus}{C}\ldots\ldots{}^{\ominus}Cl$$

(III)

for which a more nucleophilic monomer which can only favor positively charged or electro-free-radical route is obtained; noting also that when made to exist in Equilibrium state of existence, a cumulenic aldimine (Male) is obtained.
(Laws of Creations for Aziridines)

Rule 1161: This rule of Chemistry for **larger membered Cyclic Amines carrying H₂C= group cumulatively placed next to the N center,** states that, these can be made to undergo polymerization only positively or electro-free-radically as shown below-

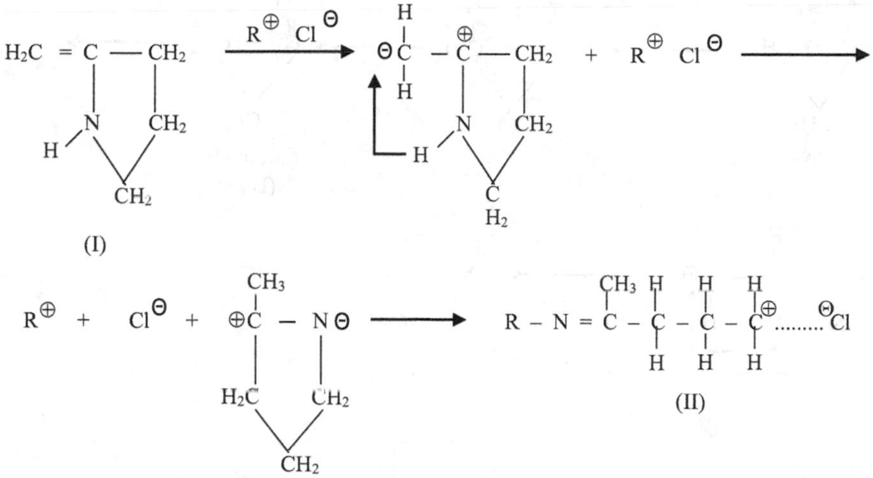

(I)

(II)

for which for the five-membered ring shown above, (II) is the favored monomer unit for the Initiation step, provided (I) is not made to exist in Equilibrium state of existence chargedly; radically same product will be obtained if made to, via resonance stabilization.
(Laws of Creations for Cyclic Amines)

Rule 1162: This rule of Chemistry for **the compounds shown below,** states that, these are no cyclic amines, but amines, (in which unlike the case of etheric counterparts where

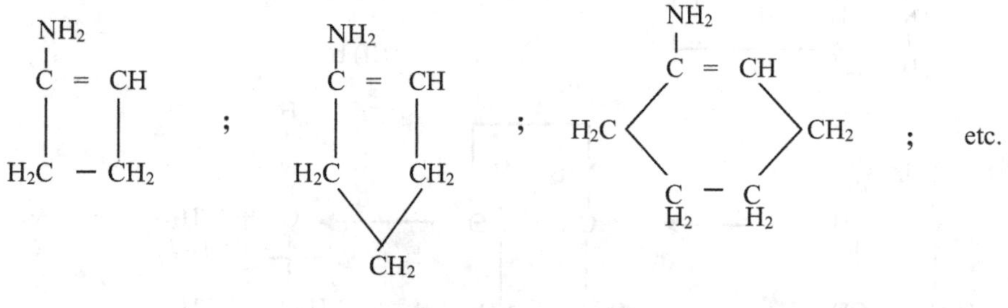

Case (a)- Existence not favored

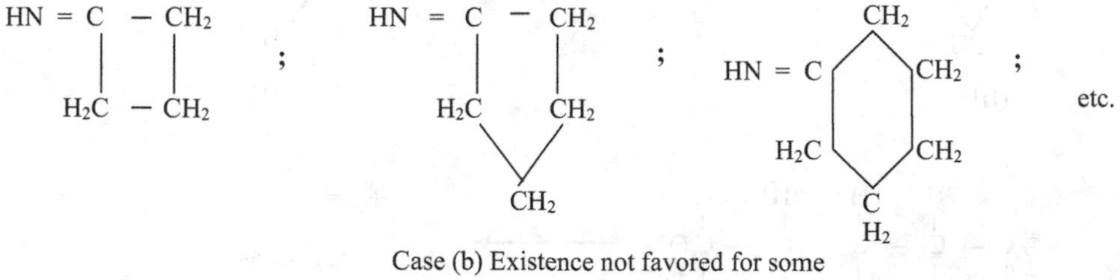

Case (b) Existence not favored for some

it is only six-membered rings and above that rearrange), here they (i.e. case (a) above) all do and therefore do not exist, because they are SELF-ACTIVATED monomers which undergo molecular rearrangement of the first kind to produce those of case (b) above which are more strained, because of the strong radical-pushing capacity of HN= group.
(Laws of Creations for Cyclic alkanes carrying NH₂ group)

Rule 1163: This rule of Chemistry for **the compounds shown below,** states that, their existences are not favored, because they molecularly rearrange to give Nitriles as shown below-

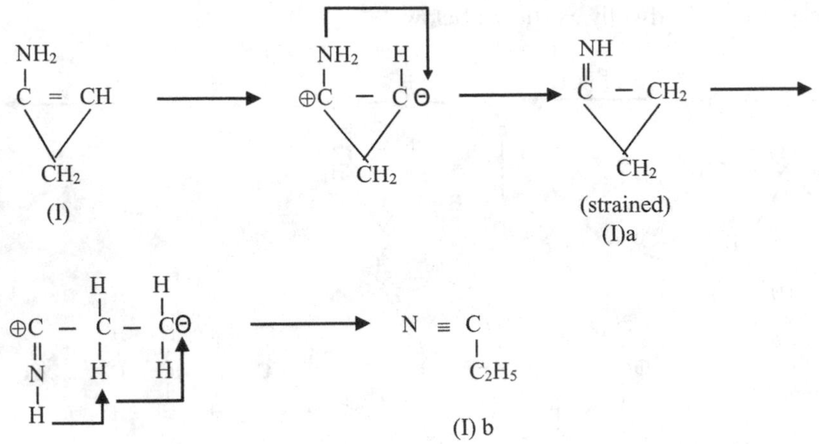

(a)

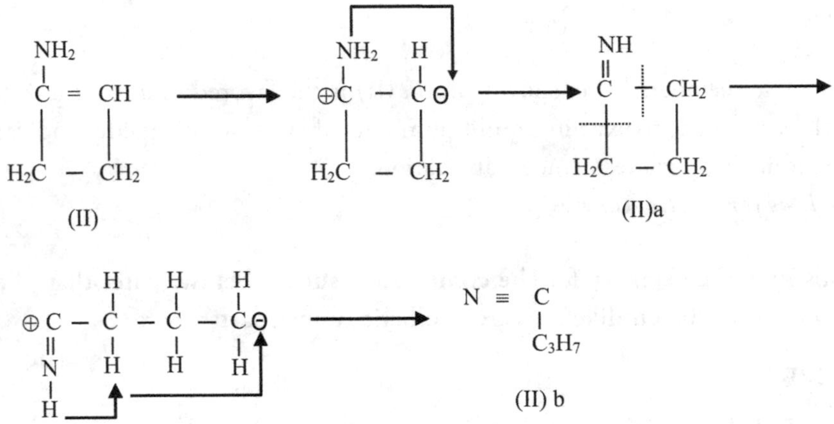

(b)

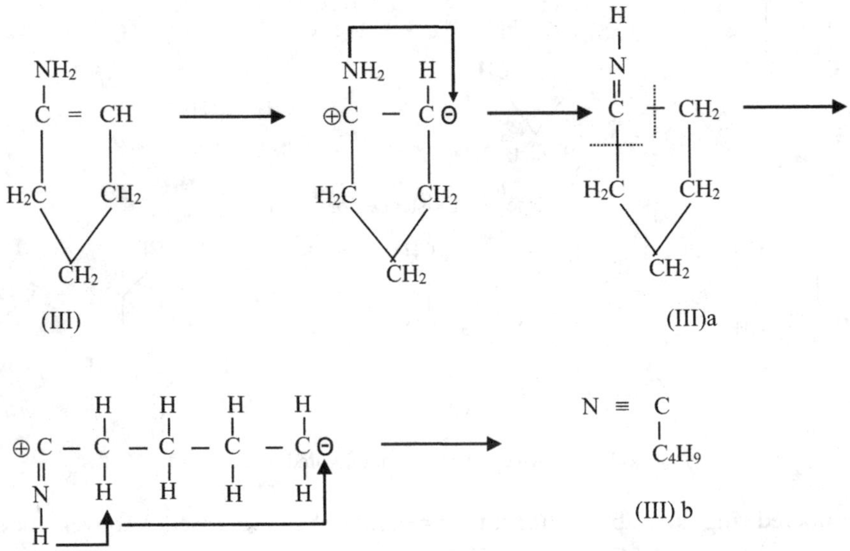

(c)

for which, based on the types of nitriles obtained after molecular rearrangement of the third kind, that is, (I)a, (II)a and (III)a from three, four and five-membered rings, it is obvious that the following is valid, **noting that these largely take place radically.**

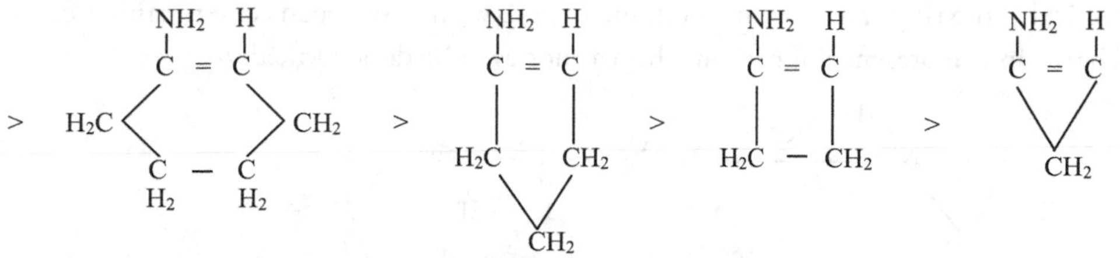

$$\text{NH}_2 \; \text{H} \qquad \text{NH}_2 \; \text{H} \qquad \text{NH}_2 \; \text{H} \qquad \text{NH}_2 \; \text{H}$$

Structures showing order of nucleophilicity of cycloalkene amines.

Order of Nucleophilicity (d)

noting vividly that the NH_2 group is greater than the OH group in radical-pushing capacity.
(Laws of Creations for Cycloalkanes with NH$_2$ group)

Rule 1164: This rule of Chemistry for **compounds shown below,** states that, these are no Cyclic amines or Cyclic amides, but a hybrid of cyclic amine and cyclic ketone, far more strained than cyclic amines and unlike larger membered rings of cyclic amines, the

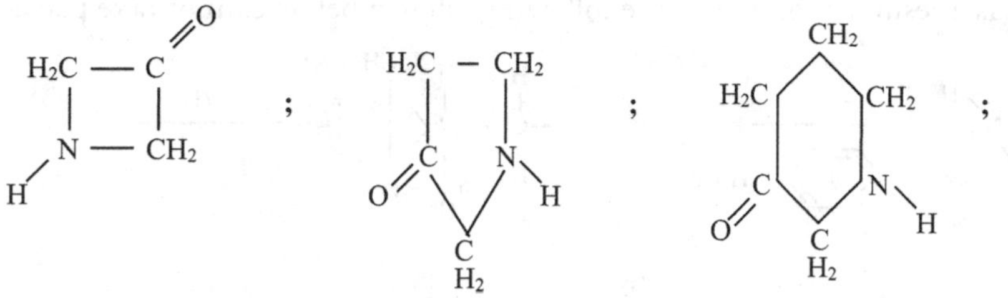

Chemical structures of cyclic amine/ketone hybrids.

(a)

hydrogen atom on the nitrogen center is more loosely bonded than in conventional larger membered cyclic amines, because the compounds above are more nucleophilic and less basic (Secondary cyclic amine) in character, for which the following is valid-

Chemical structures showing order of nucleophilicity and strain energy.

Order of Nucleophilicity and Strain Energy (b)

and unlike four, five and larger membered rings of cyclic amines, these favor positively charged or electro-free-radical polymerizations since MRSE can readily be provided for the rings which are very strained.
(Laws of Creations for Cyclic Etheric/Ketonic compound)

Rule 1165: This rule of Chemistry for **Oxaazacyclopropane a three-membered ring with two hetero atoms in the ring as two nucleophilic functional centers adjacently located,** states that, these cannot be used as functional centers for Addition polymeriza-tion for the following reasons-

(i)   The bond having the highest radical-pushing potential difference is the bond.

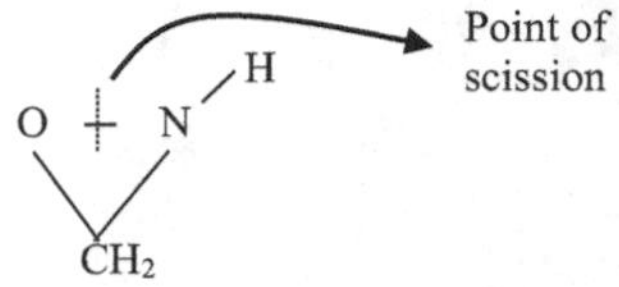

Chemical structure showing point of scission.

Point of scission

(a)

(ii) Ionically, the ring cannot be opened, since the oxygen or nitrogen centers cannot carry a positive charge in the presence of carbon; this cannot even be done radically-

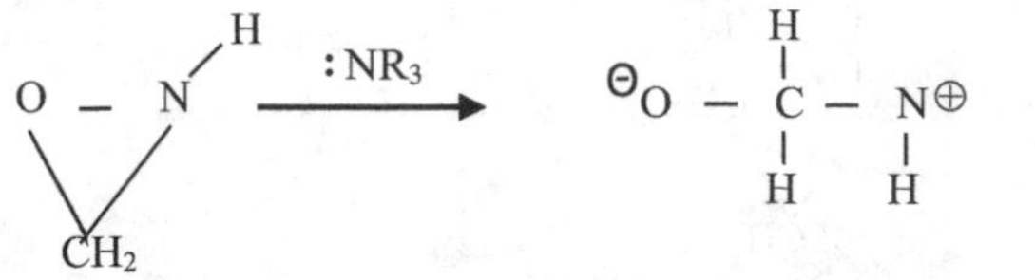

<u>Impossible activated state</u>   (b)

(iii) The compound is always in Equilibrium state of existence

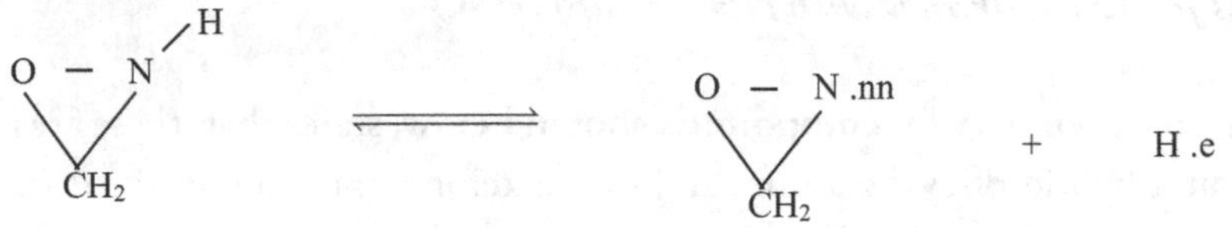

                                                    (c)

(iv) For which as a result of the above, the followings shown below cannot take place-

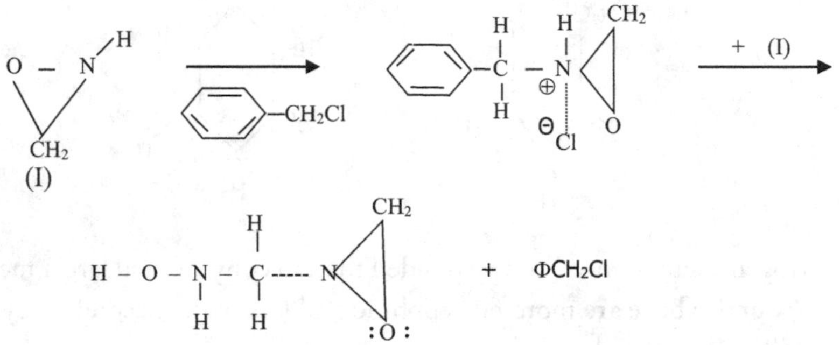

                                                    (d)

because the initiator above an Electrostatically anionically charged-paired initiator, cannot be obtained even with HCl since this compound is always in Equilibrium state of existence (iii). [See Rules 274 to 277] *(Laws of Creations for Oxaazacyclopropane)*

Rule 1166: This rule of Chemistry for **Oxaazacyclopropane**, states that, for it to be used as monomer, *it must first be suppressed (i.e., be in Stable state of existence), and if suppressed, the followings are supposed to be obtained* if indeed it has to exist as a monomer and be used as such, though not all monomers can be polymerized-
Nucleo-non-free-radically:

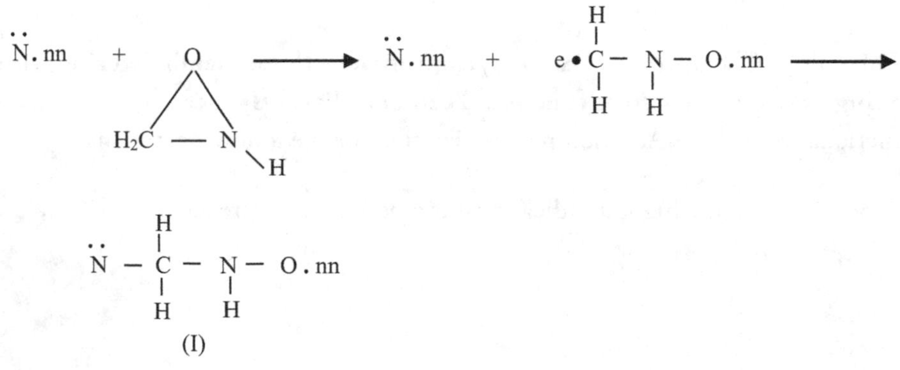

                                                    (a)

- **Electro-free-radically:**

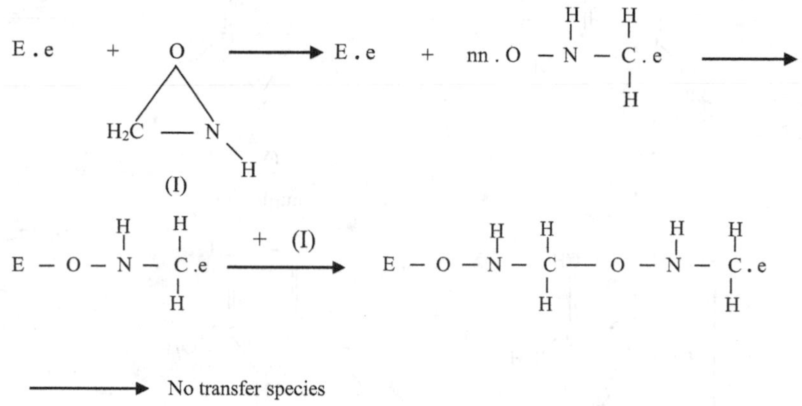

$$E.e \quad + \quad \underset{\substack{H_2C \; — \; N \\ (I) \qquad \; \backslash H}}{O} \quad \longrightarrow \quad E.e \quad + \quad nn.O \; - \; \underset{\substack{| \\ H}}{\overset{\substack{H \quad H \\ | \quad |}}{N}} \; - \; C.e \quad \longrightarrow$$

$$E \; - \; O \; - \; \underset{\substack{| \\ H}}{\overset{\substack{H \quad H \\ | \quad |}}{N}} \; - \; C.e \quad + \quad (I) \quad \longrightarrow \quad E \; - \; O \; - \; \overset{\substack{H \quad H \\ | \quad |}}{N} \; - \; C \; - \; O \; - \; \overset{\substack{H \quad H \\ | \quad |}}{N} \; - \; C.e$$

$$\longrightarrow \quad \text{No transfer species} \qquad\qquad\qquad\qquad\qquad\qquad \text{(b)}$$

for which, only the electro-free-radical route is favored by it being a strong Nucleophile, and the fact that it is not the N-O bond that is the point of scisstion, but the C-O bond, C being far more electropositive than N.
(Laws of Creations for Oxaazacyclopropane)

Rule 1167: This rule of Chemistry for **Oxaazacyclopropane,** states that, by the nature of the instantaneous opening of the ring, it seems that when N and O are adjacently placed in a ring, the following is valid.

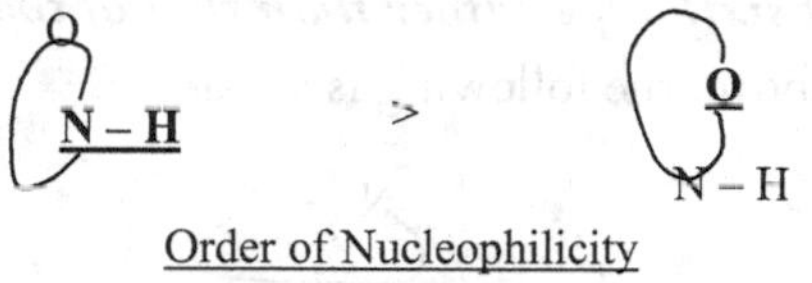

Order of Nucleophilicity

(Laws of Creations for Oxaazacyclopropane)

Rule 1168: This rule of Chemistry for **Oxaazacyclopropane, Cyclic Ether and Amine,** states that, between them *since the number of nucleophilic or electrophilic centers whether adjacently located in a ring or not is additive and is a measure of their strong or weak characters,* the followings are valid for them-

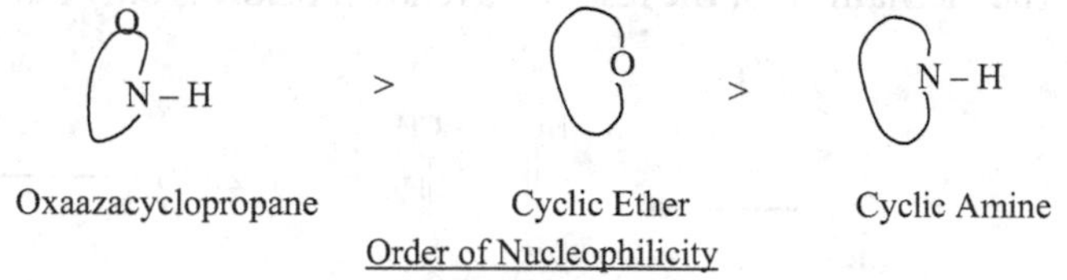

Oxaazacyclopropane   Cyclic Ether   Cyclic Amine
Order of Nucleophilicity

(Laws of Creations for Oxaazacyclopropane)

Rule 1169: This rule of Chemistry for **Succinimide,** states that, this compound cannot be used as a monomer, because it is always in *Activated/Equilibrium state of existence all the time* as shown below-

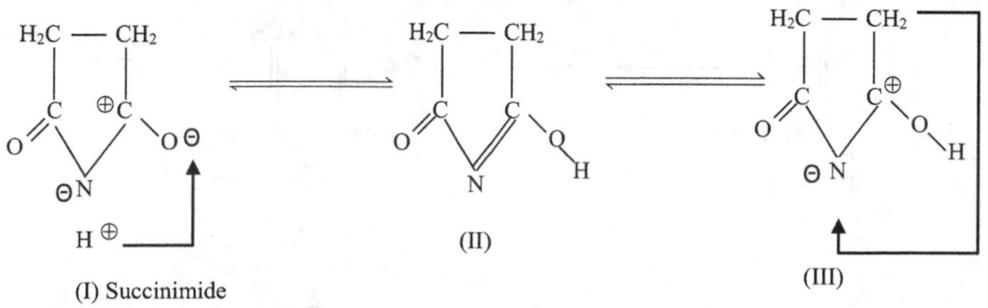

(I) Succinimide   (II)   (III)

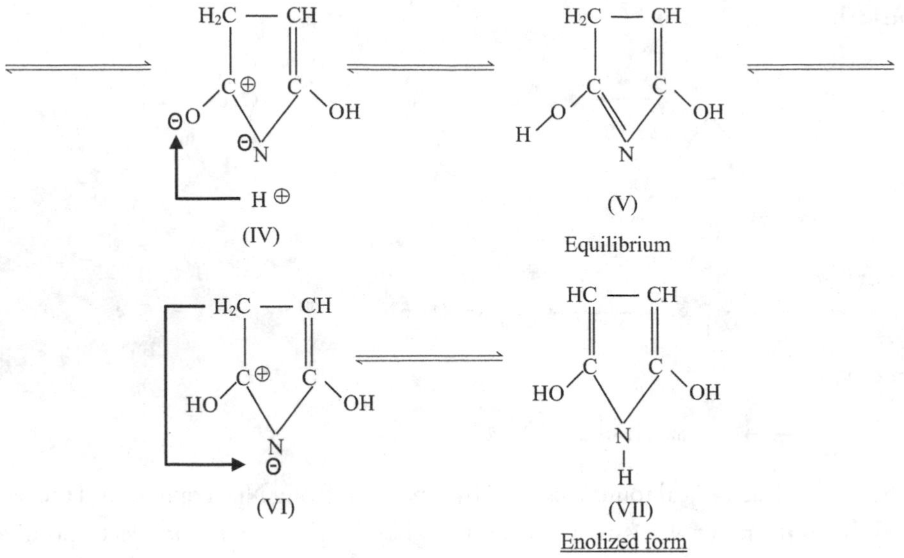

(IV)

(V)
Equilibrium

(VI)

(VII)
Enolized form

i)

for which though shown chargedly above, it can only take place radically, with it being made to undergo *Enolization/Molecular rearrangement phenomena*, enolization taking place via Activated/Equilibrium state of existence and molecular rearrangement of the first kind taking place via Activation of C=N center, *for which it is transfer species of the first kind of sixth type rather than that of the first type from OH group that is involved;* noting that from the above, the following is valid-

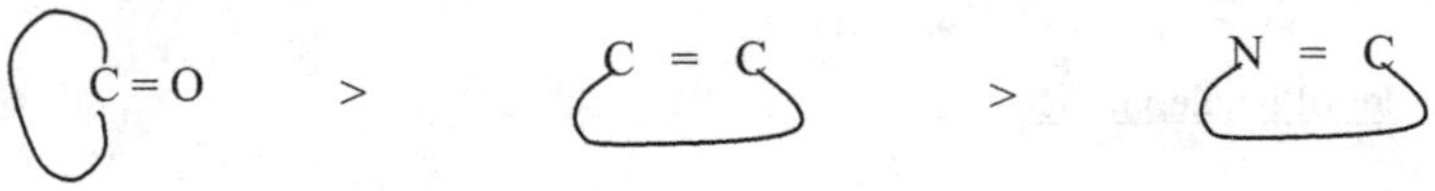

Order of Nucleophilicity

ii)

(Laws of Creations for Succinimide

Rule 1170: This rule of Chemistry for **Pyrrole**, states that, when *Succinimide is heated with Zn dust,* pyrrole is produced, for which the mechanism of the reaction as shown below is only radical in character, since Zn is a non-ionic metal-

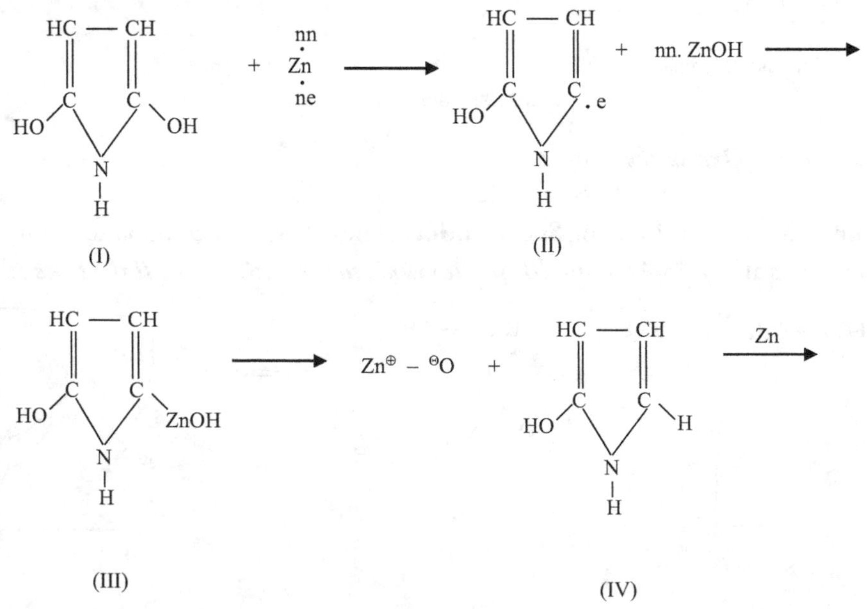

(I)

(II)

(III)

(IV)

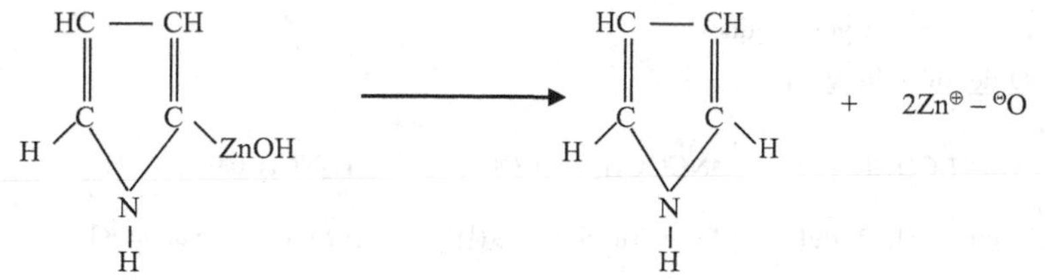

(V)   Pyrrole

a four stage Equilibrium mechanism system, with no form of distillation; noting the structure of Zinc dust which could not suppress the Activated/Equilibrium state of existence of the succinimide, but that of the enolized form, from which (I) was used and ZnO were released in the second and fourth stages along with heat.

(Laws of Creations for Succinimide)

Rule 1171: This rule of Chemistry for **Succinimide,** states that, when it undergoes Enolization/Molecular rearrangement phenomena, the followings are valid from the rearrangements-

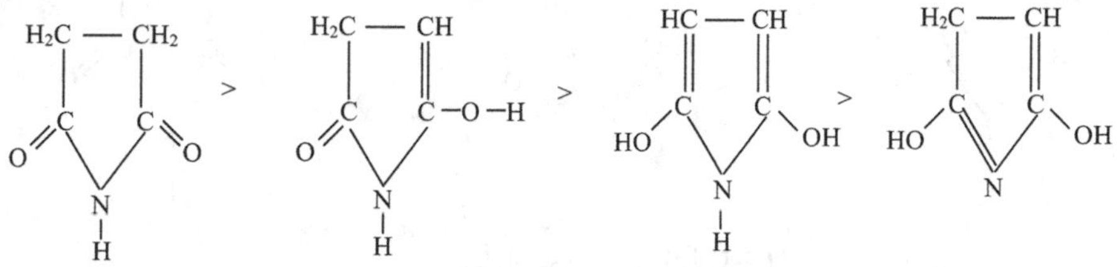

Order of Nucleophilicity

(a)

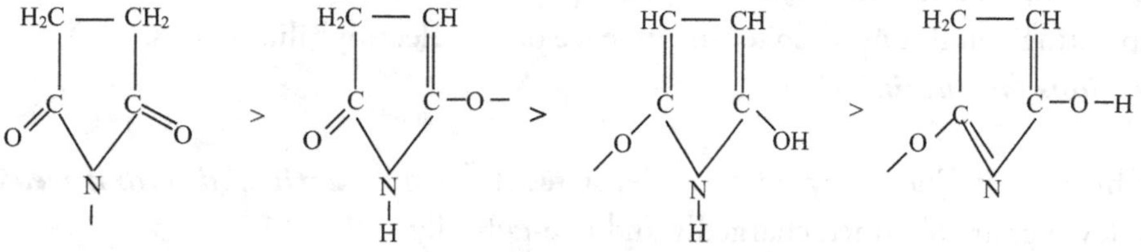

Order of Radical-pushing capacity

(b)

(Laws of Creations for Succinimide)

Rule 1172: This rule of Chemistry for **Succinimide,** states that, while *2-pyrrolidone, a five membered lactam* can be polymerized electro-free-radically using water, succinic-mide cannot be polymerized electro-free-radically, whereas the ring of the succinimide can indeed be opened under heat as shown below using water (same type of initiator used for 2-pyrrolidone); the reasons being the followings-

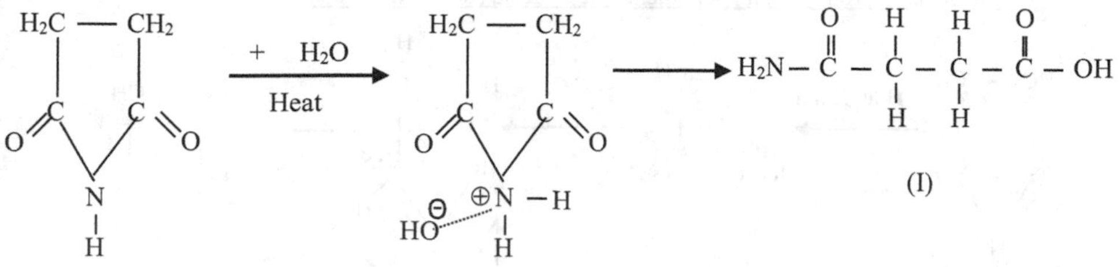

(I)

(a)

Succinimide   >   Cyclic amides

Order of Nucleophilicity

(b)

299

Succunimide > Cyclic amides

Order of Acidity

(c)

$$HOOCCH_2CH_2COOH \quad > \quad H_2NCOCH_2CH_2COOH \quad > \quad H_2NCH_2CH_2COOH$$

[From Succinic anhydride]　　　[From Succinimide (I)]　　　[From Cyclic amides]

Order of Acidity

(d)

with succinic anhydride and succinimide having two Ys and one X centers, while Cyclic amide and cyclic esters have one Y and one X centers.
(Laws of Creations for Succinimide)

Rule 1173: This rule of Chemistry for **Succinimide,** states that, in view of the presence of O = group adjacently located to a N center in place of 2H atoms in 2-pyrrolidone, succinimide is far less strained than 2-pyrrolidone, for if it was separately or distantly located as shown in (A), below, then it will be more strained.

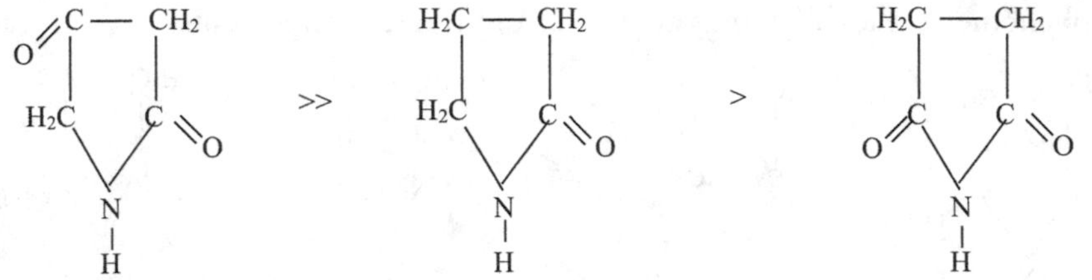

Order of strain energy in ring

for which as one of the reasons, succinimides are not popularly known to undergo anionic polymerization under mild operating conditions, despite the presence of two electrophilic centers.
(Laws of Creations for Succinimides)

Rule 1174: This rule of Chemistry for **Pyrrole,** states that, when ***activated with a weak activating source,*** the followings are obtained chargedly and free-radically-

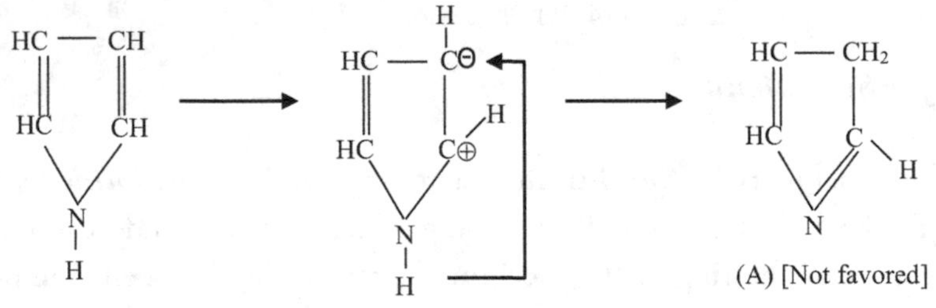

Molecular rearrangement phenomenon of 1st Kind (Chargedly)-Incomplete

(a)

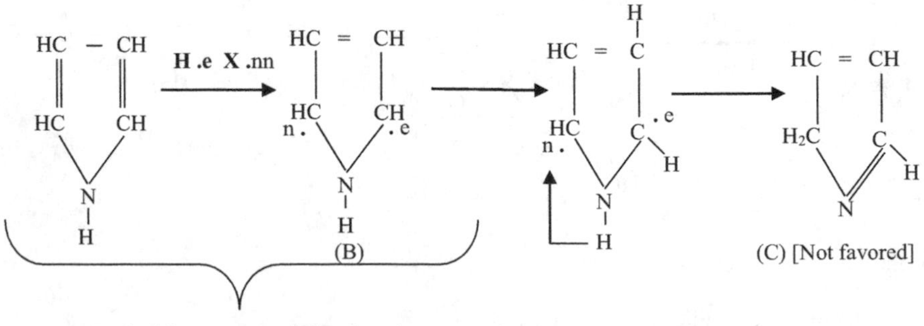

Discrete Radical /resonance stabilization phenomena (Favored)

(b)

for which the presence of (B) via resonance stabilization (that which cannot be obtained chargedly) is worthy of note, and so also are (A) and (C) that which cannot be obtained chargedly or radically, being less nucleophilic than pyrrole.
(Laws of Creations for Pyrrole)

Rule 1175: This rule of Chemistry for **Pyrrole**, states that, when *kept in Equilibrium state of existence*, it like phenol and aniline undergo what is called *Continuous Closed-loop Electro-radicalization phenomena*, that in which the radical carried by them when H is held can be resonance stabilized as shown below-

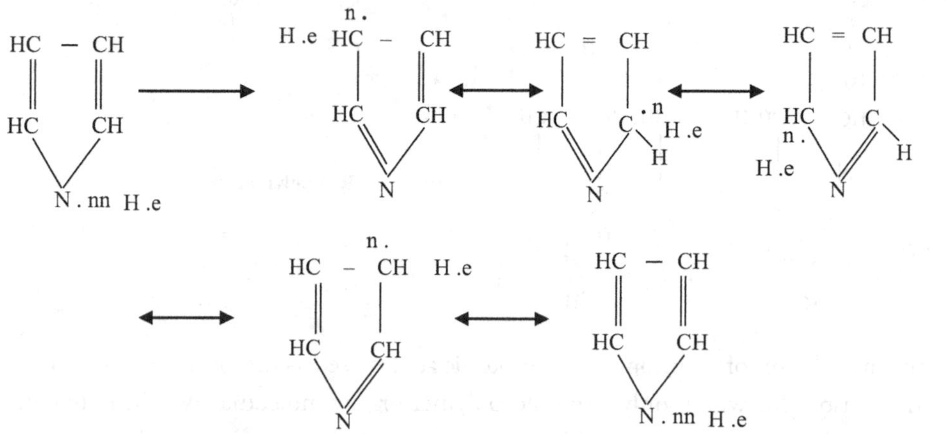

Continuous closed-loop Electro radicalization phenomena (Favored) (a)

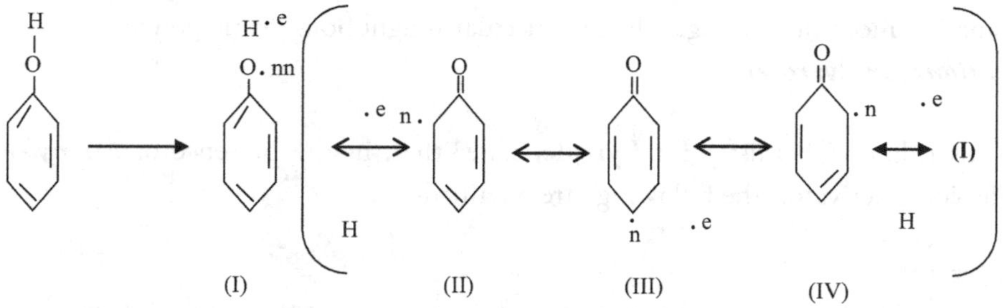

Continuous closed-loop Electro-radicalization phenomena (favored) (b)

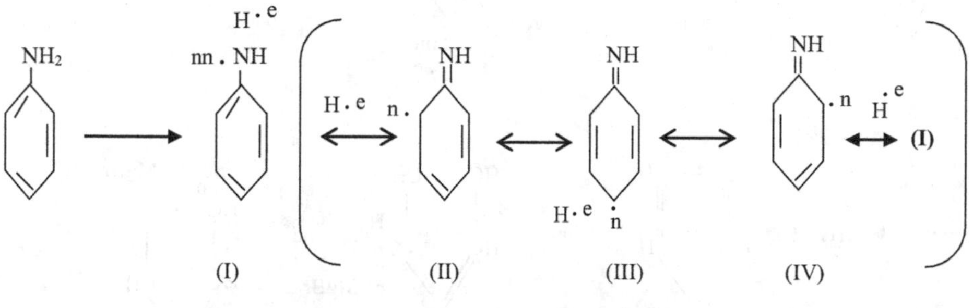

Continuous closed-loop Electro-radicalization phenomena (favored) (c)

noting that with pyrrole, one has to move twice round the ring to come back to its starting point (like in benzene) and also the 2-,or 5- and 3- or 4- positions available for replacements, while with phenol and aniline only one round movement was required, with o- and p- positions available for replacements with radical-pushing groups.
(Laws of Creations for Pyrrole)

Rule 1176: This rule of Chemistry for **Pyrrole,** states that, when used as a monomer, this can only be done electro-free-radically, the route natural to it, since when activated it must undergo resonance stabilization in order to provide centers for addition and since the ring has no point of scission, due to steric limitations, only polymeric resins can be produced as shown below-

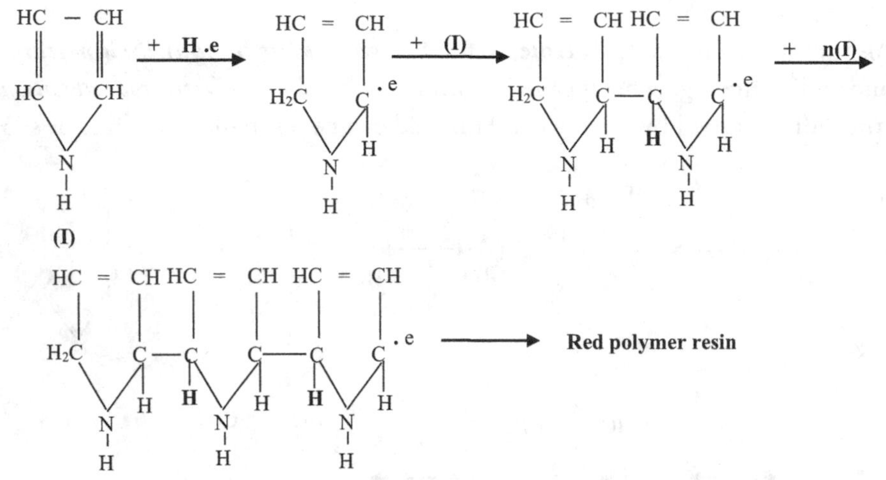

noting that the mechanism of addition when a protonic acid is used as initiator is Equilibrium mechanism via backward addition, for which only either no polymer or low molecular weight polymers (resins) are produced when sufficient time is given for it and noting again that the rings are indeed syndiotactically placed, not isotactically as shown above; for which when an initiator from NaCN is used, the mechanism will be Combination mechanism, to give large molecular weight non-resinic polymers.

(Laws of Creations for Pyrrole)

Rule 1177: This rule of Chemistry for **Pyrrole,** states that, in the presence of *Methyl-magnesium bromide,* a non-ionic activator, the followings are obtained-

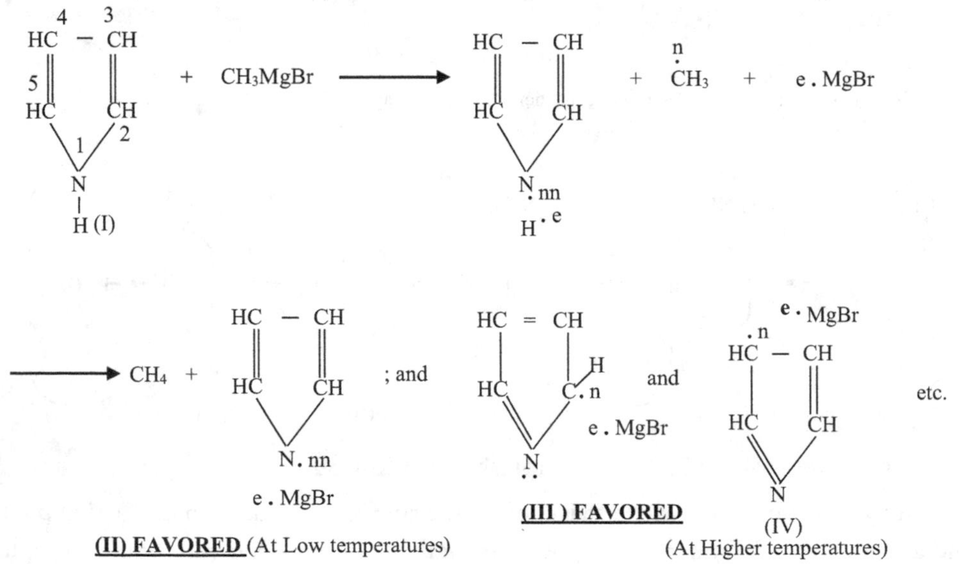

for which, how the positions can be used under different operating conditions have been well defined with the use of this activator called Grignard's reagent whose full structure has not fully been shown above.

(Laws of Creations for Pyrrole)

Rule 1178: This rule of Chemistry for **Pyrrole,** states that, after reaction with *Methyl magnesium bromide* with evolution of methane to give the bromomagnesium amide in Equilibrium state of existence, at 0°C, addition of alkyl halides or acyl chlorides gives N-alkyl derivative or N-acyl derivative respectively as shown below-

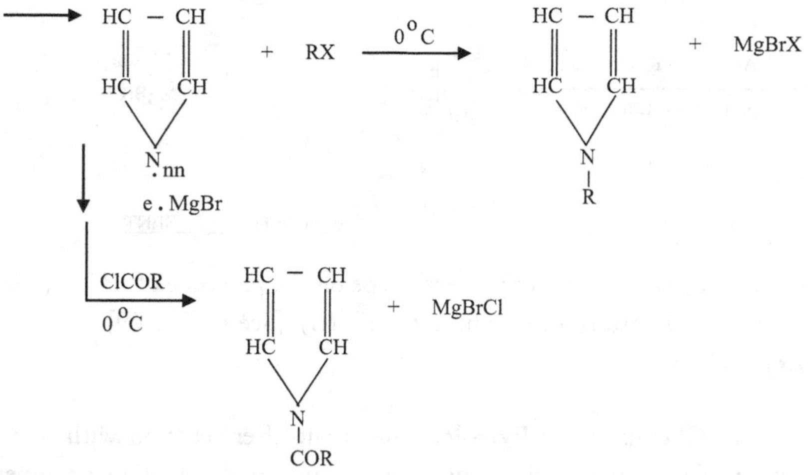

HC — CH
HC   CH
N nn
e . MgBr   + RX → (0°C) → HC — CH / HC CH / N / R   + MgBrX

ClCOR (0°C) → HC — CH / HC CH / N / COR   + MgBrCl

NO ACTIVATION

a one stage Equilibrium mechanism systems for both of them.
(Laws of Creations for Pyrrole)

Rule 1179: This rule of Chemistry for **Pyrrole,** states that, after reaction with *Methyl magnesium bromide* with evolution of methane to give the bromomagnesium amide in Equilibrium state of existence at higher temperatures, addition of alkyl halides or acyl chlorides gives 2-substituted alkyl- or acyl pyrroles as shown below-

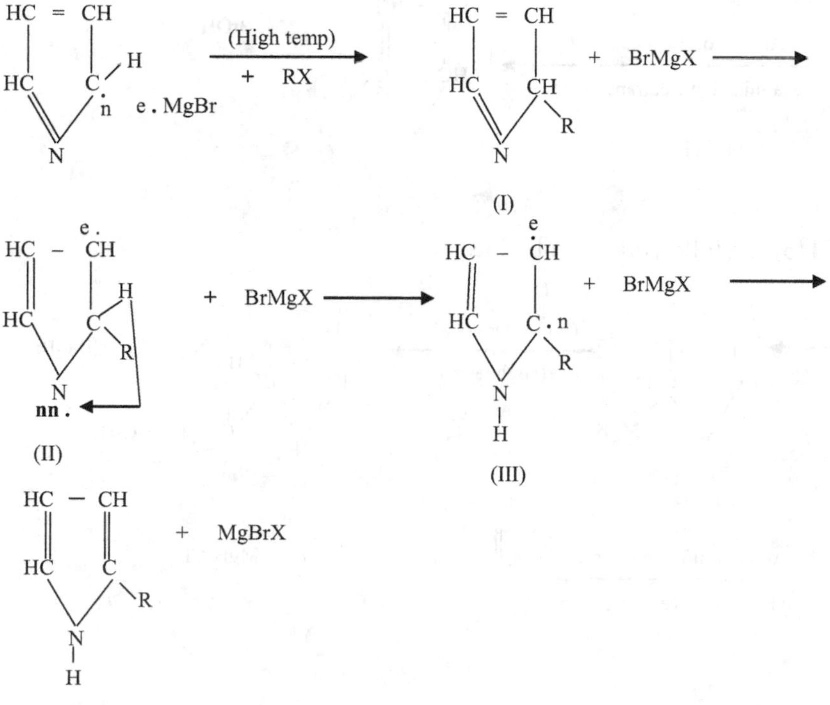

HC = CH / HC C H / n e . MgBr → (High temp) + RX → HC = CH / HC CH / R / N   + BrMgX →   (I)

HC — CH e. / HC C H R / N nn . (II)   + BrMgX → HC — CH e. / HC C . n R / N H   (III)   + BrMgX →

HC — CH / HC C R / N H   + MgBrX   (IV)

ACTIVATION PRESENT   (a)

303

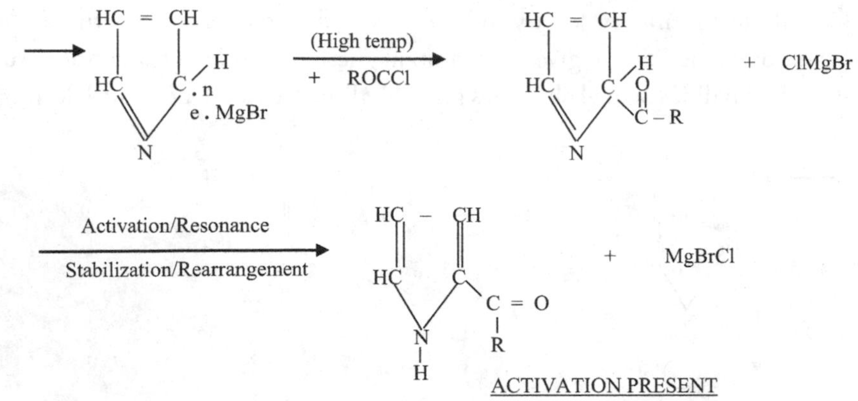

ACTIVATION PRESENT

(b)

for without the molecular rearrangement of the sixth type on (I), presence of (IV) would be impossible, noting that (I) when activated, resonance stabilized to give (II). [See Rule 1169]

(Laws of Creations for Pyrrole)

Rule 1180: This rule of Chemistry for **Pyrrole,** states that, after reaction with methyl magnesium bromide with evolution of methane to give bromomagnesium amide, at higher temperature, Carbon dioxide gives the 2-carboxylic acid; ethyl chloroformate gives ethyl-2 pyrrolecarbo-xylate, and ethyl formate gives 2-pyrrolecarboxaldehyde as shown below all via Equilibrium mechanism-

For CO_2 and water or CH_4, the followings are to be expected-

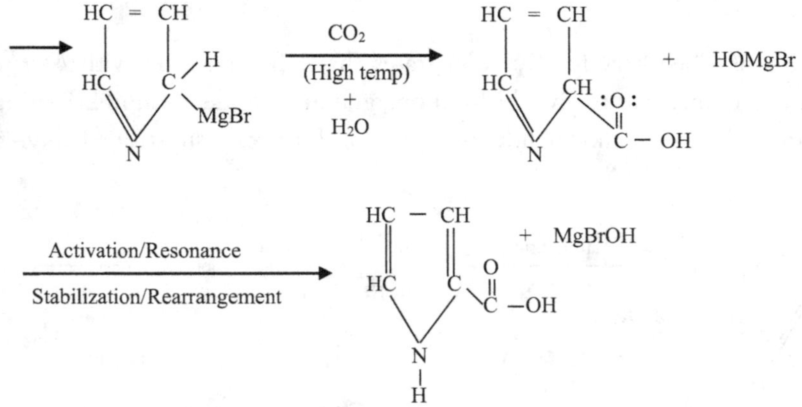

(a)

For $ClCOOC2H5$, the followings are obtained.

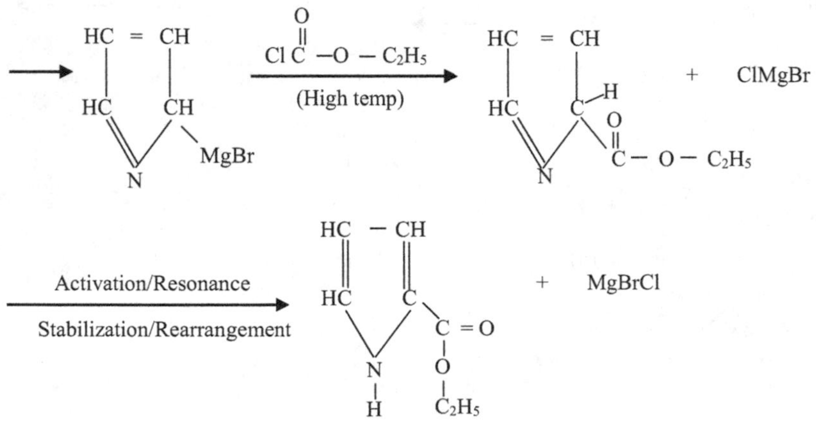

(b)

304

With HCOOC2H5, the followings are obtained.

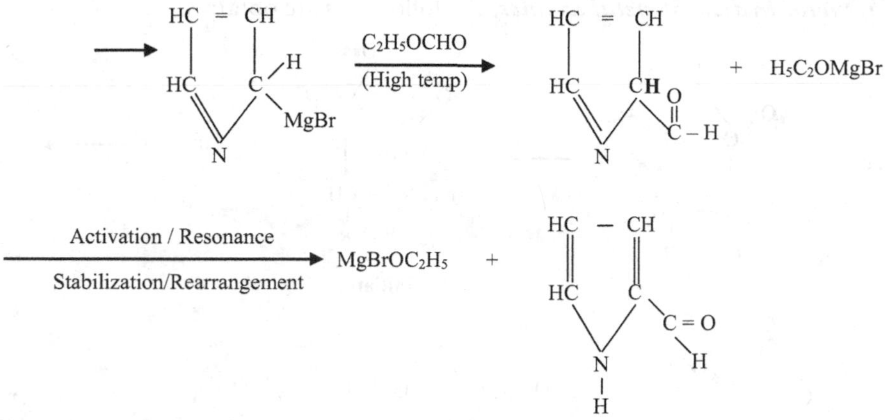

(c)

from which one can observe the unique and dual functions of Grignard's reagent here.
(Laws of Creations for Pyrroles)

Rule 1181: This rule of Chemistry for **Pyrrole,** states that, from the character exhibited by it, the followings are valid –

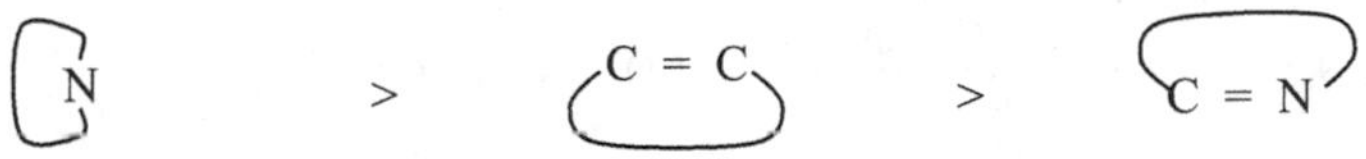

Order of Nucleophilicity   [See Rule 1169]

(Laws of Creations for Pyrrole)

Rule 1182: This rule of Chemistry for **Carboanhydrides (N-carboxyl - α - amino acid anhydrides) also referred to as oxazolidine - 2, 5- diones,** states that, unlike lactams, carboanhydrides have two hetero atoms or two nucleophilic functional centers –  and  and two different types of electrophilic centers, for which therefore, the followings are valid-

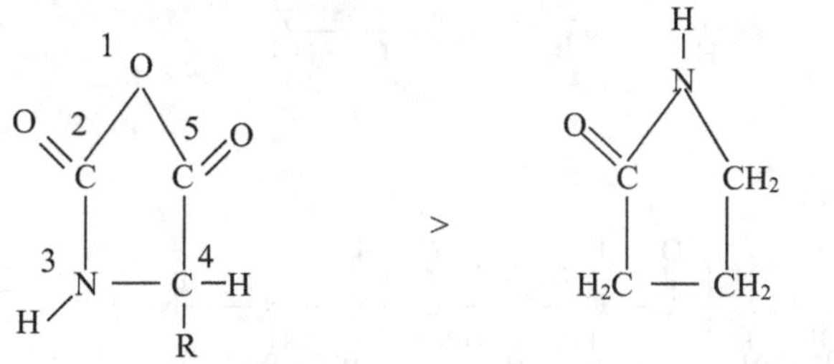

N - Carboxyl - α - amino acid anhydride   **γ - butyrolactam**

Order of nucleophilicity and electrophilicity

(a)

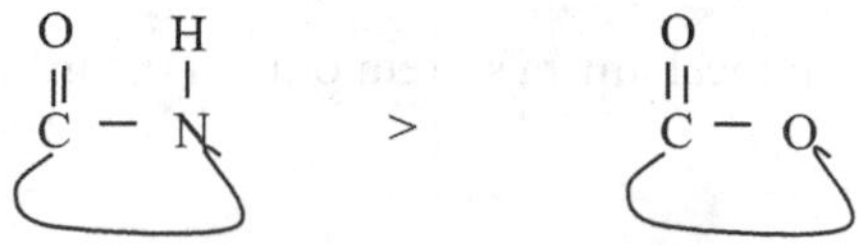

Order of electrophilicity

(b)

(Laws of Creations for Carboanhydrides)

305

Rule 1183: This rule of Chemistry for **Carboanhydrides,** states that, anionically, using "so called" strong bases such as *Na, metal hydrides, metal amides,* the followings are obtained-

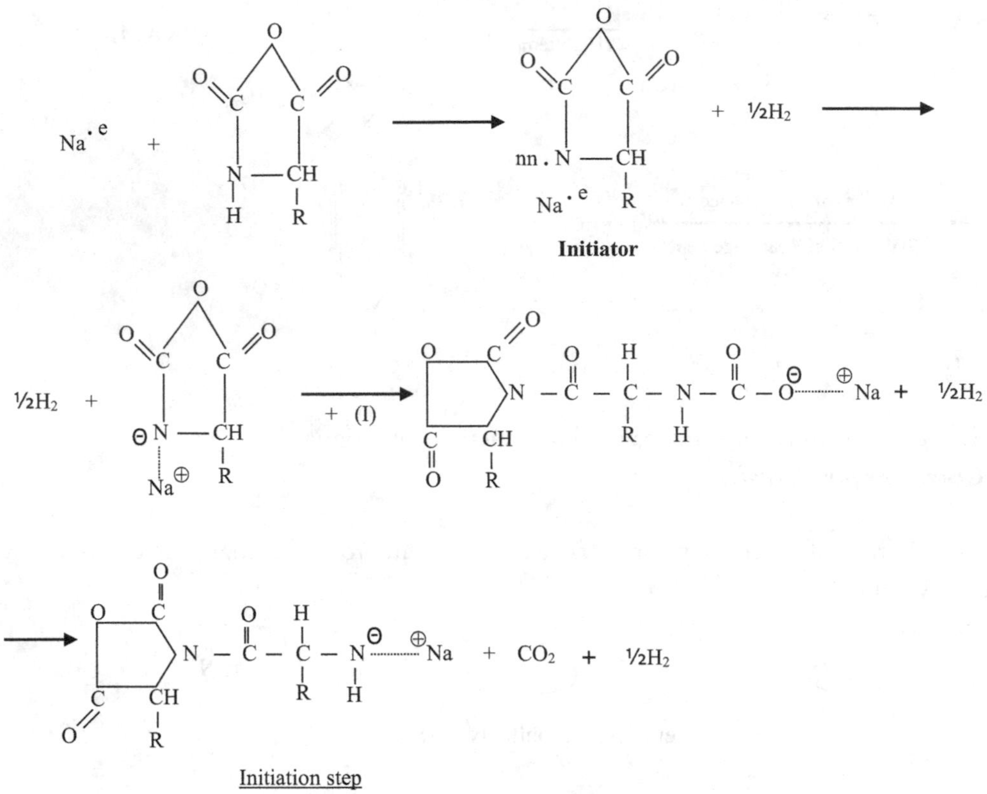

Initiator

Initiation step

(a)

for which the release of CO_2 continues throughout the propagation step as shown below, to produce **Nylon 2** anionically with a terminal ring after termination using a foreign agent; that is, the monomer is part of the imitator and the carrier of the chain-.

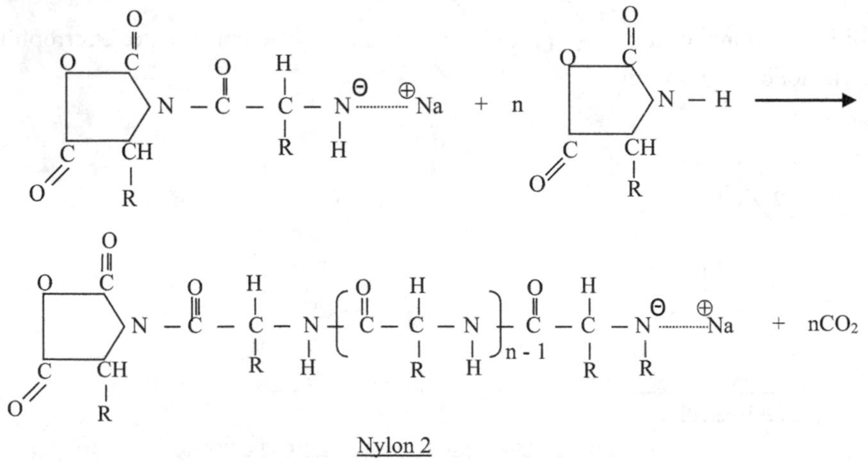

Nylon 2

(b)

noting how H_2 was released in a one stage Equilibrium mechanism system only in the initiator preparation step.

(Laws of Creations for Carboanhydrides)

Rule 1184: This rule of Chemistry for **Carboanhydrides,** states that, when weak bases such as ***non-metallic amine are involved,*** the route is "cationic" i.e., electro-free-radical for which higher polymerization temperatures will be desired for addition of more energy to the ring-

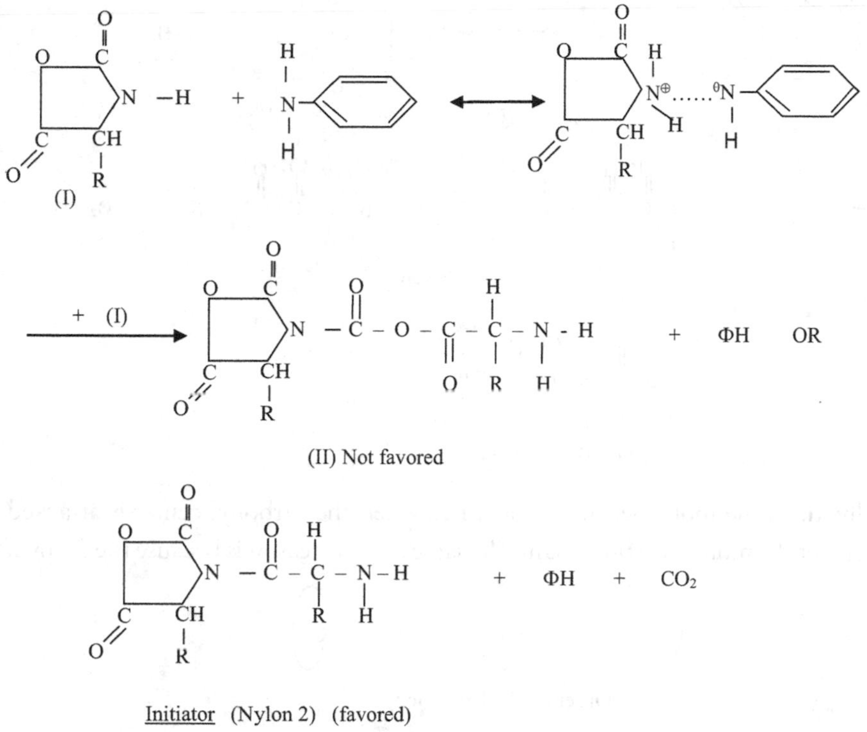

(I)

(II) Not favored

Initiator (Nylon 2) (favored)

Where Φ is a phenyl-NH- group . (a)

noting that when the H electro-free-radical species peaks up the monomer, CO_2 is released before adding to the backwardly growing center with ΦH being only released at the end of propagation step and not during initiation as shown above or during propaga- tion, while CO_2 is released in both initiation and propagation steps

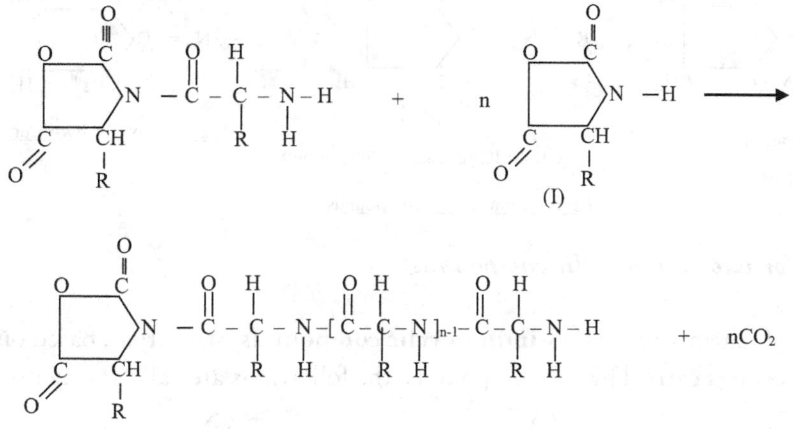

(I)

+ nCO_2

(b)

noting that only primary and secondary amines can be used as source of initiators electro-free-radically when amines are involved and this is only possible when the initiator when prepared remains in Equilibrium state of existence, otherwise if in Stable state of existence, then only the anionic route will be favored.
(Laws of Creations for Carboanhydrides)

Rule 1185: This rule of Chemistry for **the type of carboanhydride shown below wherein H₂C= group is cumulatively placed on the ring,** states that, the followings are obtained when used as a monomer-.

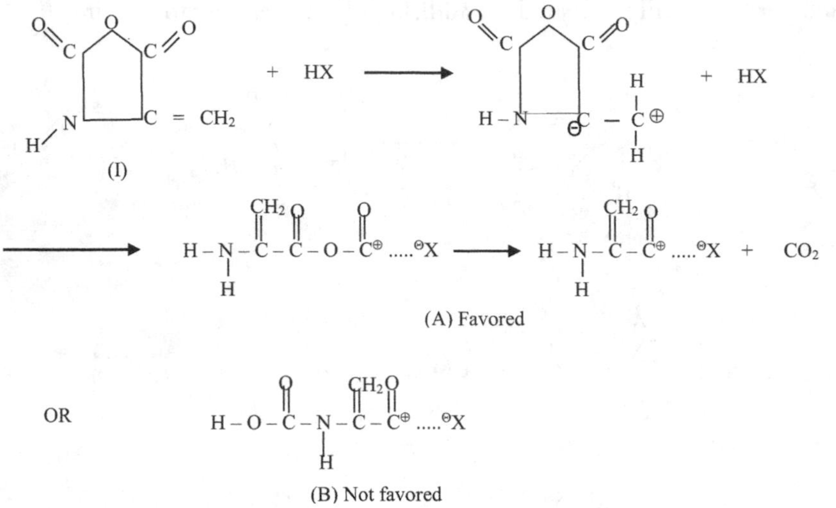

(A) Favored

OR

(B) Not favored

(a)

and since anionically, the same monomer unit is obtained when the carbonyl center is attacked, then the Electrophilic character of the monomer still remains the same and the reason is because the following is valid.

Order of Nucleophilicity

(b)

(Laws of Creations for Carboanhydrides)

Rule 1186: This rule of Chemistry for **Exo-imino-cyclic compounds,** states that, these are cyclic Electrophilic compounds visibly carrying R-N= group cumulatively placed in between either two hetero atoms or one hetero and one non-hetero atom centers in the rings, three examples of which are shown below-

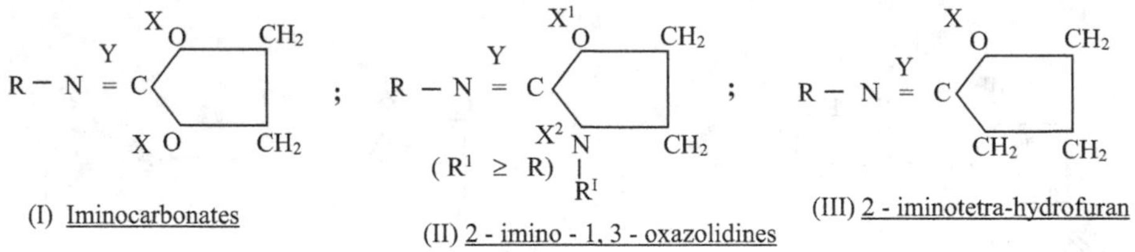

(I) Iminocarbonates ; (II) 2 - imino - 1, 3 - oxazolidines ($R^1 \geq R$) ; (III) 2 - iminotetra-hydrofuran

(Exo - imino-cyclic compounds)

(Laws of Creations for Exo-imino-cyclic compounds)

Rule 1187: This rule of Chemistry for **Exo-imino-cyclic compounds**, states that, based on the types of activation/functional centers carried by the compounds, the followings are valid for them-

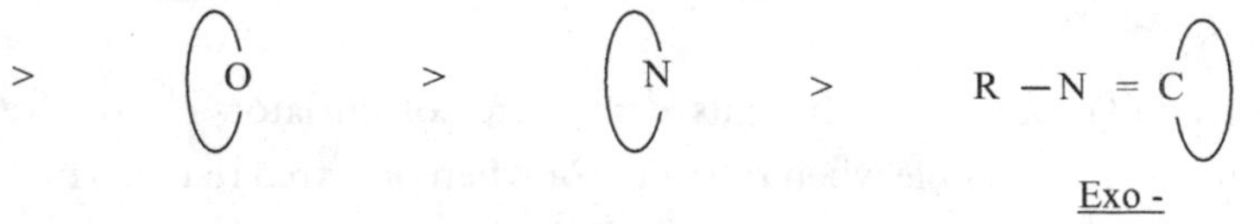

Exo -

Order of Nucleophilicity (R is a radical-pushing group)

(Laws of Creations for Exo-imino-cyclic compounds)

308

Rule 1188: This rule of Chemistry for **Endo-imino-cyclic compounds,** states that, these are cyclic Nucleophilic compounds visibly carrying C=N activation center next located to an O center in the rings, three examples of which are shown below-

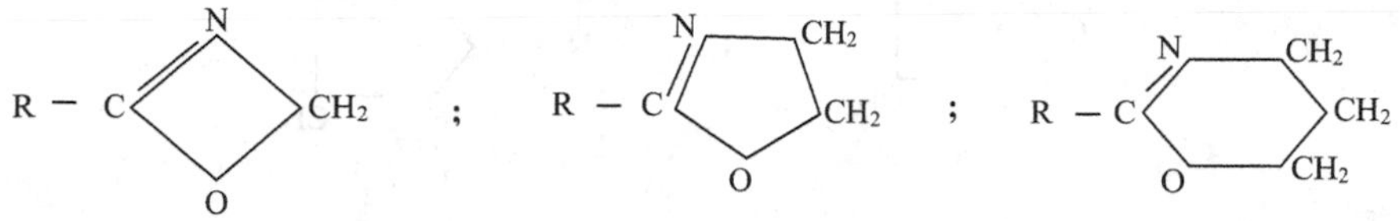

(Endo - imino cyclic ethers)

(Laws of Creations for Endo-imino cyclic compounds)

Rule 1189: This rule of Chemistry for **Endo-imino-cyclic compounds,** states that, based on the types of activation/functional centers carried by these unique ether compounds, the followings are valid for them-

Order of Nucleophilicity

(Laws of Creations for Endo-imino-cyclic compounds)

Rule 1190: This rule of Chemistry for **Exo-Iminocarbonates,** states that, when the R group on the N center is changed to H, the compound cannot be used as a monomer as shown below-

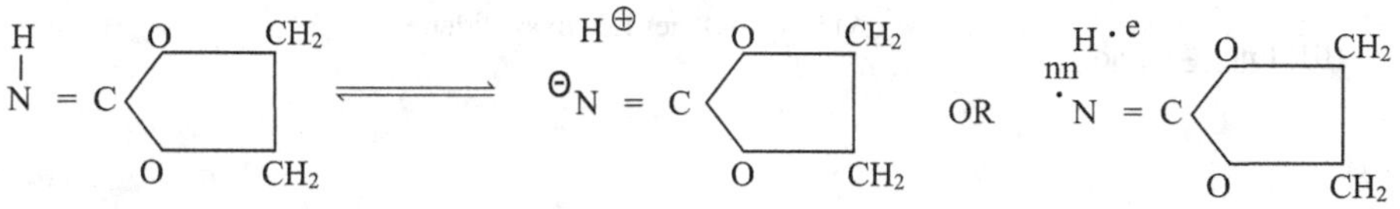

(I) An Iminocarbonate

(II) Favored existence

(a)

for which like monomers with N = C activation centers, where the hydrogen atom on the nitrogen center must be replaced with an alkylane or other suitable groups before they can fully be used as a monomer, the same applies here as shown below-

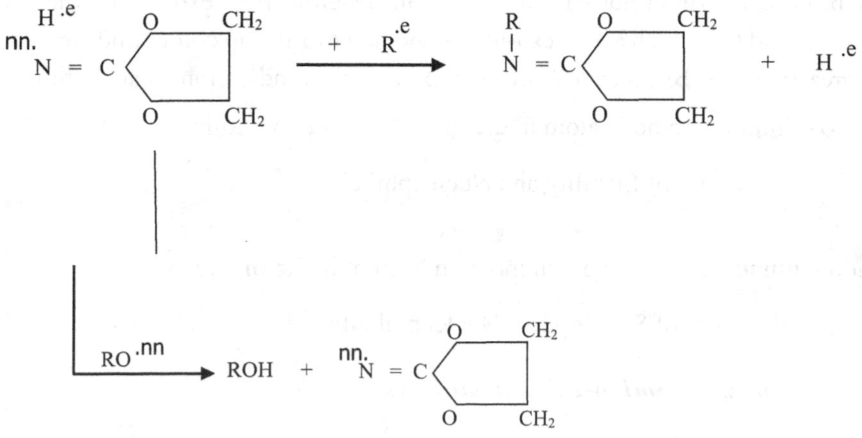

(b)

(Laws of Creations for Iminocarbonates)

309

Rule 1191: This rule of Chemistry for **Exo 2-imino-1,3-oxazolidines,** states that, when one of alkylane groups (R and R^1) is changed to H, the followings are obtained-

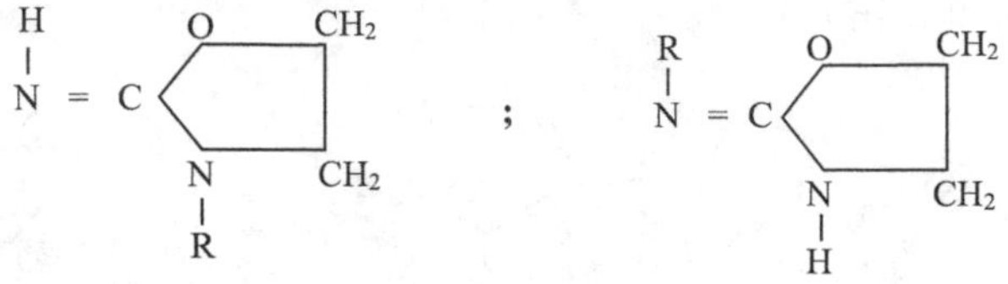

(I)   ;   (II)   (a)

in which for (I), the hydrogen atom must be replaced with an alkylane group of less capacity than R, before it can fully be used as a monomer; and for (II), the followings take place when weakly activated-

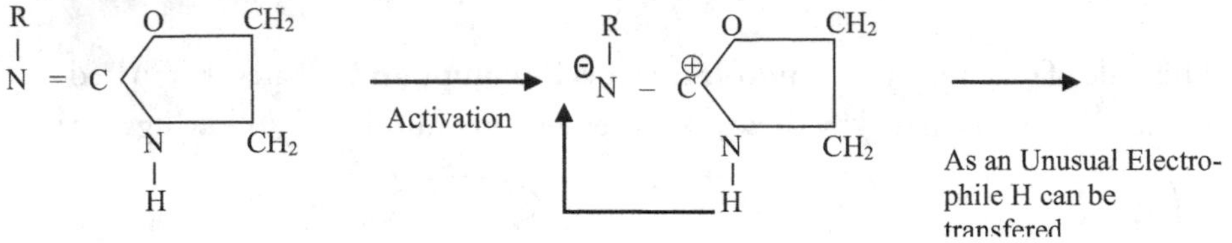

(II) <u>Exo - imino</u>

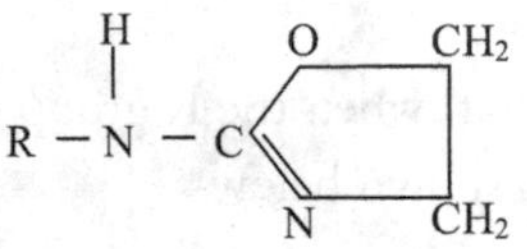

As an Unusual Electro-phile H can be transfered

(III) <u>Endo - imino</u>   ;

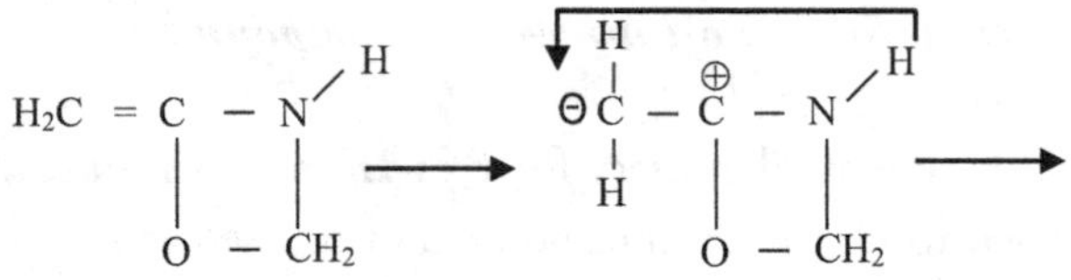

(A) 2-methylene 1,3- oxazolidene

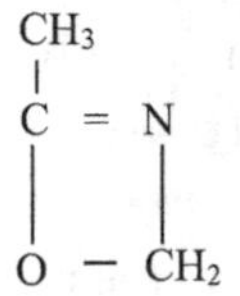

(B) Endo-imino cyclic ether   (b)

for which it can be observed that Endo- umino is more nucleophilic than exo-imino when N is in the ring, and between the (II) and (A) which are an exo-imino and non exo-imino compounds respectively, only (A) can molecularly rearrange to (B), an endo-imino compound, clear indications that the followings are valid-

Exo – imino with no N atom in the ring   >   Endo – imino

Order of Stability and Nucleophilicity   (c)

Endo – imino   >   Exo – imino with N atom in the ring

Order of Stability and Nucleophilicity   (d)

(Laws of Creations for Exo 2-imino-1,3-oxazolidenes)

Rule 1192: This rule of Chemistry for **Exo-imino-cyclic compounds,** states that, when used as monomers, since if these are said to be Nucleophiles, the followings are obtained "cationically" or indeed electro-free-radically for them-

For Iminocarbonates

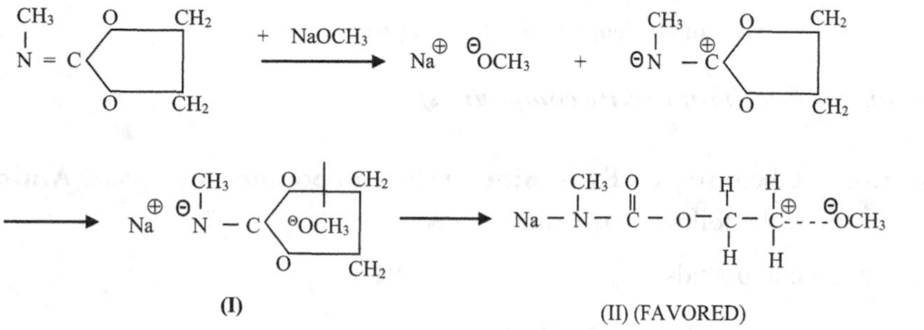

(I)   (II) (FAVORED)   (a)

For 2-imino-1,3- oxazolidenes

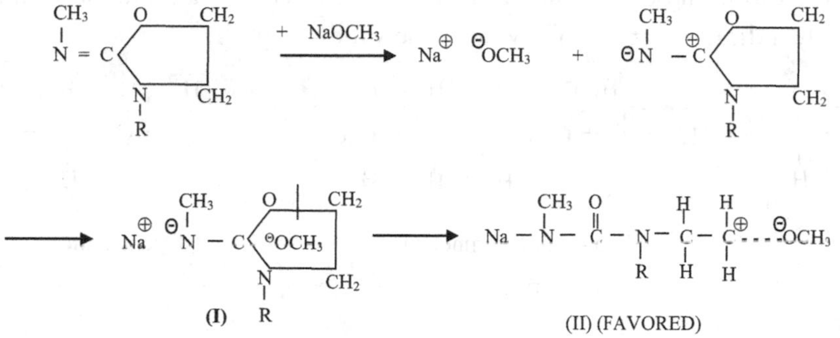

(I)   R   (II) (FAVORED)   (b)

For 2-iminotetra-hydrofuran

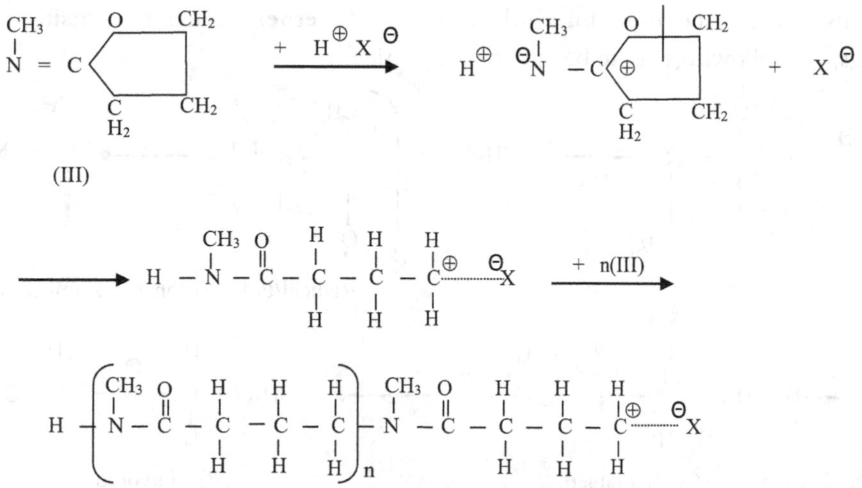

(III)

No transfer species   (c)

noting that none of the reactions above can take place chargedly as shown above, but only radically, and the three exo-imino-cyclic compounds are weak Electrophiles because there is no transfer species electro-free-radically, while anionically (the route natural to the monomers) the rings can be opened via transfer species of the eleventh type; noting that *such rings cannot readily be opened instantaneously (otherwise cumulenic aldimines are obtained,).*

(Laws of Creations for Exo-imino-cyclic compounds)

311

Rule 1193: This rule of Chemistry for **Exo-imino-cyclic compounds,** states that, between the three members, based on what is carried by them, the followings are valid.

Iminocarbonates > 2-imino-1,3-oxazolidenes > 2-iminotetrahydrofuran

(I)　　　　　　　　　　**(II)**　　　　　　　　　　**(III)**

Order of Nucleophilicity/Electrophilicity

(Laws of Creations for Exo-imino-cyclic compounds)

Rule 1194: This rule of Chemistry for **Exo-imino-cyclic compounds and Cyclic Amines,** states that, between the two families, the following is valid-.

Exo - imino compounds　　　　>　　　　Cyclic amines

<u>Order of Nucleophilicity</u>　　　　　　　　　　　　　　　(a)

and the fact that the exo-imino-compounds favor the electro-free-radical, anionic or nucleo-non-free-radical routes, clearly indicates that the following is also valid –

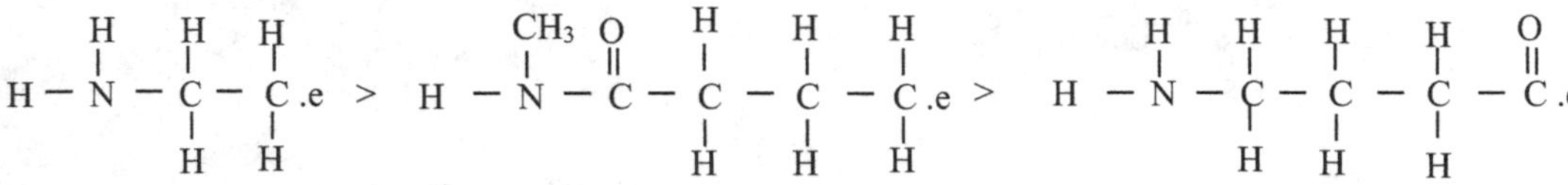

$$
\underset{\text{Cyclic amines}}{H-\overset{\overset{\displaystyle H}{|}}{N}-\overset{\overset{\displaystyle H}{|}}{\underset{\underset{\displaystyle H}{|}}{C}}-\overset{\overset{\displaystyle H}{|}}{\underset{\underset{\displaystyle H}{|}}{C}}.e} \;>\; \underset{\text{Exo - imino compounds}}{H-\overset{\overset{\displaystyle CH_3}{|}}{N}-\overset{\overset{\displaystyle O}{\|}}{C}-\overset{\overset{\displaystyle H}{|}}{\underset{\underset{\displaystyle H}{|}}{C}}-\overset{\overset{\displaystyle H}{|}}{\underset{\underset{\displaystyle H}{|}}{C}}-\overset{\overset{\displaystyle H}{|}}{\underset{\underset{\displaystyle H}{|}}{C}}.e} \;>\; \underset{\text{Cyclic amides}}{H-\overset{\overset{\displaystyle H}{|}}{N}-\overset{\overset{\displaystyle H}{|}}{\underset{\underset{\displaystyle H}{|}}{C}}-\overset{\overset{\displaystyle H}{|}}{\underset{\underset{\displaystyle H}{|}}{C}}-\overset{\overset{\displaystyle H}{|}}{C}-\overset{\overset{\displaystyle O}{\|}}{C}.e}
$$

Cyclic amines　　>　　Exo - imino compounds　　>　　Cyclic amides

<u>Order of Basicity</u>　　　　　　　　　　　　　　　(b)

(Laws of Creations for Exo-imino-cyclic compounds)

Rule 1195: This rule of Chemistry for **Endo-imino-cyclic ethers**, states that, using only the four-membered rings, the followings are to be expected anionically-

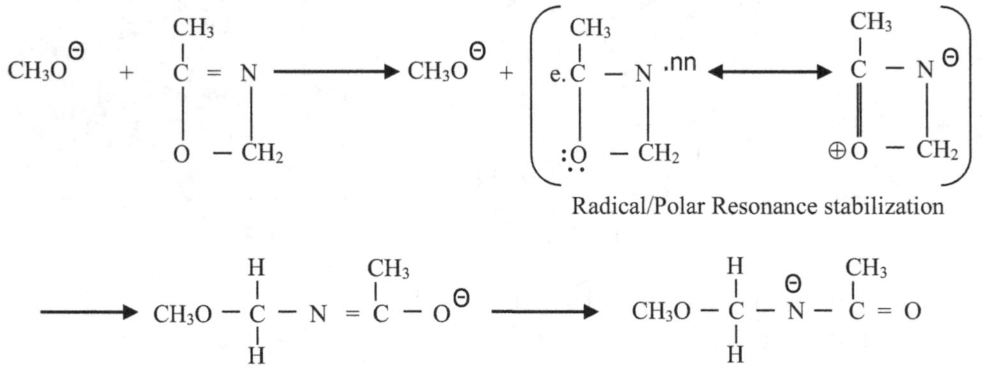

Radical/Polar Resonance stabilization

(A) <u>If strained</u>　　　　　　　　　(B) [Favored]

noting that the ring has been opened either via instantaneously opening the ring in view of its size or via abstraction of transfer species of the eleventh type from inside the ring and also that the ring when activated, can be made to undergo *Radical/polar resonance stabization phenomenon* as shown above.
(Laws of Creations for Endo-Imino-Cyclic Ethers)

Rule 1196: This rule for Chemistry for **Endo-imino-cyclic ethers,** states that, as Nuc-leophiles, when polymerized via the route natural to them (that is, positively or electro-free-radically), the followings are obtained using $NaOCH_3$ as initiator only electro-free-radically-

(A) FAVORED

OR

(B) NOT FAVORED

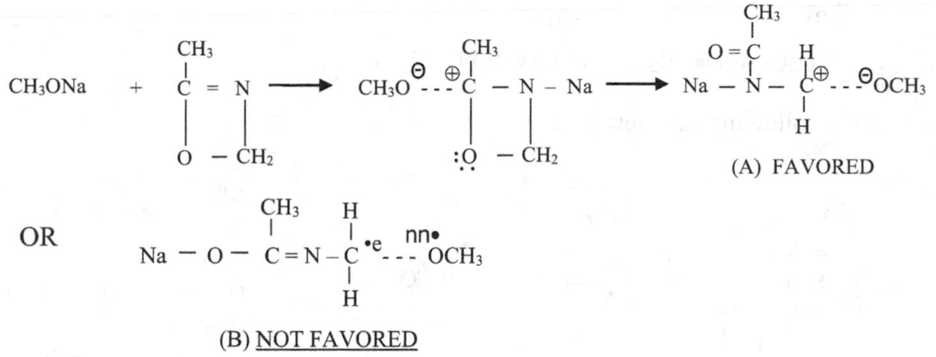

noting that the same monomer unit is obtained anionically and electro-free-radically and not cationically as shown above, since a cation cannot carry a chain during propagation, but positively, i.e., use of non-ionic metallic or non-metallic centers.

(Laws of Creations for Endo-imino-cyclic compounds)

Rule 1197: This rule of Chemistry for **Exo- with N in the ring and Endo- imino cyclic compounds,** states that, when placed side by side, the followings are valid-

Exo – imino (2-imino 1,3-oxazolidenes)   <   Endo -

Order of nucleophilicity   [See Rule 1191]

(Laws of Creations for Exo and Endo-imino-cyclic- compounds)

Rule 1198: This rule of Chemistry for **Exo- and Endo-imino-cyclic compounds,** states that, when the exo- is 2-imino 1,3-oxazolidenes with R^1 equal to R and the R group in the endo is replaced with NR_2, the followings are obtained when the rings are placed side by side in the absence of rearrangement-

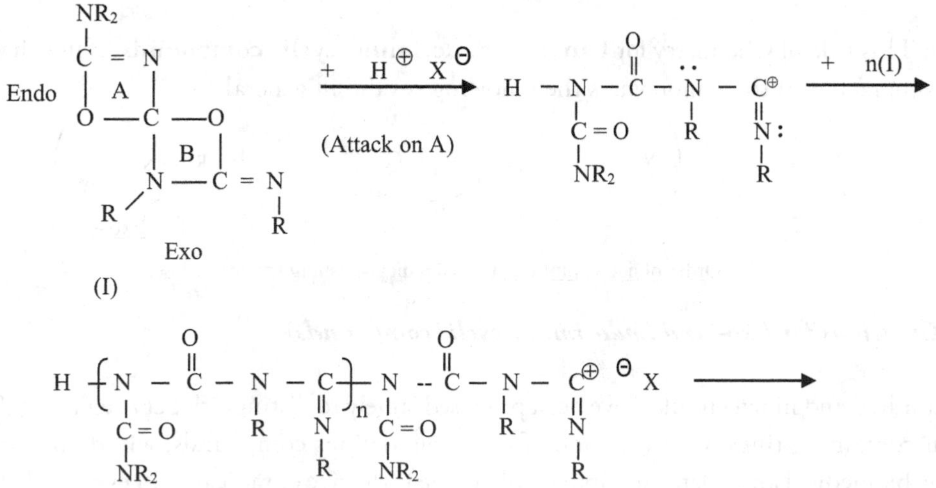

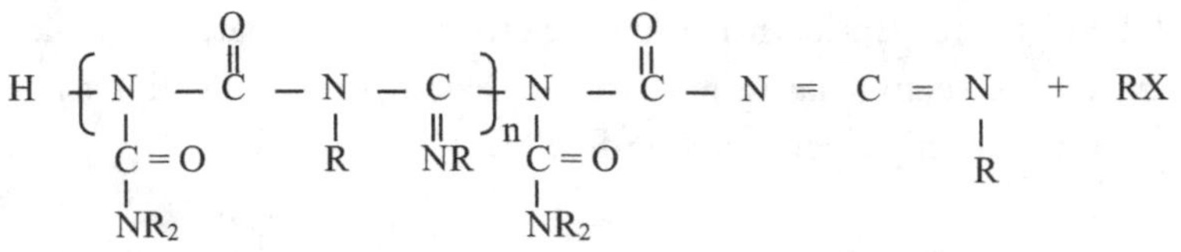

$$H \left\{ N - \overset{\overset{\displaystyle O}{\|}}{C} - N - C \right\}_n N - \overset{\overset{\displaystyle O}{\|}}{C} - N = C = N + RX$$

(with substituents $C=O$, NR_2, R, NR, $C=O$, NR_2, R as drawn)

Transfer species of 2nd kind [FAVORED only radically]   (a)

while anionically, the followings are obtained-

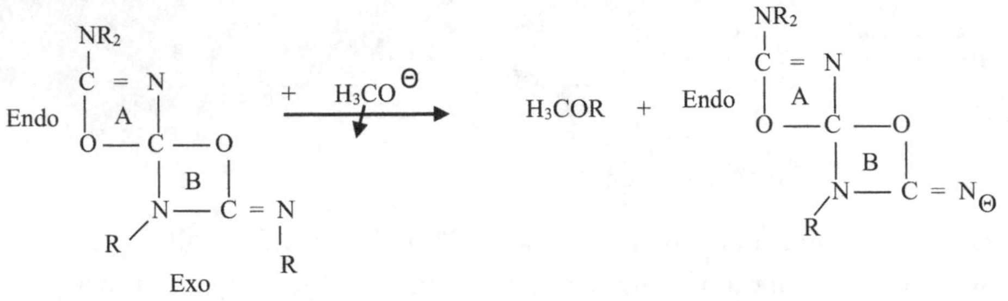

Endo (I) structure with NR_2, $C=N$, A, $O-C-O$, B, $N-C=N$, R, Exo

$+ H_3CO^{\ominus}$

$\longrightarrow$

H_3COR + Endo structure with NR_2, $C=N$, A, $O-C-O$, B, $N-C=N^{\ominus}$, R   (b)

noting that the wrong transfer species was used above, and anionically initiation is indeed favored to give the same monomer unit as obtained electro-free-radically and If (B) was allowed to molecularly rearrange, the followings are obtained "cationically"-

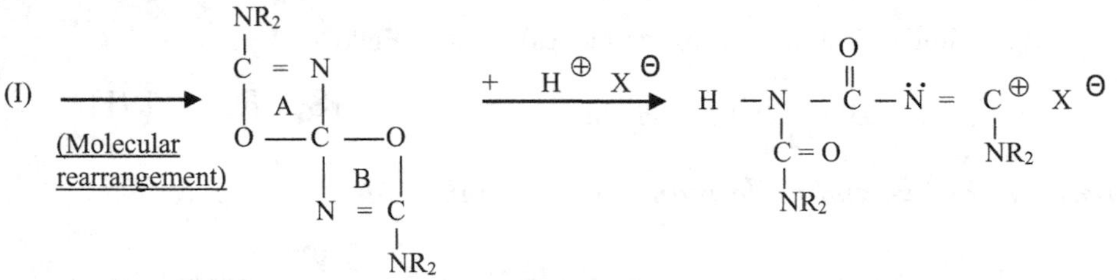

(I) $\longrightarrow$ (Molecular rearrangement) structure with NR_2, $C=N$, A, $O-C-O$, B, $N=C$, NR_2

$+ H^{\oplus} X^{\ominus}$

$H - N - \overset{\overset{\displaystyle O}{\|}}{C} - \overset{..}{N} = C^{\oplus} X^{\ominus}$

(with $C=O$, NR_2, NR_2)   (c)

while anoinically, initiation is not favored.

(Laws of Creations for Exo- and Endo-imino-cyclic compounds)

Rule 1199: This rule of Chemistry for **Exo- and Endo-imino-cyclic compounds,** states that, based on the unique characters of these monomers the following is valid in general-

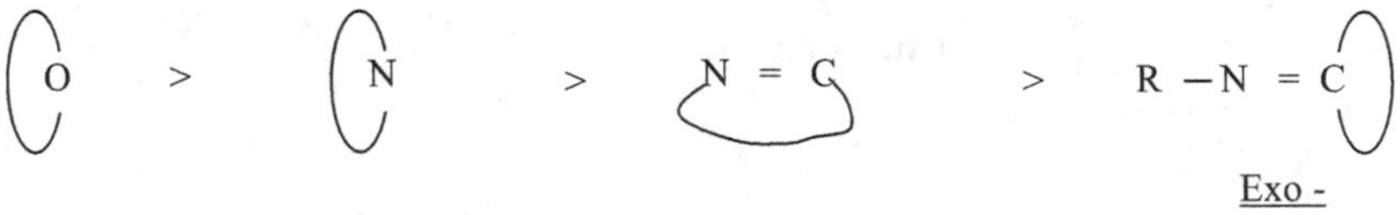

$$\left(O \right) > \left(N \right) > \left(N = C \right) > \left(R - N = C \right)$$

Exo -

Order of nucleophilicity (R is a radical-pushing group)

(Laws of Creations for Exo- and Endo-imino-cyclic compounds)

One hundred and nineteen rules have been proposed largely to distinguish between the different types of nitrogen containing rings. Consideration of these foundations compounds, will definitely open new doors in the biomedical and related disciplines where the influence of radicals will eventually be found to be very dominant. From all the considerations so far, there is no doubt that one has moved Science and all associated disciplines from the domain of application of empirical rules, rules of the thumb, trial and error methods, etc. to reality. Though, one has tried to show to some extent, how these New Frontiers were developed, very little has been shown in view of the complexity of the existence of so many erroneous possibilities in chemical systems of Present-day Science. Those possibilities have been eliminated by applications of foundation rules which have previously been proposed in steps and new definitions and concepts, since in every natural creations, there must be an order.

References

1. N. Yoda and A. Miyake, J. Polymer Sci. 43 : 117 (1960).
2. N. Okata, ibid., A1 : 3151 (1963).
3. R. C. P. Cubbon, MaKromol. Chem., 80 : 44 (1964).
4. J. Brandrup and E. H. Immergut (eds.), "Polymer Handbook" pp. 11 - 363, Interscience Publishers John Wiley & Sons, Inc., New York, 1966.
5. K. Dachs and E. Schwatz, Angew Chen. (Intern. Ed), 1 : 430 (1962).
6. G. Odian, "Principles of Polymer Systems", McGraw-Hill Book Company, (1970), pgs. 479 - 494.
7. F. Rodriguez, "Principles of Polymer Systems", McGraw-Hill Book Company, (1970), pg. 72.
8. C. R. Noller, "Textbook of Organic Chemistry", W. B. Saunder Company, (1966), pgs. 617 - 618.
9. R. A. Patsiga, J. MaCromol. Sci. - Revs. Macromol. Chem., C1(2) : 223 (1967).
10. E. H. Mottus, R. M. Hedrick, and J. M. Butler, Polymer Preprints, 9(1) : 390 (1968).
11. G. D. Jones, D. C. MacWilliams, and N. A Braxtor, J. Org. Chem., 30 : 1994 (1965).
12. G. D. Jones, Nitrogen Compounds, in P. H. Plesch (ed.), "The chemistry of Cationic Polymerization", Chap. 14, Pergamon Press Ltd., Oxford, (1963).
13. G. Odian, "Principles of Polymer Systems", McGraw-Hill Book Company, (1970), pgs. 499 - 503.
14. C. R. Noller, "Textbook of Organic Chemistry", W. B. Saunders Company, (1966), pg. 611.
15. C. R. Noller, ibid., pgs. 629 - 631.
16. T. Yasumoto, J. Polymer Sci., A3 : 3301 (1965).
17. H. Tani and T. Konomi, ibid., A - 1(4) : 301 (1966).
18. R. P. Scelia, S. E. Schonfield, and L. G. Denaruma, J. Appl. Polymer Sci., 8 : 1363 (1964).
19. M. Szwarc, Pure and Appl. Chem., 12(1) : 127(1966) ; Forsch. Hochpolymer - Forsch. (Advan. Polymer Sci.), 4 : 1 (1965).
20. V. M. Rothe, G. Reinisch, W. Jaeger, and I. Schopor, MaKromol. Chem., 54 : 183 (1962).
21. V. S. Doubravsky, and F. Geleji, ibid., 113 : 270 (1968).
22. V. Kagiya, H. Kishimoto, S. Narisawa, and K. Fukui, J. Polymer Sci., A3 : 145 (1965).
23. G. M. Burnet, J. N. Hay, and A. J. MacArthur, Polymerization of Caprolactam, in "The Chemistry of Polymerization Process", pp. 139 - 156, Monograph 20, Society of Chemical Industry, London, 1966.
24. P. H. Hermans and D. Heikens, J. Polymer Sci., 44 : 429, 437 (1960).
25. A. J. Amass and J. N. Hays, MaKromol. Chem., 103 : 244 (1967).
26. J. N. Hay, J. Polymer Sci., B5 : 577 (1967).
27. A. V. Tobolsky, J. Polymer Sci., 25 : 220 (1957).
28. A.V. Tobolsky and A. Eisenberg, J. Am. Chem. Soc., 81(1959); 82 : 289 (1960).
29. F. Wiloth, Z. Physik. Chem., N. F. (Frankfurt), 4 : 66 (1955).

30. T. Kagiya, S. Narisawa, T. Ichida, K. Fukui, H Yokota, and M. Kondo, J. Polymer Sci., A-1(4) : 293 (1966).

31. T. Kagiya, I. Maruta, T. Ichida, S. Narisawa, and K. Fukui, J. Polymer Sci., A-1(5) : 1645(1967).

32. T. Kagiya, S. Narisawa, K. Manabe and K. Fukui, J. Polymer Sci., B3 : 617 (1965).

33. C. R. Noller, "Textbook of Organic Chemistry", W. B. Saunders Company, (1966), pgs. 503 - 510.

34. R. B. Seymour, "Introduction to Polymer Chemistry", International Student Edition, McGraw-Hill Book Company, (1971), pg. 134.

35. W. Cooper, D. R. Morgan, and R. T. Wragg, Eur. Polym. J., 5(1) : 71(1969).

36. M. Terbojevich, G. Pizziolo, E. Peggion, A. Cosani, and E. Scoffone, J. Am, Chem. Soc., 89 : 2733 (1967).

37. T. Mukaiyama, T. Fujisawa, H. Nohira, and T. Hijugaji, J. Org. Chem. 27 : 3337 (1962).

38. A. Levy and M. Litt, J. Polymer Sci., A-1(6) : 57 (1968).

39. J. Hine, "Physical Organic Chemistry", International Student Edition, McGraw-Hill Book Company, (1962), pgs. 513 - 514.

40. J. F. Kerwin, G. E. Uuyot, R. C. Fuson, and C. L. Zirkle, J. Am. Chem. Soc., 69, 2961 (1947) : E. M. Schutz and J. M. Sprague, J. Am. Chem. Soc., 70, 2760 (1948).

Problems

13.1   List and define all the deferent types of resonance stabilization phenomena so far identified.

13.2   Shown below are three monomers -

(i)
$$H_2C = CH_2$$
with N and O structure

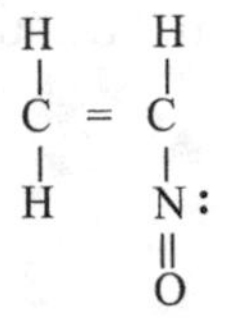

(ii)   Structure with $CH_2 - CH_2$, H_2C, $C=O$, N, H, $C=C$, H, H

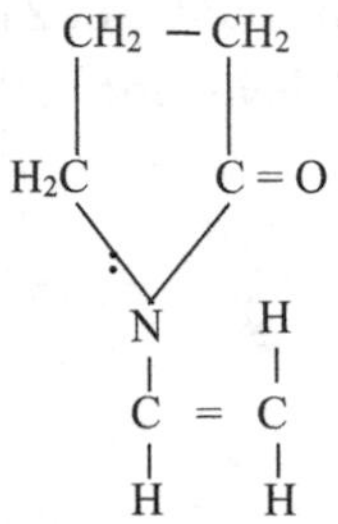

(iii)   Structure with $H_2C - CH_2$, N, H, $C=C$, H, H

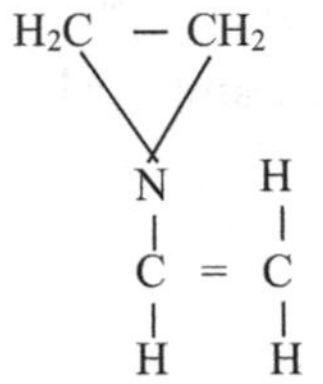

(a)   Based on the New Frontiers, distinguish between (i), (ii) and (iii).

(b)   Why is the ethene group not loosely bonded to the nitrogen center in (ii) and (iii)?

13.3   (a)   Shown below are two monomers -

(i) $H_2C - C=O$, N, H structure

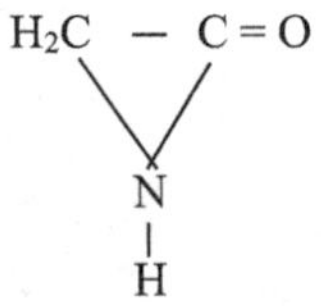

(ii) $H_2C - C$ with $=O$ and O structure

α - Lactam   α - Lactones

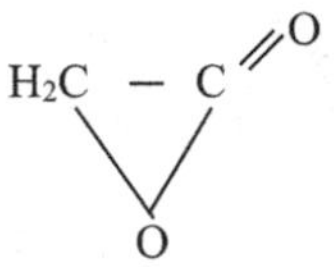

Lactones are said to be less strained than lactams, Hence (ii) exists while (i) is not known to exist. Explain why this is so and what can be done to make (i) exist.

(b)   Why does succinimide shown below not presently favor being popularly used as a monomer, whereas γ - butyro lactams does?

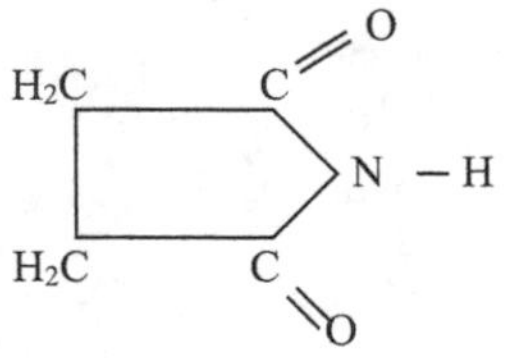

Succinimide structure with H_2C, $C=O$, N—H, H_2C, C, O

versus

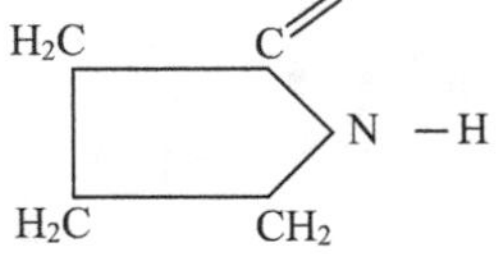

γ - butyrolactam structure with H_2C, $C=O$, N—H, H_2C, CH_2

Succinimide   γ - butyrolactam

(c)   Why also does succinic anhydride not presently favor being used as a monomer whereas maleic anhydrides does.

318

13.4   Shown below is a nitrile oxide compound -

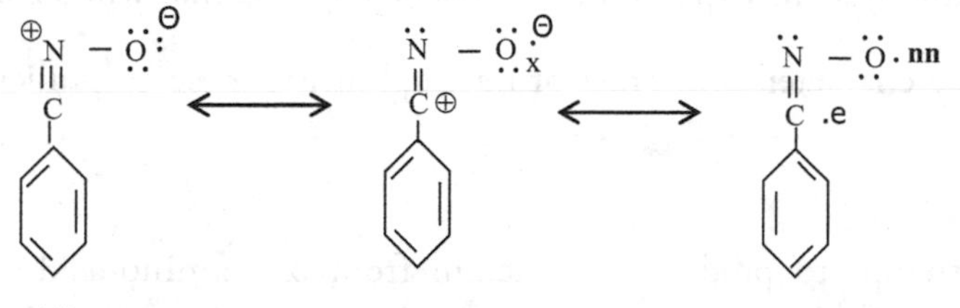

(A)   (I)   (II)

(a) What type of resonance stabilization is indicated above for (A)?

(b) Is the existence of (I), (II) favored? Explain.

(c) Can the compound favor any existence without carrying charges or radicals?

(d) Can the compound be used as a monomer radically or chargedly? Explain.

13.5   (a)   A specific example of a 1, 3- dipolar addition is the reaction of maleic anhydride with the nitric oxide in Q 13.4 to produce a bicyclic ring[39].
Explain the mechanism.

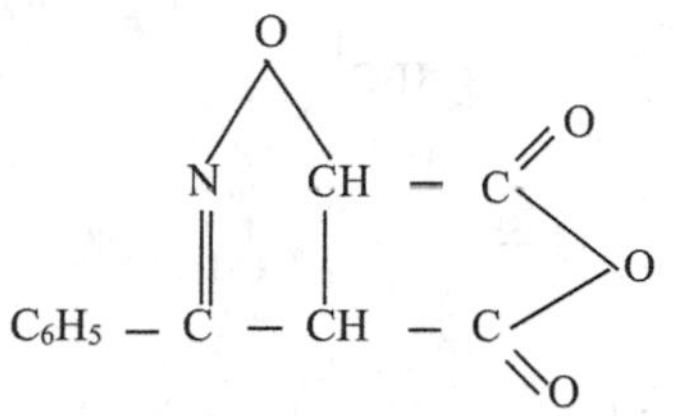

(b)   Compare the compound (nitric oxide) above with that of diazomethane in terms of resonance stabilization phenomenon.

13.6.   Shown below are multi-center types of reactions[39].

(i)

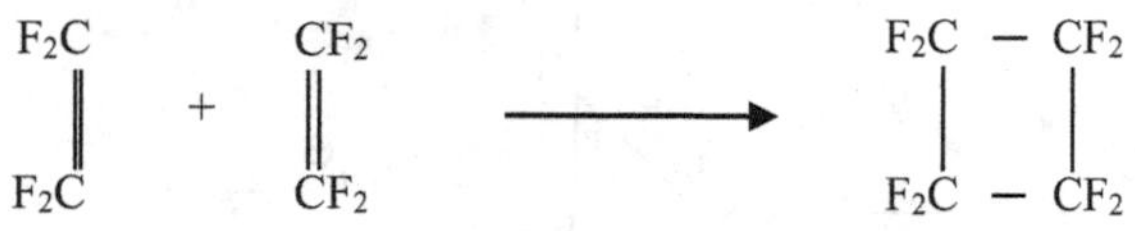

(ii)

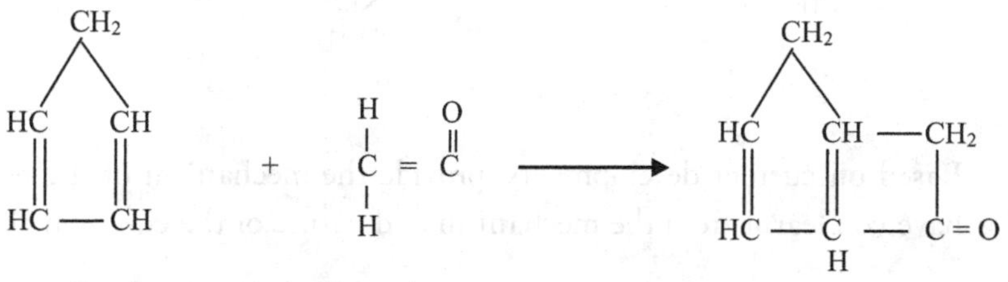

(Cyclopentadiene)   (Ketene)

(iii)

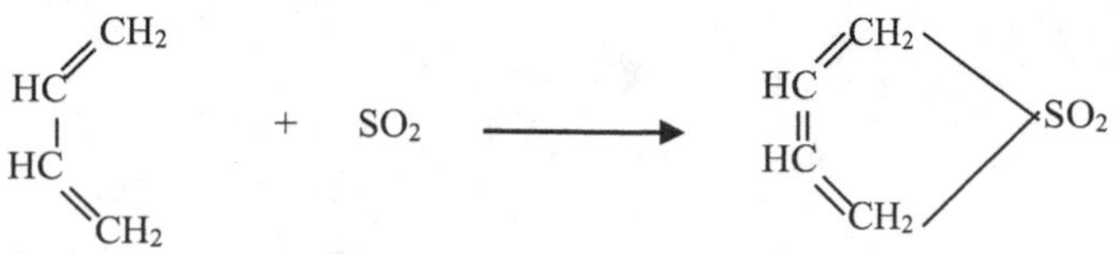

(a) Provide the mechanisms of the reactions above.

(b) Which of the cases above will favor the existence of linear copolymers if an initiator is present in the system? Explain the types of copolymers that will be obtained.

(c) Identify the characters and order of nucleophilicity or electrophilicity of the monomers above.

13.7 (a) In an attempt to produce α - lactam from α - amino-acid ($NH_2 CH_2COOH$), 2,5-Dioxopiperazine six-membered ring is obtained when heated at 170°C in glycerol. Explain why this is so and the mechanism of the reaction.

(b) Several workers[40] have reported the rearrangement of 1-chloro-2-dialkyl-aminopropane to 1-dialkylamino-2-chloropropanes using S_N mechanisms and participation of neighboring groups. What is the meaning of this?

(c) Shown below is a ring expansion reaction.

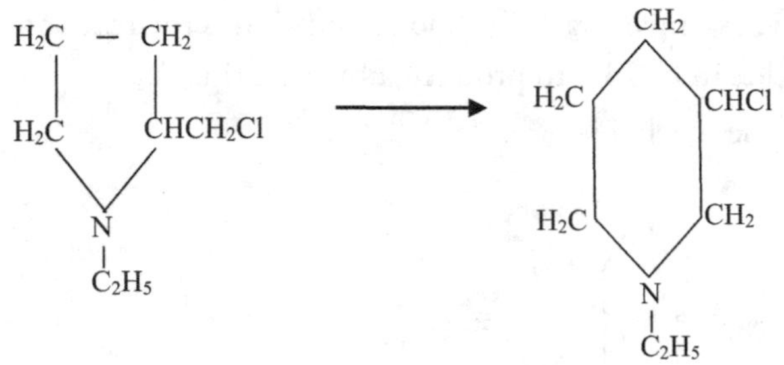

Based on the current development, provide the mechanisms for the reaction.

13.8 (a) Shown below is the isomerization of an amino primary chloride to an amino secondary chloride.

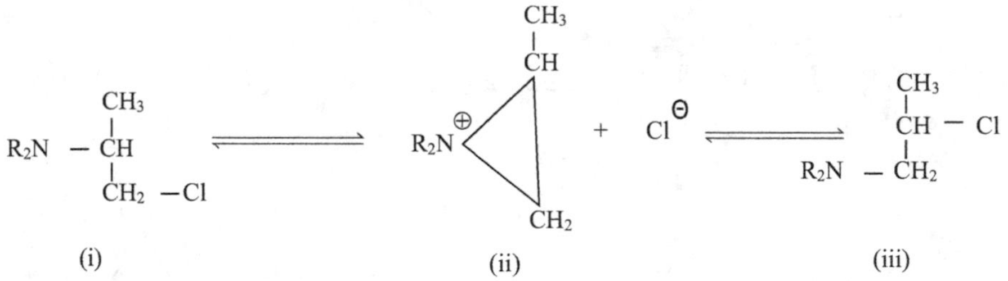

Based on current developments, provide the mechanism of the reaction above. What have you learnt from the mechanism and nature of the equation above?

(b) Below is the cyclic dimerization of methyl-bis-b-chloroethylamine in aqueous acetone solution. The reaction is said to involve several steps with comparable rate constants.

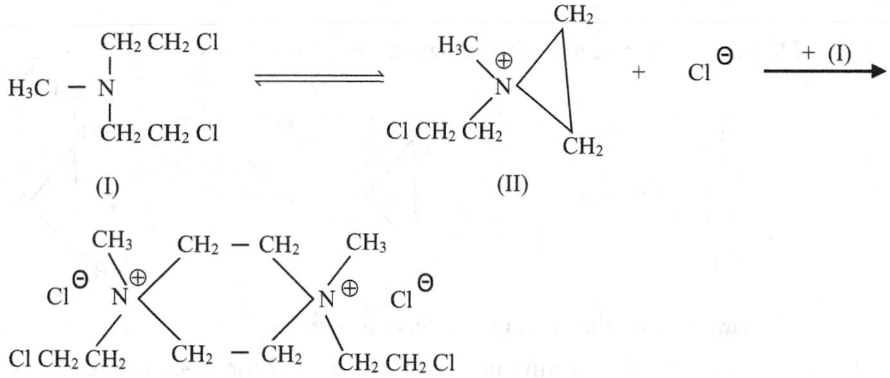

$$
H_3C - N \begin{array}{l} CH_2\ CH_2\ Cl \\ \\ CH_2\ CH_2\ Cl \end{array} \rightleftharpoons \quad H_3C - \overset{\oplus}{N} \overset{CH_2}{\underset{CH_2}{\big<}} \quad + \quad Cl^{\ominus} \quad \xrightarrow{\ +\ (I)\ }
$$

(I)   (II)

(III)

$$
Cl^{\ominus}\ \overset{CH_3}{\underset{Cl\ CH_2\ CH_2}{N^{\oplus} \!\!\big<}} \overset{CH_2-CH_2}{\underset{CH_2-CH_2}{\big>}} \overset{CH_3}{\underset{CH_2\ CH_2\ Cl}{N^{\oplus}}}\ Cl^{\ominus}
$$

Based on current development, provide the mechanism of the reaction.

(c) Shown below is a charged compound

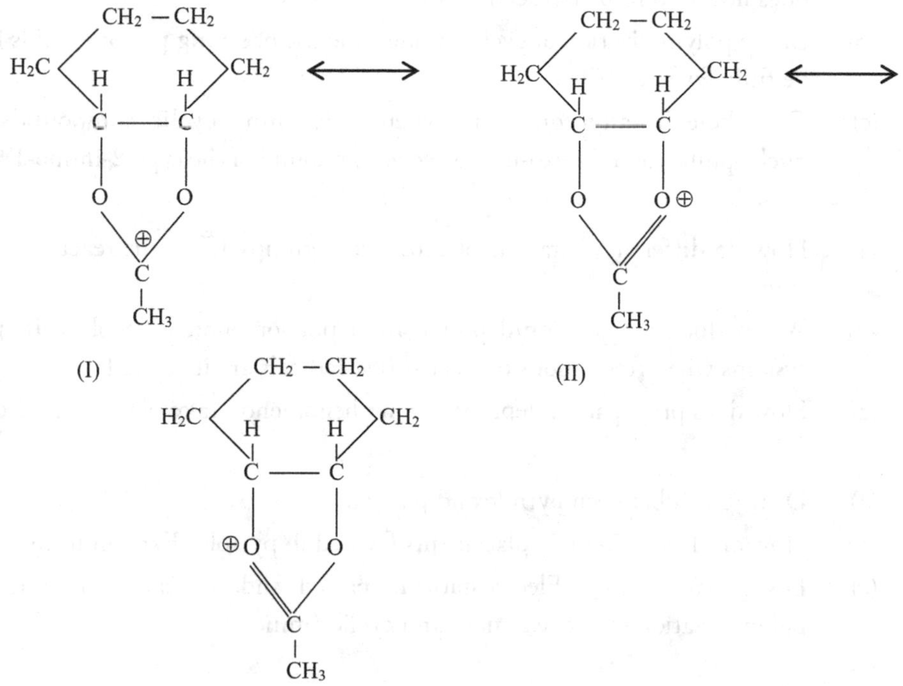

(I) ⟷ (II) ⟷

(III)

(i) Does the type of resonance stabilization phenomenon indicated above exist? If it does, what is the type? Can the positive charge be isolatedly placed?

(ii) What type of products will be obtained when CH_3COOH is added to (II) or (III)?

(iii) What types of products will be obtained when H_2O and C_2H_5OH are added to (I) or (II)? Explain noting that C_2H_5O is weaker than OH in providing electrostatic forces.

13.9   (a)   Distinguish between cyclic amides and carboanhydrides.

         (b)   Distinguish between aziridine and imino-cyclic compounds?

13.10   (a)   Distinguish between the three aziridines shown below

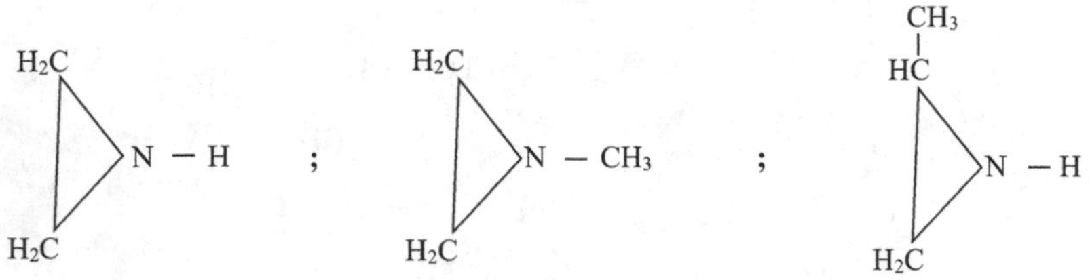

               in terms of how they can be used as monomers.

         (b)   Why does tertiary amine not favor being used for so-called "cationic" polymerization of cyclic amides.

13.11.   (a)   In the "cationic" polymerization of aziridine, cyclic amides etc., the H atom on nitrogen does not need to be replaced. Why?

         (b)   In the polymerization of cyclic amides, the use of strong protonic acids is not known to be fully favored. Why?

         (c)   Can there be any conditions when endo-imino cyclic compounds become more nucleophilic than the exo-imino cyclic compound of the type-2-imino-1,3-oxazolidenes?.

13.12   (a)   How do different characters of substituent groups affect the reactivity or opening of a ring?

         (b)   When does propagation/depropagation phenomenon take place in poly-merization systems when free cations or electro-free-radicals are invol-ved?

         (c)   How does propagation/depropagation phenomenon take place in aziri-dine?

13.13   (a)   Distinguish between pyrrole and phenol.

         (b)   How are 1- and 2- or 5- placements favored in pyrrole? Explain using examples.

         (c)   Discuss the use of Electrostatically-paired and Ionically-paired initiators for the polymerization of cyclic amides and cyclic amines.

13.14   (a)   Anionic polymerization of three membered rings that favor Instantaneous opening, is usually limited to producing a relatively low molecular weight polymer. Provide the reasons for this occurrence.

         (b)   The "cationic" polymerization of lactams require higher temperature poly-merization, than their anionic polymerizations. Provide the reasons for this.

         (c)   In the "cationic" polymerization of iminocarbonates carbon dioxide cannot be released in the process. Discuss the reasons why this is so.

13.15   (a)   When can transfer species in imino cyclic compounds be made present?

   (b)   Distinguish between the functional centers in imino cyclic compounds.

13.16   Shown below are three compounds -

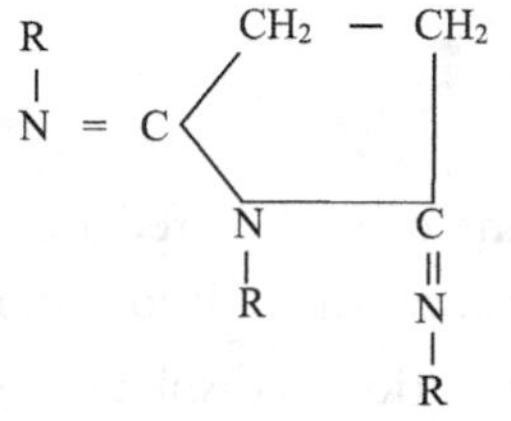

(I) 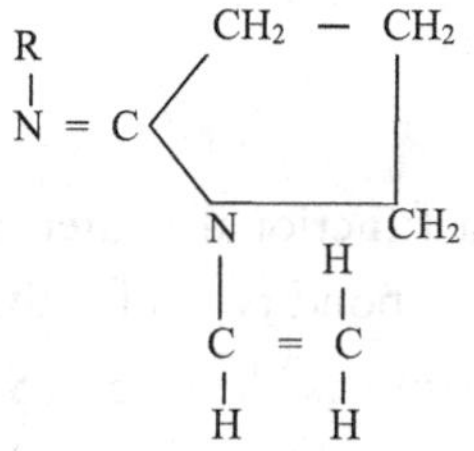   (II) 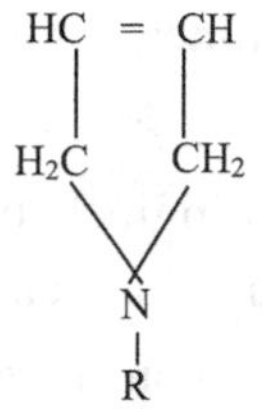   (III)

   (a)   Which of the compounds above favor Radical/polar resonance stabiliza-tion phenomena?

   (b)   For those that can be polymerized, identify the routes favored by them and their monomer units.

   (c)   Distinguish between (III) above and pyrrole.

13.17   Shown below are two compounds -

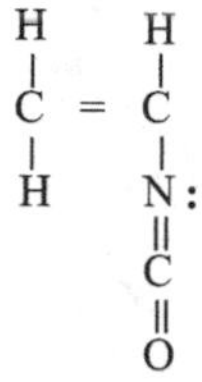

(I) 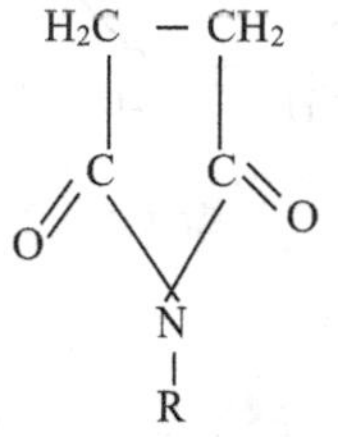   (II)

   (a)   Can any undergo resonance stabilization phenomena? Explain.

   (b)   Based on your experience so far with cyclic amides, how can (II) be polymerized anionically or nucleo-non-free-radically and electro-free-radically?

13.18.   (a)   Compare (II) of Q 13.17 with succinimide.

   (b)   Why can't oxaazacyclopropane ring be opened instantaneously chargedly? Under what conditions can it be used as a monomer?

   (c)   Why is the hydrogen atom not loosely bonded ionically to the nitrogen center?

   (d)   Why is oxaazacyclopropane more nucleophilic than ethylenimine? Which is less strained?

13.19   (a)   List the different methods by which the characters of monomers or compounds can be determined.

   (b)   Under what conditions are alternating copolymers obtained between

   (i)   Aziridines (with radical-pushing group) and carbon monoxide

   (ii)   Propylene and a nitroso compound (R - N = O)

   (d)   Show the monomer units obtained for the copolymers of (b).

13.20 (a) Cyclic acetals can be copolymerized with alkenes, but not cyclic ethers or cyclic esters. If it is true, why is this so?

(b) Can endo - imino and exo-imino cyclic compounds be made to favor copolymerization with alkenes?

(c) Under what conditions will cyclic amides and amines favor copolymeri-zation with lactones, cyclic ethers?

13.21 (a) Distinguish between a functional center, a vacant orbital/paired unbonded type of activation center, and a p-bond type of activation center in Addition monomers.

(b) Based on current developments, do S_{N1} and S_{N2} substitution or displace-ment mechanisms or E_1 and E_2 elimination mechanisms exist? Explain.

(c) The reaction below is said to be Favorskii rearrangement, that is, a reaction of a-halo ketones with hydroxide (or alkoxide) ions to give salts (or esters) of acids with a somewhat rearranged carbon skeleton but the same number of carbon atoms.

Based on the New Frontiers, explain the mechanism of the reaction above.

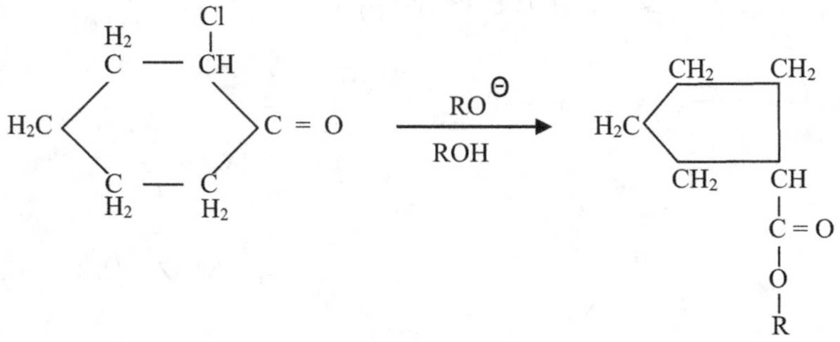

13.22 (a) Identify all the possible types of functional center in oxygen and nitrogen containing rings and state the order of their characters.

(b) Based on the sources of rays - gamma, beta- etc. types of rays, should they have any character? Explain.

(c) What are "invisible forces"?

13.23 (a) Distinguish between molecular rearrangements of the first and sixth types.

(b) Explain how CO_2 is released anionically and cationically during polymerization of carboanhydrides.

(c) What is or are the conditions that favor the existence of Radical/polar resonance stabilization phenomena?

13.24. (a) Distinguish between Radical/polar resonance stabilization phenomenon and others.

(b) Based on the influence of electrostatic forces, explain the mechanism of the polymerization of trioxanes and cyclic formals.

CHAPTER 14

TRANSFER OF TANSFER SPECIES IN CYCLOSILOXANES, CYCLIC SULPHIDES AND RELATED COMPOUNDS AND INORGANIC OR SEMI-INORGANIC CYLIC MONOMERS

14.0 Introduction

Silicon unlike carbon atom is very unique. It is unique in the sense that it is more metallic in character. It is a non-transition and non-ionic metal. If it had been ionic, it should have only been quadrivalent cationically; a situation in which the last shell is completely emptied. However, this is not possible in view of the very strong tendency of the four radicals in the last shell to undergo hybridization all the time.

With consideration of silicon ringed monomers, mono-cyclic sulphides, and some inorganic/semi-inorganic cyclic monomers, this will end full analysis of ring opening monomers, for which the rules proposed, will find complete applications to most types of ring opening and non-opening monomers or compounds. Silicon containing ringed monomers to be considered for study include cyclic siloxanes. Two well-known members of interest are the cyclic tetramers- octamethylcyclotetrasiloxane and 1,1,3,3-tetramethyl- 1,3 – disilacyclo-butane, both shown below.

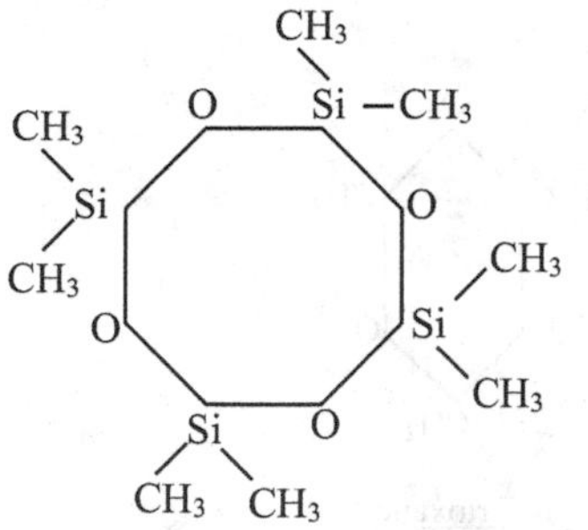

(I) Octamethylcyclotetrasiloxane

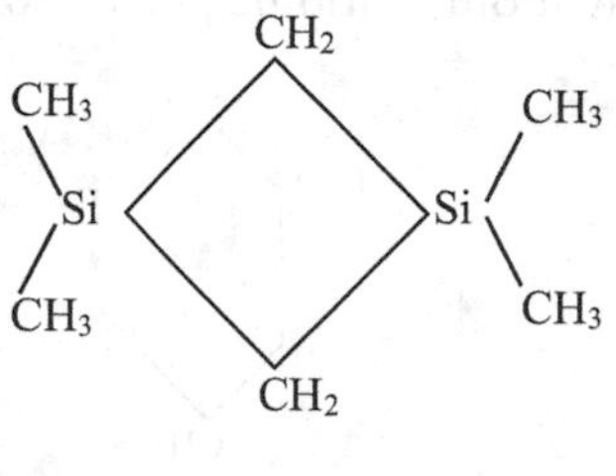

(II) 1,1,3,3 - tetramethyl- 1,3 - disilacyclobutane

14.1

(I) is known to undergo both anionic and cationic polymerizations[1-4]. Looking at (I), there seems to exist one type of visible functional center, (INSERT IMAGE), which can only be used positively or electro-free-radically. (I) an eight-membered ring may not seem to possess large strain energy (SE) for instantaneous opening of the ring. In the absence of any other visible functional center, the question is - how does the monomer undergo anionic attack? If instantaneous opening of the ring was assumed possible anionically, then the anionic route will be favored as shown below.

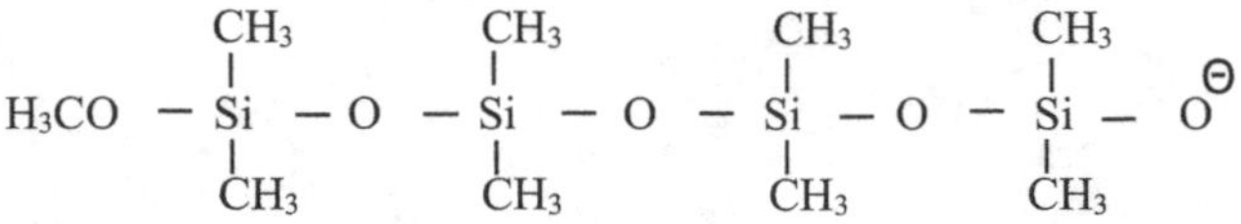

$$H_3CO - \underset{\underset{CH_3}{|}}{\overset{\overset{CH_3}{|}}{Si}} - O - \underset{\underset{CH_3}{|}}{\overset{\overset{CH_3}{|}}{Si}} - O - \underset{\underset{CH_3}{|}}{\overset{\overset{CH_3}{|}}{Si}} - O - \underset{\underset{CH_3}{|}}{\overset{\overset{CH_3}{|}}{Si}} - O^{\ominus}$$

14.2

Transfer species of the first kind cannot be rejected positively or electro-free-radically unlike all other C – C containing monomers, because Si is more electropositive than H while C is less electropositive than H. On the other hand, the formation of a double bond is not possible with silicon, since silicon unlike carbon is a metal. ***With metals in general, at best polar bonds are formed.*** Note that even then, all the four radicals in the last shell are fully hybridized.

(II) is known to readily undergo only the "cationic" route, using acid catalysis[5]. Without doubt, (II) seems to possess SE close to the MRSE, since there is no visible functional center on the monomer. The silicon centers like the carbon centers, do not have paired unbonded radicals in their last shell which "cations" can be attracted to. ***Without instantaneous opening of ring, no route will be favored.***

Amongst the sulfur containing cyclic monomers, cyclic sulfides below, will be considered. These are analogous to cyclic ethers.

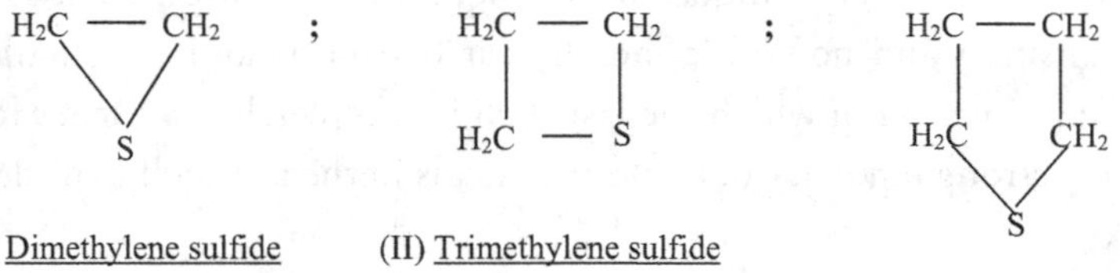

(I) Dimethylene sulfide   (II) Trimethylene sulfide

(III) Tetramethylene sulfide   14.3

Like cyclic ethers, only the three membered cyclic sulfides favor both anionic[6] and cationic homo-routes[1,7], an indication of the presence of close to MRSE in this ring. All the other cyclic sulfides favor only the cationic homopolymerization route where possible.

Analogous to trioxanes (cyclic acetals) are trithianes the cyclic trimer to thioform-aldehyde, which are known to favor only cationic polymerizations.

Trithiane   versus   Trioxane   14.4

Of particular interest are cyclic disulfides and rhombic sulfur, monomers which are distinctively known to favor radical polymerizations[8].

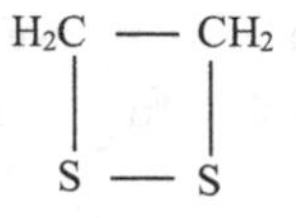

(I) A cyclic disulfide 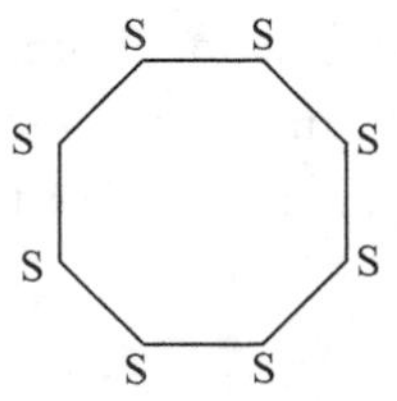   (II) Rhombic sulphur   14.5

If the monomers have to favor radical homopolymerization, then it will be none other than the non-free-radical routes, since the sulphur atom under such conditions cannot carry a free-radical. While the possibility of (I) undergoing non-free-radical route may be obvious, if the SE is close to the MRSE of the ring, that of (II) is not obvious, for which a new driving force of opening of such rings may become obvious.

There has been over the years great amount of interest in the synthesis of inorganic and semi-inorganic polymers. Their use as monomers has been delayed in view of the types of present-day methods used in understanding mechanisms of reactions. Some of the polymers studied have been obtained by ring opening polymerizations[9]. Of interest will be the cyclic trimmer of dichlorophosphazene (phosphonitrilic chloride) and cyclosilazanes such as hexamethylcyclo-trisilazane.

(I) Dichlorophosphazene or phosphonitrilic chloride

(II) Hexamethylcyclotrisilazane

14.6

(I) is particularly considered, because the structure indicated above, by which it is generally known is not the true structure of the monomer, since the phosphorus center like the nitrogen center cannot carry more than eight radicals in the last shell. Before beginning the considerations of the characters of these monomers, there is need to recall the past, in order to distinguish between silicon and carbon atoms.

14.1 The Silicon Atom[10]

Over the years, two aspects of the chemistry of silicon have been of interest: (a) the preparation and properties of silicon analogs of carbon compounds and (b) the preparation and properties of organosilicon compounds in which carbon is bound to silicon. Being next to carbon in the Periodic Table, tetracovalent compounds with tetrahedral bond angles like in carbon, are favored for most silicon compounds. Though silicon has been said to possess a coordination number of six, the 3d orbitals are never involved, since there are no 3d orbitals in its last shell.

The differences between the properties of silicon and carbon compounds arise not from the availability of d-orbitals (which indeed is not available), but because the valence radicals are further from the nucleus in silicon than in carbon. Consequently, the valence radicals are held less strongly by the nucleus, making silicon to be more electropositive than carbon. Secondly, there is more room for atoms or groups to form bond with silicon. Thirdly and most importantly, overlap of p-orbitals with p-orbitals of other elements is not possible because of full hybridization in silicon, with the result that ordinary double bonds do not form. In fact only electronegative elements have the tendency of forming double or triple polar bonds and anions.

As already indicated, silicon is electropositive. Though it is tetravalent, it cannot form a tetravalent cation, a condition in which the last shell is emptied of radicals.

$$SiCl_4 \quad \nrightarrow \quad Si^{4+} \quad + \quad 4:\overset{\cdot\cdot}{\underset{\cdot\cdot}{Cl}} x^{\ominus}$$

14.7

The reaction above is favored only if the four radicals in silicon do not undergo any form of hybridization. But in view of the non-existence of double bonds for silicon containing compounds, the radicals seem to very strongly undergo hybridization phenomenon. Hence in general, silicon cannot carry ionic charges-cations. For electropositive metallic elements which favor ionic bond formations, there is no hybridization of the radicals in the last shell. Only excitation of the radicals exists. Since double bonds cannot be formed with Si, triple bonds cannot also be formed.

While silicon hydrides are said to be analogs of alkanes, the silicon-hydrogen bonds are much more reactive than the carbon-hydrogen bonds, because the hydrogen atoms in silicon are all loosely covalently radically bonded to the silicon center. When crude magnesium silicide, prepared by heating silica ($:Si^{\ominus} - {}^{\ominus}O$) with Mg, reacts with mineral acids, a mixture containing 40% silane SiH_4, 30% disilane Si_2H_6, 15% trisilane Si_3H_8, 10% tetrasilane Si_4H_{10} and 5% mixture of higher silanes, is obtained. The mechanisms of the reactions involved are all radical in character. The silicon center seems to favor carrying electro-free-radical, in the absence of more electropositive metals in particular ionic metals. Hence, tetravalent silicon compounds are more vulnerable to nucleophilic attack than carbon compounds. Thus, although SiH_4 is stable to aqueous acids (i.e., insoluble in for example HCl), they are hydrolyzed with the evolution of hydrogen when boiled with strong bases.

$$
\begin{array}{l}
\underset{\underset{H}{|}}{\overset{\overset{H}{|}}{H-Si-H}} \quad + \quad H_2O \quad + \quad 2Na^{\cdot e} \ OH^{\cdot nn} \longrightarrow \\[3em]
\underset{\underset{H}{|}}{\overset{\overset{OH}{|}}{H-Si-OH}} \quad + \quad H_2O \quad + \quad 2\,NaH \longrightarrow \\[3em]
\underset{\underset{H}{|}}{\overset{\overset{ONa}{|}}{H-Si-ONa}} \quad + \quad H_2O \quad + \quad 2H_2 \longrightarrow
\end{array}
$$

$$
\underset{\underset{ONa}{|}}{\overset{\overset{OH}{|}}{H-Si-ONa}} \; + \; 3H_2 \longrightarrow \underset{\underset{ONa}{|}}{\overset{\overset{OH}{|}}{Si^{\oplus}-O^{\ominus}}} \; + \; NaH \; + \; 3H_2
$$

(II)   (III)

$$
\longrightarrow \underset{\underset{ONa}{|}}{\overset{\overset{O^{\ominus}}{\overset{|\oplus}{}}}{Si-ONa}} \; + \; 4H_2
$$

Na_2SiO_3

(IV)   14.8

The hydrogen molecule is formed between the reaction of a nucleo-free-radical $H^{\cdot n}$ (an element called hydride) and an electro-free-radical $H^{\cdot e}$ (the atom). It can be observed here that the hydrogen element is what is loosely covalently bonded to the silicon center. This is the opposite of carbon. As opposed to what has been thought to be the case in the past, hydride ions do not exist under normal or abnormal conditions. The hydrogen nucleo-free-radical is displaced one at a time using NaOH. The reaction which is an Equilibrium mechanism system has at least eight stages. Since, silicon cannot form a double bond with the oxygen or any other atom or element including itself, polar bonds are formed between the Si and O centers, Si being strongly electropositive and O strongly electronegative. Through the mechanisms of the reactions, one can identify the Equilibrium state of Existence of some compounds. That of H_2 was suppressed in the last step of some stages due to the presence of Na.

The presence of radical-pushing groups are said to always lead to a decrease in reactivity. Despite the fact that silicon is more electropositive than carbon, it has been noted that there is no evidence that "siliconium ions" $R_3Si^{\text{Å}}$, analogous to "carbonium ions", $R_3C^{\text{Å}}$, are intermediates in the reaction of silicon compounds, and there is no kinetic evidence for reactions of the so-called S_N1 type. Like the silicon hydrides, silicon tetrachlorides, $SiCl_4$, undergoes entirely free-radical reactions.

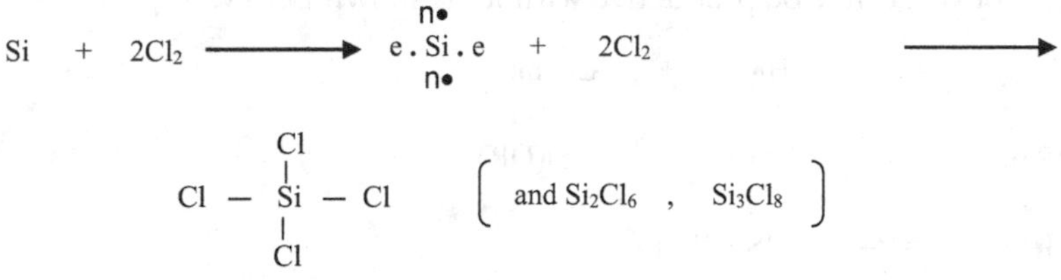

$$Si \;+\; 2Cl_2 \longrightarrow \overset{n\bullet}{\underset{n\bullet}{e\,.\,Si\,.\,e}} \;+\; 2Cl_2 \longrightarrow$$

$$\underset{\underset{Cl}{|}}{\overset{\overset{Cl}{|}}{Cl-Si-Cl}} \qquad \Big[\; \text{and } Si_2Cl_6 \;,\quad Si_3Cl_8 \;\Big] \hspace{3cm} 14.9$$

This is more than a two stage Equilibrium mechanism system. The formation of higher members, is a result of reactions between two or three silicon centers by activation, with excessive presence of Cl_2 molecules in the system all in Stable state of existence. The silicon shown above is one of the elements of Si obtained by activation of the ground state silicon.

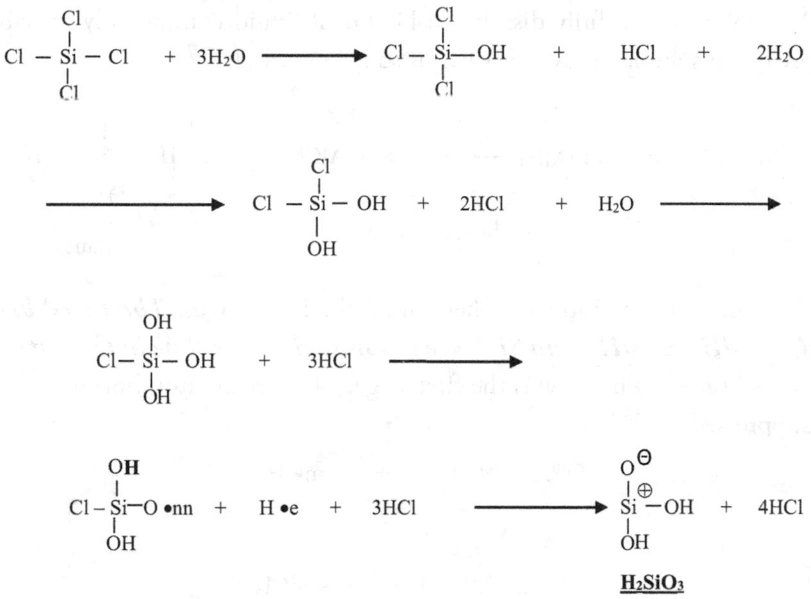

$$\underset{\underset{Cl}{|}}{\overset{\overset{Cl}{|}}{Cl-Si-Cl}} \;+\; 3H_2O \longrightarrow \underset{\underset{Cl}{|}}{\overset{\overset{Cl}{|}}{Cl-Si-OH}} \;+\; HCl \;+\; 2H_2O$$

$$\longrightarrow \underset{\underset{OH}{|}}{\overset{\overset{Cl}{|}}{Cl-Si-OH}} \;+\; 2HCl \;+\; H_2O \longrightarrow$$

$$\underset{\underset{OH}{|}}{\overset{\overset{OH}{|}}{Cl-Si-OH}} \;+\; 3HCl \longrightarrow$$

$$\underset{\underset{OH}{|}}{\overset{\overset{\mathbf{OH}}{|}}{Cl-Si-O\bullet nn}} \;+\; H\bullet e \;+\; 3HCl \longrightarrow \underset{\underset{OH}{|}}{\overset{\overset{O^{\ominus}}{|^{\oplus}}}{Si-OH}} \;+\; 4HCl$$

$$\mathbf{\underline{H_2SiO_3}} \hspace{3cm} 14.10$$

It was H_2O that commenced the reaction to form $Cl^{\ominus}......^{\oplus}OH_3$ (Dilute HCl), water and the silicon compound in the first two stages via Equilibrium mechanism. This continued to the third stage with no water anymore in the system. The seventh stage which is the last stage ended with release of the last HCl followed by deactivation.

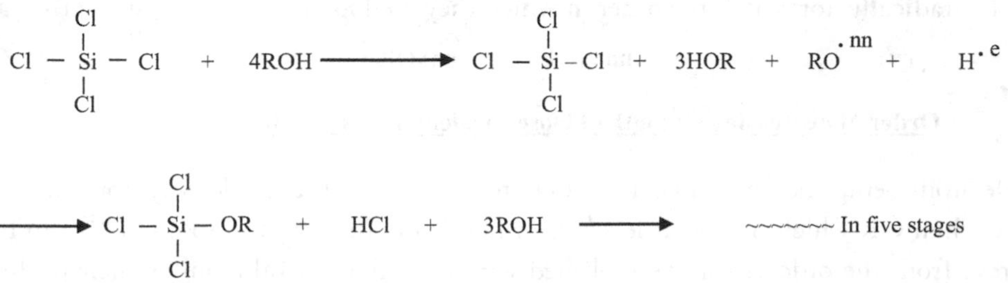

$$\underset{\underset{Cl}{|}}{\overset{\overset{Cl}{|}}{Cl-Si-Cl}} \;+\; 4ROH \longrightarrow \underset{\underset{Cl}{|}}{\overset{\overset{Cl}{|}}{Cl-Si-Cl}} \;+\; 3HOR \;+\; RO^{\bullet\,nn} \;+\; H^{\bullet e}$$

$$\longrightarrow \underset{\underset{Cl}{|}}{\overset{\overset{Cl}{|}}{Cl-Si-OR}} \;+\; HCl \;+\; 3ROH \longrightarrow \quad \sim\!\sim\!\sim\sim \text{ In five stages}$$

$$\longrightarrow \quad Si(OR)_4 \quad + \quad 4HCl$$

(Orthosilicates)

14.11a

This is an eight stage Equilibrium mechanism system. Notice that the Equilibrium state of Existence of $SiCl_4$ is suppressed by the alcohol which is acidic, by HCl and even water which is acidic. Water is alcoholic, basic, and acidic, since H is a metal. $SiCl_4$ is a stable salt insoluble in HCl. The orthosilicate formed is so stable that it can readily be hydrolyzed by water to give $Si(OH)_4$ and ROH. Yet the HCl, one of the products above cannot be productive with it, as shown below.

Stage 1: $\quad HCl \quad \rightleftharpoons \quad H \bullet e \quad + \quad Cl \bullet nn$

$\qquad H \bullet e \quad + \quad Si(OR)_4 \quad \rightleftharpoons \quad ROH \quad + \quad e \bullet Si(OR)_3$

$\qquad (OR)_3Si \bullet e \quad + \quad Cl \bullet nn \quad \rightleftharpoons \quad ClSi(OR)_3$

[Reactive, stable and soluble]

14.11b

Since the last step in the stage above is a double half headed arrow instead of one single headed arrow, the stage is unproductive, clear indication that $Si(OR)$ is soluble in HCl, noting that dissolution or miscibilization is different from solubilization. For while HCl is polar/ ionic, $ClSi(OR)_3$ is polar/non-ionic, for which $ClSi(OR)_3$ if solid cannot fully dissolve in HCl or if liquid cannot fully miscibilize in HCl. Notice that, while $SiCl_4$ is insoluble in HCl, $Si(OR)_4$ is soluble in HCl.

$$Cl - \underset{\underset{Cl}{|}}{\overset{\overset{Cl}{|}}{Si}} - Cl \quad + \quad LiAlH_4 \quad \longrightarrow \quad LiAlCl_4 \quad + \quad H - \underset{\underset{H}{|}}{\overset{\overset{H}{|}}{Si}} - H$$

Radical reactions

(Silane)   14.12a

It is the Al center that does all the abstraction here in all the four stages. ***The Equilibrium states of existence of Cl_3SiH, Cl_2SiH_2, $ClSiH_3$, and SiH_4 are suppressed all the time in the last step of the four stages by the presence of Li.*** Shown below is the first stage, wherein the Equilibrium state of Existence of $SiCl_4$ is as usual suppressed.

Stage 1: $\quad LiAlH_4 \quad \rightleftharpoons \quad Li^{\oplus} \ldots \ldots \; {}^{\theta}\underset{\bullet e}{Al}H_3 \quad + \quad n \bullet H$

(A)

$\quad (A) \quad + \quad SiCl_4 \quad \rightleftharpoons \quad Li^{\oplus} \ldots \ldots {}^{\ominus}AlH_3Cl \quad + \quad e \bullet SiCl_3$

$Cl_3Si \bullet e \quad + \quad n \bullet H \quad \longrightarrow \quad Cl_3SiH$

14.12b

Overall Equation: $\quad LiAlH_4 \quad + \quad SiCl_4 \quad \longrightarrow \quad LiAlH_3Cl \quad + \quad Cl_3SiH$

14.12c

From the reactions above, it is obvious that all substituent groups on silicon are loosely bonded to different degrees nucleo-radically, for which the order in which they are loosely bonded in strength is as follows—

$$n.C_2H_5 \quad > \quad n.CH_3 \quad > \quad H \bullet n \quad > \quad nn.NH_2 \quad > \quad nn \bullet OR \quad > \quad nn.OH \quad > \quad nn.Cl \quad >$$

Order of decreasing strength of loose covalent bonds on silicon

14.13

The chloride atom being the only radical-pulling group above, is the less loosely bonded species to the silicon center than OR which in turn is less loosely bonded than CH_3 and so on. This can be observed to be different from the order already established with C as the central atom as their pushing/pulling capacities and is therefore particularly valid for any operating conditions with Si. Under Equilibrium state of existence, with C, H is held electro-free-radically, unlike with Si.

Similar types of reactions are favored by trichlorosilane as is obvious from the first stage above. Trisubstituted silanes add to carbon-carbon double and triple bonds in the presence of platinum catalyst or free-radical initiators. With platinum catalyst, the reactions are also free-radical in character, with the Pt and the free-radical initiator unable to polymerize the acetylene, after activating it.

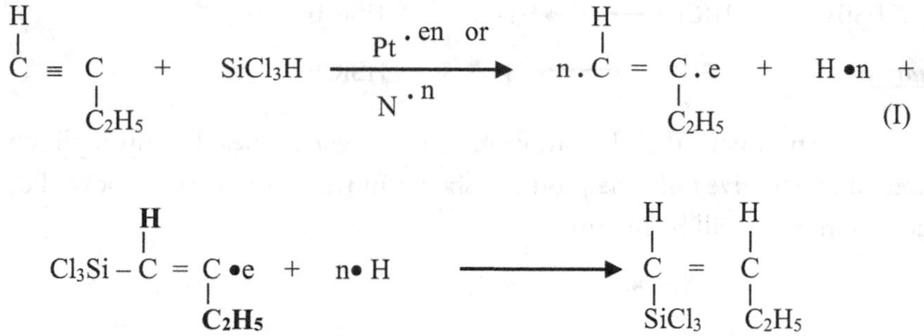

$$
\begin{array}{c}
\overset{\displaystyle H}{\underset{\displaystyle C_2H_5}{\overset{|}{\underset{|}{C}}}} \equiv C \quad + \quad SiCl_3H \quad \xrightarrow[N\,\cdot n]{\;Pt\,\cdot en\ or\;} \quad n\,.\,\overset{\displaystyle H}{\underset{\displaystyle C_2H_5}{\overset{|}{\underset{|}{C}}}} = C\,.\,e \quad + \quad H \bullet n \quad + \quad e\,.\,SiCl_3 \qquad (I)
\end{array}
$$

$$
\begin{array}{c}
\overset{\displaystyle \mathbf{H}}{\underset{\displaystyle \mathbf{C_2H_5}}{\overset{|}{\underset{|}{Cl_3Si - C}}}} = C \bullet e \quad + \quad n\bullet H \quad \longrightarrow \quad \overset{\displaystyle H}{\underset{\displaystyle SiCl_3}{\overset{|}{\underset{|}{C}}}} = \overset{\displaystyle H}{\underset{\displaystyle C_2H_5}{\overset{|}{\underset{|}{C}}}}
\end{array}
$$

<u>Addition of trichlorosilane to acetylenes using Pt - catalyst and free-radical catalyst</u>   14.14

Since Cl is less loosely bonded than H, it is (I) that is favored, and that is the finger-print of $HSiCl_3$, i.e. its Equilibrium State of Existence. It is the $Cl_3Si\bullet e$ that adds to the activated acetylene which is a Nucleophile, for which the equation is radically balanced. If $Cl\bullet nn$ has been the component held in Equilibrium state of Existence, the product obtained would have been different. If the free-radical initiator is a Nucleo-free-radical generating one, it would not have favored the initiation step, because of presence of transfer species of the first kind of the first type on the C_2H_5 group carried by the acetylene. The Pt which is carrying an electro-non-free-radical cannot also favor the initiation of the monomer, because the equation will also not be radically balanced. However, the Pt being very polar (because of the presence of four paired unbonded radicals in the last shell) is a passive catalyst in its ability not only to suppress the Equilibrium state of existence of the acetylene, but also activate it. There is need to show why the Equilibrium state of existence of the $HSiCl_3$ is as shown above in Equation 14.14, despite the order shown in Equation 14.13.

Trichlorosilane (silicochloroform) first prepared by Woehler in 1857, results along with silicon tetrachloride from the action of dry hydrogen chloride on silicon.

$$Si \quad + \quad 3HCl \quad \longrightarrow \quad HSiCl_3 \quad + \quad H_2 \ (and\ SiCl_4) \qquad 14.15$$

14.15

<u>Stage 1:</u> $\quad e \bullet \overset{\bullet n}{\underset{\bullet n}{Si}} \bullet e \quad + \quad HCl \quad \rightleftharpoons \quad Cl - \overset{\bullet n}{\underset{\bullet e}{Si}} \bullet n \quad + \quad e\bullet H$

$$\longrightarrow \quad Cl - \overset{\bullet n}{\underset{\bullet e}{Si}} - H$$

$$(A)$$

<u>Overall Equation:</u> $Si \; + \quad HCl \quad \longrightarrow \quad (A)$ 14.16a

14.16b

<u>Stage 2:</u> $\quad (A) \quad + \quad HCl \quad \rightleftharpoons \quad Cl - \overset{\bullet n}{Si}\,n - H \quad + \quad e\bullet H$

$$\underset{\displaystyle Cl}{|}$$

$$\longrightarrow \quad Cl_2SiH_2$$

<u>Overall Equation'</u> $\quad (A) \quad + \quad HCl \quad \longrightarrow \quad Cl_2SiH_2$ 14.17a

<u>Overall overall equation:</u> $\quad Si \quad + \quad 2HCl \quad \longrightarrow \quad Cl_2SiH_2$ 14.17b

14.17c

Stage 3:

$$Cl_2SiH_2 \rightleftharpoons HCl_2Si \bullet e + n \bullet H$$

$$HCl \rightleftharpoons H \bullet e + nn \bullet Cl$$

$$H \bullet e + n \bullet H \rightleftharpoons H_2$$

$$HCl_2Si \bullet e + nn \bullet Cl \longrightarrow HSiCl_3 \qquad \text{14.18a}$$

Overall Equation: $Cl_2SiH_2 + 3HCl \longrightarrow H_2 + HSiCl_3 \qquad$ 14.18b

Overall overall equation: $Si + 3HCl \longrightarrow H_2 + HSiCl_3 \qquad$ 14.18c

If $SiCl_4$ has to be one of the products, then Equation 14.15 no longer applies. It only applies based on the molar ratios indicated above to give only the products shown in the last equation above. For $SiCl_4$ to be a part of the product, more HCl will be required.

Stage 4:

$$HSiCl_3 \rightleftharpoons Cl_3Si \bullet e + n \bullet H$$

$$HCl \rightleftharpoons H \bullet e + nn \bullet Cl$$

$$H \bullet e + n \bullet H \rightleftharpoons H_2$$

$$Cl_3Si \bullet e + nn \bullet Cl \longrightarrow SiCl_4 \qquad \text{14.19a}$$

Overall Equation: $HSiCl_3 + HCl \longrightarrow H_2 + SiCl_4 \qquad$ 14.19b

Overall overall equation: $Si + 4HCl \longrightarrow 2H_2 + SiCl_4 \qquad$ 14.20

Evidently the final overall equation does not reflect the presence of $HSiCl_3$ as one of the products. Hence the final overall equation is as follows.

Overall overall equation: $2Si + 7HCl \longrightarrow 3H_2 + SiCl_4 + HSiCl_3 \qquad$ 14.21

This is an Equilibrium mechanism system with 7 stages either in series or in series and parallel as shown below I Figure 14.1.

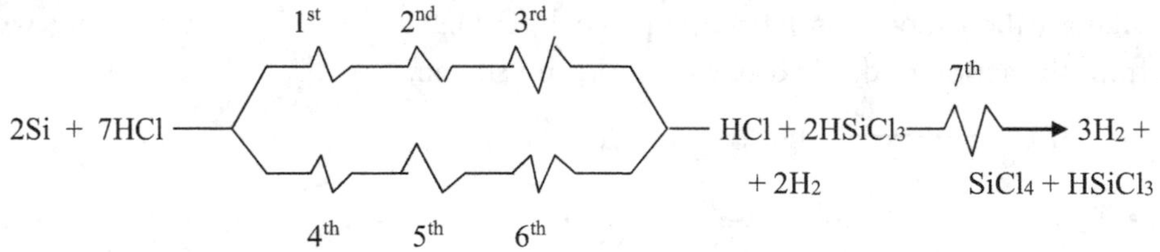

Figure 14.1 Schematic representation of the reaction of Equation 14.21

All along so far, we have only been encountering stages taking place in series. For the first time is a case of stages in series and parallel making the kinetic modeling different from what has been known in the past and or present. In fact in general, based on what has been seen so far, wherein the real mechanisms of all reactions are being systematically provided in a unique order, there is no doubt that kinetic modeling of these systems is going to be very, different, unique and simple. The reason why the conversion of Si to $SiHCl_3$ took place in parallel, is because of the molar ratios involved (Mathematics) and the fact that Nature is Orderliness wherein chemical reactions take place on molar basis based on Avogadro's constant number and concept.

The first compound to be prepared that contained a carbon-silicon bond was tetraethylsilane, b.pt 153^0C. It was made by Friedel and Grafts in 1863 by the reaction of ethyl zinc and silicon tetrachloride.

$$SiCl_4 \quad + \quad Zn(C_2H_5)_2 \quad \xrightarrow{160°C} \quad Cl - \underset{\underset{Cl}{|}}{\overset{\overset{Cl}{|}}{Si}} - Cl \quad +$$

$$en \bullet ZnC_2H_5 \quad + \quad n \bullet C_2H_5 \quad \longrightarrow \quad H_5C_2 - \underset{\underset{Cl}{|}}{\overset{\overset{Cl}{|}}{Si}} - Cl \quad + \quad ClZnC_2H_5 \quad \longrightarrow$$

$$ZnCl_2 \quad + \quad Cl - \underset{\underset{C_2H_5}{|}}{\overset{\overset{C_2H_5}{|}}{Si}} - Cl \quad \xrightarrow{+ \mathbf{Zn(C_2H_5)_2}} \quad H_5C_2 - \underset{\underset{C_2H_5}{|}}{\overset{\overset{C_2H_5}{|}}{Si}} - C_2H_5 \quad + \quad 2ZnCl_2$$

$$14.22$$

This is a four stage Equilibrium mechanism system in which only $Zn(C_2H_5)_2$ is in Equilibrium state of Existence while $SiCl_4$ is stable. From the stages, one can easily identify the finger prints (i.e. Equilibrium state of existence of the compounds formed).

$$ClZnC_2H_5 \quad \underset{\text{Existence}}{\overset{\text{Equilibrium State of}}{\rightleftharpoons}} \quad ClZn \bullet en \quad + \quad n \bullet C_2H_5$$

$$Cl_3SiC_2H_5 \quad \underset{\text{Existence}}{\overset{\text{Equilibrium State of}}{\rightleftharpoons}} \quad C_2H_5Cl_2Si \bullet e \quad + \quad nn \bullet Cl$$

$$\vdots$$

$$ClSi(C_2H_5)_3 \quad \rightleftharpoons \quad (C_2H_5)_3Si \bullet e \quad + \quad nn \bullet Cl \qquad 14.23$$

All the reactions so far, can be observed to be radical in character. Silicon tetrachloride reacts radically with Grignard reagents (the real structures of which are yet to be shown), to give a mixture of the compounds having from one to four chloride atoms replaced by hydrocarbon groups.

$$Cl - \underset{\underset{Cl}{|}}{\overset{\overset{Cl}{|}}{Si}} - Cl \quad + \quad RMgX \quad \longrightarrow \quad Cl - \underset{\underset{Cl}{|}}{\overset{\overset{Cl}{|}}{Si}} - Cl \quad + \quad R^{\bullet n} \quad + \quad e \cdot MgX$$

$$\longrightarrow Cl - \underset{\underset{Cl}{|}}{\overset{\overset{Cl}{|}}{Si}} - R \quad + \quad MgXCl \quad \xrightarrow{RMgX} \quad R - \underset{\underset{Cl}{|}}{\overset{\overset{Cl}{|}}{Si}} - R \quad + \quad 2MgXCl$$

$$\xrightarrow{2RMgX} 4MgXCl \quad + \quad R - \underset{\underset{R}{|}}{\overset{\overset{R}{|}}{Si}} - R$$

$$14.24$$

By regulating conditions, it is possible to obtain good yields of the desired products.

$$H - \underset{\underset{Cl}{|}}{\overset{\overset{Cl}{|}}{Si}} - Cl \quad + \quad RMgX \quad \longrightarrow \quad H - \underset{\underset{Cl}{|}}{\overset{\overset{Cl}{|}}{Si}} - Cl \quad + \quad R^{\bullet n} \quad + \quad e \cdot MgX$$

$$\longrightarrow H - \underset{\underset{Cl}{|}}{\overset{\overset{Cl}{|}}{Si}} - R \quad + \quad MgXCl \quad \xrightarrow{2RMgX} \quad H - \underset{\underset{R}{|}}{\overset{\overset{R}{|}}{Si}} - R \quad + \quad 3MgXCl$$

$$14.25$$

Notice herein in all the four stages, the Equilibrium state of existence of the silane is suppressed by the presence of Mg which is more electropositive than Silicon, with abstraction done by the magnesium center.

Trialkylchlorosilanes, R_3SiCl, are silicon analogs of tertiary alkyl chlorides; yet it is not possible to eliminate hydrogen chloride to form an unsaturated compound with a carbon-silicon double bond. Reaction with anhydrous ammonia or amines gives the trialkylsilyl-amine.

$$R - \underset{\underset{R}{|}}{\overset{\overset{CH_3}{|}}{Si}} - Cl \longrightarrow R - \underset{\underset{R}{|}}{\overset{\overset{H}{|}}{Si}} = \underset{\overset{}{H}}{\overset{\overset{H}{|}}{C}} + HCl$$

(Existence not favored)　　　　　　　　　14.26

Herein, the R groups are not greater than CH_3 in capacity, otherwise, the R group will be the provider of transfer species. However, none is favored, clear indication that transfer species (H·e) cannot be rejected from a growing chain electro-free-radically when Si is present.

$$R - \underset{\underset{R}{|}}{\overset{\overset{CH_3}{|}}{Si}} - Cl + 2NH_3 \longrightarrow R - \underset{\underset{R}{|}}{\overset{\overset{CH_3}{|}}{Si}} - Cl + H^{\cdot e} + nn \cdot NH_2 + \ddot{N}H_3$$

$$\longrightarrow R - \underset{\underset{R}{|}}{\overset{\overset{CH_3}{|}}{Si}} - NH_2 + H^{\cdot e} + Cl^{\cdot nn} + \ddot{N}H_3 \longrightarrow$$

$$H_4N \overset{\oplus}{\text{------}} \overset{\ominus}{Cl} + R - \underset{\underset{R}{|}}{\overset{\overset{CH_3}{|}}{Si}} - NH_2$$

$\underline{\text{trialkylsilamine}}$　　　　　　　　　14.27

Note that the alkylane groups are not replaced by $^{nn} \cdot NH_2$ group. The need for the extra mole of NH_3 does not arise since the trialkylsilamine is insoluble in HCl. It only helps to isolate it. Further reaction of the silylamine with trialkylsilyl chloride gives the disilylamine (disilazone)

$$R - \underset{\underset{R}{|}}{\overset{\overset{R}{|}}{Si}} - Cl + R - \underset{\underset{R}{|}}{\overset{\overset{R}{|}}{Si}} - NH_2 + NH_3 \longrightarrow H \bullet e + R - \underset{\underset{R}{|}}{\overset{\overset{R}{|}}{Si}} - \overset{\overset{H}{|}}{N} \cdot nn$$

(I)　　　　　　　　(II)

$$+ R - \underset{\underset{R}{|}}{\overset{\overset{R}{|}}{Si}} - Cl + NH_3 \longrightarrow R - \underset{\underset{R}{|}}{\overset{\overset{R}{|}}{Si}} - NH - \underset{\underset{R}{|}}{\overset{\overset{R}{|}}{Si}} - R$$

$$+ H^{\cdot e} Cl^{\cdot nn} + NH_3 \longrightarrow R - \underset{\underset{R}{|}}{\overset{\overset{R}{|}}{Si}} - \overset{\overset{H}{|}}{N} - \underset{\underset{R}{|}}{\overset{\overset{R}{|}}{Si}} - R + H_4N \overset{\oplus}{\text{------}} Cl^{\ominus}$$

(III) Disilylamine　　　　　　　　　14.28

Though the ionic version of the reaction may seem possible in writing, it does not exist, since the silicon center cannot carry an ionic charge. Ionic actions in some cases only exist during electrostatic addition via movement of electro-radical to form an imaginary bond. *In the presence of silicon compounds, ionic groups such as, amino, and imino etc., isolate their ionic identities; while in the presence of strong ionic centers, they isolate their radical identities. This is what equally takes place when monomers are activated.*

Hydrolysis of trialkylsilylamines or of trialkylchlorosilanes gives the silanol.

$$R - \underset{\underset{R}{|}}{\overset{\overset{R}{|}}{Si}} - NH_2 \ + \ H_2O \ \longrightarrow \ R - \underset{\underset{R}{|}}{\overset{\overset{R}{|}}{Si}} - NH_2 \ + \ \overset{nn}{\cdot}OH \ + \ H \cdot e \ \longrightarrow$$

$$R - \underset{\underset{R}{|}}{\overset{\overset{R}{|}}{Si}} - OH \ + \ NH_3 \tag{14.29}$$

This clearly indicates that NH_2 is less loosely bonded than OH group is; that is, OH is more nucleo-radical in character than NH_2 group, noting that their bonds ($Si - NH_2$, $Si - OH$) are not affected when they exist in Equilibrium state of existence.

$$R - \underset{\underset{R}{|}}{\overset{\overset{R}{|}}{Si}} - Cl \ + \ H_2O \ + \ NH_3 \ \longrightarrow \ R - \underset{\underset{R}{|}}{\overset{\overset{R}{|}}{Si}} - Cl \ + \ H \overset{e}{\cdot} + \ HO \overset{nn}{\cdot}$$

$$+ \ NH_3 \ \longrightarrow \ R - \underset{\underset{R}{|}}{\overset{\overset{R}{|}}{Si}} - OH \ + \ NH_4Cl \tag{14.30}$$

By the reactions above, though not clearly obvious, Cl is greater than OH which in turn is greater than NH_2 in nucleo-non-free-radical capacity (See Equation 14.13).

These silanols loose water easily and form the oxides known as hexalkyldisiloxanes.

$$R - \underset{\underset{R}{|}}{\overset{\overset{R}{|}}{Si}} - OH \ + \ HO - \underset{\underset{R}{|}}{\overset{\overset{R}{|}}{Si}} - R \ \longrightarrow \ R - \underset{\underset{R}{|}}{\overset{\overset{R}{|}}{Si}} - O \cdot nn \ +$$

(Active fraction)   (stable fraction)

$$e \cdot \underset{\underset{R}{|}}{\overset{\overset{R}{|}}{Si}} - R \ + \ H_2O \ \longrightarrow \ R - \underset{\underset{R}{|}}{\overset{\overset{R}{|}}{Si}} - O - \underset{\underset{R}{|}}{\overset{\overset{R}{|}}{Si}} - R \ + \ H_2O \tag{14.31}$$

Thus, it is impossible to obtain unsaturated compounds from silanols by the loss of water. That is, Si cannot carry a double or polar bond with C, not even with itself.

Hydrolysis of dialkyldichlorosilanes, R_2SiCl_2, gives diols, which are insoluble in water but soluble in aqueous alkali.

$$Cl - \underset{\underset{R}{|}}{\overset{\overset{R}{|}}{Si}} - Cl \ + \ 2H_2O \ \longrightarrow \ 2HO \overset{nn}{\cdot} \ + \ 2H \overset{e}{\cdot} \ + \ Cl - \underset{\underset{R}{|}}{\overset{\overset{R}{|}}{Si}} - Cl \ \xrightarrow{\text{(2 stages)}}$$

$$HO - \underset{\underset{R}{|}}{\overset{\overset{R}{|}}{Si}} - OH \ + \ 2HCl \ \xrightarrow[\text{(2 Stages)}]{2NaOH} \ NaO - \underset{\underset{R}{|}}{\overset{\overset{R}{|}}{Si}} - ONa \ + \ 2H_2O \ + \ 2HCl \tag{14.32}$$

The mechanisms for the insolubility in water and solubility in NaOH or aqueous NaOH are shown below.

Stage 1:

$$HO - \underset{\underset{R}{|}}{\overset{\overset{R}{|}}{Si}} - OH \rightleftharpoons H \bullet e \;+\; nn\bullet O - \underset{\underset{R}{|}}{\overset{\overset{R}{|}}{Si}} - OH$$

$$(A)$$

$$H_2O \rightleftharpoons H \bullet e \;+\; nn\bullet OH$$

$$(A) \;+\; e\bullet H \rightleftharpoons (HO)_2SiR_2$$

$$H \bullet e \;+\; nn\bullet OH \rightleftharpoons H_2O$$

(Reactive stable and Insoluble)   14.33

Stage 1:

$$NaOH \rightleftharpoons Na\bullet e \;+\; nn\bullet OH$$

$$Na\bullet e \;+\; HO - \underset{\underset{R}{|}}{\overset{\overset{R}{|}}{Si}} - OH \rightleftharpoons HO - \underset{\underset{R}{|}}{\overset{\overset{R}{|}}{Si}} - ONa \;+\; e\bullet H$$

$$H \bullet e \;+\; nn\bullet OH \rightleftharpoons H_2O$$

(Reactive, stable and soluble)   14.34

When insoluble in water, the same products as reactants are obtained, all held in equilibrium; yet it will miscibilize in water both being polar/ionic. When soluble in NaOH, notice that different products from the reactants are obtained, but all held in Equilibrium (See Equation 14.11b). Yet, the NaOH will dissolve in the diol, both being polar/ionic. However, based on the last part of Equation 14.32, products are obtained from the reaction between the diol and NaOH. This was made possible because of the presence of HCl in the system, either by suppressing the last step of Equation 14.34 or as shown below.

Stage 1:

$$NaOH \rightleftharpoons Na\bullet e \;+\; nn\bullet OH$$

$$HCl \rightleftharpoons H \bullet e \;+\; nn\bullet Cl$$

$$H \bullet e \;+\; nn\bullet OH \rightleftharpoons H_2O$$

$$Na\bullet e \;+\; HO - \underset{\underset{R}{|}}{\overset{\overset{R}{|}}{Si}} - OH \rightleftharpoons NaO - \underset{\underset{R}{|}}{\overset{\overset{R}{|}}{Si}} - OH \;+\; e\bullet H$$

$$H \bullet e \;+\; H_2O \rightleftharpoons en\bullet \underset{\underset{H}{|}}{\overset{\overset{H}{|}}{O}} - H$$

$$(A)$$

$$Cl\bullet nn \;+\; (A) \longrightarrow Cl^{\ominus}........^{\oplus}OH_3$$

Dilute hydrochloric acid   14.35a

Overall Equation: $NaOH + HCl + (HO)_2SiR_2 \longrightarrow NaO(OH)SiR_2 \;+\; ClOH_3$ 14.35b

The second stage follows in a similar manner. One can observe how unique NATURE can be. Note however, that the first two steps in Equations 14.34 and 14.35a have not been fully represented, because as is already known, Heat must be released in those stages. NaOH cannot exist in Equilibrium state of existence in the presence of water, the silanol and HCl. Nevertheless, the stages remain the same, since the oxidizing oxygen product attacks the H_2 formed to give $HO \cdot nn$ and $H \cdot e$.

It is important to distinguish between the radical and ionic reactions. As a matter of fact, no Equilibrium mechanism reaction can take place ionically, because when they do, no products can be obtained. Yet there are polar/ionic compounds. A ring with functional centers can only be opened cationically or with positive charges. The diols are useful condensation monomers which readily loose water to form linear, cross-linked and cyclic polymers, which are also called silicones or polysilicones. The condensation reaction is radical in character as will become clear in subsequent Volumes.

Methylvinyldichlorosilane (Vinylmethyldichlorosilane), ***used for rubbers of low compression set,*** is made by the addition of methyldichlorosilane to acetylene in the presence of either peroxides or platinum catalyst.

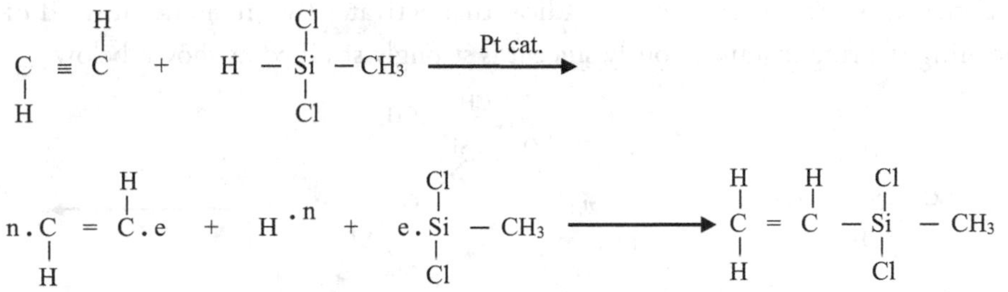

$$14.36$$

With acetylene being involved, there is no doubt that the route is free-radical in character. In fact acetylene cannot be activated chargedly due to electrodynamics forces of repulsion. (3 - Cyanopropyl) methyl dichlorosilane, ***used for oil resistant rubbers,*** is made in the same way from allyl cyanide, $H_2C=CHCH_2CN$.

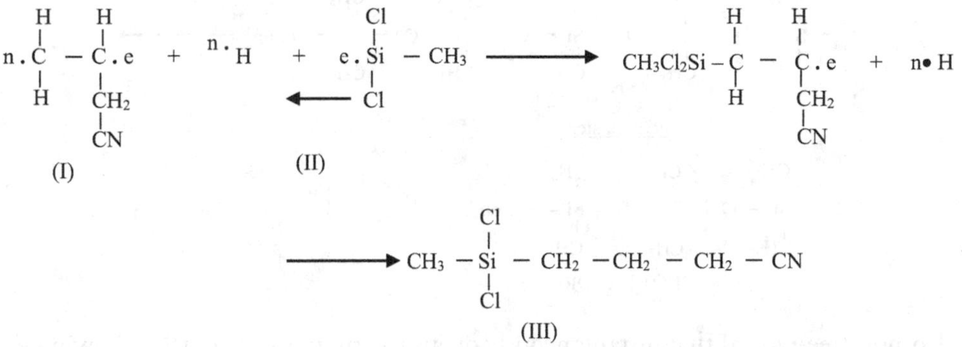

$$14.37$$

If the route had been ionic or charged, the activation of (I) would have been impossible as already stated into law. From the reaction above, CH_2X groups, are radical-pushing groups free-radically, where X is either a non-free-radical or free-radical-pulling group in general as already identified. In general, it can be observed most importantly that silicon unlike carbon, from all the reactions above, cannot carry real charges.

14.1.1 Cyclosiloxanes

14.1.1.1. "Nucleo-non-free-radical polymerization"

Beginning with (I) of Equation 14.1, the anionic polymerization of the cyclic monomer is said to be initiated by **alkali metal oxides (e.g. Na$_2$O) and hydroxides (e.g. NaOH), silanolates such as potassium trimethylsilanolate, (CH$_3$)SiOK and other bases**[1,2]. While all the catalysts just mentioned, can produce ionic and radical initiators, it is the radical initiations that are involved. For example, considering (CH$_3$)$_3$SiOK, the followings are obtained.

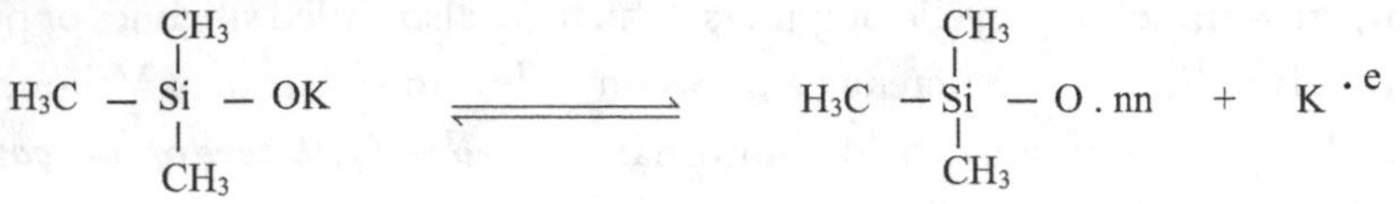

$$H_3C - \underset{\underset{CH_3}{|}}{\overset{\overset{CH_3}{|}}{Si}} - OK \quad \rightleftharpoons \quad H_3C - \underset{\underset{CH_3}{|}}{\overset{\overset{CH_3}{|}}{Si}} - O \cdot nn \quad + \quad K^{\cdot e}$$

(I)   14.38

It is "assumed" that it is the nucleo-non-free-radical that activates the monomer instead of an anionic initiator by opening the ring instantaneously since it is strongly strained as shown below.

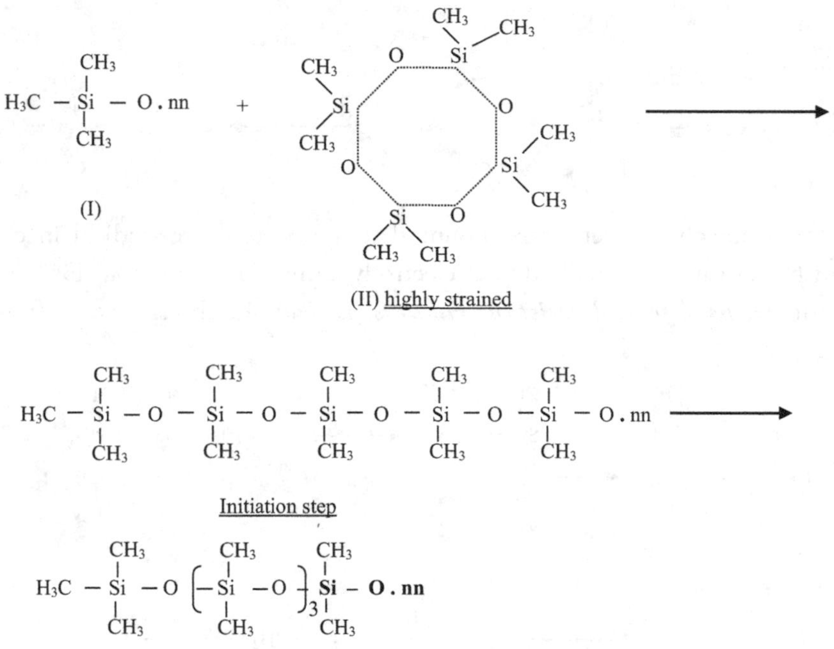

$$H_3C - \underset{\underset{CH_3}{|}}{\overset{\overset{CH_3}{|}}{Si}} - O \cdot nn \quad + \quad \text{(cyclosiloxane ring)} \quad \longrightarrow$$

(I)   (II) <u>highly strained</u>

$$H_3C - \underset{\underset{CH_3}{|}}{\overset{\overset{CH_3}{|}}{Si}} - O - \underset{\underset{CH_3}{|}}{\overset{\overset{CH_3}{|}}{Si}} - O - \underset{\underset{CH_3}{|}}{\overset{\overset{CH_3}{|}}{Si}} - O - \underset{\underset{CH_3}{|}}{\overset{\overset{CH_3}{|}}{Si}} - O - \underset{\underset{CH_3}{|}}{\overset{\overset{CH_3}{|}}{Si}} - O \cdot nn \longrightarrow$$

<u>Initiation step</u>

$$H_3C - \underset{\underset{CH_3}{|}}{\overset{\overset{CH_3}{|}}{Si}} - O \left[- \underset{\underset{CH_3}{|}}{\overset{\overset{CH_3}{|}}{Si}} - O \right]_3 \underset{\underset{CH_3}{|}}{\overset{\overset{CH_3}{|}}{Si}} - O \cdot nn$$

NOT FAVORED   14.39

It is the nucleo-non-free-radical that instantaneously opens the ring in view of the following -

(i) The weak bonds existing between the silicon and oxygen center, -Si-O-; hence the dotted lines shown for (II) above to indicate the weakness of the covalent bond. In view of the presence of weak bonds, therefore, the ring is still fairly strained.

(ii) The need to achieve the MRSE for such a ring does not arise, since the Si - O bonds along the rings are loosely bonded. The strain energy in the ring is not altered by the addition of the MRSE which is quite small.

It was for these reasons that the heat of polymerization, ΔH, for the reaction of such a monomer has been observed to be nearly zero and, amazingly, the changes in entropy, ΔS is positive by about 1.6cal/

338

mole-°K. In order words, the reaction is moving from order to disorder, an unusual case observed also with cyclic octamers of sulfur and selenium[11]. This is not to be expected. This is said to be clear indication of the increased or high degree of flexibility of the linear polymer chains, resulting in greater degrees of freedom in the linear polymer compound of the cyclic monomer[8]. Propagation reactions follow in the same manner. While the reason above looks valid, one should not forget the great influence of the initiator on the growing polymer chain she is carrying. It can be observed that the polymerization of the monomer is not anionic ion-paired, *but Electrostatically anionic-paired i.e. that in which the K center is bonded to the O center in the ring to give it a positive charge which is electrostatically bonded to the anionic center.* Anionically or nucleo-free or non-free-radically, there is no functional center in the ring. Only nucleophilic functional centers exist in the ring. The use of nucleo-free-radical is not favored, since the equation will not be radically balanced in the initiation step.

$$N \cdot^n \; + \; (II) \xrightarrow[\text{opening}]{\text{Instantaneous}}$$

$$\left[: NR_3 \right]$$

$$N - \underset{\underset{CH_3}{|}}{\overset{\overset{CH_3}{|}}{Si}} - O - \underset{\underset{CH_3}{|}}{\overset{\overset{CH_3}{|}}{Si}} - O - \underset{\underset{CH_3}{|}}{\overset{\overset{CH_3}{|}}{Si}} - O - \underset{\underset{CH_3}{|}}{\overset{\overset{CH_3}{|}}{Si}} - O \cdot nn$$

Not radically balanced

14.40

Nucleo-non-free-radically, there is no transfer species,

When Covalently charged-paired initiators are involved, polymerization is not favored negatively, since the equation will not be chargedly balanced and the monomer is a Nucleophile. When other electrostatic types of initiators are involved, polymerization will not be favored if the ring is opened instantaneously and the initiator is in stable state of existence. Electro-free-radically, it will be favored if the initiator is in Equilibrium state of existence.

(I) Imaginary

(II)

Not Favored

14.41

In (I) above, NR_3 has been used as cocatalyst and ROH as the catalyst. (I) exists in Equili-brium state of existence and uses the functional center to open the ring with backward addition electro-free-radically.

$$H_9C_4{}^{\ominus} \cdots {}^{\oplus}Li \; + \; (II) \longrightarrow \text{No reaction negatively}$$

(III) Real

14.42

In general, it can be observed that, while imaginary anionic polymerization of the monomer may be possible if the ring is opened instantaneously, real anionic polymerization is not possible. The monomer being Nucleophilic will largely favor only electro-free-radical poly-merization which inherently is cationic.

Indeed, all the above were shown to see the possibility of using the said initiator and others for the purpose of exploration. How can a nucleophilic monomer favor anionic or nucleo-non-free-radical route when the ring is so large in size and the K• e or the H•e have not been fully isolated? These initiators will largely exist in Equilibrium state of Existence in such a way that the route becomes electro-free-radical in character as shown below for NaONa.

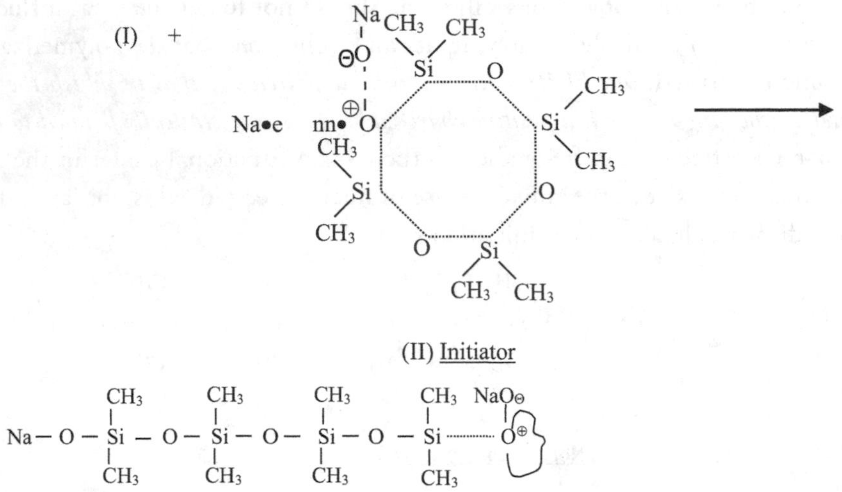

(II) Initiator

(III) Initiation step   14.43

One can observe that if the anionic route (NaO$^\ominus$) had been favored, Na will still be carrying the chain. But, it not favored. When $(CH_3)_3SiOK$ is indeed used, the route is electro-free-radical with K carrying the chain and the initiator will be like the case above. When NaOH is used, the route is also the same with the chain carried by Na. When LiC_4H_9 is used, the route will be favored with Li carrying the chain. When radicals are carried, will the center remain paired? Is it the charges carried by the centers that make pairing possible? However, notice that the chains above are carried by ionic metals, Na, K and even H. This is an additional reason to support the so-called observations above, that in which ΔH is almost equal to zero and ΔS is positive. It is for this reason, the sub-title was placed in inverted commas above. By this method of addition, there is no release of transfer species, but depropagation.

14.1.1.2. "Cationic ion-paired" and Free-radical Polymerizations.

Cationically, the situation is different. Cationic polymerization of cyclic siloxanes are said to have been carried out with protonic and Lewis acids. When free cationic initiators (protonic acids) are involved, provided the counter-ion of the initiator is less nucleophilic than the monomers, which is the case, the followings are obtained.

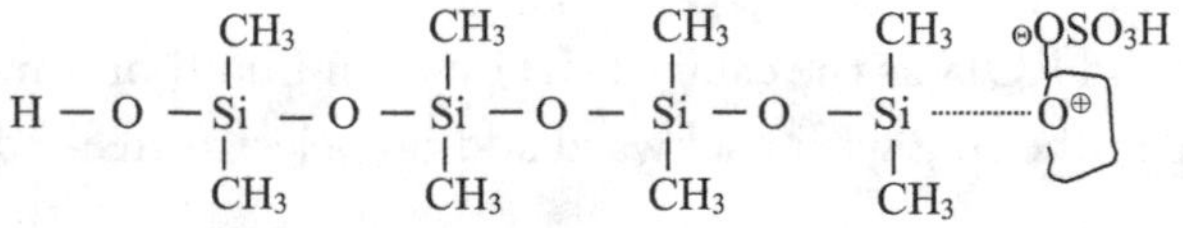

(III) Initiation step   14.44

It is important to note that the silicon center is not allowed to carry any charge. On the other hand, no evidence exist for the actual existence of the siliconium ion. One monomer can be used for the initiation preparation step, to favor the full existence of an electrostatic anionic-paired initiator which if used electro-free-radically, must exist in Equilibrium state of Existence. When in this state, H is held electro-free-radically, but the ring is opened cationically. The route is ELECTRO-FREE-RADICAL POLYMERIZATION and not Cationic as highlighted in the sub-title, whether the ring is opened instantaneously or not. Unlike cyclic ethers or all oxygen containing cyclic carbon-monomers, where presence of "carbonium ion" is favored for their growing polymer chains, presence of "siliconium ion" is not favored during Addition polymerization. The silicon center can only carry imaginary (Electrostatic and polar) charges. Polymerization can be observed to take place electro-free-radically for this monomer via backward addition. One can imagine the exact opposite of what was thought to be the case for so many years.

When complex electrostatically positively charged-paired initiators are involved, the followings are supposed to be obtained.

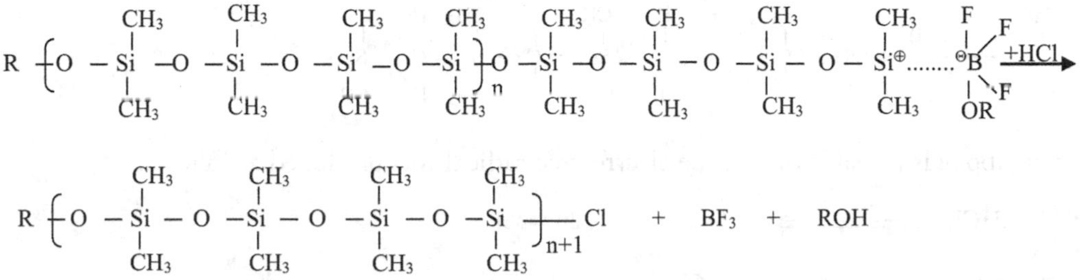

$$R \left(O - \underset{\underset{CH_3}{|}}{\overset{\overset{CH_3}{|}}{Si}} - O - \underset{\underset{CH_3}{|}}{\overset{\overset{CH_3}{|}}{Si}} - O - \underset{\underset{CH_3}{|}}{\overset{\overset{CH_3}{|}}{Si}} - O - \underset{\underset{CH_3}{|}}{\overset{\overset{CH_3}{|}}{Si}} \right)_n O - \underset{\underset{CH_3}{|}}{\overset{\overset{CH_3}{|}}{Si}} - O - \underset{\underset{CH_3}{|}}{\overset{\overset{CH_3}{|}}{Si}} - O - \underset{\underset{CH_3}{|}}{\overset{\overset{CH_3}{|}}{Si}} - O - \underset{\underset{CH_3}{|}}{\overset{\overset{CH_3}{|}}{Si^{\oplus}}} \cdots \overset{\ominus}{B} \underset{OR}{\overset{\overset{F}{|}}{\underset{}{}}}\overset{F}{\diagdown}\, \xrightarrow{+HCl}$$

$$R \left(O - \underset{\underset{CH_3}{|}}{\overset{\overset{CH_3}{|}}{Si}} - O - \underset{\underset{CH_3}{|}}{\overset{\overset{CH_3}{|}}{Si}} - O - \underset{\underset{CH_3}{|}}{\overset{\overset{CH_3}{|}}{Si}} - O - \underset{\underset{CH_3}{|}}{\overset{\overset{CH_3}{|}}{Si}} \right)_{n+1} Cl \quad + \quad BF_3 \quad + \quad ROH$$

NOT FAVORED

14.45

Whether the optimum chain length has been reached or not, living polymer chains will always be obtained since transfer species of the first kind of the first type cannot be released to form a dead terminal polar bond polymer. A stronger protonic acid than the counter-ion of the initiator or stronger anionic center is required to terminate the chain, as shown above if indeed favored. Propagation/depropagation phenomenon will not take place even if polymerization temperature is high. But electro-free-radically via backward addition if the temperature is above the Ceiling temperature. What indeed we have been calling electro-free-radical route is "inherently cationic", but not cationic. Note also that the positive charge above on the electrostatic bond is real and active. It is the negative charge that is imaginary. Hence, it is indeed a COMPLEX bond. Hence the reaction above is said not to be favored, since the silicon center cannot carry a real charge.

For (II) of (Equation of 14.1), 1,1,3,3-Tetramethyl-1,3- disilacyclobutane, acid catalyzed polymerization is said to proceed via the intermediate shown below.

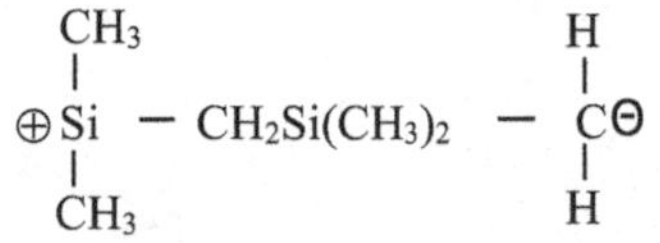

$$\underset{\underset{CH_3}{|}}{\overset{\overset{CH_3}{|}}{\oplus Si}} - CH_2Si(CH_3)_2 - \underset{\underset{H}{|}}{\overset{\overset{H}{|}}{C\ominus}}$$

14.46

Though the Si - C bonds are not strongly covalent, the Si - C bond is stronger than the Si - O bond. Si cannot carry ionic or covalent charges, but only imaginary charges. Because of the size, *the ring here is more strained than the first case above.* When the acids are involved, the followings are obtained.

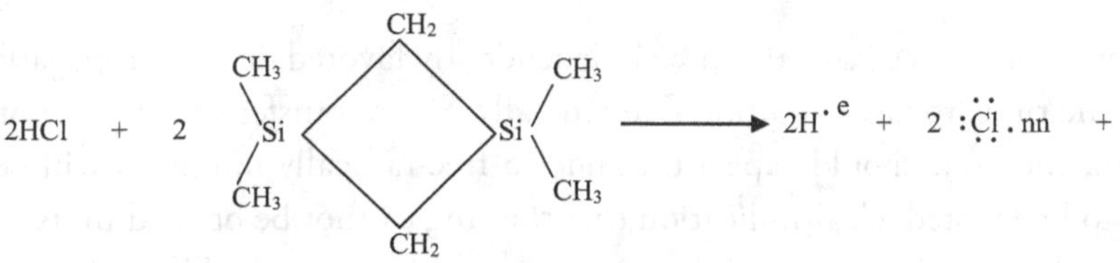

$$2HCl \;+\; 2 \;\; \underset{\underset{CH_3}{|}}{\overset{\overset{CH_3}{|}}{Si}}\diamondsuit\underset{\underset{CH_3}{|}}{\overset{\overset{CH_3}{|}}{Si}} \;\; \overset{CH_2}{\underset{CH_2}{}} \longrightarrow 2H^{\cdot\, e} \;+\; 2 \;:\ddot{C}l \cdot nn \;+$$

341

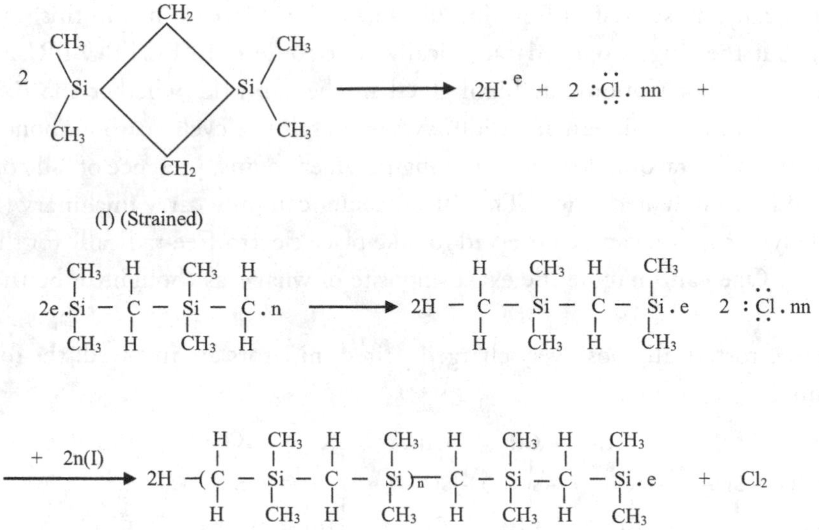

$$2 \quad \underset{CH_3}{\overset{CH_3}{\underset{|}{Si}}} \overset{CH_2}{\diamond} \underset{CH_3}{\overset{CH_3}{\underset{|}{Si}}} \quad \longrightarrow \quad 2H^{\cdot e} + 2 :\overset{..}{\underset{..}{Cl}} . \; nn \quad +$$

(I) (Strained)

$$2e . \underset{CH_3}{\overset{CH_3}{\underset{|}{Si}}} - \underset{H}{\overset{H}{\underset{|}{C}}} - \underset{CH_3}{\overset{CH_3}{\underset{|}{Si}}} - \underset{H}{\overset{H}{\underset{|}{C}}} . n \quad \longrightarrow \quad 2H - \underset{H}{\overset{H}{\underset{|}{C}}} - \underset{CH_3}{\overset{CH_3}{\underset{|}{Si}}} - \underset{H}{\overset{H}{\underset{|}{C}}} - \underset{CH_3}{\overset{CH_3}{\underset{|}{Si}}} . e \quad 2 :\overset{..}{\underset{..}{Cl}} . nn$$

$$+ 2n(I) \quad \longrightarrow \quad 2H - \left(\underset{H}{\overset{H}{\underset{|}{C}}} - \underset{CH_3}{\overset{CH_3}{\underset{|}{Si}}} - \underset{H}{\overset{H}{\underset{|}{C}}} - \underset{CH_3}{\overset{CH_3}{\underset{|}{Si}}} \right)_{\overline{n}} \underset{H}{\overset{H}{\underset{|}{C}}} - \underset{CH_3}{\overset{CH_3}{\underset{|}{Si}}} - \underset{H}{\overset{H}{\underset{|}{C}}} - \underset{CH_3}{\overset{CH_3}{\underset{|}{Si}}} . e \quad + \quad Cl_2 \qquad 14.47$$

The reaction above is possible only if the electro-free-radical was produced as follows.

Stage 1: $\quad$ 2HCl $\quad \rightleftharpoons \quad$ 2H •e $\quad + \quad$ 2nn• Cl

Cl •nn + nn• Cl $\quad \longrightarrow \quad$ Cl$_2$

Overall Equation: $\quad$ 2HCl $\quad \xrightarrow{Heat} \quad$ 2H •e + Cl$_2$ $\qquad$ 14.48

If this was not the case, then a stable product would have been obtained, because Cl.nn has not been isolated. Chargedly, the stage above is impossible. However, the ring is opened instantaneously by the paired unbonded radicals on the chlorine center. This can only activate the monomer, but not initiate it. When the ring is opened instantaneously, based on the types of radicals carried by it, only the electro-free-radical initiators can be used. With the use of the acid, the route is electro-free-radical to produce a living polymer chain in the absence of foreign terminating agents. Electro-free-radicals and nucleo-free-radicals cannot open the ring. Paired unbonded radicals on the Nucleo-non-free- and electro-non-free-radical carrying species can open the ring, and the radicals cannot polymerize it. Once the ring is instantaneously opened by other means, nucleo-free-radicals cannot polymerize the monomers as shown below.

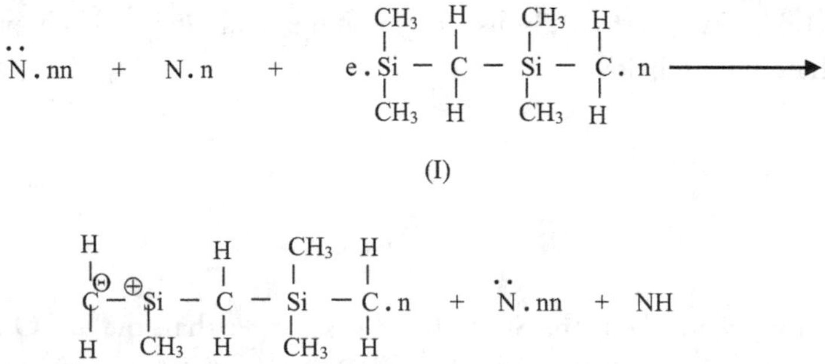

$$\overset{..}{N} . nn + N . n + e . \underset{CH_3}{\overset{CH_3}{\underset{|}{Si}}} - \underset{H}{\overset{H}{\underset{|}{C}}} - \underset{CH_3}{\overset{CH_3}{\underset{|}{Si}}} - \underset{H}{\overset{H}{\underset{|}{C}}} . n \quad \longrightarrow$$

(I)

$$\underset{H}{\overset{H}{\underset{|}{\overset{|}{C}}}}{}^{\ominus} - \overset{\oplus}{\underset{CH_3}{\underset{|}{Si}}} - \underset{H}{\overset{H}{\underset{|}{C}}} - \underset{CH_3}{\overset{CH_3}{\underset{|}{Si}}} - \underset{H}{\overset{H}{\underset{|}{C}}} . n \quad + \quad \overset{..}{N} . nn \quad + \quad NH$$

Transfer species of the first kind of the first type NOT FAVORED $\qquad$ 14.49

If the presence of the nucleo-non-free-radical can be independently favored during propagation, then the ring will be opened all the time, to favor addition free-radically. Since transfer species cannot be released from the growing chain, then one should expect that nucleo-free-radically initiation will be favored. It is however not known to be favored, clear indication that the ring cannot be opened in its presence, but remain closed. On the other hand, the route is not natural to the monomer. Since there is no site for addition of monomers along the electro-free-radical growing chain, branching reactions will not occur. Unlike the first type of silicon monomers considered above, the favored electro-free-radical route is with forward addition. Like the first case, the growing polymer chains when made to grow forwardly cannot terminate itself even when starved. A dead terminal polar bonded polymer as shown below cannot be obtained.

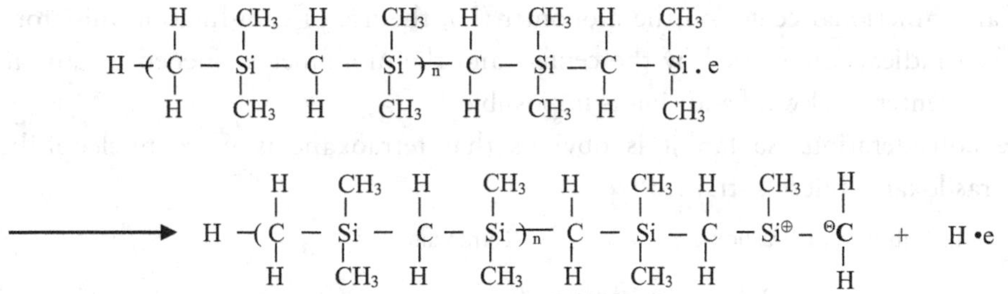

$$H + C - Si - C - Si \xrightarrow{}_n C - Si - C - Si . e$$

$$\longrightarrow H - C - Si - C - Si \xrightarrow{}_n C - Si - C - Si^{\oplus} - {}^{\ominus}C + H \cdot e$$

Dead terminal polar bonded polymer Not favored   14.50

From the considerations above, it can be observed that the oxygen containing cyclic silicon monomers (I) are more nucleophilic than their cycloalkane analogs ((II) of Equation 14.1) due to the absence of functional centers in (II). (II) is more strained than (I).

Cyclo silicanes (II)   <   Oxygen containing cyclic silicon monomers (I)

(Order of Nucleophilicity)   14.51

When the rings are of the same size the order above remains the same.

When so-called "cationic and anionic ion-paired initiators" are involved for (II), polymerization is not favored.

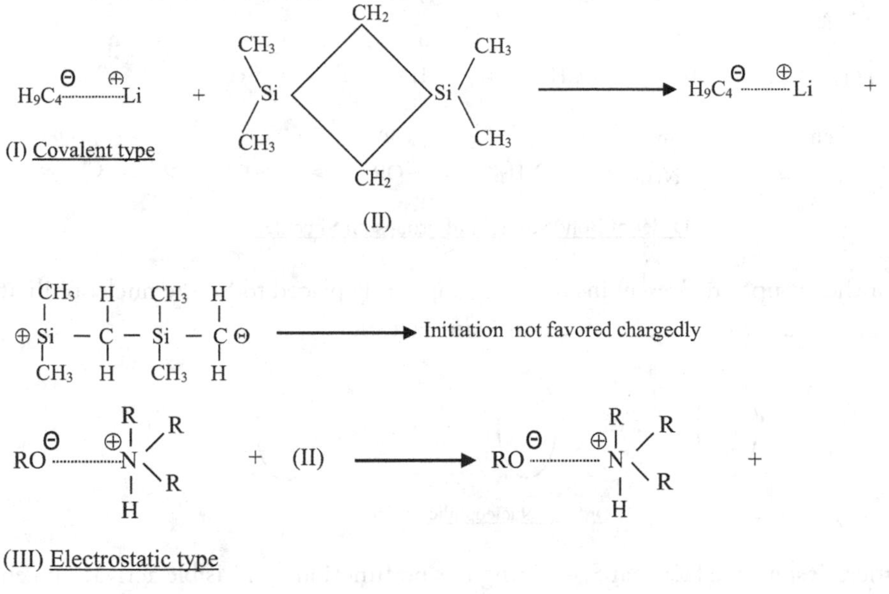

$$H_9C_4 \overset{\ominus}{\cdots\cdots} \overset{\oplus}{Li} + (II) \longrightarrow H_9C_4 \overset{\ominus}{\cdots\cdots} \overset{\oplus}{Li} +$$

(I) Covalent type

(II)

$$\overset{\oplus}{Si} - C - Si - C\overset{\ominus}{} \longrightarrow \text{Initiation not favored chargedly}$$ 14.52

$$RO \overset{\ominus}{\cdots\cdots} \overset{\oplus}{N} \overset{R}{\underset{R}{<}} + (II) \longrightarrow RO \overset{\ominus}{\cdots\cdots} \overset{\oplus}{N} \overset{R}{\underset{R}{<}} +$$

(III) Electrostatic type

$$e . Si - C - Si - C . n \longrightarrow$$ Propagation not favored if Initiator can exist in Equilibrium state of existence.   14.53

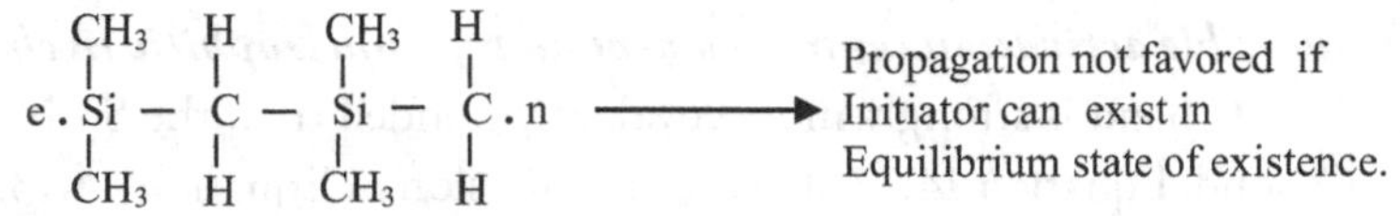

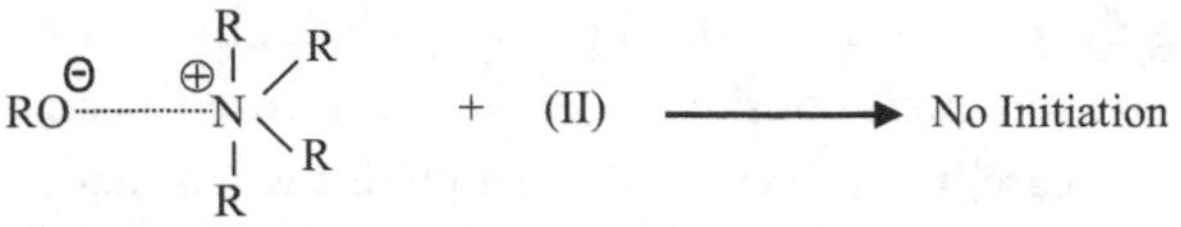

$$RO^{\ominus} \cdots \overset{\overset{\displaystyle R}{\overset{|}{N^{\oplus}}}}{\underset{\displaystyle R}{\overset{|}{}}} \overset{\displaystyle R}{\underset{\displaystyle R}{\diagup}} \quad + \quad (II) \quad \longrightarrow \quad \text{No Initiation}$$

(IV) <u>Electrostatic type</u>

14.54

In the absence of any functional center on the monomer (II), the use of coordination initiators is not possible, unless when radicals are carried by the centers and the monomer is opened instantaneously. Without a functional center, backward addition is impossible.

Based on the considerations so far, it is obvious that tetraoxane is more nucleophilic than octamethylcyclotetrasiloxane, but less strained.

Octamethylcyclotetrasiloxane   >   Tetraoxanes

<u>Order of Strain energy</u>

14.55

1,1,3,3 - tetramethyl - 1,3 - disilacyclobutane favors only the electro-free route and not the positively charged route, via instantaneously opening of the ring. Cyclobutane can favor both routes. When one H atom in cyclobutane is changed to CH_3, both electro-free-radical and positively charged routes are favored that which is not the case with silicon. However, since disilacyclobutane looks more strained than cyclo-butanes, the following is valid.

Cycloalkanes   >   Disilacyclobutane

<u>Order of Nucleophilicity</u>

14.56

Based on the analysis so far, the general order in which groups are loosely bonded to the silicon center are as shown below. This is valid for non-ringed and ring compounds, [See Equation 14.13].

$$\overset{n}{\overset{\bullet}{C_3H_7}} > \overset{n}{\overset{\bullet}{C_2H_5}} > \overset{n}{\overset{\bullet}{CH_3}} > \overset{n}{\overset{\bullet}{H}} > \overset{n}{\overset{\bullet}{C_2F_5}} > \overset{n}{\overset{\bullet}{CF_3}} >$$

$$\overset{nn}{\overset{\bullet}{NR_2}} > \overset{nn}{\overset{\bullet}{NHR}} > \overset{nn}{\overset{\bullet}{NH_2}} > \overset{nn}{\overset{\bullet}{OR}} > \overset{nn}{\overset{\bullet}{OH}} > \overset{nn}{\overset{\bullet}{Cl}}$$

<u>Order of Bond strength of groups on Si center</u>

14.57

When some of the groups are present inside a ring adjacently placed to Si, the nucleophilicity of the ring and what the groups are carrying, become dominant factors.

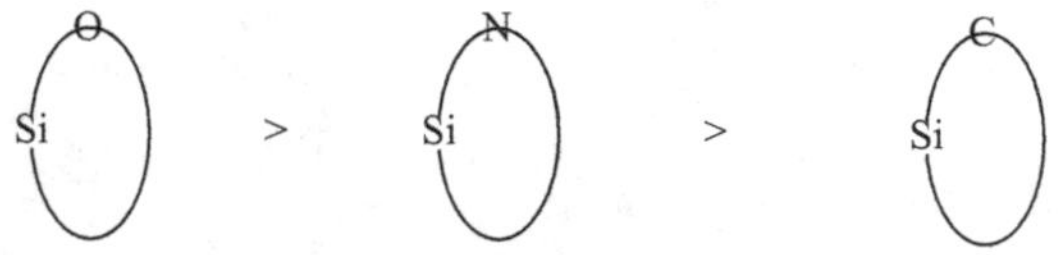

<u>Order of Nucleophilicity</u>

14.58

This is the order, despite the fact that Si – C ring has no functional or visible activation center. The fact that HCl cannot be released from $ClSi(CH_3)_3$, but can from $ClC(CH_3)_3$ means that polar bond between Si and C does not exist. ***The invisible activation center is a π-bond type nucleophilic in character and not a polar-bond type.*** The Si - C bond is strongly more covalently bonded than the S - N bond which in turn is stronger than Si – O bond. Equation 14.57 above is an extension of Equation 14.13. The reason why the nucleo-free-radical center on the alkylane groups is weaker than the nucleo-non-free-radical center in Cl atom, is largely because those on the carbon centers are more likely to be hybridized, removing the possibility of the carbon centers ever carrying anionic charges. The order of the nucleo-non-free-radical capacities of the non-free-radical pushing groups is observed to be the same as the order of their radical-pushing capacities when they carry electro-radicals, but the reverse when they carry nucleo-radicals. Based on the analysis above, one should know what to expect of the order of free radical-pulling groups such as COOR, CONH$_2$ etc. when they are nucleo-free-radical in character. They should be stronger than any of the alkylane groups, making it far more difficult to displace COOR groups from a silicon center.

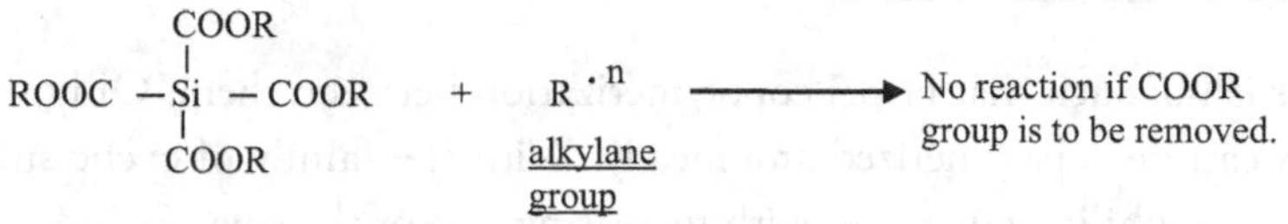

$$ROOC - \underset{\underset{COOR}{|}}{\overset{\overset{COOR}{|}}{Si}} - COOR \quad + \quad R \cdot^n \longrightarrow \begin{array}{l} \text{No reaction if COOR} \\ \text{group is to be removed.} \end{array}$$

$$\underset{\text{alkylane group}}{}$$

14.59

The bond between Si and COOR cannot easily be broken in one step.

14.2 Sulphur Containing Ringed Monomers

14.2.1 Cyclic Sulphides

Their methods of polymerization are analogous to those of cyclic ethers. The three- membered ring has strain energy close to the minimum required strain energy to favor instantaneous opening of the ring. Hence, it can be polymerized anionically and cationically as shown below.

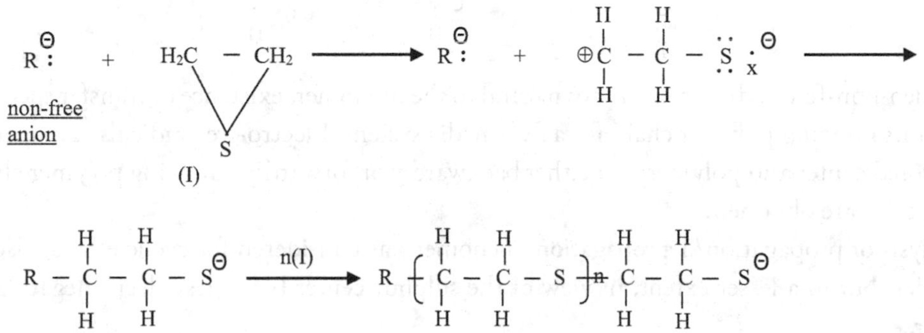

$$R \overset{\ominus}{:} \quad + \quad \underset{\underset{S}{\diagdown\diagup}}{H_2C - CH_2} \longrightarrow R \overset{\ominus}{:} \quad + \quad \oplus\overset{\overset{\text{II}}{|}}{\underset{\underset{H}{|}}{C}} - \overset{\overset{\text{II}}{|}}{\underset{\underset{H}{|}}{C}} - \overset{..}{\underset{..}{S}} :^{\ominus}_{x} \longrightarrow$$

non-free
anion

(I)

$$R - \overset{\overset{H}{|}}{\underset{\underset{H}{|}}{C}} - \overset{\overset{H}{|}}{\underset{\underset{H}{|}}{C}} - S^{\ominus} \xrightarrow{n(I)} R \left\{ \overset{\overset{H}{|}}{\underset{\underset{H}{|}}{C}} - \overset{\overset{H}{|}}{\underset{\underset{H}{|}}{C}} - S \right\}_n \overset{\overset{H}{|}}{\underset{\underset{H}{|}}{C}} - \overset{\overset{H}{|}}{\underset{\underset{H}{|}}{C}} - S^{\ominus}$$

14.60

This takes place only if the counter-charged center is isolately placed or not present in the system. This is more better accomplished with the use of nucleo-non-free-radicals.

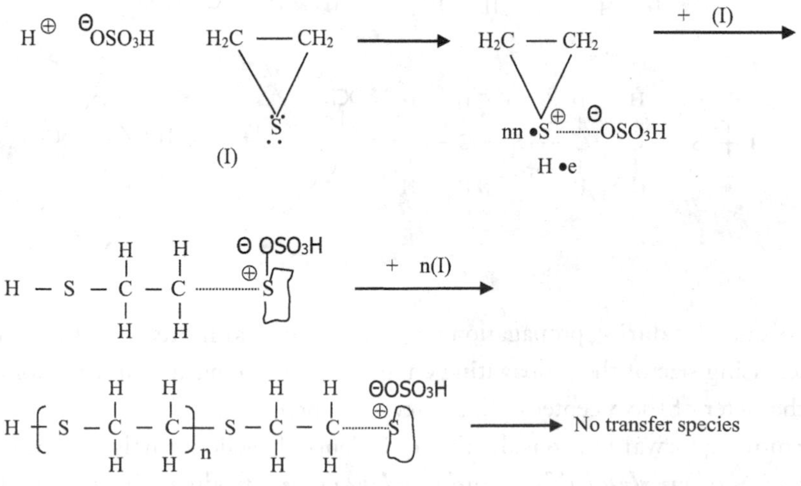

$$H^{\oplus} \quad ^{\ominus}OSO_3H \quad \underset{\underset{S}{\diagdown\diagup}}{H_2C - CH_2} \longrightarrow \underset{\underset{nn \bullet S^{\oplus} \cdots \cdots ^{\ominus}OSO_3H}{\diagdown\diagup}}{H_2C - CH_2} \xrightarrow{+ \ (I)}$$

(I)

H •e

$$H - S - \overset{\overset{H}{|}}{\underset{\underset{H}{|}}{C}} - \overset{\overset{H}{|}}{\underset{\underset{H}{|}}{C}} \cdots \overset{\ominus \; OSO_3H}{\underset{}{\overset{\oplus}{S}}} \xrightarrow{+ \ n(I)}$$

$$H \left\{ S - \overset{\overset{H}{|}}{\underset{\underset{H}{|}}{C}} - \overset{\overset{H}{|}}{\underset{\underset{H}{|}}{C}} \right\}_n S - \overset{\overset{H}{|}}{\underset{\underset{H}{|}}{C}} - \overset{\overset{H}{|}}{\underset{\underset{H}{|}}{C}} \cdots \overset{\ominus OSO_3H}{\underset{}{\overset{\oplus}{S}}} \longrightarrow \text{No transfer species}$$

14.61

For larger membered rings, only the cationic route can be involved if the MRSE can be provided for the ring. Some four-membered rings have close to the MRSE to favor anionic route.

Since the sulphur atom is more electropositive than the oxygen atom, cyclic ethers are more nucleophilic than cyclic sulfides.

$$\text{Cyclic ethers} \quad > \quad \text{Cyclic sulfides}$$

Order of Nucleophilicity   14.62

The difference in the order is not such that favors copolymerization between them. Only their three or some four-membered rings can be copolymerized anionically. With the family of cyclic sulphides, like cyclic ethers, the order of nucleophilicity increases with increasing size of the ring.

Radically, only the three and some four-membered rings can be homopolymerized. Using nucleo-non-free-radicals, the followings are obtained.

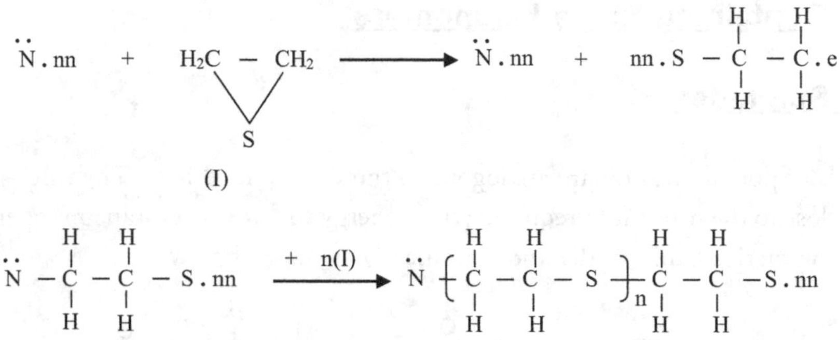

(I)

14.63

Since the nucleo-non-free-radical route is not natural to the monomer, existence of transfer species cannot be favored for its growing polymer chain for a free-media system. Electro-free-radicals can open the ring via the functional center and polymerize it either backwardly or forwardly. Growing polymer chains with no transfer species are obtained.

The analysis of propagation/depropagation phenomena as considered for cyclic ethers, also apply to cyclic sulphides, but to a lesser extent, in view of the sulphur center being less electronegative than the oxygen center.

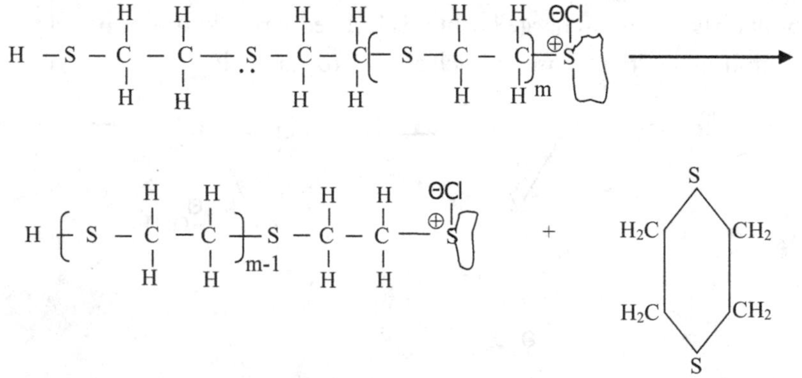

14.64

This will readily occur later during propagation rather than early far more so forwardly than backwardly, in view of the increasing size of the propagating chain in the route natural to the monomer and the less electronegative character of the S center than an oxygen center.

Now, before moving forward to consider the equivalents of cyclic formals e.g. trithiane, there is need to reconsider the cases of **vinylepisulfides and vinylepoxides** introduced in chapter 10 of this Volume, when carbon-carbon ringed monomers were considered. It was said that vinylepisulfide undergoes ring-opening rearrangement polymeri-zation under the influence of Friedel Crafts halides, but not under anionic conditions with Et_2Zn. H_2O. Unfortunately, $Et_2Zn.H_2O$ cannot provide anions, since zinc is a non-ionic metal. Secondly, as was clearly indicated in Chapter 10, the sulfur center can never carry positive charges of any type in the presence of carbon centers in the ring. Thirdly and most importantly, the Zn center carries only electro-non-free-radicals when not activated. Thus, for both the oxygen and sulfur containing vinyl monomers, the followings are to be expected.

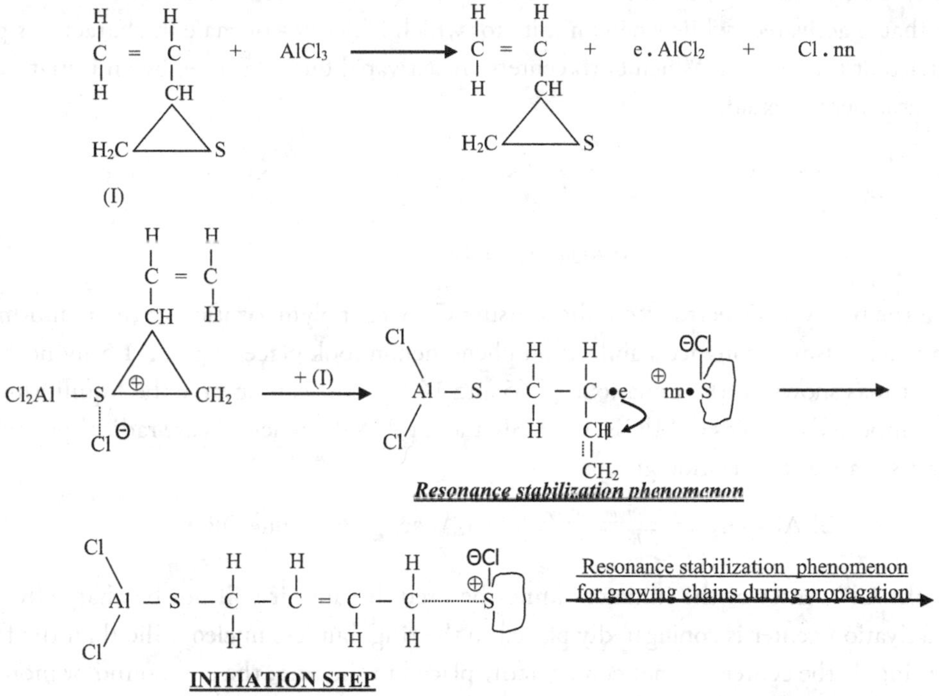
(I)
Resonance stabilization phenomenon
Resonance stabilization phenomenon for growing chains during propagation
INITIATION STEP

14.65

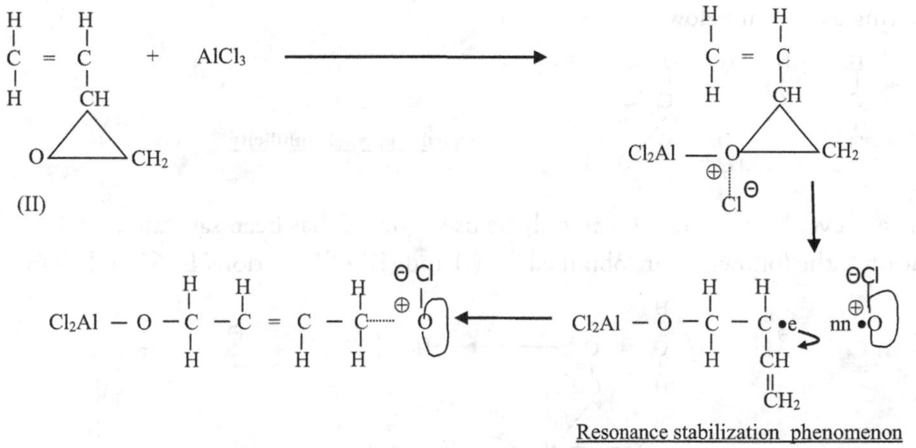
(II)
Resonance stabilization phenomenon

14.66

The reaction above is more favored with (II) than with (I) since the oxygen center is more electronegative than the sulphur center. The reason why the oxygen or sulfur center in the ring are first attacked nucleophilically is because, looking at the monomers with the eye of the needle, these are Electrophiles, i.e. MALES and not Nucleophiles as shown below. They have the X and Y centers conjugatedly placed, similar to those of acrylamide, acrylonitrile, acrolein and many others in these families of Aliphatic non-ringed compounds.

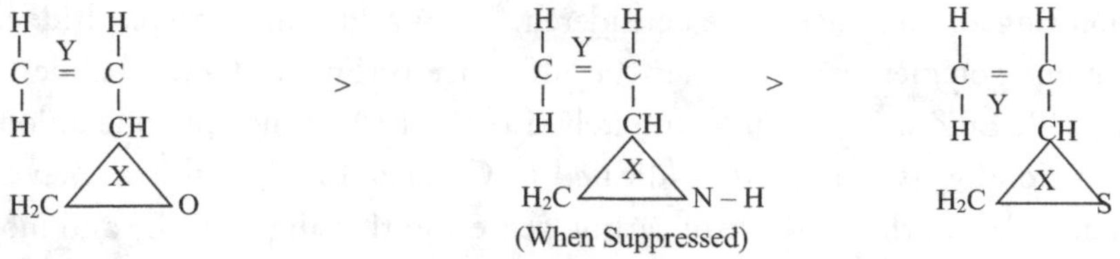

$$
\begin{array}{ccc}
\underset{\substack{| \\ H}}{\overset{\substack{H \\ |}}{C}} \overset{Y}{=} \underset{\substack{| \\ CH}}{\overset{\substack{H \\ |}}{C}} & > & \underset{\substack{| \\ H}}{\overset{\substack{H \\ |}}{C}} \overset{\substack{Y \\ ||}}{=} \underset{\substack{| \\ CH}}{\overset{\substack{H \\ |}}{C}} & > & \underset{\substack{| \\ H}}{\overset{\substack{H \\ |}}{C}} = \underset{\substack{| \\ CH}}{\overset{\substack{H \\ |}}{C}}
\end{array}
$$

(When Suppressed)

Order of Nucleophilicity   14.67

The order shown below is valid, because when an initiator nucleo- or female in character is present, it is the Y center that is activated, while when an initiator which is electro- or male in character is present, it is the X center that is activated. When both centers are activated one at a time, by an initiator, then the male monomer is heterosexual.

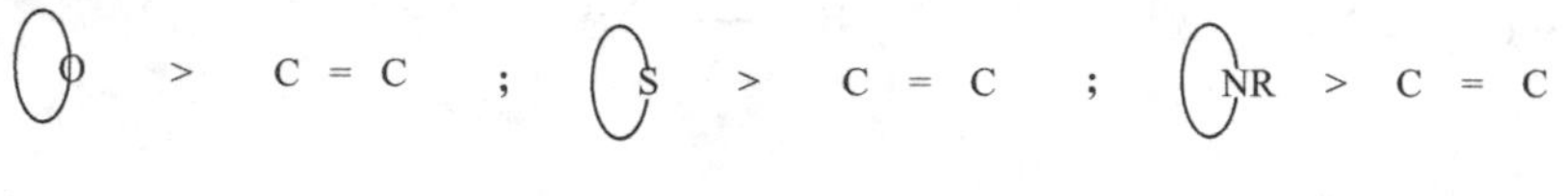

$$
\overset{\curvearrowright}{O} \quad > \quad C = C \quad ; \quad \overset{\curvearrowright}{S} \quad > \quad C = C \quad ; \quad \overset{\curvearrowright}{NR} \quad > \quad C = C
$$

Order of Nucleophilicity   14.68

Thus, during the backward electro-free-radical (using $Cl_2Al\bullet e$) polymerization of the mono-mers using Equilibrium mechanism, resonance stabilization phenomenon took place to give a 1,5-mono-form from a 1,3-mono-form as shown in the Equations 14.65 and 14.66. One can observe the Equilibrium state of existence of compounds such as Cl_2AlOR, Cl_2AlSR and Cl_2AlNR_2, where Rs are radical-pushing groups in particular resonance stabilization groups.

$$
Cl_2Al - OR \underset{\textit{Existence}}{\overset{\textit{Equilibrium State of}}{\rightleftharpoons}} \quad Cl_2Al \bullet e \quad + \quad nn\bullet\, OR
$$

14.69

Anionically, the $Cl^{\ominus}$ center of the initiator cannot be used, because it will not be chargedly balanced. The C = C activation center is conjugatedly placed to the ring, but less nucleophilic than the functional center in the ring. If the center was not conjugatedly placed to the ring, the compound or monomer will become Nucleophilic and allylic with the C = C still remaining less nucleophilic than the functional center in the ring as shown below.

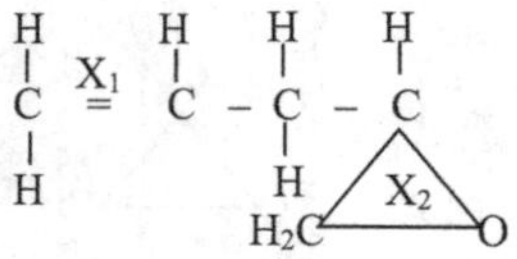

$$
\underset{\substack{| \\ H}}{\overset{\substack{H \\ |}}{C}} \overset{X_1}{=} \underset{\substack{| \\ H}}{\overset{\substack{H \\ |}}{C}} - \underset{\substack{| \\ H_2C}}{\overset{\substack{H \\ |}}{C}} - \overset{\substack{H \\ |}}{C} \qquad\qquad X_2 \quad > \quad X_1
$$

Order of Nucleophilicity   14.70

X_2 will rarely or never be activated. It can only be used once X_1 has been saturated.

"Anionically", the followings are obtained for (I) and (II) of Equations 14.65 and 14.66 respectively.

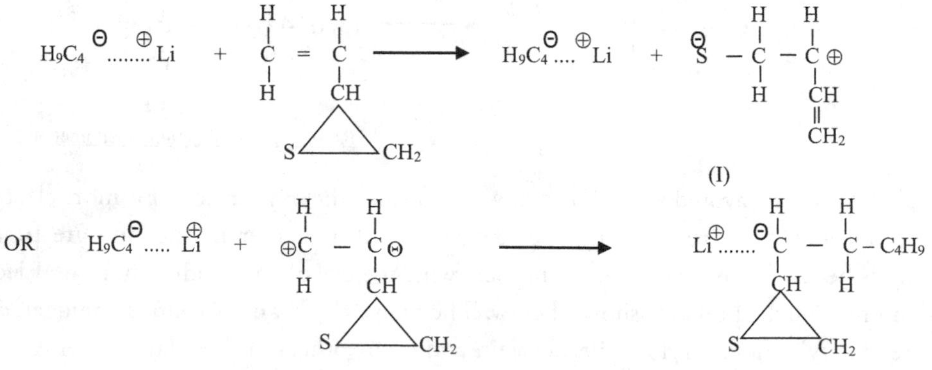

$$
H_9C_4 \overset{\ominus}{}\,......\,\overset{\oplus}{Li} \quad + \quad \underset{\substack{| \\ H}}{\overset{\substack{H \\ |}}{C}} = \underset{\substack{| \\ CH}}{\overset{\substack{H \\ |}}{C}} \longrightarrow H_9C_4 \overset{\ominus}{}\,....\,\overset{\oplus}{Li} \quad + \quad \overset{\ominus}{S} - \underset{\substack{| \\ H}}{\overset{\substack{H \\ |}}{C}} - \underset{\substack{|| \\ CH_2}}{\overset{\substack{H \\ |}}{C}}\overset{\oplus}{}
$$

(I)

$$
OR \quad H_9C_4 \overset{\ominus}{}\,.....\,\overset{\oplus}{Li} \quad + \quad \overset{\oplus}{} \underset{\substack{| \\ H}}{\overset{\substack{H \\ |}}{C}} - \underset{\substack{| \\ CH}}{\overset{\substack{H \\ |}}{C}} \overset{\ominus}{} \longrightarrow \overset{\oplus}{Li}\,.......\,\underset{\substack{| \\ CH}}{\overset{\substack{H \\ |}}{C}} - \underset{\substack{| \\ H}}{\overset{\substack{H \\ |}}{C}} - C_4H_9
$$

(II)   Favored   14.71

Anionically using $H_3CO^{\ominus}.....^{\oplus}Na$, it is the less nucleophilic center (Y) that is activated without initiation being favored, since the equation will not be chargedly balanced. When $H_9C_4^{\ominus}.....^{\oplus}Li$ is used, initiation is favored since the monomer is not resonance stabilized and the equation is chargedly balanced. H cannot be abstracted as a transfer species. The ring is never opened instantaneously under such conditions, that is, it is (II) that is favored above, since the center activated is the C = C center.

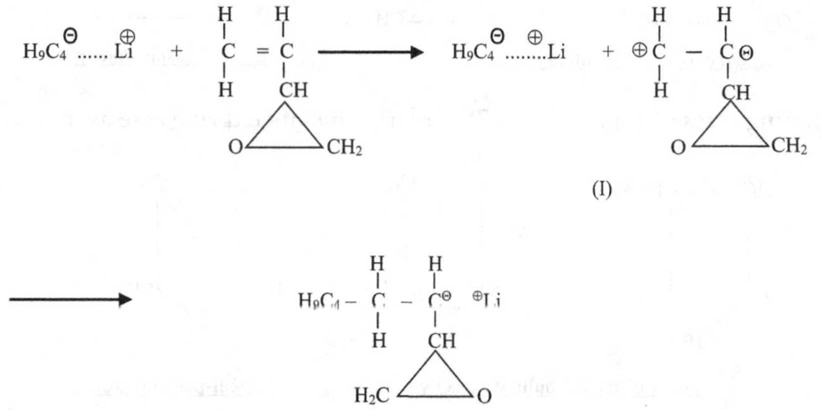

$$H_9C_4^{\ominus} ...... {}^{\oplus}\overset{\circ}{Li} \quad + \quad \begin{matrix} H & H \\ | & | \\ C & = & C \\ | & | \\ H & CH \end{matrix} \quad \longrightarrow \quad H_9C_4^{\ominus} .... {}^{\oplus}Li \quad + \quad {}^{\oplus}\begin{matrix} H & H \\ | & | \\ C & - & C {}^{\ominus} \\ | & | \\ H & CH \end{matrix}$$

$$O \triangle CH_2 \qquad\qquad O \triangle CH_2$$

(I)

$$\longrightarrow \qquad H_9C_4 - \begin{matrix} H & H \\ | & | \\ C & - & C^{\ominus} \quad {}^{\oplus}Li \\ | & | \\ H & CH \end{matrix}$$

$$H_2C \triangle O$$

14.72

One can observe a unique case of monomer, one in which the male center is aliphatic and the female center is in the ring carrying a functional center. This is unlike the cases of for example lactams or non-ringed Electrophiles which cannot be resonance stabilized. It is when the ring is opened, that the resonance stabilization is visibly seen. But, when not opened, it cannot be seen. But it is there, invisible because the π-bond in the ring is invisible; yet it is there and cannot therefore be resonance stabilized. Thus, the monomer units obtained under such conditions are not the same, just like with non-ringed Electrophilic monomers.

However, when the ring is opened instantaneously radically, the followings are obtained.

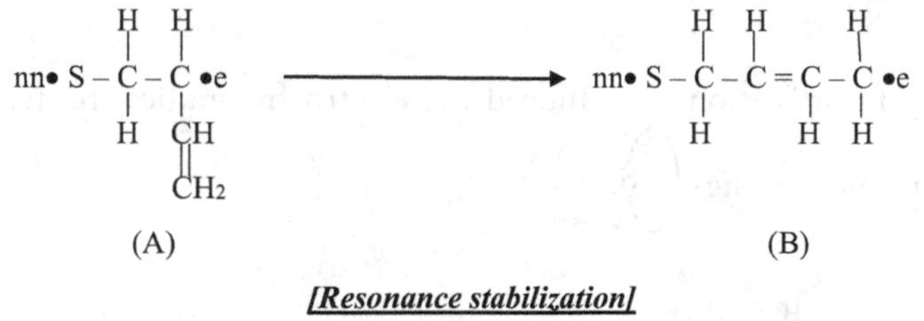

$$nn\bullet S - \begin{matrix} H & H \\ | & | \\ C & - & C \\ | & | \\ H & CH \\ & & || \\ & & CH_2 \end{matrix} \bullet e \qquad \longrightarrow \qquad nn\bullet S - \begin{matrix} H & H & & H \\ | & | & & | \\ C & - & C & = & C & - & C \\ | & | & & | \\ H & H & & H \end{matrix} \bullet e$$

(A)   (B)

[Resonance stabilization]   14.73

Resonance stabilization can only take place radically and not chargedly since charges cannot move alone. When the ring is opened instantaneously using a radical generating initiator, (B) is obtained after resonance stabilization. Thus, only nucleo-non-free-radicals and electro-free-radicals can be used to polymerize the monomer, clear indication that the monomer is bisexual. It is only under this condition that the same monomer unit can be obtained via both routes. When the ring is not opened, it can be polymerized nucleo-free-radically being natural to it. It can also be polymerized electro-free-radically. Transfer species become present with them as shown below.

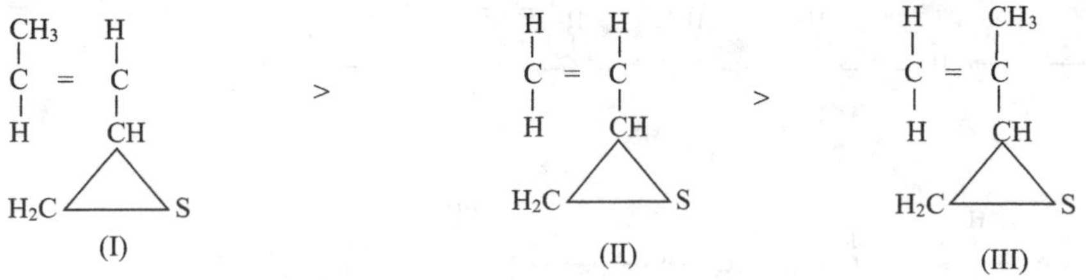

$$\begin{matrix} CH_3 & H \\ | & | \\ C & = & C \\ | & | \\ H & CH \end{matrix} \qquad > \qquad \begin{matrix} H & H \\ | & | \\ C & = & C \\ | & | \\ H & CH \end{matrix} \qquad > \qquad \begin{matrix} H & CH_3 \\ | & | \\ C & = & C \\ | & | \\ H & CH \end{matrix}$$

$$H_2C \triangle S \qquad\qquad H_2C \triangle S \qquad\qquad H_2C \triangle S$$

(I)   (II)   (III)

Order of Nucleophilicity of some family members   14.74

Only (I) has transfer species-the CH_3 group. The CH_3 in (III) does not need to be shielded. Thus, it can finally be observed that the vinylepisulphides, vinylepoxides and the likes are weak electrophiles, for which the followings are additionally valid.

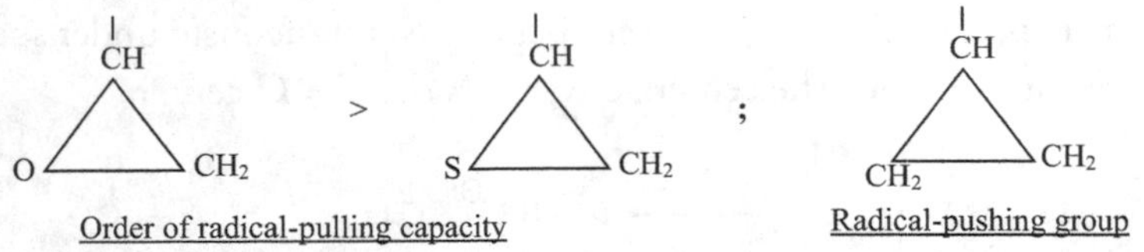

Order of radical-pulling capacity   Radical-pushing group   14.75

The corresponding cases of Equations 14.75 for larger membered rings are as follows

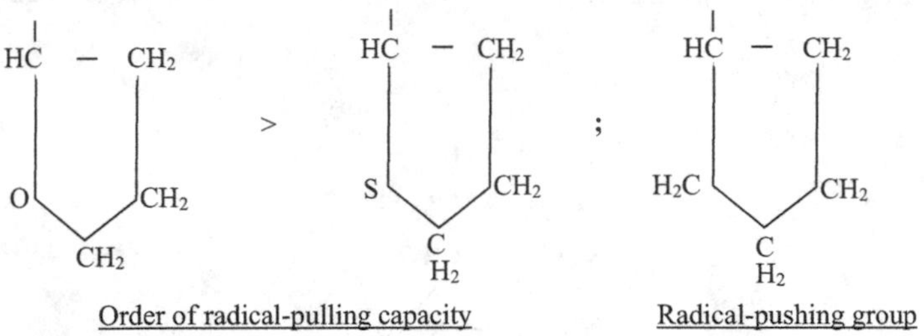

Order of radical-pulling capacity   Radical-pushing group   14.76

Note that, the carbon center carrying radical- pushing groups such as the resonance stabilization group ($-CH=CH_2$) must be adjacently located to the nucleophilic center. What is dominating the character of these monomers is the ring, which when not opened, is an Electrophile and when opened, it becomes a Nucleophile, unlike the case where both nucleo-philic and electrophilic centers are located on the ring.

Based on the developments so far and the selective use of pertinent data from the past, one can observe how the ideas being developed fall in place with what we see every day.

14.2.2 Trithiane

Like trioxane, this will only favor "cationic" or indeed the electro-free-radical route, since there is only one type of functional center in the ring- .

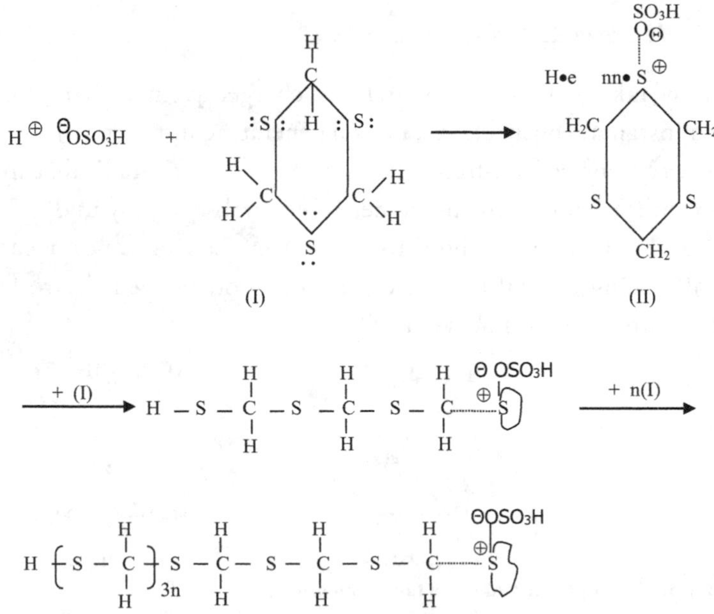

(I)   (II)

14.77

Like trioxane, there will exist propagation/depropagation phenomena, but to a lesser extent. The route is electro-free-radical in character with the polymer chain growing backwardly. The "sulfonium ion" shown above ((II)), favors transient existence only when the MRSE is present in the ring and not when the SE in the ring is less than MRSE. When the MRSE cannot be produced, (II) remains in the system as the initiator in Equilibrium state of existence without opening.

Copolymerization between trioxane and trithiane has been accomplished[12]. This can only be possible electro-free-radically.

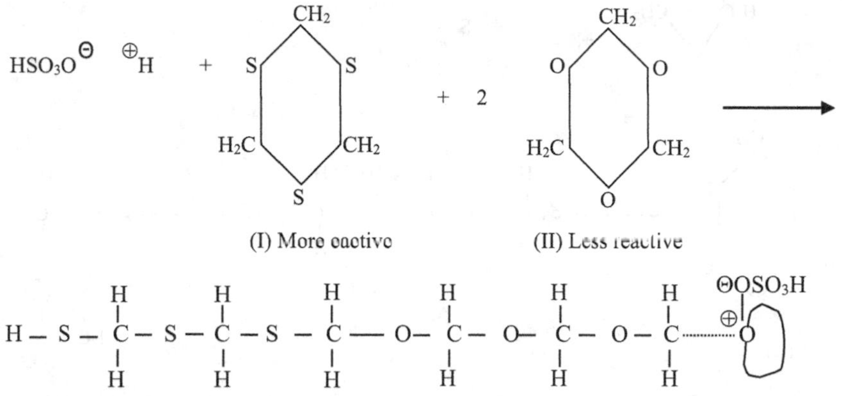

(I) More eactive   (II) Less reactive

Random copolymer with less (II)

14.78

Random copolymers are obtained with less of trioxane along the chain than trithiane, trioxane being more nucleophilic.

Trioxane   >   Trithiane

<u>Order of nucleophilicity</u>

14.79

Since trithiane or sulfur containing rings are less nucleophilic than trioxane or oxygen containing rings, then they are more strained.

Sulfur containing rings   >   Oxygen containing rings

Order of strain energy

14.80

Hence more four-membered rings of sulfur containing cyclic sulfides will have more SE than four-membered oxygen containing cyclic ethers.

It may therefore not be surprising to note that the corresponding cases of cyclic formals for sulfur containing monomers are not commonly known, if the strain energy in the rings is an important factor.

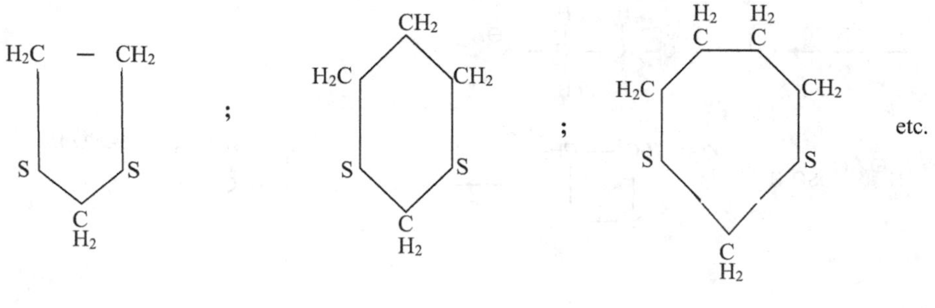

14.81

However, their existence not being favored is because they are self-activated monomers at S.T.P., for which they decompose radically to produce thiocarbonyl and cyclic sulfide compounds.

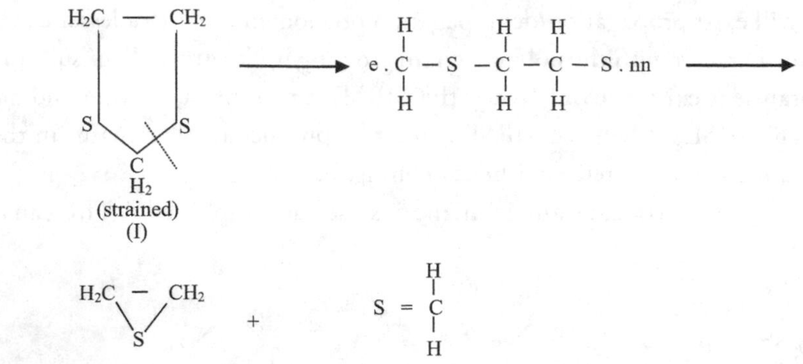

$$H_2C \text{ --- } CH_2 \quad\longrightarrow\quad e.\overset{H}{\underset{H}{C}} - S - \overset{H}{\underset{H}{C}} - \overset{H}{\underset{H}{C}} - S.nn \quad\longrightarrow$$

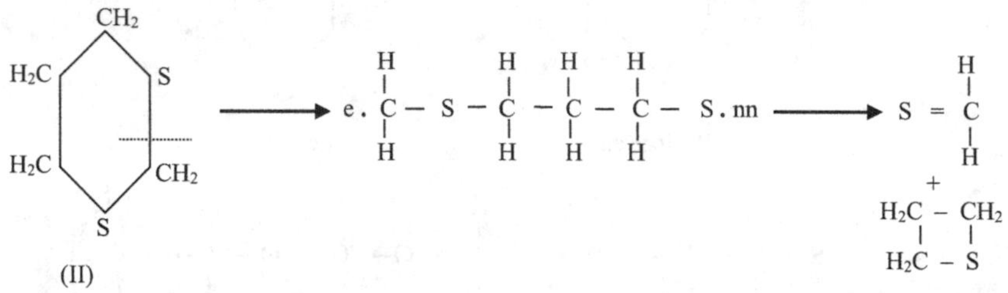

(strained) (I)

$$H_2C \text{ --- } CH_2 \quad + \quad S = \overset{H}{\underset{H}{C}} \qquad\qquad\qquad 14.82$$

$$e.\overset{H}{\underset{H}{C}} - S - \overset{H}{\underset{H}{C}} - \overset{H}{\underset{H}{C}} - \overset{H}{\underset{H}{C}} - S.nn \longrightarrow S = \overset{H}{\underset{H}{C}} + \underset{H_2C - S}{H_2C - CH_2} \qquad 14.83$$

(II)

The reactions above do also apply to their four-membered case which is analogous to 1,3 – dioxane.

The corresponding cases of cyclic esters for sulfur containing rings are not popularly known also. These are Electrophiles.

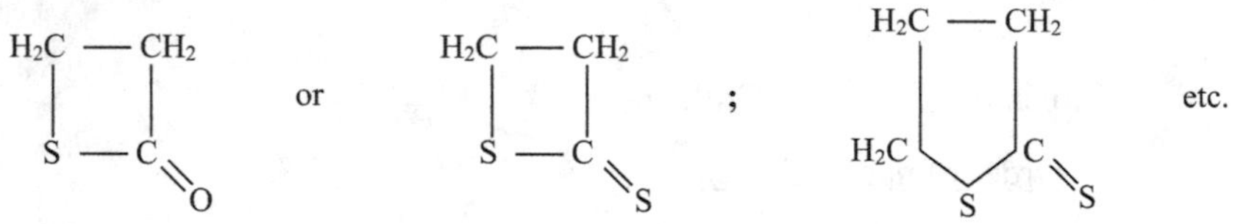

$$\begin{matrix} H_2C \text{ --- } CH_2 \\ | \qquad | \\ S \text{ --- } C \\ \quad\searrow O \end{matrix} \quad or \quad \begin{matrix} H_2C \text{ --- } CH_2 \\ | \qquad | \\ S \text{ --- } C \\ \quad\searrow S \end{matrix} \quad ; \quad \begin{matrix} H_2C \text{ --- } CH_2 \\ | \qquad | \\ H_2C \qquad C \\ S \quad\searrow S \end{matrix} \quad etc. \qquad 14.84$$

Whether cyclic thioesters exists or not, they are less nucleophilic than cyclic esters.

$$\text{Cyclic thioesters} \quad < \quad \text{Cyclic esters}$$

<u>Order of Nucleophilicity</u>   14.85

They will readily favor polymerization, since the rings are more strained than those of cyclic esters.

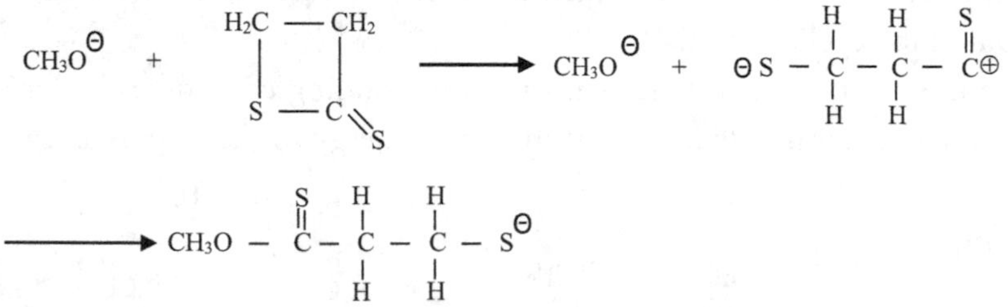

$$CH_3O^{\ominus} \quad + \quad \begin{matrix} H_2C \text{ --- } CH_2 \\ | \qquad | \\ S \text{ --- } C \\ \quad\searrow S \end{matrix} \quad\longrightarrow\quad CH_3O^{\ominus} \quad + \quad {}^{\ominus}S - \overset{H}{\underset{H}{C}} - \overset{H}{\underset{H}{C}} - \overset{S}{\overset{\|}{C}}{}^{\oplus}$$

$$\longrightarrow \quad CH_3O - \overset{S}{\overset{\|}{C}} - \overset{H}{\underset{H}{C}} - \overset{H}{\underset{H}{C}} - S^{\ominus} \qquad\qquad 14.86$$

$$H^{\oplus} \quad {}^{\ominus}OSO_3H \quad + \quad \begin{matrix} H_2C \text{ --- } CH_2 \\ | \qquad | \\ S \text{ --- } C \\ \quad\searrow S \end{matrix} \quad\longrightarrow\quad H - S - \overset{H}{\underset{H}{C}} - \overset{H}{\underset{H}{C}} - \overset{S}{\overset{\|}{C}} \overset{{}^{\ominus}OSO_3H}{\underset{\oplus}{S}} \qquad 14.87$$

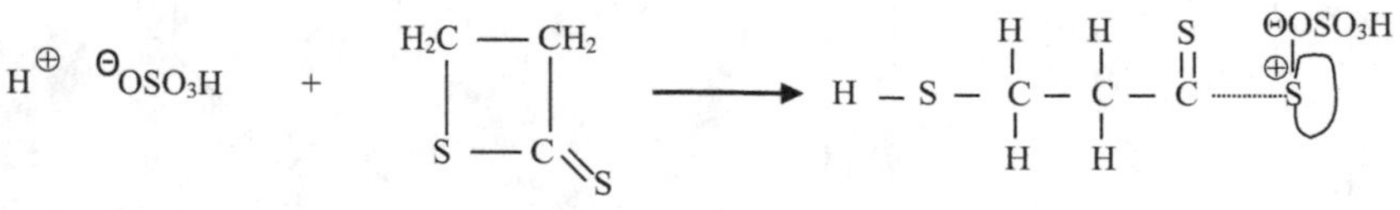

Instantaneous opening of the ring is favored, since it is more strained than the corresponding cases in cyclic esters. In the last reaction above, the initiator combination is the protonic acid/ episulfide. If the monomer is instantaneously opened radically, the following below is obtained.

$$
\begin{array}{c}
H_2C - CH_2 \\
| \qquad | \\
S - C\!\!\diagdown\!\!_S
\end{array}
\xrightarrow{\hspace{1cm}}
nn\cdot S - \overset{\overset{\displaystyle H}{|}}{C} - \overset{\overset{\displaystyle H}{|}}{\underset{\underset{\displaystyle H}{|}}{C}} - \overset{\overset{\displaystyle S}{\|}}{C}\cdot e
\xrightarrow{\hspace{1cm}}
S = C = \overset{\overset{\displaystyle H}{|}}{\underset{\underset{\underset{\displaystyle H}{|}}{S}}{\underset{\underset{\displaystyle |}{CH_2}}{C}}}
$$

14.88

They could also break down to a thioketene and thioformaldehyde. However, there is no doubt that some of the first members of the family of these monomers or compounds favor self-activated existence at STP. Presence of one single sulfur atom in a ring seem to favor the existence of more SE than presence of one single oxygen atom in a ring.

Now, consider introducing $H_2C =$ group into the rings - cyclic sulphides, cyclic thio acetals and cyclic thio esters.

(I) Cyclic sulphide

$$
\xrightarrow{\hspace{1cm}}
H - \overset{\overset{\displaystyle H}{|}}{\underset{\underset{\displaystyle H}{|}}{C}} - \overset{\overset{\displaystyle S}{\|}}{C} - \overset{\overset{\displaystyle H}{|}}{\underset{\underset{\displaystyle H}{|}}{C}} - \overset{\overset{\displaystyle H}{|}}{\underset{\underset{\displaystyle H}{|}}{C}} - \overset{\overset{\displaystyle H}{|}}{C} \cdots \overset{\ominus OSO_3H}{\overset{\oplus|}{S}\!\!\triangleleft}
$$

$$
\xrightarrow[+ \; n(I)]{\hspace{1cm}}
H \left\{ \overset{\overset{\displaystyle H}{|}}{\underset{\underset{\displaystyle H}{|}}{C}} - \overset{\overset{\displaystyle S}{\|}}{C} - \overset{\overset{\displaystyle H}{|}}{\underset{\underset{\displaystyle H}{|}}{C}} - \overset{\overset{\displaystyle H}{|}}{\underset{\underset{\displaystyle H}{|}}{C}} - \overset{\overset{\displaystyle H}{|}}{\underset{\underset{\displaystyle H}{|}}{C}} \right\}_n \overset{\overset{\displaystyle H}{|}}{\underset{\underset{\displaystyle H}{|}}{C}} - \overset{\overset{\displaystyle S}{\|}}{C} - \overset{\overset{\displaystyle H}{|}}{\underset{\underset{\displaystyle H}{|}}{C}} - \overset{\overset{\displaystyle H}{|}}{\underset{\underset{\displaystyle H}{|}}{C}} - \overset{\overset{\displaystyle H}{|}}{C} \cdots \overset{\ominus OSO_3H}{\overset{\oplus|}{S}\!\!\triangleleft}
$$

14.89

Thus, like oxygen containing monomers, the followings are valid.

(I)   (II)

Order of nucleophilicity

14.90

The order in (II) above is due to the fact that $\left(\!\!\begin{array}{c}\\O\\\end{array}\!\!\right) > \left(\!\!\begin{array}{c}\\S\\\end{array}\!\!\right)$ in capacity.

(I)

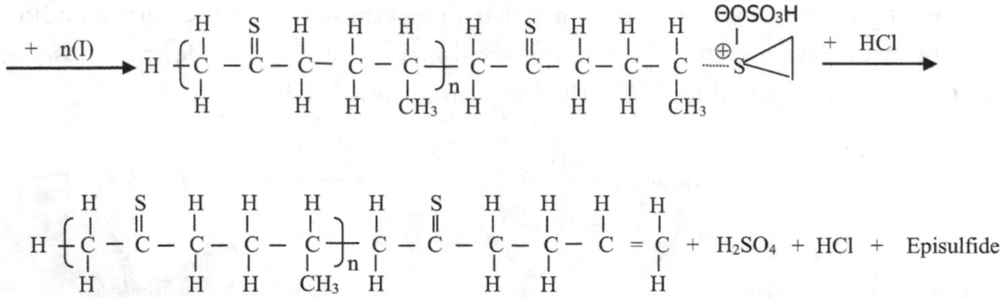

$$\xrightarrow{\quad + \ n(I) \quad} H\left\{\overset{H}{\underset{H}{C}}-\overset{S}{\overset{\|}{C}}-\overset{H}{\underset{H}{C}}-\overset{H}{\underset{CH_3}{C}}-\overset{H}{\underset{H}{C}}\right\}_n \overset{H}{\underset{H}{C}}-\overset{S}{\overset{\|}{C}}-\overset{H}{\underset{H}{C}}-\overset{H}{\underset{CH_3}{C}}\cdots\overset{\oplus}{\underset{\ominus OSO_3H}{S}}\!\!\triangleleft \xrightarrow{\quad + \ HCl \quad}$$

$$H\left\{\overset{H}{\underset{H}{C}}-\overset{S}{\overset{\|}{C}}-\overset{H}{\underset{H}{C}}-\overset{H}{\underset{CH_3}{C}}-\overset{H}{\underset{H}{C}}\right\}_n \overset{H}{\underset{H}{C}}-\overset{S}{\overset{\|}{C}}-\overset{H}{\underset{H}{C}}-\overset{H}{\underset{H}{C}}-\overset{H}{\underset{H}{C}}=\overset{H}{\underset{H}{C}} + H_2SO_4 + HCl + \text{Episulfide}$$

Transfer species of first kind of the first type   14.91

Note that this is backward addition, wherein the possibility of the chain existing in Equilibrium state of existence after the Initiation step is remote. Hence it better for these cases to use forward addition mechanism such as the use of just Na•e or e•TiCl₃ or other suitable electro-free-radical generating catalyst.

With two S atoms placed on the ring, the following are to be expected.

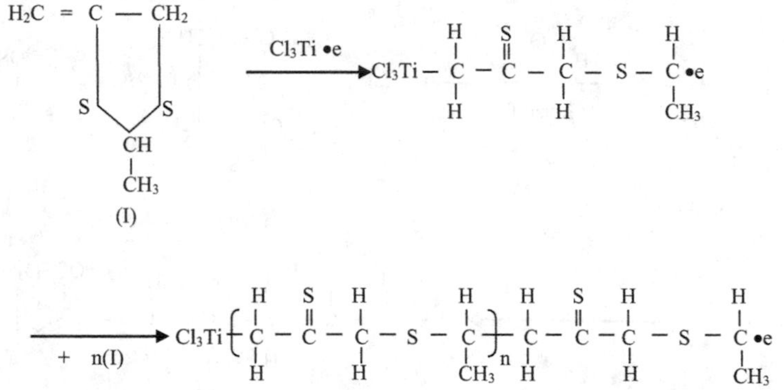

$$\xrightarrow{\quad + \ n(I) \quad} Cl_3Ti\left\{\overset{H}{\underset{H}{C}}-\overset{S}{\overset{\|}{C}}-\overset{H}{\underset{H}{C}}-S-\overset{H}{\underset{CH_3}{C}}\right\}_n \overset{H}{\underset{H}{C}}-\overset{S}{\overset{\|}{C}}-\overset{H}{\underset{H}{C}}-S-\overset{H}{\underset{CH_3}{C}}\bullet e \qquad 14.92$$

Like 2-methyl-4-methylene -1,3-dioxolane, (I) above also has transfer species of the first kind of the first type for its growing polymer chain. In general the followings are valid.

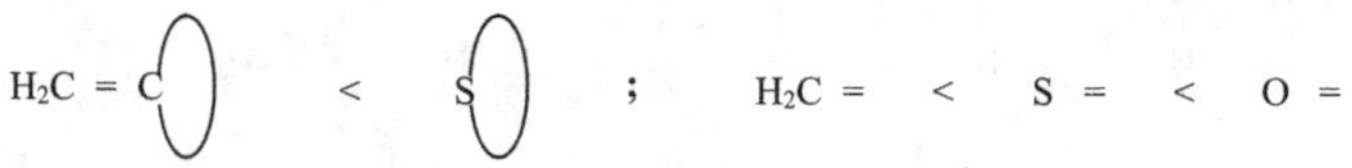

Order of Nucleophilicity and radical-pushing capacity   14.93

With trithiane, the followings are to be expected.

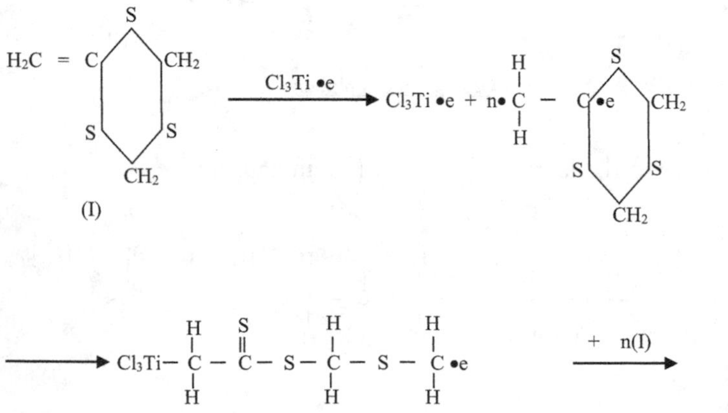

$$\xrightarrow{\qquad} Cl_3Ti-\overset{H}{\underset{H}{C}}-\overset{S}{\overset{\|}{C}}-S-\overset{H}{\underset{H}{C}}-S-\overset{H}{\underset{H}{C}}\bullet e \qquad \xrightarrow{\quad + \ n(I) \quad}$$

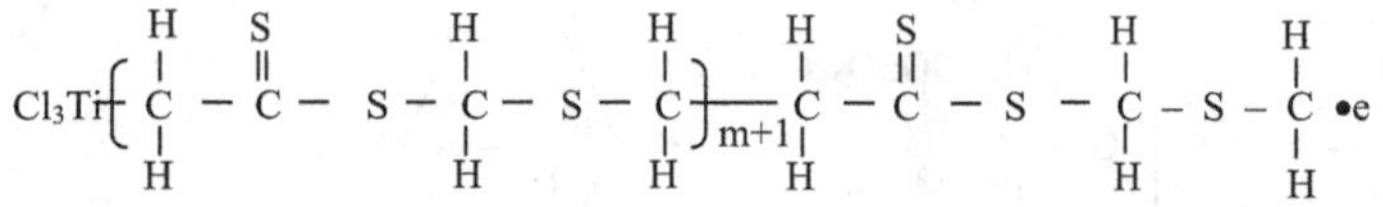

$$Cl_3Ti\left[\begin{array}{c} H \\ | \\ C \\ | \\ H \end{array} - \begin{array}{c} S \\ || \\ C \end{array} - S - \begin{array}{c} H \\ | \\ C \\ | \\ H \end{array} - S - \begin{array}{c} H \\ | \\ C \\ | \\ H \end{array}\right]_{m+1}\begin{array}{c} H \\ | \\ C \\ | \\ H \end{array} - \begin{array}{c} S \\ || \\ C \end{array} - S - \begin{array}{c} H \\ | \\ C \\ | \\ H \end{array} - S - \begin{array}{c} H \\ | \\ C \\ | \\ H \end{array} \cdot e \qquad 14.94$$

Thus, it can be observed that for cyclic acetals, cyclic thio acetals, oxanes, thiooxanes etc., for every addition of monomer, only thioformaldehyde or formaldehyde (and not thioacet-aldehyde, acetaldehyde or thioacetone, acetone) is released, *since the receiving center of the electro-free radical when thioformaldehydes are released is of stronger capacity than when it was at the end of the growing chain.* Hence, monomers such as shown below are not popularly known to be useful as such, despite the fact that they are either strained or do not exist.

$$\qquad\qquad\qquad (I) \qquad\qquad\qquad\qquad\qquad\qquad (II) \qquad\qquad\qquad\qquad 14.95$$

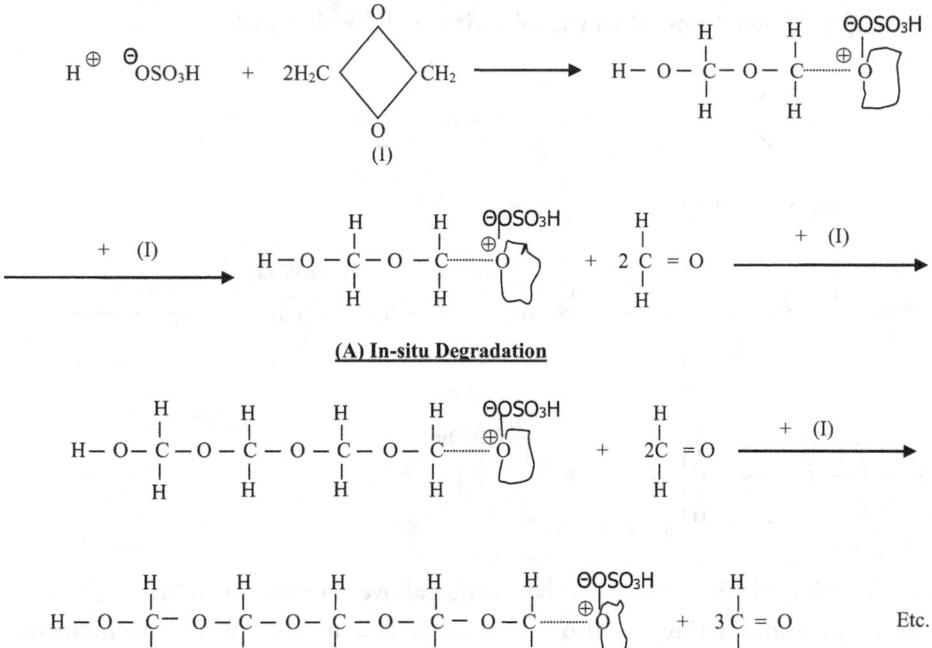

(A) In-situ Degradation

$$\qquad\qquad\qquad\qquad\qquad\qquad\qquad\qquad 14.96$$

The equilibrium state of existence of the terminal end of the growing chain is stronger here than when trioxane or tetraoxane or (II) above is involved. Hence, in the process of addition, formaldehyde is lost in a stage before moving to the stage of another addition. This is not depropagation, but in-situ degradation. The formaldehyde formed as a result of higher temperature of polymerization will subsequently be added to the chain. For the growing polymer chain, the followings may also take place.

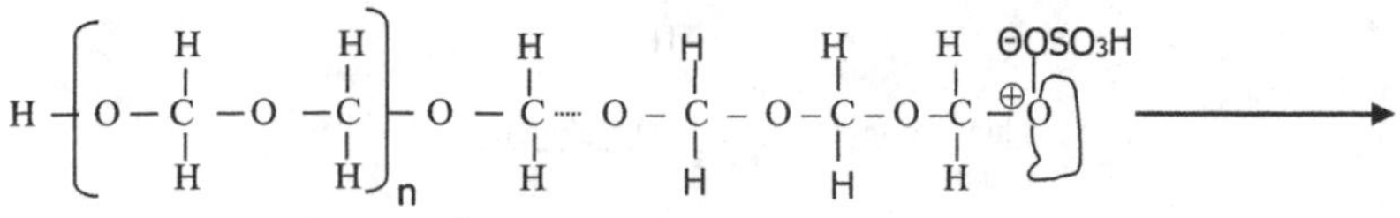

$$(A)$$

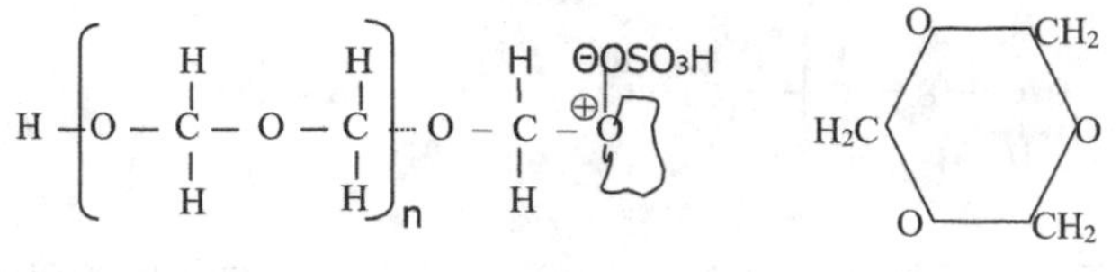

$$H \left[O - \underset{\underset{H}{|}}{\overset{\overset{H}{|}}{C}} - O - \underset{\underset{H}{|}}{\overset{\overset{H}{|}}{C}} \right]_n O - \underset{\underset{H}{|}}{\overset{\overset{H}{|}}{C}} - \overset{\oplus}{\underset{\ominus}{O}} \text{OSO}_3H \qquad \underset{O}{\overset{O}{\underset{\diagdown}{\diagup}}} \overset{CH_2}{\underset{CH_2}{\diagdown}}_O$$

Depropagation   14.97

Thus, during depropagation at temperatures above the ceiling point of the monomer, trioxane ring is more readily formed than the dioxane ring because of the better distribution of the SE in trioxane than in dioxane, noting that the SE is the same for members of the same family.

For (II) of Equation 14.95, the situation is different. Because it is more strained, it will however favor a self-activated state of existence as shown below at S.T.P and break down to give thioformaldehyde,

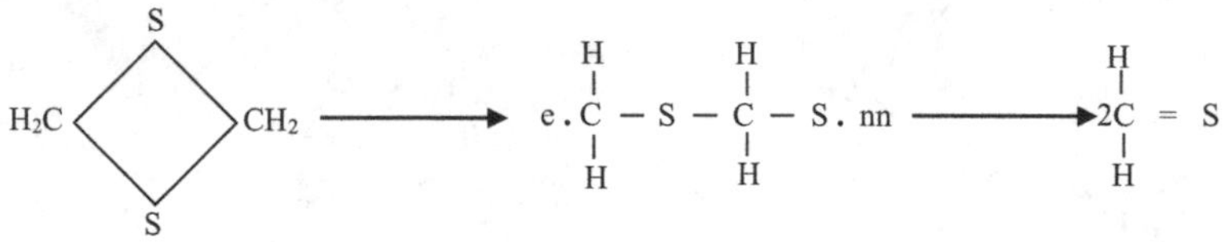

$$H_2C \overset{S}{\underset{S}{\diamondsuit}} CH_2 \longrightarrow e.\underset{\underset{H}{|}}{\overset{\overset{H}{|}}{C}} - S - \underset{\underset{H}{|}}{\overset{\overset{H}{|}}{C}} - S.\,nn \longrightarrow 2\underset{\underset{H}{|}}{\overset{\overset{H}{|}}{C}} = S$$

14.98

Imagine if the non-existing four-membered ring was made to exist at very low temperatures, and made to carry $H_2C=$ group as shown below, then the followings will be obtained.

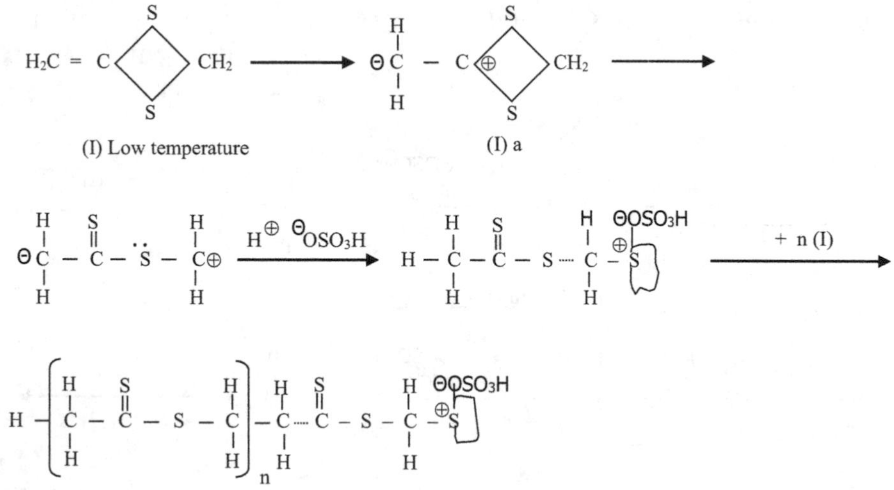

$$H_2C = C \overset{S}{\underset{S}{\diamondsuit}} CH_2 \longrightarrow \ominus \underset{\underset{H}{|}}{\overset{\overset{H}{|}}{C}} - C \overset{S}{\underset{S}{\overset{\oplus}{\diamondsuit}}} CH_2 \longrightarrow$$

(I) Low temperature   (I) a

$$\ominus \underset{\underset{H}{|}}{\overset{\overset{H}{|}}{C}} - \overset{\overset{S}{\parallel}}{\underset{..}{C}} - \overset{..}{\underset{..}{S}} - \overset{\overset{H}{|}}{\underset{H}{|}}{C}\oplus \xrightarrow{H^\oplus \;\ominus OSO_3H} H - \underset{\underset{H}{|}}{\overset{\overset{H}{|}}{C}} - \overset{\overset{S}{\parallel}}{C} - S - \underset{\underset{H}{|}}{\overset{\overset{H}{|}}{C}} - \overset{\oplus}{\underset{\ominus}{S}}\text{OSO}_3H \xrightarrow{+ n\,(I)}$$

$$H \left[\underset{\underset{H}{|}}{\overset{\overset{H}{|}}{C}} - \overset{\overset{S}{\parallel}}{C} - S - \underset{\underset{H}{|}}{\overset{\overset{H}{|}}{C}} \right]_n \underset{\underset{H}{|}}{\overset{\overset{H}{|}}{C}} - \overset{\overset{S}{\parallel}}{C} - S - \underset{\underset{H}{|}}{\overset{\overset{H}{|}}{C}} - \overset{\oplus}{\underset{\ominus}{S}}\text{OSO}_3H$$

14.99

Unlike for (I) of Equation 14.95 which favors the reaction above via forward addition, (II) may not favor the reactions above. Instead, the ring will also be opened instantaneously. When instantaneous opening of the ring is favored, (I)a above decomposes as follows when activated.

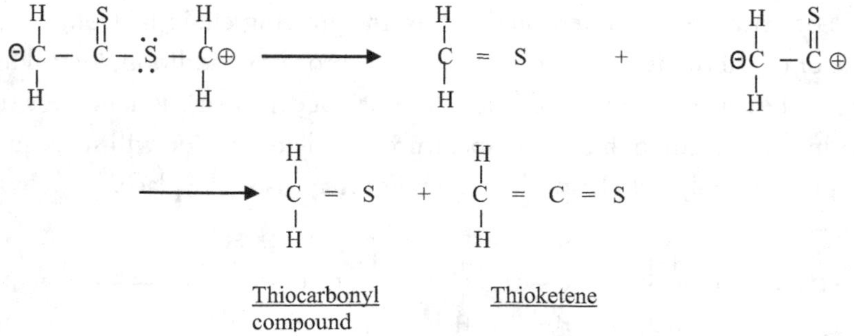

$$\ominus \underset{\underset{H}{|}}{\overset{\overset{H}{|}}{C}} - \overset{\overset{S}{\parallel}}{\underset{..}{C}} - \overset{..}{\underset{..}{S}} - \underset{\underset{H}{|}}{\overset{\overset{H}{|}}{C}}\oplus \longrightarrow \underset{\underset{H}{|}}{\overset{\overset{H}{|}}{C}} = S \quad + \quad \ominus \underset{\underset{H}{|}}{\overset{\overset{H}{|}}{C}} - \overset{\overset{S}{\parallel}}{C}\oplus$$

$$\longrightarrow \underset{\underset{H}{|}}{\overset{\overset{H}{|}}{C}} = S \quad + \quad \underset{\underset{H}{|}}{\overset{\overset{H}{|}}{C}} = C = S$$

<u>Thiocarbonyl</u>   <u>Thioketene</u>   14.100
compound

While 1,3-thio-dioxane will decompose to produce a thiocarbonyl compound, the case above will produce a thiocarbonyl compound and thioketene.

With trithiane, the followings are to be expected chargedly or radically, if the ring is so strained

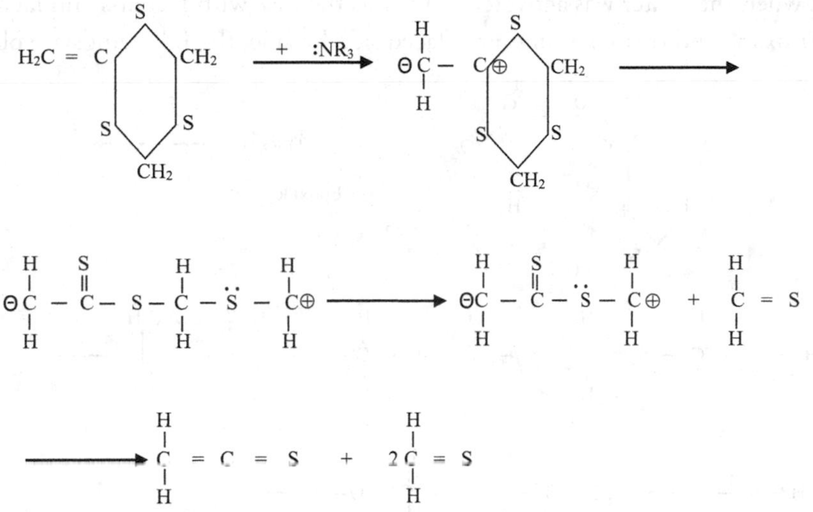

$$14.101$$

For thioesters, the followings are to be expected.

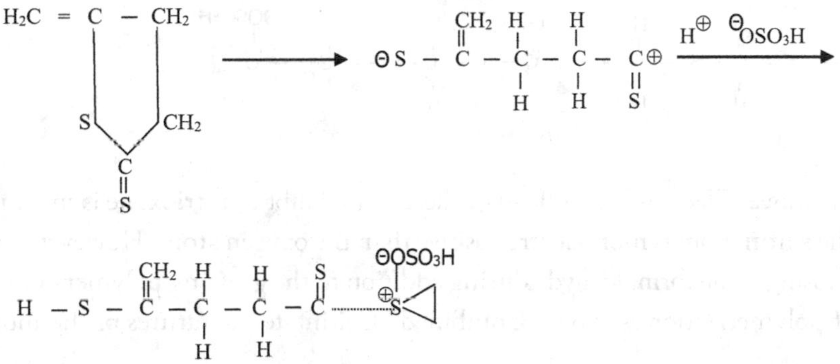

$$14.102$$

Both the electro-free-radical and anionic or nucleo-non-free-radical routes will be favored by the monomer above despite the presence of $H_2C=$ group where placed. The same monomer units are obtained electro-free-radically and anionically as shown below for which the followings are valid.

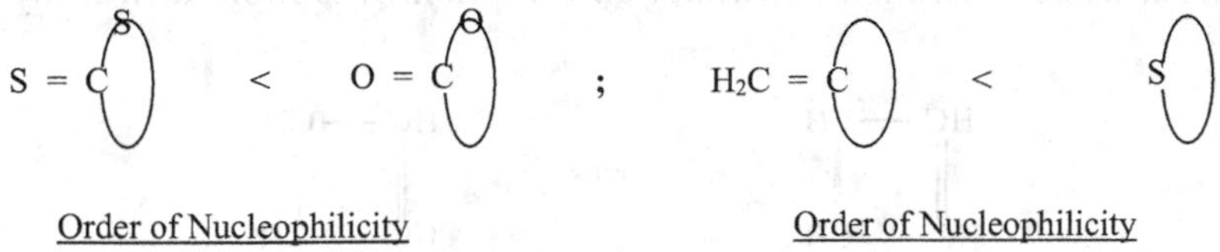

Order of Nucleophilicity   Order of Nucleophilicity

$$14.103$$

(A)

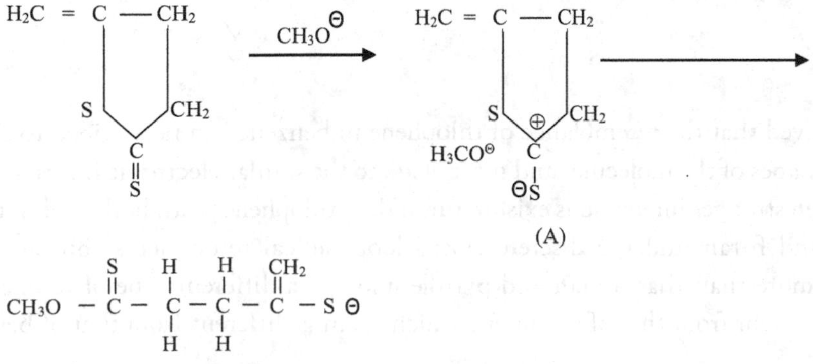

(II) FAVORED

$$14.104$$

This is the monomer unit also obtained, if the S center had been used instead of the $H_2C=$ center used in Equation 14.102 when the center was activated. This was the case with lactams and lactones.

Now when trioxane and trithiane rings are placed side-by-side, the followings are obtained.

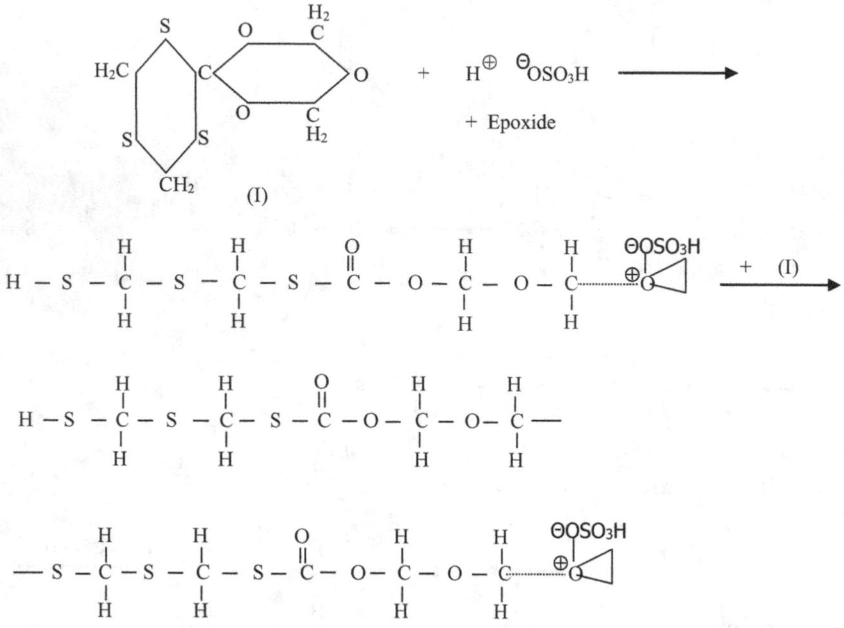

$$14.105$$

From the reaction above (Electro-free-radically), there is no doubt that trioxane is more nucleophilic than trithiane, since the sulfur atom is more electropositive than the oxygen atom. However, one cannot rule out the possibility of losing some formaldehyde during addition to the growing polymer chain (depropagation) if temperature of polymerization is above "Combined" Ceiling temperatures of the monomers.

14.2.3 Thiophene and Furan

Though thiophene cannot be used as a ring-opening monomer, because of its structural resemblance to pyrrole and furan, its consideration is without an exception. It is more identically similar to furan, than pyrrole.

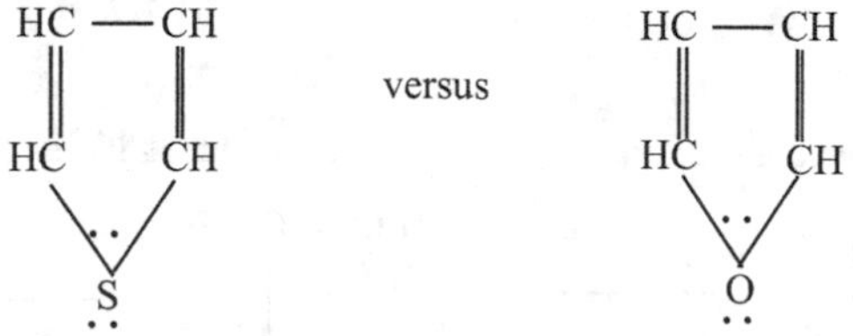

$$14.106$$

It has been observed that the resemblance of thiophene to benzene can be ascribed to similar molecular weight, similar shapes of the molecules and most of all, to the similar electronic interactions[13]. It has been observed that even stronger interactions exist in furan than thiophene, with both yet less than in pyrrole[13]. While pyrrole and furan undergo discrete closed loop radical resonance stabilization phenomenon, thiophene does more than that. Furan and pyrrole undergo a different type of resonance stabilization phenomenon, different from that of thiophene which again is different from that of benzene.

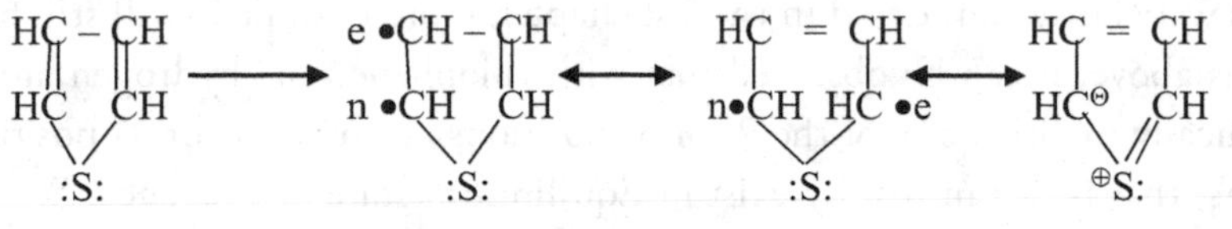

(A)

Free-radical-Polar resonance stabilization phenomenon.   14.107a

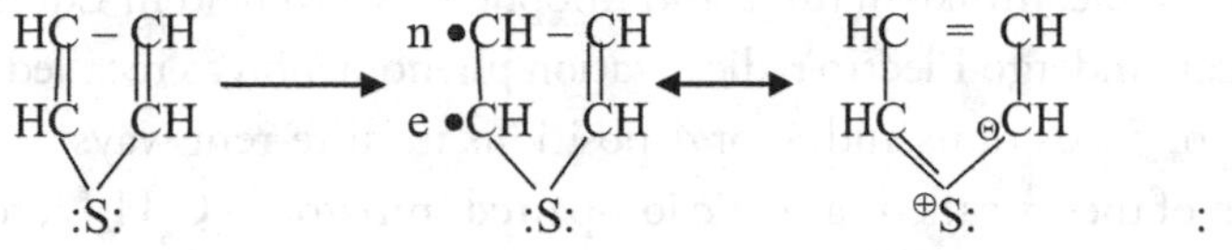

(B)

Free-radical-Polar resonance stabilization phenomenon.   14.107b

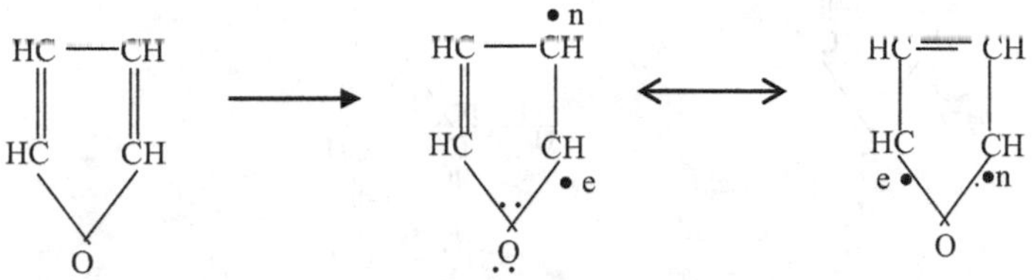

Discrete resonance stabilization phenomenon.   14.108a

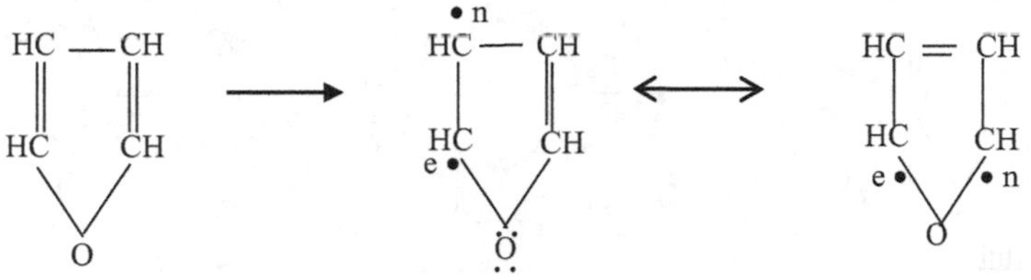

Discrete resonance stabilization phenomenon.   14.108b

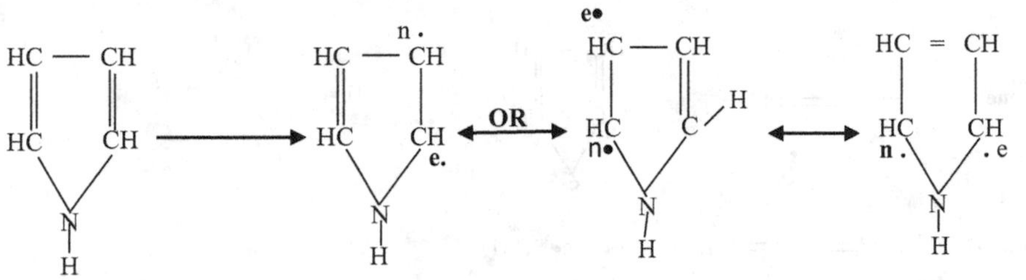

(I)   (I)   (II)

Discrete resonance stabilization phenomenon   14.109

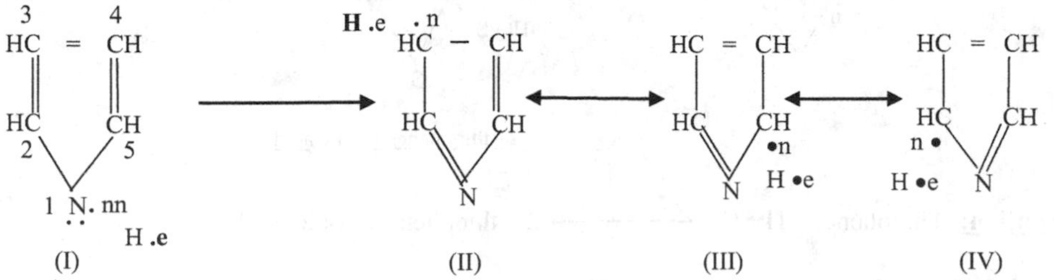

(I)   (II)   (III)   (IV)

Continuous Electroradicalization phenomenon   14.110

359

Since these compounds were fairly considered in the last chapter, more than little will still be done here. From the considerations above, it can be observed that with thiophene, one hydrogen atom is loosely bonded electro-free-radically to either any of the 2- or 5- positions or any of 3- or 4- positions one at a time without activations, that is, it can readily exist in Equilibrium state of existence. With furan and pyrrole, only one H atom in the 2- or 5-position is loosely held in Equilibrium state of Existence without activation. The 3,- or 4- positions cannot be used, because they are internally located while 2- or 5-positions are externally located. In pyrrole, unlike in furan and thiophene, when held in Equilibrium state of Existence from the N center, it can undergo Electro radicalization phenomenon as recalled in Equation 14.110 above, making available 2- or 5-positions and 3- or 4-positions in diffe-rent ways.

Now, considering the addition of the so-called "anionic ion-paired initiator" - C_4H_9Li, to thiophene, the followings are obtained.

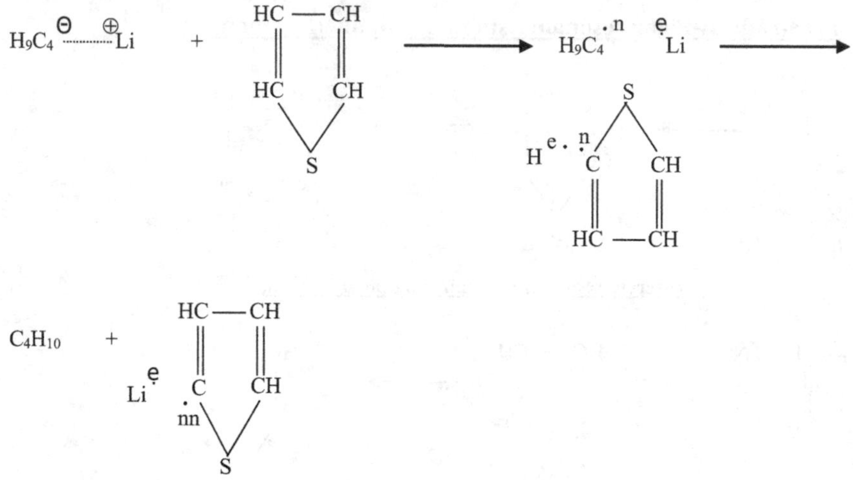

$$H_9C_4 \overset{\ominus}{-\!-\!-} \overset{\oplus}{Li} \quad + \quad \text{[thiophene ring]} \quad \longrightarrow \quad H_9C_4 \overset{.n}{} \overset{\underline{e}}{Li} \quad \longrightarrow$$

$$C_4H_{10} \quad + \quad \text{[2-thienyllithium ring structure]}$$

2 - thienyllithium

14.111

When thiophene is attacked by H_2SO_4 (concentrated) at room temperature preferably in an inert solvent, the followings occur.

Stage 1:

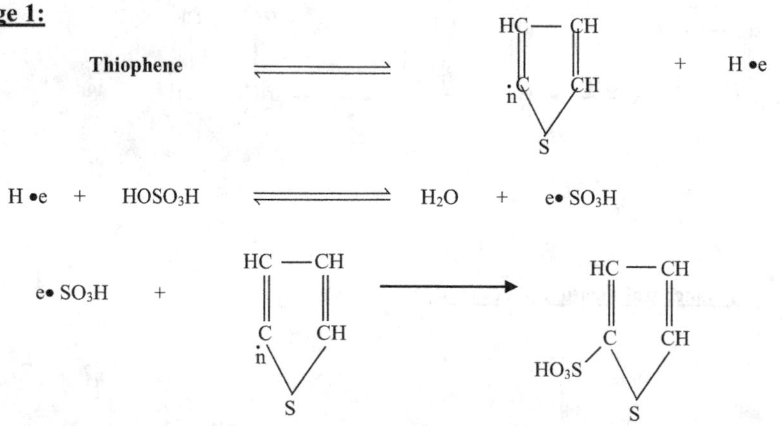

$$\textbf{Thiophene} \quad \rightleftharpoons \quad \text{[ring]} \quad + \quad H \bullet e$$

$$H \bullet e \quad + \quad HOSO_3H \quad \rightleftharpoons \quad H_2O \quad + \quad e\bullet SO_3H$$

$$e\bullet SO_3H \quad + \quad \text{[ring]} \quad \longrightarrow \quad \text{[2-thiophenesulfonic acid ring]}$$

2 - thiophenesulfonic acid

14.112

Overall Equation: Thiophene + H_2SO_4 $\longrightarrow$ 2 – thiophenesulfonic acid

14.113

Ionically or chargedly, the reaction is impossible. In the same manner, nitric acid combines with it to give 2-nitrothiophene. What is shocking to note is that thiophene is so strong as to suppress the Equilibrium state of Existence of these concentrated acids. The mechanism is a one stage Equilibrium mechanism, noting that all the three major mechanisms-Equilibrium, Combination and Decomposition take place in stages, just as all our operations on earth take place in stages either in series or parallel or both.   .

With chlorine molecule, the followings are obtained.

$$Cl_2 \; + \; \underset{S}{\overset{HC \!-\! CH}{\underset{HC \quad CH}{\diagdown \diagup}}} \longrightarrow Cl_2 \; + \; \underset{S}{\overset{HC \!-\! CH}{\underset{e \cdot H \; \overset{..}{C} \quad CH}{\diagdown \diagup}}} \longrightarrow$$

$$\underset{S}{\overset{HC \!-\! CH}{\underset{H \diagup C \qquad C \diagdown Cl}{\diagdown \diagup}}} \quad + \quad HCl$$

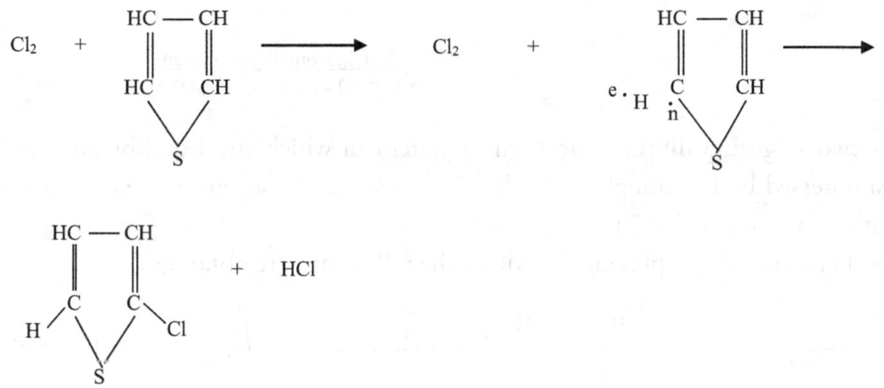

2 - Chlorothiophene(2 - thienyl chloride)

14.114

It is a one stage Equilibrium mechanism system in which the chlorine molecule is in Stable state of Existence.

When thiophene combines with NH_4Cl/formaldehyde, the followings are obtained.

$$\underset{H}{\overset{H}{C}} = O \; + \; \overset{\oplus}{NH_4} \; \overset{\ominus}{Cl} \; + \; \underset{S}{\overset{HC \!-\! CH}{\underset{HC \quad \overset{..}{C}}{\diagdown \diagup}}} \; H \cdot e \longrightarrow$$

$$\underline{\text{Non-ionic bonds}}$$

$$HO - \underset{H}{\overset{H}{C}} . e \; + \; NH_4Cl \; + \; \underset{S}{\overset{HC \!-\! CH}{\underset{HC \quad C \bullet n}{\diagdown \diagup}}} \longrightarrow \underset{S}{\overset{HC \!-\! CH}{\underset{HC \quad C - CH_2 - OH}{\diagdown \diagup}}}$$

$$H \bullet e \; + \; nn \bullet NH_3Cl \longrightarrow \underset{S}{\overset{HC \!-\! CH}{\underset{HC \quad C - \overset{H}{\underset{H}{C}} - \overset{\oplus}{N}\overset{\ominus}{H_3Cl}}{\diagdown \diagup}}} \quad + \quad H_2O$$

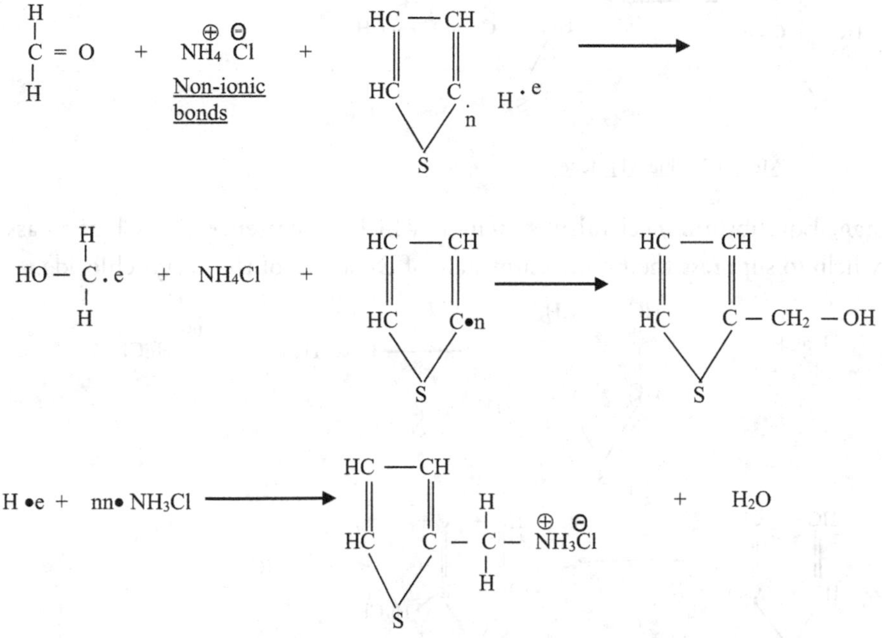

2 - Thenylammonium chloride

14.115

This is a two-stage Equilibrium mechanism system which the reader is asked to identify.

Other similar reactions undergone include the followings.

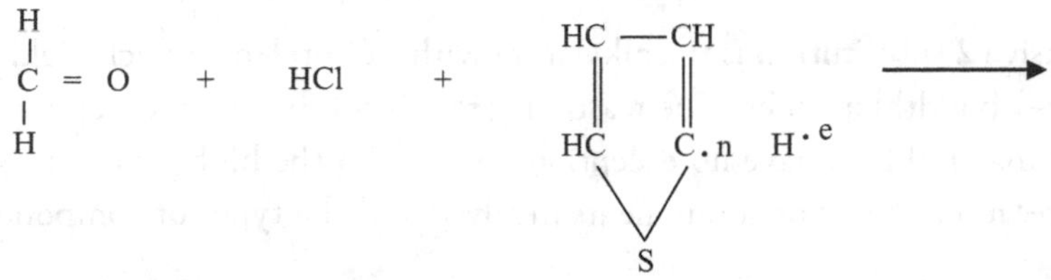

$$\underset{H}{\overset{H}{C}} = O \; + \; HCl \; + \; \underset{S}{\overset{HC \!-\! CH}{\underset{HC \quad C.n}{\diagdown \diagup}}} \; H \cdot e \longrightarrow$$

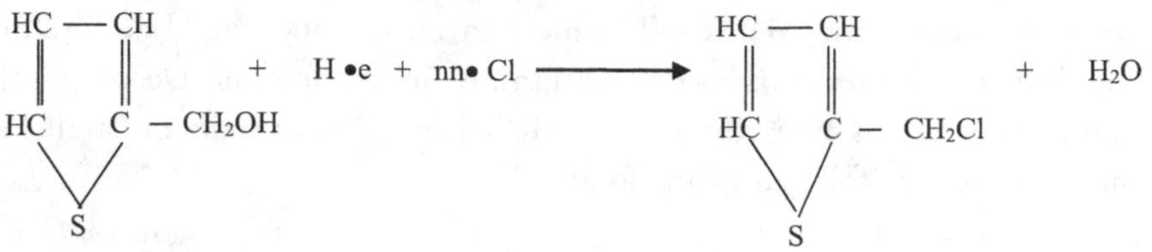

$$2 - \text{Chloromethylthiophene}$$
(2 - thenyl chloride)   14.116

This is also a two-stage Equilibrium mechanism system in which the Equilibrium state of Existence of HCl was suppressed by the thiophene in the first stage. In the second stage the HCl was no longer suppressed with no thiophene left in the system.

With acetyl chloride in the presence of $AlCl_3$, the followings are obtained.

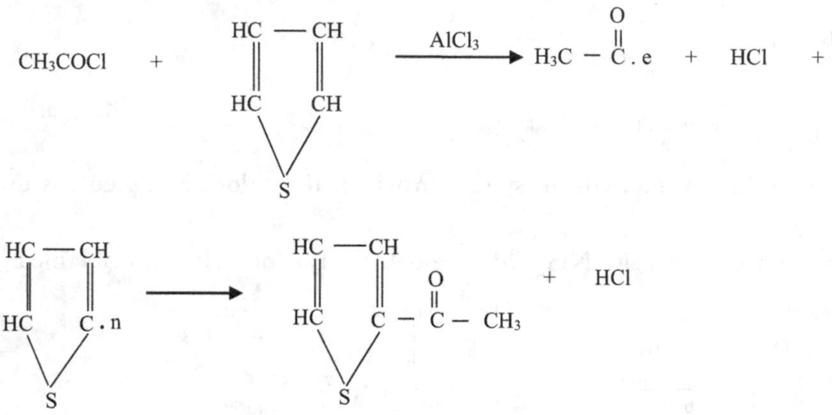

Methyl 2 - thienyl ketene   14.117

This is a one-stage Equilibrium mechanism system, in which the presence of $AlCl_3$ as a passive catalyst is to additionally help to suppress the Equilibrium state of Existence of the acetyl chloride.

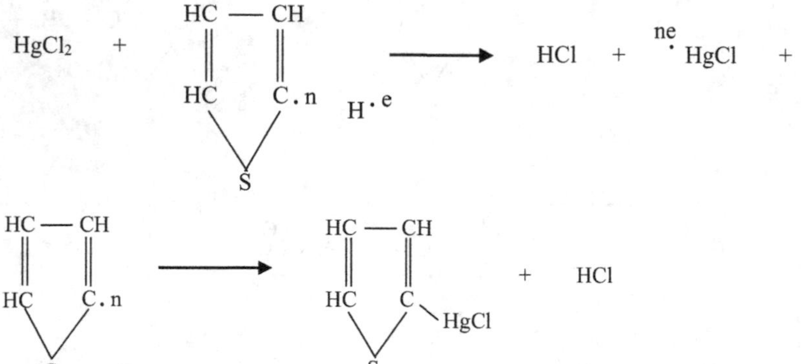

2–Chloromercurithiophene   (2-thienylmercuric chloride)   14.118

This again is a one-stage Equilibrium mechanism system in which the Equilibrium state of Existence of $HgCl_2$ is suppressed.

An exception to exclusive 2 substitution is the alkylation with isobutylene, which yields about equal amounts of the 2 - and 3 - t butylthiophenes. The reason for this is not being an exception, because the laws of NATURE unlike material laws have no exception, but due to the higher operating conditions- which include temperature, molar ratios of components involved and the types of components.

$$2 \; \substack{HC-CH \\ \| \quad \| \\ HC \quad CH \\ \diagdown_S\diagup} \quad + \quad 2 \; \substack{H \quad CH_3 \\ | \quad | \\ C = C \\ | \quad | \\ H \quad CH_3} \quad + \quad H_2SO_4 \quad \xrightarrow{60-65°C}$$

$$(75\%)$$

$$\substack{HC-C \cdot n \; H \cdot e \\ \| \quad \| \\ HC \quad CH \\ \diagdown_S\diagup} \quad + \quad \substack{HC-CH \\ \| \quad \| \\ HC \quad C \cdot n \\ \diagdown_S\diagup {}_{H \cdot e}} \quad + \quad 2 \; n. \; \substack{H \quad CH_3 \\ | \quad | \\ C = C \cdot e \\ | \quad | \\ H \quad CH_3} \quad \xrightarrow{H_2SO_4}$$

$$2 \; (CH_3)_3 \; C \cdot e \quad + \quad \substack{HC-C \cdot n \\ \| \quad \| \\ HC \quad CH \\ \diagdown_S\diagup} \quad + \quad \substack{HC-CH \\ \| \quad \| \\ HC \quad C \cdot n \\ \diagdown_S\diagup} \quad + \quad H_2SO_4$$

$$\xrightarrow{\qquad} \quad \substack{HC-C{}^{\diagup C(CH_3)_3} \\ \| \quad \| \\ HC \quad CH \\ \diagdown_S\diagup} \quad + \quad \substack{HC-CH \\ \| \quad \| \\ HC \quad C \\ \diagdown_S\diagup {}_{C(CH_3)_3}} \quad + \quad H_2SO_4$$

$$14.119$$

The presence of the H_2SO_4 which is suppressed by the thiophene is probably to keep the thiophene in dual equilibrium state of existence and or assist in activating the alkene. The mechanism above is a one-stage Equilibrium mechanism system, two of them in parallel. It can be observed that all the displacement reactions are largely radical in character.

Unlike in benzene, where all the groups carried are internally located, with thiophene and furan the groups carried are not all internally located to the hetero center, in view of the different types of resonance stabilization phenomena favored by them and the types of monomers or compounds. Hence, when a second group is to be introduced into the ring, the situation is different. If the first group present in the ring is in the 2 - position, and that group is less radical-pushing than H, then the followings are obtained.

$$\substack{HC-CH \\ \| \quad \| \\ HC \quad C \\ \diagdown_{:S}\diagup {}_{NO_2}} \quad \xrightarrow{\qquad} \quad \substack{HC \; - \; CH^{\bullet e} \\ \| \quad \\ HC \quad C \cdot n \\ \diagdown_S\diagup {}_{NO_2}} \quad \longleftrightarrow \quad \substack{HC = CH \\ | \quad | \\ HC \quad C^{\ominus} \\ \diagdown_{S\oplus}\diagup {}_{NO_2}}$$

(I) $\quad\quad$ *(**H > NO₂ in capacity**)* $\quad\quad$ (I)a $\quad\quad$ 14.120

If the group is more radical-pushing than H, then the followings are obtained.

$$\substack{HC-CH \\ \| \quad \| \\ HC \quad C \\ \diagdown_{S:}\diagup {}_{C(CH_3)_3}} \quad \xrightarrow{\qquad} \quad \substack{HC \; - \; CH^{\bullet e} \\ \| \quad \\ HC \quad C \\ {}_{n\bullet}\diagdown_S\diagup {}_{C(CH_3)_3}} \quad \longleftrightarrow \quad \substack{HC = CH \\ | \quad | \\ HC \quad C \\ {}^{\ominus}\diagdown_{S \oplus}\diagup {}_{C(CH_3)_3}}$$

(II) $\quad\quad$ *(**C(CH₃)₃ > H in capacity**)* $\quad\quad$ (II)a $\quad\quad$ 14.121

When the H atom in the 5-position of (I)a is held in Equilibrium state of Existence, the followings are obtained.

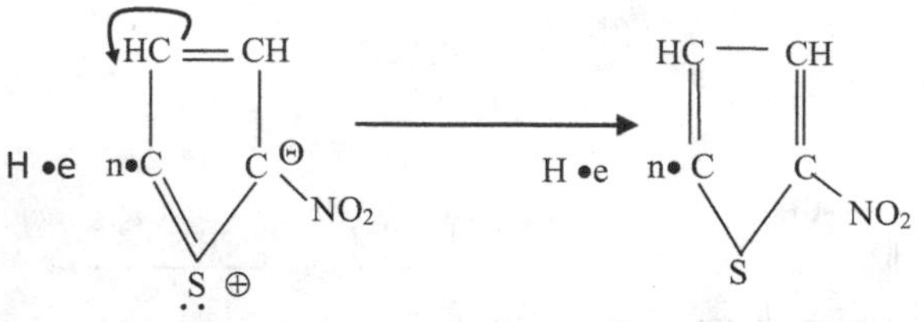

$$(H \; > \; NO_2 \; in \; capacity)$$

(I)a

14.122

When the H atom in 5-position of (II)a is held in Equilibrium state of Existence, the followings are obtained. While the first case was instantaneous, this case allowed for deactivation.

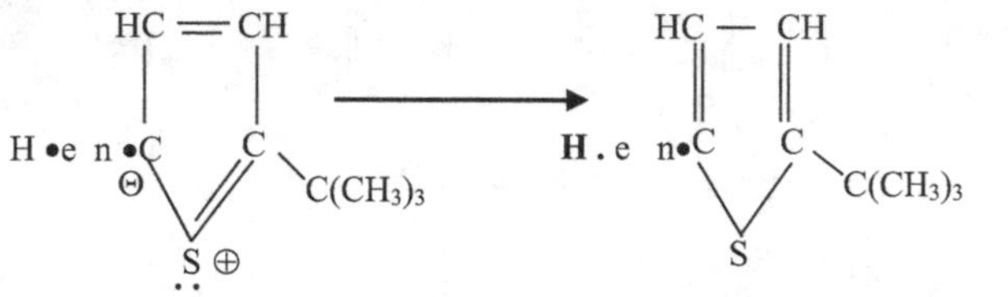

(II)a $\qquad$ $(C(CH_3)_3 \; > \; H \; in \; capacity)$

14.123

Shockingly enough, another $-C(CH_3)_3$ can be placed on the 5-position when the group first used is of equal or greater capacity than the group. One can observe why there has been great need for determining the capacities of groups carried by centers, their characters and much more, because the world of Chemistry is a complex world wherein one can see all the disciplines in humanity in totality. The need arises more so, ***because chemicals were in existence before all creations including the universe in which we leave in.***

Equation 14.122 also applies to radical-pulling free-radical centered groups as shown below; the reason being that C and S are equally electropositive.

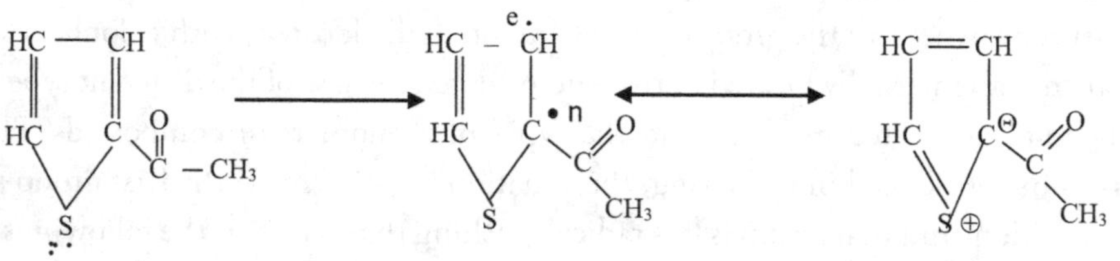

(I) $\qquad\qquad\qquad\qquad\qquad\qquad\qquad\qquad$ (I)a $\qquad$ 14.124

Thus, with radical-pulling groups in the 2 - position, the hydrogen atom in the 5 - position is loosely bonded to the ring. All above favor polar-radical resonance stabilization pheno-menon in view of the driving forces favoring its occurrence. Note how all above favor this type of resonance stabilization. The H atom in the 5- position remains loosely bonded radically, based on the boundary nature of C and S and the operating conditions.

When allylic groups are involved, the monomer is also resonance stabilized and one H atom is loosely bonded.

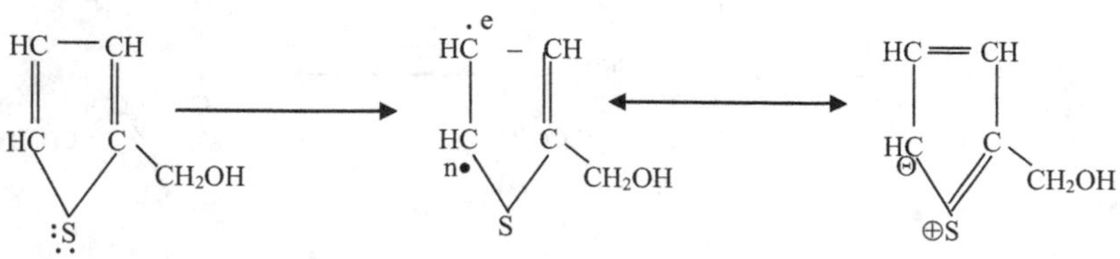

(II) $\qquad\qquad\qquad\qquad\qquad\qquad\qquad\qquad\qquad\qquad\qquad\qquad$ 14.125

Thus, when groups are radical-pushing groups of greater capacity than H, then the 5- position can also be used. Using the case of nitration, the second is introduced in one stage as follows.

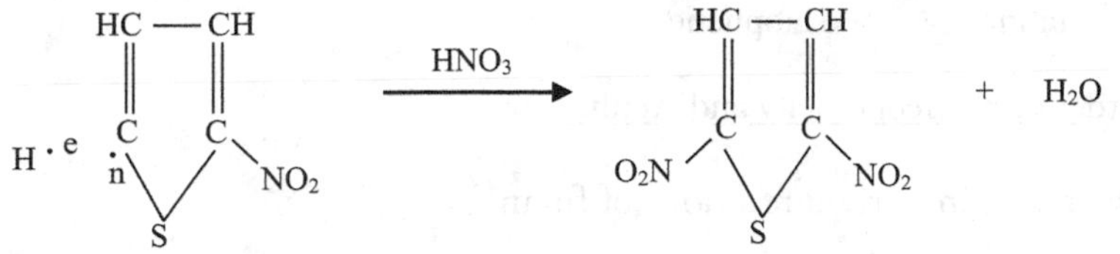

HC —— CH
    C   C
H·e n NO_2
    S

$\xrightarrow{HNO_3}$

HC —— CH
    C   C
O_2N NO_2
    S

+ H_2O

<u>2, 5 - dinitrothiophene</u>   14.126

Not all these reactions can be favored by furan, since O is less electropositive than C and the paired unbonded radicals on O are too strongly held to the O center.

Now, consider when the groups are initially in the 3 - or 4 - position.

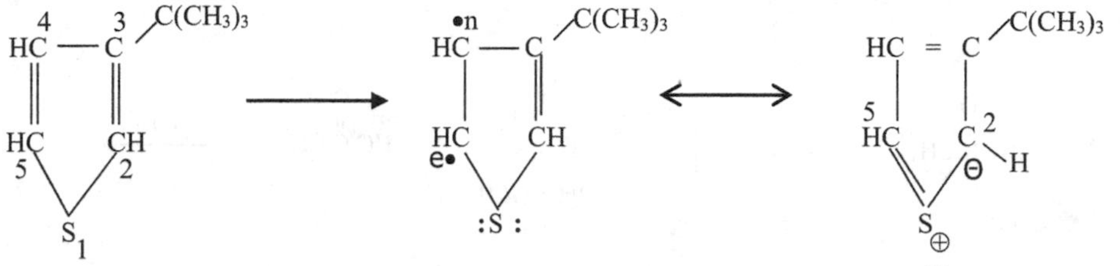

(I)   <u>Polarly/Radically resonance stabilized - 2 or 5 - position</u>   14.127

When the H atom in 2- or 5- position is held in Equilibrium state of existence, the 2- or 5- positions are made available for substituent groups, thus giving mixtures of isomers-2-/3- filled positions and 3-/5- filled positions all obtained in parallel Equilibrium mechanism system.

Using a group with less capacity than H on the 3- or 4- position, the followings are to be expected.

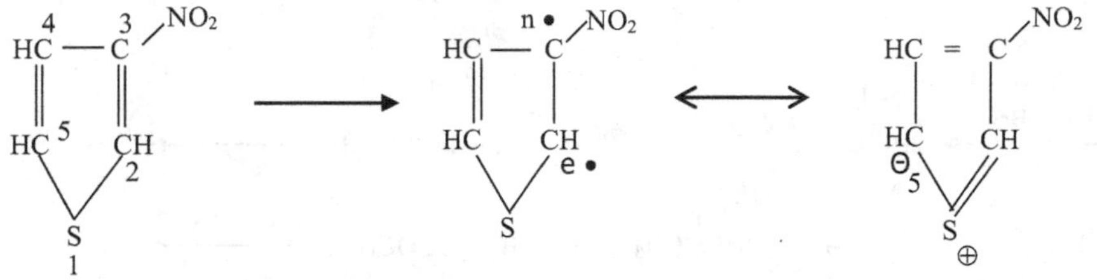

(II)   <u>Polarly/Radically resonance stabilized – 2 or 5 - position</u>   14.128

When the radical-pushing group is of greater capacity than H, displacement can take place in 2- or 5- positions. Also with radical-pushing groups of lesser capacity than H, displacement takes place in the same 2- or 5- position.

With radical-pulling groups, the followings are obtained.

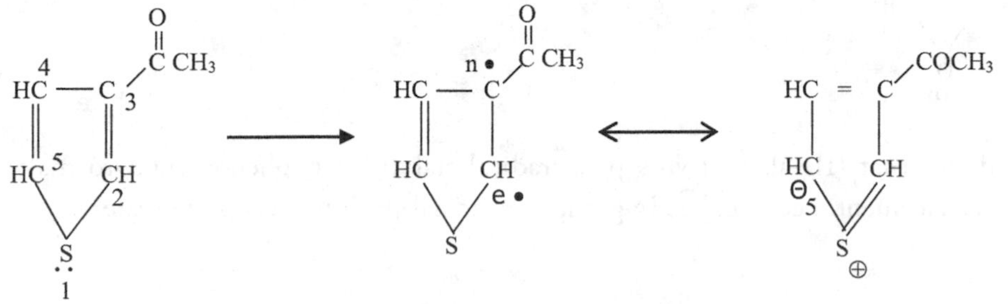

(II)   <u>Polarly/Radically resonance stabilized - 2 or 5 - position</u>   14.129

Sunny N.E. Omorodion

Despite the greater reactivity of thiophene than furan, the followings remain valid since O (3.5) is far more electronegative than S (2.5).

Furan > Thiophene

Order of Nucleophilicity and Acidity 14.130

Shown below are the major important reactions of furan.

(A)

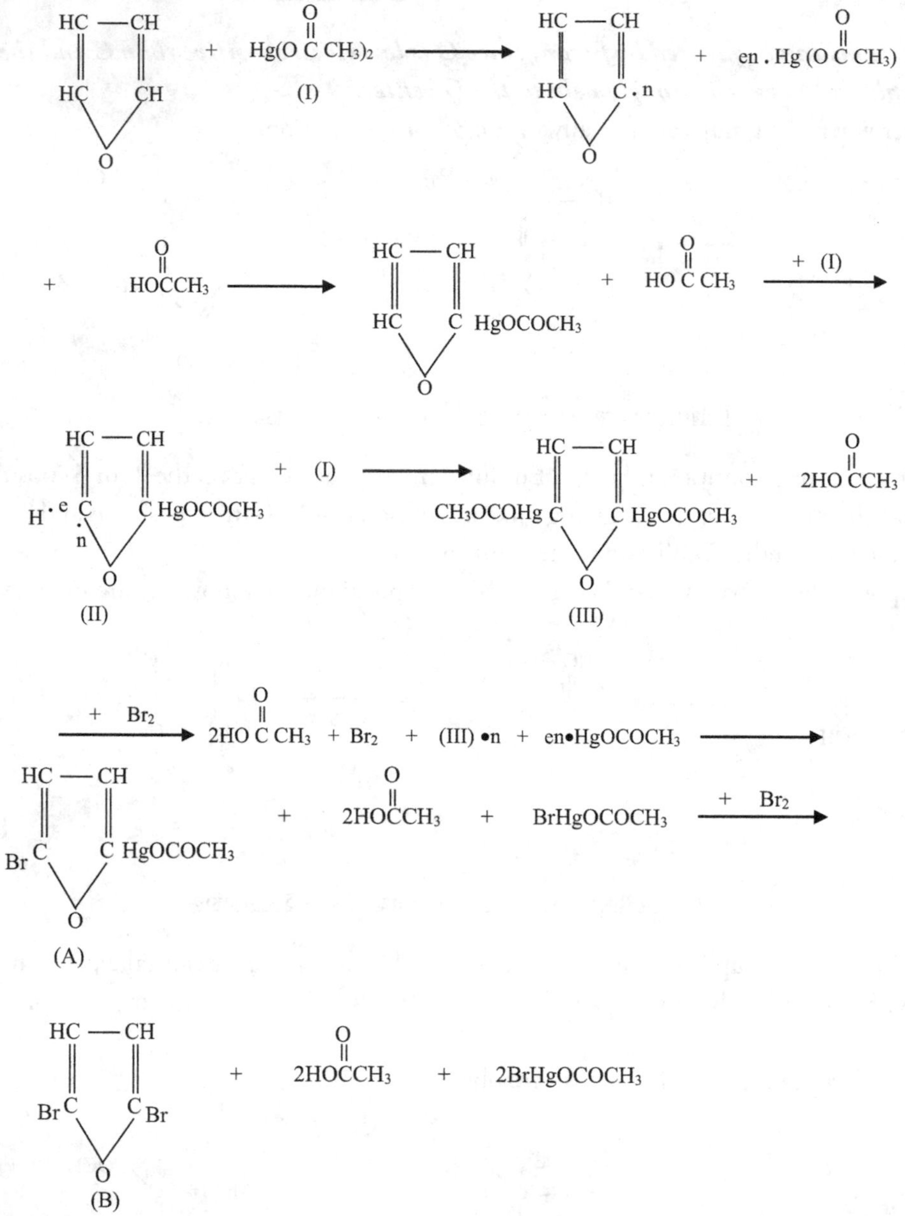

There is no doubt that (II) above favors polar/radical stabilization phenomenon to make 5-position available for displacement, because the Hg group is a radical-pushing group of greater capacity than H.

14.131

366

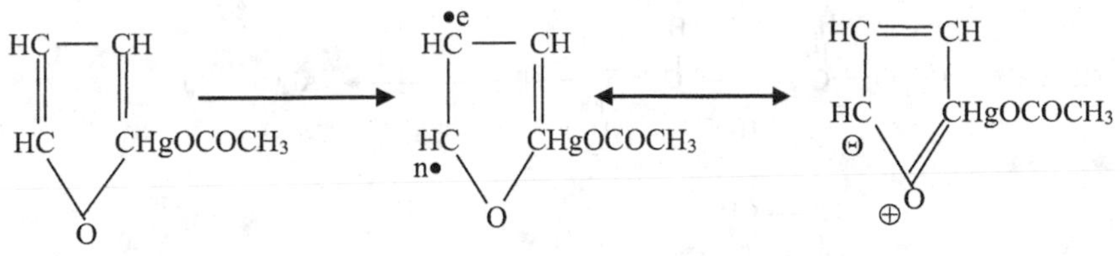

HC — CH $\qquad$ •e HC — CH $\qquad$ HC = CH

HC CHgOCOCH₃ $\quad$ HC CHgOCOCH₃ $\quad$ HC CHgOCOCH₃

O $\qquad$ n• O $\qquad$ O

(II) $\qquad\qquad\qquad\qquad\qquad\qquad$ (III)

<u>**Polar/Radical resonance stabilization in FURAN**</u> $\qquad$ 14.132

It is (III) that makes it possible for the 5- position to be the point where H is held in Equilibrium state of Existence when 2-position carries a group with no transfer species of far greater capacity than H.

One can observe the great significance of resonance stabilization phenomenon in ringed compounds, the order of suppression of components, Equilibrium states of existence of some compound involved in the reactions and influence of species when placed in different locations on the ring. ***While those placed in 2- or 5- positions are externally located in Furan and pyrrole and those in 3- or 4- positions are internally located in furan and pyrrole, those located in any of the positions are all internally located in thiophene.***

(B)

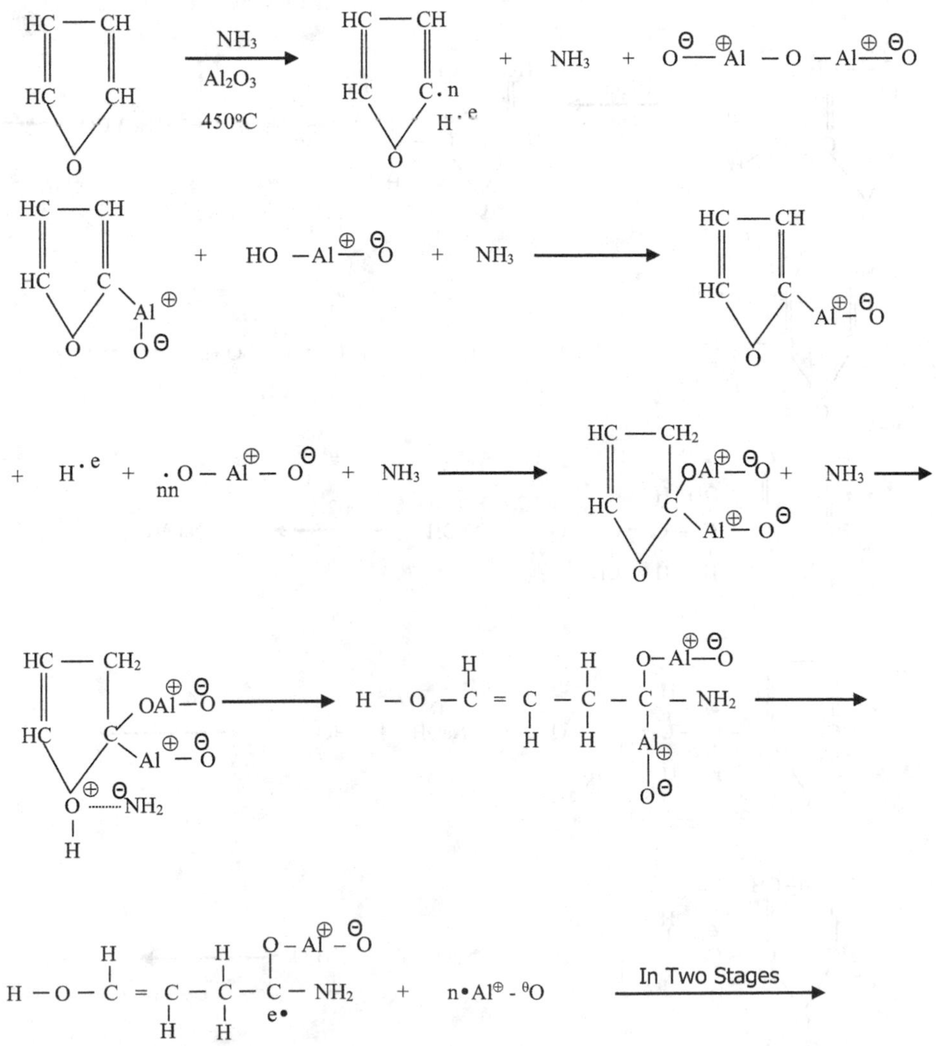

HC — CH $\quad$ NH₃ $\quad$ HC — CH $\quad$ + NH₃ + O⁻—Al—O—Al—O

HC CH $\quad$ Al₂O₃ $\quad$ HC C·n

O $\quad$ 450°C $\quad$ H·e

HC — CH + HO —Al—O + NH₃ ⟶ HC — CH

HC C $\qquad\qquad\qquad\qquad$ HC C—Al—O

O Al⁺ O⁻ $\qquad\qquad\qquad$ O

+ H·e + ·O — Al — O⁻ + NH₃ ⟶ HC — CH₂ OAl—O + NH₃ ⟶

nn $\qquad\qquad\qquad\qquad\qquad$ HC C Al—O⁻

O

HC — CH₂ OAl—O ⟶ H — O — C = C — C — C — NH₂ ⟶

HC C Al—O $\qquad$ H H O-Al-O

O⁺ O⁻NH₂ $\qquad\qquad\qquad$ Al⁺

H $\qquad\qquad\qquad\qquad\qquad$ O⁻

H — O — C = C — C — C — NH₂ + n•Al⁺-⁰O $\qquad$ **In Two Stages** ⟶

H H e•

367

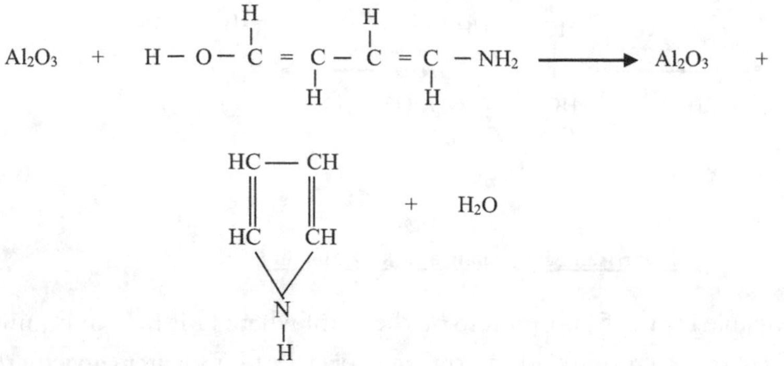

$$Al_2O_3 \;+\; H-O-\underset{\underset{H}{|}}{\overset{\overset{H}{|}}{C}}=C-\underset{\underset{H}{|}}{\overset{\overset{H}{|}}{C}}=C-NH_2 \longrightarrow Al_2O_3 \;+$$

$$\begin{array}{c} HC\!\!-\!\!CH \\[2pt] \| \quad\; \| \\[2pt] HC \quad CH \\[2pt] \diagdown \!\! N \!\! \diagup \\[2pt] | \\ H \end{array} \;+\; H_2O$$

14.133

It can be observed that most of these reactions are radical in character. Aluminum is a non-ionic metal. Without Al_2O_3, the opening of the ring would be impossible. The Al_2O_3 strongly partakes in the reaction to the end where it is released. Its presence in 2-position reduced the nucleophilicity of the C = C bond to the minimum. The existence of the pyrrole ring is readily favored, though it is more strained than the furan ring.

That the oxygen analogue of (II) of Equation 14.125 is not polarly resonance stabilized when the group carried is a radical-pulling group is supported by similar types of reaction favored by it. **Furfural** indeed is also not polarly resonance stabilized, based on the following reactions.

(A)

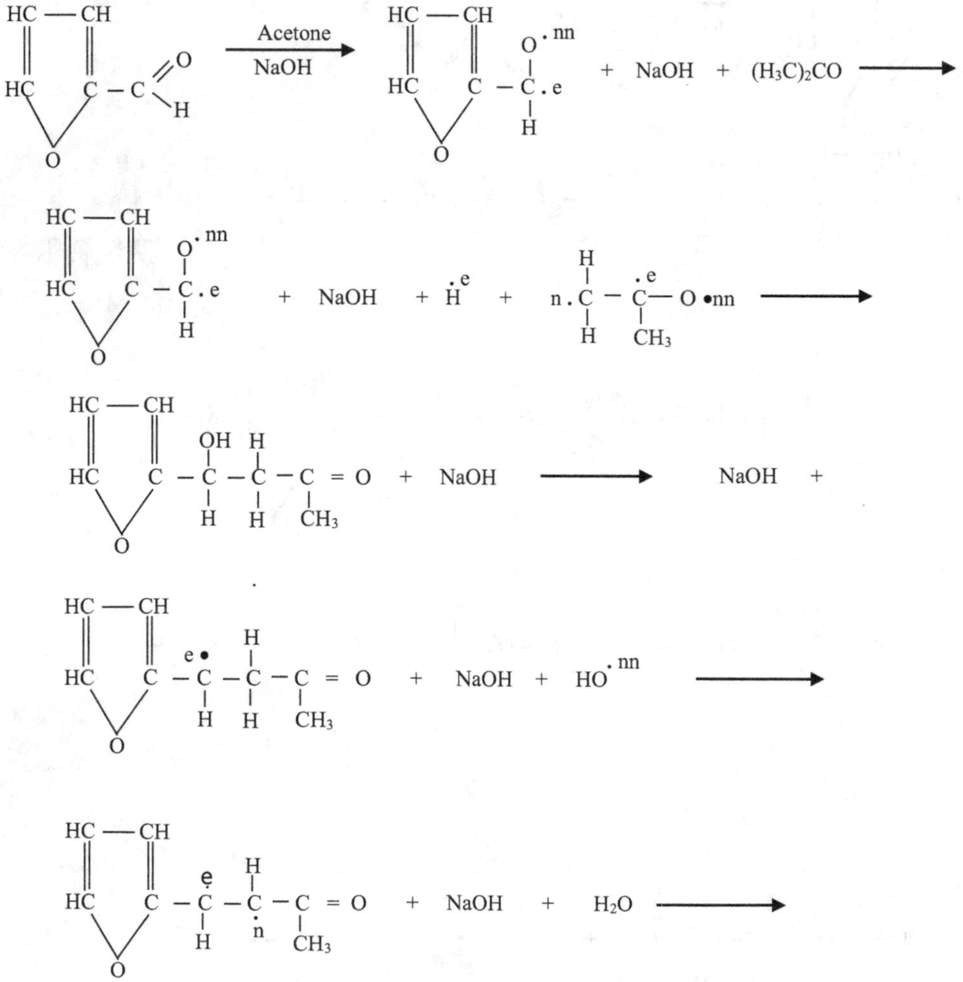

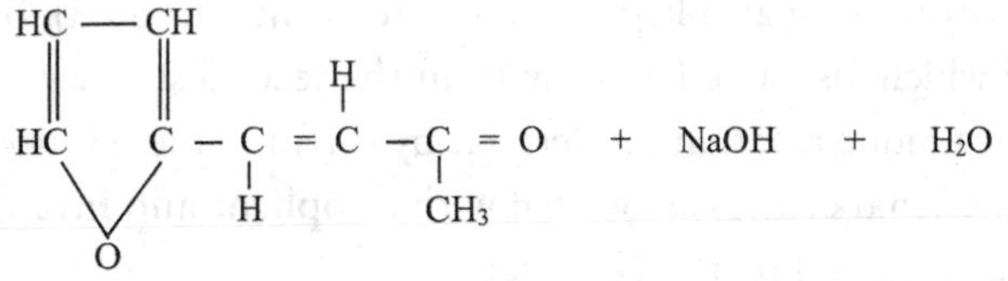

HC —— CH
HC   C — C = C — C = O   +   NaOH   +   H$_2$O
        | | |
        O   H   CH$_3$
       (with H above the second C)

Furfurylideneacetone   14.134

(B)

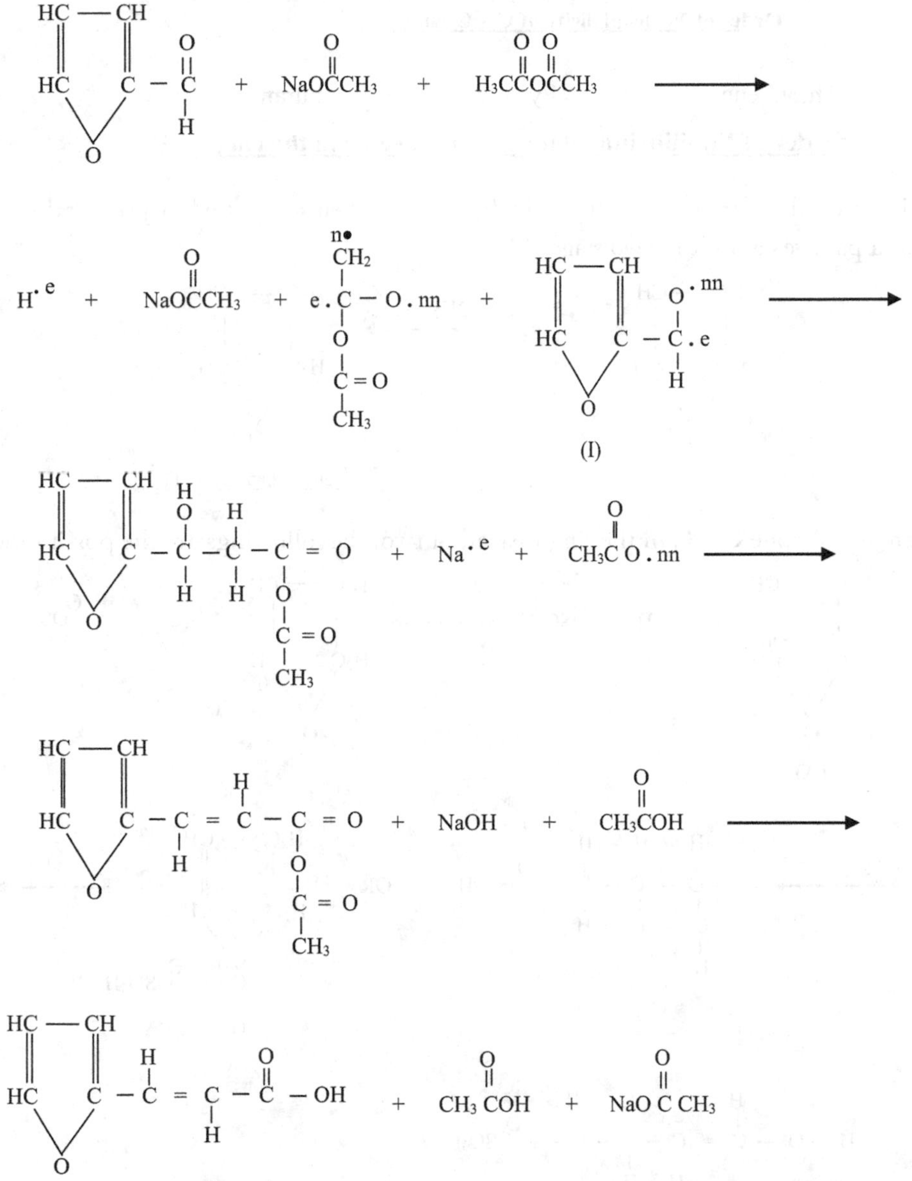

HC —— CH
HC   C — C   +   NaOCCH$_3$ +   H$_3$CCOCCH$_3$ ⟶
        | ‖   ‖   ‖ ‖
        O   O   O   O O
            H

H$\cdot^e$ +   NaOCCH$_3$ +   e.C — O.nn   +   (furan structure) O.nn   ⟶
                    ‖ | |
                    O   CH$_2$ (n•)   C.e
                              | H
                              O
                              |
                              C = O
                              |
                              CH$_3$
                                                            (I)

HC —— CH   H
HC   C — C — C — C = O   +   Na$\cdot^e$ +   CH$_3$CO.nn   ⟶
        | | | | ‖
        O   H   H   O   O
            O |
            H   C = O
                    |
                    CH$_3$

HC —— CH
HC   C — C = C — C = O   +   NaOH   +   CH$_3$COH   ⟶
        | | | ‖
        O   H   O   O
                    |
                    C = O
                    |
                    CH$_3$

HC —— CH
HC   C — C = C — C — OH   +   CH$_3$COH   +   NaO C CH$_3$
        | | ‖   ‖   ‖   ‖
        O   H | O   O   O

Furylacrylic acid   14.135

All these are Equilibrium mechanism systems with two, three, four or more stages. There are rules guiding the existence of a productive stage. There are also rules guiding the existence of a non-reactive, non-productive stage. There are also rules guiding the existence of a reactive non-productive stage- those which define Solubility and Insolubility (Not Dissolution and Miscibilization and their reverses). Just like the case with acetone, the acetic anhydride was kept in Equilibrium/Activated state of Existence

(Enolization, but with another center on furan also kept activated adjacently located), by the presence of the sodium components (passively) which also acted actively from the second stage.

Before continuing to identify the unique reactions undergone by furfural, there is need at this point in time to identify the new types of functional centers associated with thiophene and furan, but not pyrrole. Pyrrole was excluded because of the H carried by the N center.

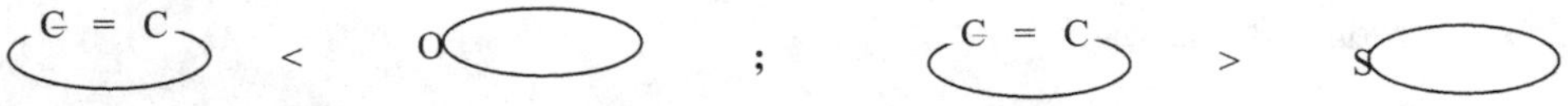

Order of Nucleophilicity of C = C center   14.136

Thiophene   >   Pyrrole   >   Furan

Order of Equilibrium State of Existence from the ring   14.137

It is for this reason that furan can readily be hydrogenated when suppressed to give tetrahydrofuran (THF) using Ni as a passive catalyst in two stages.

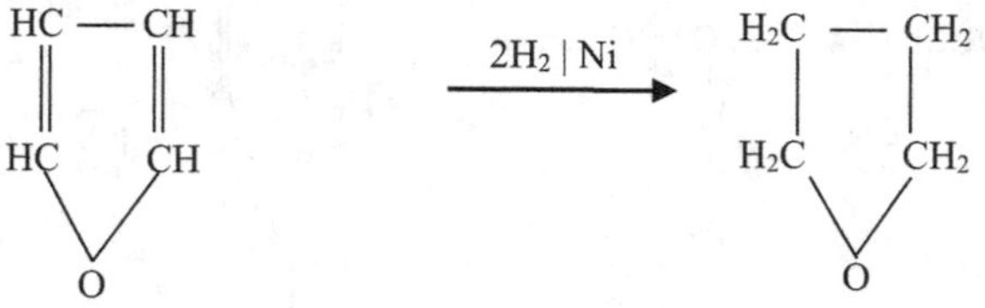

Tetrahydrofuran (THF)   14.138

When there is only one C = C in the ring instead of two, the followings are the possibilities.

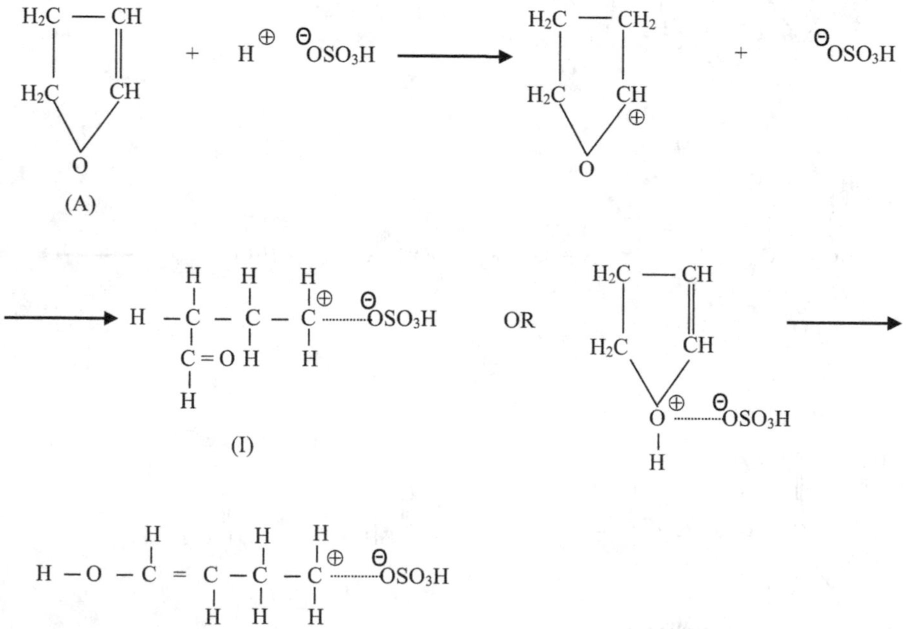

14.139

Since furan can be fully hydrogenated, it is (I) that is favored and not (II), for which the followings are valid as already shown in Equation 14.136.

$$\overset{O}{\bigcirc} \quad > \quad \overset{C\,=\,C}{\underset{\smile}{}} \quad ; \quad \overset{C\,=\,C}{\underset{\smile}{}} \quad > \quad \overset{S}{\bigcirc}$$

<table>
<tr><td>Order of nucleophilicity</td><td>Order of nucleophilicity</td><td>14.140</td></tr>
</table>

With pyrrole, the followings are also valid.

$$\overset{N-H}{\bigcirc} \quad > \quad \overset{C\,=\,C}{\underset{\smile}{}}$$

Order of Nucleophilicity   14.141

Pyrrole can also be hydrogenated to give pyrrolidine.

$$\begin{array}{c} HC - CH \\ \| \quad\quad \| \\ HC \quad CII \\ \diagdown \diagup \\ N \\ | \\ H \end{array} \quad \xrightarrow[\textbf{IN TWO STAGES}]{+\ 2H_2/Pt} \quad \begin{array}{c} H_2C - CH_2 \\ | \quad\quad | \\ H_2C \quad CH_2 \\ \diagdown \diagup \\ N \\ | \\ H \end{array}$$

PYRROLIDINE   14.142

This implies that cyclic ether is less nucleophilic than (A) of Equation 14.139 above, and in turn less nucleophilic than furan. Furan is more stained than (A) which in turn is also more strained than the cyclic ether. This is particularly obvious when four membered ring of (A) is considered.

$$\begin{array}{c} H_2C - CH_2 \\ | \quad\quad | \\ H_2C \quad CH_2 \\ \diagdown \diagup \\ O \end{array} \quad < \quad \begin{array}{c} H_2C - CH \\ | \quad\quad \| \\ H_2C \quad CH \\ \diagdown \diagup \\ O \end{array} \quad < \quad \begin{array}{c} HC - CH \\ \| \quad\quad \| \\ HC \quad CH \\ \diagdown \diagup \\ O \end{array}$$

Order of nucleophilcity   14.143

$$\begin{array}{c} HC - CH_2 \\ \| \quad\quad | \\ HC - O \end{array} \quad \longrightarrow \quad \begin{array}{c} \overset{\ominus}{HC} - CH_2 \\ | \quad\quad | \\ HC = \overset{\oplus}{O} \end{array} \quad \longrightarrow \quad \begin{array}{c} H \quad\quad H \\ | \quad\quad | \\ n\bullet C - C\bullet e \\ | \quad\quad | \\ C = O \ H \\ | \\ H \end{array}$$

(strained)   (Weak Electrophile)   14.144

For the four-membered ring, when activated or polarly stabilized, instantaneous opening of the ring is favored to produce a monomer which is a weak electrophile. This is only possible with the first member of the family which is highly stained to favor any existence. The fact that thiophene favors polar/radical resonance stabilization phenomenon, clearly shows that the following is also valid.

$$\begin{array}{c} C = C \\ | \quad\quad \diagdown \\ | \quad\quad\quad S \\ C = C \diagup \end{array} > \begin{array}{c} C = C \\ | \quad\quad \diagdown \\ | \quad\quad\quad N-H \\ C = C \diagup \end{array} > \begin{array}{c} C = C \\ | \quad\quad \diagdown \\ | \quad\quad\quad O \\ C = C \diagup \end{array} > \begin{array}{c} C = C \\ | \quad\quad \diagdown \\ | \quad\quad\quad CH_2 \\ C = C \diagup \end{array}$$

(Cyclo - dienes)

Order of Resonance stabilization capacity   14.145

Now, coming back to the reactions of furfural, the followings are obtained for some important cases of interest.

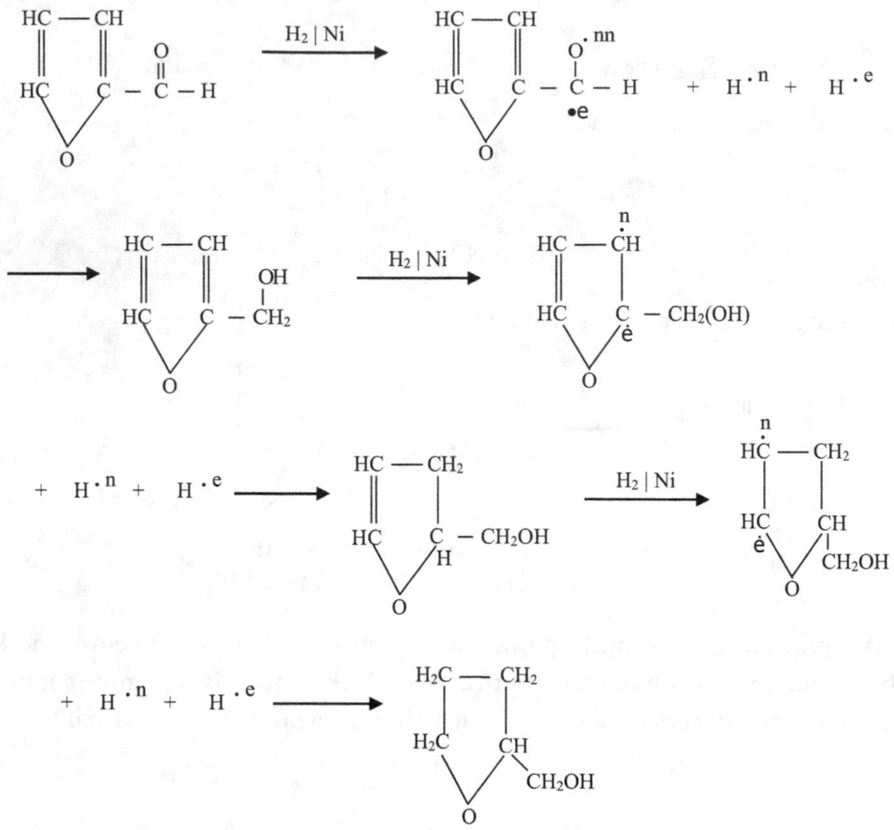

Tetrahydrofurfuryl alcohol

14.146

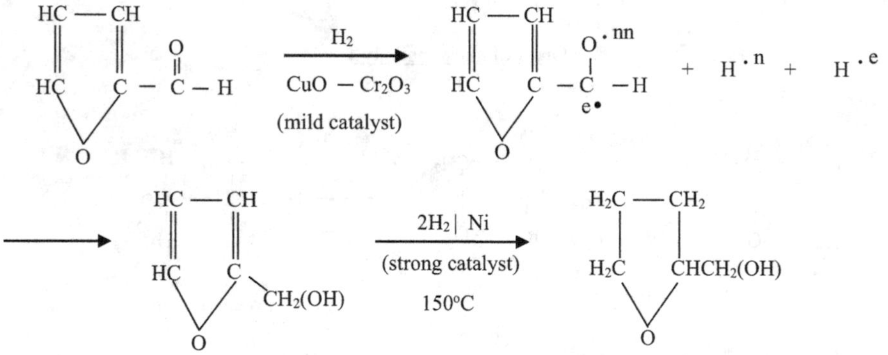

Furfuryl
alcohol

Tetrahydrofurfuryl
alcohol

14.147

All the reactions can largely be observed to be radical in character whether there is polar resonance stability or not.

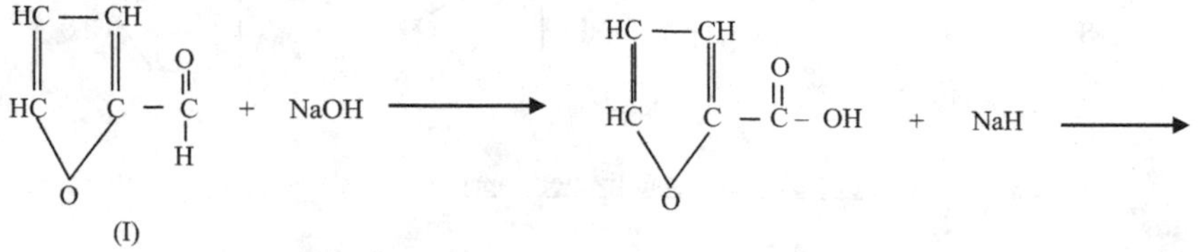

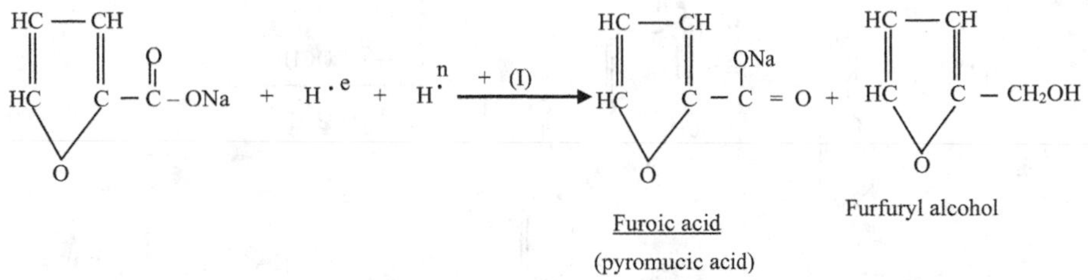

Furoic acid
(pyromucic acid)

Furfuryl alcohol

14.148

In the first stage, the acid and NaH were formed. In the second stage, the sodium salt of the acid was formed in such a way that hydrogenation of (I) was done to give furfuryl alcohol.

It can thus be noted that Perkins, Cannizzaro and so many other reaction mechanisms do not indeed exist, since only one mechanism cuts across chemical systems, following the laws of nature as will finally be shown in the series. While few may for example identify the involvement of hydrides, none of them knows what a hydride is and so many things too countless to list. This has been the major problem in humanity-lack of understanding of how NATURE operates. Even what NATURE is, no body and I repeat again no body knows. But they including the author all think they know. Our life is full of countless so-called mysteries, all of which are not, because they all have very simple explanations which we always wave aside with the hand with so much simplicity as if known. The only mystery in humanity is THE ALMIGHTY INFINITE GOD.

Of particular interest is the case where the ring of furfural is opened in the presence of aniline and acetic acid, because the manner by which it is opened has never been known. Furfural is said to be detected by the brilliant color that it gives with aniline in the presence of acetic acid. Two moles of aniline are said to condense with one of furfural with opening of the ring.

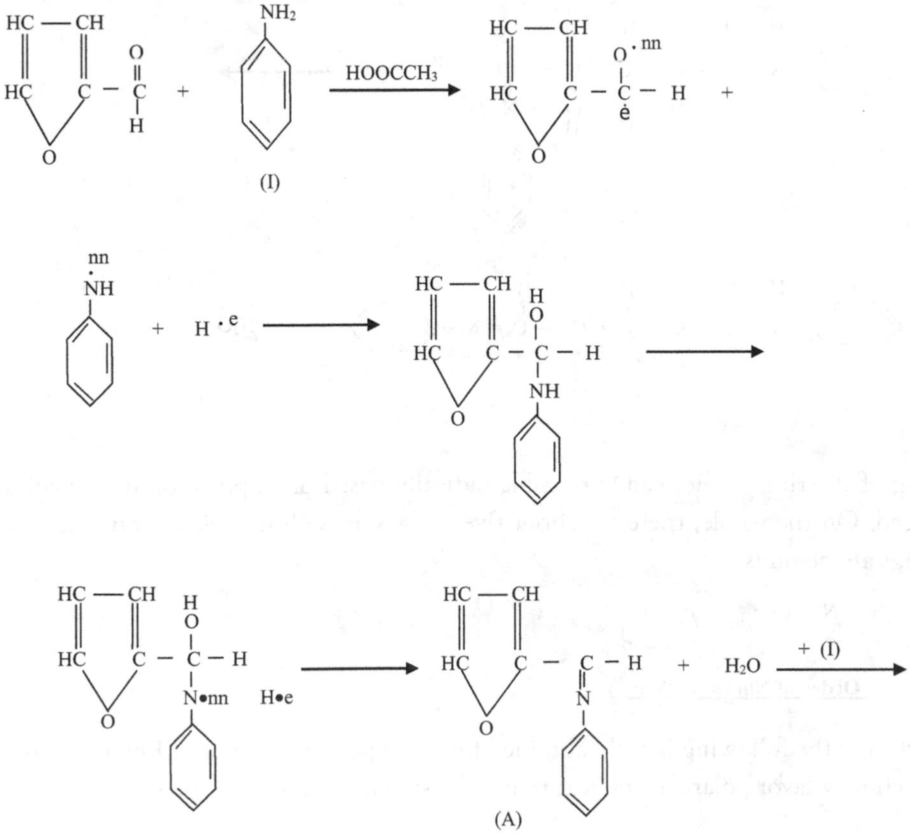

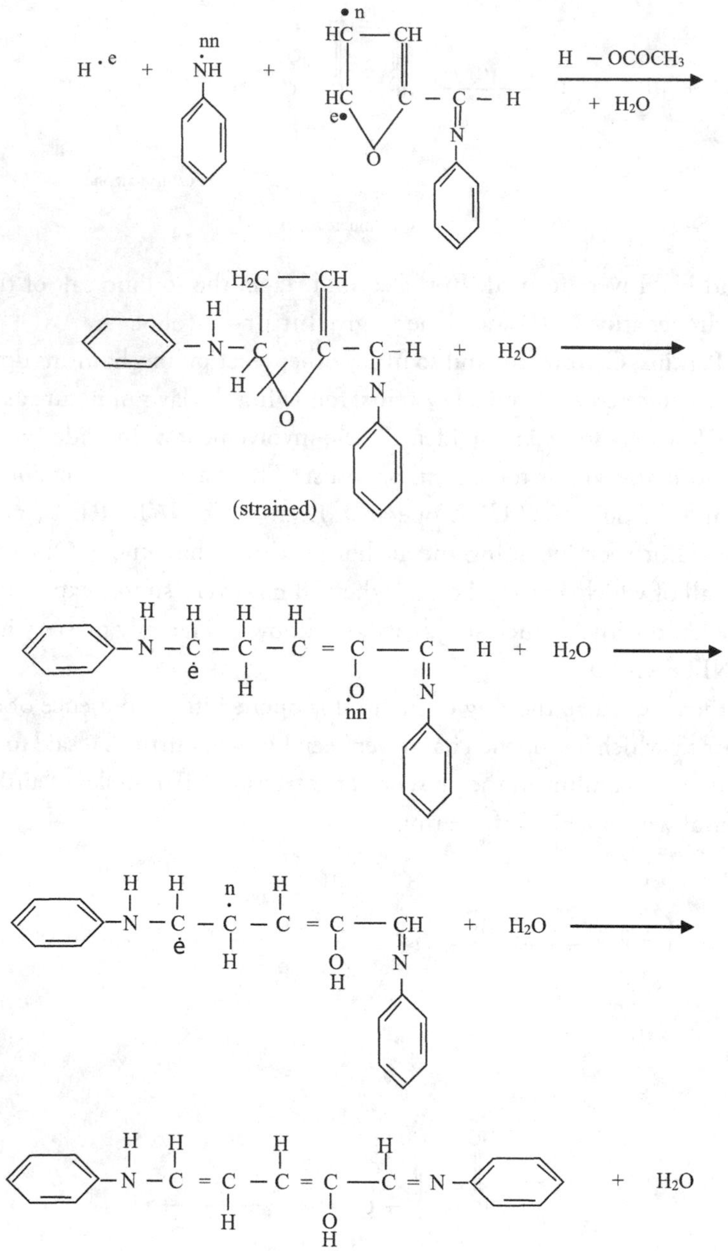

The opening of the ring is one readily possible radically based on application of the rules which have been proposed. On the whole, there are about five or six stages all in series. From the reactions above, the followings are obvious-

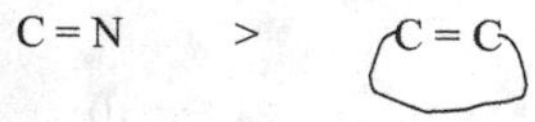

$$C = N \quad > \quad C = C \qquad\qquad \textbf{14.150}$$

Order of Nucleophilicity

Free-radically the following is truly obtained for the type of monomer of Equation 14.148- furfuryl alcohol, which may favor polar/free-radical resonance stabilization pheno-menon.

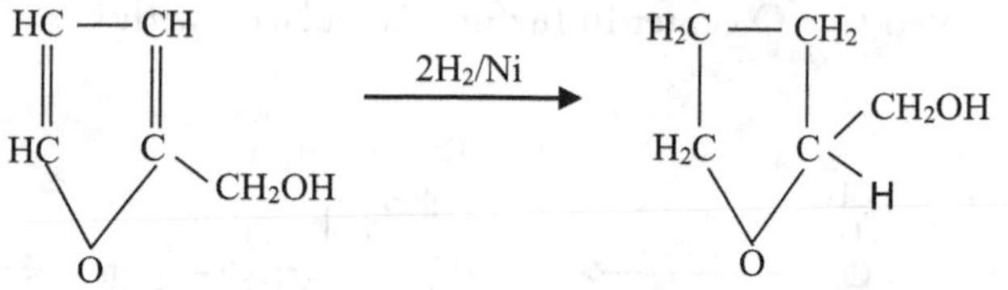

$$HC - CH \quad \xrightarrow{2H_2/Ni} \quad H_2C - CH_2$$

Tetrahydrofurfuryl alcohol

14.151

Free-radically, it is the activation center not carrying the allylic group that is first activated. Tetrahydrofurfuryl alcohol is obtained. The tetrahydrofurfuryl alcohol can react with ammonia at 500°C to give pyridine.

$$\xrightarrow{NH_3/500^0C}$$

(STRAINED)

$$HO - CH_2 - CH_2 - CH_2 - \overset{\square e}{\underset{H}{C}} - CH_2 - \underset{\square nn}{N} H \quad + \quad H_2O \quad \longrightarrow$$

$$HO - CH_2 - CH_2 - CH_2 - CH = CH - NH_2 \quad + \quad H_2O \quad \longrightarrow$$

$$e\bullet CH_2 - CH_2 - CH_2 - CH = CH - \overset{H}{N} \bullet nn \quad + \quad 2H_2O \quad \longrightarrow$$

$$+ \quad 2H_2O \quad \longrightarrow \quad + \quad H_2 \quad + \quad 2H_2O$$

$$\longrightarrow \quad + \quad 2H_2 \quad + \quad 2H_2O$$

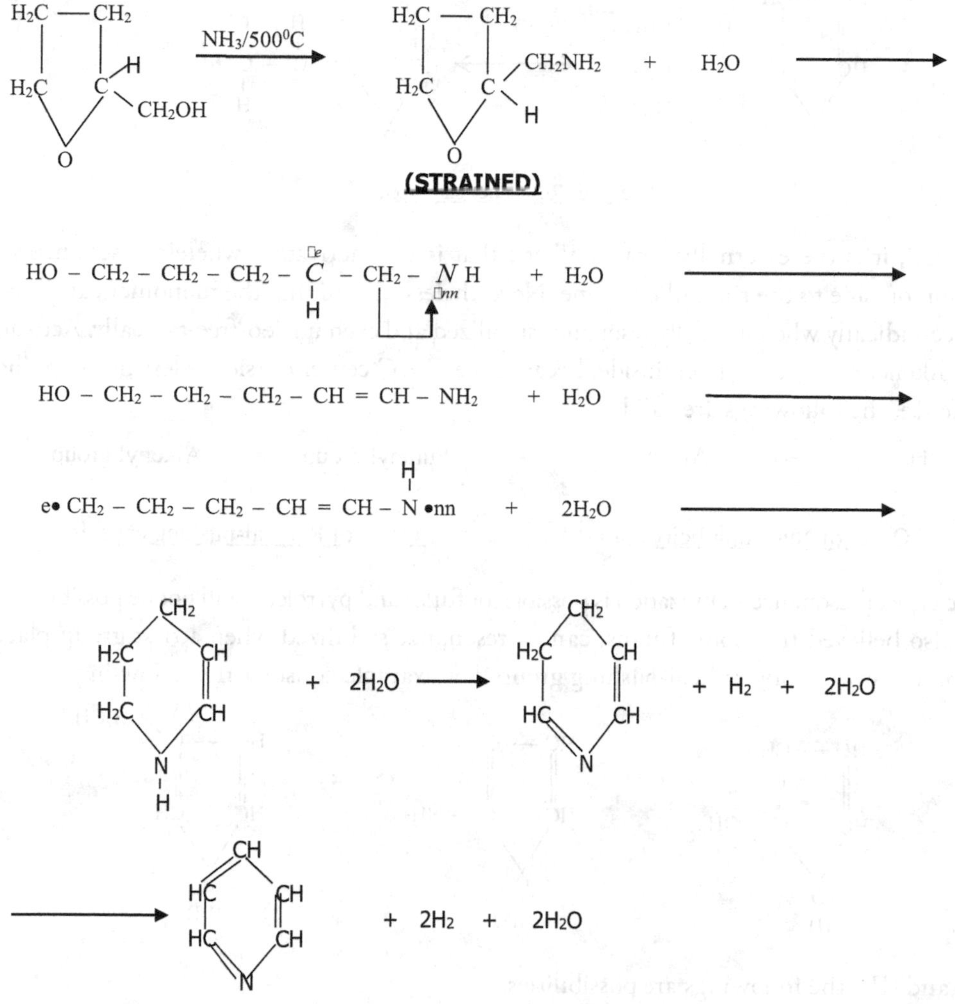

Pyridine

14.152

On the whole, there are six stages. The question is- why at such a high temperature? The reason for the question is that such a reaction can still take place at very low temperature-Surface reaction. Indeed, the high temperature was necessitated by the Cracking done to release the last H_2 molecule. To keep some components under Equilibrium state of Existence in the presence of other unknown components can call for very harsh operating conditions.

Looking very closely at furfural itself, it is an Electrophile where the C = O center is the X center and the C = C center adjacently located to it is the Y center. Hence, when hydrogenated, hydrogenation started from the C = O center despite the fact that it is more nucleophilic than the C = C center. Thus, Furfural unlike Furan is an Electrophile.

When the externally located C = O center in furfural is replaced with C = C center, the followings are obtained.

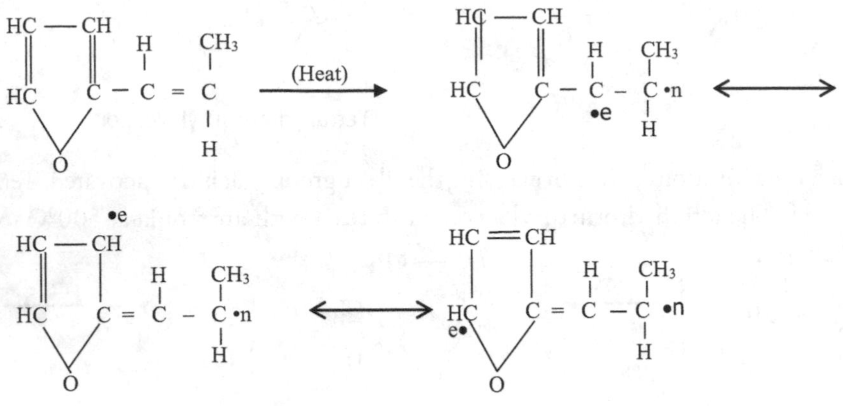

$$\text{(A)} \underline{2, 4, 6\text{- Resonance stabilized}} \qquad\qquad 14.153$$

Above in (A), it is the externally located alkene that is first activated, wherein resonance stabilization moved from outside to the ring, like a triene. Nevertheless, notice that the monomer can be polymerized electro-free-radically when it is fully resonance stabilized and even nucleo-free-radically. Activation started from the alkene, rather than from inside, because the C = C center outside is less nucleophilic than that inside. Hence, the followings are valid.

$$\text{Furan} \quad > \quad \text{Alkenes} \qquad ; \qquad \text{Furanyl group} \quad > \quad \text{Alkenyl group}$$

$$\underline{\text{Order of Nucleophilicity}} \qquad\qquad \underline{\text{Order of Radical-pushing capacity}} \qquad 14.154$$

While the type of resonance stabilization is possible for furan and pyrrole, it will not be possible for thiophene.

It is also believed that some furans, can be resonance stabilized when a first group placed in 2- or 5- position is a very strong radical-pushing group. For example consider the follow-ings.

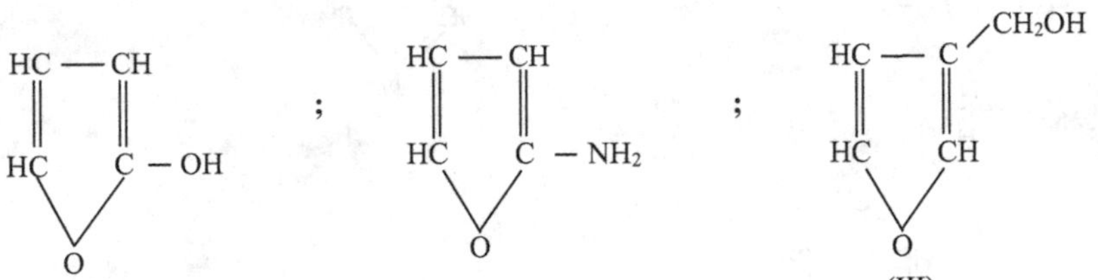

$$\text{(I)} \qquad\qquad \text{(II)} \qquad\qquad \text{(III)} \qquad\qquad 14.155$$

With (I) and (II), the followings are possibilities.

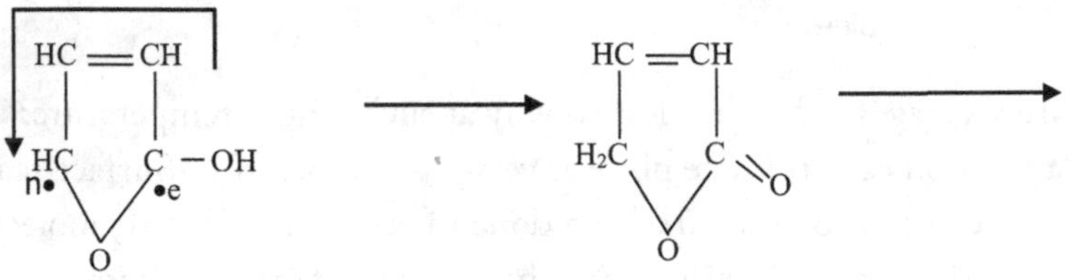

$$\text{(I)} \qquad\qquad\qquad\qquad \text{(Strained)}$$

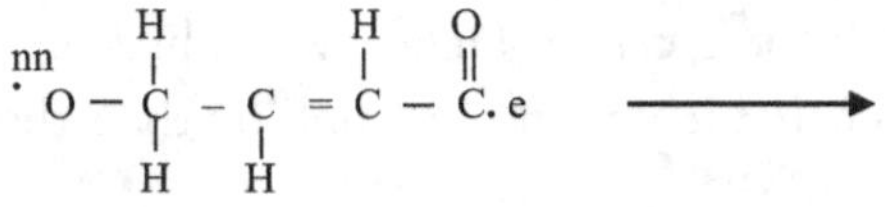

$$14.156$$

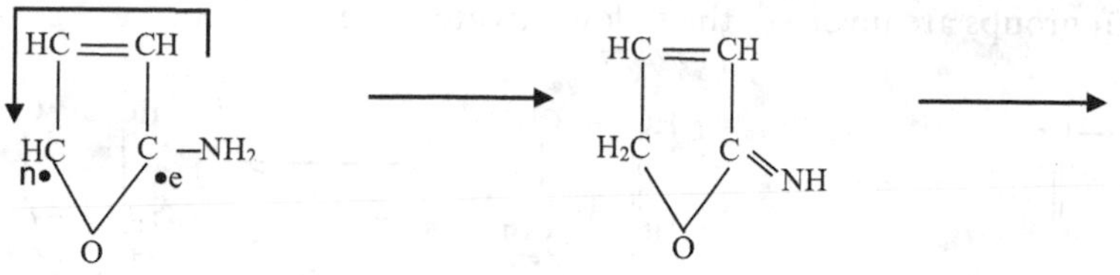

$$\text{(II)}$$

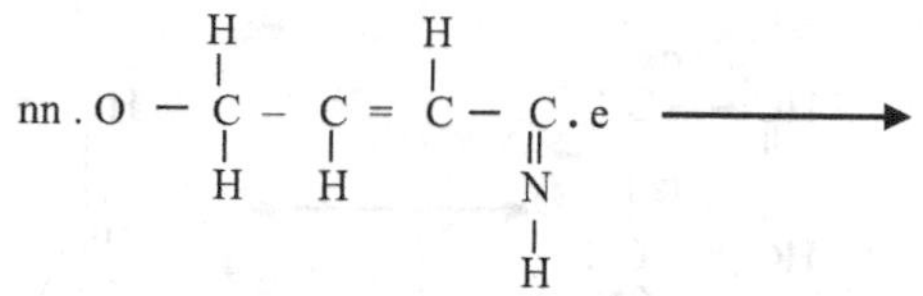

$$\text{14.157}$$

When molecular rearrangement is allowed to take place, the ring can readily be opened. The groups can only be resonance stabilized If the electro-free-radical carried by their C centers are strong enough to abstract a nucleo-non-free-radical from the oxygen center like the case in Equation 14.132 where the group carried is -HgOCOCH$_3$. It is believed that since the radical pushing capacities of –OH and –NH$_2$ groups are high (still far more than =CH$_2$ group), polar/free-radical resonance stabilization may be favored, unless when suppressed (See Equation 14.133, wherein HOAl$^{\oplus}$ - $^{\ominus}$O was a suppressor of the ring carrying -Al$^{\oplus}$ - $^{\ominus}$O in the 2-position).

Unlike the first two cases, (III) above can be resonance stabilized, if polar/radical stabilization could take place. In resonance stabilization, it is the electro-free- or non-free-radical that moves to grab a nucleo-free- or non-free-radical from an adjacently located p-bond after activation and vice visa. In Electroradicalization, it is the electro-free- or non-free-radical that moves from a p-bond to an adjacently located nucleo-free- or non-free-radical and vice visa after existence in Equilibrium state of Existence. This has been the case all along so far. If the electro-free-radical shown below can grab the nucleo-non-free-radical on the O center, then full resonance stabilization (i.e. polar/radical resonance stabilization) will be favored. However, this is not the case here because the radicals are too strongly held to the O center. But with S, it is possible. On the other hand, the electro-radical on the oxygen center is the one to move to grab the nucleo-free-radical, unlike the first two cases.

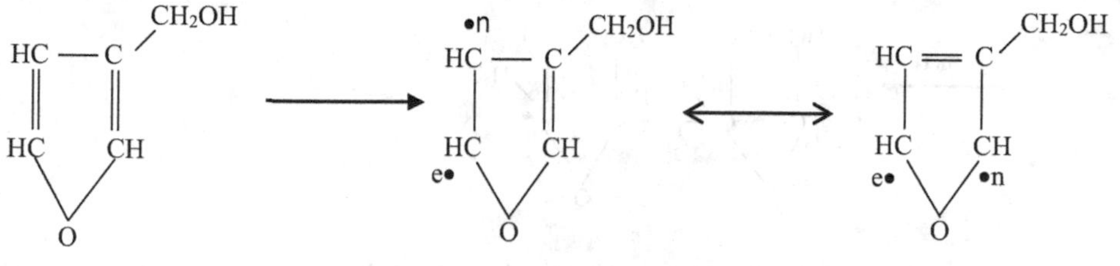

$$\text{(III)}$$

$$\text{14.158}$$

Here, because it is discretely resonance stabilized, the H atom in the 2- position cannot be held in Equilibrium state of existence after the first substitution. Thus, (III) cannot be resonance stabilized polarly/radically, unlike (I) and (II).

When halogen groups are involved, the followings are to be expected.

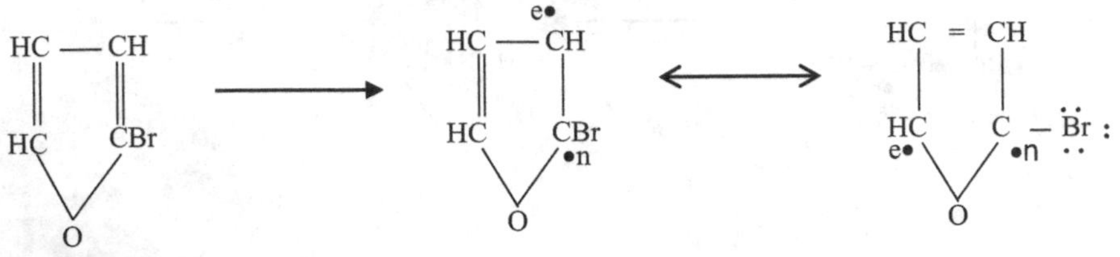

(IV)

14.159

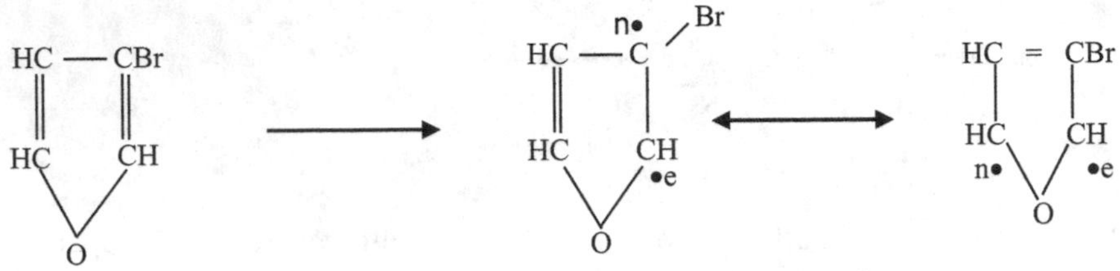

(VI)

14.160

If the resonance stabilization was like the case in thiophene, then with the group placed in the 2-position, the H atom in the 5-position would have been held in equilibrium state of existence.

Furan like pyrrole, reacts with mineral acids to form dark brown polymer. Thiophene should similarly favor reaction with mineral acids to form another colored type of polymer, but for the fact that it favors polar/radical resonance stabilization readily. The reaction here however can be that in which no acid is added to the activated monomers unlike what was done with pyrrole. Instead of the situation where each monomer is activated one after the other for each addition, here all the monomers are kept in an activated state of existence due to the presence of the HCl.

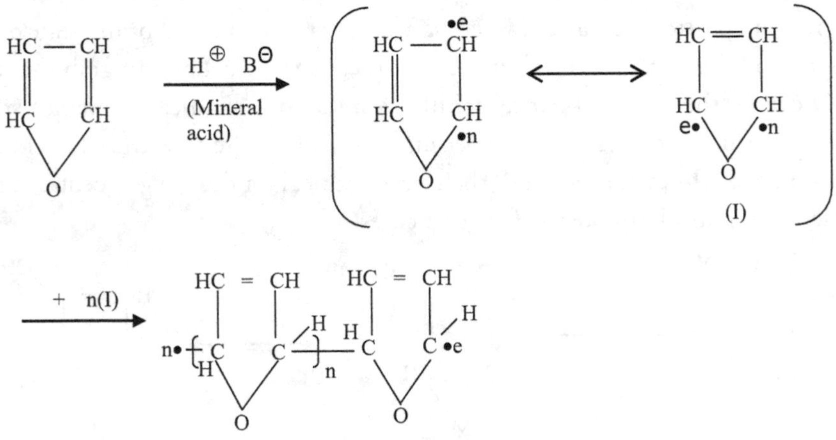

(I)

14.161

From time to time the acid could finally add to close both ends of the chain, beginning from the growing center which is electro-free-radical end.

When furfural is oxidized using $KMnO_4$ or O_2 (air) with Fe_2O_3-Ag_2O cat, the followings are obtained.

Stage 1:

$$O_2 \rightleftharpoons 2 \; en\bullet \; O \; \bullet nn$$

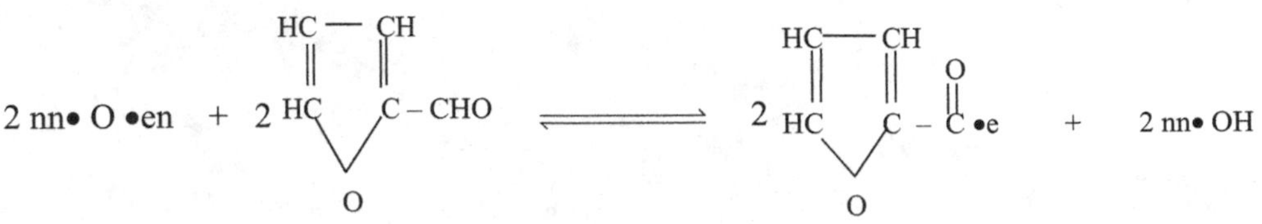

378

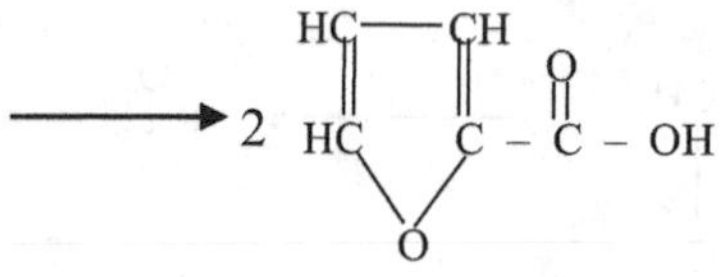

$$\longrightarrow 2 \quad \text{HC} \overset{\displaystyle \text{HC} \text{—} \text{CH}}{\underset{\displaystyle \text{O}}{\bigtriangleup}} \text{C} - \overset{\text{O}}{\text{C}} - \text{OH}$$

Furioc acid (Pyromucic acid)   14.162a

Overall Equation: O_2 + 2Furfural $\longrightarrow$ 2 Furoic acid   14.162b

How $KMnO_4$ produces the oxidizing oxygen molecule will be shown downstream. For the use of molecular O_2 from air using metallic oxides saturated with oxygen, makes the molecular oxygen to break into two parts inside the metallic pores.

$$O_2 \text{ (AIR)} \quad \underset{Cat.}{\overset{Fe_2O_3-Ag_2O}{\rightleftharpoons}} \quad 2en\bullet O \bullet nn$$

[Molecular Oxygen]   [Oxidizing Oxygen molecule]   14.163

Stage 1:

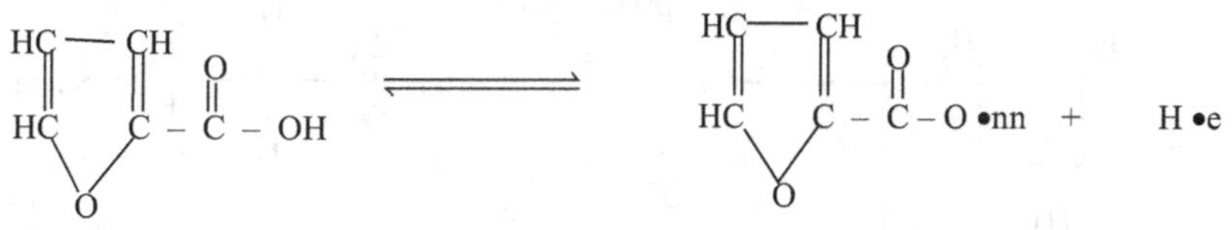

$$\text{HC} \overset{\displaystyle \text{HC} \text{—} \text{CH}}{\underset{\displaystyle \text{O}}{\bigtriangleup}} \text{C} - \overset{\text{O}}{\text{C}} - \text{OH} \quad \rightleftharpoons \quad \text{HC} \overset{\displaystyle \text{HC} \text{—} \text{CH}}{\underset{\displaystyle \text{O}}{\bigtriangleup}} \text{C} - \overset{\text{O}}{\text{C}} - O \bullet nn \quad + \quad H \bullet e$$

Furioc acid (Pyromucic acid)   (A)

$$(A) \quad \rightleftharpoons \quad \text{HC} \overset{\displaystyle \text{HC} \text{—} \text{CH}}{\underset{\displaystyle \text{O}}{\bigtriangleup}} \text{C} \bullet n \quad + \quad e\bullet \overset{\text{O}}{\text{C}} - O \bullet nn$$

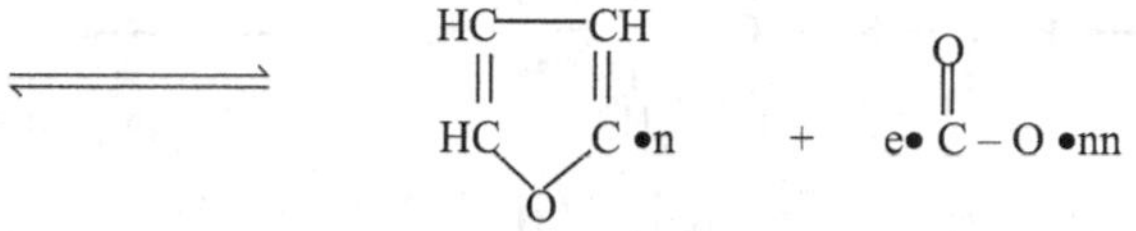

(B)   (C)

H $\bullet$e   +   (B) $\rightleftharpoons$ FURAN

(C) $\xrightarrow{\text{Deactivation}}$ CO_2 + ΔH 14.164a

Overall Equation: Furoic acid $\longrightarrow$ Furan + CO_2 14.164b

The furoic acid when heated loses CO_2 in a single stage as shown above. Any stage which cannot be interpreted mathematically (the Natural language of Communication) without making any necessary and unnecessary assumptions such as "rate determining step" is no stage and absolute nonsense.

14.2.4 Cyclic Disulphides and Rhombic Sulphur

14.2.4.1 Cyclic Disulphides

Beginning with a four-membered ring, one should expect it to possess close to the MRSE, for which the followings should be expected if the ring exists.

$$R\overset{\ominus}{:} \quad + \quad \underset{\underset{\displaystyle S - S}{|\qquad\quad|}}{H_2C - CH_2} \quad \longrightarrow \quad R\overset{\ominus}{:} \quad \overset{\ominus}{S} - S - \underset{\underset{\displaystyle H}{|}}{\overset{\overset{\displaystyle H}{|}}{C}} - \underset{\underset{\displaystyle H}{|}}{\overset{\overset{\displaystyle H}{|}}{C}}\oplus \quad OR$$

(A) strained   (I)

$$\overset{\ominus}{S} - \underset{\underset{\displaystyle H}{|}}{\overset{\overset{\displaystyle H}{|}}{C}} - \underset{\underset{\displaystyle H}{|}}{\overset{\overset{\displaystyle H}{|}}{C}} - S\overset{\oplus}{} \quad \longrightarrow \quad \text{No reaction}$$

(II)   14.165

None of (I) and (II) is favored, since the S - S bond the weakest bond in the ring will always remain the point of scission in the ring and since the S atom may not carry a positive charge in the presence of a carbon atom. The O – O (33-35 kcal/mol) bond is far weaker than S – S (51-54 kcal/mol) bond. It is for this reason the O - O corresponding case of (A) above does not exist. The S - S bond can only be cleaved homolytically.

$$\overset{..}{N}\cdot nn \quad + \quad \underset{\underset{\displaystyle S - S}{|\qquad\quad|}}{H_2C - CH_2} \quad \longrightarrow \quad \overset{..}{N}\cdot nn \quad + \quad en.\overset{..}{\underset{..}{S}} - \underset{\underset{\displaystyle H}{|}}{\overset{\overset{\displaystyle H}{|}}{C}} - \underset{\underset{\displaystyle H}{|}}{\overset{\overset{\displaystyle H}{|}}{C}} - \overset{..}{\underset{..}{S}}.nn$$

(I)

$$\longrightarrow \quad \overset{..}{N} - S - \underset{\underset{\displaystyle H}{|}}{\overset{\overset{\displaystyle H}{|}}{C}} - \underset{\underset{\displaystyle H}{|}}{\overset{\overset{\displaystyle H}{|}}{C}} - \overset{..}{\underset{..}{S}}.nn \quad \overset{+ \ n(I)}{\longrightarrow}$$

$$\overset{..}{N}\left\{ S - \underset{\underset{\displaystyle H}{|}}{\overset{\overset{\displaystyle H}{|}}{C}} - \underset{\underset{\displaystyle H}{|}}{\overset{\overset{\displaystyle H}{|}}{C}} - S \right\}_n S - \underset{\underset{\displaystyle H}{|}}{\overset{\overset{\displaystyle H}{|}}{C}} - \underset{\underset{\displaystyle H}{|}}{\overset{\overset{\displaystyle H}{|}}{C}} - S.nn \quad \longrightarrow$$

$$\overset{..}{N} - S - \underset{\underset{\displaystyle H}{|}}{\overset{\overset{\displaystyle H}{|}}{C}} - \underset{\underset{\displaystyle H}{|}}{\overset{\overset{\displaystyle H}{|}}{C}} - S - S - \underset{\underset{\displaystyle H}{|}}{\overset{\overset{\displaystyle H}{|}}{C}} - \underset{\underset{\displaystyle H}{|}}{\overset{}{C}} = S \quad + \quad H\cdot n$$

<u>Transfer species of 2nd first kind</u>   14.166

All will be shown downstream and already highlighted none of above is favored. Electro-non-free-radically, the followings are supposed to be obtained.

$$\overset{..}{E}\cdot en \quad + \quad \underset{\underset{\displaystyle S - S}{|\qquad\quad|}}{H_2C - CH_2} \quad \longrightarrow \quad \overset{..}{E}\cdot en \quad + \quad nn.S - \underset{\underset{\displaystyle H}{|}}{\overset{\overset{\displaystyle H}{|}}{C}} - \underset{\underset{\displaystyle H}{|}}{\overset{\overset{\displaystyle H}{|}}{C}} - S.en$$

(I)

$$\longrightarrow \quad \overset{..}{E} - S - \underset{\underset{\displaystyle H}{|}}{\overset{\overset{\displaystyle H}{|}}{C}} - \underset{\underset{\displaystyle H}{|}}{\overset{\overset{\displaystyle H}{|}}{C}} - S.en \quad \overset{+ \ n(I)}{\longrightarrow}$$

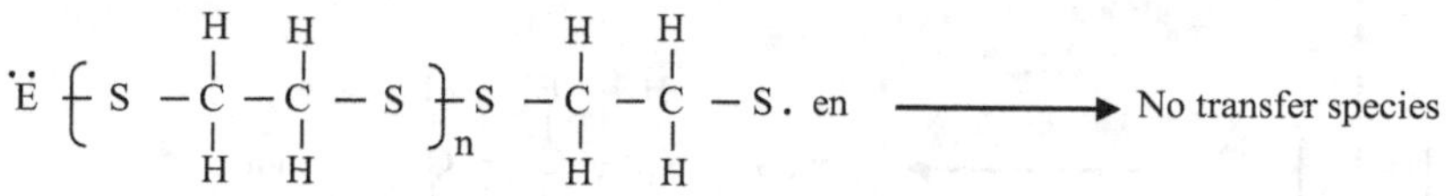

$$\ddot{E} \left\{ S - \underset{\underset{H}{|}}{\overset{\overset{H}{|}}{C}} - \underset{\underset{H}{|}}{\overset{\overset{H}{|}}{C}} - S \right\}_n S - \underset{\underset{H}{|}}{\overset{\overset{H}{|}}{C}} - \underset{\underset{H}{|}}{\overset{\overset{H}{|}}{C}} - S \,.\, en \longrightarrow \text{No transfer species}$$

14.167

As will be shown very shortly, this is also not favored. Instantaneous opening of the ring via the S - S bond is favored (and not S - C bond) because of the following reasons -

(i)   In the first case, C and S atoms have the same electronegativity; so that the radical-pushing potential difference between the C - S and S - S bonds are essentially the same. Even if they are not, the second reason below is very important.

(ii) The C - S (62-65 kcal/mol) bond strength (free and non-free centers) are far greater than the S - S bond strength (two non-free centers).

Hence, the point of scission instantaneously is more favored via the S - S bond than the C - S bond. By the nature of the monomer, it seems to belong to full non-free-radical monomer family. Secondly, the monomer is symmetric, in the sense that the two carbon centers carry same groups (hydrogen atoms). Just like the symmetric cases below where instantaneous opening is favored, the same applies to indeed the entire family of cyclic disulfides, where instantaneous opening of the ring is favored.

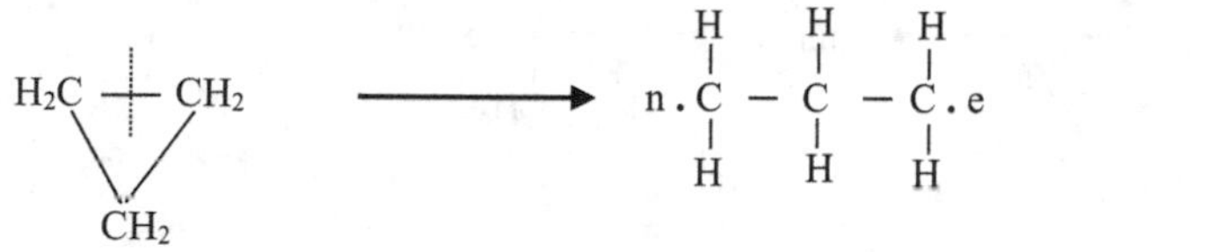

Full free - monomer

14.168

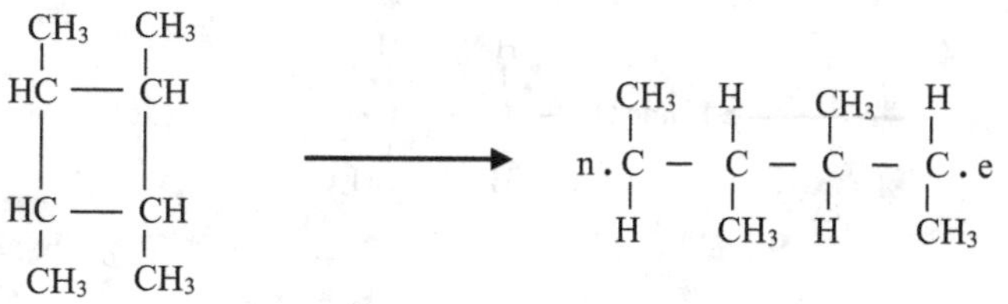

Full free - monomer

14.169

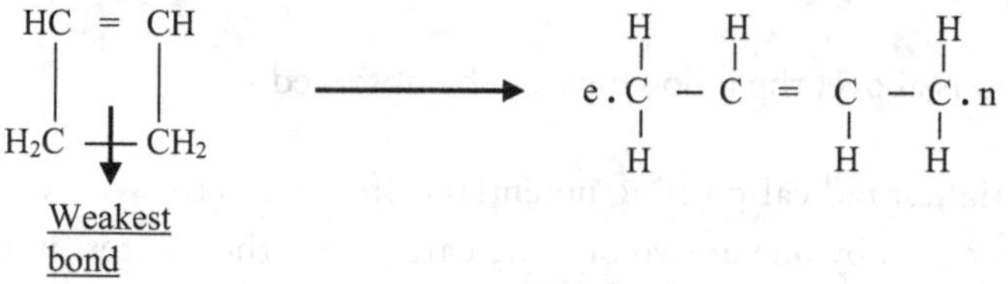

Full free - monomer

14.170

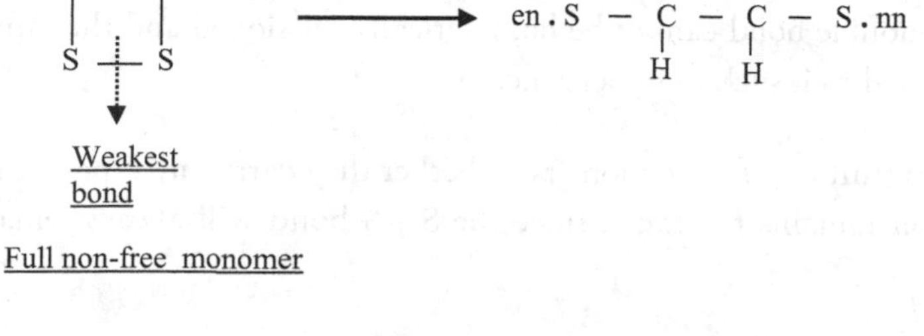

Full non-free monomer

14.171

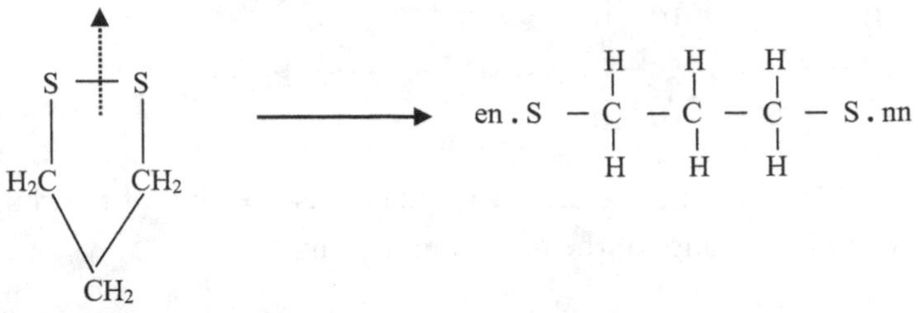

$$S \text{---} S$$

$$H_2C \quad\quad CH_2$$

$$CH_2$$

Full non-free monomer

$$\longrightarrow \quad en \,.\, S \; - \; \underset{\underset{H}{|}}{\overset{\overset{H}{|}}{C}} \; - \; \underset{\underset{H}{|}}{\overset{\overset{H}{|}}{C}} \; - \; \underset{\underset{H}{|}}{\overset{\overset{H}{|}}{C}} \; - \; S \,.\, nn$$

14.172

With non-symmetric cases, the followings are obtained.

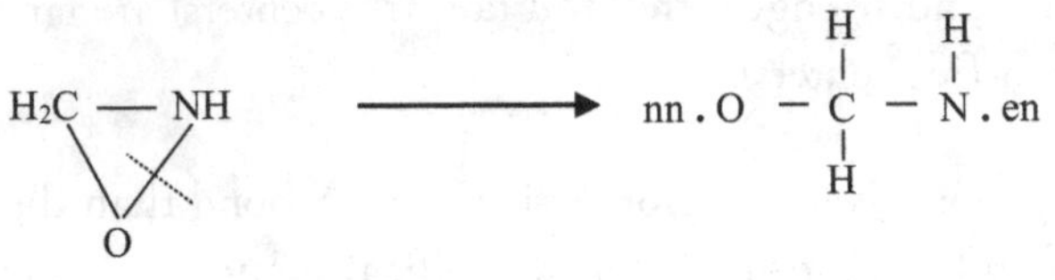

$$H_2C \text{------} NH$$
$$O$$

Weakest bond

$$\longrightarrow \quad nn \,.\, O \; - \; \underset{\underset{H}{|}}{\overset{\overset{H}{|}}{C}} \; - \; \overset{\overset{H}{|}}{N} \,.\, en$$

14.173

Full non - free monomer

$$\overset{CH_3}{\underset{|}{HC}} \text{------} CH_2$$
$$O$$

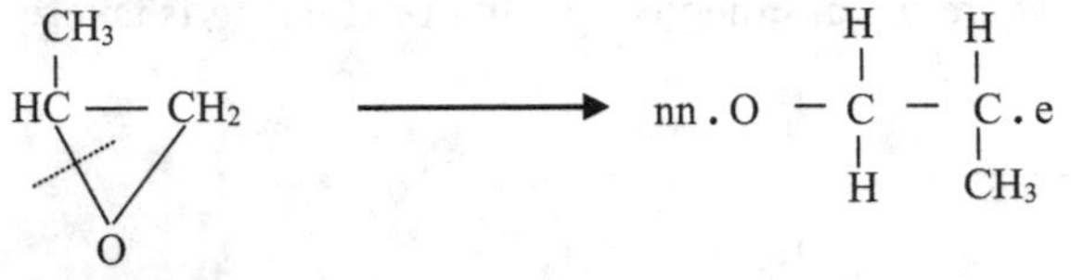

Weakest bond

$$\longrightarrow \quad nn \,.\, O \; - \; \underset{\underset{H}{|}}{\overset{\overset{H}{|}}{C}} \; - \; \underset{\underset{CH_3}{|}}{\overset{\overset{H}{|}}{C}} \,.\, e$$

14.174

Half free monomer

$$\overset{CH_2Cl}{\underset{|}{HC}} \text{------} CH_2$$
$$O$$

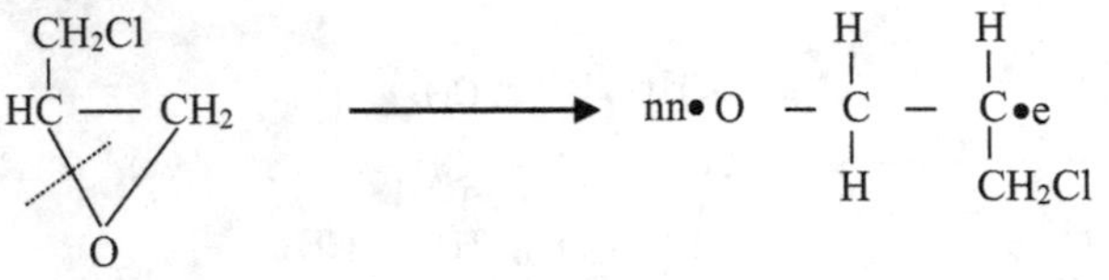

Weakest bond

$$\longrightarrow \quad nn \bullet O \; - \; \underset{\underset{H}{|}}{\overset{\overset{H}{|}}{C}} \; - \; \underset{\underset{CH_2Cl}{|}}{\overset{\overset{H}{|}}{C}} \bullet e$$

14.175

Half free monomer

On the basis of the considerations above, the followings can be observed -

(i)   The bond with the highest radical-pushing potential difference is the weakest bond in a ring.

(ii) The types of groups carried by one or two or more carbon or other centers in the ring is a next factor in determining the weakness of a bond.

(iii) The main character of the monomer also determine point of scission in the ring (half and full free monomers).

(iv) Existence of point(s) of scission in the ring; for some rings are well strained with no point of scission. A C=C double bond cannot be homolytically scissioned and the same applies to the single bond attached to it such as in benzene.

Thus, with cyclic disulfides (full non-free monomers), whether they carry any type of substituent group or not, the point of scission remains the same, since the S - S bond will always remain the weakest.

The reason for the above development is because, a brilliant student may ask the question - as shown diagrammatically below -

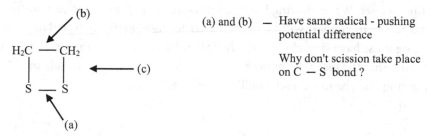

(a) and (b) — Have same radical - pushing potential difference

Why don't scission take place on C – S bond ?

14.176

without the realization that the following is valid and the centers forming the bonds are different in character as shown below.

C – C (83)   >   C – S (65)   >   S – S (51)

bond   bond   bond

(two free centers)   (one free and one non-free center)   (two non-free centers)

Order of Bond strength   14.177

It should be noted that the bonds in question are covalent σ bonds (not ionic bonds).

This is no doubt that when single bonds such as O – O (bond strength of 33 kcal/ mole), S – S (bond strength of 51kcal/mol), N – N (bond strength of 38kcal/mol) exist in a ring, the ring will be strongly strained in the following order –

O – O   >   N – N   >   S – S

(I)   (II)   (III)

Order of **Strains** in the rings of same size   14.178

to the point where their favored existences are in the reverse order above.

With cyclic disulfides whose existences are favored, and others whose existences are not favored at STP ((I) and some of (II) above), there seems to be no usable functional centers free-radically and chargedly.

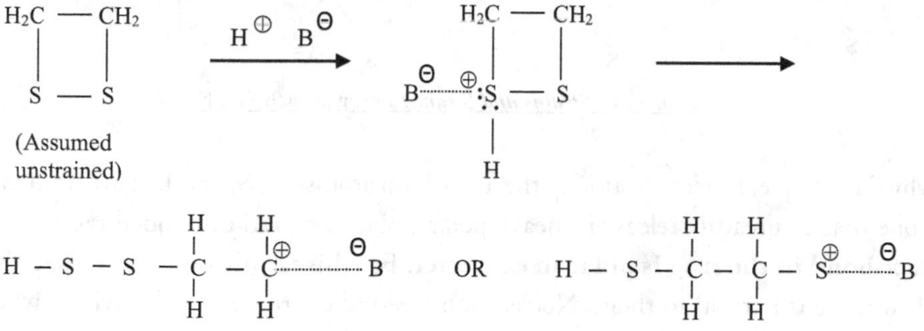

H_2C — CH_2 | S — S (Assumed unstrained) $H^{\oplus}$ $B^{\ominus}$ → H_2C — CH_2 | $B^{\ominus}$ $^{\oplus}:S$ — S | H →

$H - S - S - \overset{\overset{H}{|}}{\underset{\underset{H}{|}}{C}} - \overset{\overset{H}{|}}{\underset{\underset{H}{|}}{\overset{\oplus}{C}}} \cdots B^{\ominus}$ OR $H - S - \overset{\overset{H}{|}}{\underset{\underset{H}{|}}{C}} - \overset{\overset{H}{|}}{\underset{\underset{H}{|}}{C}} - \overset{\oplus}{S} \cdots B^{\ominus}$

(I)   (II)   14.179

In general when functional centers are involved, after the natural initiator has attacked, the weakest bond in the ring is that scissioned. Hence, (I) above is not favored. (II) is not favored, since the S atom cannot carry a positive ionic or covalent charge in the presence of C. Radically, electro-free-radicals will not also add since the equation will not be radically balanced whether free-media-wise or not unless when the mechanism is Equilibrium. With electro-non-free-radical route, the route natural to it, the route will not be favored via the functional center on the monomer even after losing something as will be shown downstream in this chapter. When the ring is opened instantaneously, then nucleo-non-free-radical route, (route not natural to the monomer) will be favored after losing something to balance the equation. In order words the ring must have very close to the MRSE, which will exist up to ring sizes larger than or equal to eight, noting that the more the number of similar hetero atoms placed side by side in a ring, the less strained is the ring and the more nucleophilic is the ring. That is–

H2 H2
C C
H2C
CH2
H2C
CH2
S S
>
CH2 CH2
H2
CH2
H2C
S
S S
>~~~~~~>
S S
S
S
S
S
S S

A - Cyclic disulphide A cyclic trisulfide Rhombic sulfur

Order of **Strain Energy in different families** 14.180

H2C — CH2
S — S
>
H2C — CH2
H2C
S
S
>
CH2
H2C
CH2
H2C
S
S
>

Order of **Strains** in the same family 14.181

It is believed that rings where there are more than eight sulfur atoms placed side by side can exist, because three-, four- and probably five- membered rings are not known to exist, noting that the following is therefore valid for them.

S S
<
S S
<
S
<
S
<
etc.

Order of ***Strains and Strain Energy*** in each family 14.182

The reason why S is unique, is because among the Transition atoms- C, S, and I, apart from other qualities, it is the only one that can readily release its heavy polar cloud (2 paired unbonded radicals) inside a ring. **The invisible π-bond in the ring is polar in character.** For this family of monomers, only the electro-non-free-radical route is natural to them. Nucleo-non-free-radical route is only favored when the ring is opened instantaneously and this is what takes place in general when polymerized.

S S
S
S
S
S
S S
nn• S – S – S – S – S – S – S – S •en OR ⊖S – S2⊕ – S – S – S – S – S – S⊖

14.183

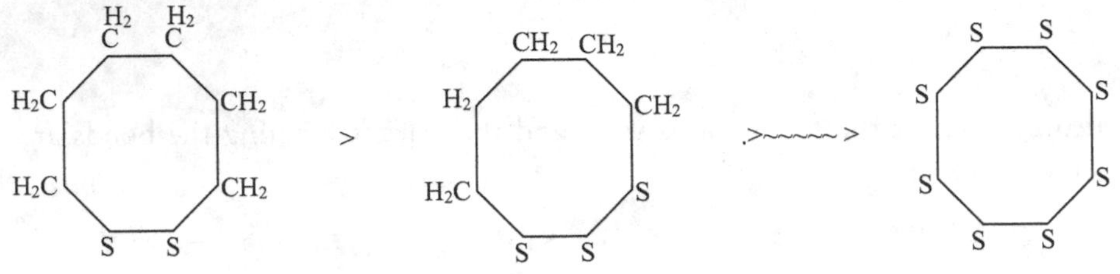

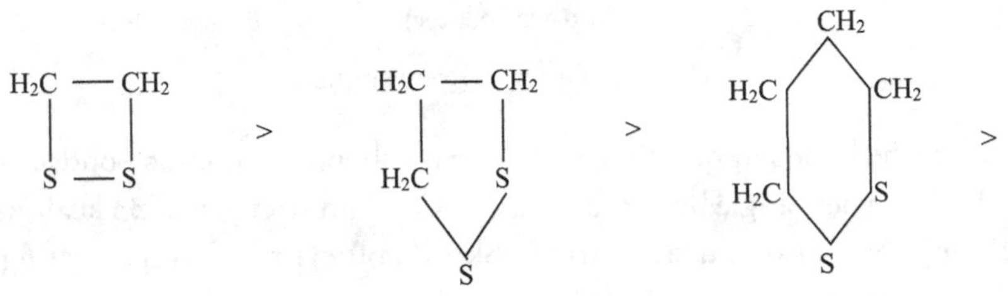

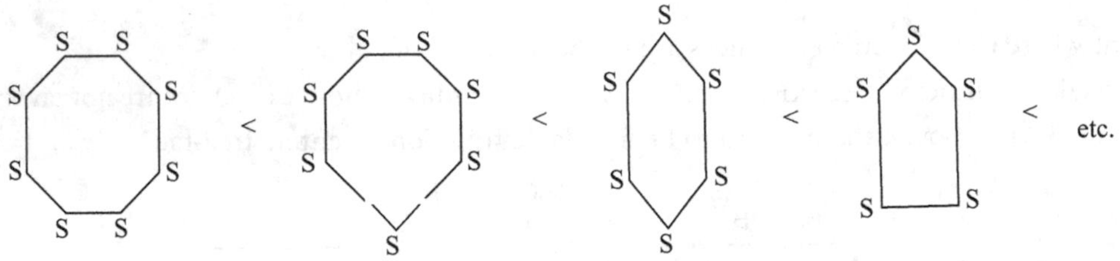

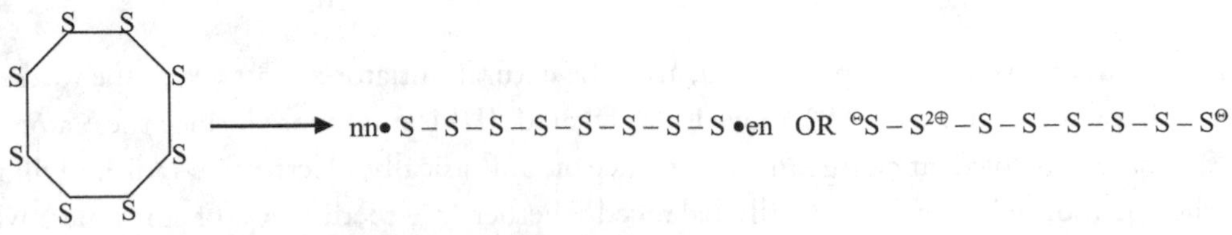

If rhombic sulfur is decomposed, what are called sulfurizing sulfur analogous to oxidizing oxygen is produced. The affinity of this for H is less than that from oxygen.

$$\text{Rhombic Sulfur} \xrightarrow{\text{Heat}} 8 \text{ en} \bullet S \bullet nn \quad ; \quad O_2 \xrightarrow{Fe_2O_3/Ag_2O} 2 \text{ en} \bullet O \bullet nn$$

$$\underline{\textit{Sulfurizing Sulfur}} \qquad\qquad\qquad \underline{\textit{Oxidizing Oxygen}}$$

14.184

The fact that the number of functional centers carried side by side in a ring, is a true measure of the level of nucleophilicity, clearly indicates that the following is valid.

Cyclic tetrasulphides   >   Cyclic trisulphides   .>   Cyclic disulphides   >   Cyclic sulphides

<u>Order of Nucleophilicity</u>

14.185

Where instantaneous opening of the rings of any member of cyclic disulfides are favored, they exhibit both nucleophilic and electrophilic characters. Consider placing a cyclic disulfide ring side by side to a cyclic sulfide ring.

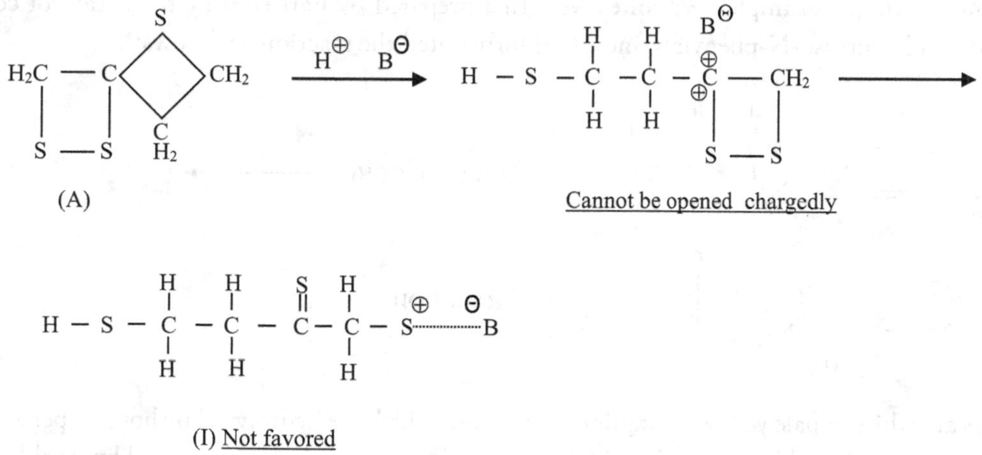

(A)   Cannot be opened chargedly

(I) <u>Not favored</u>   14.186

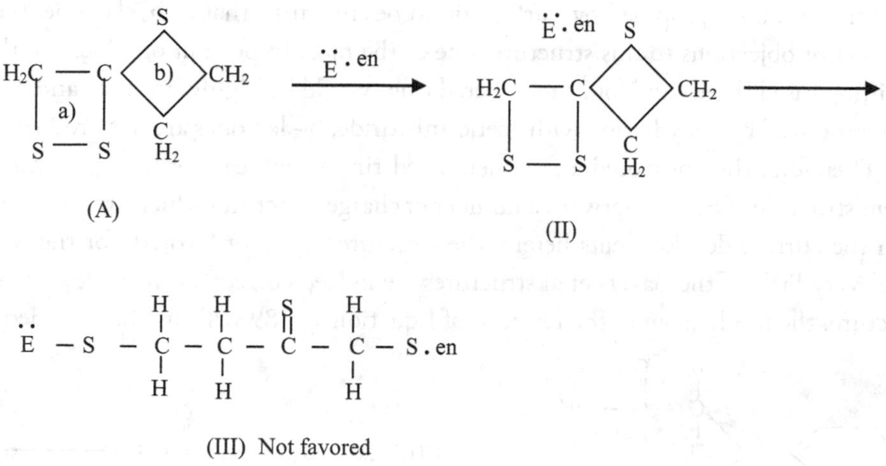

(A)   (II)

(III) Not favored   14.187

Above, note that the b) was first opened, if it is less nucleophilic than a). Cationically, anionically and nucleo-free-radically, polymerization of (A) is not favored. It is not favored non-free-radically even if the disulfide ring is opened instantaneously. The followings are obtained electro-non-free-radically if a) is less nucleophilic than b).

$$\overset{..}{E} - S - \underset{H}{\overset{H}{\underset{|}{\overset{|}{C}}}} - \overset{S}{\overset{||}{C}} - \underset{H}{\overset{H}{\underset{|}{\overset{|}{C}}}} - \underset{H}{\overset{H}{\underset{|}{\overset{|}{C}}}} - S.en$$

(IV) Not favored   14.188

(I) above is not favored since the sulfur center cannot carry free cation of the ionic type. It is the electron-non-free-radical that can add to (II) to favor the existence of (III), if cyclic sulphide is less nucleophilic than cyclic disulphide, unlike the case of (IV). Though it is (IV) that looks favored when the functional center is used via a) taken to be less nucleophilic, it is not favored, until something is lost. A nucleo-non-free-radical cannot be added to (II), if one of the rings is not opened instantaneously.

In Equation 14.173 and in the last chapter, it was stated that for the oxaazacyclo-propane ring, the weakest bond is the N - O bond, instead of the C - O bond (where in N - O bond, two non-free centers are involved), since N - O, O – O, S – S types of bonds are weak. That though N - O bond is weak, C – O bond could become weaker depending on what the C center is carrying. This is illustrated with the following unique example. **Sydnones** were first prepared by Earl et al by the action of cold acetic anhydride on N-nitroso-N-phenylglycines. Earl formulated the reaction as follows[14].

                                                                          14.189

Sydnones are white or pale yellow crystalline compounds which are hydrolyzed by hot five percent sodium hydroxide to the original N-nitroso-N-arylglycine, and by moderately con-centrated hydrochloric acid to an acrylhydrazine, formic acid and carbon dioxide.

The structure (I) above proposed by Earl is said to be similar to that of b - lactone, but Baker et al offered a number of objections to this structure, one of the most important of which, in the fact that a system containing fused three - and four - membered rings would be highly strained, and consequently is unlikely to be produced by dehydration with acetic anhydride; b - lactones are not produced under these conditions[14]. Therefore, they proposed a five-membered ring which cannot be represented by any one purely covalent structure. They put forward a number of charge structures which are resonance stabilized.

Based on the current developments herein, the structure (I) is not favored, for the reasons already offered above. Very little of the Bakers et al structures are indeed correct. Nevertheless before proposing the true structure, the mechanism of the reaction of Equation 14.189 will first be provided.

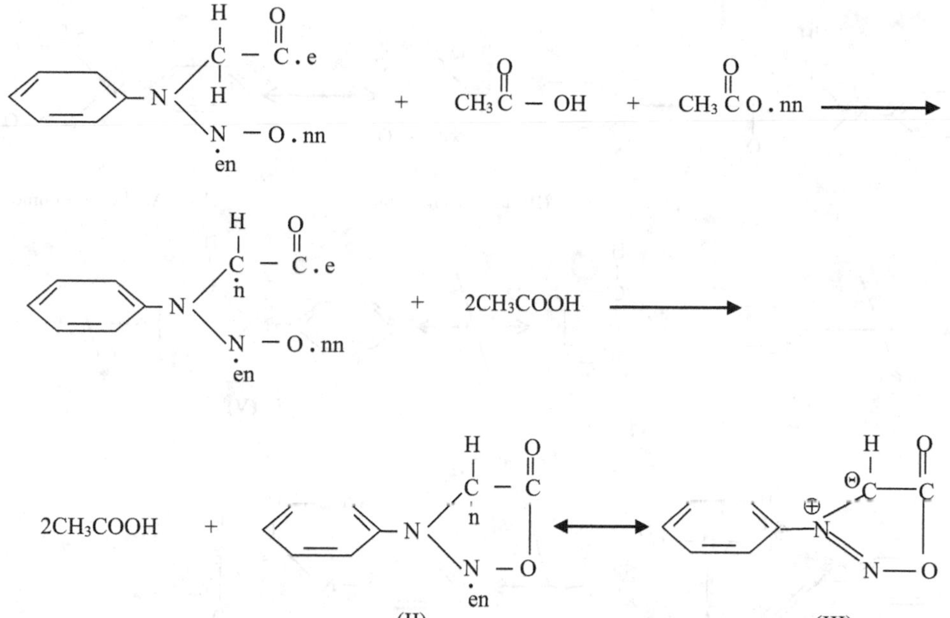

$$(II) \longleftrightarrow (III) \qquad 14.190a$$

Shockingly enough, the reaction above, (I) down to (III), is just a one- or two-stage Equili-brium system. The acetic anhydride was in Equilibrium state of existence to commence the reaction. (I) could also be the one to commence the reaction to first form acetic acid in one stage. It is indeed the one since it is an acid.

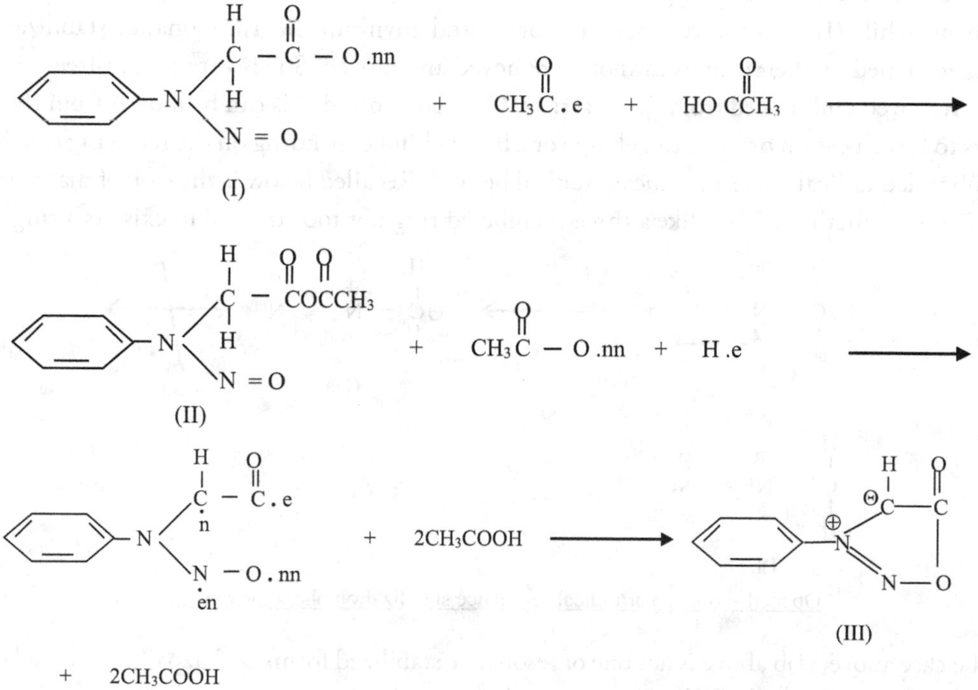

$$(I) \qquad (II) \qquad (III) \qquad 14.190b$$

It is Equation 14.190b that is favored as opposed to Equation 14.190a. This is a two stage Equilibrium mechanism system. (III) is indeed what is obtained. It is a self-activated monomer or compound. Thus, like diazoalkanes, it favors polar/radical resonance stabiliza-tion phenomenon, but of the closed-loop type and discrete in character. It is discrete, because based on the mechanism of resonance, it does not go beyond (III). It is the standing electro-radical that moves to grab a nucleo-free or non-free-radical either from an adjacently located

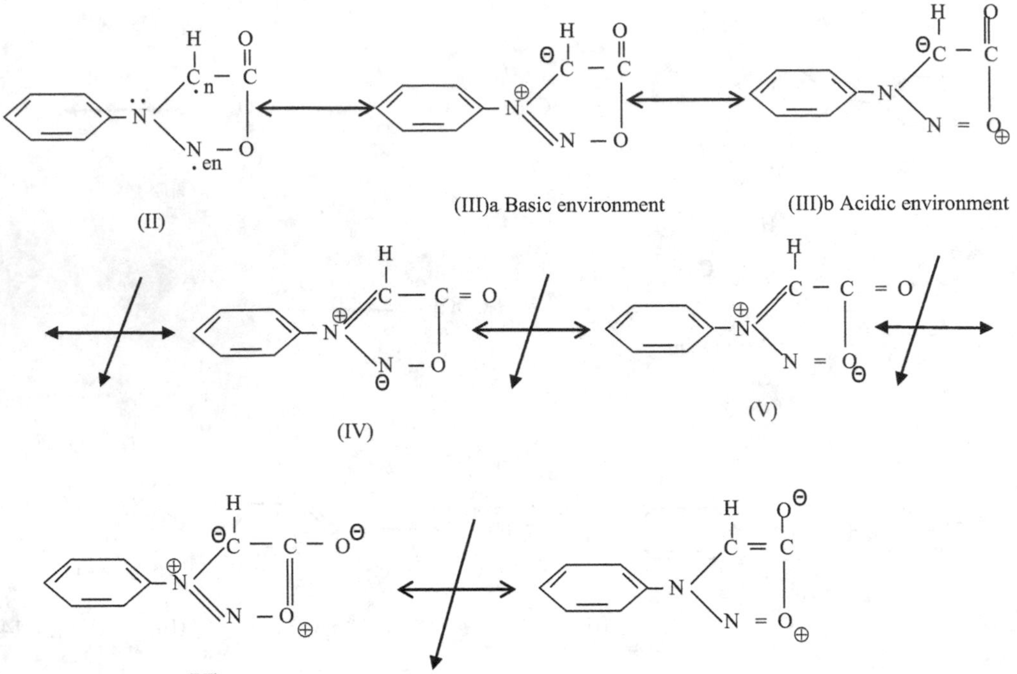

(II) (III)a Basic environment (III)b Acidic environment

(IV) (V)

(VI) (VII) 14.191

π-bond or paired unbonded radicals, for which the existence of (IV), (V), (VI), and (VII) are not possible. (III)b is favored, because the Electrophilic character of the monomer still remains. It is favored in an acidic environment, while (III)a is favored in a basic or neutral environment. In resonance stabilization, note that charges carried by their centers cannot be removed and moved. So also are nucleo-free- or non-free-radical be removed and moved from their carriers. Only electro-radicals can be moved from their centers or carriers to leave a cation or a positive charge or a hole behind depending on the receiving center or from paired unbonded radicals to leave a nucleo-radical behind. Recalled below is the case of diazomethane in Chapter 7. Diazo-methane looks like a three-membered ring not too strained to exist as a ring

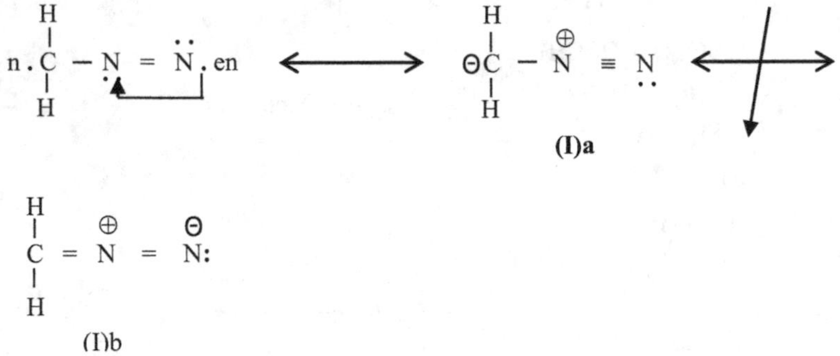

(I)a

(I)b

Opened - loop polar/radical resonance stabilization phenomenon 14.192

Unlike the case above, (I)b above is not one of resonance stabilized forms of diazoalkanes based on a basic environment. (II), (III)a and (III)b of Equation 14.191 are the only forms of sydnone.

When the sydnone is hydrolyzed with hot 5% NaOH, the following are possibilities.

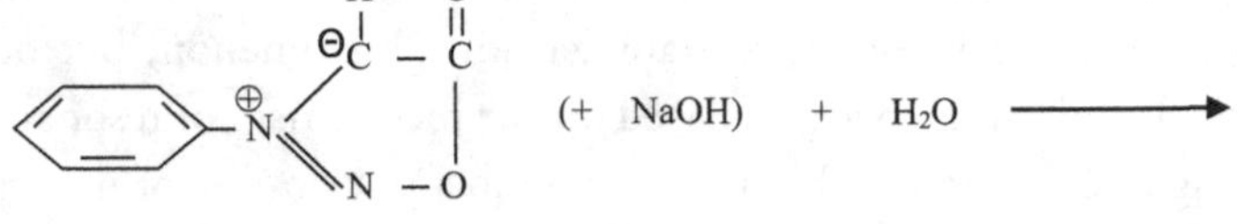

$(+ \text{ NaOH}) + H_2O \longrightarrow$

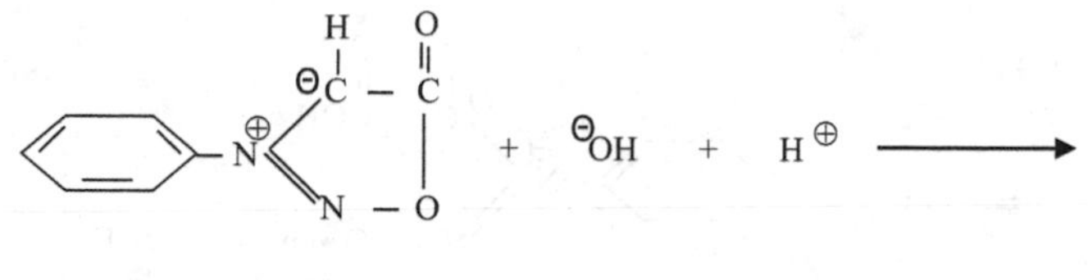

(III)

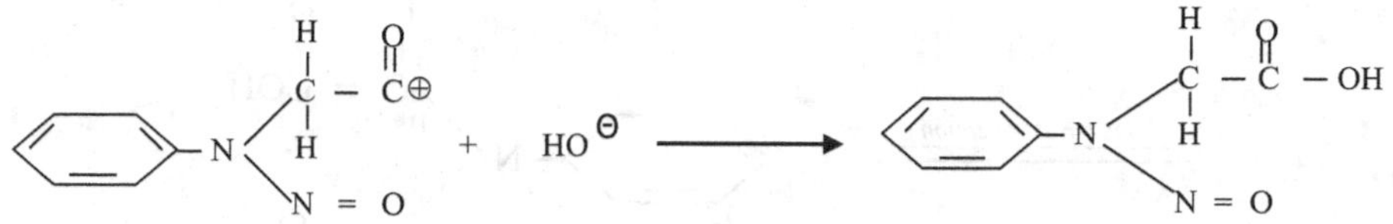

14.193

It was the polar bond that was used above just as can be done with diazoalkanes, noting that while above there are other π-bond and functional activation centers, in diazoalkanes there are none besides the polar center. So the route above cannot be said to be favored yet, though the sydnone is already in its activated state of existence. The ring was open Instantaneously instead of release of a molecule like in diazomethane.

However, worthy of note is that the sydnone is an Electrophile, i.e., a Male since it looks like a lactone with the following centers $O = C$. As such, when attacked nucleo-non-free-radically, the followings are obtained.

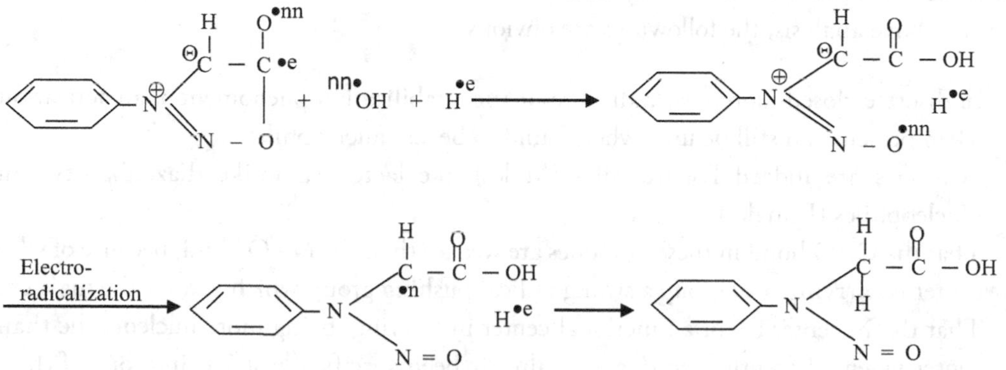

14.194

This could not be done chargedly, because charges cannot move and be moved during Electro radicalization or Resonance stabilization. Electro-free-radically, if the same monomer unit is obtained, then Equation 14.194 is the route. Electro-free-radically, the followings are obtained.

Stage 1: $H_2O \rightleftharpoons H\bullet e + nn\bullet OH$

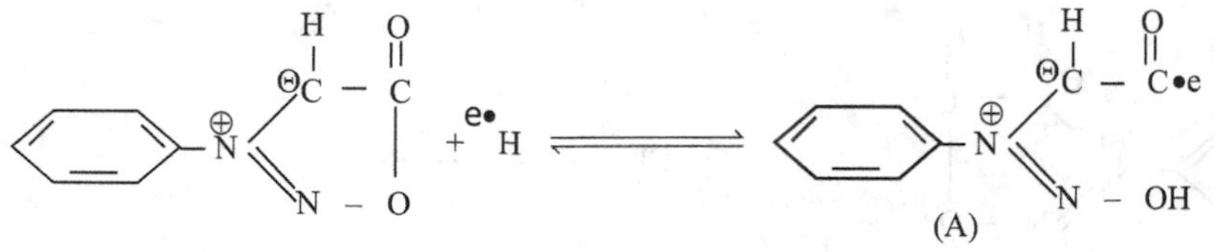

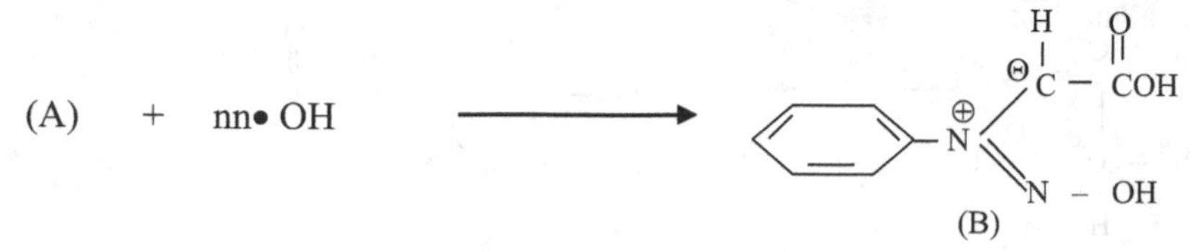

(A)

(B)

14.195

Stage 2:

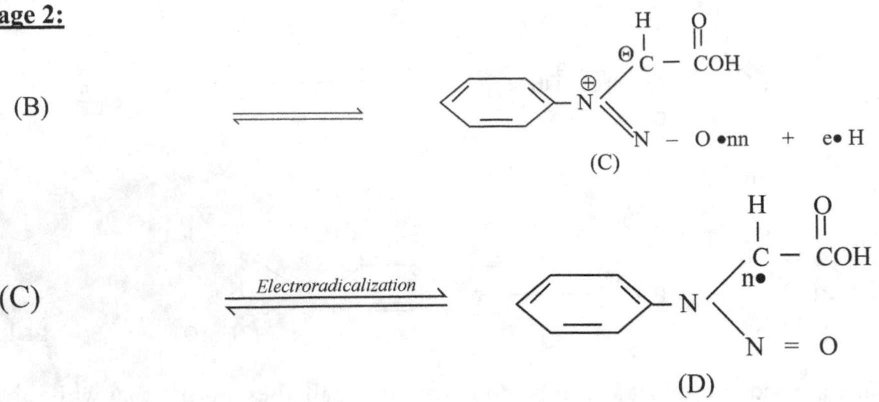

(B) $\Longrightarrow$

(C)

(C) $\xrightarrow{\text{Electroradicalization}}$

(D)

$$H \bullet e \ + \ (D) \ \longrightarrow \ \text{N-nitroso-N-phenylglycine} \qquad\qquad 14.196$$

If NaOH was involved, it would have been both passively and actively, because water may be too weak to exist in equilibrium state of existence. NaOH was not involved because of the very low concentration in which what was used anionically in Equation 14.194 is $HO^{\ominus}....^{\oplus}OH_2Na$. The initiator must have been too weak to exist in Equilibrium state of existence. The question is- can this take place chargedly? It can only take place when the bond is radically paired.

From the above analysis, the followings are obvious -

(i)   In discrete closed-loop polar/radical resonance stabilization phenomenon, an activation center when present can still be used when found to be less nucleophilic.

(ii) Sydnones are indeed Electrophiles (Males) like lactones, unlike diazoalkanes which are Nucleophiles (Females)

(iii) That the C - O bond in these sydnones are weaker than the N - O bond, because of what the C center is carrying- =O group a strong radical pushing group which gives it the male character.

(iv) That the N center is not a functional center in the ring, being more nucleophilic than the O center when adjacently placed, as has already been specifically stated into one of the rules or laws of Nature in Chapter 13.

(v)   That with the use of the eye of the needle, no laws has been contravened, because the laws of Nature have no exception.

When sydnones are attacked by moderately concentrated HCl, the followings are the true mechanisms of the reactions which up to now cannot be explained.

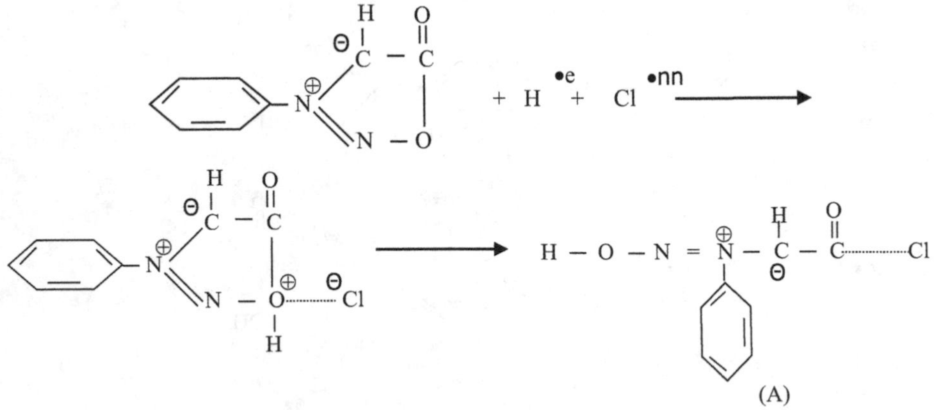

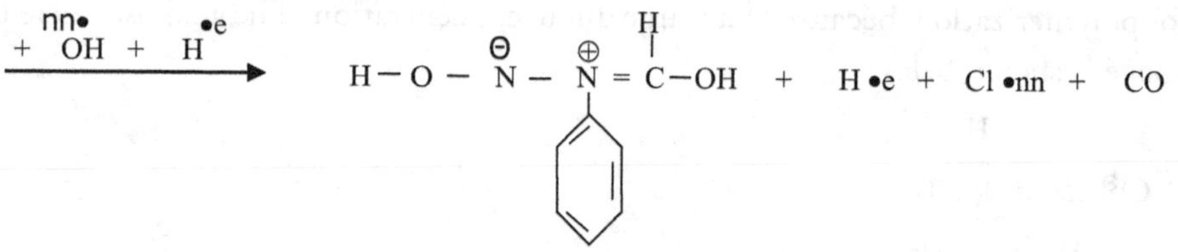

$$nn\bullet \quad \bullet e$$
$$+ \ OH \ + \ H \longrightarrow$$

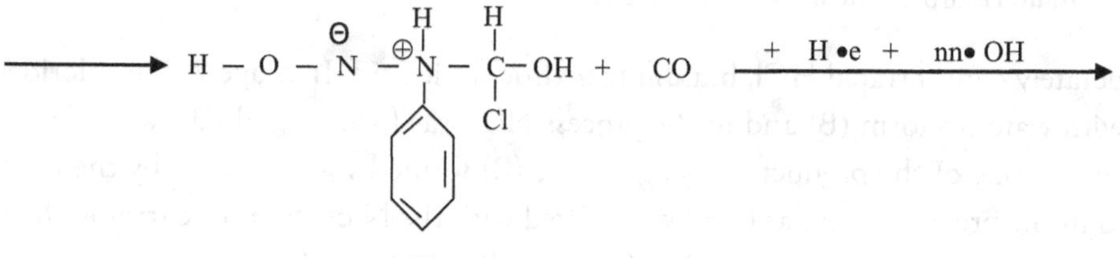

$$H-O-\overset{\ominus}{N}-\underset{\underset{\text{(B)}}{\overset{|}{C_6H_5}}}{\overset{\oplus}{N}}=\overset{\overset{H}{|}}{C}-OH \quad + \quad H\bullet e \ + \ Cl\bullet nn \ + \ CO$$

$$\longrightarrow \ H-O-\overset{\ominus}{N}-\overset{\overset{H}{|}}{\underset{\underset{\text{(C)}}{\overset{|}{C_6H_5}}}{\overset{\oplus}{N}}}-\overset{\overset{H}{|}}{\underset{\underset{Cl}{|}}{C}}-OH \ + \ CO \quad \overset{+ \ H\bullet e \ + \ nn\bullet OH}{\longrightarrow}$$

(D)
$$\overset{\overset{\overset{H}{|}}{\overset{O}{\|}}}{\underset{\underset{\overset{|}{H-C-OH}}{\overset{|}{N}}}{\underset{\overset{|}{OH}}{}}}\text{(phenyl)}-\overset{\ominus}{\underset{\oplus}{N}}\cdots H \ + \ HCl \ + \ CO \longrightarrow$$

(E)
$$\overset{\overset{\overset{OH}{|}}{\overset{\ominus}{N}}}{\text{(phenyl)}}-\overset{\oplus}{\underset{\underset{H}{|}}{N}}\cdots H \ + \ CO \ + \ HCl$$
$$+ \ H-\overset{\overset{O}{\|}}{C}-OH$$

$$\longrightarrow \ HCl \ + \ \overset{\overset{\overset{OH}{|}}{\overset{\ominus}{N}}}{\text{(phenyl)}}-\overset{\oplus}{\underset{\underset{H}{|}}{N}}\bullet nn \ + \ H\bullet e \ + \ CO \ + \ HCOOH \longrightarrow$$

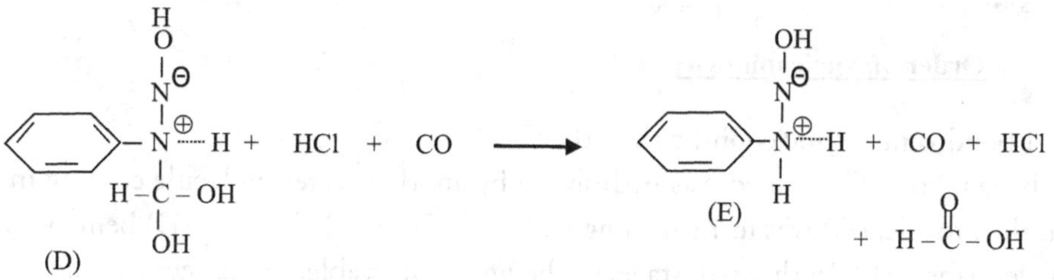

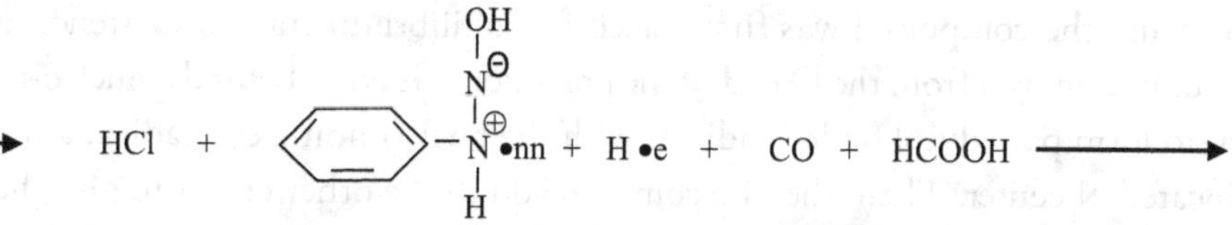

ELECTRO/POLAR RADICALIZATION OF THE SECOND KIND

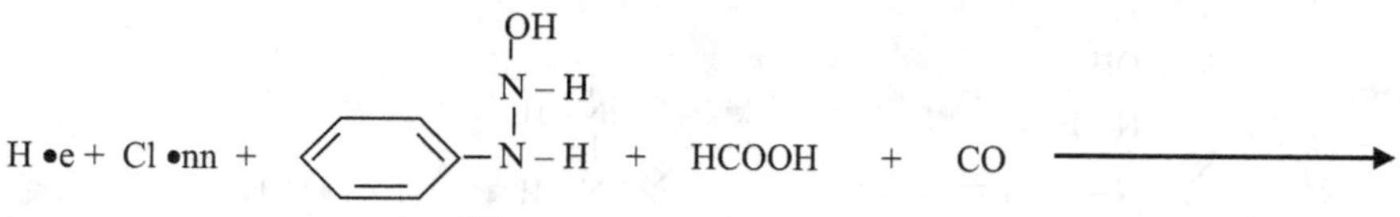

$$H\bullet e + Cl\bullet nn + \ \overset{\overset{\overset{OH}{|}}{\overset{N-H}{|}}}{\text{(phenyl)}}-\overset{\underset{\underset{\text{(F)}}{}}{|}}{N}-H \ + \ HCOOH \ + \ CO \longrightarrow$$

$$HCl \ + \ \underset{\text{arylhydrazine}}{\overset{\overset{H}{|}}{\text{(phenyl)}}-N-NH_2} \ + \ CO_2 \ + \ \underset{\text{Formic acid}}{HCOOH} \qquad 14.197$$

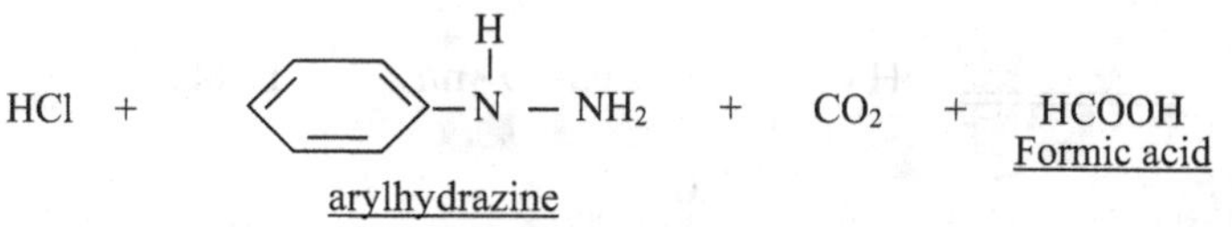

Overall Equation: $2H_2O \ + \ HCl \ + \ C_6H_5N_2OCHCO \longrightarrow C_6H_5N_2H_3 \ + \ CO_2$
$$+ \ HCl \ + \ HCOOH \qquad 14.198$$

Thus, one can observe how the arylhydrazine, formic acid and carbon dioxide were produced. The ring was opened electro-free-radically above by H•e and not by Cl•nn because of the type of acid used. With the acid in Equilibrium state of existence, (A) a chloride was formed in Stage 1 with the opening of the ring. Whether with Cl•nn or H•e, the same monomer unit will be obtained. The acid here could not be used as an initiator for polymerization, because of its large dilute concentration. The acid used here based on the mechanism above is shown below.

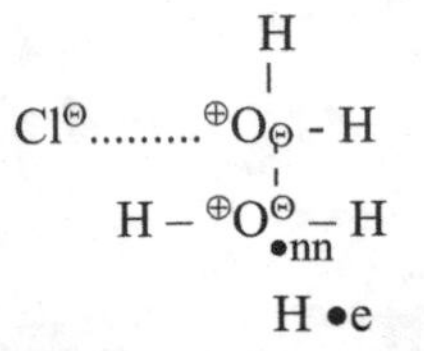

$$
\begin{array}{c}
\text{H} \\
|\\
\text{Cl}^{\ominus}\ldots\ldots\ldots{}^{\oplus}\text{O}_{\ominus} - \text{H} \\
|\\
\text{H} - {}^{\oplus}\text{O}^{\ominus}_{\bullet nn} - \text{H} \\
\\
\text{H} \bullet \text{e}
\end{array}
$$

(Dilute Hydrochloric acid – 2H₂O/HCl)   14.199

This is not a moderately concentrated HCl, but dilute hydrochloric acid. In Stage 2, the chloride formed was hydrolyzed with water to form (B) and in the process HCl was formed and CO was released. With the presence of HCl as one of the products in Stage 2, the (B) formed was activated by the HCl to give (C) a chloride and in the process the polar bond was shifted with the N center now carrying the negative charge. Based on the activation center activated in (B), the following is valid.

$$
\overset{|}{-}\text{N}^{\ominus} - {}^{\oplus}\overset{|}{\text{N}} - \qquad > \qquad \text{C} = \text{N}
$$

Order of Nucleophilicity   14.200

That is, the polar bond is more nucleophilic than the C = N bond.

In the fourth stage, the (C) formed was hydrolyzed by another water molecule coming in as one of the reactants. In the process, (D) was formed along with HCl. In the fifth stage, (D) being very unstable loses formaldehyde to form (E). In the sixth stage, (E) being also unstable, rearranges via a new tautomeric method herein called *Electro/polar radicalization phenomenon of the Second kind of the first type.* It is so called, because the compound was first placed in Equilibrium state of existence, after which an electro-non-free-radical moved from the Q end of the polar bond, leaving behind a nucleo-non-free-radical on the N center, to form paired unbonded radicals with the nucleo-non-free –radical already waiting on the adjacently located N center. Then, the H·e comes to add to the other center to give the (F) above. In the seventh stage, the HCl still present as one of the products comes to attack the (F) formed as follows.

Stage 7:   HCl $\rightleftharpoons$ H •e   +   nn• Cl

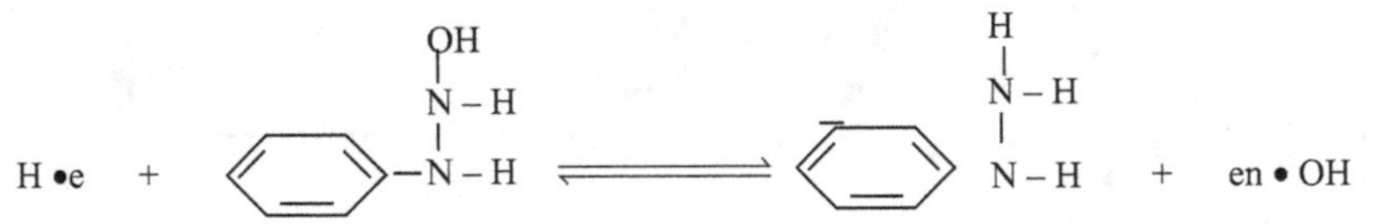

(H) $\xrightarrow{\text{Deactivation}}$ CO₂   +   Heat   14.201

Overall Equation: HCl + (F) + CO $\longrightarrow$ CO₂ + HCl + (G)   14.202

There are six steps in this single stage. Without the CO, the stage would not have been a stage. It would have been an Equilibrium Solubilization stage in which (F) is said to be soluble in HCl, i.e., a stable, reactive and unproductive stage. The need to show this last stage was important so that one can see how oxidizing oxygen was produced as an intermediate product not to be moved into another stage, but to be used in-situ in the stage where it appeared and disappeared, making it look as if it was not there. One can see the great wonders of NATURE too much to comprehend. One can see why there is need to look at every reaction so that one can see how NATURE operates, for without it, one is in complete Darkness. For one thing, we all think that we see the Light without realizing that we have only just begun. We have since antiquity been collecting abundance of data in all disciplines without having full explanation for any one of them and the main reason is because we do not know how NATURE operates, that which required a new definition of an ATOM.

14.2.4.2 Rhombic Sulphur

As has been shown, this is virtually the largest size of sulfur - sulfur containing compounds that has been known to exist. The strain energy in the ring is great and close to the MRSE. This is largely provided by sixteen paired unbonded radicals inside and outside the ring. ***They are not there for nothing, because "nothing is there for nothing".*** For thiophene, there are two paired bonded (i.e. 2p-bonds) and two paired unbonded radicals, while for benzene, there are three paired bonded radicals (3p-bonds). Octatetraene with four paired bonded radicals (4p-bonds) has only transient existence and it can readily molecularly rearrange to styrene. While paired bonded ones (p-bonds) are not free when not activated with plenty energy content, the paired unbonded ones are free with also some energy content but not as much as those in p-bonds. They both are very important, for they play very important roles in humanity. Thus, rhombic sulfur has a large number of invisible p-bonds in the ring, almost close to that in benzene ring, the difference being that while benzene has no point of scission, rhombic sulfur has eight points of scission. How can three-membered ring with six paired unbonded radicals exist? It is like having more than one but less than two p-bonds in cyclo-propane (Real name is cyclo-propene) ring. ***In fact, it is only for "this family" (Only S containing rings) of rings that the Strain energy for the family of rings is not the same.*** That is, while the stain energy for all cyclo-alkane rings is the same at about 25 kcals/mol, the strain energy in cyclo-sulfur rings are not the same, ***because as we move from three-membered ring to four membered ring we have moved from one family to another family since the number of paired unbonded radicals has increased.*** The strain energy in cyclo-alkane rings cannot be the same as the strain energy in cyclo-alkene rings. This applies in going from three-membered ring to four-membered ring (See Equation 14.182).

They readily can disintegrate to give at the end another element of sulfur as has already been shown. The possibility of forming smaller ring may not be possible because they will be so strained. During polymerization of a three-membered ring via a means where depropa-gation takes place, the growing chain does not depropagate to form the same three-membered ring, but a larger more difficult to unzip ring at the operating conditions for opening the three-membered ring. Though the sulfur centers can carry ionic positive charges in the absence of C which is equi-electropositive with her, rhombic sulfur cannot be opened instantaneously chargedly due to electrostatic forces of repulsion. ***On the other hand, only one of the sulfur functional centers in the ring can be used, and this cannot be done nucleo-non-free-radically***

and nucleo-free-radically. Hence, the ring can only be opened instantaneously radically as shown below. It may however be done electro-non-free-radically.

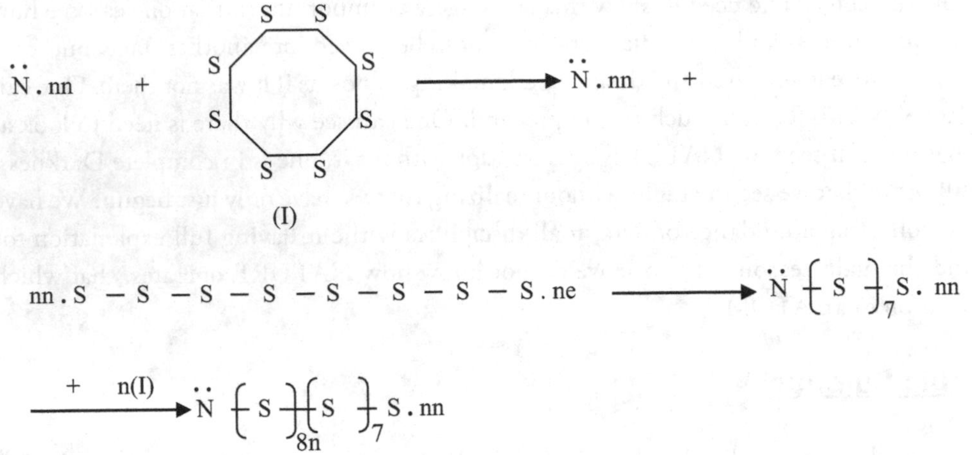

$$\ddot{N}.nn \quad + \quad \text{(I)} \quad \longrightarrow \quad \ddot{N}.nn \quad +$$

$$nn.S - S - S - S - S - S - S - S.ne \quad \longrightarrow \quad \ddot{N} \left\{ S \right\}_7 S.nn$$

$$\xrightarrow{\; + \quad n(I)\;} \quad \ddot{N} \left\{ S \right\}_{8n} \left\{ S \right\}_7 S.nn \qquad\qquad 14.203$$

$$\ddot{E}.en \quad + \quad \text{(I)} \quad \longrightarrow \quad \ddot{E}.en \quad + \quad nn.S \left\{ S \right\}_6 S.en$$

$$\longrightarrow \quad \ddot{E} \left\{ S \right\}_7 S.en \quad \xrightarrow{\; n(I)\;} \quad E \left\{ S \right\}_{8n} \left\{ S \right\}_7 S.en \qquad\qquad 14.204$$

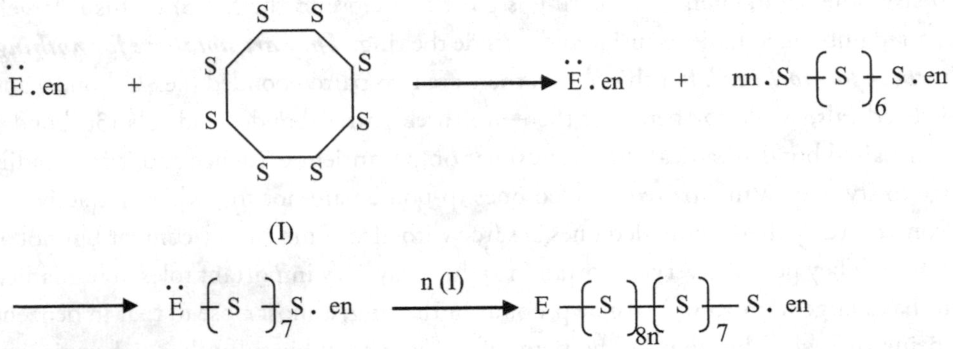

Thus, the monomer like oxaazacyclopropane, cyclic disulfides, is a full non-free-radical monomer and all are both nucleophilic and electrophilic in character. Despite that, the followings are valid for them.

$$\text{Oxaazacyclopropane} \quad > \quad \text{Rhombic sulfur} \quad > \quad \text{Cyclic disulfide}$$

<u>Order of Nucleophilicty</u>   14.205

Only nucleo-non-free and electro-non-free-radicals can be used to polymerize the monomer- rhombic sulfur. Nucleo-free- radically they cannot be polymerized. Electro-free-radically, they can be polymerized via Equilibrium mechanism, that is, Backward addition. Via Equilibrium mechanism, they can only add in the first stage after which no addition takes place again by Combination mechanism, because the equation will not be radically balanced.

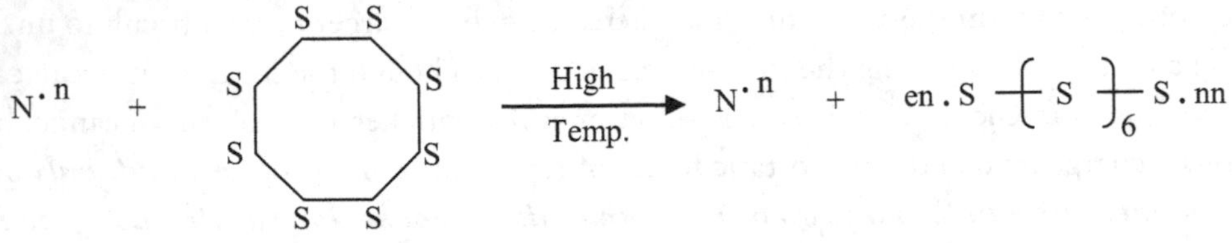

$$N^{\cdot n} \quad + \quad \text{(I)} \quad \xrightarrow[\text{Temp.}]{\text{High}} \quad N^{\cdot n} \quad + \quad en.S \left\{ S \right\}_6 S.nn$$

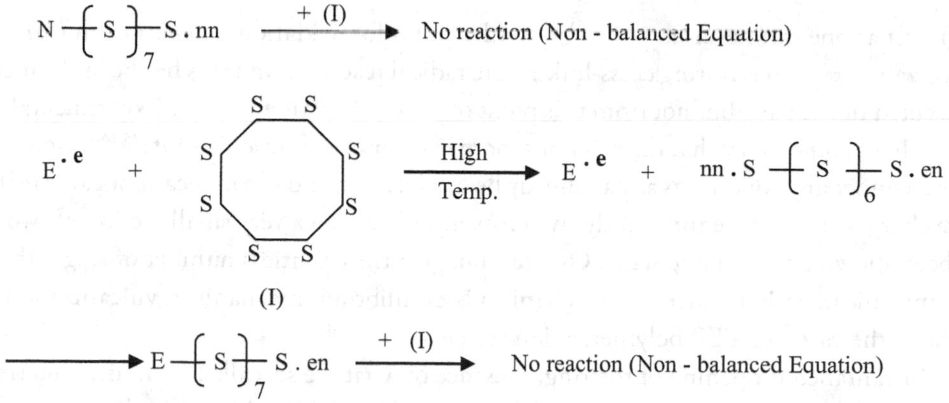

$$N\!-\!\left(\!S\!\right)_7\!\!-\!S . nn \xrightarrow{\quad +\ (I)\quad} \text{No reaction (Non - balanced Equation)}$$

14.206

$$E\cdot^e + \text{(S}_8\text{ ring)} \xrightarrow[\text{Temp.}]{\text{High}} E\cdot^e + nn . S\!-\!\left(\!S\!\right)_6\!\!-\!S . en$$

$$\xrightarrow{\qquad} E\!-\!\left(\!S\!\right)_7\!\!-\!S . en \xrightarrow{\quad +\ (I)\quad} \text{No reaction (Non - balanced Equation)}$$

14.207

Cross-linking between two unsaturated polymer chains such as the linear chains from 1,3-butadiene using rhombic sulfur can be explained as follows.

Stage 1:

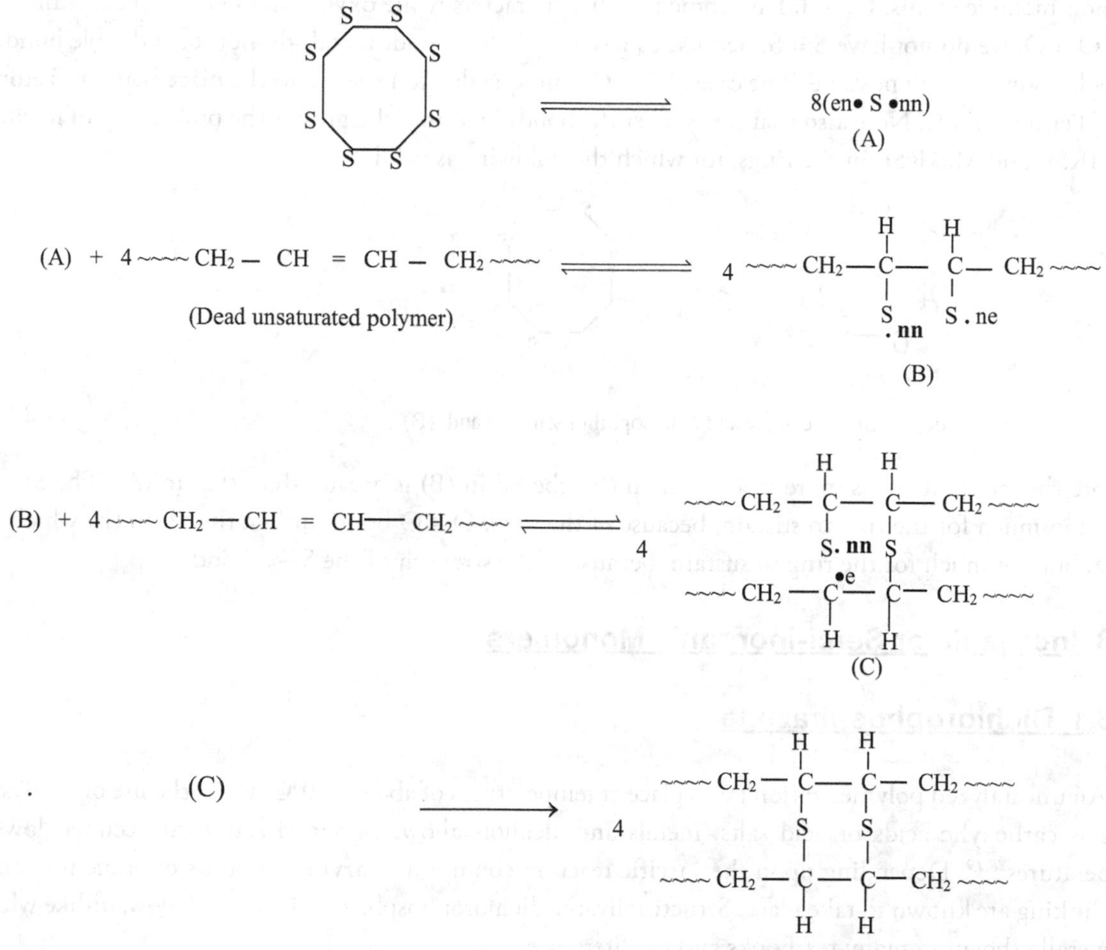

$$\text{(S}_8\text{ ring)} \rightleftharpoons 8(en\bullet S \bullet nn)$$

(A)

$$\text{(A)} + 4 \sim\!\!\sim CH_2 - CH = CH - CH_2 \sim\!\!\sim \rightleftharpoons 4 \sim\!\!\sim CH_2\!-\!\underset{\underset{.\,nn}{S}}{\overset{H}{\underset{|}{C}}}\!-\!\underset{\underset{S\,.\,ne}{|}}{\overset{H}{\underset{|}{C}}}\!-\!CH_2 \sim\!\!\sim$$

(Dead unsaturated polymer)

(B)

$$\text{(B)} + 4 \sim\!\!\sim CH_2 - CH = CH - CH_2 \sim\!\!\sim \rightleftharpoons 4 \quad \begin{array}{c} \sim\!\!\sim CH_2 - \overset{H}{\underset{|}{C}} - \overset{H}{\underset{|}{C}} - CH_2 \sim\!\!\sim \\ \underset{\bullet e}{S.\,nn} \quad S \\ \sim\!\!\sim CH_2 - \overset{}{\underset{|}{C}} - \overset{}{\underset{|}{C}} - CH_2 \sim\!\!\sim \\ H \quad H \end{array}$$

(C)

$$\text{(C)} \xrightarrow{\qquad\qquad} 4 \quad \begin{array}{c} \sim\!\!\sim CH_2 - \overset{H}{\underset{|}{C}} - \overset{H}{\underset{|}{C}} - CH_2 \sim\!\!\sim \\ S \quad S \\ \sim\!\!\sim CH_2 - \overset{}{\underset{|}{C}} - \overset{}{\underset{|}{C}} - CH_2 \sim\!\!\sim \\ H \quad H \end{array}$$

Cross-linked network

14.208

Notice that cross-linking can never take place via Combination mechanisms. It can only take place via Equilibrium mechanism as shown above. To cross-link using rhombic sulfur without opening the ring instantaneously, is impossible. One cannot use $nn\bullet S \bullet nn$ (the atom) for cross-linking, but $en\bullet S \bullet nn$

(an element), just as one can never use nn• O •nn (the atom) for oxidation, but en• O •nn (an element). Sulfur vulcanization reactions during cross-linking are radical reactions and this has been the mechanism generally accepted in the past, but not from this point of view as explained above. Experimental evidence in recent times have come to say that the reactions more likely proceed by ionic route[15,16,17]. Ionic presence is never favored under any conditions as has already been explained and a center cannot carry real positive and negative charges at the same time. Only two cross-links between a very small section of two polymer chains has been shown above in one stage. One can imagine the countless number of stages that will be involved during vulcanization. Since, the mechanism is Equilibrium mechanism, vulcanization is a long process just as is the case in STEP polymerization systems.

Without instantaneous opening of the ring, presence of what are so-called "diradical intermediates" whose real name is di-non-free-radicals, would be impossible. While rhombic sulfur exists, rhombic oxygen or any ring with O - O bond along the ring does not exist, in view of the presence of MaxRSE in their rings provided by the paired unbonded radicals, the weaker O - O bond with respect to S - S bond and the fact that S unlike O is among the atoms that form the boundary between metallic atoms and non-metallic atoms. C, S, I, have some metallic characters while oxygen does not. Hence, while we have O = O, we do not have S = S, because as has already been said, metals do not form double bonds, but polar bonds where possible. The case of C = C is unique due to its size and the placement of C atom in the Periodic Table. Note also that the weaker the bonds in a ring, the greater the probability of having the MRSE and MaxRSE in the rings, for which the following is valid.

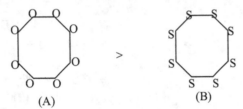

$$14.209$$

Order of Strain energy and Nucleophilicity in (A) and (B)

Despite the fact that (A) is more strained than (B), the SE in (B) is greater than that in (A). The SE in (A) is too much for the ring to sustain, because of the weak O – O bond, unlike the SE in (B) which is larger, but not much for the ring to sustain, because of the strength of the S – S bond.

14.3 Inorganic or Semi-inorganic Monomers

14.3.1 Dichlorophosphazene

An uncatalyzed polymerization takes place at temperatures of about 250°C, while the use of catalysts such as carboxylic acids or acid salts, metals and alcohols allows polymeri-zation to occur at lower temperatures[18,19]. Depending upon the specific reaction conditions, varying amounts of branching and crosslinking are known to take place. Structurally, the dichlorophosphazene is shown below, unlike what is generally shown in many textbooks and the literature.

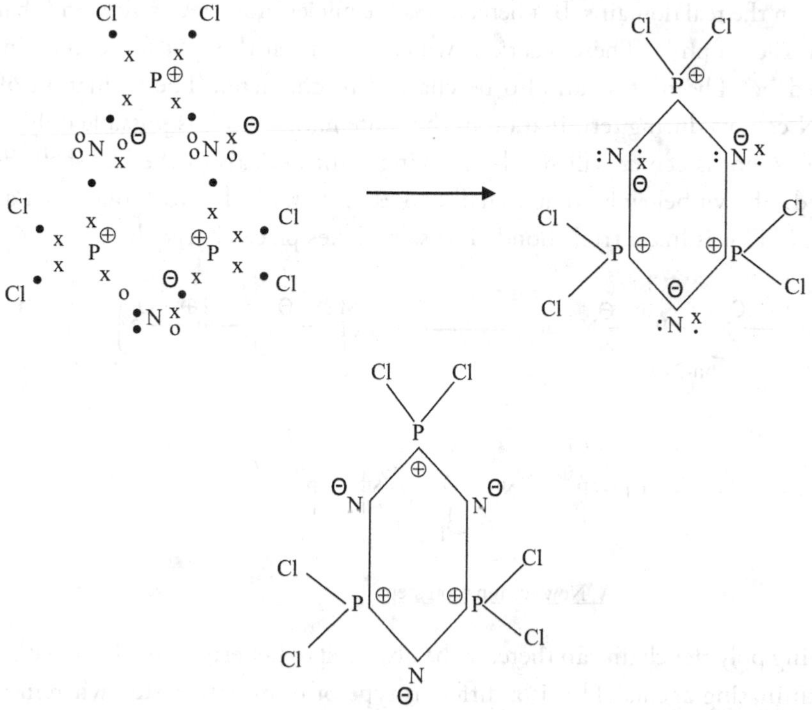

Real structure of Dichlorophosphazene

14.210

There are three polar bonds and six covalent bonds in the ring. Since according to one of the boundary laws wherein P like N cannot carry more than eight radicals in the last shell, the P center cannot carry a double bond. It is not resonance stabilized in the absence of a visible electro-radical. All the last shells of the atoms have a maximum of eight radicals. In view of the polar character of the bonds, the ring can be opened instantaneously, when possible radically and chargedly via the single bond without polar character. The opening is enhanced by the fact that, with three polar bonds and three paired unbonded radicals, the ring is well strained but far less than that provided by the three π-bonds in Benzene. While benzene has no point of scission, this has three points of scission. Hence, at high temperatures of about 250°C, the followings are obtained.

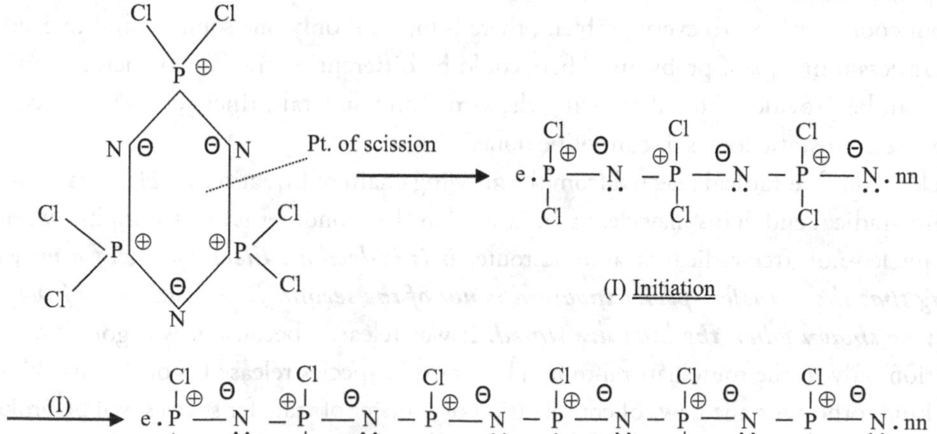

(I) Initiation

14.211

The growing polymer chain can grow only from one side nucleo-non-free-radically in the absence of foreign agents. Usually, it is the electro-radical or positive charge carrier that diffuses all the time to commence all reactions in the real domains. But here, it was the nucleo-non-free-radical end that diffused, the monomer being an Electrophile. These reactions which have been thought to be ionic in character all the time can indeed be. The routes can also be charged in character. The formation of a double bond between P and N centers during termination in the route natural to it is possible only radically or chargedly, since the phosphorus center will still be carrying eight radicals in the last shell. But a triple bond cannot be formed. Shown below is what actually takes place radically, with one polar and double bonds between P and N. This is like a triple bond. The same takes place chargedly.

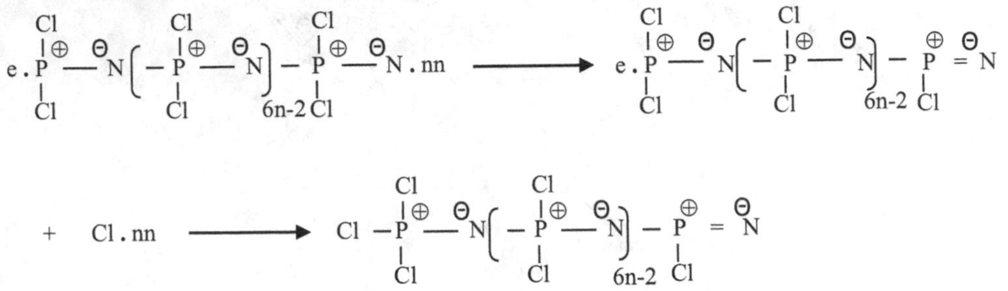

$$e \cdot \overset{\overset{\text{Cl}}{|}}{\underset{\underset{\text{Cl}}{|}}{P}} \overset{\oplus}{\underset{}{}} \overset{\ominus}{N} \left(- \overset{\overset{\text{Cl}}{|}}{\underset{\underset{\text{Cl}}{|}}{P}} \overset{\oplus}{\underset{}{}} \overset{\ominus}{N} \right) - \overset{\overset{\text{Cl}}{|}}{\underset{\underset{6n-2 \text{ Cl}}{|}}{P}} \overset{\oplus}{\underset{}{}} \overset{\ominus}{N} \cdot nn \longrightarrow e \cdot \overset{\overset{\text{Cl}}{|}}{\underset{\underset{\text{Cl}}{|}}{P}} \overset{\oplus}{\underset{}{}} \overset{\ominus}{N} \left(- \overset{\overset{\text{Cl}}{|}}{\underset{\underset{\text{Cl}}{|}}{P}} \overset{\oplus}{\underset{}{}} \overset{\ominus}{N} \right) - \overset{\oplus}{\underset{\underset{6n-2 \text{ Cl}}{|}}{P}} = N$$

$$+ \quad Cl \cdot nn \longrightarrow Cl - \overset{\overset{\text{Cl}}{|}}{\underset{\underset{\text{Cl}}{|}}{P}} \overset{\oplus}{\underset{}{}} \overset{\ominus}{N} \left(- \overset{\overset{\text{Cl}}{|}}{\underset{\underset{\text{Cl}}{|}}{P}} \overset{\oplus}{\underset{}{}} \overset{\ominus}{N} \right) - \overset{\oplus}{\underset{\underset{6n-2 \text{ Cl}}{|}}{P}} = \overset{\ominus}{N}$$

<u>A New termination step</u>   14.212

This new type of growing polymer chain can therefore be observed to be terminated from within, without the use of external terminating agents. This is a different type of termination step wherein no transfer species was released or added to kill the chain unlike all the cases we have seen so far. It may likely take place with some other monomers we have encountered so far without knowing or applying it. For example, it may be possible to polymerize a monomer such as propene without the use of a real initiator as shown below.

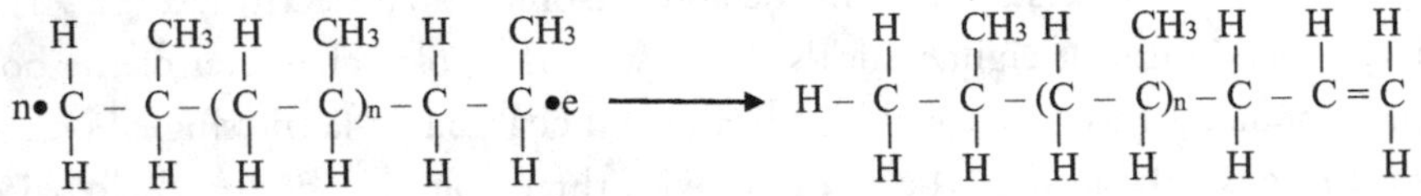

$$n \bullet \overset{\overset{\text{H}}{|}}{\underset{\underset{\text{H}}{|}}{C}} - \overset{\overset{\text{CH}_3}{|}}{\underset{\underset{\text{H}}{|}}{C}} - (\overset{\overset{\text{H}}{|}}{\underset{\underset{\text{H}}{|}}{C}} - \overset{\overset{\text{CH}_3}{|}}{\underset{\underset{\text{H}}{|}}{C}})_n - \overset{\overset{\text{H}}{|}}{\underset{\underset{\text{H}}{|}}{C}} - \overset{\overset{\text{CH}_3}{|}}{\underset{\underset{\text{H}}{|}}{C}} \bullet e \longrightarrow H - \overset{\overset{\text{H}}{|}}{\underset{\underset{\text{H}}{|}}{C}} - \overset{\overset{\text{CH}_3}{|}}{\underset{\underset{\text{H}}{|}}{C}} - (\overset{\overset{\text{H}}{|}}{\underset{\underset{\text{H}}{|}}{C}} - \overset{\overset{\text{CH}_3}{|}}{\underset{\underset{\text{H}}{|}}{C}})_n - \overset{\overset{\text{H}}{|}}{\underset{\underset{\text{H}}{|}}{C}} - \overset{\overset{\text{H}}{|}}{\underset{\underset{\text{H}}{|}}{C}} = \overset{\overset{\text{H}}{|}}{\underset{\underset{\text{H}}{|}}{C}}$$

<u>Termination by Self</u>   14.213

The termination above has been called ***Termination by Self***, since nothing was lost from the growing chain at the point of termination. The only problem above, is how do we get the regular placement of CH_3 group without coordination? To every problem, there is one and only one solution and to every solution there are countless numbers of problems. There could be different methods by which the one and only one solution can be provided, but all applying the same fundamental principles. Without coordination or influence of electrostatic forces, it cannot be done.

The nucleo-non-free-radical rejected from the growing chain of Equation 14.212 is used to terminate the electro-free-radical end. This may clearly indicate that the monomer is an electrophile whose natural route is the nucleo-non-free-radical or anionic route. ***It is indeed an Electrophile of a very different kind, noting that the transfer species involved is not of the second first kind, but of the first kind as will fully be shown when the laws are stated.*** It was released because it was going to be used for self-termination only in the route natural to it. That transfer species released is of the first kind and not second first kind, otherwise the Laws of conservation of transfer of transfer species will be broken. In the route not natural to the monomer, but natural to a center (i.e., the electro-free-radical or positively charged route), there seems to be no initiation. With the type of growing chain involved, it is not surprising to note why branch formations and cross-linking reactions take place. How these are formed will be fully explained herein and in subsequent volumes.

In view of the polar character of the monomer, ionically, there are six functional centers in the ring. These are as follows -

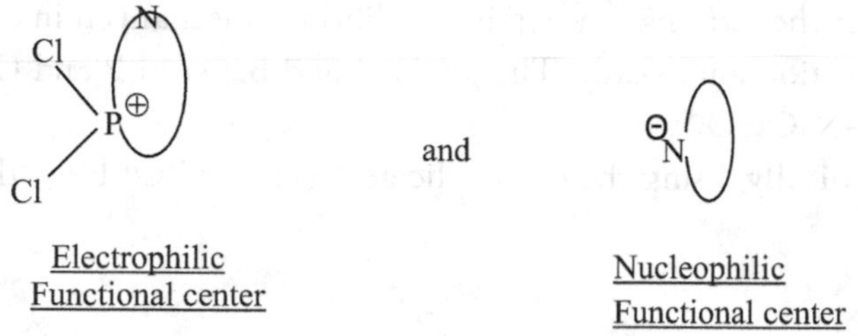

Electrophilic
Functional center

and

Nucleophilic
Functional center

14.214

The electrophilic center is polar in character and imaginary. The same almost applies to the nucleophilic center which is but real. From all observations so far, when polar bonds are attacked, either a stable molecule is released or a ring must be opened where possible. Chargedly, when the ring is opened, one end of the N or P will be carrying two charges at the same time whether the ring is opened instantaneously or not. This is only possible when one of the charges is polar or electrostatic in character. However, it is believed that, functional centers are usable one at a time, just like in rhombic sulfur where all the centers are functional centers. This ring like the case of rhombic sulfur can be opened instantaneously, radically. Chargedly rhombic sulfur cannot be opened. So far, it looks as if only Electro-free-radical, positively charged routes, and anionic routes will be favored by this monomer which has no four-membered ring because of presence of MaxRSE. The ring cannot exist, just like the case with the lower members of rhombic sulfur.

When carboxylic acid or acid salts, such sodium benzoate are involved, the route is said to be anionic, for which the mechanisms proposed looks questionable.

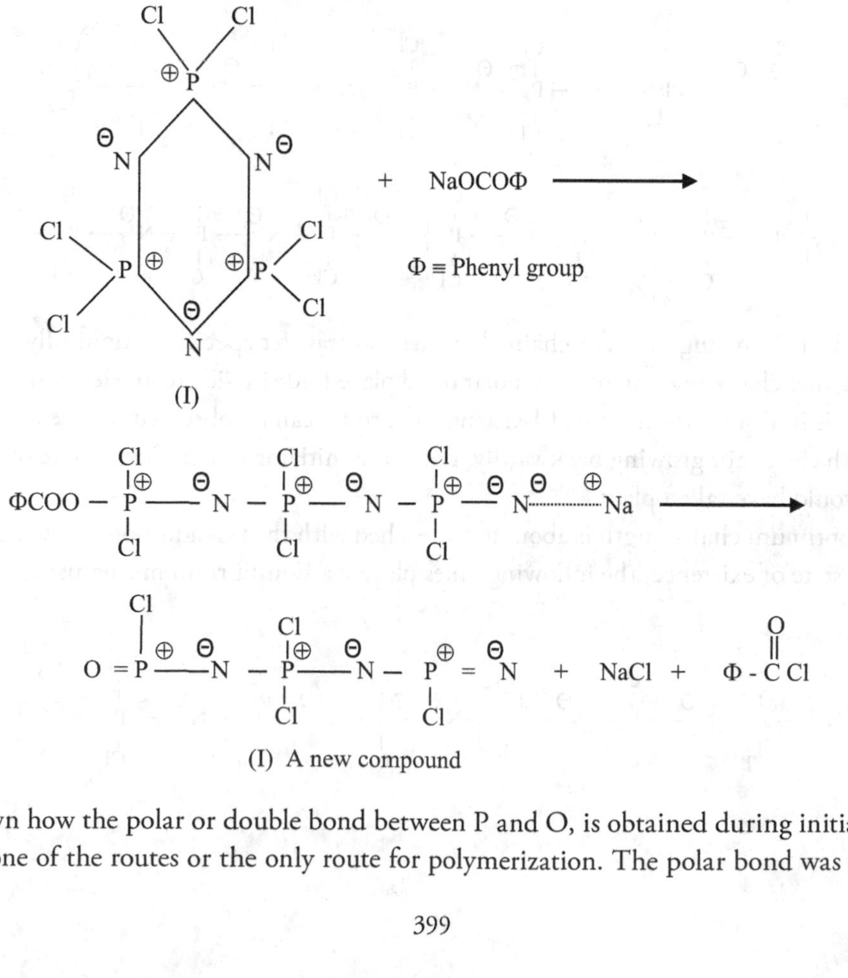

$\Phi \equiv$ Phenyl group

(I)

(I) A new compound

14.215

One has shown how the polar or double bond between P and O, is obtained during initiation. This can therefore be one of the routes or the only route for polymerization. The polar bond was obtained when

$Na^{\oplus}$ (from NaCl coming from the first stage), abstracts Cl externally located to leave a positive charge on the P center. The presence of the positive charge drove away $^{\oplus}CO\Phi$, followed by deactivation to form the double bond. The mechanism for the termination step is Equilibrium mechanism in one stage after Combination mechanism for propagation anionically. The double bond between P and O was formed because of the type of initiator used-NaOCOΦ.

"Cationically" or electro-free-radically, using the carboxylic acid (HOOCR º HB), the followings are to be expected.

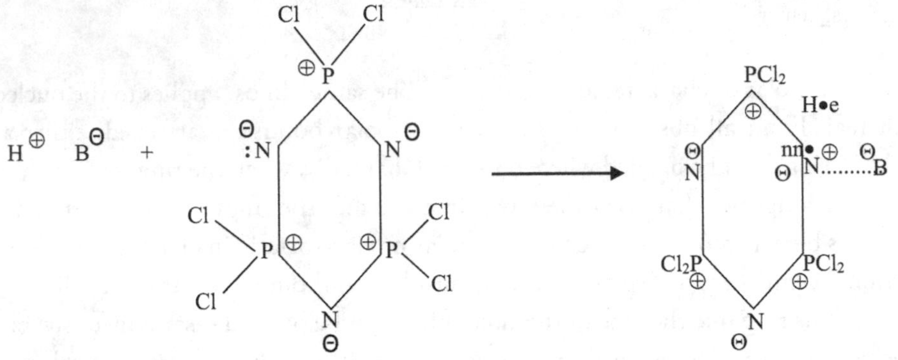

(I)   (II) Initiator

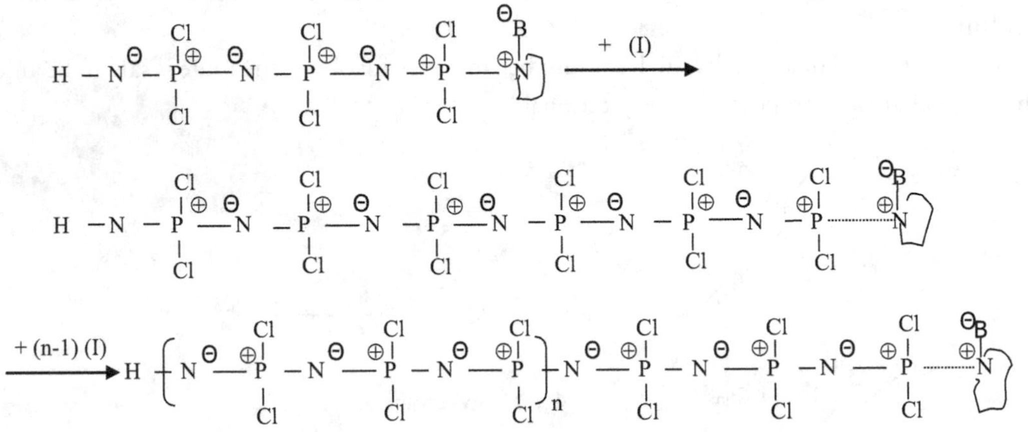

14.216

As has been said, the growing polymer chain above has no transfer species "cationically". At the end of the growing polymer chain, one can observe polar bond placed side by side to an electrostatic bond. Since the initiator (II) is in Equilibrium state of Existence, the route can be observed to be electro-free-radical in character with the chain growing backwardly. For if the initiator was in Stable state of existence, the anionic route would have taken place.

When the optimum chain length is about to be reached with the last addition and the growing chain in Equilibrium state of existence, the following takes place via Equilibrium mechanism.

Stage 1:

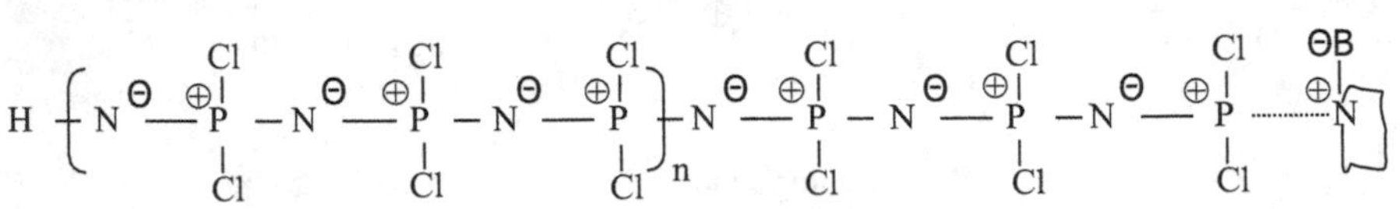

$$\Longrightarrow \; H\bullet e \; + \; nn\bullet N \overset{\ominus}{-} \overset{\oplus}{\underset{|}{P}} - N \overset{\ominus}{-} \overset{\oplus}{\underset{|}{P}} - N \overset{\ominus}{-} \overset{\oplus}{\underset{|}{P}} \left[N \overset{\ominus}{-} \overset{\oplus}{\underset{|}{P}} - N \overset{\ominus}{-} \overset{\oplus}{\underset{|}{P}} - N \overset{\ominus}{-} \overset{\oplus}{\underset{|}{P}} \right]_n \cdots \overset{\oplus}{N}$$

(A)

$$(A) \Longrightarrow Cl\bullet nn \; + \; nn\bullet N \overset{\ominus}{\underset{e\bullet}{-}} \overset{\oplus}{\underset{|}{P}} - N \overset{\ominus}{-} \overset{\oplus}{\underset{|}{P}} - N \overset{\ominus}{-} \overset{\oplus}{\underset{|}{P}} \left[N \overset{\ominus}{-} \overset{\oplus}{\underset{|}{P}} - N \overset{\ominus}{-} \overset{\oplus}{\underset{|}{P}} - N \overset{\ominus}{-} \overset{\oplus}{\underset{|}{P}} \right]_n \cdots \overset{\oplus}{N}$$

(B)

$$H\bullet e \; + \; Cl\bullet nn \; \Longrightarrow \; HCl$$

$$(D) \longrightarrow \overset{\ominus}{N} - \overset{\oplus}{P} \quad \overset{\ominus}{N} \; \overset{\oplus}{\underset{|}{P}} - N \overset{\ominus}{-} \overset{\oplus}{\underset{|}{P}} \left[N \overset{\ominus}{-} \overset{\oplus}{\underset{|}{P}} - N \overset{\ominus}{-} \overset{\oplus}{\underset{|}{P}} - N \overset{\ominus}{-} \overset{\oplus}{\underset{|}{P}} \right]_n \cdots \overset{\oplus}{N}$$

(C) 14.217

Stage 2: $HCl \; \longrightarrow \; H\bullet e \; + \; Cl\bullet nn$

$$H\bullet e \; + \; (C) \; \Longrightarrow \; HB \; + \; \text{Monomer} \; (N) \; +$$

$$\overset{\ominus}{N} = \overset{\oplus}{\underset{|}{P}} - N \overset{\ominus}{-} \overset{\oplus}{\underset{|}{P}} \left[N \overset{\ominus}{-} \overset{\oplus}{\underset{|}{P}} - N \overset{\ominus}{-} \overset{\oplus}{\underset{|}{P}} - N \overset{\ominus}{-} \overset{\oplus}{\underset{|}{P}} \right]_n N \overset{\ominus}{-} \overset{\oplus}{\underset{|}{P}} \bullet e$$

(D)

$$(D) \; + \; Cl\bullet nn \; \longrightarrow \; \overset{\ominus}{N} = \overset{\oplus}{\underset{|}{P}} - N \overset{\ominus}{-} \overset{\oplus}{\underset{|}{P}} \left[N \overset{\ominus}{-} \overset{\oplus}{\underset{|}{P}} - N \overset{\ominus}{-} \overset{\oplus}{\underset{|}{P}} - N \overset{\ominus}{-} \overset{\oplus}{\underset{|}{P}} \right]_n N \overset{\ominus}{-} \overset{\oplus}{\underset{|}{P}} - Cl$$

14.218

In two stages, both ends of the polymer chain were killed from within- Termination by Self. This is for the chain growing backwardly. If it was a chain growing forwardly electro-free-radically such as using Na metal, this would not have been possible. Again, observe that the termination step is an Equilibrium mechanism system.

When carboxylic salts, such sodium benzoate are involved, the route was said to be anionic in character with no carrier from the initiator for the chain. However, unlike the case above in Equation 14.215, the chain is growing forwardly, since the initiator is an **Ionically charged-paired initiator with two active centers.** It is the anionic center that is used, since the monomer is an Electrophile. Uniquely enough, when positively charged or electro-free-radical initiator when present alone in the system, is used, the same monomer unit is obtained, and this is to be expected for an Electrophile, whether the monomer is opened instantaneously or not. However, when the ring is not opened instantaneously, the nucleo-non-free-radical route will not take place, since there is no electro-free-radical visibly present in the ring in the absence of a π-bond type of activation center. The same too applies anionically, since the polar positive charge is imaginary. So far one can observe the very unique character of this monomer or compound-one releasing transfer species of the second first kind in the route natural to it. This will contravene one of the Laws or Rules already put in place, that which is not possible. For exploratory purposes, recalling the reaction of Equation 14.215, the followings are obtained

$$
\text{(I)} \quad + \quad NaOCO\Phi \quad \longrightarrow
$$

$$\Phi \equiv \text{Phenyl group}$$

$$
\Phi COO - \underset{Cl}{\overset{Cl}{\underset{|}{P}}} \oplus \overset{\ominus}{N} - \underset{Cl}{\overset{Cl}{\underset{|}{P}}} \oplus \overset{\ominus}{N} - \underset{Cl}{\overset{Cl}{\underset{|}{P}}} \oplus \overset{\ominus}{N} \cdots \overset{\oplus}{Na} \quad \xrightarrow{+ \ (I)}
$$

$$
\Phi COO \left\{ \underset{Cl}{\overset{Cl}{\underset{|}{P}}} \oplus \overset{\ominus}{N} - \underset{Cl}{\overset{Cl}{\underset{|}{P}}} \oplus \overset{\ominus}{N} - \underset{Cl}{\overset{Cl}{\underset{|}{P}}} \oplus \overset{\ominus}{N} \right\} \underset{Cl}{\overset{Cl}{\underset{|}{P}}} \oplus \overset{\ominus}{N} - \underset{Cl}{\overset{Cl}{\underset{|}{P}}} \oplus \overset{\ominus}{N} - \underset{Cl}{\overset{Cl}{\underset{|}{P}}} \oplus \overset{\ominus}{N} \cdots \overset{\oplus}{Na} \quad \xrightarrow{+ \ (n-1) \ (I)}
$$

$$
\Phi COO \left\{ \underset{Cl}{\overset{Cl}{\underset{|}{P}}} \oplus \overset{\ominus}{N} - \underset{Cl}{\overset{Cl}{\underset{|}{P}}} \oplus \overset{\ominus}{N} - \underset{Cl}{\overset{Cl}{\underset{|}{P}}} \oplus \overset{\ominus}{N} \right\}_n \underset{Cl}{\overset{Cl}{\underset{|}{P}}} \oplus \overset{\ominus}{N} - \underset{Cl}{\overset{Cl}{\underset{|}{P}}} \oplus \overset{\ominus}{N} - \underset{Cl}{\overset{Cl}{\underset{|}{P}}} \oplus \overset{\ominus}{N} \overset{\oplus}{Na}
$$

$$14.219$$

Note the presence of covalent or ionic and polar charges placed side by side. The reason as has already been stated into law, is because polar and electrostatic charges do not repel or attract, because they are imaginary charges. To kill the growing polymer chain above, either we use external agents or let the chain grow optimally or let the chain release $Cl^{\ominus}$ by starvation and form NaCl. In the one stage, the NaCl comes and removes $^{\ominus}Cl$ from the other end to form NaCl and release $^{\oplus}CO\Phi$ which adds to the $^{\ominus}Cl$ to form $ClCO\Phi$. Finally, deactivation follows to form the double bond between P and O. This again is termination by Self. One can observe the presence of different types of dead polymers based on the type of initiator used and other operating conditions.

Thus, for the inorganic or semi-inorganic monomer in question, both radical (Electro-free-radical) and ionic (Anionic) or charged (Positively charged,) routes are observed to be favored. Radically and chargedly, the monomer is a half free-radical monomer. The P center can carry positive covalent charges in the presence of positive polar charges. The N center can also carry anionic charges in the presence of negative polar charges. It can be observed that the behavior of this monomer is completely different from any so far encountered. In view of the fact that much heat have to be applied in opening these rings in the absence of other forces, clearly indicates that the MRSE is moderately high for them.

Branching in the chains of polymers can only take place when dead polymer chains are present in the system, since a living polymer chain which has not reached its optimum chain length cannot be adding and branching at the same time or simultaneously. Branching unlike Cross-linking can take place via Equilibrium and Combination mechanisms. Cross-linking largely takes place only via Equilibrium mechanism. Dead chains are not commonly present when initiators (i.e. so-called catalyzed polymerization) are used, because sometimes to kill these chains, external agents may be required if one is not patient enough to let the chain kill itself when all are in place. They are only present when the chains are allowed to grow to their optimum length. Even then, some will still be living. However, since there are indeed no usable branching sites when initiators are used, the chain cannot form branches during polymerization. Branches can be introduced after polymerization or when there is a dead polymer along with ionic species and monomers in the system. Any attempt to use the polar bond as branching site, as we already know, the chain breaks into two linear chains. Branching and cross-linking can take place in both catalyzed and uncatalyzed system when the ring is opened instantaneously. In the uncatalyzed system for example, one should expect a branching site to be formed between its linear living polymer chain and a linear dead chain. However this may or may not be the case as shown below.

Stage 1:

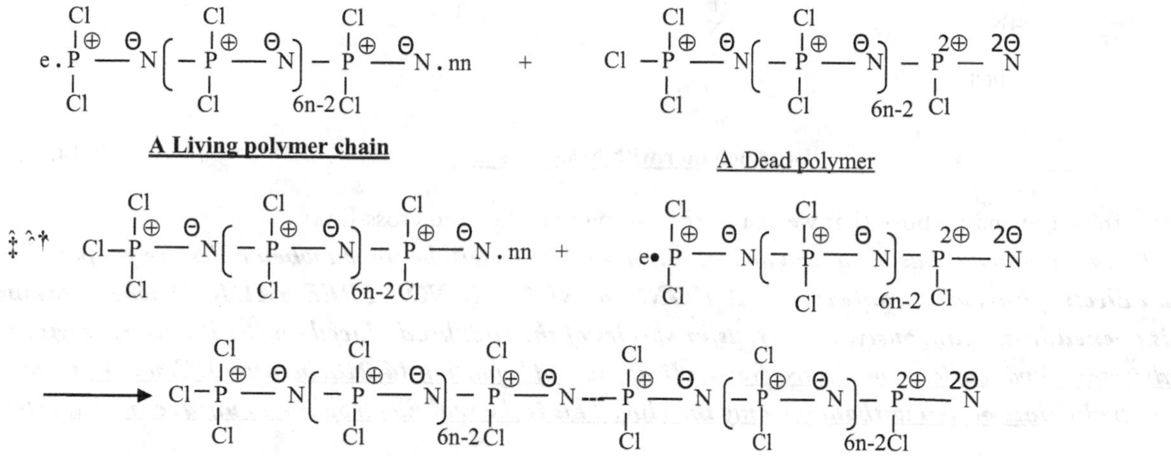

A Living polymer chain   **A Dead polymer**

Linear dead polymer   14.220

A linear dead chain is obtained in the reaction between a chain living from both ends and a dead polymer. If the Cl abstracted had been from inside, then a dead branched chain would have been obtained. It is however not the one abstracted, but the one carrying more of Cl. If indeed the monomer is heated at temperature as high as 250^0C, then some fractions of the monomer must be kept in Activated/Equilibrium State of existence as shown below.

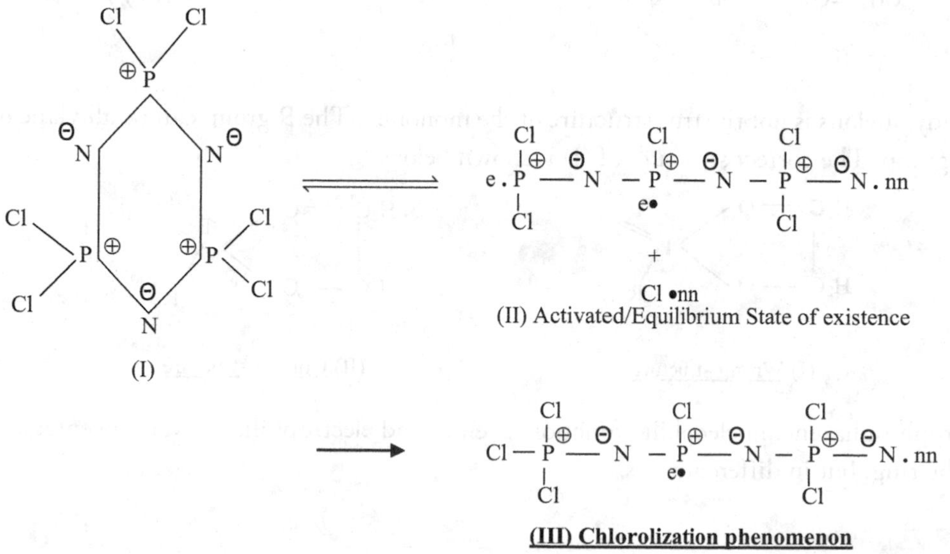

(I)

(II) Activated/Equilibrium State of existence

(III) Chlorolization phenomenon   14.221

403

Thus, the major source of branching and cross-linking in this system is (III) above. Like in Enolization and Amidization, the monomer is first kept in Activated/Equilibrium state of existence. The nucleo-non-free-radical species held in Equilibrium state of existence is grabbed by the old electro-free-radical end leaving a new electro-free-radical center. But unlike Enolization and Amidization, what is finally formed [(III) above] cannot deactivate because of the presence of MaxRSE if a four-membered was to be formed. The phenomenon above has been called herein ***Chlorolization phenomenon***, in line with Enolization and Amidization. When a living polymer chain or another monomer adds to it, one knows exactly what to expect- a linear chain with $-N^{\ominus} - {}^{\oplus}PCl_2 -$ group placed along the chain between two polar bonds. The electro-free- radicals formed along the chain become the sites for branch formation.

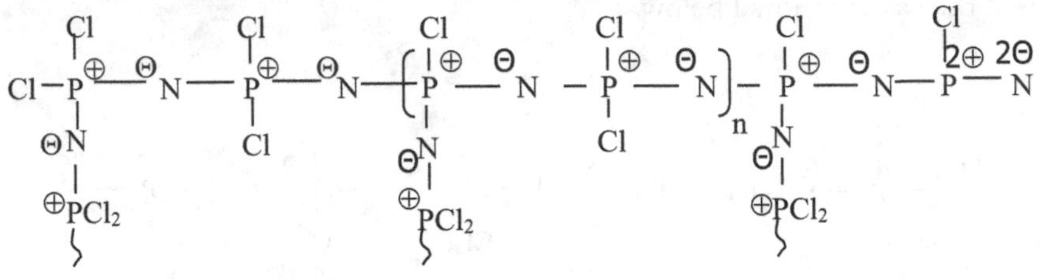

Dead polymer with branching sites   14.222

It is from the chain above that we can start forming branches and cross-links.

Based on what has been seen above, this monomer as will be shown when rules are proposed, is an Electrophile, whose only route is ANIONIC or NUCLEO-NON-FREE-RADICAL when the ring is opened instantaneously, with transfer species of the first kind placed on the P center. It cannot therefore favor cationic or electro-free-radical route of Equation 14.216, because HCl will be formed in the Initiation step without opening the ring. This is the real Electrophile so unique in character.

14.3.2 Cyclic Phosphonates and Phosphites

Low molecular weight aliphatic polyesters are produced by several methods, including ring opening polymerization of cyclic phosphonates such as shown below[20].

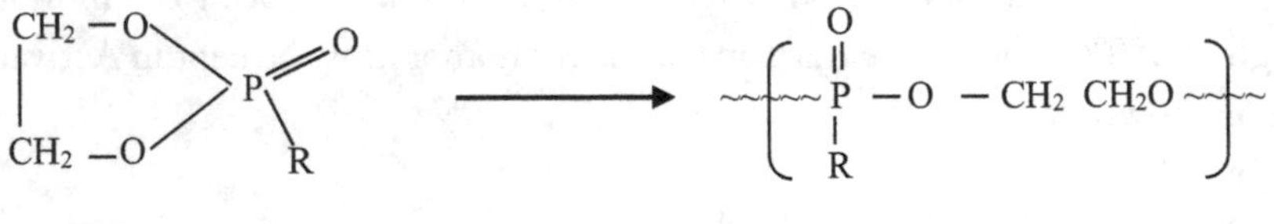

(I)   14.223

(I) as is already obvious is not the true structure of the monomer. The R group can be alkylane or aryl or any type of group. The correct structure of (I) is shown below.

(I) Wrong structure   ;   (II) Correct structure   14.224

Like dichlorophosphazene, nucleophilic functional center and electrophilic activation center seem to be present in the ring, but in different ways.

<u>Nucleophilic functional center</u>   <u>Electrophilic activation center</u>   14.225

Beginning with the use of a protonic acid, the followings are to be expected.

$$
\underset{(I)}{
\begin{array}{c}
H_2C - O \\
| \\
H_2C - O
\end{array}
> P^{\oplus} \underset{R}{\overset{O^{\ominus}}{<}}
}
\quad + \quad H^{\oplus} \quad B^{\ominus} \quad \longrightarrow \quad
\begin{array}{c}
H_2C - O \\
| \\
H_2C - O
\end{array}
> P^{\oplus} \underset{R}{\overset{O^{\ominus}}{<}}
\begin{array}{c}
\ominus B \\
\oplus \\
H
\end{array}
$$

$$
\xrightarrow{\text{(I)}} \quad H - O - \overset{\overset{H}{|}}{\underset{\underset{H}{|}}{C}} - \overset{\overset{H}{|}}{\underset{\underset{H}{|}}{C}} - O - \overset{\overset{O^{\ominus}}{|}}{\underset{\underset{R}{|}}{P^{\oplus}}} \cdots \overset{\oplus}{O} \; \ominus B \quad + \quad n\,(I) \quad \longrightarrow
$$

$$
H \left[O - \overset{\overset{H}{|}}{\underset{\underset{H}{|}}{C}} - \overset{\overset{H}{|}}{\underset{\underset{H}{|}}{C}} - O - \overset{\overset{O^{\ominus}}{|}}{\underset{\underset{R}{|}}{P^{\oplus}}} \right]_n O - \overset{\overset{H}{|}}{\underset{\underset{H}{|}}{C}} - \overset{\overset{H}{|}}{\underset{\underset{H}{|}}{C}} - O - \overset{\overset{O^{\ominus}}{|}}{\underset{\underset{R}{|}}{P^{\oplus}}} \cdots \overset{\oplus}{O} \; \ominus B
$$

(A)   14.226

Notice that the route is electro-free-radical route after starting cationically. It was the functional center that was involved above. Using the polar center cationically, the following are obtained.

$$
\underset{(I)}{
\begin{array}{c}
H_2C - O \\
| \\
H_2C - O
\end{array}
> \overset{\oplus}{\underset{R}{P}} \overset{O^{\ominus}}{<}
}
\quad + \quad H^{\oplus} \quad B^{\ominus} \quad \longrightarrow \quad
\begin{array}{c}
H - O \\
\\
R
\end{array}
> \overset{\oplus}{P} \underset{O - CH_2}{\overset{O - CH_2}{<}}
$$

$$
\longrightarrow \quad H - O - \overset{\overset{O^{\ominus}}{|}}{\underset{\underset{R}{|}}{P^{\oplus}}} - O - \overset{\overset{H}{|}}{\underset{\underset{H}{|}}{C}} - \overset{\overset{H}{|}}{\underset{\underset{H}{|}}{C}} \cdots \overset{\oplus}{O} \; \ominus B \quad + \quad n(I) \quad \longrightarrow
$$

$$
H \left[O - \overset{\overset{O^{\ominus}}{|}}{\underset{\underset{R}{|}}{P^{\oplus}}} - O - \overset{\overset{H}{|}}{\underset{\underset{H}{|}}{C}} - \overset{\overset{H}{|}}{\underset{\underset{H}{|}}{C}} \right]_n O - \overset{\overset{O^{\ominus}}{|}}{\underset{\underset{R}{|}}{P^{\oplus}}} - O - \overset{\overset{H}{|}}{\underset{\underset{H}{|}}{C}} - \overset{\overset{H}{|}}{\underset{\underset{H}{|}}{C}} \cdots \overset{\oplus}{O} \; \ominus B
$$

(B)   14.227

(B) may look similar to (A) above, but it is not. It is (A) that is the monomer unit. This is supported by the monomer unit obtained anionically.

$$\begin{array}{c} H_2C-O \\   \searrow P^{\oplus} \\ H_2C-O \diagup R \end{array} \overset{\overset{\ominus}{O}}{\diagdown} + \overset{\oplus}{Na} \overset{\ominus}{OCH_3} \longrightarrow$$

$$\overset{\overset{O}{\underset{|}{\parallel}}{\ominus}}{CH_3O - \underset{\underset{R}{|}}{P^{\oplus}} - O - \overset{\overset{H}{|}}{\underset{\underset{H}{|}}{C}} - \overset{\overset{H}{|}}{\underset{\underset{H}{|}}{C}} - \overset{\ominus}{O}\text{---}\overset{\oplus}{Na}} \overset{+ n(I)}{\longrightarrow}$$

$$CH_3O \left\{ \begin{array}{c} \overset{O}{\underset{\parallel}{}}{}^{\ominus} \\ P^{\oplus} \\ \underset{R}{|} \end{array} - O - \overset{\overset{H}{|}}{\underset{\underset{H}{|}}{C}} - \overset{\overset{H}{|}}{\underset{\underset{H}{|}}{C}} - O \right\}_n \begin{array}{c} \overset{O}{\underset{\parallel}{}}{}^{\ominus} \\ P^{\oplus} \\ \underset{R}{|} \end{array} - O - \overset{\overset{H}{|}}{\underset{\underset{H}{|}}{C}} - \overset{\overset{H}{|}}{\underset{\underset{H}{|}}{C}} - \overset{\ominus}{O}\text{---}\overset{\oplus}{Na}$$

(C)

14.228

Since (A) and (C) have the same monomer unit, the following is therefore valid.

$$\begin{array}{c} \overset{\ominus}{O} \\ \diagdown \\ P^{\oplus} \\ \diagup \\ R \end{array} > \overset{O}{\bigcirc}$$

Order of Nucleophilicity

14.229

For if R is Cl, OH or NH$_2$, the anionic route may no longer be favored and the electro-free-radical route for a chain growing forwardly will have transfer species of the first kind of the first type. The monomer then behaves as a Nucleophile.

$$H \left\{ O - \overset{\overset{H}{|}}{\underset{\underset{H}{|}}{C}} - \overset{\overset{H}{|}}{\underset{\underset{H}{|}}{C}} - O - \overset{\overset{O}{\underset{\parallel}{}}{}^{\ominus}}{\underset{\underset{OH}{|}}{P^{\oplus}}} \right\}_n O - \overset{\overset{H}{|}}{\underset{\underset{H}{|}}{C}} - \overset{\overset{H}{|}}{\underset{\underset{H}{|}}{C}} - O - \overset{\overset{\ominus}{O}}{\underset{\underset{OH}{|}}{P^{\oplus}}}{\bullet}e \longrightarrow\!\!\!\!/$$

$$H \left\{ O - \overset{\overset{H}{|}}{\underset{\underset{H}{|}}{C}} - \overset{\overset{H}{|}}{\underset{\underset{H}{|}}{C}} - O - \overset{\overset{O}{\underset{\parallel}{}}{}^{\ominus}}{\underset{\underset{OH}{|}}{P^{\oplus}}} \right\}_n O - \overset{\overset{H}{|}}{\underset{\underset{H}{|}}{C}} - \overset{\overset{H}{|}}{\underset{\underset{H}{|}}{C}} - O - \overset{\overset{\ominus}{O}}{\underset{\underset{}{|}}{P^{\oplus}}} = O + H\bullet e$$

NOT FAVORED

14.230

As it seems, the transfer species cannot be released, because P and H are equi-electropositive for which reason, H cannot be released as an atom, except when an operating condition is made to have large presence of H atoms in the system for these two equi-electropositive atoms, that which is not possible. Under the conditions above, H can only be abstracted as a hydride just like the case with Si. When R is –OH, -NH$_2$ types of group or indeed any group, the monomer remains an Electrophile. It readily undergoes both nucleo-non-free-radical, anionic, electro-free-radical and positively charged routes.

The low molecular weight polyesters observed, is due to the presence of O atoms along the chain and

406

other factors such as the method of polymerization, the types of initiator used and the temperature of polymerization with respect to strong presence of depropadation. When the chain is growing backwardly, depropagation may take place if the temperature of polymerization is above the Ceiling temperature of the monomer. However, when other factors are put in place, the presence of low molecular weight products can be envisaged as follows.

$$
HO \left[\overset{\overset{O^{\ominus}}{|}}{\underset{R}{P^{\oplus}}} - O - \overset{H}{\underset{H}{C}} - \overset{H}{\underset{H}{C}} - O \right]_x \overset{\overset{O^{\ominus}}{|}}{\underset{R}{P^{\oplus}}} - O - \overset{H}{\underset{H}{C}} = \overset{H}{\underset{H}{C}} - O \left[\overset{\overset{O^{\ominus}}{|}}{\underset{R}{P^{\oplus}}} - O - \overset{H}{\underset{H}{C}} - \overset{H}{\underset{H}{C}} - O \right]_s Y \quad +
$$

$$
CH_3O \left[\overset{\overset{O^{\ominus}}{|}}{\underset{R}{P^{\oplus}}} - O - \overset{H}{\underset{H}{C}} - \overset{H}{\underset{H}{C}} - O \right]_y \overset{\overset{O^{\ominus}}{|}}{\underset{R}{P^{\oplus}}} - O - \overset{H}{\underset{H}{C}} - \overset{H}{\underset{H}{C}} - O^{\ominus} \cdots \cdots Na \longrightarrow
$$

$$
HO \left[\overset{\overset{O^{\ominus}}{|}}{\underset{R}{P^{\oplus}}} - O - \overset{H}{\underset{H}{C}} - \overset{H}{\underset{H}{C}} - O \right]_x \overset{\overset{O^{\ominus}}{|}}{\underset{R}{P^{\oplus}}} \left[O - \overset{H}{\underset{H}{C}} - \overset{H}{\underset{H}{C}} - O - \overset{\overset{O^{\ominus}}{|}}{\underset{R}{P^{\oplus}}} \right]_{y+1} OCH_3
$$

$$
Y \left[O - \overset{H}{\underset{H}{C}} - \overset{H}{\underset{H}{C}} - O - \overset{\overset{O^{\ominus}}{|}}{\underset{R}{P^{\oplus}}} \right]_s O - \overset{H}{\underset{H}{C}} - \overset{H}{\underset{H}{C}} - O^{\ominus} \cdots \cdots Na \qquad \qquad 14.231
$$

$$
CH_3O \left[\overset{\overset{O^{\ominus}}{|}}{\underset{R}{P^{\oplus}}} - O - \overset{H}{\underset{H}{C}} - \overset{H}{\underset{H}{C}} - O \right]_y \overset{\overset{O^{\ominus}}{|}}{\underset{R}{P^{\oplus}}} - O - \overset{H}{\underset{H}{C}} - \overset{H}{\underset{H}{C}} - O^{\ominus} \cdots \cdots Na \quad +
$$

$$
CH_3O \left[\overset{\overset{O^{\ominus}}{|}}{\underset{R}{P^{\oplus}}} - O - \overset{H}{\underset{H}{C}} - \overset{H}{\underset{H}{C}} - O \right]_x \overset{\overset{O^{\ominus}}{|}}{\underset{R}{P^{\oplus}}} - O - \overset{H}{\underset{H}{C}} - \overset{H}{\underset{H}{C}} - O^{\ominus} \cdots \cdots Na \longrightarrow
$$

Living Dead Polymer   (x >> y)

$$
CH_3O \left[\overset{\overset{O^{\ominus}}{|}}{\underset{R}{P^{\oplus}}} - O - \overset{H}{\underset{H}{C}} - \overset{H}{\underset{H}{C}} - O \right]_{y+1} \overset{\overset{O^{\ominus}}{|}}{\underset{R}{P^{\oplus}}} \left[O - \overset{H}{\underset{H}{C}} - \overset{H}{\underset{H}{C}} - O - \overset{\overset{O^{\ominus}}{|}}{\underset{R}{P^{\oplus}}} \right]_x OCH_3 \quad +
$$

$$
Na - O - \overset{H}{\underset{H}{C}} - \overset{H}{\underset{H}{C}} - O - \overset{\overset{O^{\ominus}}{|}}{\underset{R}{P^{\oplus}}} - O - \overset{H}{\underset{H}{C}} - \overset{H}{\underset{H}{C}} - O^{\ominus} \cdots \cdots Na
$$

$$
14.232
$$

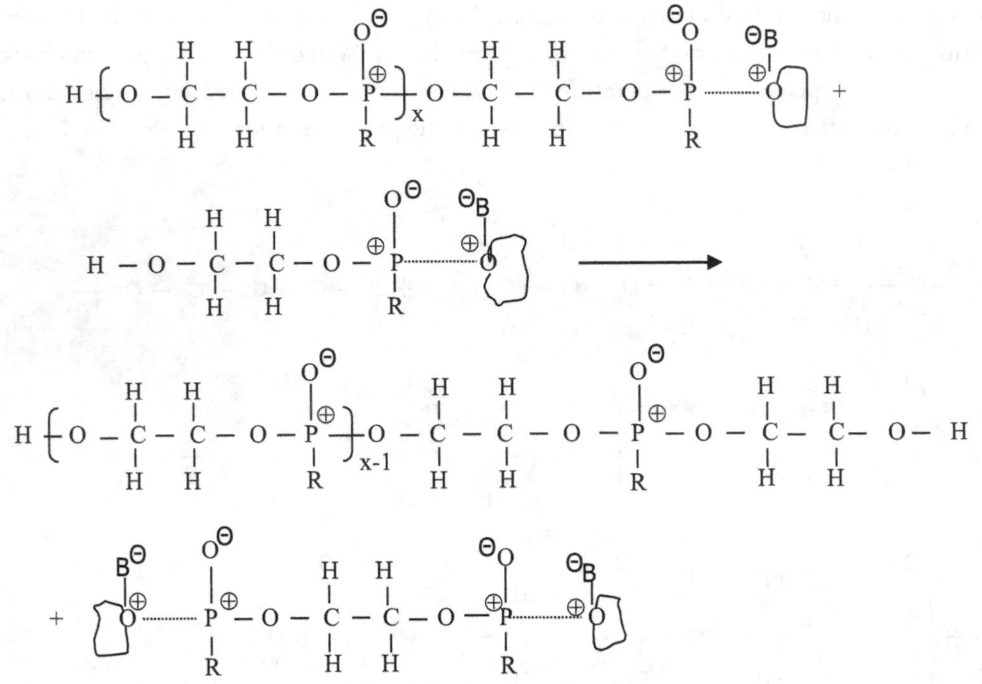

$$14.233$$

These are Backbiting reactions. The chains are increased and decreased in size throughout the entire course of polymerization, making growth of the chain an unstable process. Dead polymers some of which still look living is produced for every addition between two polymeric chains.

Aliphatic polyphosphonates with high molecular weights are produced by ring opening polymerization of **cyclic phosphites,** which is said to be accompanied by *an Arbuzov rearrangement*[20].

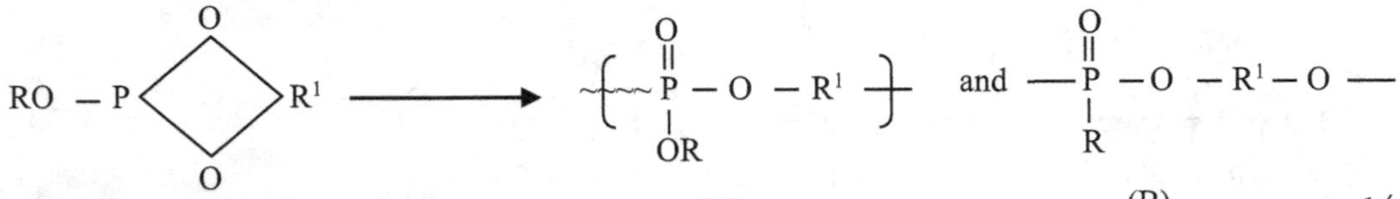

$$R^1 \equiv -(CH_2)_n- \ ; n = 1,2,3..... \qquad \text{(A)} \qquad \text{(B)} \qquad 14.234$$

As it seems, (A) looks like the monomer unit from the phosphites, while (B) is that from the phosphonates as already shown above.

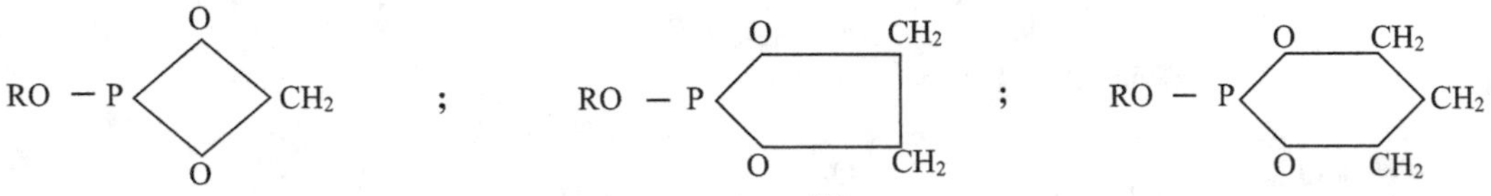

(I) ; (II) ; (III)

Cyclic phosphites   14.235

It is (II) above that can be identified with the five-membered cyclic phosphonate after undergoing a particular type of molecular rearrangement.

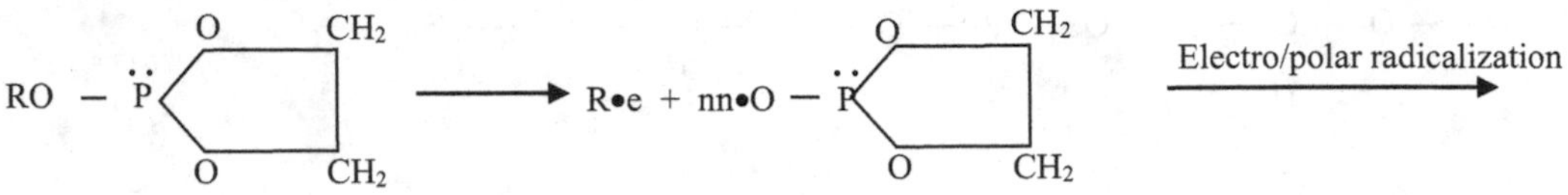

Electro/polar radicalization

$$R\bullet e + nn\bullet O - \overset{..}{P}$$

408

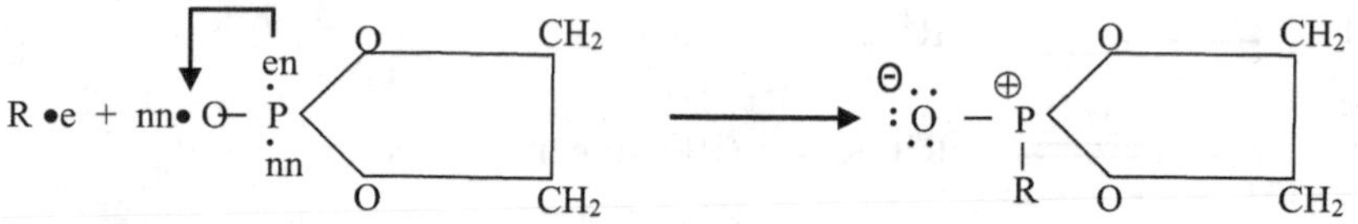

Electropolaradicalization phenomenon 14.236

The so-called Arbuzov rearrangement is indeed **Electropolaradicalization of the Second kind of the second type**, because instead of movement of an electro-radical from a π- bond adjacently located to a sitting nucleo-radical formed after the compound has existed in Equilibrium state of existence (First type of Electroradicalization), the movement is from paired unbonded radicals adjacently located to •nn to instantaneously form a polar bond instead of another π-bond. After this, the component held in Equilibrium state of Existence now adds to the new center to give the five-membered cyclic phosphonate. (III) above in Equation 14.235 will also undergo the same type of rearrangement. With (I), this may not be possible if the size of R group is large where R^1 is $- CH_2 -$ {See Equation 14.234].

The larger the size of the ring, the more favored is the molecular rearrangement of this type, since the following is valid.

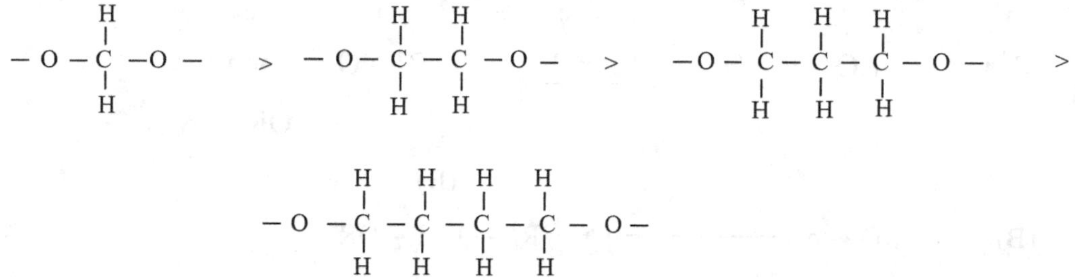

$$-O-\overset{\overset{\textstyle H}{|}}{\underset{\underset{\textstyle H}{|}}{C}}-O- \quad > \quad -O-\overset{\overset{\textstyle H}{|}}{\underset{\underset{\textstyle H}{|}}{C}}-\overset{\overset{\textstyle H}{|}}{\underset{\underset{\textstyle H}{|}}{C}}-O- \quad > \quad -O-\overset{\overset{\textstyle H}{|}}{\underset{\underset{\textstyle H}{|}}{C}}-\overset{\overset{\textstyle H}{|}}{\underset{\underset{\textstyle H}{|}}{C}}-\overset{\overset{\textstyle H}{|}}{\underset{\underset{\textstyle H}{|}}{C}}-O- \quad >$$

$$-O-\overset{\overset{\textstyle H}{|}}{\underset{\underset{\textstyle H}{|}}{C}}-\overset{\overset{\textstyle H}{|}}{\underset{\underset{\textstyle H}{|}}{C}}-\overset{\overset{\textstyle H}{|}}{\underset{\underset{\textstyle H}{|}}{C}}-\overset{\overset{\textstyle H}{|}}{\underset{\underset{\textstyle H}{|}}{C}}-O-$$

Order of radical-pushing capacity of radical-pushing
di-etheric groups 14.237

Also, this rearrangement is more enhanced with the presence of a corresponding alkyl halide or another alkyl halide. Shown below are two cases from organophosphorus compounds.[21]

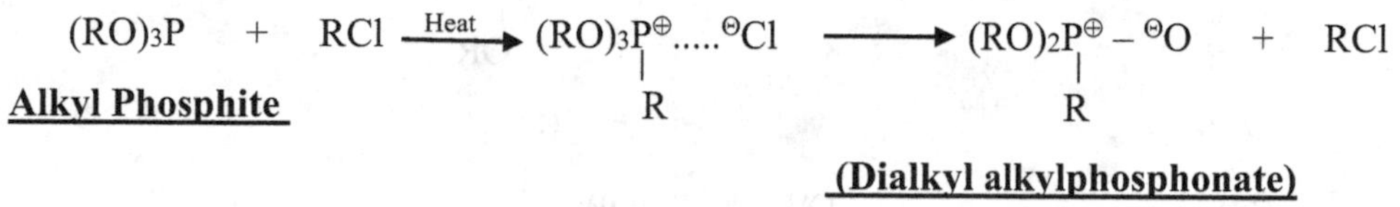

$$(RO)_3P \quad + \quad RCl \xrightarrow{\text{Heat}} (RO)_3\overset{\oplus}{\underset{R}{P}}.....\overset{\ominus}{C}l \longrightarrow (RO)_2\overset{\oplus}{\underset{R}{P}}-\overset{\ominus}{O} \quad + \quad RCl$$

Alkyl Phosphite

(Dialkyl alkylphosphonate)

14.238a

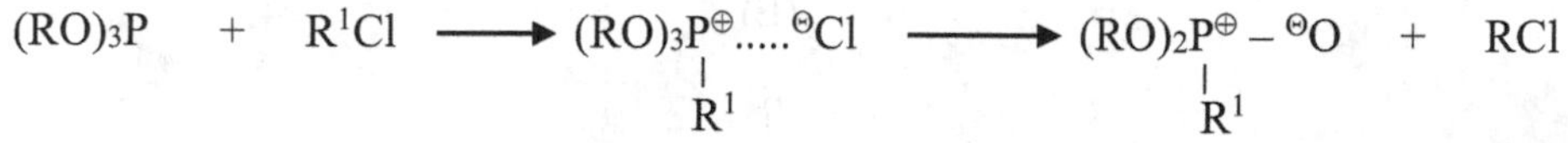

$$(RO)_3P \quad + \quad R^1Cl \longrightarrow (RO)_3\overset{\oplus}{\underset{R^1}{P}}.....\overset{\ominus}{C}l \longrightarrow (RO)_2\overset{\oplus}{\underset{R^1}{P}}-\overset{\ominus}{O} \quad + \quad RCl$$

14.238b

Note t osphonates formed is carrying R^1. This obviously is more than expected, since here the alkyl halide is taking part in the reaction and not just a catalyst of enhancement.

Stage 1: $R^1Cl \;\rightleftharpoons\; R^1 \bullet e \;+\; Cl \bullet nn$

$R^1 \bullet e \;+\; (RO)_3P \;\rightleftharpoons\; R^1OR \;+\; (RO)_2P \bullet en$

$(RO)_2P \bullet en \;\rightleftharpoons\; R \bullet e \;+\; en \bullet P - O \bullet nn$
$$\qquad\qquad\qquad\qquad\qquad\qquad\qquad\qquad |$$
$$\qquad\qquad\qquad\qquad\qquad\qquad\qquad\quad OR$$

$$(A)$$

$R \bullet e \;+\; nn \bullet Cl \;\rightleftharpoons\; RCl$

$(A) \;\longrightarrow\; RO - :P = O \;+\; Heat$ \hfill 14.239a

Overall Equation: $R^1Cl \;+\; (RO)_3P \longrightarrow R^1OR \;+\; RCl \;+\; RO - P = O$

$$\qquad\qquad\qquad\qquad\qquad\qquad\qquad\qquad\qquad +\; Heat$$ \hfill 14.239b

Stage 2: $R^1OR \;\rightleftharpoons\; R^1 \bullet e \;+\; nn \bullet OR$

$R^1 \bullet e \;+\; RO - P = O \;\xrightarrow{\;Activation\;}\; R^1 - O - P \bullet en$
$$\qquad\qquad\qquad\qquad\qquad\qquad\qquad\qquad\qquad\quad |$$
$$\qquad\qquad\qquad\qquad\qquad\qquad\qquad\qquad\quad OR$$

$$(B)$$

$(B) \;+\; RO \bullet nn \;\longrightarrow\; R^1 - O - P - OR$
$$\qquad\qquad\qquad\qquad\qquad\qquad\qquad\qquad |$$
$$\qquad\qquad\qquad\qquad\qquad\qquad\qquad OR$$

$$(C)$$ \hfill 14.240a

Overall Equation: $R^1OR \;+\; RO - P = O \;\longrightarrow\; (C)$ \hfill 14.240b

Stage 3: $(C) \;\rightleftharpoons\; R^1 \bullet e \;+\; nn \bullet O - \overset{..}{P} - OR$
$$\qquad\qquad\qquad\qquad\qquad\qquad\qquad\qquad\qquad\qquad |$$
$$\qquad\qquad\qquad\qquad\qquad\qquad\qquad\qquad\qquad OR$$

$$(D)$$

$(D) \;\rightleftharpoons\; O^{\ominus} - {}^{\oplus}\overset{\bullet n}{P} - OR$
$$\qquad\qquad\qquad\qquad\qquad\qquad\qquad |$$
$$\qquad\qquad\qquad\qquad\qquad\qquad OR$$

$$(E)$$

$$\qquad\qquad\qquad\qquad\qquad\qquad\qquad R^1$$
$$\qquad\qquad\qquad\qquad\qquad\qquad\qquad |$$
$R^1 \bullet e \;+\; (E) \;\longrightarrow\; O^{\ominus} - {}^{\oplus}P - OR$
$$\qquad\qquad\qquad\qquad\qquad\qquad\qquad |$$
$$\qquad\qquad\qquad\qquad\qquad\qquad OR$$ \hfill 14.241a

Overall Equation: $R^1Cl \;+\; (RO)_3P \;\longrightarrow\; RCl \;+\; (RO)_2P^{\oplus} - {}^{\ominus}O$
$$\qquad\qquad\qquad\qquad\qquad\qquad\qquad\qquad\qquad\qquad\qquad\qquad |$$
$$\qquad\qquad\qquad\qquad\qquad\qquad\qquad\qquad\qquad\qquad\qquad R^1$$ \hfill 14.241b

Thus, it can be seen that RCl is active with the Electropolaradicalization taking place only in the second step of the last stage- the third stage. For the replacement above to have taken place, $\underline{\mathbf{R} \geq \mathbf{R^1}}$ in radical-pushing capacity, i.e., half of what is inside the ring must be greater than or equal to what is being held outside. For if R^1 was greater than R, no replacement will take place. For example when HCl is used in place of R^1Cl the stages above will be favored. The ***so-called S_N2 addition which is shown in Equations 14.238a and 14.238b, is not,*** but Electrostatic Addition which takes place so that the boundary laws in the shell are not contravened. If Electrostatic addition had taken place before the rearrangement, then the phosphonate cannot be obtained as shown below.

Stage 1:

$$R^1Cl \;\rightleftharpoons\; R^1 \;+\; nn\bullet Cl$$

$$R^1\bullet e \;\rightleftharpoons\; R^{1\oplus} \;+\; \bullet e$$

$$e\bullet \;+\; Cl\bullet nn \;\rightleftharpoons\; Cl^{\ominus}$$

$$R^{1\oplus} \;+\; (RO)_3P \;\rightleftharpoons\; \underset{\underset{R^1}{|}}{(RO)_3P^{\oplus}}$$

$$(A$$

$$Cl^{\ominus} \;+\; (A) \;\longrightarrow\; \underset{\underset{R^1}{|}}{Cl^{\ominus}......^{\oplus}P(RO)_3}$$

$$(B) \hspace{4cm} 14.242a$$

How can (B) molecularly rearrange when the Equilibrium state of existence of (B) is as follows? This cannot even be done via Decomposition mechanism.

$$(B) \;\rightleftharpoons\; \underset{\underset{\bullet n}{}}{Cl^{\ominus}.......^{\oplus}\overset{RO\diagdown\;\diagup OR}{P}} - OR \;+\; e\bullet R^1 \hspace{2cm} 14.242b$$

Based on what we see in Equations 14.238 and 14.239, one can observe the types of number of possibilities that exist when the mechanisms of how chemical reactions take place are not understood. ***To understand the mechanisms of how chemical reactions take place is to understand how NATURE operates, because Chemistry is the study of the LAWS of NATURE in the real and imaginary domains. Physics is the study of the FORCES of NATURE in the real and imaginary domains and the FORCES in question are the SENSES-Feelings, Sight, Sound, Smell and Taste. In order words, we have ten SENSES. Mathematics is the NATURAL language of COMMUNICATION in the real and imaginary domains.***

Thus, when the molecular rearrangement is favored, different products will be obtained. For monomers such as shown below, however, both anionic and positively charged routes will also be favored, once MRSE can be provided for the ring.

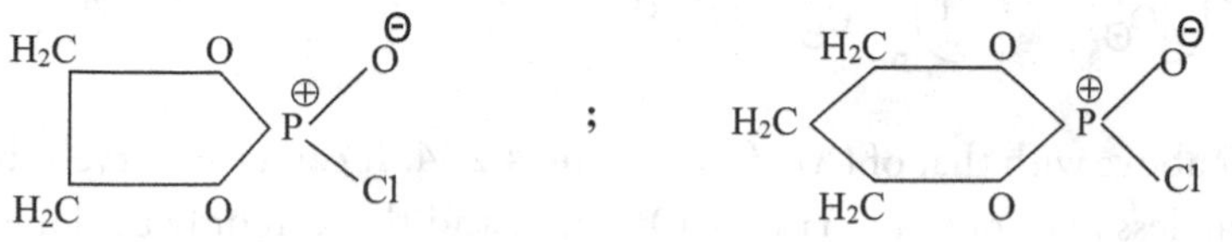

$$14.243$$

411

The Cl atom is not a transfer species since the polar bond is not involved. But it can be a branching site. Hence the electrophilic activation center of Equation 14.225 can also be as follows.

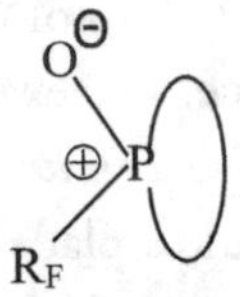

$$
\underset{R_F}{\overset{O^{\ominus}}{\underset{\oplus}{\bigodot}}}P \qquad \text{(where } R_F \text{ is a radical-pulling group)}
$$

<u>Electrophilic Activation Center</u>   14.244

The four-membered ring of cyclic phosphites will favor rearrangement via *Electropola-radicalization phenomenon of the First kind of the second type if R is large.* But when R is replaced with H, it will not favor it.

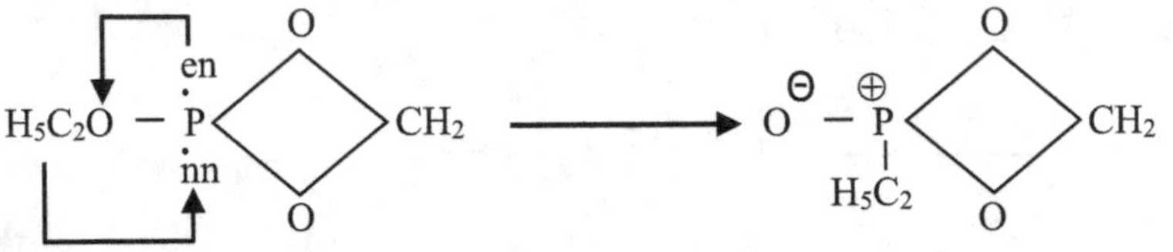

$$
H_5C_2O - \overset{en}{\underset{nn}{P}} \diamondsuit CH_2 \longrightarrow O^{\ominus} - \overset{\oplus}{\underset{H_5C_2}{P}} \diamondsuit CH_2
$$

14.245

$$
HO - \overset{en}{\underset{nn}{P}} \diamondsuit CH_2 \longrightarrow \text{Not favored}
$$

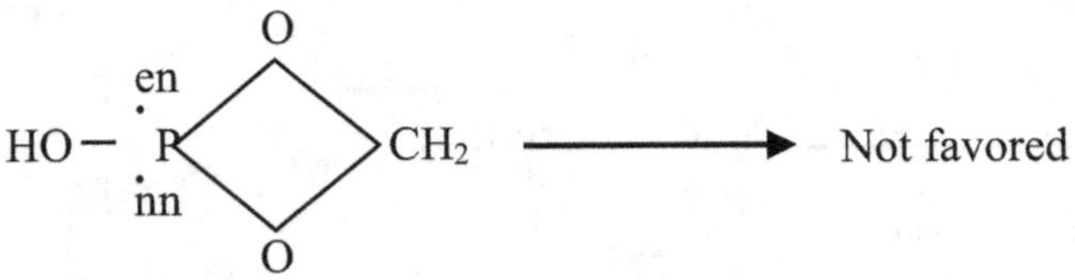

(more stable)   14.246

When protonic acid is used as initiator, the followings are obtained.

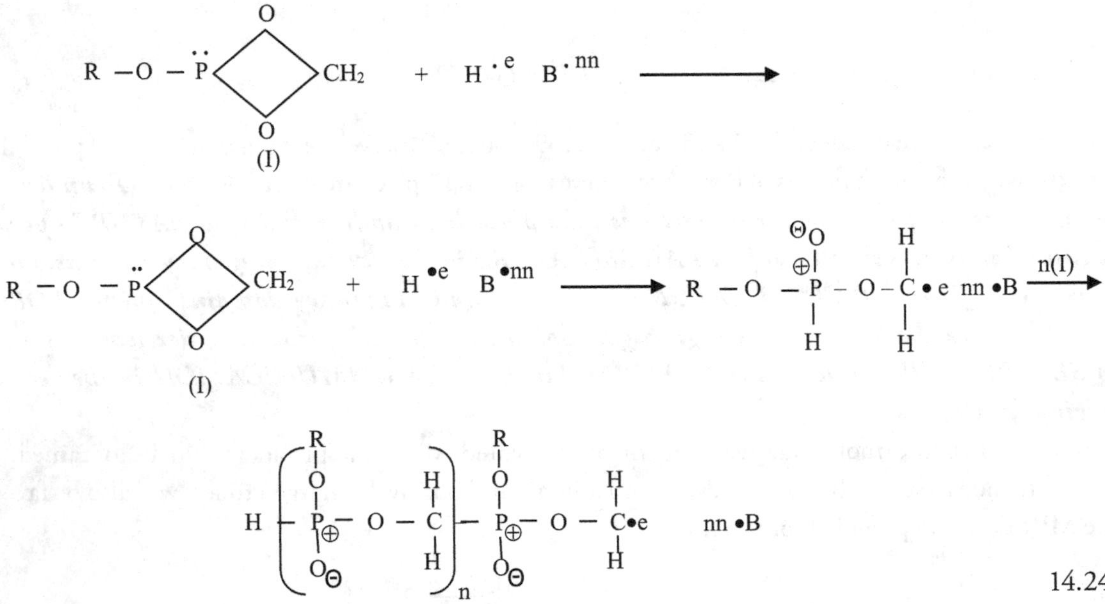

$$
R - O - \overset{..}{P} \diamondsuit CH_2 \quad + \quad H \cdot^e \quad B \cdot^{nn} \longrightarrow
$$
(I)

$$
R - O - \overset{..}{P} \diamondsuit CH_2 \quad + \quad H \cdot^{e} \quad B \cdot^{nn} \longrightarrow R - O - \overset{\overset{\ominus}{O}}{\underset{H}{\overset{\oplus}{P}}} - O - \overset{H}{\underset{H}{C}} \cdot e \;\; nn \cdot B \xrightarrow{n(I)}
$$
(I)

$$
H \left(\begin{array}{c} R \\ | \\ O \\ | \\ \overset{\oplus}{P} - O - \overset{H}{\underset{H}{C}} \\ | \\ O_{\ominus} \end{array} \right)_n \begin{array}{c} R \\ | \\ O \\ | \\ \overset{\oplus}{P} - O - \overset{H}{\underset{H}{C}} \cdot e \\ | \\ O_{\ominus} \end{array} \quad nn \cdot B
$$

14.247

Compare the monomer unit above with that of (A) of Equation 14.234. It can be observed that the center attacked is the P center being less nucleophilic than the O center, and this can only be done electro-free-radically by adding to the nucleo-free-radical on the P center to form a bond and instantaneously open the ring, since the H could not sit on the P center due to boundary laws to form an electrostatic bond.

412

The electro-free-radical left on the P center then forms a polar bond with O. ***Worthy of note is that propagation continued with paired radical centers, since radicals cannot repel and attract. Radicals have identities.*** Thus, while phosphites are Nucleophiles when there is no molecular rearrangement, phosphonates are Electrophiles and the route can never be charged. Since addition is forward and paired, depropagation cannot take place to loose formaldehyde.

Though it cannot be opened chargedly, it can however still undergo charged reactions as shown below without opening of the ring, using NaOH.

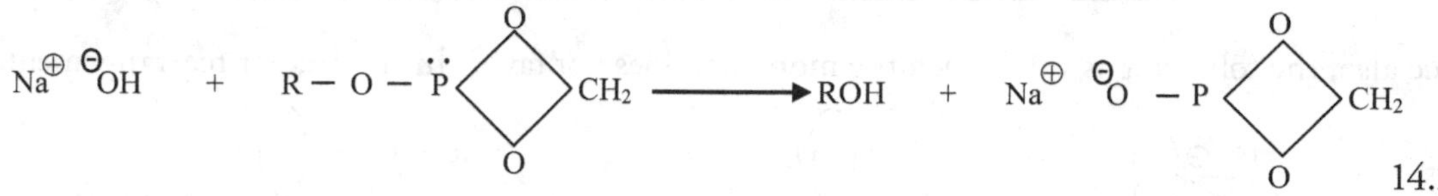

$$Na^{\oplus} \, {}^{\ominus}OH \quad + \quad R-O-\overset{..}{P}\diamondsuit CH_2 \longrightarrow ROH \quad + \quad Na^{\oplus} \, {}^{\ominus}O - P\diamondsuit CH_2$$

$$14.248$$

For the purpose of exploration, consider the positively charged polymerization of this monomer, using BF_3, in the absence of the current developments.

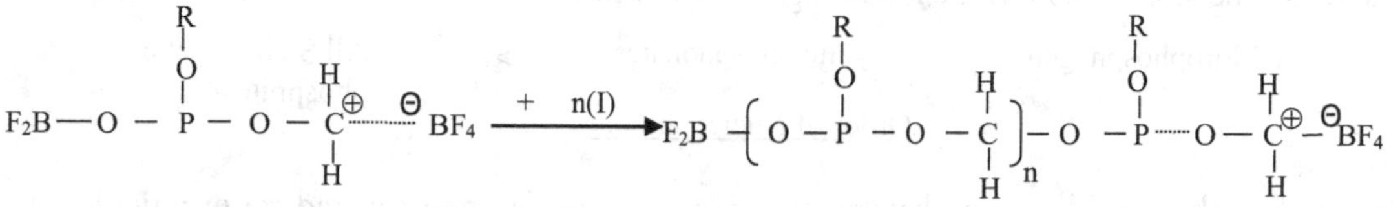

$$F_2B-O-\overset{\overset{R}{\overset{|}{O}}}{\underset{}{P}}-O-\overset{\overset{H}{|}}{\underset{\underset{H}{|}}{C}}{}^{\oplus}\,{}^{\ominus}BF_4 \quad + \quad n(I) \longrightarrow F_2B\left[O-\overset{\overset{R}{\overset{|}{O}}}{\underset{}{P}}-O-\overset{\overset{H}{|}}{\underset{\underset{H}{|}}{C}}\right]_n-O-\overset{\overset{R}{\overset{|}{O}}}{\underset{}{P}}-O-\overset{\overset{H}{|}}{\underset{\underset{H}{|}}{C}}{}^{\oplus}{}^{\ominus}BF_4$$

IMPOSSIBLE REACTION

$$14.249$$

If the initiator had been ferrous chloride ($FeCl_2$), then transfer species of the first kind of the first type would have been released from the growing polymer chain to give a terminal P=O double bond, if the iron center had a receiving center, as shown below only radically.

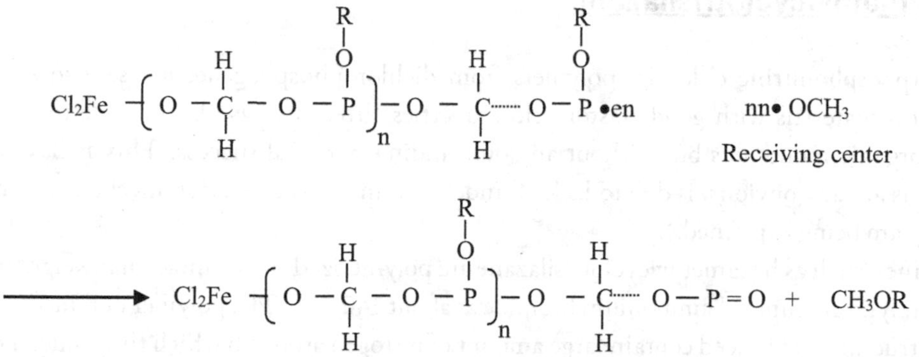

$$Cl_2Fe\left[O-\overset{\overset{H}{|}}{\underset{\underset{H}{|}}{C}}-O-\overset{\overset{R}{\overset{|}{O}}}{\underset{}{P}}\right]_n-O-\overset{\overset{H}{|}}{\underset{\underset{H}{|}}{C}}-O-\overset{\overset{R}{\overset{|}{O}}}{\underset{}{P}}\bullet en \qquad nn\bullet OCH_3$$

Receiving center

$$\longrightarrow Cl_2Fe\left[O-\overset{\overset{H}{|}}{\underset{\underset{H}{|}}{C}}-O-\overset{\overset{R}{\overset{|}{O}}}{\underset{}{P}}\right]_n-O-\overset{\overset{H}{|}}{\underset{\underset{H}{|}}{C}}-O-P=O \quad + \quad CH_3OR$$

IMPOSSIBLE REACTION

$$14.250$$

Based on Equation 14.237, only the four-membered phosphite will not molecularly rearrange when R is H or CH_3, because of the capacity of the di-etheric group shown below in the ring alone, half the capacity of which is less than what is externally located to the P center (a more electro-positive element than O). The phosphite for R equal to H is more in stable state of existence than in Equilibrium state of existence at low and high operating conditions.

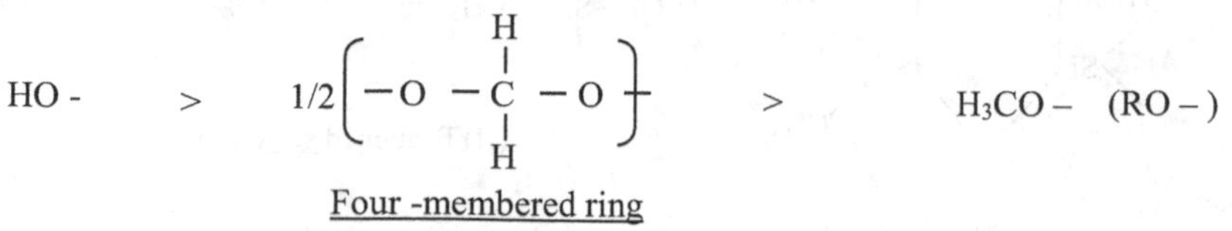

$$HO- \quad > \quad \frac{1}{2}\left[-O-\overset{\overset{H}{|}}{\underset{\underset{H}{|}}{C}}-O-\right] \quad > \quad H_3CO- \quad (RO-)$$

<u>Four -membered ring</u>

$$14.251$$

where R is measured by the number of C atoms in the central ring. For example for the four-membered ring, when R is H rearrangement will not take place. When R is CH_3, rearrangement may take place. But when R is C_2H_5, rearrangement will fully take place. Hence, the following is valid for phosphites of any ring size when RO is less than or equal to half of what is in the ring.

Cyclic phosphites $\xrightarrow[\text{rearranged}]{\text{molecularly}}$ Cyclic phosphonates

Valid for Cyclic phosphites when $[\text{H-}(CH_2)_n\text{O-}] \leq \frac{1}{2}(\text{-O-}(CH_2)_n\text{-O-}]$ 14.252

Hence also, the following is valid when the monomer does not favor this molecular rearrangement.

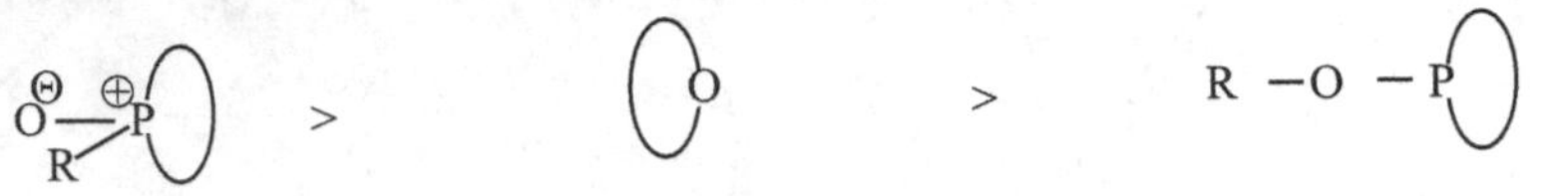

Order of Nucleophilicity   14.253

Based on the analysis so far, the followings are also valid-.

Dichlorophosphagene	>	Cyclic phosphonates	>	All Stable cyclic phosphites

Order of Nucleophilicity   14.254

despite the fact that Dichlorophosphagene favors only the nucleo-non-free-radical or anionic route (A complete Electrophile), Cyclic phosponates favor both nucleo-non-free-radical or anionic and electro-free-radical routes (An Electrophile), while Stable Cyclic phosphites favor only the electro-free-radical route (A complete Nucleophile).

14.3.3 Hexamethylcyclotrisilazane

Since poly(phosphonitrilic chloride) polymers from dichlorophosphagenes are said to show great promise as useful materials with good elastomeric properties, attempts have been made to polymerize other cyclic inorganic monomers but without any outstanding practical success. This as has been said in the past and is already obvious, is due to lack of understanding of the reaction mechanisms involved, which presently are being explained.

Cyclosilazanes, such as hexamethylcyclotrisilazane are polymerized to low molecular weight products by the use of catalytic amounts of ammonium bromide at about $140°C$[22]. The polymer obtained is said not to be linear in structure and indeed contain large amount of nitrogen atoms to which three silicon moieties are attached. The polymers are therefore said to contain cyclic and branch structures as shown below.

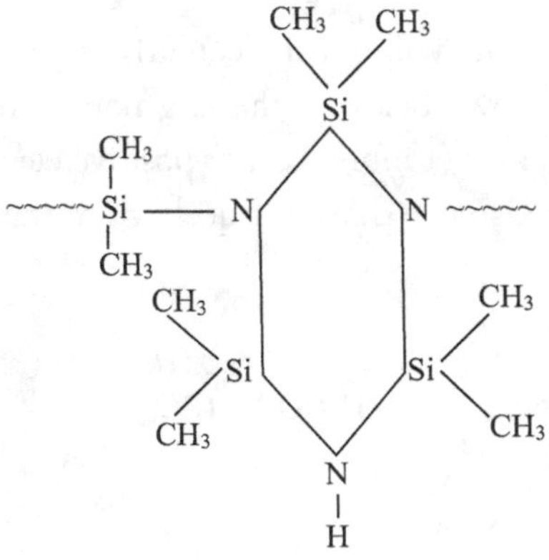

(I) Cyclic structure 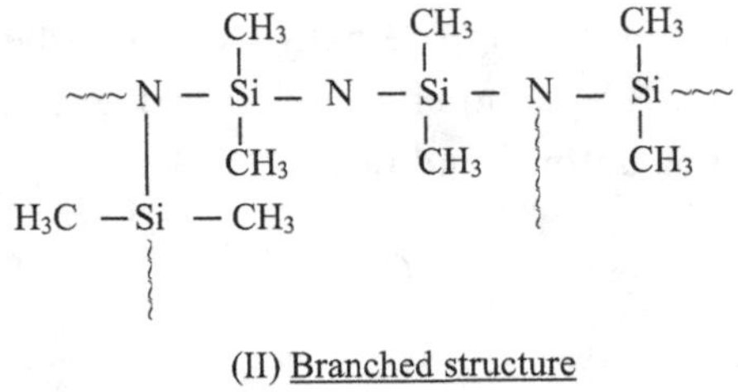   (II) Branched structure   14.255

414

Looking at the monomer itself, the Si - N bond is weak. Secondly the hydrogen atoms on the nitrogen centers are loosely bonded radically one at a time. There are three of them as shown below. With the presence of three paired unbonded radicals on the N center, unique character of Si, the ring is bound to be strained, It has three functional centers, Nucleophilic in character

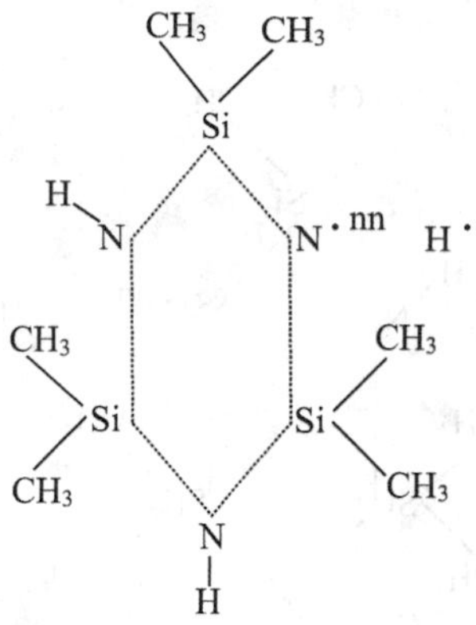

$$
\begin{array}{c}
CH_3 \quad CH_3 \\
Si \\
H-N \qquad N \cdot nn \quad H \cdot e \\
CH_3 \qquad CH_3 \\
Si \qquad Si \\
CH_3 \qquad CH_3 \\
N \\
H
\end{array}
$$

14.256

The radical bonding of hydrogen instead of ionic bonding is due to the presence of silicon in the ring, which as is now obvious does not favor ionic or covalent charged existence. With the use of the ammonium bromide, the followings are obtained.

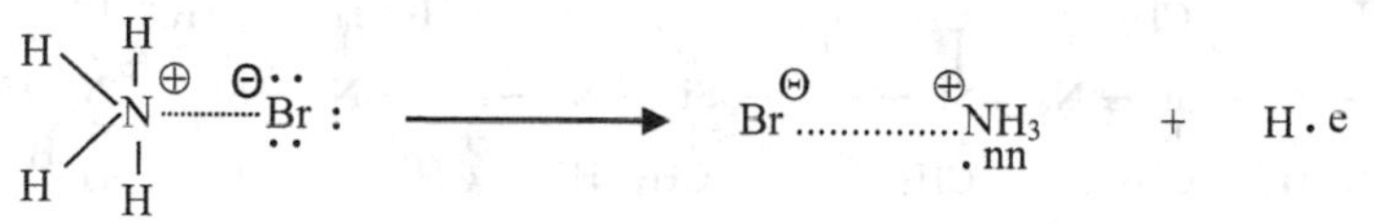

$$
\underset{H}{\overset{H}{\underset{|}{\overset{|}{N^{\oplus}}}}} \cdots Br^{\ominus} \colon \longrightarrow Br^{\ominus} \cdots NH_3^{\oplus} \cdot nn \;+\; H \cdot e
$$

14.257

$$
H \cdot e \;+\; H-N \qquad N-H \quad \longrightarrow
$$

(with the ring structure containing Si, CH$_3$ groups)

(III)

$$
\begin{array}{l}
\underset{H}{\overset{H}{N}} - \underset{CH_3}{\overset{CH_3}{Si}} - \underset{H}{\overset{H}{N}} - \underset{CH_3}{\overset{CH_3}{Si}} - \underset{H}{\overset{H}{N}} - \underset{CH_3}{\overset{CH_3}{Si}} \overset{\oplus}{-} \overset{\ominus Br}{\underset{H}{\overset{|}{N}}} - H \qquad +\; x(III) \longrightarrow
\end{array}
$$

$$
H \left[N - \underset{CH_3}{\overset{CH_3}{Si}} - N - \underset{CH_3}{\overset{CH_3}{Si}} - N - \underset{CH_3}{\overset{CH_3}{Si}} \right]_x N - \underset{CH_3}{\overset{CH_3}{Si}} - N - \underset{CH_3}{\overset{CH_3}{Si}} - N - \underset{CH_3}{\overset{CH_3}{Si}} \overset{\oplus}{-} \overset{\ominus Br}{N} - H
$$

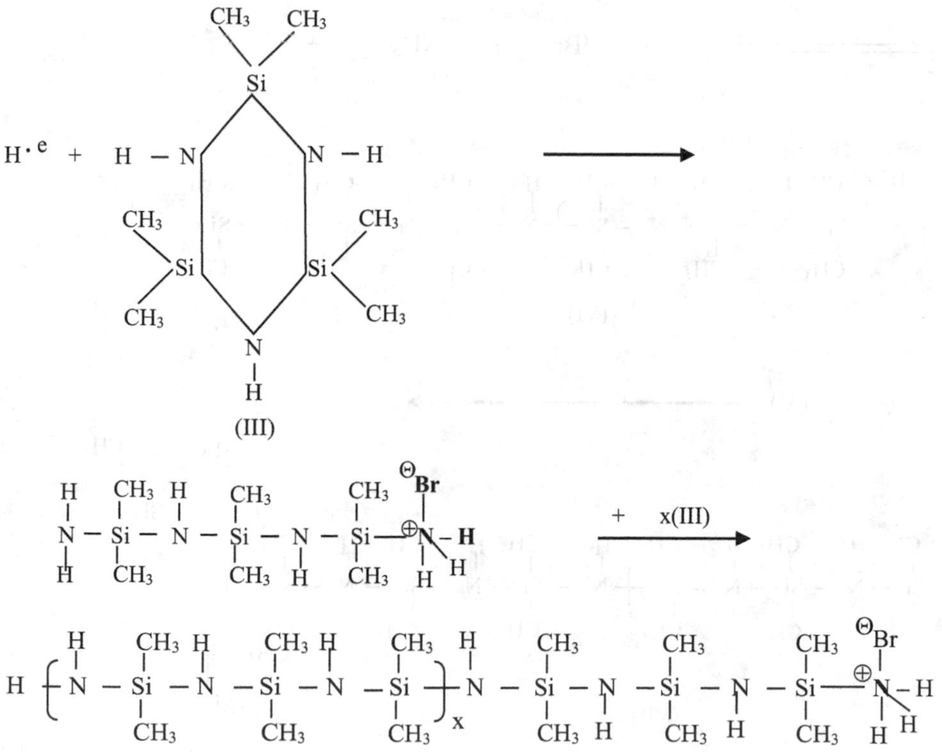

(IV)

14.258

415

Notice that the route is electro-free-radical route, with the chain growing backwardly by Equilibrium mechanism. Addition continued until at a particular point in time, instead of the monomer remaining in Stable state of Existence, it now was put in Equilibrium state of Existence, for which the following shown below took place.

Stage 1:

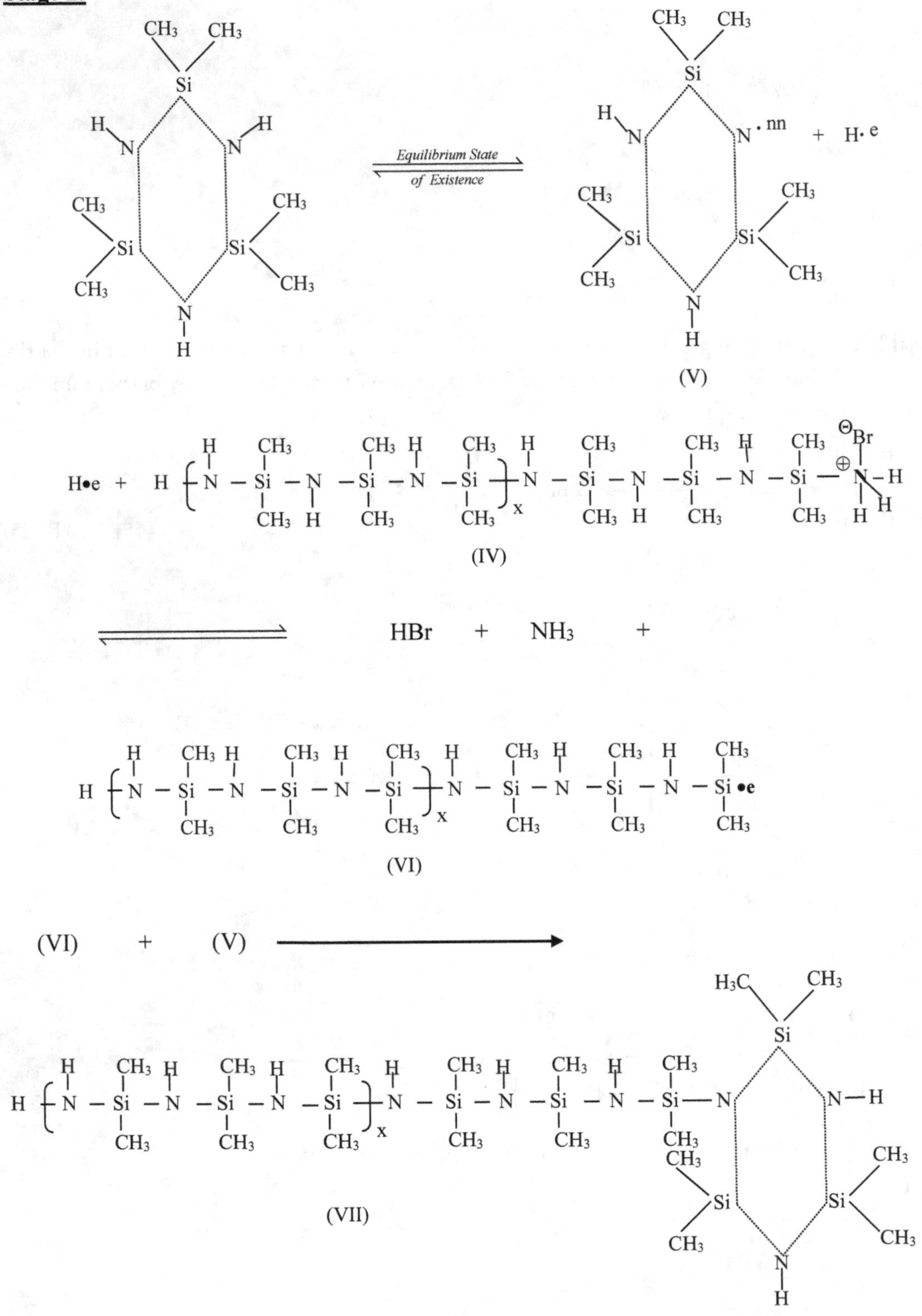

(V)

(IV)

$$\rightleftharpoons \quad HBr \quad + \quad NH_3 \quad +$$

(VI)

$$(VI) \quad + \quad (V) \quad \longrightarrow$$

(VII)

14.259

416

(VII) was now formed with the monomer placed at the terminal of the chain. In the next stage, (VII) was forced to exist in Equilibrium state of existence from the other N center to begin the same stage above to produce the chain shown below.

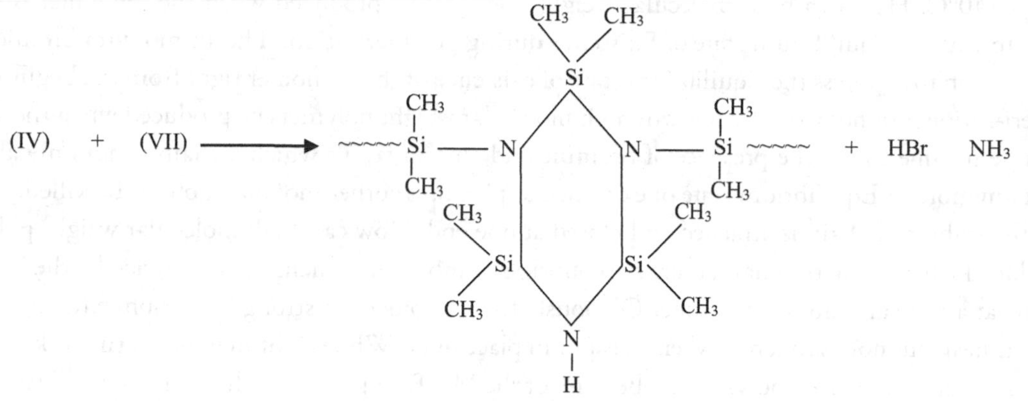

$$(IV) \quad + \quad (VII) \longrightarrow$$

(I) of Equation 14 . 255

14.260

In the same manner, branches are also formed. These happen when the chains have reached their optimum length based on the polymerization temperature and the type of polymer. The dead polymer now first exists in Equilibrium state of existence from one of the N centers inside the ring. The H•e now held comes to kill a growing polymer chain as shown in the second step of Stage 1 of Equation 14.259. The (VI) formed now comes to add to the N center along the chain to form a branch. The HBr formed when a chain is disengaged from the terminal carrying the coordinated initiator, can also be used to kill another growing chain from the same terminal. With the formation of the first branch, other branches begin to grow along the chain to give what is shown below.

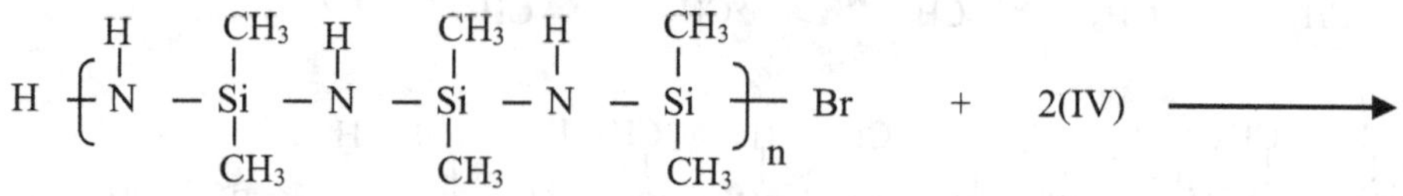

Dead polymer with optimum chain length

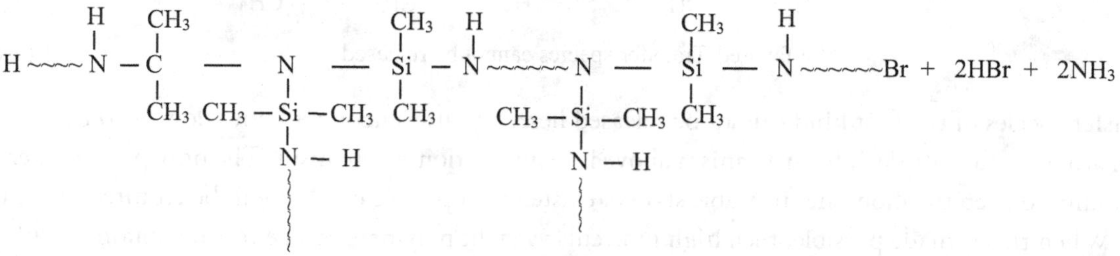

(II) of Equation 14 . 255

14.261

Though, it was not intended to go into the concepts of branching, cross-looking and other formations in these systems at this stage of development, the need sometimes arises, in order to produce convincing evidence of the mechanisms of these reactions being systematically provided. If all reactions were ionic, products can never be produced. Other charges {Covalent, Polar, and electrostatic] we see are more radical in character than ionic. In fact, they are all radical in character. It is important to note that the route above is an electro-free-radical route favored only by H•e via the functional center. ***The ring which has the ability to exist in Equilibrium state of existence during propagation cannot instantaneously be opened by the forces carried by Br and N centers. This is major reason why low molecular weight polymers are obtained***

How can high molecular weight polymers be produced when in the first case the mechanism of polymerization is Equilibrium mechanism as we have for Step polymerization systems? In the case of backward addition, depropagation can readily take place particularly at the temperature of polymerization which is 140°C. How can high molecular weight polymers be produced when the monomer is strong enough to exist in Equilibrium state of Existence during polymerization? The ammonium bromide was strong enough to suppress the Equilibrium state of existence of the monomer right from the beginning of polymeri-zation, but not to end. How can high molecular weight polymers be produced when the system from time to time shows the presence of bromine and ammonia, for which a chain is not only allowed to exist anymore in Equilibrium state of existence to pick up another monomer, but to be killed? At this point, the stabilized chain is attacked and closed at one end. How can high molecular weight polymers be produced when along the chain, there are countless numbers of branching sites carried by the N center used one at a time and not all at a time? Obviously the monomer is a strong Nucleophile (more so than cyclic amines), but not as much as when O is put in place of N. While eight-membered ring is known for O, six-membered rings are known for N, because of the MaxRSE present in the six-membered ring for O and less SE required for the existence of the eight-membered ring for N. One should therefore expect the possibility of instantaneous opening of the N containing ring to be less than that of the O containing ring.

Because of the type of initiator used above (NH_4Br), wherein the route is electro-free-radical route, with the chain growing backwardly, there is no transfer species to be released from the growing chain. But if the chain is made to grow forwardly, then we see a transfer species which cannot be released. If metals such as sodium is used, it will displace H from the centers to form H_2 and Na silico-amide instead of acting as an initiator. If H_2 can be decomposed to give $2H\cdot$ e, the followings are supposed to be obtained.

Not favored-Transfer species cannot be released   14.262

Transfer species of the first kind cannot be released here, because silicon will not allow H to be released as an atom, but a hydride. The mechanism above is Combination mechanism. The only problem here, is the ability to keep the monomer in Stable state of existence. A passive catalyst will be required to achieve this. When this is made possible, then high molecular weight polymers can be readily obtained with the use of a good initiator.

14.4 Concluding Observations

There is no doubt that the invisible p-bonds present in ringed compounds have male (Y) and female (X) characters. These characters are determined by what the ring is carrying and these characters are imaginary. What we see in the rings are real. There are different types of singly placed rings which can be used as monomers. Considering rings where there are no externally located π-bond activation centers or polar bonds cumulatively placed to the ring, we have the followings-

i)   Rings that do not have functional centers or π-bond activation centers.
ii) Rings that have only one functional center.
iii) Rings that have two different functional centers placed side by side.
iv) Rings that have isolatedly placed two or more same or different functional centers.
v)   Rings that have two or more same functional centers placed side by side.
vi) Rings that have two or more functional centers conjugatedly placed.
vii) Rings that have only one activation center.
viii) Rings that have two or more activation centers conjugatedly placed.
ix) Rings that have one or two activation centers and one functional center conjugatedly placed.
x)   Rings that have only two activation centers conjugatedly or isolatedly placed.
xi) Rings that have three or four activation centers conjugatedly placed.
xii) Rings that have three polar activation centers conjugatedly placed.

All the rings above are nucleophilic (i.e. Females) in character. The same applies to their invisible π- or polar bonds. Where invisible polar bonds exist, the rings can only be opened instantaneously (At high temps) if the ring has points of scission or functional center(s) (At low temps). Such is the case with rhombic sulfur (case v) above. Where visible polar bonds exists all along the ring, the rings can only be opened instantaneously (At high temps) or via the negatively charged center (At low temps). Such is the case with dichlorophosphazene or phosphonitrilic chloride (case xii) above. Where no functional or activation centers exist in the ring, the rings can only be opened instantaneously if the MRSE can be provided for the ring. A ring is like a SPRING in Physics. It is only in CHEMISTRY, one can see Mathema-tics, Physics, Engineering, Social Sciences, Applied Medical Sciences, Arts including RELIGION; and indeed all disciplines come into play. When we talk of rings, all the above is just but the beginning of the beginning. There are fused rings, conjugatedly placed rings, side by side rings and so on.

So far, four types of charges have clearly been identified and these are the main existing charges.

(i)   Positive and negative ionic charges.
(ii) Positive and negative electrostatic charges (Two types)
(iii) Positive and negative polar charges (Two types)
(iv) Positive and negative covalent charges.

While the charges in (i) can be isolatedly placed, the charges in (ii), (iii) and (iv) cannot be isolatedly placed. (i) and (iv) are real while (ii) and (iii) are imaginary or indeed complex. Because they are imaginary, they do not undergo electrodynamic forces of repulsion or attraction. While ionic charges and their carriers can be moved or can move, ionic charges cannot be removed from their carriers. Covalent, polar and electrostatic charges and their carriers cannot be moved alone without the partner. So also, can the charges not be removed from their carriers.

Like charges, there are also male and female radicals, based on the second law of Nature- The Law of Duality. Radical do not undergo electro-dynamic forces of repulsion and attraction, because they have identities. The male and female radicals can attract themselves, but the male and male radicals or female and female radicals do not repel themselves under any operating conditions. They can combine together to form stable molecules when carried or a sigma (σ-bond) [not π-bond] under specific operating conditions. Male and female radicals can also combine together to form σ-bond under

Equilibrium operating conditions in the absence or presence of π- or polar bonds. ***They when carried, only "communicate" between themselves. Ionic charges when paired, covalent, electrostatic, and polar charges also "communicate" between themselves.*** Radicals can be isolated. While only the male radicals and their carriers can be moved, female radical and their carriers cannot be moved or cannot move. They only move when their last shell is filled and/or when the male counterpart or partner is not in its immediate environment. While only the male radical can be removed from its carrier to form a HOLE, or a CATION, or a POSITIVE CHARGE, depending on the receiving center, the female radicals cannot be removed from its carrier.

So far, one has observed the existence of different types of monomers for ringed and non-ringed monomers.

(i) Full free covalently charged monomers.

(ii) Full free-radical monomers.

(iii) Full non-free-radical monomers.

(iv) Full non-free covalently charged monomers

(v) Half free-covalently charged monomers.

(vi) Half free-radical monomers.

Based on the foundation on which today's Chemistry has been laid, balancing of Equations were thought to be limited to stoichiometry of Chemical equation alone, i.e. balancing of the number of different atoms of components on both left and right hand side of chemical equations. It is far more than that, because now we can see that ***based on the new classifications for radicals, bonds and charges, mechanisms of chemical and polymeric reactions, state of existence, compounds***, chemical equations must be radically and chargedly balanced under Combination mechanism. While these are not important for chemical reactions which take place largely via Equilibrium mechanisms, it is very important for Addition polymerization reactions which take place largely via Decomposition and Combination mechanisms. When a compound on the left hand side of the equation decomposes not by abstraction, but by external forces such as heating to give radicals on the other side of the Equation, radical balancing does not arise. There is need to know all these things in order to know how NATURE operates. In polymerization systems, there are free-media initiators and paired-media initiators. We have seen examples of all the different types of paired-media initiators in the chapters so far in particular Chapter 13. All paired-media initiators carry charges and some carry radicals. Two radical carrying species may or may not be able to form paired centers. When paired-media initiators are used under Equilibrium mechanism conditions, the addition of monomers is backwardly and radically and the balancing of radicals does arise during initiation or grabbing of the monomer. But when the paired centers are used under Combination mechanism, radicals must be balanced. What we mean by balancing of radicals and charges is that one cannot have free-radicals or charges on the left-hand side of the equation and have non-free-radicals or charges on the right-hand side. For example, the positive charges carried by $:Fe^{2\oplus}$ are non-free and non-ionic because the last shell is not empty, while the charges carried $Mg^{2\oplus}$, $Na^{\oplus}$ are free and ionic because the last shell is empty. The positive charge carried by $Cl\text{-}Mg^{\oplus}$ is free and non-ionic because the last shell is not empty. For it to become ionic the last shell must be emptied of radicals. The positive charge carried by $H_3C^{\oplus}$ is free and non-ionic because the last shell of the carbon center is not empty. The negative charge carried by $:\overset{\square}{\underset{\square}{Cl}}{}^{\ominus}$ is non-free and ionic because of presence of paired unbonded radicals in the last shell (Polarity) and the fact that the shell is full after receiving a male radical from another center.

The same applies to $H_2 \overset{\square}{N}{}^{\ominus}$. The more the numbers of paired unbonded radicals on a center, the more polar is the center and components carrying them. $H_4C_9{}^{\ominus}$ is free and non-ionic because of the absence of paired unbonded radicals in the last shell of C. When all these are known, then we will start beginning to understand the mechanisms of polymerization reactions, the GREATEST of all the sub-disciplines or branches of Chemistry. It is from it for the first time one saw the existence of MALE and FEMALE compounds, just as exists with charges and radicals. One will appreciate all the above with some examples.

Consider the polymerization of formaldehyde (Half-free covalently charged monomer) and ethene (Full-free covalently charged monomer) using t-butyllithium. The monomers are Females. Therefore their natural route is free-positively charged route for this initiator. Li^A is free and ionic, paired to a free and non-ionic center. Hence the bond is covalent and real.

14.263

14.264

The initiation step is not chargedly balanced and therefore not favored. Before, we thought that the route is "anionic" for which the initiator was named "Anionic ion-paired initiator" universally. Interesting enough above, the carrier is $Li^{\oplus}$ a cation, and the positive charge carried on the monomer is covalent. This is only possible under Equilibrium mechanism and not under Combination mechanism conditions. The monomer comes in, gets activated and add, but cannot propagate. The Li center can therefore only be used electro-free-radically and free-medially (Not paired-medially), i.e., in the absence of C_4H_9. For example, aldehydes, ketones, propene (propylene), butenes and propylene oxide can be polymerized electro-free-radically by only the Li center when present alone. But because NATURE abhors a vacuum, the same initiator can be used while paired chargedly on ethylene only and not formaldehyde with $H_9C_4{}^{\ominus}$ as the carrier of the chain. These are first members of their families.

When a male monomer is in the vicinity of this initiator, the followings are obtained.

14.265

After communication between the active centers, initiation step is favored with the monomer carried by $H_9C_4{}^{\ominus}$, i.e., the route is negatively charged and not anionic. Because the initiator is dual in character, like the Z/N types of initiators, hence they were classified as Covalently charged-paired initiators. However, unlike Z/N types of initiators, the duality is limited, because the Li center cannot be used cationically, but only electro-free-radically.

When Electrostatically anionic-paired initiator of the type shown below is used for the same monomers above, one will think that the routes shown below are favored.

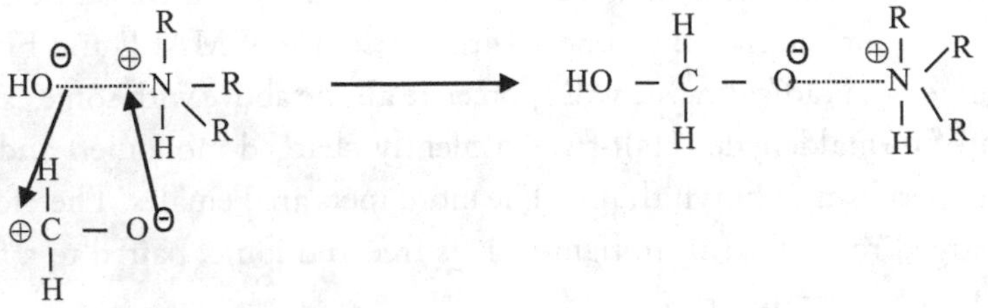

Favored only if initiator cannot exist in Equilibrium State of Existence   14.266

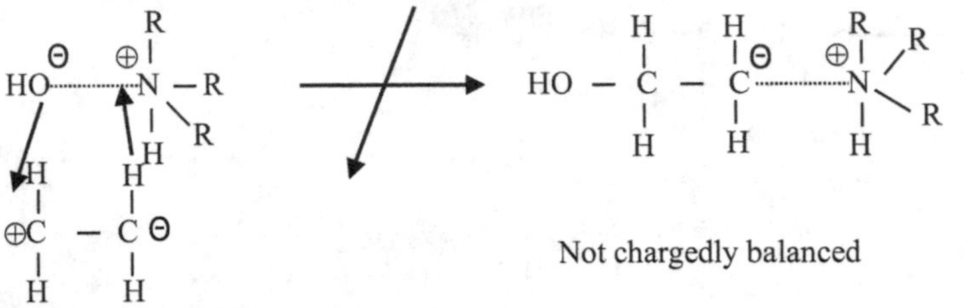

Not chargedly balanced   14.267

Nature abhors a vacuum, non-linearity, differentiation and so many things. Yet they exist with us because of the Law of Duality. For any to exist, they must first co-exist. How can you know what is sweet when there is no bitterness or what is evil when there is no good or what is plus whern there is no minus and so on? Anionic route above in the first equation should not be favored because the monomer is a Nucleophile. If the initiator can exist in Equilibrium state of existence, then electro-free-radical route will be favored with the chain growing backwardly. But if initiator cannot exist in equilibrium state of existence, and NATURE abhors a vacuum, the route shown in the first equation above will become favored. With the second reaction, whether the initiator is placed in Stable state of existence or not, the polymerization of the monomer will be impossible. Electro-free-radically, the growing chain cannot readily exist in Equilibrium state of existence for continued addition. It may if made to behave like a normal alkane.

Full non-free-radical monomers cannot be used by LiC_4H_9 initiator for the following reasons.

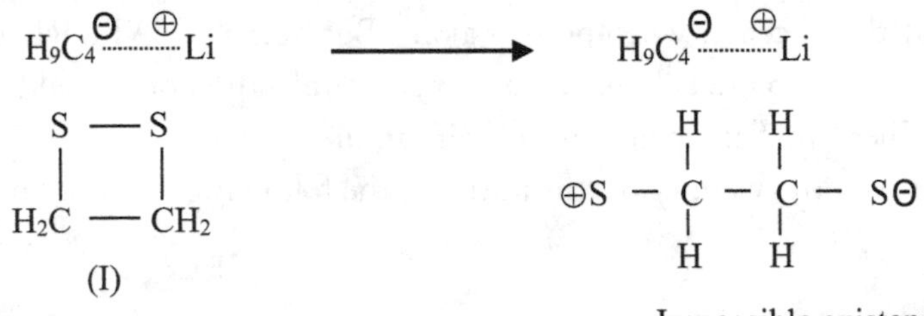

Impossible existence for $S^{\oplus}$ 14.268

Cationically or anionically, the monomer cannot be polymerized since the equation will not be chargedly balanced. The S center cannot carry a non-free positive charge in the presence of C and H, when the possiblity of releasing a stable compound exists. Something must be released along with so much energy, and this can only be done radically. When the second initiator is used, the followings are obtained.

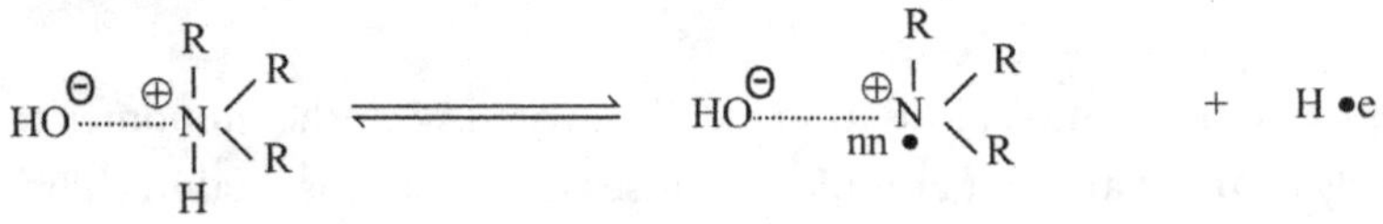

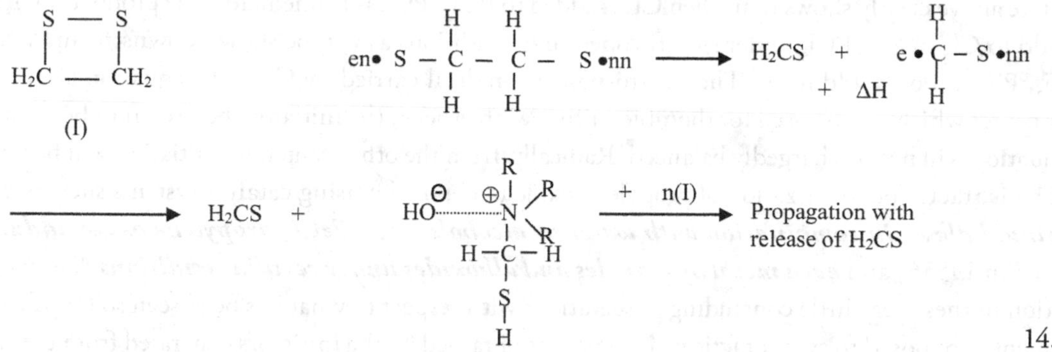

$$\text{(I)}$$

$$14.269$$

Thus, one can see that the four-membered disulfide if it exists like the other higher members in the family is Half free-radical monomer and not Full non-free-radical monomer. Electro-free-radically the monomer can be polymerized with backward addition for this initiator and no need for the equation being radically balanced. When the monomer is opened instant-aneously or via the functional center under Equilibrium conditions radically, thioformal-dehyde is released instantaneously along with large amount of energy to leave behind an electro-free-radical on the C center. For every addition of monomer, during propagation thioformaldehyde must be released. The monomer units are as follows and unlike the types shown in the literature and textbooks as already previously shown.

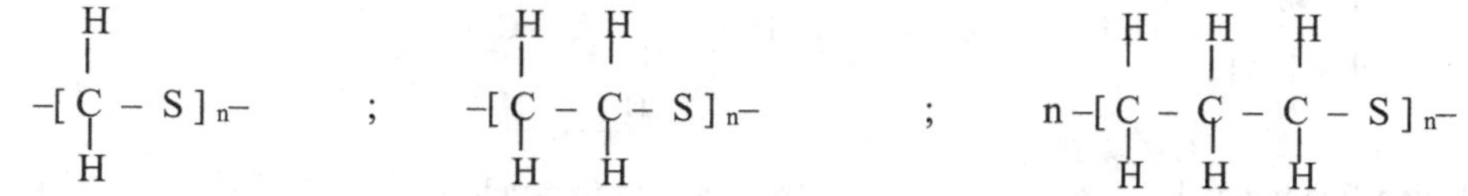

4-membered ring   **5-membered ring**   **6-membered ring**   14.270

Now consider the use of Phosphorus pentachloride for the two monomers above.

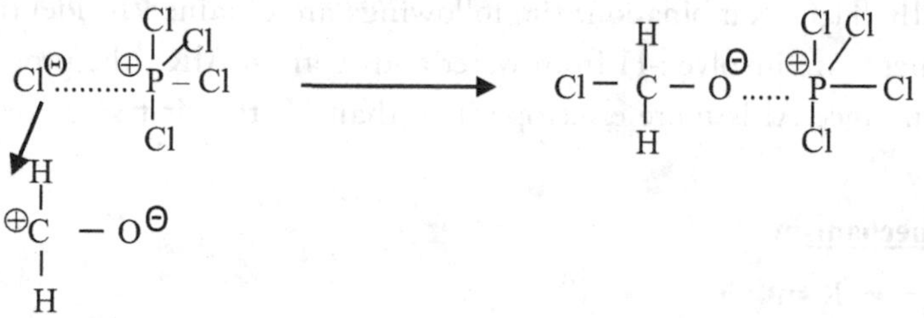

$$14.271$$

NATURE abhors a vacuum. The monomer is a Nucleophile, the first member in the family of Aldehydes. Its natural route is electro-free-radical or cationic or free-positively charged route, routes which cannot be provided by the initiator above. While formaldehyde will favor being polymerized with this initiator, the other members in the family of Aldehydes such as acetaldehyde will not favor polymerization because of presence of transfer species of the first kind of the first type and the fact that they are more Nucleophilic (i.e. more Female) than formaldehyde the first member of the family. The initiator above has only one active center on the electrostatic bond, that is the negative center which is anionic. The positive center cannot be used, since it is imaginary. ELECTROSTATIC and POLAR bonds are the one that show that our world is complex. Another center which is on the P center is as shown below.

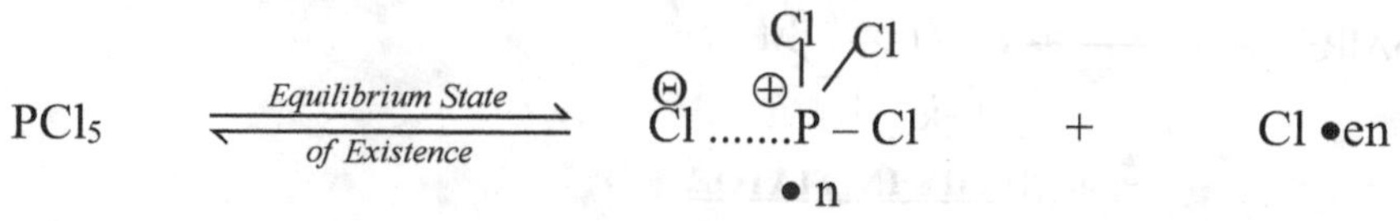

$$PCl_5 \quad \underset{\text{of Existence}}{\overset{\text{Equilibrium State}}{\rightleftharpoons}} \quad \overset{\ominus}{Cl}\text{.......}\overset{\oplus}{P} - Cl \quad + \quad Cl \bullet en$$

$$14.272$$

The state above clearly shows that when Cl_2 is added to PCl_3, PCl_5 is immediately the product. PCl_3 could also add to Cl_2 to give PCl_5 under specific operating conditions as will be shown downstream. With PF_3 and F_2, PF_3 can only add to F_2. The electro-non-free-radical carried by Cl in the equation above, is the active center which can be used for rhombic sulfur. With ethene, the initiator above cannot be used, since the equation will not be chargedly balanced. Radically, from the other center, no initiation will be possible.

The **isotactic** polymerization of propylene oxide (an epoxide) using catalyst systems such as ***zinc or aluminum alkyls in combination with water or alcohols***[23,24], a ***FeCl₃-propylene oxide adduct*** first reported in 1955[25], and ***even metal hydroxides and alkoxides under certain conditions***[26], is worthy of attention in these very little concluding observations with respect to what has been seen so far. If isotactic placement as opposed to syndiotactic placement was obtained by the initiators generated from them, then paired media initiating system must be involved. Secondly, either there is only one reservoir present or one side of the horizontal plane of addition is sterically hindered. The monomer being Nucleophilic with transfer species, the natural route must be either electro-free-radical or positively charged.

[Nucleo-free-radical]   [Nucleo-non-free-radical]   14.273

Anionically or with free negative charge, the same as above is obtained, that is there is no initiation, the route being unnatural to it.

For the first combination-AlR_3/H_2O combination, the followings are obtained in identifying the initiator. Since two metallic elements are involved-H from water and Al from AlR_3, the process begins with Decomposition mechanism. Since Al is more electropositive than H, the first stage begins with aluminum trialkyl.

Decomposition mechanism

Stage 1: $AlR_3 \longrightarrow R \bullet n + e\bullet AlR_2$

$R_2Al \bullet e + H_2O \longrightarrow R_2Al - OH + H \bullet e$

$H \bullet e + n\bullet R \rightleftharpoons RH$ 14.274a

Overall Equation: $AlR_3 + H_2O \xrightarrow{HEAT} R_2AlOH + RH$ 14.274b

Equilibrium mechanism

Stage 1: $AlR_3 \rightleftharpoons R \bullet e + n\bullet AlR_2$

$R \bullet e + R_2AlOH \rightleftharpoons ROH + e\bullet AlR_2$

(I) **THE INITIATOR**   14.274a

424

Overall Equation: AlR_3 + R_2AlOH ———————➤ ROH + (I) 14.275b

Overall overall Equation: $2AlR_3$ + H_2O ————➤ ROH + (I) + RH 14.275c

It is in the same manner Z/N initiators are obtained from their combinations such as $Al_3R_3/TiCl_4$ combination, so also is the case above. Note that the exact ratio of AlR_3/H_2O combination is 2 : 1 For the Z/N combination, for isotactic placement the ratio of Al to Ti is also 2 to 1. This is not just a matter of coincidence but the rule. Notice that the initiator is dual in character, that is, there are two active centers; one for a male and the other for a female. The same initiator can also be used for propene, butenes, (vinyl chloride), and so many. When used on vinyl chloride, the same initiator above are radically paired. When used for propylene oxide, the monomer stays all the time in the reservoir carried by the negative center being a Female to obtain the followings-

Isotactic poly (propylene oxide) 14.276

For the AlR_3/R^1OH combination, the followings are similarly obtained.

Decomposition mechanism

Stage 1: AlR_3 ——————➤ R •n + e• AlR_2

R_2Al •e + R^1OH ——————➤ R_2Al – OH + R^1•e

R^1•e + n• R ⇌ RR^1 14.277a

Overall Equation: AlR_3 + R^1OH —— HEAT ——➤ R_2AlOH + RR^1 14.277b

Stage 2 is the same as above. The same initiator (I) is obtained. ***Al metal*** has remained trivalent.

For ZnR_2/H_2O combination, the situation is slightly different, because based on the RADICAL Configuration (No longer Electronic Configuration, since electrons reside inside the nucleus of an atom- in fact, one has started preparing the Electronic configuration in the nucleus for all existing known Atoms) of atoms in the Periodic Table. Zn is highly polar with six paired unbonded radicals and three vacant orbitals (in 4p) in the last shell. How can one reduce the two vacant orbitals in ZnR_2 to one and how can one remove the paired unbonded radicals so that zinc will now be made to carry electro-free or nucleo-free-radicals instead of electro-non-free or nucleo-non-free-radicals? One of the six paired unbonded radicals has been used for bonding to the two alkyl groups via excitation, leaving five. With the advantage offered by the presence of two vacant orbitals, one can begin to reduce the numbers of vacant orbitals.

Equilibrium Mechanism

Stage 1: R_2Zn ⇌ R – Zn •en + R •n

(A)

$$(A) \quad + \quad R \bullet n \quad \longrightarrow \quad RZn - ZnR_3$$
$$(B) \qquad\qquad 14.278a$$

Overall Equation: $2ZnR_2 \longrightarrow RZn - ZnR_3$ $\qquad$ 14.278b

A fraction of the ZnR_2 was kept in Equilibrium state of existence while another fraction was kept in Stable state of existence. Because of the presence of the vacant orbitals, the ZnR_2 in Stable state of existence was activated to give (A) which finally combined with $R \bullet n$ to give (B), reducing the number of vacant orbitals from 2 to 1 and the number of paired unbonded radicals from 5 to 4 on that stable Zn center. With focus still on that stable Zn center, the second stage follows as shown below.

Stage 2: $\qquad ZnR_2 \quad \rightleftharpoons \quad R_2Zn\text{-}\square$

$$R_2Zn\text{-}\square \quad + \quad (D) \quad \xrightarrow{\textit{Dative Bonding}} \quad
\begin{array}{c} ZnR_2 \\ \uparrow \\ RZn -\overset{..}{Zn} - R \\ \diagup \quad \diagdown \\ R \qquad R \end{array}$$
$$(D) \qquad\qquad 14.279a$$

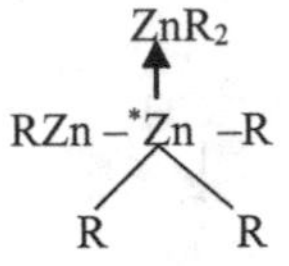

Overall Equation: $\quad 3ZnR_2 \longrightarrow (D)$ $\qquad$ 14.279b

So far, notice that we have consumed 3 moles of ZnR_2. The stable central Zn element has been marked with asterisk above. It is the one carrying the load. On that Zn center, there are three paired unbonded radicals left and one vacant orbital. The next step is to bond with the three paired unbonded radicals, so that Zn center can be made to carry free-radicals.

Stage 3: $\qquad ZnR_2 \quad \rightleftharpoons \quad R_2Zn\text{-}\square$

$$R_2Zn\text{-}\square \quad + \quad (D) \quad \xrightarrow{\textit{Dative Bonding}} \quad
\begin{array}{c} RZn \qquad ZnR_2 \\ \nwarrow \quad \nearrow \\ RZn -\overset{*}{Zn} - R \\ \diagup \quad \diagdown \\ R \qquad R \end{array}$$
$$14.280$$

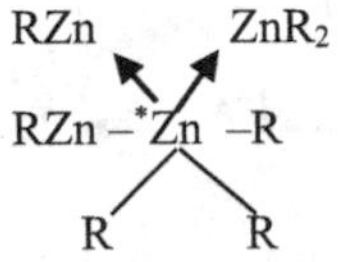

Note that no activation could further take place anymore despite the presence of one vacant orbital, because of the operating conditions. The ZnR_2 could no longer exist in Equilibrium state of existence being suppressed by the presence of (D). Instead, the (D) was datively bonded to ZnR_2 as shown above. With two paired unbonded radical now left, these are now bonded in the last two stages (Stages 4 and 5) to give the followings.

Stages 4 & 5:

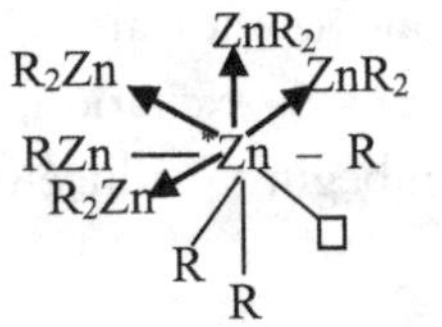

$$\begin{array}{c} ZnR_2 \\ R_2Zn \nwarrow \uparrow \nearrow ZnR_2 \\ RZn \longrightarrow \overset{*}{Zn} - R \\ R_2Zn \swarrow \diagup \big| \diagdown \square \\ R \big| \\ R \end{array}$$
$$(F) \qquad\qquad 14.281$$

Another reason why this had to be done instead of forming additional bonds is because the last shell of Zn based on the Period she belongs to in the Periodic Table, cannot carry more than eighteen (18) radicals in the last shell. Though the dative bonds formed in (F) above are chargeless, there are now 16 radicals in the last shell of Zn. With the presence of the dative bonds, Zn can now carry free-radicals and no more non-free radicals. Everything required to do the job are all in the system in the ratios desired as we shall very shortly show, Mathematics being their natural language of communication.

Overall Equations: $6ZnR_2 \longrightarrow$ (F)   14.282

Decomposition Mechanism

Stage 1:

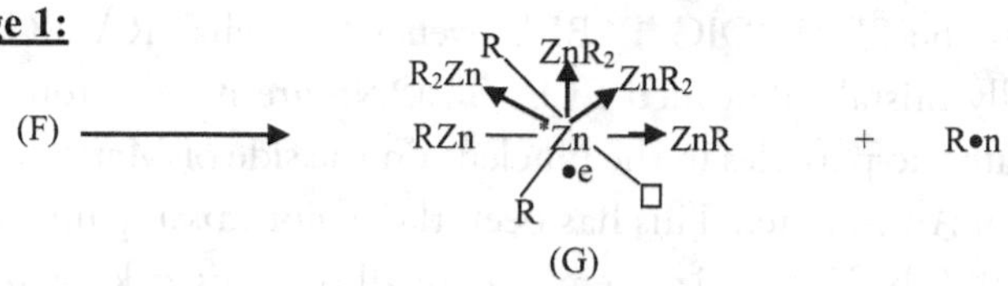

(F) $\longrightarrow$ + R•n

(G)

(G) + $H_2O \longrightarrow$ (G) – OH + H •e

(H)

H•e + R •n $\rightleftharpoons$ RH   14.283a

Overall Equation: (F) + $H_2O \longrightarrow$ RH + (H)   14.283b

Equilibrium Mechanism

Stage 1:

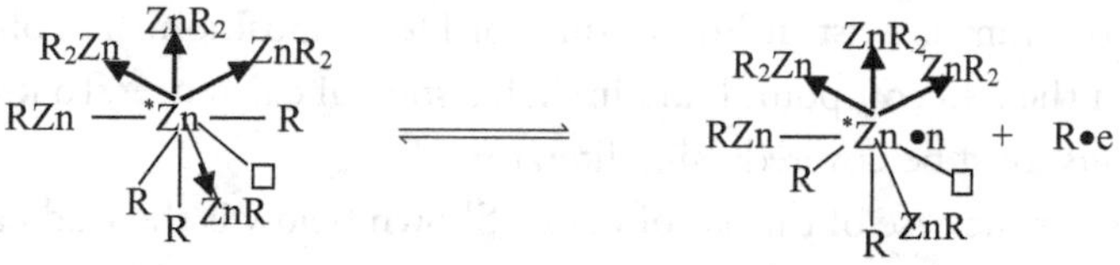

$\rightleftharpoons$ + R•e

(F)   (I)

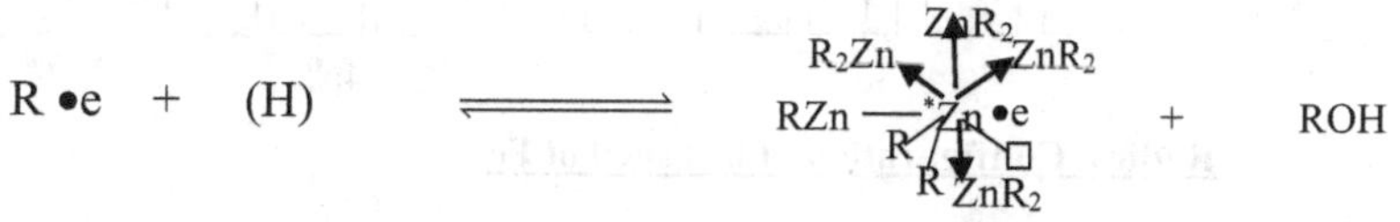

R •e + (H) $\rightleftharpoons$ + ROH

(J)

(J) + (I) $\longrightarrow$

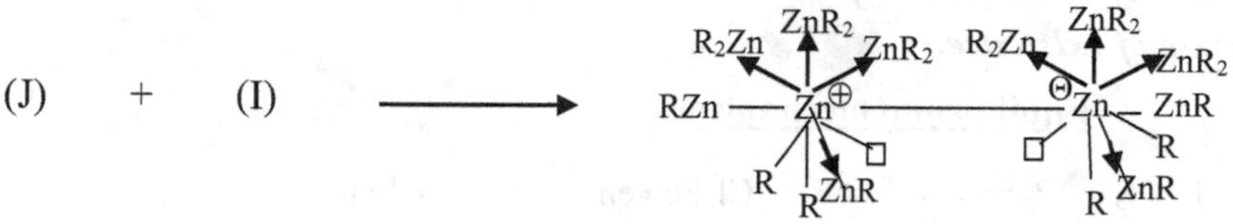

(II) THE INITIATOR   14.284

Overall overall Equation: $12ZnR_2$ + $H_2O \longrightarrow$ RH + ROH + (II)   14.285

(II) is the initiator of Z/N type. This is like the first case above- very unique set of initiators. Notice that the metals involved are Non-ionic metals. The initiator is a Covalently charged-paired initiator, since both centers are active- one for Males and other for Females. Note the ratio of the combination- ZnR_2/H_2O used. It is 12 to 1. Without using this ratio (Mathematics), the initiator can never be obtained. Only isotactic placements can be obtained with these initiators. When used with propylene oxide, isotactic poly (propylene oxide) is formed. Zn which is naturally known to be divalent, can be observed to be also tetravalent and hexavalent under certain conditions without breaking the Boundary laws. Based on the mechanisms which we have been observing since Volume (I), one is bound to ask wonderful questions. Why is there so much ORDERLINESS where no LAW is broken? Do these species 'COMMUNICATE" between themselves? What is the essence of OPERATING CONDITIONS? Are these species LIVING and living in a different world of their own? The questions are too countless to list, but to end with WHAT IS THE FORCE BEHIND ALL THESE GREAT WONDERS WHICH ARE INCOMPREHENSIBLE? That is the ONE AND ONLY MISTERY in humanity.

The greatest foundation for humanity is the PERIODIC TABLE even when what RADICALS are, had not been known. What is universally mistakenly called ELECTRONS are no electrons, but radicals. The electron is one of the eight sub-atomic-particles in the Nucleus on the side of Matter, while POSITRON, its mirror image is on the side of Anti-matter. This has been the major missing link since antiquity that which contains more than 50,000 NOBEL prizes, embracing all disciplines known and yet unknown. The award of NOBEL prizes in our world today is the second greatest achievement with respect to advancement of KNOWLEDGE, after the PERIODIC TABLE, but also with same type of major missing link.

The Equilibrium state of Existence of compounds is very important, because they are fixed and these are their FINGER-PRINTS, just like humans have finger-prints- the thumbs. One has already started preparing an encyclopedia for the finger prints of compounds. For without knowing them, one cannot get the real mechanisms of all reactions, since before chemical reactions can take place between two components, one of them or both must exist in Equilibrium or Decomposition state of existence. How can a reaction take place when the two compounds are in Stable state of existence? To know these finger prints, many chemical reactions must be covered using literature data.

Now, we will finally consider the case of the use of $FeCl_3$. Shown below is the Radical configuration of the last shell of Fe.

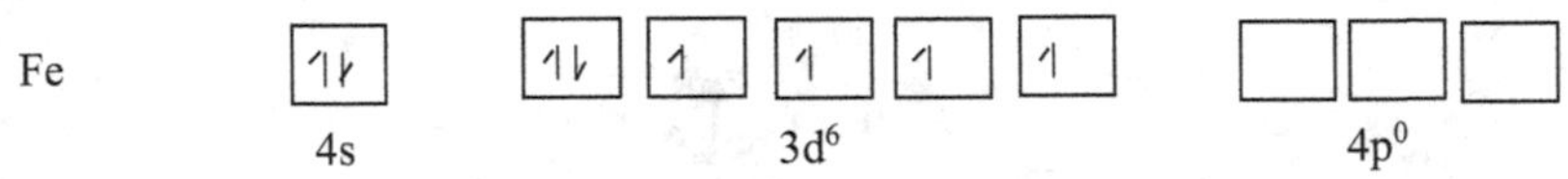

Fe   4s   3d⁶   4p⁰

Radical Configuration of last shell of Fe   14.286

↑ is male and ↓ is f ere are two paired unbonded radicals, an electro-n ing with three chlorine atoms. *With the presence of electro-non –free-radical, $FeCl_3$ can still be activated and can still exist in Equilibrium state of existence.*

Equilibrium mechanism

Stage 1: $FeCl_3$ $\rightleftharpoons$ $Cl_2Fe \bullet en$ + $Cl \bullet nn$

$Cl_2Fe \bullet en$ + $FeCl_3$ $\xrightarrow{Activation}$ $Cl_2Fe - \overset{Cl \diagdown \diagup Cl}{\underset{Cl}{Fe}} \bullet en$

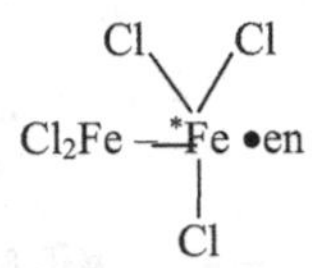

                                                          (A)

(A) + $Cl \bullet nn$ $\longrightarrow$ $Cl_2Fe -^*FeCl_4$

                                              (B)   14.287a

Overall Equation: $2FeCl_3$ $\longrightarrow$ (B)   14.287b

Note that the central Fe atom has been highlighted. This was the fraction that was kept in Stable state of existence. On that Fe, the number of paired unbonded radicals has been reduced to one, leaving behind two vacant orbitals. That center is again activated in stage 2, noting that the last shell of Fe can accommodate only eighteen (18) radicals based on one of the Boundary laws already stated.

Stage 2: $\qquad$ $FeCl_3 \rightleftharpoons Cl_2Fe\bullet en + Cl\bullet nn$

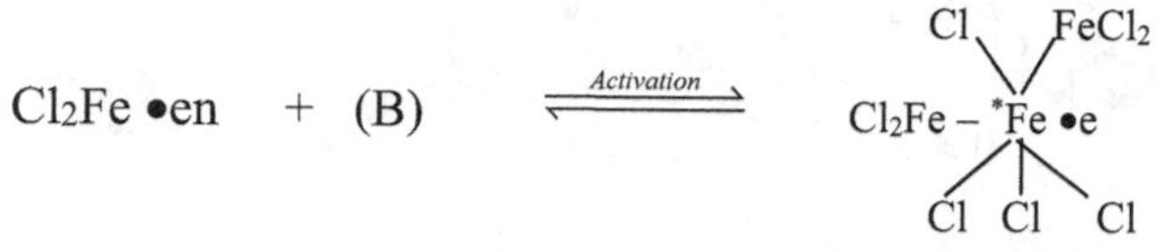

$Cl_2Fe\bullet en + (B) \xrightarrow{Activation}$ (structure labeled (Fe^*))

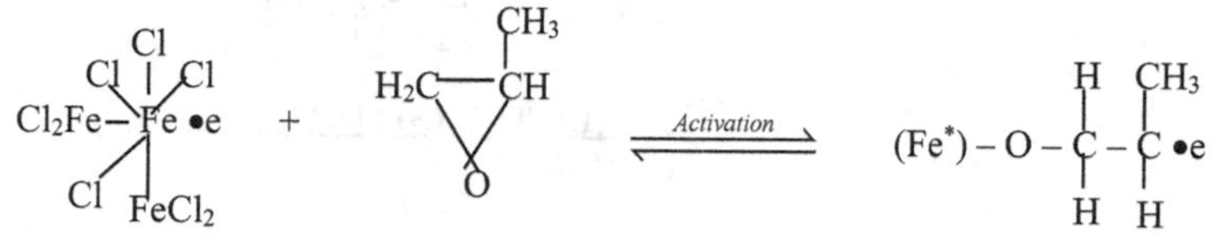

(structures with Activation step producing $(Fe^*)-O-\underset{H}{\overset{H}{C}}-\underset{H}{\overset{CH_3}{C}}\bullet e$)

$\rightleftharpoons (Fe^*)-O-CH_2-\overset{\bullet e}{CH}-CH_2 + H\bullet e$ (E), with $\bullet n$

$H\bullet e + Cl\bullet nn \rightleftharpoons HCl$

$(E) \xrightarrow{Deactivation} (Fe^*)-O-CH_2-CH=CH_2 + Heat$ (F) $\qquad$ 14.288a

Overall Equation: $3FeCl_3 + H_3CCH_2CHO \longrightarrow HCl + (F)$ $\qquad$ 14.288b

Note that the second step above is radically balanced, because the activation is taking place under equilibrium conditions (neither here nor there), the mechanism being Equilibrium mec-hanism. In the step, propylene oxide was next activated and in the process transfer species of the first kind of the first type was released to form HCl and finally (F). ***This clearly indicates that H is greater than -OFeCl_2 group in radical-pushing capacity, i.e., -FeCl_2 is less than H in radical-pushing capacity.*** At the end, the central Fe atom now has one vacant orbital and one electro-free-radical, since no paired unbonded radicals exist in the shell any more. The activation of the propylene oxide was necessitated by something else, as will shortly be explained.

Equilibrium Mechanism

Stage 3: $\qquad$ $HCl \rightleftharpoons H\bullet e + Cl\bullet nn$

$H\bullet e + (F) \rightleftharpoons HO-CH_2-CH=CH_2 + (Fe^*)$

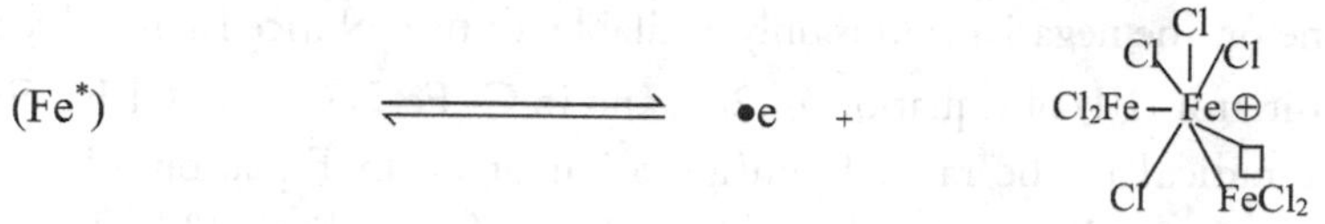

$(Fe^*) \rightleftharpoons \bullet e +$ (structure labeled $(Fe^\oplus)$)

$$•e \quad + \quad nn• \, Cl \quad \rightleftharpoons \quad Cl^{\ominus}$$

$$Cl^{\ominus} \quad + \quad (B) \quad \rightleftharpoons \quad Cl_2Fe-\overset{Cl}{\underset{Cl \quad Cl}{\overset{Cl \,\,\, Cl}{Fe}}}{\ominus} \, \square$$

$$(Fe^{\ominus})$$

$$(Fe^{\oplus}) \quad + \quad (Fe^{\ominus}) \quad \longrightarrow \quad Cl_2Fe-\overset{Cl \,\,\, Cl}{\underset{Cl \quad FeCl_2}{Fe}}\overset{\oplus}{}- \,\, \square \,\, \overset{\ominus}{....}Fe-FeCl_2$$

(III) THE INITIATOR   14.289

Overall Equation: $HCl \,\, + \,\, (F) \,\, + \,\, (B) \longrightarrow HOCH_2CH=CH_2 \,\, + \,\, (III)$ 14.290a

Overall overall Equation $6FeCl_3 + H_3CCH_2CHO \longrightarrow (III) + HOCH_2CH=CH_2$ 14.290b

The Equilibrium state of existence of (B) $[Cl_2Fe - FeCl_4]$ and $Cl_2Fe - FeCl_5 - FeCl_2$ are as follows.

$$Cl_2Fe - FeCl_4 \quad \underset{\textit{of Existence}}{\overset{\textit{Equilibrium State}}{\rightleftharpoons}} \quad Cl•nn \,\, + \,\, en• \, Fe-\overset{Cl \,\,\, Cl}{\underset{Cl \quad Cl}{Fe}}-Cl$$

2FeCl₃   14.291

$$Cl_2Fe - FeCl_5 - FeCl_2 \quad \underset{\textit{of Existence}}{\overset{\textit{Equilibrium State}}{\rightleftharpoons}} \quad Cl•nn \,\, + \,\, Cl_2Fe-\overset{Cl \,\,\, Cl}{\underset{Cl \quad FeCl_2}{Fe}}•e$$

3FeCl₃   14.292

It is no surprise therefore why the third step of Equation 14.287a is what it is, that is, that with a single right double headed arrow for Combination [Not Equilibrium or Decomposition States of existences]. It is also no surprise why propylene was involved in the third step of Stage 2 of Equation 14. 288a, otherwise there would have been no productive stage. The stage would have been "stable, reactive and insoluble". That is, $FeCl_3$ and $Cl_2Fe - FeCl_4$ would be insoluble in themselves. (III) is the real initiator, which unlike the others is carrying an imaginary bond-electrostatic bond. The initiator is Electrostatically positively charged-paired initiator, because only the positive center is active and the positive charge is not ionic but free. The negative center is not active, i.e., it cannot be used. Note that, both centers carry reservoirs for monomers, of which the one on the negative end is only available for use. Notice first and foremost, that in $(Fe^{\oplus})$ and $(Fe^{\ominus})$ of the Initiator (III) of Equation 14.289, and in $Cl_2Fe\text{-}FeCl_5$ and $Cl_2Fe\text{--}FeCl_5\text{--}FeCl_2$, the fourth electro-non-free-radical in the radical configuration of Fe in Equation 14.286, remained untouched throughout the stages to the end, but became an electro-free-radical. This clearly shows why Fe is not known to be tetravalent, despite the presence of that single radical in the last shell. Though in character, it is divalent and trivalent, it can be pentavalent and heptavalent, but not tetravalent, hexavalent or octavalent. But when coordinated, it can be made tetravalent as will be shown downstream. *One can clearly observe the wonders of NATURE. All the observations which have been highlighted above, are just but a grain of sand, because how can one continue to live in a world filled with IGNORANCE, when in NATURE Believing is seeing. In the Physical side of our world it is also true that Seeing is believing. Hence, in our world which is both Physical and Natural, no human is fully mentally balanced including the Author.*

14.5 Proposition of Rules of Chemistry and Concluding Remarks

This concludes all considerations so far on initiation of ringed monomers or compounds. Cyclic silicon containing monomers, cyclic sulfur containing monomers, inorganic or semi-inorganic cyclic monomers were considered. For the first time, silicon containing monomers have been distinguished from carbon containing monomers. Like Carbon, Silicon as a center cannot carry ionic charges, i.e., it is a non-ionic metal. Unlike Carbon, Silicon cannot carry real charges. It can however carry polar and electrostatic charges which all have radical characters. However, silicon compounds mainly undergo radical reactions. Silicon is more electropositive than H, while Carbon is less electropositive than H. Hence Alkanes are very different from Silanes. For the first time, the true chemistry of silicon containing compounds and monomers such as cyclosil**o**xanes (e.g., octamethyl-cyclotetrasil**o**xane), cyclosil**a**zanes (e.g., hexamethylcyclotrisil**a**zane), were provided. Why and how polymer chains which contain cyclic and branched structures along the chain during polymerization of hexamethylcyclotrisil**a**zane were explained, from which one can begin to know how high molecular weight polymers can be obtained from them.

For the first time, rings containing S - S, Si - O bonds etc., single bonds which are relatively weak have been distinguished from other types of rings. When such bonds exist in a ring, it has been shown, based on literature data, why their small sizes cannot exist and much more. The chemistry of cyclic sulfides, cyclic disulfides, cyclic trimmer of thioformal-dehyde and rhombic sulfur were also for the first time provided. The chemistry of thiophenes, furans and related monomers or compounds have been provided.

For the first time, a new type of electroradicalization phenomenon associated with cyclic phosphites has been identified. It is so unique, because an electro-radical is made to move from paired unbonded radicals instead of bonded one, i.e., p-bond, to grab a sitting nucleo-radical to form a polar bond instead of a p-bond. The points of scission in rings, instantaneously or via functional centers have been identified to be the weakest bond in all Nucleophilic rings all the time, but different in some Electrophilic rings such as in Sydnones. For the first time the real structures of many compounds including sydnones were provided or confirmed.

From the beginning of time, it has always been thought that, only ionic, and dative charges exist. Not even covalent charges were generally known to exist. It has been shown that there are indeed four types- two real charges and two imaginary ones. These are used to form their bonds-ionic, covalent, electrostatic and polar bonds. There is also an additional bond, the Dative bond which do not carry charges, because no electro-radicals have been moved when they are formed. ***Charges can only be formed when electro-radicals move or are moved.*** How they favor electrostatic/electrodynamic forces of repulsion and attraction between themselves and with paired unbonded radicals, have been identified. Since one has begun to distinguish between different types of bonds and charges in chemical systems, one will commence stating laws relevant to the driving forces favoring their existences.

For the first time, the nature of an inorganic monomer such as the cyclic trimmer of di-chlorophosphazene or phosphonitrilic chloride were provided. In the process a new type of molecular rearrangement called Chlorolization analogous to Enolization and Imidization (or Amidization) was identified. So also was another type of Termination step identified and this was called Termination by self.

So far, one can observe that the world of Chemistry is too much to comprehend to the point where *the Department of Chemistry should be made a Faculty of Chemistry with many sub-disciplines as Departments in the faculty under the School of Natural Sciences. Physics and Mathematics will also be faculties in that School. This School will be the only School in any tertiary institution, while the others, will be Colleges with their respective Faculties and Departments.*

Rule 1200: This rule of Chemistry for **Movement of electro-radicals alone,** states that, their movement can be for *discrete or continuous applications,* for which when discrete, bonds and charges are formed and when continuous many applications too countless to list are made possible.
(Laws of Creations for Electro-radicals)

Rule 1201: This rule of Chemistry for **Movement of Radicals and Charges alone,** states that, since only electro-radicals can be removed and move, all phenomena associated with it such as Resonance stabilization, Transmission of current, Mass communication can never take place chargedly.
(Laws of Creations for Engineering)

Rule 1202: This rule of Chemistry for **Movement of species which carry negative charges,** states that, apart from their ability to diffuse only in the *absence of a positive charge* in the whole system or when paired to a positive center, they can also diffuse to the Central atom of a species which has a vacant orbital to *form a Covalent bond and place an imaginary negative charge on the Central atom of the species* only in the *presence of a positive charge for Electrostatic bond formation.*
(Laws of Creations-for Positive Electrostatic bond formation)

Rule 1203: This rule of Chemistry for **Movement of species which carry positive charges,** states that, apart from being the only one to always diffuse and begin a chemical reaction in a charged environment and in the presence or absence of negative species, and diffuse when paired to a negative center, they can also diffuse alone to a Central atom of a species which has paired unbonded radicals to *form a Covalent bond and place an imaginary positive charge on the central atom* of the species in the presence of a negative charge for Electrostatic bond formation.
(Laws of Creations-for Negative Electrostatic bond formation)

Rule 1204: This rule of Chemistry for **Movement of species which carry electro-radicals,** states that, apart from being the only one to always diffuse and begin a chemical reaction in a radical environment and in the presence or absence of nucleo-radicals, due to limitation placed on the boundaries of atoms, *they can be made to diffuse in addition to form an imaginary bond such as in dimeric KCN, only when vacant orbitals exist on the Central atom.*
(Laws of Creations for Boundaries)

Rule 1205: This rule of Chemistry for **Movement of electro-radical carrying species,** states that, when two of them carried by two different centers are present in the system, it is the center which is more electropositive that is the first to diffuse.
(Laws of Creations for movement of radicals)

Rule 1206: This rule of Chemistry for **Movement of species which carry nucleo-radicals,** states that, apart from their ability to diffuse only in the absence of electro-radicals in the whole system or when

paired, due to limitations placed on the boundaries of atoms, *they can be made to diffuse to form an imaginary bond only when vacant orbitals exist in the Central atom.*
(Laws of Creations for Boundaries)

Rule 1207: This rule of Chemistry for **Polar bond formation,** states that, they cannot be formed in the absence of a Sigma (σ) bond and when formed their presence is either due to the limitations placed on the boundaries of atoms, for which an electro-radical is donated to another center carrying a nucleo-radical to yield positive and negative charges which *forms an imaginary double bond* or due to the fact that *one or two metallic centers are involved.*
(Laws of Creations for Boundaries)

Rule 1208: This rule of Chemistry for **Equilibrium State of Existence (Finger-Prints) of compounds,** states that, when favored based on the operating conditions, this can only be done *radically and not ionically,* since ions are creations from movement of electro-free-radical and not all compounds are ionic in character; for which Ionic Equilibrium state of existence exists only for Ionic/Polar compounds in an ionic environment and since this is the only state of existence favored by them, no productive stage can be obtained ionically.
(Laws of Creations for Equilibrium State of Existence)

Rule 1209: This rule of Chemistry for **Ringed and non-ringed monomers that carry at least one hetero atom such as cyclic ethers and aldehydes,** states that, while for non-ringed monomers, the route natural to them, (when Electrostatically paired initiators that favor Equilibrium state of existence are used), is Electro-free-radical; for ringed monomers, the route is also Electro-free-radical, since the ring is instantaneously opened once the electro-free radical has touched the functional center and provide the MRSE for the ring without breaking the boundary laws.
(Laws of Creations for Boundaries-Ringed monomers with functional centers)

Rule 1210: This rule of Chemistry for **Resonance Stabilization phenomenon**, states that, in general, where two or more π-bond types of Activation centers are conjugatedly or cumulatively placed, the systems must be resonance stabilized more so conjugatedly than cumulatively; for which imaginary links are built into the overall systems.
(Laws of Creations for Resonance Stabilization Phenomenon)

Rule 1211: This rule of Chemistry for **Resonance Stabilization phenomenon,** states that, the only time an electro-non-free-radical can stay on elements like O and N in the presence of more electropositive elements such as C and H on the same compound without release of energy and a stable molecule, is only when the electro-non-free-radical is resonance stabi-lized, such as in Sydnones, Diazoalkanes and p-Benzoquinones.
(Laws of Creations for Resonance Stabilization Phenomenon)

Rule 1212: This rule of Chemistry for **Electroradicalization phenomenon,** states that, in general, there are **two types of the first kind** of Electroradicalization based on what happens when a compound with a π-bond conjugatedly placed to group(s) is put in Equilibrium state of Existence-Type I- that in which a sitting electro-radical (visible) moves to grab a nucleo-radical from a conjugatedly placed π-bond to form another π-bond instantaneously leaving an electro-radical on another center to which what was held in Equilibrium state of existence adds to it to form another compound;

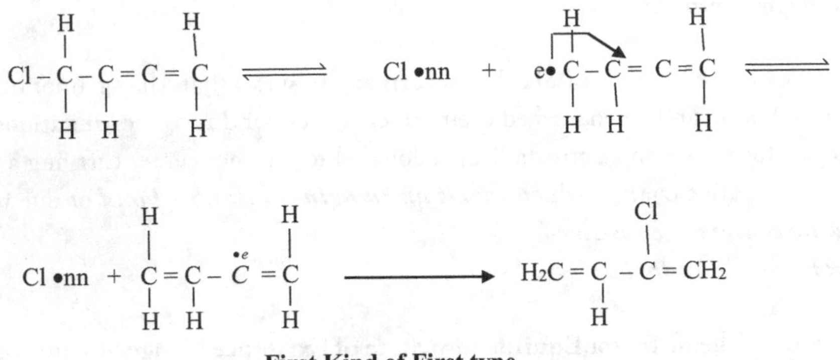

First Kind of First type

and Type II- that in which an electro-radical (invisible) moves from a conjugatedly placed π-bond to grab a sitting nucleo-radical to form another p-bond instantaneously leaving a nucleo-radical on another center to which what was held in Equilibrium state of existence adds to form another compound-

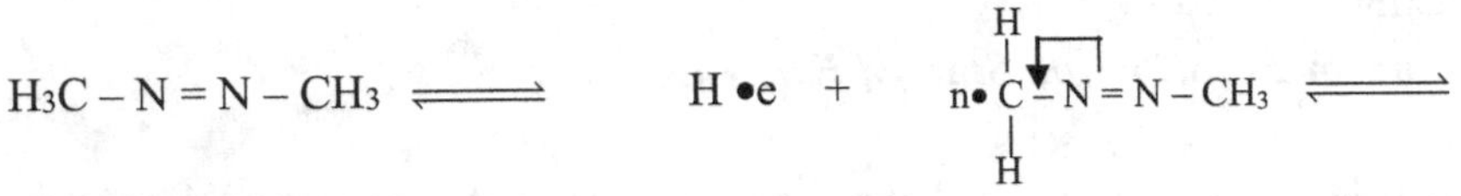

(Laws of Creations-for Electroradicalization phenomenon)

Rule 1213: This rule of Chemistry for **Electropolaradicalization phenomenon,** states that, this form of molecular rearrangement *so-called Arbuzov rearrangement,* takes place when a compound where the central atom has paired unbonded radicals is placed in Equilibrium state of existence, *an electro-radical moves from the paired unbonded radicals to grab the adjacently located sitting nucleo-radical to form a polar bond leaving a nucleo-radical on the central atom to which what was held in Equilibrium state of existence adds to form another compound with polar bond* as shown below-

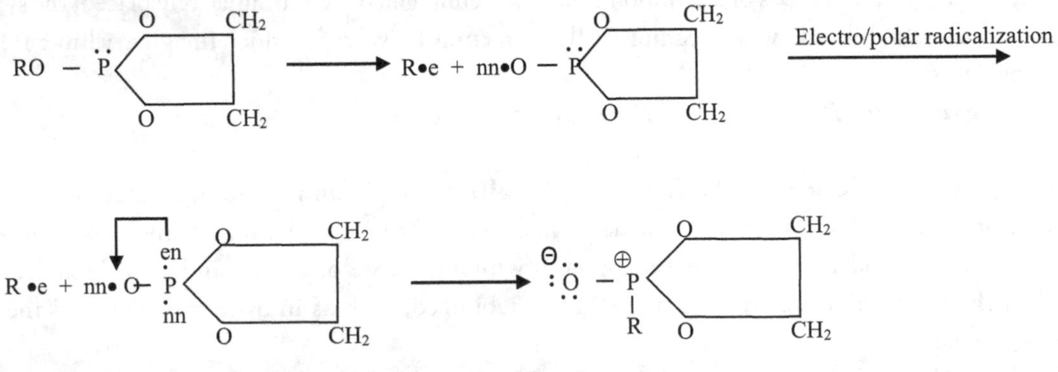

Electropolaradicalization phenomenon

(Laws of Creations-for Electropolaradicalization phenomenon)

Rule 1214: This rule of Chemistry for **Chlorolization phenomenon,** states that, just like Enolization and Amidization phenomena, when a chlorinated unsaturated compound is put in Activated/Equilibrium state of existence wherein Cl is held as an atom, the electro-radical center which was present on activation

434

adds to the chlorine atom to form a stable group leaving a new electro-radical center where the chlorine atom departed from, to form new activation centers as shown below-

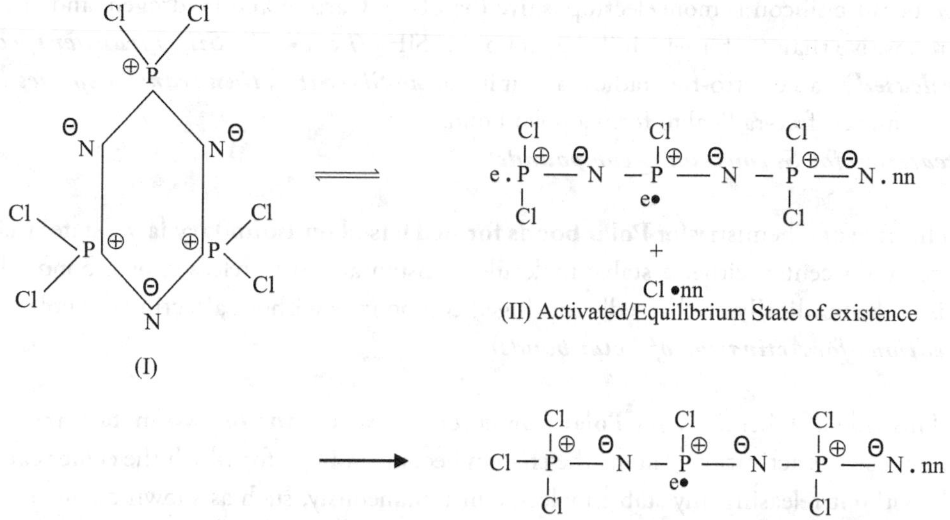

(I)

(II) Activated/Equilibrium State of existence

(III) Chlorolization phenomenon

for which unlike Enolization and Amidization, what is finally formed [(III) above] cannot deactivate because of the presence of MaxRSE if a four-membered was to be formed.

(Laws of Creations-for Chlorolization)

Rule 1215: This rule of Chemistry for **Termination Step in polymerization systems,** states that, **Termination by Self** is that which takes place *when a monomer either in self-activated state or activated in the absence of an initiator,* grows in chains to form living polymers from both ends, which when terminated by release of transfer species of the first kind only from one end, the other end diffuses to add to it if what was released is a nucleo-radical or a negative charge; for if what was released was an electro-radical or a positive charge, it diffuses to kill the other end, noting that the dead polymer is a dead terminal double or triple or polar bond polymer as shown below-

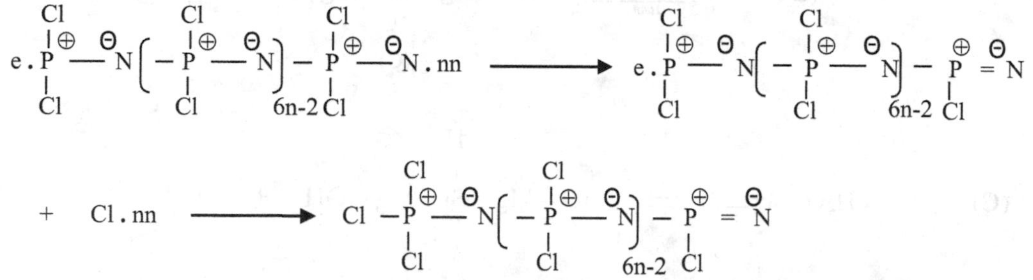

Termination by Self (i)

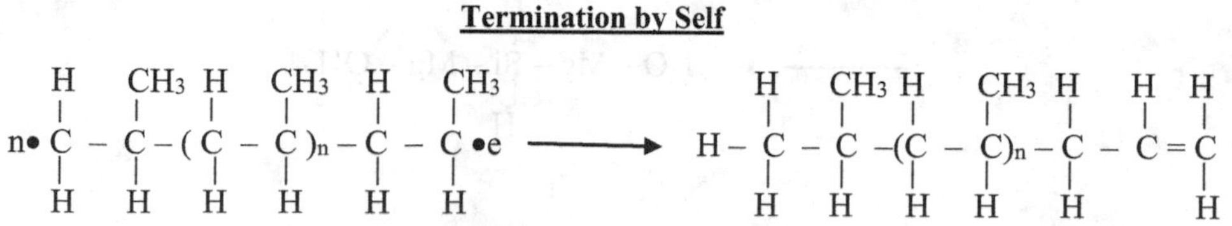

Termination by Self (ii)

(Laws of Creations-for Termination by Self)

Rule 1216: This rule of Chemistry for **Silicon center carrying an alkylane group such as CH$_3$,** states that, unlike Carbon center, *transfer species of the first kind of the first type cannot be released from its growing polymer chain*, because silicon is more electropositive than both Carbon and Hydrogen, and also that H is more electropositive than C; for which if the group was SiH$_3$ *(i.e. e• Si – SiH$_3$) transfer species (H) will not be released* as an electro-free-radical and if it was *n• Si – SiH$_3$, then transfer species (H) will be released* as a nucleo-free-radical to form a polar bond.
(Laws of Creations- for Si containing compounds)

Rule 1217: This rule of Chemistry for **Polar bonds formed based on Boundary laws,** states that, when used as an activation center, either a stable molecule is instantaneously released or a bond adjacently located to it is broken radically or chargedly for ringed compounds without affecting the polar bond.
(Laws of Creations-for Activation of Polar bonds)

Rule 1218: This rule of Chemistry for **Polar bonds formed when one or two metals are involved,** states that, when used as activation centers, the charges become radical, for which the center can only be used radically without releasing any stable molecule instantaneously, such as shown below for reaction between magnesium silicide and water or aqueous acid on heating.

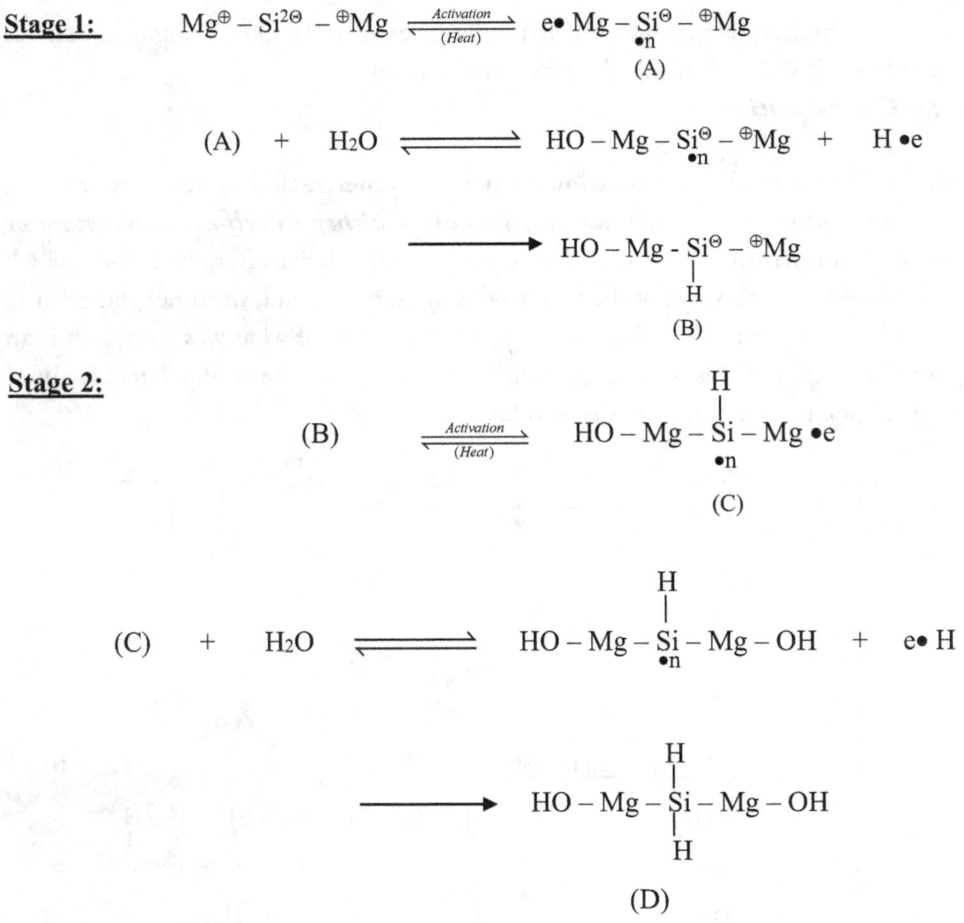

Stage 1:

$$Mg^{\oplus} - Si^{2\ominus} - {}^{\oplus}Mg \xrightarrow[\text{(Heat)}]{\text{Activation}} e\bullet\, Mg - \underset{\bullet n}{Si^{\ominus}} - {}^{\oplus}Mg$$
$$(A)$$

$$(A) \quad + \quad H_2O \rightleftharpoons HO - Mg - \underset{\bullet n}{Si^{\ominus}} - {}^{\oplus}Mg \quad + \quad H\bullet e$$

$$\longrightarrow \quad HO - Mg - \underset{|}{Si^{\ominus}} - {}^{\oplus}Mg$$
$$H$$
$$(B)$$

Stage 2:

$$(B) \quad \xrightarrow[\text{(Heat)}]{\text{Activation}} \quad HO - Mg - \underset{\bullet n}{\overset{\overset{\displaystyle H}{|}}{Si}} - Mg \bullet e$$
$$(C)$$

$$(C) \quad + \quad H_2O \rightleftharpoons HO - Mg - \underset{\bullet n}{\overset{\overset{\displaystyle H}{|}}{Si}} - Mg - OH \quad + \quad e\bullet H$$

$$\longrightarrow \quad HO - Mg - \underset{\overset{|}{H}}{\overset{\overset{\displaystyle H}{|}}{Si}} - Mg - OH$$
$$(D)$$

Stage 3:

$$(D) \rightleftharpoons HO - Mg \bullet e + n \bullet \overset{\overset{H}{|}}{\underset{\underset{H}{|}}{Si}} - Mg - OH$$

$$(E)$$

$$HO - Mg \bullet e + H_2O \rightleftharpoons HO - Mg - OH + H \bullet e$$

$$H \bullet e + (E) \longrightarrow H_3Si - Mg - OH$$

Stage 4: $\quad H_3Si - Mg - OH \rightleftharpoons H_3Si \bullet n + e \bullet Mg - OH$

$$HO - Mg \bullet e + H_2O \rightleftharpoons HO - Mg - OH + H \bullet e$$

$$H \bullet e + n \bullet SiH_3 \longrightarrow SiH_4$$

$$\text{Silane}$$

Overall Equation: $\quad Mg_2Si + 4H_2O \longrightarrow 2Mg(OH)_2 + SiH_4$

(Laws of Creations for Activation of Polar Bonds)

Rule 1219: This rule of Chemistry for **Paired unbonded radicals carried by some central atoms,** states that, these can be used during activation either when there is a vacant orbital on the central atom or when presence of an electrostatic bond can be favored along the chain as shown below in the copolymerization of alkenes ($H_2C=CHR$) with phosphines (PR^1_3) to pro-duce alternating placements-

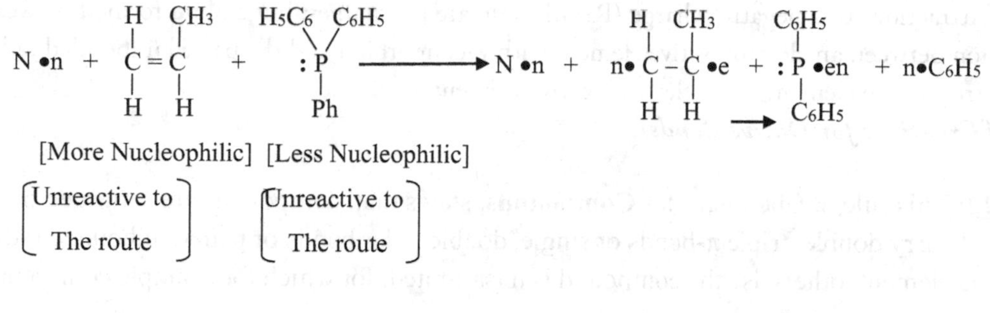

[More Nucleophilic] [Less Nucleophilic]

$$\begin{bmatrix} \text{Unreactive to} \\ \text{The route} \end{bmatrix} \quad \begin{bmatrix} \text{Unreactive to} \\ \text{The route} \end{bmatrix}$$

(A)[Couple]

UNIQUE POLYMER CHAIN   **INITIATION STEP**

for which the nucleo-free-radical which is unreactive to both monomers waites for them to form a couple via diffusion of the more female monomer to the less female monomer kept in Equilibrium state of existence due to limitations imposed by the Boundary laws and adds to the nucleo-radical on the paired unbonded radicals and forms an Electrostatic bond simulta-neously, to form a couple, followed by addition of the initiator to the couple to complete the Initiation step; noting that the same will apply when the phenyl group is replaced with alkyl groups.

(Laws of Creations-for Activation of paired unbonded radicals)

Rule 1220: This rule of Chemistry for **Silenes (Disilenes) and Silynes (Disilynes),** states that, unlike alkenes, these and their members do exist not with double or triple bonds, but with polar bonds as shown below; for which one can observe that it is in the self –activated state and unstable, making them difficult to use as monomers, since as Nucleophiles (Only X center), their natural route will become the nucleo-free-radical route as opposed to the electro-free-radical when C is used in place of Si, making C so unique to Life and Living.

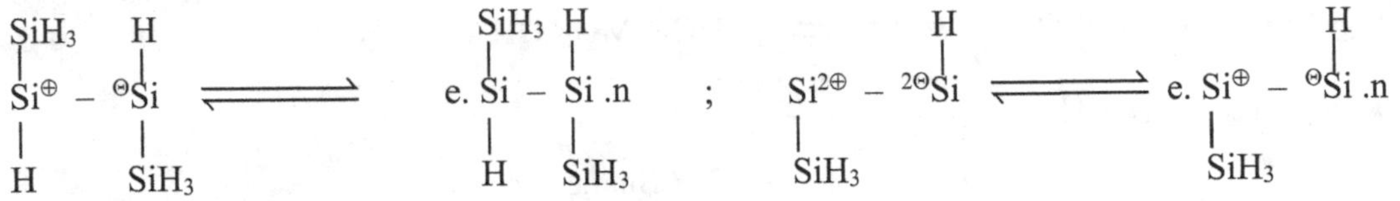

A SILENE ; A SILYNE

(Laws of Creations for Silenes, Silynes)

Rule 1221: This rule of Chemistry for **Bonds,** states that, of all bonds known to exist-**Real bonds** consisting of Ionic, Covalent and Dative bonds and **Imaginary bonds** consisting of Polar bonds and Electrostatic bonds, *only the Dative bonds do not carry charges,* since no electro-radical is moved or transferred and paired unbonded radicals [though behave like negative charges in the presence of a positive (Attraction) or a negative charge (Repulsion)], are no charges; but a bond formed by weak force of attraction between an electropositive element with vacant orbital and the paired unbonded radicals on a less electropositive element or an electronegative element.

(Laws of Creations for Dative Bonds)

Rule 1222: This rule of Chemistry for **Compounds,** states that, compounds are saturated only when they do not carry double, triple π-bonds or single, double polar bonds or paired unbonded radicals on a non-ionic element, otherwise the compound is unsaturated, for which for example compounds such as phosphines are unsaturated-

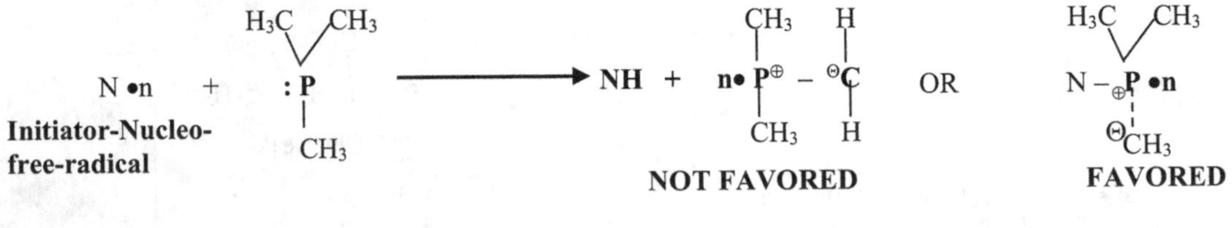

Rule 1223: This rule of Chemistry for All **Atoms with respect to their established valences which is very much in order,** states that, when paired unbonded radicals are present in the last shell of the atom, the atom can go beyond its established valence state as long as the boundary laws are not contravened for example in $HClO_4$, Chlorine atom which is naturally monovalent with only one bond carried by it, is tetravalent as shown below with the Chlorine element carrying not more than eight radicals in the last shell-

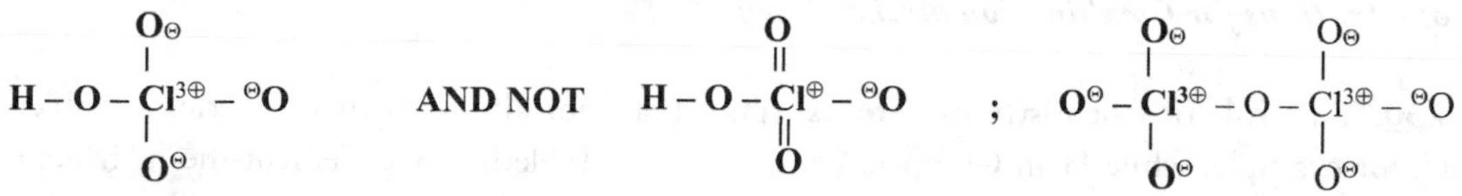

$$H - O - Cl^{3\oplus} - {}^{\ominus}O \qquad \textbf{AND NOT} \qquad H - O - Cl^{\oplus} - {}^{\ominus}O \quad ; \quad {}^{\ominus}O - Cl^{3\oplus} - O - Cl^{3\oplus} - {}^{\ominus}O$$

Perchloric acid   **Chlorine heptoxide**

noting that the overall valency state of the Cl atom still remains at -1, i.e., monovalent.
(Laws of Creations for Valence States)

Rule 1224: This rule of Chemistry for **All Chemical and Polymeric reactions that take place via Equilibrium mechanism,** states that, before any chemical reaction between two compounds can take place, one or both of them must exist in Equilibrium or Activated State of existence, otherwise if both are in Stable state of existence, no chemical reactions can take place.
(Law of Creations for Existence of Chemical Reactions by Equilibrium mechanism)

Rule 1225: This rule of Chemistry for **All Chemical and Polymeric reactions that take place via Decomposition mechanism,** states that, before any chemical reaction between two compounds can take place, one of them must be in Decomposition state of Existence, while the other is in Stable state of existence, otherwise no chemical reactions can take place.
(Laws of Creations for Existence of Chemical Reactions by Decomposition mechanism).

Rule 1226: This rule of Chemistry for **All Chemical and Polymeric reactions that take place via Decomposition mechanism,** states that, when chemical reactions take place between two compounds, it is *the more Electropositive central atom of one of the molecules* that commences the stage; clear indication of one of the greatest influences of Electro-positivity/Electronegativity of atoms in the Periodic Table.
(Laws of Creations for Existence of Chemical Reactions by Decomposition mechanism)

Rule 1227: This rule of Chemistry for **Chemical reactions that take place via Decomposition mechanism,** states that, when the central atoms of the two molecules reacting are equi-electropositive, it is the compound in larger molar concentration that commences the stage, otherwise if they are of the same concentration, no reaction takes place; clear indication of the great influence of molar concentrations in all systems.
(Laws of Creations for Existence of Chemical Reactions by Decomposition mechanism)

Rule 1228: This rule of Chemistry for **All Chemical reactions that take place via Equilibrium mechanism,** states that, in chemical reactions involving three or more compounds, they take place two or three at a time in stage(s) to give final products which can never react with themselves under the operating conditions; for which if two of the compounds are either soluble or insoluble in themselves, the presence of the third compound is to provide a productive stage.
(Laws of Creations for Existence of Chemical Reactions)

Rule 1229: This rule of Chemistry for **Chemical and Polymeric reactions,** states that, never is there a time chemical reactions take place via Combination mechanism; only polymeric reactions take place via

Combination mechanism in the absence of any terminating agent, because in chemical reactions, what are called Initiators are not involved.

(Laws of Creations for Combination mechanisms)

Rule 1230: This rule of Chemistry for **Atoms,** states that, not all atoms in the Periodic Table have elements, for example, while H in Group Ia of the Periodic Table has one element-the hydride, other members in the family do not have elements- their element is the atom.

(Laws of Creations for Elements/Atoms)

Rule 1231: This rule of Chemistry for **Paired unbonded radicals present on some atoms,** states that, these are like anionic or negative charges but of weaker capacity, since anionic charge is coming from two central atoms, while paired unbonded radicals is carried by one central atom.

(Laws of Creations for Paired Unbonded Radicals)

Rule 1232: This rule of Chemistry for **Paired unbonded radicals present on some atoms,** states that, based on limitations imposed by Boundary laws, while paired unbonded radicals will not repel themselves or anions or negative charges when present on the same central atom, they will repel anions or negative charges which are adjacently located on another central atom as shown below-

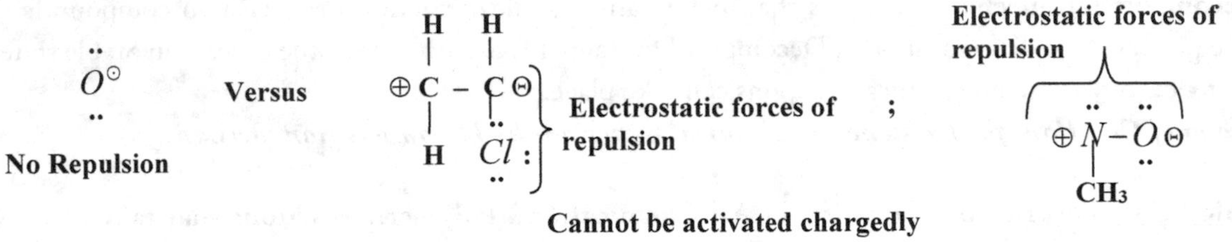

Cannot be activated chargedly

(Laws of Creations for Paired Unbonded Radicals)

Rule 1233: This rule of Chemistry, for **Paired unbounded radicals,** states that, just like radicals, when paired unbonded radicals are adjacently located, no electrostatic forces of repulsion takes place as it does between paired unbonded radicals and a negative charge; otherwise a compound such as $FeCl_3$ will not exist.

(Laws of Creations for Paired unbounded radicals)

Rule 1234: This rule of Chemistry for **Paired unbonded radicals present on some atoms,** states that, when specially located on a monomer, during propagation when polymerized particularly in the route not natural to the monomer, due to what is herein called **Electro-dynamic forces of repulsion,** the monomers are syndiotactically placed along the chain as shown below using vinyl chloride (A Nucleophile), noting that the natural route of the monomer is electro-free-radical route, being a weak Nucleophile.

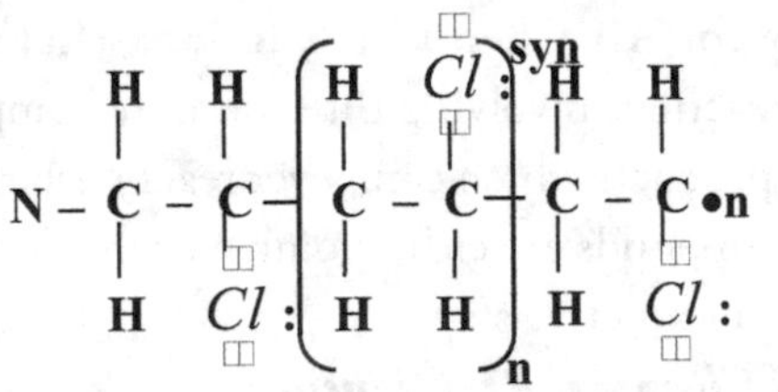

Electro-dynamic forces of repulsion

(Laws of Creations for Paired unbonded radicals)

Rule 1235: This rule of Chemistry for **Paired bonded radicals called π-bonds,** states that, a negative charge or anion when it exists cannot be adjacently located to it due to electrostatic forces of repulsion, because p-bond can create positive and negative charges, electro- and nucleo-radicals on activation, for which compounds which carry triple bonds cannot be activated as shown below.

$$\overset{\displaystyle CH_3}{\underset{\displaystyle CH_3}{{}^{\Theta}C = C^{\oplus}}} \qquad ; \qquad {}^{\Theta}\overset{..}{N} = \overset{..}{N}^{\oplus}$$

Electrostatic forces of repulsion.
[They cannot be activated chargedly)

(Laws of Creations for Paired Bonded Radicals)

Rule 1236: This rule of Chemistry for **Paired unbonded and paired bonded radicals,** states that, these when placed side by side do not repel themselves as shown below.

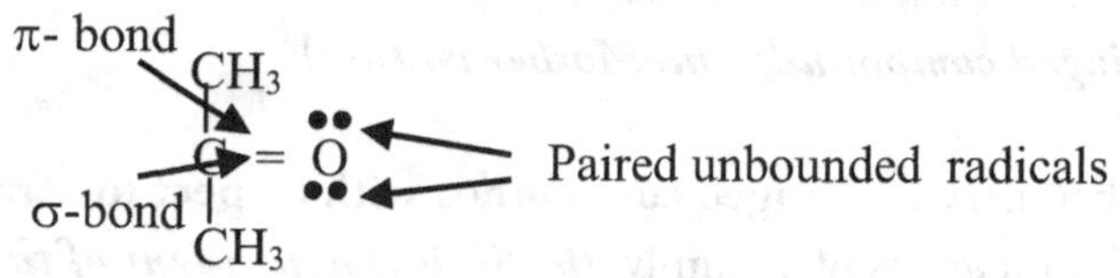

$$\begin{array}{c} \pi\text{- bond} \\ \underset{\sigma\text{-bond}}{\overset{CH_3}{\underset{CH_3}{C}}} = \overset{..}{\underset{..}{O}} \leftarrow \text{Paired unbounded radicals} \end{array}$$

(Laws of Creations for Paired Bonded and Unbonded Radicals)

Rule 1237: This rule of Chemistry for **Compounds,** states that, Equilibrium, Decomposition, and Combination States of Existences take place only via σ- bond expressed only radically, while only those with ionic bonds can be expressed radically and ionically.
(Laws of Creations for States of Existence)

Rule 1238: This rule of Chemistry for **Paired unbonded radicals,** states that, the number of paired unbonded radicals present on an atom and a molecule is *a measure of its polar character and its ability to passively suppress the Equilibrium state of Existence of some compounds*; for the more the number, the greater the polarity and the greater the ability to suppress the Equilibrium State of existence of some compounds passively just as anions do.
(Laws of Creations for Paired Unbonded Radicals)

Rule 1239: This rule of Chemistry for **Ionic metals,** states that, when present during a chemical reaction, they in view of their strong electropositivity have the ability amongst other unique abilities, to suppress the Equilibrium state of Existence of all compounds where H is present, since there is hierarchy **in Equilibrium states of existence of compounds,** just as there is in Life.
(Laws of Creations for Ionic metals)

Rule 1240: This rule of Chemistry for **Nucleophilic (Female) and Electrophilic (Male) compounds,** states that, while a Nucleophile (X) can exist without carrying a Y center, an Electrophile (X and Y) cannot exist without carrying an X center adjacently located to a Y center.
(Laws of Creations –MOTHER NATURE & Woo-MAN)

Rule 1241: This rule of Chemistry for **Ringed and non-ringed monomers or compounds,** states that, just as the number of nucleophilic centers in a compound *is additive and a measure of the order of nucleophilicity of the compound,* so also it applies to the number of electrophilic centers in a compound, noting however that all activation centers are Nucleo-philic in character.
(Laws of Creations for Unsaturated Compounds/Mother Nature)

Rule 1242: This rule of Chemistry for **Ringed monomers,** states that, the character of strain energies in rings which are said to be invisible π-bonds, is determined by the characters of visible functional and π-bond activation centers present in the ring, *for which the strain energy in the ring can be said to be the difference between the total bond energies in the ring and the total bond energies of its non-ringed isomeric monomer or counterpart.*
(Laws of Creations for Mathematics/Physics)

Rule 1243: This rule of Chemistry for **Carbon Ringed monomers or compounds,** states that, in general the strain energy in any ring whether the monomer or compound is Nucleophilic or not is just a C=C π-bond, Nucleophilic or Electrophilic in character.
(Laws of Creations for Ringed compounds and Mother Nature)

Rule 1244: This rule of Chemistry for **Ringed compounds with respect to Strain Energy (SE),** states that, for a series of particular members of a Family, *the SE is independent of the size of the rings, but dependent on the types of elements and groups carried by them,* for which for a particular family wherein the elements and the groups carried are the same, as shown below for cycloalkanes; the SE is fixed at for example 25 kcal/mol for all Cyclo-alkanes-

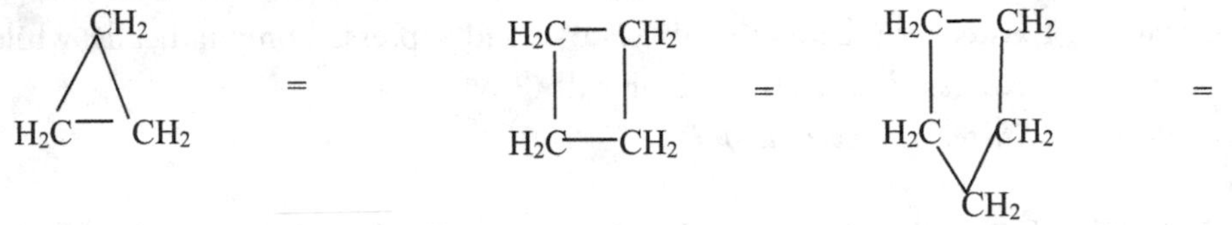

SE = 25 kcal/mol

noting that because of this, the smaller the size of the ring, the more stained is the ring, for which in order to open the smallest sized ring, very small energy will be required to reach what has been called the Minimum Required Strain Energy (MRSE), the energy required increasing in size with increasing size of ring.
(Laws of Creations for Physics/Mathematics)

Rule 1245: This rule of Chemistry for **Compounds,** states that, when Silicon compounds are present with other compounds, ionic groups such as, amino, and imino, OH, etc., isolate their ionic identities; just as what equally takes place when monomers are activated radically and chargedly; *making it look as if atoms and compounds know when charges and radicals are around.*
(Laws of Creations for Identities of Compounds)

Rule 1246: This rule of Chemistry for **Si and C atoms,** states that, when they are activated from their ground state, the followings are obtained; for which for silicon, activated silicon is obtained and for carbon, a carbon element which is what is known as Activated carbon is obtained, noting that the arrows used to indicate the male and female characters of the radicals have different sizes based on the electronegativity of the atom.

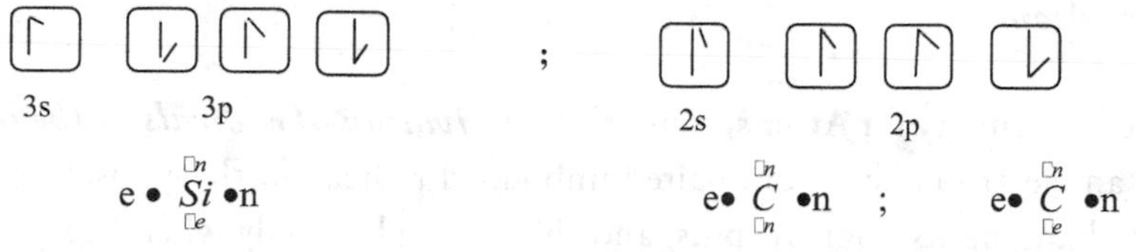

e • *Si* •n

Activated Silicon

e• *C* •n ; e• *C* •n

(A) Activated Carbon   **(B) Activated Carbon**

Black

(Laws of Creations for Si and C atoms)

Rule 1247: This rule of Chemistry for **Silicon carbide (SiC),** states that, the composition of silicon carbide is $C(Si)_4$ wherein C is centrally placed with Si at the corners of a hexagon all connected together with a single bond, for which the presence of more silicons, makes the carbide as strong as or stronger than diamond.
(Laws of Creations for Si and C atoms)

Rule 1248: This rule of Chemistry for **Silanes and Alkanes,** states that, their Equilibrium states of existence as shown below clearly indicates that Si is more electropositive than H while C is less electropositive than H; for which the same applies to other cases which can form hydrides with H.

$$SiH_4 \quad \rightleftharpoons \quad e• \, SiH_3 \; + \; H •n \quad ; \quad CH_4 \quad \rightleftharpoons \quad n• \, CH_3 \; + \; H•e$$

(Laws of Creations for Si and C atoms)

Rule 1249: This rule of Chemistry for **Tertiary Alkyl Chlorides (R_3CCl) and Tertiary Alkyl-ChloroSilanes (R_3SiCl),** states that, the fact that it is possible to eliminate HCl from tertiary alkyl chlorides to form an unsaturated compound with carbon-carbon double bond when heated and not possible with tertiary alkylchlorosilanes to form carbon-silicon double bond, does imply that there is no form of carbon-silicon "double bond"; noting that H cannot be released as an atom being less electropositive than Si, and cannot also be released as a hydride from a C center adjacently located to a Si center, whether silicon is carrying an electro-free-radical or not.
(Laws of Creations for Si and C atoms) [See Rule 1216]

Rule 1250: This rule of Chemistry for **Atoms,** states that, *all paired unbonded radicals in a sub-atomic-orbital of the last shell of an atom are of opposite spins, with none of parallel spin, not even for the cases of all Group IIa atoms in the Periodic Table* – for this is not a case of two like poles attracting, like two hydrogen atoms attracting to form H_2 molecule; for which by the nature of Excitation phenomenon, no Hybridization (i.e., mixing of radicals in s-, p-, and d- orbitals to keep them at the same energy levels) can take place.
(Laws of Creations for Atoms)

Rule 1251: This rule of Chemistry for **Atoms,** states that, *Excitation of radicals in the last shell of an atom* is movement of an electro-radical from paired unbonded radicals with opposite spins to another vacant orbital thereby changing the electronegativity of the atom from its natural state to another electronegativity; *for which the need to reverse or not to reverse the spin of the radical may arise in the process, depending on the type of another atom or central atom in its neighborhood or vicinity.* *(Laws of Creations for Atoms)*

Rule 1252: This rule of Chemistry for **Atoms,** states that, *Activation of radicals in the last shell of an atom* is movement of an electro-radical from paired unbonded radicals with opposite spins to another vacant orbital without changing its original spins, and this takes place only when heat or its equivalent is applied.
(Laws of Creations for Atoms)

Rule 1253: This rule of Chemistry for **Atoms,** states that, there are atoms which *when acti-vated or excited from their ground states, give the same element,* for example, while Mg, Si give the same element when activated or excited, C and Zn give two different elements when excited and when activated, for which one of Zn is Activated Zn (Called Zinc dust) and one of C is Activated C.
(Laws of Creations for Atoms)

Rule 1254: This rule of Chemistry for **Atoms,** states that, when more than one radical is present in the p- or d- or f- orbitals in their Ground state, they do not have to be of the same spin (i.e. parallel spin), (for example Si) *since the character of their spins is determined by the Group/Period to which they belong to in the Periodic Table and the physical state of the atom at STP.*
(Laws of Creations for Atoms)

Rule 1255: This rule Chemistry for **Atoms,** states that, *Excitation and Activation* can only take place *when vacant orbitals and paired unbonded radicals are present* in the last shell of the atom.
(Laws of Creations for Atoms)

Rule 1256: This rule of Chemistry for **Atoms,** states that, when there is still room to bond (i.e., presence of single radicals) in the orbitals of the last shell and paired unbonded radicals are present but with no vacant orbital, no Excitation and Activation can take place, for which the valence state of the atom remains the same, for example N, O, F and some family members (Groups Vb, VIb and VIIb); *for which the need for hybridization does not arise.*
(Laws of Creations for Atoms)

Rule 1257: This rule of Chemistry for **Atoms,** states that, when many single radicals more than the identified valence state of the atom are present in the last shell such as in Iron (Fe), never is there a time the extra single radical is involved during chemical reactions even when the valence state of the atom has increased beyond its natural valence state; for which the only time such extra single radical is involved is when the compound is a Coordination compound in which the atom is the Central atom.
(Laws of Creations for Atoms)

Rule 1258: This rule of Chemistry for **Atoms,** states that, when a single radical exists in the s-orbital with presence of paired-unbonded radicals in lower orbitals and vacant orbitals in the last shell of the atom, such as Cu in Group Ib of the Periodic Table, *such an atom cannot be excited but only activated,* for which such an atom is monovalent without activation, or divalent when activated, leaving a single radical (female) untouched throughout the entire course of a chemical reaction and such atom like some in the

Group-Silver (Ag) cannot be hybridized as shown below for Cu where one has also shown the difference between CuCl and Cu_2Cl_2, noting that the latter cannot be obtained via excitation.

$$Cu \quad \begin{matrix} 4s^1 \\ 3d^{10} \\ 4p^0 \end{matrix} \quad \underset{Heat}{\overset{Activation}{\rightleftarrows}} \quad \begin{matrix} 4s^1 \\ 3d^9 \\ 4p^1 \end{matrix}$$

$$OR$$

$$2CuCl \quad \xrightarrow[Heat]{Activation} \quad Cl - Cu^\oplus - {}^\ominus Cu - Cl$$

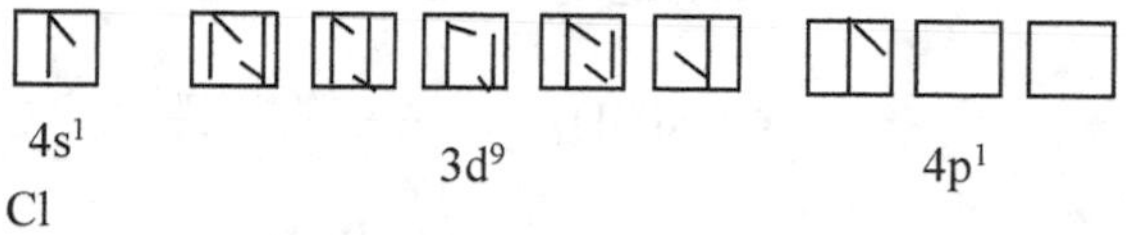

(Laws of Creations for Atom)

Rule 1259: This rule of Chemistry for **Atoms,** states that, ***Hybridization of orbitals*** takes place only when there are more than three radicals singly placed in s- and p- or d-orbitals in the last shell of the atom, ***after Excitation and not after Activation and when the radicals are of the same spin and quantum number.***
(Laws of Creations for Hybridization)

Rule 1260: This rule of Chemistry for **Atoms,** states that, without the existence of radicals in the last shell of an atom (i.e., the Boundary) the existence of our world would have been impossible; for which when no radicals exist outside the nucleus of an atom, the atom is said to be a ZERO ATOM, the foundation of all creations. [See Rule 1]
(Laws of Creations for Humanity).

Rule 1261: This rule of Chemistry for **Passive catalysts,** states that, though these catalysts do not partake chemically in chemical reactions when present, they serve countless numbers of functions, ***such as ability to keep some compounds which are very active in stable state of existence either in full or in part***, ability to make some compounds be in Equilibrium state of existence, ***ability to keep some stubborn monomers in activated state of existence***, ability to keep some compounds in Activated/Equilibrium state of existence, and so on, for which the ability to provide these functions are visibly carried by them internally and externally.

(Laws of Creations for Passive Catalysts)

Rule 1262: This rule of Chemistry for **Some Metallic Compounds,** states that, their Equili-brium state of existence are as follows-.

$$(H_5C_2)_2Zn \quad \underset{of\ Existence}{\overset{Equilibrium\ State}{\rightleftarrows}} \quad H_5C_2Zn \bullet en \ + \ n\bullet C_2H_5$$

$$H_5C_2ZnOH \quad \underset{of\ Existence}{\overset{Equilibrium\ State}{\rightleftarrows}} \quad HOZn \bullet en \ + \ n\bullet C_2H_5$$

$$ClZnC_2H_5 \quad \underset{Existence}{\overset{Equilibrium\ State\ of}{\rightleftarrows}} \quad ClZn \bullet en \ \ + \ n\bullet C_2H_5$$

$$(H_5C_2)_2Mg \quad \underset{of\ Existence}{\overset{Equilibrium\ State}{\rightleftarrows}} \quad H_5C_2Mg \bullet e \ + \ n\bullet C_2H_5$$

$$ClMgC_2H_5 \quad \underset{of\ Existence}{\overset{Equilibrium\ State}{\rightleftarrows}} \quad ClMg \bullet e \ \ + \ n\bullet C_2H_5$$

$$Si(C_2H_5)_4 \quad \underset{of\ Existence}{\overset{Equilibrium\ State}{\rightleftarrows}} \quad (H_5C_2)_3Si \bullet e \ + \ n\bullet C_2H_5$$

$$Cl_3SiC_2H_5 \quad \underset{Existence}{\overset{Equilibrium\ State\ of}{\rightleftarrows}} \quad C_2H_5Cl_2Si \bullet e \ + \ nn\bullet Cl$$

$$Cl_2Si(C_2H_5)_2 \underset{Existence}{\overset{Equilibrium\ State\ of}{\rightleftharpoons}} (C_2H_5)_2ClSi \bullet e \ + \ nn \bullet Cl$$

$$ClSi(C_2H_5)_3 \underset{Existence}{\overset{Equilibrium\ State\ of}{\rightleftharpoons}} (C_2H_5)_3Si \bullet e \ + \ nn \bullet Cl$$

$$Cl_3SiNH_2 \underset{of\ Existence}{\overset{Equilibrium\ State}{\rightleftharpoons}} Cl_3SiHN \bullet nn \ + \ e \bullet H$$

$$Al(C_2H_5)_3 \underset{of\ Existence}{\overset{Equilibrium\ State}{\rightleftharpoons}} (C_2H_5)_2Al \bullet n \ + \ e \bullet C_2H_\%$$

$$(H_5C_2)_2 AlCl \underset{of\ Existence}{\overset{Equilibrium\ State}{\rightleftharpoons}} (C_2H_5)_2Al \bullet e \ + \ nn \bullet Cl$$

$$(H_5C_2)AlCl_2 \underset{of\ Existence}{\overset{Equilibrium\ State}{\rightleftharpoons}} (H_5C_2) ClAl \bullet e \ + \ nn \bullet Cl$$

$$ROAlCl_2 \underset{of\ Existence}{\overset{Equilibrium\ State}{\rightleftharpoons}} Cl_2Al \bullet e \ + \ nn \bullet OR$$

$$RSAlCl_2 \underset{of\ Existence}{\overset{Equilibrium\ State}{\rightleftharpoons}} Cl_2Al \bullet e \ + \ nn \bullet SR$$

$$R_2AlOOH \underset{of\ Existence}{\overset{Equilibrium\ State}{\rightleftharpoons}} R_2AlO \bullet en \ + \ nn \bullet OH$$

$$Ti(C_2H_5)_4 \underset{of\ Existence}{\overset{Equilibrium\ State}{\rightleftharpoons}} (C_2H_5)_3Ti \bullet e \ + \ n \bullet C_2H_5$$

$$ClTi(C_2H_5)_3 \underset{of\ Existence}{\overset{Equilibrium\ State}{\rightleftharpoons}} Cl(C_2H_5)_2Ti \bullet e \ + \ n \bullet C_2H_5$$

$$Cl_2Ti(C_2H_5)_2 \underset{of\ Existence}{\overset{Equilibrium\ State}{\rightleftharpoons}} Cl_2(C_2H_5)Ti \bullet e \ + \ n \bullet C_2H_5$$

$$Cl_3TiC_2H_5 \underset{of\ Existence}{\overset{Equilibrium\ State}{\rightleftharpoons}} Cl_3Ti \bullet e \ + \ n \bullet C_2H_5$$

$$Cl_2V - V(C_2H_5)_4 \underset{of\ Existence}{\overset{Equilibrium\ State}{\rightleftharpoons}} Cl_2V - (C_2H_5)_3V \bullet e \ + \ n \bullet C_2H_5$$

$$Cl_2V - VCl(C_2H_5)_3 \underset{of\ Existence}{\overset{Equilibrium\ State}{\rightleftharpoons}} Cl_2V - Cl(C_2H_5)_2V \bullet e \ + \ n \bullet C_2H_5$$

$$\vdots$$

$$Cl_2V - VCl_3C_2H_5 \underset{of\ Existence}{\overset{Equilibrium\ State}{\rightleftharpoons}} Cl_2V - Cl_3V \bullet e \ + \ n \bullet C_2H_5$$

(*Laws of Creations for Finger-Prints of Some Metallic Compounds*)

Rule 1263: This rule of Addition for **Silicon containing compounds,** states that, in view of the radical nature of all its displacement reactions, the followings is the order in which groups are loosely bonded to the silicon center or the order of their nucleo-radical strength –

$$\overset{n}{\underset{\bullet}{C_3H_7}} > \overset{n}{\underset{\bullet}{C_2H_5}} > \overset{n}{\underset{\bullet}{CH_3}} > \overset{n}{\underset{\bullet}{H}} > \overset{n}{\underset{\bullet}{C_2F_5}} > \overset{n}{\underset{\bullet}{CF_3}} >$$

$$\overset{nn}{\underset{\bullet}{NR_2}} > \overset{nn}{\underset{\bullet}{NHR}} > \overset{nn}{\underset{\bullet}{NH_2}} > \overset{nn}{\underset{\bullet}{OR}} > \overset{nn}{\underset{\bullet}{OH}} > \overset{nn}{\underset{\bullet}{Cl}}$$

Order of Bond strength of groups on Si center

(*Laws of Creations for Groups on Silicon*)

Rule 1264: This rule of Chemistry for **Octamethylcyclotetrasiloxane one of the members of Cyclosiloxanes,** states that, this monomer is a strong Nucleophile, since it like the other members has one type of functional center ◯ , four of them in the ring.

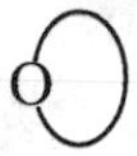

(Laws of Creations for Cyclosiloxanes)

Rule 1265: This rule of Chemistry for **Cyclosiloxanes,** states that, when alkali metal oxides and hydroxides, silanolates such as potassium trimethylsilanolate $((CH_3)_3SiOK)$ and other bases are used as source of initiators, the monomer is involved and the only route of polymerization is *Electro-free-radical (Backward Addition) route in which the metals are the carriers of the chain,* noting that these initiators will largely exist in Equilibrium state of Existence as shown below using NaONa;

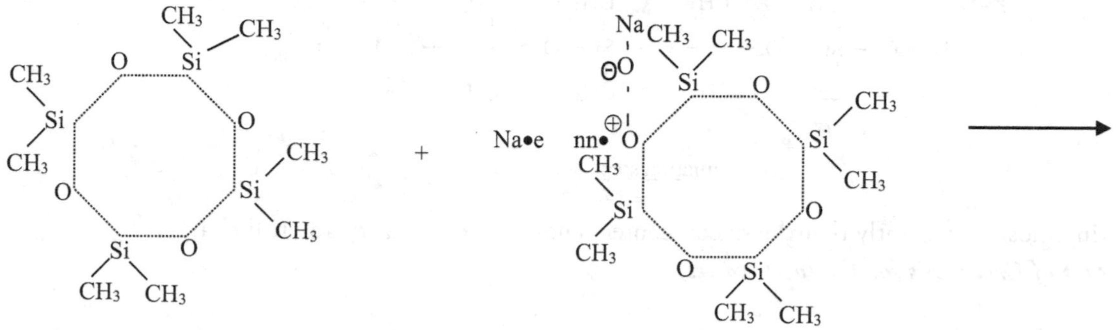

(I) highly strained

(II) Initiator

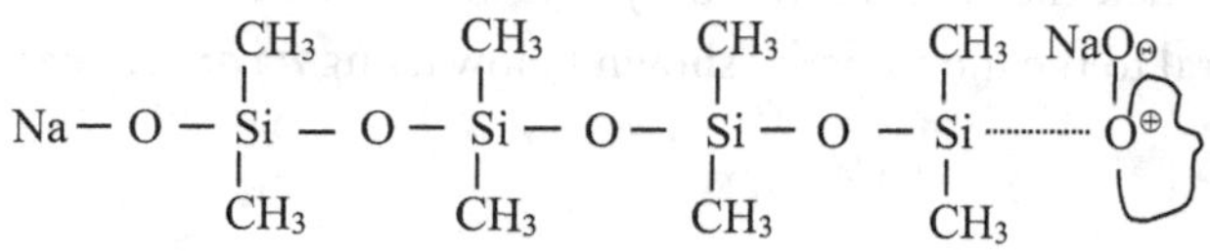

(III) Initiation step

for which when $(CH_3)_3SiOK$ is used, K will be the carrier of the chain and the when NaOH is used, Na will also be the carrier of the chain, the route still remaining electro-free-radical.

(Laws of Creations for Cyclosiloxanes)

Rule 1266: This rule of Chemistry for **Cyclosiloxanes,** states that, when protonic acids are used as source of initiators, the monomer is involved and the route of polymerization is *Electro-free-radical route (Backward Addition) in which H a metal, is the carrier of the chain,* and the chain like many others where ionic metals are involved as carriers is a Living chain, as shown below-

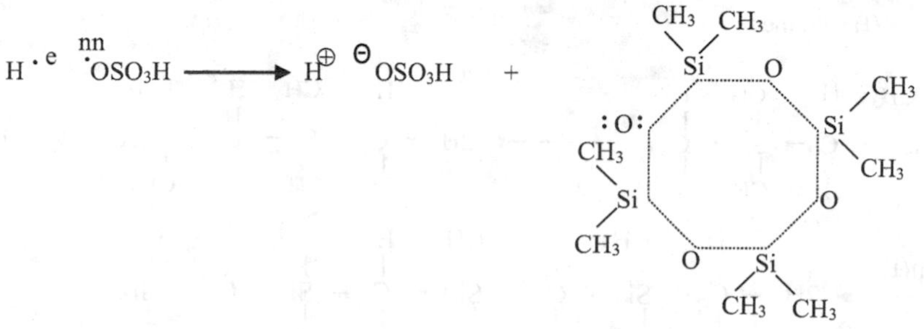

(I)

447

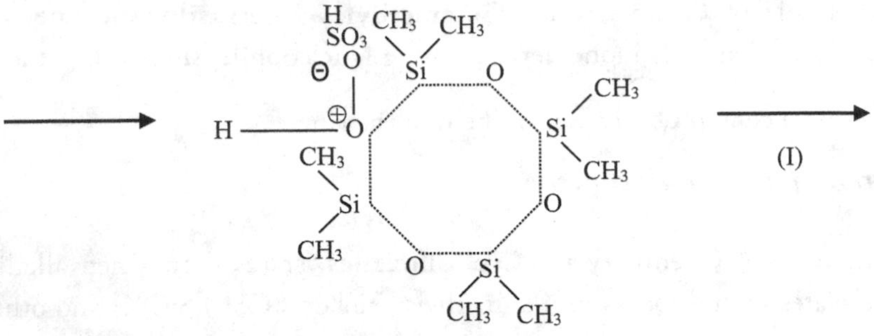

(II) Initiator

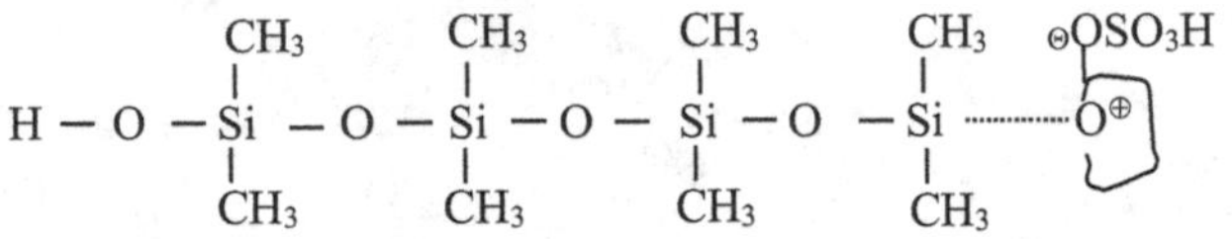

(III) Initiation step

noting most importantly that the silicon center is not allowed to carry any real charge.
(Laws of Creations for Cyclosiloxanes)

Rule 1267: This rule of Chemistry for **1,1,3,3,-tetramethyl-1,3-disilacyclobutane,** states that, this four-membered cyclic monomer a Nucleophile with no functional center and no visible activation center, can only be opened instantaneously, for which the routes favored by it is only the *Electro-free-radical Forward Addition route,* the routes natural to the monomer as shown below using *a concentrated acid,* in a Non-polar/Non-ionic solvent-

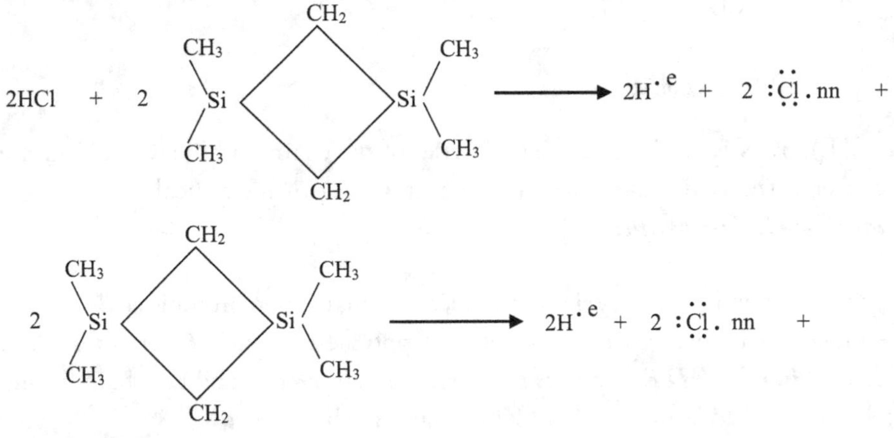

(I) (Strained)

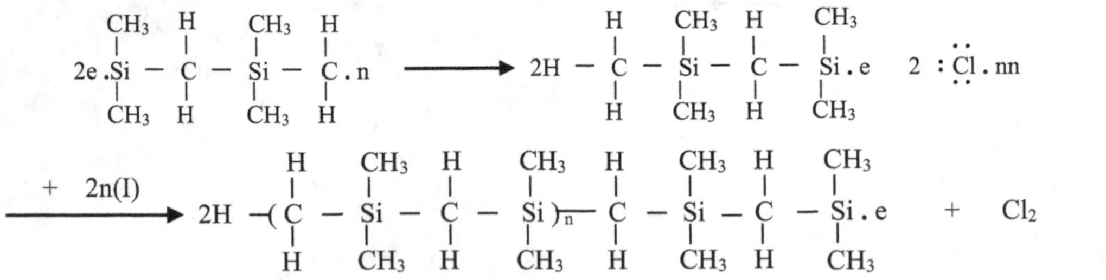

(a)

448

noting that t llows from the acid, just as Na •e is obtained from NaCN with release of cyanogen.

Stage 1: $2HCl$ ⇌ $2H •e$ + $2nn• Cl$

   $Cl •nn + nn• Cl$ ⟶ Cl_2

Overall Equation: $2HCl$ $\xrightarrow{Heat}$ $2H •e$ + Cl_2 (b)

(Laws of Creations for Si- C Chain Polymer)

Rule 1268: This rule of Chemistry for **Cyclic Si –O and Si – C monomers,** states that, based on what is carried by them, the following is valid-

Cyclic Si-O monomer   >   Cyclic Si-C monomer

(Order of Nucleophilicity)

(Laws of Creations for Cyclic Si-C and Si-O monomers)

Rule 1269: This rule of Chemistry for **Tetraoxane and Octamethylcyclotetrasiloxane,** states that, while the former is more Nucleophilic than the latter, it is less strained-

Octamethylcyclotetrasiloxane   >   Tetraoxanes

Order of Strain energy

(Laws of Creations for C-O and Si-O rings)

Rule 1270: This rule of Chemistry for **Cyclic Si – C and C – C four-membered rings,** states that, while 1,1,3,3-tetramethyl-1,3-disilacyclobutane favors only the electro-free-radical route with no transfer species, via instantaneously opening of the ring and Cyclobutane favors both electro-free-radical and positively charged routes, nucleo-free-radical and negatively charged routes also via instantaneous opening of the ring and when one H atom in cyclo butane is changed to CH_3, transfer species are made present, and since the disilacyclobutane looks more strained than the cyclo butane, the following is valid in general-

Cyclobutanes   >   Disilacyclobutanes

Order of Nucleophilicity

(Laws of Creations for Si-C and C-C Ringed Monomers)

Rule 1271: This rule of Chemistry for **O, N, S, C containing Silicon cyclic compounds,** states that, when adjacently placed to Si, the nucleophilicity of the ring and what the groups are carrying, become dominant factors for which the following is valid-.

Order of Nucleophilicity

noting that Si – C ring has no functional center and that Si - C bond is strongly more covalently bonded than the Si – S bond which in turn is stronger than Si – N bond which in turn is stronger than Si – O bond.

(*Laws of Creations for Si/O, Si/N, Si/S, Si/C cyclic compounds*)

Rule 1272: This rule of Chemistry for **Cyclic sulfides,** states that, these monomers/com-pounds like cyclic ethers are Nucleophiles, wherein because sulphur atom is more electro-positive than the oxygen atom, cyclic ethers are more Nucleophilic than cyclic sulfides-

Cyclic ethers   >   Cyclic sulfides

Order of Nucleophilicity

(*Laws of Creations for Cyclic Sulfides*)

Rule 1273: This rule of Chemistry for **Cyclic sulfides,** states that, being Nucleophiles, their natural routes are the Electro-free-radical (and positively charged) routes, for which only the fisrt member of three-membered ring can be made to additionally favor the nucleo-non-free-radical (and anionic) routes, since the ring can be opened instantaneously as shown below-

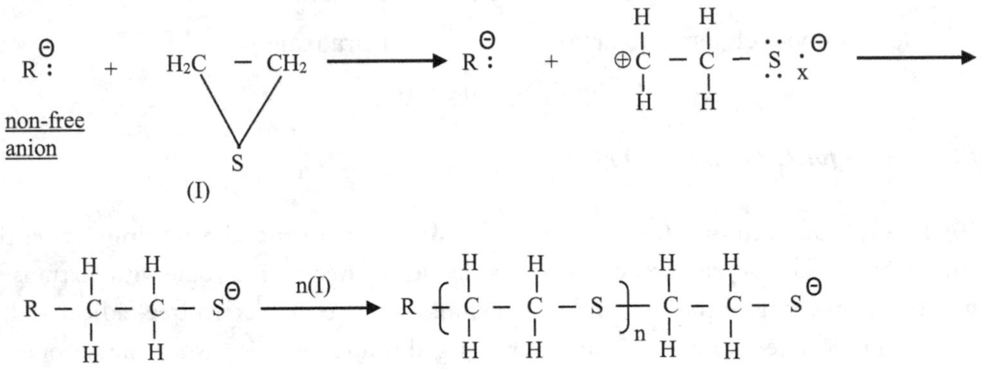

(a)

noting that this can only take place if the counter-charged center is isolately placed or electrostatically paired and stable or not present in the system.

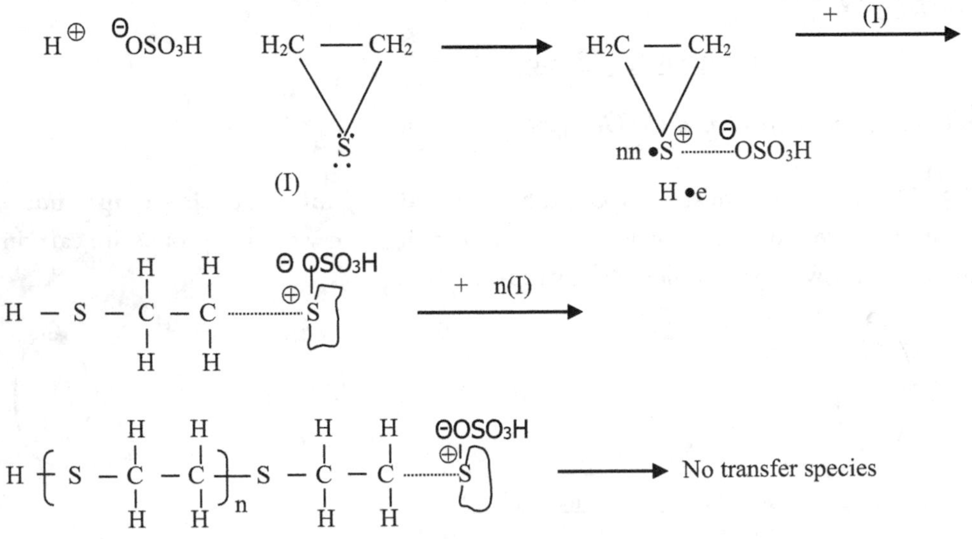

(b)

(*Laws of Creations for Cyclic Sulfides*)

Rule 1274: This rule of Chemistry for **Propagation/Depropagation phenomenon of Cyclic sulfides,** states that they like cyclic ethers undergo this phenomenon to a lesser extent only electro-free-radically as shown below for the three-membered ring-

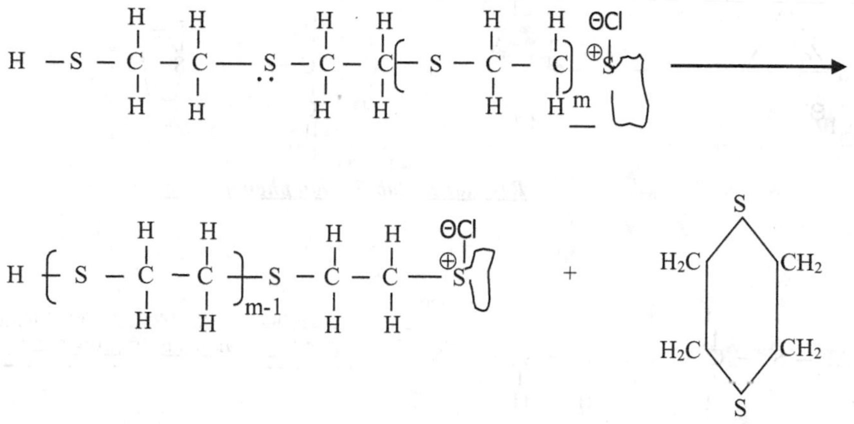

noting that this will readily occur later during propagation rather than early, in view of the increasing size of the propagating chain in the route natural to the monomer and the less electronegative character of the S center than an oxygen center.

(Laws of Creations for Cyclic sulfides)

Rule 1275: This rule of Chemistry for **Vinylepisulfide and Vinylepoxide,** states that, when their rings are opened instantaneously using a radical generating initiator, the followings are obtained using vinylepisulfide-

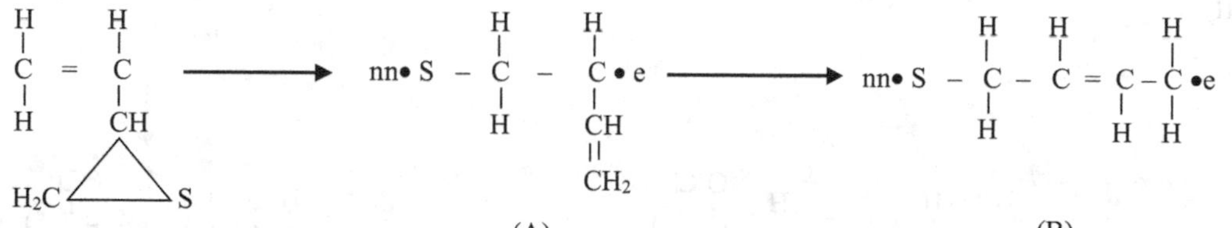

[Resonance stabilization]

for which (B) is obtained after resonance stabilization from 1,3- mono-form to 1,5- mono-form (Half non-free to Half-non-free mono-forms), noting that only nucleo-non-free-radicals and electro-free-radicals can be used for their polymerizations.

(Laws of Creations for Vinylepisulfides and Vinylepoxides)

Rule 1276: This rule of Chemistry for **Vinylepisulfides and Vinylepoxide,** states that, *when AlCl$_3$ is used as initiator,* the followings are to be expected when the initiator is in Equilibrium state of existence-

Vinylepisulfide

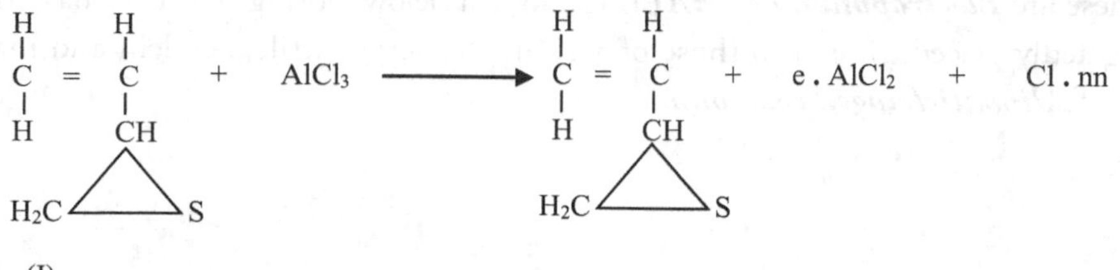

451

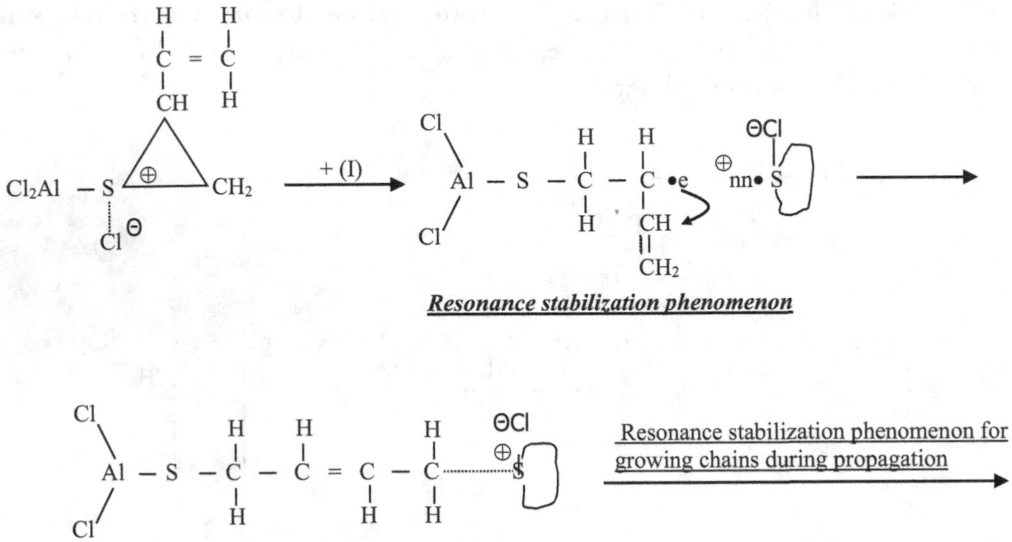

Resonance stabilization phenomenon

Resonance stabilization phenomenon for growing chains during propagation

INITIATION STEP

Vinylepoxide

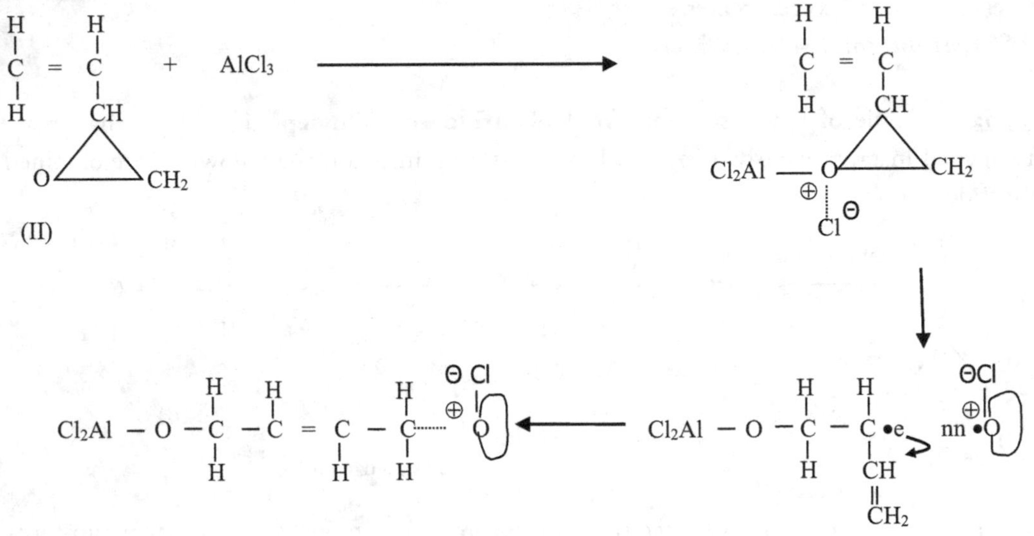

(II)

Resonance stabilization phenomenon

noting that, the reaction above is more favored with (II) than with (I) since the oxygen center is more electronegative than the sulphur center and that resonance stabilization takes place for every addition of monomer to the growing chain.

(Laws of Creations for Vinylepisulfides and Vinylepoxides)

<u>**Rule 1277:**</u> This rule of Chemistry for **Vinylepisulfides and Vinylepoxides,** states that, looking at the monomers, these are *Electrophiles, i.e. MALES* as shown below, noting that they have their X and Y centers conjugatedly placed, similar to those of acrylamide, acrylonitrile, acrolein and many others in these families of *Aliphatic/ringed compounds.*

$$
\begin{array}{ccc}
\overset{\displaystyle H}{\underset{\displaystyle H}{C}} - Y - \overset{\displaystyle H}{\underset{\displaystyle CH}{C}} & & \overset{\displaystyle H}{\underset{\displaystyle H}{C}} \equiv Y \equiv \overset{\displaystyle H}{\underset{\displaystyle CH}{C}} & & \overset{\displaystyle H}{\underset{\displaystyle H}{C}} = \overset{\displaystyle H}{\underset{\displaystyle CH}{C}} \\
H_2C \overset{X}{\triangle} O & > & H_2C \overset{X}{\triangle} N-H & > & H_2C \overset{X}{\triangle} S \\
& & \text{(When Suppressed)} &
\end{array}
$$

Order of Nucleophilicity

(Laws of Creations for Vinylepisulfides and Vinylepoxides)

Rule 1278: This rule of Chemistry for **Vinylepisulfides and Vinylepoxide,** states that, ***when NaOCH₃ is used as initiator,*** the followings are to be expected-

$$
H_3CO^{\ominus} \ldots \overset{\oplus}{Na} \quad | \quad \overset{H}{\underset{H}{C}} = \overset{H}{\underset{CH}{C}} \underset{S\triangle CH_2}{\;} \longrightarrow H_3CO^{\ominus} \ldots Na \; + \; \overset{\oplus H}{\underset{H}{C}} - \overset{H \ominus}{\underset{CH}{C}}_{S\triangle CH_2} \longrightarrow
$$

$$
H_3CO - \overset{H}{\underset{H}{C}} - \overset{H}{\underset{\ominus}{C}} - \overset{H \oplus}{\underset{H}{C}} - \overset{H}{\underset{\;}{C}} \cdots \overset{\ominus}{S} \cdots \overset{\oplus}{-} Na \longrightarrow H_3CO - \overset{H}{\underset{H}{C}} - \overset{H}{\underset{\;}{C}} = \overset{H}{\underset{\;}{C}} - \overset{H}{\underset{\;}{C}} \cdots \overset{\ominus}{S} \cdots \overset{\oplus}{-} Na
$$

INITIATION STEP

noting that the monomer has been polymerized anionically or nucleo-non-free-radically without resonance stabilizing, because the route is natural to them and monomer unit is the same as that obtained with the use of positive charge; for if the ring was not opened, then the route will not be favored.

(Laws of Creations for Vinylepisulfides and Vinylepoxides)

Rule 1279: This rule of Chemistry for **Vinylepisulfides and Vinylepoxides,** states that, ***when LiC₄H₉ is used as initiator,*** the followings are to be expected for them-

Vinylepisulfide

$$
H_9C_4{}^{\ominus} \ldots \overset{\oplus}{Li} \; + \; \overset{H}{\underset{H}{C}} = \overset{H}{\underset{CH}{C}}_{S\triangle CH_2} \longrightarrow H_9C_4{}^{\ominus} \ldots \overset{\oplus}{Li} \; + \; \overset{\ominus}{S} - \overset{H}{\underset{CH\;\|\;CH_2}{C}} - \overset{H}{\underset{\;}{C}} \oplus
$$

(I)

$$
OR \quad H_9C_4{}^{\ominus} \ldots \overset{\oplus}{Li} \; + \; \overset{\oplus H}{\underset{H}{C}} - \overset{H \ominus}{\underset{CH}{C}}_{S\triangle CH_2} \longrightarrow H_9C_4 - \overset{H}{\underset{H}{C}} - \overset{H \ominus \oplus}{\underset{CH}{C}} \ldots Li_{S\triangle CH_2}
$$

(II)   **Favored**   (i)

453

Vinylepoxide.

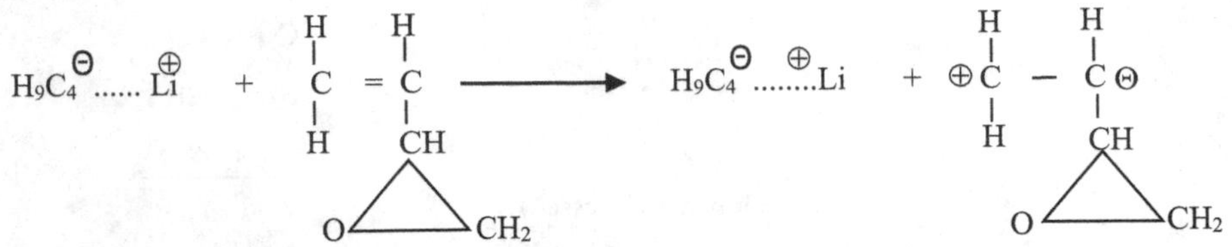

$$H_9C_4 \overset{\ominus}{......} \overset{\oplus}{Li} \quad + \quad \underset{\underset{\underset{O \diagup\!\!\diagdown CH_2}{|}}{\overset{|}{CH}}}{\overset{H}{\underset{|}{C}}} = \underset{\overset{|}{H}}{\overset{H}{C}} \quad \longrightarrow \quad H_9C_4 \overset{\ominus}{........} \overset{\oplus}{Li} \quad + \quad \oplus\overset{H}{\underset{\overset{|}{H}}{C}} - \underset{\underset{O \diagup\!\!\diagdown CH_2}{\overset{|}{CH}}}{\overset{H}{\underset{|}{C}}}\ominus$$

(I)

$$\longrightarrow \quad H_9C_4 - \underset{\overset{|}{H}}{\overset{\overset{H}{|}}{C}} - \underset{\underset{H_2C \diagup\!\!\diagdown O}{\overset{|}{CH}}}{\overset{H}{\underset{|}{C}}}\ominus \overset{}{........} \overset{\oplus}{Li}$$

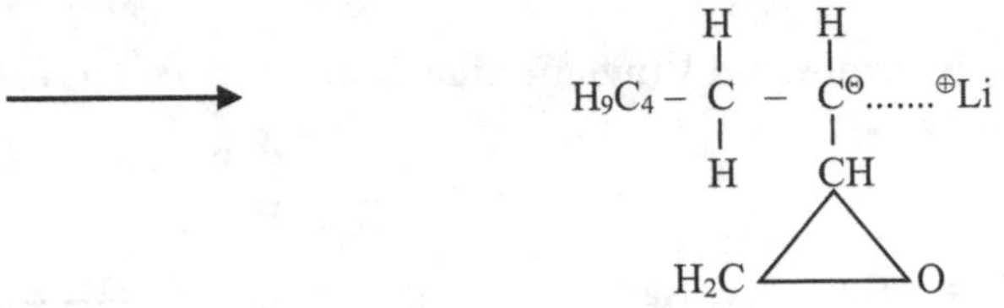

(ii)

for which, if the ring is opened instantaneously as shown in (I) above, the male (Y) center cannot be used; noting also that nucleo-free-radically polymerization is favored, when the ring is not opened in the absence of steric limitations.
(Laws of Creations for Vinylepisulfides and Vinylepoxides)

Rule 1280: This rule of Chemistry for **Vinyleposulfides and Vinylepoxides,** states that, these are unique sets of monomers, one in which the male center (Y) is aliphatic and the female center is in a ring carrying a functional center (X), unlike the case of (i) ringed Electrophiles with X and Y centers in the ring (e.g. Lactams), or (ii) non-ringed Electrophiles (e.g. ketenes, acrylamide,) or (iii) ringed Electrophiles without functional centers (e.g. Benzo-quinone) which is the only one that can be resonance stabilized, but like (i) can instantaneously be opened, yet still different; for which common to all of them, is the adjacent or cumulative placement of their X and Y centers, that which sends a message to Medical Scientists.
(Laws of Creations for Electrophiles)

Rule 1281: This rule of Chemistry for **C = C activation center and cyclic ether or sulfide ring conjugatedly placed (Episulfides and Epoxides),** states that, the C = C center (Y) is less nucleophilic than the functional center (X) in the ring, while when they are isolatedly placed, the compound or monomer becomes a Nucleophile and allylic with the C = C still remaining less nucleophilic than the functional center in the ring as shown below-

$$\bigcirc\!\!O \quad > \quad C = C \quad ; \quad \bigcirc\!\!S \quad > \quad C = C \quad ; \quad \bigcirc\!\!NR \quad > \quad C = C$$

Order of Nucleophilicity

(a)

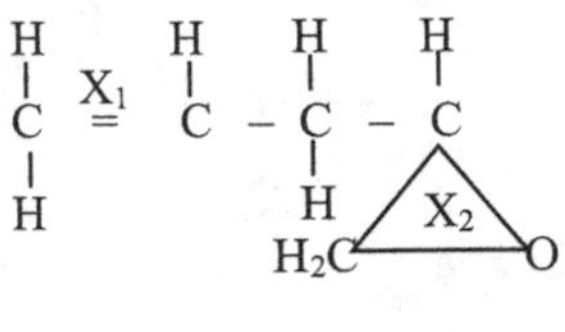

$$\underset{\overset{|}{H}}{\overset{\overset{H}{|}}{C}} \overset{X_1}{=} \underset{\underset{H_2C \diagup\!\!\diagdown_{X_2} O}{\overset{|}{H}}}{\overset{H}{\underset{|}{C}}} - \underset{\overset{|}{H}}{\overset{\overset{H}{|}}{C}} - \overset{H}{\underset{|}{C}} \qquad\qquad X_2 \quad > \quad X_1$$

Order of Nucleophilicity

(b)

for which X_2 will never be activated until X_1 has been saturated.
(Laws of Creations for C=C and Cyclic ethers/sulfides)

454

Rule 1282: This rule of Chemistry for **Vinylepisulfides shown below,** states that, whether the ring is opened instantaneously or not, transfer species become present with them as shown below-

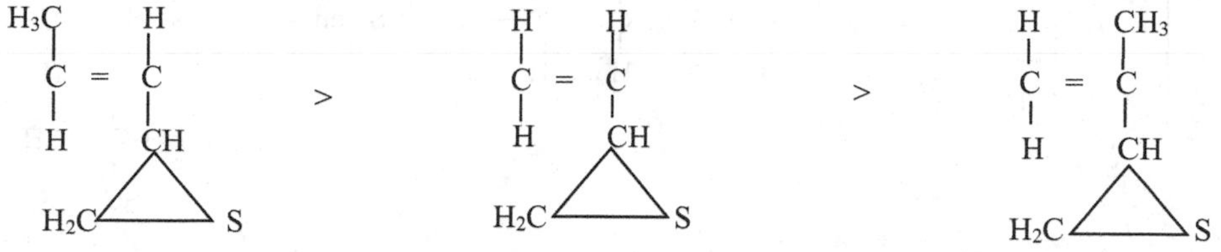

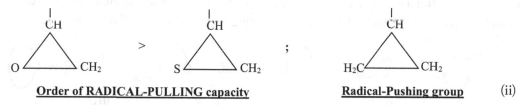

(I) (II) (III)

Order of Nucleophilicity   (i)

for which, only (I) has transfer species-from the CH_3 group, and the CH_3 in (III) is shielded and (II) is the first member with no transfer species like (III); noting the followings are additionally valid-

Order of RADICAL-PULLING capacity   ;   **Radical-Pushing group**   (ii)

(Laws of Creations for Vinylepisulfides and Vinylepoxides)

Rule 1283: This rule of Chemistry for **Trithianes, Cyclic thioformals, Cyclic thioesters, and all corresponding cases with O,** states that, these like their O containing counterparts favor the same routes in particular electro-free-radical and in addition nucleo-non-free-radical for the thioesters; *noting that these are more strained and less Nucleophilic than their O counterparts* and being more strained makes it possible for some of them not to exist, such as the cyclic thioformals-

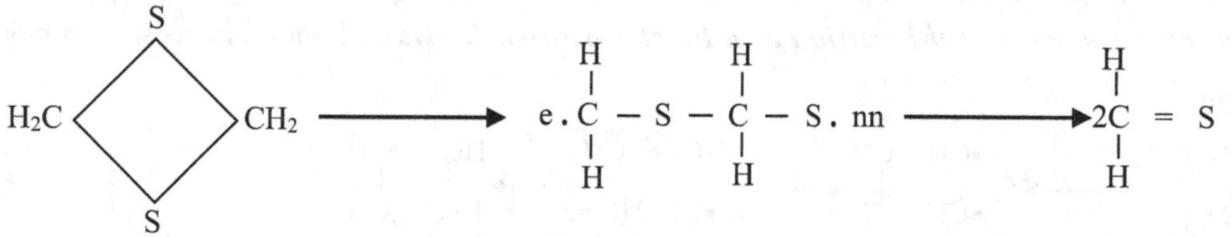

$$e.\overset{\overset{\displaystyle H}{|}}{\underset{\underset{\displaystyle H}{|}}{C}} - S - \overset{\overset{\displaystyle H}{|}}{\underset{\underset{\displaystyle H}{|}}{C}} - S.nn \longrightarrow 2\overset{\overset{\displaystyle H}{|}}{\underset{\underset{\displaystyle H}{|}}{C}} = S$$

(a)

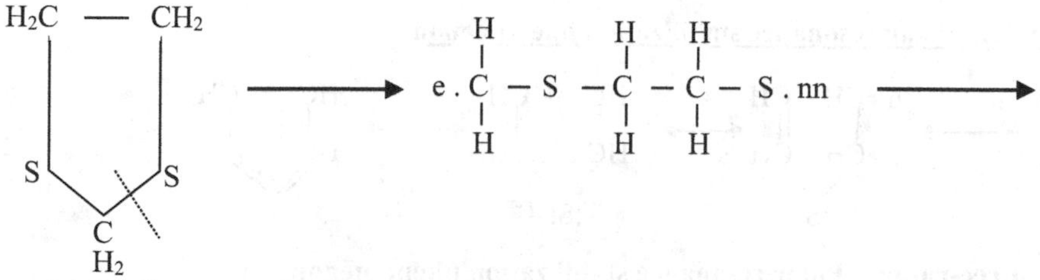

$$e.\overset{\overset{\displaystyle H}{|}}{\underset{\underset{\displaystyle H}{|}}{C}} - S - \overset{\overset{\displaystyle H}{|}}{\underset{\underset{\displaystyle H}{|}}{C}} - \overset{\overset{\displaystyle H}{|}}{\underset{\underset{\displaystyle H}{|}}{C}} - S.nn \longrightarrow$$

(I) (strained)

$$\quad + \quad S = \overset{\overset{\displaystyle H}{|}}{\underset{\underset{\displaystyle H}{|}}{C}}$$

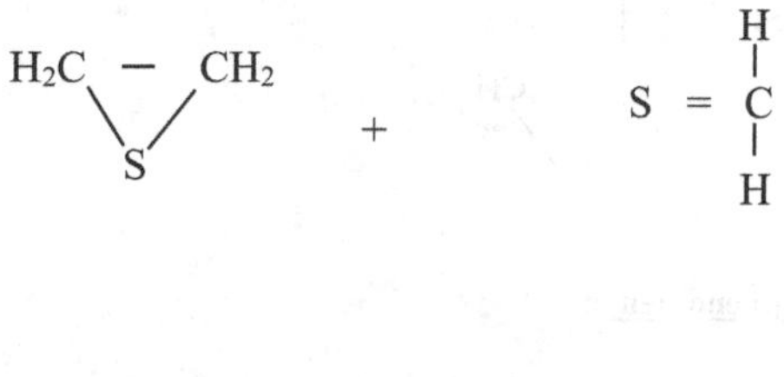

(b)

455

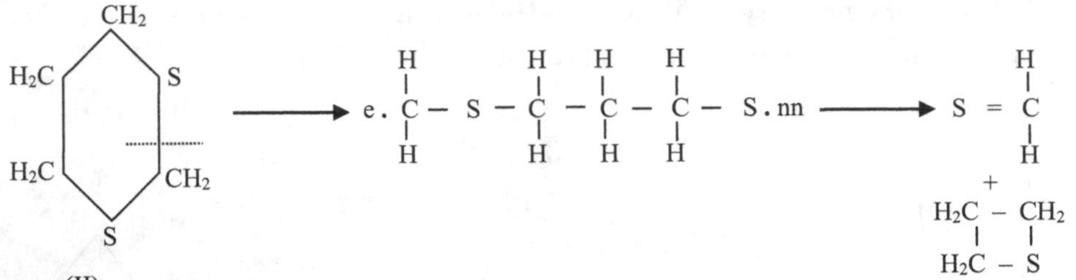

$$e . \overset{H}{\underset{H}{C}} - S - \overset{H}{\underset{H}{C}} - \overset{H}{\underset{H}{C}} - \overset{H}{\underset{H}{C}} - S . nn \longrightarrow \begin{matrix} H \\ | \\ S = C \\ | \\ H \\ + \\ H_2C - CH_2 \\ | \quad\quad | \\ H_2C - S \end{matrix}$$

(II)   (c)

(Laws of Creations for S containing Rings)

<u>**Rule 1284:**</u> This rule of Chemistry for **Cyclic sulfides, Trithianes, Cyclic thioformals, Cyclic thioacetals,** states that, *when $H_2C=$ group is adjacently located to a functional center*, the functional center ceases to function, but not the Y center if present, because the new center becomes the least nucleophilic center; for which like oxygen containing monomers, the following is valid-

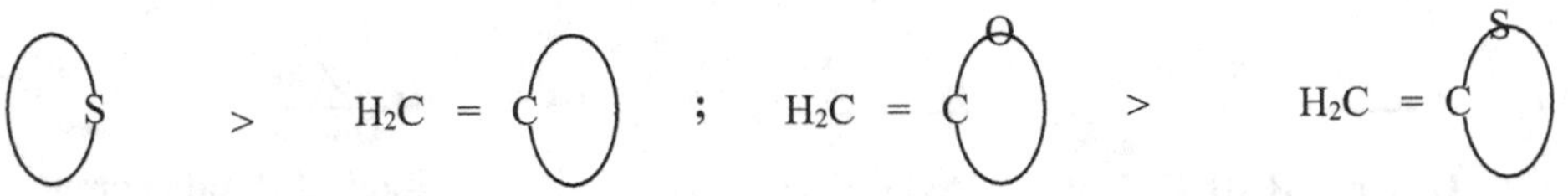

(ring with S) > $H_2C = C$ (ring) ; $H_2C = C$ (ring with O) > $H_2C = C$ (ring with S)

(I)   (II)

<u>Order of Nucleophilicity</u>

for if the Y center is present, the monomer (Cyclic thioesters) still remains electrophilic.

(Laws of Creations for S containing rings)

<u>**Rule 1285:**</u> This rule of Chemistry for the **Family members of Pyrrole, Furan and Thiophene,** states that, while Thiophene undergoes what is called *Free-radical/Polar resonance stabilization* phenomenon, Furan undergoes *Discrete resonance stabilization* phenomenon and Pyrrole undergoes both *Discrete resonance stabilization and Electro-radicalization* phenomena as shown below-

<u>**Thiophene**</u>

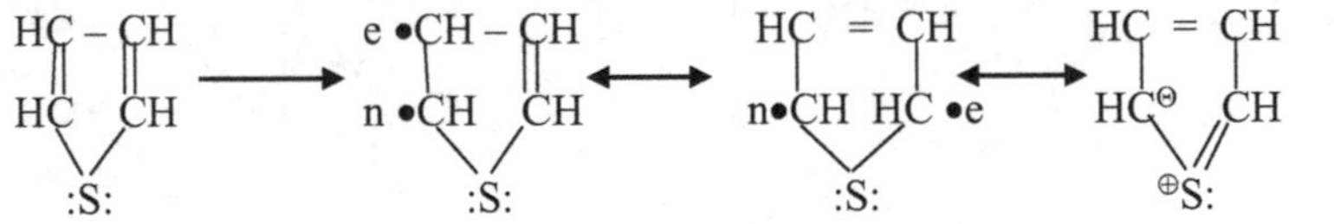

<u>Free-radical/Polar resonance stabilization phenomenon:</u>   (i)

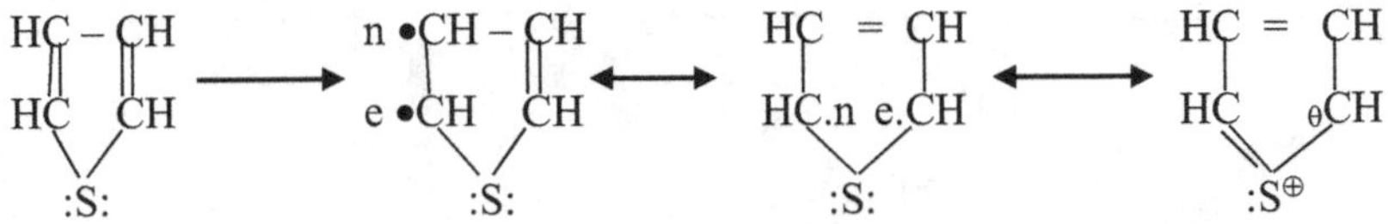

<u>Free-radical-Polar resonance stabilization phenomenon:</u>   (ii)

<u>**Furan**</u>

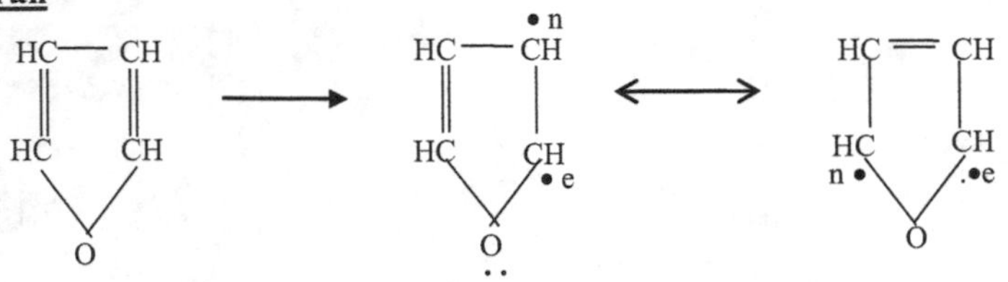

<u>Discrete resonance stabilization phenomenon</u>   (iii)

456

Pyrrole

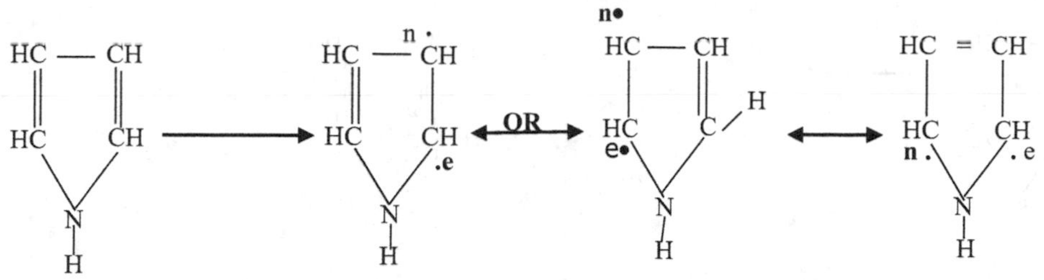

HC — CH / HC — CH / N / H → HC — ^{n}· CH / HC — CH ·e / N / H ⟷ **OR** ⟷ $^{n•}$ HC — CH / HC e• — C ⁄ H / N / H ⟷ HC = CH / HC — CH n·— ·e / N / H

(I)   (I)   (II)

Discrete resonance stabilization phenomenon   (iv)

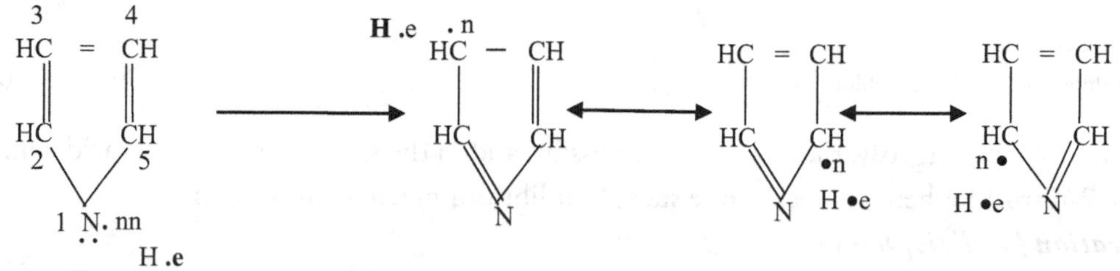

3   4 / HC = CH / HC CH / 2   5 / 1 N· nn / ·· / H ·e / (I) → **H** ·e ·^{n} HC — CH / HC CH / N / (II) ⟷ HC = CH / HC CH / •n / N   H •e / (III) ⟷ HC = CH / n• HC CH / H •e   N / (IV)

Continuous Electroradicalization phenomenon   (v)

from which, it can be observed that, **with thiophene**, one hydrogen atom is loosely bonded electro-free-radically at either any of the 2- or 5- positions or any of 3- or 4- positions one at a time without activation, that is, when energized like in Benzene or made to exist in Equilibrium state of existence; **with furan and pyrrole**, only one H atom in the 2- or 5-position is loosely held in Equilibrium state of Existence without activation, noting that the 3,- or 4- positions cannot be used, because they are internally located while 2- or 5- positions are externally located and in addition, with pyrrole, unlike in furan and thiophene, when held in Equilibrium state of Existence from the N center, it can undergo Electroradicalization phenomenon making available 2- or 5-positions and 3- or 4-positions in different ways.

(Laws of Creations for Thiophene, Furan and Pyrrole)

Rule 1286: This rule of Chemistry for **Thiophene,** states that, its Equilibrium state of Existence is so strong that it can suppress those of Concentrated H_2SO_4, Concentrated HCl, $H_2C=O$, Cl_2, H_3CCOCl, Hg_2Cl_2 and more, and can activate C=C and some other bonds; for which when this is done, the 2- or 5- position is replaced with a group as shown below-

Stage 1:

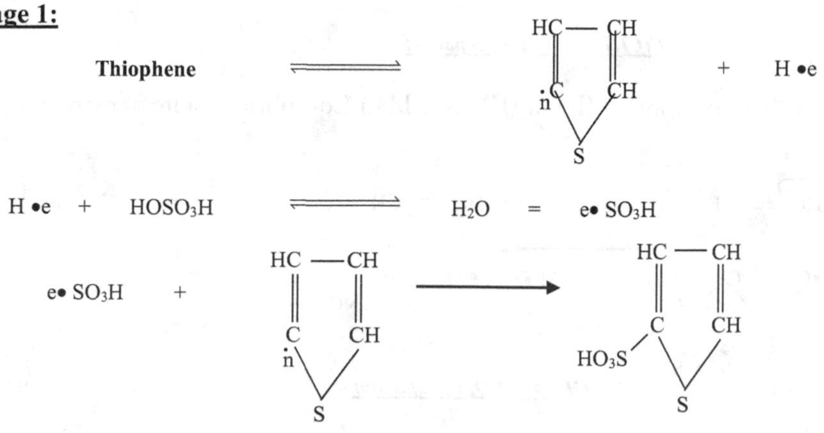

Thiophene ⇌ HC — CH / $\overset{..}{n}$C CH / S + H •e

H •e + HOSO$_3$H ⇌ H$_2$O = e• SO$_3$H

e• SO$_3$H + HC — CH / $\overset{..}{n}$C CH / S → HC — CH / C CH / HO$_3$S S

2 - thiophenesulfonic acid   (i)

457

Overall Equation: Thiophene + H_2SO_4 $\longrightarrow$ 2 – thiophenesulfonic acid　　　　(ii)

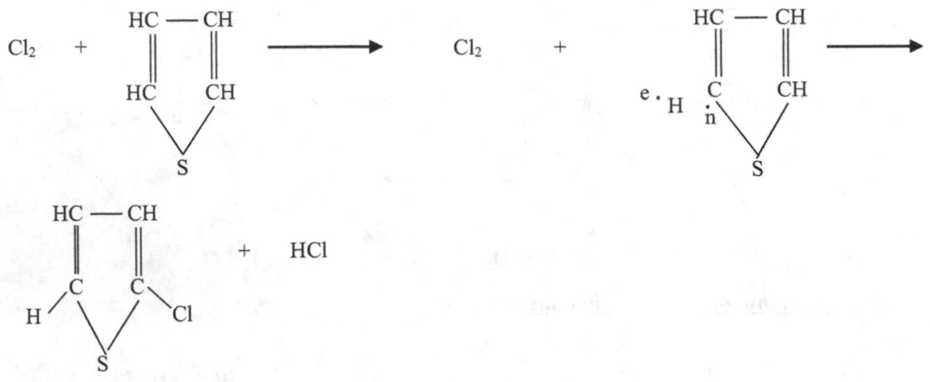

Cl_2　+　[thiophene structure: HC—CH, HC, CH, S]　$\longrightarrow$　Cl_2　+　[thiophene structure with e·H, C, :n, CH, S]　$\longrightarrow$

[structure: HC—CH, C, C, H, Cl, S]　+　HCl

　2 - Chlorothiophene(2 - thienyl chloride)　　　　(iii)

noting that ionically or chargedly, the reaction is impossible and in the same manner, nitric acid combines with it to give 2-nitrothiophene, all being one stage Equilibrium mechanism system.
(Laws of Creation for Thiophene)

Rule 1287: This rule of Chemistry for **Thiophene,** states that, when a first group has been placed in the 2-position, the entrance of the second group to the 5-position is independent of the capacity of radical-pushing or pulling group unlike the case of benzene as shown below-

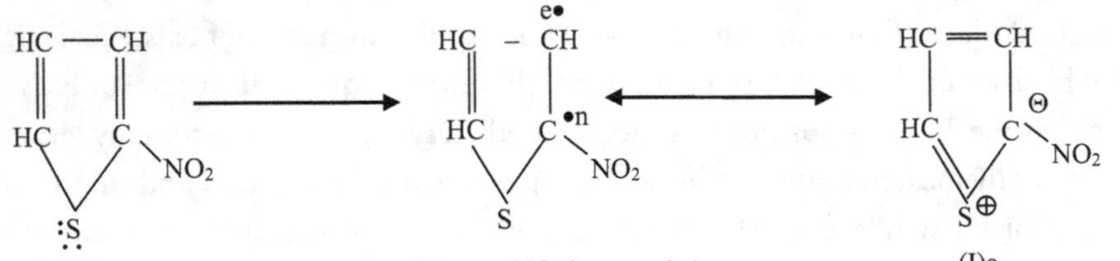

[structure (I): HC—CH, HC, C, NO_2, :S]　$\longrightarrow$　[structure with e•, HC–CH, HC, C•n, NO_2, S]　$\longleftrightarrow$　[structure (I)a: HC=CH, HC, C$^\ominus$, NO_2, S$^\oplus$]

(I)　　　　*(H > NO_2 in capacity)*　　　　(I)a　　　　(i)

[structure (II): HC—CH, HC, C, $C(CH_3)_3$, S:]　$\longrightarrow$　[structure with •e, HC–CH, HC, C, n•, $C(CH_3)_3$, S]　$\longleftrightarrow$　[structure (II)a: HC=CH, HC$^\ominus$, C, $C(CH_3)_3$, S$^\oplus$]

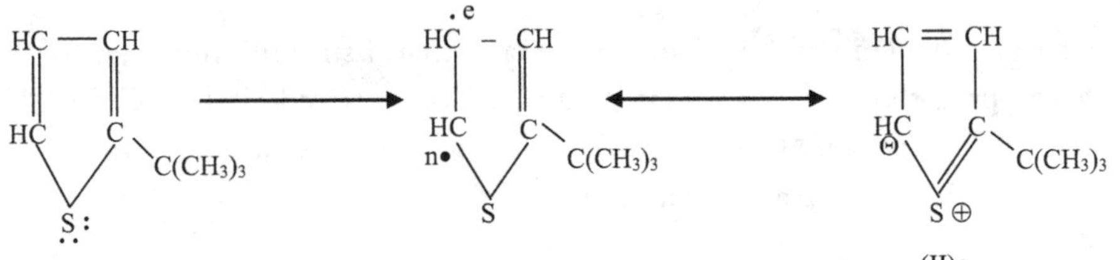

(II)　　　　*(C(CH₃)₃ > H in capacity)*　　　　(II)a　　　　(ii)

for which when the H atom in the 5-position of (I)a or (II)a is held in Equilibrium state of existence, the followings are obtained-

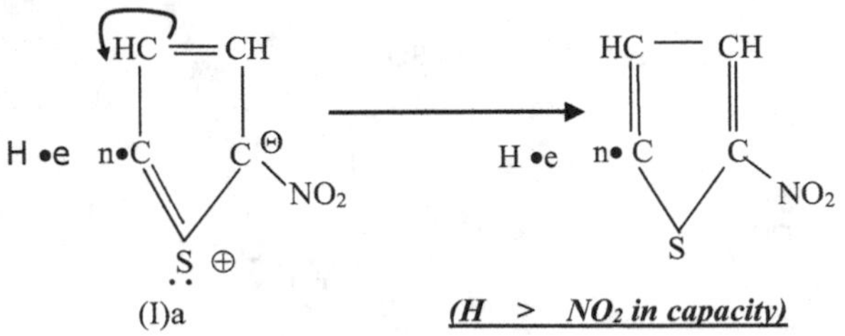

H •e　n•C　[structure: HC=CH, C$^\ominus$, NO_2, S$^\oplus$]　$\longrightarrow$　H •e　n• C　[structure: HC—CH, C, NO_2, S]

(I)a　　　　*(H > NO_2 in capacity)*　　　　(iii)

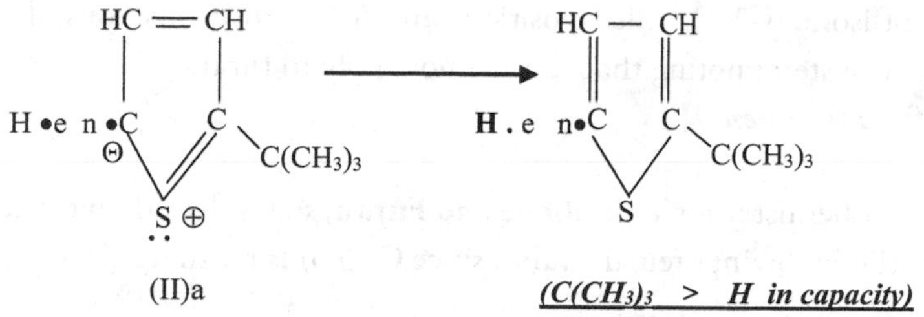

(II)a   *(C(CH₃)₃ > H in capacity)*   (iv)

noting that this will also almost apply to furan.
(Laws of Creations for Thiophene)

Rule 1288: This rule of Chemistry for **Thiophene and Furan,** states that, though they both contain the same number of paired unbonded radicals, not all reactions favored by thiophene can be favored by furan, since O is less electropositive than S and the paired unbonded radicals on O are too strongly held to the O center which is smaller in size and the fact that C and S are equi-electropositive.
(Laws of Creations for Furan and Thiophene)

Rule 1289: This rule of Chemistry for **Thiophene,** states that, when a group is initially in the 3- or 4– position, whether the group is of greater capacity than H or not, via Polar/Radical resonance phenomenon, the hydrogen atom in the 2- or 5- position is loosely bonded as shown below-

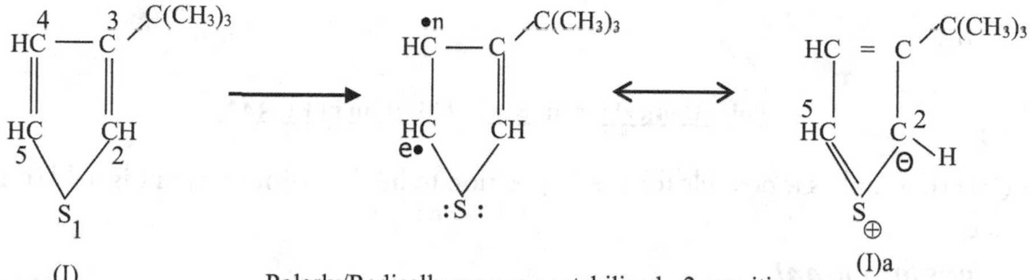

(I)   Polarly/Radically resonance stabilized - 2 - position   (I)a   (i)

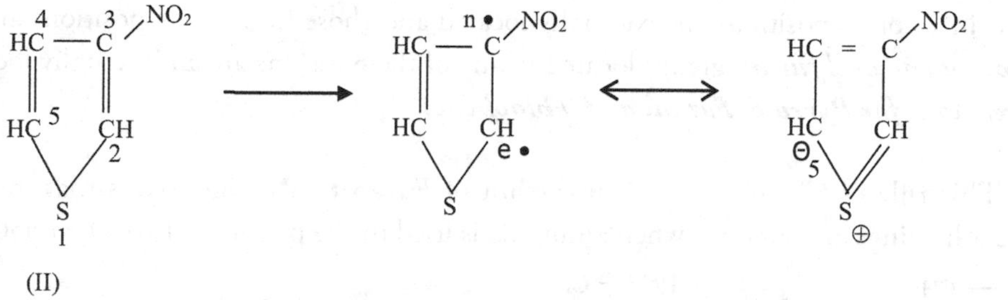

(II)   Polarly/Radically resonance stabilized – 5 - position   (ii)

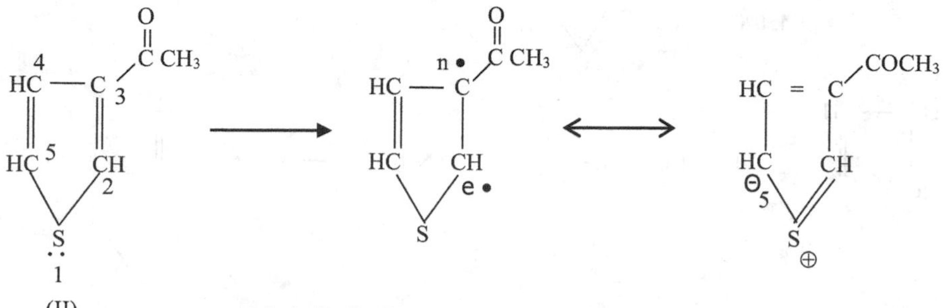

(II)   Polarly/Radically resonance stabilized - 5 - position   (iii)

459

thus giving mixtures of isomers-2-/3- filled positions and 5-/3- filled positions all obtained in parallel Equilibrium mechanism system; noting that this cannot apply to furan.
(Laws of Creations for Thiophene)

Rule 1290: This rule of Chemistry for **Thiophene and Furan,** states that, despite the greater reactivity of thiophene than furan, the followings remain valid, since O (3.5) is far more electronegative than S (2.5)-

Furan   >   Thiophene

Order of Nucleophilicity and Acidity

(Laws of Creations for Thiophene and Furan)

Rule 1291: This rule of Chemistry for **Furan,** states that, when a strong radical pushing group carried by a metallic center is in 2-position, the hydrogen atom in the 5-position is loosely bonded via Polar/Radical resonance stabilization phenomenon as shown below-

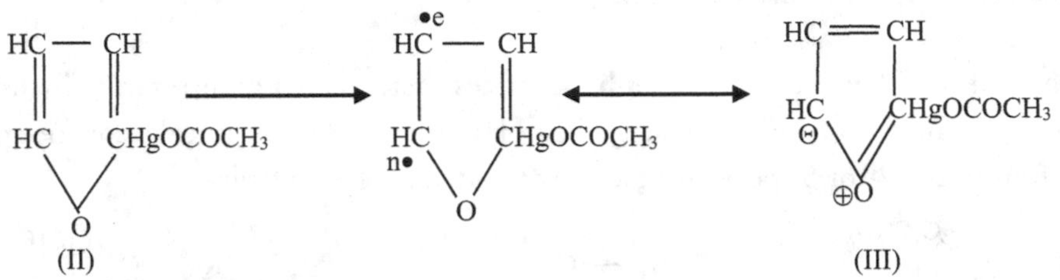

Polar/Radical resonance stabilization in FURAN

in which, it is (III) that makes it possible for the 5- position to be the point where H is held in Equilibrium state of Existence.
(Laws of Creations for Furan)

Rule 1292: This rule of Chemistry for **Pyrrole, Furan and Thiophene,** states that, while in *pyrrole,* groups placed in 2- or 5- positions are externally located and those in 3- or 4- positions are internally located, in ***thiophene and furan,*** groups located in any of the positions are all internally located.
(Laws of Creations for Pyrrole, Furan, and Thiophene)

Rule 1293: This rule of Chemistry for **Conversion of Furan to Pyrrole,** states that, in one of the methods, the followings are obtained when ammonia is used in the presence of Al_2O_3 at 450°C-

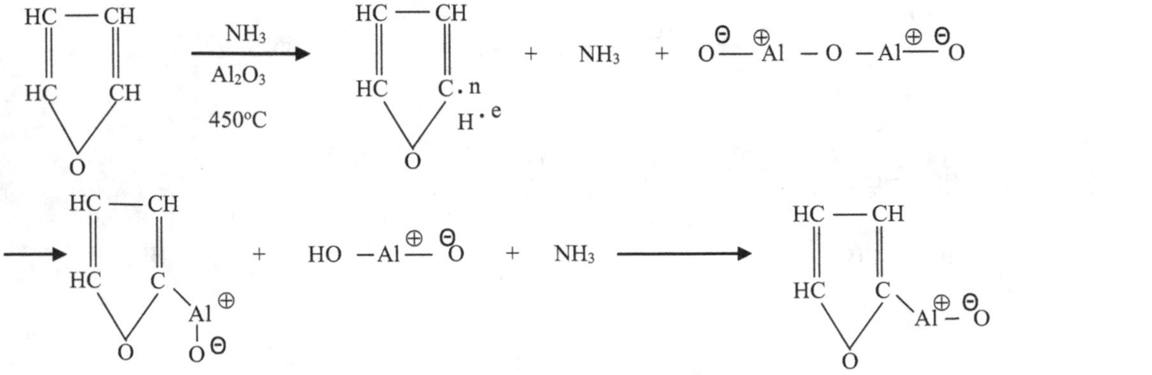

$$+ \quad H \cdot^e \quad + \quad \underset{nn}{\cdot}O - \overset{\oplus}{Al} - O^{\ominus} \quad + \quad NH_3 \quad \longrightarrow \quad \begin{array}{c} HC - CH_2 \\ \| \quad | \quad \overset{\oplus}{OAl} - O^{\ominus} \\ HC \quad C \\ \quad \backslash \quad \overset{\oplus}{Al} - O^{\ominus} \\ O \end{array} \quad + \quad NH_3 \quad \longrightarrow$$

$$\begin{array}{c} HC - CH_2 \\ \| \quad | \quad \overset{\oplus}{OAl} - O^{\ominus} \\ HC \quad C \\ \quad \backslash \quad \overset{\oplus}{Al} - \overset{\ominus}{O} \\ O \overset{\oplus}{} \overset{\ominus}{\cdots} NH_2 \\ H \end{array} \quad \longrightarrow \quad \begin{array}{c} H \\ | \\ H - O - C = C - \overset{|}{\underset{|}{C}} - \overset{O - \overset{\oplus}{Al} - \overset{\ominus}{O}}{\underset{\underset{O^{\ominus}}{\overset{\oplus}{Al}}}{C}} - NH_2 \\ \quad H \quad H \end{array} \quad \longrightarrow$$

$$\begin{array}{c} H \\ | \\ H - O - C = C - \overset{|}{\underset{|}{C}} - \overset{O - \overset{\oplus}{Al} - \overset{\ominus}{O}}{\underset{e\bullet}{C}} - NH_2 \\ \quad H \quad H \end{array} \quad + \quad n \bullet Al^{\oplus} - {}^{\theta}O \qquad \xrightarrow{\text{In Two Stages}}$$

$$Al_2O_3 \quad + \quad \begin{array}{c} H \quad H \\ | \quad | \\ H - O - C = C - C = C - NH_2 \\ | \quad | \\ H \quad H \end{array} \quad \longrightarrow \quad Al_2O_3 \quad +$$

$$\begin{array}{c} HC - CH \\ \| \quad \| \\ HC \quad CH \\ \backslash \quad / \\ N \\ | \\ H \end{array} \quad + \quad H_2O$$

for which, one of the double bonds was first saturated with Al_2O_3 in two stages, followed by opening of the ring with the ammonia in one stage; after which the Al_2O_3 was recovered in two stages; and the product obtained closed itself with the release of water, bringing the total number of stages to six stages, noting that never is there a time one can bypass a stage.

(Laws of Creations for Opening of Furan)

Rule 1294: This rule of Chemistry for **Furan, Pyrrole and Thiophene,** states that, in view of the fact that *furan and pyrrole can be hydrogenated and thiophene cannot,* clearly indicates that the followings are valid-

$$\text{(O)} \quad > \quad \overset{\frown}{\underset{\smile}{C = C}} \qquad ; \qquad \overset{\frown}{\underset{\smile}{C = C}} \quad > \quad \text{(S)}$$

Order of Nucleophilicity   Order of Nucleophilicity

$$\underset{N-H}{\bigcirc} \quad > \quad \underset{C=C}{\bigcirc}$$

<u>Order of Nucleophilicity</u>

(Laws of Creations for Furan, Pyrrole and Thiophene)

Rule 1295: This rule of Chemistry for **Furan and Cyclic ethers,** states that, the following is valid-

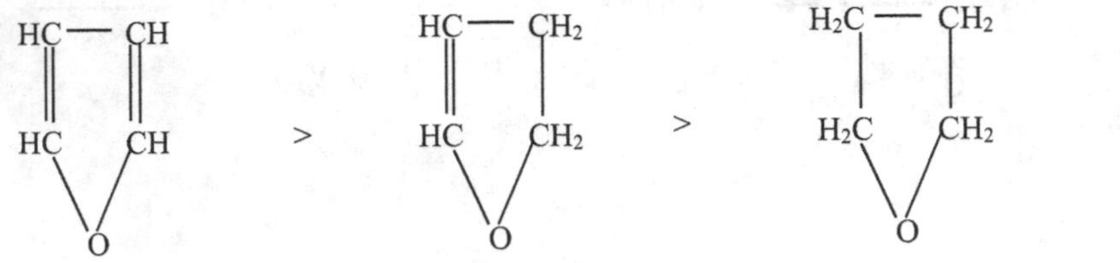

(A) Furan $>$ (B) Dihydrofuran $>$ (C) Tetrahydrofuran

<u>**Order of Nucleophilicity and Strain Energy**</u>   (i)

for which it is only for the four-membered ring of (B), when activated or radically/polarly stabilized, instantaneous opening of the ring is favored to produce a monomer which is a **weak Electrophile (Acrolein)** as shown below-

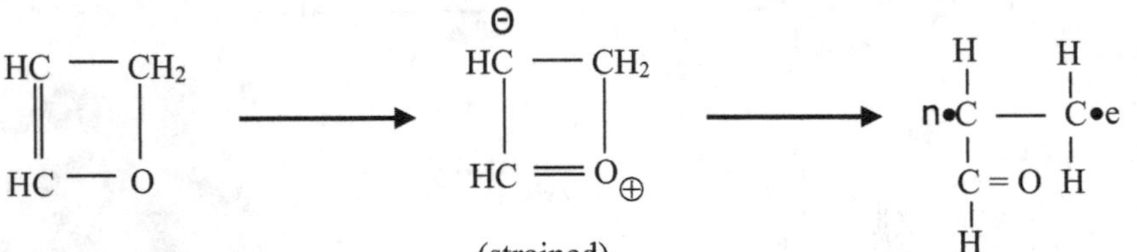

(strained)   (Weak Electrophile)   (ii)

noting that this is only possible with the first member of this family which is highly strained to favor any existence, while all the other larger sized members of this family (B) are Nucleophiles.

(Laws of Creations for C/O rings)

Rule 1296: This rule of Chemistry for **Furan, Pyrrole and Thiophene,** states that, the fact that thiophene favors polar/radical resonance stabilization phenomenon, clearly shows that the following is valid-

$$\underset{C=C}{\overset{C=C}{\Big|}}\!\!\!> S \quad > \quad \underset{C=C}{\overset{C=C}{\Big|}}\!\!\!> O \quad > \quad \underset{C=C}{\overset{C=C}{\Big|}}\!\!\!> N-H \quad > \quad \underset{C=C}{\overset{C=C}{\Big|}}\!\!\!> CH_2$$

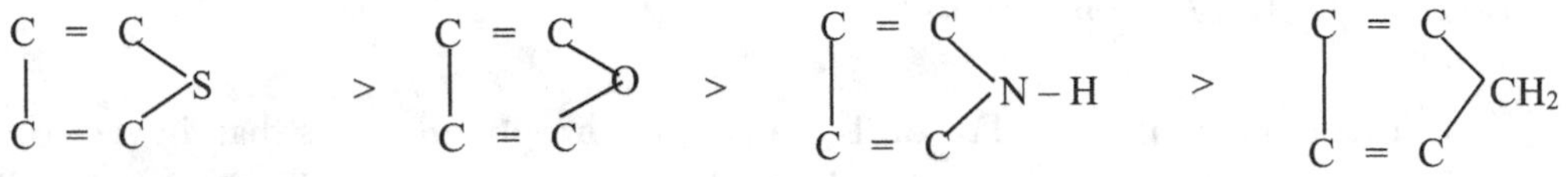

(Cyclo - dienes)

<u>Order of resonance stabilization capacity</u>

(Laws of Creations for Thiophene, Furan and Pyrrole)

462

Rule 1297: This rule of Chemistry for **Furfural,** state that, based on the types of reactions favored by it as shown below for one of them- *the so-called Cannizzaro reaction-*

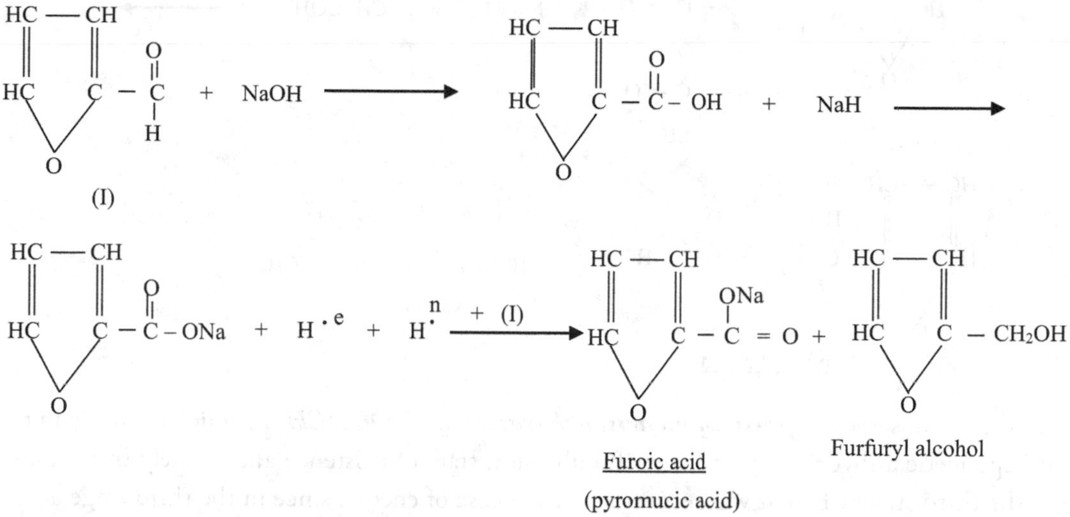

(I)

Furoic acid
(pyromucic acid)

Furfuryl alcohol

wherein in the first stage NaH and furoic acid were formed when the NaOH existed in Equilibrium state of existence to attack the furfural, and in the second stage the NaH was indirectly used to hydrogenate another furfural, which if it had been absent Stage 2 would not have taken place.
(Laws of Creations for Furfural)

Rule 1298: This rule of Chemistry for **Furfural,** states that, based on the types of reactions favored by it as shown below for one of them- *so-called Perkin reaction-*

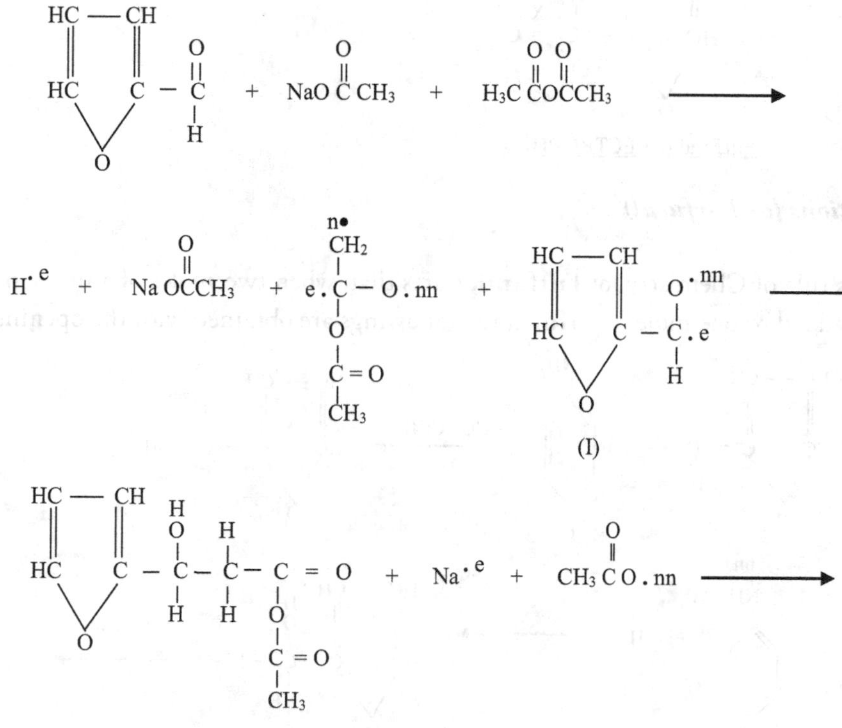

(I)

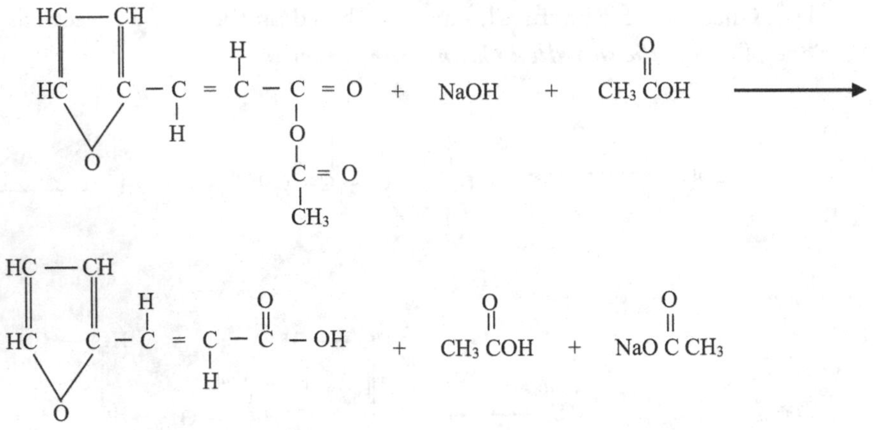

Sunny N.E. Omorodion

Furylacrylic acid

wherein one can observe **the passive and active character of NaOCOCH₃** in which passively in the first stage it kept acetic anhydride in Activated/Equilibrium state of existence and actively from the second stage to the third stage where it was recovered with release of energy, since in the third stage acetic acid commence the attack on NaOH to be released again along with sodium acetate.

(Laws of Creations for Furfural)

Rule 1299: This rule of Chemistry for **Furfural,** states that, though the C=O center externally located is more nucleophilic than the other activation centers, it was the center first attacked for example during hydrogenation, because furfural unlike furan is an Electrophile as shown below-

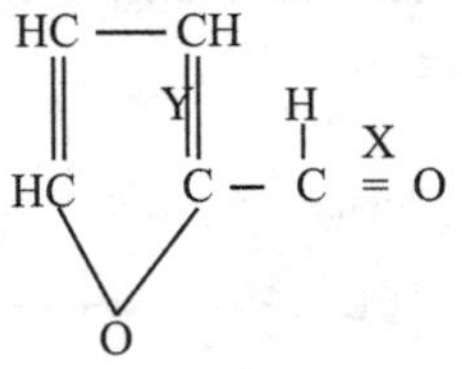

Furfural (ELECTROPHILE)

(Laws for Creations for Furfural)

Rule 1300: This rule of Chemistry for **Furfural,** states that, when two moles of aniline in the presence of acetic acid, is added to one mole of furfural, the followings are obtained with the opening of the ring-

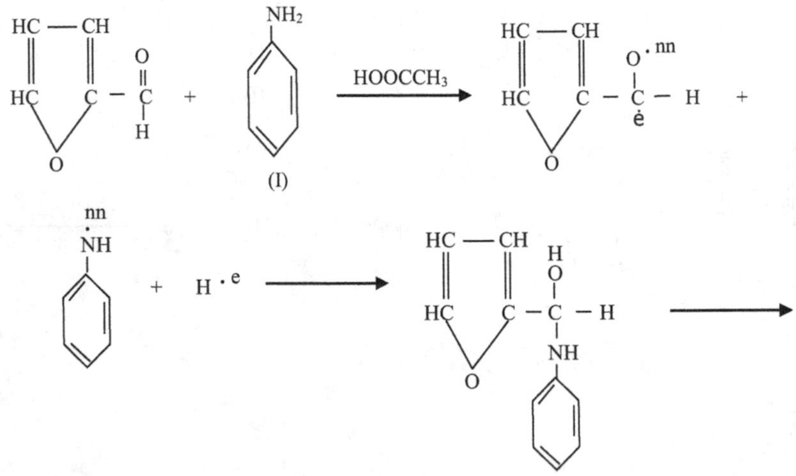

464

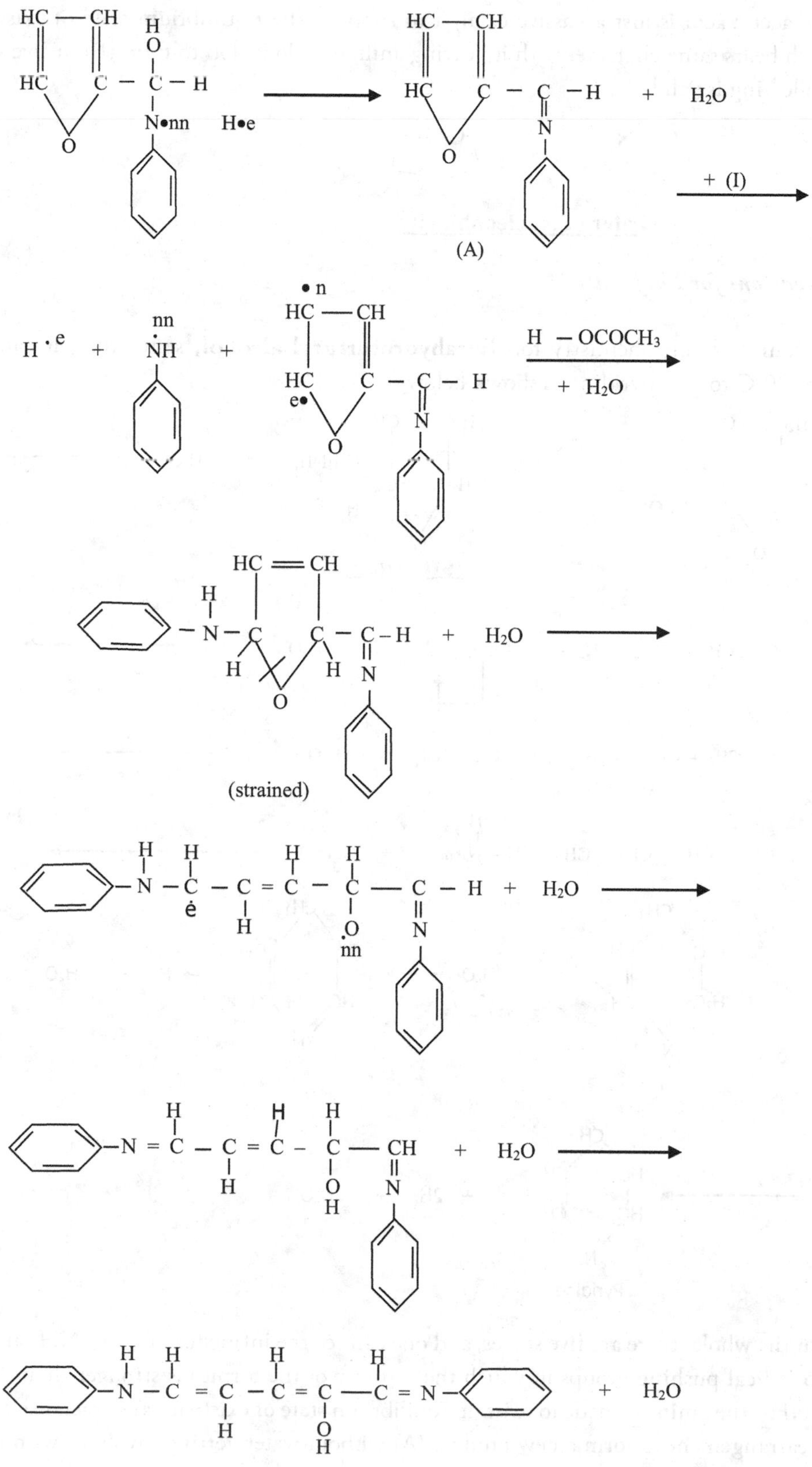

(A)

+ (I)

$H - OCOCH_3$
+ H_2O

(strained)

+ H_2O

+ H_2O

+ H_2O

+ H_2O

(B)

(i)

in which the acetic acid is just a passive catalyst to suppress the Equilibrium state of existence of the furfural which bears same character with it, leaving aniline to do its job to form (B) in five stages; from which the following is valid-

$$C = N \quad\quad > \quad\quad C = C$$

Order of Nucleophilicity (ii)

(Laws of Creations for Furfural)

Rule 1301: This rule of Chemistry for **Tetrahydrofurfuryl alcohol,** states that, it can react with ammonia at 500^0C to give pyridine as shown below-

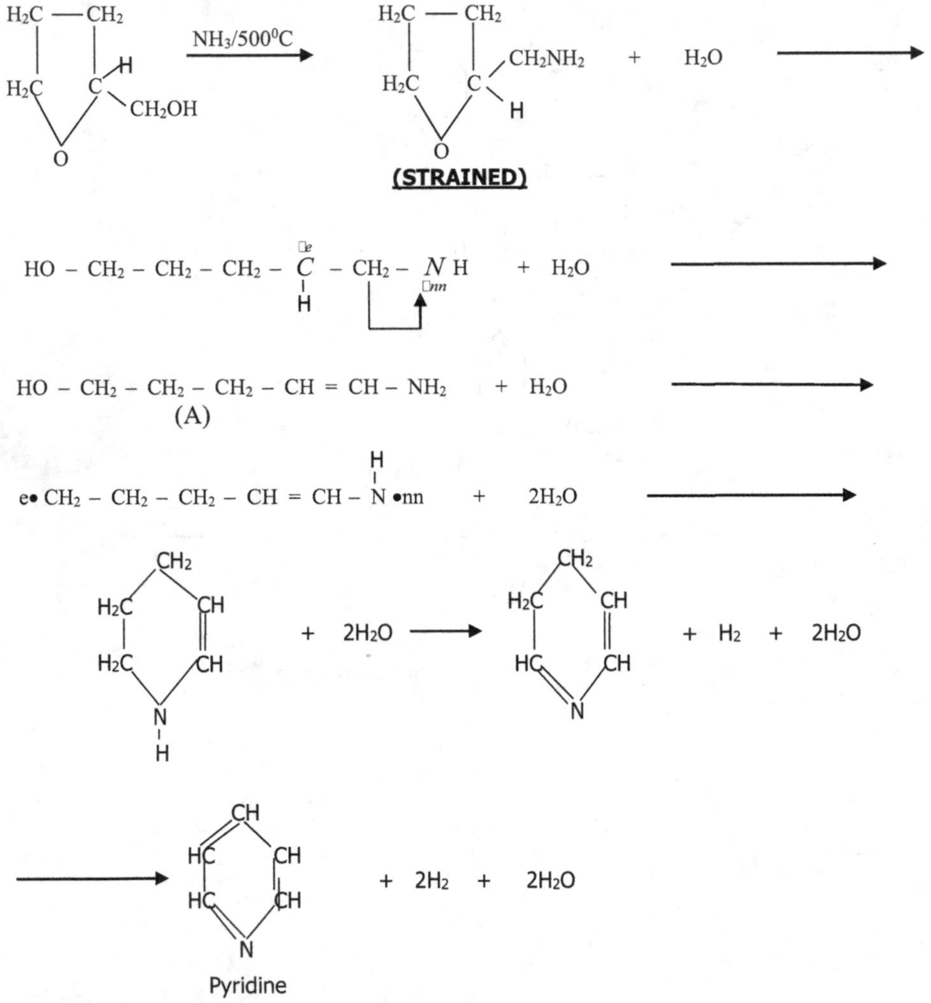

(STRAINED)

(A)

Pyridine

for which on the whole, there are five stages, and one can see the influence of $-CH_2NH_2$ and $-CH_2OH$ groups, two radical pushing groups in which the capacity of the former is stronger than the latter and with $-CH_2NH_2$, the amine is made to exist in Equilibrium state of existence and open itself, followed by molecular rearrangement to form a new product (A) without water getting involved; with the same (A)

existing in Equilibrium state of existence to form a ring by itself, for which the use of the high temperature was necessitated by the cracking done to release the two H_2 molecules using water as an active catalyst.
(Laws of Creations for Tetrahydrofurfuryl alcohol to Pyridine)

Rule 1302: This rule of Chemistry for **Furfural,** states that, when the externally located C = O center in furfural is replaced with C = C center, the followings are obtained-

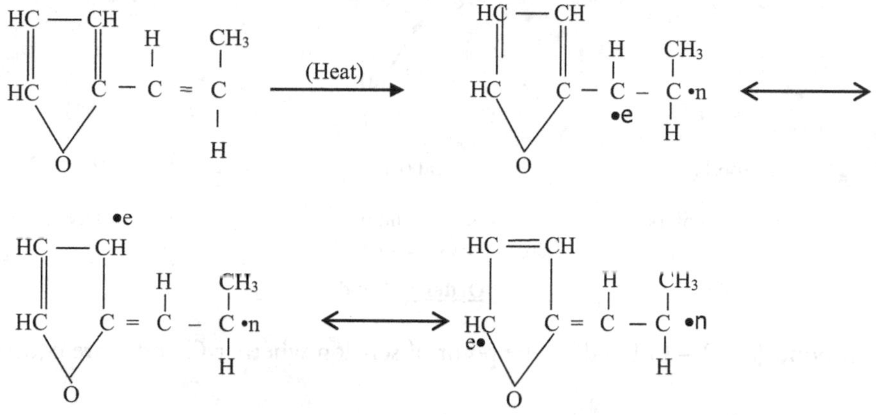

(A) 2, 4, 6- Resonance stabilized

for which the monomer is resonance stabilized as shown above and seen to favor the electro-free-radical route being a Nucleophile as well as the nucleo-free-radical route; noting that the following is valid-

Furan > Alkenes ; Furanyl group > Alkenyl group

Order of Nucleophilicity   Order of Radical-pushing capacity

noting that the following still remains valid-

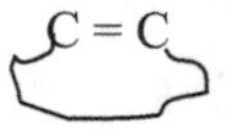

C = C > Alkenes (C = C)

Order of Nucleophilicity

and the same will apply to pyrrole, but not to thiophene.
(Laws of Creations for Furan and Pyrrole)

Rule 1303: This rule of Chemistry for **The use of molecular O_2 from air as Oxidizing oxygen without the use of Oxidizing agents such as $KMnO_4$,** states that, when metallic oxides saturated with oxygen is used, it makes the molecular oxygen to break into two parts inside the metallic pores as shown below-

$$O_2 \text{ (AIR)} \quad \underset{Cat.}{\overset{Fe_2O_3 - Ag_2O}{\rightleftharpoons}} \quad 2en\bullet \ O \bullet nn$$

[Molecular Oxygen]   [Oxidizing Oxygen molecule]

noting that one of the metals must be a transition metallic oxide which contain pores. {Surface Chemistry}
(Laws of Creations for Oxidizing oxygen)

Rule 1304: This rule of Chemistry for **C-C, C-S, and S-S single bonds in rings,** states that, the following is valid for them-

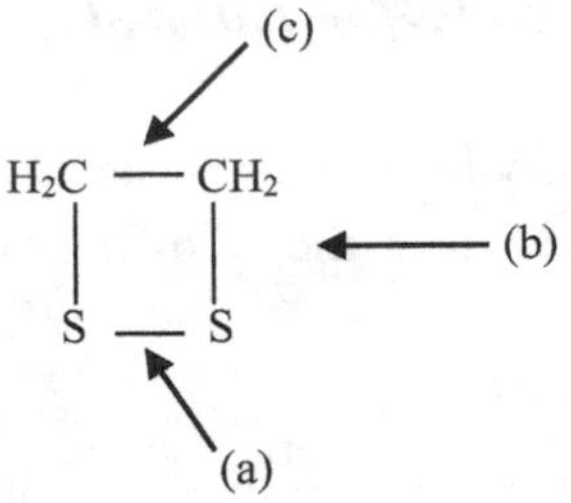

$$H_2C — CH_2$$

(with bond (c) pointing to the $H_2C—CH_2$ bond, bond (b) pointing to the right side, and bond (a) pointing to the $S—S$ bond)

$$\begin{array}{ccccc}
C — C\,(83) & > & C — S\,(65) & > & S — S\,(51) \\
\text{Bond (c)} & & \text{Bond (b)} & & \text{Bond (a)} \\
\text{(two free centers)} & & \text{(one free and one} & & \text{(two non-free} \\
& & \text{non-free center)} & & \text{centers)}
\end{array}$$

Order of Bond strength

for which S – S bond (or O – O bond) is the point of scission whether C and S are equi-electropositive or not.

(Laws of Creations for Bonds)

Rule 1305: This rule of Chemistry for **O-O, N-N, and S-S single bonds in rings whether symmetrically placed or not,** states that, when they [O - O (bond strength of 33 kcal/mole), S - S (bond strength of 51kcal /mol), N – N (bond strength of 38kcal/mol)] exist in a ring, the ring will be strongly strained in the following order –

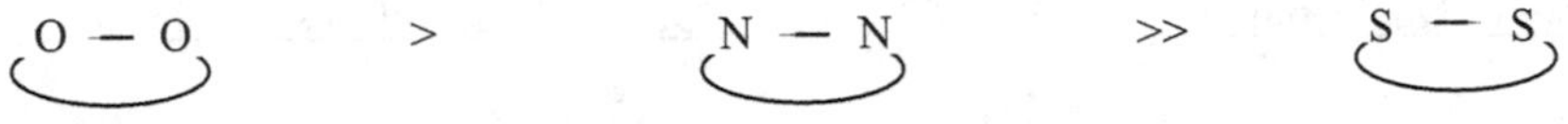

$$\begin{array}{ccccc}
O — O & > & N — N & \gg & S — S \\
\text{(I)} & & \text{(II)} & & \text{(III)}
\end{array}$$

Order of Strains in the rings of same size

to the point where their favored existences are in the reverse order above.

(Laws of Creations for Bonds)

Rule 1306: This rule of Chemistry for **Sulfur containing rings,** states that, the more the number of similar hetero atoms placed side by side in a ring, the less strained is the ring and the more nucleophilic is the ring, that is–

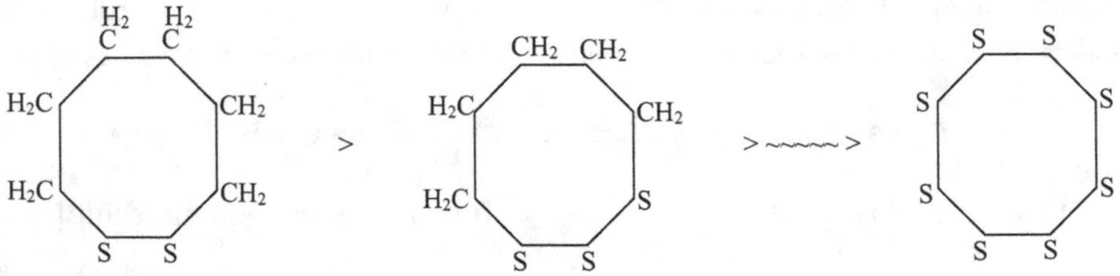

A - Cyclic disulphide   >   A Cyclic trisulfide   > ~~~~ >   Rhombic sulfur

Order of Strain Energy

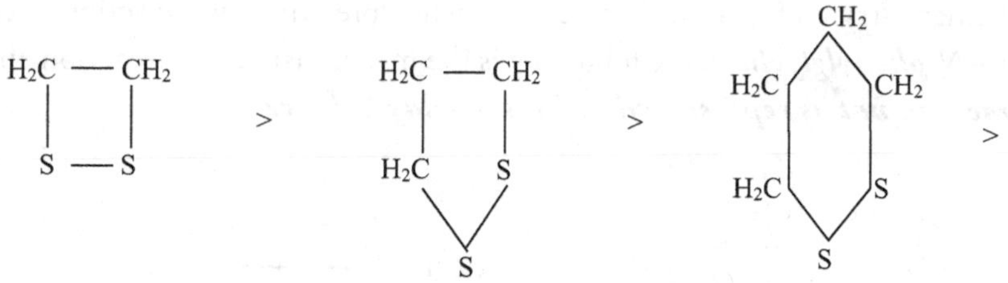

Order of Strains in the family

(Laws of Creations for S containing rings)

Rule 1307: This rule of Chemistry for **Only Sulfur containing rings,** states that, only eight sulfur atoms placed side by side can exist, while three-, four-, five- six- and seven-membered rings are not known to exist, because the following is valid for them-

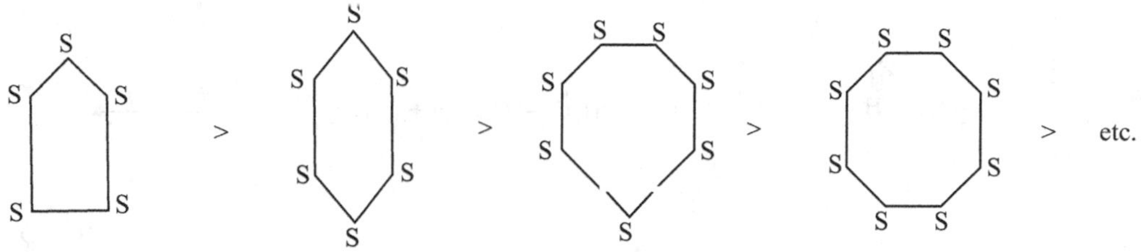

Order of Strains and Strain Energy in the family

noting that S is so unique because it is among the Transition atoms- C, S, and I and it is the only group of families where the SE is different for different sizes, since for every addition of S, two paired unbonded radicals are added to the ring and a new family is formed.

(Laws of Creations for Rhombic sulfur)

Rule 1308: This rule of Chemistry for **Rhombic sulfur,** states that, when decomposed what are called sulfurizing sulfur analogous to oxidizing oxygen is produced for which the affinity of this for H is less than that from oxygen-.

$$\text{Rhombic Sulfur} \xrightarrow{\text{Heat}} 8 \text{ en} \bullet S \bullet nn \quad ; \quad O_2 \xrightarrow[\text{AIR}]{Fe_2O_3/Ag_2O} 2 \text{ en} \bullet O \bullet nn$$

$$\underline{\textit{Sulfurizing Sulfur}} \qquad\qquad \underline{\textit{Oxidizing Oxygen}}$$

(Laws of Creations for Rhombic sulfur)

Rule 1309: This rule of Chemistry for **Cyclic sulfides, disulfides, trisulfides and tetrasulfides,** state that, since in general the number of functional centers carried in a ring, is a measure of the level of nucleophilicity, the following is valid-

Cyclic tetrasulphides   >   Cyclic trisulphides   >   Cyclic disulphide   >   Cyclic sulphides

Order of Nucleophilicity

(Laws of Creations for S containing rings)

Rule 1310: This rule of Chemistry for **Sydnones,** states that, when prepared by the action of *cold acetic anhydride on N-nitroso-N-phenylglycines,* the following is the mechanism of the reaction and *the real structure of the Sydnone and not as represented in Present-day Science-*

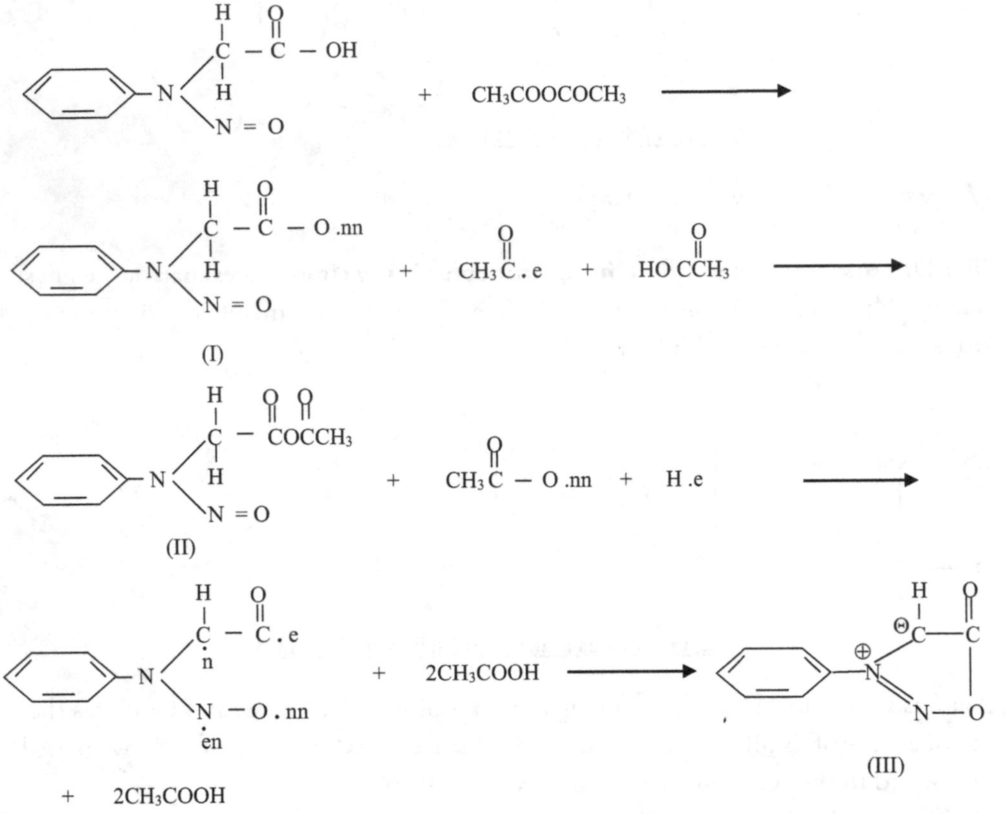

noting that this is a two stage Equilibrium mechanism system in which (III) is indeed the structure of the sydnone.

(Laws of Creations for Sydnones)

Rule 1311: This rule of Chemistry for **Sydnones,** states that, these are self-activated monomers which like diazomethanes favor polar/radical resonance stabilization phenome-none of the *closed-loop type and discrete in character* since (I) does not go beyond (II) as shown below-

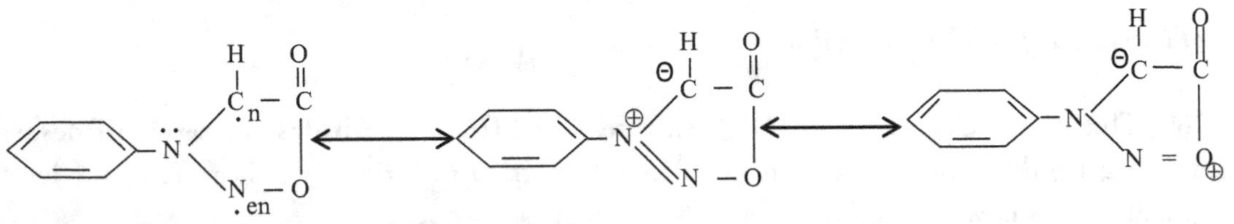

(I) Neutral environment   (II)a Basic environment   (II)b Acidic environment

Radical/Polar Resonance Stabilization phenomenon

(Laws of Creations for Sydnones)

Rule 1312: This rule of Chemistry for **Sydnones,** states that, these are *Electrophiles (i.e., Males)* which favor *discrete polar/radical resonance stabilization* different from that favored by *thiophene,*

since **thiophene** unlike sydnones has to be first activated and its resonance stabilization is **continuous in character.**
(Laws of Creations for Sydnones)

Rule 1313: This rule of Chemistry for **Sydnones,** states that, as **Electrophiles,** when attacked nucleo-non-free-radically, the route natural to them, the followings are obtained when paired or isolatedly placed-

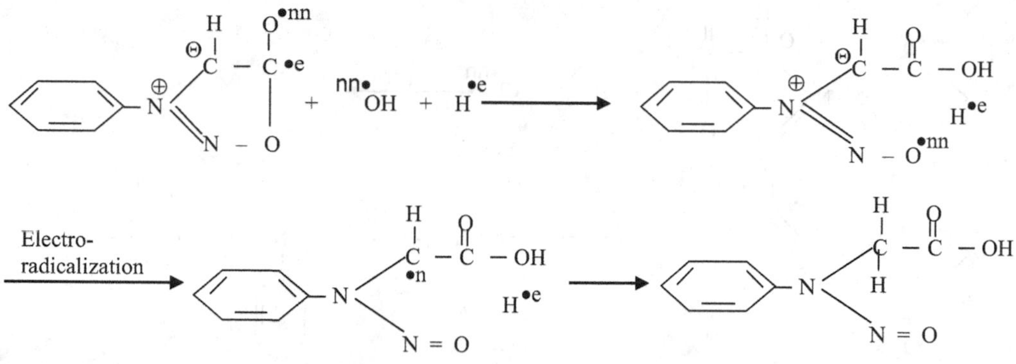

(i)

noting that this could not be done chargedly, because charges cannot move and be moved during Electroradicalization or Resonance stabilization and electro-free-radically, the followings are obtained-

Stage 1:

$$H_2O \rightleftharpoons H\bullet e + nn\bullet OH$$

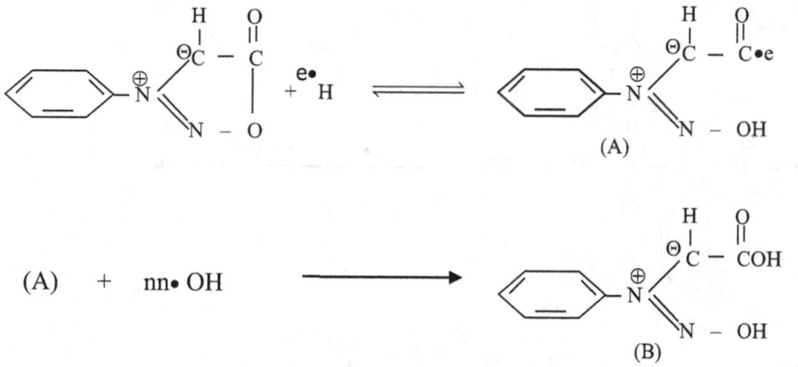

(ii)

Stage 2:

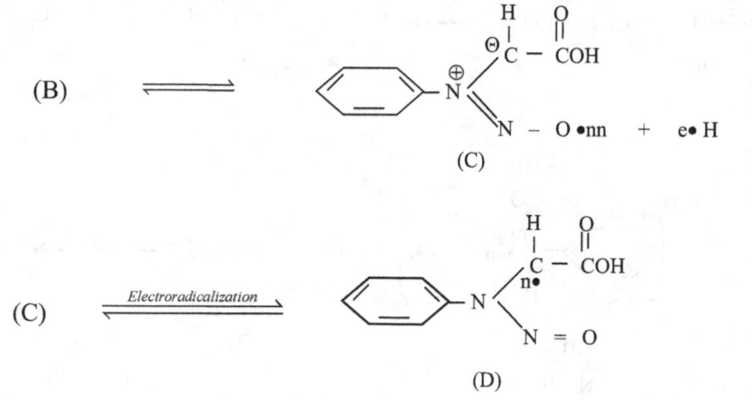

(iii)

$$H\bullet e + (D) \longrightarrow \text{N-nitroso-N-phenylglycine}$$

(iii)

471

noting that the same N-nitroso-N-phenylglycine obtained clearly indicates that Sydnones are indeed Electrophiles (Males) like lactones, unlike diazoalkanes which are Nucleophiles (Females).
(Laws of Creations for Sydnones)

Rule 1314: This rule of Chemistry for **Sydnones,** states that, when they are attacked by *moderately concentrated HCl,* the followings are the true mechanisms of the reactions,

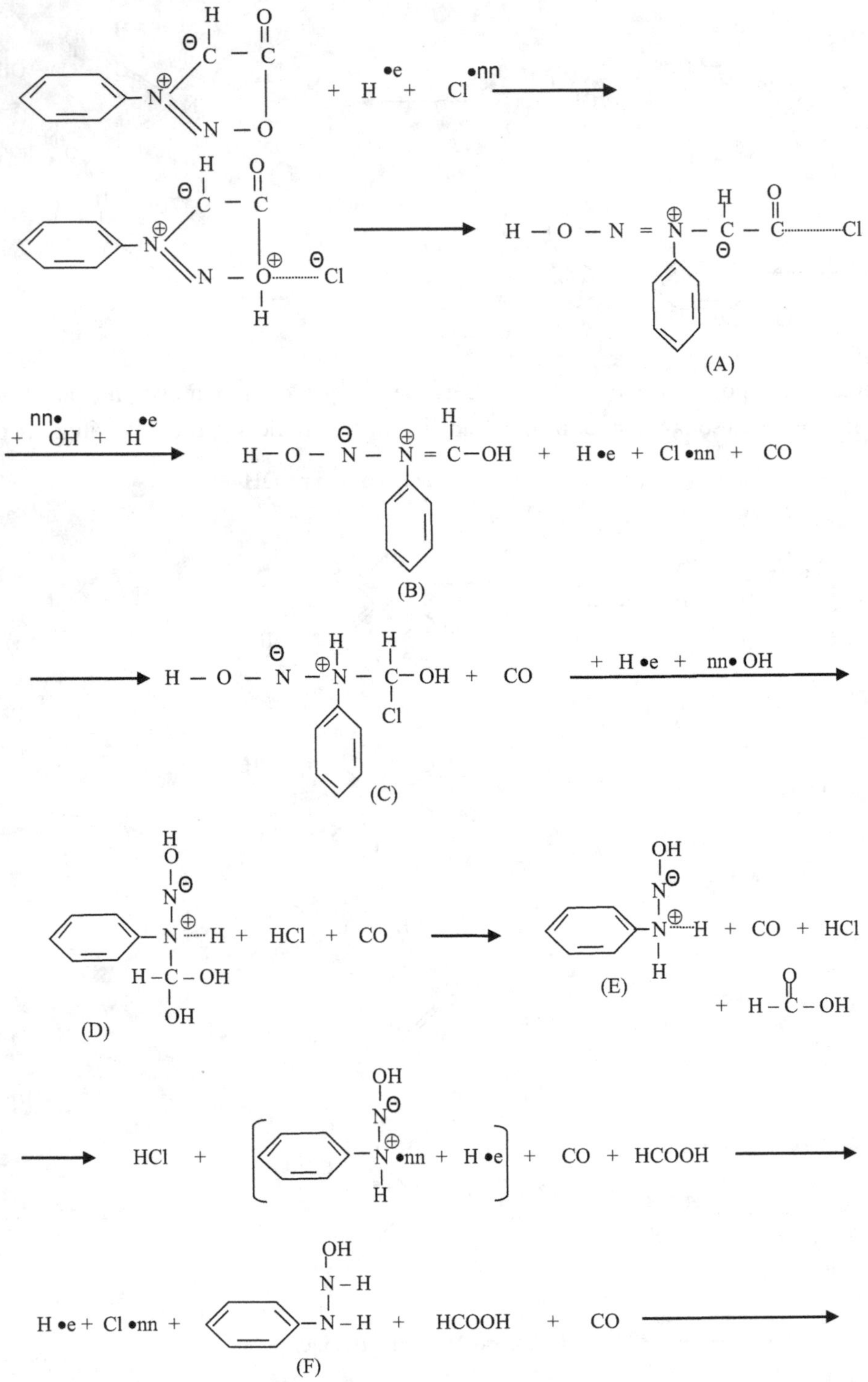

(A)

(B)

(C)

(D)

(E)

(F)

ELECTRO/POLAR RADICALIZATION

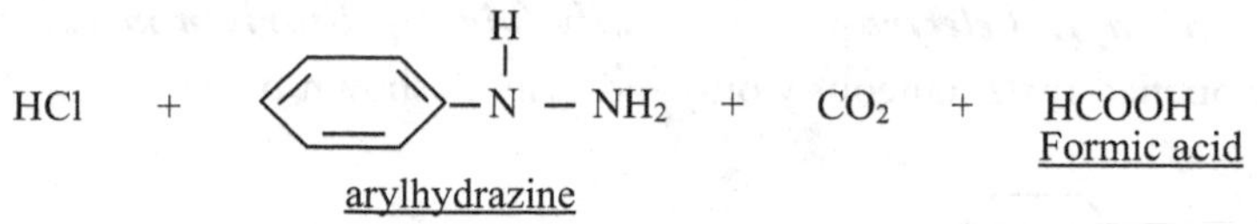

$$HCl \quad + \quad [\text{arylhydrazine ring}]-N-NH_2 \quad + \quad CO_2 \quad + \quad \underset{\text{Formic acid}}{HCOOH}$$

$$\overset{|}{H}$$

arylhydrazine

(i)

Overall Equation: $2H_2O + HCl + C_6H_5N_2OCHCO \longrightarrow C_6H_5N_2H_4 + CO_2$

$$+ HCl + HCOOH$$

(ii)

a seven stage Equilibrium mechanism system, from which, one can observe how the arylhydrazine, formic acid and carbon dioxide are produced, noting a new form of Electroradicalization phenomenon herein called ***ELECTRO/POLAR RADICALIZATION of the Second kind of the first type*** since it involves *movement of an electro-radical <u>from the negative charge of a polar bond to add to a sitting nucleo-non-free-radical on the center carrying the positive charge to form paired unbonded radicals on the carrier and leave a nucleo-non-free-radical on the donor center to which the H atom held in equilibrium adds to form a new compound without polar bond [From (E) to (F) above]*</u>, all in six stages from the beginning to (F), and finally in the seventh stage HCl attacked (F) to release oxidizing oxygen used to oxidize the CO with release of Heat and formation of product as shown below- in the seventh stage, the HCl still present as one of the products comes to attack the (F) formed as follows-

Stage 7: $\quad HCl \quad \rightleftharpoons \quad H \bullet e \quad + \quad nn \bullet Cl$

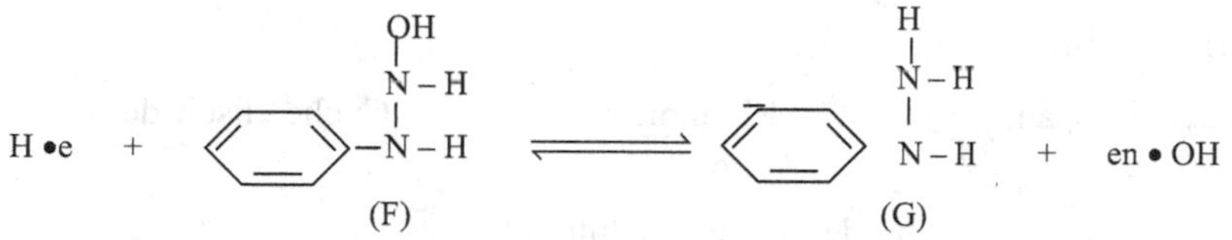

$$H \bullet e \quad + \quad [\text{ring (F)}] \quad \rightleftharpoons \quad [\text{ring (G)}] \quad + \quad en \bullet OH$$

(F) (G)

$$HO \bullet en \quad \rightleftharpoons \quad H \bullet e \quad + \quad en \bullet O \bullet nn \quad + \quad Heat$$

$$nn \bullet O \bullet en \quad + \quad e \bullet \overset{O}{C} \bullet n \quad \rightleftharpoons \quad nn \bullet O - \overset{O}{C} \bullet e$$

(H)

$$H \bullet e \quad + \quad nn \bullet Cl \quad \rightleftharpoons \quad HCl$$

(H) $\quad \xrightarrow{\textit{Deactivation}} \quad$ **CO$_2$** + **Heat**

(iii)

Overall Equation: **HCl** + **(F)** + **CO** $\longrightarrow$ **CO$_2$** + **HCl** + **(G)**

(iv)

from which it can be observed that the following is valid-

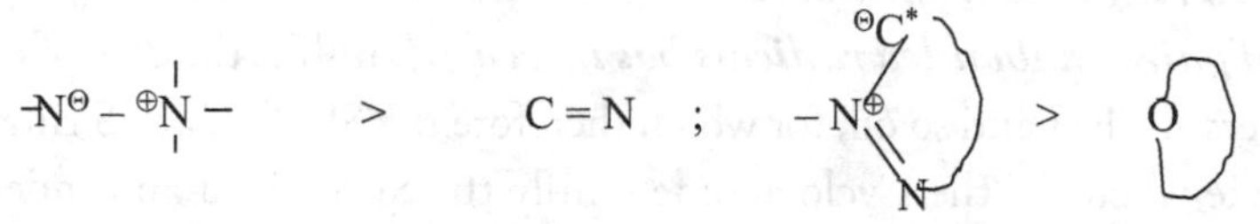

$$-N^{\ominus} - \overset{|}{\underset{|}{N}}^{\oplus} - \quad > \quad C=N \quad ; \quad -N^{\oplus} \quad > \quad [\text{ring}]$$

(Polar bond)

Order of Nucleophilicity

(v)

(Laws of Creations for Sydnones)

Rule 1315: This rule of Chemistry for **Rhombic sulfur,** states that, when used as a monomer *any one of the sulfur functional centers in the ring can be used, and this can be done electro-non-free-radically* *(via Combination mechanism and electro-free-radically (via Equilibrium mechanism),* and most favorably the ring can be opened instantaneously only radically as shown below-

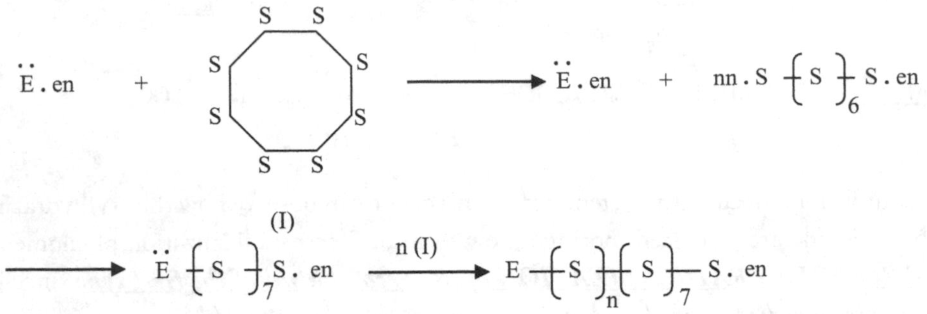

$$\ddot{E}\cdot en \quad + \quad \text{(I)} \quad \longrightarrow \quad \ddot{E}\cdot en \quad + \quad nn\cdot S\left\{S\right\}_6 S\cdot en$$

$$\longrightarrow \quad \ddot{E}\left\{S\right\}_7 S\cdot en \quad \xrightarrow{n\,(I)} \quad E\left\{S\right\}_n\left\{S\right\}_7 S\cdot en$$

noting that polymerization cannot be favored chargedly due to electrostatic forces of repulsion, and when paired initiators are used, this can only be done electro-free-radically via Backward addition using for example HCl and electro-non-free-radically using for example $FeCl_3$ via Forward and Backward additions, all depending on the operating conditions.
(Laws of Creations for Rhombic sulfur)

Rule 1316: This rule of Chemistry for **Rhombic sulfur,** states that, this monomer like *oxaazacyclopropane, cyclic disulfides (all Nucleophiles), are full non-free-radical compounds stable and unstable,* for which the followings are valid-.

Oxaazacyclopropane   >   Rhombic sulfur   >   Cyclic disulfide

Order of Nucleophilicity

(Laws of Creations for Rhombic Sulfur)

Rule 1317: This rule of Chemistry for **Rhombic sulfur,** states that, when compared with benzene, it has *sixteen paired unbonded radicals with eight points of scission,* while benzene ring has *three conjugatedly placed π-bonds with no point of scission,* clear indication of presence of large SE in rhombic sulfur for which it can be seen why smaller sized rings of S atoms cannot therefore exist.
(Laws of Creations for Rhombic sulfur)

Rule 1318: This rule of Chemistry for **Rhombic sulfur,** states that, Rhombic sulfur is the only member of its family, just like the other sized S rings are the only single members of their families, *because as we move from three-membered ring to four membered ring we have moved from one family to another family since the number of paired unbonded radicals has increased,* unlike the case of other families of Cyclo alkanes, alkenes, ethers, amines and so on; for which therefore the SEs for each S ringed compounds is fixed and different, unlike where for the Cyclo alkane family the SE is fixed and independent of the size of the ring.
(Laws of Creations for Rhombic sulfur)

Rule 1319: This rule of Chemistry for **Sulfur molecule,** states that, unlike Oxygen where O_2 exists, with S, a double bond cannot exist between two S atoms to form S_2, because of the physical state of the atom, the size, its valence states compared to O, and its unique personality yet to be identified (part metallic and part non-metallic); for at best S_2 can only exist as shown below-

$$en\square S - S\square nn \longrightarrow \overset{\oplus}{S} - S \quad (Polar\ bond)$$

making it look as if it is activated all the time and noting that polar charges do not repel or attract and this can also be used in a similar manner like sulfurizing sulfur (en .S. nn).
(Laws of Creations for Sulfur atom)

Rule 1320: This rule of Chemistry for **Rhombic sulfur,** states that, the SE in the ring is not an invisible π-bond since S cannot form double bonds with itself, but an invisible polar bond as shown below-

$$en\square S - S - S - S - S - S - S - S\square nn \longrightarrow \overset{2\oplus}{S} - S - S - S - S - S - S - S$$

Octasulfurzone

(Laws of Creations for Rhombic sulfur)

Rule 1321: This rule of Chemistry for **Rhombic sulfur,** states that, while rhombic sulfur exists, rhombic oxygen or any ring with O - O bond along the ring does not exist, because the following is valid-.

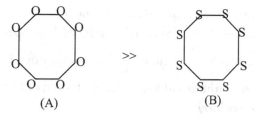

(A) >> (B)

Order of Strain and Nucleophilicity in (A) and (B)

noting that despite the fact that (A) has the MaxRSE, (B) does not; yet the SE is (B) is greater than that in (A) quantitatively, the SE in (A) being too much for the ring with weak O – O bond to sustain, unlike the SE in (B) which is not much for the ring to sustain, because of the strength of the S – S bond.
(Laws of Creations for Rhombic sulfur)

Rule 1322: This rule of Chemistry for **Rhombic sulfur**, states that, when used as *a Cross – linking agent* between two unsaturated polymer chains such as the linear chains from 1,3-butadiene, the followings are obtained-

 Stage 1:

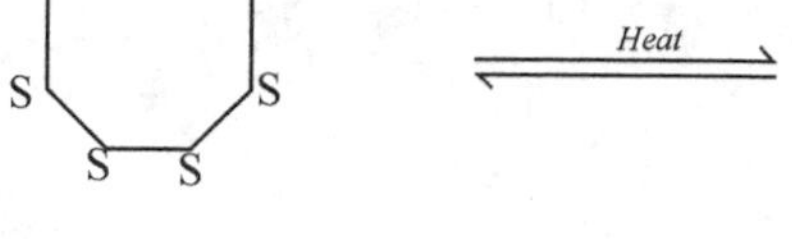

$$\xrightleftharpoons{Heat} \quad 8(en\bullet S \bullet nn)$$
(A)

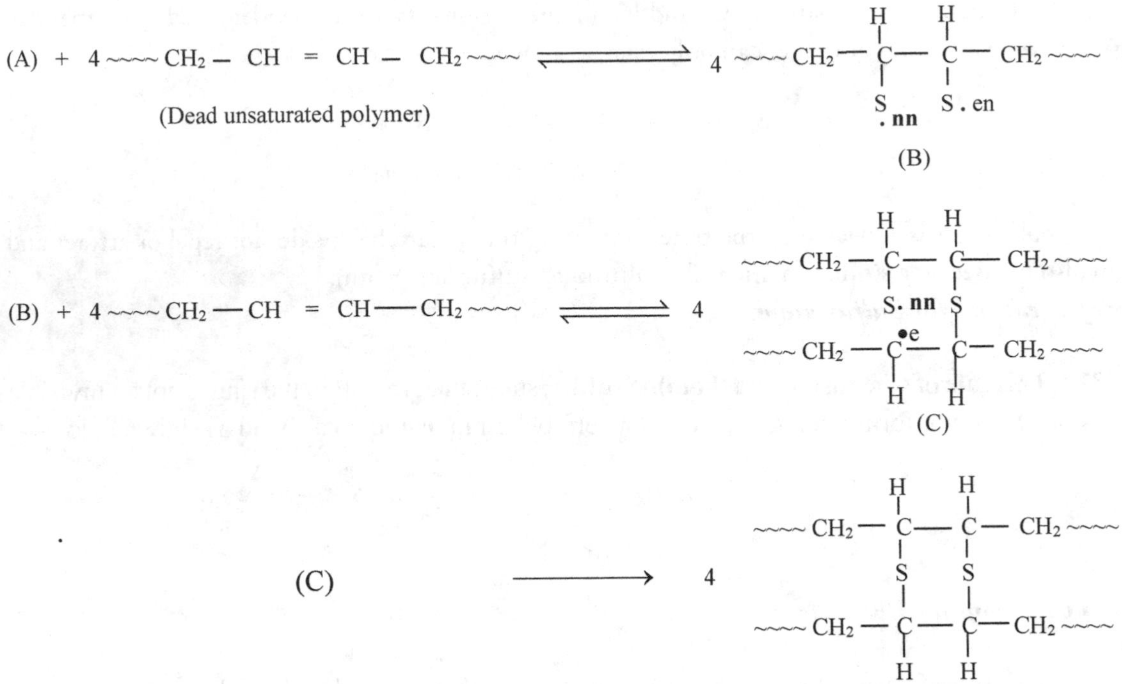

(A) + 4 ～～ CH_2 — CH = CH — CH_2 ～～ $\rightleftharpoons$ 4 ～～ CH_2—C—C—CH_2 ～～

(with H, H above the two C's, and S·nn, S·en below)

(Dead unsaturated polymer)

(B)

(B) + 4 ～～ CH_2 — CH = CH — CH_2 ～～ $\rightleftharpoons$ 4

～～ CH_2—C—C—CH_2 ～～ (H, H above; S·nn, S below)

～～ CH_2—C—C—CH_2 ～～ (•e; H, H below)

(C)

(C) $\longrightarrow$ 4

～～ CH_2—C—C—CH_2 ～～ (H, H above; S, S below)

～～ CH_2—C—C—CH_2 ～～ (H, H below)

Cross-linked network

noting that ***cross-linking can never take place via Combination mechanisms but via Equilibrium mechanism*** as shown above and to cross-link using rhombic sulfur without opening the ring instantaneously, is impossible and one cannot use $nn\bullet S \bullet nn$ (the atom) for cross-linking, but $en\bullet S \bullet nn$ (an element), just as one can never use $nn\bullet O \bullet nn$ (the atom) for oxidation, but $en\bullet O \bullet nn$ (an element).
(Laws of Creations for Rhombic sulfur)

Rule 1323: This rule of Chemistry for **Dichlorophosphazene,** states that, this is ***a semi-inorganic monomer/compound*** whose real structure is as shown below, unlike what is generally shown to be the case-

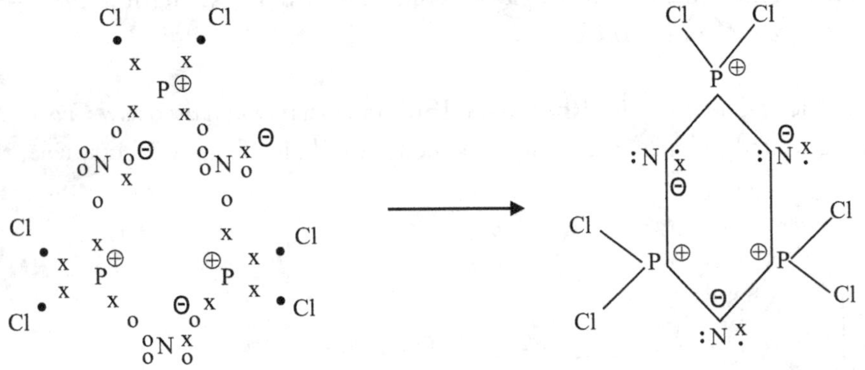

Real structures of Dichlorophosphazene

noting that, there are ***three polar bonds, three paired unbonded radicals and six covalent bonds*** in the ring and since according to one of the boundary laws wherein P like N cannot carry more than eight radicals in the last shell, the P center cannot carry a double bond here.
(Laws of Creations for Dichlorophosphazene)

Rule 1324: This rule of Chemistry for **Dichlorophosphazene,** states that, in view of the polar character of the monomer, chargedly, there are six functional centers in the ring-

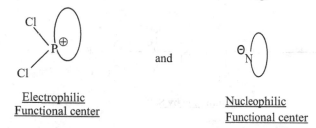

Electrophilic
Functional center

and

Nucleophilic
Functional center

all polar in character, and since when polar bonds are attacked chargedly, either a stable molecule is released in non-ringed compounds or a ring must be opened where possible; for which this is clear indication that ***this monomer is an Electrophile of a very different kind,*** since the nucleophilic center cannot be used electro-free-radically or cationically in view of presence of transfer species of the first kind, while the electrophilic center cannot be used nucleo-non-free-radically or anionically, unless when the ring is opened instantaneously.
(Laws of Creations for Dichlorophosphagene)

Rule 1325: This rule of Chemistry for **Dichlorophosphaxene,** states that, this semi-inorganic monomer which cannot undergo resonance stabilization, is a ring with three polar bonds and three paired unbonded radicals making it well strained with three points of scission (but less strained than that provided by the three p-bonds in Benzene with no point of scission), clearly indicates that the ring can instantaneously be opened, and when opened instantaneously, this can be done either radically or chargedly via the single bond.
(Laws of Creations for Dichlorophosphazene)

Rule 1326: This rule of Chemistry for **Dichlorophosphazene,** states that, when heated at temperatures far below 250°C, the followings take place-

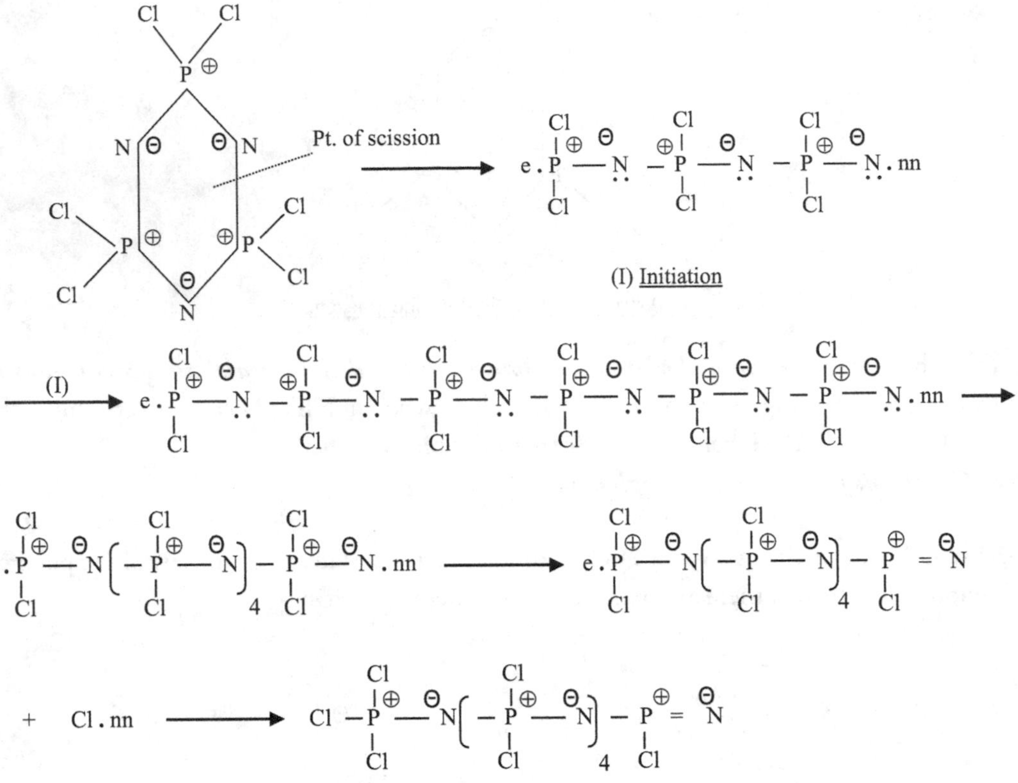

$$
\text{(I) Initiation}
$$

$$
\text{Termination by Self}
$$

for which, it can be observed that the monomer was only kept in Activated state of existence after instantaneous opening of the ring, growing from the nucleo-non-free-radical (being an electrophile) end in chains and finally killing itself after the optimum chain length has been reached to form dead terminal triple polar bond polymer with presence of a double bond; noting that the new termination step has been called **TERMINATION BY SELF. [See Rule 1215]**
(Laws of Creations for Dichlorophosphazene)

Rule 1327: This rule of Chemistry for **Dichlorophosphagene,** states that, if heated at temperature as high as 250°C, then some fractions of the monomer must be kept in Activated/ Equilibrium State of existence as shown below-

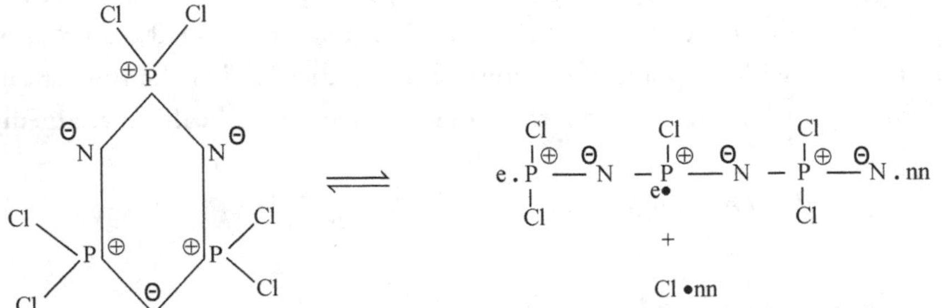

$$
\text{(II) Activated/Equilibrium State of existence}
$$

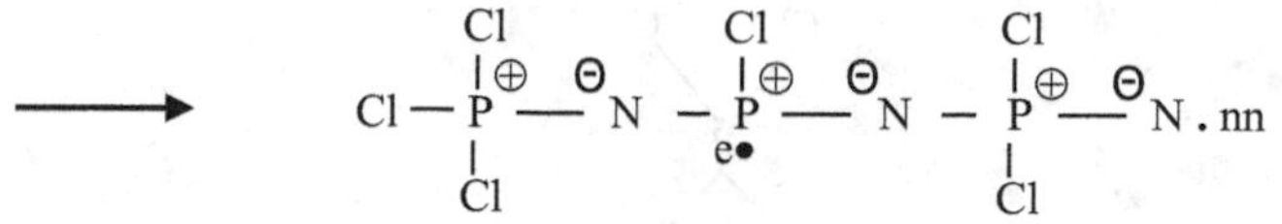

$$Cl - \overset{\overset{\textstyle Cl}{|}}{\underset{\underset{\textstyle Cl}{|}}{\overset{\oplus}{P}}} - \overset{\ominus}{N} - \underset{e\bullet}{\overset{\overset{\textstyle Cl}{|}}{\overset{\oplus}{P}}} - \overset{\ominus}{N} - \overset{\overset{\textstyle Cl}{|}}{\underset{\underset{\textstyle Cl}{|}}{\overset{\oplus}{P}}} - \overset{\ominus}{N} . nn$$

(III) Chlorolization phenomenon   i)

in which, it can be observed that the monomer is kept in Activated/Equilibrium state of existence, with the old electro-free-radical end diffusing to grab the element held in Equili-brium state of existence and leaving a new electro-free-radical center, which unlike Enolization and Amidization, what is finally formed [(III) above] cannot deactivate (because of the presence of MaxRSE if a four-membered ring was to be formed) and this new pheno-menon above has been called [See Rule 1214] called **CHLOROLIZATION PHENO-MENON**, in line with Enolization and Amidization; noting that for the living polymer chain when terminated by self, a dead terminal triple polar bond polymer with branching and cross-linking sites as shown below is obtained-

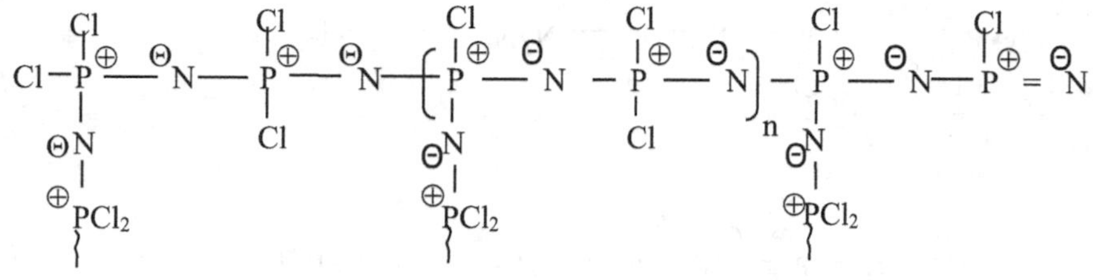

Dead polymer with branching sites   ii)

that which can only be used when in Equilibrium state of existence or when attacked anionically by a growing chain.

(Laws of Creations for Dichlorophosphagene)

Rule 1328: This rule of Chemistry for **Dichlorophosphazene**, states that, when *Carboxylic acid (HOOCR ≡ HB) is used as a source of initiator,* the electro-free-radical route or cationic routes cannot be favored whether the ring is opened instantaneously or not as shown below-

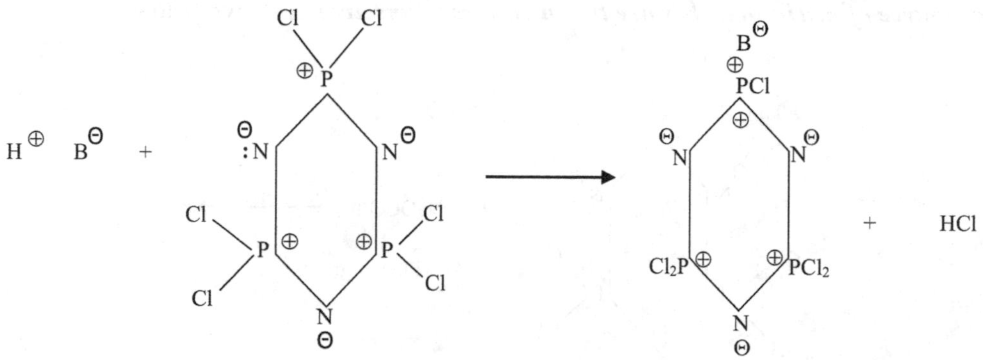

(I)   **No Initiation**   i)

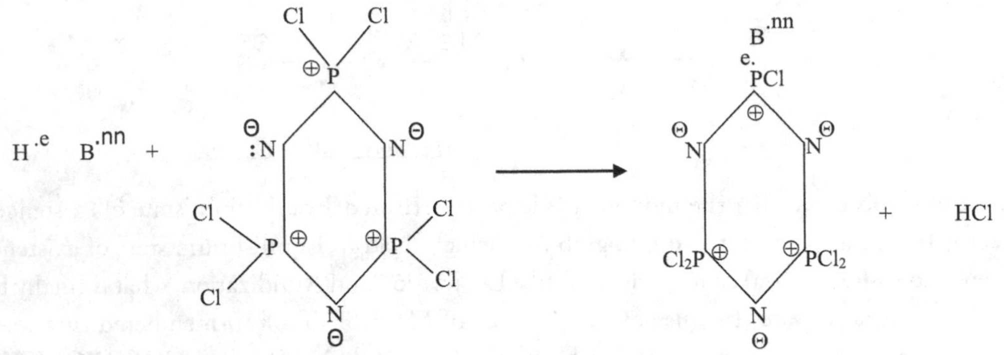

$$H \cdot^e \quad B^{\cdot nn} \; + \; \text{(I)} \quad \longrightarrow \quad \text{(No Initiation)} \; + \; HCl \qquad \text{ii)}$$

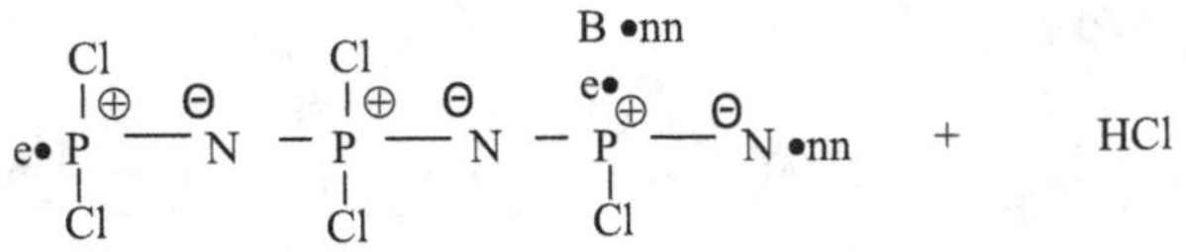

No Initiation

$$\text{OR} \quad e\bullet P \text{—} N - P \text{—} N - P \text{—} N \bullet nn \; + \; HCl \qquad \text{iii)}$$

$$\text{OR} \quad {}^{\oplus}P \text{—} N - P \text{—} N - P \text{—} N^{\ominus} \; + \; HCl \qquad \text{iv)}$$

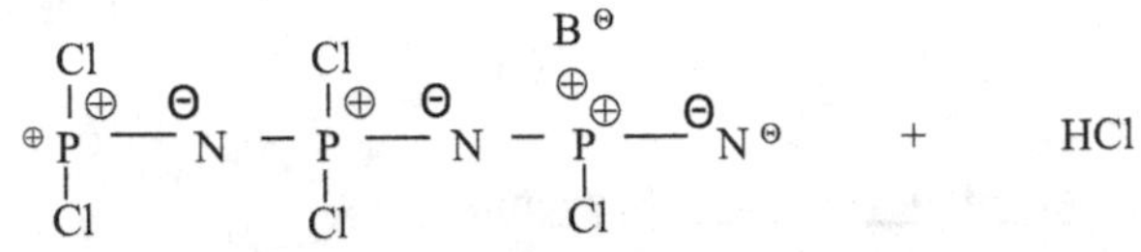

No Initiation-Radically or Chargedly

where the transfer species involved is of the **first kind and second first type** as opposed to first kind (Natural route) of the first type or second first kind (Unnatural route); noting that this is the transfer species rejected from the anionic growing polymer chain during propagation and termination when starved (the route natural to it).

(Laws of Creation for Dichlorophosphazene)

Rule 1329: This rule of Chemistry for **Dichlorophosphazene,** states that, when *carboxylic acid salts are used as source of initiators, they are the initiators themselves* as shown below-

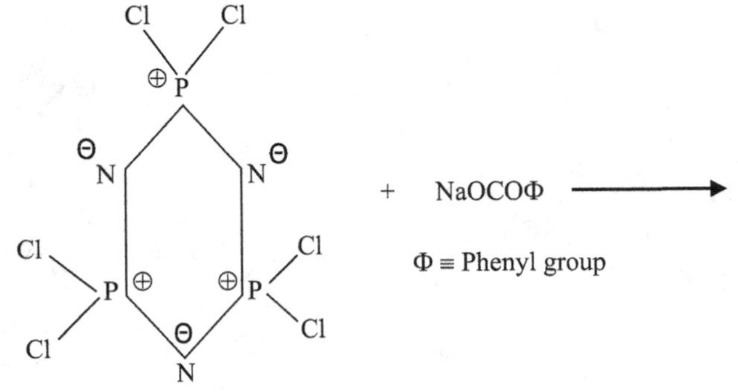

$$\text{(I)} \; + \; NaOCO\Phi \quad \longrightarrow$$

$$\Phi \equiv \text{Phenyl group}$$

$$\Phi COO - P \text{—} N - P \text{—} N - P \text{—} N^{\ominus} \cdots {}^{\oplus}Na \quad \xrightarrow{\; + \; \text{(I)}\;}$$

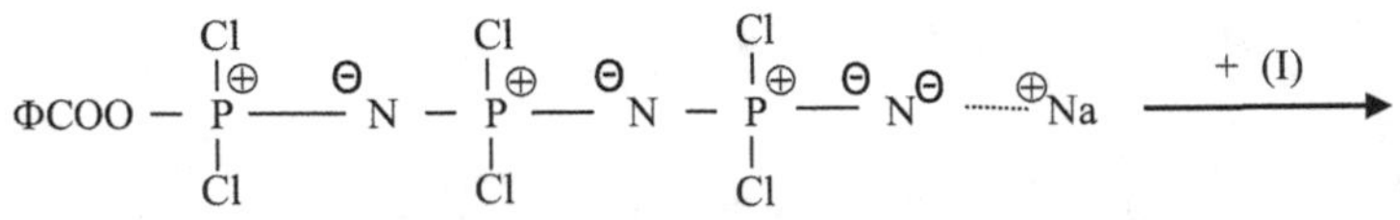

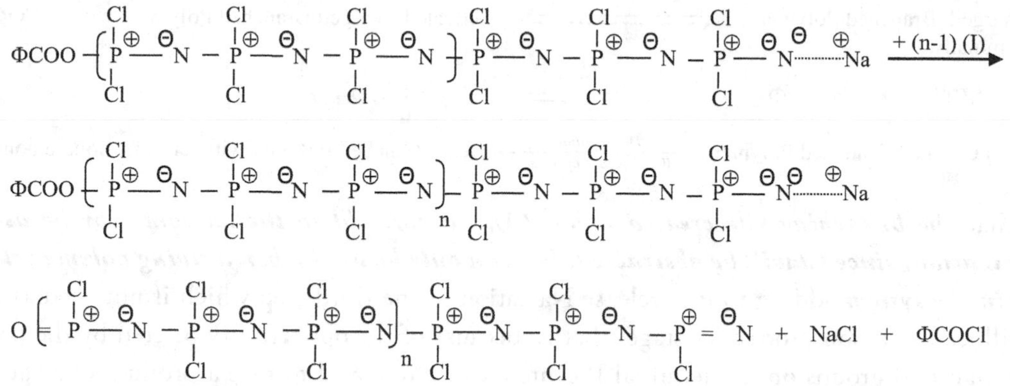

$$\Phi COO \left[\begin{array}{c} Cl \\ | \oplus \\ P \\ | \\ Cl \end{array} \!\!-\!\! \ominus N - \begin{array}{c} Cl \\ | \oplus \\ P \\ | \\ Cl \end{array} \!\!-\!\! \ominus N - \begin{array}{c} Cl \\ | \oplus \\ P \\ | \\ Cl \end{array} \!\!-\!\! \ominus N \right] \begin{array}{c} Cl \\ | \oplus \\ P \\ | \\ Cl \end{array} \!\!-\!\! \ominus N - \begin{array}{c} Cl \\ | \oplus \\ P \\ | \\ Cl \end{array} \!\!-\!\! \ominus N - \begin{array}{c} Cl \\ | \oplus \\ P \\ | \\ Cl \end{array} \!\!-\!\! \ominus N \cdots\cdots \overset{\oplus}{Na} \quad + (n-1)\,(I) \longrightarrow$$

$$\Phi COO \left\{ \begin{array}{c} Cl \\ | \oplus \\ P \\ | \\ Cl \end{array} \!\!-\!\! \ominus N - \begin{array}{c} Cl \\ | \oplus \\ P \\ | \\ Cl \end{array} \!\!-\!\! \ominus N - \begin{array}{c} Cl \\ | \oplus \\ P \\ | \\ Cl \end{array} \!\!-\!\! \ominus N \right\}_n \begin{array}{c} Cl \\ | \oplus \\ P \\ | \\ Cl \end{array} \!\!-\!\! \ominus N - \begin{array}{c} Cl \\ | \oplus \\ P \\ | \\ Cl \end{array} \!\!-\!\! \ominus N - \begin{array}{c} Cl \\ | \oplus \\ P \\ | \\ Cl \end{array} \!\!-\!\! \ominus N \cdots\cdots \overset{\oplus}{Na}$$

$$O \left[\begin{array}{c} Cl \\ | \oplus \\ P \\ | \\ \end{array} \!\!-\!\! \ominus N - \begin{array}{c} Cl \\ | \oplus \\ P \\ | \\ Cl \end{array} \!\!-\!\! \ominus N - \begin{array}{c} Cl \\ | \oplus \\ P \\ | \\ Cl \end{array} \!\!-\!\! \ominus N \right]_n - \begin{array}{c} Cl \\ | \oplus \\ P \\ | \\ Cl \end{array} \!\!-\!\! \ominus N - \begin{array}{c} Cl \\ | \oplus \\ P \\ | \\ Cl \end{array} \!\!-\!\! \ominus N - \overset{\oplus}{P} = \ominus N \quad + \quad NaCl \quad + \quad \Phi COCl$$

noting that the route here is anionic, since the monomer is an Electrophile, and when starved, it terminates itself to give a dead terminal triple polar bond polymer and in the process, Na closes the second terminal to form a double or polar bond, NaCl and $\Phi COCl$ in a one stage Equilibrium mechanism system during termination step; noting that the initiator could also be paired radically making the route being a nucleo-non-free-radical route.

(Laws of Creations for Dichlorophosphazene)

Rule 1330: This rule of Chemistry for **Branch formations,** states that, Branching in the chains of polymers can only take place when dead polymer chains are present in the system, since a living polymer which has not reached its optimum chain length cannot be adding and branching at the same time or simultaneously and ***Branching can either take place via Equilibrium or Combination mechanisms, depending on the type of monomer, while cross-linking takes place only via Equilibrium mechanism;*** for which with the case of dichlorophosphazene when catalyzed using $NaOCO\Phi$, one should expect a branching site to be formed with the presence of NaCl and $\Phi COCl$ when a chain is killed as shown below-

Stage 1:

$$NaCl \quad \rightleftharpoons \quad \overset{\smile}{Na}{}^{\oplus} \quad + \quad Cl^{\ominus}$$

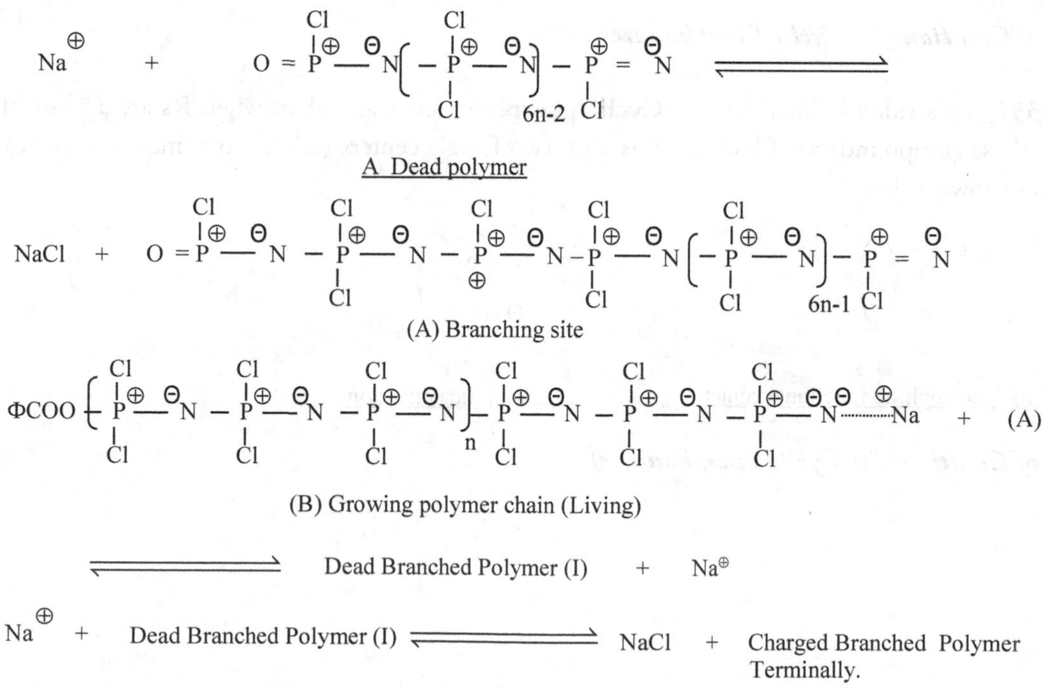

$$\overset{\oplus}{Na} \quad + \quad O = \overset{\oplus}{\underset{Cl}{P}} \!-\! \ominus N \left[- \overset{\overset{Cl}{|}\oplus}{\underset{\underset{Cl}{|}}{P}} \!-\! \ominus N \right]_{6n-2} - \overset{\oplus}{P} = \ominus N \quad \rightleftharpoons$$

A Dead polymer

$$NaCl \quad + \quad O = \overset{\overset{Cl}{|}\oplus}{\underset{\underset{Cl}{|}}{P}} \!-\! \ominus N - \overset{\overset{Cl}{|}\oplus}{\underset{\underset{Cl}{|}}{P}} \!-\! \ominus N - \overset{\overset{Cl}{|}\oplus}{\underset{\oplus}{P}} \!-\! \ominus N - \overset{\overset{Cl}{|}\oplus}{P} \!-\! \ominus N \left(- \overset{\overset{Cl}{|}\oplus}{\underset{\underset{Cl}{|}}{P}} \!-\! \ominus N \right)_{6n-1} - \overset{\oplus}{\underset{Cl}{P}} = \ominus N$$

(A) Branching site

$$\Phi COO \left\{ \begin{array}{c} Cl \\ | \oplus \\ P \\ | \\ Cl \end{array} \!\!-\!\! \ominus N - \begin{array}{c} Cl \\ | \oplus \\ P \\ | \\ Cl \end{array} \!\!-\!\! \ominus N - \begin{array}{c} Cl \\ | \oplus \\ P \\ | \\ Cl \end{array} \!\!-\!\! \ominus N \right\}_n \begin{array}{c} Cl \\ | \oplus \\ P \\ | \\ Cl \end{array} \!\!-\!\! \ominus N - \begin{array}{c} Cl \\ | \oplus \\ P \\ | \\ Cl \end{array} \!\!-\!\! \ominus N - \begin{array}{c} Cl \\ | \oplus \\ P \\ | \\ Cl \end{array} \!\!-\!\! \ominus N \cdots\cdots \overset{\oplus}{Na} \quad + \quad (A)$$

(B) Growing polymer chain (Living)

$$\rightleftharpoons \quad \text{Dead Branched Polymer (I)} \quad + \quad Na^{\oplus}$$

$$\overset{\oplus}{Na} \quad + \quad \text{Dead Branched Polymer (I)} \quad \rightleftharpoons \quad NaCl \quad + \quad \text{Charged Branched Polymer Terminally.}$$

Charged Branched Polymer $\rightleftharpoons$ Activated Charged Branched Polymer + e.COΦ
terminally

$Cl^\ominus$ + e.COΦ $\rightleftharpoons$ ClCOΦ

Activated Charged Branched Polymer $\xrightarrow[\text{With release of Heat}]{\text{Deactivation}}$ Dead Branched Polymer with terminal double bond

noting that *the branching site created above (A), cannot add to the monomer or be used for polymerization, since Cl will be abstracted, but can only be used when a living polymer chain is present in the system* adding to it to release Na cation in the third step which if not used as shown above will lead to a non-productive stage; that is, because of the opportunity offered by the presence of more than two groups on the terminal P center, one is removed to give a productive stage where NaCl which started the stage was released along with ClCOΦ and Dead branched terminally double and triple bonded polymer; noting also that this can only take place radically and not chargedly as shown above.

(Laws of Creations for Branching)

Rule 1331: This rule of Chemistry for **Cyclic phosphonates,** states that, the real structure of these compounds/monomers is as shown below-

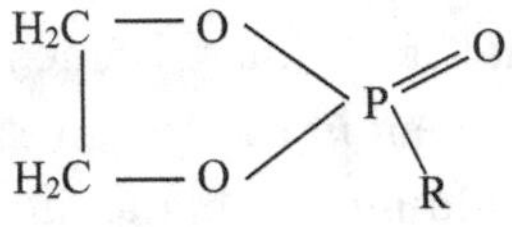

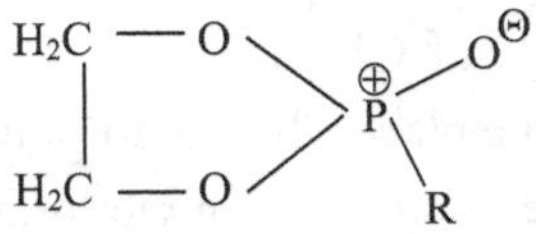

(I) <u>Wrong structure</u> ; (II) <u>Correct structure</u>

where the R group is alkyl or phenyl (aryl) or any type of group including H; wherein the P center cannot be allowed to carry more than 8 radicals in the last shell as shown in (II) as against 10 radicals in (I); noting that larger sizes of this can exist with respect to increasing number of presence of – CH$_2$ - groups.

(Laws of Creations for Cyclic Phosphonates)

Rule 1332: This rule of Chemistry for **Cyclic phosphonates,** states that, when Rs are aryl or phenyl groups, these compounds are **Electrophiles** with two female centers (X) and one male center (Y) (i.e., Males) as shown below-

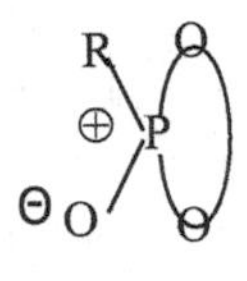

R $\equiv$ Aryl groups

<u>Nucleophilic functional center</u>   <u>Electrophilic activation center</u>

(Laws of Creations for Cyclic Phosphonates)

Rule 1333: This rule of Chemistry for **Cyclic phosphonates,** states that, when Rs are CF$_3$ types of groups, halogens, CONH$_2$, COOR types of groups and more, these compounds are *Electrophiles* with two female centers (X) and one male center (Y) (i.e., Males) as shown below-

;

R ≡ Radical-pulling groups
OR
Radical-pushing groups of
lower capacity than H

<u>Nucleophilic functional center</u>   <u>Electrophilic activation center</u>

(Laws of Creations for Cyclic Phosphonates)

Rule 1334: This rule of Chemistry for **Cyclic phosphonates,** states that, when the Rs are alkylane groups or radical-pushing groups of greater capacity than H, these compounds are Electrophiles since no transfer species exists nucleo-radically or negatively; for if transfer species exists nucleo-radically or negatively, then the positively charged or electro-free-radical routes will not be favored, with however two female centers (X) and one male center (Y) (i.e., Males) as shown below-

R ≡ Alkylane groups

<u>Electrophilic activation center</u>   <u>Nucleophilic functional center</u>

noting however that because P and H are equi-electropositive, H cannot be abstracted as an atom, and therefore no transfer species exists.

(Laws of Creations for Cyclic Phosphonates)

Rule 1335: This rule of Chemistry for **Cyclic phosphonates,** states that, when protonic acid is used as one of the sources of initiators, the followings are to be expected-

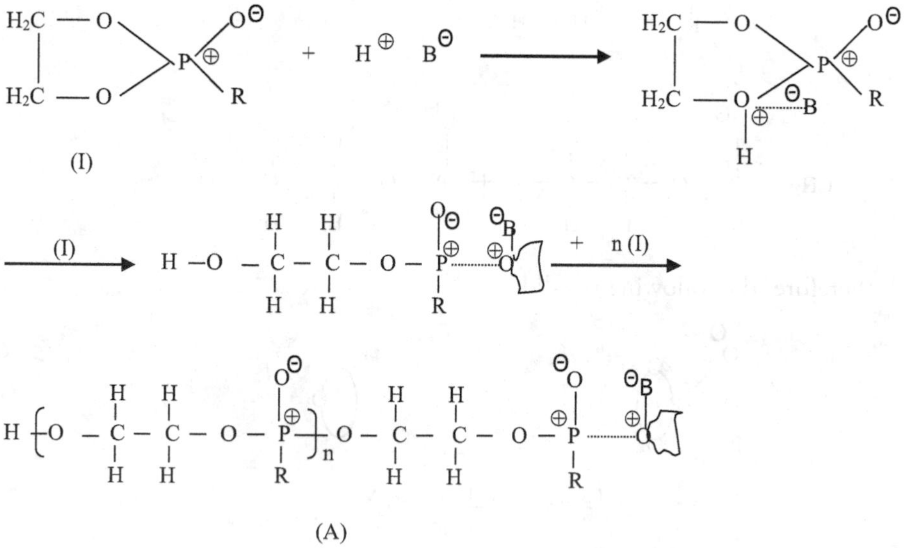

(A)   i)

noting that the route is ***electro-free-radical route*** after starting cationically, with the chain growing backwardly, for if it was growing forwardly, the followings are obtained-

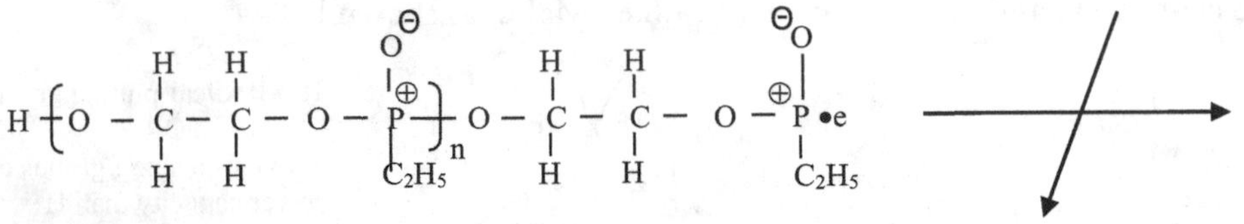

$$H \left[O - \overset{\overset{\displaystyle H}{|}}{\underset{\underset{\displaystyle H}{|}}{C}} - \overset{\overset{\displaystyle H}{|}}{\underset{\underset{\displaystyle H}{|}}{C}} - O - \overset{\overset{\displaystyle \overset{\ominus}{O}}{\|}}{\underset{\underset{\displaystyle C_2H_5}{|}}{\overset{\oplus}{P}}} \right]_n O - \overset{\overset{\displaystyle H}{|}}{\underset{\underset{\displaystyle H}{|}}{C}} - \overset{\overset{\displaystyle H}{|}}{\underset{\underset{\displaystyle H}{|}}{C}} - O - \overset{\overset{\displaystyle \overset{\ominus}{O}}{\|}}{\underset{\underset{\displaystyle C_2H_5}{|}}{\overset{\oplus}{P}}} \bullet e \quad \nrightarrow$$

(B)

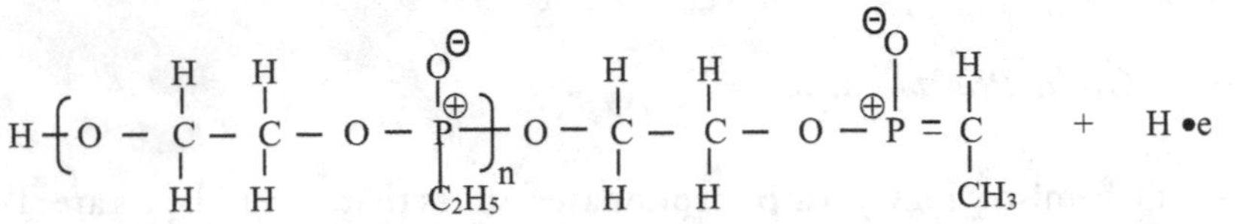

$$H \left[O - \overset{\overset{\displaystyle H}{|}}{\underset{\underset{\displaystyle H}{|}}{C}} - \overset{\overset{\displaystyle H}{|}}{\underset{\underset{\displaystyle H}{|}}{C}} - O - \overset{\overset{\displaystyle \overset{\ominus}{O}}{\|}}{\underset{\underset{\displaystyle C_2H_5}{|}}{\overset{\oplus}{P}}} \right]_n O - \overset{\overset{\displaystyle H}{|}}{\underset{\underset{\displaystyle H}{|}}{C}} - \overset{\overset{\displaystyle H}{|}}{\underset{\underset{\displaystyle H}{|}}{C}} - O - \overset{\overset{\displaystyle \overset{\ominus}{O}}{\|}}{\overset{\oplus}{P}} = \overset{\overset{\displaystyle H}{|}}{\underset{\underset{\displaystyle CH_3}{|}}{C}} \quad + \quad H \bullet e \qquad \text{ii)}$$

that is, that in which a dead terminal double bond polymer cannot be produced when the optimum chain length has been reached since H cannot be removed as an atom, but as a hydride.
(Laws of Creation for Cyclic Phosphonates)

<u>**Rule 1336:**</u> This rule of Chemistry for **Cyclic phosphonates,** states that, the fact that these are Electrophiles is supported by the fact that monomer units obtained ***anionically or nucleo-non-free-radically via the P center only when the ring is opened instantaneously as shown below is the same as that obtained electro-free-radically via the functional center shown in Rule 1335 -***

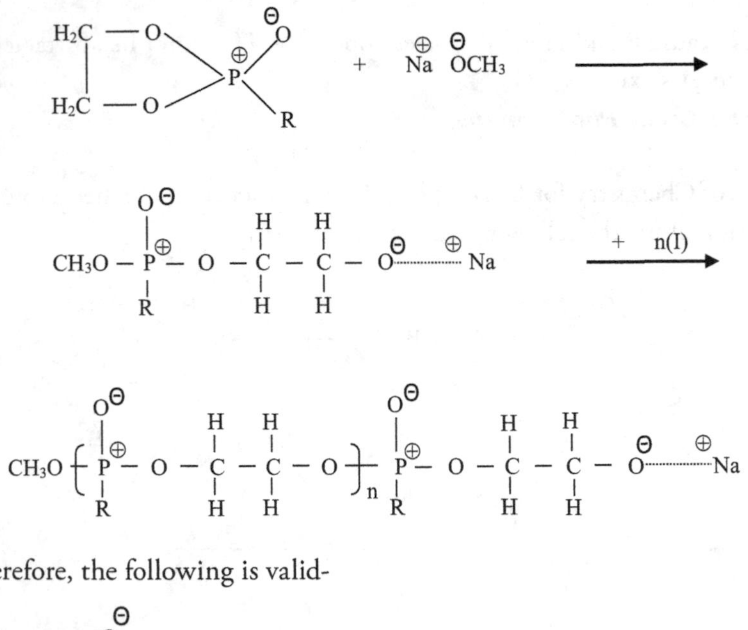

$$\begin{array}{c} H_2C - O \\ | \qquad \quad \searrow \\ H_2C - O \end{array} \overset{\overset{\ominus}{O}}{\underset{R}{\overset{\oplus}{P}}} \qquad + \quad \overset{\oplus}{Na} \; \overset{\ominus}{O}CH_3 \qquad \longrightarrow$$

$$CH_3O - \overset{\overset{\displaystyle \overset{\ominus}{O}}{\|}}{\underset{\underset{\displaystyle R}{|}}{\overset{\oplus}{P}}} - O - \overset{\overset{\displaystyle H}{|}}{\underset{\underset{\displaystyle H}{|}}{C}} - \overset{\overset{\displaystyle H}{|}}{\underset{\underset{\displaystyle H}{|}}{C}} - \overset{\ominus}{O} \cdots \overset{\oplus}{Na} \qquad \xrightarrow{\quad + \; n(I) \quad}$$

$$CH_3O \left[\overset{\overset{\displaystyle \overset{\ominus}{O}}{\|}}{\underset{\underset{\displaystyle R}{|}}{\overset{\oplus}{P}}} - O - \overset{\overset{\displaystyle H}{|}}{\underset{\underset{\displaystyle H}{|}}{C}} - \overset{\overset{\displaystyle H}{|}}{\underset{\underset{\displaystyle H}{|}}{C}} - O \right]_n \overset{\overset{\displaystyle \overset{\ominus}{O}}{\|}}{\underset{\underset{\displaystyle R}{|}}{\overset{\oplus}{P}}} - O - \overset{\overset{\displaystyle H}{|}}{\underset{\underset{\displaystyle H}{|}}{C}} - \overset{\overset{\displaystyle H}{|}}{\underset{\underset{\displaystyle H}{|}}{C}} - \overset{\ominus}{O} \cdots \overset{\oplus}{Na}$$

and for which therefore, the following is valid-

$$\overset{\overset{\ominus}{O}}{\underset{R}{\overset{\oplus}{P}}} \qquad > \qquad O$$

<u>Order of Nucleophilicity</u>

(Laws of Creations for Cyclic Phosphonates)

Rule 1337: This rule of Chemistry for **Cyclic phosphites,** states that, these are *Nucleophiles with three functional centers,* two of which are the same, the first three members of which are shown below-

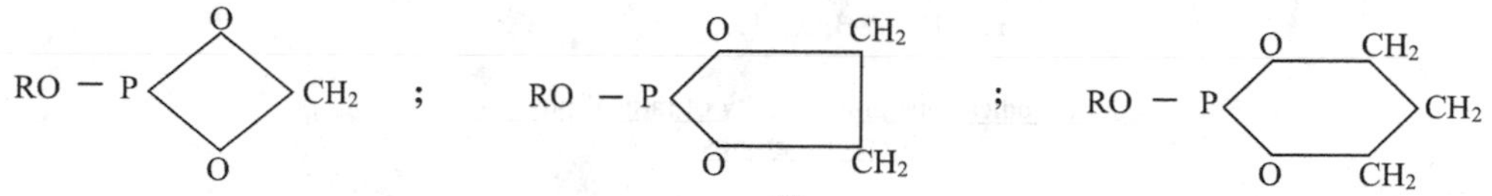

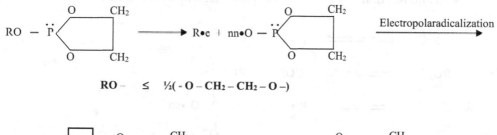

(I)   (II)

(III)

Cyclic phosphites

noting that the following is valid for this family-

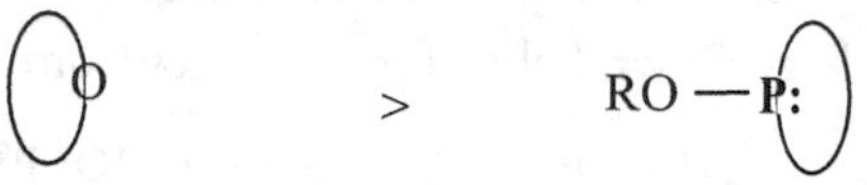

RO — P:

>

Order of Nucleophilicity

(Laws of Creation for Cyclic Phosphites)

Rule 1338: This rule of Chemistry for **Cyclic phosphites,** states that, when certain conditions specified below are met, they undergo *so-called Arbuzov rearrangement* shown below for the five-membered ring-

RO — P̈   ⟶   R•e + nn•O — P̈   Electropolaradicalization ⟶

$$RO- \quad \leq \quad ½(-O-CH_2-CH_2-O-)$$

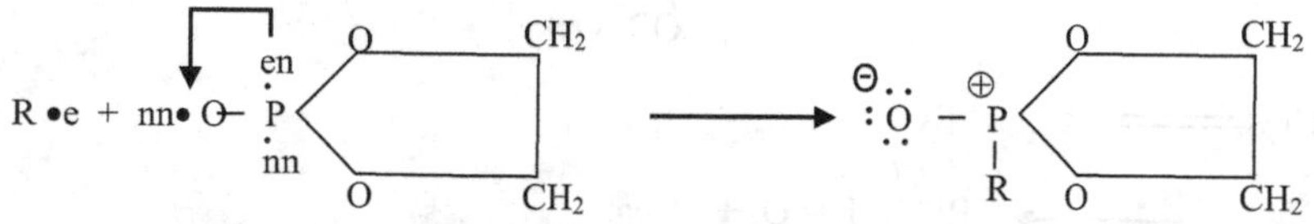

R•e + nn•O— P   ⟶   :O⁻ — P

Electropolaradicalization phenomenon

for which the so-called Arbuzov rearrangement is indeed **Electropolaradicalization of the first type** i.e., the movement of an electro-radical from paired unbonded radicals adjacently located to an •nn center to instantaneously form a polar bond, after which, the component held in Equilibrium state of Existence now adds to the new center to give the five-membered cyclic phosphonate; noting that the fraction held in Equilibrium state of existence depends on the operating conditions. [See Rule 1213]

(Laws of Creations for Cyclic Phosphites/Electropolaradicalization phenomenon)

Rule 1339: This rule of Chemistry for **Cyclic phosphites,** states that, Electropolaradicali-zation will largely be favored, the larger the size of the ring, since the following is valid-

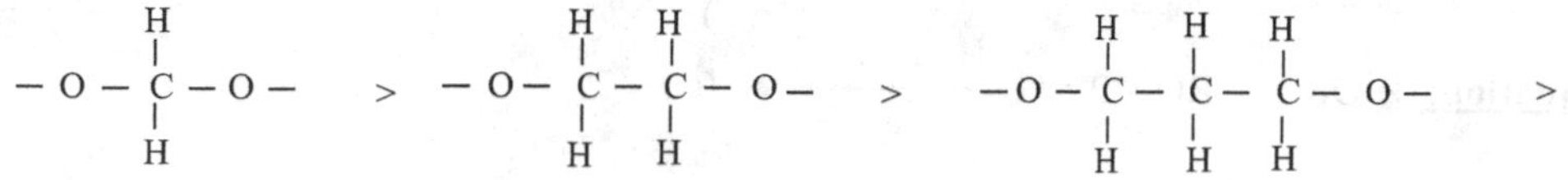

$$-O-\underset{H}{\overset{H}{C}}-O- \quad > \quad -O-\underset{H}{\overset{H}{C}}-\underset{H}{\overset{H}{C}}-O- \quad > \quad -O-\underset{H}{\overset{H}{C}}-\underset{H}{\overset{H}{C}}-\underset{H}{\overset{H}{C}}-O- \quad >$$

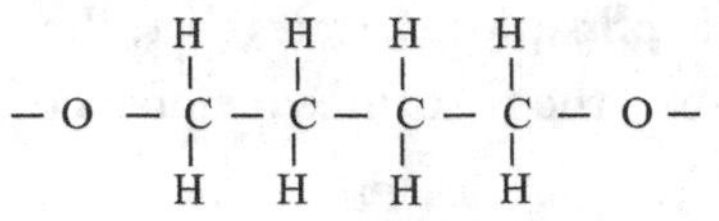

$$-O-\overset{\overset{\displaystyle H}{|}}{\underset{\underset{\displaystyle H}{|}}{C}}-\overset{\overset{\displaystyle H}{|}}{\underset{\underset{\displaystyle H}{|}}{C}}-\overset{\overset{\displaystyle H}{|}}{\underset{\underset{\displaystyle H}{|}}{C}}-\overset{\overset{\displaystyle H}{|}}{\underset{\underset{\displaystyle H}{|}}{C}}-O-$$

<u>Order of radical-pushing capacity of radical-pushing
di-etheric groups</u>

(Laws of Creations for Cyclic Phosphites)

<u>**Rule 1340:**</u> This rule of Chemistry for **Electropolaradicalization phenomenon in alkyl phosphite,** states that, the rearrangement to phosphonates is more enhanced with the presence of a corresponding alkyl halide or another alkyl halide as shown below for two cases from organophosphorus compounds-

$$(RO)_3P \quad + \quad RCl \xrightarrow{\text{Heat}} (RO)_3\overset{\oplus}{\underset{\underset{\displaystyle R}{|}}{P}}.....\overset{\ominus}{}Cl \longrightarrow (RO)_2\overset{\oplus}{\underset{\underset{\displaystyle R}{|}}{P}}-\overset{\ominus}{}O \quad + \quad RCl$$

Alkyl Phosphite

<u>**(Dialkyl alkylphosphonate)**</u> (i)

$$(RO)_3P \quad + \quad R^1Cl \longrightarrow (RO)_3\overset{\oplus}{\underset{\underset{\displaystyle R^1}{|}}{P}}.....\overset{\ominus}{}Cl \longrightarrow (RO)_2\overset{\oplus}{\underset{\underset{\displaystyle R^1}{|}}{P}}-\overset{\ominus}{}O \quad + \quad RCl$$

(ii)

noting that in the second reaction above, 95% of the phosphonates formed is carrying R^1 as shown below-

<u>**Stage 1:**</u> $\quad\quad R^1Cl \rightleftharpoons R^1{\bullet}e \quad + \quad Cl{\bullet}nn$

$R^1{\bullet}e \; + \; (RO)_3P \rightleftharpoons R^1OR \; + \; (RO)_2P\;{\bullet}en$

$(RO)_2P\;{\bullet}en \rightleftharpoons R\;{\bullet}e \; + \; en{\bullet}\,\underset{\underset{\displaystyle OR\;\;(A)}{|}}{P}-O\;{\bullet}nn$

$R\;{\bullet}e \; + \; nn{\bullet}Cl \rightleftharpoons RCl$

$(A) \longrightarrow RO-:P=O \; + \; Heat$ (iii)

<u>**Overall Equation:**</u> $R^1Cl \; + \; (RO)_3P \longrightarrow R^1OR \; + \; RCl \; + \; RO-P=O$

$+ \; Heat$ (iv)

<u>**Stage 2:**</u> $\quad\quad R^1OR \rightleftharpoons R^1{\bullet}e \; + \; nn{\bullet}OR$

$R^1{\bullet}e \; + \; RO-P=O \rightleftharpoons R^1-\underset{\underset{\displaystyle OR}{|}}{P}\;{\bullet}en$

(B)

$(B) \; + \; RO\;{\bullet}nn \longrightarrow R^1-O-\underset{\underset{\displaystyle OR}{|}}{P}-OR$

(C) (v)

<u>**Overall Equation:**</u> $R^1OR \; + \; RO-P=O \longrightarrow (C)$ (vi)

Stage 3: (C) $\rightleftharpoons$ $R^1 \bullet e$ + $nn \bullet \overset{\bullet\bullet}{O} - \overset{\bullet\bullet}{P} - OR$
$$| \atop OR$$
(D)

(D) $\rightleftharpoons$ $O^{\ominus} - {}^{\oplus}\overset{\bullet n}{P} - OR$
$$| \atop OR$$
(E)

$R^1 \bullet e$ + (E) $\longrightarrow$ $O^{\ominus} - {}^{\oplus}\overset{R^1}{\underset{OR}{P}} - OR$ (vii)

Overall Equation: R^1Cl + $(RO)_3P$ $\longrightarrow$ RCl + $(RO)_2P^{\oplus} - {}^{\ominus}O$
$$\qquad\qquad\qquad\qquad\qquad | \atop R^1$$

Phosphite $\qquad\qquad$ Phosphonate $\qquad$ (viii)

from which it can be seen that RCl is active when present with the Electropolaradicalization taking place only in the second step of the last stage- the third stage and for the replacement above to have taken place, based on the first step in Stage 3, **R≥ R^1** in radical-pushing capacity, i.e., half of what is inside the ring must be greater than or equal to what is being held outside; for when all the R^{1}Cl cannot exist in Equilibrium state of existence as shown in Stage 1, the fraction in stable state of existence that which cannot be used accounts for the 5%.
(Laws of Creations for Organo-alkyl phosphite)

Rule 1341: This rule of Chemistry for the **Four-membered phosphite,** states that, Electropolaradicalization will not take place, i.e., the phosphite is more in stable state of existence than in Equilibrium state of existence at low and high operating conditions, since

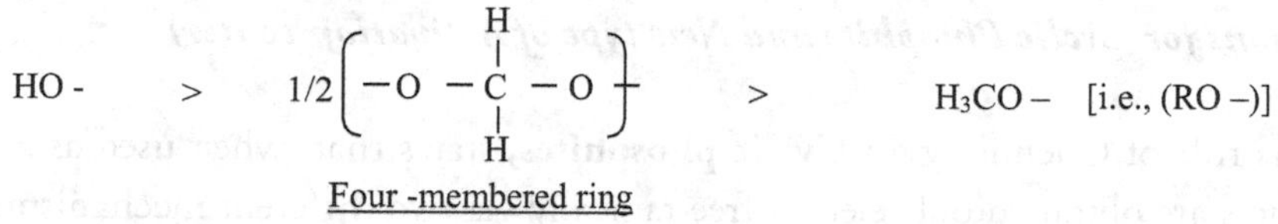

$$HO- \quad > \quad \frac{1}{2}\left(-O - \overset{\overset{H}{|}}{\underset{\underset{H}{|}}{C}} - O \right) \quad > \quad H_3CO- \quad [\text{i.e., } (RO-)]$$
$$\underline{\text{Four -membered ring}}$$

for which when R is H, rearrangement will not take place, but when R is CH_3 or C_2H_5, rearrangement will take place as shown below,

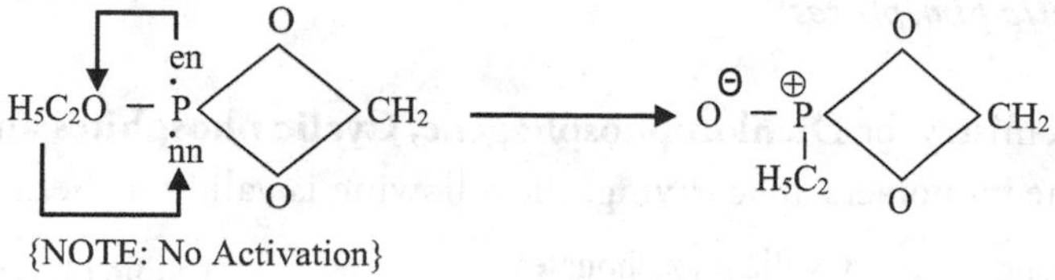

{NOTE: No Activation} (i)

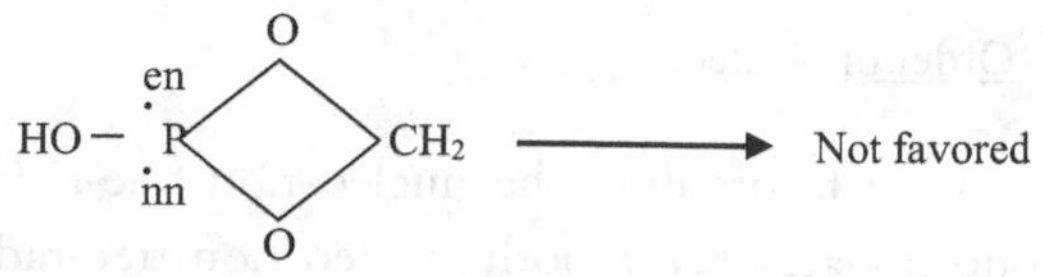

(more stable) (ii)

487

and therefore the following is valid for phosphites of any size when RO is less than or equal to half of what is in the ring-

Cyclic phosphites $\xrightarrow[\text{rearranged}]{\text{molecularly}}$ Cyclic phosphonates

Valid for Cyclic phosphites when [H-(CH₂)ₙO-) ≤ ½(-O-(CH₂)₂ₙ –O-]   (iii)

(Laws of Creations for Cyclic Phosphite)

Rule 1342: This rule of Chemistry for **four-membered ring of Cyclic phosphite,** states that, when protonic acid is used as source of initiator, the followings are obtained-

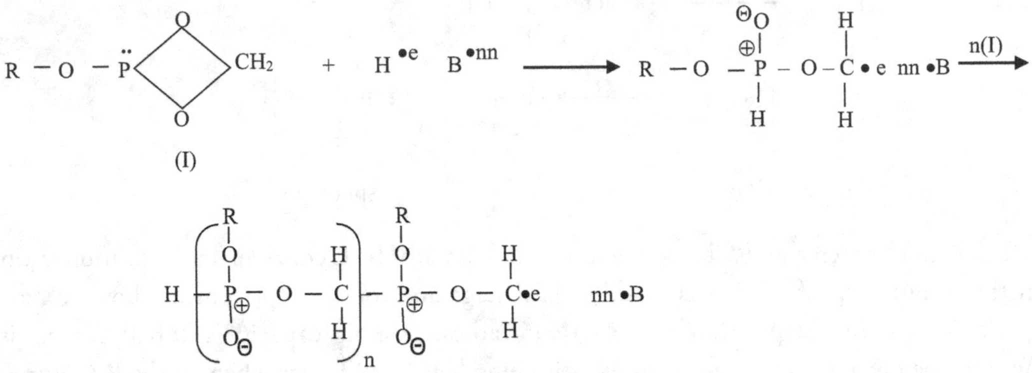

wherein, it can be observed that the center attacked is the P center being less nucleophilic than the O center, and this could only be done electro-free-radically by adding to the nucleo-free-radical on the P center to form a bond and instantaneously open the ring, since the H could not sit on the P center due to boundary laws to form an electrostatic bond (the electro-free-radical left on the P center forms a polar bond with O), *noting that propagation continued with paired radical centers, via Combination mechanism* and also while phosphites are Nucleophiles when there is no molecular rearrangement, phosphonates are Electrophiles and that the route for phosphites can never be charged.
(Laws of Creations for Cyclic Phosphite and New type of Activation center)

Rule 1343: This rule of Chemistry for **Cyclic phosphites,** states that, when used as a monomer, two polymeric products are obtained only electro-free-radically via two different mechanisms, based on the operating conditions, particularly when R is CH₃, since a fraction of the monomer is in Stable state of existence (the fraction that gives polyphosphites) and the remaining fraction is in Equilibrium state of existence (the fraction that gives polyphosponates).
(Laws of Creations for Cyclic phosphites)

Rule 1344: This rule of Chemistry for **Dichlorophosphagene, Cyclic phosphites and phosphonates,** states that, based on what the monomers are carrying, the following is valid for them-

Dichlorophosphagene   >   Cyclic phosphonates   >   Cyclic phosphite

Order of Nucleophilicity   i)

despite the fact that Dichlorophosphagene favors only the nucleo-non-free-radical or anionic route (A complete Electrophile), Cyclic phosponates favor both nucleo-non-free-radical or anionic and electro-free-radical or positively charged routes (An Electrophile), while Stable Cyclic phosphites favor only the electro-free-radical route (A complete Nucleophile), for which the followings are valid.

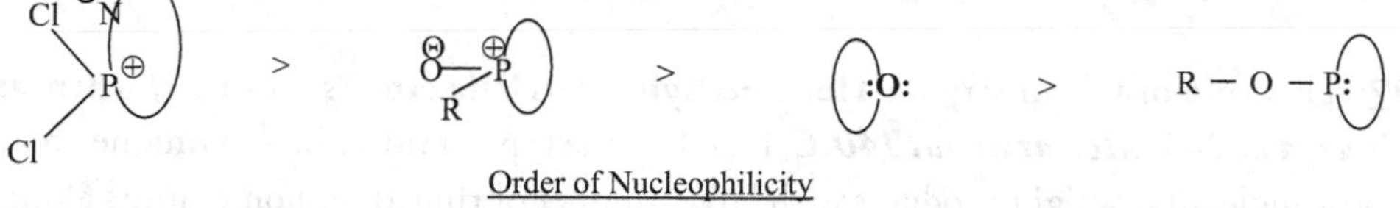

Order of Nucleophilicity

ii)

(Laws of Creations for Cyclic Phosphite)

Rule 1345: This rule of Chemistry for **Organo-phosphite compounds,** states that, based on their characters, the followings are the Equilibrium states of existence of some compounds-

$$H_5C_2OCH_3 \rightleftharpoons H_3C \bullet e + nn \bullet O - C_2H_5$$

$$HO - P(OR)_2 \rightleftharpoons H \bullet e + nn \bullet O - P(OR)_2$$

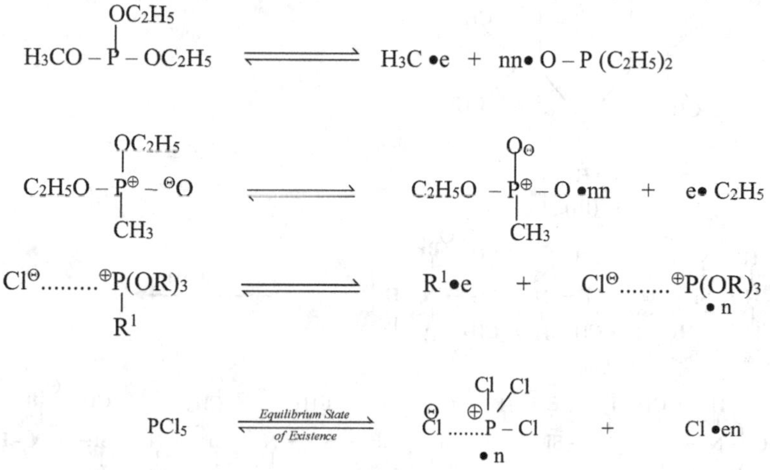

$$H_3CO - \overset{\overset{\displaystyle OC_2H_5}{|}}{P} - OC_2H_5 \rightleftharpoons H_3C \bullet e + nn \bullet O - P(C_2H_5)_2$$

$$C_2H_5O - \overset{\overset{\displaystyle OC_2H_5}{|}}{\underset{\underset{\displaystyle CH_3}{|}}{P^{\oplus}}} - {}^{\ominus}O \rightleftharpoons C_2H_5O - \overset{\overset{\displaystyle O^{\ominus}}{|}}{\underset{\underset{\displaystyle CH_3}{|}}{P^{\oplus}}} - O \bullet nn + e \bullet C_2H_5$$

$$Cl^{\ominus} .... {}^{\oplus} \overset{}{\underset{\underset{\displaystyle R^1}{|}}{P}}(OR)_3 \rightleftharpoons R^1 \bullet e + Cl^{\ominus} .... {}^{\oplus}P(OR)_3 \\ \bullet n$$

$$PCl_5 \xrightarrow[\text{of Existence}]{\text{Equilibrium State}} \overset{\ominus}{Cl} .... {}^{\oplus} \overset{\overset{\displaystyle Cl \diagup Cl}{|}}{P} - Cl + Cl \bullet en \\ \bullet n$$

(Laws of Creations for Organo phosphite compounds)

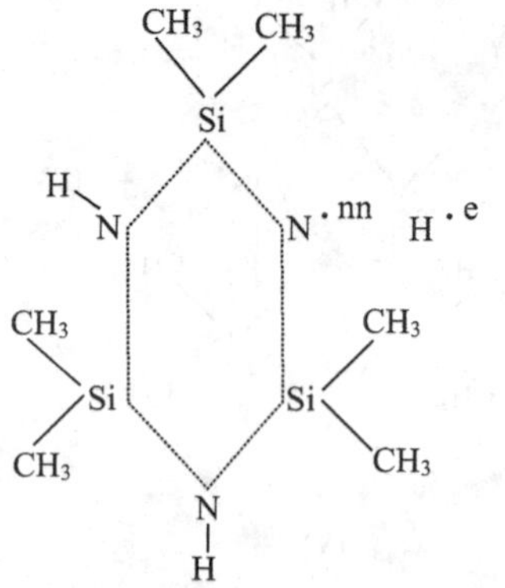

Rule 1346: This rule of Chemistry for **Hexamethylcyclotrisilazane,** states that, from a look at the structure of the monomer, it is a Nucleophile and with three functional centers due to presence of three paired unbonded radicals on the N centers, three weak Si – N bonds, hydrogen atoms on the N center loosely bonded one at a time, the ring is bound to be strained-noting the radical bonding of hydrogen instead of ionic bonding due to the presence of silicon in the ring, which as is now obvious does not favor ionic existence.

(Laws of Creations for Hexamethylcyclotrisilazane)

Rule 1347: This rule of Chemistry for **Hexamethylcyclotrisilazane,** states that, when *ammonium bromide is used as initiator at about 140⁰C*, based on the type of initiator and unique character of the monomer, low molecular weight products with large number of ringed monomer units along the chain, many branches and cross-linking are said to be obtained and this can partly be explained as follows-

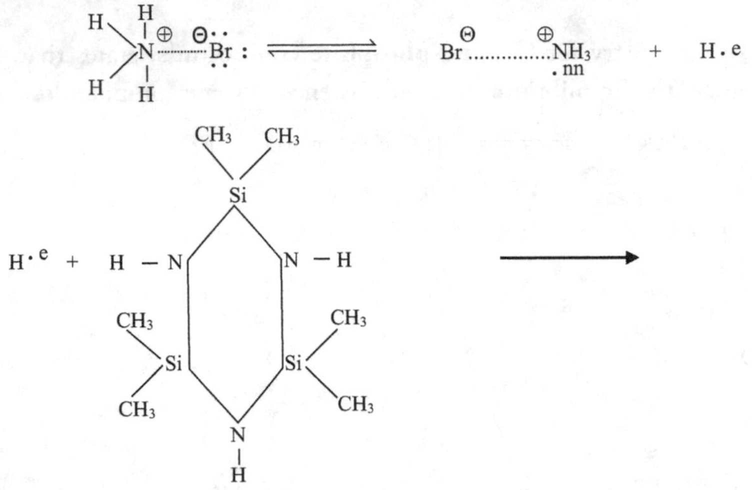

(III)

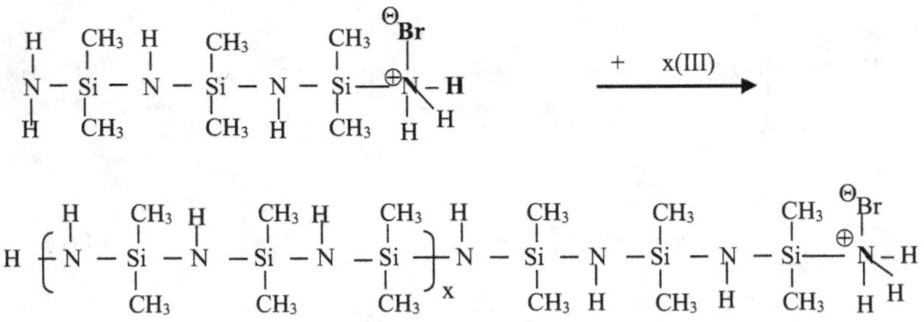

(IV)

noting that the route is electro-free-radical route, with the chain growing backwardly via Equilibrium mechanism, until at a particular point in time, instead of the monomer remaining in **Stable state of Existence**, it now was put in **Equilibrium state of Existence**, for which the following shown below was made to take place-

Stage 1:

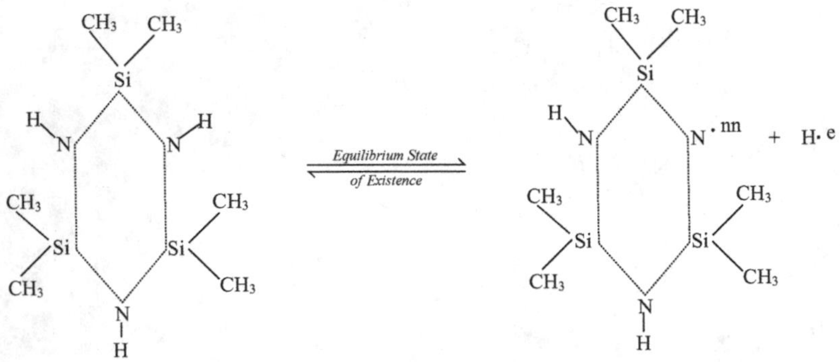

(V)

490

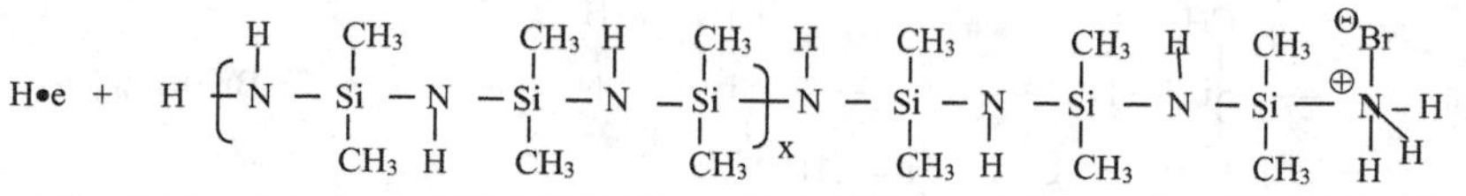

$$H\bullet e \; + \; H \left(N - \underset{\underset{CH_3}{|}}{\overset{\overset{H}{|}}{Si}} - N - \underset{\underset{H}{|}}{\overset{\overset{CH_3}{|}}{Si}} - N - \underset{\underset{CH_3}{|}}{\overset{\overset{H}{|}}{Si}} \right)_x N - \underset{\underset{H}{|}}{\overset{\overset{CH_3}{|}}{Si}} - N - \underset{\underset{CH_3}{|}}{\overset{\overset{H}{|}}{Si}} - N - \underset{\underset{H}{|}}{\overset{\overset{CH_3}{|}}{Si}} \overset{\oplus}{-} \underset{\underset{H}{|}}{\overset{\overset{\ominus Br}{}}{N}} - H$$

(IV)

$$\Longleftarrow \qquad HBr \;\; + \;\; NH_3 \;\; +$$

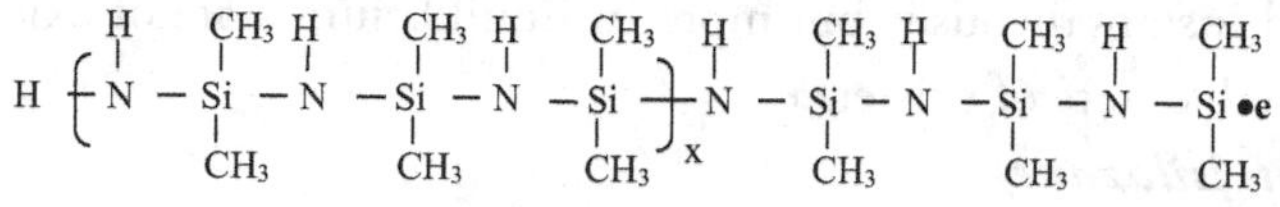

$$H \left(N - \underset{\underset{CH_3}{|}}{\overset{\overset{H}{|}}{Si}} - N - \underset{\underset{CH_3}{|}}{\overset{\overset{CH_3 \; H}{| \; \;}}{Si}} - N - \underset{\underset{CH_3}{|}}{\overset{\overset{H}{|}}{Si}} \right)_x N - \underset{\underset{CH_3}{|}}{\overset{\overset{CH_3 \; H}{| \; \;}}{Si}} - N - \underset{\underset{CH_3}{|}}{\overset{\overset{H}{|}}{Si}} - N - \underset{\underset{CH_3}{|}}{\overset{\overset{CH_3}{|}}{Si}} \bullet e$$

(VI)

(VI) + (V) $\longrightarrow$

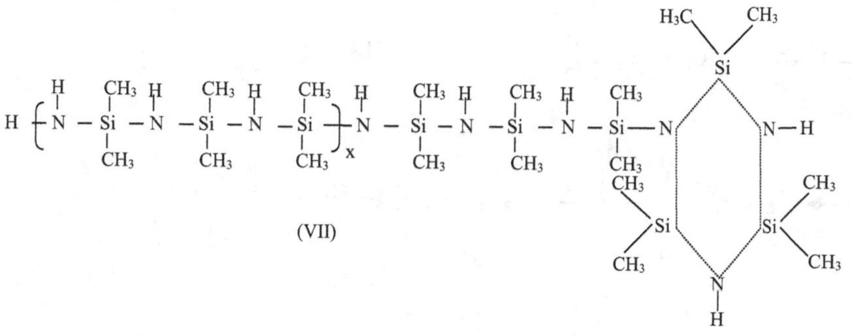

(VII)

and with (VII) now formed with the monomer placed at the terminal of the chain, in the next stage, (VII) was forced to exist in Equilibrium state of existence from the other N center to begin the same stage above to produce the chain shown below-

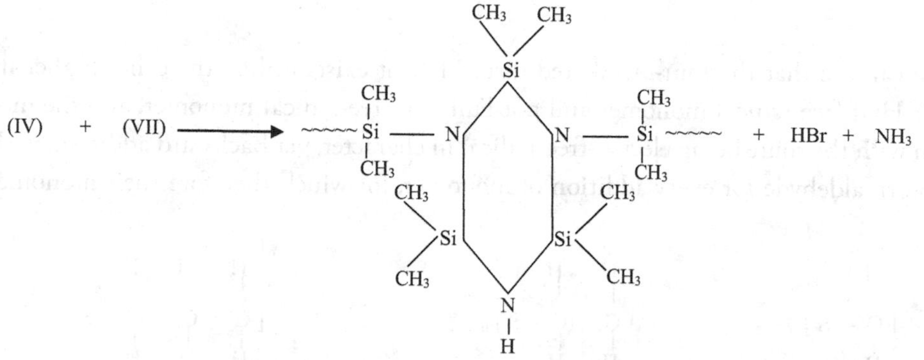

(IV) + (VII) $\longrightarrow$ ... $+ \; HBr \; + \; NH_3$

and in the same manner, branches are also formed when the chains have reached their optimum length based on the polymerization temperature and the type of polymer and with the formation of the first branch, other branches begin to grow along the chain to give what is shown below-

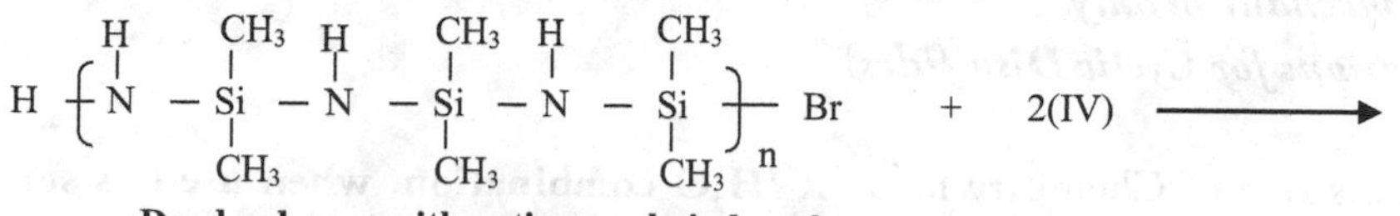

$$H \left(N - \underset{\underset{CH_3}{|}}{\overset{\overset{H}{|}}{Si}} - N - \underset{\underset{CH_3}{|}}{\overset{\overset{CH_3 \; H}{| \; \;}}{Si}} - N - \underset{\underset{CH_3}{|}}{\overset{\overset{CH_3 \; H}{| \; \;}}{Si}} \right)_n Br \;\; + \;\; 2(IV) \longrightarrow$$

Dead polymer with optimum chain length

491

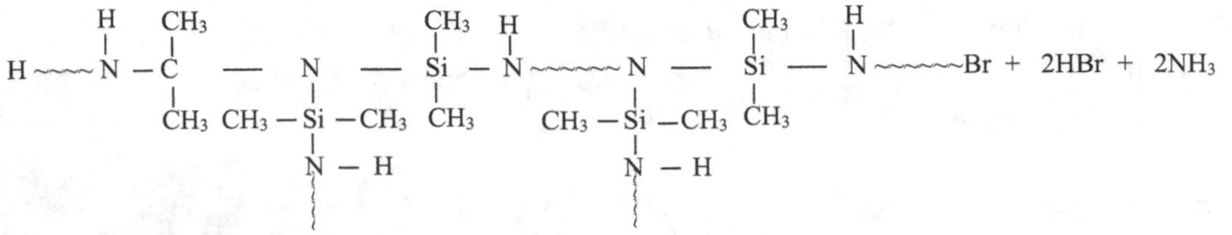

$$H \sim\sim N - \overset{\overset{\displaystyle H}{|}}{\underset{\underset{\displaystyle CH_3}{|}}{C}} - N - \overset{\overset{\displaystyle CH_3}{|}}{\underset{\underset{\displaystyle CH_3}{|}}{Si}} - N \sim\sim N - \overset{\overset{\displaystyle CH_3}{|}}{\underset{\underset{\displaystyle CH_3}{|}}{Si}} - \overset{\overset{\displaystyle H}{|}}{\underset{\underset{\displaystyle CH_3}{|}}{N}} \sim\sim Br + 2HBr + 2NH_3$$

for never is there a time the ring is opened instantaneously, but more in Equilibrium state of existence (source of low molecular weight) than in Stable state of existence.
(Laws of Creations for Hexamethylcyclotrisilazane)

Rule 1348: This rule of Chemistry for **Cyclic Disulfides,** states that, when *trialkyl ammonium hydroxide is used as initiator,* the followings are obtained-

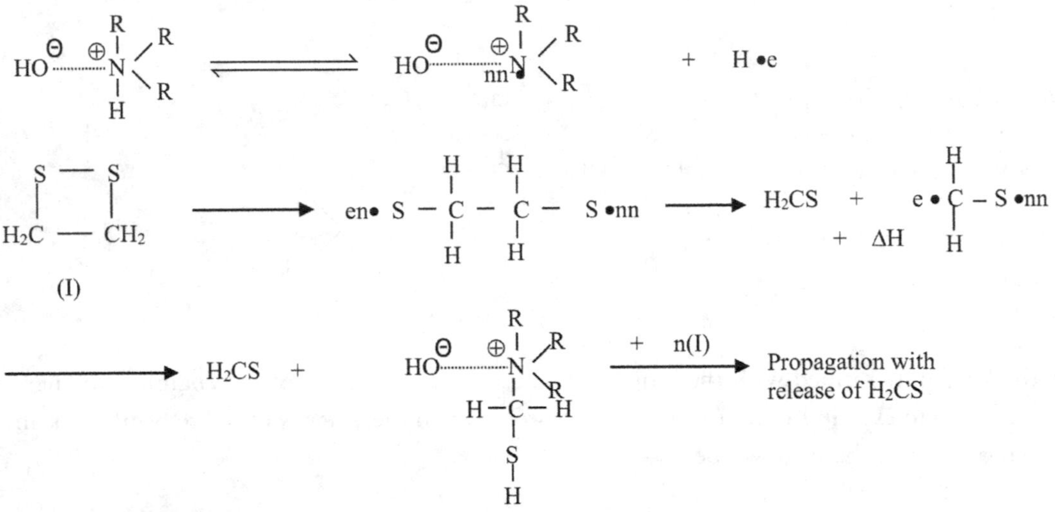

wherein one can see that the four-membered disulfide if it exists unlike the other higher members in the family is Half free-radical monomer and not Full non-free-radical monomer, and the mechanism is Equilibrium with the route being electro-free-radical in character, via Backward addition, with release of carbon thioformaldehyde for every addition of monomer, for which therefore their monomer units are as follows.

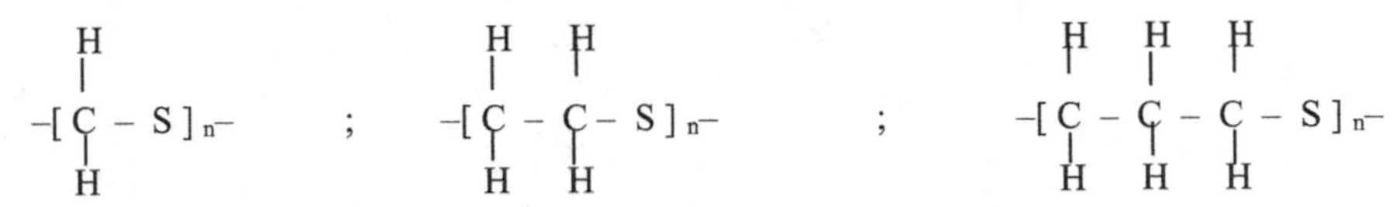

4-membered ring ; **5-membered ring** ; **6-membered ring**

for with or without instantaneous opening of the ring, polymerization will still take place via Equilibrium mechanism only.
(Laws of Creations for Cyclic Disulfides)

Rule 1349: This rule of Chemistry for **AlR₃/H₂O combination, when used as source of initiator for Isotactic placement,** states that, *since two metallic elements are involved-H (Ionic) from water and Al (Non-ionic) from AlR₃,* the process of preparation of the initiator begins with Decomposition mechanism and Al being more electropositive than H, the first stage in Decomposition mechanism begins with aluminum trialkyl-

<u>**Decomposition mechanism**</u>

Stage 1: $AlR_3 \longrightarrow R \bullet n \;+\; e\bullet AlR_2$

$R_2Al \bullet e \;+\; H_2O \longrightarrow R_2Al-OH \;+\; H\bullet e$

$H\bullet e \;+\; n\bullet R \rightleftharpoons RH$ (i)

<u>**Overall Equation:**</u> $AlR_3 \;+\; H_2O \xrightarrow{\text{HEAT}} R_2AlOH \;+\; RH$ (ii)

clear indication that H is a gaseous metal and then followed next by Equilibrium mechanism-

<u>**Equilibrium mechanism**</u>

Stage 1: $AlR_3 \rightleftharpoons R\bullet e \;+\; n\bullet AlR_2$

$R\bullet e \;+\; R_2AlOH \rightleftharpoons ROH \;+\; e\bullet AlR_2$

$R_2Al\bullet e \;+\; n\bullet AlR_2 \longrightarrow$

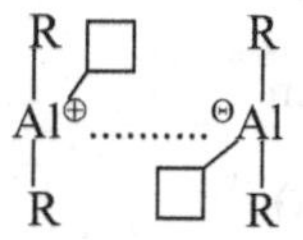

(I) **THE INITIATOR**   (iii)

<u>**Overall Equation:**</u> $AlR_3 \;+\; R_2AlOH \longrightarrow ROH \;+\; (I)$ (iv)

<u>**Overall overall Equation:**</u> $2AlR_3 \;+\; H_2O \longrightarrow ROH \;+\; (I) \;+\; RH$ (v)

for just as Z/N initiators are obtained from their combinations such as $Al_3R_3/TiCl_4$ combination, so also is the case above, noting that the exact ratio of AlR_3/H_2O combination is 2 : 1, and that the initiator (I) above is dual in character, that is, there are two active centers; one for a male and the other for a female and when used on ***propylene oxide,*** the monomer stays all the time in the reservoir carried by the negative center being a Female to obtain the followings-

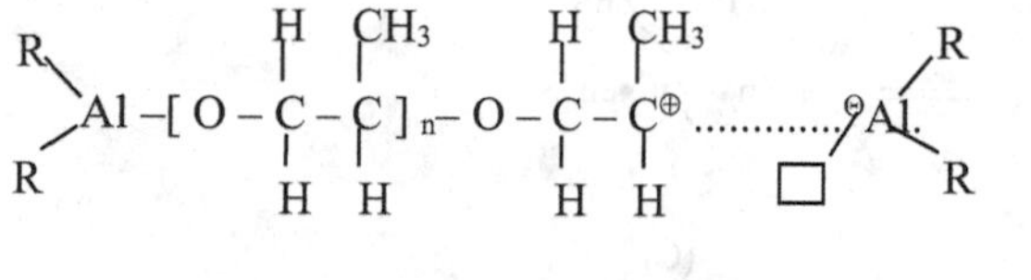

<u>**Isotactic poly (propylene oxide**</u>

(Laws of Creations for AlR$_3$/H$_2$O Combination)

<u>**Rule 1350:**</u> This rule of Chemistry for the **AlR$_3$/R^{1}OH combination when used as source of initiator for Isotactic placement,** states that, the same type of initiator as obtained for AlR_3/H_2O combination is also obtained, with only the following difference in the preparation step-

<u>**Decomposition mechanism**</u>

Stage 1: $AlR_3 \longrightarrow R \bullet n \;+\; e\bullet AlR_2$

$R_2Al \bullet e \;+\; R^1OH \longrightarrow R_2Al-OH \;+\; R^1\bullet e$

$$R^1 \bullet e \;+\; n \bullet R \;\rightleftharpoons\; RR^1$$

Overall Equation: $AlR_3 \;+\; R^1OH \xrightarrow{\text{HEAT}} R_2AlOH \;+\; RR^1$

(Laws of Creations for AlR₃/R¹OH Combination)

Rule 1351: This rule of Chemistry for **ZnR₂/H₂O combination when used as source of initiator for Isotactic placement,** states that, based on the **Radical Configuration (Not Electronic configuration as used in Present-day Science)** of Zn metal, the first step is to make the center carry free-radicals, as opposed to non-free-radicals carried by them and this is the advantage offered by the presence of two vacant orbitals, for which one begins with reduction of the numbers of vacant orbitals and paired unbonded radicals as follows, that which is called **Depolarization** as will be fully highlighted downstream-.

<u>**Equilibrium Mechanism**</u>

Stage 1: $R_2Zn \;\rightleftharpoons\; R - Zn \bullet en \;+\; R \bullet n$

$$RZn \bullet e \;+\; R_2Zn \;\xrightleftharpoons{\text{Activation}}\; \begin{array}{c} R \\ | \\ RZn - Zn^* \bullet en \\ | \\ R \;\; (A) \end{array}$$

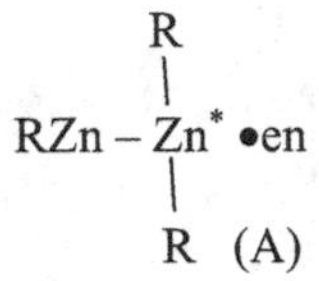

$$(A) \;+\; R \bullet n \longrightarrow RZn - Zn^*R_3$$
$$(B) \hspace{4cm} (i)$$

Overall Equation: $2ZnR_2 \longrightarrow RZn - Zn^*R_3$ \hfill (ii)

and with focus on the stable Zn center (Zn*), the Central atom, the second stage shown below will not take place, since no vacant orbital will be left to be used-

(Stage 2:) $R_2Zn \;\rightleftharpoons\; R - Zn \bullet en \;+\; R \bullet n$

$$RZn \bullet e \;+\; (B) \;\xrightleftharpoons{\text{Activation}}\; \begin{array}{c} R \;\diagdown \;\;\diagup\; ZnR \\ RZn - Zn^* \bullet en \\ \diagup \;\;\;\; \diagdown \\ R \;\;\;\;\; R \end{array}$$
$$(C)$$

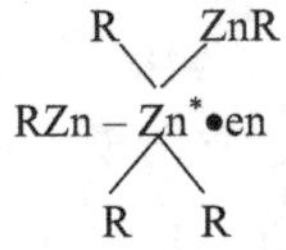

$$(C) \;+\; R \bullet n \longrightarrow (RZn) - {}^*ZnR_4 - (ZnR)$$
$$(D)\ \text{NOT FAVORED} \hspace{3cm} (iii)$$

Overall Equation: $3ZnR_2 \longrightarrow (D)$ \hfill (iv)

noting the consumption of 3 moles of ZnR₂, in which in place of (Stage 2) is that the paired unbonded radicals are shield with ZnR₂ to replace (D) with (RZn - *ZnR₃ → ZnR₂ with the stable central Zn element marked with asterisk above carrying the load and the next stage replacing (Stage 2) above follows as shown below-

Stage 2: $ZnR_2 \;\rightleftharpoons\; R_2Zn\text{-}\square$

$$R_2Zn\text{-}\square \;+\; (B) \;\xrightarrow{\text{Dative Bonding}}\; \begin{array}{c} R_2Zn \\ \nwarrow \\ RZn - {}^*Zn - R \\ \diagup \;\;\;\;\; \diagdown \\ R \;\;\;\;\;\;\; R \end{array}$$
$$(E) \hspace{5cm} (v)$$

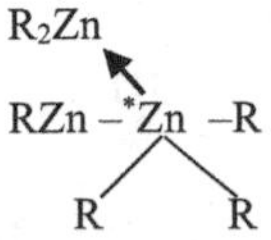

noting that no activation could further take place anymore after Stage 1 despite the presence of one vacant orbital, because of the operating conditions i.e., the ZnR_2 could no longer exist in Equilibrium state of existence being suppressed by the presence of (B), leaving behind three paired unbonded radicals, which were isolated in the same manner above in the last three stages (Stages 3, 4 and 5) to give the followings.

Stages 3, 4 & 5:

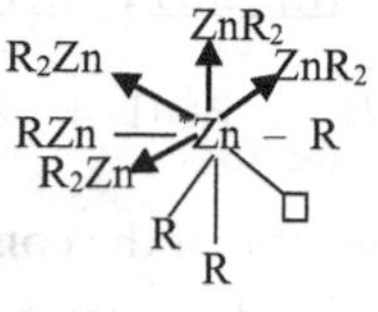

(F)

Overall Equation: $6ZnR_2 \longrightarrow$ (F)   (vi)

(Laws of Creation for Zn metal)

Rule 1352: This rule of Chemistry for **ZnR_2/H_2O combination when used as source of initiator for Isotactic placement,** states that, based on the Radical Configuration of Zn metal, after the first step of making Zn center carry free-radicals, as opposed to non-free-radicals carried them, for which six moles of ZnR_2 were consumed, the next step follows as shown below continuing from Rule 1151-

Decomposition Mechanism

Stage 1:

(G)

(F) $\longrightarrow$

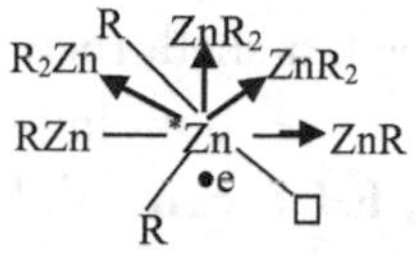

$+$ R•n

(G) $+$ H_2O $\longrightarrow$ (G) $-$ OH $+$ H •e

(H)

H•e $+$ R •n $\rightleftharpoons$ RH   (i)

Overall Equation: (F) $+$ H_2O $\longrightarrow$ RH $+$ (H)   (ii)

and this is followed next by the third step-

Equilibrium Mechanism

Stage 1:

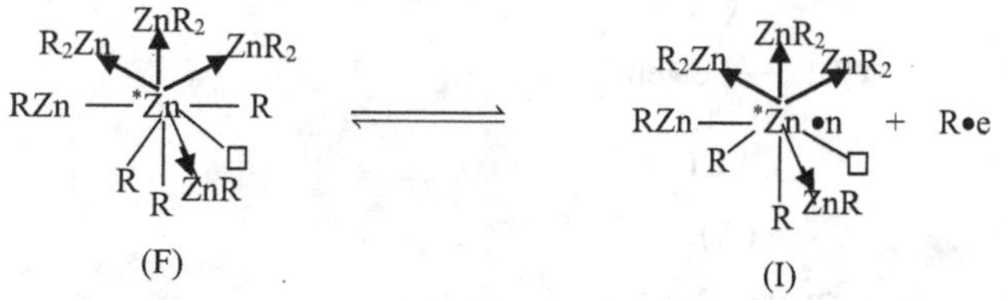

(F)   (I)

R •e $+$ (H) $\rightleftharpoons$

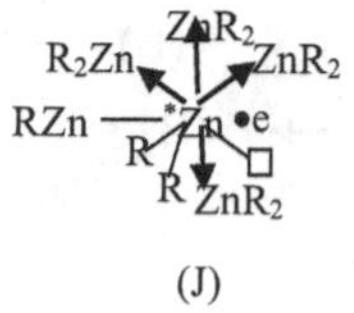

$+$ ROH

(J)

$$(J) \quad + \quad (I) \quad \longrightarrow$$

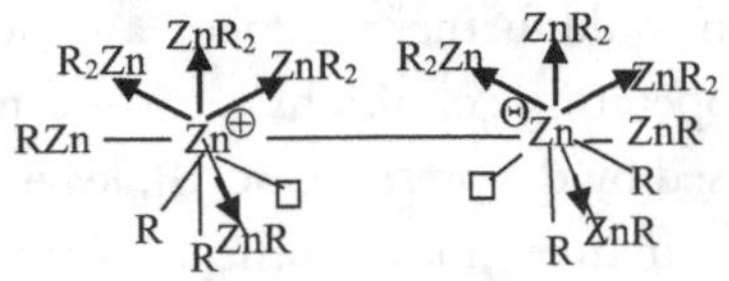

$$\underline{\textbf{(II) THE INITIATOR}} \hspace{3cm} \text{(iii)}$$

Overall overall Equation: $12ZnR_2 + H_2O \longrightarrow RH + ROH + (II)$ $\hspace{1cm}$ (iv)

for which (II) is the initiator of Z/N type, noting that the ratio of the combination-ZnR_2/H_2O used is 12 to 1 and the fact that Zn which is naturally known to be **divalent**, can be observed to be also **tetravalent** under certain conditions without breaking the Boundary laws noting that the initiator is dual in character radically or chargedly paired.
(Laws of Creations for ZnR$_2$H$_2$O Combination)

<u>**Rule 1353:**</u> This rule of Chemistry for **the use of FeCl$_3$ for many applications such as a source for initiators,** states that, based on the Radical configuration of the last shell of Fe shown below,

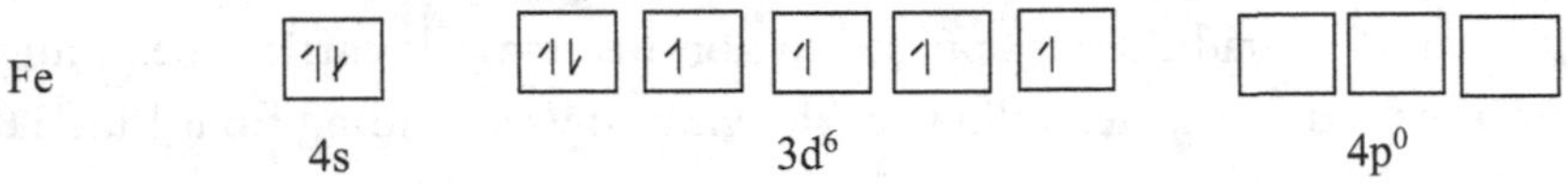

Radical Configuration of last shell of Fe

wherein | is male and | is female of opposite spins and in the last shell, there are two paired unbonded radicals, an electro-non-free-radical and three vacant orbitals after bonding with three chlorine atoms, **and despite the presence of an electro-non –free-radical, FeCl$_3$ can still be activated and can still exist in Equilibrium state of existence and therefore used for so many applications.**
(Laws of Creations for Fe metal)

<u>**Rule 1354:**</u> This rule of Chemistry for **use of FeCl$_3$ as a source of initiators for Isotactic placement,** states that, the first step is making the Fe center carry free-radicals as opposed to the non-free-radicals carried by her and this is one of the ways by which it is done in the presence of an adduct-

Equilibrium mechanism

Stage 1: $\hspace{1cm}$ FeCl$_3$ $\rightleftharpoons$ $\hspace{1cm}$ Cl$_2$Fe •en $\hspace{0.5cm}$ + $\hspace{0.5cm}$ Cl•nn

$$Cl_2Fe \cdot en \; + \; FeCl_3 \; \underset{}{\overset{Activation}{\rightleftharpoons}} \; Cl_2Fe \overset{\overset{\displaystyle Cl \quad Cl}{\diagdown \diagup}}{\underset{\underset{\displaystyle Cl}{|}}{-}} {}^*Fe \cdot en$$

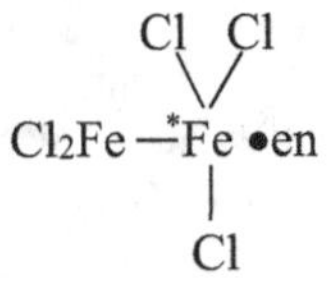

$$(A)$$

$$(A) \; + \; Cl \cdot nn \; \longrightarrow \; Cl_2Fe \; {}^{-*}FeCl_4$$

$$(B) \hspace{5cm} \text{(i)}$$

Overall Equation: $\hspace{0.3cm}$ 2FeCl$_3$ $\longrightarrow$ $\hspace{1cm}$ (B) $\hspace{4cm}$ (ii) noting that the central Fe atom highlighted, has the number of paired unbonded radicals reduced to one, leaving behind two vacant orbitals, and that center is again activated in stage 2, noting that the last shell of Fe can only accommodate only eighteen (18) radicals based on one of the Boundary laws already stated-

Stage 2: $FeCl_3$ ⇌ Cl_2Fe •en + Cl•nn

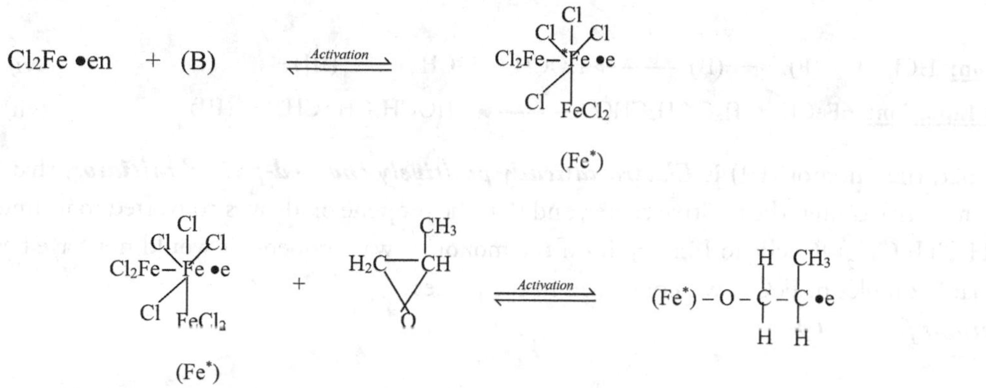

Cl_2Fe •en + (B) $\xrightarrow{Activation}$ $Cl_2Fe-\overset{\overset{\displaystyle Cl}{\underset{\displaystyle |}{Cl|Cl}}}{\underset{\overset{\displaystyle |}{Cl}}{Fe}}$•e

(Fe^*)

$Cl_2Fe-\overset{\overset{\displaystyle Cl}{\underset{\displaystyle |}{Cl|Cl}}}{\underset{\overset{\displaystyle |}{Cl}}{Fe}}$ ••e   + $H_2C-\overset{\overset{\displaystyle CH_3}{|}}{CH}$ (epoxide) $\xrightarrow{Activation}$ $(Fe^*)-O-\overset{\overset{\displaystyle H}{|}}{\underset{\underset{\displaystyle H}{|}}{C}}-\overset{\overset{\displaystyle CH_3}{|}}{\underset{\underset{\displaystyle H}{|}}{C}}$••e

(Fe^*)

⇌ $(Fe^*)-O-CH_2-\overset{\overset{\displaystyle ••e}{}}{\underset{\underset{\displaystyle •n}{}}{CH}}-CH_2$ + H••e

(E)

H••e   + Cl •nn ⇌ HCl

(E) $\xrightarrow{Deactivation}$ $(Fe^*)-O-CH_2-CH=CH_2$ + Heat   (iii)

(F)

Overall Equation: $3FeCl_3$ + H_3CCH_2CHO ⟶ HCl + (F)   (iv)

noting the involvement of the monomer such as propylene oxide in order to make the stage productive with the production of (F) where the central Fe atom now has one vacant orbital and one electro-free-radical and this was now followed by the last stage for the production of the Initiator (III) below-

Equilibrium Mechanism

Stage 3: HCl ⇌ H •e + Cl •nn

H••e + (F) ⇌ $HO-CH_2-CH=CH_2$ + (Fe^*)

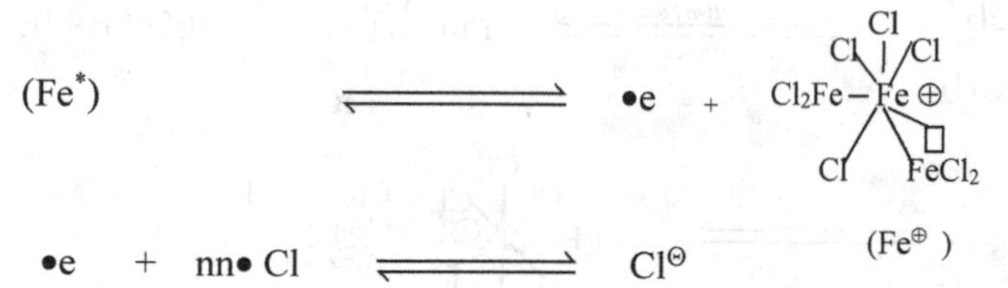

(Fe^*) ⇌ •e + $Cl_2Fe-\overset{\overset{\displaystyle Cl}{\underset{\displaystyle |}{Cl|Cl}}}{\underset{\overset{\displaystyle |}{Cl}}{Fe}} \oplus \square$

$(Fe^\oplus)$

•e + nn• Cl ⇌ $Cl^\ominus$

$Cl^\ominus$ + (B) ⇌ $Cl_2Fe-\overset{\overset{\displaystyle Cl}{\underset{\displaystyle |}{Cl|Cl}}}{\underset{\overset{\displaystyle }{Cl \quad Cl}}{Fe}} \ominus \square$

$(Fe^\ominus)$

497

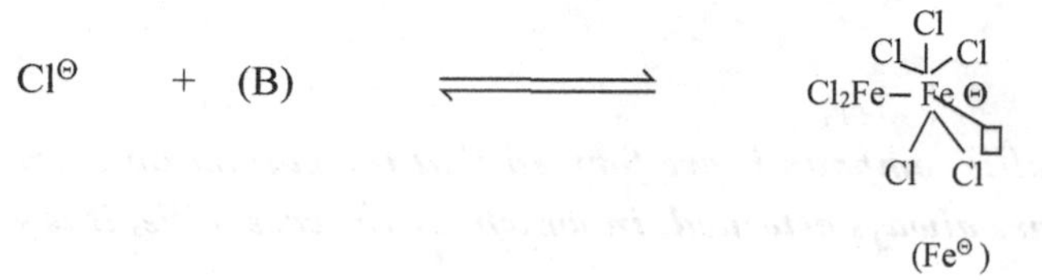

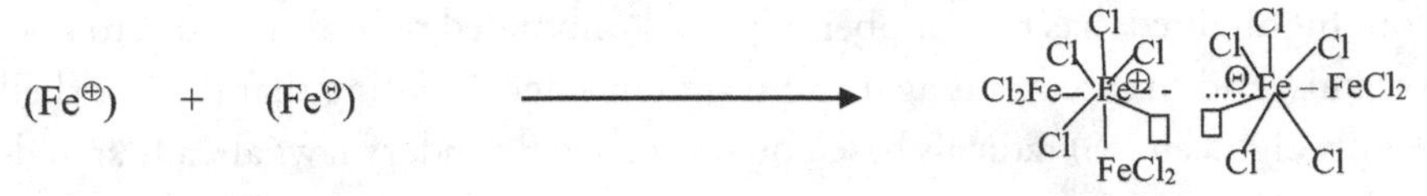

$$(Fe^{\oplus}) \quad + \quad (Fe^{\ominus}) \quad \longrightarrow \quad$$

(III) THE INITIATOR (v)

Overall Equation: $HCl \quad + \quad (F) \quad + \quad (B) \longrightarrow HOCH_2CH=CH_2 \quad + \quad (III)$ (vi)

Overall overall Equation: $6FeCl_3 + H_3C\,CH_2CHO \longrightarrow HOCH_2CH=CH_2 + (III)$ (vii)

noting finally that, the initiator (III) is *Electrostatically positively charged-paired initiator,* that is, that with only one active center, the positive center, and that the propene oxide was converted to its linear isomer ($HOCH_2CH=CH_2$) [Cyclic to Linear], for if the monomer was propene, it would not have been possible, and that Six moles of $FeCl_3$ was consumed in the process.
(Laws of Creations for FeCl₃)

Rule 1355: This rule of Chemistry for **Compounds,** states that, the Equilibrium state of Existence of some Bimetallic and trimetalic Iron compounds are as follows-

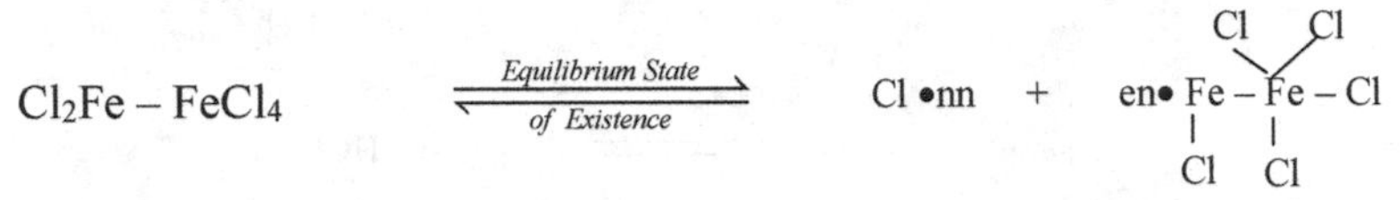

$$Cl_2Fe - FeCl_4 \quad \underset{\text{of Existence}}{\overset{\text{Equilibrium State}}{\rightleftharpoons}} \quad Cl\bullet nn \quad + \quad$$

2FeCl₃

$$Cl_2Fe - FeCl_5 - FeCl_2 \quad \underset{\text{of Existence}}{\overset{\text{Equilibrium State}}{\rightleftharpoons}} \quad Cl\bullet nn \quad +$$

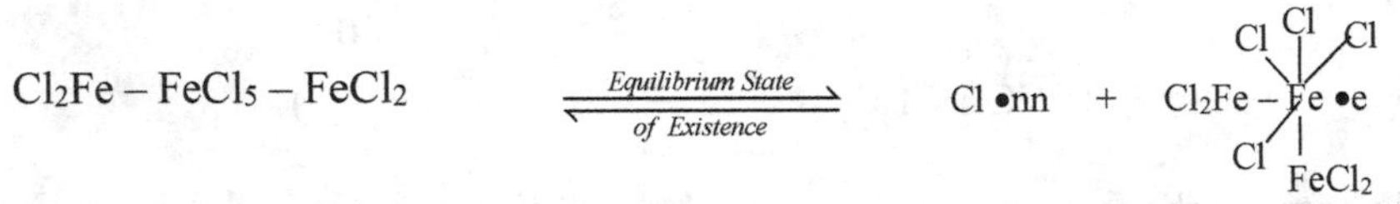

3FeCl₃

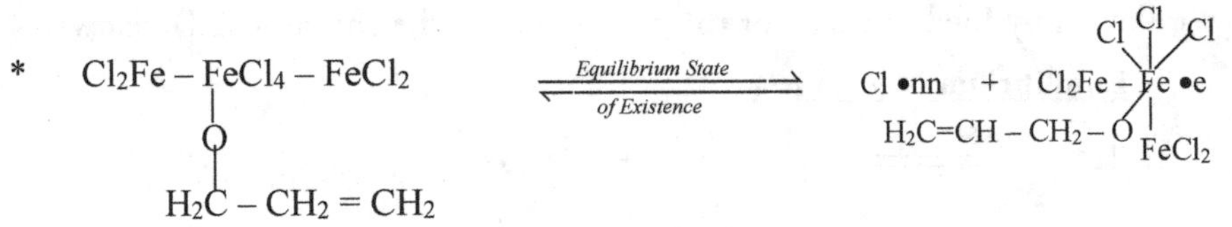

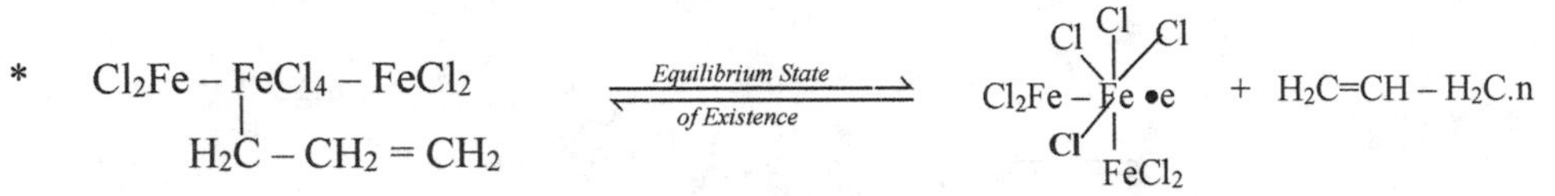

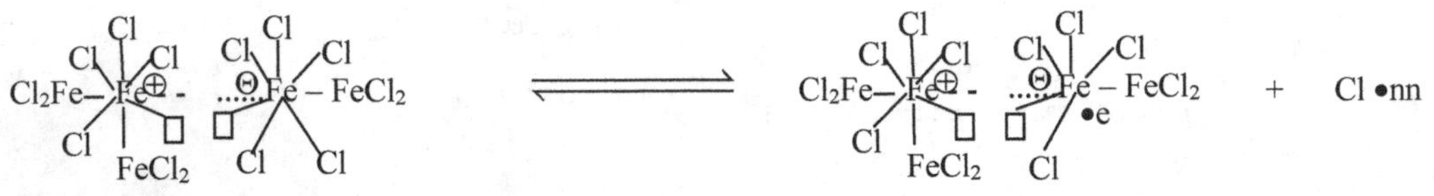

THE INITIATOR

noting that it is when bimetallic or trimetallic compounds are formed that the maximum overall valency state of the metallic Central atom are always attained, in which for the case of Fe, it is +7, all covalent bonds.
(Laws of Creations for Fe Compounds)

Rule 1356: This rule of Chemistry for **Compounds,** states that, the Equilibrium state of Existence of some Bimetallic and trimetalic compounds are as follows-

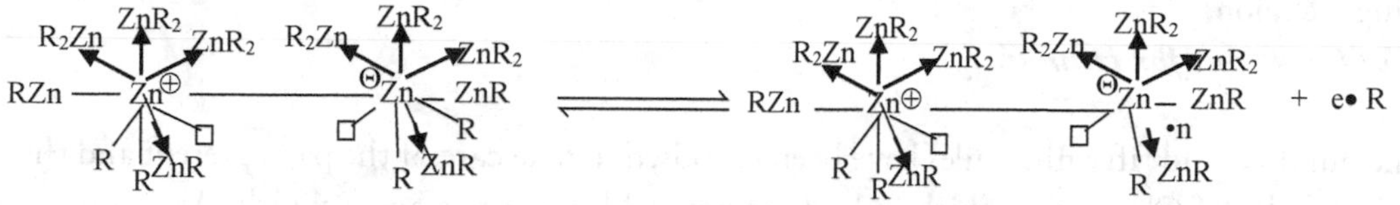

(II) THE INITIATOR

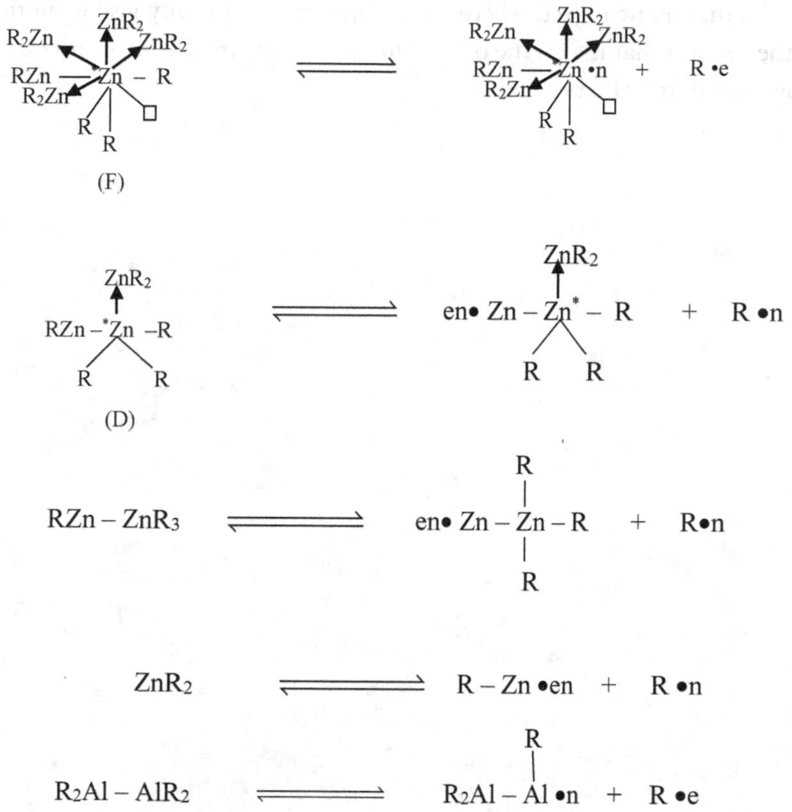

(F)

(D)

$$RZn - ZnR_3 \rightleftharpoons en\bullet Zn - \underset{\underset{R}{|}}{\overset{\overset{R}{|}}{Zn}} - R \quad + \quad R\bullet n$$

$$ZnR_2 \rightleftharpoons R - Zn \bullet en \quad + \quad R \bullet n$$

$$R_2Al - AlR_2 \rightleftharpoons R_2Al - \underset{}{\overset{\overset{R}{|}}{Al}} \bullet n \quad + \quad R \bullet e$$

noting that the maximum oveall valency state of the Central for Zn when bimetallic compounds are formed is +4 while that for Al is +3.
(Laws of Creations for Metallic Compounds)

Rule 1357: This rule of Chemistry for **FeCl$_3$ when used as source of initiators,** states that, throughout the stages of preparation, the fourth electro-non-free-radical in the radical configuration of Fe, remained untouched throughout the stages to the end, but became an electro-free-radical, clearly shows that though Fe has not shown its tetravalent character, it does exist when the compounds are Coordination compounds; when indeed so far, presence of its **divalent, trivalent, pentavalent** and **heptavalent characters** have been encountered.
(Laws of Creations for Fe)

Rule 1358: This rule of Chemistry for **Humanity,** states that, the greatest foundation ever laid since antiquity is the PERIODIC TABLE, from which one can see that ATOMS and their ELEMENTS which are the building blocks of everything in our world are LIVING SYSTEMS, for they can FEEL,

SEE, SMELL, TASTE, HEAR and communicate with themselves via the applications of the laws of Nature in Mathematics and Physics, all of which are inside CHEMISTRY, the mother of all disciplines including Religion.
(Laws of Creations for Humanity)

One hundred and fifty nine rules have been proposed to take care of the past, present and the future. To produce rules of Chemistry which indeed are natural laws, is not an easy task. As can be observed, reactions don't just take place just like that. They take place following the natural laws. Candidly speaking, there is no inanimate object, whether the object maintains any visible motion or not. It is the invisible part of the motion that makes the object animate. There is "life" in every object, so long as they are build-ups from atoms and their elements.

References

1. R. A. Parsiga, J. Macromol. Sci. - Rvs. Macromol. Chem., C1(2) : 223 (1967).

2. A. J. Barry and H. N.. Beck, Silicone Polymers, in F. G. A. Stone and W. A. G. Graham (eds), "Inorganic Polymers, Chap. 5, Academic Press, Inc., New York, 1962.

3. C. L. Lee and O. K. Johannson, J. Polymer Sci., A - 1 (4) : 3013 (1966).

4. T. C. Kendrick, J. Chem. Soc., 1965 : 2027.

5. G. Levin and J. B. Carmichael, J. Polymer Sci., A - 1 (6) : 1(1968).

6. W. Cooper, D. R. Morgan, and R. T. Wragg, Eur. Polym. J., 5(1) : 71(1969).

7. J. K. Stille and J. A Empen, J. Polymer Sci., A - 1(5) : 273 (1967).

8. G. Odian, "Principles of Polymer Systems", McGraw-Hill Book Company, (1970), pgs. 495 - 501.

9. H. R. Allcock, "Heteroatom Ring Systems and Polymers", Chaps. 6 and 8, Academic Press, Inc., New York, 1967.

10. C. R. Noller, "Textbook of Organic Chemistry", W. B. Saunders Company, (1966), pgs. 275 - 279.

11. J. Brandrup and E. H. Immergut (eds.), "Polymer Handbook", pp. 11 - 363, Interscience Publishers, John Wiley & Sons, Inc., New York, 1966.

12. J. B. Lando and P. Frayer, J. Polymer Sci., B6 : 285 (1968).

13. C. R. Noller, "Textbook of Organic Chemistry", W. B. Saunders Company, (1966), pgs. 504 - 514.

14. I. L. Finar, "Organic Chemistry", English Language Book Society and Longmans, Greens & Co Ltd., (1964), Vol Two, pgs. 435 - 436.

15. L. Bateman (ed.), "The Chemistry and Physics of Rubber like Substances, "John Wiley & Sons, Inc., New York, 1963.

16. L. Bateman, C. G. Moore, M. Porter, J. Chem. Soc., 1958 : 12866.

17. L. Bateman, C. G. Moore, M. Porter, and B. Saville, Chemistry of Vulcanization, in L. Bateman (ed.), op. Cit., Chap 15.

18. J. R. MacCallum and A. Wemick, J. Polymer. Sci.,A-1(5): 3061 (1967).

19. N. L. Paddock, Quart. Rev. (London), 18: 168 (1964).

20. (a) Friedman, L. (Union Carbide Corporation). USA Patent 3194795, 1965. (b) Frieman, L. (Union Carbide Corporation). USA Patent 3318855, 1967. (c) Friedman, L. (Pure Chemicals Limited). FR patent 1375760, 1963.

21. C. R. Noller, "Textbook of Organic Chemistry", W. B. Saunders Company, (1966), pgs. 272 - 274.

22. C. R. Kruger and E. G. Rochow, J. Polymer Sci., A2:3179 (1964).

23. T. Tsuruta, S. Inoue, and K. Tsubaki, Macromol. Chem., 111:236 (1968).

24.   T. Saegusa, H. Imai, and S. Matsumoto, J. Polymer Sci., A-1(6): 549 (1968).

25.   M. E. Pruitt and J. M. Baggett, U. S. Pat. 2,706,182 to Dow Chemical Co. (April, 1955).

26.   T. Tsuruta, Stereospecific Polymerization of Epoxides, A. D. Ketley (ed), op. cit., vol. 2, chap. 4 of A. D. Ketley (ed.), "The stereochemistry of Macromolecules", Marcel Dekker, Inc., New York, 1967.

Problems

14.1. Carbon and silicon are close neighbors of Group. IVA elements in the Periodic Table in which their last shells can carry a maximum of eight radicals. Distinguish between carbon and silicon as source of ringed Addition monomers.

14.2. (a) What types of hybridizations are favored by carbon and silicon atoms?

(b) Give a series of reactions for the conversion of -

(i) Tri - n - propylchlorosilane to hexa - n - propyldisiloxanes.

(ii) Silane to sodium salt of silicon by hydrolysis using a strong base - NaOH.

by providing the mechanisms.

14.3 Shown below is pyrazolidine obtained from pyrazole.

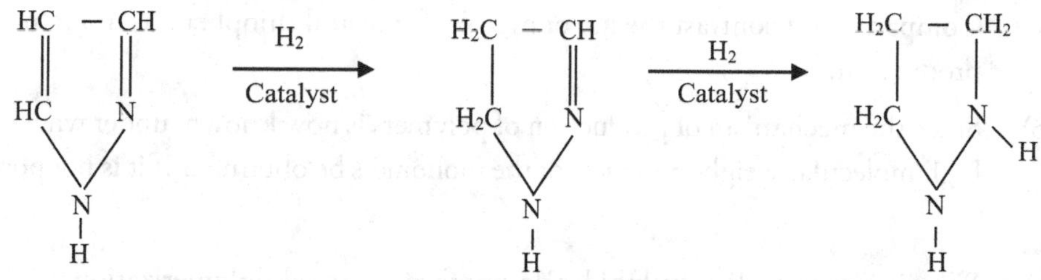

(I) Pyrazole   (II) Pyrazoline   (III) Pyrazolidine

(i) Explain the mechanism of the reactions above where possible. Check if (II) is the real structure of pyrazoline if pyrozoline is indeed the product.

(ii) Identify any phenomenon in (I) and (II).

(iii) In what manner can the ring in (III) be opened.

14.4 (a) Compare (III) of Q. 14.3 with cyclic amines.

(b) Shown below is a nitroethylene monomer.

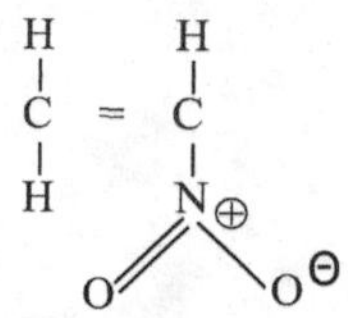

(i) Does the monomer favor any type of resonance stabilization phenomenon? Explain.

(ii) Identify the types of activation centers in the monomer and show the order of their nucleophilicity.

(iii) Show how the monomer can be polymerized.

14.5 Give a series of reactions for the conversion of -

(a) Thiophene to equal amounts of 2 - and 3 - t - butyl thiophenes.

(b) Thiophene to 2, 5 - dinitrothiophene.

(c) Furan to 2, 5 - dibromofuran.

by providing the mechanism.

14.6 (a) Distinguish between the types of resonance stabilization phenomena undergone by -

(i) Benzene

(ii) Pyrrole

(iii) Furan

(iv) Thiophene.

(b) What is polar/radical resonance stabilization phenomenon? Compare with that of free radical resonance stabilization phenomenon.

14.7 (a) Compared and contrast the use of pyrrole, furan and thiophene as monomers for polymer production.

(b) Since the mechanism of production of polymers is now known, under what conditions can high molecular weight polymers of the monomers be obtained? If it is not possible, explain.

14.8 (a) Why is it that cyclic disulphides do not favor charged polymerizations? Explain.

(b) Why is it that cyclic disulphides and rhombic sulphur exist at S.T.P, but the corresponding cases for oxygen containing rings do not exist?

(c) Distinguish between the three rings shown below, in terms of character, their use as monomers and favored existence.

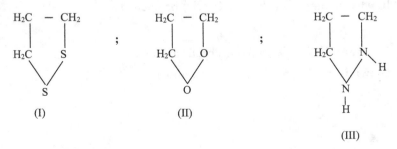

(I) ; (II) ; (III)

14.9　(a)　In the absence of initiators, when heat is involved as initiator, what goes on for different types of monomers?

(b)　Explain the mechanism of initiation and propagation of dichlorophosphazene when heat is involved as initiator.

(c)　What is the significance of carrying out some reactions under "inert conditions", based on the New Frontiers?

14.10.　(a)　For many year, only ionic types of charges (cations and anions) were known to exist. Based on the New Frontiers, how many types of charges exist?

(b)　Distinguish between the different types of charges by providing the new classifications using examples.

14.11.　(a)　What is basis of the existences of different types of charges?

(b)　What are the driving forces favoring their existence to start with?

(c)　On the basis of electrostatic forces of repulsion and attraction, distinguish between the different types of charges.

14.12.　(a)　How can dichlorophosphazene be made to favor nucleo-non-free-radical, anionic routes? Can they be made to undergo electro-free-radical and positively charged routes?

(b)　Is oxaazacyclopropane more nucleophilic than cyclic disulphides and rhombic sulphur? Explain.

14.13　Shown below is pyridine.

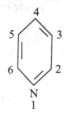

(a)　Identify the types and strength of activation centers in the ring.

(b)　Show if there is any type of resonance stabilization phenomenon favored by pyridine.

(c)　Amino group can be introduced directly into pyridine by heating with sodium amide to give the sodium salt of 2-aminopyridine (Chichibabin reaction). Addition of water gives the free amine.

(i)   Is the mechanism of the so-called Chichibabin reaction above ionically possible? Explain. If the answer is negative, propose a new mechanism for the reaction.

(ii) Why the need adding water above to give 2- aminopyridine when it already exist above?

(iii) Why is the 3- or 4- or other positions not involved in the displacement reaction?

14.14. (a) Pyridine reacts with peroxy acids or with 30 per cent hydrogen peroxide in acetic acid solution to give good yields of pyridine N - oxide.

(b) 4 - aminopyridine can be obtained from the nitro- N - oxide by reduction with iron and acetic acid.

Based on the New Frontiers, provide the mechanisms of both reactions above. Can Fe form hydrides with H? Explain.

14.15 (a) Explain the mechanism by which the a and g acids of pyridine shown below loose CO_2 when heated to form pyridine.

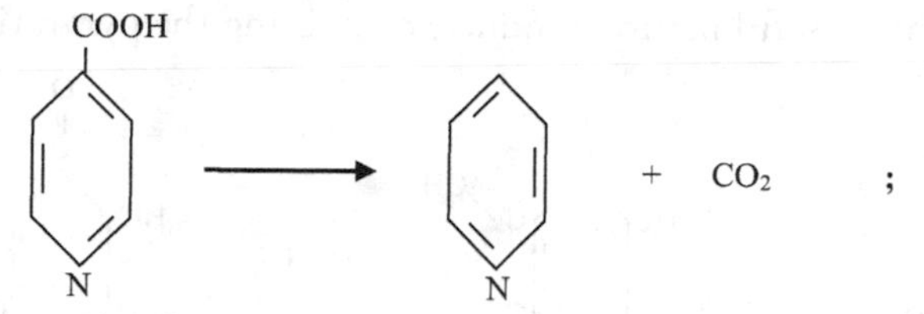

COOH

$\longrightarrow$

$+$ CO_2 ;

γ - picoline acid

COOH

$\longrightarrow$

$+$ CO_2

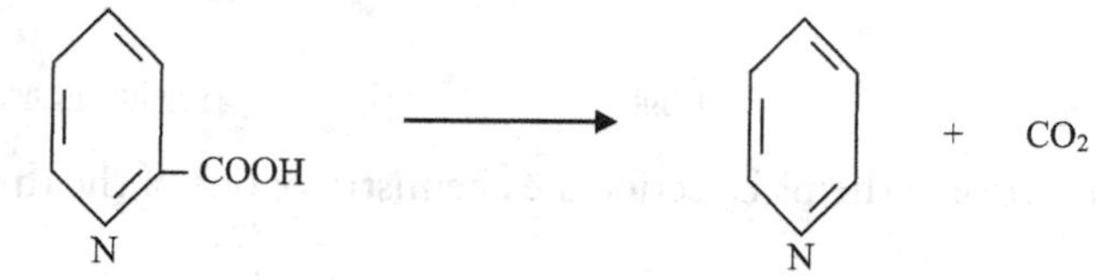

α - picolinic acid

(b) Nicotinic acid is obtained by the reaction indicated below from nicotine, that is, by oxidation of nicotine.

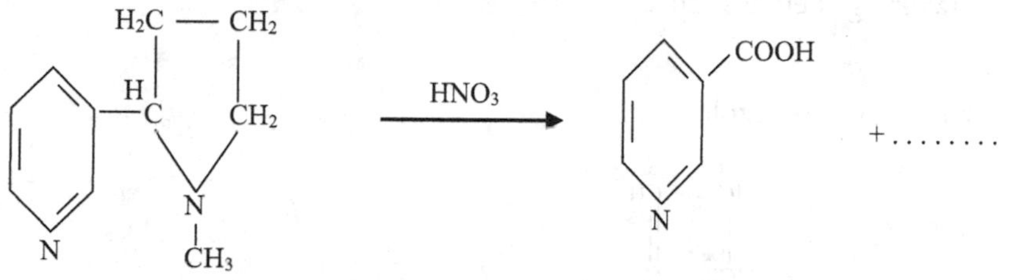

H_2C — CH_2
H
C CH_2
N
CH_3

$\xrightarrow{HNO_3}$

COOH

$+ \ldots \ldots$

Nicotine

Nicotinic acid

Explain the mechanism of the reaction above and complete the reaction.

14.16 (a) α and γ - picolines are said to condense with aromatic aldehydes to give benzylidene derivatives, but β - picoline does not

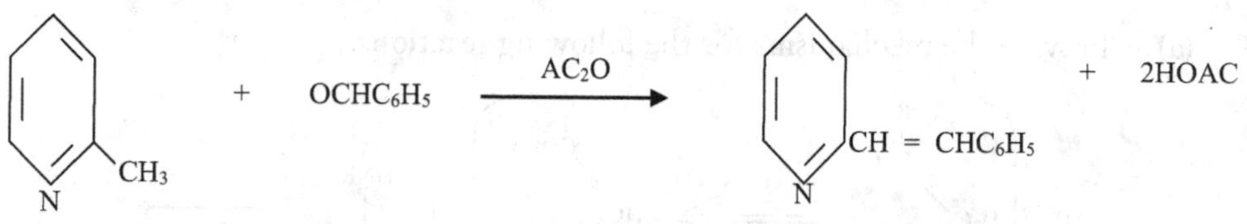

CH_3
N

$+$ $OCHC_6H_5$ $\xrightarrow{AC_2O}$

$CH = CHC_6H_5$
N

$+$ $2HOAC$

Benzylidene - α - picoline

CH_3
N

$+$ $OCHC_6H_5$ $\xrightarrow{AC_2O}$

$CH = CHC_6H_5$
N

$+$ $2HOAC$

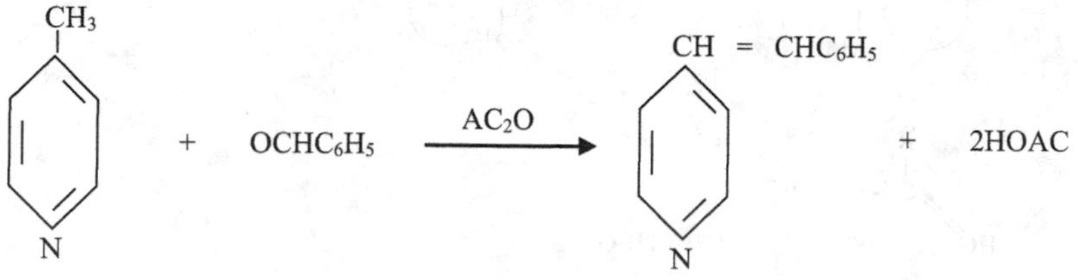

Benzylidene - γ - picoline

507

Explain the observations and the mechanism of the reactions. Why does β - picoline not favor the reaction? (Resonance stabilization phenomenon).

(b)   Shown below are three useful natural products containing the pyran ring.

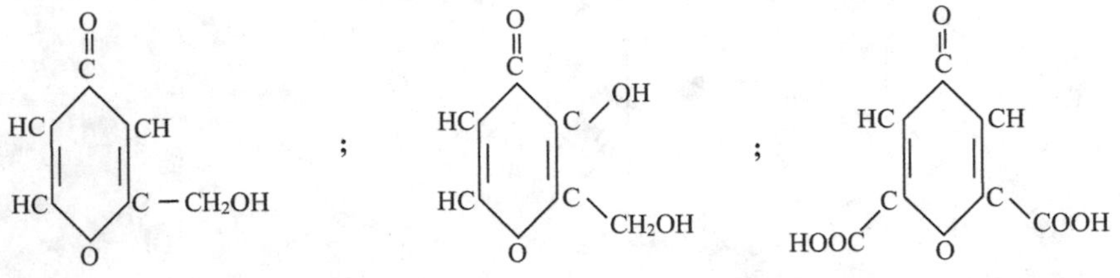

Kojic acid   Maltol   Chelidonic acid

What is or are common between them? Describe the chemistry of one of the three compounds.

14.17. (a)   Distinguish between continuous closed-loop polar/radical, discrete closed-loop polar/radical and opened-loop polar/radical resonance stabilization phenomena.

(b)   What are Electroradicalization of the first and second types? What are the driving forces favoring their existence?

14.18. Shown below is pyrazole

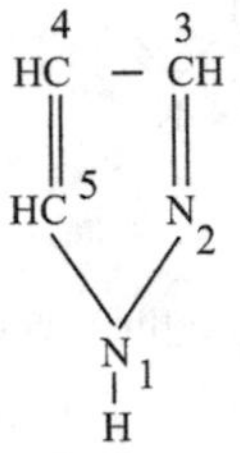

The type of resonance stabilization phenomenon that takes place with it is said to be not yet known. It is said to exhibit aromatic properties, e.g. it is readily halogenated, nitrated and sulphonated; the group enters only at the 4 - position. Based on the New Frontiers, identify the type of resonance and provide the chemistry of pyrazole.

14.19. (a)   Provide the mechanisms for the following reactions.

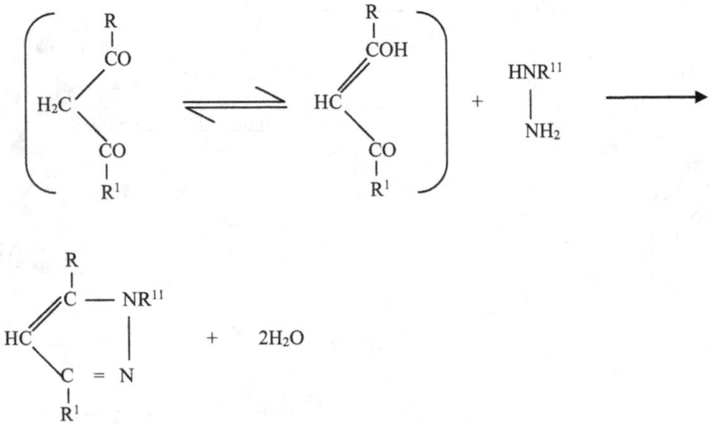

(I)

(ii)

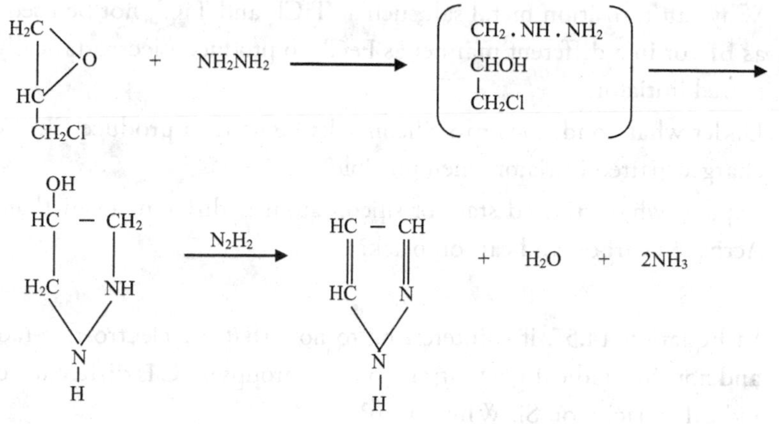

$$H_2C \underset{HC}{\overset{}{\diagdown}} O + NH_2NH_2 \longrightarrow \left(\begin{array}{c} CH_2.NH.NH_2 \\ | \\ CHOH \\ | \\ CH_2Cl \end{array} \right) \longrightarrow$$

with CH_2Cl attached below HC.

$$\begin{array}{c} OH \\ | \\ HC - CH_2 \\ | \quad\quad | \\ H_2C \quad NH \\ \diagdown \quad / \\ N \\ | \\ H \end{array} \xrightarrow{N_2H_2} \begin{array}{c} HC - CH \\ \| \quad\quad \| \\ HC \quad N \\ \diagdown \quad / \\ N \\ | \\ H \end{array} + H_2O + 2NH_3$$

(II)

(b) Are (I) and (II) resonance stabilized?

14.20. (a) Provide the mechanism for the reactions below.

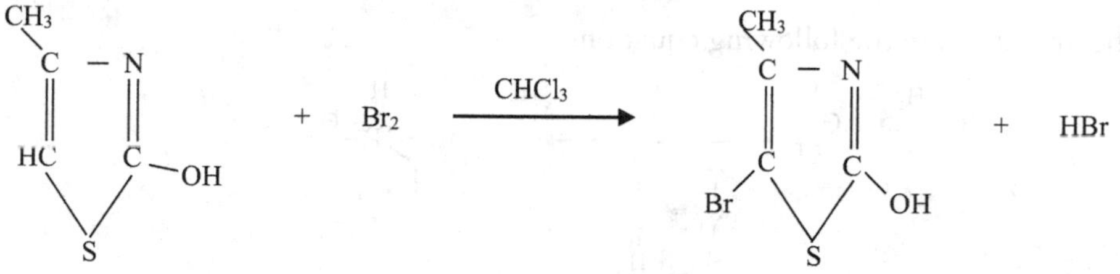

$$\begin{array}{c} CH_3 \\ | \\ C - N \\ \| \quad\quad \| \\ HC \quad C \\ \diagdown \quad \diagdown OH \\ S \end{array} + Br_2 \xrightarrow{CHCl_3} \begin{array}{c} CH_3 \\ | \\ C - N \\ \| \quad\quad \| \\ C \quad C \\ \diagup \diagdown \quad \diagdown OH \\ Br \quad S \end{array} + HBr$$

5 - bromo - 2 - hydroxy - 4 - methylthiazole

(b)

$$\begin{array}{c} CH_2.NH_2 \\ | \\ CH_2Br \end{array} + \begin{array}{c} NH \\ | \\ C.R \\ HS \diagup \end{array} \longrightarrow \begin{array}{c} H_2C - N \\ | \quad\quad \| \\ H_2C \quad C \\ \diagdown \quad \diagdown R \\ S \end{array} + NH_4Br$$

β-halogenoamines   Thioamodes   (I) Thiazolines

(c) (I) above is said to favor ring opening by the action of acid, e.g.

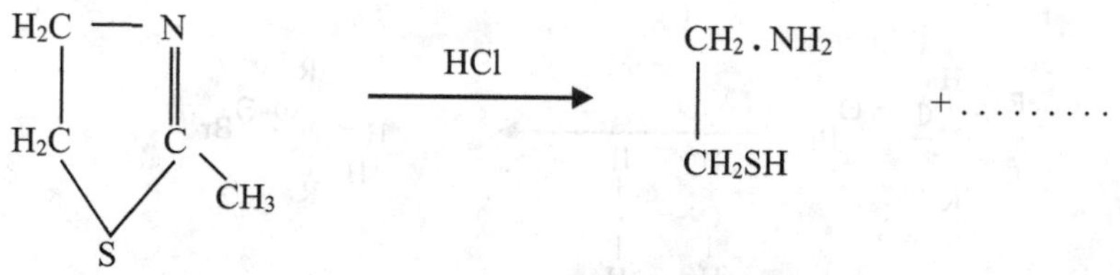

$$\begin{array}{c} H_2C - N \\ | \quad\quad \| \\ H_2C \quad C \\ \diagdown \quad \diagdown CH_3 \\ S \end{array} \xrightarrow{HCl} \begin{array}{c} CH_2.NH_2 \\ | \\ CH_2SH \end{array} + \ldots\ldots$$

2 - methylthiazoline   2 - aminoethanethiol

Complete and explain the mechanism of the reaction above.

14.21. (a) Why can transition metal salt such as $TiCl_3$ and $TiCl_4$ not be used in a similar manner as BF_3 or in a different manner as $FeCl_3$ to produce Electrostatically positively charged-paired initiator?

   (b) Under what conditions can silicon salts be used to produce Electrostatically positively charged-paired initiator when possible?

   (c) Explain why activated state of silicon atom is different from that of carbon? What is Activated carbon and carbon black?

14.22. (a) In Equation 14.57, it is interesting to note that the electro-free-radical capacity of free and non-free radical-pushing and pulling groups on C is different from their nucleo-free-radical capacity on Si. Why is it so?

   (b) Why is it that Si cannot be used in the same manner as C to give another form of "life"? Do you think it is used as such in another universe where there is Life?

   (c) Can you readily explain why per fluoro propylene, $CF_2 = CF(CF_3)$, cannot readily favor nucleo-free-radical and electro-free-radical polymerization?

14.23. Shown below are the following equations

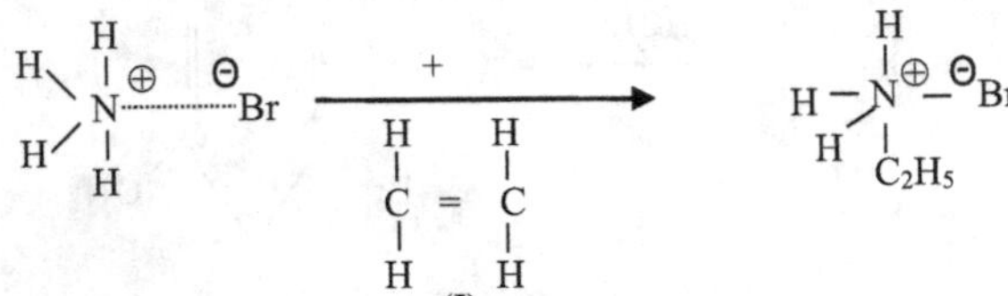

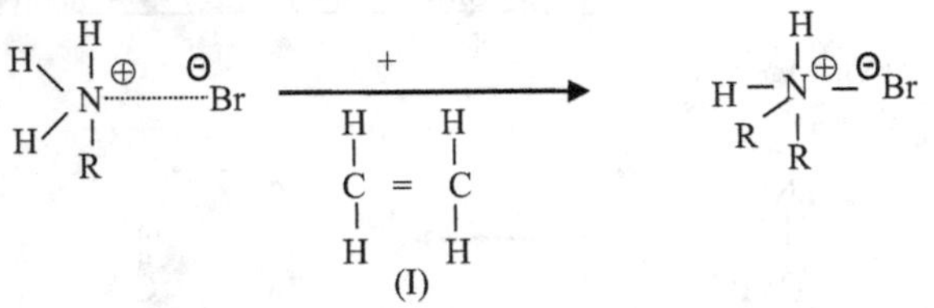

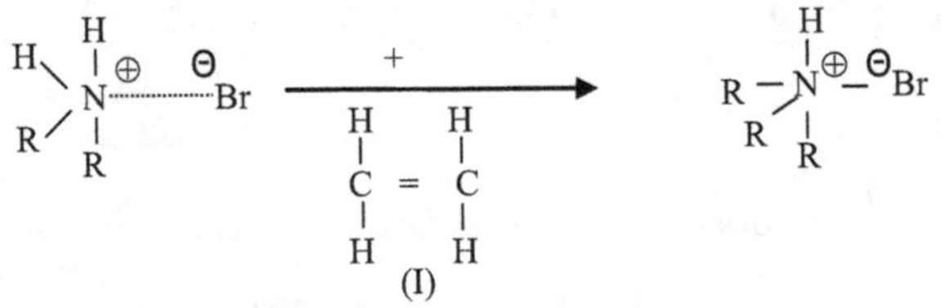

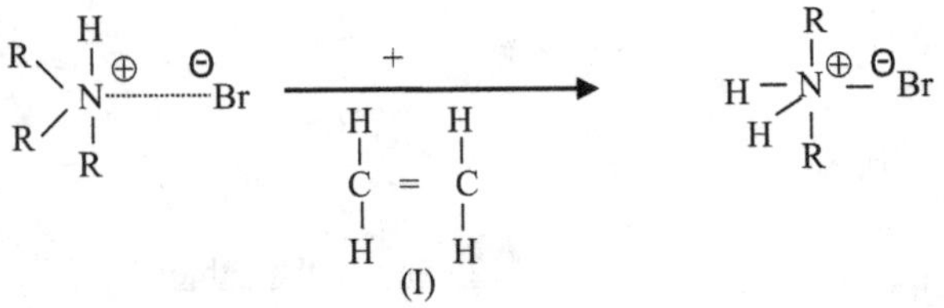

a) Describe the relevance of the equations above in terms of the suitability of the use of primary, secondary, tertiary amines as catalyst or cocatalyst for cationic and anionic polymerization. Provide the mechanism for the reactions above.

b) Why can the initiator not be used for the polymerization of ethene? What is the significance of these reactions philosophically?

14.24.   In nature, males are known to carry both the male and female hormones, while females carry only the female hormones. In nature, females have been observed to be far more than males in population regardless the effect of wars etc. Can you explain this correlation in line with what exist in chemical systems. You must go into details.

14.25.   After identifying the followings, -
   (i)   Radicals   (v)   Polar charges
   (ii)   Ionic charges   (vi)   Dative bonds
   (iii)   Covalent charges   (vii) Different types of activation centers
   (iv)   Electrostatic charges.

   provide the new classifications for Radicals, Charges, Bonds and Compounds.

14.26.   a)   Carbon disulfide are unique monomers. Show how the members of this family can be polymerized?

   b)   Why is Rhombic sulfur very unique? Under what conditions can it be used as monomer and cross-linking agents?

   c)   Can Rhombic sulfur react with H_2 to produce hydrogen sulfide?

   d)   Compare HOOH, HOSH and HSSH whether they exist at STP or not.

CHAPTER FIFTEEN

TRANSFER OF TRANSFER SPECIES IN RING - FORMING AND POLYMERIC MONOMERS

15.0. Introduction

With Addition monomers, there are essentially three major groups -

(i)   Those that add inter-molecularly.
(ii)   Those that add intra-molecularly and
(iii)   Those that add inter-/intra-molecularly.

Those that add *inter-molecularly* have been the cases largely considered. These include all the Addition monomers of alkenes, acetylenes, nitriles, aldehydes, ketones, aldimines, ketimines, isocyanates, cumulenes, ketenes etc. and all the ringed monomers considered.

Those that add *intra-molecularly* only exist amongst polymeric monomers. These are polymers which have activation centers externally located. Those that favor *inter-/intra-molecular additions* are linear or ringed or non-ringed monomers with at least two activation centers suitably distantly placed. There are three types of monomers that add inter- and intra- molecularly in terms of the types of polymeric products obtained. These include -

(i)   Inter-/intra-/inter-/intra-molecular alternating regular and random cyclo-co-polymerization.
(ii)   Inter-/intra-/inter-/intra-molecular alternating cyclo-copolymerization.
(iii)   Inter-/intra-/intra-molecular alternating and random copolymerization.

The three cases above involve the use of reactive initiators. There are also three other types where initiators are not involved. These are called-

(iv)   Di - inter- molecular cyclo-homo-polymerization.
(v)   Intra-molecular cyclo-homo polymerization.
(vi)   Equilibrium Intra-molecular Pseudo-Step cyclo-co-polymerization.

In all these cases, one monomer is involved. In (iv) however, two monomers can be used. The third, fifth and sixth cases involve the use of gigantic monomers, i.e., unsaturated polymeric monomers. Where inter-/intra-molecular alternating cyclo-polymerizations and di-inter-molecular polymerizations are involved, there can be cross-links or no cross-links in the polymeric products.

When intra-molecular addition takes place along with inter, rings are formed along the polymeric chain. Based on factors favoring the existence of a ring size, as already fully identified, *only the existence of special sizes can be favored along the chain, depending on the route of polymerization and type of ring to be formed. The rings formed must be such which do not possess the MRSE by itself in particular or if they carry functional centers, the presence of free electro-free-radicals or cations via the route of polymerization should be disallowed.* For example, existence of three- or some four- membered rings along the chain is not common, because of the favored presence of MRSE in such small rings. Existence of five- or six - membered rings along the chain one after the other in an alternating manner will strongly be favored. Existence of larger membered rings one after the other in an alternating manner is obviously possible, **all largely dependent on the amount of strain energy (SE) for the family of the monomer, conformational stability or instability, and steric limitation.** While there are countless numbers of conformations for a particular compound, there is only one configuration. Hence, *the extent of cyclization, which is measured as the percent of monomer units which are cyclized, varies from one monomer to the other in terms of character, the mode of polymerization, the distance between the active centers and immediate activation center on the monomer during propagation and other factors which will become clear at the end of this chapter.*

Examples of ring- forming monomers via Inter-/Intra- addition, to be considered for present study, include "**unconjugated dienes**" such as diallyl maleate, triallyl cyanurate, divinyl benzene, 1,6- heptadiene, diallyl quaternary ammonium salts, 1,4- etheric-, thiocarbonyl-, methylene-dienes, and **di-aldehydes** such as o-phthaldehyde, Glutaraldehyde and its 3-methyl and 3-phenyl derivatives, succinaldehyde, malealdehyde cyclohexene-cis-4,5 dicarboxyaldehyde and its methyl derivatives, adipaldehyde, and **dialkyl aldimines** some of which are shown below.

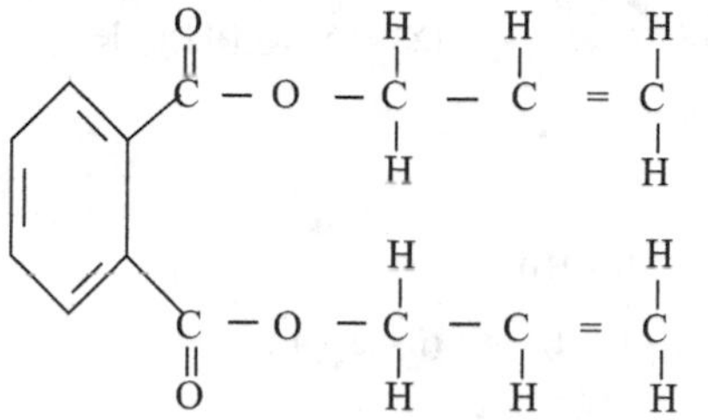

(I) <u>diallyl phthalate</u>

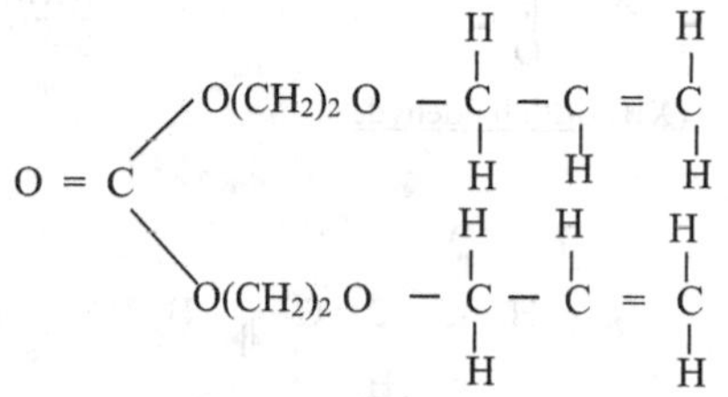

(II) <u>diethylene glycol bisallyl carbonate</u>

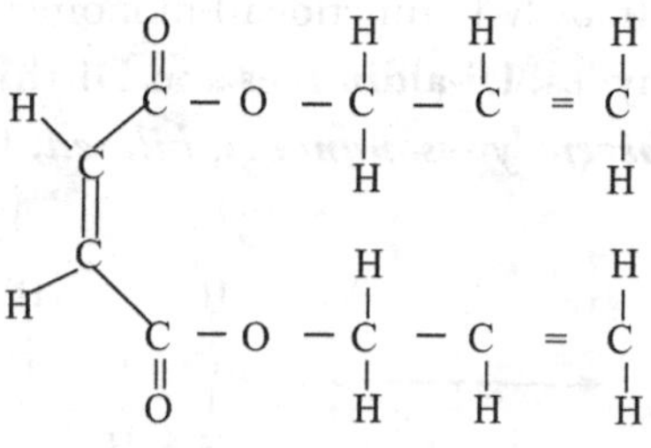

(III) <u>diallyl maleate</u>   15.1

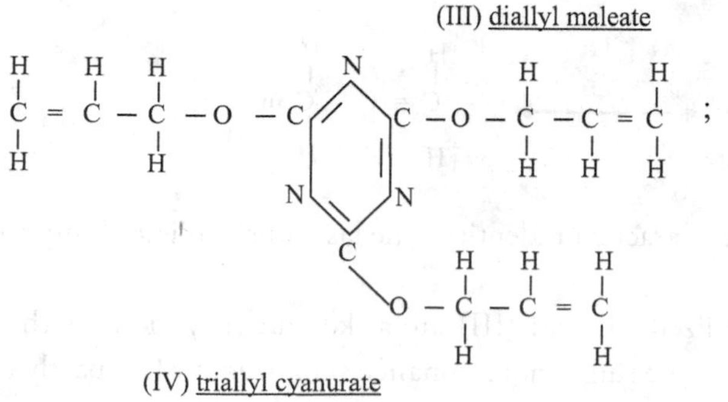

(IV) <u>triallyl cyanurate</u>

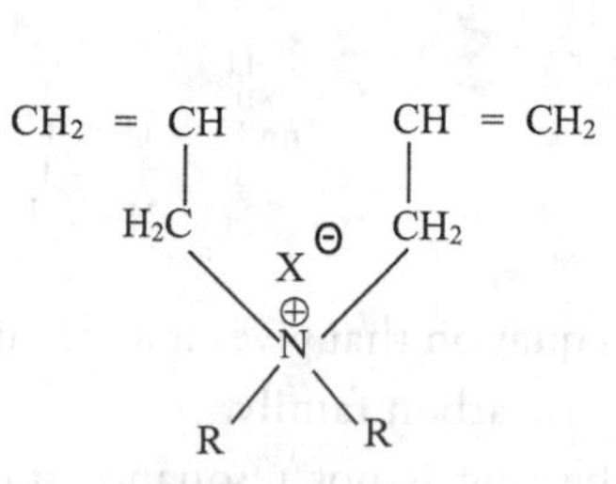

(V) <u>diallyl quaternary ammonium salts</u>

15.2

513

$$CH_2 = \overset{\overset{\displaystyle H}{|}}{C} - \!\!\!\bigcirc\!\!\!- \overset{\overset{\displaystyle H}{|}}{C} = CH_2 \quad ; \quad CH_2 = CH - \overset{\overset{\displaystyle H}{|}}{\underset{\underset{\displaystyle H}{|}}{C}} - \overset{\overset{\displaystyle H}{|}}{\underset{\underset{\displaystyle H}{|}}{C}} - \overset{\overset{\displaystyle H}{|}}{\underset{\underset{\displaystyle H}{|}}{C}} - CH = CH_2$$

(VI) <u>divinyl benzene</u>　　　　　　　　　　(VII) <u>1, 6 - heptadiene</u>　　　15.3a

$$\overset{\overset{\displaystyle H}{|}}{\underset{\underset{\displaystyle H}{|}}{C}} = \overset{\overset{\displaystyle H}{|}}{C} - \overset{\overset{\displaystyle O}{\|}}{C} - O - \overset{\overset{\displaystyle O}{\|}}{C} - \overset{\overset{\displaystyle H}{|}}{C} = \overset{\overset{\displaystyle H}{|}}{\underset{\underset{\displaystyle H}{|}}{C}} \quad ; \quad \overset{\overset{\displaystyle H}{|}}{\underset{\underset{\displaystyle H}{|}}{C}} = \overset{\overset{\displaystyle H}{|}}{C} - \overset{\overset{\displaystyle H}{|}}{\underset{\underset{\displaystyle H}{|}}{C}} - \overset{\overset{\displaystyle H}{|}}{\underset{\underset{\displaystyle H}{|}}{C}} = \overset{\overset{\displaystyle H}{|}}{\underset{\underset{\displaystyle H}{|}}{C}}$$

(VIII) <u>acrylic anhydrides</u>　　　　　　　　(IX) <u>1, 4 - diene</u>　　　15.3b

$$\overset{\overset{\displaystyle H}{|}}{\underset{\underset{\displaystyle H}{|}}{C}} = \overset{\overset{\displaystyle H}{|}}{C} - \overset{\overset{\displaystyle H}{|}}{\underset{\underset{\displaystyle O}{\|}}{C}} - \overset{\overset{\displaystyle H}{|}}{C} = \overset{\overset{\displaystyle H}{|}}{\underset{\underset{\displaystyle H}{|}}{C}} \quad ; \quad \overset{\overset{\displaystyle H}{|}}{\underset{\underset{\displaystyle H}{|}}{C}} = \overset{\overset{\displaystyle H}{|}}{C} - O - \overset{\overset{\displaystyle H}{|}}{C} = \overset{\overset{\displaystyle H}{|}}{\underset{\underset{\displaystyle H}{|}}{C}}$$

(X) <u>1, 4 - carbonyl - diene</u>　　　　　　(XI) <u>1, 4 - etheric diene</u>　　　15.4

$$\bigcirc\!\!\!\!\!\overset{\overset{\displaystyle O}{\|}}{\underset{\underset{\displaystyle O}{\|}}{\overset{\displaystyle C-H}{C-H}}} \quad ; \quad H - \overset{\overset{\displaystyle O}{\|}}{C} - (CH_2)_3 - \overset{\overset{\displaystyle O}{\|}}{C} - H \quad ; \quad H - \overset{\overset{\displaystyle O}{\|}}{C} - (CH_2)_2 - \overset{\overset{\displaystyle O}{\|}}{C} - H$$

　　　　　　　　　　　　　　(XIII) <u>Glutaldehyde</u>　　　　(XIV) <u>Succinaldehyde</u>

(XII) <u>o-Phthaldehyde</u>

$$H - \overset{\overset{\displaystyle O}{\|}}{C} - \overset{\overset{\displaystyle H}{|}}{C} = \overset{\overset{\displaystyle H}{|}}{\underset{\underset{\displaystyle O}{\|}}{C}} - \overset{\displaystyle H}{C} - H \quad ; \quad H - \overset{\overset{\displaystyle N(CH_3)}{\|}}{C} - (CH_2)_n - \overset{\overset{\displaystyle N(CH_3)}{\|}}{C} - H \quad (n = 2,3,4)$$

　　　　　　　　　　　　　　　　　(XVI) <u>Di-methylaldimines</u>

(XV) <u>Malealdehyde</u>　　　　　　　　　　　　　　　　　　15.5

All these are polyfunctional unsaturated (mostly Difunctional) monomers – **Di-allyls, Di-Vinyls, Di-acrylic, Unconjugated Dienes, Di-aldehydes, Di-aldimines and Di-thioalde-hydes**. What is unique about *allylic groups is that they are inherently resonance stabilized,* for which reason, they cannot be activated chargedly.

$$e\bullet\overset{\overset{\displaystyle H}{|}}{\underset{\underset{\displaystyle H}{|}}{C}} - \overset{\overset{\displaystyle H}{|}}{\underset{\underset{\displaystyle H}{|}}{C}} = \overset{\overset{\displaystyle H}{|}}{\underset{\underset{\displaystyle H}{|}}{C}} \quad \longleftrightarrow \quad \overset{\overset{\displaystyle H}{|}}{\underset{\underset{\displaystyle H}{|}}{C}} = \overset{\overset{\displaystyle H}{|}}{\underset{\underset{\displaystyle H}{|}}{C}} - \overset{\overset{\displaystyle H}{|}}{\underset{\underset{\displaystyle H}{|}}{C}}\bullet e \qquad 15.6a$$

$$n\bullet\overset{\overset{\displaystyle H}{|}}{\underset{\underset{\displaystyle H}{|}}{C}} - \overset{\overset{\displaystyle H}{|}}{\underset{\underset{\displaystyle H}{|}}{C}} = \overset{\overset{\displaystyle H}{|}}{\underset{\underset{\displaystyle H}{|}}{C}} \quad \longleftrightarrow \quad \overset{\overset{\displaystyle H}{|}}{\underset{\underset{\displaystyle H}{|}}{C}} = \overset{\overset{\displaystyle H}{|}}{\underset{\underset{\displaystyle H}{|}}{C}} - \overset{\overset{\displaystyle H}{|}}{\underset{\underset{\displaystyle H}{|}}{C}}\bullet n \qquad 15.6b$$

It is the first equation that gives it its allylic character or identity. The last equation is still important with respect to hydrocarbon family.

In (I), the ring is not resonance stabilized. (I) and (III) are alike and they along with (XII) are Electrophiles of different characters. In (IV), the ring is not resonance stabilized. (VI) is partly resonance stabilized as already shown and favors independent activation of the externally located activation centers free-radically and chargedly (when not resonance stabilized). All the others are not resonance stabilized.

(I), (II), (III), (IV), (V) and the rest except (VI) largely favor inter-/intra- molecular cyclohomopolymerization, in view of the sizes of their rings when formed. However, (VI) can still be used as a cross-linking agent. Hence, most or all of them are largely known to be commercially useful or important in forming *highly cross-linked thermosetting products.*

Diallyl quaternary ammonium salts (V), have been reported to give soluble, uncross linked polymers with little or no residual unsaturation during free-radical polymerization[1,2]. In order words, the products are completely linear with cyclic structures in the backbones alternatingly placed. Also, 1,4 - dienes lead to products which have little residual unsaturation and are not appreciably cross-linked[3]. Extensive works which have been done by Butler and Marvel and their coworkers, as well as others,[1-8] *have shown that the size of the ring structure which can be formed, determines whether inter- and intra- molecular alternating or random placement is the predominant reaction for a particular monomer.* The extent of cyclization have been found to generally increase with size in the order: 5- and 6- membered rings appreciably greater than 7- membered rings and the latter somewhat greater than larger sized rings. When rings of 5 or 6 can be formed, the polymerization is said to proceed almost exclusively by Inter-/Intra- molecular alternating arrangement. The extent of cyclization decreases quite sharply as one goes to ring sizes of seven or more atoms. This indeed and much more above are not true.

Di- inter- molecular homopolymerizations largely apply to conjugated dienes, such as 1,3 - dienes which have been known to yield relatively low molecular weight polymers, when complex catalysts such as $TiCl_4$ - $Al(C_2H_5)_2Cl$ or $TiCl_4$ - $C_2H_5AlCl_2$ at high Ti/Al ratios are used[9-11]. For the type of compounds and the ratio involved, Ziegler - Natta initiator cannot be obtained. Gaylord and coworkers postulated a Diels-Alder type "charge transfer mechanism" to describe the almost exclusive formation of cyclopolymers[10,11]. These are impossible situations, since charges cannot be transferred. Few polymeric monomers exist which result in cyclization via pyrolysis (Intra-addition) using the pendant functional groups, that is, external activation center. A notable example is polyacrylonitrile[12].

Based on the experiences gathered from all the above mentioned cases, Di-aldehydes and Di-alkyl aldimines will be considered almost at the end on their own merits. (XIII), (XIV) and (XVI) cannot produce cross-linked network polymers during polymerization, but (XII) and (XV) can. While the former are Nucleophiles, the latter are Electrophiles. They all favor Inter/Intra/ Inter/Intra- molecular additions

15.1. Inter/Intra Molecular Alternating Regular and Random Cyclocopolymerization

Monomers to be considered under this category include *diallyl phthalate, diethylene glycol bisallyl carbonate, triallyl cyanurate and divinyl benzene. Divinyl benzene does not actually belong to this group.* It was however included for the purpose of mis-representation of families as currently exists in general everywhere. The routes found during cyclization will also be identified.

15.1.1. Diallyl Phthalate

For ring-forming monomers in general, the existence of cyclization depends on the following variables -

(i)   Presence and location of activation centers, for which at least two must be present on the monomers.

(ii)   The distance between the activation centers.

(iii) The degree of strain energy that is to be present in the ring when formed, since the less the strain energy or the absence of any point of scission when MRSE or more are present in the ring, the greater the probability of forming a ring.

(iv)   The types of functional centers present in the ring to be formed.

(v)   The types of imitators or kinetic route involved.

Now, considering the monomer in question, since the symmetric electrophilic benzene ring cannot be opened (no point of scission), and since the C = O center is more nucleophilic than any center in the ring, activation can only take place from the C = C centers externally located, being the least nucleophilic centers. When the monomer is activated chargedly, the followings are to be expected.

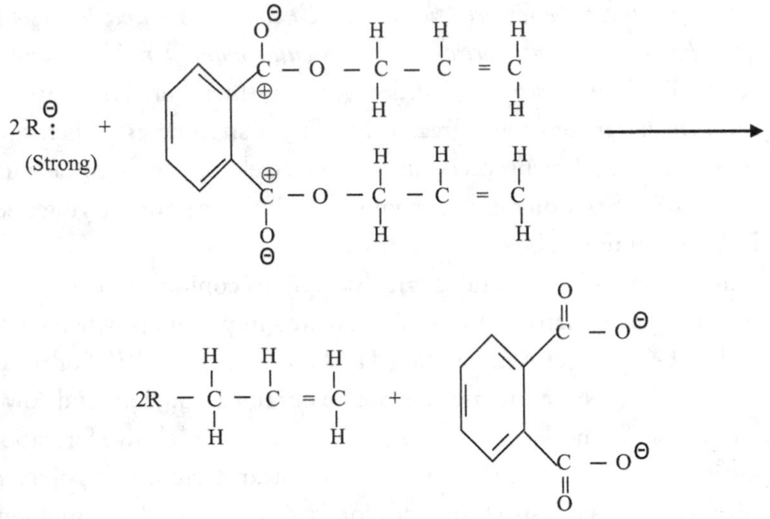

$$2\,R\overset{\ominus}{:}\quad + \qquad\qquad\qquad\qquad\qquad \longrightarrow$$

(Strong)

$$2R - \overset{\overset{\displaystyle H}{|}}{\underset{\underset{\displaystyle H}{|}}{C}} - \overset{\overset{\displaystyle H}{|}}{\underset{\underset{\displaystyle H}{|}}{C}} = \overset{\overset{\displaystyle H}{|}}{\underset{}{C}} \quad + $$

Transfer species of 1st kind of first type   15.7a

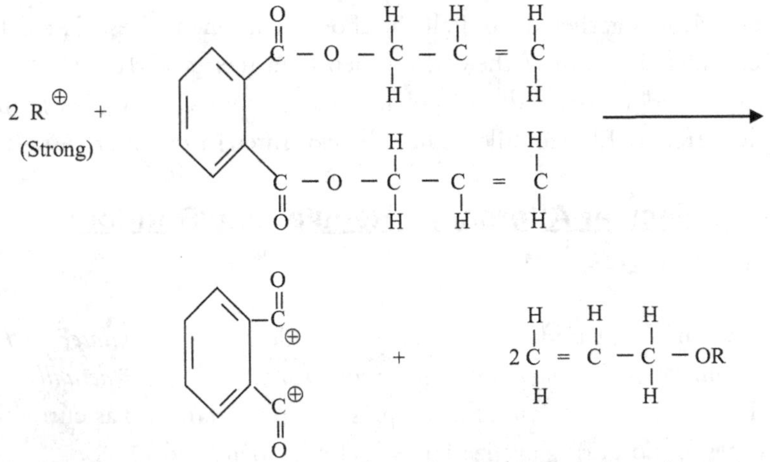

$$2\,R\overset{\oplus}{}\quad +\qquad\qquad\qquad\qquad\qquad \longrightarrow$$

(Strong)

$$+\qquad 2\,\overset{\overset{\displaystyle H}{|}}{\underset{\underset{\displaystyle H}{|}}{C}} = \overset{\overset{\displaystyle H}{|}}{\underset{}{C}} - \overset{\overset{\displaystyle H}{|}}{\underset{\underset{\displaystyle H}{|}}{C}} - OR$$

Transfer species of 2nd 1st kind of second type   15.7b

Notice that the radical pulling groups are ortho- placed on the ring, making the ring not resonance stabilized. If they were para-placed cyclization may not be favored because of the size of the ring. Note that the centers are activated one at a time, not indeed as shown above. The center activated above in the first equation is the wrong center. It was used exploratively. Chargedly, the monomer cannot be polymerized. In view of the fact that the C = C center is carrying radical-pushing allylic groups, chargedly the C = C centers cannot be activated. With activation of only C = O center, it is important to note also the types of transfer species involved "anionically and cationically". Only the radical route will favor the use of the compound as a monomer.

Free-radically, the C = C activation center is strong nucleophilically, but less so than the C = O center which cannot also favor any radical route, due to transfer species of the first kind of first type and second first kind of second type.

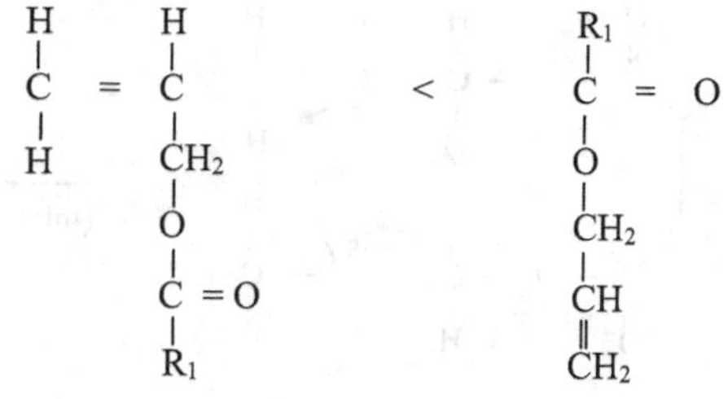

Order of nucleophilicity   15.8

Hence the followings are valid radically.

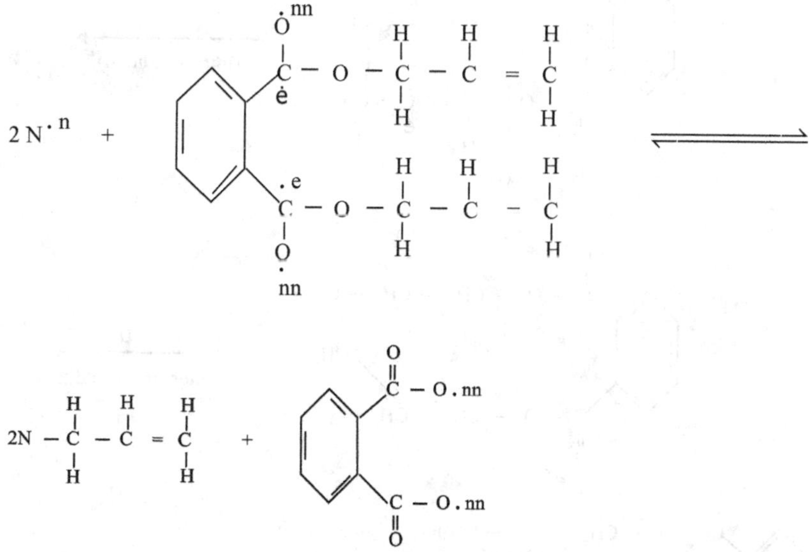

                                                                                             15.9

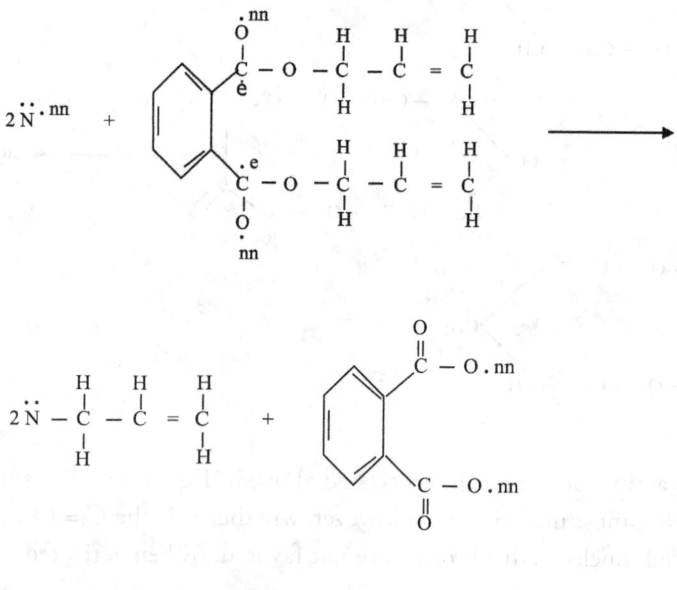

                                                                                             15.10

Indeed, the externally located C = C center being less nucleophilic, is the center first activated. The C = O is only first activated when carried by an Electrophile (i.e., Male) in the presence of an electro-free-radical or positive charge or cation. The C = C (Y) center in the benzene ring is the one adjacently located to the C = O center. The C = C center externally located is not a Y center since there is no C = O adjacently located to it.

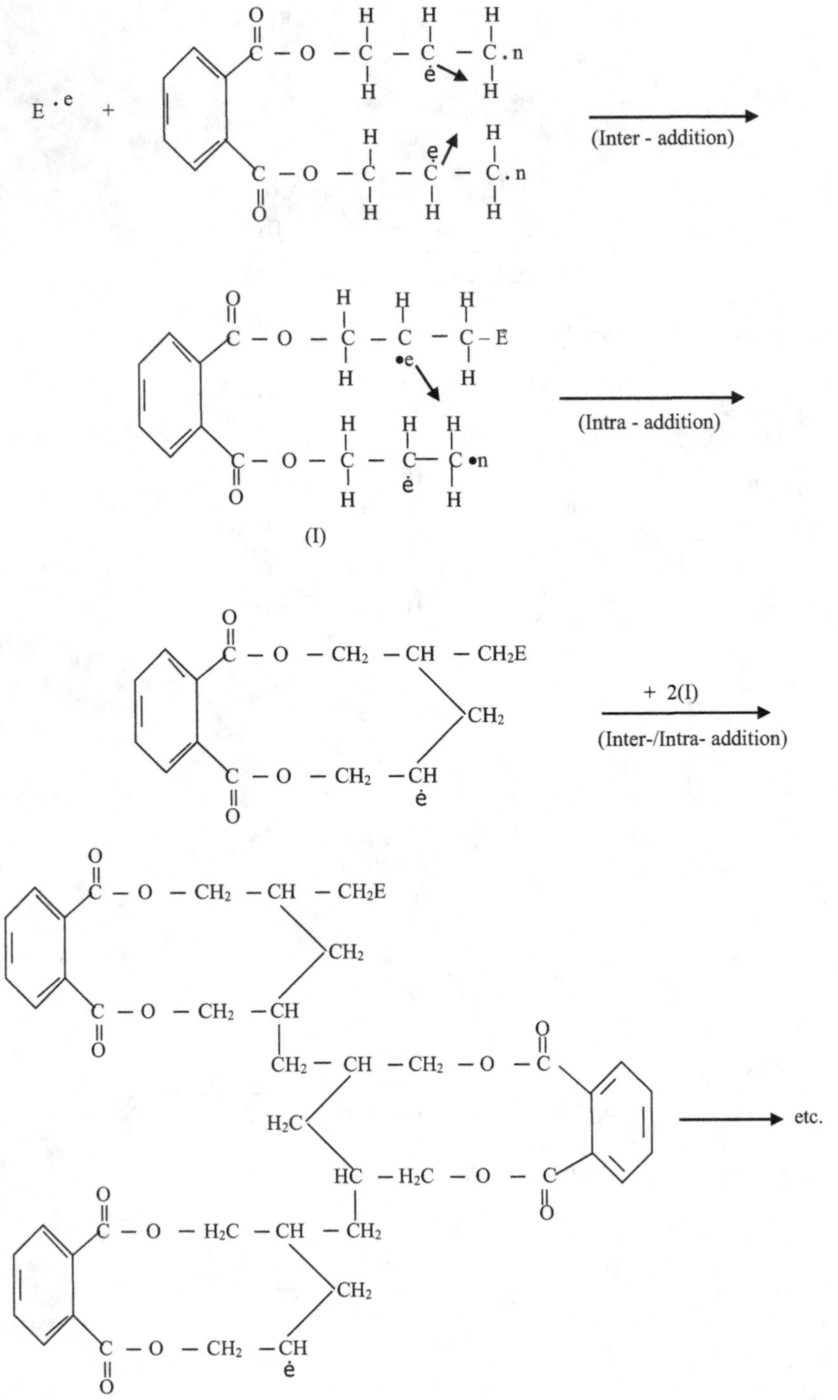

15.11

It can be observed that the activation center first activated above in Equations 15.9 and 15.10 is the C = O which should not be, being most nucleophilic. However, whether it is the C = O center or the allylic C = C center that is activated, nucleo-radical routes are not favored. When activated, it is the externally located C = C centers that are first activated being the least nucleophilic of all the centers – 2C = O, 2Y, 2 C = C and one other in the ring. With no transfer species for that center electro-free-radically, hence the reactions above. The chain is growing vertically instead of horizontally, although one can change the orientation to make it grow horizontally. One can imagine how sterically hindered the propagation step is, in view of the size of the ring. It is when there are no C = C centers present in the system or in the vicinity, that the ⬭O functional centers (and C =O centers) along the chain gets activated electro-free-radically, to favor presence of cross-linked or branched networks if –

(i) Paired initiating centers are present in the system during polymerization. For example, if HCl is present, such centers can be created.

(ii) The eleven-membered ring is strained enough to be opened via the ⬭O functional center in the ring if the operating conditions are higher than expected.

In the absence of charged-paired initiator and other factors, no cross-linking or branching from within the chain can be favored via opening of a ring. As a matter of fact, the ring cannot be opened, because the weakest nucleophilic center is the C = C bond in benzene ring distantly located to the unsaturated eleven-membered ring. Secondly, the growing polymer chain above has transfer species of the first kind of the sixth (originally first) type to reject, since it is the same (first type) involved in the opposite route when the same activation center is involved.

Thus, the presence of unsaturated electrophilic eleven-membered rings vertically placed alternatingly along the chain is only favored when –

(i) The electro-free-radical initiator is initially strong.

(ii) The two activation centers which are part of substituent groups adjacently located, come close together during conformational changes taking place during polymerization.

When (i) and (ii) are not favored, the followings are possibilities.

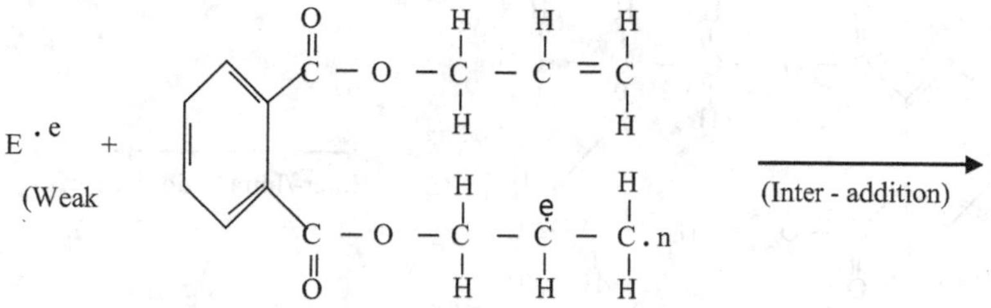

(I)

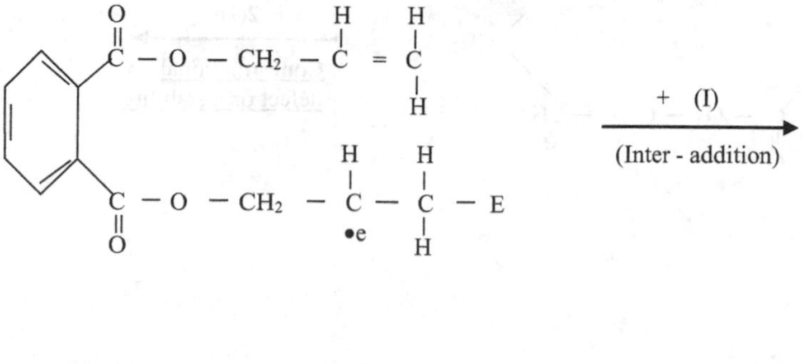

519

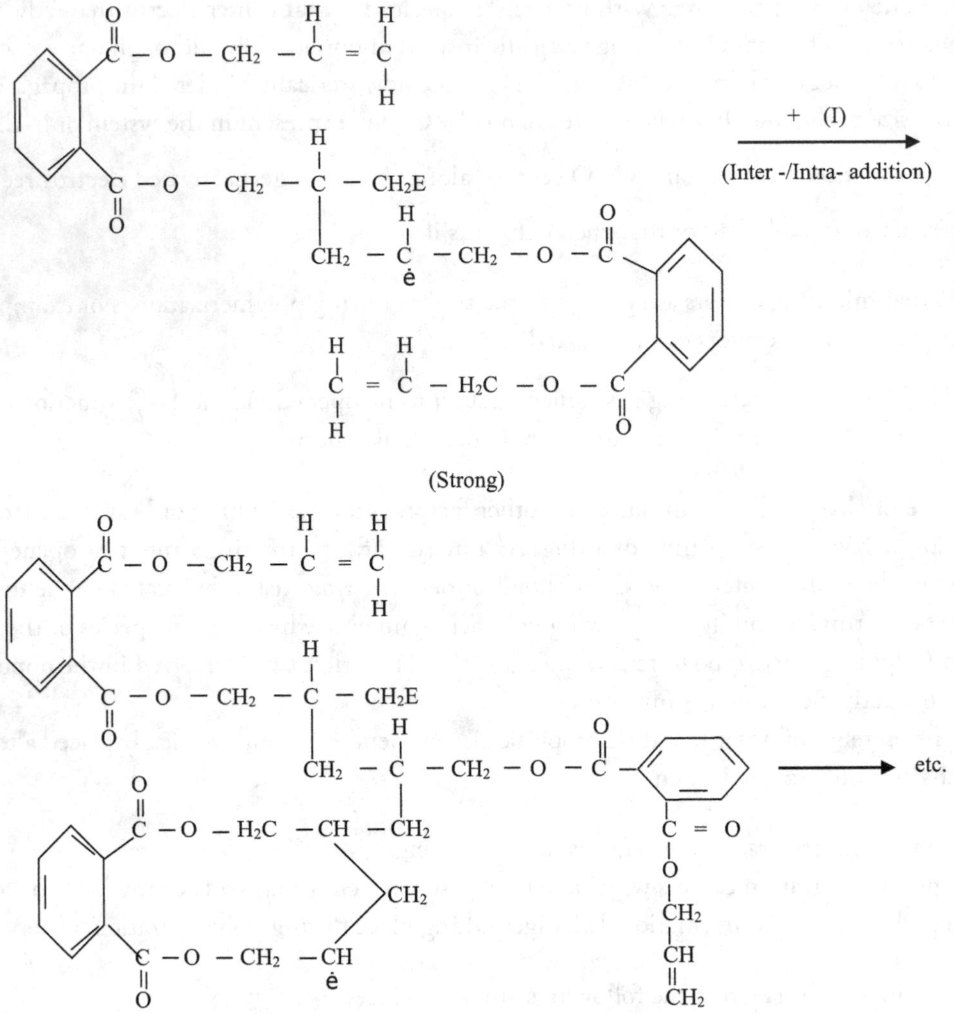

(Strong)

etc.

15.12a

$E \cdot^e$ +

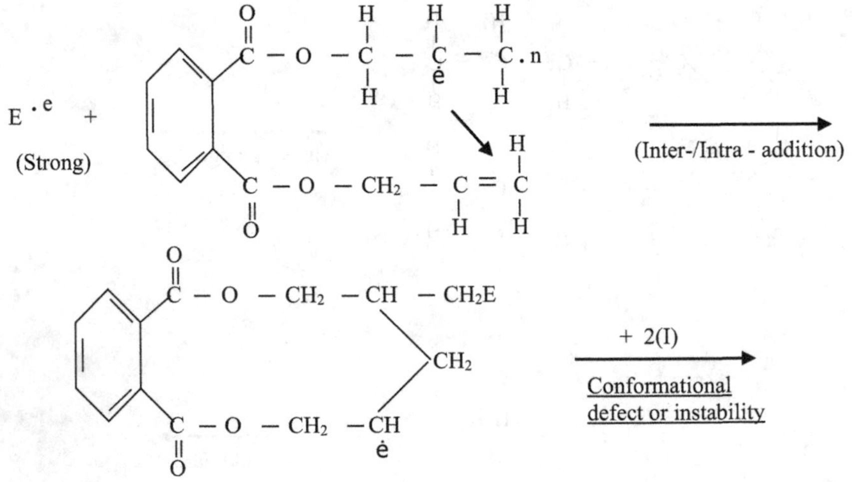

(Strong)

(Inter-/Intra - addition)

+ 2(I)

Conformational
defect or instability

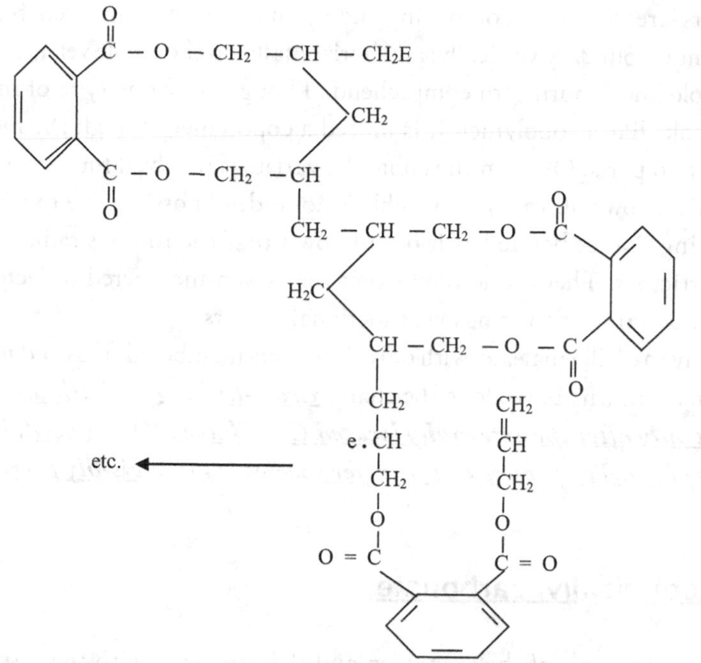

15.12b

It is important to note that it is only when one external activation center has been activated and the initiator adds to activate the second center, that the eleven-membered ring is first formed to give the active growing center. When the two activation centers do not come close enough, presence of a ring may not be favored after inter addition takes place. On the other hand, when weak initiators are initially involved, presence of rings will also not be favored. Hence, the C = C centers present along the chain under such conditions when rings are not formed, will favor their use for cross-linking reactions between chains or for branching. However, this is not where cross-linked structures are obtained. The vertical growth can also be horizontally placed. The externally located C= C bond in the benzene ring, the weakest nucleophilic center, gets activated to add to another benzene ring to form fused six-four-six-membered rings **via intra-molecular addition** horizontally along a chain. When vertically placed, they become the cross-links. The four-membered fused ring favors existence, because the groups carried apart from H, are ringed radical-pulling groups which cannot be abstracted. If the groups fused to the four-membered ring were radical-pushing groups, the ring will not exist as will be shown downstream.

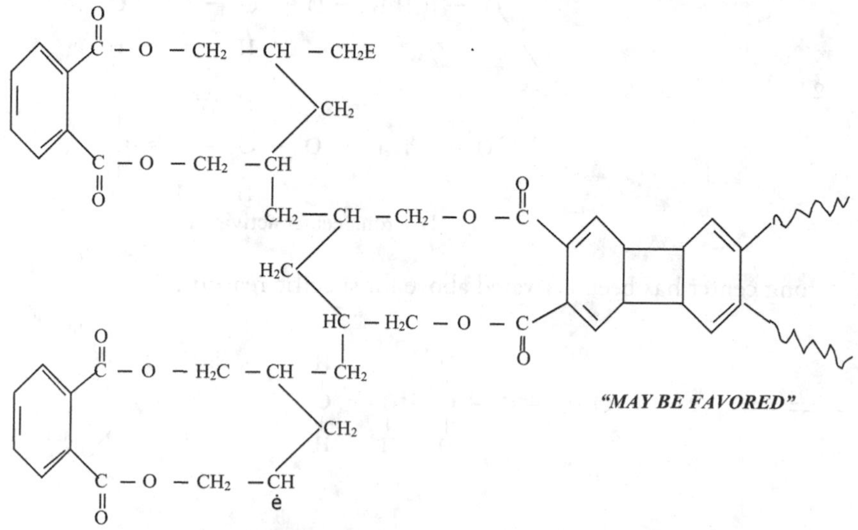

"MAY BE FAVORED"

15.13

Note that the monomer units are complex, comprising methylene, eleven-membered fused with six membered ring with one common boundary vertically; and horizontally on the whole vertical or horizontal network. The network is simple and beautiful to comprehend. Though only one type of monomer was used, the polymer obtained looks like a copolymer. It is indeed a copolymer, though the monomer unit is the whole monomer, but in two parts. One can thus note the variables involved for formation of cross-links for this monomer which as shown above to be possible is not indeed possible due to transfer species inside the eleven membered ring. Never before has it been known that the route is radical in character and electro-free-radical in particular. There is no doubt that the eleven-membered di-lactone ring will never have the MRSE to favor opening of the rings via functional centers.

The same as above will apply to diallyl maleate, with only the eleven-membered rings and methethylene, along the chain with no fusion as shown above along the chain. ***Cross-linking for both cases takes place not during propagation, but only after the externally located C = C have all been used. However, for the two cases, it cannot take place. The presence of transfer species is more visibly present in diallyl maleate than in the case above.***

15.1.2. Diethylene Glycol bisallyl carbonate

Based on the charged activation state of the monomer, and the similarity with the case above, none of the charged routes will favor cyclization of the monomer.

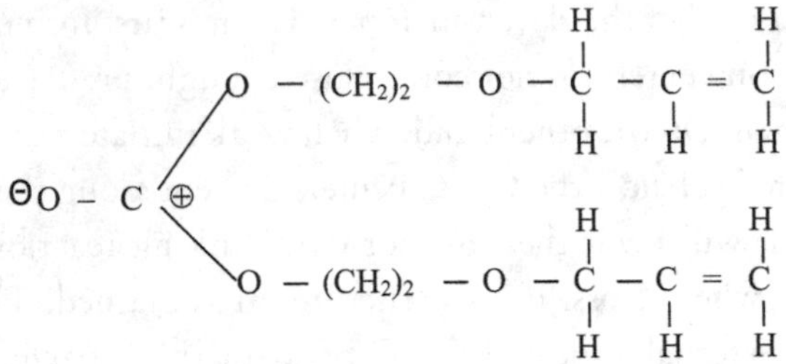

$$\ominus O - C \overset{\oplus}{\underset{\Big\backslash}{\Big/}} \quad \begin{array}{l} O - (CH_2)_2 - O - \overset{\overset{\displaystyle H}{|}}{\underset{\underset{\displaystyle H}{|}}{C}} - \overset{\overset{\displaystyle H}{|}}{\underset{\underset{\displaystyle H}{|}}{C}} = \overset{\overset{\displaystyle H}{|}}{\underset{\underset{\displaystyle H}{|}}{C}} \\[2em] O - (CH_2)_2 - O - \overset{\overset{\displaystyle H}{|}}{\underset{\underset{\displaystyle H}{|}}{C}} - \overset{\overset{\displaystyle H}{|}}{\underset{\underset{\displaystyle H}{|}}{C}} = \overset{\overset{\displaystyle H}{|}}{\underset{\underset{\displaystyle H}{|}}{C}} \end{array} \qquad 15.14$$

Note that, in general it has already long been established that C = O center is more Nucleophilic than C = C center, noting that the monomer here is a Nucleophile.

Free-radically, the followings are obtained.

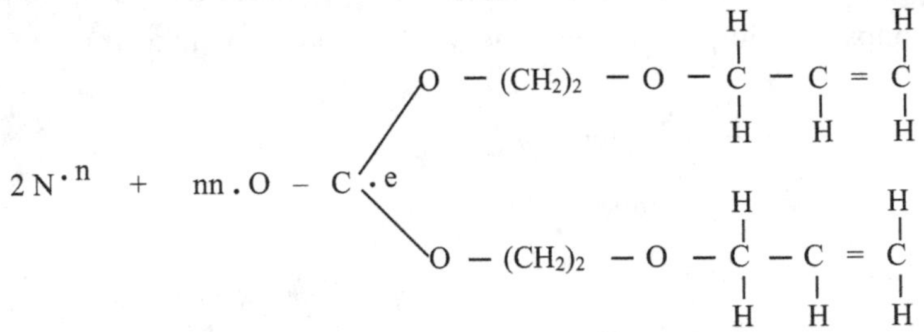

$$2 N \cdot^n + nn \cdot O - C \overset{\cdot e}{\underset{\Big\backslash}{\Big/}} \quad \begin{array}{l} O - (CH_2)_2 - O - \overset{\overset{\displaystyle H}{|}}{\underset{\underset{\displaystyle H}{|}}{C}} - \overset{\overset{\displaystyle H}{|}}{\underset{\underset{\displaystyle H}{|}}{C}} = \overset{\overset{\displaystyle H}{|}}{\underset{\underset{\displaystyle H}{|}}{C}} \\[2em] O - (CH_2)_2 - O - \overset{\overset{\displaystyle H}{|}}{\underset{\underset{\displaystyle H}{|}}{C}} - \overset{\overset{\displaystyle H}{|}}{\underset{\underset{\displaystyle H}{|}}{C}} = \overset{\overset{\displaystyle H}{|}}{\underset{\underset{\displaystyle H}{|}}{C}} \end{array}$$

(I) Wrong center activated.

Notice that the wrong center has been activated above for specific reasons.

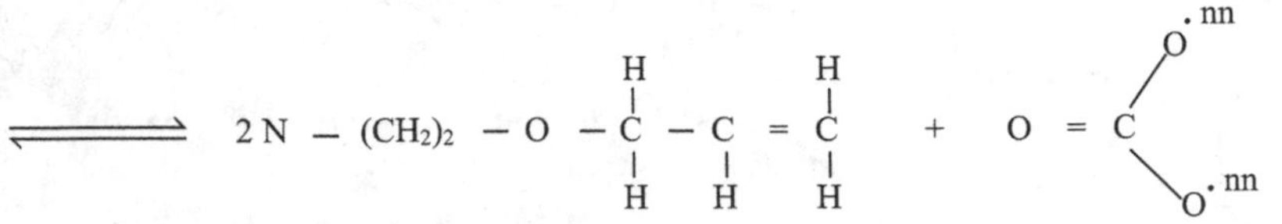

$$\Longleftarrow \quad 2 N - (CH_2)_2 - O - \overset{\overset{\displaystyle H}{|}}{\underset{\underset{\displaystyle H}{|}}{C}} - \overset{\overset{\displaystyle H}{|}}{\underset{\underset{\displaystyle H}{|}}{C}} = \overset{\overset{\displaystyle H}{|}}{\underset{\underset{\displaystyle H}{|}}{C}} \quad + \quad O = C \overset{\cdot nn}{\underset{\cdot nn}{\overset{\Big/ O}{\Big\backslash O}}} \qquad 15.15$$

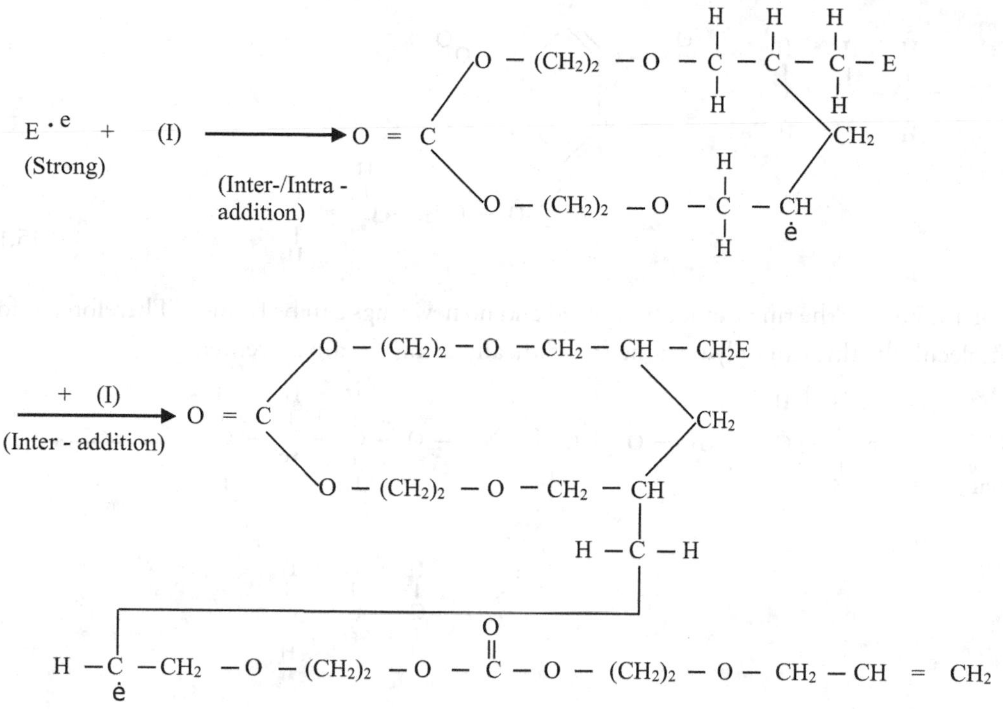

$$
\begin{array}{c}
\text{E} \cdot^e \ + \ \text{(I)} \ \xrightarrow[\text{addition)}]{\text{(Inter-/Intra -}} \\
\text{(Strong)}
\end{array}
$$

$$
\xrightarrow[\text{(Inter - addition)}]{+ \ \text{(I)}}
$$

15.16

The analysis is similar to that of diallyl phthalate in almost all respects, except that fourteen - membered rings are favored here along with methylene and the possibility of having cross-links via C = O center placed on the ring is remote and in fact impossible, because of the presence of four functional centers in the ring. For the size of the ring, it is far less strained, making it difficult to open. The chain above can grow and form rings in the absence of conformational instability and steric limitations, but cannot form cross-links or branches, unless when the rings are randomly placed, that which takes place when the strength of the initiator is weak.

15.1.3. Triallyl Cyanurate

On the compound (IV) of Equation 15.5, only the activation centers in the ring can be involved chargedly, since the C = C centers allylic in character can only be activated radically. The ring is not resonance stabilized chargedly. When attacked anionically, the followings are obtained.

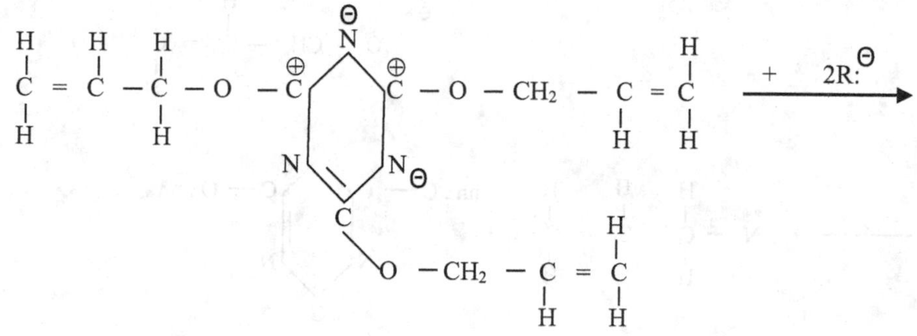

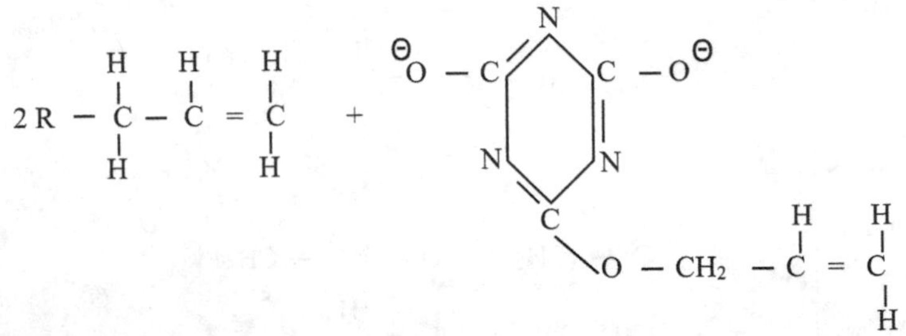

$$2R - \underset{\underset{H}{|}}{\overset{\overset{H}{|}}{C}} - \underset{\underset{H}{|}}{\overset{\overset{H}{|}}{C}} = \underset{\underset{H}{|}}{\overset{\overset{H}{|}}{C}} \quad + \quad \cdots$$

15.17

Cationically or anionically the ring cannot be opened and no new rings can be formed. Therefore, to form rings intra- molecularly, this can only be done free-radically using the allylic centers.

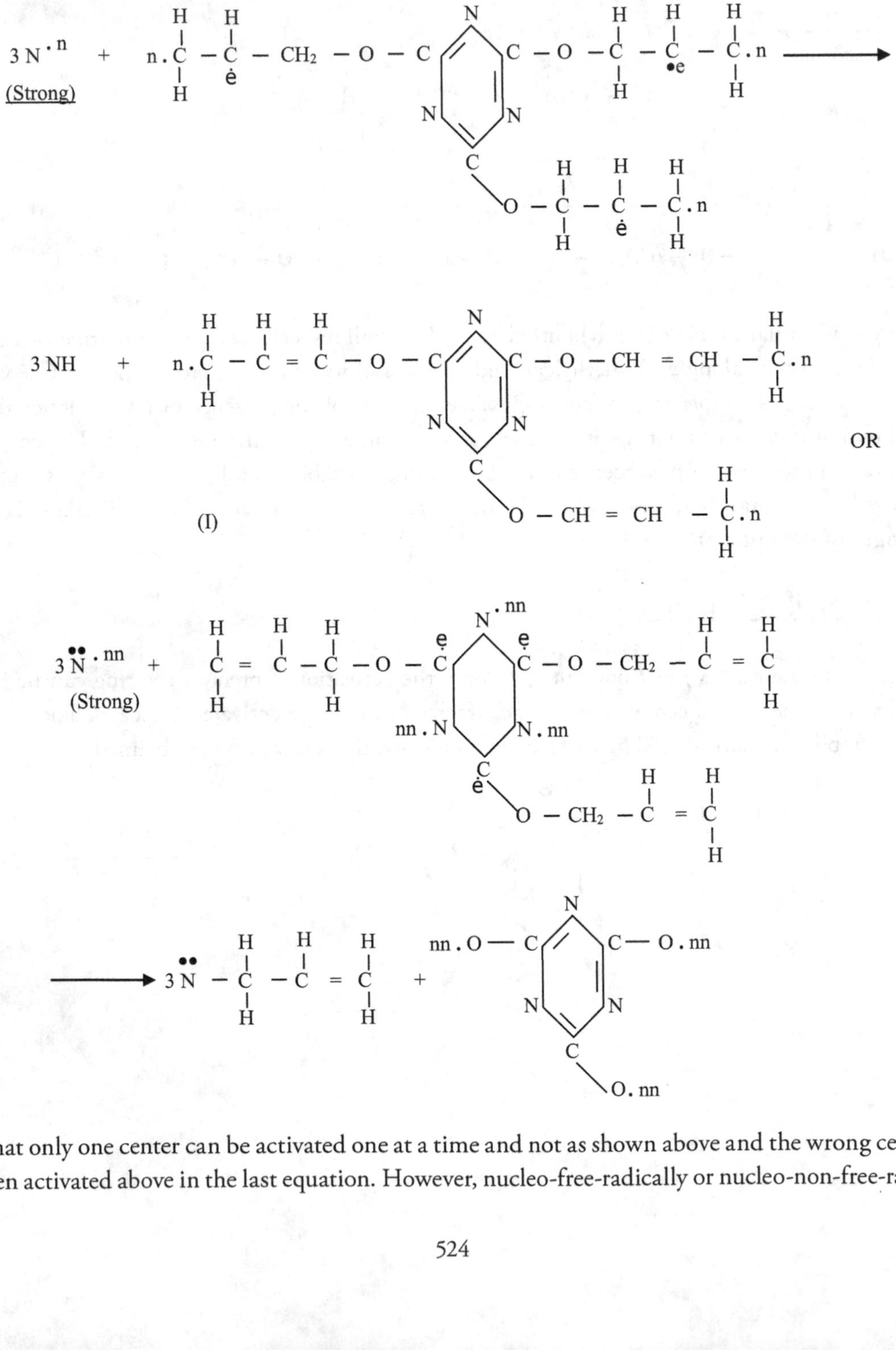

$$3N\cdot^{n} \quad + \quad n\cdot\overset{\overset{H}{|}}{\underset{\underset{H}{|}}{C}} - \underset{\dot{e}}{\overset{\overset{H}{|}}{C}} - CH_2 - O - C \cdots \longrightarrow$$

(Strong)

$$3NH \quad + \quad n\cdot\overset{\overset{H}{|}}{\underset{\underset{H}{|}}{C}} - \overset{\overset{H}{|}}{C} = \overset{\overset{H}{|}}{C} - O - C \cdots \qquad OR$$

(I)

$$3\overset{\bullet\bullet}{N}\cdot^{nn} \quad + \quad \overset{\overset{H}{|}}{\underset{\underset{H}{|}}{C}} = \overset{\overset{H}{|}}{C} - \overset{\overset{H}{|}}{C} - O - \overset{\dot{e}}{C} \cdots$$

(Strong)

$$\longrightarrow \quad 3\overset{\bullet\bullet}{N} - \overset{\overset{H}{|}}{\underset{\underset{H}{|}}{C}} - \overset{\overset{H}{|}}{C} = \overset{\overset{H}{|}}{C} \quad + \quad \cdots$$

15.18

Note that only one center can be activated one at a time and not as shown above and the wrong center has also been activated above in the last equation. However, nucleo-free-radically or nucleo-non-free-radically, initiation cannot be favored. Nevertheless, it is (I) that is first activated one at a time, an indication that the following is valid for the monomer only radically.

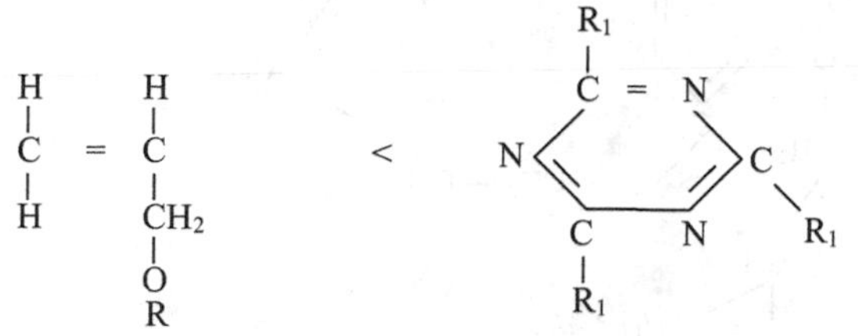

(where $R_1 \equiv OCH_2 - CH = CH_2$ and R is the R_2 in (II))

(I)   (II) $(R_1 - R_2)$

Order of Nucleophilicity (radically)   15.19

Electro-free-radically, it will be the C = C activation center that will largely be involved as shown below.

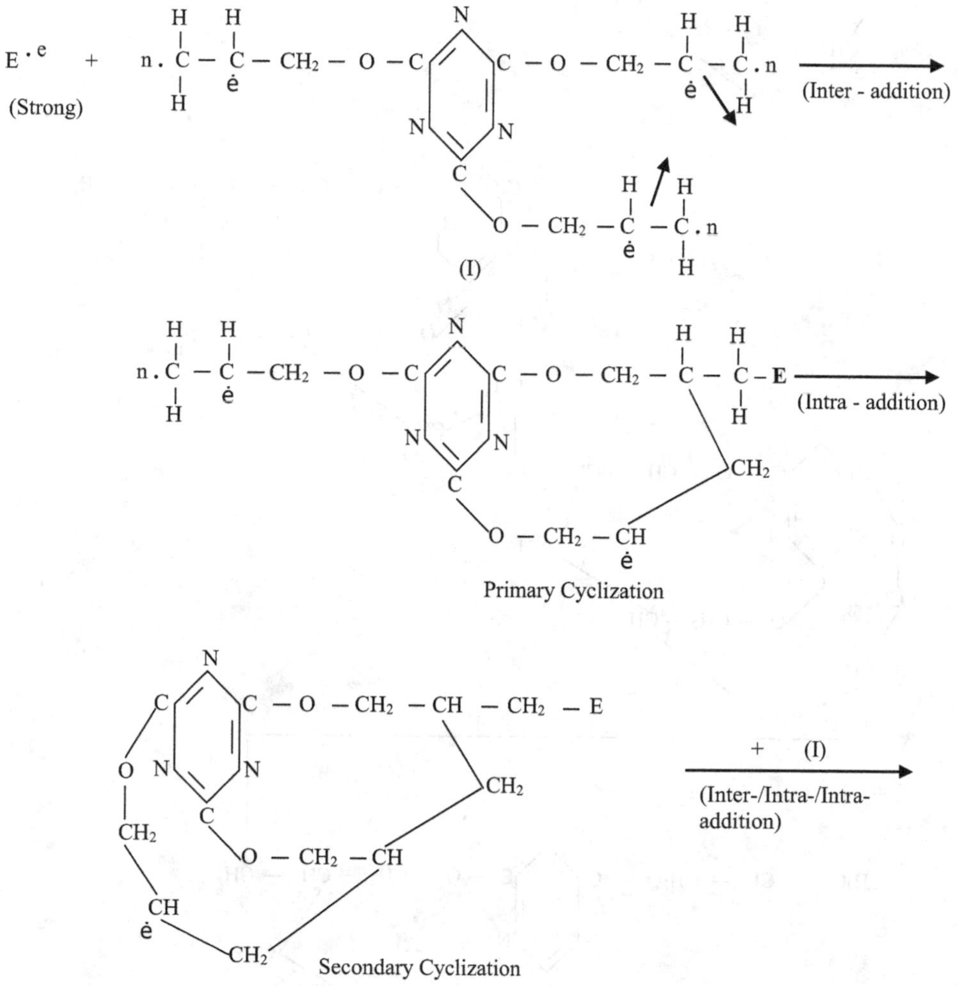

(Inter - addition)

(I)

(Intra - addition)

Primary Cyclization

+   (I)

(Inter-/Intra-/Intra-addition)

Secondary Cyclization

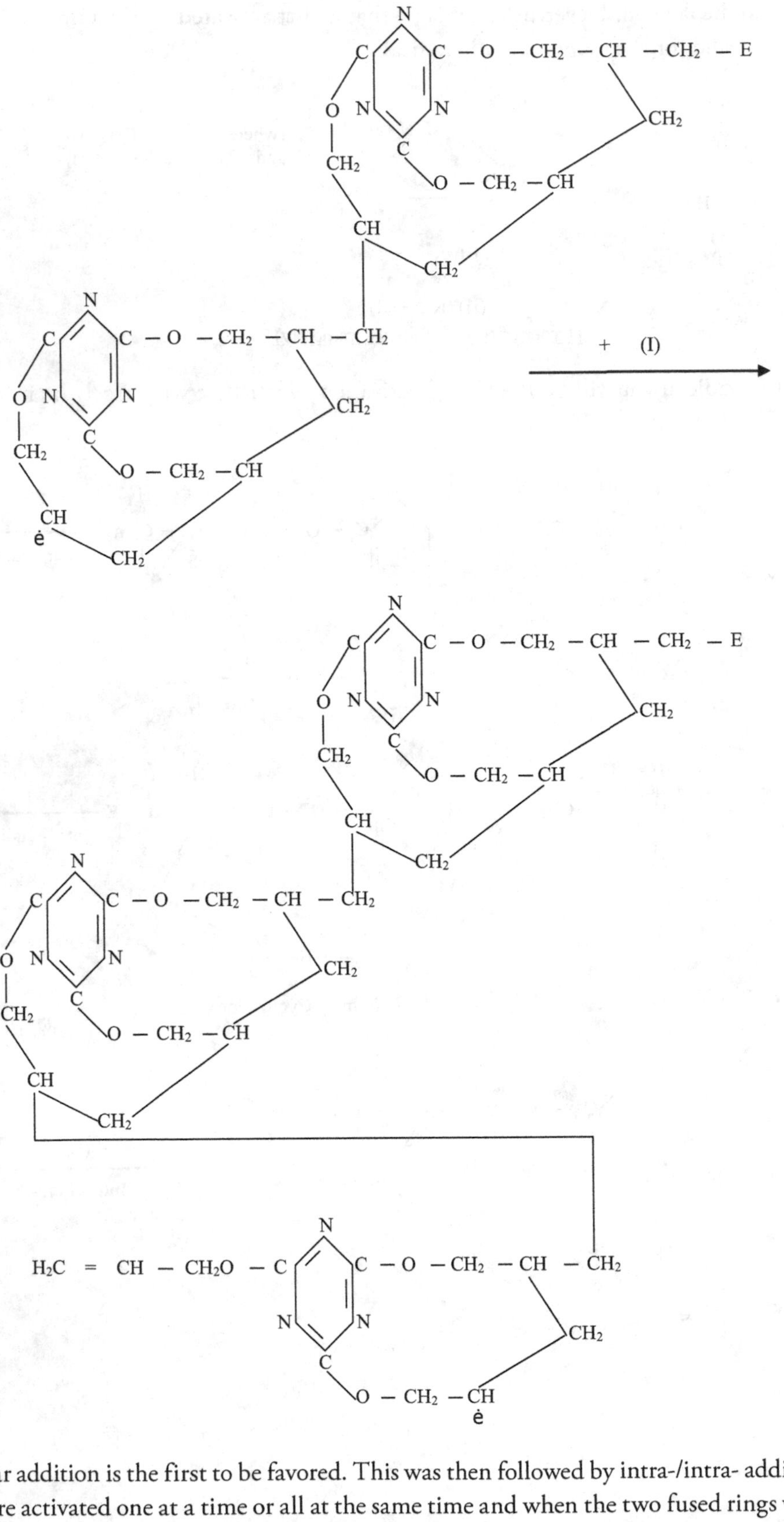

$$+ \quad (I) \longrightarrow$$

15.20

Inter-molecular addition is the first to be favored. This was then followed by intra-/intra- addition wherein the centers were activated one at a time or all at the same time and when the two fused rings were formed, an active center was left behind to continue the growth of the chain. Since the externally located activation

526

centers are of the same capacity, they could all be activated at the same time. All these take place by Combination mechanism. They can never take place via Backward addition, since the initiator cannot be held in Equilibrium state of Existence.

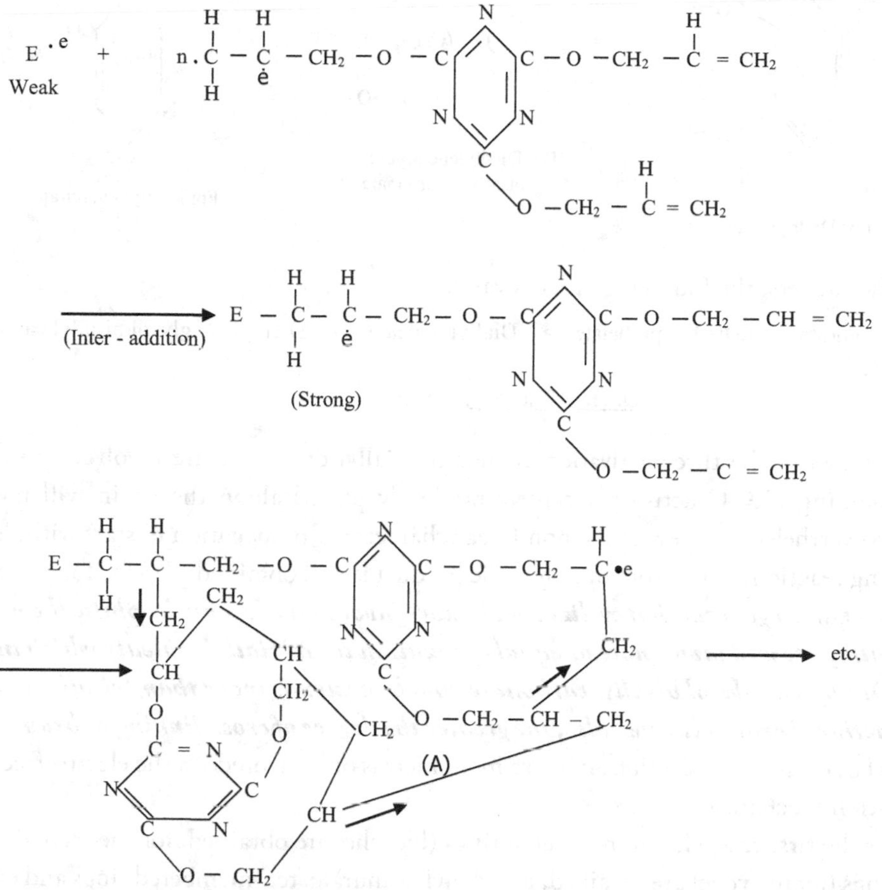

$$
\begin{aligned}
&E \cdot^e \ + \ n \cdot \overset{H}{\underset{H}{C}} - \overset{H}{\underset{\dot{e}}{C}} - CH_2 - O - C \underset{\underset{C}{N}}{\overset{N}{\diagdown}} C - O - CH_2 - \overset{H}{C} = CH_2 \\
&\text{Weak} \qquad\qquad\qquad\qquad\qquad\qquad\qquad\qquad\qquad\qquad\qquad\qquad O - CH_2 - \overset{H}{C} = CH_2
\end{aligned}
$$

$$
\xrightarrow[\text{(Inter - addition)}]{} E - \overset{H}{\underset{H}{C}} - \overset{H}{\underset{\dot{e}}{C}} - CH_2 - O - C \underset{\underset{C}{N}}{\overset{N}{\diagdown}} C - O - CH_2 - CH = CH_2
$$

(Strong)

$$O - CH_2 - C = CH_2$$

(A)

etc.

15.21

One can observe the strong non-linear character of addition of the monomers. After inter-molecular addition has taken place when the monomer is activated, this is then followed by another inter-molecular addition to another monomer where intra-molecular addition takes place in part or in full. This is just one of the possibilities that may arise due to conformational instability, thus leading to randomness. Imagine if what is shown above takes place, i.e. (A), that in which the electro-free-radical from the second monomer diffuses to the unused end of the first monomer, the network obtained would contain 16-membered ring. At the end, the chain is made to grow linearly. Above from two monomers, six fused rings were formed. In general, inter-molecular addition takes place before intra- molecular addition, when initiators are involved. ***When alternating placement of rings cannot be favored either due to conformational instability or the use of weak initiators or steric limitations, then presence of crosslinks or branches cannot be ruled out.***

Looking at the three ring forming monomers considered so far, there is something common with them. Their central molecular species carry two or three similar unsaturated groups. With only one group, no cyclization can take place. All the groups carried are terminally located, and allylic in character ($H_2C = CH - CH_2 -$). These are like STEP monomers where with only one functional group, no Step polymerization can take place. For these members of monomers, the carriers of these groups is what determines the character of the monomer. Diallyl phthalate and maleate are Electrophiles, yet strongly nucleophilic. All the others are Nucleophiles and their carriers are the same for the Electrophiles as shown below.

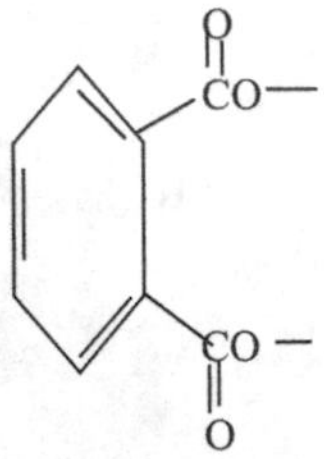

For Diethyl phthalate 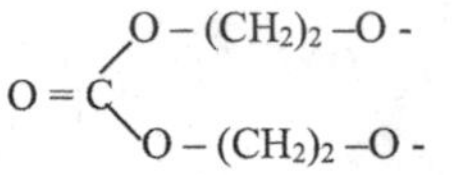 ; For Diethylene glycol bisallyl carbonate 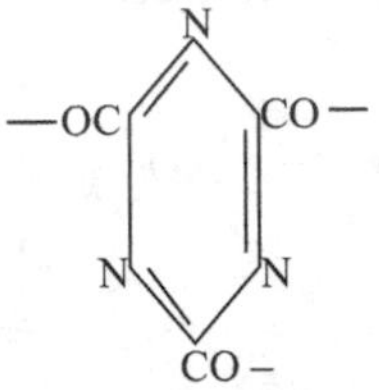 ; ForTriallyl cyanurate

15.22

Looking at the carriers, the following order is valid.

Tri-allyl Cyanurate > Diethyl phthalate > Diallyl maleate > Diethylene glycol bisallyl carbonate

<u>Order of Nucleophilicity</u>   15.23

If only one or two of the three activation centers in triallyl cyanurate are involved most of the time, then the remaining C = C activation centers randomly placed along the chain will favor cross-link formations. Nevertheless, in view of the non-linear character of propagation to start with, in the absence of cross-linking reactions, a network of polymeric product is still obtained. For the four monomers so far considered, it seems *in general that cyclization is more favored the less nucleophilic the monomer; that is, Diallyl phthalate will favor more or equal cyclization than Diallyl maleate which in turn will be more than Diethylene glycol bisallyl carbonate which in turn is more than tri-allyl cyanurate. The more the functional groups carried also, the greater the degree of cross-linking or branch formations.* In general so far, cyclization reactions in these monomers is only favored via the electro-free-radical route via Combination mechanism.

While for the first case, eleven-membered rings (bicyclic) are obtained, for the second case fourteen-membered rings (monocyclic) are obtained, for triallyl cyanurate, ten-membered rings and even higher are obtained - either bicyclic or tricyclic or more, depending of the strength of the initiator. In the absence of conformational instability, alternating placement of rings and methylene will strongly be favored during propagation, since the strength of the active center is increasing for every inter - molecular addition of monomer. It is important to note that during intra-molecular addition, it is the electro-free-radical center that diffuses to the other activation center all the time.

15.1.4. Divinyl Benzenes

Based on the para-placements of two similar alkenyl or monoene groups on benzene ring, divinyl benzene also a strong nucleophile has been shown to favor a unique resonance stabilization phenomenon, that in which one external group is not involved. Due to electrodynamic forces of repulsion, charged activation of the externally located activation centers at the same time is not possible. Here, since the activation centers are of the same capacity, can they be activated at the same time? Due to electrodynamic forces of repulsion, they cannot be activated chargedly if resonance stabilized. Yes, they can be activated chargedly, since no resonance stabilization will take place. This will be seen clearly when laws are being stated downstream. As a food for thought, consider the following cases.

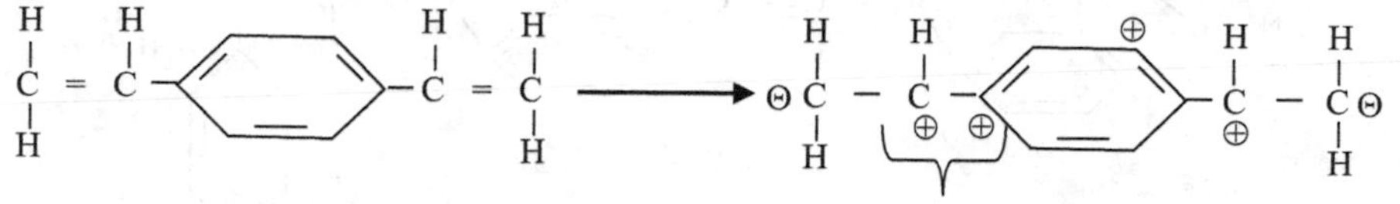

(A)
Electrodynamic
forces of repulsion

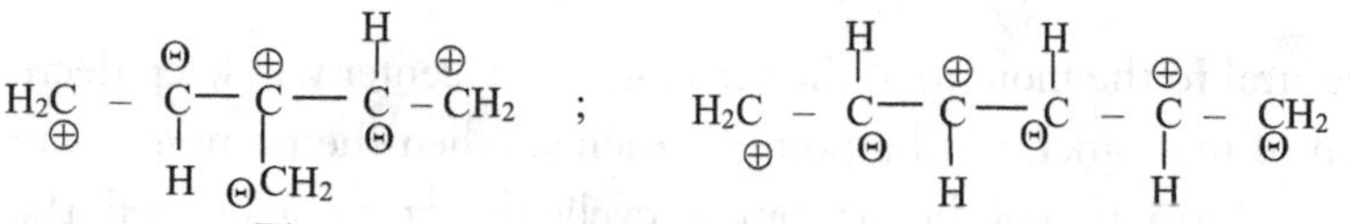

(B)   (C)

No Repulsion   No Repulsion   15.24

(B) is like (A), but different. (B) is not like (C). (C) is almost like (A), but still with differences. Above, it was assumed that resonance stabilization is favored chargedly.

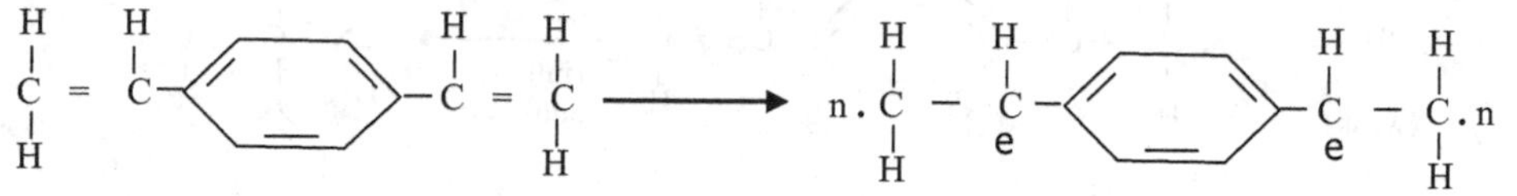

Favored State   15.25

However, cyclization reaction will only be considered herein free-radically.

Free - radically, the followings are obtained.

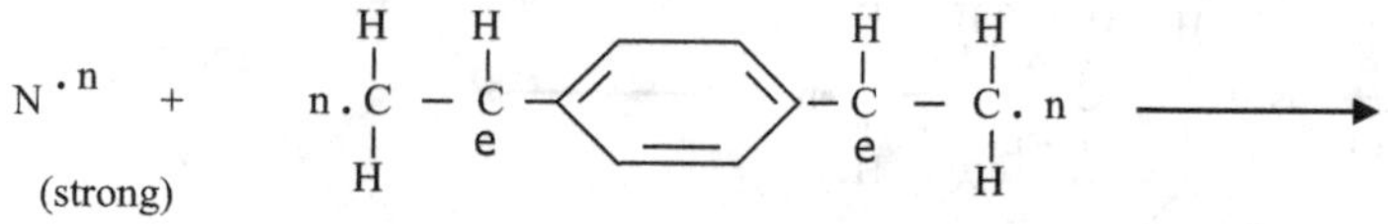

(I)

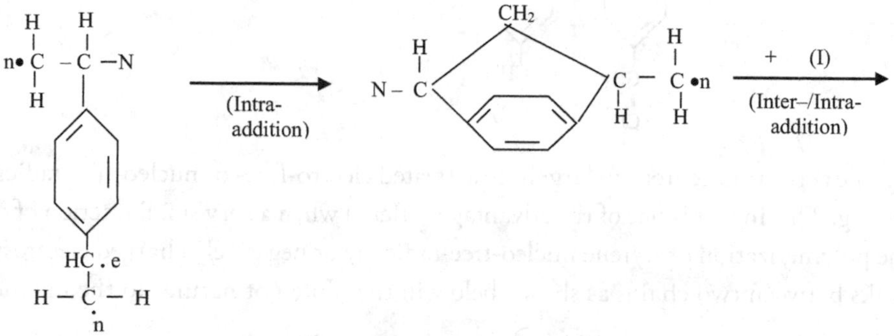

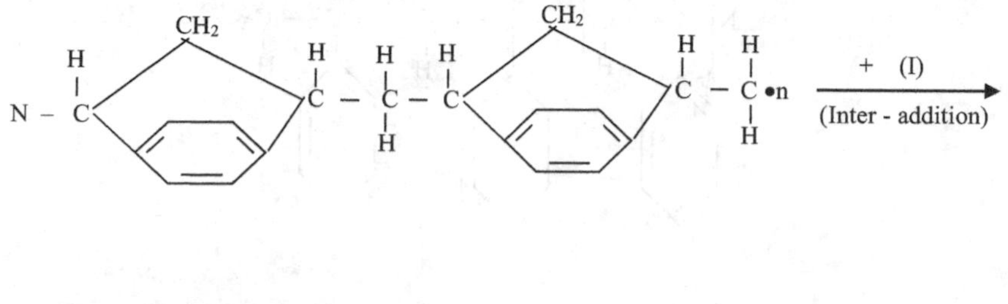

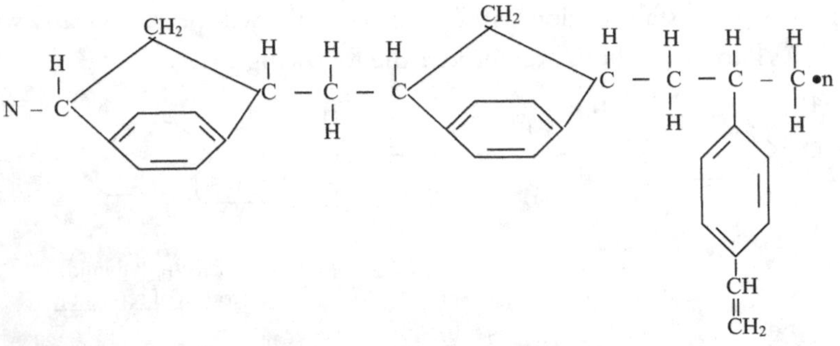

Less intra - addition

15.26

Since the route is not natural to the monomer, the active growing center will keep decreasing in strength with increasing addition of monomer, until a point is reached when the monomer can add no more to the chain. Based on the conformation of the monomers, cyclic rings may be favored, the limitation being the para-placement of the groups (i.e., distance between the centers and the route in particular. Inter-molecular addition will largely be favored with little presence of rings along the chain nucleo-free-radically.

However, it is generally believed that, in view of the strong nucleophilicity of the activation

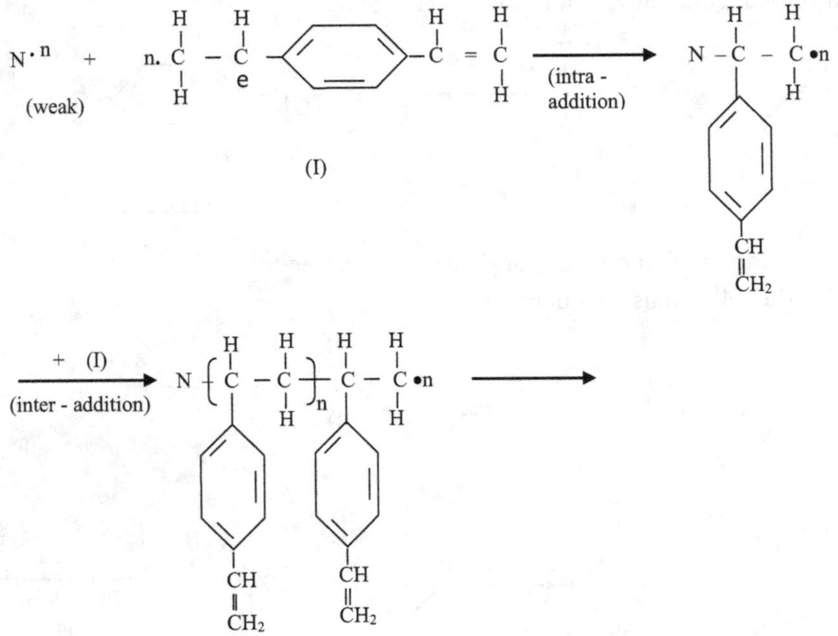

15.27

centers, only one activation center can largely be activated electro-free- or nucleo-free-radically, whether it is weak or strong. This in fact is one of the advantages offered when a very small fraction of divinylbenzene is used in the polymerization of styrene nucleo-free-radically or negatively charged electrostatic centers to give cross-links between two chains as shown below in the route not natural to the monomers.

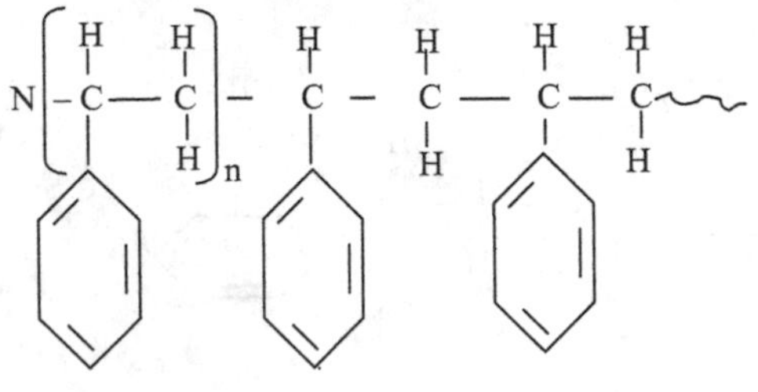

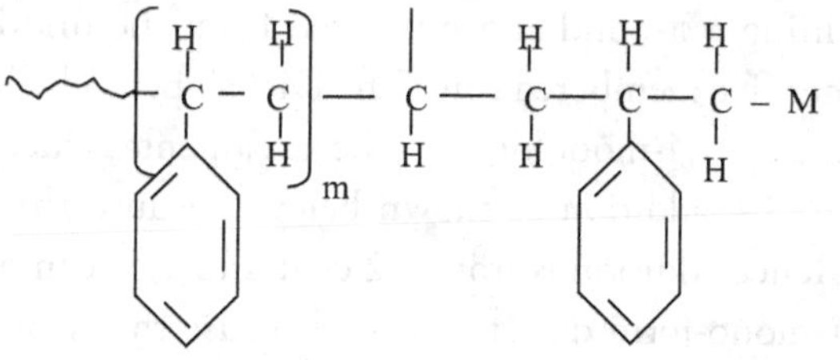

Cross-linked Polystyrene/Divinylbenzene   15.28a

This can only be done in a specific order

Electro-free-radically, the route natural to the monomers, the followings is obtained.

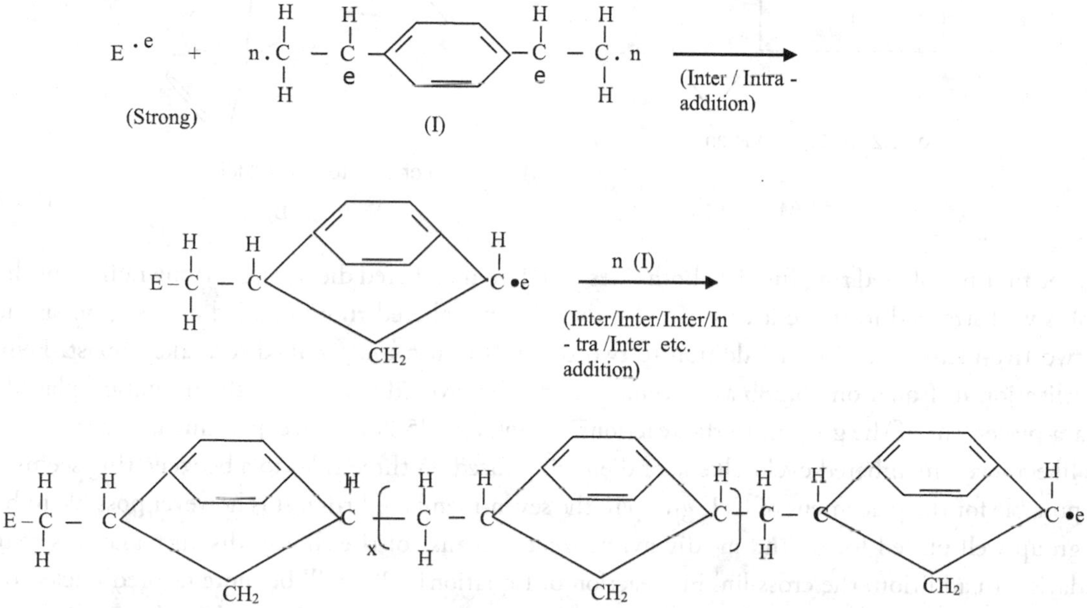

$$E \cdot^e \quad + \quad n. \underset{H}{\overset{H}{C}} - \underset{e}{C} \text{—} \langle \rangle \text{—} \underset{H}{\overset{H}{C}} - \underset{H}{\overset{H}{C}}. n \quad \xrightarrow{\quad}$$

(Strong)   (I)   (Inter / Intra - addition)

$$E - \underset{H}{\overset{H}{C}} - \overset{H}{C} \langle\rangle \overset{H}{C} \bullet e \quad \xrightarrow{\quad n\ (I)\quad}$$

(Inter/Inter/Inter/In - tra /Inter etc. addition)

$$E - \underset{H}{\overset{H}{C}} - \overset{H}{C} \langle\rangle \overset{H}{C} \{ \underset{H}{\overset{H}{C}} - \overset{H}{C} \langle\rangle \overset{H}{C} \}_x \underset{H}{\overset{H}{C}} - \overset{H}{C} \langle\rangle \overset{H}{C} \bullet e$$

NOT FAVORED (WITH LIMITATIONS)   15.28b

If the fused rings can exist despite the limited amount of strain energy in the rings, then more rings will be favored, since the strength of the active center is increasing for every addition of monomer. Here we will have six-membered ring fused with a seven-membered ring with *three common boundaries*. There are rules guiding the number of boundaries that can exist between two rings when fused. For example, six-membered ring (Hexene) can be fused with a five- membered ring (Pentene) with three common boundaries and another one with one common boundary as shown below. Cyclopentadiene dimerizes slowly on standing to dicyclo-pentadiene, m.pt. 33^0C, which dissociates at its boiling point, 170^0C, to the monomer.

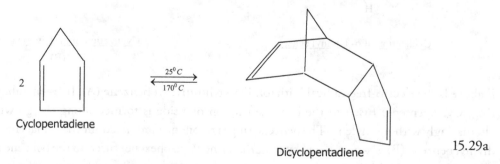

Cyclopentadiene   Dicyclopentadiene   15.29a

The six-membered ring containing a π-bond above is fused from the middle to a five-membered ring and on another side with another five membered ring. The Dicyclopentadiene obtained is 1,2-monoform addition of the cyclopentadiene to 1,4-monoform of the cyclopentadiene. It was not 1,2-addition to 1,2-addition or I,4- addition to 1,4-addition as shown below, because the rings would have been too strained to make them exist. Hence addition is from 1,2 center to 1,4- center. The order of either1,4- to 1,2-mono-forms or 1,2- to 1,4-mono-form does not arise, because the two monomers are the same and of the same nucleophilicity.

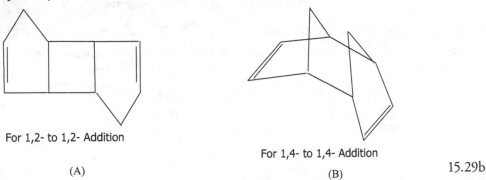

For 1,2- to 1,2- Addition

(A)

For 1,4- to 1,4- Addition

(B)

15.29b

In (A) the four-membered ring fused on both sides with five-membered diene ring (Strong radical-pushing group) is well strained to make it exist. In (B), the eight-membered ring bounded or fused on six sides with two five-membered ring in addition to two p-bonds is also too strained to make it exist. Hence, the cyclization in Equation 15.28b as it seems may not be favored apart from the limitation placed by the para-placement of the groups. In the reactions of Equation 15.28b above, placement of a ring which looks like a seven-membered cyclo-alkene or diene ring fused on three sides to a benzene ring seems not to be possible for the placement of CH_2 group in the seven-membered ring. It is however possible to have CH_2 group well placed for o- and m- divinyl benzene, because of the shorter distance and less shared boundaries. In addition, the cross-linking reaction of Equation15.28a will be more favored nucleo-free-radically than electro-free-radically since ***Cyclization seems to be more favored in the natural route than the unnatural route in general.***

While p-divinyl benzene has been shown not to favor cyclization, it is still very useful as a cross-linking agent. The o-divinyl benzene and m-divinyl benzene are well known to favor cyclization as shown below. This is only possible electro-free-radically.

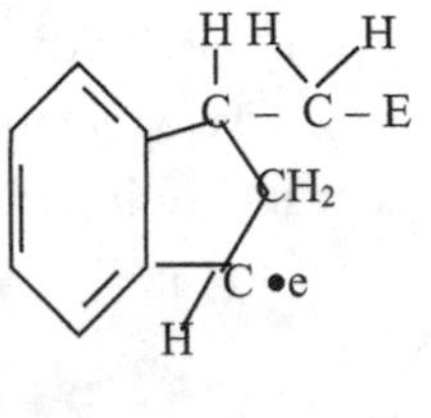

Cyclization of o- Divinyl benzene

(A)

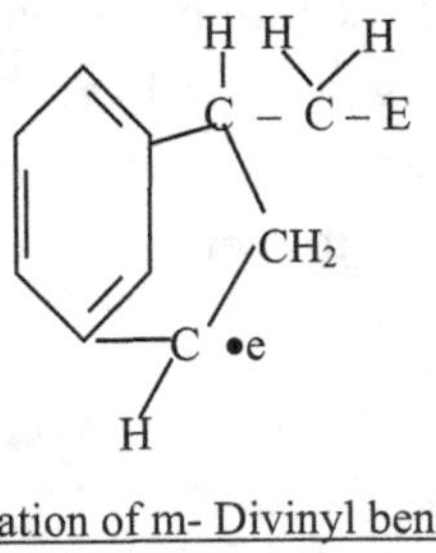

Cyclization of m- Divinyl benzene

(B)

15.30

E above is the electro-free-radical initiator. For ortho-divinyl benzene (A), five –membered hydrocarbon ring (Cyclopentene) fused to the benzene ring on one side is formed along the growing chain placed alternatingly with methylene. The benzene ring is resonance stabilized for the o-placement. For the meta-divinyl benzene (B), six-membered hydrocarbon ring (Cyclohexene) fused to the benzene ring on two sides with a common double bond is formed. The benzene ring is not resonance stabilized. Looking at the types of resonance stabilization for the para-, ortho- and meta- placements and the strain energies in the rings, ortho-placed divinyl benzene favors more cyclization than the meta-placed divinyl benzene followed by p-placed divinyl benzene which in fact does not. From the benzene ring, cross-links between two chains cannot be made possible; not even from the meta-placed ring between two chains intra- molecularly, because the four-membered ring will be too strained to exist as will fully become obvious downstream.

15.2. Inter-/Intra - Molecular Alternating Linear Cyclo-homo and co-polymerization

Monomers to be considered under this category include diallyl quaternary ammonium salts, 1,6-heptadiene, acrylic anhydrides, 1,4- dienes [(V), (VII), (VIII), and (IX) of Equations 15.2 to 15.4] and more.

15.2.1. Diallyl Quaternary Ammonium Salts

Unlike its poly-allyl counterparts considered in the last section, diallyl quaternary ammonium salt are readily known to undergo full cyclization reactions to favor an alternating inter/intra molecular polymerization without cross-links, for several reasons which include -

(i) The size of the ring formed.
(ii) The limited presence of activation centers of different types, e.g. C=O, C=C.
(iii) The proximity of the activation centers based on the structural configuration of a nitrogen center.

The monomer is *an electrostatic type of monomer in view of the presence of an electrostatic bond on the nitrogen center.* This makes it more nucleophilic in character, despite the fact that the electrostatic bond is electrophilic in character useful only for ringed Electrophiles. The presence of two allyl groups instead of one, is again worthy of note.

When the monomer is activated chargedly, the following is to be expected.

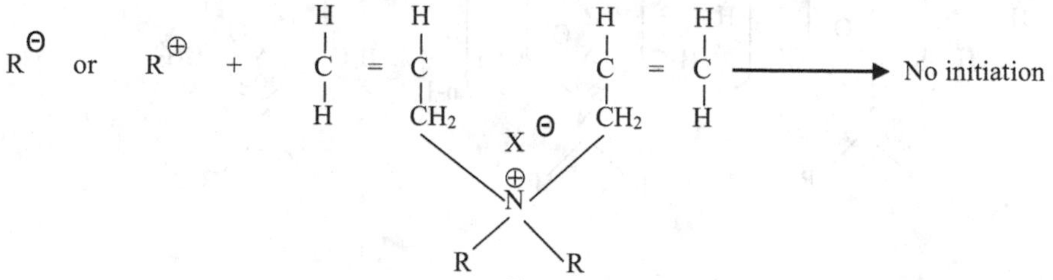

$$(I) \qquad\qquad 15.31$$

Since the $C = C$ activation centers are carrying unique allylic groups, they cannot be activated chargedly. Hence, only free-radical routes can be considered.

Free-radically, the followings are obtained.

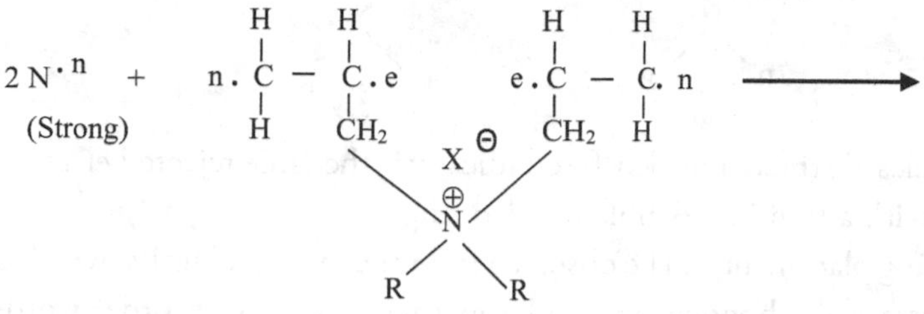

$$(I)$$

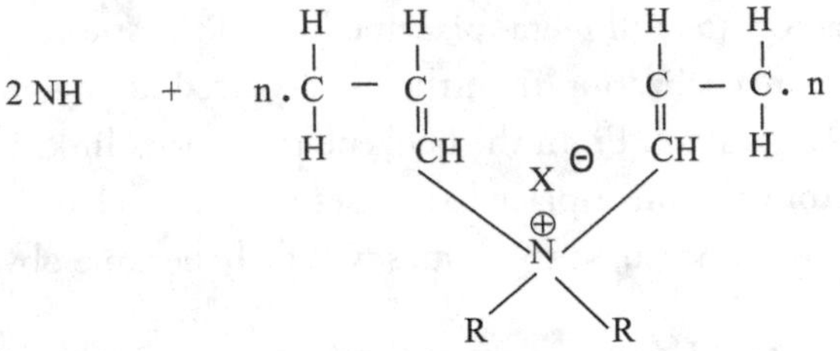

$$2 \text{ NH} \quad + \quad \underset{\text{CH}}{\overset{H \quad H}{n \cdot \overset{|}{C} - \overset{|}{\underset{\|}{C}}}} \qquad \underset{\text{CH}}{\overset{H \quad H}{\overset{|}{C} - \overset{|}{C} \cdot n}}$$

<u>Transfer species of 1st kind of 1st type</u>

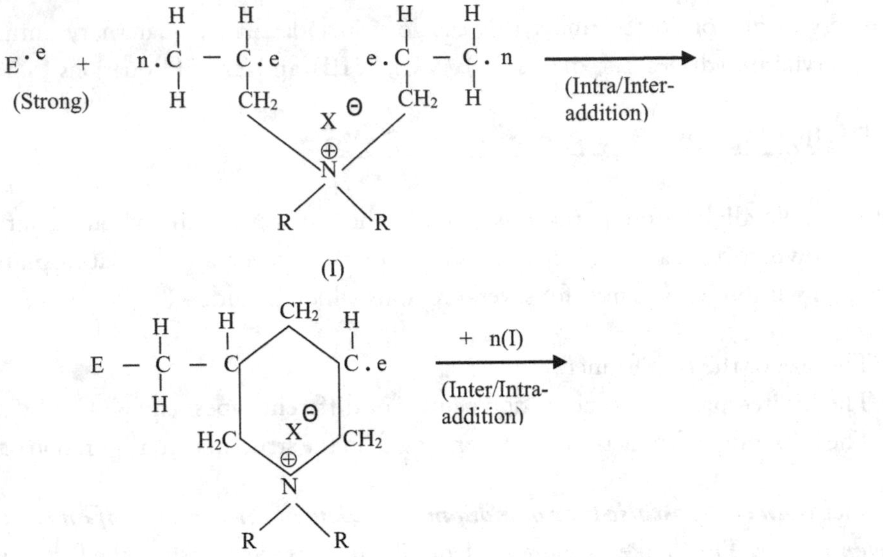

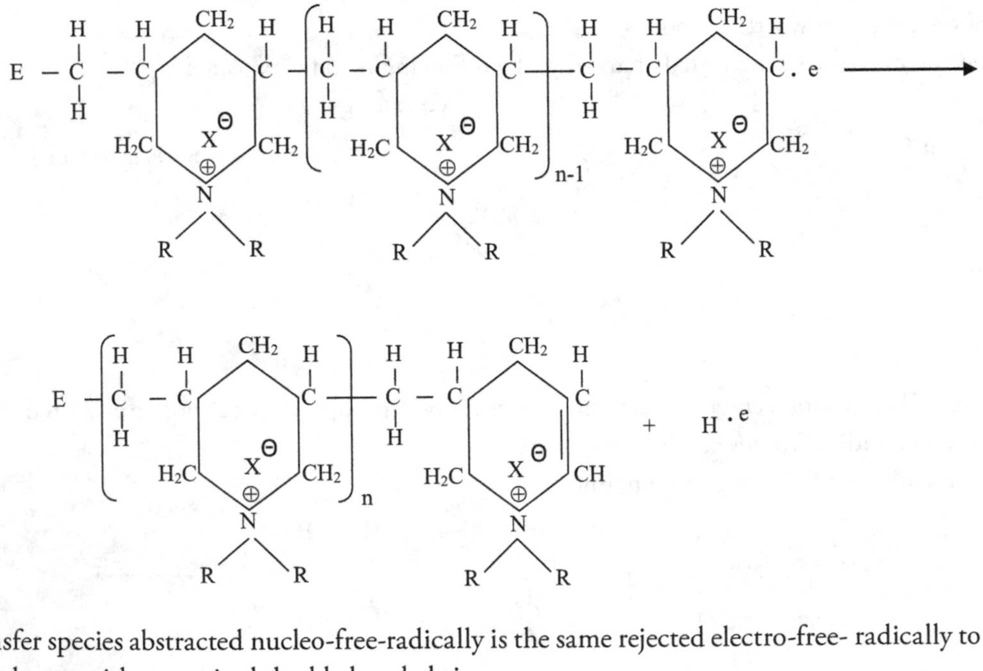

$$E \cdot^{e} \; + \; \underset{\text{(Strong)}}{} \quad n \cdot \overset{H}{\underset{H}{\overset{|}{C}}} - \overset{H}{\underset{CH_2}{\overset{|}{C}}} \cdot e \qquad e \cdot \overset{H}{\underset{CH_2}{\overset{|}{C}}} - \overset{H}{\underset{H}{\overset{|}{C}}} \cdot n \qquad \xrightarrow[\text{addition)}]{\text{(Intra/Inter-}}$$

(I) $\qquad\qquad$ 15.32

$$E - \overset{H}{\underset{H}{\overset{|}{C}}} - \overset{H}{C} \overset{CH_2}{\underset{}{}} \overset{H}{C} \cdot e \qquad \xrightarrow[\text{(Inter/Intra-addition)}]{+ \; n(I)}$$

The transfer species abstracted nucleo-free-radically is the same rejected electro-free- radically to produce a dead polymer with a terminal double bonded ring.

The alternating placement can be observed from the monomer units, which consists of a six-membered saturated electrostatically bonded ring and a methylene group. In order words, a six-membered ringed

15.33

cyclic tertiary amine is alternatingly placed to activated carbene monomer (Methylene). Like all other cases considered so far, cyclization reactions are largely favored electro-free-radically.

15.2.2. 1, 6 - heptadiene

Like all the cases above so far, only one monomer unit is required to form a cyclic ring, in view of the distance between the two activation centers. This distance is what dictates the size of ring to be formed. Note that while the groups carried by $-CH_2-CH_2-CH_2-$ (The Carrier) are monoene in character, the groups carried by $-CH_2-$ (The carrier), are allylic in character. **Which is the case?** Though which is the case should become obvious by now, this will be found very shortly. *There are no allylic groups here.*

When 1, 6 - heptadiene is activated chargedly, the followings are to be expected.

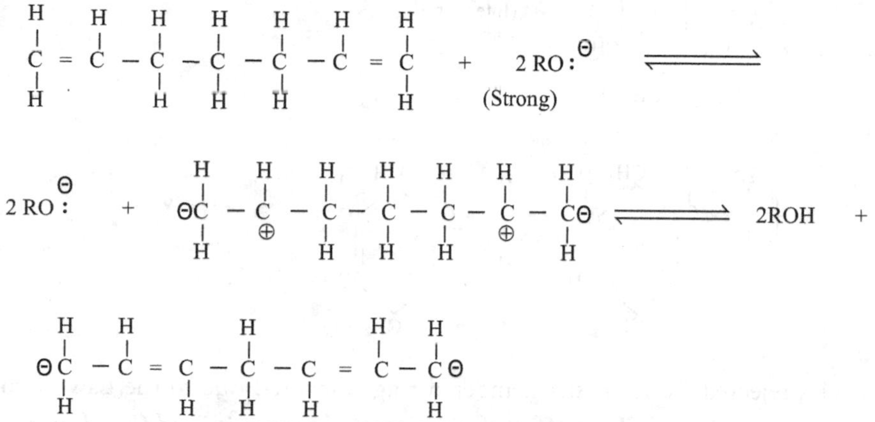

$$C = C - C - C - C - C = C \quad + \quad 2\,RO:^{\ominus} \quad \rightleftharpoons$$
(Strong)

$$2\,RO:^{\ominus} \quad + \quad {}^{\ominus}C - C - C - C - C - C - C^{\ominus} \quad \rightleftharpoons \quad 2ROH \quad +$$

$$^{\ominus}C - C = C - C - C = C - C^{\ominus} \qquad \qquad 15.34$$

Note that, the abstraction of transfer species is done one at a time and not as shown above. The reaction above is favored only under Equilibrium mechanism condition, but not under Combination mechanism condition, because the equation is not chargedly balanced. However chargedly negatively, the route is not favored.

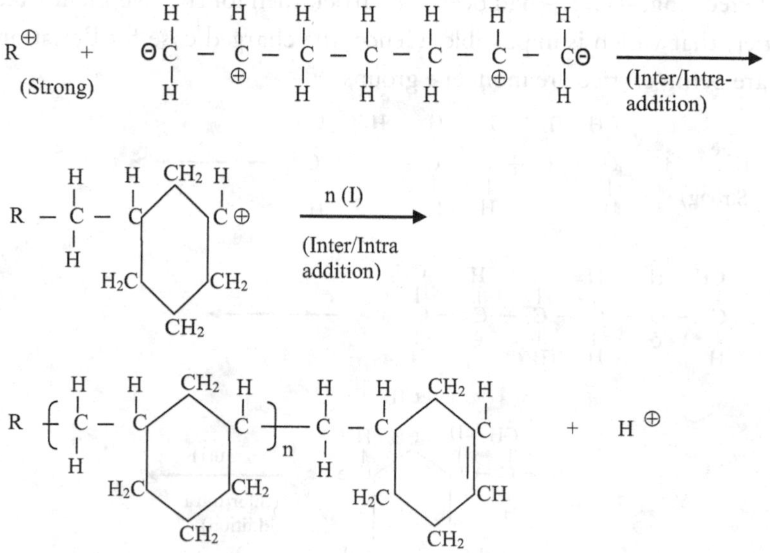

$$R^{\oplus} \quad + \quad {}^{\ominus}C - C - C - C - C - C - C^{\ominus} \quad \xrightarrow[\text{addition})]{\text{(Inter/Intra-}}$$
(Strong)

$$R - C - C \langle \text{ring} \rangle C^{\oplus} \quad \xrightarrow[\text{addition}]{n\,(I) \; \text{(Inter/Intra}} $$

$$R \left\{ C - C \langle \text{ring} \rangle C \right\}_n C - C \langle \text{ring} \rangle C \quad + \quad H^{\oplus} \qquad 15.35$$

Thus, it can be observed here that, the free-cationic or positively charged route (e.g. BF_3/ROR) favors full cyclization to produce alternating copolymer of a six- membered cycloalkane combined with a carbene. Charged activation of all the activation centers is favored due to absence of resonance stabilization in the monomers, absence of allylic groups and the fact that they are of the same capacity. Clearly, if the group carried had been allylic, the reactions above would preferably take place electro-free-radically.

Free-radically, only electro-free- radicals will polymerize the monomer to produce similar products.

$$E \cdot^e + n \cdot \overset{H}{\underset{H}{C}} - \overset{H}{\underset{\dot{e}}{C}} - \overset{H}{\underset{H}{C}} - \overset{H}{\underset{H}{C}} - \overset{H}{\underset{H}{C}} - \overset{H}{\underset{\dot{e}}{C}} - \overset{H}{\underset{H}{C}} \cdot n \longrightarrow \text{(Inter/Intra-addition)}$$

(I)

$$E - \overset{H}{\underset{H}{C}} - \overset{H}{\underset{}{C}} \langle \text{ring with } CH_2, CH_2, CH_2, H_2C \rangle \overset{H}{\underset{}{C}} \cdot e \xrightarrow{+ \; n\,(I)} \text{(Inter/Intra - addition)}$$

$$E \left[\overset{H}{\underset{H}{C}} - C \langle \text{ring } CH_2/H_2C/CH_2 \rangle C \right]_n \overset{H}{\underset{H}{C}} - C \langle \text{ring } CH_2/H_2C/CH/CH_2 \rangle C + H \cdot^e$$

$$\tag{15.36}$$

The transfer species rejected from the six-membered ring is favored due to the Law of conservation of transfer of transfer species. ***This will not affect chain growth, because it is adding directly to a standing monomer (not to a center on a linear chain).*** The existence of the six-membered ring is favored due to the fact that the strain energy for the family is small for its size and the absence of adequate means of introducing MRSE into the ring apart from instantaneous opening. The same will apply to 1,5-hexadiene and its radical-pushing substituted members, where five- membered rings are obtained along the chain. If for the six-membered ring, $- CH_2 -$ had been the carrier, then for the five-membered ring, there would have been no carrier; that which is impossible. Hence, the charged case for Equation 15.35 would take place, since what are being carried are monoene groups.

$$E \cdot^e + \overset{CH_3}{\underset{H}{C}} = \overset{H}{\underset{}{C}} - \overset{H}{\underset{H}{C}} - \overset{H}{\underset{H}{C}} - \overset{H}{\underset{}{C}} = \overset{CH_3}{\underset{H}{C}} \longrightarrow E \cdot^e +$$

$$n \cdot \overset{CH_3}{\underset{H}{C}} - \overset{H}{\underset{\dot{e}}{C}} - \overset{H}{\underset{H}{C}} - \overset{H}{\underset{H}{C}} - \overset{H}{\underset{\dot{e}}{C}} - \overset{CH_3}{\underset{H}{C}} \cdot n \longrightarrow$$

$$E - \overset{CH_3}{\underset{H}{C}} - \overset{H}{\underset{}{C}} \langle \text{5-ring with } CH_3-CH, H, H_2C, CH_2 \rangle C \cdot e \xrightarrow{+ \; n(I)} \text{(Inter/Intra addition)}$$

$$\tag{15.37a}$$

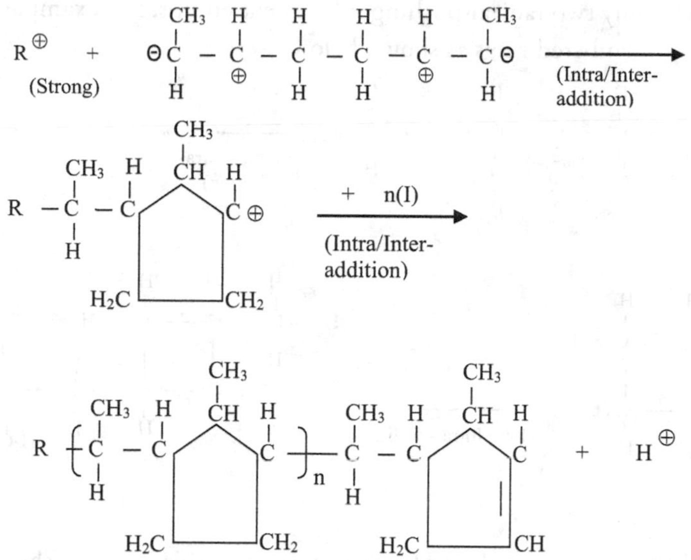

$$
R^{\oplus} \;+\; {}^{\ominus}\!\!\overset{CH_3}{\underset{H}{C}} - \overset{H}{\underset{\oplus}{C}} - \overset{H}{\underset{H}{C}} - \overset{H}{\underset{H}{C}} - \overset{H}{\underset{\oplus}{C}} - \overset{CH_3}{\underset{H}{C}}{}^{\ominus} \longrightarrow
$$
(Strong) \hspace{4cm} (Intra/Inter-addition)

(ring structure with CH₃, CH groups, H₂C—CH₂) $+ \; n(I) \longrightarrow$ (Intra/Inter-addition)

(polymer structure) $+ \; H^{\oplus}$ \hspace{2cm} 15.37b

It can be observed that both the electro-free-radical and positively charged routes are favored here, the monomer being Nucleophilic. Even if only one CH_3 group is externally located, the electro-free-radical route will still favor cyclization as shown below, but under higher and more difficult operating conditions.

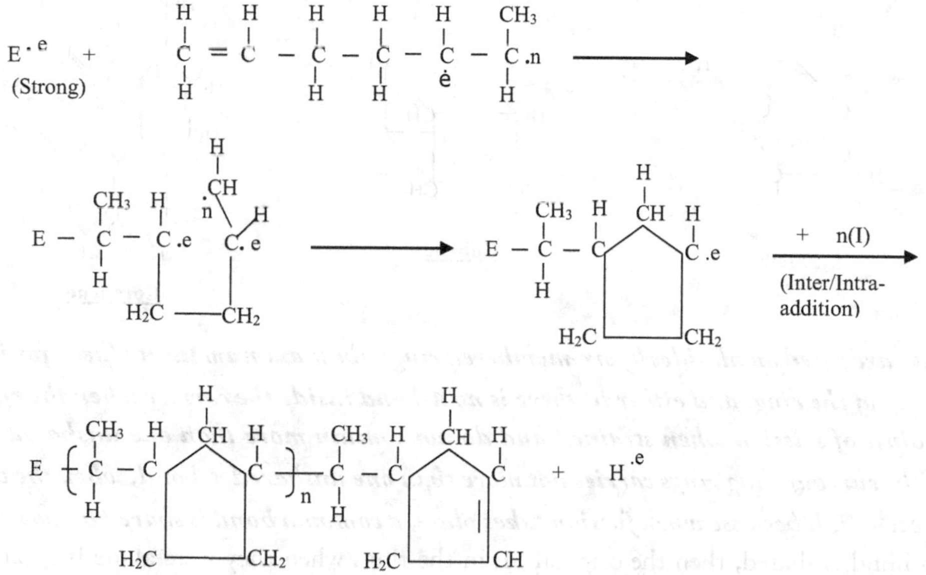

$$
E\cdot^{e} \;+\; \overset{H}{\underset{H}{C}} = \overset{H}{\underset{H}{C}} - \overset{H}{\underset{H}{C}} - \overset{H}{\underset{H}{C}} - \overset{H}{\underset{\dot{e}}{C}} - \overset{CH_3}{\underset{H}{C}}.n \longrightarrow
$$
(Strong)

(ring intermediate structures) $+ \; n(I) \longrightarrow$ (Inter/Intra-addition)

(polymer structure) $+ \; H^{\cdot e}$ \hspace{2cm} 15.38

Note that the two groups carried do not have to be exactly the same, except that when they are different it is the less nucleophilic center that is first activated instead of both at the same time. *Above, it was the less nucleophilic center that was first activated in the absence of resonance stabilization and if the source of activation was strong as used above, the new active center may be strong enough to activate the second activation center. Otherwise, cyclization may never take place. Hence in general, the two groups are usually chosen to be the same. If the initiator is weak, molecular rearrangement will take place, thereby preventing cyclization, if the transfer species is H and not -CH=CH₂.*

Existence of four-membered rings along the chain may be favored if MRSE is not present in such rings which are carrying only two radical-pushing substituent groups. For example 1; 4 - pentadiene may favor existence of four - membered rings as shown below.

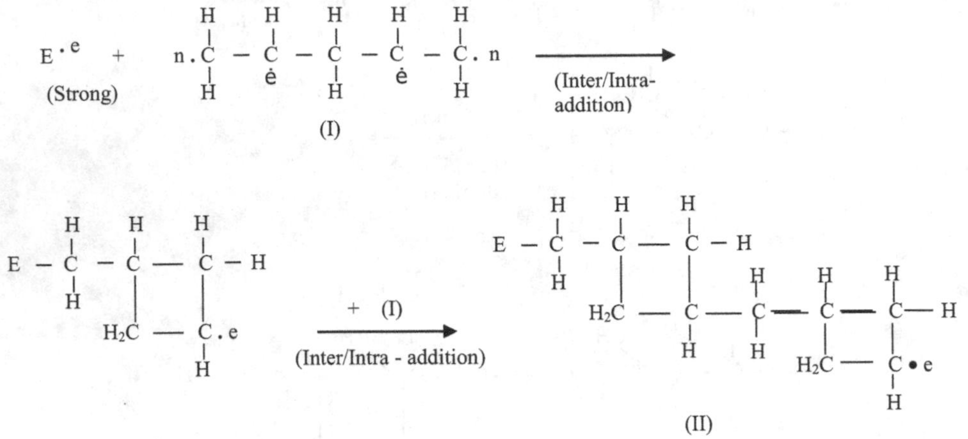

$$\text{(I)}$$

$$15.39$$

Fused four-membered ring to a six-membered hexene ring on two sides exists as shown below for α-Pinene. In Cubane, all the four-membered rings are fused on all sides. In coro-nene, a cyclohexane ring is fused on all sides just like the four membered ring in Cubane, as also shown below.

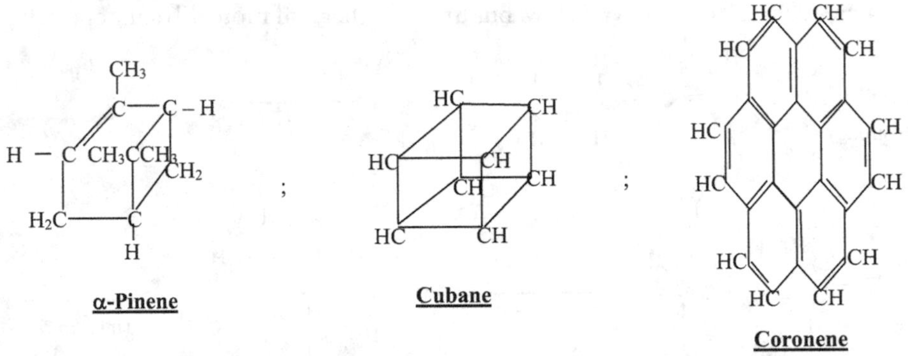

α-Pinene ; **Cubane** ; **Coronene** 15.40a

When Rings are fused on all sides by six-membered rings, the maximum limit placed for its existence are i) the size of the ring, and either ii) there is no π-bond inside the carrier when the rings carried have no points of scission when strained and do not contain more than one unshared π-bond, or iii) one of the surrounding rings carries not more than one unshared π-bond, when the central ring carries one π-bond, because when fussion takes place, a common bond is shared by two rings. When a common bond is shared, then the original SE in the rings when they were alone is greatly increased when fused. Therefore, increase in SE in rings when fused, depends on the followings-

i)   The size of the ring to be fused with. The smaller the ring, the more difficult it is for the ring to exist.

ii) The groups carried by the surrounding rings, for if they are radical-pushing groups, the more difficult it is for the carrier to exist.

iii) When two similar rings are fused together, the lesser the SE in the rings than when it was alone. Such is the case with benzene, naphthalene, and anthracene. The SE in naphthalene is less than the sum of the SE in two benzene rings. That is, while the presence of one common boundary decreases the strain energies in the fused rings, this is not the case with two or more common boundaries.

iv)   Is it true that the more the number of unshared π-bonds in the rings surrounding the central ring, the greater the possibility of the non-existence of the central ring? Coronene known to exist does when the bonds are rearranged via resonance stabilization. However, the (A) shown below is not known to exist. The reason is because of (i) above. While (A) like the corresponding five membered case are not known to exist, coronene exists. They all satisfy the same conditions, except the size. (B) shown below is also not known to exist. Two benzene rings should have been able to combine together under certain operating conditions to give (B). As it seems, it is so strained that it breaks down to benzene. If the four membered ring is so strained due to the two benzene ring fused together, then how can (A) exist? They cannot be dehydrogenated. One can observe the unique significance of the numbers four, five and six with respect to rings in general.

(A)   (B)

NOT KNOWN TO EXIST   15.40b

v)   The more the number of boundaries shared by two rings where possible, the lesser their favored existence, if their sizes are too close.

The (II) of Equation 15.39 obtained, will not be strained to the point where the four-membered ring can no longer exist as propagation continues. However, it is ones belief that the existence of the four-membered ring will be fully favored under low operating conditions, because *how (I) of Equation 15.39 is involved to produce larger sized rings in preference to full existence of four-membered rings alone along the chain will shortly be considered.* If one of the H atoms externally located on both sides and/or the two in the center are replaced with a radical-pulling group carrying no transfer species such as -Cl (not –COOR), cyclization will very readily take place as shown below.

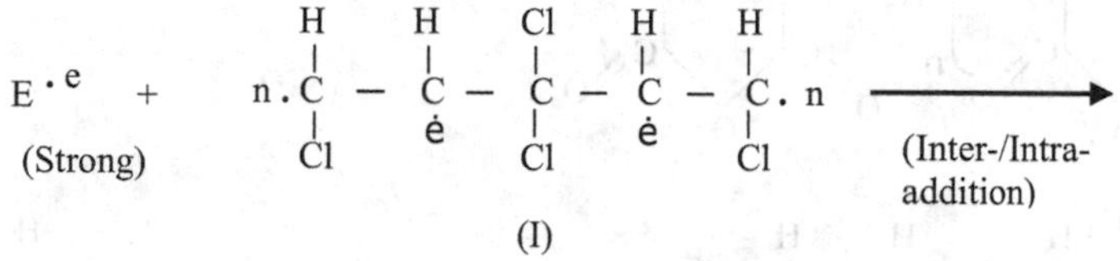

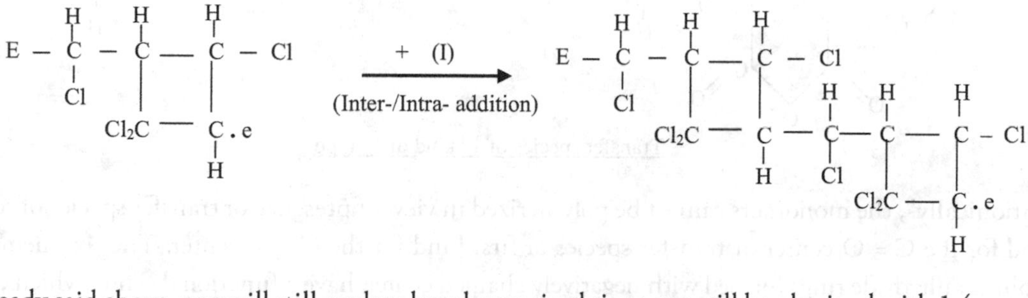

As already said above, one will still explore how larger sized rings can still be obtained with 1,4-pentadiene.

15.2.3. Acrylic Anhydrides and Pimelic Acids

With these groups (Acrylic anhydrides), there are four activation centers of two types - C = O and C = C.

Chargedly and radically, the C = O center will not favor any attack. Thus, only the C = C centers can be used to favor existence of rings along the chain. On the other hand, C = O center is nucleophilic, while the C = C is electrophilic. In order words, acrylic anhydrides are Males (i.e., Electrophiles). When activated chargedly, the followings are obtained.

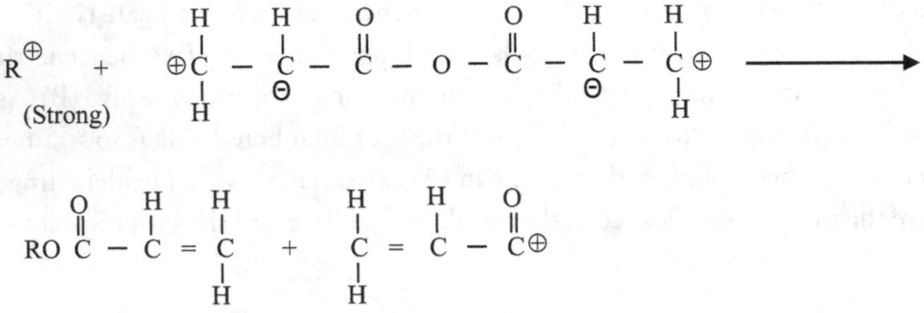

$$R^{\oplus} + {}^{\oplus}C - C - C - O - C - C - C^{\oplus} \longrightarrow$$

(Strong)

$$RO\,C - C = C \quad + \quad C = C - C^{\oplus}$$

<u>Transfer species of 1st kind of 1st type</u>

15.42

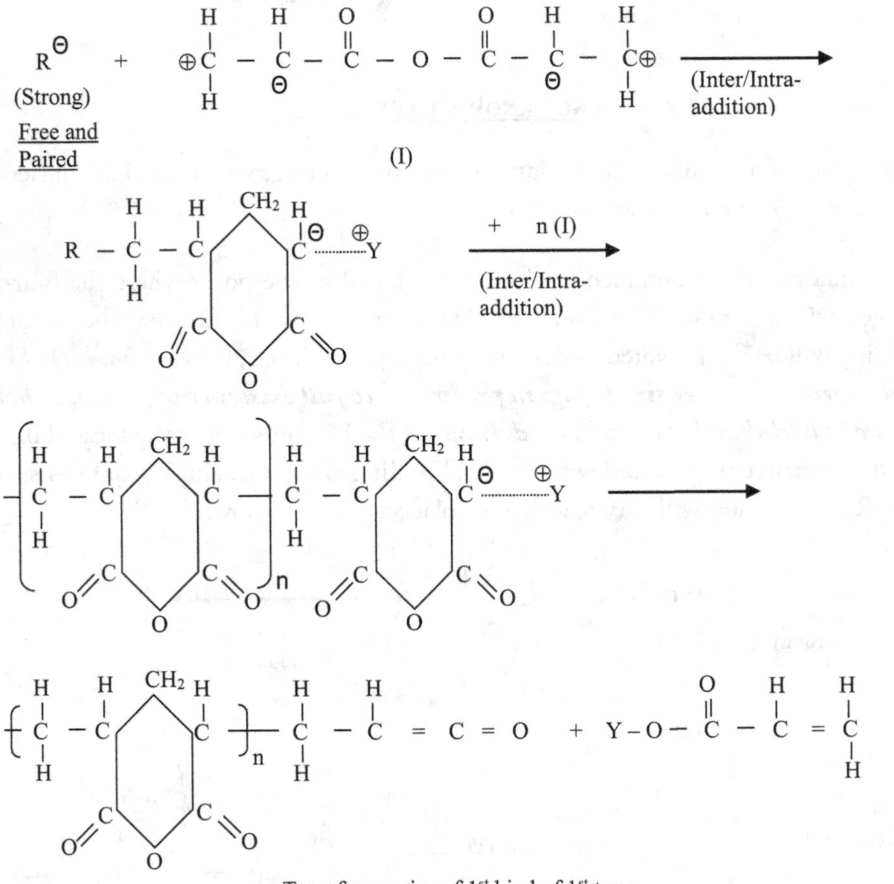

$$R^{\ominus} + {}^{\oplus}C - C - C - O - C - C - C^{\oplus} \longrightarrow$$

(Strong)

(Inter/Intra-addition)

<u>Free and Paired</u>

(I)

$$R - C - C \cdots C \cdots Y \qquad + \quad n\,(I) \longrightarrow$$

(Inter/Intra-addition)

$$R \left[C - C \cdots C \right]_n C - C \cdots C \cdots Y \longrightarrow$$

$$R \left\{ C - C \cdots C \right\}_n C - C = C = O \quad + \quad Y - O - C - C = C$$

<u>Transfer species of 1st kind of 1st type</u>

15.43

"Cationically", the monomers cannot be polymerized in view of presence of transfer species of second first kind for the C = O center or transfer species of first kind for the C = C center. The six–membered Electrophilic anhydride rings formed with negatively charged center, have a functional center which can be used "cationically" and activation center which can only be used anionically (not with negatively charged center as used above-e.g., LiC_4H_9). Since the MRSE cannot be provided for the rings, the anionic route is only favored for special types of ring sizes where the MRSE can be provided for this type of monomer which is largely electrophilic in character. In fact no other size can exist beyond this size. On the other hand, the initiator above (E.g. $H_9C_4{}^{\ominus}.....{}^{\oplus}Li$), cannot be used to open the rings. For the first time, one is encountering a case where cyclization can take place with Covalently charged-paired initiator. It is the transfer species that prevented "cationic" initiation that is rejected anionically. During termination, with increasing intensity for formation of rings, a transfer species as shown above may not indeed be released when the optimum chain length has not been reached.

Radically, for the initiation of the monomers, the nucleo-free-radical route is favored via the $C = C$ center, while the electro-free-radical route is not favored due to the presence of $C = O$ activation center. Hence, the followings are to be expected.

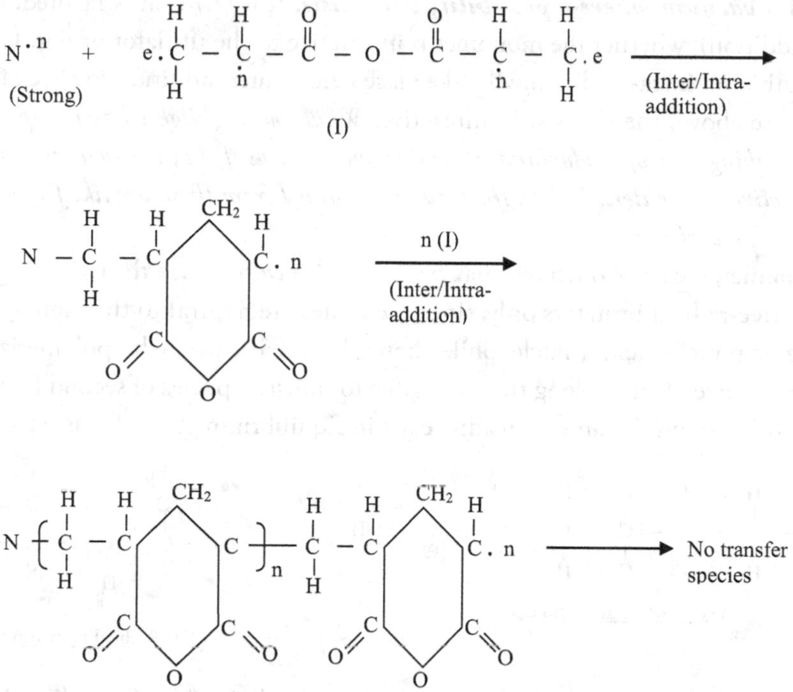

$$N \cdot^n \; (Strong) \; + \; e \cdot \overset{H}{\underset{H}{C}} - \overset{H}{\underset{\overset{\cdot}{C}}{C}}_n - \overset{O}{\overset{\|}{C}} - O - \overset{O}{\overset{\|}{C}} - \overset{H}{\underset{\overset{\cdot}{C}}{C}}_n - \overset{H}{\underset{H}{C}} \cdot e \; \longrightarrow \; \text{(Inter/Intra-addition)}$$

(I)

(Inter/Intra-addition) n (I)

15.44

There is no transfer species in the absence of a receiving center, though the route is natural to the monomer. Electro-free-radically, the followings are obtained.

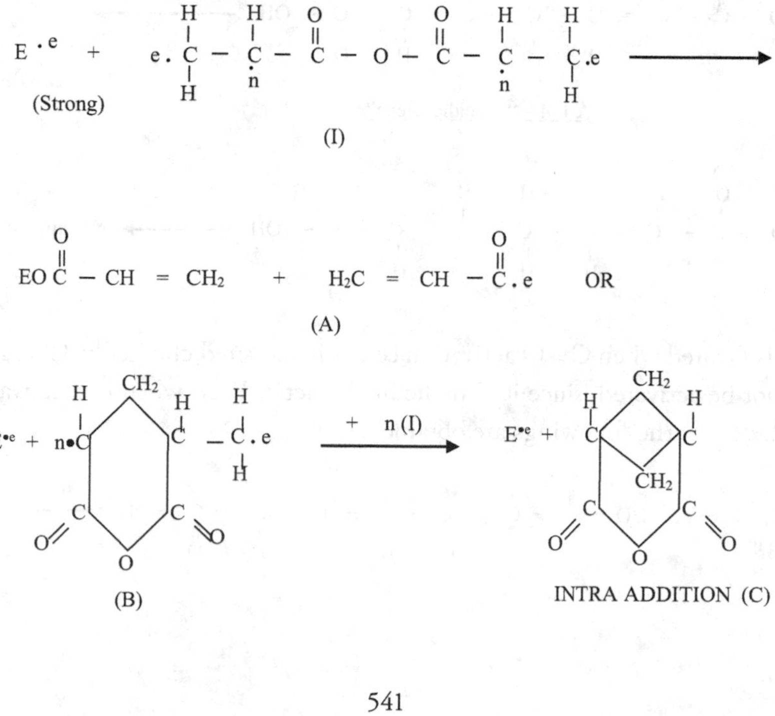

$$E \cdot^e \; (Strong) \; + \; e \cdot \overset{H}{\underset{H}{C}} - \overset{H}{\underset{\overset{\cdot}{C}}{C}}_n - \overset{O}{\overset{\|}{C}} - O - \overset{O}{\overset{\|}{C}} - \overset{H}{\underset{\overset{\cdot}{C}}{C}}_n - \overset{H}{\underset{H}{C}} \cdot e \; \longrightarrow$$

(I)

$$EO \overset{O}{\overset{\|}{C}} - CH = CH_2 \; + \; H_2C = CH - \overset{O}{\overset{\|}{C}} \cdot e \quad OR$$

(A)

(B)   INTRA ADDITION (C)   15.45

Sunny N.E. Omorodion

It is believed that (B) is not favored, since the transfer species will not allow for its initiation and since the initiator is present. If intra- molecular addition had taken place before inter molecular addition as shown above, then a stable four-membered ring fused to a six-membered ring (C) would have been obtained. But this is not possible, since (B) cannot be obtained. *The case (B) above where intra- was allowed to take place before inter-addition is very much like the case where two monomers which are unreactive to the initiator, are allowed to form a COUPLE to see if the couple formed is now reactive to the initiator; for which if reactive alternating placement will then be obtained, because NATURE abhors a vacuum where a possibility still exists.* It is (A) that is favored, (i.e., intra- takes place after inter- addition), whether the monomer is unreactive to the initiator or not. If the monomer is unreactive to the initiator, intra- addition will take place before inter- addition to check if something can be done. For the case above, the (B) is still unreactive. *What was highlighted above applies, for one, two, or more monomers; taking note of the fact that these monomers are as if they are two monomers, because they carry two or more activation centers. Hence, the products obtained from them are like from two monomers- a ring or fused rings and a carbene.*

With this monomer, it can be observed that cyclization is favored with the use of negatively charged-paired and nucleo-free-radical imitators only, since the routes are natural to the monomer.

Now, looking at pimelic acid a nucleophile shown below, it cannot be polymerized chargedly or radically to favor existence of rings along the chain, due to transfer species of second first kind and of the first kind. On the other hand, it can very readily exist in Equilibrium state of existence.

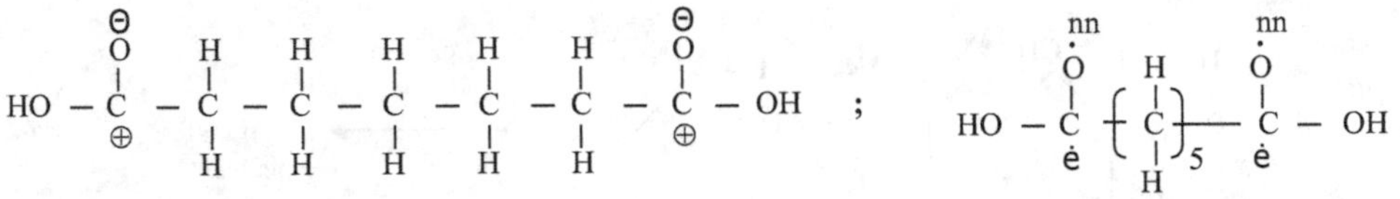

(I) <u>Charged activation</u>

(II) <u>Radical activation</u>   15.46

For the unsaturated dicarboxylic acid which can be said to be a derivative of pimelic acid, the followings are obtained chargedly, noting that this is an electrophile.

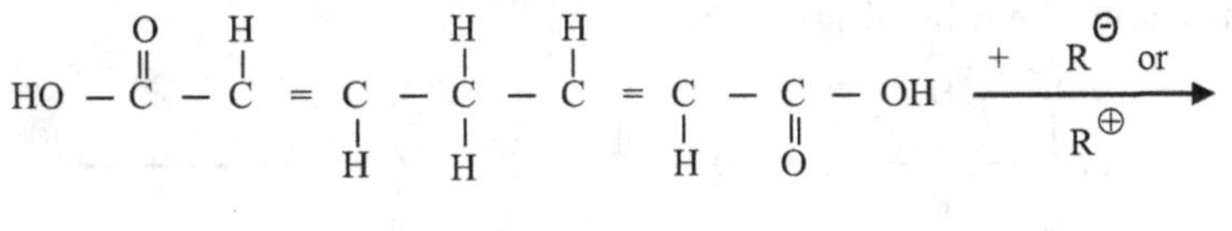

<u>A 1, 4 – "Diacidic Diene"</u>

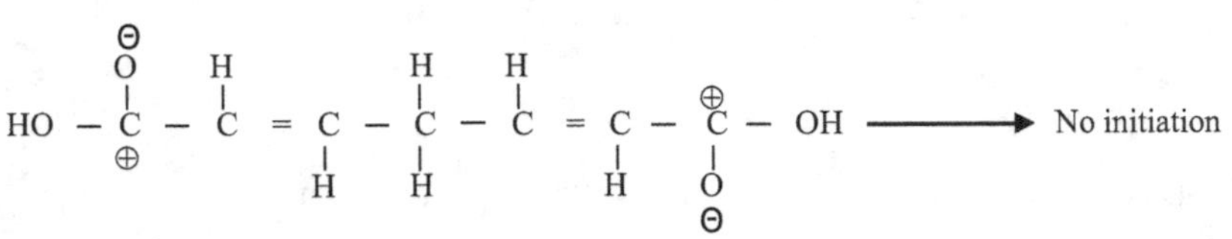

15.47

No charged route is favored when C = O activation center is activated chargedly. Chargedly however, the C = C center cannot be activated, since it is allylic in character. Even when it is activated chargedly, no route is favored. Radically the followings are obtained.

542

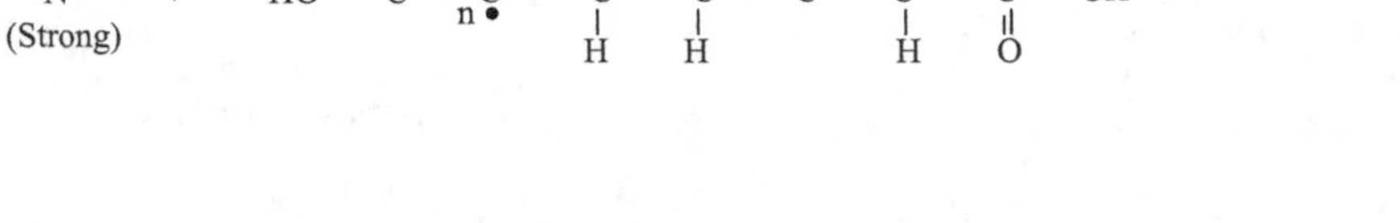

$$NH + HO - \overset{\overset{O}{\|}}{C} - \overset{\overset{H}{|}}{C} = \overset{H}{C} - \overset{\overset{\bullet n}{}}{C} - C = C - \overset{\overset{O}{\|}}{C} - OH \qquad 15.48$$

$$E \cdot e + HO - \overset{\overset{O}{\|}}{C} - \underset{n}{\overset{H}{C}} - \overset{e.}{C} - \overset{H}{\underset{H}{C}} - \underset{\grave{e}}{\overset{H}{C}} - \overset{n}{\underset{\cdot}{C}} - \overset{\overset{O}{\|}}{C} - OH \xrightarrow[\text{addition)}]{\text{(Intra}}$$

(I)

(II)   15.49

Indeed, the ring above cannot be formed because of the presence of the same transfer species –OH. It was assumed to be formed in order to show that whether intra- or inter- addition takes place first or not, the route cannot be favored electro-free-radically. ***What first takes place is Inter- addition before Intra-addition in general.*** The monomer is an Electrophile wherein its natural route is nucleo-free-radical or negatively charged route. Almost like methyl sorbate, no route is favored, unless when the groups are trans-placed followed by the use of charged-paired coordinated initiators. Unlike all the other cases so far considered, pimelic acid and the unsaturated pimelic acid used above cannot favor the existence of four-membered rings along the chain, whatever the operating conditions are. Even, when groups are cis-placed and paired initiator is used, it cannot be activated chargedly, because of the presence of allylic group. Only radically Covalent paired initiators can be used to favor the existence of four-membered rings adjacently placed with an electrophilic carbene as copolymer along the chain nucleo-free-radically. (II) above will form the monomer unit along the chain. In the absence of initiator generating catalyst, but use of heat, the monomer cannot undergo cyclization, but only under the conditions specified.

15.2.4. 1, 4 - Dienes and related Compounds.

In the last but one sub-section, it was shown that 1,4-diene will cyclize to give four-membered rings all along with limitations due to the increasing amount of SE in the ring as it grows. The possibility of forming larger membered rings along the chain still exists. Beginning with 1, 4 - pentadiene, the followings are to be expected chargedly.

$$\underset{\text{(Strong)}}{R^{\ominus}} + \overset{\ominus}{\underset{H}{C}} - \overset{\overset{H}{|}}{\underset{\oplus}{C}} - \overset{\overset{H}{|}}{\underset{H}{C}} - \overset{\overset{H}{|}}{\underset{\oplus}{C}} - \overset{\overset{H}{|}}{\underset{H}{C}} \ominus \longrightarrow$$

(I)

$$RH + \overset{\overset{H}{|}}{\underset{H}{C}} = \overset{\overset{H}{|}}{C} - \underset{\ominus}{\overset{H}{C}} - \overset{\overset{H}{|}}{C} = \overset{\overset{H}{|}}{\underset{H}{C}} \qquad 15.50$$

543

Negatively or nucleo-free-radically, it cannot be initiated. If the second activation center is not activated at the same time with the other, the transfer species still remains the same and not $-CH=CH_2$. Positively or cationically and electro-free-radically, it can be initiated as already shown in Equation 15.39.

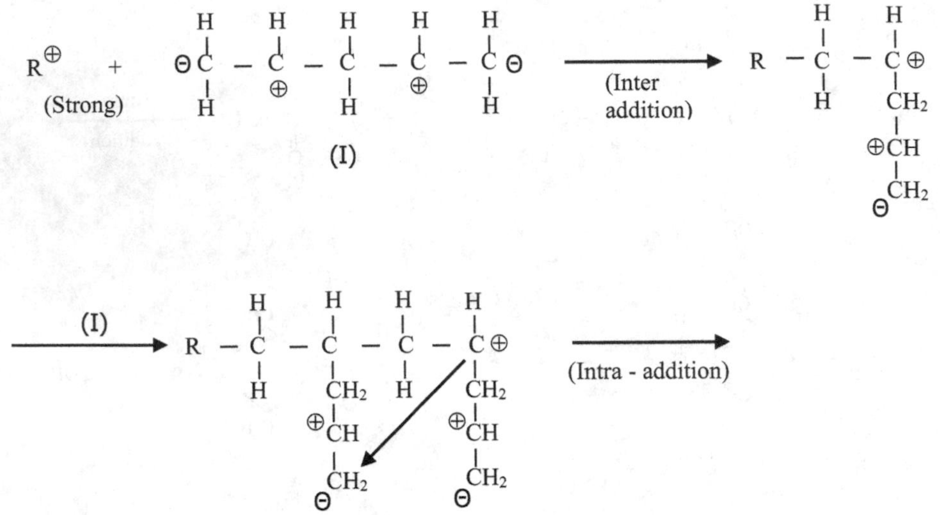

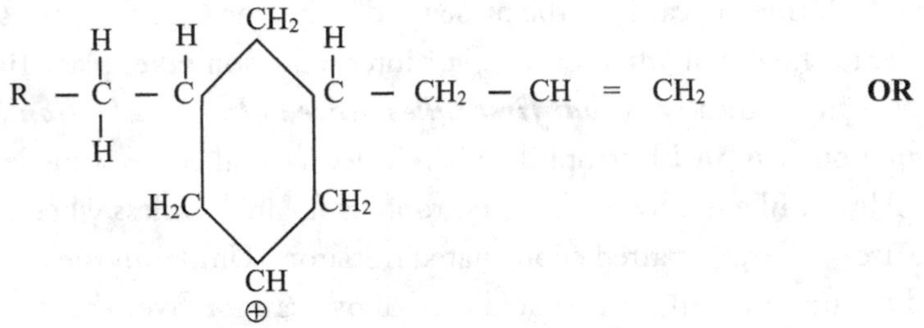

(I) Primary cyclization

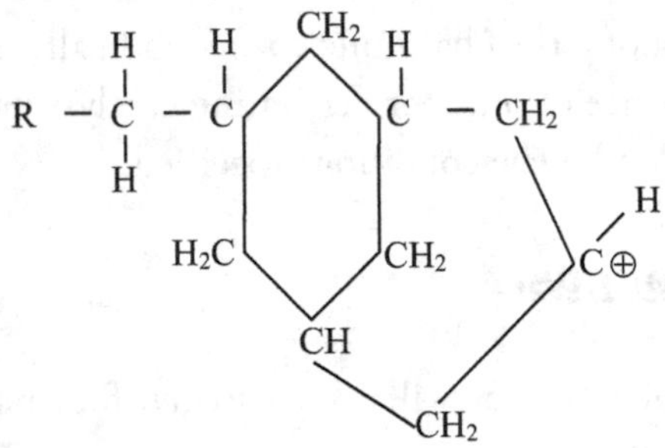

(II) Primary and Secondary cyclizations

15.51

Notice that the second ring is more difficult to obtain than the first ring, because once the first ring is formed, for the positive center to diffuse to the negative end is not easy. Positively, two monomers units are required to favor primary or secondary cyclizations. With primary cyclization, a monocyclic ring is obtained, while with secondary cyclization, a bicyclic ring is obtained with two common boundaries. The rings are six-membered cyclohexane rings. The existence of two fused eight - membered rings with six common boundaries and few points of scission, while the active center is not involved, is not possible, whether due to possible presence of MRSE in their rings and steric limitations or not.

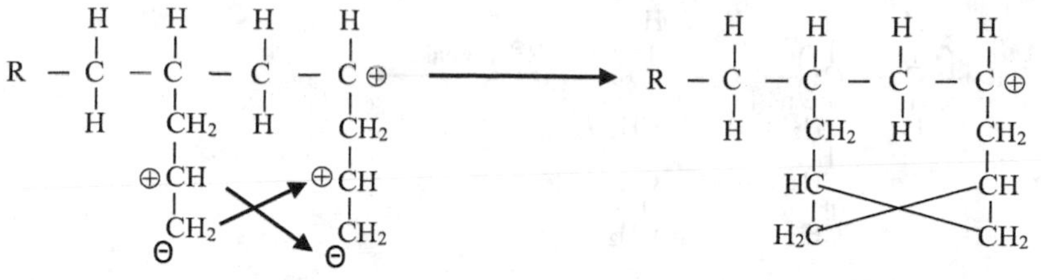

$$R-\overset{\overset{\displaystyle H}{|}}{\underset{\underset{\displaystyle H}{|}}{C}}-\overset{\overset{\displaystyle H}{|}}{\underset{\underset{\displaystyle CH_2}{|}}{C}}-\overset{\overset{\displaystyle H}{|}}{\underset{\underset{\displaystyle H}{|}}{C}}-\overset{\overset{\displaystyle H}{|}}{\underset{\underset{\displaystyle CH_2}{|}}{C}}\oplus \longrightarrow R-\overset{\overset{\displaystyle H}{|}}{\underset{\underset{\displaystyle H}{|}}{C}}-\overset{\overset{\displaystyle H}{|}}{\underset{\underset{\displaystyle CH_2}{|}}{C}}-\overset{\overset{\displaystyle H}{|}}{\underset{\underset{\displaystyle H}{|}}{C}}-\overset{\overset{\displaystyle H}{|}}{\underset{\underset{\displaystyle CH_2}{|}}{C}}\oplus$$

Not favored (strained)　　　15.52

Existence of secondary or primary cyclization will depend largely on the strength of the active center or initiator. For the growing polymer chains of (I) and (II) of Equation 15.51, the followings are obtained.

For (II)

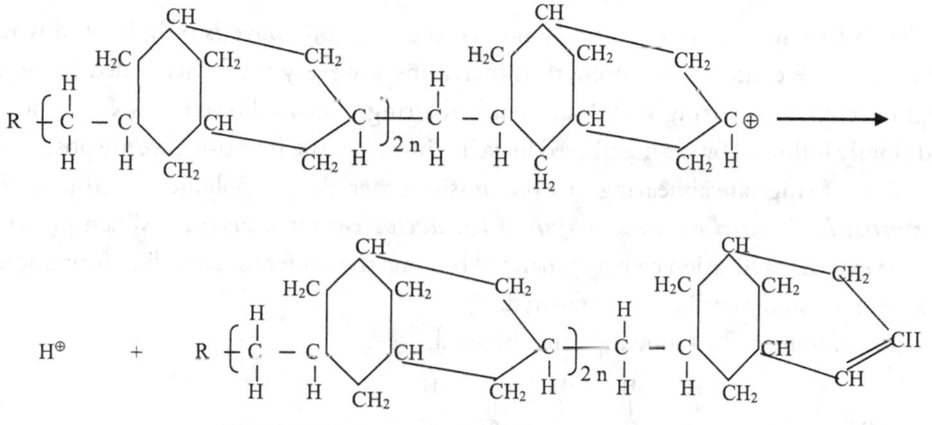

(A)　　　15.53

For (I)

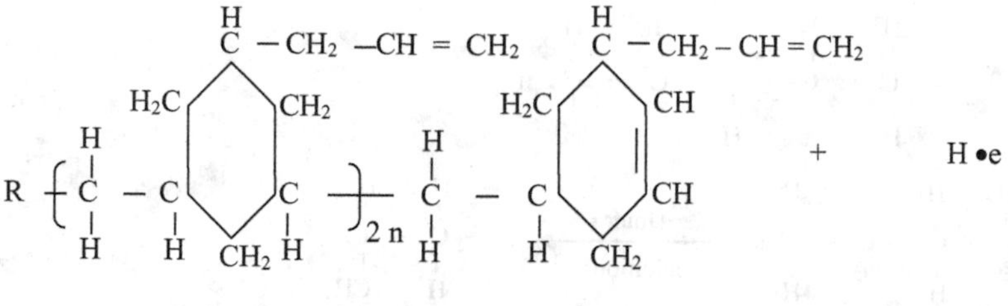

(B)　　　15.54

Similar terminal chains are obtained for monocyclization and bicyclization, since the double bond is placed inside the ring. If a weak initiator had been involved, then no cyclization will take place, wherein there is largely inter - molecular reaction.

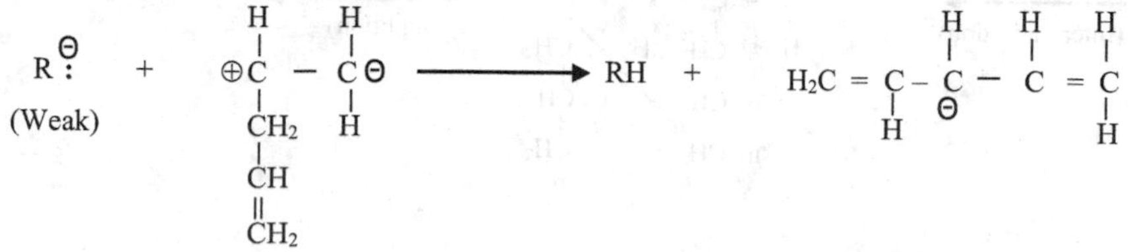

$$R\overset{\ominus}{:} \; + \; \oplus\overset{\overset{\displaystyle H}{|}}{\underset{\underset{\underset{\underset{\displaystyle CH_2}{\parallel}}{\displaystyle CH}}{|}}{\underset{\displaystyle CH_2}{C}}}-\overset{\overset{\displaystyle H}{|}}{\underset{\underset{\displaystyle H}{|}}{C}}\ominus \longrightarrow RH \; + \; H_2C=\overset{\overset{\displaystyle H}{|}}{\underset{\underset{\displaystyle H}{|}}{C}}-\overset{\ominus}{\underset{}{C}}-\overset{\overset{\displaystyle H}{|}}{C}=\overset{\overset{\displaystyle H}{|}}{\underset{\underset{\displaystyle H}{|}}{C}}$$

(Weak)　　　15.55

$$R \left\{ \begin{array}{c} H \\ | \\ C \\ | \\ H \end{array} - \begin{array}{c} H \\ | \\ C \\ | \\ CH_2 \\ | \\ CH \\ || \\ CH_2 \end{array} \right\}_{n+1} \begin{array}{c} H \\ | \\ C \\ | \\ H \end{array} - \begin{array}{c} H \\ | \\ C \oplus \\ | \\ CH_2 \\ | \\ CH \\ || \\ CH_2 \end{array} \xrightarrow{\quad R^\ominus \text{ (Weak)} \quad}$$

$$R \left\{ \begin{array}{c} H \\ | \\ C \\ | \\ H \end{array} - \begin{array}{c} H \\ | \\ C \\ | \\ CH_2 \\ | \\ CH \\ || \\ CH_2 \end{array} \right\}_n \begin{array}{c} H \\ | \\ C \\ | \\ H \end{array} - \begin{array}{c} H \\ | \\ C \\ | \\ H \end{array} = \begin{array}{c} \\ C \\ | \\ H \end{array} - CH = CH_2 \quad + \quad H^\oplus$$

15.56

Since the positively charged route is natural to the monomer, the reaction above is only favored when **n** is small after which the active center whose strength is increasing for every monomer added becomes strong enough to begin cyclization starting with four membered rings, then followed by six-membered monocyclic rings and finally followed by six-membered bicyclic rings. For the first time, we are observing a case where different sizes of rings are appearing in a progressive order along a polymeric chain. ***In the natural route, rings must be formed as the strength of the active center increases***. When primary cyclization is favored, the external activation centers generated become the center for cross-link formations. This is not possible when secondary cyclization is favored.

Free-radically for the monomer, the followings are obtained.

$$N \cdot^n \quad + \quad \begin{array}{cccccccc} H & & H & & H & & H & & H \\ | & & | & & | & & | & & | \\ n.C & - & C & - & C & - & C & - & C.n \\ | & & \dot{e} & & | & & \dot{e} & & | \\ H & & & & H & & & & H \end{array} \xrightarrow{\quad\quad}$$
(Strong)

(I)

$$NH \quad + \quad \begin{array}{cccccccc} H & & H & & & & H & & H \\ | & & | & \bullet n & & & | & & | \\ C & = & C & - & C & - & C & - & C.n \\ | & & | & & \bullet e & & | & & | \\ H & & H & & & & & & H \end{array}$$
15.57

$$E \cdot^e \quad + \quad \begin{array}{cccccccc} H & & H & & H & & H & & H \\ | & & | & & | & & | & & | \\ n.C & - & C & - & C & - & C & - & C.n \\ | & & \dot{e} & & | & & \dot{e} & & | \\ H & & & & H & & & & H \end{array} \xrightarrow[\text{addition}]{\text{(Inter -}} \quad E - \begin{array}{c} H \\ | \\ C \\ | \\ H \end{array} - \begin{array}{c} H \\ | \\ C.e \\ | \\ CH_2 \\ | \\ e.CH \\ | \\ CH_2 \\ \cdot \\ n \end{array}$$
(Strong)

(I)

$$\xrightarrow[\text{(Inter - addition)}]{+ \quad (I)} \quad E - \begin{array}{c} H \\ | \\ C \\ | \\ H \end{array} - \begin{array}{c} H \\ | \\ C \\ | \\ CH_2 \\ | \\ e.CH \\ | \\ n.CH_2 \end{array} - \begin{array}{c} H \\ | \\ C \\ | \\ H \end{array} - \begin{array}{c} H \\ | \\ C.e \\ | \\ CH_2 \\ | \\ e.CH \\ | \\ CH_2 \\ \cdot \\ n \end{array} \xrightarrow[\text{addition}]{\text{(Intra -}}$$

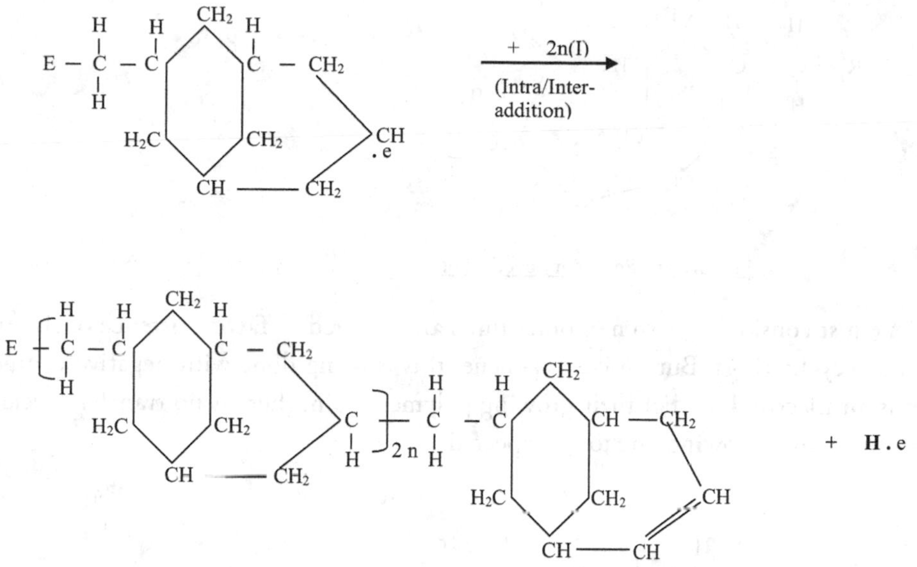

$$+ \quad \textbf{H} \cdot e$$

15.58

Like the positively charged case, two monomers units are required to favor primary and secondary cyclization. In general, as can be observed so far, mostly electro-free-radical and positively charged initiators can be used for cyclization reactions. Thus, the 1, 4 - diene like propylene, is a nucleophile which can only favor electro-free-radical and positively charged routes.

For *1,4-carbonyl diene (A Ketone- Acrolediene) an Electrophile,* the followings should be expected chargedly.

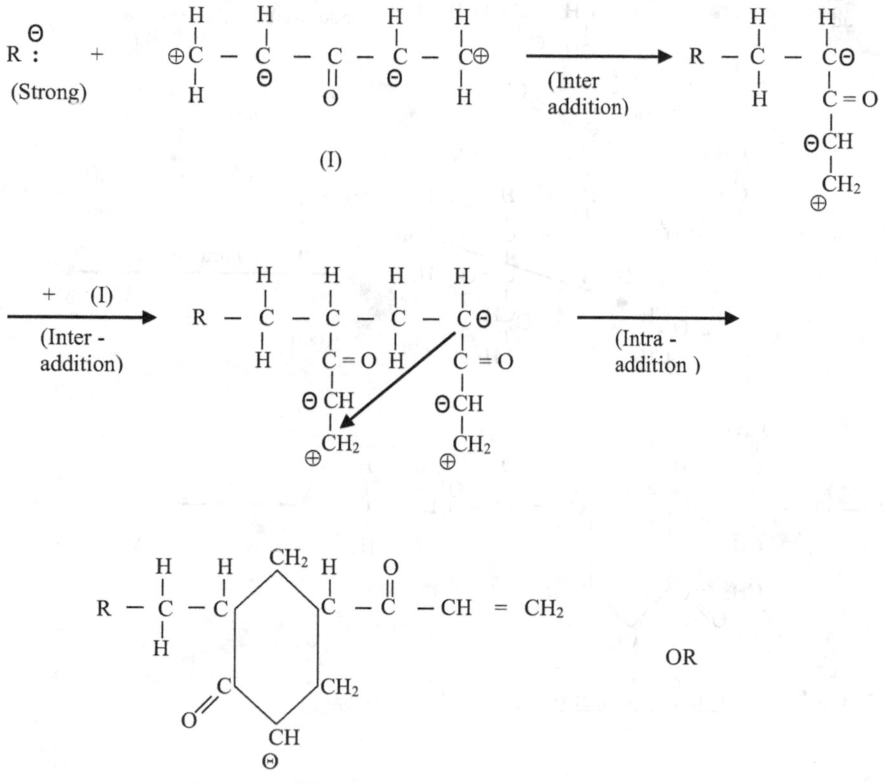

<u>Primary cyclization</u>

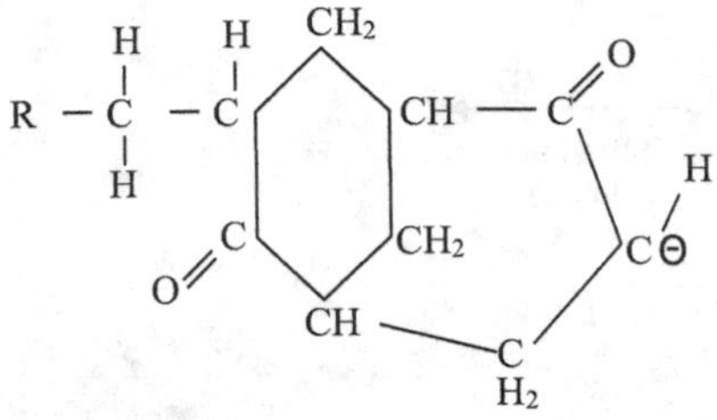

Primary and Secondary cyclization

15.59

Like 1, 4 - diene just considered, two monomer units are required to favor existence of six - membered monocyclic and bicyclic rings. But unlike 1,4-diene, this is being done with negative charges because the monomer is an Electrophile. For their growing polymer chain, there is no transfer species to reject. Therefore, positively the followings are to be expected.

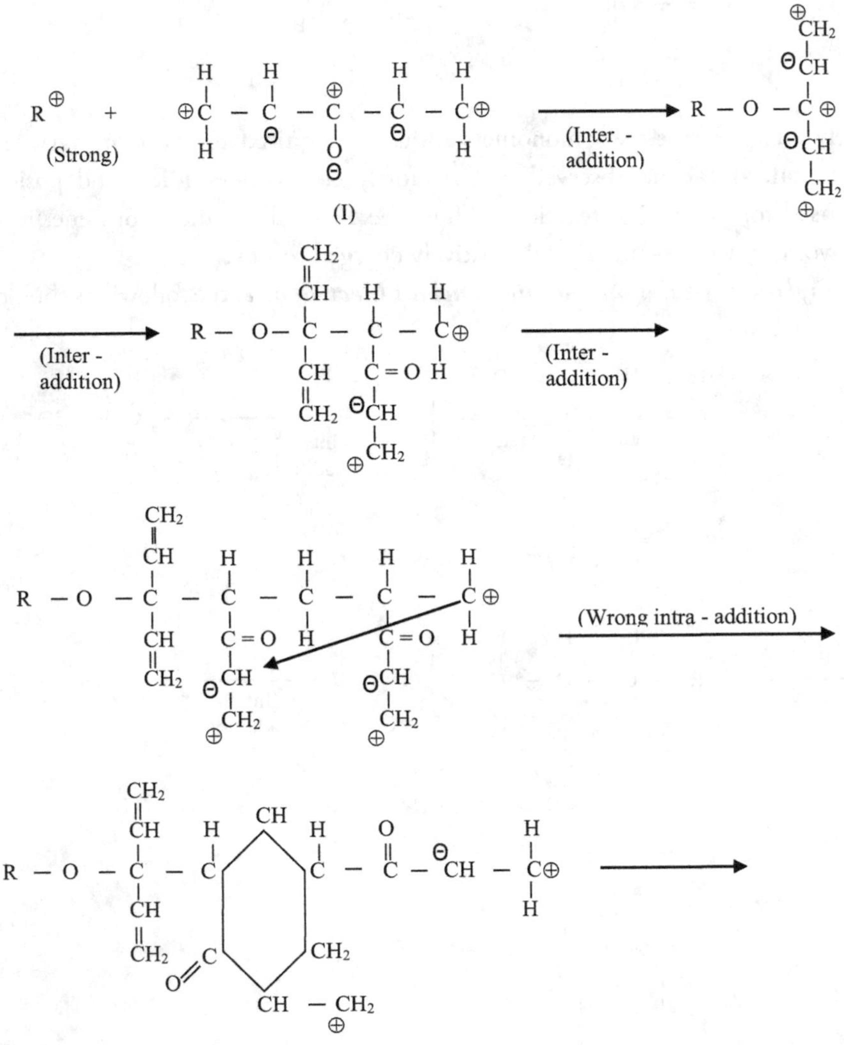

Primary cyclization

548

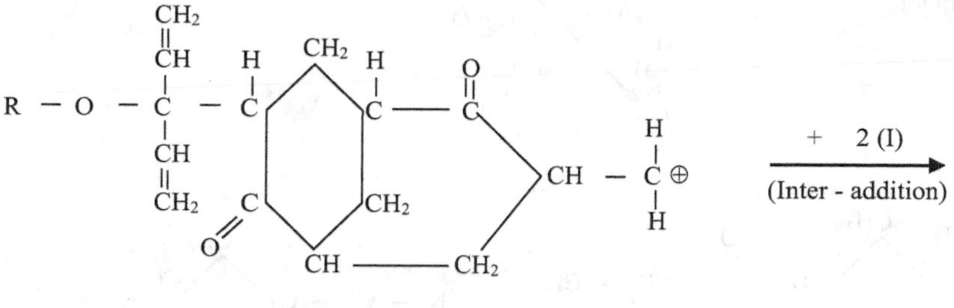

Secondary cyclization

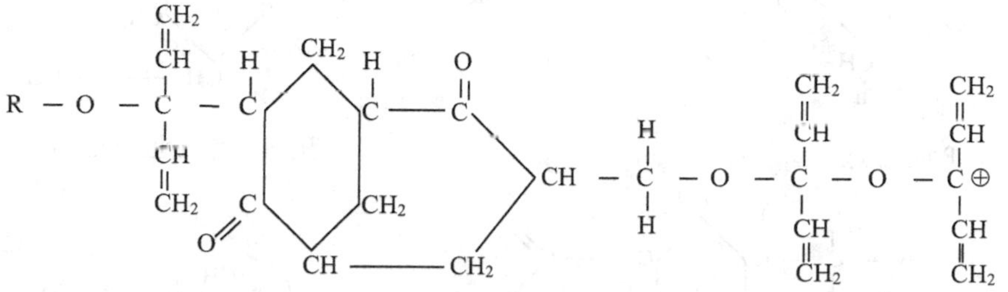

$$15.60$$

The intra-addition above is not to be expected. However, it is important to note the similarity between the present monomer and acrolein as shown below.

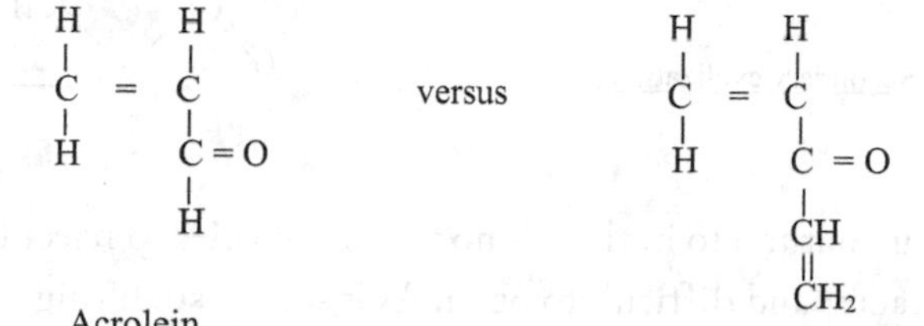

Acrolein   versus   Present monomer

$$15.61$$

With negatively charged-paired initiator, polymerization is favored only via the C = C activation center and not the C = O activation center, the monomer being an Electrophile. Positively, polymerization is favored through both centers to produce copolymers (random) with more of C = O activation center along the chain than of C = C center, since C = O center is the X center. It is more nucleophilic than C = C center. Hence positively, the existence of primary and or secondary cyclization along the ring will be very limited, if any, since its presence is only possible when two or more Y centers are placed side by side.

Free - radically, the followings are to be expected.

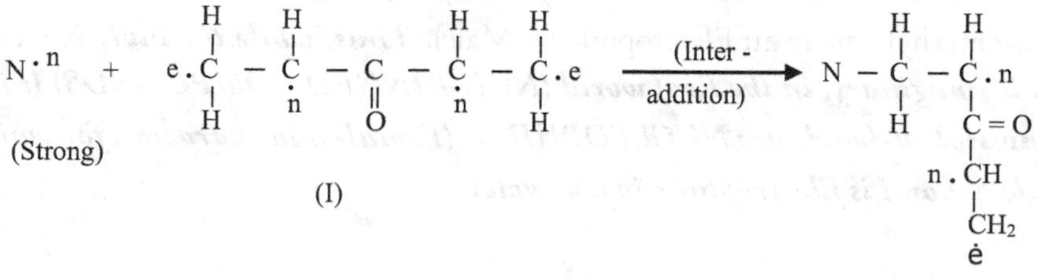

549

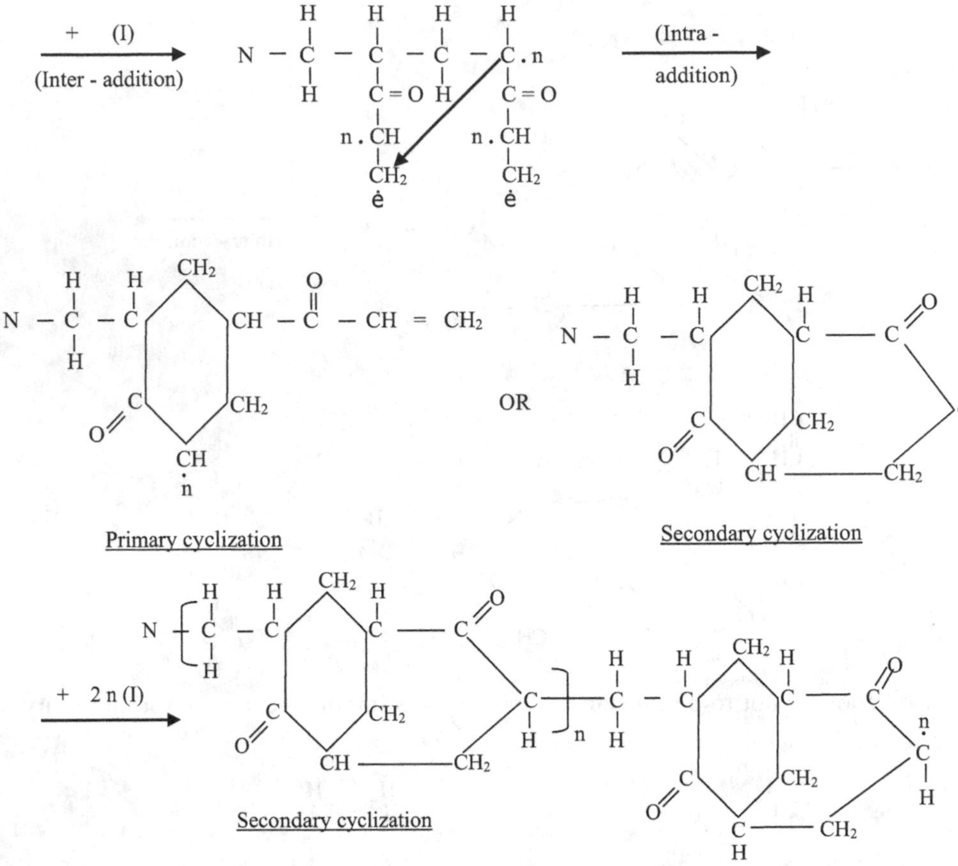

$$\xrightarrow[\text{(Inter - addition)}]{+ \quad \text{(I)}} \quad N - \overset{H}{\underset{H}{C}} - \overset{H}{\underset{\underset{n.CH}{C=O}}{C}} - \overset{H}{\underset{H}{C}} - \overset{H}{\underset{\underset{n.CH}{C=O}}{C}}.n \quad \xrightarrow[\text{addition)}]{\text{(Intra -}}$$

$$\underset{\dot{e}}{CH_2} \qquad \underset{\dot{e}}{CH_2}$$

Primary cyclization   OR   Secondary cyclization

Secondary cyclization

15.62

Nucleo-free-radically, the route natural to it, there is no transfer species to reject for this unique monomer. The rings are ketonic in character and difficult to open. As it seems, such rings almost look Electrophilic in character, wherein the Y center externally located is real while the imaginary X center is the invisible π-bond inside the ring. It is a ring which can only be opened instantaneously. For example, consider the opening of a three- or four-membered ring of this family.

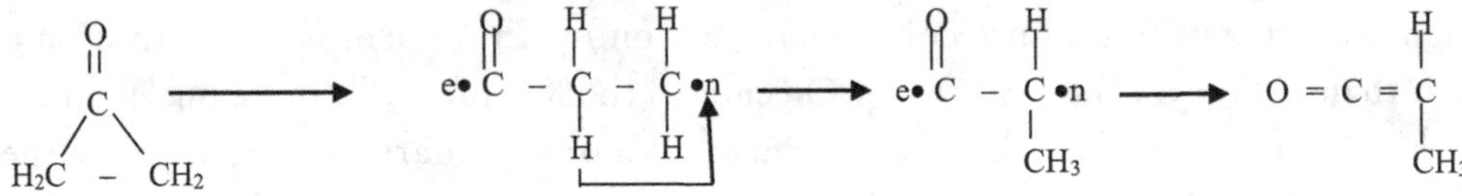

$$\underset{H_2C \ - \ CH_2}{\overset{O}{\underset{\|}{C}}} \longrightarrow e\bullet \overset{O}{\underset{\|}{C}} - \overset{H}{\underset{H}{C}} - \overset{H}{\underset{H}{C}}\bullet n \longrightarrow e\bullet \overset{O}{\underset{\|}{C}} - \overset{H}{\underset{CH_3}{C}}\bullet n \longrightarrow O = C = \overset{H}{\underset{CH_3}{C}}$$

Methyl Ketene
AN ELECTROPHILE   15.63

One can observe that when the ring is opened instantaneously, molecular rearrangement of the third kind takes place to give methyl ketene an Electrophile (A Male). ***Thus, while the invisible π-bond in the world of RINGS is Imaginary, in the Real world (NON-RINGED COMPOUNDS) it is visible and indeed all the Invisible π-bonds are NUCLEOPHILIC (Females) in character, for which there are times the invisible π-bond is Electrophilic in character.***

Using sodium cyanide (NaCN) for example as the source of initiator, the followings are obtained.

$$
\text{Na} \cdot e \;+\; e \cdot \underset{H}{\overset{H}{C}} - \underset{n}{\overset{H}{C}} - \underset{\underset{nn}{\overset{|}{O}}}{\overset{e}{C}} - \underset{n}{\overset{H}{C}} - \underset{H}{\overset{H}{C}} \cdot e \;\xrightarrow[\text{addition}]{\text{(Inter -}}\; \text{Na} - O - \underset{\substack{CH \\ \| \\ CH_2}}{\overset{\substack{CH_2 \\ \| \\ CH}}{C}} \cdot e
$$

(Strong)

(I)

$$
\xrightarrow[\substack{\text{(Inter -} \\ \text{addition}}]{+\ 2\,(I)}\; \text{Na}\left\{ O - \underset{\substack{CH \\ \| \\ CH_2}}{\overset{\substack{CH_2 \\ \| \\ CH}}{C}} \right\}_2 O - \underset{\substack{CH \\ \| \\ CH_2}}{\overset{\substack{CH_2 \\ \| \\ CH}}{C}} \cdot e
$$

15.64

$$
E \cdot e \;+\; e \cdot \underset{H}{\overset{H}{C}} - \underset{\substack{C=O \\ | \\ CH \\ \| \\ CH_2}}{\overset{H}{C}} \cdot n \quad \text{OR} \quad e \cdot \underset{H}{\overset{H}{C}} - \underset{\substack{C=O \\ | \\ n \cdot CH \\ | \\ e \cdot CH_2}}{\overset{H}{C}} \cdot n \;\longrightarrow
$$

(Weak

$$
ECH = CH_2 \;+\; e \cdot \underset{}{\overset{O}{\overset{\|}{C}}} - \underset{H}{\overset{H}{C}} = \underset{H}{\overset{H}{C}}
$$

15.65

It is important to note that, when a transfer species is in an activated form, it cannot be abstracted radically or chargedly. For the case above, with the use of a weak electro-free-radical initiator, initiation is not favored via the C=C center, because of presence of transfer species of the first kind of the first type (-CH=CH$_2$). This transfer species cannot be abstracted chargedly, because of electrodynamic forces of repulsion resulting from the double bond and the negative charge to be placed on the C center ($^\ominus$CH=CH$_2$) for the positive charge to abstract. ***This clearly indicates that during abstraction, the component to be abstracted is known by the abstractor which puts the component to be abstracted disengaged and placed in an opposite charged or radical state.*** Electro-free-radically, only the C=O can be used as shown in Equation 15.64. Recall that, one has shown that NaCN is not a so-called "Anionic ion-paired initiator" that which does not exist, but an electro-free-radical generating initiator. Even at temperatures as low as -50^0 to -40^{0}C in the presence of solvents such as tetrahydrofuran or toluene, it breaks down to give Na•e and cyanogen (N≡C-C≡N). Thus, for this monomer, cyclization is strongly favored largely with negative charges and nucleo-free-radically. Cationically or positively, cyclization cannot take place. Electro-free-radically, cyclization cannot also take place.

For *1,4 - etheric diene,* the followings will be expected chargedly.

$$
R : ^\ominus \;+\; ^\ominus\underset{H}{\overset{H}{C}} - \underset{\substack{O \\ | \\ CH \\ \| \\ CH_2}}{\overset{H}{C}} \oplus \;\longrightarrow\; RCH = CH_2 \;+\; ^\ominus O - \underset{H}{\overset{H}{C}} = \underset{H}{\overset{H}{C}}
$$

15.66

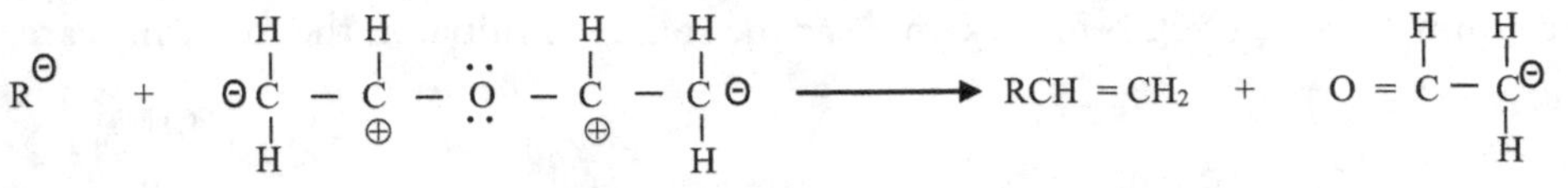

$$R^{\ominus} \quad + \quad {}^{\ominus}\!\overset{\displaystyle H}{\underset{\displaystyle H}{C}} - \overset{\displaystyle H}{\underset{\displaystyle \oplus}{C}} - \ddot{\overset{\displaystyle \cdot\cdot}{O}} - \overset{\displaystyle H}{\underset{\displaystyle \oplus}{C}} - \overset{\displaystyle H}{\underset{\displaystyle H}{C}}{}^{\ominus} \quad \longrightarrow \quad RCH = CH_2 \quad + \quad O = \overset{\displaystyle H}{C} - \overset{\displaystyle H}{\underset{\displaystyle H}{C}}{}^{\ominus}$$

15.67

Whether the anionic or negatively charged initiators are strong or weak, the transfer species must be abstracted. This monomer similar to **alkyl vinyl ethers** cannot favor the anionic or negatively charged routes. Therefore, no cyclization can take place.

Cationically or indeed positively, the following is obtained.

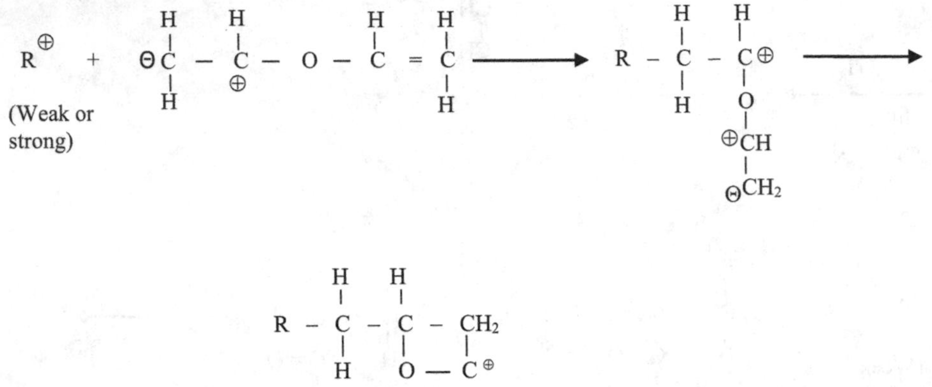

$$R^{\oplus} \quad + \quad {}^{\ominus}\!\overset{\displaystyle H}{\underset{\displaystyle H}{C}} - \overset{\displaystyle H}{\underset{\displaystyle \oplus}{C}} - O - \overset{\displaystyle H}{C} = \overset{\displaystyle H}{\underset{\displaystyle H}{C}} \quad \longrightarrow \quad R - \overset{\displaystyle H}{\underset{\displaystyle H}{C}} - \overset{\displaystyle H}{\underset{\displaystyle \underset{\underset{\ominus CH_2}{\oplus CH}}{O}}{C}}{}^{\oplus} \quad \longrightarrow$$

(Weak or strong)

$$R - \overset{\displaystyle H}{\underset{\displaystyle H}{C}} - \overset{\displaystyle H}{\underset{\displaystyle O}{C}} - \overset{\displaystyle}{\underset{\displaystyle C^{\oplus}}{CH_2}}$$

15.68

The existence of the four-membered cyclic ether with two radical-pushing groups will be favored, because the amount of strain energy in the ring is less than the MRSE for the ring. This will exist up to a particular point along the chain, after which the ring will be too strained to exist. This will then be followed with the use of two monomers as shown below.

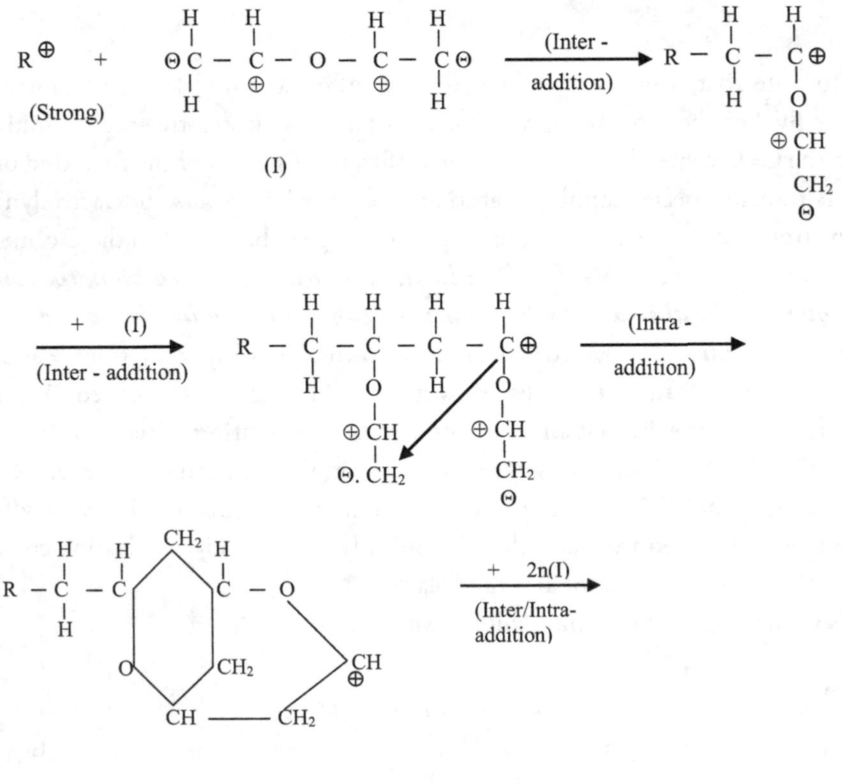

$$R^{\oplus} \quad + \quad {}^{\ominus}C\overset{\displaystyle H}{\underset{\displaystyle H}{}} - \overset{\displaystyle H}{\underset{\displaystyle \oplus}{C}} - O - \overset{\displaystyle H}{\underset{\displaystyle \oplus}{C}} - \overset{\displaystyle H}{\underset{\displaystyle H}{C}}{}^{\ominus} \quad \xrightarrow[\text{addition}]{\text{(Inter -}} \quad R - \overset{\displaystyle H}{\underset{\displaystyle H}{C}} - \overset{\displaystyle H}{\underset{\displaystyle \underset{\underset{CH_2 \; \ominus}{\oplus CH}}{O}}{C}}{}^{\oplus}$$

(Strong)

(I)

$$\xrightarrow[\text{(Inter - addition)}]{+ \quad (I)} \quad R - \overset{\displaystyle H}{\underset{\displaystyle \underset{\underset{\ominus . \, CH_2}{\oplus CH}}{H}}{C}} - \overset{\displaystyle H}{\underset{\displaystyle O}{C}} - \overset{\displaystyle H}{\underset{\displaystyle H}{C}} - \overset{\displaystyle H}{\underset{\displaystyle \underset{\underset{CH_2 \; \ominus}{\oplus CH}}{O}}{C}}{}^{\oplus} \quad \xrightarrow[\text{addition}]{\text{(Intra -}}$$

$$R - \overset{\displaystyle H}{\underset{\displaystyle H}{C}} - \overset{\displaystyle H}{C} \underset{\text{ring structure}}{\cdots} \quad \xrightarrow[\text{(Inter/Intra-addition)}]{+ \quad 2n(I)}$$

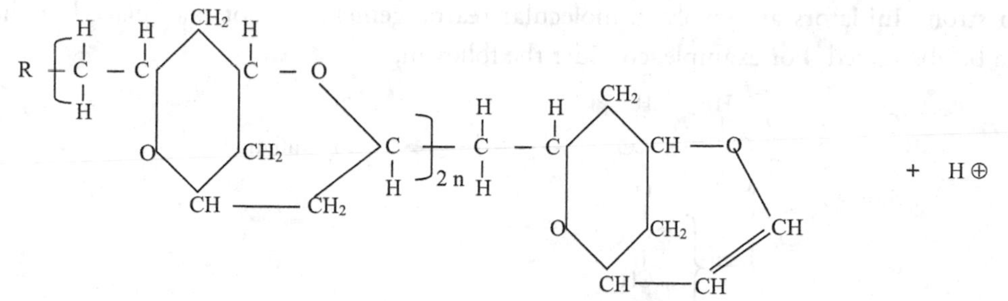

$$R \left[\begin{matrix} H & H & CH_2 & H \\ | & | & & | \\ C & - & C & \overset{}{\underset{}{C}} - O \\ | & & & \\ H & & & \end{matrix} \right]_{2n} \cdots \quad + \quad H^{\oplus}$$

15.69

Like the cases we have seen so far, two monomers units are required to favor primary and secondary cyclization. Notice that the transfer species rejected is not the same as was abstracted during initiation and this is not to be expected, because of the Laws of Conservation of transfer of transfer species. When the growth is about to reach its optimum chain length, a ring is probably never formed. It is at this point that transfer species is released to kill the chain, if it has transfer species.

Free-radically the media different from paired-media, the followings are obtained.

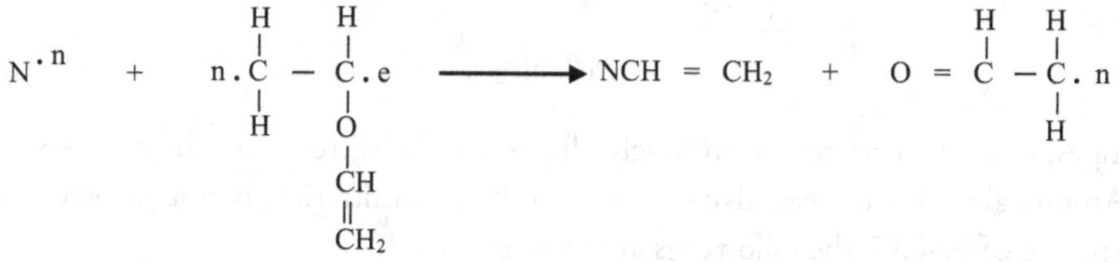

$$N^{\cdot n} \quad + \quad n \cdot \overset{H}{\underset{H}{C}} - \overset{H}{\underset{O}{C}} \cdot e \quad \longrightarrow \quad NCH = CH_2 \quad + \quad O = \overset{H}{\underset{H}{C}} - \overset{H}{C} \cdot n$$

15.70

$$:N^{\cdot nn} \quad + \quad n \cdot \overset{H}{\underset{H}{C}} - \overset{H}{\underset{\dot{e}}{C}} - O - \overset{H}{\underset{\dot{e}}{C}} - \overset{H}{\underset{H}{C}} \cdot n \quad \longrightarrow \quad \overset{nn \cdot}{O} - \overset{H}{C} = \overset{H}{\underset{H}{C}} \quad + \quad :NCH = CH_2$$

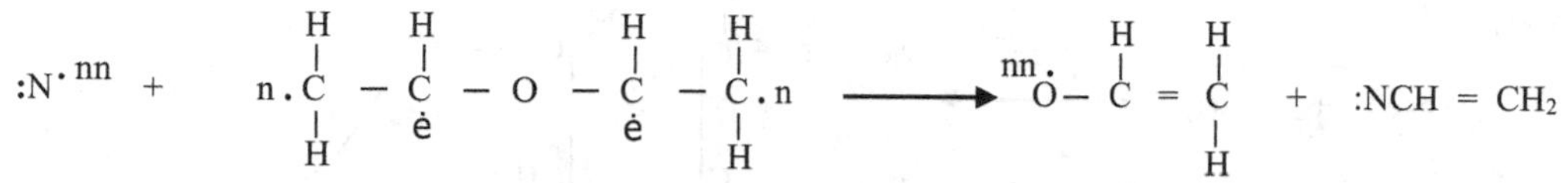

15.71

Nucleo-free- or non-free-radically, the monomer cannot be initiated and cyclization cannot take place. Electro-free-radically, cyclization reaction is favored in view of the fact that the monomer has no transfer species of any kind to be rejected.

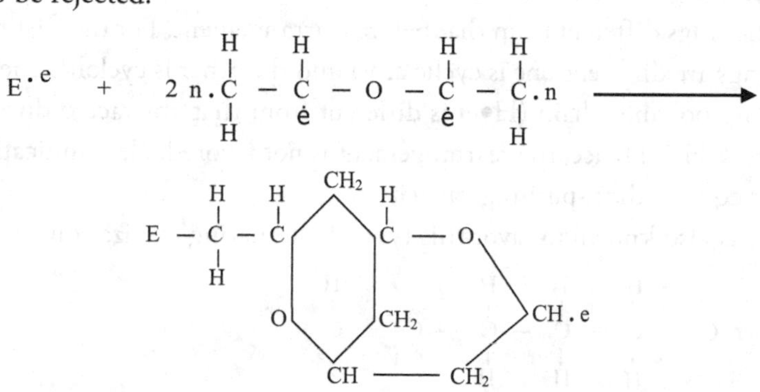

$$E^{\cdot e} \quad + \quad 2n \cdot \overset{H}{\underset{H}{C}} - \overset{H}{\underset{\dot{e}}{C}} - O - \overset{H}{\underset{\dot{e}}{C}} - \overset{H}{\underset{H}{C}} \cdot n \quad \longrightarrow$$

$$E - \overset{H}{\underset{H}{C}} - \overset{H}{C} \overset{CH_2}{\diagup} \overset{H}{\underset{}{C}} - O \diagdown CH \cdot e$$

15.72

Thus, it can be observed that cyclization for this monomer is favored positively (paired) and electro-free-radically, something which is to be expected. Bicyclic six - membered rings are produced, if cross-links are not to be obtained.

When strong initiators are involved, molecular rearrangement cannot take place but the transfer species can be abstracted. For example, consider the followings.

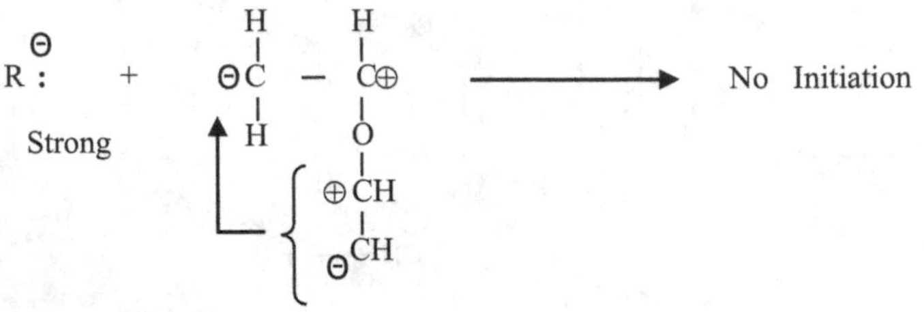

$$
R: {}^{\ominus} \quad + \quad \overset{\ominus}{C} \overset{H}{\underset{H}{|}} - \overset{\oplus}{C} \overset{H}{\underset{O}{|}} \quad \longrightarrow \quad \text{No Initiation}
$$

Strong

$$
\left\{ \begin{array}{c} \oplus CH \\ \ominus CH \end{array} \right.
$$

15.73

$$
R: {}^{\ominus} \quad + \quad \overset{\ominus}{C} \overset{H}{\underset{H}{|}} - \overset{\oplus}{C} \overset{H}{\underset{O}{|}} \quad \longrightarrow \quad R: {}^{\ominus} \quad + \quad \overset{\oplus}{C} \overset{H}{|} - O {}^{\ominus}
$$

Weak

$$
\left\{ \begin{array}{c} CH \\ \parallel \\ CH_2 \end{array} \right.
$$

$$
\begin{array}{c} CH_2 \\ | \\ CH \\ \parallel \\ CH_2 \end{array}
$$

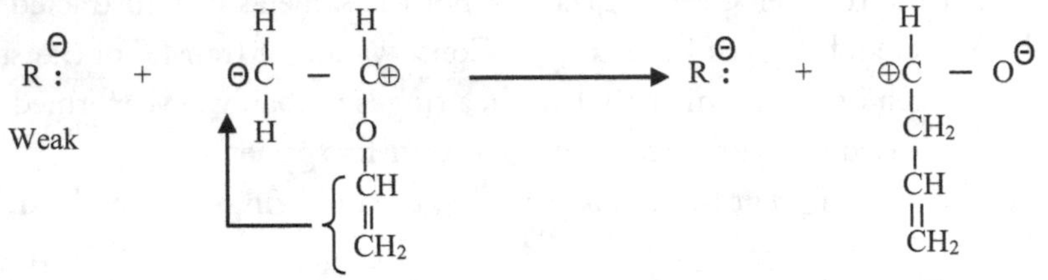

No Initiation   (I)   15.74

When it rearranges, only the cationic or positively charged and electro-free-radical routes still remains favored by it. Anionically or with negatively charged initiator, initiation is not favored. Electro-free-radically, with the use of NaCN, the followings are to be expected.

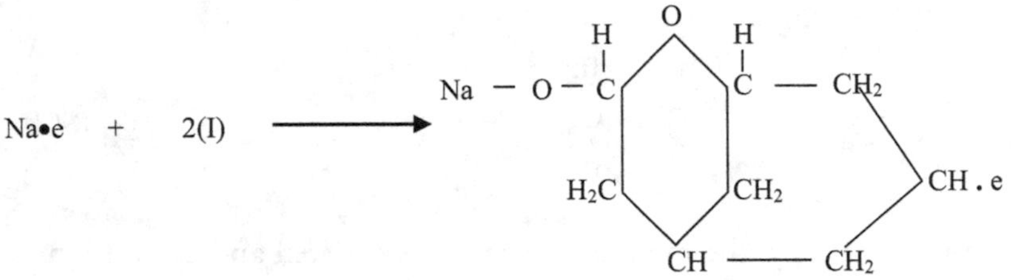

$$
Na\bullet e \quad + \quad 2(I) \quad \longrightarrow
$$

15.75

When molecular rearrangement is strongly favored, cyclization takes place to give six-membered monocyclic or bicyclic rings different from that before rearrangement. For the first time, one can observe that the two fused rings are different-one is cyclic ether and the other is cyclohexane. The transfer species rejected for its growing polymer chain (H•e) is different from that abstracted during initiation nucleo-free-radically (e• CH=CH$_2$). Hence, the rearrangement is not favored, clear indication that –CH = CH$_2$ group is of greater or equal radical-pushing capacity than H.

1, 6 - heptadiyne is also known to favor this type of alternating cyclization reaction.

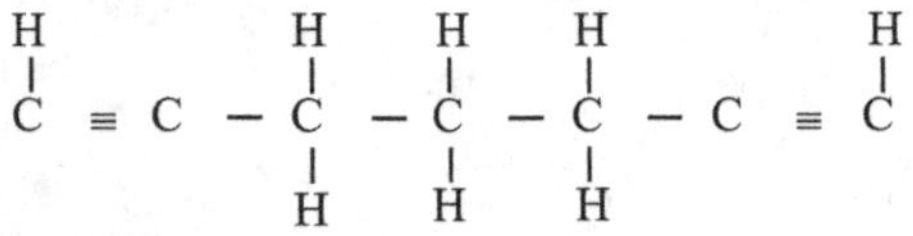

$$
\overset{H}{\underset{}{C}} \equiv C - \overset{H}{\underset{H}{C}} - \overset{H}{\underset{H}{C}} - \overset{H}{\underset{H}{C}} - C \equiv \overset{H}{\underset{}{C}}
$$

1, 6 - heptadiyne

15.76

Chargedly, the triple bond cannot be activated due to electrodynamic forces of repulsion between the π-bond and a negative charge adjacently located to it. When activated nucleo-free-radically, the followings are obtained.

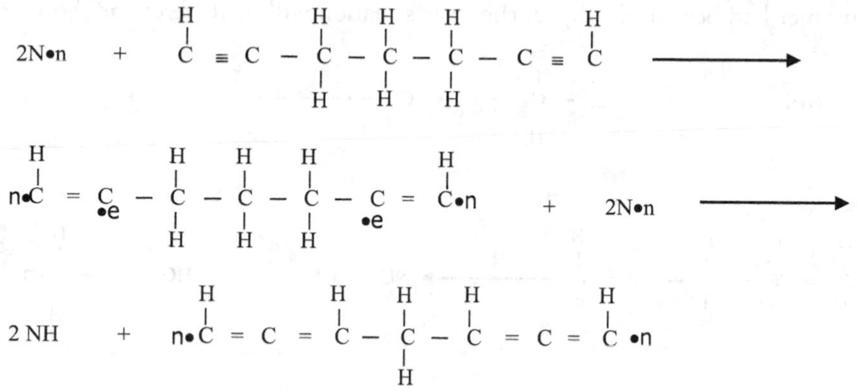

$$2N\bullet n \quad + \quad \overset{\overset{\textstyle H}{|}}{C} \equiv C - \overset{\overset{\textstyle H}{|}}{\underset{\underset{\textstyle H}{|}}{C}} - \overset{\overset{\textstyle H}{|}}{\underset{\underset{\textstyle H}{|}}{C}} - \overset{\overset{\textstyle H}{|}}{\underset{\underset{\textstyle H}{|}}{C}} - C \equiv \overset{\overset{\textstyle H}{|}}{C} \quad \longrightarrow$$

$$n\bullet \overset{\overset{\textstyle H}{|}}{C} = \underset{\underset{\textstyle \bullet e}{}}{C} - \overset{\overset{\textstyle H}{|}}{\underset{\underset{\textstyle H}{|}}{C}} - \overset{\overset{\textstyle H}{|}}{\underset{\underset{\textstyle H}{|}}{C}} - \overset{\overset{\textstyle H}{|}}{\underset{\underset{\textstyle H}{|}}{C}} - \underset{\underset{\textstyle \bullet e}{}}{C} = C\bullet n \quad + \quad 2N\bullet n \quad \longrightarrow$$

$$2\,NH \quad + \quad n\bullet \overset{\overset{\textstyle H}{|}}{C} = C = \overset{\overset{\textstyle H}{|}}{C} - \overset{\overset{\textstyle H}{|}}{\underset{\underset{\textstyle H}{|}}{C}} - \overset{\overset{\textstyle H}{|}}{C} = C = \overset{\overset{\textstyle H}{|}}{C} \bullet n$$

15.77

Electro-free-radically, the followings are obtained.

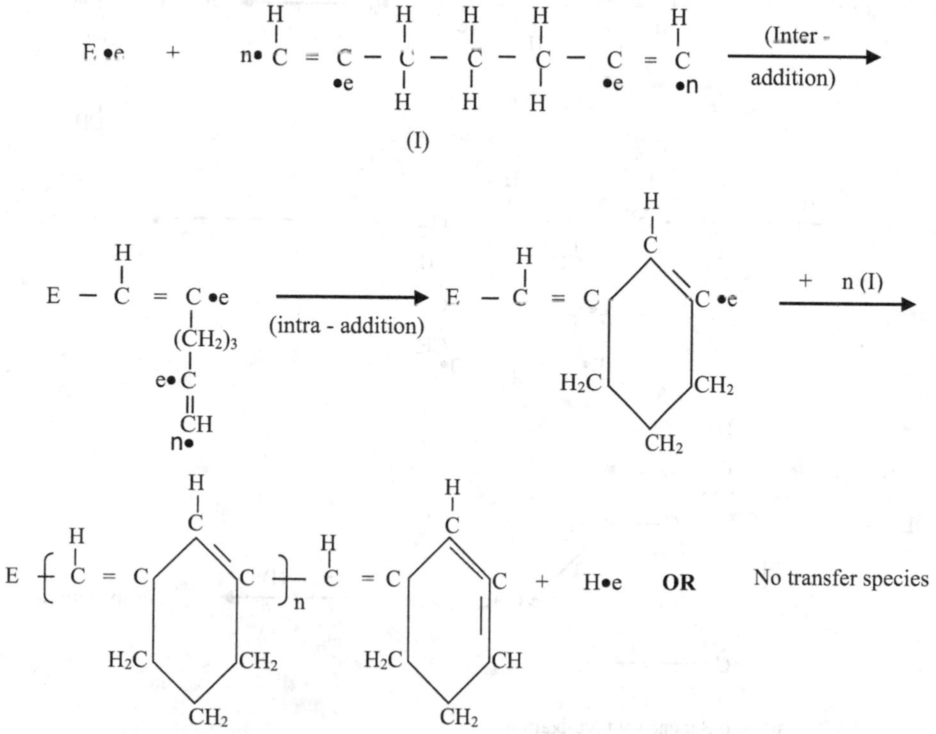

$$F \bullet e \quad + \quad n\bullet \overset{\overset{\textstyle H}{|}}{C} = \underset{\underset{\textstyle \bullet e}{}}{C} - \overset{\overset{\textstyle H}{|}}{\underset{\underset{\textstyle H}{|}}{C}} - \overset{\overset{\textstyle H}{|}}{\underset{\underset{\textstyle H}{|}}{C}} - \overset{\overset{\textstyle H}{|}}{\underset{\underset{\textstyle H}{|}}{C}} - \underset{\underset{\textstyle \bullet e}{}}{C} = \overset{\overset{\textstyle H}{|}}{\underset{\underset{\textstyle \bullet n}{}}{C}} \quad \xrightarrow{\begin{array}{c}\text{(Inter -}\\ \text{addition)}\end{array}}$$

(I)

Strained terminal ring

15.78a

If the terminal rings is going to be seriously strained when transfer species is rejected, such as the case above, then it will not be rejected. Indeed a ring is not formed at the end. A transfer species is however released to form a dead cumulenic terminal. Only one monomer unit is required to favor the existence of a six-membered cycloalkene ring along the chain. This is still a co-polymer -the ring and a sort of unactivated

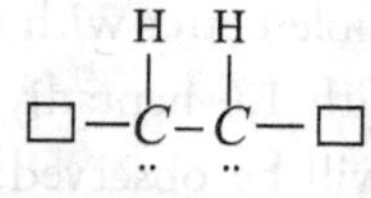

$$\square - \overset{\overset{\textstyle H}{|}}{\underset{\underset{\textstyle \cdot\cdot}{}}{C}} - \overset{\overset{\textstyle H}{|}}{\underset{\underset{\textstyle \cdot\cdot}{}}{C}} - \square$$

Unactivated Ethylyne (From Ethyne ≡ Acetylene)

15.78b

alkylyne (Half of it) as shown above. This is analogous to carbenes which when activated give alkylenes, the first member of which is methylene. When the above is activated, acetylene or ethyne is obtained, just as when carbene is activated to give methylene, ethene is obtained.

If the monomer had been 1, 4 - diyne, then the situation will be different as shown below.

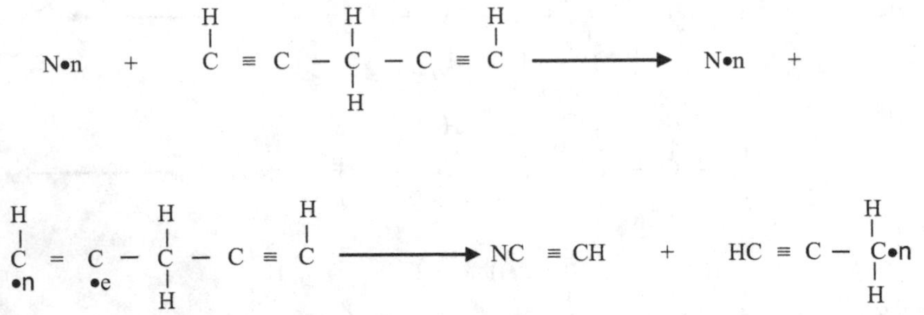

$$N\bullet n \quad + \quad \underset{\underset{H}{|}}{\overset{\overset{H}{|}}{C}} \equiv C - \underset{\underset{H}{|}}{\overset{\overset{H}{|}}{C}} - C \equiv \overset{\overset{H}{|}}{C} \quad \longrightarrow \quad N\bullet n \quad +$$

$$\underset{\bullet n}{\overset{\overset{H}{|}}{C}} = \underset{\bullet e}{C} - \underset{\underset{H}{|}}{\overset{\overset{H}{|}}{C}} - C \equiv C \quad \longrightarrow \quad NC \equiv CH \quad + \quad HC \equiv C - \underset{\underset{H}{|}}{\overset{\overset{H}{|}}{C}}\bullet n \qquad 15.79$$

$$E\bullet e \quad + \quad \underset{\bullet n}{\overset{\overset{H}{|}}{C}} = \underset{\bullet e}{C} - \underset{\underset{H}{|}}{\overset{\overset{H}{|}}{C}} - \underset{\bullet e}{C} = \overset{\overset{H}{|}}{C}\bullet n \quad \longrightarrow \quad E - \underset{\underset{\overset{|}{CH_2}}{|}}{\overset{\overset{H}{|}}{C}} = C\bullet e$$

(I)

$$+ \quad (I) \quad \longrightarrow \quad E - \overset{\overset{H}{|}}{C} = C - \overset{\overset{H}{|}}{C} = C\bullet e \quad \longrightarrow$$

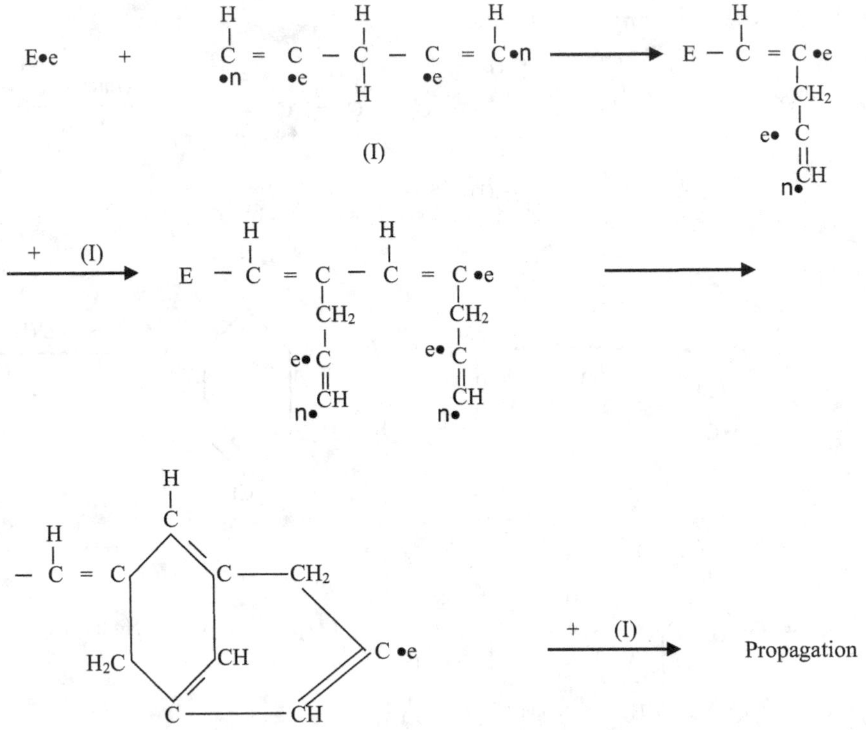

(II) Primary and Secondary Cyclization

15.80

The existence of (II) may be favored, since the six-membered diene rings carrying strong radical-pushing groups, with less than two double bonds shared between the two rings and two common boundaries, are not within the limit of MRSE. Note that, the monomer is almost identical to $HC \equiv CCH_3$. For (I) or 1,6-heptadiyne to favor being used as a monomer, one or both of the externally located hydrogen atoms on acetylene may be replaced with radical-pushing or pulling substituent groups or kept suppressed. Thus, with 1,4-pentadiyne above, cyclization may be possible better with the two radical-pulling groups (Cl) externally located, via electro-free-radical route. With 1,6-heptadiyne of Equation 15.78, very limited polymer-ization (low molecular weight products) will be observed, in view of the unstable character of acetylene. It is unstable in the sense that it is always in Equilibrium state of existence; unless when suppressed. Only the substituted 1,6-heptadiynes will favor being fully used only electro-free-radically for cyclization reactions to produce high molecular weight products.

Because of their unstable state of existence, hence for example, 1,5-hexadiyne is converted to cyclooctadecahexayne in the presence of $Cu(OAC)_2$ in pyridine[13].

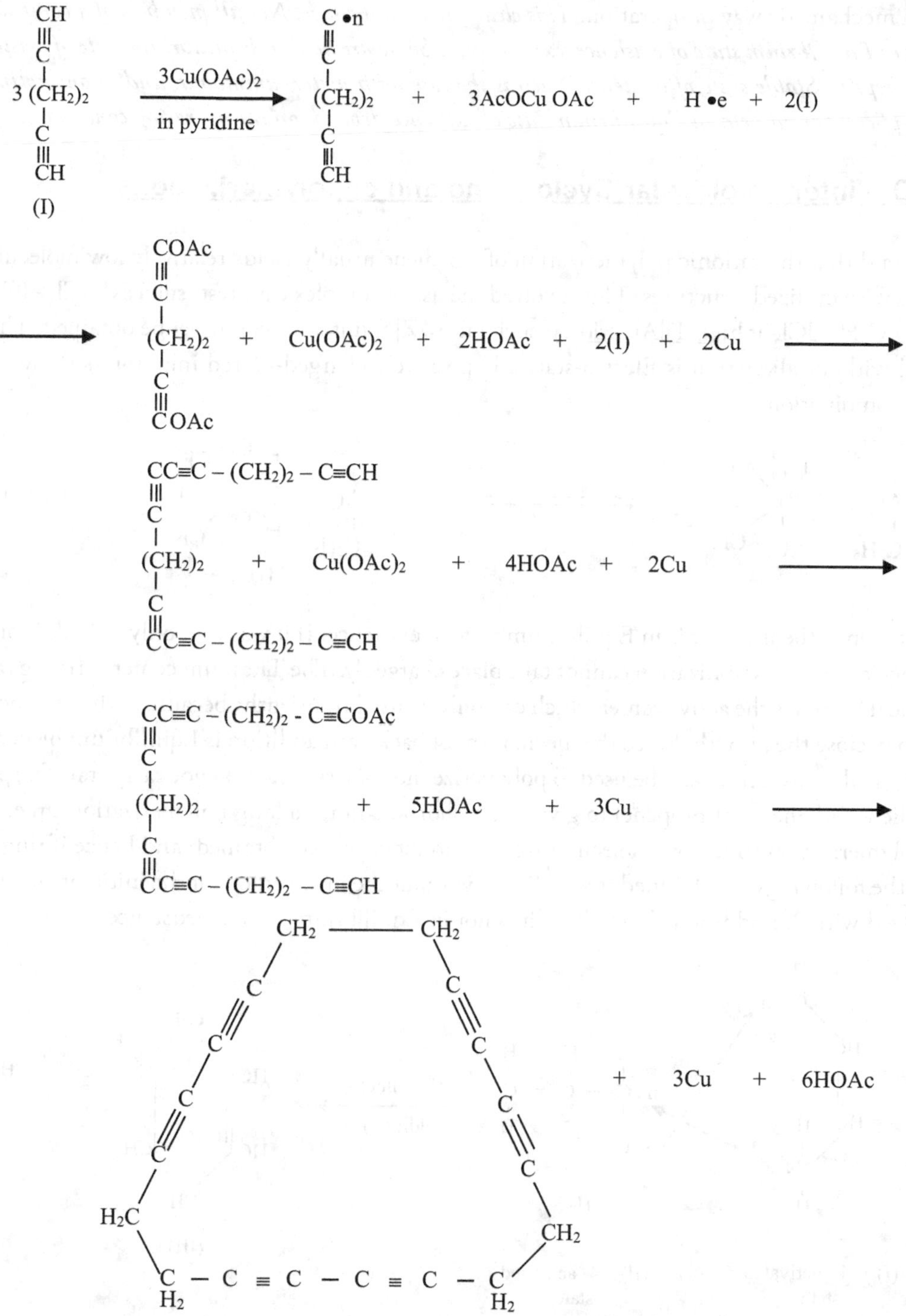

$$3 \begin{array}{c} CH \\ \parallel \\ C \\ \mid \\ (CH_2)_2 \\ \mid \\ C \\ \parallel \\ CH \\ (I) \end{array} \xrightarrow[\text{in pyridine}]{3Cu(OAc)_2} \begin{array}{c} C \bullet n \\ \parallel \\ C \\ \mid \\ (CH_2)_2 \\ \mid \\ C \\ \parallel \\ CH \end{array} + 3AcOCu\,OAc + H \bullet e + 2(I)$$

$$\longrightarrow \begin{array}{c} COAc \\ \parallel \\ C \\ \mid \\ (CH_2)_2 \\ \mid \\ C \\ \parallel \\ COAc \end{array} + Cu(OAc)_2 + 2HOAc + 2(I) + 2Cu \longrightarrow$$

$$\begin{array}{c} CC{\equiv}C - (CH_2)_2 - C{\equiv}CH \\ \parallel \\ C \\ \mid \\ (CH_2)_2 \\ \mid \\ C \\ \parallel \\ CC{\equiv}C - (CH_2)_2 - C{\equiv}CH \end{array} + Cu(OAc)_2 + 4HOAc + 2Cu \longrightarrow$$

$$\begin{array}{c} CC{\equiv}C - (CH_2)_2 - C{\equiv}COAc \\ \parallel \\ C \\ \mid \\ (CH_2)_2 \\ \mid \\ C \\ \parallel \\ CC{\equiv}C - (CH_2)_2 - C{\equiv}CH \end{array} + 5HOAc + 3Cu \longrightarrow$$

+ 3Cu + 6HOAc

15.81

The mechanism above is Equilibrium mechanism of nine stages. Note that Cu is a non-ionic metal carrying non-free-radicals. The 1,5-hexadiyne will favor cyclization to produce five membered rings, if it is substituted or suppressed and MRSE is not present in the ring of such size and content. In the reaction above, the diyne was always in Equilibrium state of existence, one at a time. One was first kept in Equilibrium state of existence from one end in two stages. In the next two stages, the same one was kept again in Equilibrium state of existence from the other end, keeping the others waiting in Stable state of existence. After finishing, the next diyne takes over and exist in Equilibrium state of existence-turn by turn in a mechanistic way of operation. *This clearly sends a message. Not all members of a compound which can exist in Equilibrium state of existence, exist as such. Some are held in Equilibrium state of existence while some are kept in Stable state of existence. Such is the case with water, ammonia, and so on; noting that the fractions of compound held in Equilibrium state of existence depends on the operating conditions.*

15.3. Di - inter - molecular Cyclo homo and copolymerizations

It is said that the cationic polymerization of 1,3-diene usually yields relatively low molecular weight polymers with cyclized structures. This involved the use of complex cata-lysts such as $TiCl_4 - (C_2H_5)_2AlCl$ or $TiCl_4 - C_2H_5AlCl_2$ at high Ti|Al ratio. At such ratios, Z|N initiators can never be obtained. The initiator obtained without alkylation is Electro-statically positively charged-paired initiator as shown below for the first combination.

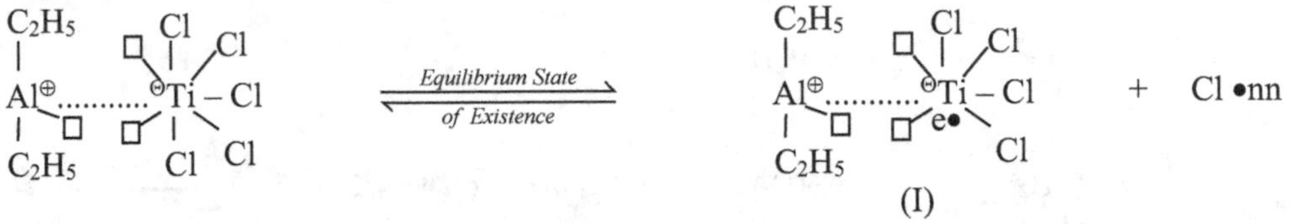

$$(I) \qquad 15.82$$

As it seems, the initiator is in Equilibrium state of existence. If it was not, only 1,2-addition will take place since resonance stabilization cannot take place chargedly. The Titanium center carrying the electro-free-radical becomes the active center which can only be used transiently, because of the presence of Cl·nn which may close the growth. Since the mechanism of backward addition is Equilibrium mechanism, the Cl.nn the real active center can be used to polymerize monomers which do not carry transfer species such as butadiene, ethene (Not propene) to give a very short chain for a long polymerization time. Hence for short polymerization time, low molecular weight products will be obtained, and hence if rings are to be formed the followings are obtained free-radically without the involvement of the initiator which will only be involved with 3,4-addition chargedly if it is not in Equilibrium state of existence.

| (I) 1, 4 - activated state | (II) 3,4- activated state | (III) |

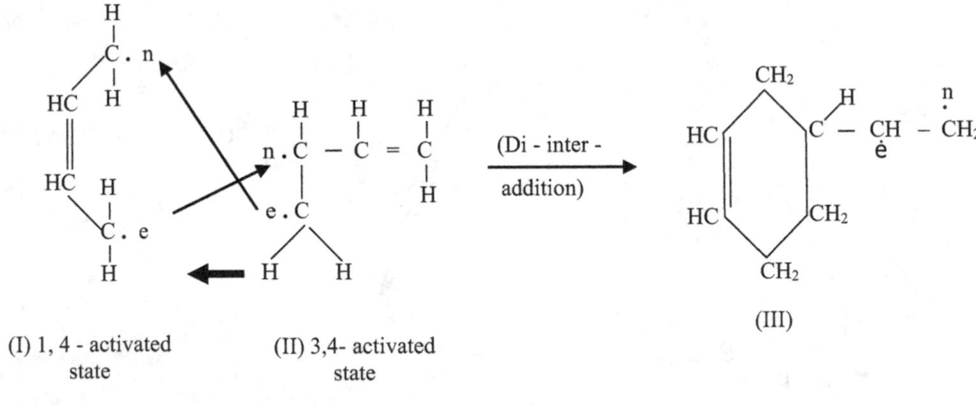

$$15.83$$

The 3,4- activated diene adds to activate a diene to a 1,4- activated diene to produce (III)- vinyl cyclohexene. This can add to another 1,4- activated diene to form another six-membered ring not fused with it. From here upwards, fused six-membered rings are obtained. The center activated is that outside the ring and not the one inside the ring, because it is less nucleophilic than the center inside the ring. Note that it is (II) that is activating a diene to give (I) above and not the other way round. It is also possible to have an eight-membered ring in the process as shown below. This is usually mistakenly said to take place with the use of a Zeigler catalyst. If it is Z/N initiator, it has to be the ones paired radically.

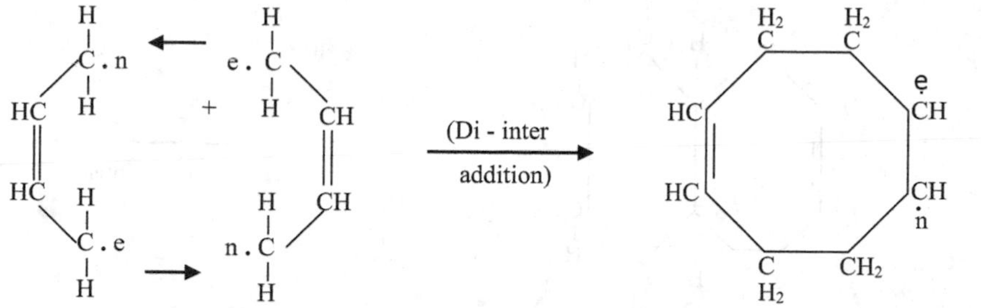

(IV) → (V)   15.84

It is hoped that the 1,5-cyclooctadiene (V), will not undergo so-called trans-annular "migration" as shown below. There is no migration here, but addition. This takes place inside a ring with more than one π-bond activation center. The p-bonds could be isolatedly placed or conjugatedly placed.

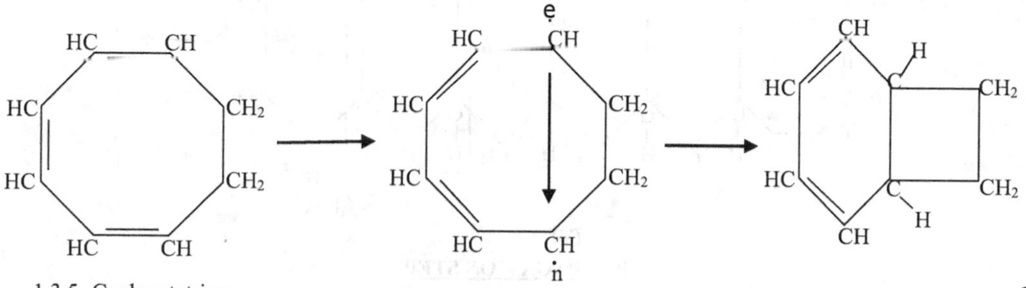

1,3,5,-Cyclooctatriene   15.85a

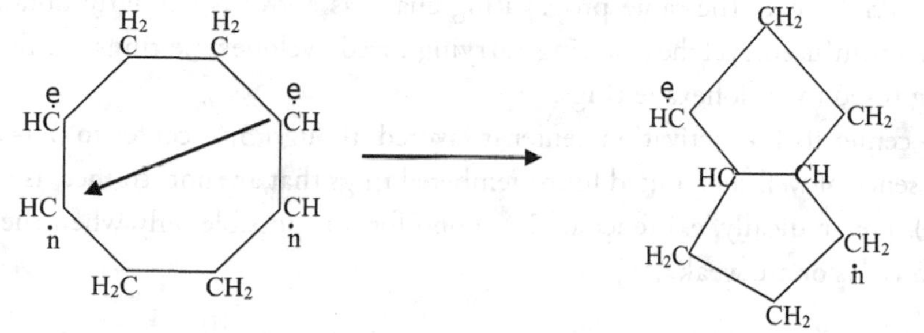

(V) → (V)a   15.85b

While the above can take place only radically, (V) can take place both radically and chargedly. One can observe how complex our world can be. When (V)a is formed, then it eventually becomes one of the monomer units along the chain.

Using (III) or (V) of Equations 15.83 or 15.84 respectively as if it is an initiator (with only one usable center- the electro-free-radical center), the followings take place.

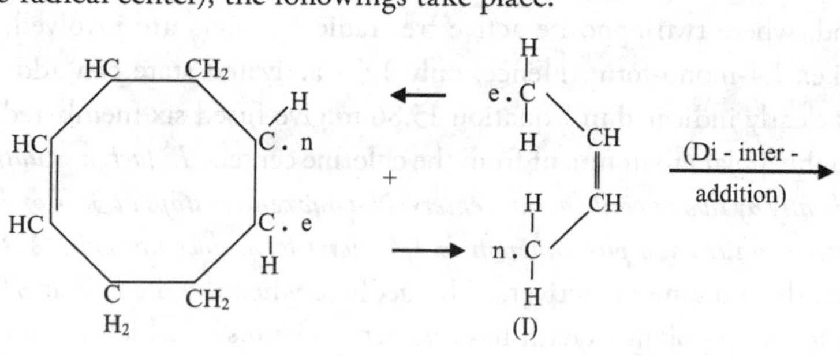

(Di - inter - addition)

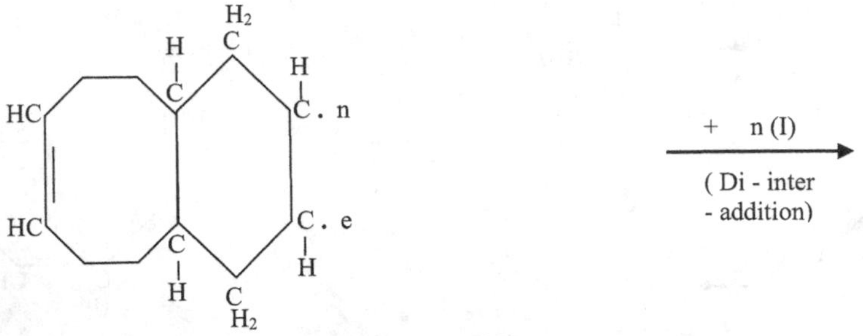

INITIATION STEP

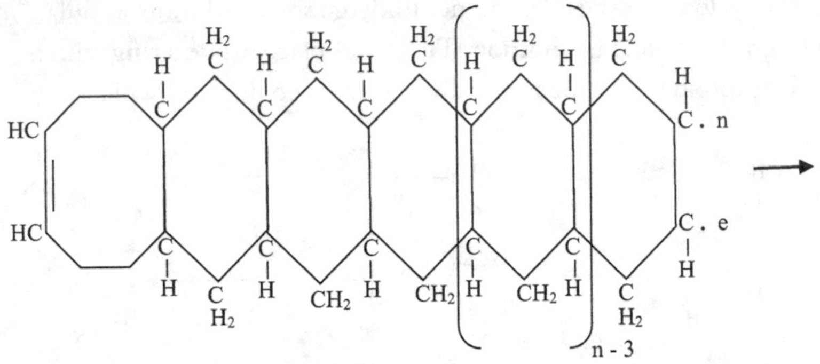

(VI)
PROPAGATION STEP

15.86

With (III)-vinyl cyclohexene, the same propagating chain as above is similarly obtained, except that (III) begins with an unfused cyclohexene ring carrying fused cyclohexane rings, while (V) begins with cyclooctaene ring fused to cyclohexane rings.

Mostly 3,4 - center to 1,4- activation center is favored, though 3,4- center to 3,4- activation center will favor the presence of well substituted four-membered rings that are not strained as shown below (See Equation 15.85a). Free-radically, existence of 3,4-mono-form is possible only when the initiator or heat or electrostatic forces is or are weak.

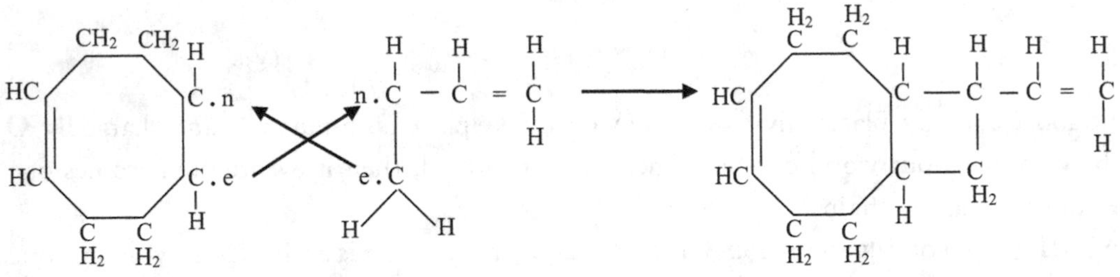

15.87

On the other hand, where two opposite active free-radical centers are involved, the activation of the 1,3-diene is full, i.e., I,4-mono-form. Hence, only 1,4 - activated state can add to (V) or its growing polymer chains as clearly indicated in Equation 15.86 to give fused six membered cyclo-alkane rings in the presence of a highly polar environment from the chlorine centers. *In fact, a situation where the monomer when activated radically by two opposite active centers (Not one center) stops at 3,4- or 1,2-mono-form without undergoing resonance stabilization phenomenon to 1,4-mono-form does not exist. 1,2- or 3,4- mono-forms can only exist when the monomer is activated chargedly or when the diene is at STP when stored (Vinyl cyclohexene).* The growing polymer chain has two active centers to which the monomer adds. In order to prevent existence of low molecular weight polymers, the active growing center must be kept activated all the time, for continued addition. Based on diffusion controlled mechanisms, and the route, the electro-free-radical center carried by the propagating chain diffuses to the activated monomers while at the same time the other electro-free-radical center on the activated monomer diffuses to the other center of the growing chain to produce six-membered ring. ***This is what makes it Di-inter addition.*** The growing polymer chain cannot reject a transfer species to close the chain. Living polymers will largely be produced in the presence of strong activating agents and in the absence of foreign agents in the system. However, when the growing chain reaches its optimum chain length, it closes by itself by Deactivation.

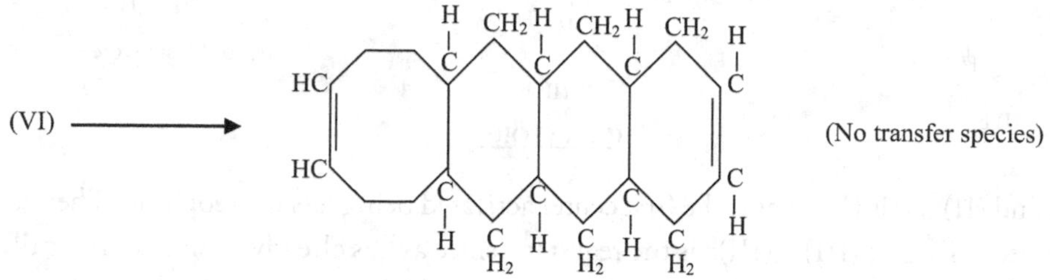

(VI) ⟶

(No transfer species)

(VII)   15.88

Without the type of catalyst combination (carrying lots of polar forces) used above, the favored existence of the reactions above is impossible. The condition here is such that the initiator obtained cannot be used radically. As already said, in addition to these cyclic products, are the 3,4-mono-forms placed along the chain with Al being the carrier. These 3,4 mono-forms form a very small percentage of the overall products, because of the great presence of free-radical polymerization in the system. When Z/N initiator is used, when charged, only 3,4- addition can take place with no existence of 1,4-mono-form along the chain. Thus, it can be observed how these reactions take place. There is nothing like a Diels-Alder type charged transfer mechanism[10,11]. All these are questionable, since they cannot be understood. ***What is meaningful is what can be understood.*** However, their ability to identify such a phenomenon which they could not understand is still commendable. Ionic charges cannot be provided by the catalyst components. From the equation above, dead terminal double bond ringed polymer (VII) is produced, without release of transfer species, since there are two active centers.

Considered above, was 1,3-butadiene. When chloroprene is involved under similar operating conditions, the followings are to be expected.

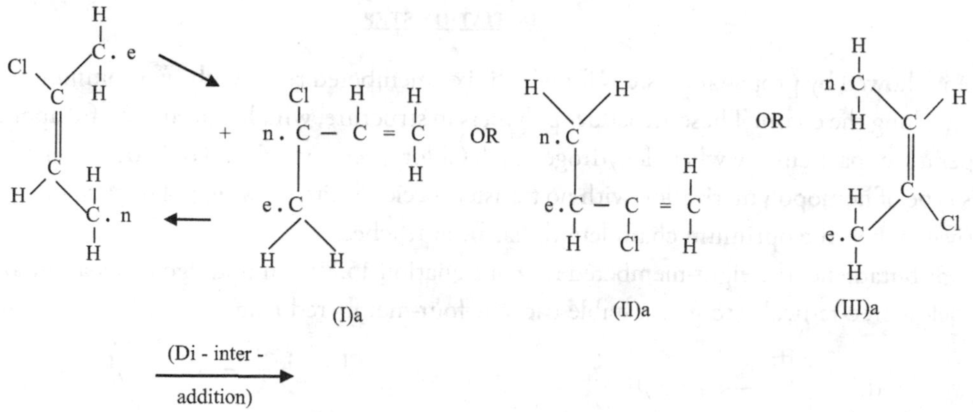

(I)a   (II)a   (III)a

(Di - inter -
addition)

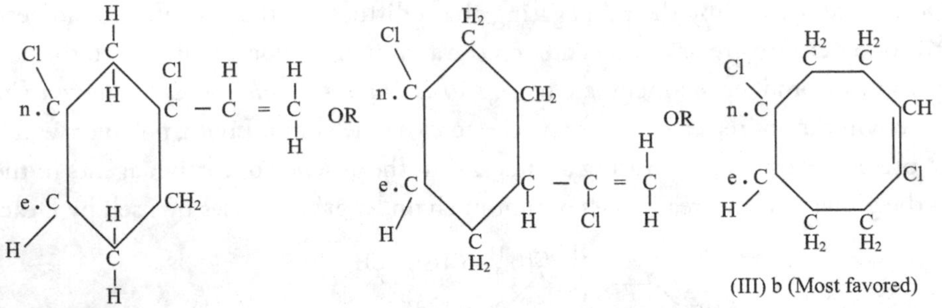

(I) b (II) b (III) b (Most favored)

INITIATORS 15.89

Between (I)a and (II)a, it is (I)a that is the first center activated being less nucleophilic. Therefore (II)a does not exist. Between (I)a and (III)a, (III)a is the real state, since as has already been said, radically 1,4-mono-form predominates over 1,2- or 3,4- mono-forms from which 1,4- monomer was derived, all being Nucleophiles. The center activated above in (I)a is 1,2-mono-form. Note that the centers activated in (I)b and (II)b are the wrong centers. When (III)b is involved, the followings are obtained in the initiation step.

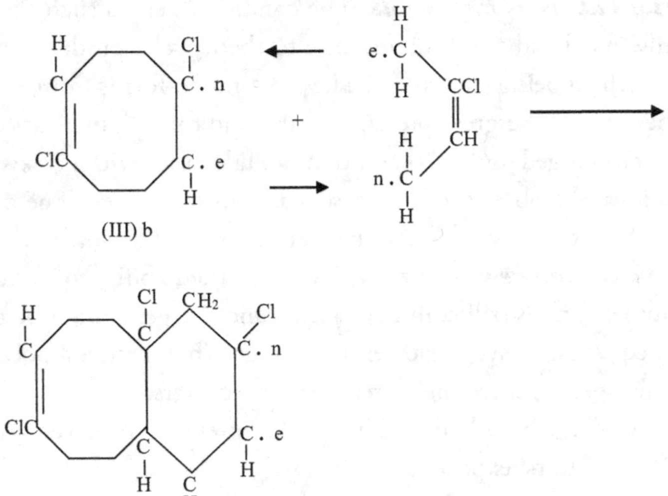

(III) b

INITIATION STEP 15.90

This is then followed by propagation steps for which six - membered rings with two common boundaries are obtained along the chain. These are ladder polymers in structure, which will find useful applications at high temperatures particularly when dehydrogenated. Chloroprene which can only be activated radically, favors this type of homopolymerization with no transfer species for its growing polymer chains. The chain will deactivate when the optimum chain length has been reached.

Like 1,3- butadiene, the eight-membered ring of Equation 15.89 can undergo trans-annular addition as shown below free-radically to give a stable tricyclic four-membered ring,

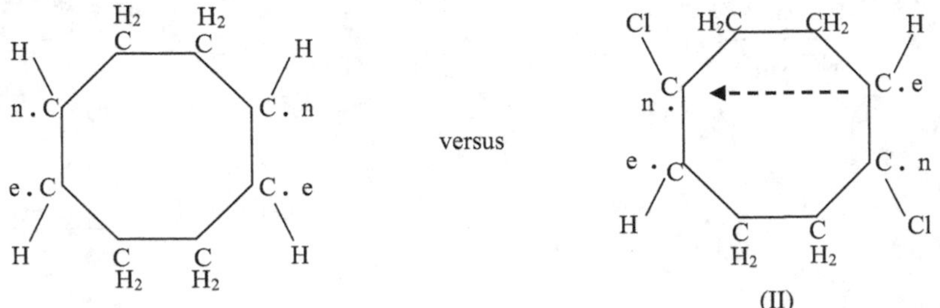

(I) versus (II) 15.91

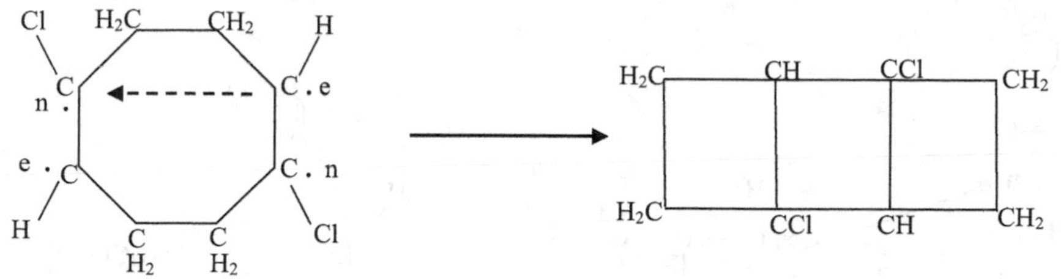

(II)

15.92

(I) rearranges to give two fused five-membered rings with two active centers too far apart to form a ring. (II) has different distribution of the active centers which conveniently add to form a four-membered tricyclic ring above without active centers.

Isoprene will also favor the same type of polymerization, in the absence of any transfer species. It is the absence of transfer species that makes it possible for them to favor nucleo-free radical polymerization when electro-free-radicals are absent in the system. But here, electro-free-radicals are present. For example, considering the case of 1,3- pentadiene, the followings are to be expected.

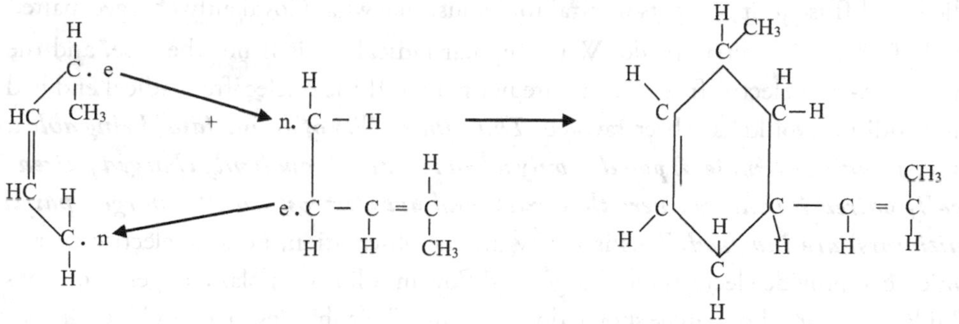

(I)

15.93

Observe that, the presence of transfer species, did not affect the addition, because it is the electro-free-radical end that diffuse all the time. Nucleo-free-radicals cannot diffuse when electro-free-radicals are present.

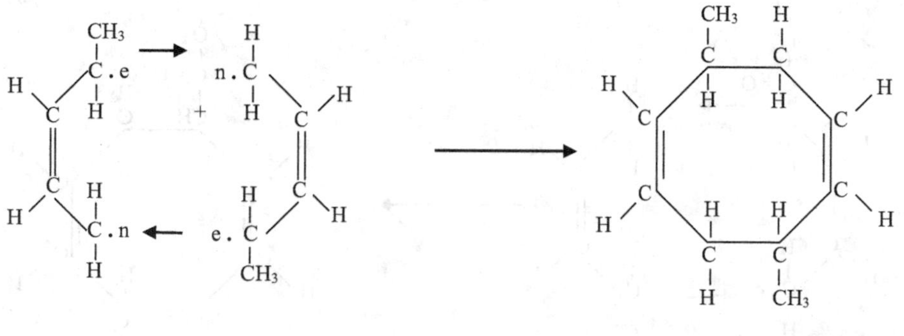

(II)

15.94

(II) above (Dimerization) is usually said to be obtained with the use of Z/N catalyst. This as can be seen is very misleading, because the catalyst is not Z/N since it cannot generate a Z/N initiator. For a 1,3-diene containing an externally located radical-pulling group, the followings are to be expected free-radically.

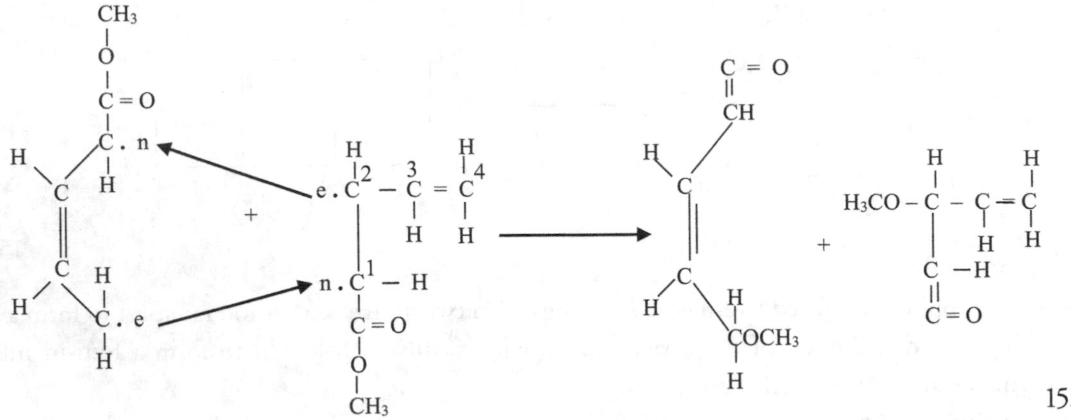

15.95

With the electro-free-radical ends being the ones to first diffuse, one can observe that no rings could be formed. Transfer species were abstracted from each diene to give two different products. However, one should expect that the monomer being an Electrophile, the radicals will be able to see it, making the nucleophilic end diffuse to it being its natural route, just like what Covalently charged-paired initiators (e.g., Z./N, $H_9C_4^{\ominus}$......$^{\oplus}$Li initiators) do. With singular radicals, this is not the case, and the reason is because the nucleo- and electro-free- radicals are not paired. If the nucleo-free-radical end had diffused, then Di-inter addition would have been favored. ***Thus, the ability of an initiator being able to identify if a monomer is male or female, is possible only when ionic and covalently charged paired initiators and radically paired initiators where they exist (and not electrostatically charged-paired or free-radical initiators) are involved.*** This is a very unique observation, because electric currents flow in solution which can provide electrostatic charges and flow in solids via polar charges as one has begun to show for fluids. That for solids will be shown downstream. Their abilities to see and identify is imaginary, because it is there inherently. One cannot rule out the possibility that the C = O may be the center of attack electro-free-radically, for which the possibility of ring formation does not exist.

If the nucleo-free-radical was the first to diffuse, because the monomer is an Electrophile, then the followings would have been obtained.

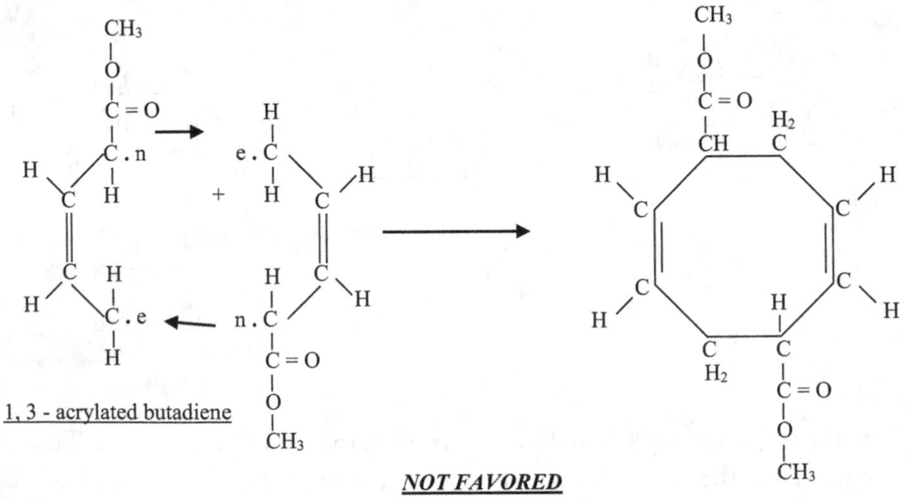

1, 3 - acrylated butadiene

NOT FAVORED

15.96

Thus, as it seems, the nucleo-free-radical cannot diffuse in the presence of an electro-free-radical when not paired. Otherwise, how does a nucleo-free-radical polymerize Nucleophilic monomers such as ethene, vinyl chloride, and more? With methyl sorbate, the situation is even worse and so clear, because both active centers have transfer species. It, like the case above, cannot undergo Di-inter cyclo molecular homopolymerization being strong Electro-philes.

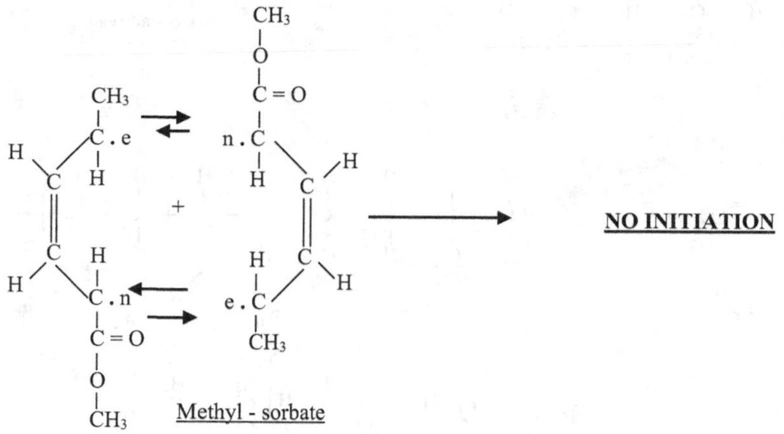

Methyl - sorbate   15.97

Only the trans-form of this monomer can be polymerized using only Covalently charged-paired initiator negatively or Electrostatically negatively charged-paired initiator and not positively as has already been shown upstream, being an electrophile.

When two different Nucleophilic monomers with or without transfer species are involved in this type of homopolymerization, random copolymers will be obtained. For example, in the copolymerization of 1,3-butadiene with chloroprene, random copolymers are obtained with more of chloroprene along the chain than 1,3-butadiene, because chloroprene is less Nucleophilic than the diene.

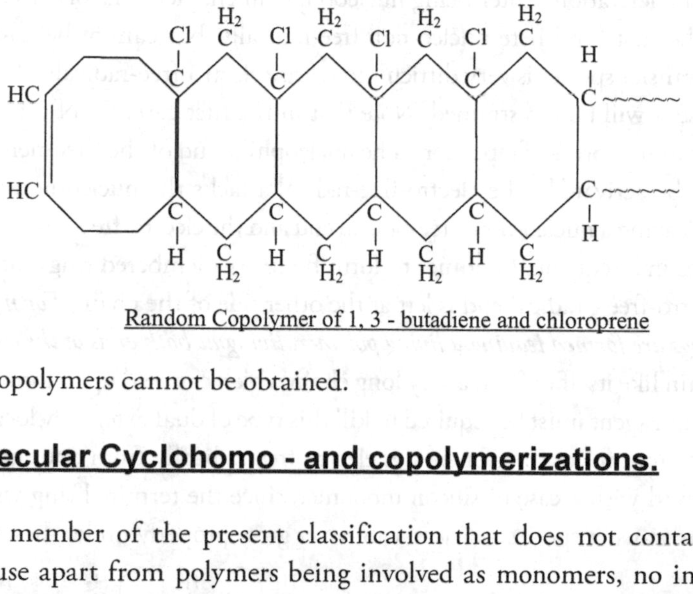

Random Copolymer of 1, 3 - butadiene and chloroprene   15.98

With Electrophiles, copolymers cannot be obtained.

15.4. Intra - molecular Cyclohomo - and copolymerizations.

This is the only member of the present classification that does not contain inter - molecular polymerization, because apart from polymers being involved as monomers, no initiators are involved. The polymers involved in this case are those which have activation centers as neighbors externally located one after the other along the chain to favor the existence of six-membered rings at certain optimum conditions. The existence of the same ring-size along the chain has to be favored on a continuous basis, otherwise in most cases the existence of the rings will not be favored along the chain, noting that no transfer species can be rejected.

The most well-known polymeric monomer which favors this type of polymerization is polyacrylonitrile[14]. **Pyrolysis** of the polymeric monomers, results in cyclization reactions as shown below in the presence of oxygen. It is the heat that activates O_2 which becomes an activating or initiating force.

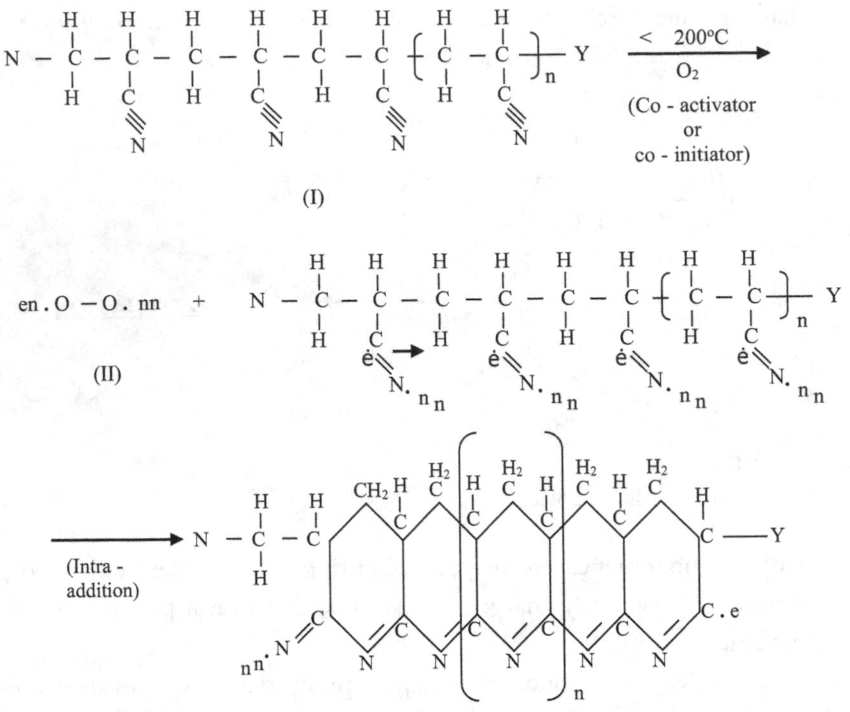

(I)

(II)

(III)

15.99

It is (II) and heat that activated (I) radically, though it cannot readily add to the centers under Combination mechanism. The $C \equiv N$ activation center being nucleophilic in character, favors electrophilic attack. There is transfer species of the first kind here nucleo-non-free-radically, but cannot be abstracted due to radical balancing. The same transfer species is very difficult to release electro-free-radically. When released, no rings can be formed, because it will be too strained. Note that in the first case, $C \equiv N$ center cannot be activated chargedly due to electrostatic forces of repulsion. The nucleophilic end of the first member of the chain being the least is the first to be activated. The electro-free-radical attacks the nucleo-non-free-radical end of the second monomer unit leaving a nucleo-non- free-radical end and the electro-free-radical growing active center. The addition continues in a sequential manner to form fused six-membered rings intra - molecularly along the chain until an electro-free - radical end is left at the other side of the chain. *For n monomer units, (n - 1) fused six-membered rings are formed leaving a living polymer, living at both ends of the chain.* In fact, the chain can add to another chain like itself to form a very long chain which can fold up to form a dead gigantic folded ring. Otherwise, a foreign agent must be required to kill this type of dual externally located active centers. On the other hand, (III) cannot reject transfer species electro-free-radically from one end and add to the other end, as has been observed with a case of silicon monomer, since the terminal ring will be very strained. In fact, any chain that has transfer species to reject, can never grow properly under these operating conditions.

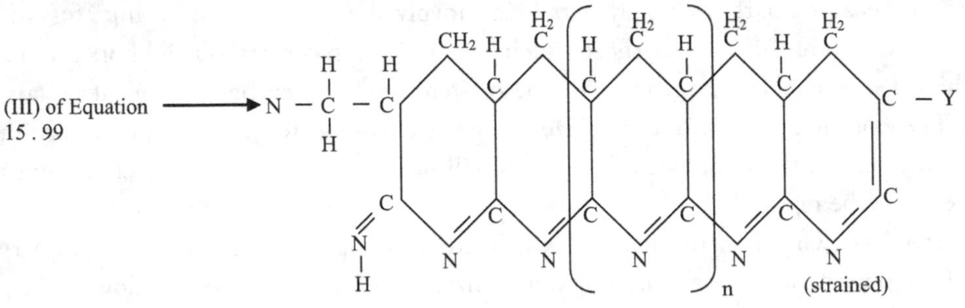

(III) of Equation
15 . 99

(IV) Not favoured

15.100

At further higher temperatures of pyrolysis, (III) above dehydrogenates to form the products shown below.

$$O_2 \quad | \xrightleftharpoons{Heat} \quad nn\bullet O - O \bullet en \xrightleftharpoons{Heat} \quad 2\,nn\bullet O \bullet en$$

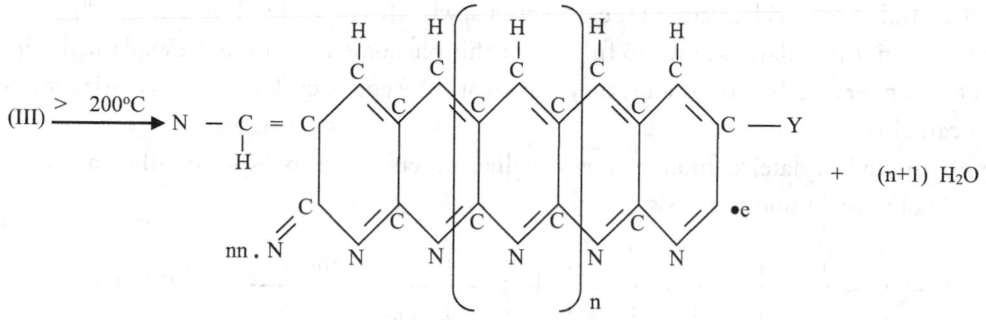

(III) $\xrightarrow{> \; 200^\circ C}$

(V) (Polyquinizarine)   15.101

+ $(n+1)\ H_2O$

The water molecules released free-radically are due to possible involvement of the O_2, via oxidation, since the possibility of the chain existing in Equilibrium state of existence from the CH_2 group is remote and may not be possible. The oxygen is most likely broken down at the operating conditions to give Oxidizing oxygen. Hydrogen is removed from CH_2 and hydrogen is released to form H_2O. **This is no longer Combination mechanism, but Equilibrium mechanism.** From the structure of (V) indicated above no hydrogen molecules can further be released and no three double bonds conjugatedly placed can exist in the ring as are previously thought to be the case. During dehydrogenation, the terminal ends of the chain can be killed. Electro-free-radicals under pyrolytic conditions may be used to also initiate the cyclohomopolymerization and when this is the case, it becomes **Inter/Intra-, intra-, intra-... molecular addition**.

Now, with polyacrolein obtained nucleo-free-radically, the followings are obtained.

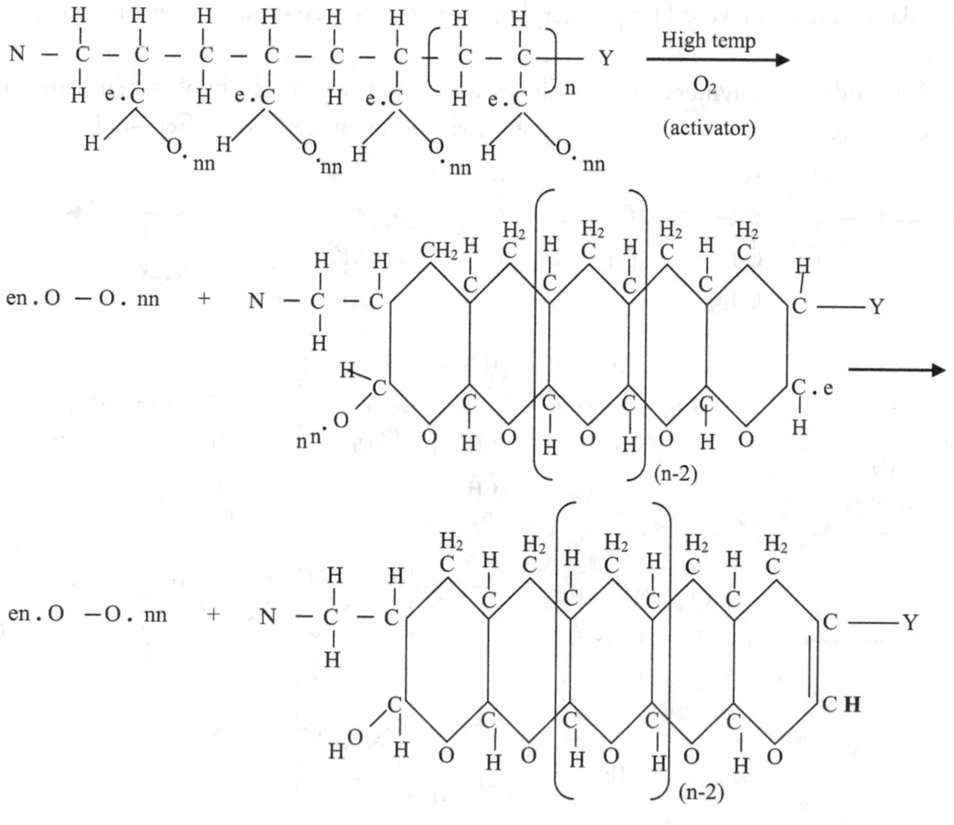

High temp $\xrightarrow{\quad}$

O_2

(activator)

en . O — O . nn   +

en . O — O . nn   +

Transfer species of 1st kind   15.102

The C = O activation center involved is nucleophilic in character. Hence it favors only electro-free-radical attack. Transfer species of the first kind of the sixth type can be rejected, based on the Laws of conservation of transfer of transfer species. This can be used to kill the other side of the chain. Because of lack of understanding of mechanisms of these reactions, why this type of polyacrolein and polymers from vinyl ketones are not popularly known to fully favor this phenomenon cannot be explained. However, it is believed that under suitable conditions, they can favor this type of cyclohomopolymerization to a great extent and can also be dehydrogenated.

For poly(methylacrylate) a monomer which like others above is Nucleophilic in character, the followings are obtained upon pyrolysis.

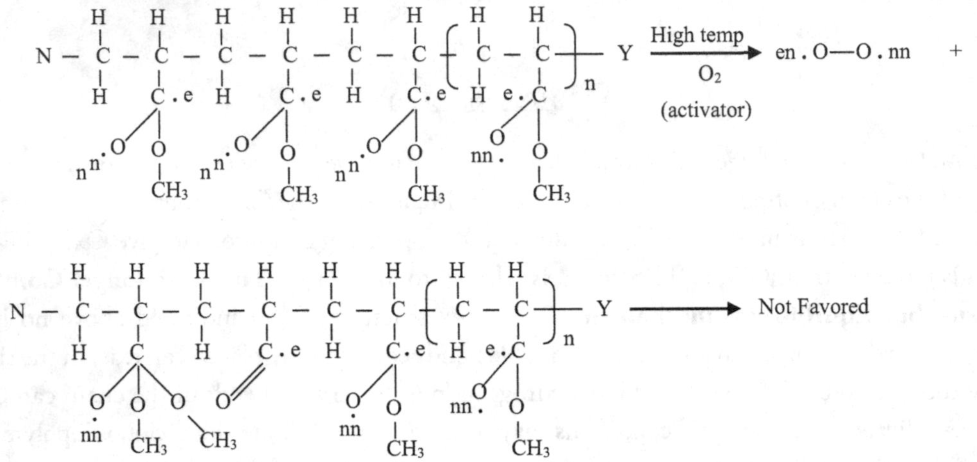

$$\text{15.103}$$

Due to presence of transfer species of the second first kind, the continuous cyclization reaction is not favored. Similar reaction is favored for polyacrylamide obtained free - radically when used as a polymeric monomer.

When 3,4- addition polymers of 1,3 - dienes which can only be fully obtained using special Z|N initiators are involved chargedly, cyclo-homopolymerization may be favored free –radically.

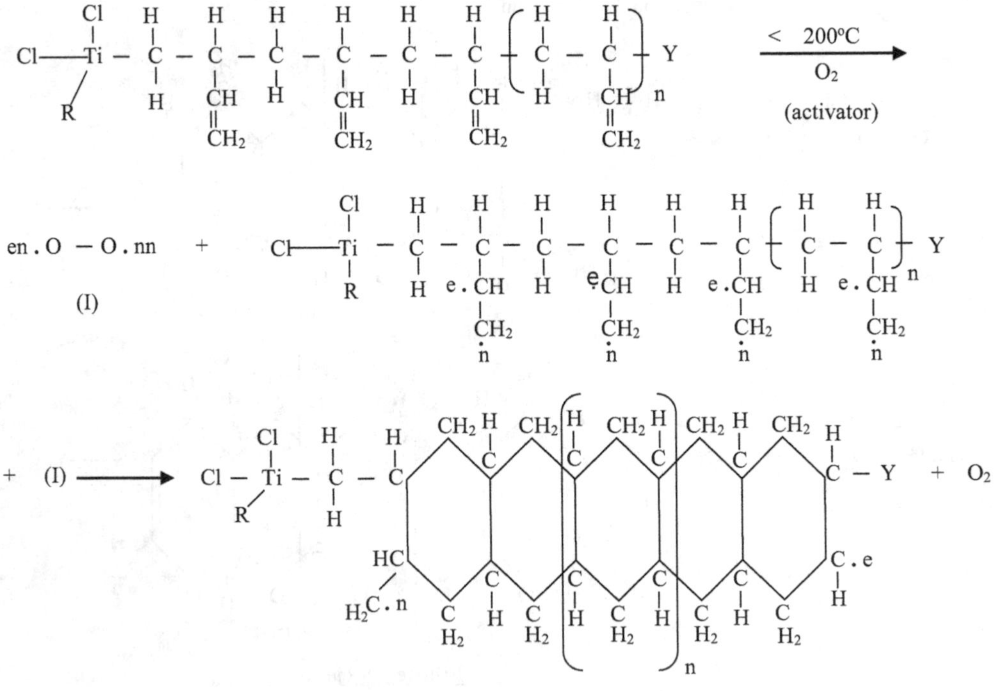

$$\text{15.104}$$

In view of the Laws of Conservation of transfer of transfer species, transfer species can be rejected here. With 1,3-pentadiene, the followings are obtained.

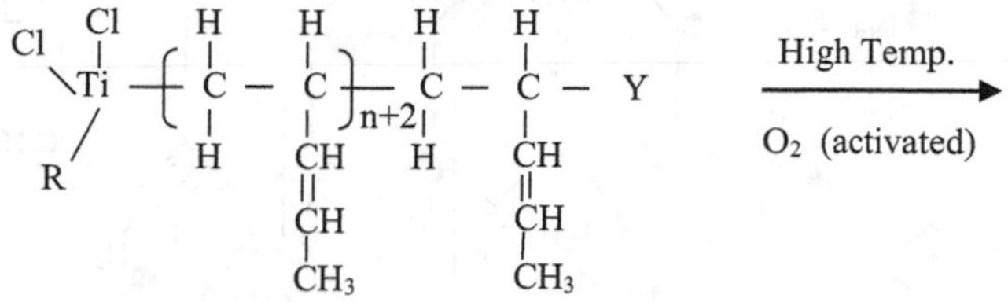

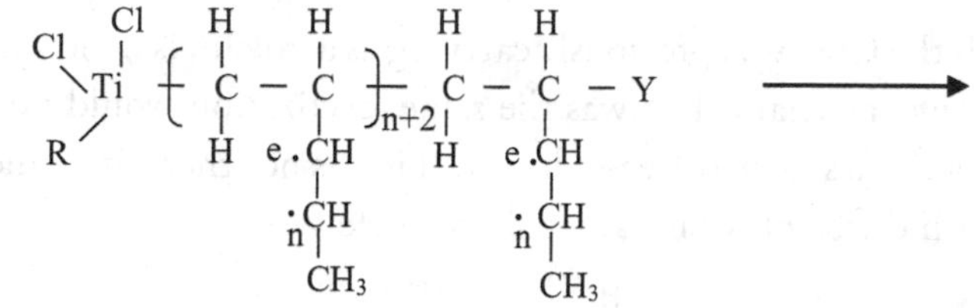

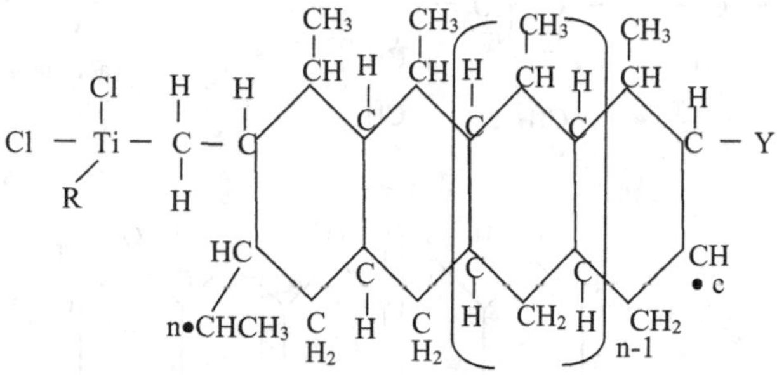

$$
\underset{R}{\overset{Cl}{\underset{|}{\overset{|}{Ti}}}} \left[C - C \right]_{n+2} C - C - Y \xrightarrow[\text{O}_2 \text{ (activated)}]{\text{High Temp.}}
$$

$$
\underset{R}{\overset{Cl}{\overset{|}{Ti}}} \left[C - C \right]_{n+2} C - C - Y \longrightarrow
$$

15.105

Presence of transfer species of the first kind of the sixth type will disturb the growth of the cyclization reactions here for 4,3- mono-form of 1,3- butadiene, 4,3- mono-form of 1,3- pentadiene and also for 4,3- mono-form of isoprene, when the centers are not close enough. All these can terminate themselves. The three monomers here are nucleophiles, and the activation centers favor electrophilic attack. ***Presence of partial cyclization is favored for these monomers, because of transfer species of the first kind of the sixth type to release, created by the time it takes for the active center to add to the next activation center.***

The cyclization reaction is not favored for chloroprene, since it cannot be polymerized using Z|N or other charged-paired initiators. It is important to note that all the pendant activation centers are nucleophilic in character.

For the case where the pendant activation center which formally was electrophilic in character, the following should be expected for polyacrolein obtained exclusively via the C = O activation center electro-free-radically using NaCN.

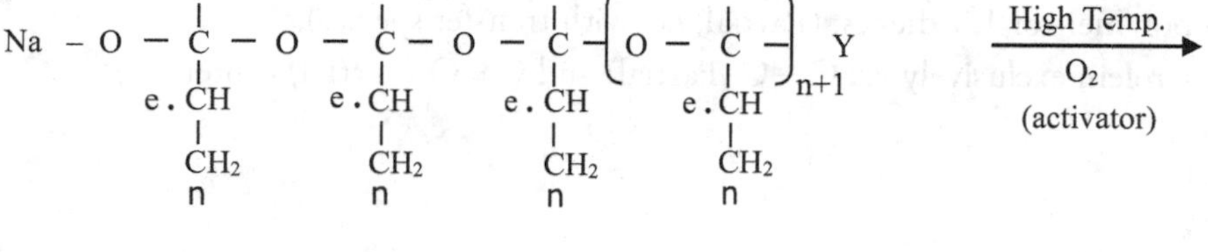

$$
Na - O - \underset{\underset{n}{\overset{|}{CH_2}}}{\overset{H}{\underset{|}{\overset{|}{C}}}} - O - \underset{\underset{n}{\overset{|}{CH_2}}}{\overset{H}{\underset{|}{\overset{|}{C}}}} - O - \underset{\underset{n}{\overset{|}{CH_2}}}{\overset{H}{\underset{|}{\overset{|}{C}}}} \left[O - \underset{\underset{n}{\overset{|}{CH_2}}}{\overset{H}{\underset{|}{\overset{|}{C}}}} \right]_{n+1} Y \xrightarrow[\text{(activator)}]{\text{High Temp.} \atop \text{O}_2}
$$

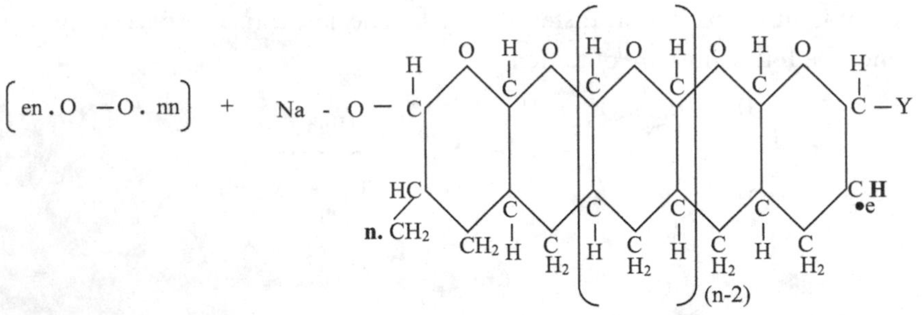

$$15.106$$

Notice that the radicals which the C=C was previously carrying as acrolein, is no longer the same, because of the groups now carried along the chain. If it was the same, cyclization would not have been favored because of presence of transfer species electro-free-radically. If it was not there, it would have been favored to give the same chain but with different terminals as shown below.

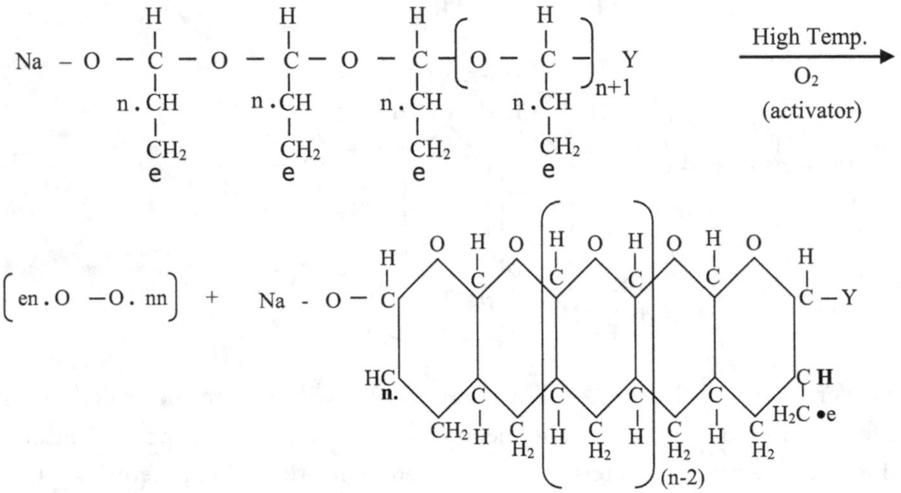

NOT FAVORED, YET LOOKS FAVORED

$$15.107$$

However, the case above does not exist, since C=C center has been wrongly activated. The real case is that shown in Equation 15.106. Note that the monomer unit is identical to that from the C = C center of Equation 15.102. When suitable conditions exist, note should be taken of the polymeric monomers which favor this type of cyclohomopolymerization. In view of the presence of transfer species of the first kind of the sixth type for the case of Equation 15.106, cyclization reactions may not be a continuous one. Nevertheless, polymeric monomers which seem to favor this type of cyclo - homopolymerization include so far-

(i)   Polyacrylonitrile - (Electro-free - radically $C \equiv N$), (Full).
(ii)   3,4- polymers of 1,3-dienes, (Partial, i.e. with transfer species).
(iii) Polyacrolein exclusively via C = C (Partial) and C = O (Partial) centers.

The presence of double and triple bonds along the chains of polymers is obviously very important in cyclopolymerization. Now, consider the case where acetylene is used to give vinylacetylene in the presence of ***Cu_2Cl_2 and NH_4Cl*** shown below.

Stage 1: $HC \equiv CH$ $\rightleftharpoons$ $H \bullet e + n \bullet C \equiv CH$ (A)

$H \bullet e + HC \equiv CH$ $\xrightleftharpoons{\text{Activation}}$ $H_2C = \overset{\overset{\displaystyle H}{|}}{C} \bullet e$ (B)

(B)   +   (A) $\longrightarrow$ $H_2C = CH - C \equiv CH$

$\qquad\qquad\qquad\qquad\qquad$ (Vinylacetylene) $\qquad\qquad\qquad$ 15.108

Free-radically, the vinylacetylene is resonance stabilized to give $e\bullet CH_2 - CH = C = HC \bullet n$. This is what will be the monomer unit along the chain when polymerized nucleo-free-radically.

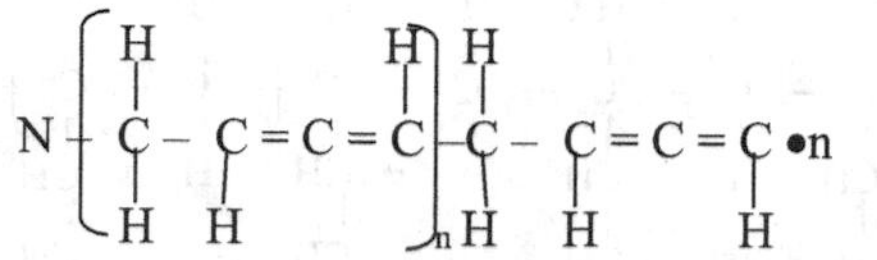

$$N \left[\overset{\overset{\displaystyle H}{|}}{\underset{\underset{\displaystyle H}{|}}{C}} - C = C = \overset{\overset{\displaystyle H}{|}}{C} \right]_n \overset{\overset{\displaystyle H}{|}}{\underset{\underset{\displaystyle H}{|}}{C}} - C = C = C \bullet n$$

$\qquad\qquad\qquad\qquad\qquad\qquad\qquad\qquad\qquad\qquad\qquad$ 15.109

Chargedly, the chain above cannot be obtained, since resonance stabilization cannot be favored. What will be obtained along the chain as pendant groups is the acetylene group. It is hoped that the H atom on acetylene side is not loosely held to it, otherwise the H atom under such condition can easily be replaced with for example CH_3 group. If no such disturbance is present as can be done by suppression, then the followings are obtained with a Z/N charged type of initiator, since the monomer can be activated chargedly, but cannot be resonance stabilized.

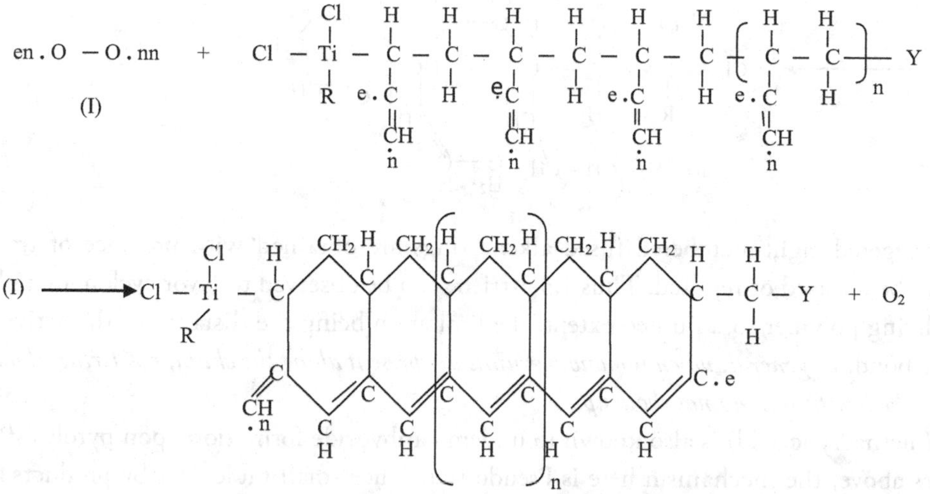

$en.O - O.nn$ + (diagram)

(I)

$\qquad\qquad\qquad\qquad\qquad\qquad\qquad\qquad\qquad\qquad\qquad\qquad\qquad\qquad$ 15.110

+ (I) $\longrightarrow$ (diagram) $+ O_2$

When it is possible to have the product, the cyclization above will be favored, provided no transfer species can be released. However, it cannot be released, because when released, a cumulenic bond inside the ring is obtained making the ring too strained to exist. Recall that we have already shown that the capacity of $H_2C = CH - \geq H > HC \equiv C -$.

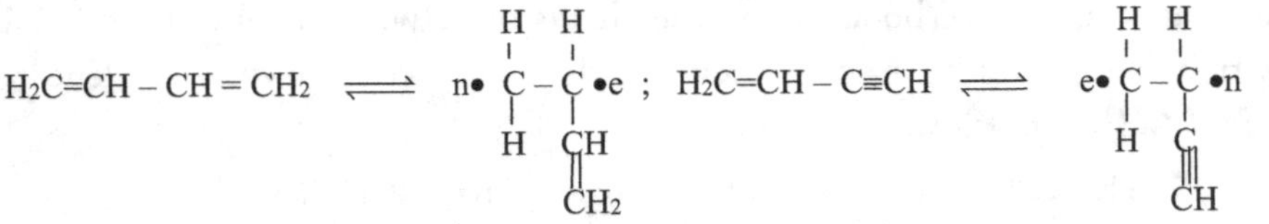

$$H_2C=CH-CH=CH_2 \rightleftharpoons n\bullet \overset{\overset{\displaystyle H}{|}}{\underset{\underset{\displaystyle CH_2}{\underset{\displaystyle \|}{CH}}}{\overset{\displaystyle H}{\underset{|}{C}}} - \overset{\overset{\displaystyle H}{|}}{C} \bullet e \quad ; \quad H_2C=CH-C\equiv CH \rightleftharpoons e\bullet \overset{\overset{\displaystyle H}{|}}{\underset{|}{C}} - \overset{\overset{\displaystyle H}{|}}{\underset{\underset{\displaystyle CH}{\underset{\displaystyle \|}{C}}}{C}} \bullet n$$

15.111

Now, let us consider the first member of 1,3,5-trienes ($H_2C=CH–CH=CH–CH=CH_2$). Can this be made to undergo cyclization? Chargedly it is not resonance stabilized. Radically, unlike divinylbenzene where resonance stabilization cannot go beyond the ring, it is discretely fully resonance stabilized, since all the activation centers are linearly conjugatedly placed. In Divinylbemzene and styrene, the resonance stabilization is continuous, coming back to where it started. When 1,2-mono-form has been obtained chargedly using Z/N type of initiator, the followings are obtained.

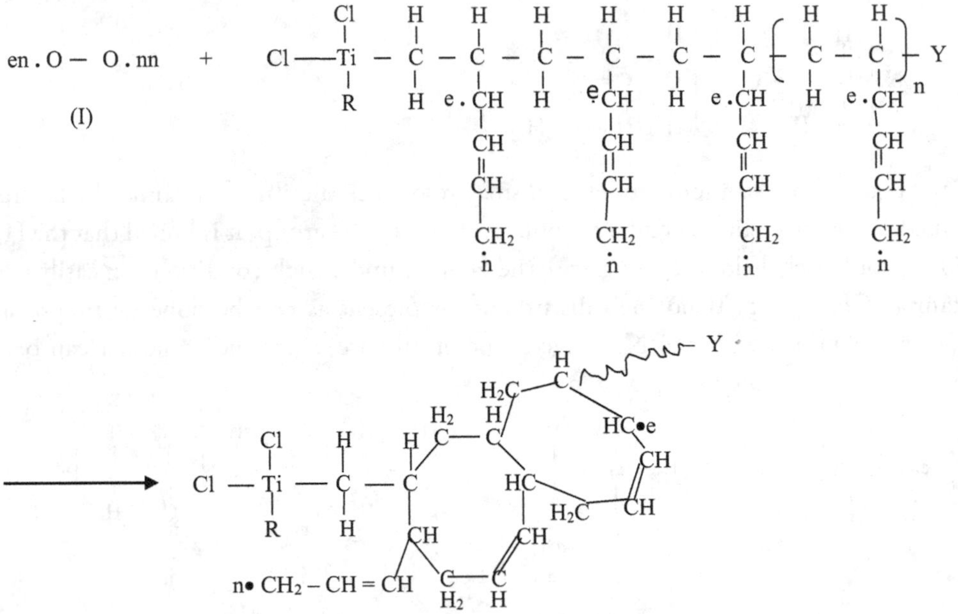

15.112

Heavily staggered eight-membered fused alkene rings are obtained with presence of transfer species available to be released or rejected. Thus 1,3,5-triene can be observed to favor cyclohomopolymerization to give a living polymer to a limited extent, the limitation being the distance of the active center from the double bond. *In general, when n monomer units are present along the chain, n - 1 rings should have been obtained if the limitations do not show up.*

Poly (methacrylic acid) is also known to undergo anhydride formation upon pyrolysis[15]. Unlike the other cases above, the mechanism here is Pseudo step, since small molecular by-products are obtained for every addition.

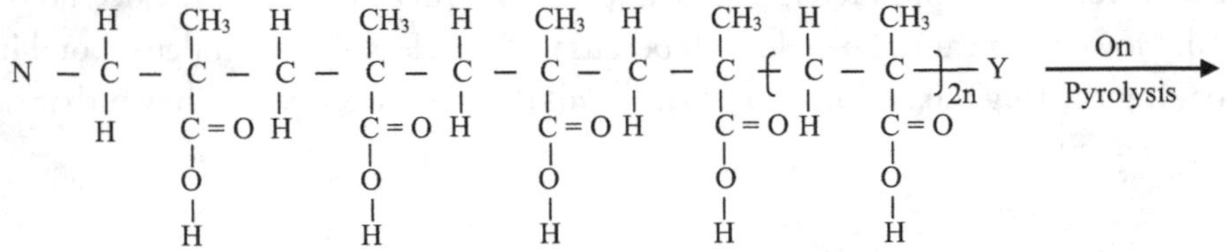

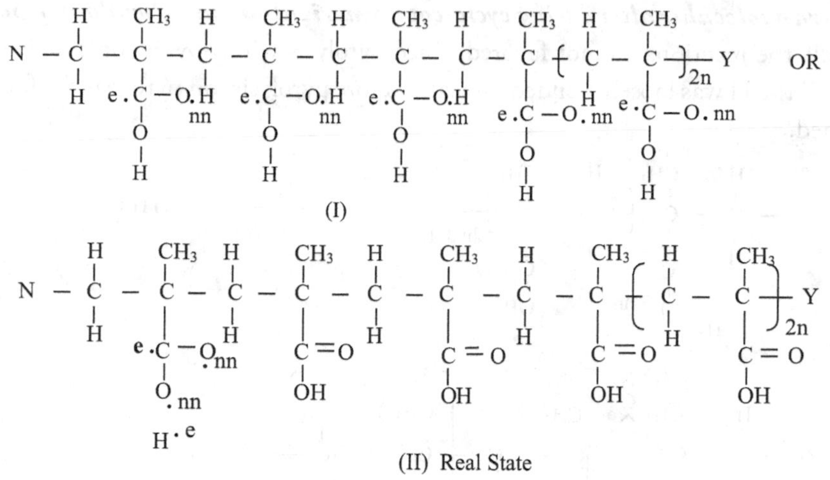

(structure I)

$$N - C - C - C - C - C - C - C \left[C - C \right]_{2n} Y \quad OR$$

(I)

(structure II)

(II) Real State

15.113

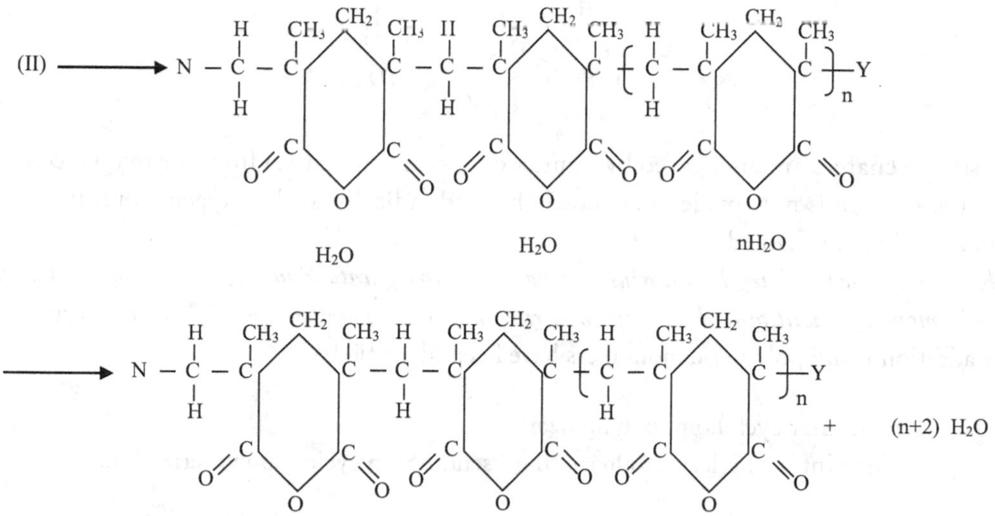

$$(II) \longrightarrow N - C - C \cdots \left[C - C \right]_n Y$$

$H_2O \qquad H_2O \qquad nH_2O$

$$\longrightarrow N - C - C \cdots \left[C - C \right]_n Y \quad + \quad (n+2)\ H_2O$$

15.114

The case of Equation 15.114 should be compared with that of Equation 15.103 where the possibility of obtaining the product above exists if the methyl group can be made to be held in Equilibrium state of existence. *When a compound is made to exist in Equilibrium state of existence, only one atom or component is held in that state one at a time if it has more than one.* Nature does not operate indiscriminately, because Nature is nothing else but ORDERLI-NESS in...... *Which H atom will first be held in Equilibrium state of existence? This is dictated by what is carrying the chain (i.e. part of the initiator) and what the chain is carrying, for which the component held in Equilibrium state of existence for this case is either from the beginning or from the end.* Such is the case with glycerol. For the case above, since the nucleo-free-radical route is natural to the monomer, the H held is from the beginning. After the first, this is then followed by the next which is the third and so on. The H of OH group is loosely bonded to oxygen either ionically or radically, while for OCH_3, CH_3 is held only radically. Due to the homolytic method of scission (Pyrolysis), it is radically loosely bonded to the oxygen center. It is (II) that is favored for methyacrylic acid. In (I), only the C = O activation center was activated with nothing held in Equilibrium state of existence (i.e. Activated/Equilibrium state of existence). After existing in Equilibrium state of existence, the $H \cdot ^e$ in (II) of Equation 15.113, removes OH group as transfer species of the second first kind, without activating the C = O center to form H_2O. The features of Pseudo–step or indeed Step polymerization kinetics can thus be observed. *Unlike the other polymeric monomers considered under this section, this is not intra-molecular cyclo-homopoly-merization, but*

Equilibrium intra-molecular alternating cyclo-copolymerization, via Pseudo step polymerization kinetics. Ionically, the reactions are not favored, since pyrolysis will provide ionic activation of C = O activation center. If the H was loosely bonded ionically before pyrolysis, on pyrolysis the followings should have been obtained.

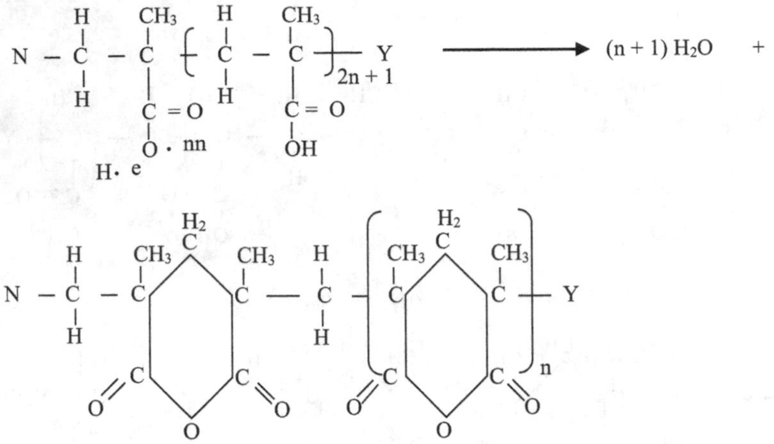

$$
\text{N} - \underset{\underset{H}{|}}{\overset{\overset{H}{|}}{C}} - \underset{\underset{\underset{O \cdot nn}{|}}{\overset{|}{C=O}}}{\overset{\overset{CH_3}{|}}{C}} \left\{ \underset{\underset{H}{|}}{\overset{\overset{H}{|}}{C}} - \underset{\underset{\underset{OH}{|}}{\overset{|}{C=O}}}{\overset{\overset{CH_3}{|}}{C}} \right\}_{2n+1} \text{Y} \longrightarrow (n+1)\,H_2O \quad +
$$

$$
\text{H} \cdot e
$$

15.115

On pyrolysis, the charges are disengaged by transfer of electro-free-radical from the oxygen center to the cation and the H atom is now made to be loosely bonded radically to the oxygen atom. It is after this, cyclization is made to be favored.

For this unique case involving Equilibrium mechanism with a giant monomer, via pseudo-step mechanisms, 2n or 2n + 1 monomers will provide n rings or n rings and one monomer unit. Thus, two types of intra-molecular addition using polymeric monomers have been identified -

(i)   Intra - molecular cyclohomopolymerization.
(ii)   Equilibrium Intra - molecular alternating Pseudo-Step cyclocopolymerization.

Some products from (i) can be further dehydrogenated during pyrolysis to give very strong products.

Finally, there is a third type to be identified. It is that which favors the existence of random cyclo-copolymerization in the presence of initiators as opposed to pyrolysis. Polymerization include natural rubber and some 1,4- poly-1,3-dienes. Considering polyiso-prene, the followings were thought to be obtained "ionically".

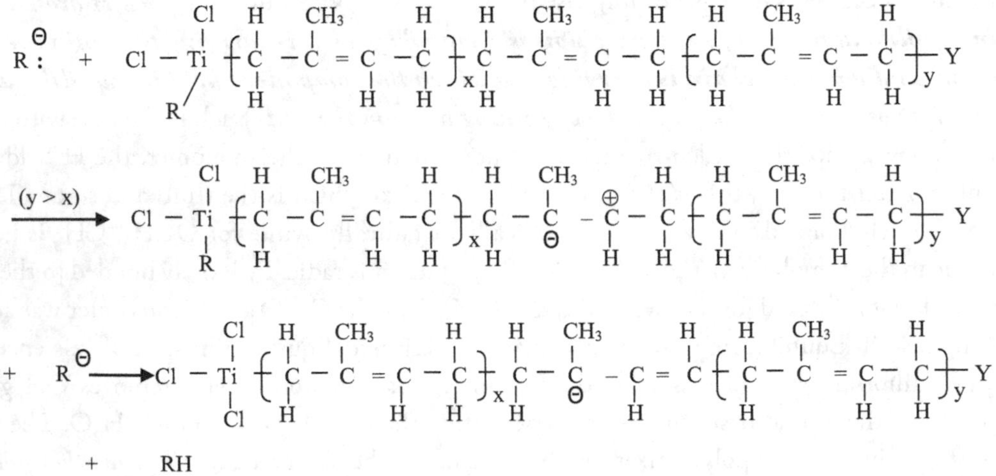

15.116

In the first case, no chemical reactions have been found to take place ionically. They all take place chargedly and radically. Secondly, resonance stabilization cannot take place chargedly.

Therefore, 1,4-mono-form cannot be obtained chargedly. The 1,4-monoform above was only obtained radically. Radically, Ti and Al centers can be paired. Pairing takes place not only when charges are formed, but also between some radical species. ***If resonance stabilization had taken place radically before initiator was prepared, covalent charges cannot be formed when the two radical carrying centers are distantly located. The same applies to electrostatic charges, but not to polar charges as shown below.*** With polar charges, this takes place only when three or more centers placed side by side carry paired unbounded radi-

$$O = \overset{\oplus}{O} - \overset{\ominus}{O} \qquad ; \qquad O = \overset{\oplus}{O} - O - O - O - O - O - O - \overset{\ominus}{O}$$

Ozone   Stable polymeric oxygen

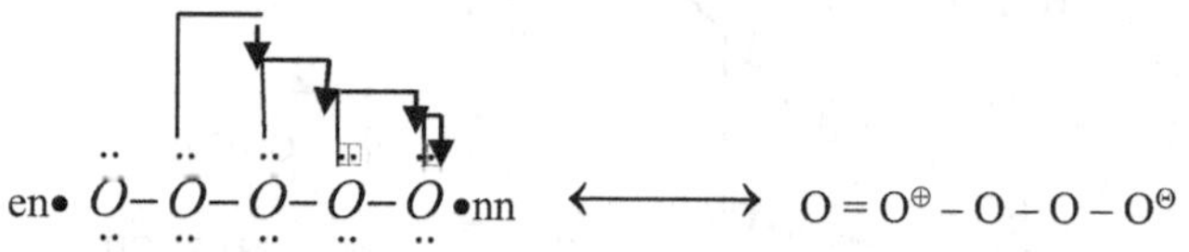

$$\text{en} \bullet \overset{..}{\underset{..}{O}} - \overset{..}{\underset{..}{O}} - \overset{..}{\underset{..}{O}} - \overset{..}{\underset{..}{O}} - \overset{..}{\underset{..}{O}} \bullet \text{nn} \quad \longleftrightarrow \quad O = O^{\oplus} - O - O - O^{\ominus} \qquad\qquad 15.117$$

cals in their last shell. It is from the last one carrying nucleo-free or non-free-radical that an electro-radical is moved from the paired unbonded radicals to form a negative charge. The movements of electro-radicals continue until the second to the last O atom where a double bond is formed and a positive charge is placed. With the movement of each electro-radical, a hole is created. Now we are beginning to get closer to showing how current flows in Solids.

Obviously, 1,4-mono-form above can only be obtained radically. In the reactions above (i.e., Equation 15.116), one center internally located was activated "indiscriminately" by a negatively charged initiator in the neighborhood. This should not be the case if there was orderliness in the operations, unless it was desired to put something in that section of the chain after doing something. Note that all the centers cannot be activated at the same time, all being of different nucleophilicity. In Nature, the center first allowed to be activated could be at the beginning or end of the chain or in the middle of the chain. But meanwhile, let us take it for granted that it is from the middle since the activation centers are along the chain internally located. The route being unnatural to it, did not favor initiation, because of presence of transfer species of the first kind of the first type, noting why "y" was said to be greater than "x". It could be the other way round, but with different transfer species.

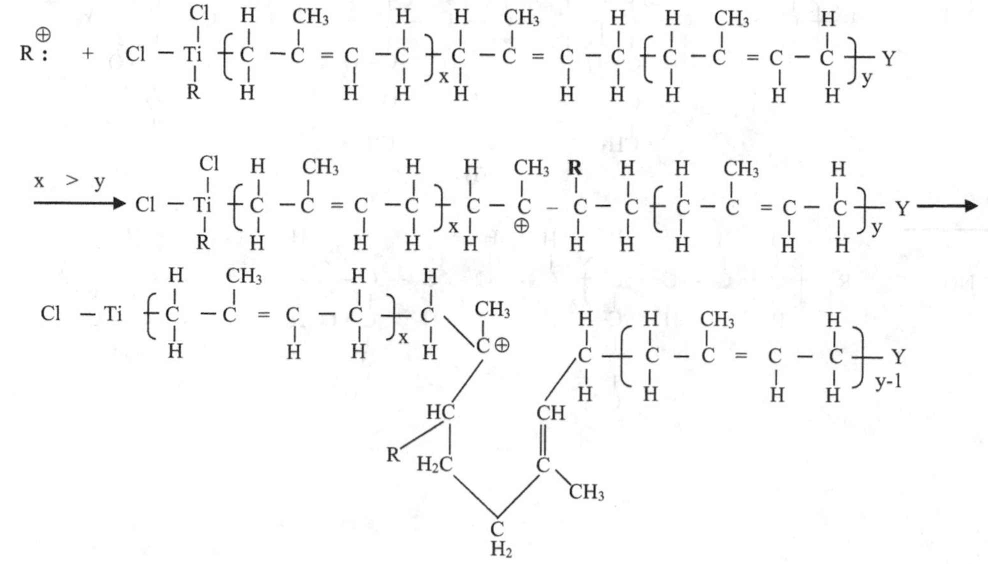

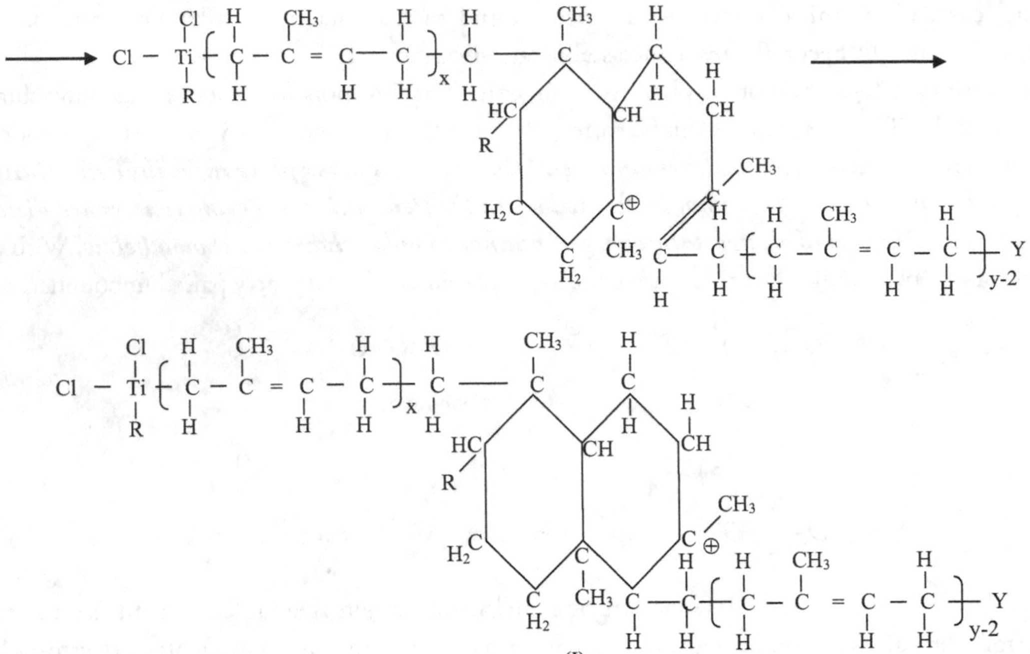

$$\text{Cl} - \underset{R}{\overset{Cl}{Ti}} \left(\overset{H}{\underset{H}{C}} - \overset{CH_3}{\underset{H}{C}} = C - \overset{H}{\underset{H}{C}} \right)_x \overset{H}{\underset{H}{C}} - C \cdots (I)$$

(I)   15.118

With continuous addition favored after activation of one of the centers by $R^{\oplus}$, six-membered alternating fused bicyclic rings moving downwardly are formed, if and only if no transfer species is released. This will be possible on a continuous basis if the activation center activated is close to the end of the chain, i.e., x > y during initiation in the absence of control. Electro-free-radically, initiation is favored due to absence of transfer species of the first kind.

In fact, all the reactions above largely take place radically.

Poly (1,3-butadiene) and some other nucleophilic polymeric monomers may favor presence of six-membered fused bicyclic fused to another bicyclic ring alternating placed diagonally along the chain down to the end of the chain, when initiated electro-free-radically, provided there is no electrophilic character in the polymeric monomer. Shown below is the polymeric product obtained when an Electrophilic monomer is polymerized nucleo-free-radically.

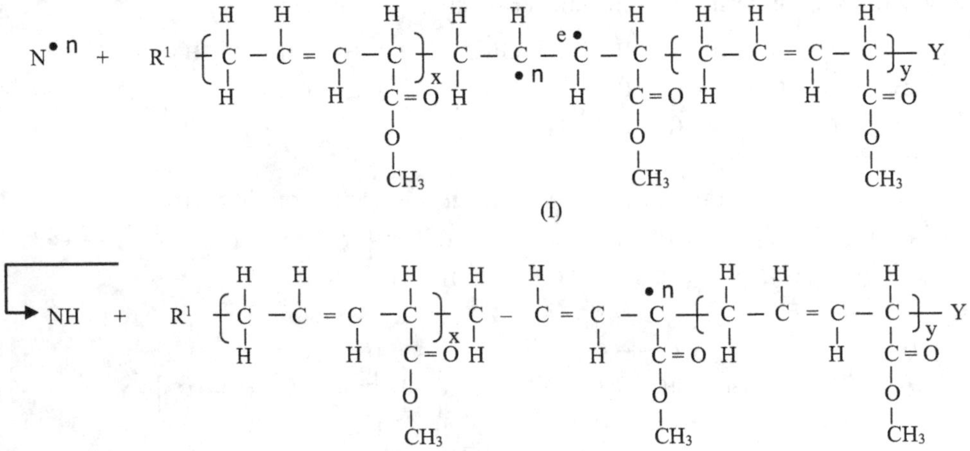

(I)

                                                                    15.119

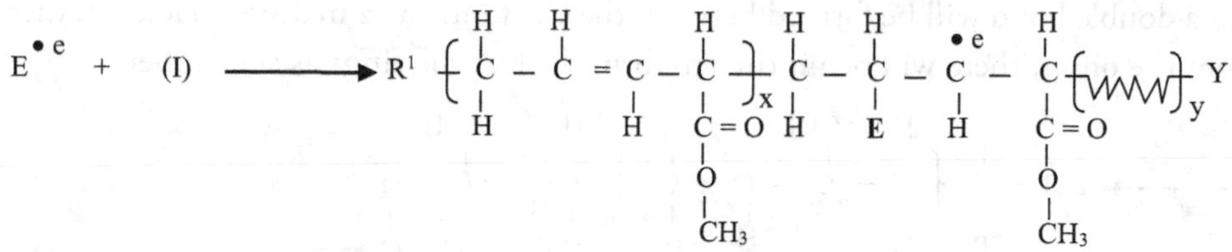

$$E^{\bullet e} + (I) \longrightarrow R^1 \left\{ \underset{H}{\overset{H}{C}} - \underset{H}{\overset{H}{C}} = C - \underset{\substack{C=O \\ | \\ O \\ | \\ CH_3}}{\overset{H}{C}} \right\}_x \underset{H}{\overset{H}{C}} - \underset{E}{\overset{H}{C}} - \underset{H}{\overset{H}{C}} - \underset{\substack{C=O \\ | \\ O \\ | \\ CH_3}}{\overset{\bullet e}{C}} \left\{ \text{www} \right\}_y Y$$

15.120

Nucleo-free-radically, initiation for mono or bicyclization is not possible. ***Chargedly, the chain cannot be activated, because it is allylic.*** Where presence of bicyclic rings are favored, two monomer units are required in the process and this is only possible electro-free-radically. For the growing polymer chain of (I) of Equation 15.118, the followings are obtained if formation of rings started from close to the beginning of the chain.

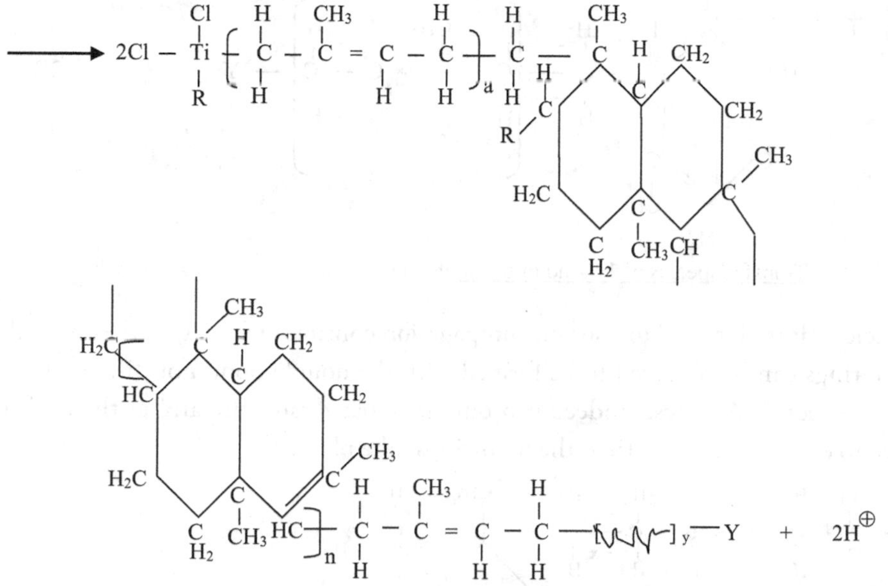

<u>Transfer species of 1st kind of the sixth type</u>

15.121

After continued propagation with formation of bicyclic rings, transfer species were released.
These reactions can only take place radically and via Combination mechanism, because apart from the fact that positive charges cannot be isolatedly placed and cations cannot be used directly, never can two hydrogen cations combine together to give H_2 molecule. This can take place with the atom or the hydride, but not with cations or positive charges.

$$H^{\oplus} + H^{\oplus} \xrightarrow{\quad\quad} H_2 \quad ; \quad H\bullet e + H\bullet e \longrightarrow H_2 \quad ; \quad H\bullet n + H\bullet n \longrightarrow H_2$$

IMPOSSIBLE

15.122

Hydrogen cation is carrying nothing. What is she going to use to form a bond? Even the positive charge itself cannot be removed leaving the carrier behind. The same also apply to $Cl^{\ominus}$. If two of them combine together, then a double bond will be formed between the Cl atoms! As a matter of fact, if transfer species is rejected from the onset, there will be no time to form the bicyclic rings as shown below.

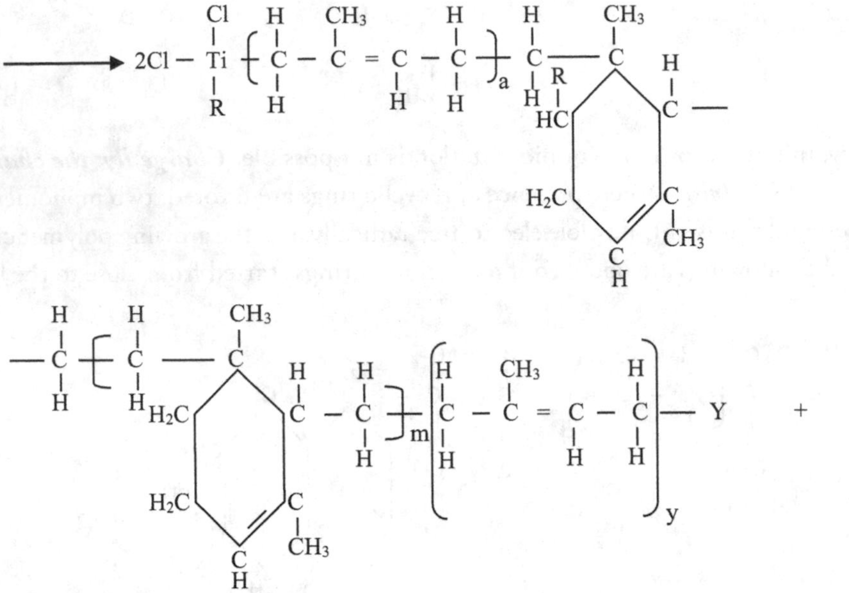

$$\longrightarrow 2Cl - Ti \left(\begin{array}{c} Cl \\ | \\ C \\ | \\ R \end{array} \begin{array}{c} H \\ | \\ C \\ | \\ H \end{array} \begin{array}{c} CH_3 \\ | \\ = C \end{array} \begin{array}{c} H \\ | \\ C \\ | \\ H \end{array} \right)_a \cdots$$

Transfer species of 1st kind of the sixth type   15.123

The transfer species released is used to continue propagation continuously to give the chain above. Thus, only monocyclic rings can be observed to be formed with the double bond not differently placed when transfer species is rejected. All these, indeed can only take place radically and at the beginning of the chain. If at close to end of the chain, then the followings take place.

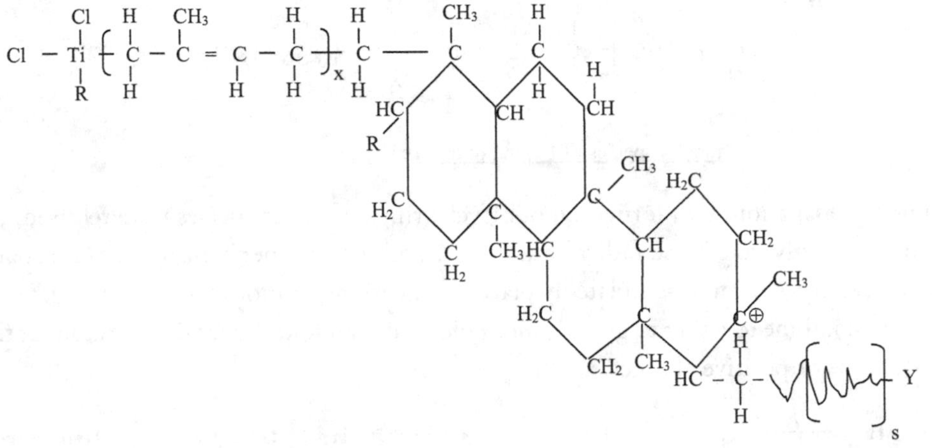

15.124

All the above, have been an exercise to show some basic fundamental principles about how transfer species are rejected in such systems. Shown below, are some important examples of some components where transfer species can and cannot be rejected.

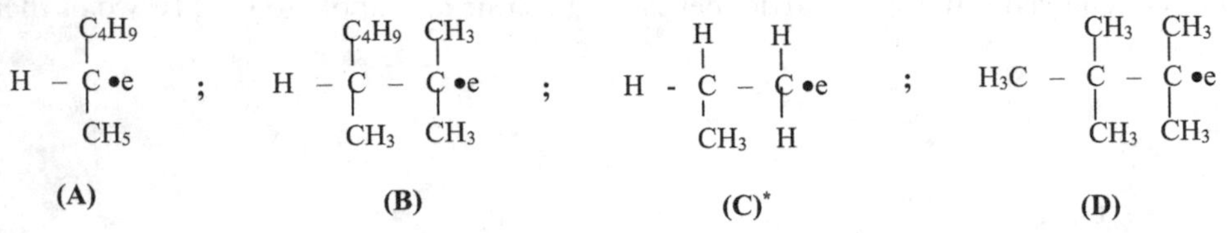

$$H - \underset{\underset{CH_5}{|}}{\overset{\overset{C_4H_9}{|}}{C}} \bullet e \quad ; \quad H - \underset{\underset{CH_3}{|}}{\overset{\overset{C_4H_9}{|}}{C}} - \underset{\underset{CH_3}{|}}{\overset{\overset{CH_3}{|}}{C}} \bullet e \quad ; \quad H - \underset{\underset{CH_3}{|}}{\overset{\overset{H}{|}}{C}} - \underset{\underset{H}{|}}{\overset{\overset{H}{|}}{C}} \bullet e \quad ; \quad H_3C - \underset{\underset{CH_3}{|}}{\overset{\overset{CH_3}{|}}{C}} - \underset{\underset{CH_3}{|}}{\overset{\overset{CH_3}{|}}{C}} \bullet e$$

            (A)   (B)   (C)*   (D)

$$
\begin{array}{cccc}
\underset{\overset{|}{CH_3}}{\overset{\overset{CH_3}{|}}{n\bullet C}} - \underset{\overset{|}{C_5H_{11}}}{\overset{\overset{H}{|}}{C}}\bullet e
& ;
& \underset{\overset{|}{CH_3}}{\overset{\overset{H}{|}}{n\bullet C}} - \underset{\overset{|}{C_4H_9}}{\overset{\overset{H}{|}}{C}}\bullet e
& ;
& \underset{\overset{|}{H}}{\overset{\overset{H}{|}}{n\bullet C}} - \underset{\overset{|}{C_4H_9}}{\overset{\overset{H}{|}}{C}}\bullet e
& ;
\end{array}
$$

(E)   (F)   (G)   (H)*   15.125

In Nature, it is the richer component that provides transfer species under certain conditions. If it cannot, the second richer component does the rejection. This has been the observation as applies to RELIGION. (A), (B), and (D) can reject in order to form a double bond when heated, but (C) cannot reject. Note that this is the case for a growing polymer chain, because of the guiding Laws of conservation of transfer of transfer species. With (D), what is going to be rejected? Is it $H\bullet e$ or $H_3C\bullet e$? The richer component that is providing transfer species does not give all it has, or the biggest of the components it is carrying, but the smallest. The rejection is H from CH_3, because it is the smallest component that is rejected whether C-H bond is stronger than C-C bond or not. (H) will reject, because of the presence of CH_3 group carried by the C center adjacently located to the center carrying the electro-free-radical. If the CH_3 group is replaced with Cl, H would more readily be rejected. If not, the H rejected will come from the other side carrying CH_2 group. With rings, the situation is slightly different, because the ring is common to both active centers when no groups other than H are present. (E), (F) and (G) look as if they have transfer species. They do, because nucleo-free-radically, the transfer species can be abstracted, since it can be rejected from its growing polymer chain. They cannot however molecularly rearrange by activation. For example, 1-hexene, 3-methyl-1-propene (an isomer of hexene), 1-pentene, can be polymerized, because transfer species can be rejected from the growing polymer chain from the second richer component to form a dead terminal double bond polymer. However, they cannot molecularly rearrange by activation. (E), (F) and (G) and similar ones, look like the monomer units in for example Equation 15.116 when activated, but not with case of Equation 15.120 which can rearrange and become unproductive. Those of Equation 15.116 cannot therefore molecularly rearrange. Secondly, the charges or radicals carried by the activation center at the beginning is the reverse of the charges or radicals carried at the terminal end of the chain. Starting from the beginning of the chain, the activation centers seem to remain the same with respect to what they are carrying until somewhere in the middle where it is reversed and this remains the same to the end of the chain. *Hence the middle monomer is the least nucleophilic center in the chain. As we move along the chain, the nucleophilicity of the activation centers keep increasing until the end or the beginning of the chain. What is carried by the active centers on the left of the middle is different from what is carried on the active centers on the right of the middle.* However for exploratory purposes, starting from the beginning of the chain, the followings are obtained.

$$15.126$$

There is no doubt that the four-membered ring will not be so strained to the point where it cannot exist, since four-membered ring like 3-methyl-1-cyclobutene which can be polymer-ized to poly(1,4-mono-form of pentadiene) exists. However, if the strain energy is greater than the minimum required energy, it will never be formed. A little after the middle, six-membered rings starts appearing along towards the end of the chain when no transfer species is released. At this end of the chain, the analogous form of (E) of Equation 15.125 is shown below, (I).

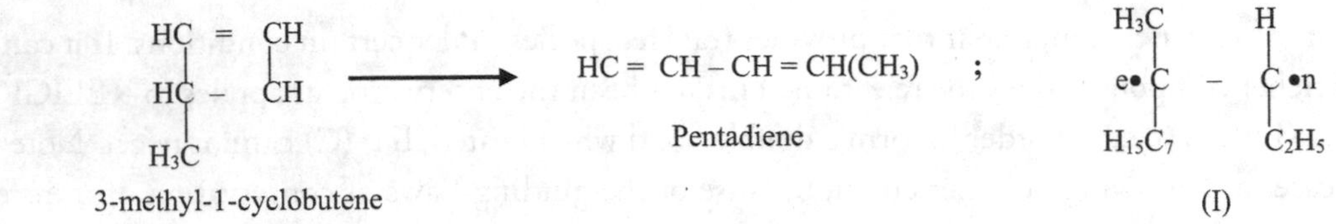

$$HC = CH - CH = CH(CH_3) \quad ; \quad \text{Pentadiene}$$

3-methyl-1-cyclobutene $\qquad$ (I) $\qquad$ 15.127

Like (E), only the positively charged, or electro-free-radical route is favored with transfer species. This does imply that there will be no transfer species for the chain when rings are formed. There must be a transition point along the chain where the change in size takes place if four-membered rings are formed. If not formed, only two fused six-membered rings diagonally fused to another two fused six-membered rings will appear along the chain. It will seem to appear that both four-membered rings and the fused six-membered rings will appear along the chain, because the four-membered ring with two heavy radical-pushing group and one small one (CH_3) will not be too strained to exist. Changing the CH_3 group in isoprene to Cl (chloroprene) will give a different sinerio, since only four-membered ring will be fully favored along the chain, unlike the case for isoprene shown below.

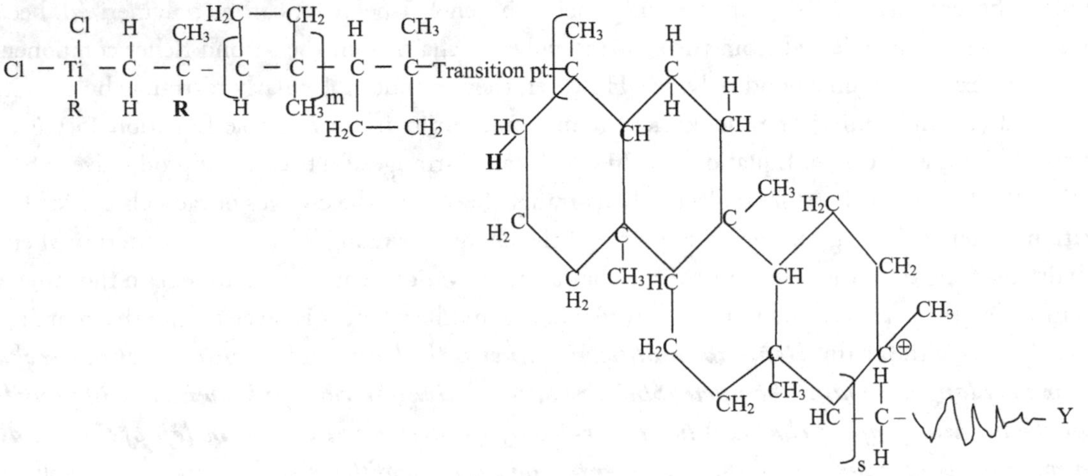

NOT POSSIBLE OR FAVORED $\qquad$ 15.128

At the transition point, the transfer species released, becomes the new initiator for the other side of the growing chain. Since we have found that the least nucleophilic center is in the middle (if odd) or close to the middle of the chain (if even), hence activation begins from the middle as shown in Equation 15.118 It starts from the middle and moves either left or right depending on what the terminals are carrying. Growth can be linearly horizontal, or linearly vertical or linearly diagonal or linearly circular and so on, since Nature also abhors non-linearity. Starting from the transition point to the left, fused six- membered ring as above are also obtained. All what are to be expected will be seen when the laws or rules are finally stated.

15.5 Cyclization of Di-aldehydes and Di-aldimines

To end this chapter, one has decided based on the experiences gathered so far, to apply to a group of family of monomers which have the features or abilities to cyclize. These were also listed in Equation 15.5 and recalled below. Note that the list is not complete, because NATURE is that where you think you know all, but only to realize that you know nothing. We including the author, are all SINNERS, living in accordance to the fact that "TO ERR IS HUMAN". That is the way it is, for "no human is an island to itself". We can only learn from our mistakes. Hence a TEACHER is no TEACHER when no assignments are provided during the course of teaching, with solutions provided thereafter by the TEACHER. Unknown is that when one goes through the marking exercise of the assignments, one learns a lot from the students. Hence not all TEACHERS are TEACHERS, just like "NOT ALL THIEVES OR CRIMINALS ARE THIEVES OR CRIMINALS. This is the first law of Nature in the field of LAW. Unknown is that every discipline must apply the first law of Nature for it to exist as a discipline. This has been the case for all ENGINEERING disciplines, Physics and Mathematics, including RELIGION, but very rarely in all other disciplines.

(XII) o-Phthaldehyde

$$H - C - (CH_2)_3 - C - H$$

(XIII) Glutaldehyde

$$H - C - (CH_2)_2 - C - H$$

(XIV) Succinaldehyde

$$H - C - C = C - C - H$$

(XV) Malealdehyde

$$H - C - (CH_2)_n - C - H \qquad (n = 2,3,4)$$

(XVI) Di-methylaldimines

(15.5)

All these are mostly difunctional monomers. The functional groups are strongly nucleophilic in character. (XII) and (XV) are Electrophiles in which the Y centers in (XII) cannot be used, because of steric limitations. The X centers cannot also singly be used, because of steric limitations. They can readily be used however to form a growing ring, because the functional groups are ortho-placed in (XII). In order words, they are not distantly located, such as being para-placed. The C=O center being strongly nucleophilic, favors largely the positively charged or electro-free-radical route. This is the only way rings can be formed. Since an initiator is going to be used for activation, the first addition is Inter-molecular addition. This is then followed by Intra-molecular addition to form ring and generate the active growing center. Chargedly, paired initiators such as $BF_3/O(C_2H_5)_2$, $Al(C_2H_5)_3/H_2O$ and even water can be used. There are some charged paired initiators that can be used electro-free-radically via backward addition. Steric limitations will not hinder their use. For the growing polymer chain, the followings are obtained using dilute hydrochloric acid as source of initiator.

Alternating bicyclic co-polymer of o-Phthaldehyde

15.129

581

The five-membered cyclic ether ring formed above via Equilibrium mechanism, looks like an electrophile. **Indeed, it is a Nucleophile.** External terminating agent will be required. If H₂O is used, the dead polymer obtained is as good as living, since the terminal(s) will be ionic in character. The Hs must be replaced with groups such as CH₃. The O along the chain makes the chain vulnerable and not as strong as when C was in its place.

The weakest nucleophilic center in the bicyclic ring is the C = C center in the ring and not the O center in the ring as has already been stated into law. It is from that center that cross-links can begin to appear as shown below.

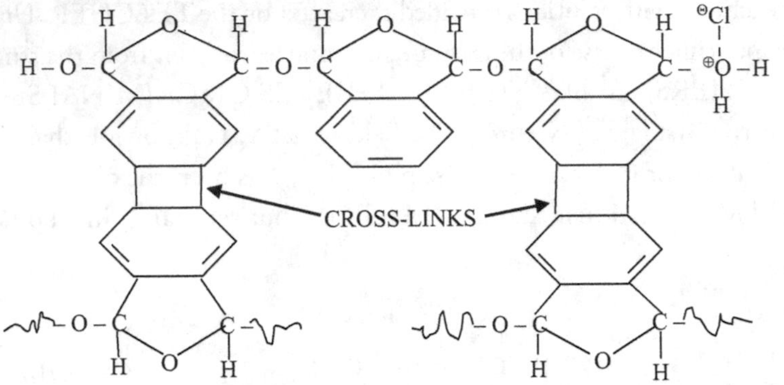

Cross-linked network of polymer from (XII)   15.130

A four-membered ring bonded on two sides by two fused rings is the cross-link between two chains. It can only take place between two dead chains. The four-membered ring is able to survive because of what the benzene ring is carrying, a ring with radical-pulling character. The monomer unit is the two fused rings and an oxygen atom.

With the second electrophile, (XV)-malealdehyde, the polymeric chain via the same route as above is as shown below.

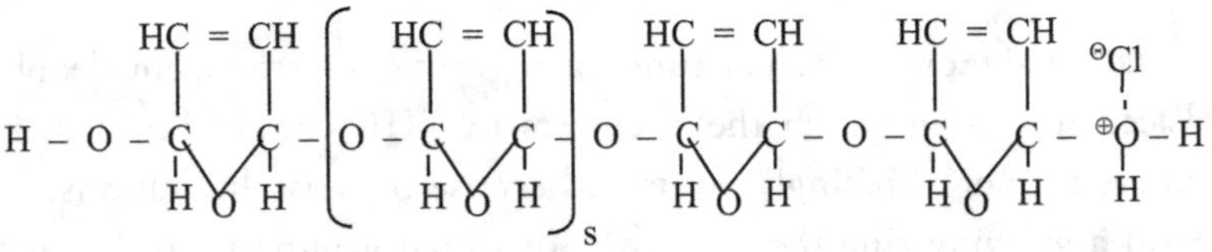

Alternating monocyclic polymer of malealdehyde   15.131

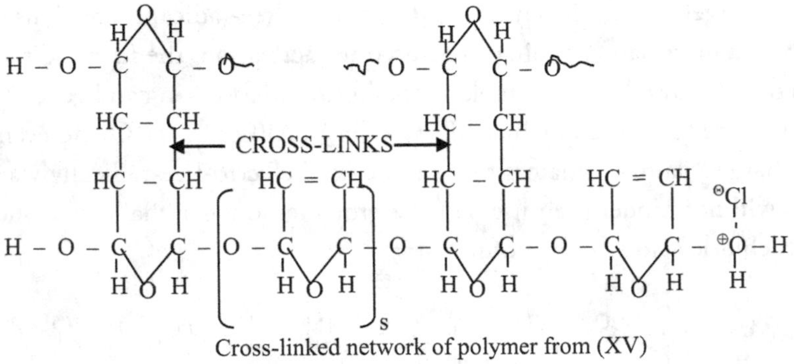

Cross-linked network of polymer from (XV)   15.132

The same four-membered ring like the case above, is the cross-link between two chains. They have been so placed to look as if so mechanically designed. Indeed, that is the way it is with Nature. The four-membered ring here is far less strained than the case above. So far, one can carefully observe with the eye of the needle how the existence of rings are favored. The presence of hetero atoms (for without which males or electrophile cannot exist) in them or adjacently located makes their existence more favored by providing less strain energy in their rings. For example calculation of the strain energies (SE) in Cyclo propene and in Cyclo acetaldehyde as shown below clearly supports this observation.

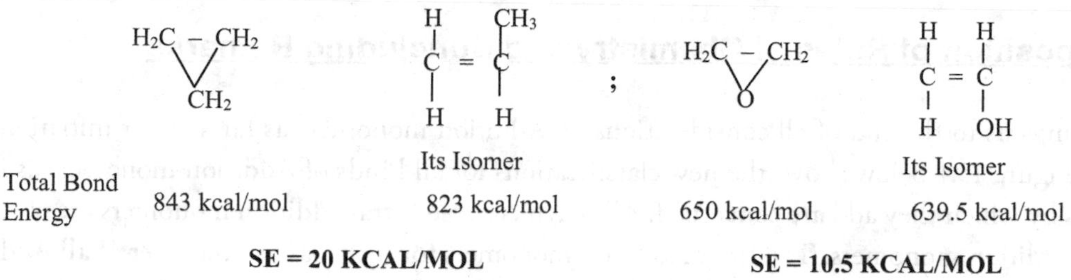

Total Bond Energy	843 kcal/mol	823 kcal/mol	650 kcal/mol	639.5 kcal/mol

SE = 20 KCAL/MOL   **SE = 10.5 KCAL/MOL**   15.133

One can see the hidden or invisible π-bond in the ring in its isomer. Like the first di-aldehyde, this is Inter-/intra-/inter/inter- molecular addition. The rings look like Cyclic ethers and at the same time look like stretched cyclic anhydrides. The initiator used above is dilute (50/50) hydrochloric acid. Note that the isomer for the three-membered ring above is vinyl alcohol and not acetaldehyde because of the presence of C=O bond. Instead of the well-established name of cyclo propane, one called it cyclo propene as has been highlighted before, because of the invisible π-bond in the ring.

For the other di-aldehydes which are Nucleophiles, the same initiators can be used. While for Adipaldehyde, seven-membered rings will be obtained along the chain, for glutaldehyde (XIII) six-membered rings will be formed, and for succinaldehyde (XIV) five-membered rings will be obtained, as shown below for glutaldehyde.

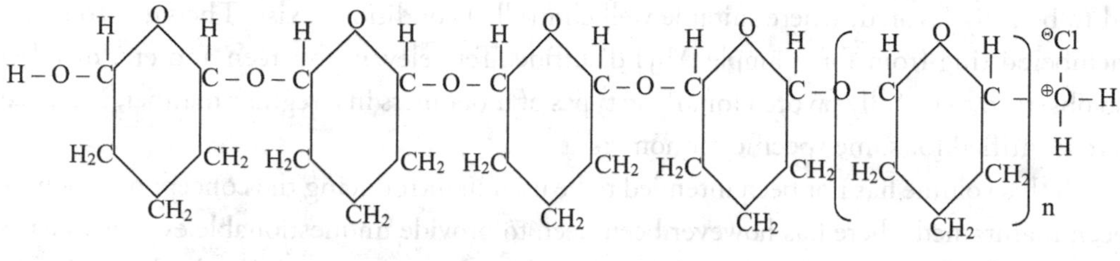

Alternating monocyclic polymer of glutaldehyde   15.134

The initiator should not be present in the system during propagation, because this may open the rings if the operating conditions are high. When this takes place as in many others, then cross-linking sites are obtained. The monomer unit has been highlighted. The seven-member-ed ring will be least strained and more difficult to open.

Finally for (XVI)-di- aldimine, unless suppressed, it will be difficult to polymerize. Hence, dimethyl aldimine was chosen. The same initiator as for aldehydes will also apply, the route being positively charged or electro-free-radical. The growing polymer chain is shown below for n equal to 3.

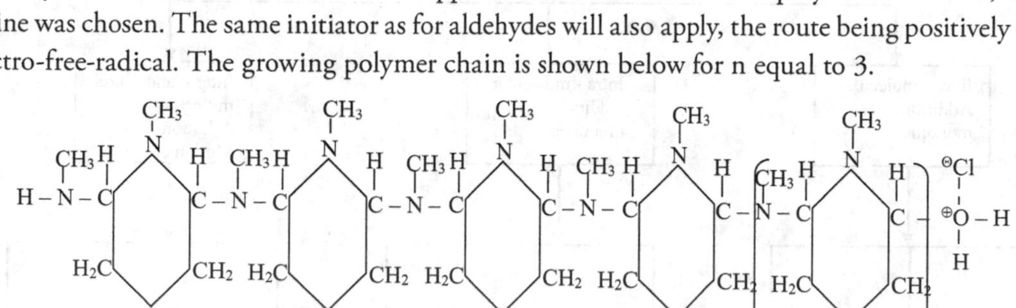

Alternating monocyclic polymer of dimethyl aldimine   15.135

The six-membered cyclic amine looks like a nucleophile. If the initiator is dilute HCl, then a secondary amine is obtained. The N centers in the ring can be used as cross-linking sites, particularly if H is put in place of CH_3 group, when the chain has stopped growing.

15.6 Proposition of Rules of Chemistry and Concluding Remarks

This brings us to the end of all considerations on Addition monomers as far as their initiations are concerned. Figure 15.1 below shows the new classifications for all kinds of Addition monomers as far as the manners by which they add are concerned. All polymerizable Intra-Addition monomers are obtained from Inter-Addition monomers. The Inter-Addition monomers form the largest member of all Addition monomers.

Without the considerations for Inter-Addition monomers, considerations of Intra - and Inter-/Intra-Addition monomers would be impossible, and without consideration of latter groups, the consideration of the former group would be incomplete.

Nevertheless, all these considerations, unfortunately are just still part of the beginning in laying the new foundations in the chemistry of atoms and compounds. When coordination initiators are being fully considered, then one would have fully completed the inorganic aspects of the chemistry of atoms and compounds.

In all the considerations with these families of monomers, molecular rearrangement phenomenon was rarely encountered in view of types of monomers and the types of polymerizations involved particularly with respect to the strength of the initiator used. Very strong activating initiators are largely involved in order to favor cyclization. In all the cases considered, six and few four and eight-membered rings were observed to be fully favored, where suitable well controlled conditions exist. The same too will apply to seven-membered ring from for example Allyl disulfide. Ten, eleven, fourteen and etc. membered rings were also observed to be fully favored for allylic types of monomers in a regular manner. Presence of fused rings were identified for some specific monomers

Though this volume has not been intended to be used in introducing the concept of copolymerization as has been maintained, there has however been need to provide unquestionable evidences based of The New Frontiers. Herein, the need has arisen based on the type of monomers involved, which are uniquely different from the conventional Inter-molecular Addition monomers.

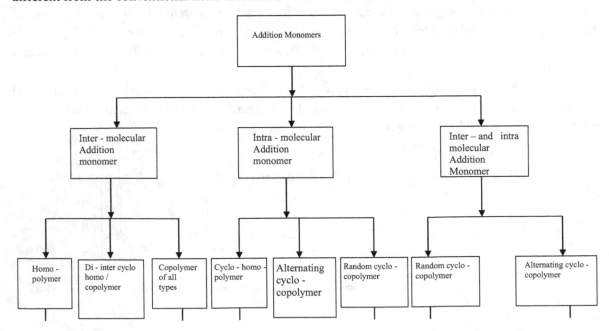

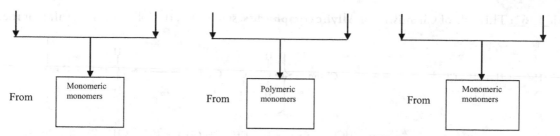

Figure 15.1. Classification of Addition monomers according to mode of Addition

Rule 1359: This rule of Chemistry for **Resonance Stabilization phenomena,** states that, there are two DYNAMIC CLASSES- DISCRETE and CONTINUOUS, for which for example, 1,3-Dienes, 1,3,5-Trienes, and higher –ienes, Furan, and Pyrrole undergo the **DISCRETE CLASS,** while Benzene, Styrene, Divinyl benzene and Thiophene undergo the **CONTINUOUS CLASS.**
(Laws of Creations for Resonance Stabilization)

Rule 1360: This rule of Chemistry for **ALLYLIC groups**, states that, these groups which are generally represented as $H_2C = CH - CH_2 -$ ***is not,*** because it is what the group is carrying that makes it allylic as shown below-

$$H_2C = CH - CH_2 - (CH_2)_n - CH_3 \quad ; \quad H_2C = CH - CH_2 - (CH_2)_n - Cl \quad ; \quad H_2C = CH - CH_2 - (CH_2)_n - OH \quad ;$$

$$H_2C = CH - CH_2 - (CH_2)_n - CH = CH_2 \quad ; \quad H_2C = CH - CH_2 - (CH_2)_n - COOH \quad ; \quad H_2C = CH - CH_2 - (CH_2)_n - NH_2$$

NON ALLYLIC COMPOUNDS (where n = 1,2,3.....) i)

$$H_2C = CH - CH_2 - Cl \quad ; \quad H_2C = CH - CH_2 - OH \quad ; \quad H_2C = CH - CH_2 - COOH \quad ; \quad H_2C = CH - CH_2 - OCOR \quad ;$$

$$H_2C = CH - CH_2 - CHO \quad ; \quad H_2C = CH - CH_2 - NH_2 \quad ; \quad H_2C = CH - CH_2 - CH = CH - COOR$$

ALLYLIC COMPOUNDS ii)

for which the groups directly carried on the CH_2 center can be radical-pushing or pulling groups and not hydrocarbon groups such as H, C_nH_{2n+1}, $CH = CH(CH_3)$, etc.
(Laws of Creations for Allylic compounds)

Rule 1361: This rule of Chemistry for **ALLYLIC groups**, states that, what is unique about them is that they are resonance stabilized, as shown below-

i)

ii)

for which as can be observed, allylic compounds cannot be activated chargedly, but only free-radically, noting that if one or more H on CH_2 group is or are replaced with only alkylane group(s) such as CH_3, the allylic character still remains; noting that it is i) that truely identifies with it, since groups carried by it are groups that carry nucleo-radicals.
(Laws of Creations for Allylic compounds)

585

Rule 1362: This rule of Chemistry for **Allylic compounds**, states that, the followings are valid for them-

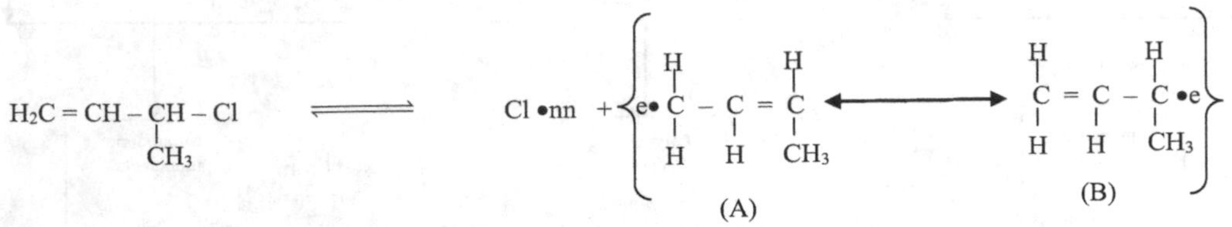

$$H_2C = CH - \underset{\underset{CH_3}{|}}{CH} - Cl \quad \rightleftharpoons \quad Cl \bullet nn \; + \left\{ \begin{array}{cc} e\bullet \underset{\underset{H}{|}}{\overset{\overset{H}{|}}{C}} - \underset{H}{\overset{H}{C}} = \underset{CH_3}{\overset{H}{C}} & \longleftrightarrow & \underset{H}{\overset{H}{C}} = \underset{H}{C} - \underset{CH_3}{\overset{H}{C}} \bullet e \\ (A) & & (B) \end{array} \right\}$$

$$\xrightarrow{+ C_2H_5OH} \quad HCl \; + \; \underset{(82\%)}{C_2H_5OCH_2 - CH = CH(CH_3)} \; + \; \underset{(18\%)}{CH_2 = CH - CH(CH_3)OC_2H_5}$$

$$CH_3CH = CH - CH_2 - \quad > \quad H_2C = CH - CH(CH_3) - \quad >> \quad H_2C = C(CH_3) - CH_2 -$$

Order of Radical-pushing capacity   i)

for which, there is more of (A) than (B), because the following is also valid-

$$CH_3CH = CH - CH_2 - \quad > \quad H_2C = CH - CH(CH_3) - \quad >> \quad H_2C = C(CH_3) - CH_2 -$$

Order of Radical-pushing capacity   ii)

(Laws of Creations for Allylic compounds)

Rule 1363: This rule of Chemistry for **a group herein called Acetyallylic group [HC≡C – CH$_2$ –]**, states that, just as the Allylic groups [H$_2$C = CH – CH$_2$ –], only specific groups similar to those that identify with allyls makes it acetyallyl, for which what is unique about it is that it is resonance stabilized as shown below-

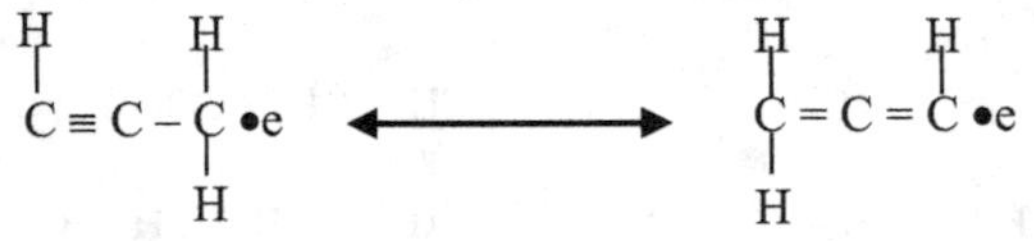

$$\underset{\underset{H}{|}}{\overset{\overset{H}{|}}{C}} \equiv C - \underset{H}{\overset{H}{C}} \bullet e \quad \longleftrightarrow \quad \underset{\underset{H}{|}}{\overset{\overset{H}{|}}{C}} = C = \overset{H}{C} \bullet e$$

$$(A) \qquad\qquad\qquad (B)$$

for which, (A) is more stable but less radical-pushing than (B), and when the Hs on CH$_2$ group are changed to only alkylane groups such as CH$_3$, the acetyallylic character still remains.

(Laws of Creations for Acetyallylic group)

Rule 1364: This rule of Chemistry for **Hydrocarbon Family Tree,** states that, after the Carbenes in the hierarchy of the families of Hydrocarbons, there is a family which is herein called **Acetylcarbenes** as shown below for their first members activated and unactivated-

CARBENES   **ACETYLCARBENES**

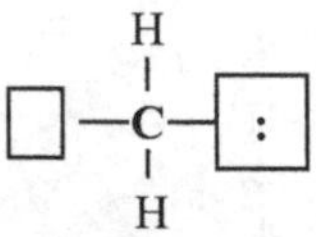

$$\square - \underset{\underset{H}{|}}{\overset{\overset{H}{|}}{C}} - \square :$$

Unactivated Methylene (From Methane)

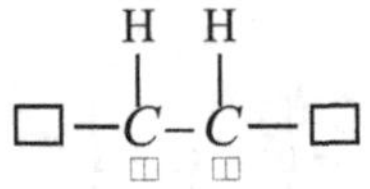

$$\square - \underset{\underset{\square}{|}}{\overset{\overset{H}{|}}{C}} - \underset{\square}{\overset{H}{C}} - \square$$

Unactivated Ethylyne (From Ethyne ≡ Acetylene)   i)

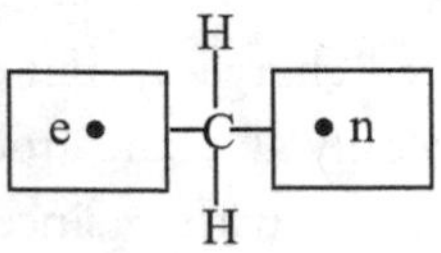

$$e \bullet \; \square - \underset{\underset{H}{|}}{\overset{\overset{H}{|}}{C}} - \square \; \bullet n$$

Methylene

$$e \bullet \; \square - \underset{\underset{n}{|}}{\overset{\overset{H}{|}}{C}} -- \underset{\square}{\overset{H}{C}} - \square$$

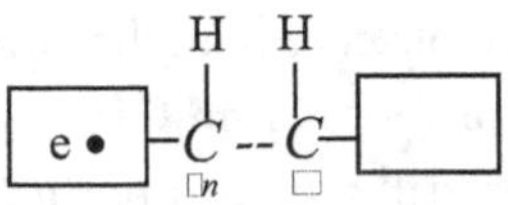

Ethylyne   ii)

noting that when the ethylyne is decomposed, it breaks down to two equal parts and when in equilibrium state of existence, it breaks down into two different parts, and on the other hand, this is analogous to carbenes which when activated give alkylenes, the first member of which is methylene and when the above **(wherein each center is activated one at a time as shown above)** is deactivated, acetylene or ethyne is obtained, just as when two methylenes combine together to give ethene.

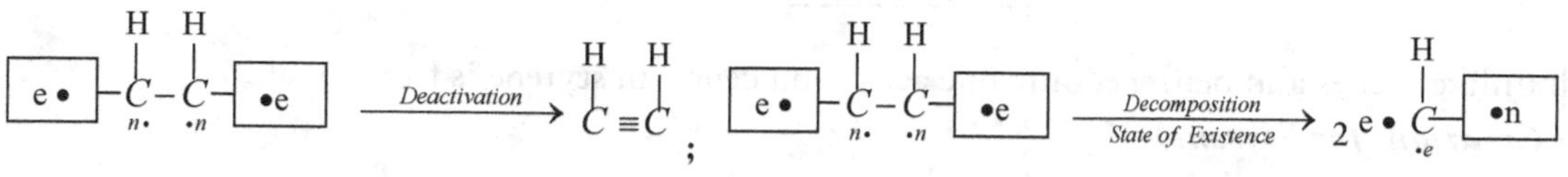

iii)

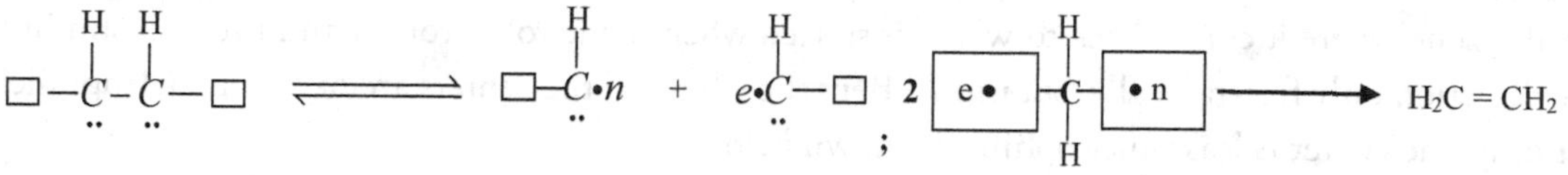

iv)

(Laws of Creations for Carbenes and AcetylCarbenes)

Rule 1365: This rule of Chemistry for **Monoene or alkenyl resonance stabilization groups,** states that, what is unique about them is that when they are used as resonance stabilization groups, no transfer species can be abstracted from them, unless when they are externally located as shown below -

$$H_2C = CH - ; \quad ClHC - CH - ; \quad H_2C = C(COH) - ; \quad H_2C = CCl - ; \quad H_2C = C(COOH) - ; \quad H_2C = C(OH) -$$

NO TRANSFER SPECIES

$$(CH_3)HC = CH - ; \quad (HOOC)HC = CH - ; \quad (HCOO)HC = CH - ; \quad (H_2N)HC = CH - ; \quad (HO)HC = CH -$$

$$(ORC)HC = CH -$$

HAVE TRANSFER SPECIES

and in addition, the group cannot carry negative charges due to electrostatic forces of repulsion.
(Laws of Creations for Mono-ene groups)

Rule 1366: This rule of Chemistry for **Vinyl groups,** states that, these groups which are generally represented as $H_2C = CH -$ **is not,** because it is what the group is carrying that makes it vinylic as shown below-

$$H_2C = CH - Cl ; \qquad H_2C = CH - OH ; \qquad H_2C = CH - NH_2 ; \qquad H_2C = CH - OCOR ;$$

$$H_2C = CH - O - CH = CH_2 \qquad ; \qquad H_2C = CH - NH - CH = CH_2$$

VINYLIC COMPOUNDS i)

$$H_2C = CH - CH_3 \qquad ; \qquad H_2C = CH - CH_2 - Cl \qquad ; \qquad H_2C = CH - CH_2 - OH ;$$

$$\left[H_2C = CH - (CO) - O - (CO) - CH = CH_2 ; \quad H_2C = CH - COOH \qquad ; \qquad H_2C = CH - COR \right]$$

(ACRYLIC COMPOUNDS)

NON VINYLIC COMPOUNDS ii)

for which the groups directly carried on the CH center can be non-free radical-pushing or pulling groups and not hydrocarbon groups such as H, C_nH_{2n+1}, $CH = CH(CH_3)$, etc. and the last group above shown under Non vinylic compounds are known as Acrylic compounds.
(Laws of Creations for Vinylic compounds)

Rule 1367: This rule of Chemistry for **Styrene**, states that, the provider of resonance stabilization, being the phenyl group clearly indicates that the following is valid-

Order of Nucleophilicity

for which unlike dienes and benzene, only one activation center in styrene is functional.
(Laws of Creations for Styrene)

Rule 1368: This rule of Chemistry for **Styrene,** states that, though its resonance stabilization is continuous like in Benzene where it comes back to where it started when made to go round and round again in the reverse direction only free-radically, but unlike Benzene where all the centers are of equal nucleophilicity, with it only one center is least nucleophilic as shown below-

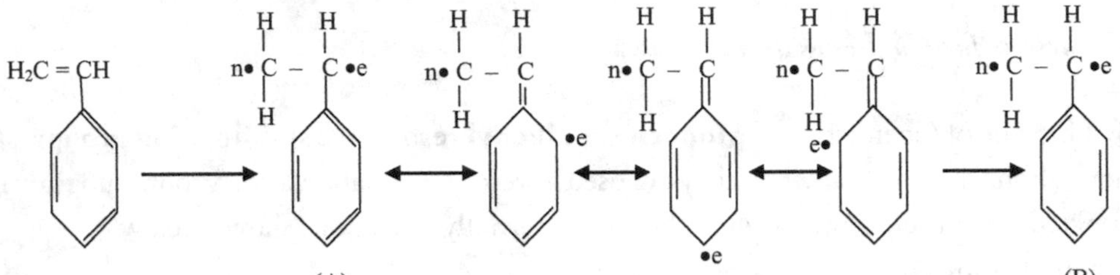

for which unlike Dienes and Benzene, (A) or (B) is the only mono-form that exists for it and the movement round the ring is only once.
(Laws of Creations for Styrene)

Rule 1369: This rule of Chemistry for **p-Divinyl benzene (3,6-Benzadiene),** states that, though its resonance stabilization is continuous as in Styrene (i.e., it comes back to where it started) only free-radically, but unlike Benzene where all the centers are of equal nucleophilicity, with it, the following is valid-

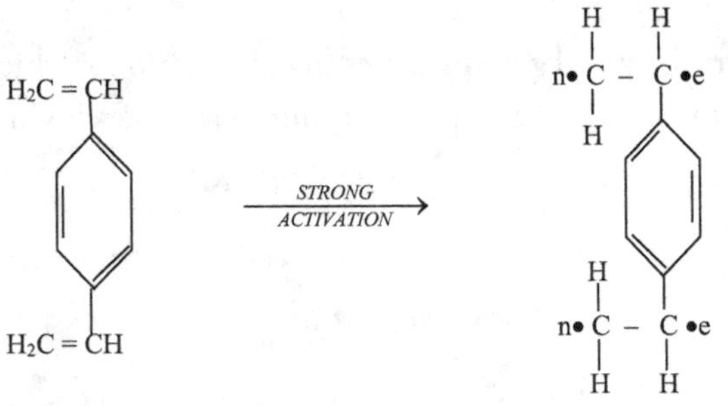

i)

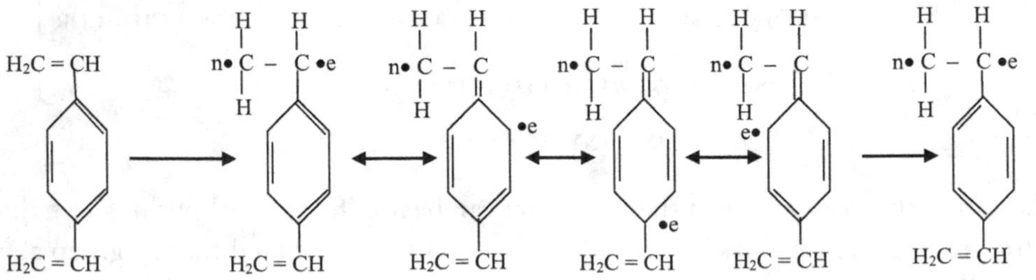

ii)

for which unlike Styrene, exists two centers of the same nucleophilic capacity and the provider of resonance stabilization for any of them one at a time is the "phenyl group" in (A), but like Styrene and unlike Dienes, (B) or (C) is the only mono-form that exist for it, with the six groups (Hs and the second vinyl group) resonance stabilized.

(Laws of Creations for p-Divinylbenzene)

Rule 1370: This rule of Chemistry for **o-Divinyl benzene (3,4-Benzadiene),** states that, though its resonance stabilization is continuous as in Styrene (i.e., it comes back to where it started) only free-radically, but unlike Benzene where all the centers are of equal nucleophilicity, with it, the following is valid-

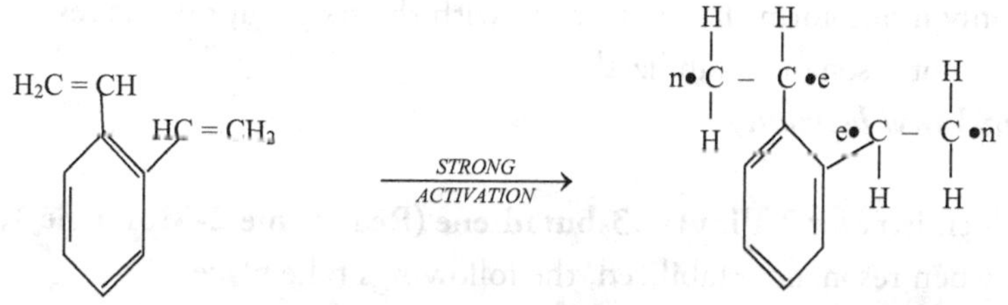

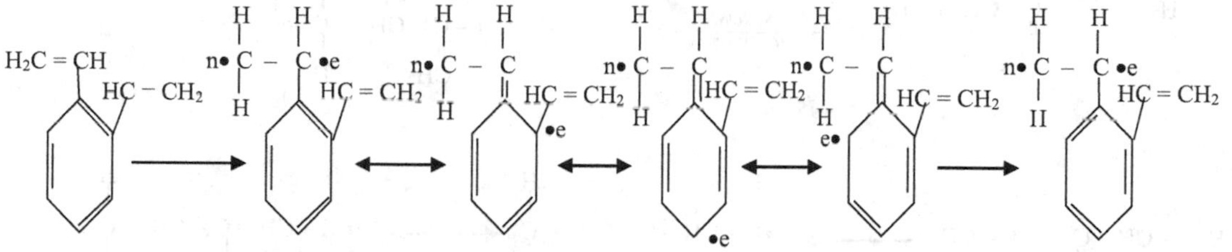

for which unlike Styrene, exists two centers of the same nucleophilic capacity and the provider of resonance stabilization for any of them one at a time is the "phenyl group" in (A), but like Styrene and unlike Dienes, (B) or (C) is the only mono-form that exist for it, with the six groups (Hs and the second vinyl group) resonance stabilized.

(Laws of Creations for o-Vinyl benzene)

Rule 1371: This rule of Chemistry for **m-Divinyl benzene (3,5-Benzadiene),** states that, though its resonance stabilization is continuous as in Styrene (i.e., it comes back to where it started) only free-radically, but unlike Benzene where all the centers are of equal nucleophilicity, with it, the following is valid-

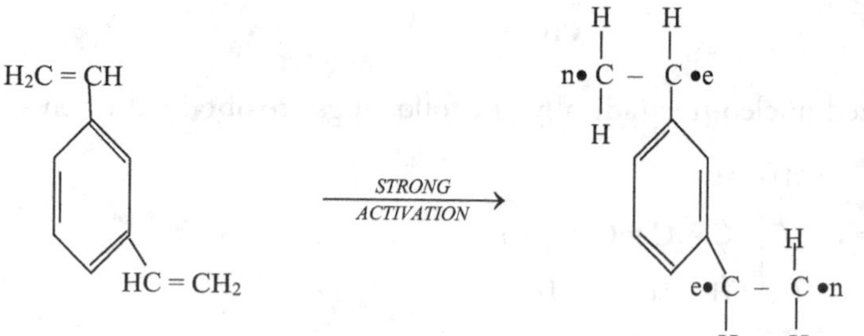

589

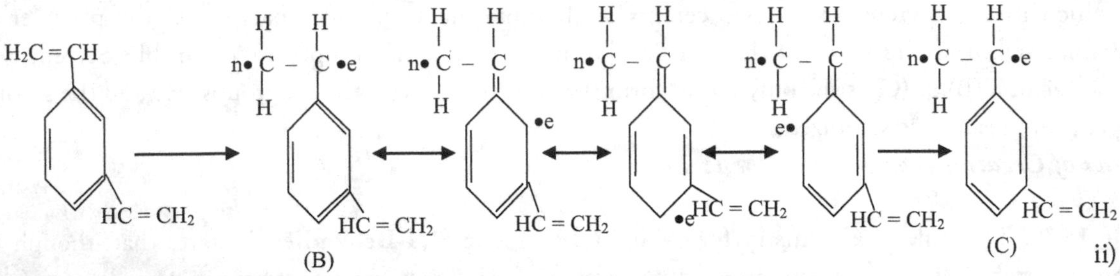

(B)   (C)   ii)

for which unlike Styrene, exists two centers of the same nucleophilic capacity and the provider of resonance stabilization for any of them one at a time is the phenyl group in (A), but like Styrene and unlike Dienes, (B) or (C) is the only mono-form that exist for it, with the six groups of Hs resonance stabilized and the second vinyl group not resonance stabilized.
(Laws of Creations for m-Vinyl benzene)

Rule 1372: This rule of Chemistry for **2-Vinyl -1,3-butadiene (Real name-2-Monoene-1,3-butadiene)** shown below, states that, when resonance stabilized, the followings take place-

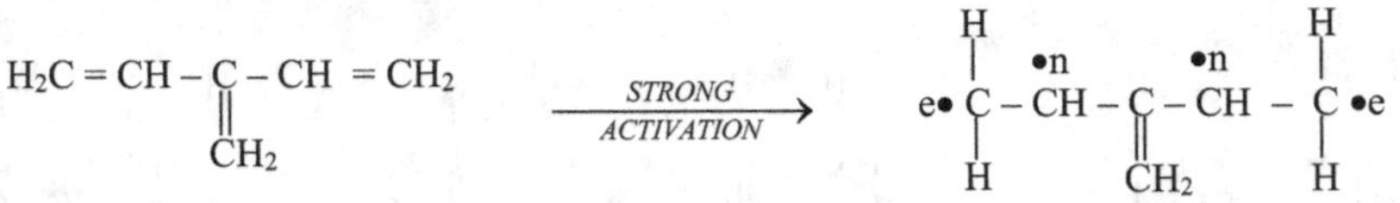

i)

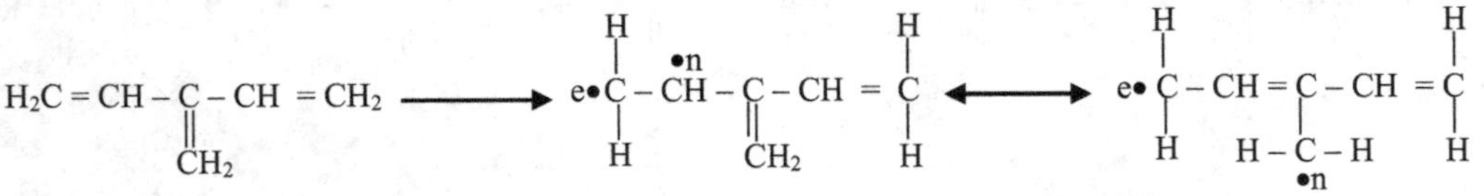

ii)

for which like divinyl benzene, there are two centers of the same nucleophilic capacity and what is providing resonance stabilization for them one at a time is the ene group in the middle and unlike divinyl benzene, the resonance stabilization provided is DISCRETE.
(Laws of Creations for 2-Vinyl 1,3-Butadiene)

Rule 1373: This rule of Chemistry for **Vinylacetylene (Ethylkene acetylene),** states that, when activated free-radically, this monomer is resonance stabilized as follows-

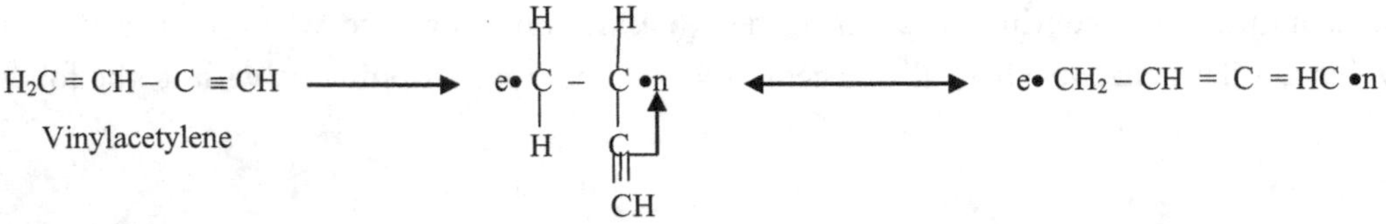

i)

for which when polymerized nucleo-free-radically, the followings are obtained when suppressed-

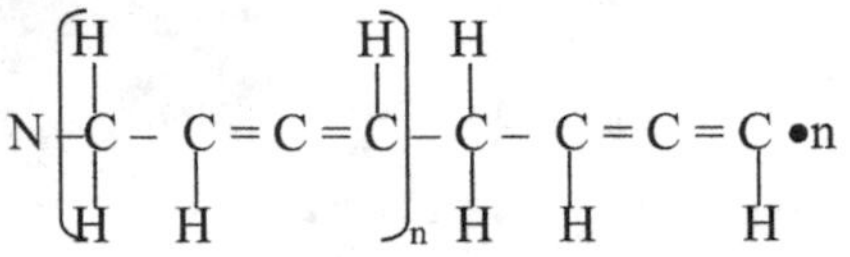

ii)

and when polymerized chargedly, the followings are similarly obtained when suppressed.

$$Cl-Ti-\left[\underset{\underset{CH}{\overset{H}{|}}}{\overset{R}{\underset{|}{C}}} - \underset{\underset{|}{H}}{\overset{H}{\underset{|}{C}}}\right]_n \underset{\underset{CH}{\overset{H}{|}}}{\overset{H}{\underset{|}{C}}} - \underset{\underset{H}{\overset{H}{|}}}{\overset{H}{\underset{|}{C}}}{}^{\oplus}.......{}^{\ominus}Al\underset{R}{\overset{R}{<}}$$

iii)

(Laws of Creations for Vinylacetylene)

Rule 1374: This rule of Chemistry for **1,3-Dienes**, states that, when activated the followings are the centers first activated and the radicals carried by the active centers-

$$H_2C = CH - CH = CH_2 \xrightarrow{Activation} n\bullet \underset{\underset{CH_2}{\overset{||}{CH}}}{\overset{\overset{H}{|}}{\underset{|}{C}}} - \underset{\underset{|}{H}}{\overset{H}{\underset{|}{C}}}\bullet e$$

4,-3-monoform of Butadiene

$$H_2C = CH - CCl = CH_2 \xrightarrow{Activation} e\bullet \underset{\underset{CH_2}{\overset{||}{CH}}}{\overset{\overset{H}{|}}{\underset{|}{C}}} - \underset{\underset{|}{H}}{\overset{Cl}{\underset{|}{C}}}\bullet n$$

1,2-monoform of Chloroprene

i)

$$H_2C = CH - C(CH_3) = CH_2 \xrightarrow{Activation} e\bullet \underset{\underset{CH_2}{\overset{||}{CCH_3}}}{\overset{\overset{H}{|}}{\underset{|}{C}}} - \underset{\underset{|}{H}}{\overset{H}{\underset{|}{C}}}\bullet n$$

4,3-monoform of Isoprene ;

$$H_2C = CH - CH = CH(CH_3) \xrightarrow{Activation} n\bullet \underset{\underset{HC(CH_3)}{\overset{||}{CH}}}{\overset{\overset{H}{|}}{\underset{|}{C}}} - \underset{\underset{|}{H}}{\overset{H}{\underset{|}{C}}}\bullet e$$

4,3-monoform of Pentadiene

ii)

$$H_2C = CH - CH = CHCl \xrightarrow{Activation} n\bullet \underset{\underset{CH_2}{\overset{||}{CH}}}{\overset{\overset{Cl}{|}}{\underset{|}{C}}} - \underset{\underset{|}{H}}{\overset{H}{\underset{|}{C}}}\bullet e$$

1,2-monoform of 1-Chloro-Butadiene

iii)

for which the followings are valid-

$$H_2C = CCl - \quad > \quad (CH_3)CH = CH - \quad > \quad H_2C = CH - \quad > \quad H_2C = C(CH_3) - > \quad ClCH = CH -$$

Pulling

Order of Radical-pushing capacity/ Nucleophilicity

iv)

(Laws of Creations for 1,3-Dienes)

Rule 1375: This rule of Chemistry for **Initiators,** states that, *Active centers* are centers carried by initiators which assist in activating and adding to monomers on a continuous basis via Combination or Equilibrium Backward mechanism, for which some initiators (free and paired) have only one active center and others (Paired) have two active centers.
(Laws of Creations for Initiators)

Rule 1376: This rule of Chemistry for **Initiators,** states that, when used on monomers with one activation center, the active center generated by it, is the light of the growing polymer chain whether the route is

natural or not; for which if the route is natural, the light keeps getting brighter and brighter for every monomer added and if not natural, the light keeps getting dimmer and dimmer for every monomer added.
(Laws of Creations for Initiators)

Rule 1377: This rule of Chemistry for **Initiators**, states that, when used on monomers with two or more activation centers isolatedly and distantly placed and are of the same nucleophilicity, the active center generated by the initiator, remains the light of the growing polymer chain whether the route is male (positive) or female (negative), for as long as it is the natural route.
(Laws of Creations for Initiators)

Rule 1378: This rule of Chemistry for **Electrophiles**, states that, when there is another activation center externally and distantly located to the X/Y centers, the least nucleophilic center in the monomer is the center externally and distantly located.
(Laws of Creations for Electrophiles)

Rule 1379: This rule of Chemistry for **Initiators**, states that, ***the ability of an initiator being able to identify if a monomer is male or female in character is possible only when ionic, covalently charged and radical paired initiators are involved and not when electrostatically charged paired or free or non- radical initiators are involved,*** and these are initiators with two active centers.
(Laws of Creations for Initiators)

Rule1380: This rule of Chemistry for **Activation of resonance stabilized monomers** such as 1,3-Dienes with Charged paired types of initiators, states that, ***if resonance stabilization had taken place radically before initiator was prepared, covalent charges cannot be formed when the two radical carrying centers are distantly located, and this is irreversible-***

IMPOSSIBLE TRANSFORMATION

i)

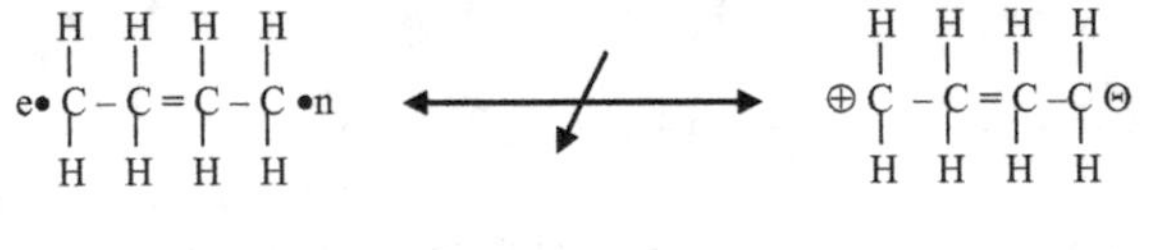

IMPOSSIBLE TRANSFORMATION

ii)

for which the same applies to electrostatic charges where possible, but not to polar charges as shown below noting that with polar charges, this takes place only when two or more centers placed side by side carry paired unbonded radicals in their last shell, for which it is

Ozone   Stable polymeric oxygen (The Ozone Layer)

iii)

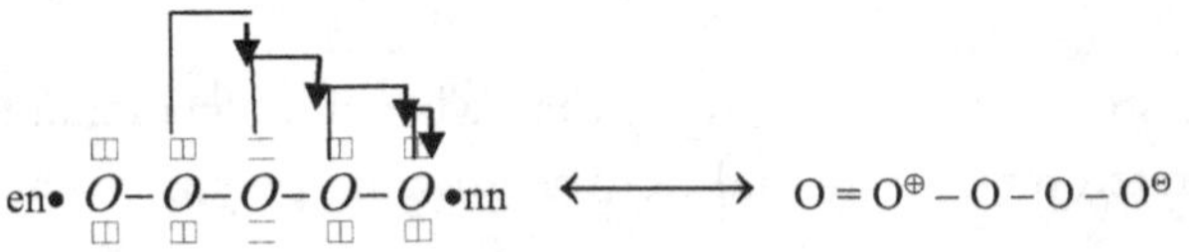

iv)

from the last one carrying nucleo-free or non-free-radical that an electro-radical is forced to move from the neighboring paired unbonded radicals to form a negative charge and the movement of electro-radicals continue until the second to the last O atom where a double bond is formed and a positive charge is placed noting that with the movement of each electro-radical, a hole is always created and the process is reversible at different operating conditions.

(Laws of Creations for Movement of Electro-radicals)

Rule 1381: This rule of Chemistry for **charges and radicals**, states that, the following reac-tions are the possible and the impossible for them-

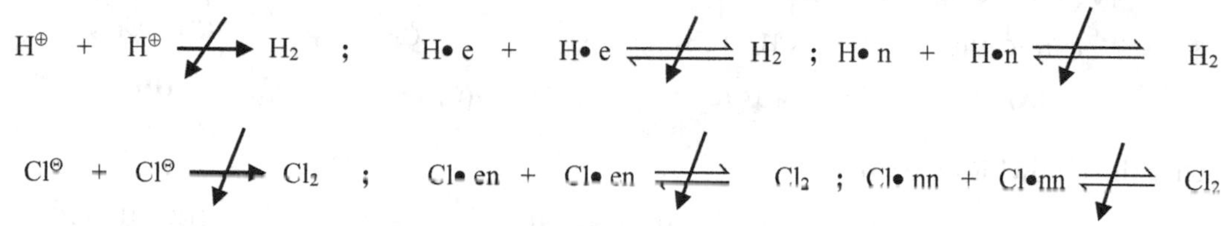

$$H^{\oplus} + H^{\oplus} \xrightarrow{\quad} H_2 \quad ; \quad H \bullet e + H \bullet e \rightleftharpoons H_2 \quad ; \quad H \bullet n + H \bullet n \rightleftharpoons H_2$$

$$Cl^{\ominus} + Cl^{\ominus} \xrightarrow{\quad} Cl_2 \quad ; \quad Cl \bullet en + Cl \bullet en \rightleftharpoons Cl_2 \quad ; \quad Cl \bullet nn + Cl \bullet nn \rightleftharpoons Cl_2$$

IMPOSSIBLE REACTIONS

i)

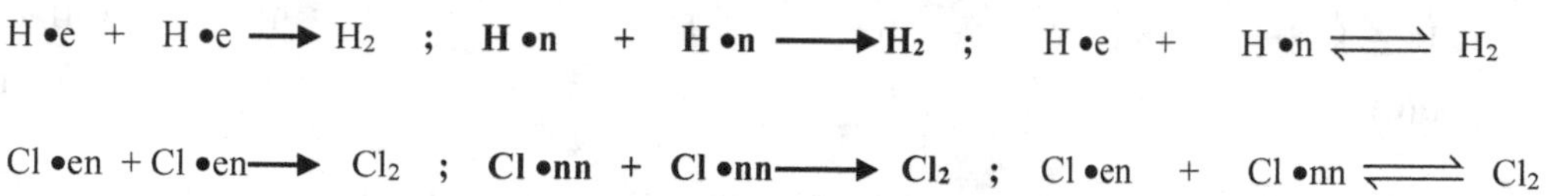

$$H \bullet e + H \bullet e \longrightarrow H_2 \quad ; \quad \mathbf{H \bullet n} + \mathbf{H \bullet n} \longrightarrow \mathbf{H_2} \quad ; \quad H \bullet e + H \bullet n \rightleftharpoons H_2$$

$$Cl \bullet en + Cl \bullet en \longrightarrow Cl_2 \quad ; \quad \mathbf{Cl \bullet nn + Cl \bullet nn} \longrightarrow Cl_2 \quad ; \quad Cl \bullet en + Cl \bullet nn \rightleftharpoons Cl_2$$

FAVORED EXISTENCE

ii)

for how can two hydrogen cations carrying nothing form a bond and how can two chlorine anions carrying four radicals form two bonds without breaking the Boundary laws and how can two males (or two females) exist together in Equilibrium to form a molecule?

(Laws of Creations for Humanity)

Rule 1382: This rule of Chemistry for **Radical carrying hydrocarbon species**, states that, while transfer species can be rejected very readily from some, for others, no transfer species can be rejected as shown below-

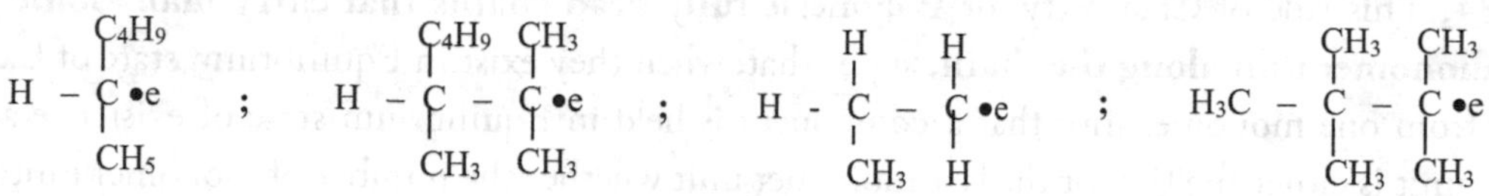

(A)   (B)   (C)   (D)

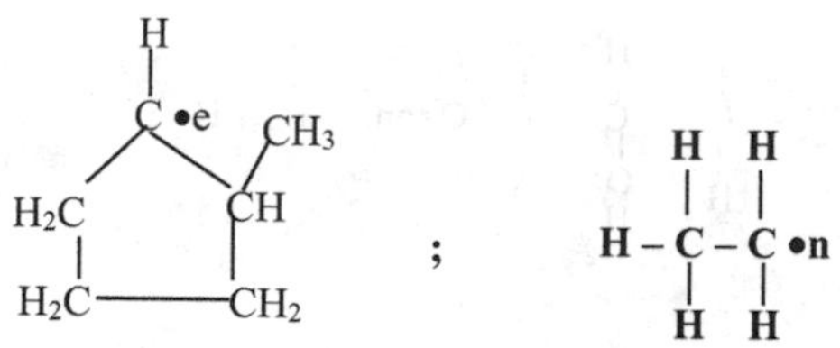

(E)   (F)

since in Nature, it is the richer component that provides transfer species under certain conditions and if it cannot, the second richer component does the rejection and the transfer species rejected is the smallest, for which (A), (B), (D) and (E) can reject in order to form a double bond when mildly heated, but (C) cannot reject and (F) can reject H as H•n (Hydride) only at very high temperatures of the order of 1000^0Cs and high pressures or in the presence of an ionic metal, noting that ability to reject has little or nothing to do

with the bond strength, but with many other laws- Laws of Conservation of transfer of transfer species, Mechanism of reaction, Balancing of Radicals and charges, operating conditions and more.
(Laws of Creations for Transfer Species)

<u>**Rule 1383:**</u> This rule of Chemistry for **Polymeric chains carrying π-bonds internally located along the chain,** states that, these gigantic monomers *are such that cannot molecularly rearrange after activation*, since they are like mono-alkenes shown below-

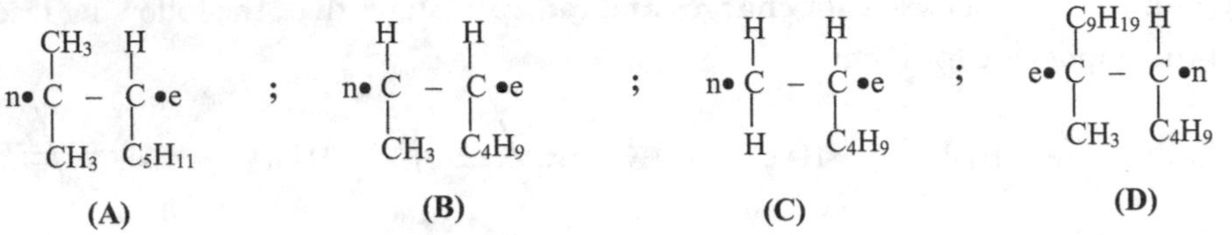

noting that the followings are valid-

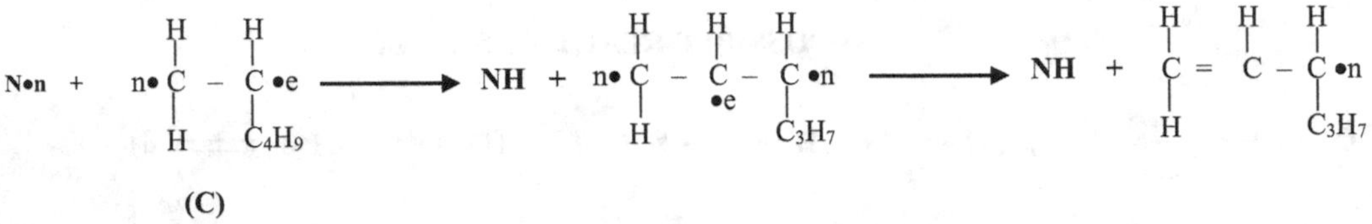

in which for the growing polymer chain, transfer species was rejected from the second richer component C_4H_9 instead of from the chain, based on the Laws of Conservation of transfer of transfer species.
(Laws of Creations for Molecular rearrangement)

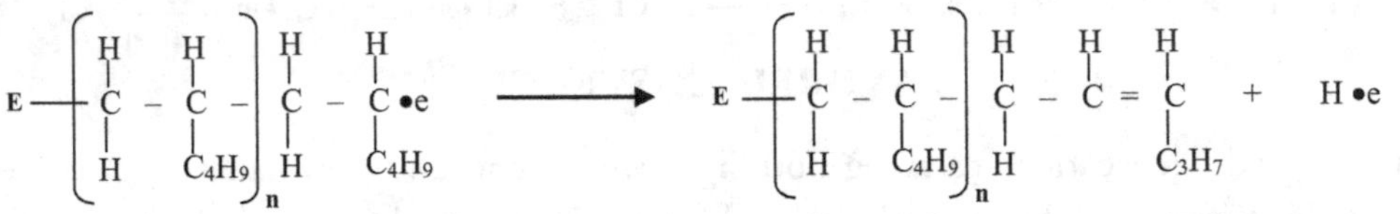

<u>**Rule 1384:**</u> This rule of Chemistry for **Polymeric fully dead chains that carry many ionic groups on each monomer unit along the chain,** states that, when they exist in Equilibrium state of Existence, it is only from one monomer unit that a component is held in Equilibrium state of existence and that monomer unit is either the first or the last monomer unit whether the number of monomer units is even or odd independent on the carrier of the chain, and never from the middle, like the case with glycerol where the terminals are carrying such ionic groups.

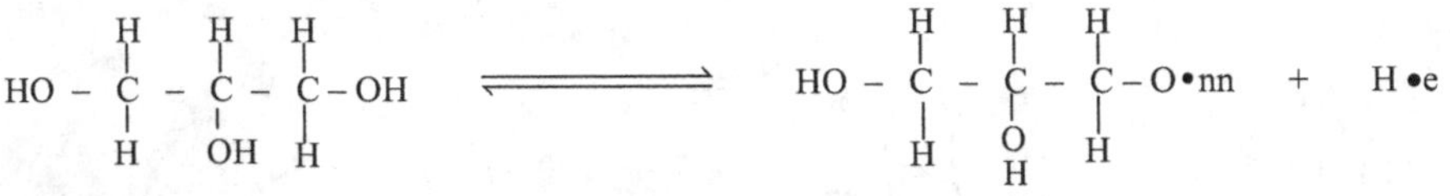

(Laws of Creations for Polymeric monomers)

<u>**Rule 1385:**</u> This rule of Chemistry for **only C/H Polymeric chains with π-bonds along the chain,** states that, since all the π-bonds are of different nucleophilicity, only one can be activated one at a time and that one is in the middle (if odd or by the middle if even) of the chain where the least nucleophilic center exists; with nucleophilicity increasing on both sides as one moves from the middle to right or to the left, depending on what is carrying the chain.
(Laws of Creations for Polymeric monomers)

Rule 1386: This rule of Chemistry for **Chloroprene**, states that, 1,4-mono-form is more largely obtained than 1,2- mono-form as their polymeric products when polymerized radically while 1,2- mono-form cannot be exclusively obtained when polymerized chargedly, since chloroprene like vinyl chloride cannot be activated chargedly.
(Laws of Creations for Chloroprene)

Rule 1387: This rule of Chemistry **for Poly (1,2- or 3,4- dienes),** states that, though all the activation centers are of different nucleophilicity, the radicals carried by the activation centers along the chain is not constant in capacity, for which the least nucleophilic center is either at the beginning or at the end of the chain depending on the route of polymerization.
(Laws of Creations for Poly(1,2- or 3,4-dienes))

Rule 1388: This rule of Chemistry for **ADDITION MONOMERS, based on the manners by which they add,** states that, there are three KINDS of Addition monomers-**INTER-,** *INTRA*- and **INTER / INTRA- ADDITION** kinds of MONOMERS, for which apart from very unique differences amongst them, INTER- do not form cyclic units along a polymeric chain, but INTRA- and INTER/INTRA-cyclization takes place during polymerization either randomly or regularly, depending on the operating conditions and the types of monomers.
(Laws of Creations for Addition Monomers)

Rule 1389: This rule of Chemistry for **States of Existences,** states that, not all compounds which can exist in Equilibrium state of existence, exist as such; there are those that are fully held in Equilibrium state of Existence, and there are those where some are kept in Stable state of existence leaving a fraction in Equilibrium state of existence such as the case with water, ammonia, and so on; noting that the fraction of compound kept in Equilibrium state of existence is always small and depends on the operating conditions.
(Laws of Creations for States of Existences)

Rule 1390: This rule of Chemistry for Vaporization of **Water,** states that, when water is boiled, while a small fraction is held in Equilibrium state of existence, a very fraction is held in Stable state of existence as shown below-

Stage 1:

$$H_2O \quad \underset{\longleftarrow}{\overset{Activation}{\longrightarrow}} \quad H \bullet e \; + \; nn \bullet OH$$

$$H \bullet e \; + \; H_2O \; \rightleftharpoons \; H_2 \; + \; en \bullet OH \; + \; \textbf{Heat}$$

$$HO \bullet en \; \rightleftharpoons \; H \bullet e \; + \; en \bullet O \bullet nn \; + \; Heat$$

$$nn \bullet O \bullet en \; + \; H_2 \; \rightleftharpoons \; HO \bullet nn \; + \; H \bullet e$$

$$\rightleftharpoons \; H_2O \; [\text{Close to Vapor state}]$$

$$H \bullet e \; + \; nn \bullet OH \; \longrightarrow \; H_2O \; [\text{Colder and Stable}] \qquad \text{(i)}$$

Overall Equation: $2H_2O \; \xrightarrow{\text{Heat}} \; H_2O$ [In vapor] $+ \; H_2O$ [Liquid-vapor] $+$ Heat (ii)

for which so many stages like the first stage take place at the same time in parallel followed by the second stage wherein the liquid water is gradually being transformed to vapor due to the heat generated in so many countless numbers of stages until a point is reached when all the liquid becomes vapor.
(Laws of Creations for Vaporization of Water)

Rule 1391: This rule of Chemistry for **NaOH/Water,** states that, when pellets of NaOH are dropped in water, the followings take place-

Stage 1:

$$LHS \qquad RHS$$
$$\underline{H_2O} \rightleftharpoons H \cdot e + nn \cdot OH$$
$$H \cdot e + \underline{NaOH} \rightleftharpoons NaO \cdot en + \underline{H_2} + Heat$$
$$NaO \cdot en \rightleftharpoons Na \cdot e + en \cdot O \cdot nn + Heat$$
$$nn \cdot O \cdot en + \underline{H_2} \rightleftharpoons HO \cdot nn + H \cdot e$$
$$\rightleftharpoons \underline{H_2O}$$
$$Na \cdot e + nn \cdot OH \longrightarrow \underline{NaOH}$$
$$(Pr \textit{oductive}) \tag{i}$$

$$\underline{Overall\ Equation}:\ NaOH + H_2O \longrightarrow NaOH + H_2O + Heat \tag{ii}$$

for which, one can observe where heat is generated, noting that what is on the Right-hand side (RHS) which is imaginary is exactly the same as what is on the Left hand side (LHS) which is real, clear indication that NaOH, dissolves in Water both being Polar/Ionic, but is insoluble in the water, noting also that the last step could be either double half headed arrows to indicate an Equilibrium Stage or a single right headed arrow as shown above for the heat generated.
(Laws of Creations for NaOH/Water combination)

Rule 1392: This rule of Chemistry for **Resonance Stabilization phenomenon,** states that, not all compounds that favor resonance stabilization through the use of one single active center get fully resonance stabilized; there are those that get fully resonance stabilized, and there are those that are partially resonance stabilized leaving a fraction not resonance stabilized, such as the case with some 1,3-dienes, noting that all these depend on the operating conditions.
(Laws of Creations for Resonance Stabilization phenomenon)

Rule 1393: This rule of Chemistry for **Resonance Stabilization phenomenon,** states that, *a situation where a monomer such as 1,3-Diene when activated radically by two opposite active centers such as in Di-Inter Addition (Not one active center) stops at 1,2-mono-form without undergoing resonance stabilization phenomenon to 1,4-mono-form does not exist; 1,2- or 3,4- mono-forms can only exist when the diene is at STP when stored (e.g., Vinyl cyclo-hexene) or when one activation center is involved.*
(Laws of Creations for Resonance Stabilization phenomenon)

Rule 1394: This rule of Chemistry for **Monomers that favor being used alone for Cycliza-tion during polymerization,** states that, these monomers are not a single monomer, but *Di- or Tri or Poly-monomers,* since they carry at least **two activation centers distantly and externally located,** for which the polymers obtained from them are mostly copolymers.
(Laws for Creations for Poly-monomers)

Rule 1395: This rule of Chemistry for **Di-Inter-Addition polymerization system,** states that, this is a method of polymerization which involves *the use of discrete type of resonance stabilization monomers such as 1,3- dienes in the absence of initiators, but presence of strong generating forces to activate*

the monomer free-radically only to generate two active centers for propagation to form six-membered cyclic fused rings along the chain via Combination mechanism, with termination by deactivation.
(Laws of Creations for Di-Inter- Addition Polymerization system)

Rule 1396: This rule of Chemistry for **Di-Inter–Addition polymerization system,** states that, what makes it Di-inter-addition is *the rules based on Diffusion controlled mechanisms,* in which the electro-free-radical center carried by the propagating chain diffuses to the nucleo-free-radical end of the activated monomers while at the same time the other electro-free-radical center on the activated monomer diffuses to nucleo-free-radical end of the growing chain to produce four, six or eight-membered rings.
(Laws of Creation for Di-Inter-Addition Polymerization system)

Rule 1397: This rule of Chemistry for **Intra- Addition polymerization system,** states that, this is a method of polymerization involving *the use of polymers with externally located double bonds along the chain, herein called <u>Polymeric monomers,</u>* to form ladder polymers electro free radically in the absence of an initiator generating "catalyst", but presence of heat or other invisible forces.
(Laws of Creations for Intra- Addition Polymerization system)

Rule 1398: This rule of Chemistry for **Inter-Addition polymerization system,** states that, this is a method of polymerization involving **the use of monomers that carry either π-bond activation centers, or vacant orbitals in the presence of paired unbonded radicals in the last shell of the central atom, or functional centers on rings or invisible π-bond inside rings,** to form linear polymers in the presence of initiator generating compound(s).
(Laws of Creations for Inter-Addition polymerization system)

Rule 1399: This rule of Chemistry for **Inter/Intra-Addition polymerization system,** states that, this is a method of polymerization involving *the use of monomers which carry at least two unsaturated functional groups distantly externally located,* to form ringed polymeric cross-linked or non-cross-linked compounds *in the presence of initiator generating compound(s)* to which inter addition is associated with, while intra addition is connected with internal addition inside the monomer to generate the active site which adds to the next monomer inter molecularly, and followed next by intra molecular addition and this continues sequentially along the chain during propagation.
(Laws of Creations for Inter/Intra-Addition polymerization system)

Rule 1400: This rule of Chemistry for **Inter/Intra-Intra-Intra-...Addition polymerization system,** states that, this is a method of polymerization involving *the use of polymers with internally or externally located π-bonds arranged along the chain, herein also called a <u>Polymeric monomer,</u>* to form fused and non-fused rings intermittently along the chain *in the presence of an initiator generating catalyst* which adds only once inter molecularly followed by continuous intra molecular addition along the chain.
(Laws of Creations for Inter/Intra-Intra-...Intra-Addition polymerization system)

Rule 1401: This rule of Chemistry for **Equilibrium Intra-Step polymerization system**, states that, this is a method of polymerization involving *the use of polymers which not only carry externally located double bonds, but also has the ability of existing in Equilibrium state of existence from the externally located center, also herein called a <u>Polymeric monomer,</u>* to form a polymer with electrophilic rings alternatingly placed along the chain **via Step polymerization method.**
(Laws of Creations for Equilibrium Intra-Step polymerization system)

Rule 1402: This rule of Chemistry for <u>**Trans annular "migration"**</u>, a phenomenon which is herein called **"Trans annular Addition"**, since there is no migration taking place, states that, this addition which takes place inside rings with more than one π-bond type of activation

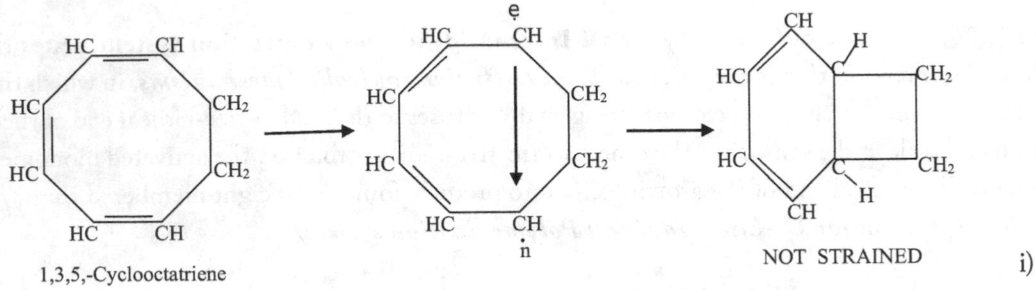

1,3,5,-Cyclooctatriene   NOT STRAINED   i)

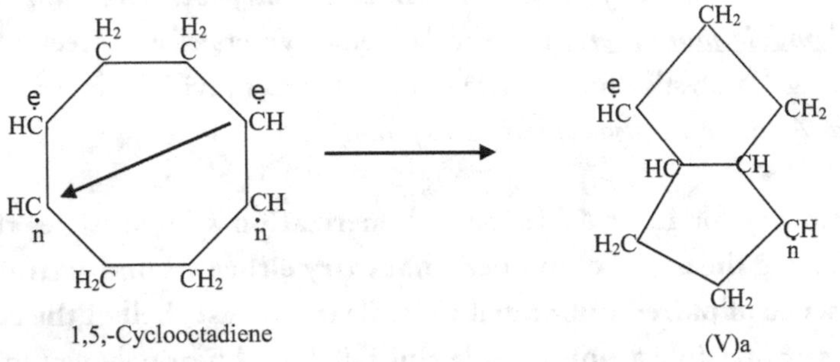

1,5,-Cyclooctadiene   (V)a   ii)

center isolatedly or conjugatedly placed (with discrete resonance stabilization), when strongly activated either only radically (1,3,5,-Cyclooctatriene), or either radically or chargedly like the case just above (1,5-Cyclooctadiene) in the presence of an initiator, produces either a stable bicyclic or tricyclic ring or a bicyclic ring with two active centers distantly located. [See Rule 959]
(Laws of Creations for Trans annular Addition)

Rule 1403: This rule of Chemistry for **Chloroprene as a monomer during Di-Inter molecular addition,** states that, the eight-membered ring formed during addition to give six-membered fused cyclohexene chain *in the absence of an initiator,* can also be made to undergo Trans-annular Addition in the presence of an activating force not to give a fused bicyclic living monomer, but a stable fused four-membered tricyclic compound as shown below.

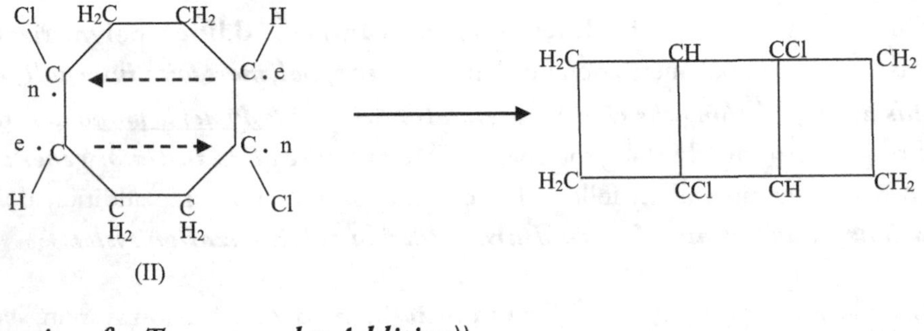

(II)

(Laws of Creations for Trans annular Addition))

Rule 1404: This rule of Chemistry for **Ringed ketonic compounds**, states that, these rings which have no functional centers, but only one type of activation center (C = O) which cannot readily be used, are unique Electrophiles where the C = O center is Y and the invisible p-bond is the X center, since when one of them is opened instantaneously as shown below,

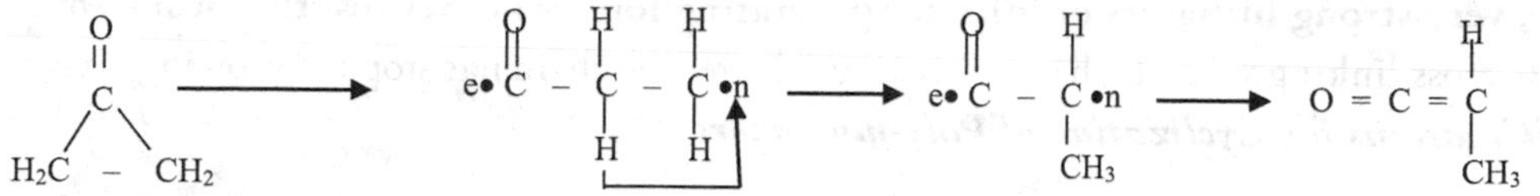

Methyl Ketene
AN ELECTROPHILE

molecular rearrangement of the third kind takes place to give methyl ketene an Electrophile (A Male); for which one can observe that *while the invisible π-bond in the world of RINGS is Imaginary, in the Real world (NON-RINGED COMPOUNDS) it is visible and indeed all the Invisible π-bonds in rings are C= C π-bonds NUCLEOPHILIC and ELECTROPHILIC in character.* [See Rule 1074] *(Laws of creations for Cyclic Ketonic compounds)*

Rule 1405: This rule of Chemistry for **Cyclization of Poly-monomers**, states that, *the first* driving force favoring the formation of rings is the poly-monomeric character of the monomer, in the sense that *two or more π-bond types of activation centers isolatedly but equally distantly placed and nucleophilic in character,* must be present along the chain of the monomer.
(Laws of Creations for Cyclization of Poly-monomers)

Rule 1406: This rule of Chemistry for **Cyclization of Poly-monomers,** states that, *the second* driving force favoring the formation of rings, is that the two or more π-bond types of activation centers must be *externally located on the chain of the monomer, the distance between them being a function of the size of ring to be formed,* for which there exists a <u>minimum and maximum distance.</u>
(Laws of Creations for Cyclization of Poly-monomers)

Rule 1407: This rule of Chemistry for **Cyclization of Poly-monomers**, states that, there are in general seven types of single π-bond type of Activation centers and one type of double π-bonds type of activation center and these are-

i) The Allylic type of Activation centers [- CH_2 – CH = CH_2]
ii) The Ene type of Activation centers [(C/H)-CH = CH_2]
iii) The Vinyl type of Activation centers [(X)-CH = CH_2]
iv) The Acrylic type of Activation centers[(CO)-CH = CH_2]
v) The Yne type of Activation centers [(C/H)-C ≡ CH]
vi) The Aldehydic type of Activation center (-(CH) = O)
vii) The Aldiminic type of Activation center (-(CH) = NR)
viii) The Thioaldehydic type of Activation center (-(CH) = S)

for which when two or more of them are carried by a Carrier, a Di or Tri- or Poly- monomer is formed.
(Laws of Creations for Cyclization of Poly-monomers)

Rule 1408: This rule of Chemistry for **Cyclization of Poly-monomers**, states that, *the third* driving force favoring the formation of rings, is that the two or more π-bond types of activation centers must be of the *same nucleophilicity.*
(Laws of Creations for Cyclization of Poly-monomers)

Rule 1409: This rule of Chemistry for **Cyclization of <u>Poly-monomers</u>,** states that, *the fourth* driving force favoring the formation of rings, is that because two or more activation centers have to be activated at the same time, **very strong initiators or initiator generating force whether reactive or not**, must be used, for if *weak,* cross-linking sites which can only be used after the chain has stopped growing, may be formed. *(Laws of Creations for Cyclization of Poly-monomers)*

Rule 1410: This rule of Chemistry for **Cyclization of <u>Poly-monomers</u>,** states that, regardless the strength of the initiator, *for as long as the route is natural to the monomer,* cyclization must take place (or rings must be formed), since the strength of the active center keeps increasing for every monomer added to the growing polymer chain. *(Laws of Creations for Cyclization of Poly-monomers)*

Rule 1411: This rule of Chemistry for **Cyclization of <u>Poly-monomers</u>,** states that, *the fifth* driving force favoring the formation of rings, is *absence of conformational instability in the system,* and this depends largely on the type of reactors used for polymerization, the type of initiator used, the manner by which the components are added, and the operating conditions. *(Laws of Creations for Cyclization of Poly-monomers)*

Rule 1412: This rule of Chemistry for **Cyclization of <u>Poly-monomers</u>,** states that, *the sixth* driving force favoring the formation of rings when the two or more π-bond types of Activation centers are isolatedly placed, is *the distance between the centers.* *(Laws of Creations for Cyclization of Poly-monomers) [See Rule 1406]*

Rule 1413: This rule of Chemistry for **Cyclization of <u>Poly-monomers</u>,** states that, *the seventh* driving force favoring the formation of rings *when non-allylic types of activation centers are involved, is the size of the ring to be formed,* for which presence of three-membered rings is absolutely impossible, because the amount of strain energy present in the ring is above the MaxRSE independent of the types of groups carried since when formed, they open up easily in view of the operating conditions. *(Laws of Creations for Cyclization of Poly-monomers)*

Rule 1414: This rule of Chemistry for **Cyclization of <u>Poly-monomers</u>,** states that, *the eight* driving force favoring the formation of rings when **non-allylic types of activation centers are involved, is again** *the size of the ring to be formed,* for which presence of <u>*six-membered ring*</u> is most largely favored where possible all the time because of its ability to accommodate large range of magnitude of SEs in their rings. *(Laws of Creations for Cyclization of Poly-monomers)*

Rule 1415: This rule of Chemistry for **Cyclization of <u>Poly-monomers</u>,** states that, *the ninth* driving force favoring the formation of rings is *the type of the ring to be formed,* for which if hetero-atom(s) exist in the ring to be formed, the lesser the SE and more favored its existence. *(Laws of Creations for Cyclization of Poly-monomers)*

Rule 1416: This rule of Chemistry for **Cyclization of <u>Poly-monomers</u>,** states that, *the tenth* driving force favoring the formation of rings *when allylic type of activation centers are involved, is again the size the ring,* for which the favored existence of large sizes of rings (from six and above) are made possible. *(Laws of Creations for Cyclization of Poly-monomers)*

Rule 1417: This rule of Chemistry for **Cyclization of <u>Poly-monomers</u>,** states that, *the eleventh* driving force favoring the formation of rings, is *the route of polymerization* for which if the route is not natural to the centers, no cyclization will take place.
(Laws of Creations for Cyclization of Poly-monomers)

Rule 1418: This rule of Chemistry for **Cyclization of <u>Poly-monomers</u>,** states that, *the twelfth* driving force favoring the formation of rings, is *the type of initiator generating compound(s) used,* for if not properly chosen and is in "excess", some rings which have been formed may be opened to favor the presence of cross-links or linear chains.
(Laws of Creations for Cyclization of Poly-monomers).

Rule 1419: This rule of Chemistry for **Cyclization of <u>Polymeric monomers</u>,** states that, *the first* driving force favoring their use as gigantic monomers for cyclization, **is *presence of externally located activation centers on all the monomer units along the chain,*** for which these can be used with or without the use of initiators, noting that these activation centers along the chain are of *different nucleophilicities,* wherein for them, the center first activated is that located either at the beginning or the end of the chain, depending on what the terminals are carrying.
(Laws of Creations for Cyclization of Polymeric monomers)

Rule 1420: This rule of Chemistry for **Cyclization of <u>Polymeric monomers</u>,** states that, *the second* driving force favoring their use as gigantic monomer for cyclization, **is *presence of internally located activation centers on all the monomer units along the chain,*** for which these can only be used with the use of initiators.
(Laws of Creations for Cyclization of Polymeric monomers)

Rule 1421: This rule of Chemistry for **Cyclization of <u>Polymeric monomers</u>,** states that, *the third* driving force favoring the formation of rings, is **the size of rings to be formed,** for which six-membered rings are largely favored, with sometimes four-membered rings for some polymeric monomers carrying in particular radical pulling groups such as Cl specially placed.
(Laws of Creations for Cyclization of Polymeric monomers)

Rule 1422: This rule of Chemistry for **Cyclization of <u>Polymeric monomers</u>,** states that, *the fourth* driving force favoring formation of rings continuously along the chain, is *absence of release of transfer species of the first kind;* for which if released when the ring is growing, partial cyclization is favored and if released when the ring is so strained, no cyclization is favored and when not released, full cyclization is favored such as the case with polyacrylo-nitrile.
(Laws of Creations for Cyclization of Polymeric monomers)

Rule 1423: This rule of Chemistry for **Cyclization of <u>Polymeric monomers</u> with activation centers externally located along the monomer units in a chain,** states that, *the fifth* driving force favoring cyclization is *when those centers are isotactically placed;* for which when they are syndiotactically placed (the distance between the activation centers becomes too large) or randomly placed, no cyclization may take place.
(Laws of Creations for Cyclization of Polymeric monomers).

Rule 1424: This rule of Chemistry for **Formation of fused rings,** states that, *when a ring is fused on all sides by six-membered rings, the maximum limit placed for its existence are as shown below-*

    i)   *The size of the ring,*

ii) ***If there is no π-bond inside the carrier, then the most unsaturated rings carried should have no points of scission when strained or should not contain more than one unshared π-bond,***

iii) ***If there is one π-bond inside the carrier, then the rings carried should not carry more than one unshared π-bond when fully unsaturated.***

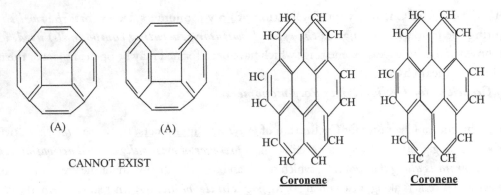

CANNOT EXIST

for which for four- (A) and five- membered inner rings, their rings are too strained to make them exist.
(Laws of Creations for Fused ring formations)

Rule 1425: This rule of Chemistry for **Formations of Fused rings**, states that, when two similar rings are fused together, the lesser the SE in the fused ring than when they were alone; such as the case with benzene, naphthalene, and anthracene where the SE in naphthalene is less than the sum of the SE in two benzene rings, that is, while the presence of one common boundary decreases the strain energies in the fused rings, this is not the case with two or more common boundaries as shown below-

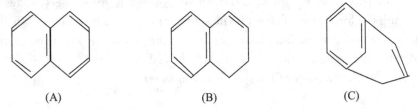

for which in (A), while two π-bonds can be placed into the second ring, in (C), this cannot be done and while in (B), there is one point of scission, in (C) there is no point of scission, noting that while in (A) and (B) there is only one common boundary, in (C) there are two common boundaries.
(Laws of Creations for Fused ring formations)

Rule 1426: This rule of Chemistry for **Formation of Fused rings,** states that, when four membered rings are formed between two polymeric chains, their existence depend on the radical pushing capacities of the groups, for which if it is a ring and more than one, the rings must be saturated, such as shown below-

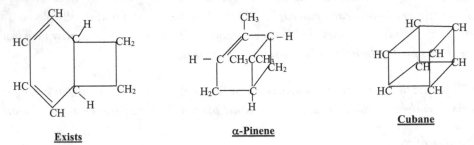

Exists　　　　　**α-Pinene**　　　　　**Cubane**

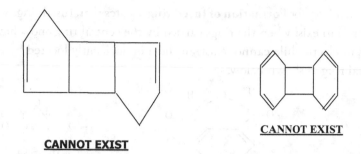

CANNOT EXIST   **CANNOT EXIST**

(Laws of Creations for Fused ring formations)

Rule 1427: This rule of Chemistry for **Formation of fused rings**, states that, increase in SE in rings when fused, depends on the size of the ring to be fused with, for the smaller the ring, the more difficult it is for the ring to exist.
(Laws of Creations for formation of rings)

Rule 1428: This rule of Chemistry for **Formation of fused rings**, states that, increase in SE of the carrier ring when formed, depends on the groups carried by the surrounding rings, for if they are *radical-pushing groups,* the more difficult it is for the carrier ring to exist.
(Laws of Creations for formation of fused rings) [See Rule 1424]

Rule 1429: This rule of Chemistry for **Formation of fused rings,** states that, the more the number of boundaries shared by two rings of the same size, the less favored their existence, because of the heavy strain involved as a result of the distance between the active centers where the bonds are overstretched beyond limit, such as the case with Divinyl benzene-

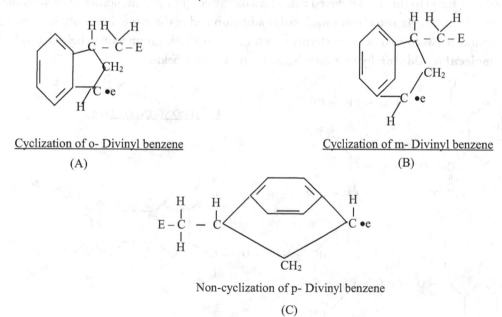

Cyclization of o- Divinyl benzene
(A)

Cyclization of m- Divinyl benzene
(B)

Non-cyclization of p- Divinyl benzene
(C)

for which while the existence of (A) and (B) are possible in the order shown, that of (C) is impossible.
(Laws of Creations for formation of fused bonds)

Rule 1430: This rule of Chemistry for **Formation of fused rings,** states that, fused rings which previously could not exist, can be made to exist when the rings carried by the central rings now have hetero atoms (for without which males or electrophile cannot exist) in them or adjacently located by provision of less strain energy in the central ring as shown below.

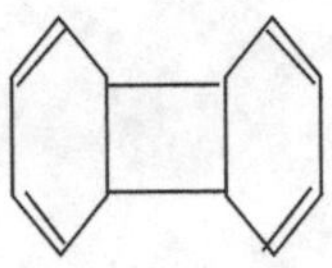

CANNOT EXIST

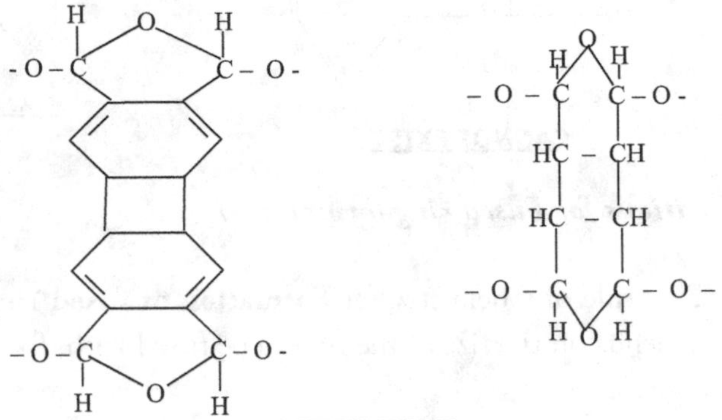

THEY EXIST

(Laws of Creations for formation of fused rings)

Rule 1431: This rule of Chemistry for **Di- or Tri- or Poly-Allylic monomers,** states that, when used for Cyclization, only Electro-free-radical generating initiator can be used as the source of initiator, since that is the only route Natural to their centers.
(Laws of Creations for Poly-Allylic monomers)

Rule 1432: This rule for Chemistry for **Diallyl phthalate,** and the likes, states that, this is **an Electrophile** which when polymerized using *a Strong initiator,* have the ability of producing copolymers of an **eleven-membered** ring fused to the six-membered ring and a methylene group as monomer units syndiotactically placed along the chain via **Inter/Intra-molecular addition** and at the same time contain some products with cross-links between two or more chains which carry no allylic group along their chains linked via **Di-Inter-molecular addition** by four membered rings as shown below.

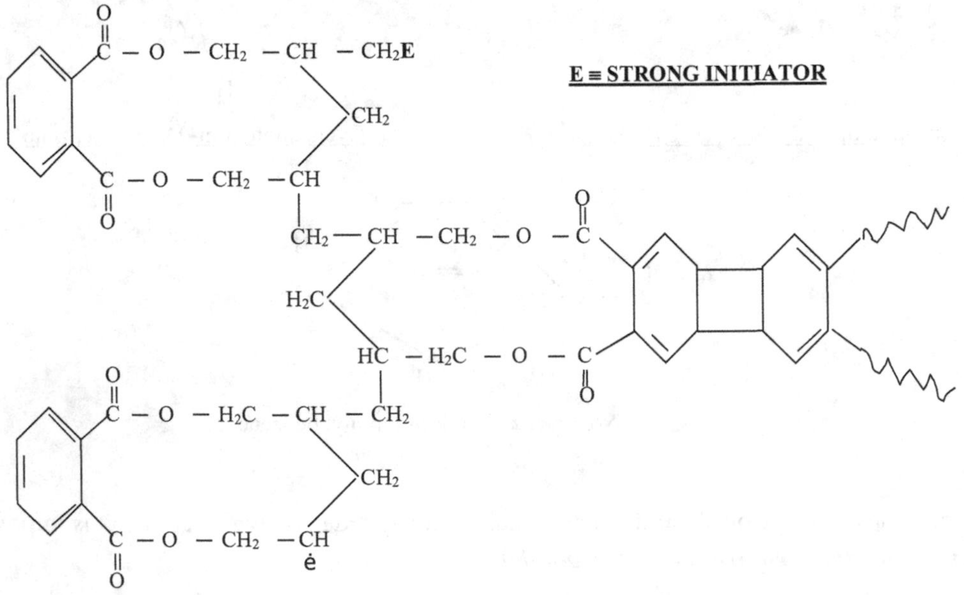

E ≡ STRONG INITIATOR

(Laws of Creations for Diallyl Phthalate)

Rule 1433: This rule of Chemistry for **Diallyl phthalate**, states that, this is **an Electrophile** which when polymerized using *a Weak initiator,* have the ability of producing copolymers of an hetero Electrophilic **eleven-membered ring fused to the six membered ring and a methylene group as monomer units** randomly placed along the chain via *Inter-/Intra-molecular addition* as shown below-

```
        O           H    H
        ||          |    |
        C — O — CH2 — C = C ——— Branching site
        |                |
       ( )               H              E ≡ WEAK INITIATOR
        |
        |
        C — O — CH2 — C — CH2E
        ||              |
        O               H
                       CH2 — C — CH2 — O — C                    etc.
        O                   |            ||
        ||                  CH2          O    ( )
        C — O — H2C — CH — CH2                |
       ( )                      \            C = O
        |                        CH2          |
        |                       /             O
        C — O — CH2 — CH                      |
        ||              ·                     CH2
        O               e                     |
                                              CH ← Branching site
                                              ||
                                              CH2
```

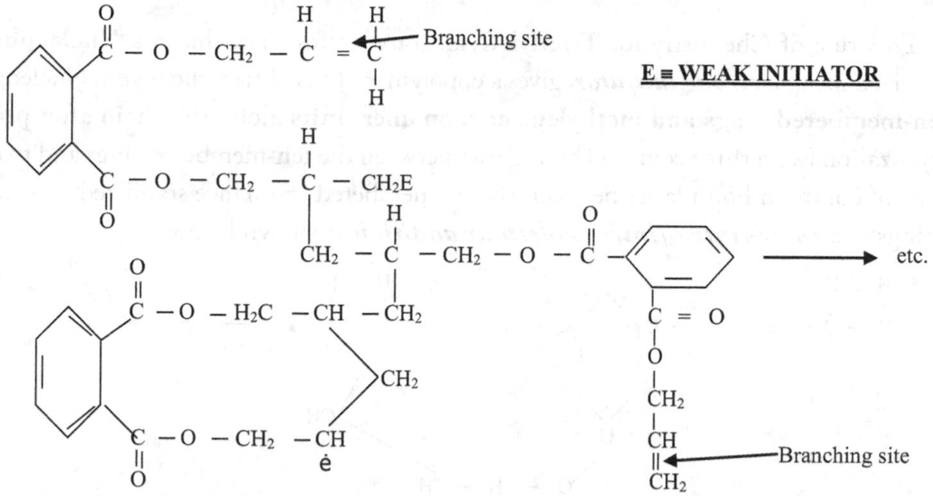

and at the same time contain some branching allylic sites which can be used either during polymerization or after polymerization to give highly branched and cross-linked networks.
(Laws of Creations for Diallyl Phthalate)

Rule 1434: This rule of Chemistry for **Diallyl Maleate**, states that, this is **an Electrophile** which when polymerized behaves exactly like Diallyl phthalate, with however no fused ring along the chain, but similar hetero Electrophilic **eleven-membered** ring alternatingly placed with methylene as monomer units syndiotactically placed or random placement with branch-ing and cross-linking sites, depending on the strength of the initiator, and all formed via *Inter-/Intra- molecular addition*.
(Laws of Creations for Diallyl Maleate)

Rule 1435: This rule of Chemistry for **Diethylene glycol bisallyl carbonate (I)**, states that, this is a **Nucleophile,** which when polymerized using *a Strong initiator,* gives a copolymer of an hetero Electrophilic **fourteen-membered ring and methylene** alternatingly place along the chain with no cross-linking site, *via Inter-/Intra- molecular addition* as shown below-

```
                    E ≡ STRONG INITIATOR
                                              H   H   H
                                              |   |   |
                             O — (CH2)2 — O — C — C — C — E
                            /                 |   |   |
                                              H       H
    E · e   +   (I) ———→  O = C                       CH2
  (strong)                    \               H      /
              (inter / intra -  O — (CH2)2 — O — C — CH
                 addition)                    |   ↓ e·
                                              H
```

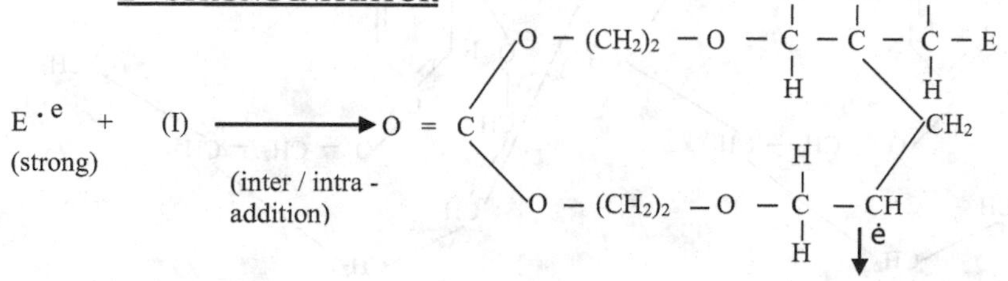

noting the presence of male and female centers which cannot be used in the ring.
(Laws of Creation for Diethylene glycol bisallyl carbonate)

Rule 1436: This rule of Chemistry for **Diethylene glycol bisallyl carbonate**, states that, this is a **Nucleophile**, which when polymerized using *a Weak initiator,* gives a copolymer of an hetero

Electrophilic **fourteen-membered ring and methylene** randomly alternatingly placed along the chain with branching and cross-linking sites along the chain and these can be used during polymerization on dead chains to give highly branched and cross-linked products.
(Laws of Creations for Diethylene glycol bisallyl carbonate)

Rule 1437: This rule of Chemistry for **Triallyl Cyanurate**, states that, this is a **Nucleophile,** which when polymerized using *a strong initiator,* gives a copolymer of fused tricyclic hetero Nucleophilic **six-, ten- and ten-membered rings and methylene as monomer units** along the chain after primary and secondary cyclization (with three common boundaries between the ten-membered rings and two different but same type of common boundaries between the six membered resonance stabilized ring and the ten membered rings) via *Inter-/Intra-/Intra-molecular addition* as shown below-

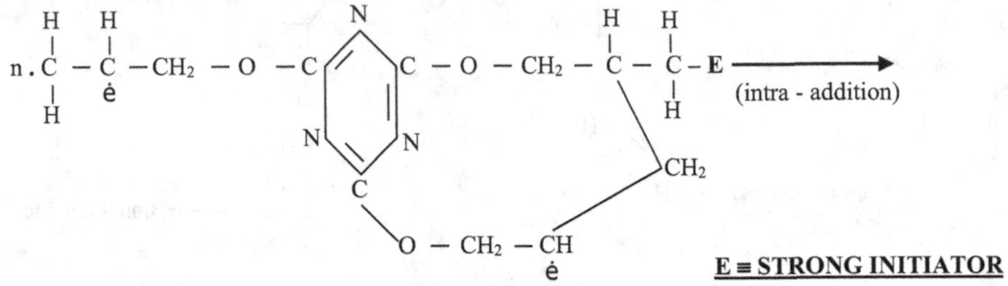

Primary Cyclization

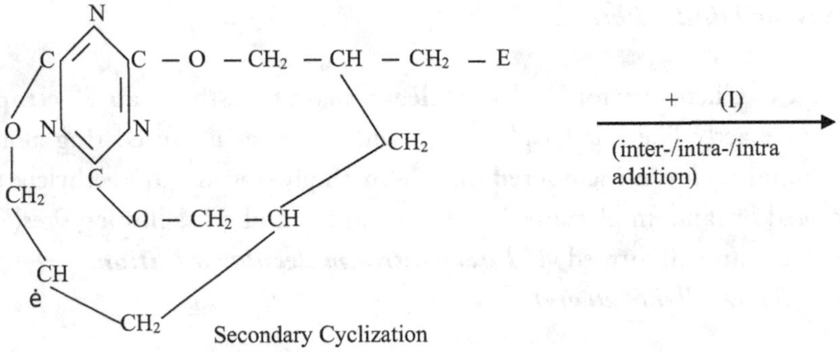

E ≡ STRONG INITIATOR

Secondary Cyclization

wherein the six-membered ring and what it is carrying are syndiotactically placed along the chain when placed horizontally as shown below-

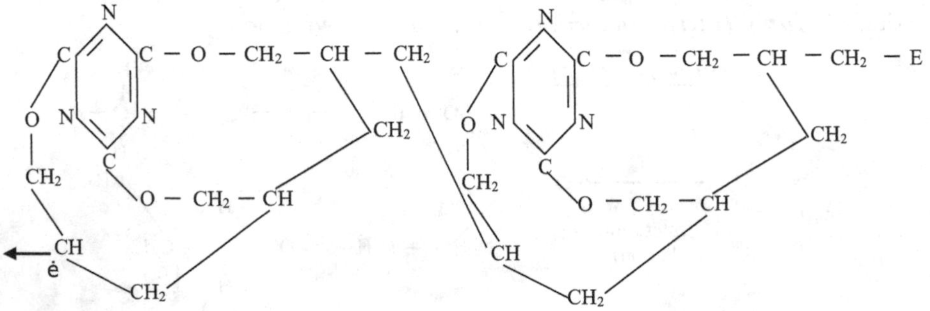

noting that *when only primary cyclization is favored that which is to be expected in view of distance between addition centers and steric limitations or a Weak initiator is involved,* highly branched and cross-linked network is obtained.
(Laws of Creations for Triallyl Cyanurate)

Rule 1438: This rule of Chemistry for **Diallyl quaternary ammonium salt**, states that, this is **an Electrostatic type of Nucleophile** in view of the presence of an electrostatic bond on the N center, which when polymerized using *a Strong initiator* via *Inter-/Intra- molecular addition,* gives a copolymer of a **six-membered electrostatic ring and methylene** alternatingly placed isotactically along the chain in the absence of electrodynamic forces of repulsion from the X (Cl) centers as shown below-

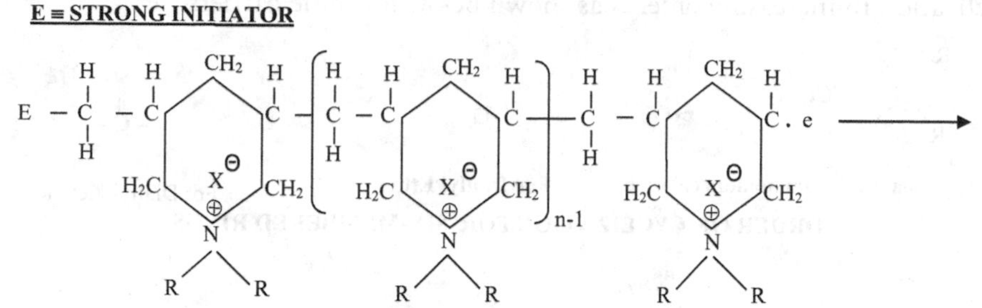

E ≡ STRONG INITIATOR

for if present, syndiotactic placement is obtained; noting that no site for branching exists on the ring; for if branching is to be desired, very weak initiators must be used

(Laws of Creations for Diallyl quaternary ammonium salt)

Rule 1439: This rule of Chemistry for **Diallyl Ketone, a Nucleophile ($H_2C=CH - CH_2 - CO - CH_2 - CH=CH_2$)**, states that, when used as a poly-monomer, this (like members in this family) can only be done when *strong electro-free-radical generating initiators* are used via *Inter-/Intra- molecular addition*, to give a copolymer of a **six-membered cyclohexanone and methylene** alternatingly placed isotactically along the chain in the absence of electro-dynamic forces of repulsion from the O center on CO centers as shown below-

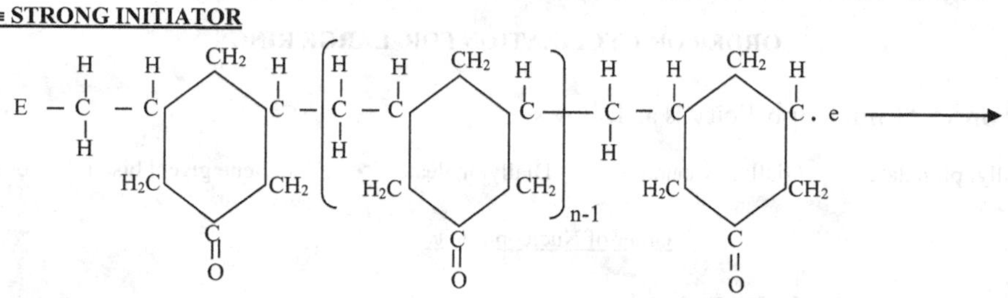

E ≡ STRONG INITIATOR

noting that site for opening of the ring exists on the ring; for if branching sites are to be desired, *very weak initiators* must be used.

(Laws of Creations for Diallyl Ketone)

Rule 1440: This rule of Chemistry for **Diallyl ether, a Nucleophile ($H_2C=CH - CH_2 - O - CH_2 - CH=CH_2$)**, states that, when used as a poly-monomer, this (like members in this family) can be done only when *strong electro-free-radical generating initiators* are used via Inter-/Intra- molecular addition, to give a copolymer of a **six-membered cyclic ether and methylene** alternatingly placed isotactically along the chain in the absence of electrodynamic forces of repulsion from the O centers in the rings as shown below-

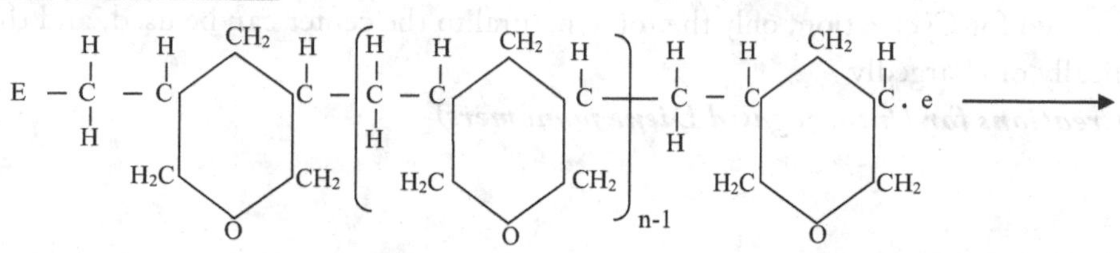

E ≡ STRONG INITIATOR

noting that site for opening of the ring exists on the ring; for if branching sites are to be desired, ***very weak initiators*** must be used.
(Laws of Creations for Diallyl Ether)

Rule 1441: This rule of Chemistry for **Allylic Poly-monomers**, states that, the order of their ability to undergo Cyclization in increasing order is as shown below for some carriers-

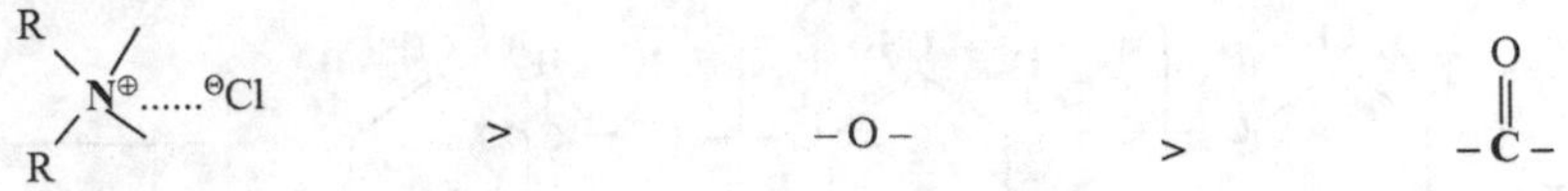

For Diallyl quaternary ammonium chloride   **For Diallyl Ether**   For Diallyl Ketone

ORDER OF CYCLIZATION FOR SIX-MEMBERED RINGS   i)

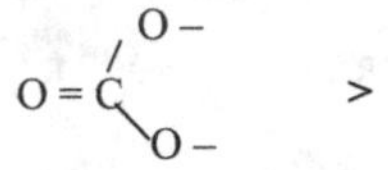

For Diallyl anhydride (8)

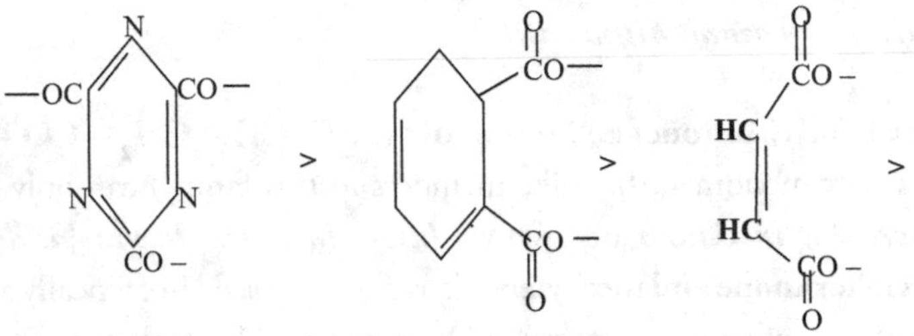

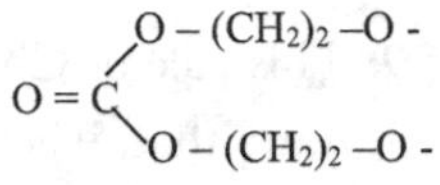

For Triallyl cyanurate (10)   For Diallyl phthalate (11)   For Diallyl maleate (11)   For Diethylene glycol bisallyl carbonate (14)

ORDER OF CYCLIZATION FOR LARGE RINGS   ii)

while their order of nucleophilicity is as follows-

Diallyl phthalate > Triallyl cyanurate > Diallyl maleate > Diethylene glycol bisallyl carbonate

Order of Nucleophilicity   iii)

(Laws of Creations for Allylic Poly-monomers)

Rule 1442: This rule of Chemistry for **Allylic Poly-monomers**, states that, the order of their ability to form branched and cross-linked networks is a function of ***the strength of the initiator,*** for if the initiator is weak, there will be more branching and cross-linking; ***the character of the carrier center,*** for if the carrier is an Electrophile, cross-links will exist; and finally ***the number for allylic centers carried by the carrier,*** for if the number is more than two there will be more branching and cross-linking.
(Laws of Creations for Allylic Poly-monomers)

Rule 1443: This rule of Chemistry for **Unconjugated Diene monomers,** states that, these are **Poly-monomers** which include *two Enes-, or two Vinyls-, or two Acrylics- types of groups (excluding allylic types),* which when used for Cyclization, only the route natural to the center can be used, and this could either be free-radically or chargedly.
(Laws of Creations for Unconjugated Diene monomers)

Rule 1444: This rule of Chemistry for **Divinyl benzenes,** states that, these are monomers which can be said to be either *a Conjugated 1,3,5,7- Tetraene (Non-symmetric) or an Unconjugated Diene (Symmetric),* based on the strength of initiator used for activation; for if the initiator is **weak or moderate,** resonance stabilization is favored only free-radically and if it is **strong,** then the two externally located activation centers are activated as Unconjugated diene, free-radically or chargedly like other unconjugated dienes.
(Laws of Creations for Divinyl benzenes)

Rule 1445: This rule of Chemistry for **p-Divinyl benzene,** states that, this is a **Nucleophile** which when polymerized using strong initiator, *cannot undergo Cyclization,* because the distance created by the para- placement of the groups, gives a seven-membered ring fused with three common boundaries to benzene ring, making it to contain more than the MaxRSE for its existence.
(Laws of Creations for p-Divinyl benzene)

Rule 1446: This rule of Chemistry for **o-Divinyl benzene,** states that, this is **Nucleophile** which when polymerized using *a strong electro-free-radical or positively charged paired initiator,* gives a copolymer of the *six-membered ring fused with cyclopentene with one common boundary and methylene as monomer units* alternatingly placed syndiotactically along the chain via *Inter/Intra-molecular addition*, with no site for cross-linking between chains intra-molecularly because the ring is still resonance stabilized with no site for branching.

E ≡ STRONG INITIATOR

Cyclization of o- Divinyl benzene

(Laws of Creations for o-Divinyl benzene)

Rule 1447: This rule of Chemistry for **m-Divinyl benzene,** states that, this is a Nucleophile, which when polymerized using *a strong electro-free-radical or positively charged paired initiator,* gives a copolymer of *six-membered ring fused with cyclohexene with two common boundaries and methylene as monomer units* syndiotactically placed along the chain via *Inter-/Intra-molecular addition,* with site on the benzene ring available for use for forming four-membered rings as cross-links between two chains intra- molecularly, but cannot be used because the four-membered rings are too strained by the presence of two benzene rings loaded with heavy radical pushing groups on two common boundaries. [See Rules 1426 and 1430]

E ≡ STRONG INITIATOR

Cyclization of m- Divinyl benzene

Impossible Cross-link between two chains

(Laws of Creations for m-Divinyl benzene)

Rule 1448: This rule of Chemistry for **Strain Energy** in ringed compounds, states that, since Strain energy has been defined as the invisible π-bond nucleophilic in character inside the ring, it can therefore be calculated by searching for its linear isomer and taking the difference between the total bond energies for the ring and its linear isomer, as shown below for Benzene-

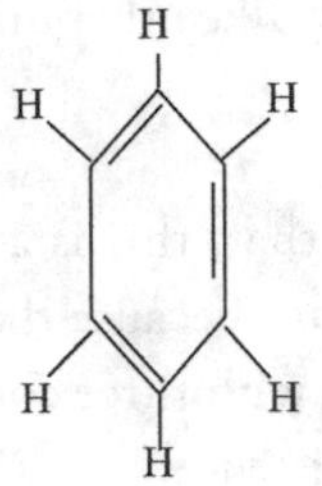

(A) C_6H_6

(B) C_6H_6

SE = Bond Energies in (A) – Bond Energies in (B)

SE in Benzene = 36 kcal

noting very carefully the invisible π-bond in the triple bond of the linear isomer which is also resonance stabilized; for which if the difference between the ring and the linear isomer is found to be negative, then the linear isomer is not the right linear isomer, for example *the linear isomer for ethylene oxide is vinyl alcohol and not acetone* because the invisible p-bond in the ring is not C=O type of bond, but C=C type of bond.
(Laws of Creations for SE in Benzene)

Rule 1449: This rule of Chemistry for **Covalent, Electrostatic and Polar Charges,** states that, while it is impossible to have an atomic center carrying both an ionic charge and a radical at the same time, with covalent, electrostatic and polar charges, it is possible, for example, en• $Fe^{3\oplus}$.
(Laws of Creations for Covalent, Electrostatic and Polar charges)

Rule 1450: This rule of Chemistry for **the compounds shown below**, states that, their Equilibrium State of existences are as follows-

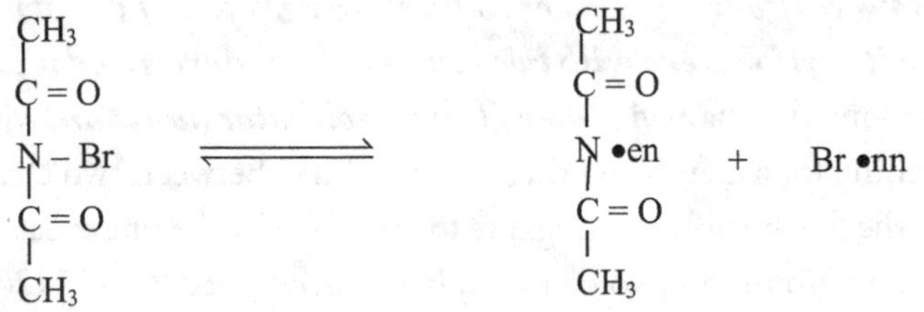

wherein N is carrying the electro-non-free-radical is the first case, despite the fact that N (3.0) is more electro-negative than Br (2.8), and the fact that Br is closer to the metallic zone of atoms in the Periodic Table than N is, and the reason is because of the type(s) of groups carried by N center, just as exists for many compounds such as AlR_3, RMgBr.
(Laws of Creations for Equilibrium State of existence of a N compound)

Rule 1451: This rule of Chemistry for **1,6-Heptadiene a Nucleophile**, states that, when used as a *Poly-monomer for cyclization,* this can only be done either with the use of positively charged paired initiator or electro-free-radically, to give alternating copolymers of an **unfused six membered cyclohexane and a methylene** when *the initiator is strong* via **Inter-/Intra- molecular addition** as shown below chargedly-

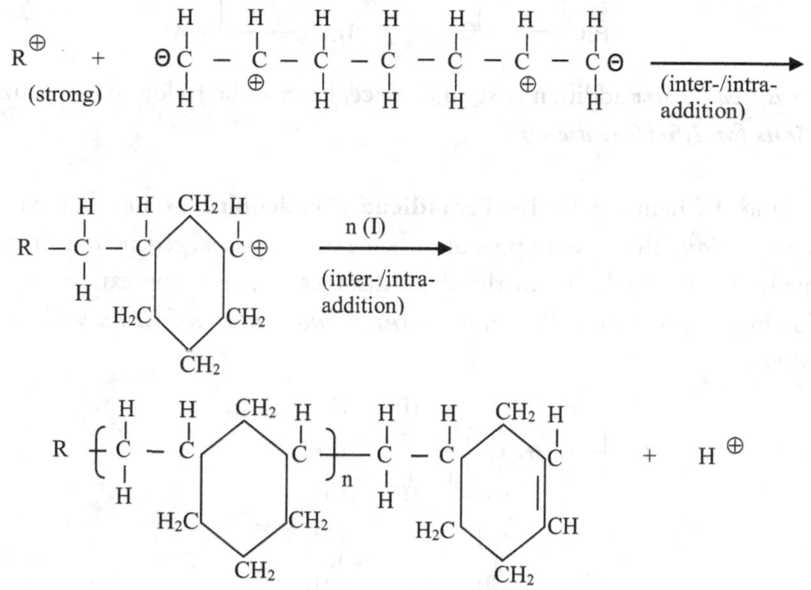

$$R^{\oplus} \quad + \quad {}^{\ominus}\overset{\overset{\displaystyle H}{|}}{\underset{\underset{\displaystyle H}{|}}{C}} - \overset{}{\underset{\displaystyle \oplus}{C}} - \overset{\overset{\displaystyle H}{|}}{\underset{\underset{\displaystyle H}{|}}{C}} - \overset{\overset{\displaystyle H}{|}}{\underset{\underset{\displaystyle H}{|}}{C}} - \overset{\overset{\displaystyle H}{|}}{\underset{\underset{\displaystyle H}{|}}{C}} - \overset{}{\underset{\displaystyle \oplus}{C}} - \overset{\overset{\displaystyle H}{|}}{\underset{\underset{\displaystyle H}{|}}{C}}{}^{\ominus} \quad \xrightarrow[\text{addition)}]{\text{(inter-/intra-}}$$

(strong)

noting that charged activation of all the activation centers is favored due to absence of charge carrying force in the monomers, absence of allylic character and the fact that they are of the same capacity; *for if the initiator is weak,* Inter-molecular addition first takes place, after which the active center becomes strong enough to begin cyclization, leaving behind branching and cross-linking sites.
(Laws of Creations for 1,6-Heptadiene)

Rule 1452: This rule of Chemistry for a **1,5 methyl carrying Hexadiene a Nucleophile**, states that, when used as a *Poly-monomer for cyclization,* only the positively charged paired initiator or electro-free-radical generating initiators can be used, to give *copolymers of five-membered cyclopentane and ethylene along the chain* when the *initiator is strong* via **Inter/Intra molecular addition** as shown below-

$$R^{\oplus} \quad + \quad {}^{\ominus}\overset{\overset{\displaystyle CH_3}{|}}{\underset{\underset{\displaystyle H}{|}}{C}} - \overset{\overset{\displaystyle H}{|}}{\underset{\displaystyle \oplus}{C}} - \overset{\overset{\displaystyle H}{|}}{\underset{\underset{\displaystyle H}{|}}{C}} - \overset{\overset{\displaystyle H}{|}}{\underset{\underset{\displaystyle H}{|}}{C}} - \overset{\overset{\displaystyle H}{|}}{\underset{\displaystyle \oplus}{C}} - \overset{\overset{\displaystyle CH_3}{|}}{\underset{\underset{\displaystyle H}{|}}{C}}{}^{\ominus} \quad \xrightarrow[\text{addition)}]{\text{(intra-/inter-}}$$

(strong)

(I)

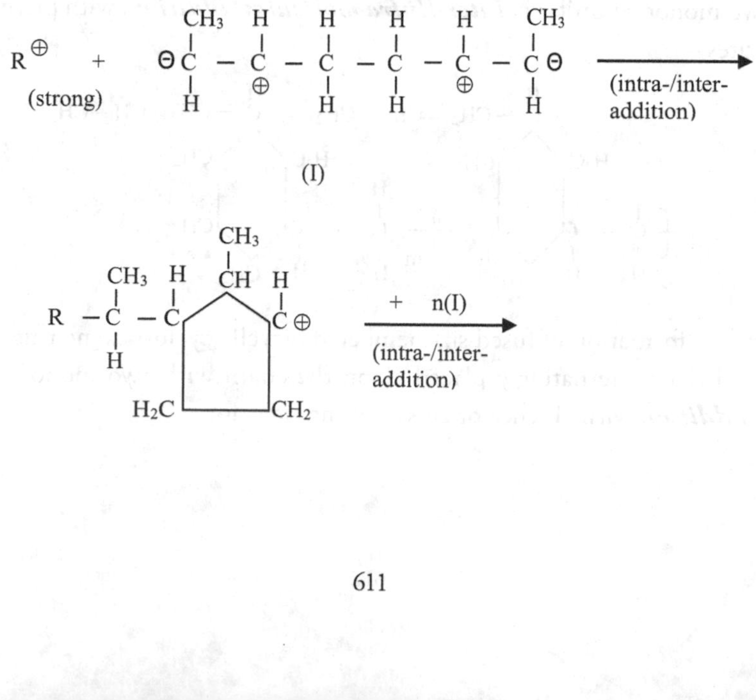

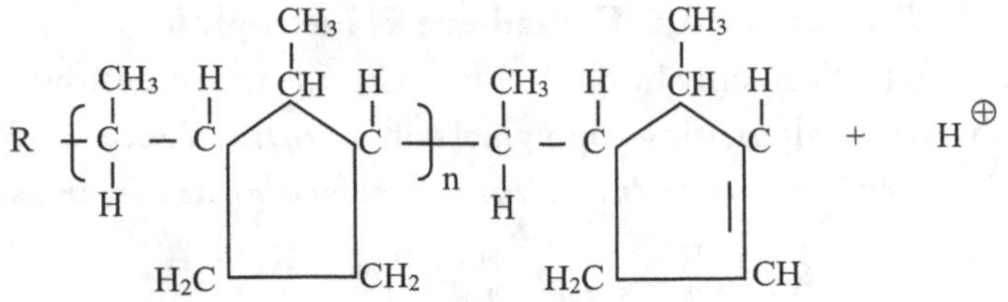

for *if the initiator is weak,* Inter addition first takes place, latter to be followed by cyclization.
(Laws of Creations for 1,5-Hexadiene)

Rule 1453: This rule of Chemistry for **1,4-Pentadiene a Nucleophile**, states that, when used as a **Poly-monomer for cyclization,** this is only possible when positively charged paired or electro-free-radical initiators are involved, for which due to the short distance between the externally located activation centers, the followings are obtained, first only by **Inter- molecular addition** with many cross-linking and branching sites -

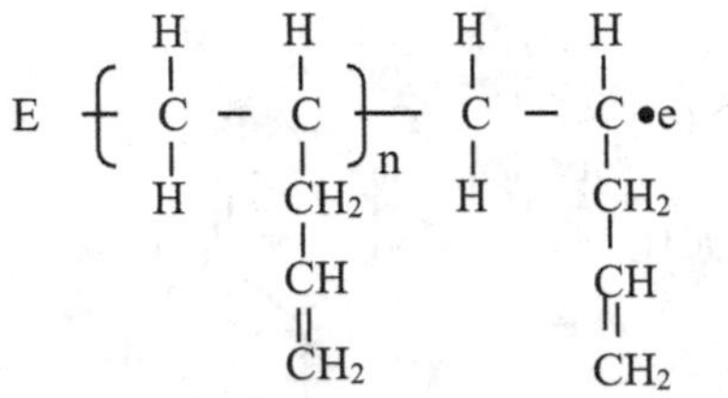

i)

then followed by presence of **four-membered rings along the chain alternating placed with methylene using only one monomer** via **Inter-/intra- molecular addition,** with no signs of cross-linking sites -

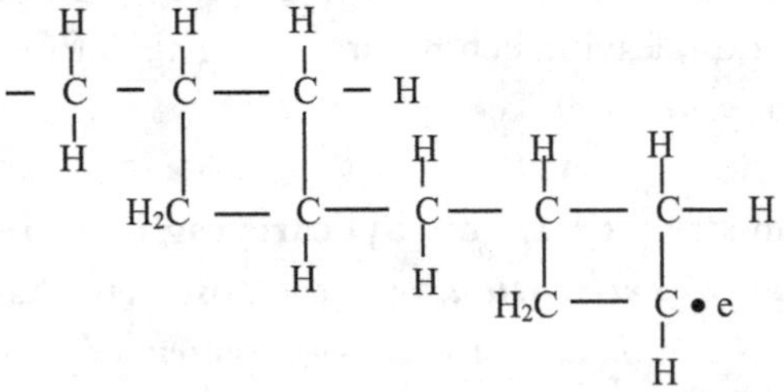

ii)

followed next by presence of six-membered unfused rings along the chain alternatingly placed with methylene using two monomer units via **Inter-/Intra molecular addition**, with presence of cross-linking and or branching sites –

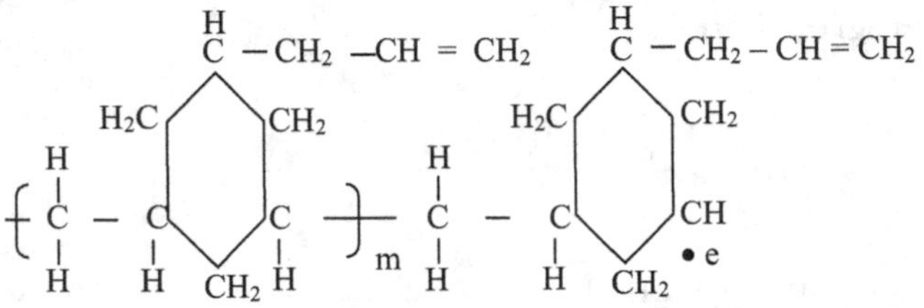

iii)

and finally followed by formation of fused six-membered bicyclic cyclo-hexane rings with two common boundaries and methylene alternatingly placed along the chain with two monomer units **via Inter-/ Intra- molecular addition** with absence of cross-linking sites; for

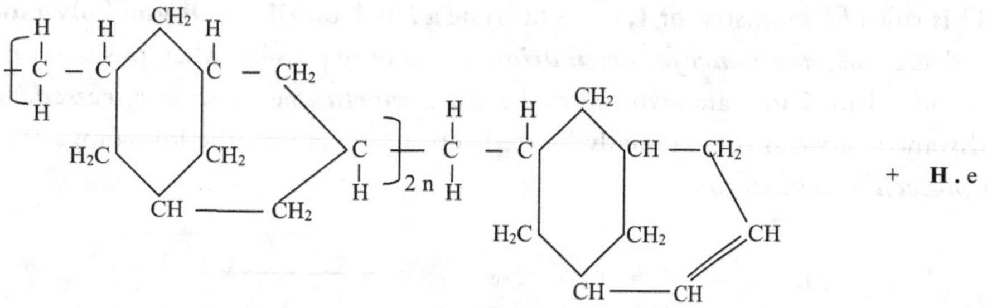

iv)

which, the presence of all these along the chain depends on the strength of the initiator, conformational stability and absence of the release of transfer species of the first kind and of the sixth type; for if the initiator is strong, only small four-membered rings and large six-membered fused bicyclic rings will appear along the chain and if weak, all the above will appear along the chain for as long as no transfer species is released, and if transfer species are released, then small molecular weight products containing rings will be obtained.
(Laws of Creations for 1,4-Pentadiene)

Rule 1454: This rule of Chemistry for **1,6-Heptadiyne a Nucleophile and Di-yne Poly-monomer,** states that, when used as a **Poly-monomer,** this can only be done electro-free-radically and ***most importantly when suppressed,*** to give copolymers of only *six-membered cyclo hexene ring with* **ethylyne** *alternatingly placed* along the chain, via ***Inter-/Intra- molecular addition*** as shown below-

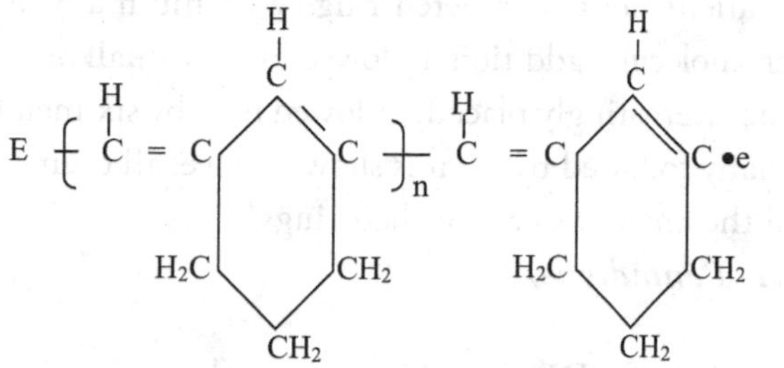

noting that transfer species is never released, because when released a cumulenic double bonded or triple bonded (after rearrangement) six-membered ring which may not be too strained to exist will be formed.
(Laws of Creations for 1,6-Heptadiyne)

Rule 1455: This rule of Chemistry for **1,5-hexadiyne a Nucleophile and a Di-yne Poly-monomer,** states that, when used as a **Poly-monomer,** this can only be done with the use of positively charged-paired initiator or electro-free-radically and ***most importantly when suppressed,*** to give a *copolymer of five-membered cyclo pentene ring and an* **ethylyne** *alternatingly placed* along the chain, via ***Inter-/Intra-molecular addition*** as shown below-

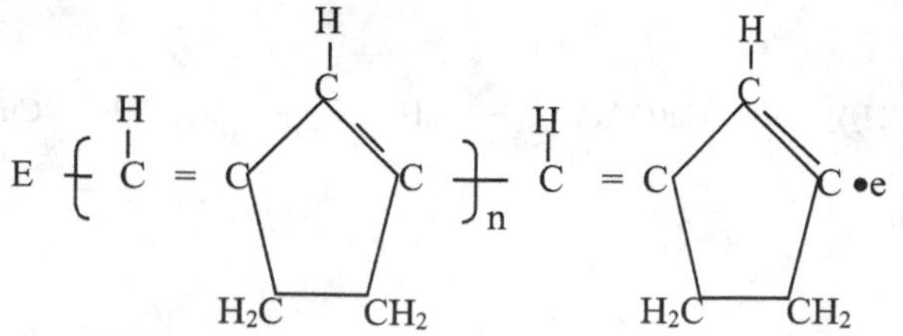

with little or no possibility to release transfer species after propagation, due to strain placed on the ring.
(Laws of Creations for 1-,5-Hexadiyne)

613

Rule 1456: This rule of Chemistry for **1,4-pentadiyne a Nucleophile, a Di-yne Poly-monomer**, states that, when used as a ***Poly-monomer for cyclization***, this is only possible when positively charged paired or electro-free-radical initiators are involved and ***most importantly when suppressed,*** for which due to the short distance between the externally located activation centers, the followings are obtained, via ***Inter-/Intra-molecular addition-***

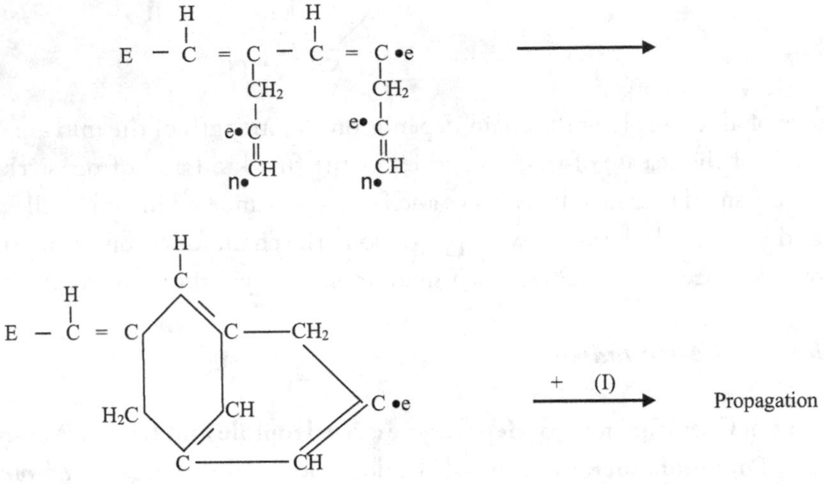

Primary and Secondary Cyclization

for which unlike 1-,4- pentadiene, four membered rings will minimally appear along the chain; with however, the first being Inter- molecular addition, followed by very small or no presence of four-membered monocyclic ring and ethylyne alternatingly placed, followed next by six-membered ring also alternatingly placed with ethylyne and finally followed by what is shown above, all their appearance depending on the strength of the initiator and the amount of SE in their rings.

(Laws of Creations for 1-,4- Pentadiyne)

Rule 1457: This rule of Chemistry for **Diynes,** states that, without suppressing them, they can never be used as monomers as shown below for 1,5-hexadiyne which is readily converted to cyclooctadecahexayne in the presence of $Cu(OAc)_2$ in pyridine.

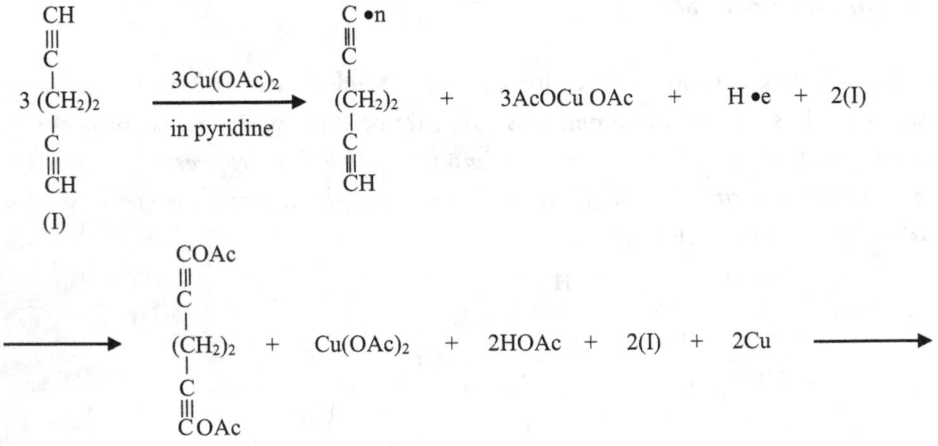

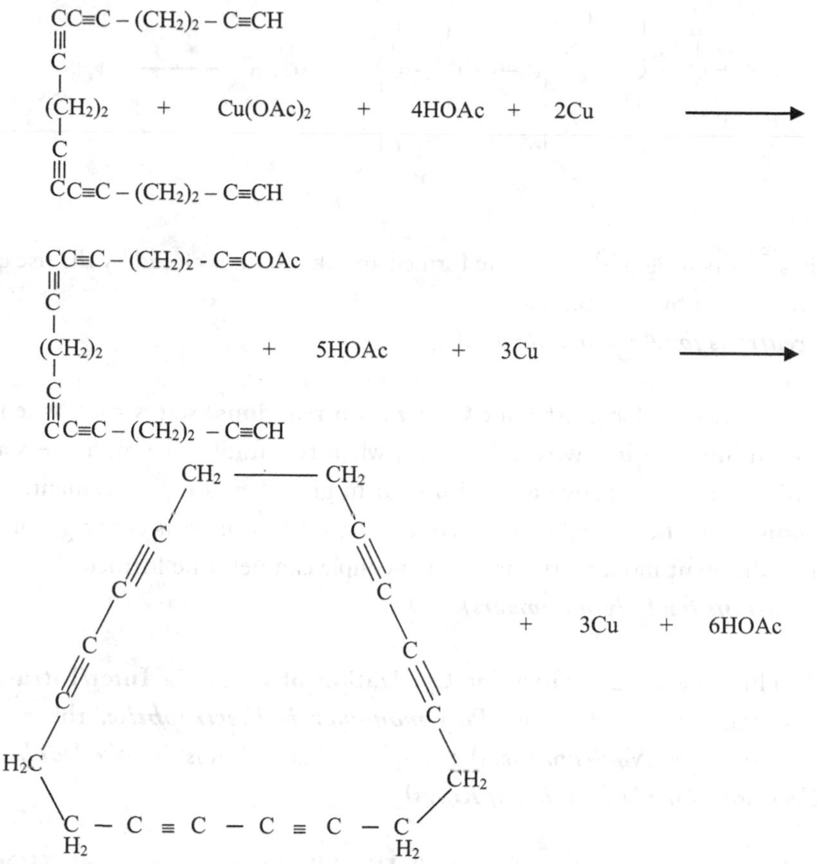

$$CC \equiv C - (CH_2)_2 - C \equiv CH$$
$$\overset{|||}{C}$$
$$|$$
$$(CH_2)_2 \quad + \quad Cu(OAc)_2 \quad + \quad 4HOAc \quad + \quad 2Cu \quad \longrightarrow$$
$$|$$
$$\overset{|||}{C}$$
$$CC \equiv C - (CH_2)_2 - C \equiv CH$$

$$CC \equiv C - (CH_2)_2 - C \equiv COAc$$
$$\overset{|||}{C}$$
$$|$$
$$(CH_2)_2 \quad + \quad 5HOAc \quad + \quad 3Cu \quad \longrightarrow$$
$$|$$
$$\overset{|||}{C}$$
$$CC \equiv C - (CH_2)_2 - C \equiv CH$$

$$+ \quad 3Cu \quad + \quad 6HOAc$$

an Equilibrium mechanism system of nine stages in which in the reaction above, the diyne was always in Equilibrium state of existence, one at a time; one was first kept in Equilibrium state of existence from one end and complete two stages and in the next two stages, the same one was kept again in Equilibrium state of existence from the other end, keeping the others waiting in Stable state of existence and after finishing, the next diyne takes over and exist in Equilibrium state of existence-turn by turn in a mechanistic way of operation, clear indication of the fact that Nature is indeed ORDERLINESS wherein exists full Control and Communication amongst components so wonderful to comprehend.
(Laws of Creations for Diynes)

Rule 1458: This rule of Chemistry for **Acrylic anhydride [$H_2C=CH-(CO)-O-(CO)-CH=CH_2$] an Electrophile, a Di-acrylic diene Poly-monomer,** states that, when used as a *Poly-monomer for cyclization,* this can only be done using *negatively charged-paired initiator or nucleo-free-radically,* to give only *six-membered unfused Electrophile ring alternatingly placed with methylene along the chain,* via Inter-/intra- molecular addition as shown below-

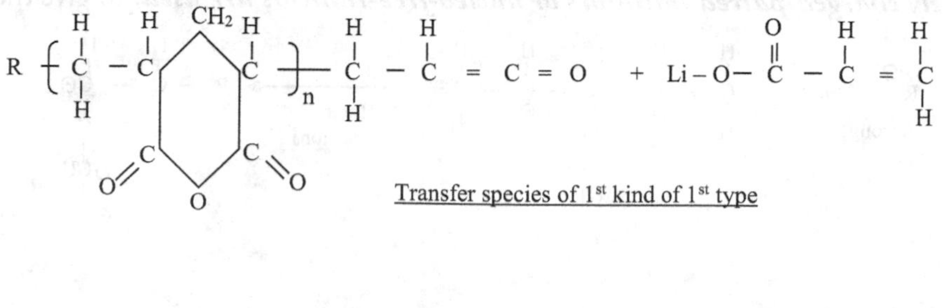

Transfer species of 1st kind of 1st type

i)

$$N \left[\begin{array}{c} H \\ | \\ C \\ | \\ H \end{array} - C \cdots \cdots C \right]_n \begin{array}{c} H \\ | \\ C \\ | \\ H \end{array} - \begin{array}{c} H \\ | \\ C \\ | \\ H \end{array} \cdots \cdots C \cdot n \longrightarrow \text{No transfer species}$$

ii)

for which sites for opening of the rings are formed and the chain is closed by the use of foreign terminating agents or termination by Combination.
(Laws of Creations for Acrylic anhydrides)

Rule 1459: This rule of Chemistry for **Cyclization reactions,** states that, when a **Poly-monomer** is unreactive to an initiator, like what takes place when two unreactive monomers are allowed to form a "Couple" which becomes reactive to the initiator to give alternating placement, this cannot take place with **Poly-monomers,** because the carriers of the activation centers is coming from one single monomer instead of two different monomers, for which a couple can never be formed.
(Laws of Creations for Poly-monomers)

Rule 1460: This rule of Chemistry for **Cyclization of rings via Inter/Intra- molecular addition mechanisms,** states that, when the *Poly-monomer is Electrophilic,* the ring obtained must be *Electrophilic* and when *Nucleophilic,* the ring also obtained must be *Nucleophilic.*
(Laws of Creations for Cyclization of Rings)

Rule 1461: This rule of Chemistry for a **1-,4- Diacidic Allylic compound (HOOC-CH = CH-CH$_2$-CH=CH-COOH),** states that, this compound which looks like methyl sorbate (H$_3$CCH = CH – CH = CHCOOH) an Electrophile, cannot be said to be a diene, but diacid and allylic in character favoring no route for cyclization-

$$HO - \overset{O}{\overset{||}{C}} - \overset{H}{\overset{|}{C}} = C - \overset{H}{\overset{|}{\underset{|}{C}}} - \overset{H}{\overset{|}{\underset{H}{C}}} = C - \overset{H}{\overset{|}{\underset{O}{C}}} - OH$$

A 1, 4 – Diacidic Allylic compound

noting what the ene group is carrying externally (COOH) in place of H.
(Laws of Creations for 1-,4-Diacidic Allylic compound)

Rule 1462: This rule of Chemistry for **1, 4- carbonyl diene (a Ketone- Acrolediene) an Electrophile, a Di-Acrylic Diene Poly-monomer,** states that, when used as a **Poly-monomer,** this is only possible when *negatively charged-paired initiators or nucleo-free-radicals are used,* to give the followings-

$$R: \overset{\ominus}{} \; + \; \oplus\overset{H}{\underset{H}{C}} - \overset{H}{\underset{\ominus}{C}} - \overset{H}{\underset{||}{C}} - \overset{H}{\underset{\ominus}{C}} - \overset{H}{\underset{H}{C}}\oplus \xrightarrow[\text{addition}]{\text{(Inter}} R - \overset{H}{\underset{H}{C}} - \overset{H}{\underset{\underset{\underset{\oplus CH_2}{\ominus CH}}{C=O}}{C}}\ominus$$

(strong)   (I)

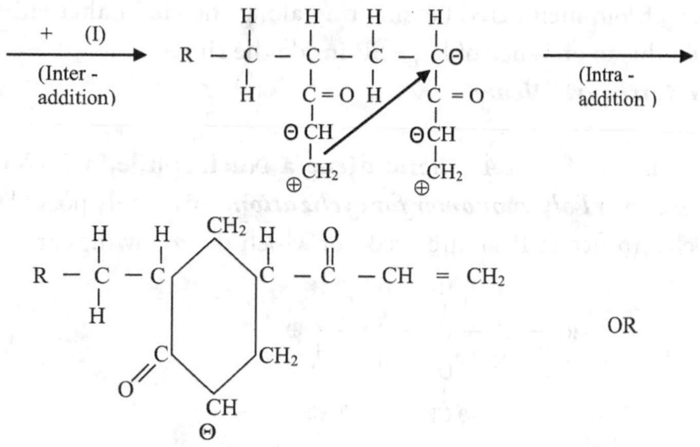

Primary cyclization

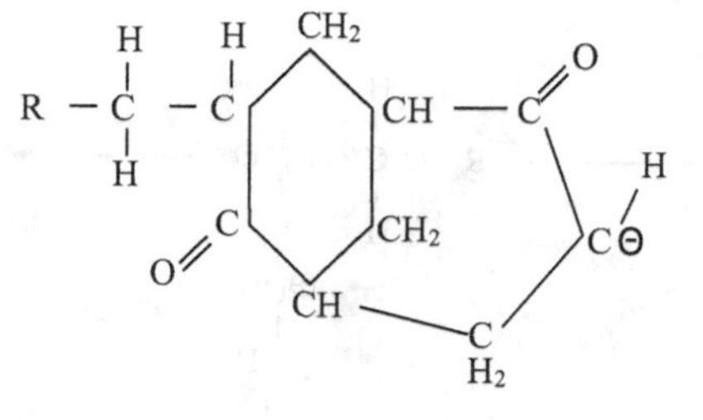

Primary and Secondary cyclization

beginning with a linear Inter- molecular addition chain small in size, followed by a cyclo-hexanone alternatingly placed with methylene, and finally followed by two fused cyclo-hexanone (if not too strained to exist) alternatingly placed with methylene via *Inter-/Intra- molecular addition;* the presence of these rings depending on the strength of the initiator, noting the absence of four-membered ring too strained to exist along the chain.

(Laws of Creations for 1-,4- Carbonyl Diene)

<u>**Rule 1463:**</u> This rule of Chemistry for **1-,4-Carbonyl Diene,** states that, this *Poly-monomer* which is almost like Acrolein, cannot cyclize when only the C= O center is involved electro-free-radically, because the externally located –CH=CH$_2$ center cannot be re-activated from behind differently as shown below to give a five-membered looking electrophilic ring, while

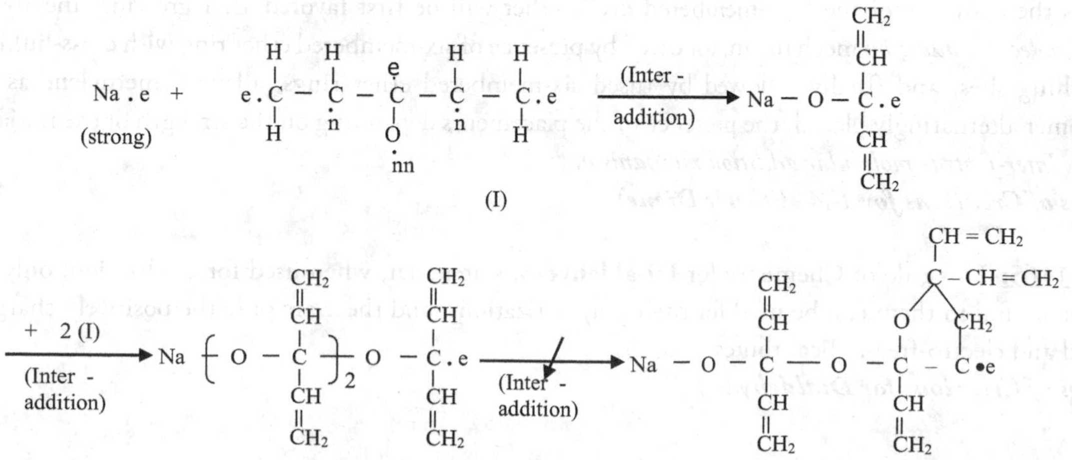

nucleo-free-radically presence of four-membered ketonic ring along the chain alternatingly placed with methylene may not be favored due to presence of large SE inside the ring.
(Laws of Creations for 1-,4-Carbonyl Diene)

Rule 1464: This rule of Chemistry for **1-,4-etheric diene a Nucleophile, a Di-Vinyl ester Poly-monomer,** states that, when used as a ***Poly-monomer for cyclization,*** this is only possible when positively charged-paired initiators or electro-free-radicals are used, for which the followings are obtained-

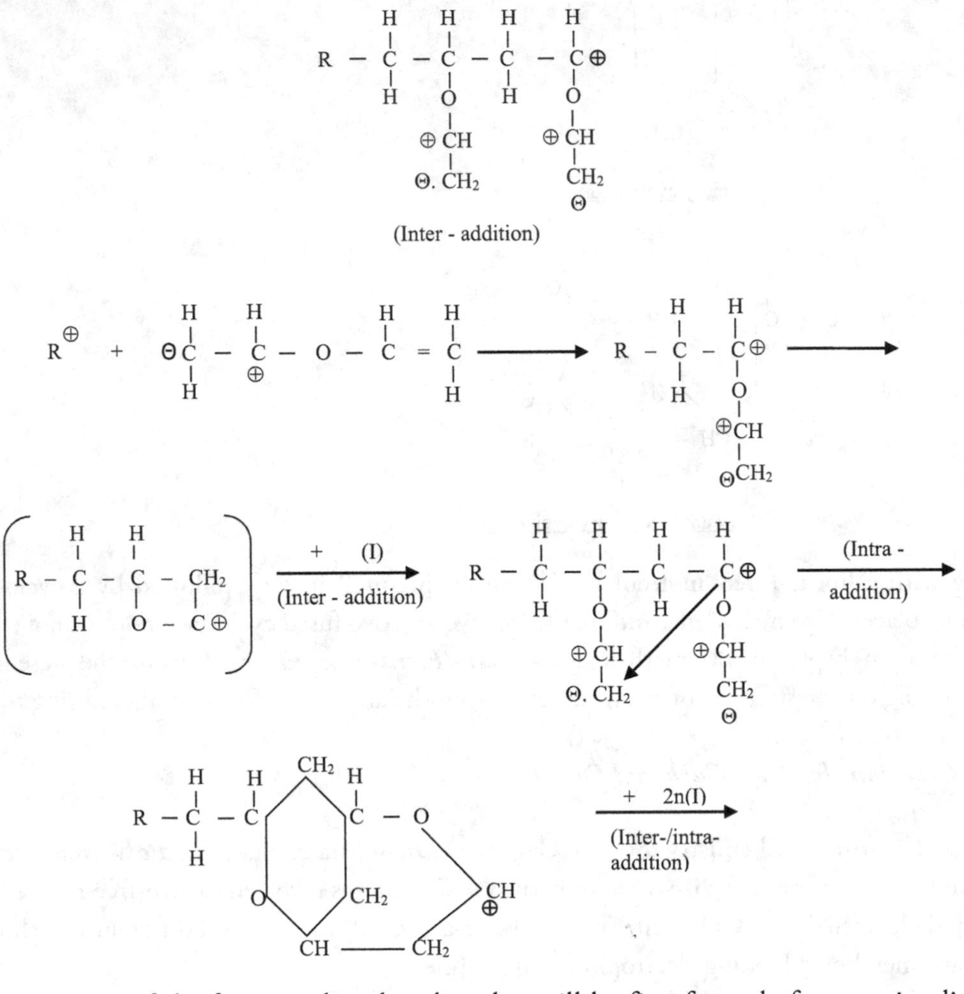

$$
R - \overset{\overset{\displaystyle H}{|}}{\underset{\underset{\displaystyle H}{|}}{C}} - \overset{\overset{\displaystyle H}{|}}{\underset{\underset{\displaystyle O}{|}}{C}} - \overset{\overset{\displaystyle H}{|}}{\underset{\underset{\displaystyle H}{|}}{C}} - \overset{\overset{\displaystyle H}{|}}{\underset{\underset{\displaystyle O}{|}}{C\oplus}}
$$

(Inter - addition)

$$
R^{\oplus} + \ominus\overset{\overset{\displaystyle H}{|}}{\underset{\underset{\displaystyle H}{|}}{C}} - \overset{\overset{\displaystyle H}{|}}{\underset{\underset{\displaystyle \oplus}{|}}{C}} - O - \overset{\overset{\displaystyle H}{|}}{C} = \overset{\overset{\displaystyle H}{|}}{\underset{\underset{\displaystyle H}{|}}{C}} \longrightarrow
$$

(Inter - addition) → (Intra - addition) →

that is the existence of the four-membered cyclic ether will be first favored after growing linearly via *Inter-molecular addition* mechanism, followed by presence of six-membered ether ring with cross-linking/branching sites, and finally followed by fused six-membered ether rings, all with methylene as co-monomer alternatingly placed; the presence of the placements depending on the strength of the initiator, all via *Inter-/Intra- molecular addition mechanism.*
(Laws of Creations for 1-,4- Etheric Diene)

Rule 1465: This rule of Chemistry for **Di-aldehydes**, states that, when used for Cyclization, only the routes natural to them can be used for their polymerizations, and these are only the positively charged-paired and electro-free-radical routes.
(Laws of Creations for Dialdehydes)

Rule 1466: This rule of Chemistry for **o-Phthaldehyde an Electrophile**, states that, when used as a *Poly-monomer for cyclization,* this is only possible when strong initiators are used, to give a copolymer of oxygen and five membered electrophilic looking type of ring fused to benzene ring on one side alternatingly placed via *Inter/Intra- molecular addition mechanisms-*

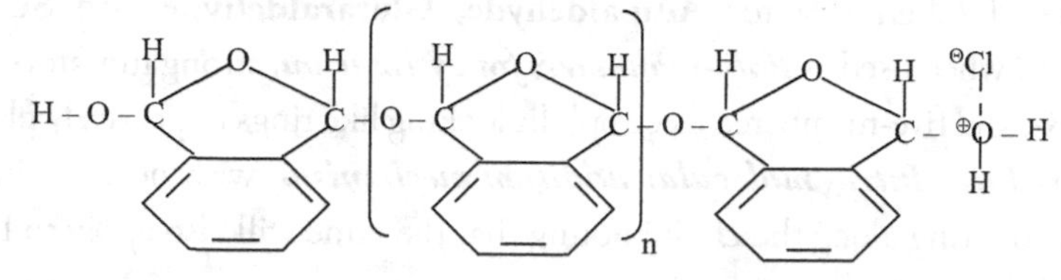

Alternating bicyclic co-polymer of o-Phthaldehyde

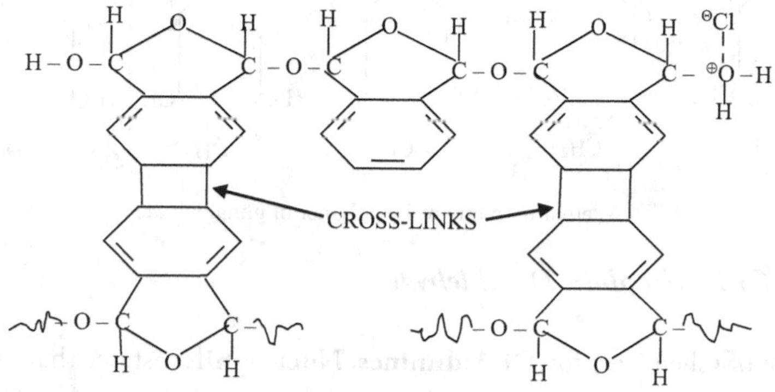

Cross-linked network of polymer

and may also form cross-links between two dead chains when suppressed via *Di-inter molecular addition mechanism if a C=C center in the ring is less nucleophilic than the functional center*; noting that while the route is natural to the center, it is not natural to the monomer whose Y center cannot be used. *(Laws of Creations for o-Phthaldehyde)*

Rule 1467: This rule of Chemistry for **Malealdehyde an Electrophile,** states that, when used as a *Poly-monomer for cyclization,* this is only possible when strong positively charged paired or electro-free-radical initiators are used, to give a copolymer of O and five-membered unsaturated electrophilic looking ring along the chain *via Inter/Intra- molecular addition mechanism.-*

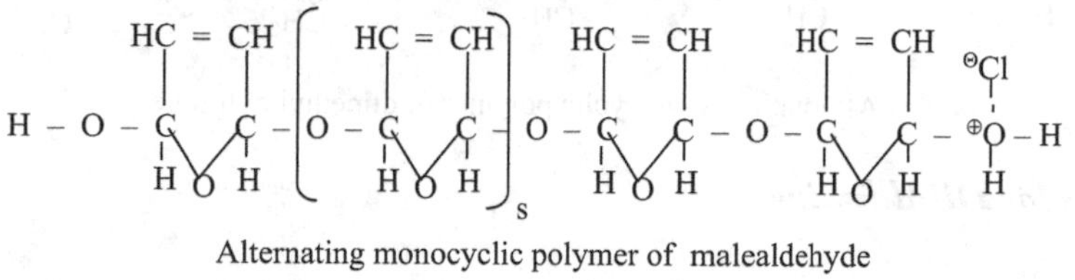

Alternating monocyclic polymer of malealdehyde

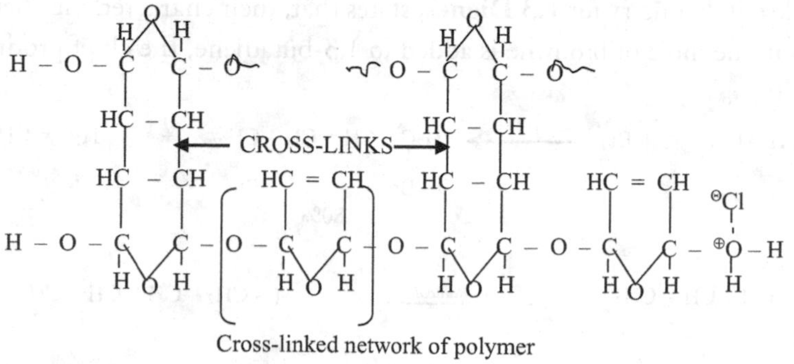

Cross-linked network of polymer

and can also form cross-links between two dead chains when suppressed via ***Di-inter molecular addition mechanism.***
(Laws of Creations for Malealdehyde)

<u>**Rule 1468:**</u> This rule of Chemistry for **Adipaldehyde, Glutaraldehyde and Succinal-dehyde, all Nucleophiles,** states that, when used as ***Poly-monomers for cyclization,*** strong initiators are required, to give copolymers of seven-, six-, and five-membered electrophilic looking like rings respectively placed along the chain alternatingly with O via ***Inter/Intra- molecular addition mechanism,*** with no cross-linking or branching sites besides the ones on the ring along the chain; noting that the same will also apply to Di –thioaldehydes.

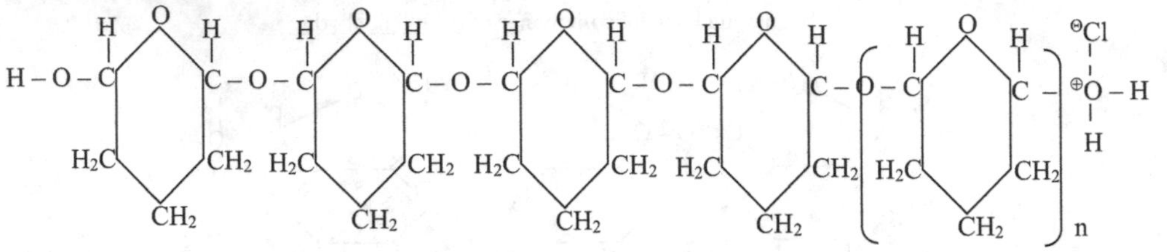

Alternating monocyclic polymer of glutaldehyde

(Laws of Creations for Nucleophilic Di-Aldehydes)

<u>**Rule 1469:**</u> This rule of Chemistry for **Di-Aldimines, Nucleophiles,** states that, when used as a ***Poly-monomer for cyclization,*** just like the Dialdehydes, this can only be done using positively charged paired or electro-free-radical initiators, to give seven-, six-, or five- membered electrophilic looking rings alternatingly placed with N along the chain via ***Inter-/Intra- molecular addition*** with no cross-linking sites besides the N center on the rings along the chain; noting that unlike Dialdehydes, the H on the N center must be changed if its Equilibrium state of existence is not suppressed.

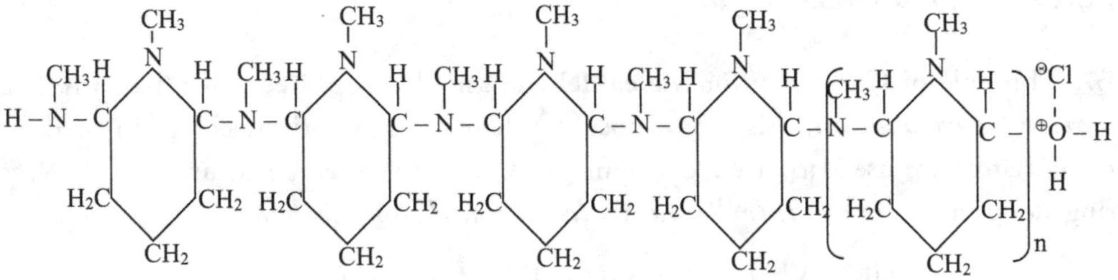

Alternating monocyclic polymer of dimethyl aldimine

(Laws of Creations for Di-Aldimines)

<u>**Rule 1470:**</u> This rule of Chemistry for **1,3 Dienes,** states that, their characteristic chemical behavior is 1, 4 addition, wherein if one mole of bromine is added to 1,3-butadiene, the chief product is 1,4-dibromo-2-butene, just at STP-

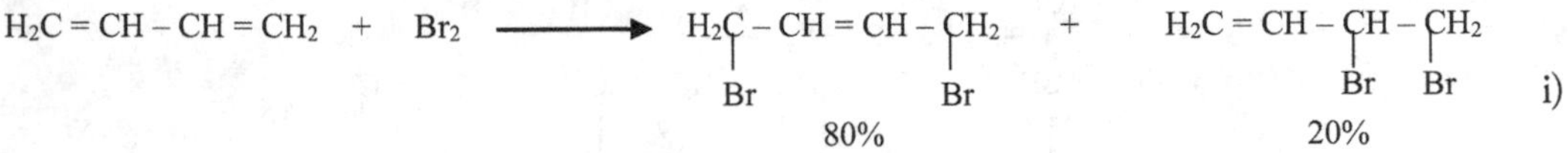

$$H_2C = CH - CH = CH_2 \ + \ Br_2 \longrightarrow \underset{80\%}{H_2\underset{Br}{C} - CH = CH - \underset{Br}{CH_2}} \ + \ \underset{20\%}{H_2C = CH - \underset{Br}{CH} - \underset{Br}{CH_2}} \qquad i)$$

$$x \ H_2C = CH - CH = CH_2 \xrightarrow[\text{Initiator}]{\textit{Free Radical}} (- CH_2 - CH = CH - CH_2 -)_x \qquad ii)$$

clear indications that 1,4-mono-form largely predominates free-radically, noting that charged-ly, it does not exist.
(Laws of Creations for 1,3-Butadiene)

Rule 1471: This rule of Chemistry for **Di-inter- molecular addition**, states that, the only route natural to them is only the electro-free-radical route, since while electro-radical carrying species can diffuse in the presence or absence of nucleo-radical carrying species, nucleo-radical carrying species cannot diffuse in the presence of electro-radical carrying species.
(Laws of Creations for Di-inter- Addition)

Rule 1472: This rule of Chemistry for **Di-inter- molecular addition,** states that, for their growing polymer chains, there is no transfer species, for which ***termination is by Deactivation*** provided either when the forces for activation of the active center can no longer be provided (One of the sources of low Molecular weight polymer) or when the optimum chain length has been reached or there is no monomer in its vicinity.
(Laws of Creations for Di-inter-Addition)

Rule 1473: This rule of Chemistry for **Di-Inter-molecular addition,** states that, ***unlike Inter- addition*** where the active center of the growing polymer chain keeps increasing in intensity with every addition of monomer in the route natural to them, with Di-Inter-addition, this does not take place, since no initiator is involved to generate a new center, i.e. for every addition of a diene, a dead ring is formed, and then re-activated to activate another diene, all via the help of the external force [Major source of low Molecular weight polymers].
(Laws of Creations for Di-Inter-Addition)

Rule 1474: This rule of Chemistry for **1,3-Butadiene,** states that, when used as a monomer for cyclization via ***Di-inter- molecular addition*** in the absence of an active initiator, but presence of a strong generating force for activation, it begins with ***the initiator preparation step***, wherein an eight-membered activated 1,5-cyclooctadiene ring is obtained as shown below-

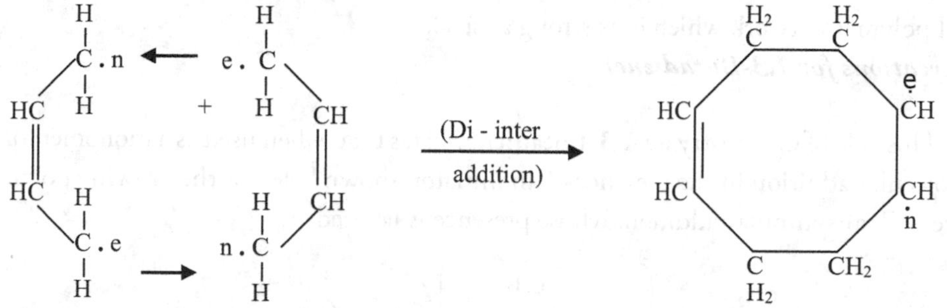

(Laws of Creations for 1,3-Butadiene)

Rule 1475: This rule of Chemistry for **1,3-Butadiene,** states that, when used as a monomer for cyclization via Di-inter molecular addition, the Initiation and Propagation steps are as follows in the absence of Trans-annular addition-

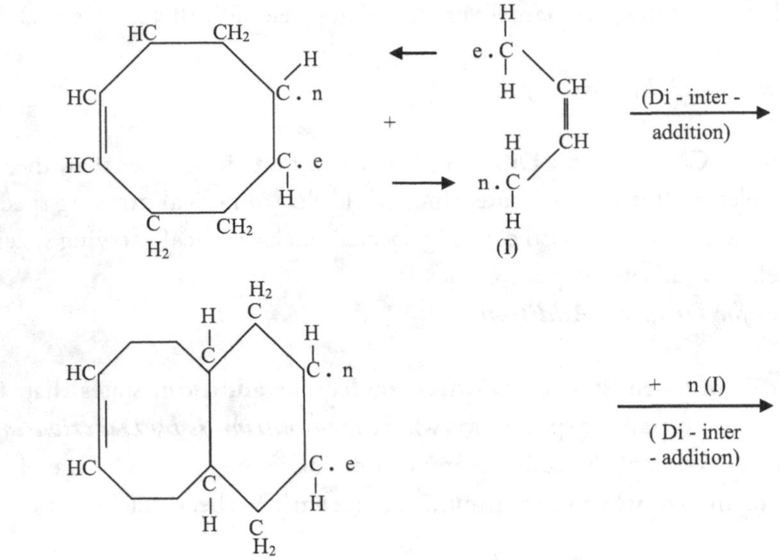

INITIATION STEP

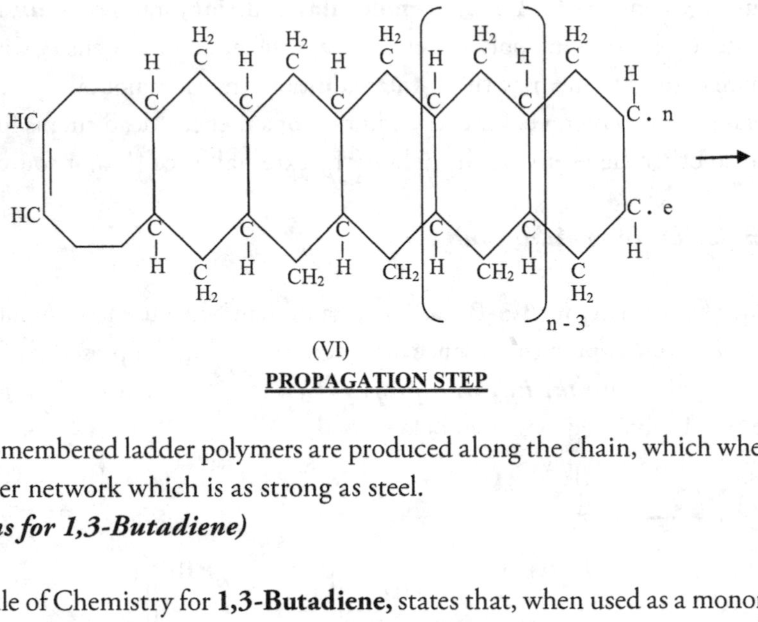

(VI)

PROPAGATION STEP

for which fused six-membered ladder polymers are produced along the chain, which when dehydrogenated gives a solid polymer network which is as strong as steel.
(Laws of Creations for 1,3-Butadiene)

Rule 1476: This rule of Chemistry for **1,3-Butadiene,** states that, when used as a monomer for cyclization via Inter molecular addition in the presence of an initiator, shown below is the growing polymer chain in the presence of Trans-annular addition, whose presence is favored

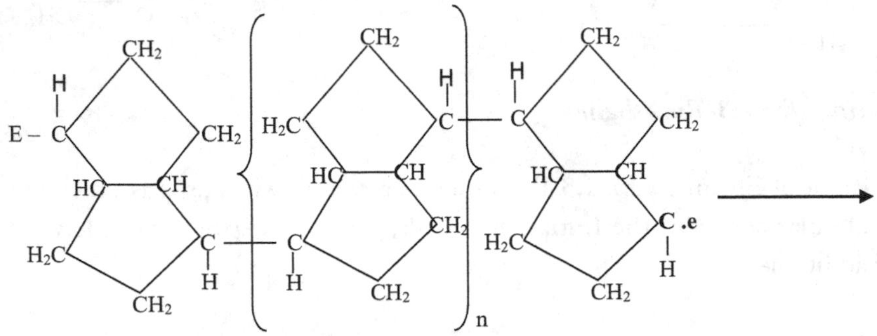

Polymer Chain after Transannular Additions

either when the initiating generating force is strong enough to activate the two activation centers in 1,5-cyclooctadiene at the same time both being of the same nucleophilicity instead of one center or conformational instability is strong enough to bring the two activation centers close together.
(Laws of Creations for 1,3-Butadiene)

Rule 1477: This rule of Chemistry for **1,3-Dienes carrying radical pushing groups internally or externally located, such as CH$_3$ (Nucleophiles),** states that, when used as a monomer for cyclization via Di-inter-molecular addition, their Initiator preparation, Initia-tion, Propagation and Termination steps are similar to that of 1,3-butadiene in the absence of Trans-annular Addition which if allowed to take place, unlike 1,3-butadiene, stable tricyclic four-membered rings are obtained as shown below for 1,3-pentadiene.

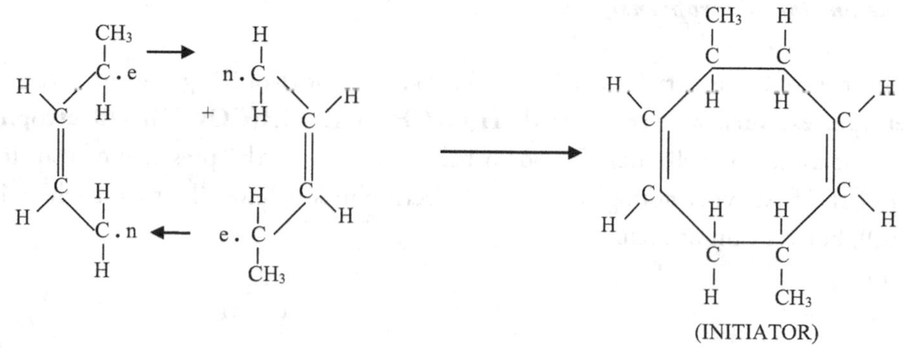

(INITIATOR)   i)

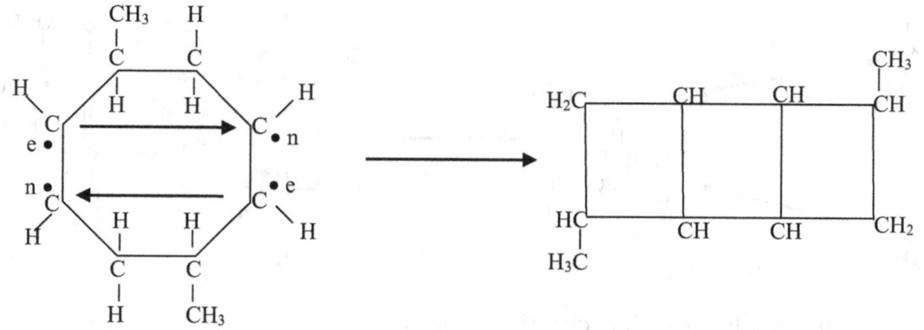

ii)

(Laws of Creations for 1,3-Radical pushing –Dienes)

Rule 1478: This rule of Chemistry for **1,3-Diene carrying radical-pulling group internally or externally located with no transfer species, such as chloroprene (Nucleophiles),** states that, it unlike the others including the externally located one which cannot be activated chargedly, and which therefore largely favors the existence of mostly 1,4- mono-form when activated radically, will be similar in character to other 1,3- Dienes with radical-pushing groups (Nucleophiles) as shown below for cyclization via Di-Inter- molecular addition-

INITIATOR

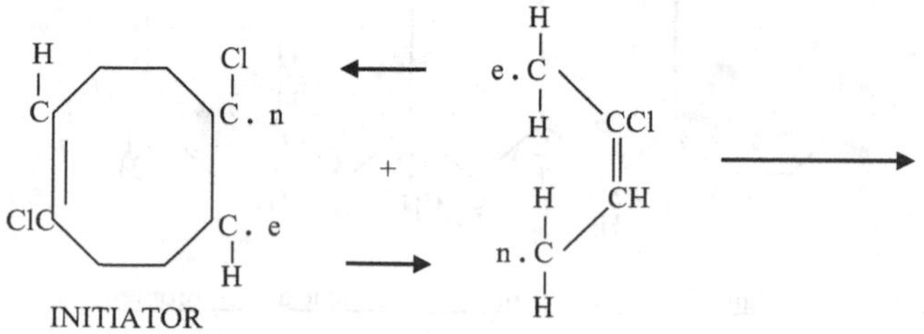

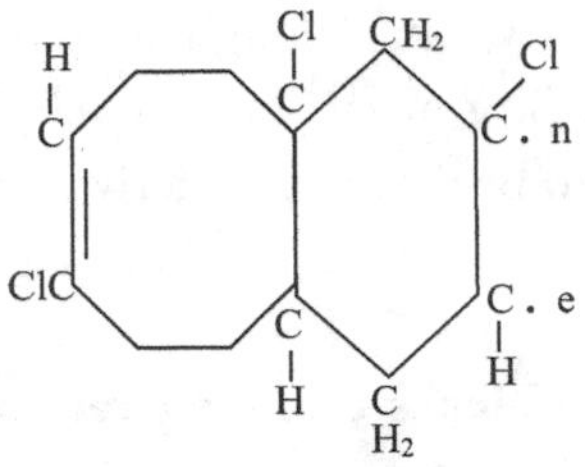

INITIATION STEP

which is then followed by propagation step wherein six- membered rings with two common boundaries ladder-like in structure are obtained and when Trans-annular addition takes place, stable tricyclic four-membered rings are obtained.

(Laws of Creations for Chloroprene)

Rule 1479: This rule of Chemistry for **1,3-Dienes carrying radical-pulling groups externally located with transfer species, such as Acrylic acid [$H_2C=CH- CH=CH(COOH)$] (Electrophiles),** states that, these cannot undergo cyclization as shown below, in view of the presence of transfer species of the first kind of the first type, noting that not all electrophiles behave like this, such as $H_2C=CH - CH=CH(CHO)$, but yet cannot cyclize-

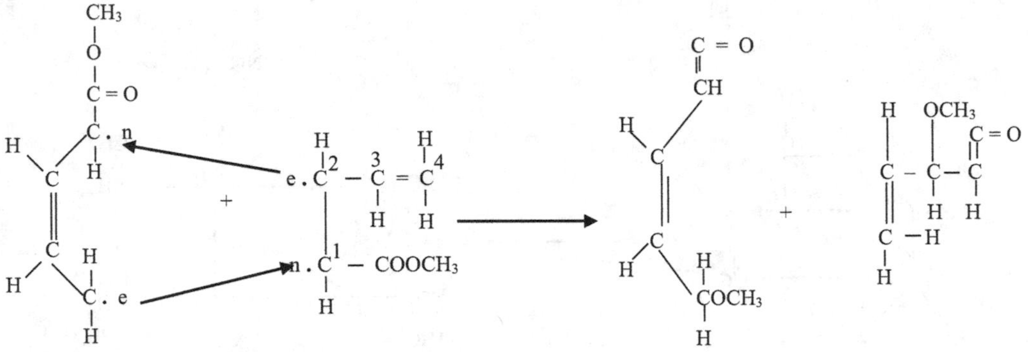

while when they are internally located, cyclization is favored.

(Laws of Creations for 1,3-Electrophilic Dienes)

Rule 1480: This rule of Chemistry for **1,3-Dienes,** states that, when two of them which can readily undergo cyclization via Di-Intra- molecular addition are present in the system, such as Butadiene and Chloroprene, random copolymers can be obtained for them either in form of blocks or together randomly placed all depending on the operating conditions.

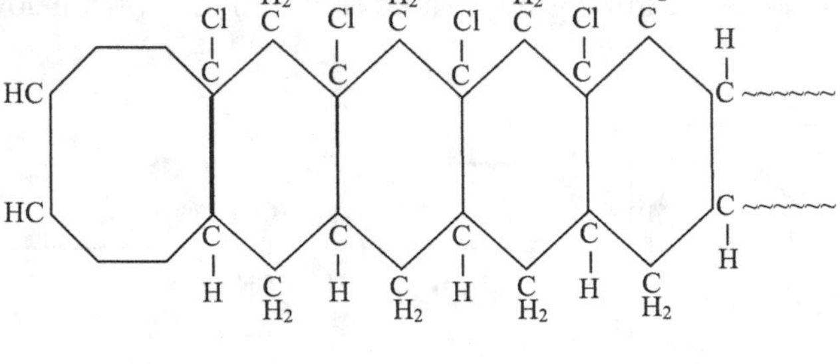

Random Copolymer of 1, 3 - butadiene and chloroprene

(Laws of Creations for Copolymerization via Di-Inter-Addition)

624

Rule 1481 This rule of Chemistry for **Intra-molecular and Inter/Intra/Intra/Intra-..mole-cular addition,** states that, when the **Polymeric monomers** are involved for Cyclization, this is only possible radically and never chargedly.
(Laws of Creations for Intra-molecular Addition)

Rule 1482 This rule of Chemistry for **Polyacrylonitrile (a product of an Electrophilic monomer obtained nucleo-free-radically),** states that, when used as a *Polymeric monomer for cyclization under pyrolytic conditions* at temperatures as high as 200^0C *in the absence of molecular oxygen,* fused ladder like gigantic polymer is obtained as shown below via *Intra-molecular addition-*

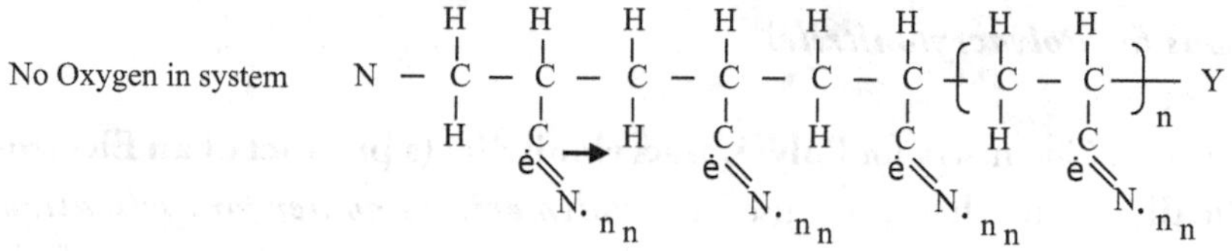

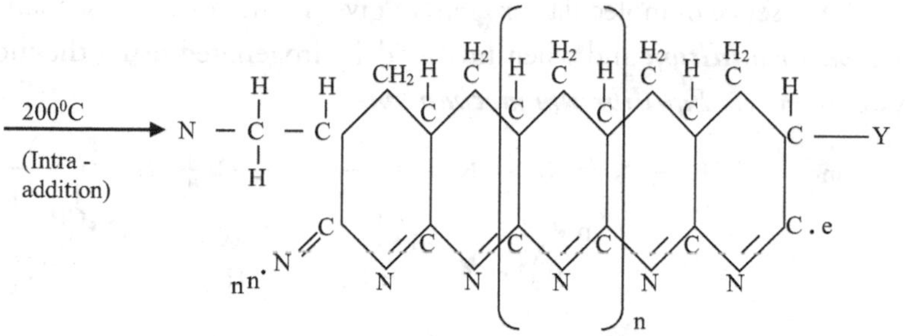

for which the living polymeric chain (A) with no possibility of releasing transfer species of the first kind, because of the limits placed by the SE, must be killed either by introducing a foreign terminating agent or letting the chains add to themselves, fold, and terminate; noting that when electro-free-radicals are used cyclization will still be favored under same operating conditions, but not nucleo-free- or non-free-radically, making it Inter-/Intra-/Tntra-.......... molecular addition.
(Laws of Creations for Polyacrylonitrile)

Rule 1483: This rule of Chemistry for **Polyacrylonitrile (a product of an Electrophilic monomer obtained nucleo-free-radically),** states that, when used as a *Polymeric monomer for Cyclization under pyrolytic conditions at higher operating conditions far greater than 200^0C in the presence of oxygen,* the oxygen breaks down as in Transition metal pores into oxidizing oxygen elements to dehydrogenate the fused ladder polymers to give a network of fully unsaturated polymer- polyquinizarine as shown below via *Intra- molecular addition mechanism* followed by oxidation via *Equilibrium mechanism-*

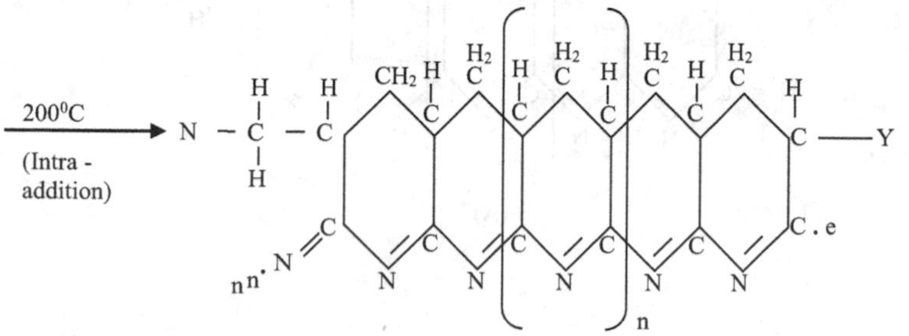

(A)

625

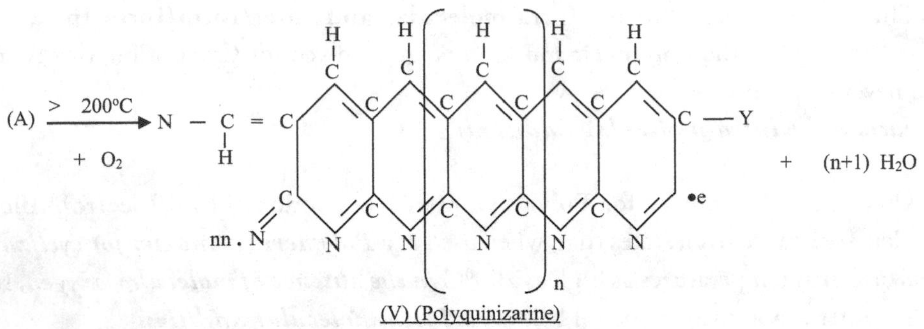

(A) $\xrightarrow[\text{+ O}_2]{\text{> 200°C}}$ ... N — C = C ... C — Y   + (n+1) H_2O

(V) (Polyquinizarine)

(Laws of Creations for Polyacrylonitrile)

Rule 1484: This rule of Chemistry for **Poly(4,3-acrylonitrile) (a product of an Electro-phile obtained electro-free-radically),** states that, when used as a *Polymeric monomer for cyclization,* due to absence of transfer species nucleo-free-radically and electro-free-radically, cyclization is fully favored under pyrolytic conditions in the absence of molecular oxygen, to give six-membered fused ladder-like gigantic polymer via *Intra-molecular addition* and when further dehydrogenated using the molecular oxygen, a stronger network is obtained via *Equilibrium mechanism-*

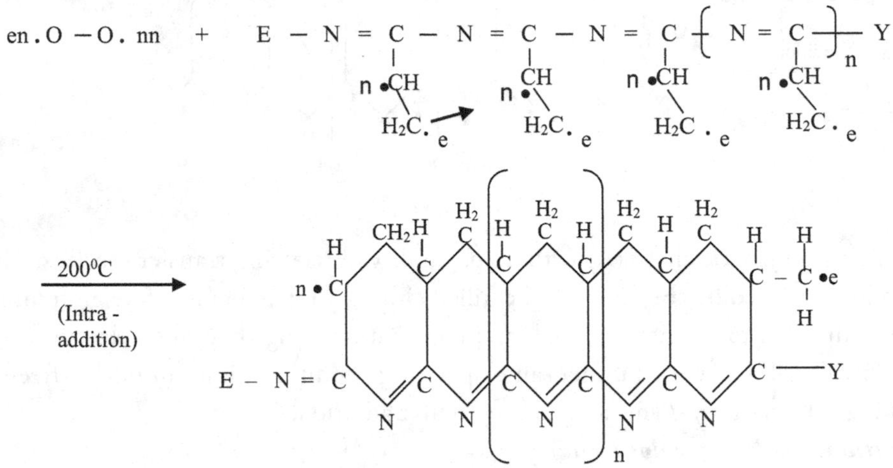

en. O — O . nn + E — N = C — N = C — N = C $\left(\!\!\begin{array}{c} N = C \end{array}\!\!\right)_n$ — Y

$\xrightarrow[\substack{\text{(Intra -} \\ \text{addition)}}]{200°C}$

(A)   i)

noting that when electro-free-radicals are used, cyclization is impossible, because the center of attack is the C =N center, the chain being Electrophilic, whereas nucleo-free-radically cyclization may be favored.

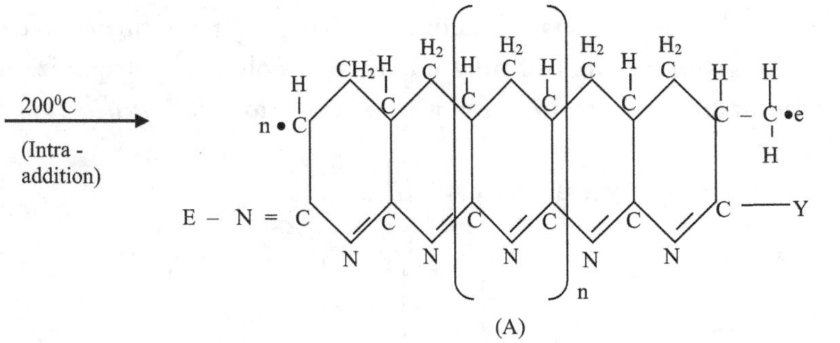

$\xrightarrow[\substack{\text{(Intra -} \\ \text{addition)}}]{200°C}$

(A)

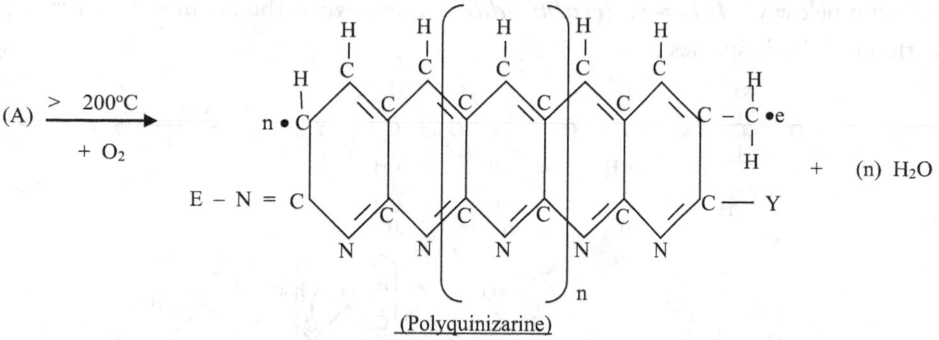

(Polyquinizarine)

ii)

(Laws of Creations for Poly(4,3 acrylonitrile))

Rule 1485: This rule of Chemistry for **Polyacrolein or polyacrylaldehyde (a product of an Electrophile obtained nucleo-free-radically),** states that, when used as a **Polymeric monomer** for cyclization, full cyclization is favored as shown below via *Intra- molecular addition-*

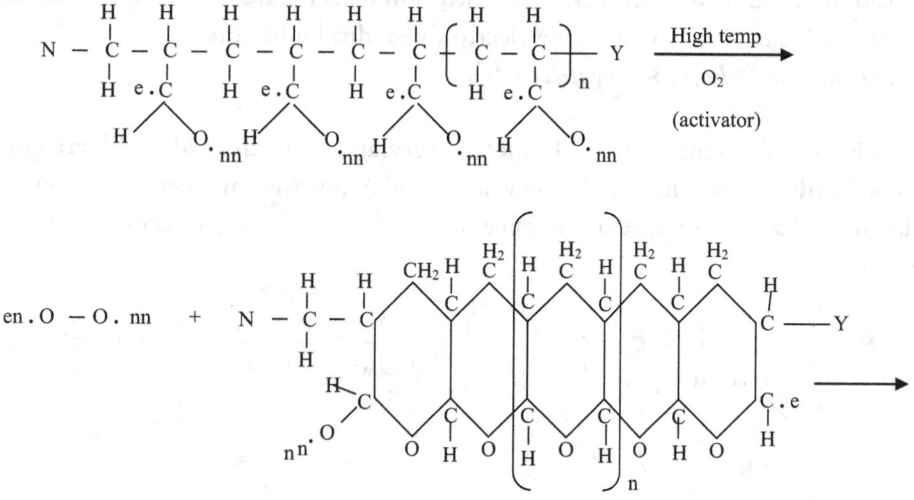

noting that when electro-free-radicals are used under similar operating conditions as used when not present, cyclization is favored, whereas nucleo-free-radically, it is not favored and when dehydrogenated, the followings are obtained-

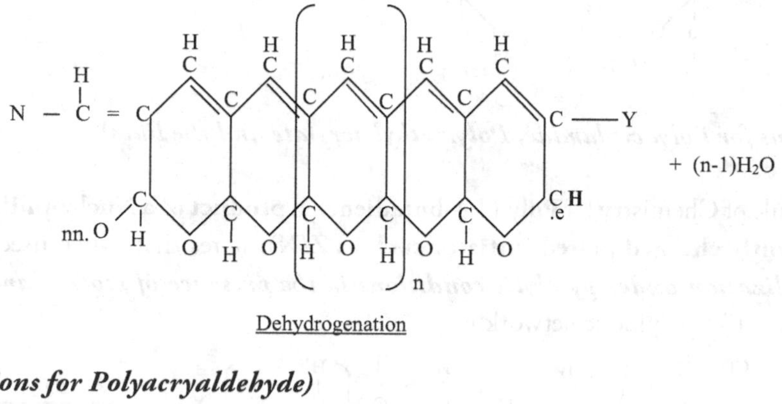

Dehydrogenation

(Laws of Creations for Polyacryaldehyde)

Rule 1486: This rule of Chemistry for **Poly(4,3-acrolein) (a product of an Electrophile obtained electro-free-radically),** states that, when used as a *Polymeric monomer for cyclization,* full cyclization is favored as shown below via ***Intra- molecular addition*** only when the Na initially carrying the chain is replaced with a non-ionic species (R) -

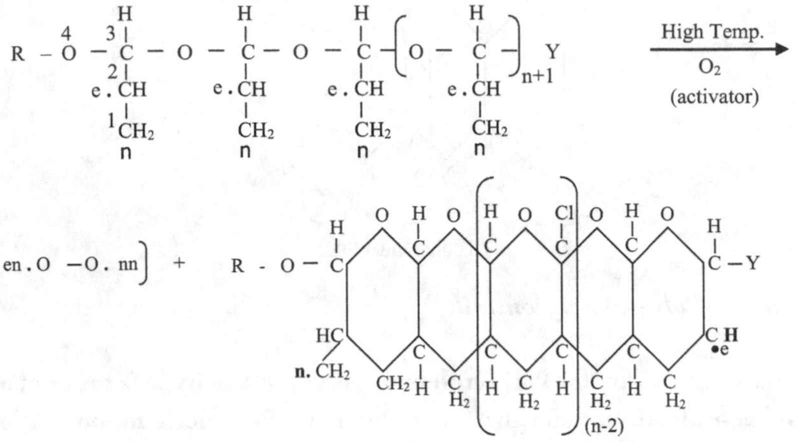

for which a dead fused ladder like network is obtained (still weak because of the functional centers carried by the ring); and this can also be done using electro-free-radical initiators.
(Laws of Creations for Poly (4,3-acryaldehyde))

<u>**Rule 1487:**</u> This rule of Chemistry for **Poly(methyl acrylate) (a product of an Electrophile obtained nucleo-free-radically),** states that, it like poly(acrylamide), poly(methyl metha-crylate), and the likes, cannot undergo cyclization, because of presence of transfer species of the second first kind, as shown below for one of them-

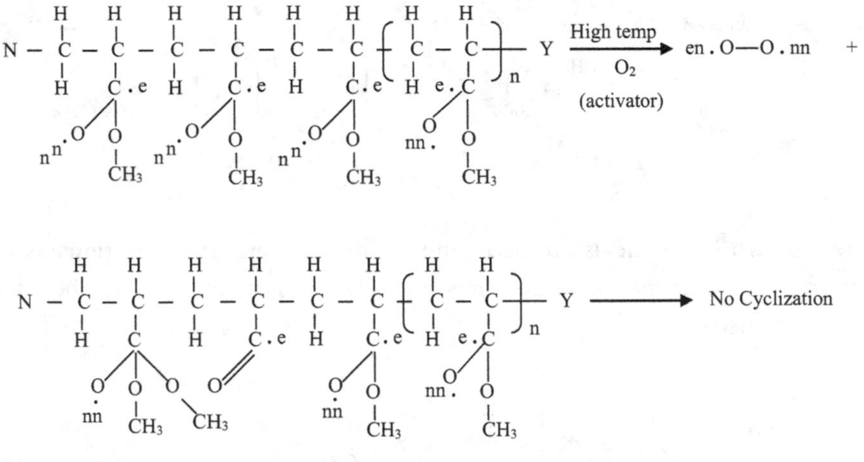

(Laws of Creations for Polyacrylamide, Polymethyl acrylate and the likes)

<u>**Rule 1488:**</u> This rule of Chemistry for **Poly (3,4-butadiene) (a product of a Nucleophile obtained from the use of Covalently charged paired initiator such as Z/N),** states that, when used as a *Polymeric monomer for cyclization under pyrolytic conditions in the presence of molecular oxygen,* a fused six-membered ladder like polymeric network is

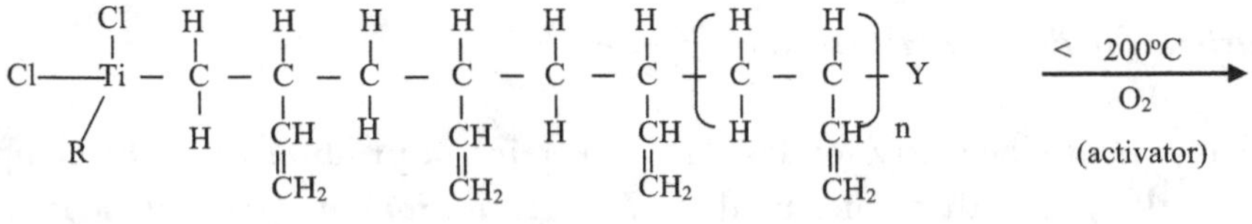

obtained as shown below *via Intra- molecular addition* in the absence of release of transfer species-

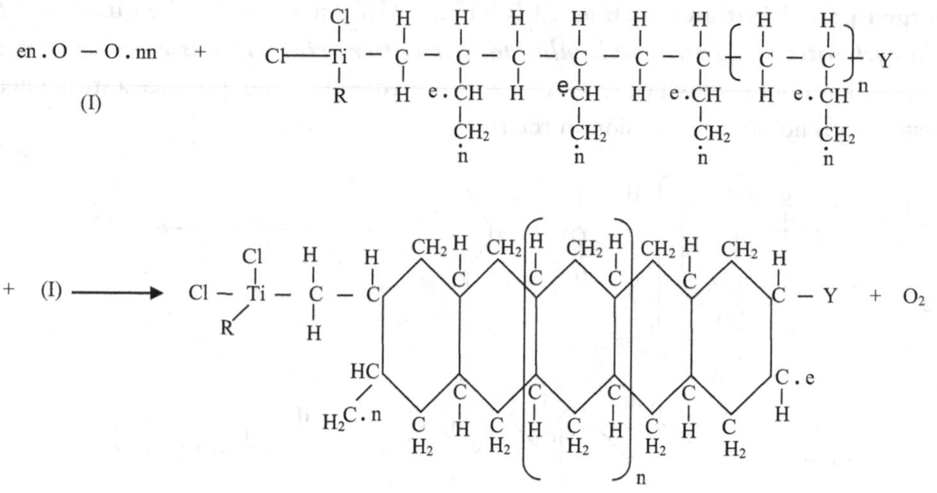

and in addition while electro-free-radicals can be used, nucleo-free-radicals cannot be used for cyclization.
(Laws of Creations for Poly(3,4-butadiene))

<u>Rule 1489:</u> This rule of Chemistry for **Poly(3,4-pentadiene) (a product of a Nucleophile obtained from the use of Covalently charged-paired initiator, such as Z/N),** states that, when used as a *Polymeric monomer for cyclization,* the same as applies to poly(3,4-butadiene) is obtained and the same will also apply to poly(isoprene) and all Nucleophilic C/H cases, for as long transfer species of the first kind of the sixth type is not released during propagation.

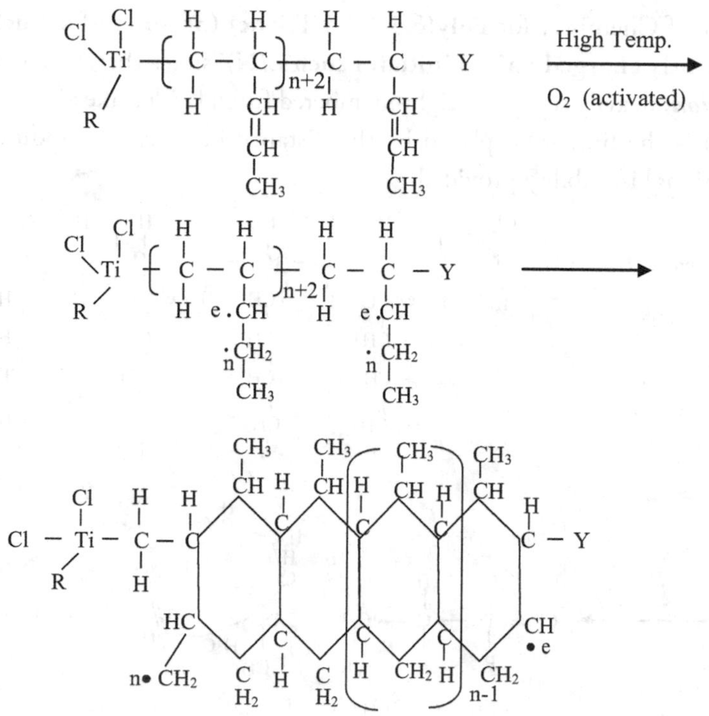

noting that when the transfer species is released, partial cyclization is favored with presence of low Molecular weight products and this can readily be prevented by replacing the transfer species.
(Laws of Creations for Nucleophilic C/H Poly (1,3-Dienes))

Rule 1490: This rule of Chemistry for **Poly(3,4-vinylacetylene) (a product of a Nucleo-phile)** obtained using a charged-paired initiator such as $Cl_2R\,Ti^{\oplus}....^{\theta}AlR_2$, states that, when used as a *Polymeric monomer for cyclization* via *Intra-molecular addition after being suppressed,* a fused six-membered ladder-like network is obtained, which when dehydrogenated will further produce a strong network ring as strong as steel with no point of scission in the rings-.

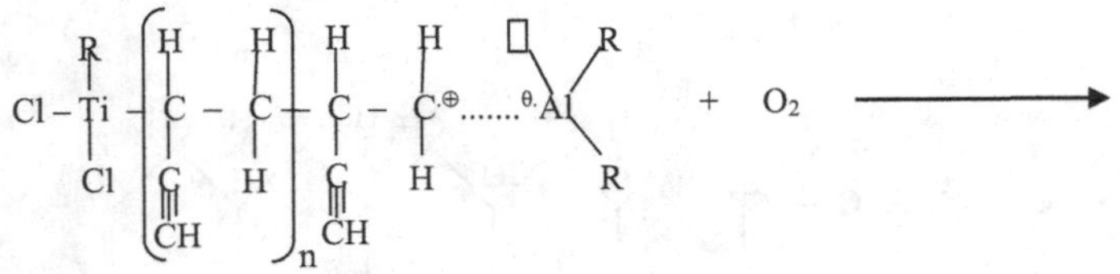

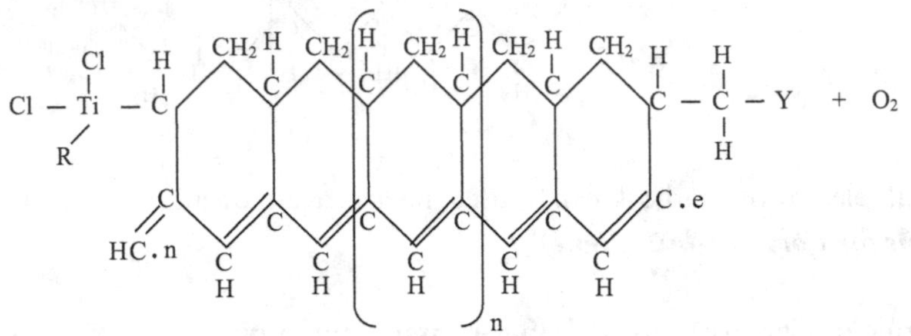

noting that no transfer can be rejected because of the large amount of strain energy present when double bonds are cumulatively placed in a ring already carrying pushing groups.
(Laws of Creations for Poly(3,4- vinylacetylene))

Rule 1491: This is rule of Chemistry for **Poly(6,5-1,3,5-Triene) (product of a Nucleophile obtained from the use of Covalently charged-paired initiator such Z/N)**, states that, when used as a *Polymeric monomer for cyclization,* heavily staggered eight-membered fused ladder like alkene rings are obtained, the existence of which is the limitation placed by the distance between the addition centers and the influence of conformational instability provided

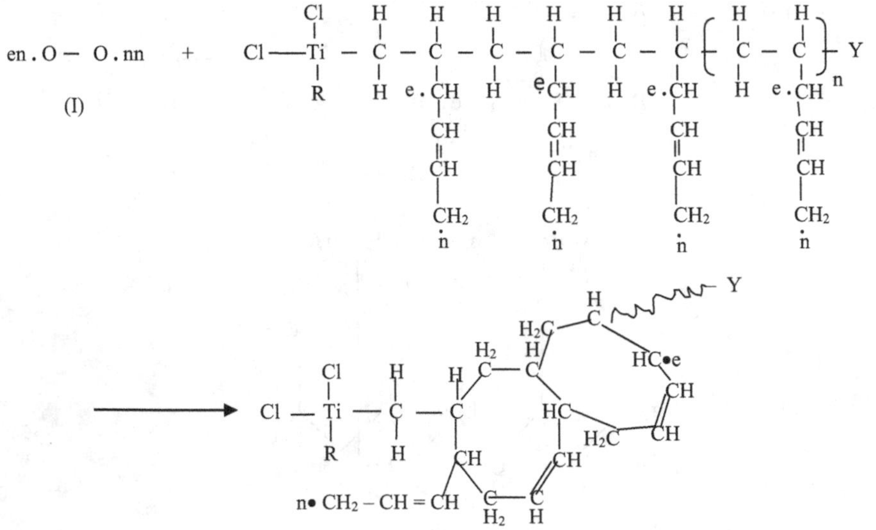

by the size of the ring; for which Cyclization may not be easily favored.
(Laws of Creations for Poly(6,5- 1,3,5-Triene))

Rule 1492: This rule of Chemistry for Equilibrium **Intra- molecular Step Polymerization system,** states that, only *Polymeric monomers that have the ability of existing in Equilibrium state of existence from special radical-pulling groups on the side chain,* can undergo this type of cyclization, and these polymeric monomers must be *Polar/Ionic* in character.
(Laws of Creations for Equilibrium Intra-molecular Step Polymerization)

Rule 1493: This rule of Chemistry for **Poly(methyl acrylic acid) (a product of an Electrophile obtained from the use of nucleo-free-radicals),** states that, when used as a *Polymeric monomer for cyclization,* copolymers of six-membered cyclic anhydride Electrophilic ring alternatingly placed with methylene along the chain is obtained along with small molecular byproducts of water as shown below-

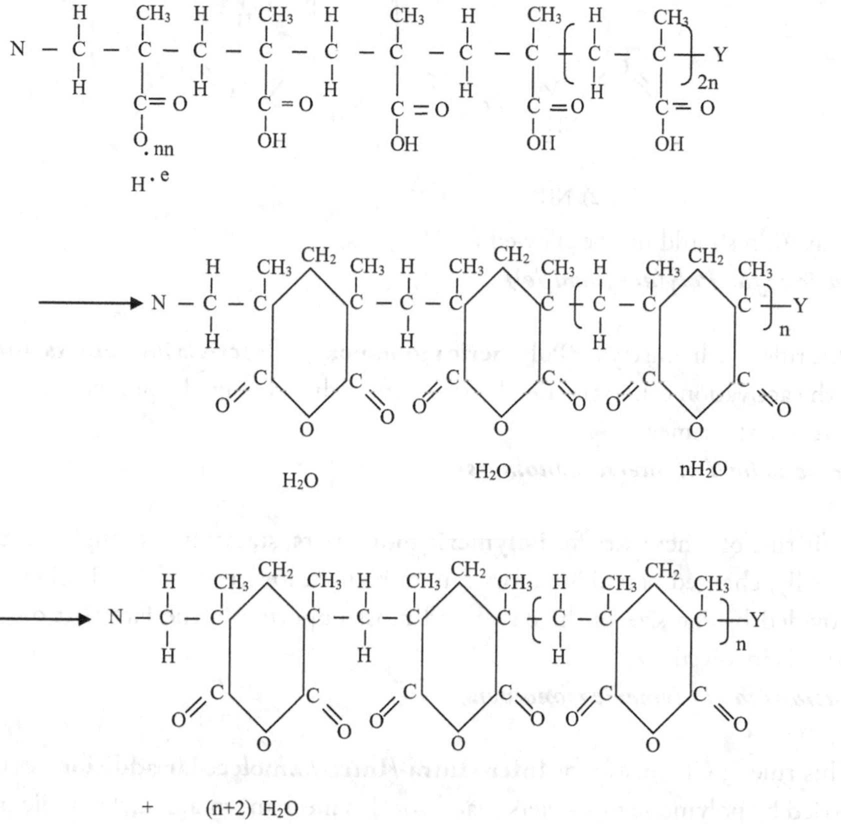

noting that Activation should not be allowed to take place.
(Laws of Creations for Poly(methyl acrylic acid))

Rule 1494: This rule of Chemistry for **Poly(acrylamide) (a product of an Electrophile obtained from the use of nucleo-free-radicals),** states that, when used as a *Polymeric monomer for cyclization,* copolymers of six-membered cyclic Electrophilic amino anhydride alternatingly placed with methylene along the chain is obtained along with small molecular byproduct of ammonia as shown below-

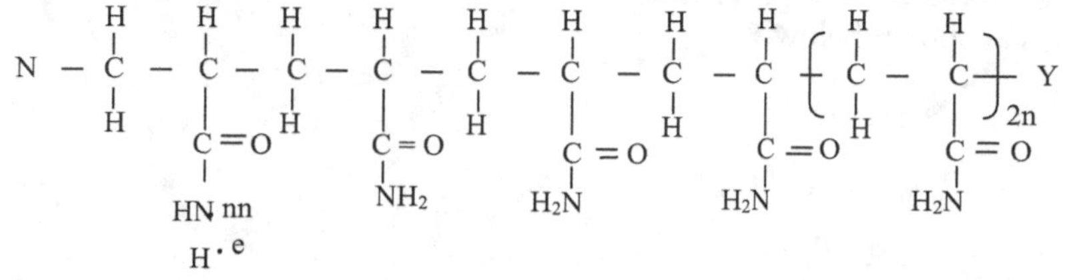

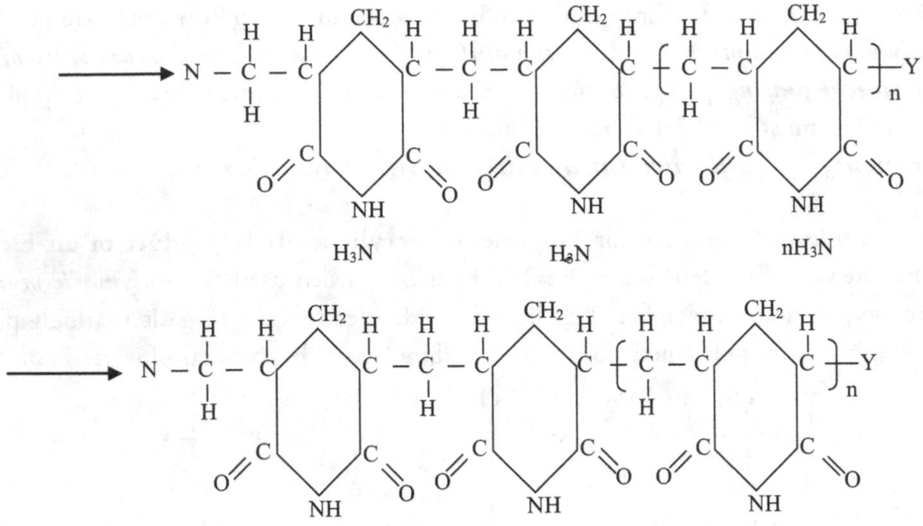

[Chemical structures of poly(acrylamide) chains shown]

$+$ $(n+2)$ NH_3

noting that Activation should not be allowed to take place.
(Laws of Creations for Poly(acrylamide))

Rule 1495: This rule of Chemistry for **Polymeric monomers with activation centers internally located**, states that, all the activation centers carried by them are of different nucleophilicities, for which only one can be activated one at a time.
(Laws of Creations for Polymeric monomers)

Rule 1496: This rule of Chemistry for **Polymeric monomers,** states that, though some of them can be activated chargedly, charged paired initiators cannot be used for their polymerization, because of steric limitations provided by the size of the chain and most importantly the fact that only one unit (The polymeric chain) is involved.
(Laws of Creations for Polymeric monomers)

Rule 1497: This rule of Chemistry for **Inter-/Intra-/Intra-/...molecular addition,** states that, since all the centers carried by polymeric monomers that favor this mechanism are nucleophilic in character, only the electro-free-radical initiators can be used for their cyclization.
(Laws of Creations for Inter-/Intra-/Intra-/.. Molecular Addition)

Rule 1498: This rule of Chemistry for **Poly(1,4-isoprene), (a product of a Nucleophile obtained free-radically),** states that, when used as a *Polymeric monomer for cyclization,* formation of rings begins from the middle of chain where the least nucleophilic center exists and moves to the right with formation of only fused six-membered rings until the right hand is filled, followed by release of transfer species which is now used to begin initiation and propagation to the left from the transition point to give same fused six-membered rings as shown below-

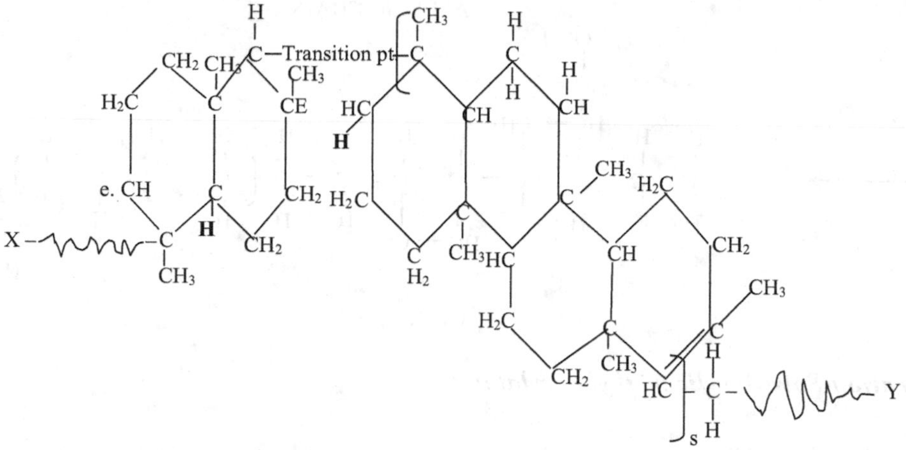

(Laws of Creations for Poly(isoprene)

Rule 1499: This rule of Chemistry for **Polychloroprene (a product of a Nucleophile obtained free-radically),** states that, when used as a *Polymeric monomer for cyclization,* four-membered cyclic rings are placed sequentially and syndiotactically along the chain, starting from the beginning or the end of the chain to the end or beginning.

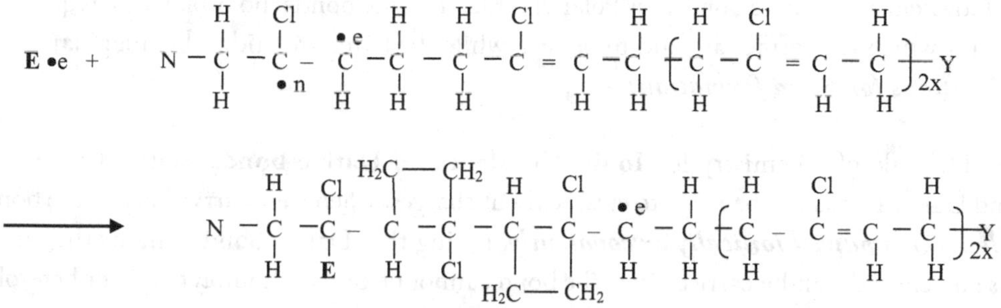

(Laws of Creations for Polychloroprene)

Rule 1500: This rule of Chemistry for **Polymers obtained from Electrophilic Diene such as $H_2C =$ CH – CH = CH(COOCH$_3$),** states that, their polymeric chains are allylic in character (and the same will apply when COOCH$_3$ is replaced with Cl (Nucleophile), CONH$_2$ etc.), for which when used as a *Polymeric monomer,* four-membered rings like the case for chloroprene are obtained along the chain.

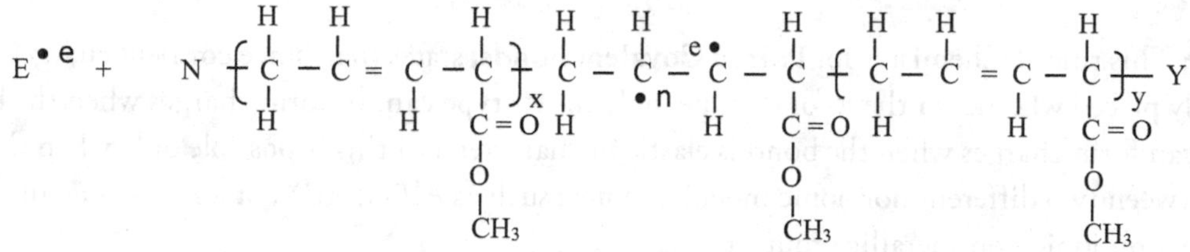

ALLYLIC CHAIN

(Laws of Creation Poly(Acrylic methyl acrylate)

Rule 1501: This rule of Chemistry for **Polymers made from Electrophilic Dienes such as $H_2C = CH – C(COOCH_3) = CH_2$,** states that, when used as a *Polymeric monomer for Inter-/Intra-/Intra-/... molecular addition*, no cyclization can be possible because of presence of transfer species of the first kind of the first type electro-free- and nucleo-free-radically.
(Laws of Creations for Poly(1,2-electrophilic diene))

Rule 1502: This rule of Chemistry for **Bond formations,** states that, while two orbitals are required to form Ionic, Covalent and Dative bonds, for Polar and Electrostatic bonds, no orbitals are required to form their bonds; for which the former are said to be real, while the latter are said to be imaginary.
(Laws of Creations for Bond formation)

Rule 1503: The rule of Chemistry for **Ionic, Covalent and Dative bonds,** states that, while Ionic, Covalent and Dative bonds can be made to exist without charges, when they carry charges, the bonds cease to exist *unless when paired ionically or covalently,* noting that Dative bond remains the same all the time because no charges can be carried since the bond cannot be broken homolytically or heterolytically.
(Laws of Creations for Ionic, Covalent and Dative Bonds)

Rule 1504: This rule of Chemistry for **Paired Ionic bonds,** states that, since ionic bonds exist only when paired, the conditions favoring the existence of Paired ionic bond is when it is formed between an ionic metal and a *bulky anionic group such as* those with large radical-pushing capacity (OCH_3, OC_2H_5, OC_3H_7, NR_2) in addition to provision of an ionic or charged environment.
(Laws of Creations for Paired Ionic Bonds)

Rule 1505: This rule of Chemistry for **Paired Covalent bonds,** states that, since covalent charges cannot be isolatedly placed whether of the π- or σ- type, only the σ- type cannot form charges when the bond is fixed, but can form charges when the bond is elastic in character and this is possible only when the bond is either between two different non-ionic metallic centers such as Al/Ti, Al/V, or an ionic metallic center and a bulky non-ionic non-metallic group such as Li/C_4H_9.
(Laws of Creations for Paired Covalent Bonds)

Rule 1506: This rule of Chemistry for **Paired Radical bonds,** states that, though radicals cannot repel and attract, and since male and female radicals exist, this unique quality can be used to create a paired radical bond only between two same or different metals such as between two non-ionic metals e.g. Al/Ti, Al/Al, Zn/Zn or between an ionic metal and a bulky anionic group e.g. Na/OCH_3 or between an

ionic metal and a non-ionic non-metallic group e.g. Na/Naphthalene, whether there is a large potential electropositive difference between the two centers or not, and a radical environment is provided only for covalent and ionic bonds.
(Laws of Creations for Paired Ionic Bonds)

Rule 1507: This rule of Chemistry for **Electrostatic bonds,** states that, unlike Ionic and Covalent bonds, these bonds in reality (Real world) do not exist, because they are charges carried by two centers in which for positively charged electrostatic bond a vacant orbital is used to form a bond on the imaginary side and for negatively charged electrostatic bond, the presence of paired unbonded radicals is used to form a bond on the imaginary side.
(Laws of Creations for Polar and Electrostatic bonds)

Rule 1508: This rule of Chemistry for **Polar bonds,** states that, there are *two types of polar bonds-* Type (I) those formed between two metals or a metal and hetero atom(s) and Type (II) those based on boundary laws, both of which are complex with real and imaginary side; for which for both of them, the imaginary side is the positive center- *a hole for both of them*.
(Laws of Creations for Polar bonds)

Rule 1509: This rule of Chemistry for **Polar, Electrostatic and Ionic positive charges,** states that, when Ionic positive charge (Cation) is created, a HOLE is left behind and the Hole is Real, while when Electrostatic (for negatively charged ones) positive charges and Polar positive charges (of Types (I) and (II)) are formed, a HOLE is left behind, while with some of those of Type (II) which carry paired unbonded radicals, these paired unbonded radicals can be used to send a signal electrically through for example a chain of SO_2 as shown below-

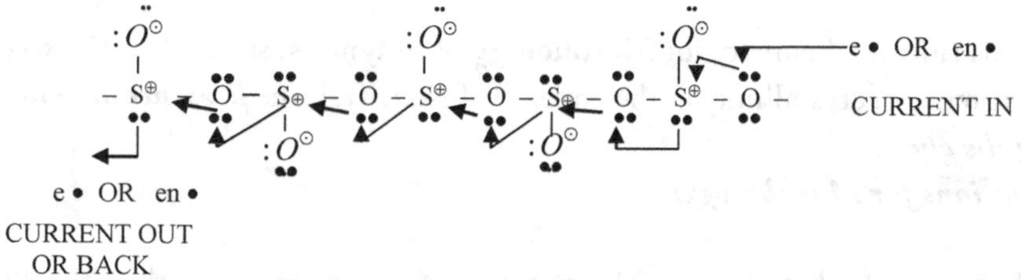

wherein as the current goes in, the electro-radical at the receiving end is released, followed by flow of radicals along the chain from hole to hole; a feat that can be done by oxygen.
(Laws of Creations for Polar, Electrostatic and Ionic Charges)

Rule 1510: This rule of Chemistry for **Stability of Bonds,** states that, only Paired Initiators with paired bonds are unstable *(that is, Elastic)* in character and these are some specific Ionic bonds, some s type of Covalent bonds (e.g., those between metals and C, Transition metals and Non-Transition Non-ionic metals), all Electrostatic "bonds" and all radical bonds; while Polar bonds are stable because of the presence of s bonds.
(Laws of Creations for Stability of Bonds)

Rule 1511: This rule of Chemistry for **Optical Activity in compounds,** states that, for a compound to be optically active, it must have *an asymmetric C center* which must carry H, and either *all three radical-pulling groups or all three radical-pushing groups or mixture of the last two for the three*

remaining groups, all arranged in an increasing order of radical-pushing capacities around the C center beginning with radical-pulling group in decreasing capacity if more than one, then followed by H and then followed by **radical-pushing groups** in increasing capacity; and they must be such that are dissymmetric in character (i.e., their mirror images are not supperposable)-a funny cross on the wheel.
(Laws of Creations for Optical Activity)

Rule 1512: This rule of Chemistry for **Morphology of polymers**, states that, since it is in reality impossible to have polymers that is hundred percent *amorphous* and hundred percent *crystalline* in general, all polymers are **visco-elastic in character** with two temperature transition domains -Glass-transition temperature (T_g) and Melting temperature (T_m) in the liquid–solid zone.
(Laws of Creations for Morphology)

Rule 1513: This rule of Chemistry for **Morphology of polymers**, states that, while the crystalline melting point (T_m) is the *melting temperature of the crystalline domain of the polymer,* the glass transition temperature (T_g) is that temperature *at which the amorphous domain of a polymer takes on the characteristic properties of the glassy state (brittleness, stiffness, and rigidity);* for which for any family of polymers, no specific values exist, but ranges of values, depending on the molecular weight averages and distribution of the polymers.
(Laws of Creations for Morphology)

Rule 1514: This rule of Chemistry for **Morphology of polymers,** states that, the *first driving force* favoring the existence of crystallinity in polymers, is *presence of tacticity or stereoregular placement* along a polymer main chain backbone.
(Laws of Creations for Morphology)

Rule 1515: This rule of Chemistry for **Morphology of polymers,** states that, the *second driving force* favoring the existence of crystallinity in the presence of tacticity, is the *presence of polar species regularly placed along the chain.*
(Laws of Creations for Morphology)

Rule 1516: This rule of Chemistry for **Morphology of polymers,** states that, the *third driving force* favoring the existence of crystallinity is the existence of secondary forces manifested by the presence of what *is herein called Ionic Electropositive/Electronegative forces of attraction* between an ionic metal such as H and a polar species (Not hydrogen bonding which does not exist) in or along the polymer main chain backbones, that which is *invisible and imaginary*.
(Laws of Creations for Morphology)

Rule 1517: This rule of Chemistry for **Morphology of polymers,** states that, the *fourth driving force* favoring the existence of crystallinity in polymers in the presence of tacticity, is the *presence of linear arrangements along the main chain backbone and absence of irregular branching or crosslinking.*
(Laws of Creations for Morphology)

Rule 1518: This rule of Chemistry for **Morphology of polymers,** states that, the *fifth driving force* favoring the existence of crystallinity in polymers in the presence of tacticity, is the *presence of free-rotational, translational secondary forces and vibrational motions of the polymer main chain backbones.*
(Laws of Creations for Morphology)

Rule 1519: This rule of Chemistry for **Morphology of polymer**, states that, the *sixth driving force* favoring the existence of crystallinity in polymers in the presence of tacticity, is *absence of specific bulky groups along the chain,* which readily favor more *conformational instability* than desired.
(Laws of Creations for Morphology)

Rule 1520: This rule of Chemistry for **Morphology of polymers**, states that, the *seventh driving force* favoring the existence of crystallinity in polymers, in the presence of tacticity, is *absence of impurities embedded inside the polymeric chains and or absence of "living" polymers* (that is polymers with loose chain ends).
(Laws of Creations for Morphology)

Rule 1521: This rule of Chemistry for **Morphology of polymers**, states that, the *eighth driving force* favoring the existence of crystallinity in polymers in the presence of tacticity, is *the reduced polydispersity of the polymer;* for which the larger the polydispersity, the lesser the crystallinity.
(Laws of Creations for Morphology)

Rule 1522: This rule of Chemistry for **Morphology for polymers**, states that, while T_m and T_g of polymers can be increased to different levels in the same directions with presence of chain entanglement forces, stiffening groups in the polymer backbone, strength of secondary forces or bonds and primary bonds, size of chain segments, T_m *can still be increased while* T_g *is decreased when there is tacticity;* for which the major driving force for existence of *amorphous character* in polymers is *atacticity.*
(Laws of Creations for Morphology)

Rule 1523: This rule of Chemistry for **Morphology of polymers**, states that, between the crystalline domain (T_m) and the amorphous domain (T_g) of a polymer, is the *semi-crystalline domain* a visco-elastic liquid favoring the presence of only *vibrational or micro-Brownian motions of the polymer main chain backbones and side chains,* and this shows up when a polymer is quenched or heated.
(Laws of Creations for Morphology)

Rule 1524: This rule of Addition for **Morphology of polymers**, states that, while for chemical compounds the freezing point (T_f) and melting point (T_m) are marked by the same transition point, for polymers the glass transition temperature (T_g) (which is analogous to T_f for compounds), and melting point (T_m) are marked by two different transition points.
(Laws of Creations for Morphology)

Rule 1525: This rule of Chemistry for **Morphology of polymers**, states that, the softening temperature (T_s) of a polymer, is *the temperature at which the polymer begins to flow when heat or pressure is applied;* and is the T_g of the polymer if the polymer is hundred percent amorphous or T_m if hundred percent crystalline or a range for semi - crystalline polymers not equal to T_m or T_g.
(Laws of Creations for Morphology)

Rule 1526: This rule of Chemistry for **Morphology of polymers**, states that, while T_f or T_m for compounds can be identified with T_m for semi - crystalline polymers, T_f or T_m of compounds cannot be identified with T_g for highly amorphous polymers.
(Laws of Creations for Morphology)

Rule 1527: This rule of Chemistry for **Functional groups,** states that, when a monomer or compound contains one functional group, it is said to be <u>monofunctional</u>; when it contains two functional groups of the same type, it is <u>di-monofunctional</u>; when it contains three functional groups of the same type, it is <u>tri-monofunctional</u>; when it contains two functional groups of different types, it is <u>di-bifunctional</u> and so on. *(Laws of Creations for Functional groups)*

Rule 1528: This rule of Chemistry for **Solubilization and Insolubilizations,** states that, while Solubilization *is a chemical reaction* between two or more components to give *different products* all held in Equilibrium state of existence, making the **reaction non-productive,** Insolubilization *is a chemical reaction* between two or more components to give *the same two or more different components all* held in Equilibrium state of existence, making the reaction non-productive; for example while sodium chloride is soluble in water (the HCl and NaOH formed cannot be seen), sodium hydroxide is insoluble in water, yet they all belong to the same family (Polar/Ionic) and dissolve in water. *(Laws of Creations for Solubilization and Insolubilization)*

Rule 1529: This rule of Chemistry for **Dissolution/Miscibilization,** states that, while Dissolution is a Solid/Liquid or Solid/Gas phase phenomenon, that involving breakdown of solid in a fluid to give a homogeneous solution, Miscibilization is a Liquid/Liquid or Liquid/Gas or Gas/Gas phase phenomenon, that involving breakdown of liquid in liquid or gas to give a homogeneous solution. *(Laws of Creations for Dissolution/Miscibilization)*

Rule 1530: This rule of Chemistry for **Dissolution/Miscibilization,** states that, this pheno-menon (which has been said to take place only when the two or more compounds involved belong to the same family for the New Classification for Compounds) *holds whether there is reaction or no reaction between them;* for example methanol and polyacrylamide belong to the same family (Polar/Ionic), but reactive with themselves and also hydrochloric acid and sodium hydroxide belong to the same family (Polar/Ionic), but reactive with themselves, both reactive to give different products not held in Equilibrium (A productive stage). *(Laws of Creations for Dissolution/Miscibilization)*

Rule 1531: This rule of Chemistry for **Sobilization/Dissolution of compounds, monomers and polymers,** states that, when they can dissolve or miscibilize in a solvent, does not mean that they can solubilize (where unproductive reactions take place) in the same solvent; for example sodium chloride dissolves in water and is also soluble in water, while acrylamide dissolves in water, but insoluble in water, so also sodium hydroxide dissolves in water, but insoluble in water. *(Law of Creations for Solubility/Dissolution)*

Rule 1532: This rule of Chemistry for **Solvents, Monomers, and Polymers,** states that, when they are all in Stable state of Existence, they cannot solubilize and insolubilize with themselves, but only dissolve/miscibilize or not dissolve/immiscibilize with themselves. *(Laws of Creations for Solvent, Monomers and Polymers)*

Rule 1533: This rule of Chemistry for **Monomers and Corresponding Polymers wherein one or both can exist in Equilibrium state of existence,** states that, while a polymer can dissolve or miscibilize in its monomer and or can solubilize or insolubilize in its monomer, it cannot chemically react productively with it, other than during polymerization when living. *(Laws of Creations for Monomers and its Polymer)*

Rule 1534: This rule of Chemistry for **Monomers and Solvents,** states that, a solvent is said to be a solvent to a monomer, *only if the monomer dissolves or miscibilizes in the solvent,* and not when it is reactive productively with it; for if it does, then the solvent is a *Non-solvent,* for example, water is a solvent for sodium chloride and sodium chloride is soluble in water, while water is a non-solvent for AlR_3, since they belong to two extreme families.
(Laws of Creations for Monomers and Solvents)

Rule 1535: This rule of Chemistry for **Polymers and Solvent,** states that, a solvent is said to be a solvent to a polymer, *only if the polymer dissolves or miscibilizes or is swollen in the solvent and reactive non-productively with it,* and not when it is reactive productively; for if it does, then the solvent is a Non-solvent, for example, *methanol is not a solvent but a non-solvent for polyacrylamide in solution (water),* because it is productively reactive with it to give methylated polyacrylamide, yet they all belong to the same family (Polar/Ionic) which makes it possible for polyacrylamide to dissolve in methanol just as it dissolves in water.
(Laws of creations for Polymers and Solvents)

Rule 1536: This rule of Chemistry for **Dissolution/Miscibility of compounds, monomers, polymers,** states that, the "*Solubility" Parameter Law* is that in which for two or more components to be dissolvable/miscible, they must have compatible "*solubility" parameters* based on ionic/non-ionic and polar/non-polar characters; for which between three compo-nents A, B and C, if A is compatible with B and B is compatible with C, then A must be compatible with C and compatibility is a measure of the polar and ionic (or di-electric) character of the compound.
(Laws of Creations for Solubility Parameter Law)

Rule 1537: This rule of Chemistry for **Polymerization of Monomers,** states that, the limiting conversion (X_{Lim}) at a particular temperature, is the maximum conversion which can be attained at temperatures below the glass transition temperatures of the polymer (T_g); for which before any treatment, the swollen polymers obtained contain the exact amount of unreactive monomer required to attain hundred percent conversion if polymerization temperatures had been at the T_g of polymer.
(Laws of Creations for Limiting Conversion)

Rule 1538: This rule of Chemistry for **Polymerization of Monomers,** states that, at the limiting conversion, the swollen polymer obtained is that in which the monomer inside is not free under isothermal conditions, because they are embedded in *the polymer sub-particles* and not in the *interstices.*
(Laws of Creations for Polymer Sub-particles during Polymerizations)

Rule 1539: This rule of Chemistry for **Classification of Polymerization Processes,** states that, a process is Single-phase homogeneous, when only one phase exists and one polymerization route is taking place only in that phase; single-phase heterogeneous, when only one phase exists and two polymerization routes are taking place only in that phase (Not common).
(Laws of Creations for Single-Phase-Polymerization)

Rule 1540: This rule of Chemistry for **Classification of Polymerization Processes,** states that, a process is Bulk single-phase, if the major components present are the monomer, initiator and polymer; for which high polymerization $(\geq T_g)$ temperatures are involved, and the phase largely present, is called *Monomer/Polymer-rich phase.*
(Laws of Creations for Bulk Single-Phase-Polymerization)

Rule 1541: This rule of Chemistry for **Classification of Polymerization Processes**, states that, a process is Bulk hetero-phase, if the major components present are the monomer, initiator and polymer; for which the polymerization temperature is *below* T_g of the polymer, and the two phases present if limiting conversion has not been reached, are called the *Monomer/Polymer rich and Monomer rich-phases* *(Laws of Creations for Bulk Hetero-Phase-Polymerization)*

Rule 1542: This rule of Chemistry for **Classification of Polymerization Processes,** states that, a process is Solution single-phase, if the major components present are the monomer, solvent, initiator and polymer; for which high polymerization $(\geq T_g)$ temperatures are involved and the phase largely present is called *Monomer/Solvent/Polymer-rich phase.* *(Laws of Creations for Solution Single-Phase-Polymerization)*

Rule 1543: This rule of Chemistry for **Classification of Polymerization Processes,** states that, a process is Solution hetero-phase, if the major components present are the monomer, solvent, initiator and polymer; for which monomer and solvent are *perfectly miscible* and the polymerization temperature is *below* T_g of polymer and the two phases present are called the *Monomer/Solvent/Polymer-rich and Monomer/ Solvent rich-phases.* *(Laws of Creations for Solution Hetero-Phase-Polymerization)*

Rule 1544: This rule of Chemistry for **Classification of Polymerization Processes,** states that, a process is Block hetero-phase, if components of Solution process are present; for which the monomer and solvent are *partially miscible* and therefore the two phases largely present are called the *Monomer/Polymer-rich (Bulk features) and Monomer/Solvent/ Polymer-rich (Solution features) phases.* *(Laws of Creations for Block Hetero-Phase Polymerization)*

Rule 1545: This rule of Chemistry for **Classification for Polymerization Processes,** states that, a process is Hetero-phase homogeneous, when more than the one phase is present and polymerization is taking place only in one phase; hetero-phase heterogeneous, when more than one phase is present and polymerization is taking place in more than one phase. *(Laws of Creations for Hetero-Phase-Homo- & Heterogeneous)*

Rule 1546: This rule of Chemistry for **Classification of Polymerization Processes,** states that, a process is Hetero-phase homogeneous interfacial Solution type, when two phases are present and polymerization is taking place at the *interface* in solution via Step polymeri-zation mechanism; for which *two immiscible miscible monomer/solvent mixtures* form the two separate phases and the polymer formed does not dissolve in the mixtures when removed continuously. *(Laws of Creations for Hetero-phase Homogeneous Interfacial Solution Polymerization)*

Rule 1547: This rule of Chemistry for **Classification of Polymerization Processes,** states that, a Heterogeneous hetero-phase Solution process is that in which two routes are taking place in one phase in a system where there are two phases or zones, when the monomers and solvents are miscible. (Not common) *(Laws of Creations for Hetero-Phase Heterogeneous Solution Polymerization)*

Rule 1548: This rule of Chemistry for **Suspension and Emulsion Polymerization process-es,** states that, these hetero-phase processes are essentially *micro- reactors suspended in a macro - reactor.* *(Laws of Creations for Micro- and Macro-Chemical Reactor System)*

Rule 1549: This rule of Chemistry for **Suspension Polymerization Systems,** states that, while an Ideal Suspension polymerization system is that which is used for the polymerization of organic-soluble polymers, an Ideal Inverse Suspension Polymerization System is that which is used for the polymerization of water-soluble polymers.
(Laws of Creations for Suspension Polymerization Systems)

Rule 1550: This rule of Chemistry for **Emulsion polymerization systems,** states that, while an Ideal Emulsion polymerization system is that which is used for the polymerization of organic-soluble polymers, an Ideal Inverse Emulsion polymerization system is that which is used for the polymerization of water-soluble polymers.
(Laws of Creations for Emulsion Polymerization Systems)

Rule 1551: This rule of Chemistry for **Suspension polymerization systems,** states that, an Ideal Suspension Polymerization System is that in which the micro-reactors which are the reservoirs for the monomer (in droplets) miscible or dissolvable with the initiator and the polymer particles, are suspended in *an aqueous dispersion medium (Polar/Ionic)* containing the suspending agent which is viscous in character and other additives; for which the initiator must be *immiscible or not dissolvable* in water.
(Laws of Creations for Ideal Suspension Polymerization Systems)

Rule 1552: This rule of Chemistry for **Suspension polymerization systems,** states that, an Ideal Inverse Suspension Polymerization System is that in which the micro-reactors which are the reservoirs for the monomer (in droplets) miscible or dissolvable with the initiator and the polymer particles, are suspended in *an organic dispersion medium (Polar/Non-ionic)* containing the suspending agents which is viscous in character and other additives; for which the initiator must be *immiscible or not dissolvable* with the organic phase.
(Laws of Creations for Ideal Inverse Suspension Polymerization Systems)

Rule 1553: This rule of Chemistry for **Emulsion polymerization systems,** states that, an Ideal Emulsion Polymerization System is that in which the micro-reactors are *single organic emulsified polymer particles* surrounded by another phase of monomers or monomer/its solvent, suspended in *an aqueous phase* in which only the *initiator and emulsifiers are miscible or dissolvable with*, but not the monomer or the polymer.
(Laws of Creations for Ideal Emulsion Polymerization Systems)

Rule 1554: This rule of Chemistry for **Emulsion polymerization systems,** states that, an Ideal Inverse Emulsion Polymerization System is that in which the micro-reactors are *single aqueous emulsified polymer particles* surrounded by another phase of monomer or monomer/its solvent, suspended in **an organic phase** in which the *initiator and emulsifiers are miscible or dissolvable with,* but not the monomer or the polymer.
(Laws of Creations for Ideal Inverse Emulsion Polymerization Systems)

Rule 1555: This rule of Chemistry for **Suspension and Emulsion polymerization systems**, states that, the two systems can be placed side by side in a process, after favoring the conditions for their existences, *provided two different initiators are involved for the separate mini-reactors.*
(Laws of Creations for Suspension and Emulsion Polymerization Systems)

Rule 1556: This rule of Chemistry for **Emulsion polymerization systems,** states that, the driving force favoring the existence of non-ideality, is *the partial miscibility* (less than 5 - 10%) of the monomer in the aqueous phase; for which the existence of *Homogeneous heterophase Solution/Emulsion polymerization system or Homogeneous single-phase Bulk/ Emulsion polymerization system* is favored.
(Laws of Creation for Non-Ideal Emulsion Polymerization Systems)

Rule 1557: This rule of Chemistry for **Equilibrium between phases during Polymeri-zation,** states that, Equilibrium can only take place between two or more phases only when there is polymerization taking place in at least one phase, for which there must be mass transfer of monomers and some components if any between two or more phases during the course of polymerization, to the point of which when mass transfer ceases, **the system is said to be in Equilibrium.**
(Laws of Creations for Equilibrium between Phases during Polymerization)

Rule 1558: This rule of Chemistry for **Suspension and Emulsion polymerization systems,** states that, while many Initiation steps take place in the micro-reactor in Suspension polymerization systems, only one Initiation step takes place outside instantaneously followed by emulsification to form the micro-reactor in Emulsion polymerization systems; for which in the second stage of polymerization in Emulsion polymerization systems, it is the monomers that diffuse into the growing polymer particle by Mass transfer in the micro - reactor, in favor of achieving Equilibrium between the two phases as opposed to diffusion controlled mechanisms where the active center diffuses to add to the activated monomer.
(Laws of Creations for Emulsion Systems)

Rule 1559: This rule of Chemistry for **Suspension and Emulsion polymerization systems,** states that, in the Ideal systems, the particle size of the polymer particles obtained is far smaller in Emulsion systems than in Suspension systems, since in Emulsion systems, only one polymer particle is involved in the micro-reactors <u>with no room for flocculation and cementing</u> of swollen polymers sub-particles, while in Suspension systems, many polymer particles are involved in the micro-reactors with favored existence of <u>flocculation and cementing of swollen polymer sub-particles</u> inside the micro-reactors.
(Laws of Creations for Particle Size in Suspension and Emulsion Systems)

Rule 1560: This rule of Addition for **Hetero-phase homogeneous/heterogeneous Bulk or Solution or Block systems and Suspension polymerization systems,** states that, the sizes of the *polymer particles* obtained in the latter systems is *far smaller* than those obtained in the former systems, since in the former, there are no micro-reactors, but a macro-reactor with room for favored existence of larger particles via <u>aggregation</u> and <u>strong cementing</u> of swollen polymer sub-particles.
(Laws of Creations for Particle Size in Bulk, Solution, Block, and Suspension Systems)

Rule 1561: This rule of Chemistry for **Emulsion polymerization systems,** states that, what distinguishes it from other systems is that while with others, monomers with or without radical transfer species of the first kind can be homopolymerized in the mini-reactor in the route natural to it, with Emulsion only monomers which favor the nucleo-free-radical route can be homopolymerized; for which during copolymerization, it unlike others, can copolymerize with a co-monomer carrying transfer species of the first kind, in the mini-reactor in the route not natural to it to give copolymers, such as propene adding to a nucleo-free-radical growing polymer chain of polyacrylamide.
(Laws of Creations for Emulsion Systems)

Rule 1562: This rule of Chemistry for **Emulsion polymerization systems,** states that, for both Ideal Emulsion and Inverse Emulsion polymerization systems, polymerizations can only take place radically, via Addition polymerization kinetics, while with Inverse, in addition to Addition polymerization, Step polymerization kinetics can also take place only radically such as in Genes preparations in Living systems.
(Laws of Creations for Emulsion Polymerization Systems)

Rule 1563: This rule of Chemistry for **Emulsion polymerization systems,** states that, Initia-tion step can take place inside the mini-reactor, *if and only if the mini-reactor is initially made to contain a "seeded monomer", that is, a single dead polymer particle with activation centers along the polymer chain, in the presence of a suitable initiator immiscible with the aqueous or organic medium-*(Grafting).
(Laws of Creations for Emulsion Systems)

Rule 1564: This rule of Chemistry for **Suspension and Emulsion polymerization systems,** states that, while the existence of the micro-reactors in Suspension system is determined by *the type and concentration of suspending agents used, the level of agitation in the system and the manner by which the ingredients are added,* the existence of micro-reactors in Emulsion systems is determined by *the type and concentration of emulsion used, concentration of initiator used and the manner by which the ingredients are added.*
(Laws of Creations for Suspension and Emulsion Polymer)

Rule 1565: This rule of Chemistry for **Configurations and Conformations of Monomers,** states that, while Configurations which are placements of groups along the Central atoms of a monomer or compound along the horizontal axis are fixed above and below the axis, Conformations of the groups due to vibrational, translational and rotational motions in the system are ever changing, below and above the horizontal axis.
(Laws of Creations for Configurations and Conformations)

Rule 1566: This rule of Chemistry for **Configurations and Placements,** states that, while Configuration applies to all chemical and polymeric compounds, Placement which is the arrangement of monomer units along a chain applies only to Polymeric systems for which there are different kinds of placements of monomer units along a chain too countless to list; noting that Placements like Configurations are fixed.
(Laws of Creations for Configurations and placements)

Rule 1567: This rule of Chemistry for **Chemical reactions between two or more com-pounds,** states that, the choice of the exact MOLAR RATIOS of the reacting components is the most important variable of great concern before desired products can be obtained when other right operating conditions are put in place.
(Laws of Creations for Chemical reactions)

Rule 1568: This rule of Chemistry for **Polymeric reactions involving the use of Initiators,** states that, the monomer involved are ADDITION monomers, for which the most important variable of great concern is the MOLAR RATIO of the Initiator with respect to the monomer (s), for which very minute concentrations of the initiator must always be used, depending on the polydispersity of the polymeric products desired.
(Laws of Creations for Polymeric reactions)

Rule 1569: This rule of Chemistry for **Initiators,** states that, there are *six types of Free-Media Initiators* and these are-

(i)   Electro-free-radical initiators- e.g. $Na.e$
(ii)   Nucleo-free-radical initiators- e.g. $H.n$
(iii)   Nucleo-non-free-radical initiators- e.g. $H_6C_5\textbf{COO}.nn$
(iv)   Electro-non-free-radical initiators- e.g. $Cl_2Fe.en$
(v)   Cationic initiators- e.g. $H^\oplus$ [which must become electro-free-radical in order to be used].
(vi)   Anionic initiators- e.g. $H_3CO^\ominus$

(Laws of Creations for Free-Media Initiator)

Rule 1570: This rule of Chemistry for **Initiators,** states that, there are *eight types of Paired- Media Initiators* and these are-

(i)   Full Free-radical-paired initiators- e.g. $Na^{.e}.......^{n.}C\equiv N$, $RClTi^e......^{n.}AlR_2$
(ii)   Half-Free-radical-paired initiators- e.g. $Na^{.e}......^{nn.}OCH_3$
(iii)   Ionically charged paired initiators- e.g. $Na^\oplus.......^\ominus OC_4H_9$
(iv)   Covalently charged paired initiators- e.g. $H_9C_4^\ominus.....^\oplus Li$, $RClTi^\oplus.....^\ominus AlR_2$
(v)   Electrostatically cationic charged-paired initiators- e.g. $Li^\oplus.....^\ominus AlH_4$
        (Only H can be used nucleo-free-radically for telomers)
(vi)   Electrostatically positively charged-paired initiators- e.g. $R^\oplus.....^\ominus BF_3(OR)$, $RClAl^\oplus.....^\ominus TiCl_5$
(vii)   Electrostatically anionically charged-paired initiators- e.g. $RO^\ominus.....^\oplus NR_3H$, $Cl^\oplus.....^\ominus N(C_4H_9)_4$, $Cl^\ominus.....^\oplus PCl_4$, $Cl^\ominus......^\oplus Cl(PCl_3)$
(viii) Electrostatically negatively charged-paired initiators- e.g.

$$H_2C^\ominus.......^\oplus N \overset{\displaystyle H}{\underset{\displaystyle \overset{|}{\underset{\displaystyle CH_3}{C=O}}}{\triangleleft}}$$

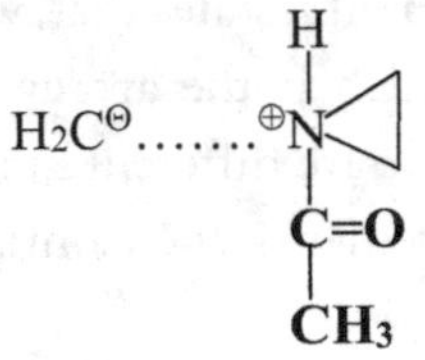

noting that some of them have two active centers, that is, they are dual in character while some have only one active center and in addition not all of them shown above have reservoirs used for stereoregular placements.
(Laws of Creations for Paired-Media Initiators)

Rule 1571: This rule of Chemistry for **Paired-Media Initiators,** states that, of all the paired types, *only some specific types of Electrostatically charged-paired initiators* can be used for Addition polymerization via Equilibrium mechanism radically in particular electro-free-radically for as long as they are made to exist in Equilibrium state of existence.
(Laws of Creations for Electrostatically charged-paired initiators)

Rule 1572: This rule of Chemistry for **Covalently charged paired initiators,** states that, though they are dual in character, not all of them can be used to polymerize all Hydrocarbon Nucleophilic monomers, such as the case with $H_9C_4{}^{\ominus}.....{}^{\oplus}Li$, since pairing cannot take place between two carbon centers as shown below-

(i) $\quad Li - CH_2 - CH_2{}^{\oplus}.....{}^{\ominus}C_4H_9 \equiv Li^{\oplus}.....{}^{\ominus}C_6H_{13} \equiv LiC_6H_{13}$

(ii) $\quad H_9C_4 - CH_2 - H_2C^{\ominus}.....{}^{\oplus}Li$

for which such initiators can only be used for the first member via the route not natural to it under harsher operating conditions (ii), because Nature abhors a vacuum when there still exist a possibility, whereas other members will not favor the route and the use of the initiator.
(Laws of Creations for Covalently charged-paired initiators)

Rule 1573: This rule of Chemistry for **Paired-Media Initiators**, states that, Pairing radically can only take place between two centers after specific condition have been satisfied only when the bond between the centers is covalent or ionic in character, for if it is electrostatic in character, based on the laws guiding the existence of these bonds, pairing cannot take place radically.
(Laws of Creations for Paired-Media Initiators)

Rule 1574: This rule of Chemistry for **Free-Media Initiators,** states that, *all radical Free-media initiators can be prepared largely via Decomposition mechanisms* with few via Equilibrium or Decomposition mechanism (e.g. NaCN), in view of the marked differences between the two mechanisms.
(Laws of Creations for Radical Free-Media initiators)

Rule 1575: This rule of Chemistry for **Paired-Media Initiators of the Electrostatic types,** states that, these are the only types where some specific ringed hetero monomers which are Polar/Ionic or Polar/Non-ionic in character can be used as part of the Initiator preparing steps to prepare the initiator before the Initiation step, for which small induction periods must be present.
(Laws of Creations for Paired-Media Initiators)

Rule 1576: This rule of Chemistry for **Paired-Media Initiators of the Z/N types,** states that these are *Covalently Charged-paired types of initiators* which are dual in character and can only be prepared between two same or different types of metallic types of compounds, only *via Decomposition and Equilibrium mechanisms if and only if the metallic centers as Central atoms are free-radical in character, otherwise before the two steps above, the non-free-radical character of the Central atoms must be made Free-radical before* Decomposi-tion mechanism can commence followed by Equilibrium mechanism.
(Laws of Creations for Z/N Initiators)

Rule 1577: This rule of Chemistry for **Paired-Media Initiators of the Z/N types,** states that, since two metallic centers are involved, and since the initiator is dual in character and since part of them are the carriers of the polymeric chains, the components involved in preparation of the Z/N initiators are NO CATALYSTS, but INITIATOR GENERATING Metallic COMPONENT COMBINATIONS.
(Laws of Creations for Z/N Initiators)

Rule 1578: This rule of Chemistry for **Paired-Media Initiators of the Z/N types,** states that, since two metallic centers are involved, any of the metallic centers can be the Carrier of the chain based on the character of the monomer, for which the metallic component carrying the chain is not called the

CATALYST (when the non-carrier is called the CO-CATALYST), but herein called the **Z/N CHAIN CARRIER**.
(Laws of Creations for Z/N Initiators).

Rule 1579: This rule of Chemistry for **Paired Media Initiators of the Z/N types,** states that, since the two initiator generating metallic components provide the carriers of the chain, a Combination such as H_2O/AlR_3, does not belong to Z/N family, because the initiator obtained from the combination $(R_2Al^{\oplus}.....^{\ominus}AlR_2)$ is that in which Al is the carrier of the chains for both types of monomers, clear indication that Z/N initiators are based on metals which are Non-ionic in character.
(Laws of Creations for Z/N Initiators)

Rule 1580: This rule of Chemistry for **Paired-Media Initiators of the Z/N types**, states that, the essence of Z/N initiators is **ALKYLATION,** that is, partial or full replacement of halogenated radical-pulling groups on a metallic component with radical-pushing alkane groups from the second metallic component.
(Laws of Creations for Z/N Initiators)

Rule 1581: This rule of Chemistry for **Paired-Media Initiators of the Z/N types,** states that, when the two metallic Central atoms of the two components are equi-electropositive, and alkylation is made to take place, it can only be *Limited,* that is partial alkylation, for which *Full alkylation* takes place only when the two metals have different electropositivities and exact molar ratios are used.
(Laws of Creations for Limited and Full Alkylation)

Rule 1582: This rule of Chemistry for **Paired-Media Initiators of the Z/N types,** states that, during *Decomposition mechanism between two metallic components*, it is the more electro-positive metallic component that commences **alkylation;** for if they are equi-electropositive, it is the more concentrated component that commences the alkylation.
(Laws of Creations for Z/N Initiators)

Rule 1583: This rule of Chemistry for **Paired-Media Initiators of the Z/N types,** states that, when the MOLAR RATIOS of the two metallic components *are the same, no Z/N initiator can be obtained;* for which to obtain Z/N initiator, the Molar ratio of the more electropositive metallic component to the less electropositive metallic component must be greater than one or indeed two if full conversion must be obtained.
(Laws of Creations for Z/N Initiators)

Rule 1584: This rule of Chemistry for **Paired-Media Initiators,** states that, when the Molar ratio of the more electro-positive metallic component to the less electro-positive metallic component *is less than one, no Z/N initiator can be obtained;* instead an Electrostatically charged-paired initiator with only one active center is obtained.
(Laws of Creations for Z/N Initiators)

Rule 1585: This rule of Chemistry for **Initiators,** states that, while few initiators exist on their own without the need for preparation, most initiators must first be prepared before polymerization can take place; for which this must always be first accomplished before monomers are added, even when the monomer will be part of the initiator.
(Laws of Creations for Initiators)

Rule 1586: This rule of Chemistry for **Free-Media-Radical Initiators,** states that, when Radical initiator generating components are involved, Nucleo-radicals are in general obtained; presence of electro-radicals made possible only when the Central atom in one of the components is a Transition metal such as Iron (Fe), for example Persulfate/Ferrous salt redox system, $FeCl_2$ or $FeCl_3$/HOOH system or combination, both of which provide H.e or the Central atom is an ionic metal such as in NaCN.
(Laws of Creations for Radical Initiators)

Rule 1587: This rule of Chemistry for **Radical Generating Initiator Components,** states that, these are too countless to list, few examples of most popularly known ones are many peroxides such as Benzoyl peroxide [n.C_6H_5], many azo-alkanes such as so-called Azo-methane [n.CH_3], Aromatic diazonium compounds and Azonitriles such as 2-Azo-bis-isobutyronitrile; all of which give nucleo-radicals, for which shown below is the mechanism for the decomposition of the last two cases-

For an Aromatic diazo compound

Electroradicalization

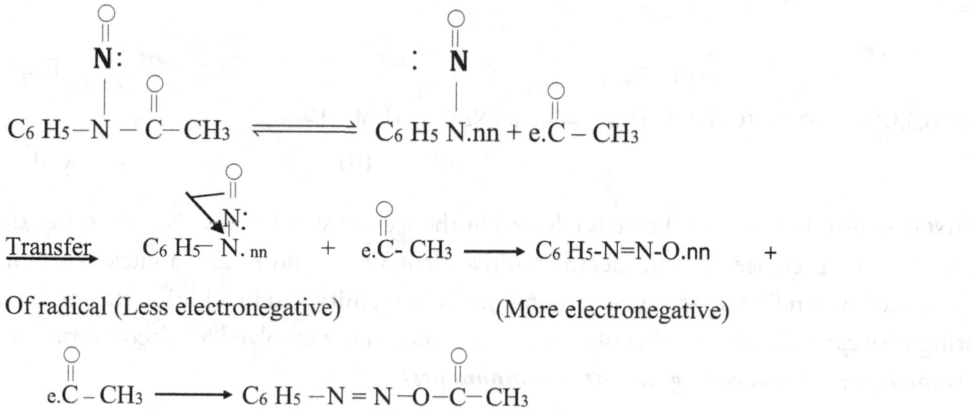

$$2C_6H_5-N=N\{O-\overset{O}{\underset{||}{C}}-CH_3\}$$

$$C_6H_5-\overset{\overset{O}{||}}{\underset{|}{N:}} - \overset{O}{\underset{||}{C}} - CH_3 \rightleftharpoons C_6H_5\ N.nn + e.\overset{O}{\underset{||}{C}} - CH_3$$

Transfer $\quad C_6H_5-\overset{\overset{O}{||}}{N}.nn \quad + \quad e.\overset{O}{\underset{||}{C}}- CH_3 \longrightarrow C_6H_5-N=N-O.nn \quad +$

Of radical (Less electronegative) $\qquad$ (More electronegative)

$$e.\overset{O}{\underset{||}{C}} - CH_3 \longrightarrow C_6H_5 -N = N-O-\overset{O}{\underset{||}{C}}-CH_3 \tag{i}$$

Decomposition mechanism

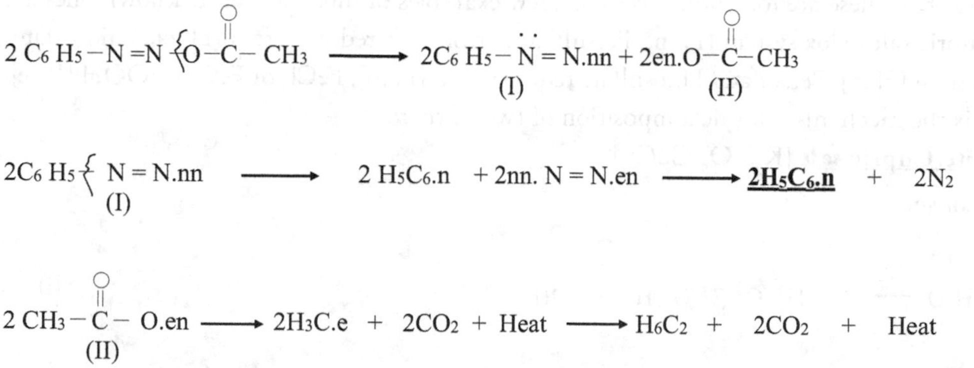

$$2\ C_6H_5-N=N\{O-\overset{O}{\underset{||}{C}}-CH_3 \longrightarrow 2C_6H_5-\overset{..}{N}=N.nn + 2en.O-\overset{O}{\underset{||}{C}}-CH_3$$
$$\qquad\qquad\qquad\qquad\qquad (I)\qquad\qquad\qquad (II) \tag{ii}$$

$$2C_6H_5\{N=N.nn \longrightarrow 2\ H_5C_6.n\ +2nn.\ N=N.en \longrightarrow \mathbf{2H_5C_6.n} \quad + \quad 2N_2$$
$$\qquad (I) \tag{iii}$$

$$2\ CH_3-\overset{O}{\underset{||}{C}}-O.en \longrightarrow 2H_3C.e\ +\ 2CO_2\ +\ Heat \longrightarrow H_6C_2\ +\ 2CO_2\ +\ Heat$$
$$\quad (II) \tag{iv}$$

Overall equation: $2H_5C_6(N=O)N(CO)CH_3 \longrightarrow 2N_2\ +\ 2CO_2\ +\ C_2H_6\ +$
$$\mathbf{2n.C_6H_5}\ +\ Heat \tag{v}$$

For 2-Azo-bis-isobutyronitrile

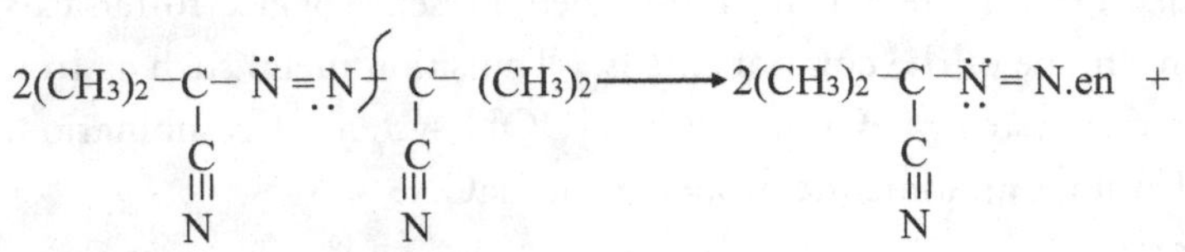

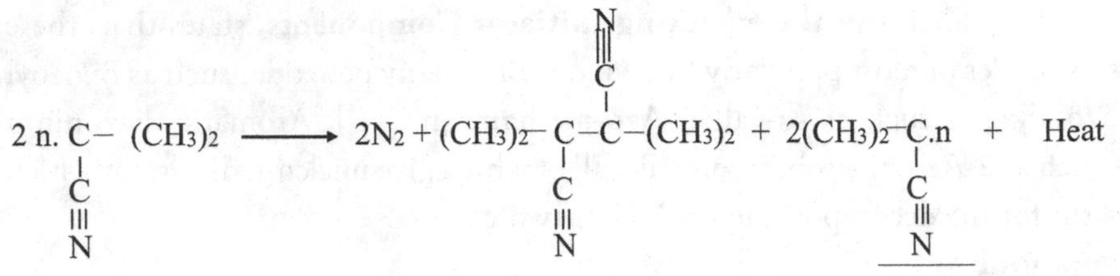

$$2(CH_3)_2-C-\ddot{N}=N+C-(CH_3)_2 \longrightarrow 2(CH_3)_2-C-\ddot{N}=N.en +$$

(with $\underset{N}{\overset{C}{\underset{|||}{|}}}$ groups shown below the carbons)

$$2 \; n.C-(CH_3)_2 \longrightarrow 2N_2 + (CH_3)_2-C-C-(CH_3)_2 + 2(CH_3)_2-C.n + \text{Heat}$$

(I)   (II)   (III)   (i)

$$(III) \; \rightleftharpoons \; 2\underline{(CH_3)_2}-C=C=N.nn$$

(IV)   (ii)

Overall equation: $2(CH_3)_2C(C\equiv N)-N=N-(C\equiv N)C(CH_3)_2 \longrightarrow 2N_2 + \text{Heat} +$

(III) + (II)   (iii)

for which like the first case heat in the second case is released in the second step because N is carrying an electro-non-free-radical in the presence of more electro-positive elements-C and H and a nucleo-free-or non-free-radical is produced depending on the type of monomer in its vicinity and in addition the product (II) is obtained, noting very carefully the number of moles of the component involved for decomposition.
(Laws of Creations for Radical generating initiator components)

Rule 1588: This rule of Chemistry for **Radical Generating Initiator compounds of the so-called Redox types,** states that, these are too countless to list, few examples of most popularly known ones are Metabisulphite/Cupric salt redox system [H .n], Persulfate/Ferrous salt redox system [H .e], Thiosulfate/cupric salt redox system [H .n], Persulfate/Thiosulfate redox system [H .n], $FeCl_2$ or $FeCl_3$/HOOH [H.e], and shown below is the mechanism for decomposition of two of them.

For Metabisulphite/Cupric salt $[K_2S_2O_5/CuCl_2]$

Equilibrium mechanism

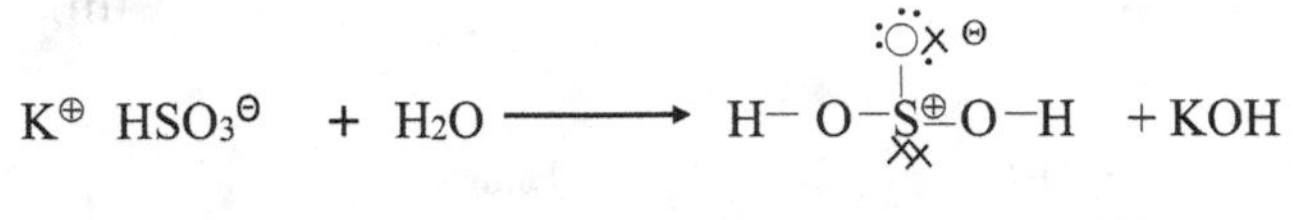

$$K^{\oplus} HSO_3^{\ominus} + H_2O \longrightarrow H-O-\overset{\overset{\ddot{O}x^{\ominus}}{|}}{\underset{xx}{S^{\oplus}}}-O-H + KOH$$

(i)

$$2H-O-\overset{\overset{\ddot{O}x^{\ominus}}{|}}{\underset{xx}{S^{\oplus}}}-O-H + {}^{4}\square\overset{\overset{3d\;Cl}{|}}{\underset{Cl}{Cu}}\overset{\square_{en}}{\diagdown} \longrightarrow HO-\overset{\overset{O^{\ominus}}{|}}{S^{\oplus}}-O-Cu-O-\overset{\overset{O^{\ominus}}{|}}{S^{\oplus}}-OH$$

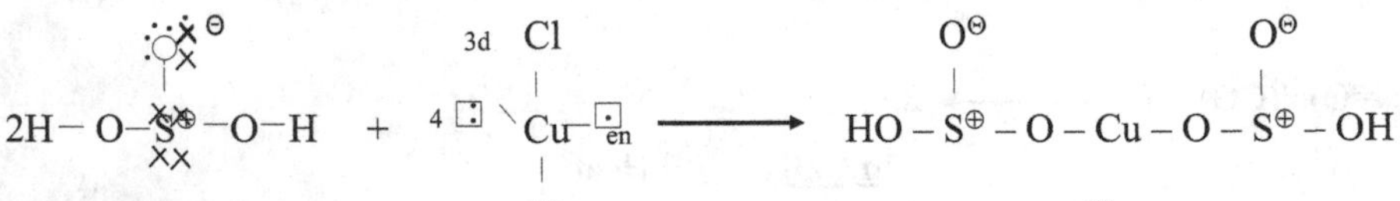

(I)

$$+ \; 2HCl$$

(ii)

$$(I) \longrightarrow HO - \overset{\overset{O^{\ominus}}{|}}{S^{\oplus}} - O - \overset{\overset{O^{\oplus}}{|}}{S^{\oplus}} - OH \quad + \quad Cu^{\oplus} - {}^{\ominus}O \qquad (iii)$$

$$CuO \quad + \quad 2HCl \longrightarrow CuCl_2 \quad + \quad H_2O \qquad (iv)$$

for which after the first hydrolysis of $K_2S_2O_5$, the followings are obtained-

Decomposition mechanism

$$2H.O - \overset{\overset{..\overset{x\ominus}{x}}{|}}{S^{\oplus}} - \overset{..}{\underset{..}{O}} - \overset{\overset{:\overset{.\ominus}{x}}{|}}{S^{\oplus}} - \overset{..}{O}H \longrightarrow 2HO - \overset{\overset{O^{\ominus}}{|}}{S^{\oplus}} - O.en \quad + \quad 2nn.\overset{\overset{O^{\ominus}}{|}}{S^{\oplus}} - OH \cdot$$

$$(III) \qquad\qquad\qquad (IV)$$

$$(III) \longrightarrow 2SO_2 \quad + \quad HOOH \; ; \quad (IV) \longrightarrow 2SO_2 \quad + \quad \underline{\textbf{2H.n}}$$

$$+ \quad Heat \qquad\qquad (v)$$

Overall equation: $4KHSO_3 + 4H_2O + 2CuCl_2 \longrightarrow 4SO_2 + 2H_2O +$

$$Heat \quad + \quad 4KOH \quad + \quad 2CuCl_2 \quad + \quad HOOH \quad + \quad \underline{\textbf{2H}\bullet\textbf{n}} \qquad (vi)$$

for which it can be observed that two nucleo-free-radicals are produced in the process along with other products (HOOH, SO_2, KOH, and H_2O) involving five stages.

For Thiosulfate/cupric salt redox system ($Na_2S_2O_3/CuCl_2$)

In an aqueous system, the followings are obtained-

Equilibrium mechanism

$$Na_2S_2O_3 \quad + \quad H_2O \longrightarrow H_2S_2O_3 \quad + \quad KOH \qquad (i) \qquad (i)$$

$$\overset{\overset{Cl}{|}}{\underset{\underset{Cl}{|}}{Cu}} + 2 H - O - S - O - S - O - H \longrightarrow HO\text{-}S\text{-}O\text{-}S\text{-}O\text{-}Cu\text{-}O\text{-}S\text{-}O\text{-}S\text{-}OH + 2HCl$$

$$(I) \qquad\qquad (ii) \qquad\qquad (ii)$$

$$(I) \longrightarrow HO\text{-}S\text{-}O\text{-}S\text{-}O\text{-}S\text{-}OH \quad + \quad Cu^{\oplus} - {}^{\ominus}O \qquad (iii)$$

$$(II)$$

$$2HCl \quad + \quad CuO \longrightarrow CuCl_2 \quad + \quad H_2O \qquad (iv)$$

Decomposition mechanism

$$(II) \longrightarrow HO\text{-}S\text{-}O.en \quad + \quad nn.S\text{-}O\text{-}S\text{-}O\text{-}S\text{-}OH$$

$$2HO\text{-}S\text{-}O.en \longrightarrow HOOH \quad + \quad 2S^{\oplus} - {}^{\ominus}O$$

$$2HO\text{-}S\text{-}O\text{-}S\text{-}O\text{-}S.nn \longrightarrow 2H.n \quad + \quad 6S^{\oplus} - {}^{\ominus}O \qquad (v)$$

Overall equation: $2CuCl_2 + 4H_2S_2O_3 \longrightarrow 2CuCl_2 + HOOH + 8SO + 2H_2O$

$$+ \quad \underline{\textbf{2H.n}} \quad + \quad Heat \qquad (vi)$$

noting that so-called redox reactions are decomposition reactions via Decomposition mechanisms
(Laws of Creations for Radical generating Initiator components)

Rule 1589 This rule of Chemistry for **Metals,** states that, based on universal data and the Group to which they belong to in the Periodic Table, unlike what is presently known, metals can be *reclassified* as-

(a) *Ionic Non-Transition metals (Groups 1A and IIA),*
(b) *Ionic Transition metals (Group IIIA),*
(c) *Non-ionic Transition metals (Groups IVA, VA, VIA and VIIA),*
(d) *Non-ionic **Transition-transition** metals (Groups VIIIA), and*
(e) *Non-ionic Non-Transition metals (Groups IB, IIB, IIIB and some of Groups IVB, VB, VIB, and VIIB).*

for which all the metals (gas, liquid and solid) are uniquely different based on their Radical Configurations (Called Electronic Configuration in present-day Science), noting that only few metals are Ionic in character.
(Laws of Creations for Metals)

Rule 1590: This rule of Chemistry for **Ringed Monomers,** states that, these monomers are *no STEP monomers, but ADDITION monomers,* for which just as some few non-ringed ADDITION monomers can be made to undergo Step polymerization kinetics, so also exists few Ringed ADDITION monomer which can be made to undergo Step polymerization kinetics; so also there exists few STEP monomers which can be made to undergo Addition polymerization kinetics.
(Laws of Creations for Ringed monomers)

Rule 1591: This rule of Chemistry for **Decomposition mechanism,** states that, it is impossi-ble to produce oxidizing oxygen as a product via Decomposition mechanism, for which it is impossible to obtain H.n from decomposition of HOOH alone.
(Laws of Creations for Oxidizing oxygen)

Rule 1592: This rule of Chemistry for **Hydrocarbons,** state that, when used as a source of Chemical energy, the only environmentally friendly one is *Methane the first member,* since unlike other members in the family which undergo combustion and oxidation during combustion, methane undergoes only combustion to give only CO_2 and H_2O both environmentally friendly, unlike the others; unless where carbonic acid is formed from CO_2 and water vapor (Acid Rain), that which has nothing to do with global warming.
(Laws of Creations for Hydrocarbons)

Rule 1593: This rule of Chemistry for **Benzene,** states that, when a benzene ring carries *triple or cumulenic bonds adjacently or conjugatedly placed* as shown below, the compound is very unstable-

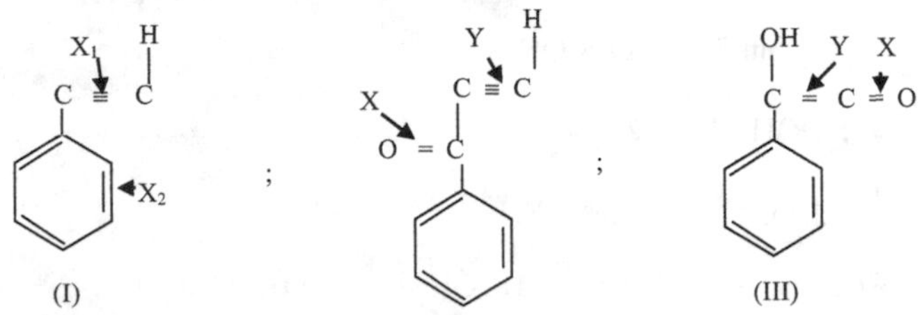

(I) is a Nucleophile, while (II) and (III) are Electrophiles, since at mild operating conditions, they readily decompose.
(Laws of Creations for Benzene)

Rule 1594: This rule of Chemistry for **an Aromatic ketene of the type** shown below, states that, being unstable, it can readily be decomposed via Equilibrium mechanism to give the products shown below in the presence or absence of oxidizing oxygen-

Stage 1:

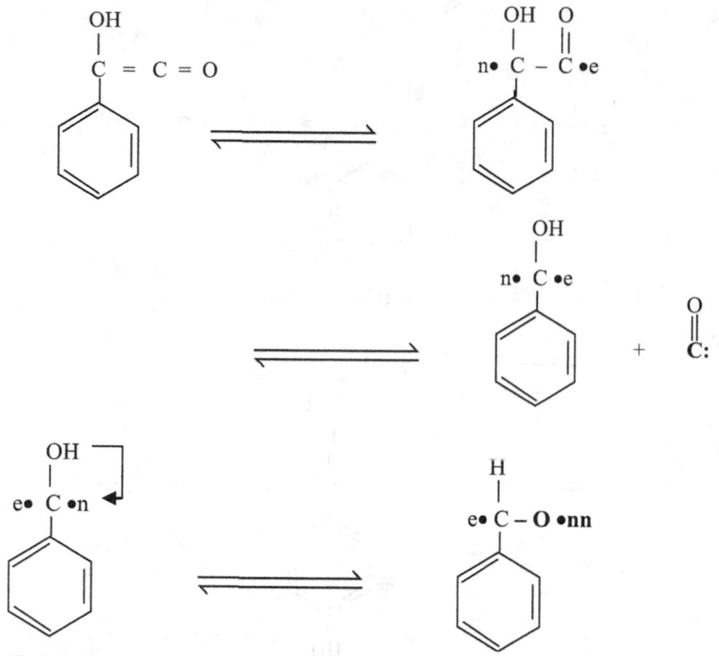

(J) A Carbene   (K)

**MOLECULAR REARRANGEMENT OF THE SECOND KIND
OF THE FIRST TYPE**

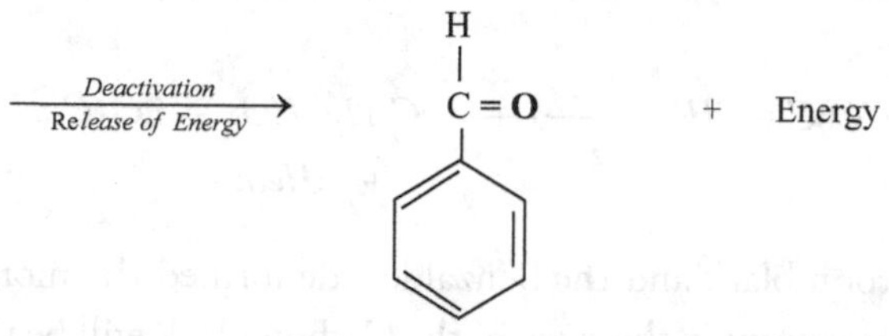

(L)   (i)

Overall equation : $2C_6H_5C(OH)C=O \longrightarrow C_6H_5CHO + CO + Heat$ (ii)

in which the aromatic carbene was made to undergo what has been called *Molecular rearrangement of the Second kind of the first type* in the third step to form benzaldehyde; for if oxidizing oxygen was present in the system, of the two compounds formed, the CO will be the first to be oxidized being less stable than the benzaldehyde; for which one knows what to expect when OH group is replaced with H, CH_3, C_2H_5, etc. type of groups
(Laws of Creations for Aromatic ketenes)

Rule 1595: This rule of Chemistry for **Phenyl carbonyl acetylene an Electrophile shown below,** states that, it can readily be decomposed via Equilibrium mechanism to give the products shown below being very unstable in the presence or absence of oxidizing oxygen molecule-

Stage 1:

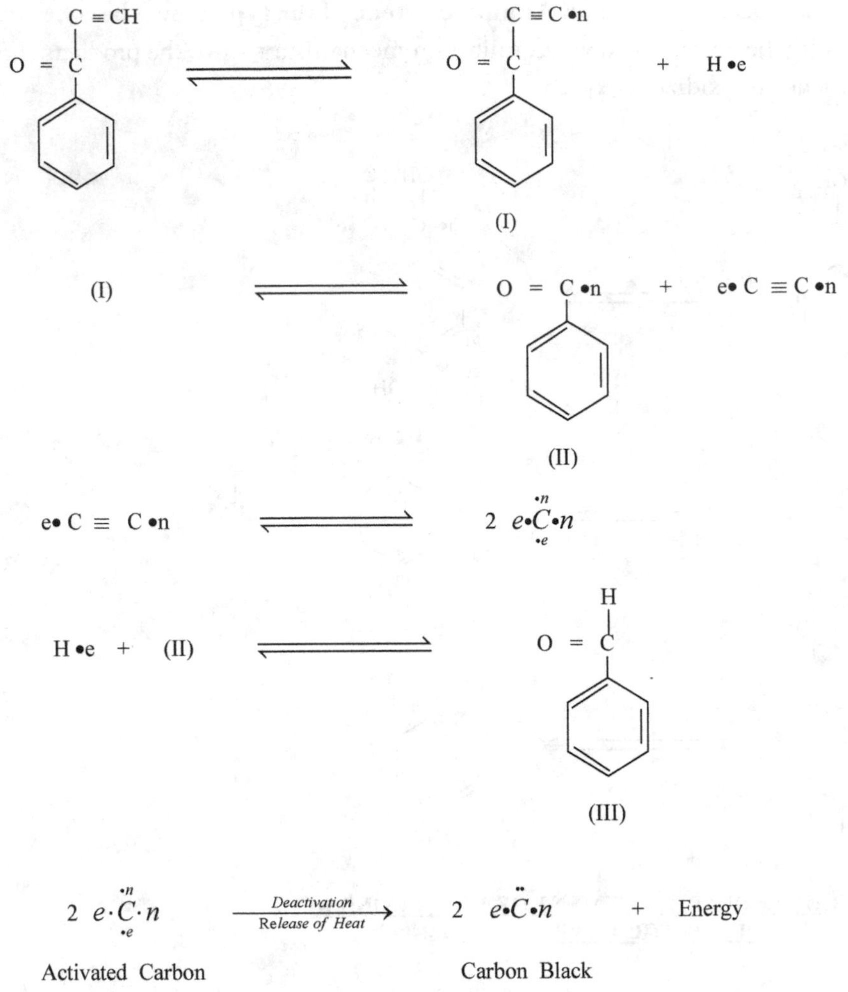

(i)

Overall Eqyation: $C_6H_5(CO)C \equiv CH \longrightarrow C_6H_5CHO + 2C\,(Carbon\;Black)$

$+ \quad Heat$ (ii)

in which between the Carbon black and the benzaldehyde formed the more stable is the benzaldehyde and if oxidizing oxygen was present in the system, the Carbon black will be the first to be oxidized before the benzaldehyde.

(Laws of Creations for Phenyl carbonyl acetylene)

Rule 1596: This rule of Chemistry for **Phenyl acetylene,** states that, being unstable, it can readily be decomposed via Equilibrium mechanism to give the products shown below even in the presence of oxidizing oxygen-

Stage 1:

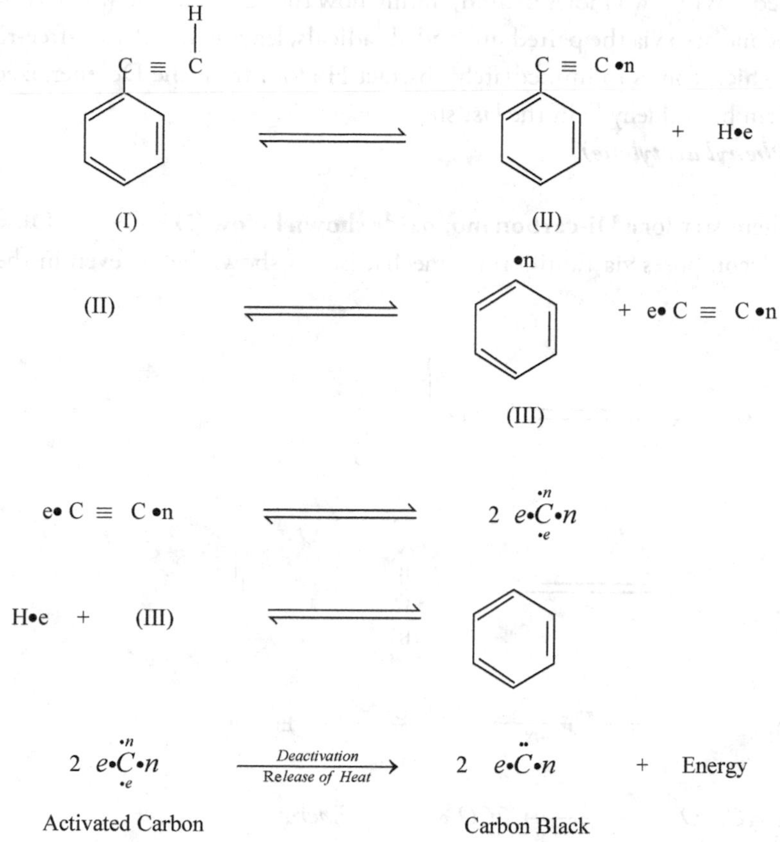

$$C \equiv CH \text{ (I)} \rightleftharpoons C \equiv C\bullet n \text{ (II)} + H\bullet e$$

(I)   (II)

(II) $\rightleftharpoons$ (III) $\quad + \quad e\bullet C \equiv C\bullet n$

(III)

$$e\bullet C \equiv C\bullet n \rightleftharpoons 2\ e\bullet\overset{\bullet n}{\underset{\bullet e}{C}}\bullet n$$

$$H\bullet e \quad + \quad \text{(III)} \rightleftharpoons$$

$$2\ e\bullet\overset{\bullet n}{\underset{\bullet e}{C}}\bullet n \xrightarrow[\text{Release of Heat}]{\text{Deactivation}} 2\ e\bullet\overset{\bullet\bullet}{C}\bullet n \quad + \quad \text{Energy}$$

Activated Carbon   Carbon Black

(i)

Overall equation: $2\,Phenyl\ Benzene \longrightarrow 2\,Benzene\ +\ 4\,Carbon\ Black$
$$+\ Energy$$

(ii)

Stage 2: $O_2 \rightleftharpoons 2nn\bullet O\bullet en$

$$2nn\bullet O\bullet en \ +\ 2\ e\bullet\overset{\bullet\bullet}{C}\bullet n \rightleftharpoons 2nn\bullet O - \overset{\bullet e}{\underset{\bullet n}{C}}\bullet e$$

$$2nn\bullet O - \overset{\bullet e}{\underset{\bullet n}{C}}\bullet e \ +\ 2 \text{ (IV)} \rightleftharpoons 2 \text{ (IV)}^{\bullet e} \ +\ 2\,O = \overset{\bullet n}{C} - H \text{ (V)}$$

(IV)   (V)

$$\longrightarrow 2 \text{ (VI)}$$

(VI)

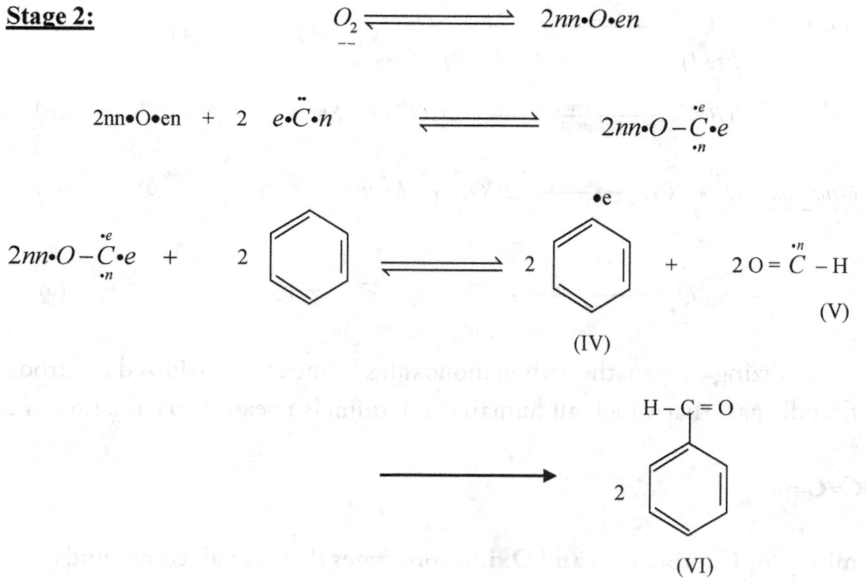

(iii)

Overall equation: $2C_6H_5C \equiv CH\ +\ O_2 \longrightarrow 2C_6H_5CHO\ +\ 2C\ +\ Heat$

(*Carbon Black*)

(iv)

for which in the presence of oxidizing oxygen, in the second stage only two moles of the Carbon black from Stage1 were consumed leaving two moles behind, noting how the carbon black was activated by the oxidizing oxygen in the second step via the paired unbonded radicals, leaving the electro-free-radical end of the Carbon monoxide which comes to immediately abstract H atom from the De-energized benzene to give (IV) and (V) to form benzaldehyde in the last step.
(*Laws of Creations for Phenyl acetylene*)

Rule 1597: This rule of Chemistry for **a Di-carbon monoxide shown below (O = C = C = O),** states that, being unstable, it readily decomposes via Equilibrium mechanism as shown below even in the presence of oxidizing oxygen-

Stage 1:

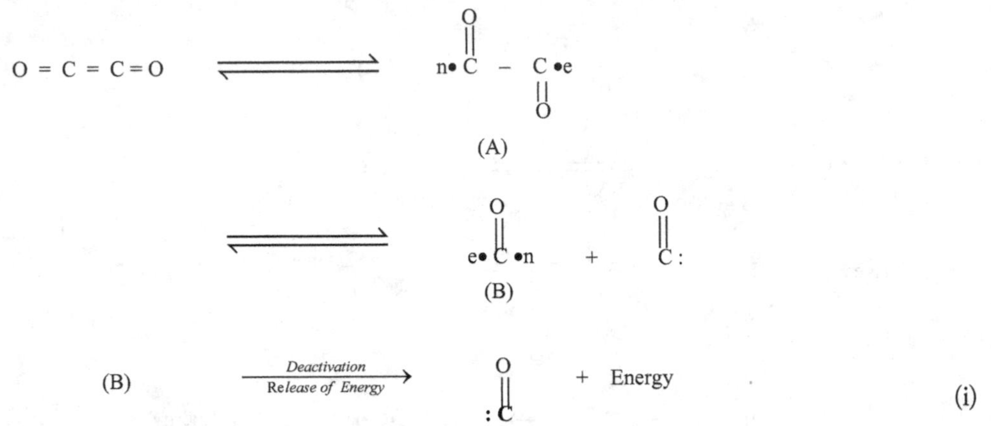

$$O = C = C = O \quad \rightleftharpoons \quad n \bullet \overset{\overset{O}{\|}}{C} - \overset{\underset{\|}{O}}{C} \bullet e$$

(A)

$$\rightleftharpoons \quad e \bullet \overset{\overset{O}{\|}}{C} \bullet n \quad + \quad \overset{\overset{O}{\|}}{C} :$$

(B)

$$(B) \quad \xrightarrow[\text{Release of Energy}]{\text{Deactivation}} \quad \overset{\overset{O}{\|}}{:C} \quad + \quad Energy \qquad (i)$$

Overall equation : $O = C = C = O \longrightarrow 2CO + Energy$ (ii)

Stage 2:

$$O_2 \xrightarrow[\text{Oxygen molecule}]{\text{Equilibruim State Existence of Oxidi sing}} 2en \cdot \ddot{O} \cdot nn$$

$$2en \cdot O \cdot nn + \underset{(Stabilized)}{2C = O} \xrightarrow{\text{Oxidation}} \underset{(B)}{2e \bullet \overset{\overset{O}{\|}}{C} - O \bullet nn}$$

$$(B) \quad \xrightarrow[\text{Release of Energy}]{\text{Deactivation}} \quad 2CO_2 + Energy \qquad (iii)$$

Overall Equation : $O_2 + 2CO \longrightarrow 2CO_2 + Energy$ (iv)

Overall equation: $O = C = C = O + O_2 \longrightarrow 2CO_2 + Energy (3)$ (v)

for which due to the presence of oxidizing oxygen, the carbon monoxides formed were oxidized to carbon dioxide an environmentally friendly gas, that which all humans and animals release every fraction of a second every day.
(*Laws of Creations for O=C=C=O*)

Rule 1598: This rule of Chemistry for **Combustion and Oxidation,** states that, not all compounds that can be combusted can be oxidized and not all compounds that can be oxidized can be combusted, since molecular and oxidizing oxygens are two different compounds of oxygen.
(*Laws of Creations for Combustion and Oxidation*)

Rule 1599: This rule of Chemistry for **Chemical and Polymeric reactions,** states that, based on the ways Chemical and Polymeric reactions have been observed to take place, one can see that in general in Nature, ***Males or Electrophiles or electro-radicals are "POLYGAMISTS"* DEPENDING ON THE OPERATING CONDITIONS;** *while Females or Nucleophiles or nucleo-radicals are* ***"PROSTITUTES"* DEPENDING ON THE OPERATING CONDITIONS,** noting the great importance of the emphasis placed on the word OPERATING CONDITIONS in all operations in our world for without it nothing takes place.
(Laws of Creations for Chemical and Polymeric reactions)

Rule 1600: This rule of Chemistry for **Halogenated hydrocarbons,** states that, their Equili-brium states of existences can be obtained from some of the examples shown below-

$$CH_3F \rightleftharpoons H\bullet e + n\bullet CH_2F \quad ; \quad CH_3Cl \rightleftharpoons H_3C\bullet e + Cl\bullet nn$$

$$F_2CH_2 \rightleftharpoons H\bullet e + n\bullet CHF_2 \quad ; \quad Cl_2CH_2 \rightleftharpoons ClH_2C\bullet e + Cl\bullet nn$$

$$F_3CH \rightleftharpoons H\bullet e + n\bullet CF_3 \quad ; \quad Cl_3CH \rightleftharpoons HCl_2C\bullet e + Cl\bullet nn \quad .$$

$$[CF_4 \rightleftharpoons F_3C\bullet e + nn\bullet F] \quad ; \quad CCl_4 \rightleftharpoons Cl_3C\bullet e + Cl\bullet nn$$

$$H_3CI \rightleftharpoons I\bullet en + n\bullet CH_3$$

noting that that of Br is identical to that of Cl, and that CF_4 is too stable to ever exist in Equilibrium state of existence.
(Laws of Creations for Halogenated Hydrocarbons)

Rule 1601: This rule of Chemistry for **Fluorinated hydrocarbons and F$_2$,** states that, these compounds are very stable to the point, where when they are made to exist in Equilibrium state of existence, never is there a time F atom in the component held, because apart from the fact that F is the smallest size of and the most electronegative of all atoms, the potential difference between the electronegativities of C/H (0.4) is very small compared to that of C/F (1.4) [that for C/Cl is 0.5], noting that the larger the potential difference, the greater the electrostatic forces of attraction.
(Laws of Creations for F$_2$ and Fluorinated Hydrocarbon Compounds)

Rule 1602: This rule of Chemistry for **Atoms,** states that, without the presence of vacant orbitals or paired unbonded radicals on some metals and non-metals, electrostatic bonds can never be formed.
(Laws of Creations for Atoms/Electrostatic bonds)

Rule 1603: This rule of Chemistry for **Chemical reactions,** states that, when chemical reactions take place via Equilibrium mechanisms, ***the transfer species or component abstracted is not the component usually held in Equilibrium state of existence, unless there is only one state of existence for the compound such NaCl or the component held is removed not as it was held, such as removing H as H.n in NH$_2$Cl, and in addition amongst metals only ionic metal or metals of more or equal electropositivity than H can displace a lower ionic metal such as H from a compound electro-free-radically; for example*** instead of H•e abstracting OH from NaOH, as it did when it abstracts OH from CH_3OH, it abstracts H from NaOH since OH is component held in Equilibrium state of existence of NaOH and H is the component held in Equilibrium state of existence of CH_3OH.
(Laws of Creations for Chemical reactions)

Rule 1604: This rule of Chemistry for **The common compounds shown below,** states that the following is the order of their Equilibrium states of existence-

$$HCl \quad >> \quad H_2O \quad > \quad NaOH \quad > \quad NaCl$$
$$\underline{ORDER \ OF \ EQUILIBRIUM \ STATE \ OF \ EXISTENCE} \tag{i}$$

$$H_2O \quad > \quad NH_3 \quad > \quad NH_2R \quad > \quad NHR_3$$
$$\underline{ORDER \ OF \ EQUILIBRIUM \ STATE \ OF \ EXISTENCE} \tag{ii}$$

(Laws of Creations for Acid, Bases and Salt)

Rule 1605: This rule of Chemistry for **Chlorination of Ammonia,** states that, when ammonia is chlorinated, the mechanism is a three stage Equilibrium mechanism system as shown below-

Stage 1:

$$Cl_2 \quad \xrightleftharpoons[Chlorine]{Equilibrium \ State \ of} \quad Cl \cdot en + nn \cdot Cl$$

$$Cl \cdot en + NH_3 \quad \xrightleftharpoons{Abstraction} \quad HCl \quad + \quad en \cdot NH_2$$

$$Cl \cdot nn + en \cdot NH_2 \quad \xrightarrow[State]{Combination} \quad Cl - NH_2 \tag{i}$$

$$\tag{ii}$$

$$\underline{Overall \ Equation}: NH_3 + Cl_2 \longrightarrow HCl + Cl - NH_2$$

Stage 2:

$$HCl \quad \xrightleftharpoons{} H \cdot e + Cl \cdot nn$$

$$H \cdot e + H_2NCl \quad \xrightleftharpoons{} H_2 + en \cdot NHCl + Heat$$

$$ClHN \cdot en + Cl \cdot nn \longrightarrow NHCl_2$$

$$\underline{Overall \ equation}: HCl + H_2NCl \longrightarrow H_2 + NHCl_2 \tag{iii}$$

$$\underline{Overall \ overall \ equation}: Cl_2 + NH_3 \longrightarrow H_2 + NHCl_2 + Heat \tag{iv}$$

Stage 3:

$$Cl_2 \quad \xrightleftharpoons[Chlorine]{Equilibrium \ State \ of} \quad Cl \cdot en + nn \cdot Cl$$

$$Cl \cdot en + NCl_2H \quad \xrightleftharpoons{Abstraction} \quad HCl \quad + \quad en \cdot NCl_2$$

$$Cl \cdot nn + en \cdot NCl_2 \quad \xrightarrow[State]{Combination} \quad Cl - NCl_2 \tag{v}$$

$$\underline{Overall \ Equation}: NH_3 + 2Cl_2 \longrightarrow HCl + NCl_3 + H_2 + Heat \tag{vi}$$

noting the abstraction of H by Cl.en, and the followings -

$$NH_3 \quad \xrightleftharpoons{} \quad H \cdot e + nn \cdot NH_2$$

$$ClNH_2 \xrightleftharpoons{} \quad H \cdot e + nn \cdot NHCl$$

$$Cl_2NH \xrightleftharpoons{} \quad H \cdot e + nn \cdot NCl_2$$

$$Cl_3N \quad \xrightleftharpoons{} \quad Cl \cdot en + nn \cdot NCl_2$$

(Laws of Creations for Chlorination of Ammonia)

Rule 1606: This rule of Chemistry for **NCl$_3$**, states that, this compound which **is said to be a** *powerful oxidizing agent is not since the compound is not carrying OXYGEN, for what* is doing the job when NCl$_3$ is used as a bleaching agent is Cl using the molecular oxygen of the air as shown below-

Stage 1:

$$NCl_3 \rightleftharpoons Cl_2N \cdot nn + Cl \cdot en$$

$$Cl \cdot en + O = O \rightleftharpoons Cl - O - O \cdot en$$
$$(AIR) \qquad\qquad (A)$$

$$(A) + Cl_2N \cdot nn \longrightarrow Cl - O - O - NCl_2$$
$$(B) \qquad\qquad\qquad\qquad \text{(i)}$$

$$\underline{Overall\ equation:} NCl_3 + O_2 \longrightarrow Cl - O - O - NCl_2 \qquad \text{(ii)}$$

Stage 2:

$$Cl - O - O - NCl_2 \rightleftharpoons Cl - O \cdot nn + en \cdot O - NCl_2$$

$$Cl_2N - O \cdot en \rightleftharpoons Cl \cdot en + Cl - N = O + Heat$$

$$Cl \cdot en + Cl - O \cdot nn \longrightarrow Cl - O - Cl \qquad\qquad \text{(iii)}$$

$$\underline{Overall\ equation:} 2Cl - O - O - NCl_2 \longrightarrow 2Cl - N = O + 2Cl - O - Cl + Heat \qquad \text{(iv)}$$

Stage 3:

$$2Cl - O - Cl \rightleftharpoons 2Cl - O \cdot en + 2nn \cdot Cl$$

$$2Cl - O \cdot en \rightleftharpoons 2Cl \cdot en + O_2 + Heat$$

$$2Cl \cdot en + 2Cl \cdot nn \xrightarrow{\substack{Suppressed\ by \\ O_2}} Cl_2 \qquad\qquad \text{(v)}$$

$$\underline{Overall\ equation:}\ 2Cl - O - Cl \longrightarrow O_2 + 2Cl_2 + Heat \qquad \text{(vi)}$$

$$\underline{Overall\ equation:} 2NCl_3 + 2O_2 \longrightarrow 2Cl - N = O + O_2 + 2Cl_2 + Heat \qquad \text{(vii)}$$

noting that the NCl$_3$ through Cl converted the molecular oxygen of the AIR to Oxidizing oxygen molecule, for which NCl$_3$ under such conditions cannot be said to be an oxidizing agent but *__a Molecular oxygen converter,__* and not even a catalyst, unlike KMnO$_4$ or HOOH which are carrying the oxidizing oxygen called *Oxidizing agents*

(Laws of Creations for NCl$_3$- a Molecular Oxygen converter)

Rule 1607: This rule of Chemistry for **Fluorination of ammonia,** states that, when ammonia is fluorinated, the mechanism is a three stage Equilibrium system as shown below-

Stage 1:

$$NH_3 \xrightleftharpoons{\substack{Equilibrium\ State\ of \\ Ammonia}} H \cdot e + nn \cdot NH_2$$

$$H \cdot e + F_2 \xrightleftharpoons{Abstraction} HF + en \cdot F + Energy$$

$$F \cdot en + nn \cdot NH_2 \xrightarrow{\substack{Combination \\ State}} F - NH_2 \qquad\qquad \text{(i)}$$

$$\underline{Overall\ Equation:}\ NH_3 + F_2 \longrightarrow HF + F - NH_2 + Energy \qquad \text{(ii)}$$

Stage 2:

$$FNH_2 \quad \underset{\text{Ammonia}}{\overset{\text{Equilibrium State of}}{\rightleftharpoons}} \quad H \cdot e + nn \cdot NHF$$

$$H \cdot e + F_2 \quad \overset{\text{Abstraction}}{\rightleftharpoons} \quad HF + en \cdot F + Energy$$

$$F \cdot en + nn \cdot NHF \quad \underset{\text{State}}{\overset{\text{Combination}}{\longrightarrow}} \quad H - NF_2 \qquad\qquad\qquad\qquad \text{(iii)}$$

$$\underline{Overall\ Equation}: NH_2F + F_2 \longrightarrow HF + H - NF_2 + Energy \qquad \text{(iv)}$$

Stage 3:

$$HNF_2 \quad \underset{\text{compound}}{\overset{\text{Equilibrium State of}}{\rightleftharpoons}} \quad H \cdot e + nn \cdot NF_2$$

$$H \cdot e + F_2 \quad \overset{\text{Abstraction}}{\rightleftharpoons} \quad HF + en \cdot F + Energy$$

$$F \cdot en + nn \cdot NF_2 \quad \overset{\text{Suppressed}}{\longrightarrow} \quad NF_3 \qquad\qquad\qquad\qquad\qquad \text{(v)}$$

$$\underline{Overall\ Equation}: NF_2H + F_2 \longrightarrow HF + NF_3 + Energy \qquad \text{(vi)}$$

$$\underline{Overall\ overall\ equation}: NH_3 + 3F_2 \longrightarrow NF_3 + 3HF + Energy \qquad \text{(vii)}$$

from which it can be seen that fluorination is more explosive than chlorination of ammonia and for which the followings are worthy of note-

$$NH_3 \quad \rightleftharpoons \quad H \cdot e + nn \cdot NH_2$$

$$FNH_2 \rightleftharpoons \quad H \cdot e + nn \cdot NHF$$

$$F_2NH \rightleftharpoons \quad H \cdot e + nn \cdot NF_2$$

$$F_3N \rightleftharpoons \quad F \cdot nn + en \cdot NF_2$$

(Laws of Creations for Fluorination of Ammonia)

Rule 1608: This rule of Chemistry for **NF$_3$ a colorless gas,** states that, when sparked with water, the following is said to be obtained-

$$3NF_3 + 3H_2O \quad \underset{(Heat\ from\ Water\ vapor)}{\overset{Sparked}{\longrightarrow}} \quad 6HF + N_2O_3 \qquad\qquad \text{(a)}$$

and this can be explained as follows

Stage 1:

$$H_2O \quad \rightleftharpoons \quad H \cdot e + nn \cdot OH$$

$$H \cdot e + NF_3 \rightleftharpoons \quad HNF_2 + en \cdot F + Heat$$

$$F \cdot en + nn \cdot OH \longrightarrow \quad HOF \qquad\qquad\qquad\qquad\qquad \text{(i)}$$

$$\underline{Overall\ equation}: H_2O + NF_3 \longrightarrow HNF_2 + HOF + Heat \qquad \text{(ii)}$$

Stage 2

$$HOF \quad \rightleftharpoons \quad H \cdot e + nn \cdot OF$$

$$FO \cdot nn \rightleftharpoons \quad F \cdot nn + en \cdot O \cdot nn + Heat$$

$$H \cdot e + nn \cdot F \rightleftharpoons \quad HF$$

$$nn \cdot O \cdot en + HNF_2 \rightleftharpoons \quad HO \cdot nn + en \cdot NF_2$$

$$\longrightarrow HONF_2 \tag{iii}$$

$\underline{Overall\ equation}:\ H_2O\ +\ NF_3\ \longrightarrow\ HF\ +\ HONF_2 \tag{iv}$

Stage 3:

$$HONF_2 \rightleftharpoons H \cdot e\ +\ nn \cdot ONF_2$$

$$H \cdot e\ +\ NF_3 \rightleftharpoons HNF_2\ +\ en \cdot F\ +\ Heat$$

$$F \cdot en\ +\ nn \cdot ONF_2 \longrightarrow FONF_2 \tag{v}$$

$\underline{Overall\ equation}:\ HONF_2\ +\ NF_3\ \longrightarrow\ HNF_2\ +\ FONF_2\ +\ Heat \tag{vi}$

Stage 4:

$$HNF_2 \rightleftharpoons H \cdot e\ +\ nn \cdot NF_2$$

$$H \cdot e\ +\ FONF_2 \rightleftharpoons HF\ +\ en \cdot ONF_2\ +\ Heat$$

$$F_2N \cdot nn\ +\ en \cdot ONF_2 \longrightarrow F_2NONF_2 \tag{vii}$$

$\underline{Overall\ equation}:\ HNF_2\ +\ FONOF_2\ \longrightarrow\ HF\ +\ F_2NONF_2\ +\ Heat \tag{viii}$

Stage 5:

$$H_2O \rightleftharpoons H \cdot e\ +\ nn \cdot OH$$

$$H \cdot e\ +\ F_2N - O - NF_2 \rightleftharpoons HF\ +\ F\overset{\cdot en}{N} - O - NF_2\ +\ Heat$$

$$F\overset{\cdot en}{N} - O - NF_2\ +\ nn \cdot OH \longrightarrow F - \underset{\underset{(A)}{|}}{\overset{\overset{OH}{|}}{N}} - O - NF_2 \tag{ix}$$

$\underline{Overall\ equation}:\ F_2N - O - NF_2\ +\ H_2O \longrightarrow F(OH)N - O - NF_2\ +\ HF \tag{x}$

Stage 6:

$$F - \underset{\underset{(A)}{|}}{\overset{\overset{OH}{|}}{N}} - O - NF_2 \rightleftharpoons H \cdot e\ +\ F - \underset{\underset{(B)}{|}}{\overset{\overset{O \cdot nn}{|}}{N}} - O - NF_2$$

$$(B) \rightleftharpoons F \cdot nn\ +\ en \cdot \underset{\underset{(C)}{|}}{\overset{\overset{O \cdot nn}{|}}{N}} - O - NF_2$$

$$H \cdot e\ +\ F \cdot nn \rightleftharpoons HF$$

$$(C) \xrightarrow[\text{Release of Heat}]{\text{Deactivation}} O = N - O - NF_2 \tag{xi}$$

$\underline{Overall\ equation}:\ F(OH)N - O - NF_2\ \longrightarrow\ HF\ +\ O = N - O - NF_2 \tag{xii}$

Stage 7:

$$H_2O \rightleftharpoons H \cdot e\ +\ nn \cdot OH$$

$$H \cdot e\ +\ F_2N - O - N = O \rightleftharpoons HF\ +\ F\overset{\cdot en}{N} - O - N = O\ +\ Heat$$

$$F\overset{\cdot en}{N} - O - NF_2\ +\ nn \cdot OH \longrightarrow F - \underset{\underset{(D)}{|}}{\overset{\overset{OH}{|}}{N}} - O - N = O \tag{xiii}$$

Overall equation $: F_2N - O - N = O + H_2O \longrightarrow F(OH)N - O - N = O + HF$ (xiv)

Stage 8:

$$\underset{\displaystyle F - \overset{\displaystyle |}{\underset{\displaystyle |}{N}} - O - N = O}{\overset{\displaystyle OH}{}} \rightleftharpoons H \cdot e + \underset{\displaystyle F - \overset{\displaystyle |}{N} - O - N = O}{\overset{\displaystyle O \cdot nn}{}}$$

$$(E)$$

$$(E) \rightleftharpoons F \cdot nn + \underset{\displaystyle en \cdot \overset{\displaystyle |}{N} - O - N = O}{\overset{\displaystyle O \cdot nn}{}}$$

$$(F)$$

$$H \cdot e + F \cdot nn \rightleftharpoons HF$$

$$(F) \xrightarrow[\text{Release of Heat}]{\text{Deactivation}} O = N - O - N = O + Heat \tag{xv}$$

Overall equation $: F(OH)N - O - N = O \longrightarrow HF + O = N - O - N = O + Heat$

(xvi)

Overall overall equation $: 2NF_3 + 3H_2O \longrightarrow 6HF + O = N - O - N = O$

$+ Heat$ (a)

an eight stage Equilibrium mechanism system wherein heat was released in the first five, seventh and the last stages, noting the order of existence of the Equilibrium states of existence of the compounds in which for example the Equilibrium state of existence of F(OH)N-O-NF$_2$ or F(OH)N-O-N=O is stronger than that of H$_2$O which in turn is greater than that of HF, with all of them less than that of HCl.
(Laws of Creations for NF$_3$)

Rule 1609: This rule of Chemistry for **some Halogenated Nitrogen compounds viz- NI$_3$.NH$_3$, NBr$_3$.6NH$_3$**, states that, their structures are as follows-

$$NBr_3 \rightleftharpoons Br_2N \cdot nn + Br \cdot en \tag{i}$$

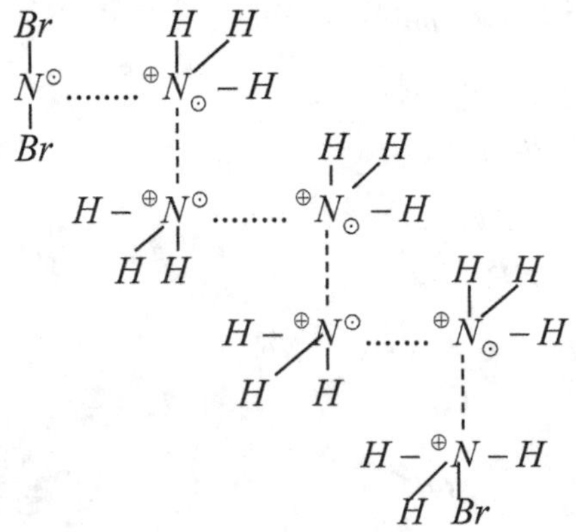

$$NBr_3.6NH_3 \tag{ii}$$

660

$$NI_3 \;\rightleftharpoons\; I_2N \cdot nn \;+\; I \cdot en \tag{i}$$

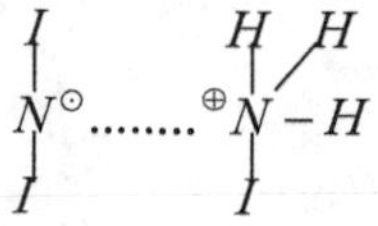

$$
\begin{array}{c}
\overset{H\;H}{\underset{I}{N^{\circleddash}} \cdots\cdots \overset{\oplus}{N} - H} \\[2pt]
\end{array}
$$

$$\underline{NI_3NH_3} \tag{ii}$$

noting that based on their structures, these can conduct electricity.
(Laws of Creations for Halogenated Nitrogen/NH$_3$ compounds)

<u>**Rule 1610:**</u> This rule of Chemistry for **Synthesis of Urea from CO$_2$ and NH$_3$**, states that, when the reaction is carried out in an aqueous media at 170-180^{0}C, in a slurry of ammonia and carbon dioxide, the followings are obtained-

(a)   <u>**Synthesis of Urea from CO$_2$ and NH$_3$**</u>

Stage 1:

$$NH_3 \;\rightleftharpoons\; H \cdot e \;+\; nn \cdot NH_2$$

$$H \cdot e \;+\; O = \overset{O}{\underset{\|}{C}} \;\rightleftharpoons\; H - O - \overset{O}{\underset{|}{C}} \cdot e$$
$$(A)$$

$$(A) \;+\; nn \cdot NH_2 \;\longrightarrow\; HO - \overset{O}{\underset{\|}{C}} - NH_2$$
$$\text{Carbamic acid} \tag{i}$$

<u>Overall equation</u>: $CO_2 \;+\; NH_3 \;\longrightarrow\; HO(CO)NH_2$ (ii)

Stage 2:

$$HO - \overset{O}{\underset{\|}{C}} - NH_2 \;\rightleftharpoons\; H \cdot e \;+\; nn \cdot O - \overset{O}{\underset{\|}{C}} - NH_2$$
$$(B)$$

$$H \cdot e \;+\; NH_3 \;\rightleftharpoons\; e \cdot NH_4$$

$$(B) \;+\; e \cdot NH_4 \;\longrightarrow\; H_2N - \overset{O}{\underset{\|}{C}} - O^{\circ} \cdots\cdots \overset{\oplus}{N}H_4$$
$$(\text{Ammonium carbamate}) \tag{iii}$$

<u>Overall equation</u>: $HO(CO)NH_2 \;+\; NH_3 \;\longrightarrow\;$ Ammonium carbamate   (iv)

Stage 3:

$$NH_3 \;\rightleftharpoons\; H \cdot e \;+\; nn \cdot NH_2$$

$$H \cdot e + H_2N - \overset{O}{\underset{\|}{C}} - O^{\circ} \cdots\cdots \overset{\oplus}{N}H_4 \;\rightleftharpoons\; H_2N - \overset{O}{\underset{\|}{C}} \cdot e \;+\; H - O^{\circ} \cdots\cdots \overset{\oplus}{N}H_4$$

$$(C)$$

$$(C) \;+\; nn\cdot NH_2 \;\longrightarrow\; H_2N-(CO)-NH_2$$
$$(Urea)$$

(v)

Overall equation: $Ammonium\ carbamate\ +\ NH_3 \longrightarrow Urea + NH_4OH$

(vi)

Overall overall equation: $3NH_3 + CO_2 \longrightarrow H_4NOH + H_2N(CO)NH_2$

(vii)

for if ammonium carbonate is one of the products instead of ammonium hydroxide, then this must have been obtained as follows-

Stage 4:

$$HO^{\circleddash}\dots\dots^{\oplus}NH_4 \;\rightleftharpoons\; H\cdot e \;+\; HO^{\circleddash}\dots\dots^{\oplus}N\underset{\cdot n}{H_3}$$

$$(C)$$

$$H\cdot e \;+\; O=C\!\!\overset{O}{\underset{}{\big\|}} \;\rightleftharpoons\; H-O-\overset{O}{\overset{\|}{C}}\cdot e$$

$$(D)$$

$$(C) \;\rightleftharpoons\; HO\cdot nn \;+\; NH_3$$

$$(D) \;+\; nn\cdot OH \;\longrightarrow\; HO-\overset{O}{\overset{\|}{C}}-OH$$
$$(Carbonic\ acid)$$

(viii)

Overall equation: $HONH_4 \;+\; CO_2 \longrightarrow HO(CO)OH \;+\; NH_3$

(ix)

Stage 5:

$$HO-\overset{O}{\overset{\|}{C}}-OH \;\rightleftharpoons\; H\cdot e \;+\; HO-\overset{O}{\overset{\|}{C}}-O\cdot nn$$

$$(E)$$

$$H\cdot e \;+\; NH_3 \;\rightleftharpoons\; e\cdot NH_4$$

$$H_4N\cdot e \;+\; (E) \;\longrightarrow\; H_4N^{\oplus}\dots\dots^{\circleddash}O-\overset{O}{\overset{\|}{C}}-OH$$
$$(Ammonium\ bicarbonate)$$

(x)

Stage 6:

$$HO-\overset{O}{\overset{\|}{C}}-O^{\circleddash}\dots\dots^{\oplus}NH_4 \;\rightleftharpoons\; H\cdot e \;+\; nn\cdot O-\overset{O}{\overset{\|}{C}}-O^{\circleddash}\dots^{\oplus}NH_4$$

$$(F)$$

$$H\cdot e \;+\; NH_3 \;\rightleftharpoons\; e\cdot NH_4$$

$$H_4N\cdot e \;+\; (F) \;\longrightarrow\; H_4N^{\oplus}\dots\dots^{\circleddash}O-\overset{O}{\overset{\|}{C}}-O^{\circleddash}\dots^{\oplus}NH_4$$
$$(Ammonium\ carbonate)$$

(xi)

Overall equation: $HONH_4 \;+\; CO_2 \;+\; NH_3 \longrightarrow (H_4N)_2CO_3$

(xii)

Overall overall equation: $4NH_3 + 2CO_2 \longrightarrow H_2N(CO)NH_2 + (H_4N)_2CO_3$

(xiii)

for which one can observe that the molar ratio of ammonia to carbon dioxide is 2 to 1, in which at the beginning ammonia was forced to existing in Equilibrium state of existence by the use of harsh operating

conditions, but with the immediate presence of Carbonic and Carbamic acids, it no longer could exist in Equilibrium state of existence anymore; hence, it was possible to produce the ammoniums noting that even the ammonium bicarbonate could not exist in Equilibrium state of existence from the ammonium side, because of the presence of the carboxylic end; no wonder urea is a very unique compound.
(Laws of Creations for Urea)

Rule 1611: This rule of Chemistry for **Acids,** states that, the names- *dilute, concentrated, and fuming acids apply more to inorganic acids than to organic acids carrying C,* because of the non-polar character of C element unlike as in N, P, S elements and unique placement of C in the Periodic Table, noting that Carbonic, Carbamic acids are no organic or inorganic acids, but hybrids of them in which all of them contain Oxygen (O) and for this reason, O is never the CENTRAL ATOM in any of them; also the presence of (CH_2) groups in carboxylic acids does a great deal with respect to the strength of the acids.
(Laws of Creations for Acids)

Rule 1612: This rule of Chemistry for **Sulphuric acids,** states that, while the structures of the dilute and concentrated acids have been shown, the followings are obtained for fuming sulphuric acid-

$$\underline{Fu\min g\ Sulphuric\ acid:}\quad 2SO_3\ +\ 2H_2O$$
$$(Fumes) \tag{a}$$

Stage 1:

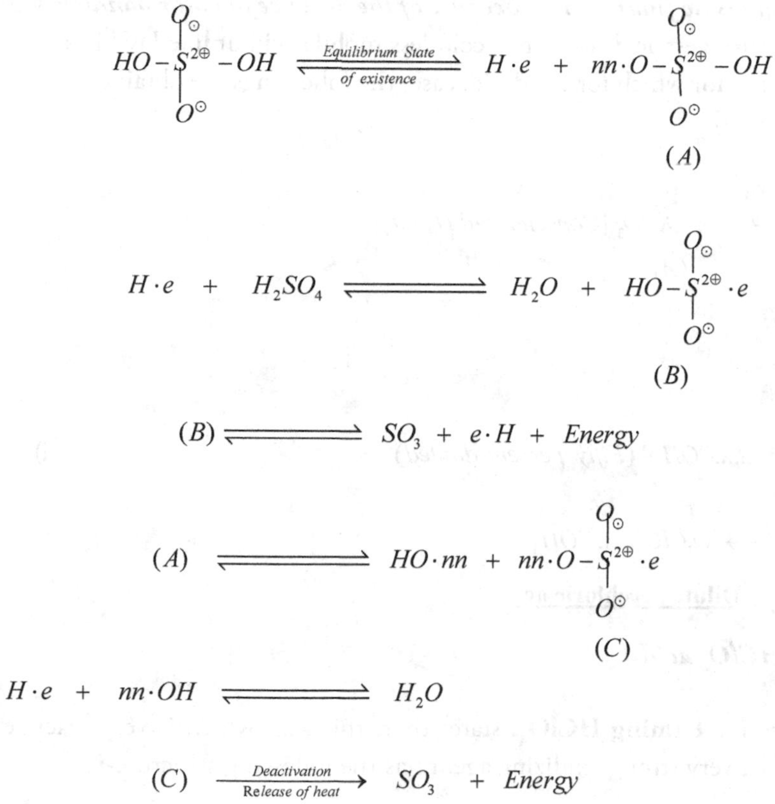

$$HO-\underset{O^\odot}{\overset{O_\odot}{S^{2\oplus}}}-OH \underset{of\ existence}{\overset{Equilibrium\ State}{\rightleftharpoons}} H\cdot e\ +\ nn\cdot O-\underset{O^\odot}{\overset{O_\odot}{S^{2\oplus}}}-OH$$
$$(A)$$

$$H\cdot e\ +\ H_2SO_4 \rightleftharpoons H_2O\ +\ HO-\underset{O^\odot}{\overset{O_\odot}{S^{2\oplus}}}\cdot e$$
$$(B)$$

$$(B) \rightleftharpoons SO_3\ +\ e\cdot H\ +\ Energy$$

$$(A) \rightleftharpoons HO\cdot nn\ +\ nn\cdot O-\underset{O^\odot}{\overset{O_\odot}{S^{2\oplus}}}\cdot e$$
$$(C)$$

$$H\cdot e\ +\ nn\cdot OH \rightleftharpoons H_2O$$

$$(C) \xrightarrow[Release\ of\ heat]{Deactivation} SO_3\ +\ Energy \tag{i}$$

$$\underline{Overall\ equation:}\ 2H_2SO_4 \longrightarrow 2H_2O\ +\ 2SO_3\ +\ Energy \tag{ii}$$

for which two moles of concentrated sulfuric acid is required to give what is called fuming sulfuric acid a mixture of two moles of water and two moles of sulfur trioxide with heat and fumes from the SO_3; noting that, the SO_3 can only provide oxidizing oxygen only under Equilibrium conditions, since when activated, it gives a non-productive stage wherein the oxidizing oxygen is held in equilibrium with SO_2, making it such that the oxidizing oxygen can only be used in situ for specific compounds as shown below to give a productive stage.

Stage 1:

$$O^{\ominus} - S^{2\oplus} - {}^{\otimes}O \xrightleftharpoons{\text{Activation}} nn \cdot \overset{..}{O} \cdot en \ + \ :S^{\oplus} - {}^{\circ}O \tag{iii}$$

[*Stable, reactive and insoluble*]

$$nn \cdot O \cdot en \ + \ :CO \xrightleftharpoons{\hspace{1cm}} nn \cdot O - \overset{\|}{C} \cdot e$$
$$\xrightarrow{\hspace{1cm}} CO_2 \ + \ Heat \tag{iv}$$

$$\underline{Overall \ equation: \ SO_3 \ + \ CO \longrightarrow \ SO_2 \ + \ CO_2 \ + \ Heat} \tag{v}$$

(Laws of Creations for Fuming Sulphuric acid)

Rule 1613: This rule of Chemistry for **HCl and HClO$_4$ acids**, states that, while there are dilute and concentrated hydrochloric acids, *there is no fuming HCl, because of the absence of polar bond(s);* with Perchloric acid (HClO$_4$) one of the strongest acids known, a colorless mobile acid unlike HCl has three parts-dilute, concentrated and fuming, for which for the dilute case, the followings are obtained.

HClO$_4$ [Concentrated perchloric acid]

Stage 1:

$$HClO_4 \xrightleftharpoons{\hspace{1cm}} H \cdot e \ + \ nn \cdot OClO_3 \ [Concentrated \ HClO_4]$$
$$(A)$$

$$H \cdot e \ + \ H_2O \xrightleftharpoons{\hspace{1cm}} en \cdot OH_3$$
$$(B)$$

$$(A) \ + \ (B) \xrightarrow{\hspace{1cm}} O_3ClO^{\ominus}........{}^{\oplus}OH_3 \ (Fifty \ percent \ diluted) \tag{i}$$

$$\underline{Overall \ equation: \ HClO_4 \ + \ H_2O \longrightarrow \ O_3ClO^{\ominus}......{}^{\oplus}OH_3} \tag{ii}$$

Dilute perchloric acid

(Laws of Creations for HCl and HClO$_4$ acids)

Rule 1614: This rule of Chemistry for **Fuming HClO$_4$,** states that, this acid which is very reactive, explosive when heated at 1 atm. and a very strong oxidizing agent has the following structure-

$$\textbf{Fuming HClO}_4 \ \ 4HClO_4 \xrightarrow{\hspace{1cm}} \underline{2Cl_2 \ + \ 2H_2O \ + \ 7O_2 \ + \ Energy} \tag{a}$$

Stage 1:

$$HClO_4 \rightleftharpoons H \cdot e + nn \cdot O - \overset{\overset{\displaystyle O_\odot}{\|}}{\underset{\underset{\displaystyle O^\odot}{|}}{Cl^{3\oplus}}} - {}^\odot O$$

$$(A)$$

$$H \cdot e + HOClO_3 \rightleftharpoons H_2O + e \cdot \overset{\overset{\displaystyle O_\odot}{\|}}{\underset{\underset{\displaystyle O^\odot}{|}}{Cl^{3\oplus}}} - {}^\odot O$$

$$(B)$$

$$(A) + (B) \longrightarrow O_3Cl - O - ClO_3$$
$$(Di - Chlorine\ Heptoxide) \tag{i}$$

$$\underline{Overall\ equation:}\ 2HClO_4 \longrightarrow O_3Cl - O - ClO_3 + H_2O \tag{ii}$$

noting that chlorine heptoxide is not the real name of O_3Cl-O-ClO_3 as used in today's Science but di-chlorine heptoxide-

Stage 2:

$$2\overset{\overset{\displaystyle O_\odot}{\|}}{\underset{\underset{\displaystyle O^\odot}{|}}{O^\odot - Cl^{3\oplus}}} - O - \overset{\overset{\displaystyle O}{\|}}{\underset{\underset{\displaystyle O}{|}}{{}^{3\oplus}Cl^\odot}} - {}^\odot O \rightleftharpoons 2\overset{\overset{\displaystyle O_\odot}{\|}}{\underset{\underset{\displaystyle O^\odot}{|}}{O^\odot - Cl^{3\oplus}}} - O \cdot en + 2n \cdot \overset{\overset{\displaystyle O_\odot}{\|}}{\underset{\underset{\displaystyle O^\odot}{|}}{Cl^{3\oplus}}} - {}^\odot O$$

$$(A) \qquad\qquad (B)$$

$$(A) \rightleftharpoons O_2 + 2O_3Cl \cdot e + Energy$$

$$2O_3Cl \cdot e + 2n \cdot ClO_3 \longrightarrow 2O_3Cl - ClO_3$$
$$[Di - Chlorine\ Hexoxide] \tag{iii}$$

$$\underline{Overall\ equation:}\ 4HClO_4 \longrightarrow O_2 + 2H_2O + 2O_3Cl - ClO_3 + Energy \tag{iv}$$

in which the di-chlorine hexoxide a dimer of chlorine trioxide, is the least explosive oxide of chlorine, but like the other oxides explodes if brought into contact with organic compounds and as it seems, it is very stable when alone, but unstable when activated-

Stage 3:

$$2\overset{\overset{\displaystyle O_\odot}{\|}}{\underset{\underset{\displaystyle O^\odot}{|}}{O^\odot - Cl^{3\oplus}}} - \overset{\overset{\displaystyle O}{\|}}{\underset{\underset{\displaystyle O}{|}}{{}^{3\oplus}Cl^\odot}} - {}^\odot O \underset{by\ Heat}{\overset{Activated}{\rightleftharpoons}} 2\overset{\overset{\displaystyle O_\odot}{\|}}{\underset{\underset{\displaystyle O^\odot}{|}}{O^\odot - Cl^{3\oplus}}} \cdot e + 2 : \overset{\overset{\displaystyle O_\odot}{\|}}{\underset{\underset{\displaystyle O^\odot}{|}}{Cl^{2\oplus}}} - O \cdot nn$$

$$(A) \qquad\qquad (B)$$

$$(A) \quad \rightleftharpoons \quad O_2 \; + \; 2 : \overset{\overset{O_\ominus}{|}}{\underset{\underset{O^\ominus}{|}}{\underset{--}{Cl^{2\oplus}}}} \cdot en \; + \; Energy$$

$$(C)$$

$$(B) \; + \; (C) \quad \longrightarrow \quad O_2Cl - O - ClO_2$$
$$(D) - Di - Chlorine \; pentoxide \qquad\qquad (v)$$

Overall equation : $4HClO_4 \longrightarrow 2H_2O + 2O_2 + 2\underset{--}{O_2Cl} - O - ClO_2 + Energy$ \qquad (vi)

the Di-chlorine pentoxide never been known to exist, is one of the intermediate products when the number of moles is four or more-

Stage 4:

$$2 : \overset{\overset{O_\ominus}{|}}{\underset{\underset{O^\ominus}{|}}{Cl^{2\oplus}}} - O - \overset{\overset{\ominus O}{|}}{\underset{\underset{\ominus O}{|}}{{}^{2\oplus}Cl}} : \quad \rightleftharpoons \quad 2 : \overset{\overset{O_\ominus}{|}}{\underset{\underset{O^\ominus}{|}}{Cl^{2\oplus}}} \cdot nn \; + \; 2 : \overset{\overset{O_\ominus}{|}}{\underset{\underset{O^\ominus}{|}}{Cl^{2\oplus}}} - O \cdot en$$
$$\qquad\qquad\qquad\qquad\qquad (A) \qquad\qquad\quad (B)$$

$$(B) \quad \rightleftharpoons \quad O_2 \; + \; 2 : \overset{\overset{O_\ominus}{|}}{\underset{\underset{O^\ominus}{|}}{\underset{--}{Cl^{2\oplus}}}} \cdot en \; + \; Energy$$

$$(C)$$

$$(A) \; + \; (C) \quad \longrightarrow \quad 2O_2Cl - ClO_2$$
$$(D) - Di - Chlorine \; tetroxide \qquad\qquad (vii)$$

Overall equation : $4HClO_4 \longrightarrow 2H_2O + 3O_2 + 2\underset{--}{O_2Cl} - ClO_2 + Energy$ \qquad (viii)

the di-chlorine tetroxide (Cl_2O_4) a dimer of chlorine dioxide (ClO_2), like di-chlorine hexoxide is stable when alone, but unstable when activated or an oxidizable compound is in its neighborhood-

Stage 5:

$$2 : \overset{\overset{O_\ominus}{|}}{\underset{\underset{O^\ominus}{|}}{Cl^{2\oplus}}} - {}^{2\oplus}\overset{\overset{\ominus O}{|}}{\underset{\underset{\ominus O}{|}}{Cl}} : \quad \underset{\text{by Heat}}{\overset{\text{Activated}}{\rightleftharpoons}} \quad 2 : \overset{\overset{O_\ominus}{|}}{\underset{\underset{O^\ominus}{|}}{Cl^{2\oplus}}} \cdot en \; + \; 2 : \overset{\overset{O_\ominus}{|}}{Cl^{2\oplus}} - O \cdot nn$$
$$\qquad\qquad\qquad\qquad\qquad (A) \qquad\qquad\quad (B)$$

$$(A) \quad \rightleftharpoons \quad O_2 \; + \; 2 : \overset{\overset{O_\ominus}{|}}{\underset{..}{Cl^{\oplus}}} \cdot en \; + \; Energy$$

$$(C)$$

$$(B) \; + \; (C) \quad \longrightarrow \quad OCl - O - ClO$$
$$(D) - Di - Chlorine \; trioxide \qquad\qquad (ix)$$

Overall equation: $4HClO_4 \longrightarrow 2H_2O + 4O_2 + 2OCl-O-ClO + Energy$ (x)

the Di-chlorine trioxide here also not been known to exist is one of the intermediates in the fuming acids in-situ preparation and since it is very unstable, hence the next stage follows-

Stage 6:

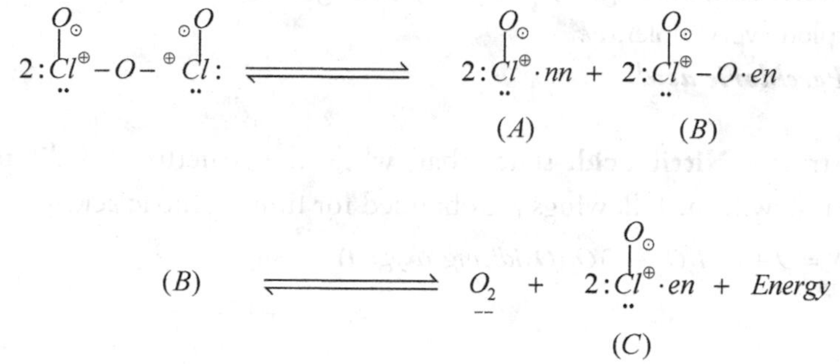

$$2:\overset{O}{\underset{..}{Cl}}{}^{\oplus}-O-\overset{O}{\underset{..}{Cl}}{}^{\oplus}: \rightleftharpoons 2:\overset{O}{\underset{..}{Cl}}{}^{\oplus}\cdot nn + 2:\overset{O}{\underset{..}{Cl}}{}^{\oplus}-O\cdot en$$

$$(A) \qquad\qquad (B)$$

$$(B) \rightleftharpoons O_2 + 2:\overset{O}{\underset{..}{Cl}}{}^{\oplus}\cdot en + Energy$$

$$(C)$$

$$(A) + (C) \longrightarrow 2OCl-ClO$$

$$(D)-Di-Chlorine\ dioxide \qquad\qquad (xi)$$

Overall equation: $4HClO_4 \longrightarrow 2H_2O + 5O_2 + 2O_2Cl-ClO_2 + Energy$ (xii)

the di-chlorine dioxide above one of the intermediate products not known to exist is very unstable and explosive in character-

Stage 7:

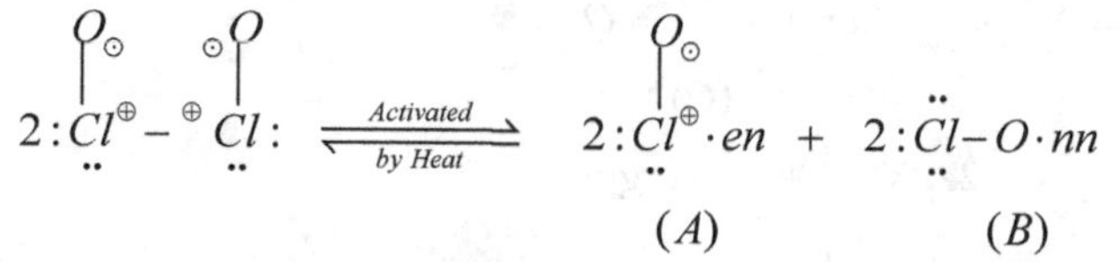

$$2:\overset{O}{\underset{..}{Cl}}{}^{\oplus}-\overset{O}{\underset{..}{Cl}}{}^{\oplus}: \underset{by\ Heat}{\overset{Activated}{\rightleftharpoons}} 2:\overset{O}{\underset{..}{Cl}}{}^{\oplus}\cdot en + 2:\overset{..}{\underset{..}{Cl}}-O\cdot nn$$

$$(A) \qquad\qquad (B)$$

$$(A) \rightleftharpoons O_2 + 2:\overset{..}{\underset{..}{Cl}}\cdot en + Energy$$

$$(C)$$

$$(B) + (C) \longrightarrow Cl-O-Cl$$

$$(D)-Di-Chlorine\ monoxide \qquad\qquad (xiii)$$

Overall equation: $4HClO_4 \longrightarrow 2H_2O + 6O_2 + 2Cl-O-Cl + Energy$ (xiv)

the di-chlorine monoxide (Present-day Science calls it chlorine hemioxide or monoxide, both of which are wrong) produced above, is known to be explosive when alone and explodes violently on contact with easily oxidized substances such as hydrogen, sulfur, ammonia, nitric oxide, etc.-

Stage 8:

$$2:\overset{..}{\underset{..}{Cl}}-O-\overset{..}{\underset{..}{Cl}}: \rightleftharpoons 2:\overset{..}{\underset{..}{Cl}}\cdot nn + 2:\overset{..}{\underset{..}{Cl}}-O\cdot en$$

$$2:\overset{..}{\underset{..}{Cl}}-O\bullet en \rightleftharpoons 2:\overset{..}{\underset{..}{Cl}}\bullet en + O_2 + Energy$$

667

$$2 : \overset{..}{Cl} \bullet en \ + \ 2 : \overset{..}{Cl} \bullet nn \xrightarrow[\text{Oxidizing oxygen}]{\text{Suppressed by}} \ 2Cl_2 \qquad \text{(xv)}$$

$$\underline{Overall \ equation} : 4HClO_4 \longrightarrow 2Cl_2 \ + \ 2H_2O \ + \ 7\underset{--}{O_2} \ + \ Energy \qquad \text{(xvi}$$

an eight stage Equilibrium system in which one can shockingly observe that Fuming Perchloric acid is indeed a very strong acid, a very strong oxidizing agent, and a very strong source of energy since all the seven stages after the first stage explode very violently.

(Laws of Creations for Fuming Perchloric acid)

Rule 1615: This rule of Chemistry for **Nitric acid,** states that, while the structures of dilute and concentrated nitric acids have been shown, the followings are obtained for fuming nitric acid-

$$\underline{Fu\min g \ nitric \ acid} : \ 2O = N - N = O \ + \ 2H_2O \ + \ 3\underset{--}{O_2}(Oxidizing \ oxygen) \qquad \text{(a)}$$

Stage 1:

$$HNO_3 \ \rightleftharpoons \ H \cdot e \ + \ nn \cdot O - NO_2$$
$$(A)$$

$$H \cdot e \ + \ HNO_3 \ \rightleftharpoons \ H_2O \ + \ e \cdot NO_2$$
$$(B)$$

$$(A) \ + \ (B) \ \longrightarrow \ O_2N - O - NO_2 \qquad \text{(i)}$$

$$\underline{overall \ equation} : \ 2HNO_3 \ \longrightarrow \ O_2N - O - NO_2 \ + \ H_2O \qquad \text{(ii)}$$

Stage 2:

$$2O = \overset{\overset{O}{\underset{\ominus}{\|}}}{N^{\oplus}} - O - \overset{\overset{\ominus}{\underset{O}{\|}}}{N} = O \ \rightleftharpoons \ 2O = \overset{\overset{O}{\underset{\ominus}{\|}}}{N^{\oplus}} \cdot n \ + \ 2en \cdot O - \overset{\overset{\ominus}{\underset{O}{\|}}}{N} = O$$
$$(C)$$

$$(C) \ \rightleftharpoons \ O_2 \ + \ 2e \cdot N\underset{--}{O_2} \ + \ Energy$$

$$2O_2N \cdot e \ + \ 2n \cdot NO_2 \ \longrightarrow \ 2O_2N - NO_2 \qquad \text{(iii)}$$

$$\underline{Overall \ equation} : 4HNO_3 \ \longrightarrow \ 2O_2N - NO_2 \ + \ O_2 \ + \ 2H_2O \ + \ Energy \qquad \text{(iv)}$$

Stage 3:

$$2O = \overset{\overset{O}{\underset{\ominus}{\|}}}{N^{\oplus}} - \overset{\overset{\ominus}{\underset{O}{\|}}}{N} = O \ \rightleftharpoons \ 2O = \overset{\overset{O}{\underset{\ominus}{\|}}}{N^{\oplus}} \cdot e \ + \ 2n \cdot \overset{\overset{\ominus}{\underset{O}{\|}}}{N} = O$$
$$(D)$$

$$(D) \ \rightleftharpoons \ O_2 \ + \ 2en \cdot \overset{..}{N} O \ + \ Energy$$

$$2n \cdot NO_2 \ \rightleftharpoons \ 2O = N - O \cdot nn$$

$$2O \overset{..}{N} \cdot en \ + \ 2nn \cdot O - N = O \ \longrightarrow \ 2O = N - O - N = O \qquad \text{(v)}$$

$$\underline{Overall \ equation} : 4HNO_3 \ \longrightarrow \ 2O = NON = O \ + \ 2\underset{--}{O_2} \ + \ 2H_2O \ + \ Energy \qquad \text{(vi)}$$

Stage 4:

$$2O = N - O - N = O \ \rightleftharpoons \ 2O = N - O \cdot en \ + \ 2nn \cdot N = O$$
$$(D)$$

$$(D) \quad \rightleftharpoons \quad O_2 + 2en\cdot NO + Energy$$

$$2ON\cdot en + 2nn\cdot NO \xrightarrow{\hspace{3cm}} 2ON-NO \hspace{3cm} \text{(vii)}$$

<u>Overall equation</u>: $4HNO_3 \xrightarrow{\hspace{1cm}} 2ON-NO + 3O_2 + 2H_2O + Energy$ \hspace{2cm} (viii)

from which one can clearly see why fuming nitric acid is a very strong oxidizing agent and observe the great significance and effect of number of moles of compounds on some compounds, and also observe the great significance of many things we have always ignored; one can observe the origins of the fumes, and one can see why these fuming acids must be stored in CLOSED SYSTEMS at very low operating conditions.
(Laws of Creations for Fuming Nitric acids)

Rule 1616: This rule of Chemistry for **Combustion and Oxidation,** states that, despite the fact that both phenomena use oxygen to provide Energy in form of Heat, the types of oxygen used are different, since while the former uses the molecular oxygen of the Air, the latter uses what is called oxidizing oxygen molecule.
(Laws of Creations for Combustion and Oxidation)

Rule 1617: This rule of Chemistry for **Molecular and Oxidizing Oxygens,** states that, while Molecular oxygen is non-cyclic linear compound with a π-bond type of Activation center sitting on top a sigma (σ) bond, Molecular oxidizing oxygen is a two membered cyclic ring too strained to exist permanently and therefore always in Equilibrium state of existence as shown below-

$$O=O \xrightarrow{\hspace{0.3cm}\text{ACTIVATION}\hspace{0.3cm}} nn\cdot O-O\cdot en \quad ; \quad O=O \xrightleftharpoons{\hspace{0.3cm}\text{ACTIVATION}\hspace{0.3cm}} nn\cdot O-O\cdot en$$

$$\text{\textit{(Molecular Oxygen)}} \hspace{6cm} \text{(i)}$$

$\xrightleftharpoons[\text{OF EXISTENCE}]{\text{EQUILIBRIUM STATE}} 2\,nn\cdot O\cdot en$

$$\text{\textit{(Oxidizing Oxygen Molecule)}} \hspace{2cm} \text{\textit{(Oxidizing Oxygen Elements)}} \hspace{2cm} \text{(ii)}$$

for which while molecular oxygen cannot exist in Equilibrium state of existence wherein an element or group is held, but can only be activated via Decomposition or Equilibrium mechanism, Oxidizing oxygen molecule cannot be activated, but can only exist in Equili-brium state of existence all the time; clear indication that while Molecular oxygen is non-poisonous, oxidizing oxygen molecule is a poison coming from the electro-non-free-radical center.
(Laws of Creations for Molecular and Oxidizing Oxygens)

Rule 1618: This rule of Chemistry for **Molecular and Oxidizing Oxygens,** states that, based on the unique differences between the two, Molecular oxygen which is in abundance can only be used when activated by a neighbor in the presence of a force such as Heat, while Oxidizing which is not in abundance (Hence the existence of Oxidizing agents) can only be used instantaneously and mathematically under Equilibrium mechanism conditions.
(Laws of Creations for Molecular and Oxidizing Oxygens)

Rule 1619: This rule of Chemistry for **Oxidizing Oxygen Molecule,** states that, these can be obtained either from Molecular oxygen using Surface Chemistry, that is, that in which specific passive catalysts are used, or from the use of Oxidizing agents such as, Hydrogen peroxide, Potassium permanganate,

Vanadium pentoxide, fuming acids and more too large to list; for which for some of them the oxidizing has to be used in-situ in a Stage or can be obtained as a product all via Equilibrium mechanisms.
(Laws of Creations for Oxidizing Oxygen Molecule)

Rule 1620: This rule of Chemistry for **Combustion of a Compound,** states that, before a compound can be combusted, the compound must either have the ability to first exist in Equilibrium state of Existence such as found with very light hydrocarbons (e.g. Methane) or must be such that can be made to exist in such a manner that it is carrying an electro-radical center such as found with Rhombic sulfur.
(Laws of Creations for Combustion of Compounds)

Rule 1621: This rule of Chemistry for **Oxidation of a Compound,** states that, before a compound can be oxidized, the compound must be such that can be kept in Stable State of existence when oxidizing oxygen is present, and secondly it must be such that carries a transfer species in particular H to abstract and since what is doing the abstraction is an electro-non-free-radical on an O center, the mechanism of the reaction is essentially carried out via Equilibrium mechanism only.
(Laws of Creations for Oxidation of Compounds)

Rule 1622: This rule of Chemistry for **Molecular and Oxidizing Oxygens,** states that, while Molecular oxygen is carrying lots of latent energy in the visible π-bond which can only be released or manifested when activated, Oxidizing oxygen molecule is also carrying lots of energy hidden inside the electro-non-free-radical carried by the Oxygen center, that which was inside the ring too strained to exist.
(Laws of Creations for Molecular and Oxidizing Oxygens)

Rule 1623: This rule of Chemistry for **Combustion of Alkane Hydrocarbon family of Compounds,** states that, beginning with the first member *Methane,* the followings are obtained-

$$Stage\,1: \quad CH_4 \underset{\substack{\text{Existence Of Methane}\\\text{At any Temperature}}}{\overset{\text{Equilibrium State Of}}{\rightleftharpoons}} H \cdot e + n \cdot CH_3$$

$$H \cdot e + O = O \underset{\substack{\text{Oxygen (Heat required)}}}{\overset{\text{Activation of}}{\rightleftharpoons}} H \cdot e + nn \cdot \ddot{O} - \ddot{O} \cdot en$$

$$\rightleftharpoons H - O - O \cdot en$$

$$(A)$$

$$(A) + n \cdot CH_3 \xrightarrow{\text{Combination Step}} H - O - O - CH_3 \qquad\qquad \text{(i)}$$

$$\underline{Overall\ Equation}: CH_4 + O_2 \longrightarrow H - O - O - CH_3$$
$$\quad\quad (Air) \qquad\qquad (B) \qquad\qquad\qquad\qquad\qquad \text{(ii)}$$

$$Stage\,2: \quad H - O - O - CH_3 \underset{\substack{\text{Existence of the peroxide}}}{\overset{\text{Equilibrium State of}}{\rightleftharpoons}} H - O \cdot nn + en \cdot OCH_3$$

$$en \cdot O - \underset{\underset{H}{|}}{\overset{\overset{H}{|}}{C}} - H \underset{\text{Release}}{\overset{\text{Energy}}{\rightleftharpoons}} H \cdot e + \quad O = \underset{\underset{H}{|}}{\overset{\overset{H}{|}}{C}} \quad + Energy\ Release$$

$$(Formaldehyde)$$

$$H \cdot e + nn \cdot OH \xrightarrow[\substack{\text{by presence of the energy Or product formed}}]{\text{Combination state sup pressed}} H_2O \qquad\qquad \text{(iii)}$$

Overall Equation: $CH_4 + O_2 \longrightarrow H_2O + H_2C = O + Energy$ Release (iv)

Stage3: $H_2C = O \; \underset{\text{Existence of formaldehyde}}{\overset{\text{Activation State of}}{\rightleftharpoons}} \; H \cdot e + n \cdot \overset{\overset{H}{|}}{C} = O$

$H \cdot e + O = O \; \underset{\text{Oxygen}}{\overset{\text{Activation of}}{\rightleftharpoons}} \; H - O - O \cdot en$

(A)

$(A) + n \cdot \overset{\overset{H}{|}}{C} = O \; \xrightarrow{\text{Combination}} \; H - O - O - \overset{\overset{H}{|}}{C} = O$

(*Another peroxide*) (v)

Overall equation: $H_2C = O \; + \; O_2 \longrightarrow HOOCH = O$ (vi)

Stage4: $H - O - O - \overset{\overset{H}{|}}{C} = O \; \underset{\text{Existence}}{\overset{\text{Equilibrium State of}}{\rightleftharpoons}} \; H - \ddot{O} \cdot nn + en \cdot O - \overset{\overset{H}{|}}{C} = O$

(C)

$(C) \; \underset{\text{Release}}{\overset{\text{Energy}}{\rightleftharpoons}} \; CO_2 + H \cdot e + Energy$

$H \cdot e + nn \cdot OH \; \xrightarrow{\text{Combination}} \; H_2O$ (vii)

Overall equation: $HOOCH = O \longrightarrow CO_2 + H_2O + Heat$ (viii)

Overall overall Equation: $CH_4 + 2O_2 \longrightarrow CO_2 + 2H_2O + Energy$ (ix)

a four stage Equilibrium mechanism system, in which Energy in form of Heat are released in the second and last stage from Equilibrium decomposition of PEROXIDES unique compounds which contain invisible energy inside them, just like the atomic sub-particles inside the Nucleus of an atom but by far much less; noting that no Oxidizing oxygen molecule or Carbon monoxide are present as one of the products, clear indication that its combustion is environmentally friendly.

(Laws of Creations for the Combustion of Methane)

<u>**Rule 1624:**</u> This rule of Chemistry for **Combustion of Alkane Hydrocarbon family of Compounds**, states that, considering the second member *Ethane,* the followings are obtained-

Stage1: $C_2H_6 \; \underset{\text{Existence at above Melting Point}}{\overset{\text{Equilibrium State of}}{\rightleftharpoons}} \; H \cdot e + n \cdot C_2H_5$

$H \cdot e + O = O \; \underset{\text{Oxygen}}{\overset{\text{Activation of}}{\rightleftharpoons}} \; H - O - O \cdot en$

$H - O - O \cdot en + n \cdot C_2H_5 \; \xrightarrow{\text{Combination State}} \; H - O - O - C_2H_5$

(*A peroxide*) (i)

Overall equation: $C_2H_6 \; + \; O_2 \longrightarrow HOOC_2H_5$ (ii)

Stage 2 :

$$H-O-O-\underset{\underset{H}{|}}{\overset{\overset{H}{|}}{C}}-\underset{\underset{H}{|}}{\overset{\overset{H}{|}}{C}}-H \quad \underset{\textit{Existence of the Peroxide}}{\overset{\textit{Equilibrium State of}}{\rightleftharpoons}} \quad H-O\cdot nn + en\cdot O-\underset{\underset{H}{|}}{\overset{\overset{H}{|}}{C}}-\underset{\underset{H}{|}}{\overset{\overset{H}{|}}{C}}-H$$

$$(A)$$

$$(A) \quad \underset{\textit{Release}}{\overset{\textit{Energy}}{\rightleftharpoons}} \quad H\cdot e + O = \underset{\underset{H}{|}}{\overset{\overset{CH_3}{|}}{C}} + \textit{Energy}$$

$$(\textit{Acetaldehyde})$$

$$H\cdot e + nn\cdot O - H \quad \overset{\textit{Combination}}{\longrightarrow} \quad H_2O \qquad \qquad \text{(iii)}$$

Overall Equation : $HOOC_2H_5 \longrightarrow H_2O + H_3CCHO + Energy$ (iv)

Stage 3 :

$$CH_3CHO \quad \underset{\textit{Existence of the Unactivated Aldehyde}}{\overset{\textit{Equilibrium state of}}{\rightleftharpoons}} \quad H\cdot e + O = \overset{\overset{CH_3}{|}}{C}\cdot n$$

$$H\cdot e + O = O \quad \underset{\textit{Oxygen}}{\overset{\textit{Activation of}}{\rightleftharpoons}} \quad H-O-O\cdot en$$

$$H-O-O\cdot en + n\cdot \overset{\overset{CH_3}{|}}{C} = O \quad \overset{\textit{Combination}}{\longrightarrow} \quad H-O-O-\overset{\overset{CH_3}{|}}{C} = O$$

$$(B) - A \ peroxide \qquad \qquad \text{(v)}$$

Overall equation : $H_3CCHO + O_2 \longrightarrow HOOC(CH_3) = O$ (vi)

Stage 4 :

$$H_3C - \overset{\overset{O}{\|}}{C} - O - O - H \quad \underset{\textit{of Existence}}{\overset{\textit{It's Equilibrium State}}{\rightleftharpoons}} \quad 2H-O\cdot en + 2nn\cdot O - \overset{\overset{O}{\|}}{C} - CH_3$$

$$2nn\cdot O - \overset{\overset{O}{\|}}{C} - CH_3 \quad \rightleftharpoons \quad 2nn\cdot O - \overset{\overset{O}{\|}}{C}\cdot e + 2n\cdot CH_3$$

$$2H - O\cdot en \quad \underset{\textit{Release}}{\overset{\textit{Energy}}{\rightleftharpoons}} \quad 2H\cdot e + \quad \underline{O_2} \quad + \textit{Energy}$$

$$(\textit{Oxidi} \sin g \ \textit{Oxygen})$$
$$\textit{molecule}$$

$$2H\cdot e + 2n\cdot CH_3 \quad \underset{\textit{Existence of CH}_4}{\overset{\textit{Equilibrium State of}}{\rightleftharpoons}} \quad 2CH_4$$

$$2e\cdot C - O\cdot nn \quad \underset{\textit{Energy Release}}{\overset{\textit{Deactivation}}{\longrightarrow}} \quad 2CO_2 + \textit{Energy}$$
$$\underset{O}{\overset{\|}{}}$$

$$\text{(vii)}$$

Overall Equation : $2C_2H_6 + 4O_2 \longrightarrow \underline{2CH_4} + 2CO_2 + 2H_2O + \quad \underline{O_2} \quad + \quad \textit{Energy}$

$$(\textit{Air}) \quad (\textit{Methane}) \qquad (\textit{Oxidi} \sin g \ \textit{Oxygen})$$
$$\textit{Molecule} \qquad \qquad \text{(viii)}$$

Stage 5 : $O_2 \underset{\text{Oxygen molecule}}{\overset{\text{Equilibruim State Existence of Oxidi}\sin g}{\rightleftharpoons}} 2en \cdot \ddot{O} \cdot nn$

$2en \cdot O \cdot nn + 2CH_4 \underset{}{\overset{\text{Oxidation}}{\rightleftharpoons}} 2H - O \cdot nn + 2e \cdot CH_3$
   (Stabilized)

$\longrightarrow 2CH_3OH$ (ix)

Overall Equation : $O_2 + 2CH_4 \longrightarrow 2CH_3OH$ (x)

Stage 6 : $H_3COH \underset{\text{of Existence}}{\overset{\text{Equilibrium State}}{\rightleftharpoons}} H \bullet e + nn \bullet OCH_3$

$H \bullet e + O = O \underset{\text{(Heat)}}{\overset{\text{Activation of } O_2}{\rightleftharpoons}} H - O - O \bullet en$
   (A)

$(A) + nn \bullet OCH_3 \longrightarrow H - O - O - O - CH_3$
   (B) (xi)

Overall equation : $H_3COH + O_2 \longrightarrow H - O - O - O - CH_3$ (xii)

Stage 7 :

$(B) \underset{\text{of existence}}{\overset{\text{Equilibrium State}}{\rightleftharpoons}} H - O - O \bullet nn + en \bullet O - \overset{\overset{\textstyle H}{|}}{\underset{\underset{\textstyle H}{|}}{C}} - H$
   (C)

$(C) \rightleftharpoons H \bullet e + H_2C = O$
   (Formaldehyde)

$H - O - O \bullet nn + H \bullet e \longrightarrow H - O - O - H$ (xiii)

Overall equation : $(B) \longrightarrow H - O - O - H + H_2C = O$ (xiv)

Stage 8 : $2H - O - O - H \underset{\text{Existence of Hyrogen peroxide}}{\overset{\text{Equilibrium State of}}{\rightleftharpoons}} 2H - O \cdot en + 2nn \cdot O - H$

$2H - O \cdot en \underset{\text{Release}}{\overset{\text{Energy}}{\rightleftharpoons}} 2H \cdot e + O_2 + Energy$

$2H \cdot e + 2nn \cdot O - H \underset{\text{presence of } O_2}{\overset{\text{Suppressed by the}}{\longrightarrow}} 2H_2O$ (xv)

Overall Equation : $2HOOH \longrightarrow 2H_2O + O_2 + Energy$ (xvi)

Stage 9 :

$O_2 \underset{\text{Oxygen molecule}}{\overset{\text{Equilibruim State Existence of Oxidi}\sin g}{\rightleftharpoons}} 2en \cdot \ddot{O} \cdot nn$

$$2en \cdot O \cdot nn + 2H_2C = O \quad \underset{}{\overset{Oxidation}{\rightleftharpoons}} \quad 2H - O \cdot nn + 2e \cdot \overset{\displaystyle H}{\underset{}{C}} = O$$

(*Stabilized*)

$$\longrightarrow \quad 2O = CHOH \qquad \text{(xvii)}$$

Overall Equation: $\quad O_2 + 2H_2C = O \longrightarrow \quad 2O = CHOH \qquad$ (xviii)

*Stage*10:

$$O = CHOH \quad \underset{of\ Existence}{\overset{Equilibrium\ State}{\rightleftharpoons}} \quad H \bullet e \quad + \quad n \bullet \overset{\displaystyle O}{\underset{}{C}} - OH$$

$$H \bullet e \quad + \quad O = O \quad \underset{(Heat)}{\overset{Activation\ of\ O_2}{\rightleftharpoons}} \quad H - O - O \bullet en$$

(*A*)

$$(A) \quad + \quad n \bullet \overset{\displaystyle O}{\underset{}{C}} - OH \quad \longrightarrow \quad H - O - O - \overset{\displaystyle O}{\underset{}{C}} - OH$$

(*B*) $\qquad$ (xix)

Overall equation: $\quad 2O = CHOH \quad + \quad 2O_2 \quad \longrightarrow \quad 2H - O - O - \overset{\displaystyle O}{\underset{}{C}} - OH \qquad$ (xx)

*Stage*11:

$$2H - O - O - \overset{\displaystyle O}{\underset{}{C}} - OH \quad \underset{of\ existence}{\overset{Equilibrium\ State}{\rightleftharpoons}} \quad 2H - O \bullet en \quad + \quad 2nn \bullet O - \overset{\displaystyle O}{\underset{}{C}} - OH$$

(*A*)

$$2H - O \bullet en \quad \rightleftharpoons \quad 2H \bullet e \quad + \quad O_2$$

$$(A) \quad \rightleftharpoons \quad 2HO \bullet nn \quad + \quad 2e \bullet \overset{\displaystyle O}{\underset{}{C}} - O \bullet nn$$

(*B*)

$$2H \bullet e \quad + \quad 2nn \bullet OH \quad \rightleftharpoons \quad 2H_2O$$

$$(B) \quad \underset{Release\ of\ Heat}{\overset{Deactivation}{\longrightarrow}} \quad 2CO_2 \quad + \quad Energy \qquad \text{(xxi)}$$

Overall equation: $2O = CHOH \quad + \quad 2O_2 \quad \longrightarrow 2H_2O + 2CO_2 + O_2 + Energy$

(xxii)

FULL COMBUSTION

Final Overall Equation: $2C_2H_6 + 8O_2 \longrightarrow 4CO_2 + 6H_2O$

$\qquad\qquad\qquad\qquad + \quad O_2$ (*Oxidizing oxygen molecular*)

$\qquad\qquad\qquad\qquad + \quad Energy$(*From five Stages*) $\qquad$ [a]

an eleven stage Equilibrium mechanism system in which Energy as a form of Heat are released in five stages with two moles of ethane as against two stages with one mole of methane, and unlike methane, oxidizing oxygen an environmentally unfriendly gas can be observed to be one of the products; for if the operating conditions had been slightly higher than the above, in place of Oxidizing oxygen, Carbon monoxide would then be released as shown below after Stage 8 above-

$$\underline{Stage\,9}: \quad O_2 \xrightleftharpoons[Oxygen\,molecule]{Equilibruim\,State\,Existence\,of\,Oxidising} 2en\cdot\ddot{O}\cdot nn$$

$$en\cdot O\cdot nn + \quad 2O=CH_2 \quad \xrightleftharpoons{Oxidation} 2H-O\cdot nn + 2e\cdot\overset{\overset{O}{\|}}{C}-H$$
$$(Stabilized) \qquad\qquad\qquad\qquad (B)$$

$$(B) \quad \xrightleftharpoons{} \quad 2H\bullet e + 2n\bullet\overset{\overset{O}{\|}}{C}\bullet e$$
$$\qquad\qquad\qquad\qquad (A)$$

$$2H\bullet e + 2nn\bullet OH \quad \xrightleftharpoons{} \quad 2H_2O$$
$$(A) \qquad \xrightarrow[release\,of\,Energy]{Deactivation} \quad :C=O \;+\; Energy \qquad\qquad (xvii)$$

$$\underline{Overall\,Equation}: O_2 + 2O=CH_2 \longrightarrow 2H_2O \;+\; CO \;+\; Energy \qquad (xviii)$$

FROM COMBUSTION (FULL)

$$\underline{Final\,Overall\,equation}: 2C_2H_6 \;+\; 6O_2 \longrightarrow 2CO_2 + 6H_2O \;+\; 2CO$$
$$+ \; Energy(From\,five\,Stages) \qquad [b]$$

noting that this equation complements the first final equation above for the combustion of ethane at high operating condition with nine stages as opposed to eleven stages when CO is not formed, for with nine stages, only 6 moles of O_2 as opposed to 8 moles are required and this would apply to higher alkanes, from which it can be seen that *while methane is the only member of the family of Alkanes that undergoes full combustion with molecular oxygen of the AIR, all the other members in the family undergo both combustion and oxidation with molecular oxygen,* and the reason is because methane is the first member of the family, something which is always unique with all first members of families of compound- Alkene, alkynes, Cyclic members, Aldehydes, Ketones, Carboxylic acids and so on.
(Laws of Creations for Combustion of Ethane)

Rule 1625: This rule of Chemistry for **Combustion of Alkane Hydrocarbon family of Compounds**, states that, if during the combustion of Ethane, the operating conditions had been very high, the situation will resemble the combustion of Diazomethane as shown below

$$\underline{Stage\,1}: \quad C_2H_6 \xrightleftharpoons[Existence\,at\,above\,Melting\,Point]{Equilibrium\,State\,of} H\cdot e + n\cdot C_2H_5$$

$$H\cdot e + O=O \quad \xrightleftharpoons[Oxygen]{Activation\,of} \quad H-O-O\cdot en$$

$$H-O-O\cdot en + n\cdot C_2H_5 \xrightarrow{Combination\,State} H-O-O-C_2H_5$$
$$(A\;peroxide) \qquad\qquad (i)$$

$$\underline{Overall\,equation}: \quad C_2H_6 \;+\; O_2 \longrightarrow HOOC_2H_5 \qquad\qquad (ii)$$

Stage 2:

$$H-O-O-\underset{\underset{H}{|}}{\overset{\overset{H}{|}}{C}}-\underset{\underset{H}{|}}{\overset{\overset{H}{|}}{C}}-H \quad \underset{\text{Existence of the Peroxide}}{\overset{\text{Equilibrium State of}}{\rightleftharpoons}} \quad H-O\cdot nn + en\cdot O-\underset{\underset{H}{|}}{\overset{\overset{H}{|}}{C}}-\underset{\underset{H}{|}}{\overset{\overset{H}{|}}{C}}-H$$

$$(A)$$

$$(A) \quad \underset{\text{Release}}{\overset{\text{Energy}}{\rightleftharpoons}} \quad H\cdot e + O=\underset{\underset{H}{|}}{\overset{\overset{CH_3}{|}}{C}} \quad + \quad Energy$$

$$(Acetaldehyde)$$

$$H\cdot e + nn\cdot O-H \quad \xrightarrow{\text{Combination}} \quad H_2O \tag{iii}$$

Overall Equation: $HOOC_2H_5 \longrightarrow H_2O + H_3CCHO + Energy$ (iv)

Stage 3:

$$CH_3CHO \quad \underset{\text{Existence of the Unactivated Aldehyde}}{\overset{\text{Equilibrium state of}}{\rightleftharpoons}} \quad H\cdot e + O=\overset{\overset{CH_3}{|}}{C}\cdot n$$

$$H\cdot e + O=O \quad \underset{\text{Oxygen}}{\overset{\text{Activation of}}{\rightleftharpoons}} \quad H-O-O\cdot en$$

$$H-O-O\cdot en + n\cdot \overset{\overset{CH_3}{|}}{C}=O \quad \xrightarrow{\text{Combination}} \quad H-O-O-\overset{\overset{CH_3}{|}}{C}=O$$

$$(B)-A\ peroxide \tag{v}$$

Overall equation: $H_3CCHO + O_2 \longrightarrow HOOC(CH_3)=O$ (vi)

Stage 4:

$$H_3C-\overset{\overset{O}{\|}}{C}-O-O-H \quad \underset{\text{of Existence}}{\overset{\text{It's Equilibrium State}}{\rightleftharpoons}} \quad 2H-O\cdot en + 2nn\cdot O-\overset{\overset{O}{\|}}{C}-CH_3$$

$$2nn\cdot O-\overset{\overset{O}{\|}}{C}-CH_3 \quad \rightleftharpoons \quad 2nn\cdot O-\overset{\overset{O}{\|}}{C}\cdot e + 2n\cdot CH_3$$

$$2H-O\cdot en \quad \underset{\text{Release}}{\overset{\text{Energy}}{\rightleftharpoons}} \quad 2H\cdot e \quad + \quad \underline{O_2} \quad + Energy$$
$$(Oxidising\ Oxygen)$$
$$molecule$$

$$2H\cdot e + 2n\cdot CH_3 \quad \underset{\text{Existence of } CH_4}{\overset{\text{Equilibrium State of}}{\rightleftharpoons}} \quad 2CH_4$$

$$2e\cdot \underset{\underset{O}{\|}}{C}-O\cdot nn \quad \underset{\text{Energy Release}}{\overset{\text{Deactivation}}{\longrightarrow}} \quad 2CO_2 \quad + \quad Energy \tag{vii}$$

Overall Equation: $2C_2H_6 + 4O_2 \longrightarrow \underline{2CH_4} + 2CO_2 + 2H_2O + \underline{O_2} + Energy$
$$(Air) \quad (Methane) \quad\quad (Oxidising\ Oxygen)$$
$$Molecule \tag{viii}$$

Stage 5: $O_2 \underset{\text{Oxygen molecule}}{\overset{\text{Equilibruim State Existence of Oxidising}}{\rightleftharpoons}} 2en \cdot \ddot{O} \cdot nn$

$2en \cdot O \cdot nn + 2CH_4 \xrightleftharpoons{\text{Oxidation}} 2H - O \cdot nn + 2e \cdot CH_3$
$\qquad\qquad$ (*Stabilized*)

$$2e \bullet CH_3 \rightleftharpoons 2H \bullet e + 2e \bullet \overset{\displaystyle H}{\underset{\displaystyle H}{\overset{|}{\underset{|}{C}}}} \bullet n$$

$\qquad\qquad\qquad\qquad\qquad$ (*A*) − *Methylene*

$2H \bullet e + 2nn \bullet OH \rightleftharpoons 2H_2O$
$\qquad$ (*A*) $\qquad \xrightarrow[\text{Release of Energy}]{\text{Deactivation}} 2 : CH_2 \, (A\,Carbene) + Energy \qquad$ (ix)

Overall Equation : $O_2 + 2CH_4 \longrightarrow 2H_2O + 2 : CH_2 + Energy \qquad$ (x)

Stage 6 :

$$: CH_2 \rightleftharpoons n \cdot \overset{\displaystyle H}{\underset{\displaystyle H}{\overset{|}{\underset{|}{C}}}} \cdot e$$

$\qquad\qquad\qquad\qquad$ (*A*)

$$(A) + O = O \rightleftharpoons n \cdot \overset{\displaystyle H}{\underset{\displaystyle H}{\overset{|}{\underset{|}{C}}}} - O - O \cdot en$$

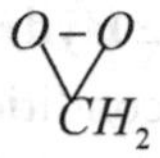

$$\longrightarrow \overset{\displaystyle O - O}{\underset{\displaystyle CH_2}{\diagdown\diagup}}$$

$\qquad\qquad\qquad\qquad$ (*B*) $\qquad\qquad\qquad\qquad$ (xi)

Overall equation : $2H_2C + 2O_2 \longrightarrow$ (*B*) $\qquad\qquad$ (xii)

Stage 7 :

$$(B) \rightleftharpoons 2nn \cdot O - \overset{\displaystyle H}{\underset{\displaystyle H}{\overset{|}{\underset{|}{C}}}} - O \cdot en$$

$$\rightleftharpoons O_2 + 2nn \cdot O - \overset{\displaystyle H}{\underset{\displaystyle H}{\overset{|}{\underset{|}{C}}}} \cdot e$$

$\qquad\qquad\qquad\qquad\qquad$ (*C*)

$(C) \xrightarrow[\text{Release of energy}]{\text{Deactivation}} 2H_2C = O + Heat \qquad\qquad$ (xii)

Overall equation : (*B*) $\longrightarrow O_2 + 2H_2C = O + Energy \qquad\qquad$ (xiv)

Stage 8 : $O_2 \underset{\text{Oxygen molecule}}{\overset{\text{Equilibruim State Existence of Oxidising}}{\rightleftharpoons}} 2en \cdot \ddot{O} \cdot nn$

$$en \cdot O \cdot nn + \quad 2O = CH_2 \quad \xrightleftharpoons{\text{Oxidation}} \quad 2H - O \cdot nn + 2e \cdot \overset{\overset{O}{\|}}{C} - H$$
$$\text{(Stabilized)} \qquad\qquad\qquad\qquad (B)$$

$$(B) \quad \rightleftharpoons \quad 2H \cdot e + 2n \cdot \overset{\overset{O}{\|}}{C} \cdot e$$
$$(A)$$

$$2H \cdot e + 2nn \cdot OH \quad \rightleftharpoons \quad 2H_2O$$

$$(A) \quad \xrightarrow[\text{release of Energy}]{\text{Deactivation}} \quad :C = O + \quad Energy \qquad\qquad (xv)$$

$$\underline{Overall\ Equation: O_2 + 2O = CH_2} \xrightarrow{} 2H_2O + CO + Energy \qquad (xvi)$$

$$\underline{Final\ Overall\ equation: 2C_2H_6 + 6O_2 \xrightarrow{} 2CO_2 + 6H_2O + 2CO}$$
$$+ Energy(From\ five\ Stages) \qquad\qquad [c]$$

an eight stage Equilibrium system wherein Energy in the form of Heat still remains released in five stages and Carbon monoxide an environmentally unfriendly gas is one of the products; noting that methylene is an intermediate as shown above, in view of the operating condition. with *the last equation above complimenting two other cases of eleven and nine stages, because of the great importance of OPERATING CONDITIONS and STAGE-WISE operations.*
(Laws of Creations for Combustion of Ethane)

Rule 1626: This rule of Chemistry for **Combustion of Alkane Hydrocarbon family of Compounds**, states that, if during the combustion of Ethane, the operating conditions is such that Enolization types of catalysts are present at high conditions, the situation is completely different as shown below-

$$Stage1: \quad C_2H_6 \xrightleftharpoons[\text{Existence at above Melting Point}]{\text{Equilibrium State of}} H \cdot e + n \cdot C_2H_5$$

$$H \cdot e + O = O \xrightleftharpoons[\text{Oxygen}]{\text{Activation of}} H - O - O \cdot en$$

$$H - O - O \cdot en + n \cdot C_2H_5 \xrightarrow{\text{Combination State}} H - O - O - C_2H_5$$
$$(A\ peroxide) \qquad\qquad (i)$$
$$\underline{Overall\ equation: \quad C_2H_6 + O_2 \xrightarrow{} HOOC_2H_5} \qquad\qquad (ii)$$

$$Stage2:$$

$$\underset{\underset{H}{|}\ \underset{H}{|}}{\overset{\overset{H}{|}\ \overset{H}{|}}{H - O - O - C - C - H}} \xrightleftharpoons[\text{Existence of the Peroxide}]{\text{Equilibrium State of}} H - O \cdot nn + en \cdot O - \underset{\underset{H}{|}\ \underset{H}{|}}{\overset{\overset{H}{|}\ \overset{H}{|}}{C - C - H}}$$
$$(A)$$

$$(A) \quad \xrightleftharpoons[\text{Release}]{\text{Energy}} \quad H \cdot e + O = \underset{\underset{H}{|}}{\overset{\overset{CH_3}{|}}{C}} + Energy$$
$$(Acetaldehyde)$$

678

$$H \cdot e + nn \cdot O - H \xrightarrow{\text{Combination}} H_2O \qquad \text{(iii)}$$

Overall Equation: $HOOC_2H_5 \longrightarrow H_2O + H_3CCHO + Energy$ \qquad (iv)

Stage 3:

$$CH_3CHO \xrightleftharpoons[\text{Existence of the Unactivated Aldehyde}]{\text{Equilibrium state of}} H \cdot e + nn \cdot O - \underset{\underset{H}{|}}{\overset{\overset{n \cdot CH_2}{|}}{C}} \cdot e$$

$$H \cdot e + O = O \xrightleftharpoons[\text{Oxygen}]{\text{Activation of}} H - O - O \cdot en$$

$$H - O - O \cdot en + e \cdot \underset{\underset{H}{|}}{\overset{\overset{n \cdot CH_2}{|}}{C}} - O \bullet nn \xrightarrow{\text{Combination}} H - O - O - \underset{\underset{H}{|}}{\overset{\overset{H}{|}}{C}} - \overset{H}{\underset{}{C}} = O$$

$$(B) - A\ peroxide \qquad \text{(v)}$$

Stage 4:

$$H - \overset{\overset{O}{\|}}{C} - \underset{\underset{H}{|}}{\overset{\overset{H}{|}}{C}} - O - O - H \xrightleftharpoons[\text{of Existence}]{\text{It's Equilibrium State}} H - O \cdot nn + en \bullet O - \underset{\underset{H}{|}}{\overset{\overset{H}{|}}{C}} - \overset{\overset{O}{\|}}{C} - H$$

$$en \cdot O - \underset{\underset{H}{|}}{\overset{\overset{H}{|}}{C}} - \overset{\overset{O}{\|}}{C} - H \xrightleftharpoons{} H - \overset{O}{\underset{}{C}} \cdot e + H_2C = O \quad + \quad Energy$$

$$(C)$$

$$(C) \xrightleftharpoons{} e \bullet \overset{\overset{O}{\|}}{C} \bullet n \quad + \quad H \bullet e$$

$$H \cdot e + nn \cdot OH \xrightleftharpoons{} H_2O$$

$$e \cdot \underset{\underset{O}{\|}}{C} \bullet n \xrightarrow{\substack{\text{Deactivation} \\ \text{Energy Release}}} CO + Energy$$

\qquad (vi)

Overall Equation: $C_2H_6 + 2O_2 \longrightarrow CO + 2H_2O + H_2C = O + Energy$ \qquad (vii)

Stage 5:

$$H_2C = O \xrightleftharpoons[\text{Existence of formaldehyde}]{\text{Activation State of}} H \cdot e + n \cdot \overset{\overset{H}{|}}{C} = O$$

$$H \cdot e + O = O \xrightleftharpoons[\text{Oxygen}]{\text{Activation of}} H - O - O \cdot en$$

$$(A)$$

$$(A) + n \cdot \overset{\overset{H}{|}}{C} = O \xrightarrow{\text{Combination}} H - O - O - \overset{\overset{H}{|}}{C} = O \qquad \text{(viii)}$$

$$(Another\ peroxide)$$

Stage 6:

$$H - O - O - \overset{\overset{H}{|}}{C} = O \xrightleftharpoons[\text{Existence}]{\text{Equilibrium State of}} H - \ddot{O} \cdot nn + en \cdot O - \overset{\overset{H}{|}}{C} = O$$

$$(C)$$

$$(C) \quad \underset{Re\,lease}{\overset{Energy}{\rightleftharpoons}} \quad CO_2 \ + \ H \cdot e \ + \ Energy$$

$$H \cdot e + nn \cdot OH \quad \xrightarrow{Combination} \quad H_2O \qquad\qquad\qquad (ix)$$

$$\underline{Overall \ Equation}: C_2H_6 + 3O_2 \ \longrightarrow \ CO \ + \ CO_2 \ + \ 2H_2O \ + \ Energy \qquad (x)$$

for which with the use of Activated/Equilibrium state of existence of acetaldehyde for the continued combustion of ethane, only six stages wherein energy is released four times in three stages were obtained, that which is quite different from the case where the Equilibrium state of existence of acetaldehyde was used; noting the complete absence of Oxidizing oxygen as an intermediate product, *clear indication that these do not take place during combustion of Alkanes.*
(Laws of Creations for Combustion of Ethane)

Rule 1627: This rule of Chemistry for **Oxidation of Alkane Hydrocarbon family of Compounds,** states that, beginning with the first member *Methane,* the followings are obtained-

Stage 1:

$$O_2 \quad \underset{Oxygen\,molecule}{\overset{Equilibruim\ State\ Existence\ of\ Oxidi\sin g}{\rightleftharpoons}} \quad 2en \cdot \ddot{O} \cdot nn$$

$$2en \cdot O \cdot nn + \ 2CH_4 \quad \underset{}{\overset{Oxidation}{\rightleftharpoons}} \ 2H - O \cdot nn \ + \ 2e \cdot CH_3$$
$$(Stabilized)$$

$$\longrightarrow \quad 2CH_3OH \qquad\qquad\qquad (i)$$

$$\underline{Overall \ Equation}: O_2 + 2CH_4 \ \longrightarrow \qquad 2CH_3OH \qquad\qquad (ii)$$

Stage 2:

$$O_2 \quad \underset{Oxygen\,molecule}{\overset{Equilibruim\ State\ Existence\ of\ Oxidi\sin g}{\rightleftharpoons}} \quad 2en \cdot \ddot{O} \cdot nn$$

$$2en \cdot O \cdot nn + \ 2CH_3OH \quad \underset{}{\overset{Oxidation}{\rightleftharpoons}} \ 2H - O \cdot nn \ + \ 2CH_3O \bullet en$$
$$(Stabilized)$$

$$2CH_3O \bullet en \quad \rightleftharpoons \quad 2H \bullet e \quad + \quad 2\overset{H}{\underset{H}{\overset{|}{\underset{|}{C}}}} = O \quad + \quad Energy$$
$$(A)$$

$$2H \bullet e \ + \ 2nn \bullet OH \ \longrightarrow \ 2H_2O \qquad\qquad\qquad (iii)$$

$$\underline{Overall \ Equation}: O_2 + 2CH_3OH \ \longrightarrow \ 2H_2O \ + \ 2H_2C = O \ + \ Energy \qquad (iv)$$

Stage 3:

$$O_2 \quad \underset{Oxygen\,molecule}{\overset{Equilibruim\ State\ Existence\ of\ Oxidi\sin g}{\rightleftharpoons}} \quad 2en \cdot \ddot{O} \cdot nn$$

$$2en \cdot O \cdot nn + \ 2O = CH_2 \quad \underset{}{\overset{Oxidation}{\rightleftharpoons}} \ 2H - O \cdot nn \ + \ 2e \cdot \overset{O}{\overset{\|}{C}} - H$$
$$(Stabilized) \qquad\qquad\qquad (B)$$

$$\longrightarrow \quad 2HO-\overset{\overset{H}{|}}{C}=O \tag{v}$$

Overall Equation: $\underset{\underline{\underline{\ }}}{O_2} + 2O = CH_2 \longrightarrow 2HO-CHO \tag{vi}$

Stage 4 : $\quad \underset{\underline{\underline{\ }}}{O_2} \xrightleftharpoons[\text{Oxygen molecule}]{\text{Equilibruim State Existence of Oxidi sin g}} 2en \cdot \overset{\cdot\cdot}{\underset{\cdot\cdot}{O}} \cdot nn$

$2en \cdot O \cdot nn + \; 2HO-CHO \quad \xrightleftharpoons{\text{Oxidation}} 2H-O \cdot nn + 2\,H\overset{\overset{O\bullet en}{|}}{C}=O \; + \; Heat$

$\qquad\qquad (Stabilized) \qquad\qquad\qquad\qquad\qquad\quad (B)$

$\qquad\qquad (B) \qquad \rightleftharpoons \qquad 2\,CO_2 \; + \; 2H \bullet e \; + \; Heat$

$\qquad\qquad\qquad\qquad\qquad\qquad (C)$

$2H \bullet e \; + \; 2nn \bullet OH \longrightarrow \qquad 2H_2O \tag{vii}$

Overall Equation: $\underset{\underline{\underline{\ }}}{O_2} + 2HO-CHO \longrightarrow 2H_2O \; + \; 2CO_2 \; + \; Energy \tag{viii}$

FROM OXIDATION (FULL)

Overall overall equation: $2CH_4 + 4O_2 \xrightarrow{\text{Oxidation}} 4H_2O + 2CO_2 + Energy(2)$

$\qquad\qquad\qquad\qquad\qquad\qquad\qquad\qquad (From\ four\ Stages)\quad [A]$

835

FROM COMBUSTION (FULL)

Overall overall equation: $2CH_4 + 4O_2 \xrightarrow{\text{Combustion}} 4H_2O + 2CO_2 + Energy(2)$

$\qquad\qquad\qquad\qquad\qquad\qquad\qquad\qquad (From\ four\ Stages)\quad [B]$

for if the operating conditions had been higher than above, instead of Stages 3 and 4 above, the followings would be obtained-

Stage 3 : $\quad \underset{\underline{\underline{\ }}}{O_2} \xrightleftharpoons[\text{Oxygen molecule}]{\text{Equilibruim State Existence of Oxidi sin g}} 2en \cdot \overset{\cdot\cdot}{\underset{\cdot\cdot}{O}} \cdot nn$

$2en \cdot O \cdot nn + \; 2O = CH_2 \quad \xrightleftharpoons{\text{Oxidation}} 2H-O \cdot nn + 2e \cdot \overset{\overset{O}{\|}}{C}-H$

$\qquad\qquad (Stabilized) \qquad\qquad\qquad\qquad\qquad (B)$

$\qquad\qquad (B) \qquad \rightleftharpoons \qquad 2H \bullet e \; + \; 2n \bullet \overset{\overset{O}{\|}}{C} \bullet e$

$\qquad\qquad\qquad\qquad\qquad\qquad\qquad (A)$

$2H \bullet e \; + \; 2nn \bullet OH \qquad \rightleftharpoons \qquad 2H_2O$

$\qquad\qquad (A) \qquad \xrightarrow[\text{release of Energy}]{\text{Deactivation}} \quad 2:C=O \; + \; Energy \tag{v}$

Overall Equation: $\underset{\underline{\underline{\ }}}{O_2} + 2O = CH_2 \longrightarrow 2H_2O \; + \; 2CO \; + \; Energy \tag{vi}$

Overall overall equation: $2CH_4 + 3\underline{O_2} \xrightarrow{\text{Oxidation}} 4H_2O + 2CO \; + \; Energy(2)$

$\qquad\qquad\qquad\qquad\qquad\qquad\qquad\qquad (From\ three\ Stages)\quad [C]$

$Stage\,4:\quad O_2 \underset{Oxygen\;molecule}{\overset{Equilibrium\;State\;Existence\;of\;Oxidising}{\rightleftharpoons}} 2en\cdot\ddot{O}\cdot nn$

$2en\cdot O\cdot nn + \underset{(Stabilized)}{2C=O} \overset{Oxidation}{\rightleftharpoons} \underset{(B)}{2\,e\bullet\overset{\overset{O}{\|}}{C}-O\bullet nn}$

$(B) \xrightarrow[Release\;of\;Energy]{Deactivation} 2CO_2 + Energy$ \hfill (vii)

$Overall\;Equation: O_2 + 2CO \longrightarrow 2CO_2 + Energy$ \hfill (viii)

$Overall\;overall\;equation: 2CH_4 + 4O_2 \xrightarrow{Oxidation} 4H_2O + 2CO_2 + Energy(3)$

$(From\;four\;Stages)\;\;[D]$

\hfill (ix)

from which it can be observed that if there is no oxidizing oxygen in the system, CO_2 will not be produced in Stage 4, and one can begin to see *the origin of Carbon monoxide both during Combustion and Oxidation, but less in Oxidation than in Combustion; noting that at higher operating conditions, more heat is produced than above and via Combustion, and with controlled oxidation, methanol can be obtained at lower operating conditions.*
(Laws of Creations for Oxidation of Methane)

Rule 1628: This rule of Chemistry for **Oxidation of Ethane,** states that, when a suitable oxidizing agent is used, the followings are obtained-

$Stage\,1:\quad O_2 \underset{Oxygen\;molecule}{\overset{Equilibrium\;State\;Existence\;of\;Oxidising}{\rightleftharpoons}} 2en\cdot\ddot{O}\cdot nn$

$2en\cdot O\cdot nn + \underset{(Stabilized)}{2C_2H_6} \overset{Oxidation}{\rightleftharpoons} 2H-O\cdot nn + 2e\cdot C_2H_5$

$2e\bullet C_2H_5 \rightleftharpoons 2H\bullet e + 2e\bullet\underset{\substack{| \\ H\;\;H}}{\overset{\substack{H\;\;H \\ |}}{C}-C}\bullet n$

(A)

$2H\bullet e + 2nn\bullet OH \rightleftharpoons 2H_2O$

$(A) \xrightarrow[Release\;of\;Energy]{Deactivation} H_2C=CH_2 + Energy$ \hfill (i)

$Overall\;Equation: O_2 + 2C_2H_6 \longrightarrow 2H_2O + 2H_2C=CH_2 + Energy$ \hfill (ii)

$Stage\,2:\quad O_2 \underset{Oxygen\;molecule}{\overset{Equilibrium\;State\;Existence\;of\;Oxidising}{\rightleftharpoons}} 2en\cdot\ddot{O}\cdot nn$

$2en\cdot O\cdot nn + \underset{(Stabilized)}{2H_2C=CH_2} \overset{Oxidation}{\rightleftharpoons} 2HO\bullet nn + \underset{(B)}{2e\bullet\overset{\overset{H}{|}}{C}=CH_2}$

$(B) \rightleftharpoons H\bullet e + e\bullet\overset{\overset{H}{|}}{C}=\underset{\underset{H}{|}}{C}\bullet n$

682

$$2H\bullet e \;+\; 2nn\bullet OH \;\rightleftharpoons\; 2H_2O$$

$$(B) \xrightarrow[\text{Release of Energy}]{\text{Deactivation}} 2HC \equiv CH \;+\; Energy \qquad\qquad\text{(iii)}$$

$$\underline{Overall\;Equation}: O_2 + 2H_2C = CH_2 \longrightarrow 2H_2O + 2HC \equiv CH + Energy \qquad\text{(iv)}$$

$\underline{Stage\,3}$

$$O_2 \xrightleftharpoons[\text{Oxygen molecule}]{\text{Equilibruim State Existence of Oxidising}} 2en\cdot\ddot{O}\cdot nn$$

$$2en\cdot O\cdot nn + \; 2HC \equiv CH \xrightleftharpoons{\text{Oxidation}} 2HO\bullet nn \;+\; 2\overset{H}{C} \equiv C\bullet e$$
$$\text{(Stabilized)} \qquad\qquad\qquad\qquad (C)$$

$$(C) \rightleftharpoons 2H\bullet e \;+\; 2e\bullet C \equiv C\bullet n$$
$$(D)$$

$$(D) \rightleftharpoons 4e\bullet\overset{\bullet n}{\underset{\bullet e}{C}}\bullet n$$
$$(E)\,Activated\;Carbon\;black$$

$$2H\bullet e \;+\; 2nn\bullet OH \;\rightleftharpoons\; 2H_2O$$

$$(E) \xrightarrow[\text{Release of Energy}]{\text{Deactivation}} 4:\overset{\bullet nn}{\underset{\bullet en}{C}} \;+\; Energy \qquad\qquad\text{(v)}$$
$$\text{(Carbon black)}$$

$$\underline{Overall\;Equation}: O_2 + 2HC \equiv CH \longrightarrow 2H_2O + 4:\overset{\bullet nn}{\underset{\bullet en}{C}} + Energy \qquad\text{(vi)}$$

$\underline{Stages\,4\,\&\,5}:\,[In\;parallel]$

$$O_2 \xrightleftharpoons[\text{Oxygen molecule}]{\text{Equilibruim State Existence of Oxidising}} 2en\cdot\ddot{O}\cdot nn$$

$$2en\cdot O\cdot nn + \; 2:\overset{\bullet n}{\underset{\bullet e}{C}} \xrightleftharpoons{\text{Oxidation}} 2nn\bullet O - \overset{\bullet e}{\underset{\bullet n}{C}}\bullet e$$
$$\text{(Stabilized)} \qquad\qquad (F)$$

$$\xrightarrow[\text{Release of Energy}]{\text{Deactivation}} 2:C = O \;+\; Energy \qquad\qquad\text{(vii)}$$

$$\underline{Overall\;Equation}: 2O_2 + 4\overset{\bullet n}{\underset{\bullet e}{C}}: \longrightarrow 4:C = O \;+\; Energy \qquad\text{(viii)}$$

FOR OXIDATION (PARTIAL)

$$\underline{Final\;Overall\;equation}: 2C_2H_6 \;+\; 5O_2 \longrightarrow 4CO + 6H_2O + Energy\text{(4)}$$
$$\text{(From five Stages)} \quad [A]$$

$$\underline{Final\;Overall\;equation}: 2CH_4 \;+\; 3O_2 \longrightarrow 2CO + 4H_2O + Energy\text{(3)}$$
$$\text{(From three Stages)} \quad [B]$$

FROM COMBUSTION (FULL)

$$\underline{Final\;Overall\;equation}: 2C_2H_6 \;+\; 6O_2 \longrightarrow 2CO_2 + 6H_2O \;+\; 2CO$$
$$+\; Energy\text{(5)}$$
$$\text{(From eight or nine Stages)} \qquad [C]$$

$\underline{Stages\,6\,\&\,7}:$

$$O_2 \xrightleftharpoons[\text{Oxygen molecule}]{\text{Equilibruim State Existence of Oxidising}} 2en\cdot\ddot{O}\cdot nn$$

$$2en \cdot O \cdot nn + 2C = O \xrightleftharpoons{Oxidation} 2e \cdot \overset{\overset{O}{\|}}{C} - O \cdot nn$$
$$\text{(Stabilized)} \qquad\qquad (B)$$

$$(B) \xrightarrow[\text{Release of Energy}]{\text{Deactivation}} 2CO_2 + Energy \qquad\qquad\qquad\qquad \text{(ix)}$$

Overall Equation 2: $O_2 + 4CO \longrightarrow 4CO_2 + Energy$ $\qquad\qquad\qquad$ (x)

FULL OXIDATION

Final Overall equation: $2C_2H_6 + 7O_2 \longrightarrow 4CO_2 + 6H_2O + Energy(5)$

$$\text{(From seven Stages)} \qquad\qquad [D]$$

a five stage Equilibrium mechanism system in which the last two stages are two in one (i.e., in parallel) for FULL OXIDATION, for when oxidation is PARTIAL, carbon monoxide an environmentally unfriendly gas appears as one of the products only in Stage 4 (two in one), taking cognizance of the fact that while in Combustion, the formation of carbon monoxide cannot be prevented in the absence of oxidizing oxygen in the system or an oxidizing agent, in Oxidation, the formation can be prevented by not starving the system with oxidizing oxygen.

(Laws of Creations for Oxidation of Ethane)

Rule 1629: This rule of Chemistry for **Oxidation of n-Propane,** states that, when a suitable oxidizing agent is used, the followings are obtained-

Stage 1 $\qquad O_2 \xrightleftharpoons[\text{Oxygen molecule}]{\text{Equilibruim State Existence of Oxidising}} 2en \cdot \ddot{O} \cdot nn$

$$2en \cdot O \cdot nn + 2C_3H_8 \xrightleftharpoons{Oxidation} 2e \cdot \overset{H}{\underset{H}{C}} - \overset{H}{\underset{H}{C}} - \overset{H}{\underset{H}{C}} - H + 2nn \cdot OH$$
$$\text{(Stabilized)}$$
$$(B)$$

$$\longrightarrow 2HO - C_3H_7 \qquad\qquad\qquad\qquad \text{(i)}$$

Overall Equation: $O_2 + 2H_8C_3 \longrightarrow 2C_3H_7OH$ $\qquad\qquad\qquad$ (ii)

worthy of note in this first stage is that unlike methane and ethane, hydrogen atom (H•e) could not be released from (B) above, because when released the carbon center from where it was released cannot carry a nucleo-free-radical to form an alkene and CH_3 group cannot be released because it is more radical pushing than H, and unlike with the case of methane and ethane, propanol can readily be produced only from n-propane regardless the operating conditions, for as long as the exact molar ratios of the two components have been used-.

Overall equation: $2CH_4 + O_2 \longrightarrow 2H_2O + Unactivated\ Methylene2(:CH_2)$

$$OR\ 2METHANOL$$
$$+ Energy \qquad\qquad [A\}$$

$$2C_2H_6 + O_2 \longrightarrow 2H_2O + 2H_2C = CH_2 + Energy \qquad [B]$$

$$2C_3H_8 + O_2 \longrightarrow 2C_3H_7OH \qquad\qquad\qquad [C]$$

Stage 2:

$$O_2 \xrightleftharpoons[\text{Oxygen molecule}]{\text{Equilibruim State Existence of Oxidising}} 2en \cdot \ddot{O} \cdot nn$$

$$2en \cdot O \cdot nn + 2C_3H_7OH \xrightleftharpoons{\text{Oxidation}} 2H - \underset{\underset{H}{|}}{\overset{\overset{H}{|}}{C}} - \underset{\underset{H}{|}}{\overset{\overset{H}{|}}{C}} - \underset{\underset{H}{|}}{\overset{\overset{H}{|}}{C}} - O \bullet en + 2nn \bullet OH$$

(*Stabilized*) $\qquad\qquad\qquad (C)$

$$(C) \xrightleftharpoons{\quad\quad} 2H_5C_2HC = O + 2H \bullet e + Energy$$

$$2H \bullet e + 2nn \bullet OH \longrightarrow 2H_2O \tag{iii}$$

$$\underline{Overall\ Equation:}\ 2O_2 + 2C_3H_8 \longrightarrow 2H_2O + \begin{array}{l} 2H_5C_2HC = O \\ + \quad Energy \end{array} \tag{iv}$$

for which the hydrogen atom removed is that from the center carrying OH group, that which is more radical-pushing than H and when the H was removed, an aldehyde was instantan-eously formed with release of energy and oxidation of the aldehyde follows-

Stage 3:

$$O_2 \xrightleftharpoons[\text{Oxygen molecule}]{\text{Equilibruim State Existence of Oxidising}} 2en \cdot \ddot{O} \cdot nn$$

$$2en \cdot O \cdot nn + 2O = CHC_2H_5 \xrightleftharpoons{\text{Oxidation}} 2H - O \cdot nn + 2e \cdot \overset{\overset{O}{\|}}{C} - C_2H_5$$

(*Stabilized*) $\qquad\qquad\qquad (D)$

$$\longrightarrow 2H_5C_2 - \overset{\overset{O}{\|}}{C} - OH$$

$\qquad\qquad\qquad\qquad (E) \tag{v}$

$$\underline{Overall\ Equation:}\ O_2 + 2O = CHC_2H_5 \longrightarrow 2H_2O + 2(E) \tag{vi}$$

Stage 4:

$$O_2 \xrightleftharpoons[\text{Oxygen molecule}]{\text{Equilibruim State Existence of Oxidising}} 2en \cdot \ddot{O} \cdot nn$$

$$2en \cdot O \cdot nn + 2(E) \xrightleftharpoons{\text{Oxidation}} HO \bullet nn + 2en \cdot O - \overset{\overset{O}{\|}}{C} - C_2H_5 \bullet e$$

(*Stabilized*) $\qquad\qquad\qquad (F)$

$$(F) \rightleftharpoons 2CO_2 + 2e \bullet C_2H_5 + Energy$$

$$2e \bullet C_2H_5 \rightleftharpoons 2n \bullet CH_2 - H_2C \bullet e + 2H \bullet e$$

$\qquad\qquad\qquad\qquad (G)$

$$2H \bullet e + 2nn \bullet OH \rightleftharpoons 2H_2O$$

$$(G) \xrightarrow{\text{Deactivation}} 2H_2C = CH_2 + Energy \tag{vii}$$

$$\underline{Overall\ Equation:}\ O_2 + 2(E) \longrightarrow 2CO_2 + 2H_2C = CH_2 + 2H_2O \\ + \quad Energy \tag{viii}$$

$$\underline{Overall\ Overall\ equation:}\ 2C_3H_8 + 4O_2 \longrightarrow 4H_2O + 2CO_2 + 2H_2C = CH_2 \\ + \quad Energy \qquad [a]$$

Stage 5:

$$O_2 \xrightleftharpoons[\text{Oxygen molecule}]{\text{Equilibruim State Existence of Oxidi sing}} 2en\cdot\ddot{O}\cdot nn$$

$$2en\cdot O\cdot nn + 2H_2C=CH_2 \xrightleftharpoons{\text{Oxidation}} 2HO\bullet nn + 2e\bullet\overset{\overset{\displaystyle H}{|}}{C}=CH_2$$
$$\text{(Stabilized)} \qquad\qquad\qquad (H)$$

$$(H) \xrightleftharpoons{} H\bullet e + e\bullet\overset{\overset{\displaystyle H}{|}}{C}=\underset{\underset{\displaystyle H}{|}}{C}\bullet n$$
$$(I)$$

$$2H\bullet e + 2nn\bullet OH \xrightleftharpoons{} 2H_2O$$

$$(I) \xrightarrow[\text{Release of Energy}]{\text{Deactivation}} 2HC\equiv CH + Energy \qquad\qquad (ix)$$
$$(J)$$

$$\underline{Overall\ Equation:}\ O_2 + 2H_2C=CH_2 \longrightarrow 2H_2O + 2HC\equiv CH + Energy \qquad (x)$$

Stage 6:

$$O_2 \xrightleftharpoons[\text{Oxygen molecule}]{\text{Equilibruim State Existence of Oxidi sing}} 2en\cdot\ddot{O}\cdot nn$$

$$2en\cdot O\cdot nn + 2HC\equiv CH \xrightleftharpoons{\text{Oxidation}} 2HO\bullet nn + 2\overset{\overset{\displaystyle H}{|}}{C}\equiv C\bullet e$$
$$\text{(Stabilized)} \qquad\qquad\qquad (K)$$

$$(K) \xrightleftharpoons{} 2H\bullet e + 2e\bullet C\equiv C\bullet n$$
$$(L)$$

$$(L) \xrightleftharpoons{} 4e\bullet\overset{\overset{\displaystyle \bullet n}{}}{C}\bullet n$$
$$(M)\ Activated\ Carbon\ Black$$

$$2H\bullet e + 2nn\bullet OH \xrightleftharpoons{} 2H_2O$$

$$(M) \xrightarrow[\text{Release of Energy}]{\text{Deactivation}} 4:\overset{\overset{\displaystyle \bullet nn}{}}{\underset{\underset{\displaystyle \bullet en}{}}{C}} + Energy \qquad\qquad (xi)$$
$$(Carbon\ black)$$

$$\underline{Overall\ Equation:}\ O_2 + 2HC\equiv CH \longrightarrow 2H_2O + 4:\overset{\overset{\displaystyle \bullet nn}{}}{\underset{\underset{\displaystyle \bullet en}{}}{C}} + Energy \qquad (xii)$$

Stage 7&8: [IN PARALLEL]

$$O_2 \xrightleftharpoons[\text{Oxygen molecule}]{\text{Equilibruim State Existence of Oxidi sing}} 2en\cdot\ddot{O}$$

$$2en\cdot O\cdot nn + 2:\overset{\overset{\displaystyle \bullet n}{}}{\underset{\underset{\displaystyle \bullet e}{}}{C}} \xrightleftharpoons{\text{Oxidation}} 2nn\bullet O-\overset{\overset{\displaystyle \bullet e}{}}{\underset{\underset{\displaystyle \bullet n}{}}{C}}\bullet e$$
$$\text{(Stabilized)} \qquad\qquad\qquad (N)$$

$$\xrightarrow[\text{Release of Energy}]{\text{Deactivation}} 2:C=O + Energy \qquad\qquad (xiii)$$

$$\underline{Overall\ Equation:}\ 2O_2 + 4\overset{\overset{\displaystyle \bullet n}{}}{\underset{\underset{\displaystyle \bullet e}{}}{C}}: \longrightarrow 4:C=O + Energy \qquad\qquad (xiv)$$

Stage 9&10: [IN PARALLEL]

$$O_2 \xrightleftharpoons[\text{Oxygen molecule}]{\text{Equilibruim State Existence of Oxidi sing}} 2en\cdot\ddot{O}\cdot nn$$

$$2en \cdot O \cdot nn + \ 2C = O \ \underset{}{\overset{Oxidation}{\rightleftharpoons}} \ 2e \bullet \overset{\overset{O}{\|}}{C} - O \bullet nn$$

$$\text{(Stabilized)} \qquad\qquad\qquad (O)$$

$$(O) \quad \xrightarrow[\substack{Release\ of\ Energy}]{\substack{Deactivation}} \quad 2CO_2 \ + \ Energy \qquad (xv) \qquad\qquad (xv)$$

$$\underline{Overall\ Equation}:\ O_2 + 2CO \ \longrightarrow \ 2CO_2 \ + \ Energy \qquad (xvi) \qquad\qquad (xvi)$$

$$\underline{Final\ Overall\ Equation}:\ 2C_3H_8 \ + \ 10O_2 \ \longrightarrow 6CO_2 + 8H_2O \ + \ Energy(9)$$

$$\text{(From ten Stages)} \quad [b]$$

an eight stage Equilibrium mechanism system in which the last two stages are two in one (i.e., in parallel) for FULL OXIDATION, for when oxidation is PARTIAL, carbon monoxide an environmentally unfriendly gas appears only in Stage 7 (two in one).
(Laws of Creations for Oxidation of n-Propane)

Rule 1630: This rule of Chemistry for **Oxidation of Alkane Hydrocarbons,** states that, based on the overall equation obtained for them as shown below for propane, ethane and methane-

$$\underline{Final\ Overall\ equation}:\ 2C_3H_8 \ + \ 10O_2 \ \longrightarrow 6CO_2 \ + \ 8H_2O \ + \ Energy(9)$$

$$\text{(From ten Stages)} \qquad\qquad (a)$$

$$\underline{Final\ Overall\ equation}:\ 2C_2H_6 \ + \ 7O_2 \ \longrightarrow 4CO_2 + 6H_2O \ + \ Energy(5)$$

$$\text{(From seven Stages)} \qquad\qquad (b)$$

$$\underline{Final\ overall\ equation}:\ 2CH_4 \ + \ 4O_2 \ \xrightarrow{Oxidation} 4H_2O \ + \ 2CO_2 \ + \ Energy(4)$$

$$\text{(From four Stages)} \qquad\qquad (c)$$

from which the General overall equation is as follows

GENERAL OVERALL EQUATION

$$2C_nH_{2n+2} \ + \ (3n+1)O_2 \xrightarrow{OXIDATION} 2nCO_2 \ + \ 2(n+1)H_2O \ + \ HEAT \qquad [A]$$

and not the type shown below and used universally both for Combustion and Oxidation.

$$C_nH_{2n+2} \ + \ \frac{3n+1}{2}O_2 \ \longrightarrow \ nCO_2 \ + \ (n+1)H_2O \qquad [B]$$

noting that the general equation for Combustion which contains CO is different from the case above for Oxidation, and that the "Molecular oxygen" shown in the Equation in the absence of a passive Oxygen catalyst, is not Molecular oxygen, but Oxidizing oxygen molecule as distinguished above in [A].
(Laws of Creations for Oxidation of Alkane Hydrocarbons)

Rule 1631: This rule of Chemistry for **Combustion of Benzene,** states that, when combusted, the followings are obtained-

Stage 1:

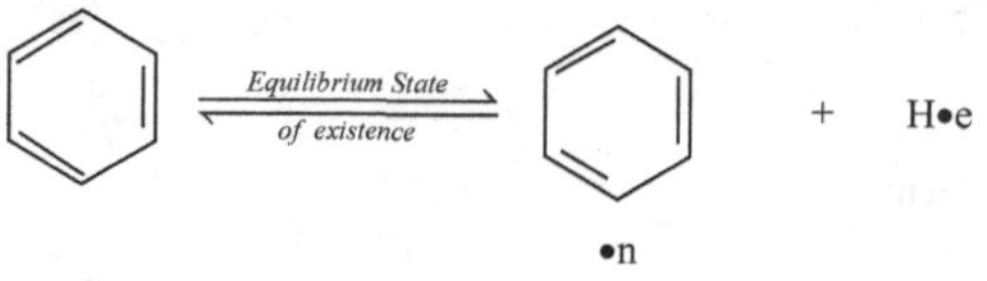

$$\bullet n$$

$$(A)$$

$$H\bullet e \quad + \quad O=O \quad \rightleftharpoons \quad H-O-O\bullet en$$

(B)

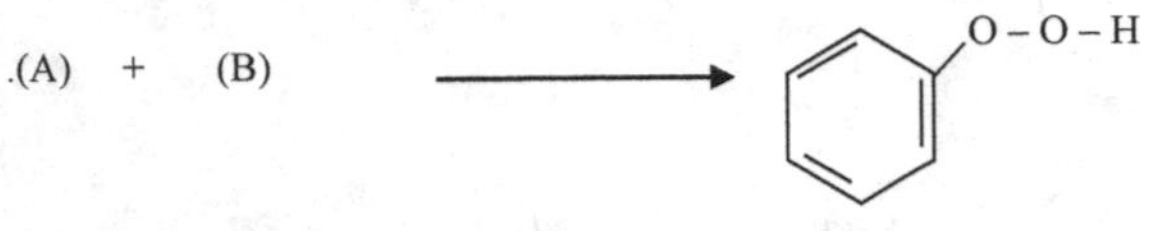

.(A) + (B) ⟶

O–O–H

(C)

(i)

Stage 2:

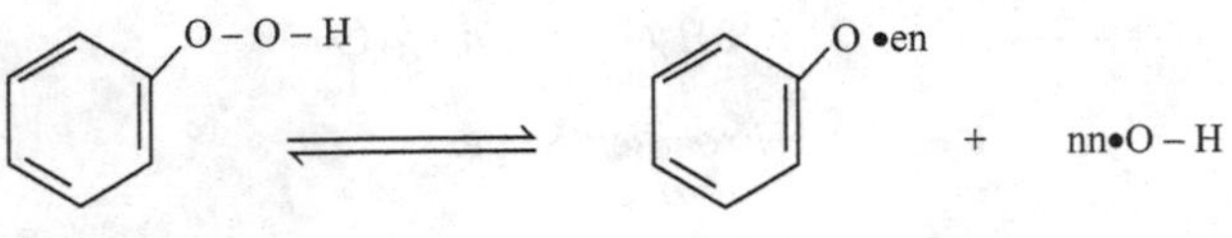

O–O–H

(C) ⇌ O•en

(D) + nn•O–H

(D) ⇌

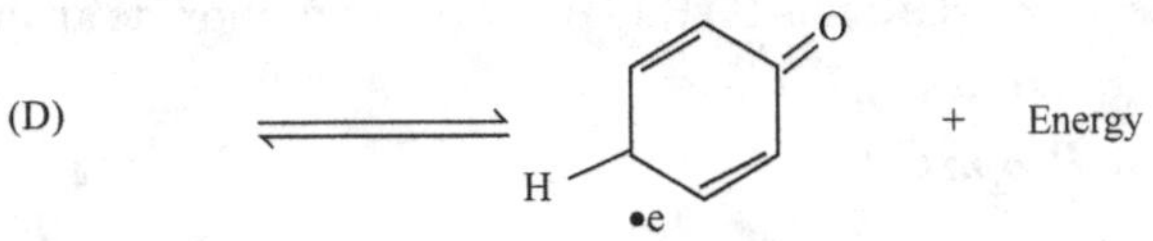

(E) + Energy

(E) + nn•OH ⟶

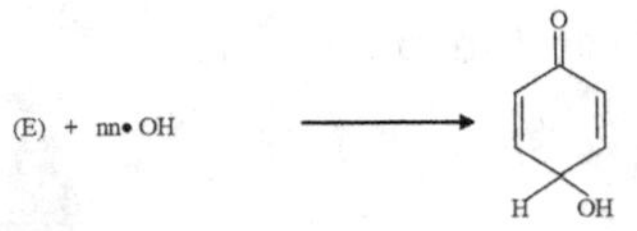

(F) An Acid

(ii)

Overall equation: $H_6C_6 + O_2 \longrightarrow$ (F) + Energy

(iii)

Stage 3:

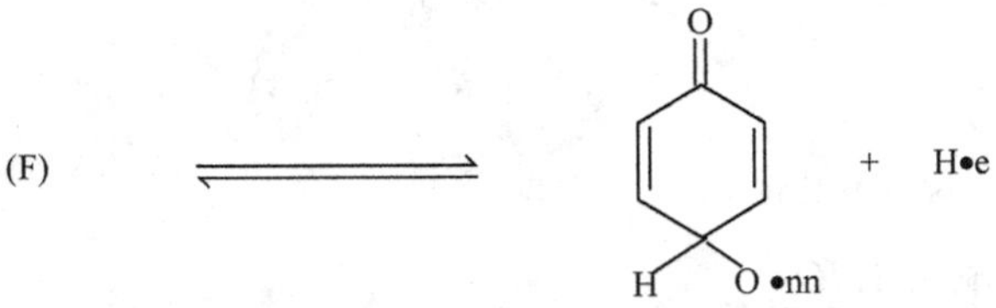

(F) ⇌

+ H•e

(G)

$$H\bullet e \quad + \quad O=O \quad \rightleftharpoons \quad H-O-O\bullet en$$

(B)

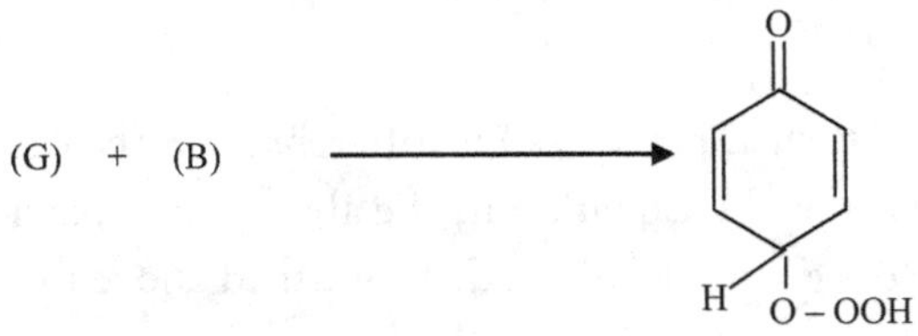

(G) + (B) ⟶

(H)

(iv)

Stage 4:

(H) ⇌

+ nn•OOH

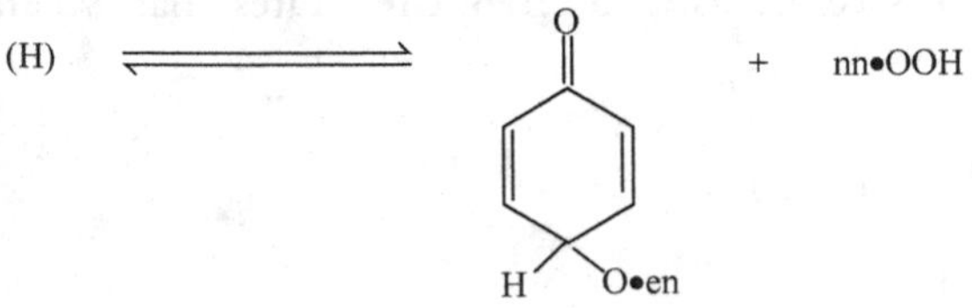

(I)

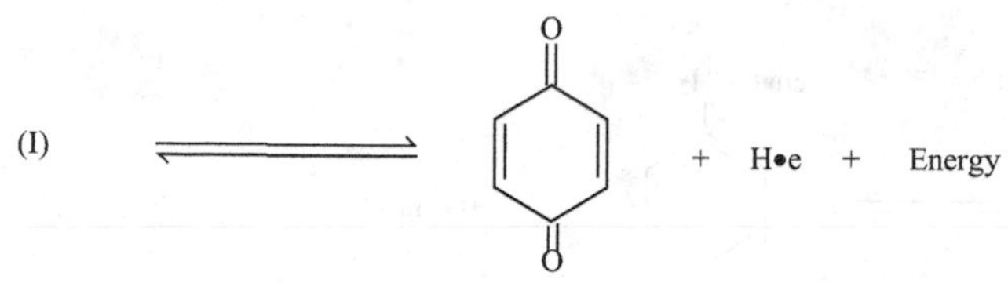

(I)

Aromatic Carbon dioxide

$$H\bullet e \quad + \quad nn\bullet OOH \longrightarrow \quad HOOH \quad \text{(v)}$$

Stage 5:

$$2H-O-O-H \xrightleftharpoons[\text{Existence of Hyrogen peroxide}]{\text{Equilibrium State of}} 2H-O\cdot en + 2nn\cdot O-H$$

$$2H-O\cdot en \xrightleftharpoons[\text{Release}]{\text{Energy}} 2H\cdot e + \underline{O_2} + Energy$$

$$2H\cdot e + 2nn\cdot O-H \xrightarrow[\text{presence of } O_2]{\text{Suppressed by the}} 2H_2O \quad \text{(vi)}$$

$$\underline{Overall\ Equation}: 2HOOH \longrightarrow 2H_2O + \underline{O_2} + Energy \quad \text{(vii)}$$

Overall equation: $2H_6C_6 + 4O_2 \longrightarrow$ 2Benzoquinone + Energy

$$+ \quad 2H_2O \quad + \quad \underline{O_2} \quad \text{(viii)}$$

a five stage Equilibrium mechanism system, in which energy was released from three stages more than that released from methane, noting the presence of the real poisonous Carbon dioxide aromatic in character (p-Benzoquinone), carcinogenic in character and presence of another environmentally unfriendly gas-oxidizing oxygen; for which their use as FUEL in an OPENED system should cease.
(Laws of Creations for Combustion of Benzene)

Rule 1632: This rule of Chemistry for **Combustion of Toluene an Alipharomatic Compound,** states that, when combusted, the followings are obtained-

Stage 1:

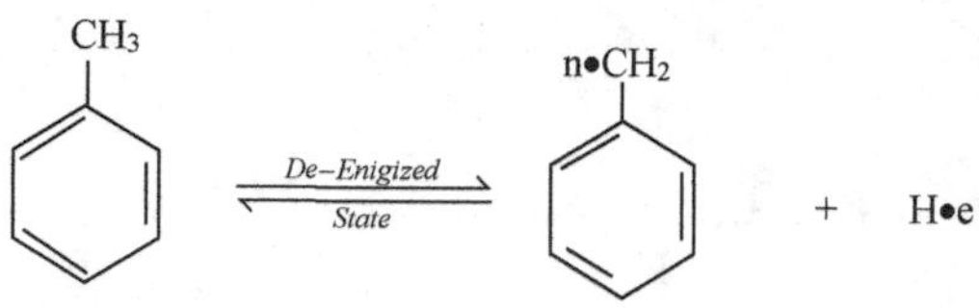

(A)

$$H\bullet e + O=O \quad \rightleftharpoons \quad H-O-O\bullet en$$

(B)

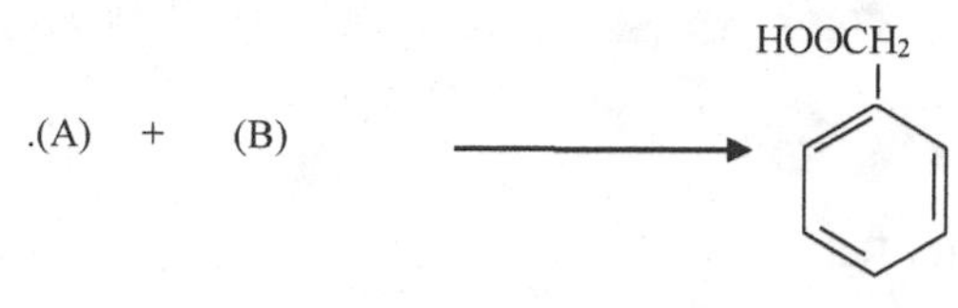

$$.(A) \quad + \quad (B) \longrightarrow$$

(C) A peroxide

(i)

689

Stage 2:

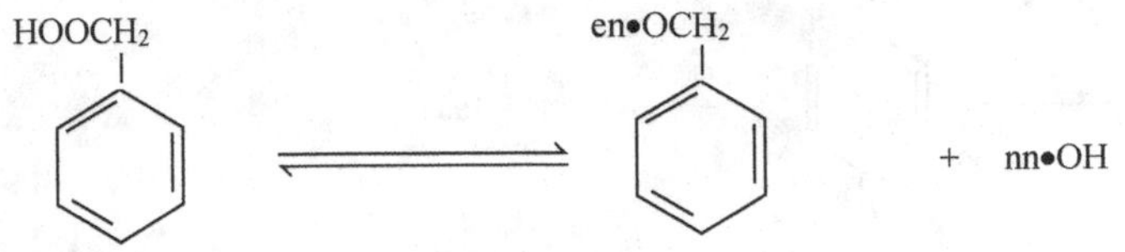

HOOCH$_2$ — (C) A peroxide ⇌ en•OCH$_2$ (D) + nn•OH

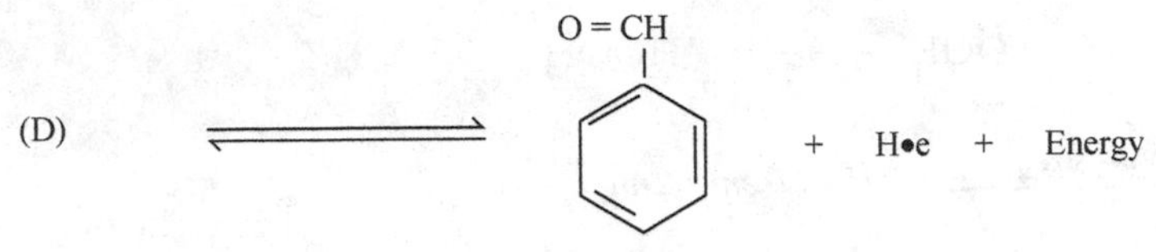

(D) ⇌ O=CH (E) + H•e + Energy

$$H•e \quad + \quad nn•OH \quad \longrightarrow \quad H_2O \qquad\qquad \text{(ii)}$$

Overall equation: $H_3CC_6H_5 + O_2 \longrightarrow H_2O + (E) + Energy$ (iii)

Stage 3:

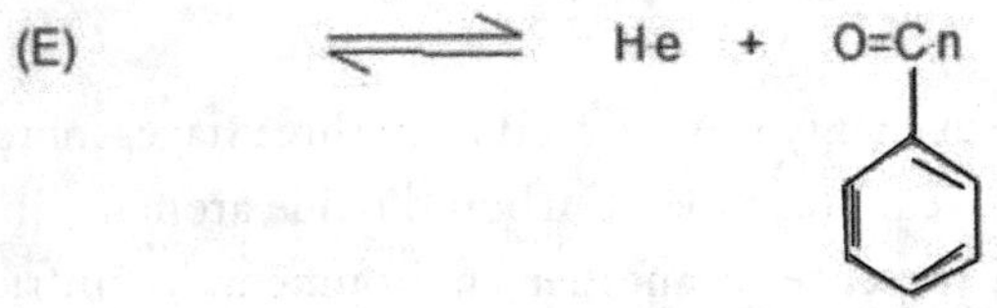

(E) ⇌ H•e + O=Cn (F)

$$H•e + O=O \rightleftharpoons H—O—O•en$$

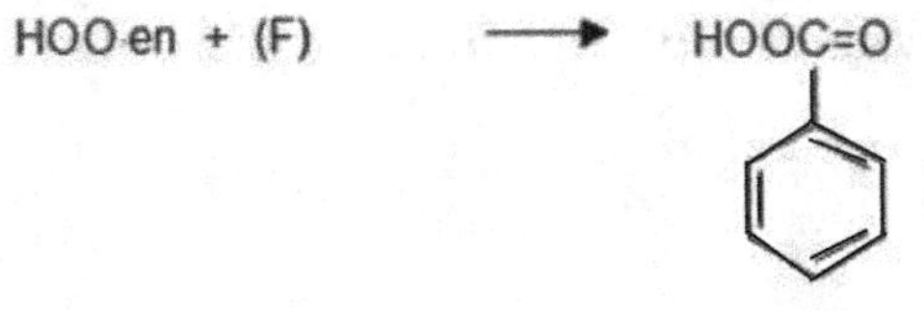

$$HOO•en + (F) \longrightarrow HOOC=O \quad (G) \qquad\qquad \text{(iv)}$$

Overall Equation: $2C_6H_5CH_3 + 4 O_2 \longrightarrow 2H_2O + 2 (G)$ (v)

Stage 4:

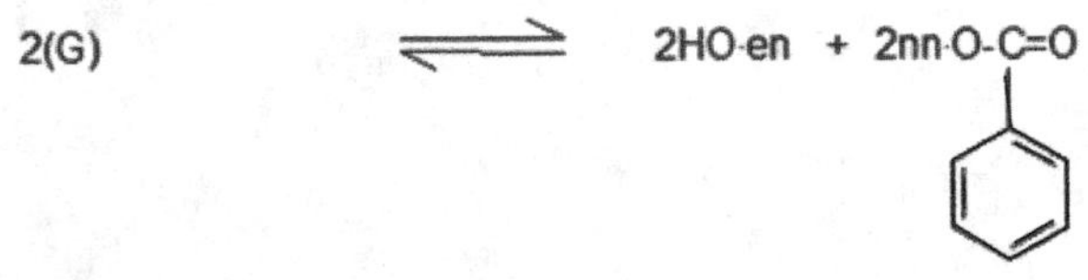

2(G) ⇌ 2HO•en + 2nn•O-C=O (H)

$$2 HO•en \rightleftharpoons 2H•e + \underline{O_2} + Energy\ Release$$

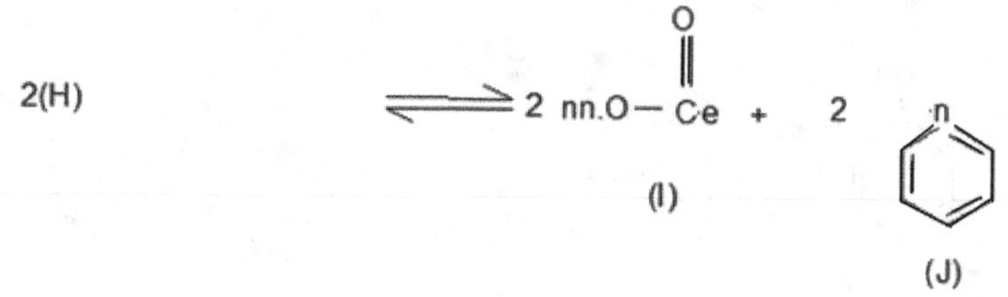

$$2(H) \rightleftharpoons 2 \; nn.O-\overset{\overset{\displaystyle O}{\|}}{Ce} + 2 \quad \text{(J)}$$

(I)

$$2(J) + 2H \cdot e \rightleftharpoons 2 C_6H_6$$

$$2(I) \longrightarrow 2CO_2 + \text{Energy Release} \qquad \text{(vi)}$$

Overall Equation: $2C_6H_5CH_3 + 4O_2 \longrightarrow 2C_6H_6 + 2CO_2 + 2H_2O$

(Air)

$$+ \underline{O_2} + \text{Energy}$$
(Oxidizing Oxygen) (vii)

(a) Oxidation of Benzene
Stage 5:

$$O_2 \rightleftharpoons 2nn \cdot O \cdot en$$

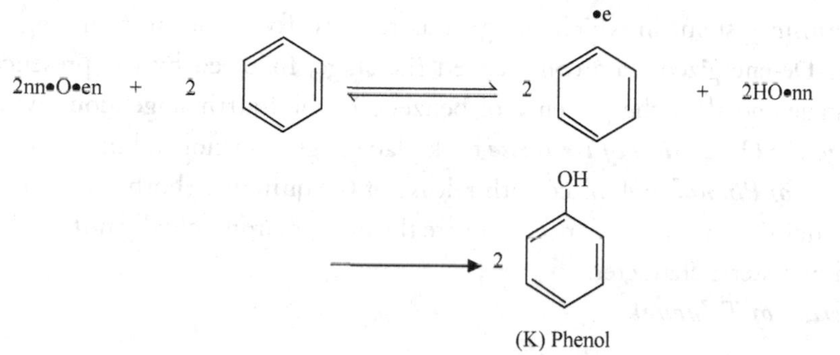

$$2nn \cdot O \cdot en + 2 \quad \rightleftharpoons \quad 2 \quad + \quad 2HO \cdot nn$$

$$\longrightarrow 2 \quad \text{(K) Phenol} \qquad \text{(viii)}$$

Overall equation: $O_2 + 2C_6H_6 \longrightarrow 2HOC_6H_5$ (ix)

(b) Combustion of phenol
Stage 6:

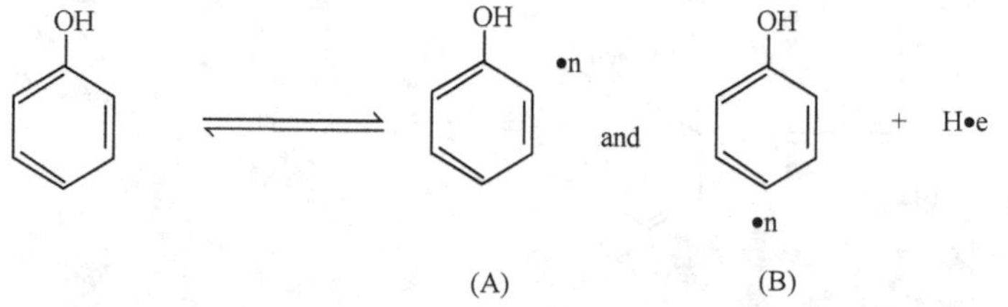

(A) and (B) + H•e

$$H \cdot e + O = O \rightleftharpoons H - O - O \cdot en$$
(C)

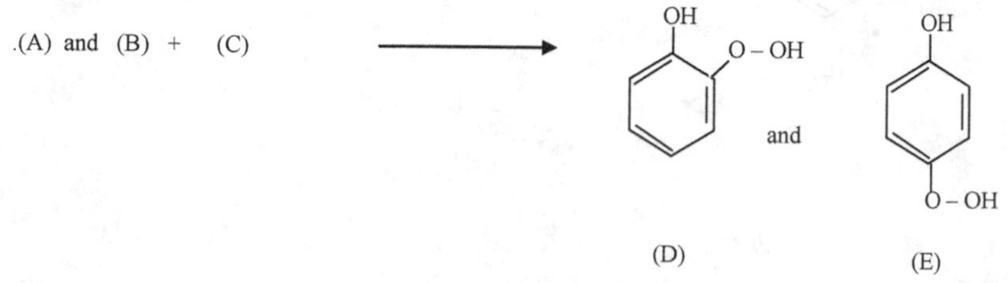

.(A) and (B) + (C) $\longrightarrow$

(D) and (E) (x)

691

Stage 7:

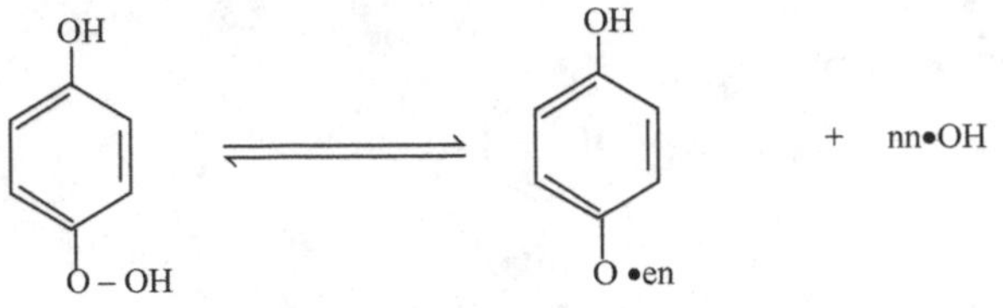

(E)   (F)

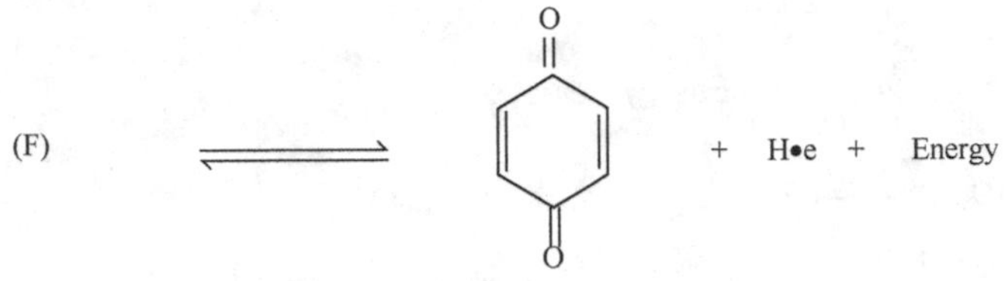

(F) $\rightleftharpoons$ (G) $+$ H•e $+$ Energy

H•e $+$ nn• OH $\longrightarrow$ H_2O (xi)

Final Overall equation : $2C_6H_5CH_3 + 6O_2 \longrightarrow 2CO_2 + 4H_2O$
(AIR) $+ 2C_6H_4O_2 + Energy(5)$
(G)   (xii)

a seven stage Equilibrium mechanism system, in which energy was released five times in four stages, that in which the toluene in the De-energized state commenced the stage, followed by the presence of Benzaldehyde in the second stage and then the presence of benzene in the fourth stage along with oxidizing oxygen; in the fifth stage *the Oxidation of Benzene* took place to give phenol and in the sixth and seventh stages the *Combustion of Phenol* took place with release of the quinones (both ortho- and para-), water and more energy- the quinones in particular the para are the only environmentally unfriendly products in view of its linear and non-steric character.
(Laws of Creations for Combustion of Toluene)

Rule 1633: This rule of Chemistry for **Oxidation of Phenol,** states that, when subjected to the this phenomenon, the followings are obtained-

Stage 1: $O_2 \rightleftharpoons 2nn{\cdot}O{\cdot}en$

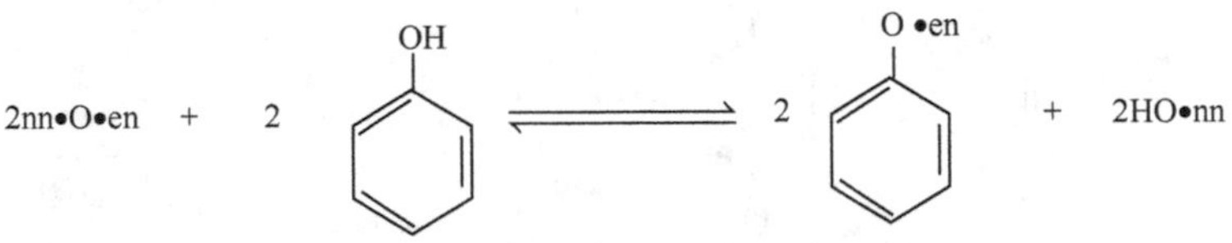

$2nn{\cdot}O{\cdot}en + 2$ (phenol) $\rightleftharpoons 2$ (A) $+ 2HO{\cdot}nn$

(A) $\rightleftharpoons 2$ (B) $+ Energy$

692

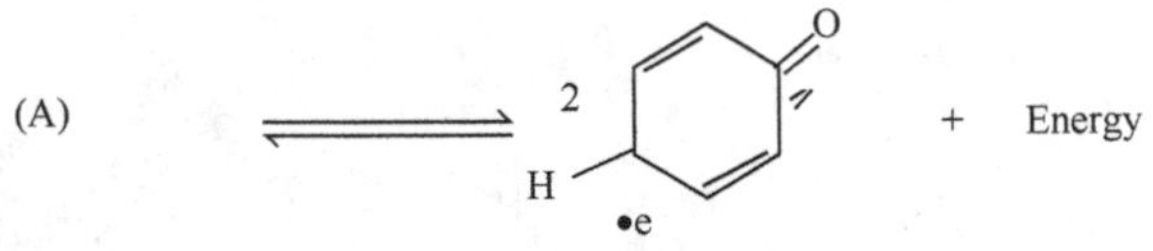

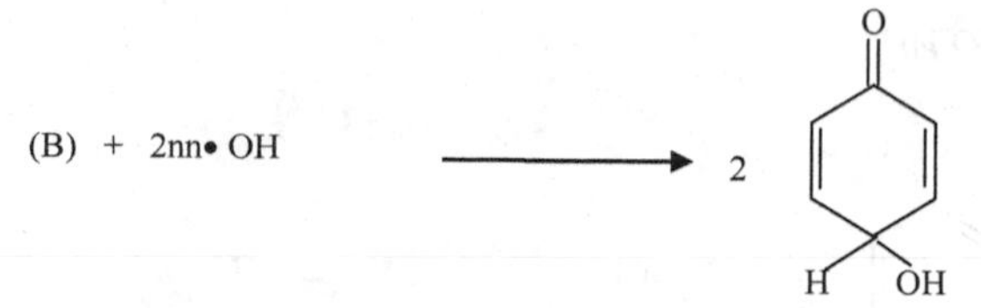

(B) + 2nn• OH $\longrightarrow$ 2

(C) An Acid   (i)

Overall equation: $2H_5C_6OH$ + $\underline{O_2}$ $\longrightarrow$ (C) + Energy   (ii)

Stage 2: $O_2 \rightleftharpoons 2nn{\cdot}O{\cdot}en$

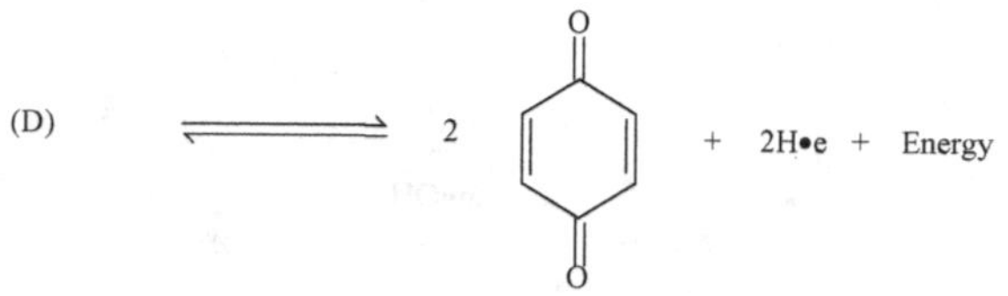

2nn•O•en + (C) $\rightleftharpoons$ 2

+ 2 HO•nn

(D) $\rightleftharpoons$ 2

+ 2H•e + Energy

2H•e + 2nn•OH $\longrightarrow$ $2H_2O$ (iii)

Overall equation: $2C_6H_5OH$ + $2\underline{O_2}$ $\longrightarrow$ $2C_6H_4O_2$ + $2H_2O$

+ *Energy*   (iv)

for which two stages are required for the FULL OXIDATION of phenol, just like for its combustion, both giving the same products- water, benzoquinones, but unlike with combustion (See Rule 1633), energy is released here in the two stages, clear indication of more release of Energy via Oxidation than via Combustion.

(Laws of Creations for Oxidation of Phenol)

Rule 1634: This rule of Chemistry for **Oxidation of Toluene an Alipharomatic Compound,** states that, when oxidized, the followings are obtained-

Stage 1: $O_2 \rightleftharpoons 2nn{\cdot}O{\cdot}en$

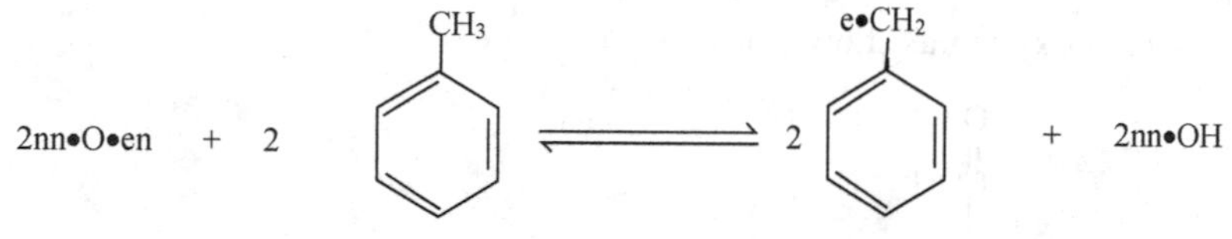

2nn•O•en + 2

$\rightleftharpoons$ 2

+ 2nn•OH

(A)

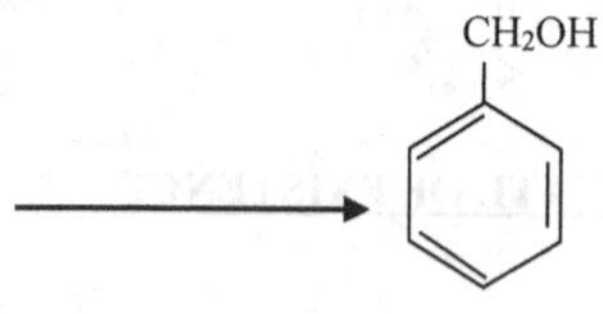

(B)   (i)

693

Stage 2:
$$O_2 \rightleftharpoons 2nn\square O\square en$$

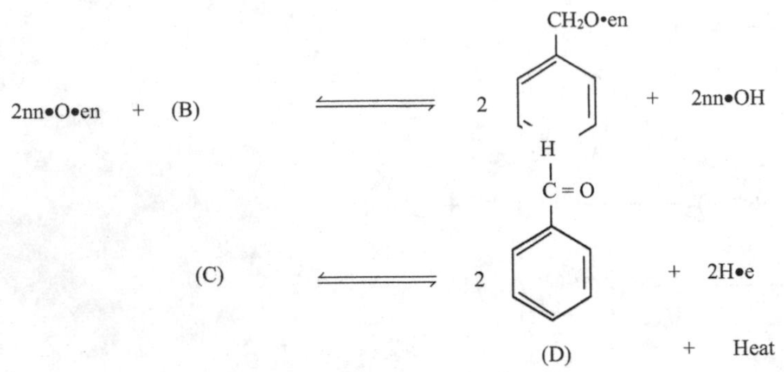

2nn•O•en + (B) $\rightleftharpoons$ 2 [structure: CH₂O•en substituted ring] + 2nn•OH

(C) $\rightleftharpoons$ 2 [structure: H C=O substituted benzene ring] + 2H•e

(D) + Heat

2H•e + 2HO •nn $\longrightarrow$ 2H$_2$O

(ii)

Stage 3:
$$O_2 \rightleftharpoons 2nn\square O\square en$$

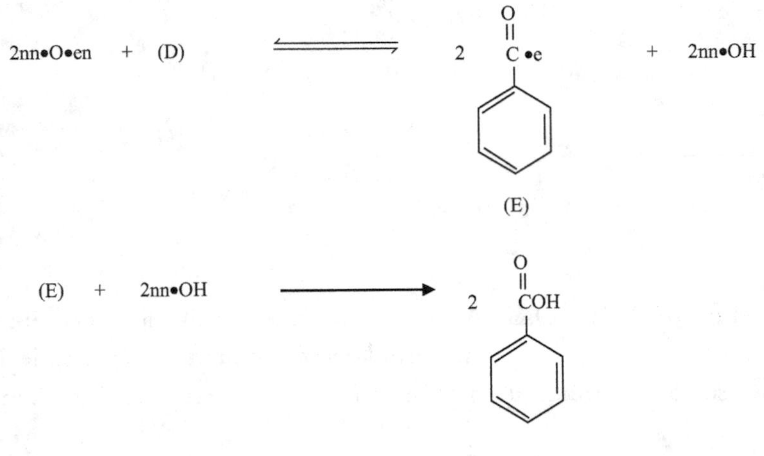

2nn•O•en + (D) $\rightleftharpoons$ 2 [structure: C•e with =O on benzene ring] + 2nn•OH

(E)

(E) + 2nn•OH $\longrightarrow$ 2 [structure: COH with =O on benzene ring]

(F)

(iii)

Overall equation: $2C_6H_5CH_3 + 3O_2 \longrightarrow 2C_6H_5COOH + 2H_2O + Energy$(1)

(iv)

a three stage Equilibrium mechanism system at FULL OXIDATION, in which Energy is released only in one stage, noting that even in the presence of excess Oxidizing oxygen in the system, oxidation of the carboxylic acid (F) formed is impossible, because the carboxylic acid is always in the Energized state of existence in the presence of the oxidizing oxygen as shown below-

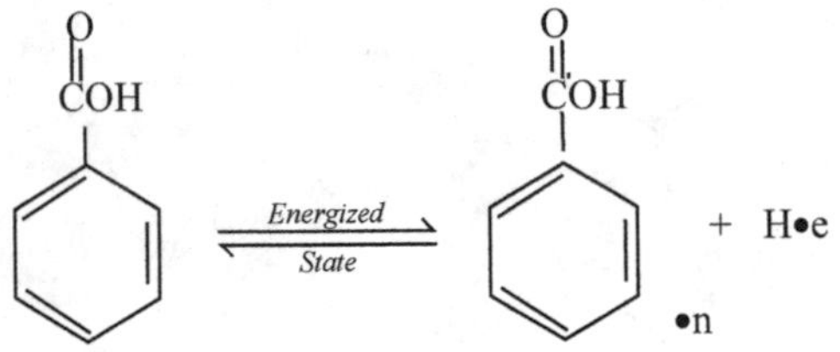

[structure: benzoic acid] $\underset{State}{\overset{Energized}{\rightleftharpoons}}$ [structure: benzoic acid radical •n] + H•e

ENERGIZED EQUILIBRIUM STATE OF EXISTENCE

(A)

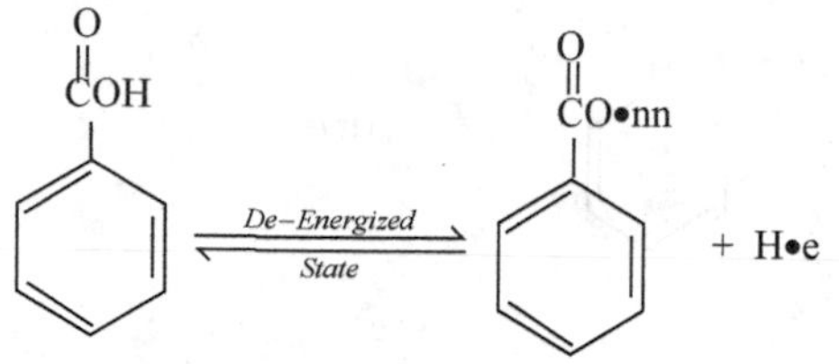

<u>**DE-ENERGIZED EQUILIBRIUM STATE OF EXISTENCE**</u>   (B)

and the same will not apply if it was in the De-energized Equilibrium state of existence, for benzene and carbon dioxide would have been formed in the absence of oxidizing oxygen in the system; nothing that the movement is from-

Toluene ⟶ **Benzyl Alcohol** ⟶ **Benzaldehyde** ⟶ **Benzoic acid**

Stage 1   **Stage 2**   **Stage 3**

with no enol present as an intermediate; thus in general benzoic acid is always in the Energized Equilibrium state of existence in the presence of oxidizing oxygen.

(Laws of Creations for Oxidation of Toluene)

<u>**Rule 1635:**</u> This rule of Chemistry for **Benzoic acid,** states that, if the acid can be kept in Stable state of existence in the presence of Oxidizing oxygen, then the followings are to be expected-

<u>**Stage 1:**</u> $O_2 \rightleftharpoons 2nn \cdot O \cdot en$

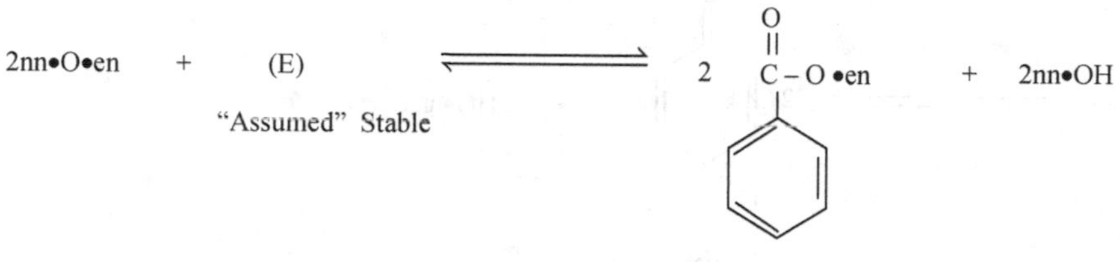

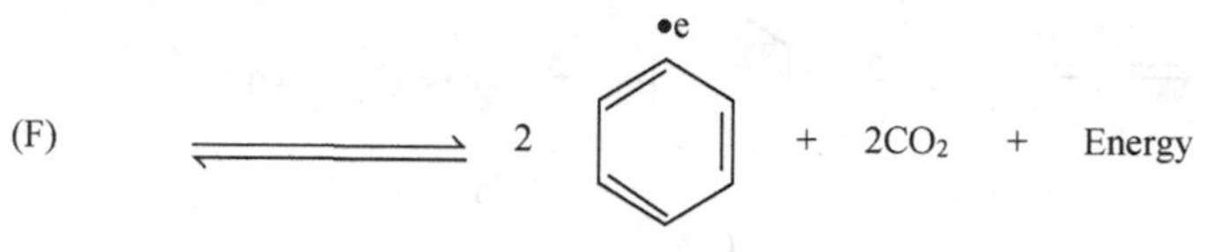

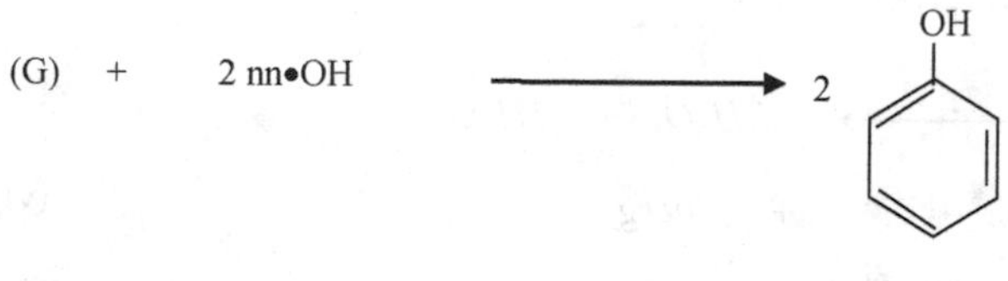

(i)

Overall equation: $C_6H_5COOH + O_2 \longrightarrow C_6H_5OH + CO_2 + Energy$ (ii)

<u>**Stage 2:**</u>

$O_2 \rightleftharpoons 2nn \cdot O \cdot en$

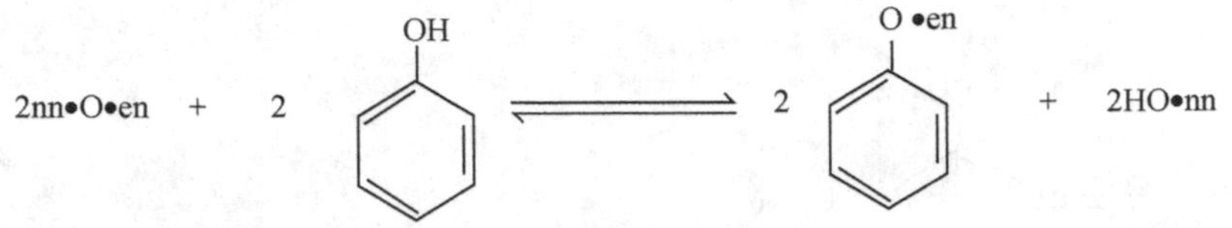

$$2nn \bullet O \bullet en \ + \ 2 \ \text{(phenol, OH)} \ \rightleftharpoons \ 2 \ \text{(A, O} \bullet en) \ + \ 2HO \bullet nn$$

(A)

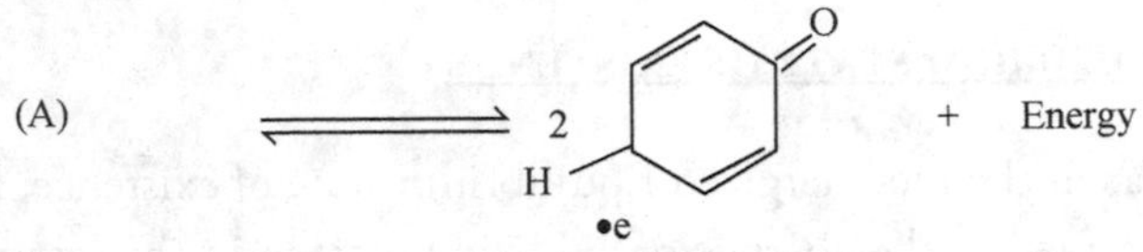

$$(A) \ \rightleftharpoons \ 2 \ \text{(B)} \ + \ \text{Energy}$$

(B)

$$(B) \ + \ 2nn \bullet OH \ \longrightarrow \ 2 \ \text{(C)} \qquad \text{(iii)}$$

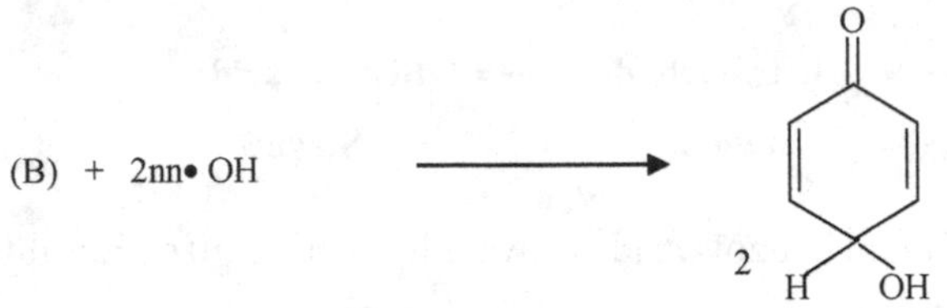

(C) An Acid

Overall equation: $2H_5C_6OH \ + \ \underline{O_2} \ \longrightarrow \ (C) \ + \ Energy$ (iv)

Stage 3: $\underline{O_2} \ \rightleftharpoons \ 2nn \bullet O \bullet en$

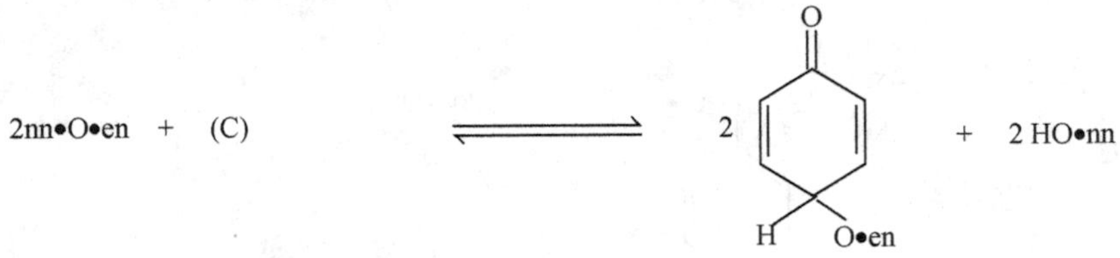

$$2nn \bullet O \bullet en \ + \ (C) \ \rightleftharpoons \ 2 \ \text{(D)} \ + \ 2 HO \bullet nn$$

(D)

$$(D) \ \rightleftharpoons \ 2 \ \text{(quinone)} \ + \ 2H \bullet e \ + \ Energy$$

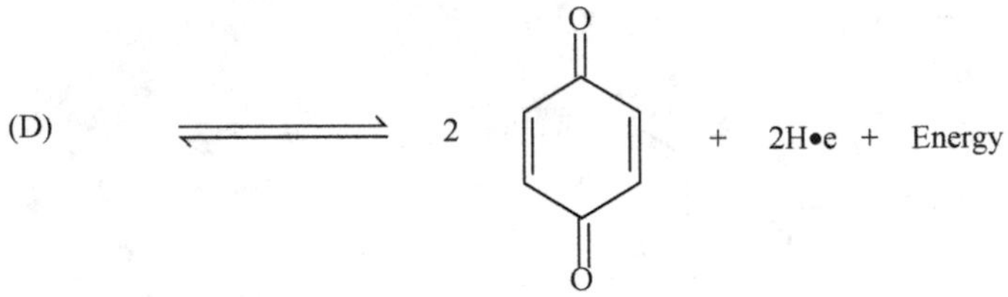

$$2H \bullet e \ + \ 2nn \bullet OH \ \longrightarrow \ 2H_2O \qquad \text{(v)}$$

Overall equation: $2C_6H_5OH \ + \ 2\underline{O_2} \ \longrightarrow \ 2C_6H_4O_2 \ + \ 2H_2O$
$$+ \ Energy \qquad \text{(vi)}$$

Overall equation: $2C_6H_5COOH \ + \ 3\underline{O_2} \longrightarrow 2C_6H_4O_2 \ + \ 2H_2O \ + \ Energy(3)$
$$+ \ 2CO_2 \qquad \text{(vii)}$$

Overall equation: $2C_6H_5CH_3 \ + \ 6\underline{O_2} \longrightarrow 2C_6H_4O_2 \ + \ 4H_2O \ + \ Energy(4)$
$$+ \ 2CO_2 \qquad \text{(viii)}$$

in which worthy of note is that when the carboxylic acid was oxidized, so much energy was generated in the three-stage Equilibrium mechanism system in every stage, none of which is via de-activation, for which one can see that the same product cannot be obtained for them via Combustion and Oxidation and why the benzoic acid cannot be oxidized.

(Laws of Creations for Forced Oxidation of Benzoic acid).

Rule 1636: This rule of Chemistry for **Oxidation of Ethyl benzene,** states that, when oxidized, the followings are obtained *when the operating condition is high-*

Stage 1:

$$O_2 \rightleftharpoons 2nn \cdot O \cdot en$$

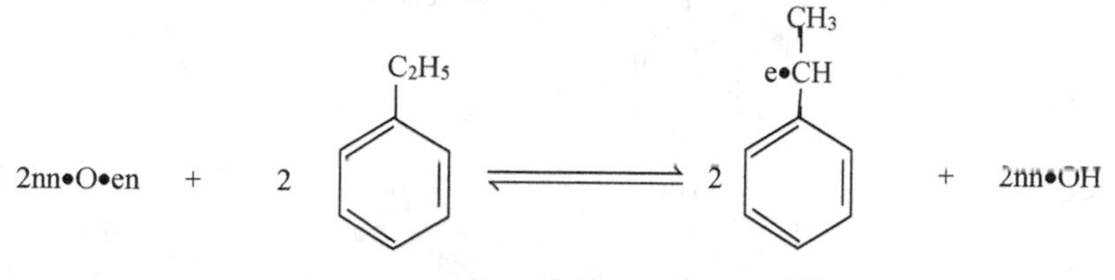

$$2nn \cdot O \cdot en \; + \; 2\; \text{[C}_2\text{H}_5\text{ benzene]} \;\rightleftharpoons\; 2\; \text{[CH}_3\text{, e}\cdot\text{CH benzene]} \; + \; 2nn \cdot OH$$

(A)

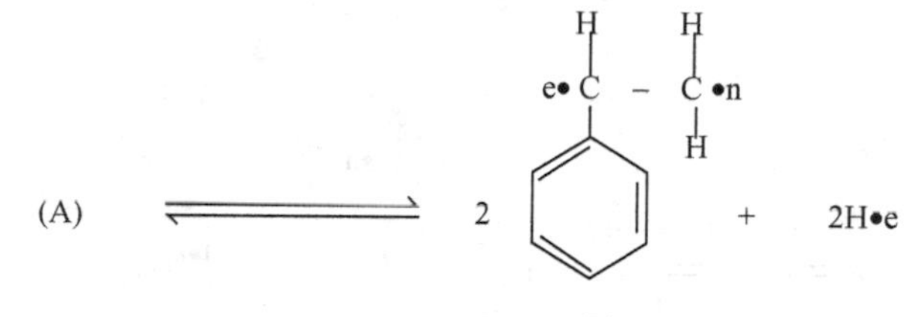

$$(A) \;\rightleftharpoons\; 2\; \text{[e}\cdot\text{C}H - \text{C}H\text{, H}\cdot\text{n benzene]} \; + \; 2H\cdot e$$

(B)

$$2H\cdot e \; + \; 2HO \cdot nn \;\rightleftharpoons\; 2H_2O$$

$$(B) \xrightarrow[\text{Release of Energy}]{\text{Deactivation}} 2\; \text{[C}H = C H_2 \text{ benzene]} \; + \; \text{Energy}$$

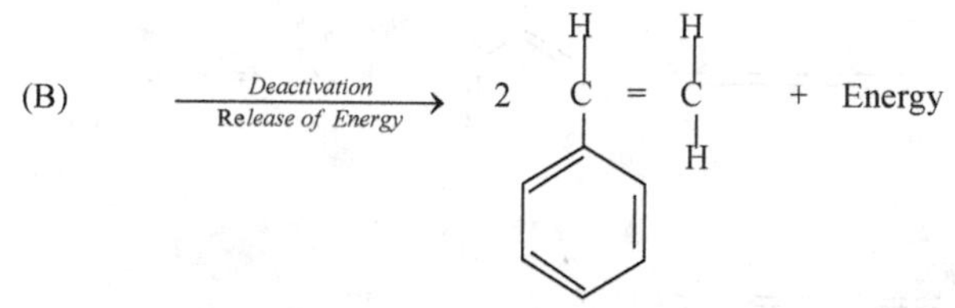

(C)   (i)

$$\underline{Overall\ equation}: 2C_6H_5C_2H_5 \; + \; O_2 \longrightarrow 2H_2C = CHC_6H_5 \; + \; 2H_2O$$

$$+ \; Energy$$ (ii)

Stage 2:

$$O_2 \rightleftharpoons 2nn \cdot O \cdot en$$

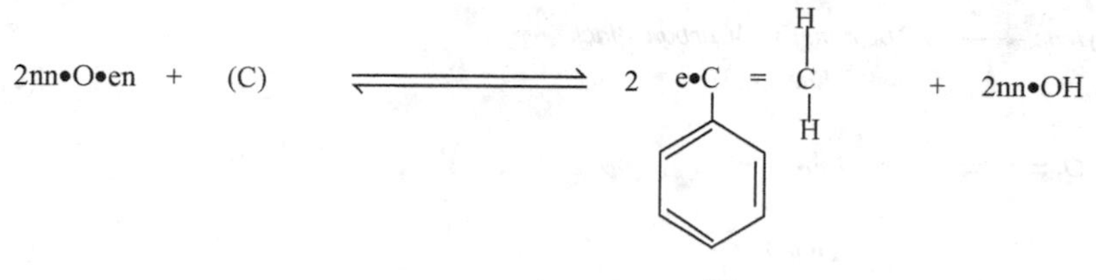

$$2nn \cdot O \cdot en \; + \; (C) \;\rightleftharpoons\; 2\; \text{[e}\cdot C = C H \text{, H benzene]} \; + \; 2nn \cdot OH$$

(D)

697

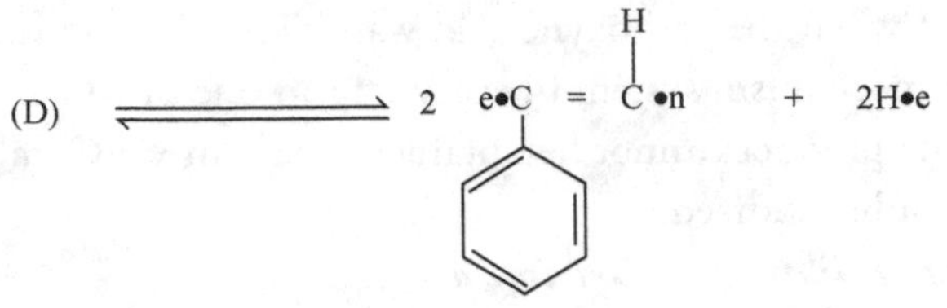

$$(D) \rightleftharpoons 2 \quad e{\bullet}C = \overset{\overset{H}{|}}{C}{\bullet}n \quad + \quad 2H{\bullet}e$$

(E)

$$2H{\bullet}e \quad + \quad 2HO{\bullet}nn \quad \rightleftharpoons \quad 2H_2O$$

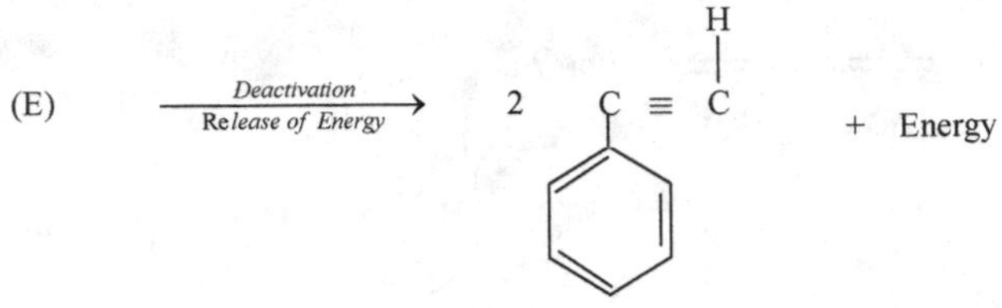

$$(E) \quad \xrightarrow[\text{Release of Energy}]{\text{Deactivation}} \quad 2 \quad C \equiv \overset{\overset{H}{|}}{C} \quad + \quad Energy$$

(F)

(iii)

Overall equation: $2C_6H_5C_2H_5 + 2\underset{--}{O_2} \longrightarrow 2HC \equiv CC_6H_5 + 4H_2O + Energy$

(iv)

Stage 3:

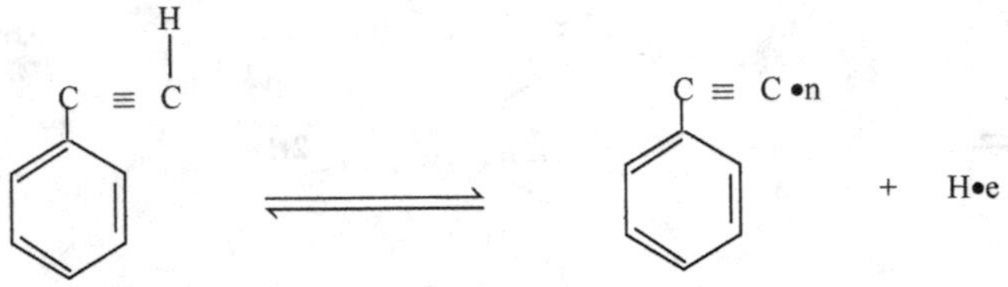

$$\overset{\overset{H}{|}}{C} \equiv C \quad (F) \quad \rightleftharpoons \quad C \equiv C{\bullet}n \quad (F) \quad + \quad H{\bullet}e$$

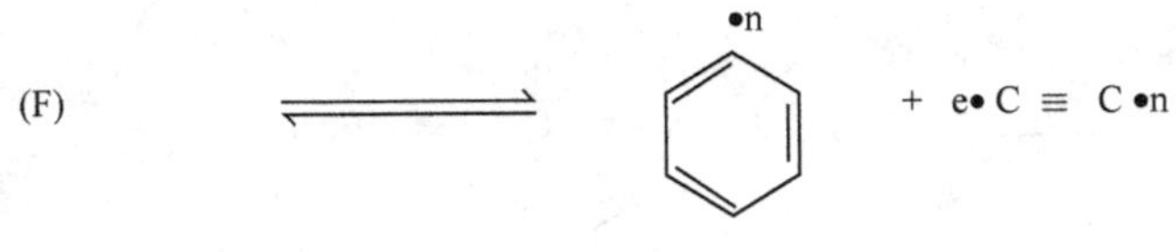

$$(F) \quad \rightleftharpoons \quad \overset{{\bullet}n}{\bigcirc} \quad + \quad e{\bullet} C \equiv C{\bullet}n$$

(F)

$$H{\bullet}e \quad + \quad (F) \quad \rightleftharpoons \quad \bigcirc$$

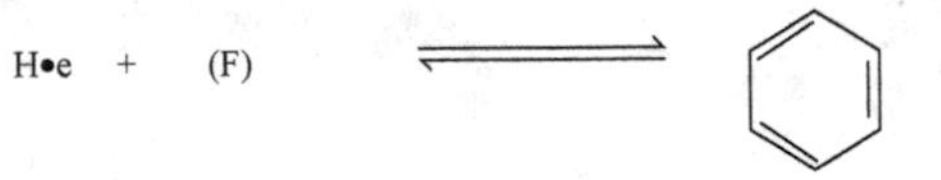

$$e{\bullet} C \equiv C{\bullet}n \quad \rightleftharpoons \quad 2 \; e{\bullet}\overset{\overset{{\bullet}n}{}}{\underset{{\bullet}e}{C}}{\bullet}n$$

$$2 \; e{\bullet}\overset{\overset{{\bullet}n}{}}{\underset{{\bullet}e}{C}}{\bullet}n \quad \xrightarrow[\text{Release of Heat}]{\text{Deactivation}} \quad 2 \; e{\bullet}\overset{\overset{{\bullet}{\bullet}}{}}{C}{\bullet}n \quad + \quad Energy$$

Activated Carbon   Carbon Black

(v)

Overall equation: $2Phenyl\ Acetylene \longrightarrow 2Benzene + 4Carbon\ Black + Energy\,(3)$

(vi)

Stage 4:

$$O_2 \rightleftharpoons 2nn{\bullet}O{\bullet}en$$

$$2nn{\bullet}O{\bullet}en \quad + \quad 2 \; e{\bullet}\overset{\overset{{\bullet}{\bullet}}{}}{C}{\bullet}n \quad \rightleftharpoons \quad 2nn{\bullet}O - \overset{\overset{{\bullet}e}{}}{\underset{{\bullet}n}{C}}{\bullet}e$$

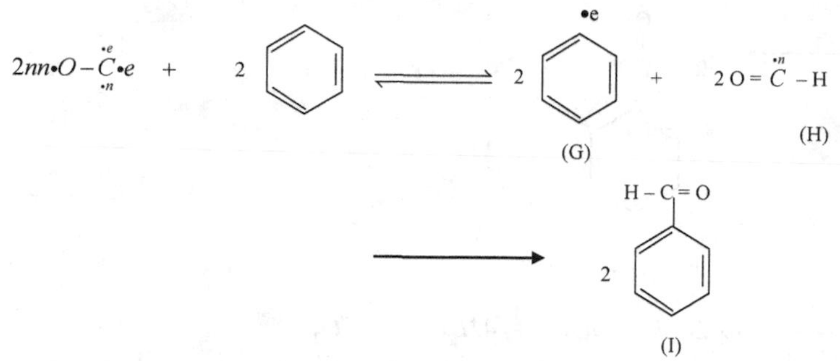

$$2nn \cdot O - \overset{\cdot e}{\underset{\cdot n}{C}} \cdot e \ + \ 2 \ \bigcirc \ \rightleftharpoons \ 2 \ \bigcirc^{\bullet e} \ + \ 2 \ O = \overset{\cdot n}{C} - H$$

(G)   (H)

$$\xrightarrow{\hspace{2cm}} \quad 2 \ \underset{(I)}{\bigcirc^{H-C=O}}$$

(vii)

Overall equation: $2C_6H_5C_2H_5 + 3\underset{--}{O_2} \longrightarrow 2C_6H_5CHO + 2C + 4H_2O$

(Carbon Black)

$+ \ Energy(3)$

(viii)

Stage 5:

$$\underset{--}{O_2} \underset{Oxygen\ molecule}{\overset{Equilibruim\ State\ Existence\ of\ Oxidising}{\rightleftharpoons}} 2en \cdot \ddot{O} \cdot nn$$

$$2en \cdot O \cdot nn + \ 2 : \overset{\cdot n}{\underset{\cdot e}{C}} \ \overset{Oxidation}{\rightleftharpoons} \ 2\,nn \cdot O - \overset{\cdot e}{\underset{\cdot n}{C}} \cdot e$$

(Stabilized)   (F)

$$\xrightarrow[Release\ of\ Energy]{Deactivation} \quad 2 : C = O \ + \ Energy$$

(ix)

Overall Equation: $\underset{--}{O_2} + 2\overset{\cdot n}{\underset{\cdot e}{C}} : \longrightarrow \ 2 : C = O \ + \ Energy$

(x)

Stage 6:

$$\underset{--}{O_2} \underset{Oxygen\ molecule}{\overset{Equilibruim\ State\ Existence\ of\ Oxidising}{\rightleftharpoons}} 2en \cdot \ddot{O} \cdot nn$$

$$2en \cdot O \cdot nn + \ 2C = O \ \overset{Oxidation}{\rightleftharpoons} \ 2e \cdot \overset{O}{\overset{\|}{C}} - O \cdot nn$$

(Stabilized)   (B)

$$(B) \quad \xrightarrow[Release\ of\ Energy]{Deactivation} \quad 2CO_2 \ + \ Energy$$

(xi)

Overall Equation $\quad \underset{--}{O_2} + 2CO \longrightarrow 2CO_2 \ + \ Energy$

(xii)

Overall equation: $2C_6H_5C_2H_5 + 5\underset{--}{O_2} \longrightarrow 2C_6H_5CHO + 2CO_2 + 4H_2O$

$+ \ Energy(5)$

(xiii)

Stage 7:

$$\underset{--}{O_2} \rightleftharpoons \ 2nn \cdot O \cdot en$$

(v)

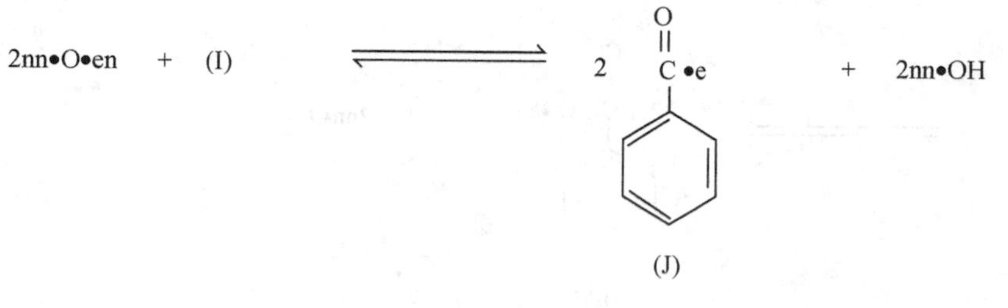

$$2nn \bullet O \bullet en \ + \ (I) \ \rightleftharpoons \ 2 \ \underset{(J)}{\bigcirc^{\overset{O}{\overset{\|}{C} \bullet e}}} \ + \ 2nn \bullet OH$$

(vi)

699

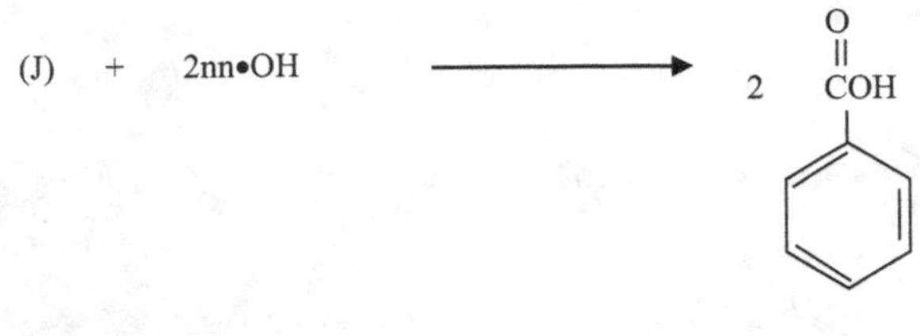

$$\text{(J)} \quad + \quad 2nn\bullet OH \quad \longrightarrow \quad 2 \; \underset{\text{(K)}}{\overset{\overset{\displaystyle O}{\overset{\|}{COH}}}{\bigcirc\!\!\!\bigcirc}}$$

(xiv)

<u>*Overall equation*</u>: $2C_6H_5C_2H_5 \; + \; 6\underline{O_2} \longrightarrow 2C_6H_5COOH \; + \; 4H_2O \; + \; 2CO_2$

$+ \quad Energy(5)$

(xv)

a seven stage Equilibrium mechanism system at FULL OXIDATION, with no possibility of formation of Alcohol, Ketone, Enol, but only Styrene, benzene, benzaldehyde and the carboxylic acid, because of the high operating conditions.
(Laws of Creations for Oxidation of Ethyl benzene)

Rule 1637: This rule of Chemistry for **Oxidation of Ethyl benzene,** states that, *when oxidized, the followings are obtained when the operating condition is mild-*

<u>Stage 1:</u> $\qquad\qquad O_2 \rightleftharpoons 2nn\bullet O\bullet en$

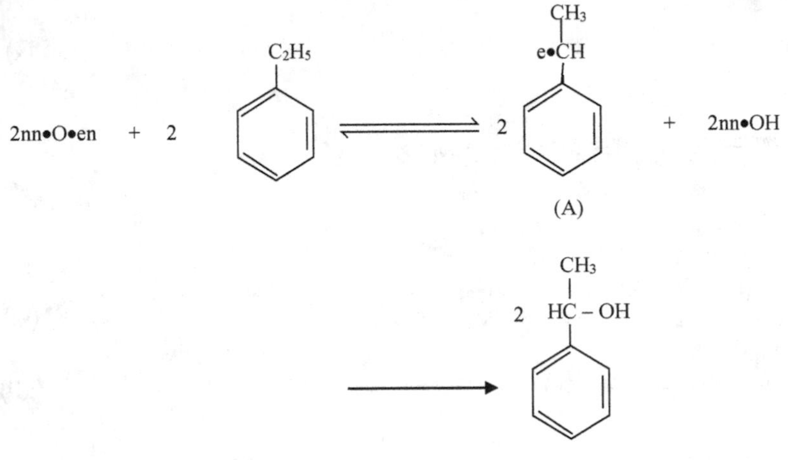

$$2nn\bullet O\bullet en \quad + \quad 2 \; \underset{C_2H_5}{\bigcirc\!\!\!\bigcirc} \quad \rightleftharpoons \quad 2 \; \underset{\overset{e\bullet CH}{|}}{\underset{CH_3}{\bigcirc\!\!\!\bigcirc}} \quad + \quad 2nn\bullet OH$$

(A)

$$\longrightarrow \quad 2 \; \underset{\text{(B)}}{\underset{HC-OH}{\overset{CH_3}{\overset{|}{\bigcirc\!\!\!\bigcirc}}}}$$

(i)

<u>*Overall equation*</u>: $2C_6H_5C_2H_5 \; + \; \underline{O_2} \longrightarrow 2C_6H_5CH(OH)CH_3$

(ii)

<u>Stage 2:</u> $\qquad\qquad O_2 \rightleftharpoons 2nn\bullet O\bullet en$

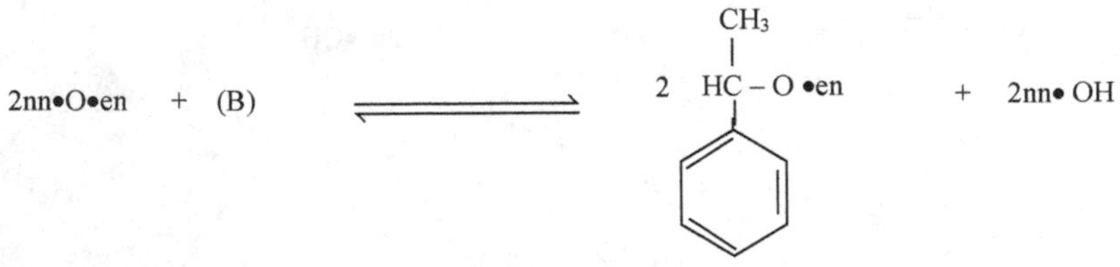

$$2nn\bullet O\bullet en \quad + \quad \text{(B)} \quad \rightleftharpoons \quad 2 \; \underset{\text{(B)}}{\underset{HC-O\bullet en}{\overset{CH_3}{\overset{|}{\bigcirc\!\!\!\bigcirc}}}} \quad + \quad 2nn\bullet OH$$

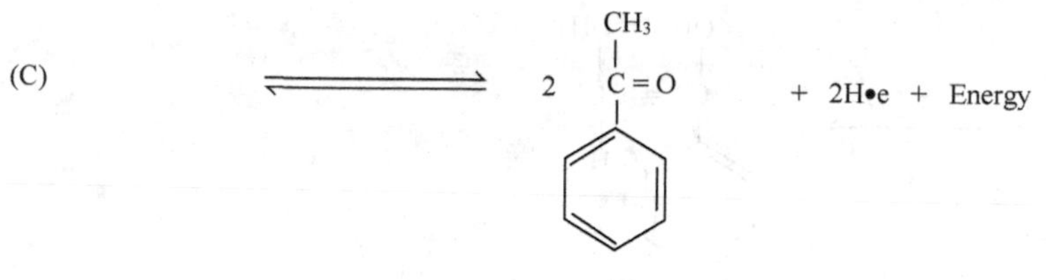

(C) $\rightleftharpoons$ 2
$$\underset{\text{(D)}}{\overset{\displaystyle CH_3 \atop \displaystyle | \atop \displaystyle C=O}{\bigcirc}}$$
    $+ \ 2H{\bullet}e \ + \ \text{Energy}$

$2H{\bullet}e \ + \ 2HO{\bullet}nn \longrightarrow 2H_2O$ (iii)

Overall equation : $2C_6H_5C_2H_5 \ + \ 2\underline{O_2} \longrightarrow 2C_6H_5COCH_3 \ + \ 2H_2O \ + \ Energy$ (iv)

Stage 3:

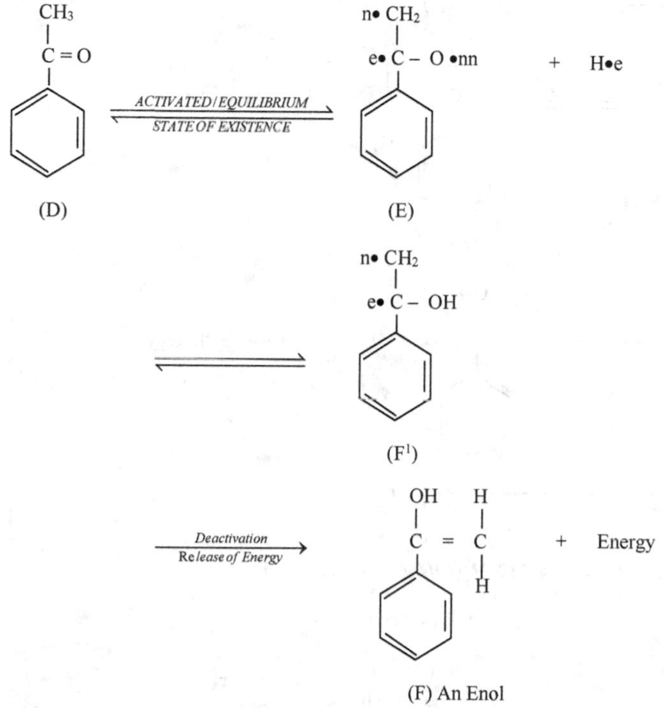

$$\underset{\text{(D)}}{\overset{\displaystyle CH_3 \atop \displaystyle | \atop \displaystyle C=O}{\bigcirc}} \quad \underset{\text{STATE OF EXISTENCE}}{\overset{\text{ACTIVATED / EQUILIBRIUM}}{\rightleftharpoons}} \quad \underset{\text{(E)}}{\overset{\displaystyle n{\bullet}CH_2 \atop \displaystyle | \atop \displaystyle e{\bullet}C-O{\bullet}nn}{\bigcirc}} \quad + \quad H{\bullet}e$$

$$\rightleftharpoons \quad \underset{\text{(F}^1\text{)}}{\overset{\displaystyle n{\bullet}CH_2 \atop \displaystyle | \atop \displaystyle e{\bullet}C-OH}{\bigcirc}}$$

$$\underset{\text{Release of Energy}}{\overset{\text{Deactivation}}{\longrightarrow}} \quad \underset{\text{(F) An Enol}}{\overset{\displaystyle OH \quad H \atop \displaystyle | \quad\quad | \atop \displaystyle C \ = \ C \atop \displaystyle \bigcirc \quad\quad | \atop \displaystyle H}{}} \quad + \quad \text{Energy}$$
     (v)

Overall equation : $C_6H_5COCH_3 \ \xrightarrow[\text{or Base catalyst}]{Acid} \ C_6H_5(OH)C=CH_2$ (vi)

Overall equation : $2C_6H_5C_2H_5 \ + \ 2\underline{O_2} \ \xrightarrow[\text{or Based catalyzed}]{Acid} \ 2C_6H_5(OH)C=CH_2 \ + \ 2H_2O$
$$+ \ Energy\,(2)$$
     (vii)

Stage 4: $\underline{O_2} \rightleftharpoons 2nn{\bullet}O{\bullet}en$

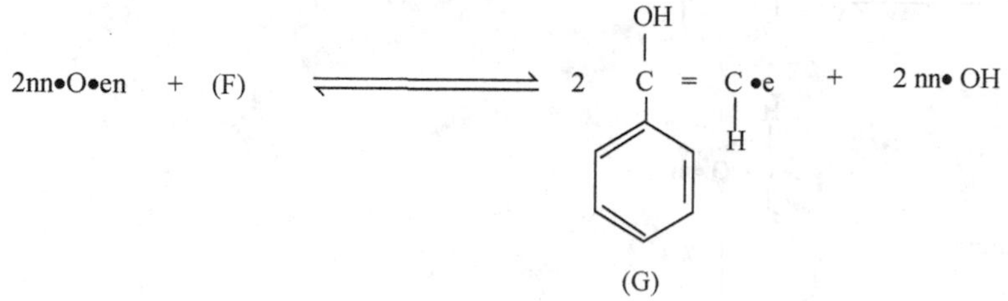

$2nn{\bullet}O{\bullet}en \ + \ \text{(F)} \quad \rightleftharpoons \quad 2$
$$\underset{\text{(G)}}{\overset{\displaystyle OH \atop \displaystyle | \atop \displaystyle C \ = \ C{\bullet}e \atop \displaystyle \bigcirc \quad | \atop \displaystyle H}{}} \quad + \quad 2\,nn{\bullet}\,OH$$

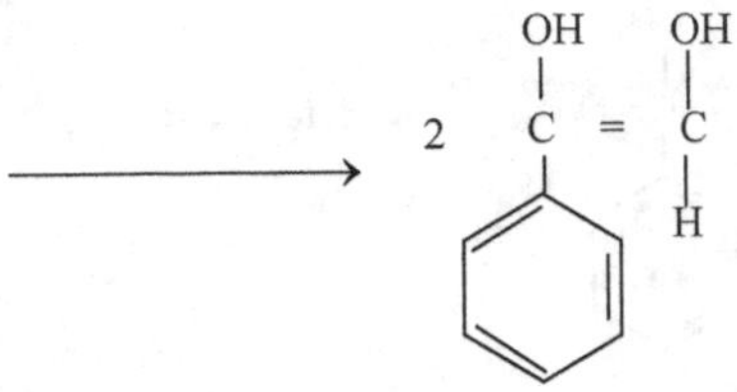

$$2 \quad \underset{\underset{\underset{C_6H_5}{|}}{|}}{\overset{OH}{\underset{|}{C}}} = \underset{\underset{H}{|}}{\overset{OH}{\underset{|}{C}}}$$

(H)   (viii)

Overall equation: $2C_6H_5C_2H_5 + 3\underline{O_2} \longrightarrow 2C_6H_5(OH)C = CH(OH) + 2H_2O$

$+ \; Energy(2)$ (ix)

Stage 5: $\underline{O_2} \rightleftharpoons 2nn{\cdot}O{\cdot}en$

$2nn{\cdot}O{\cdot}en \; + \; (H) \rightleftharpoons 2 \; \underset{C_6H_5}{\overset{OH}{\underset{|}{C}}} = \underset{\underset{H}{|}}{\overset{O{\cdot}en}{\underset{|}{C}}} \; + \; 2\,nn{\cdot}OH$

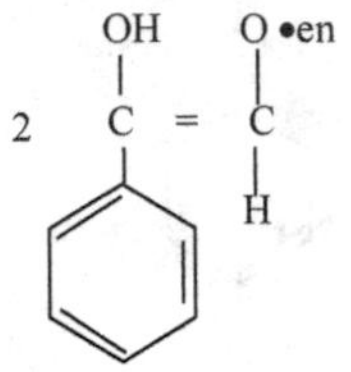

(I)

$\rightleftharpoons 2 \; \underset{C_6H_5}{\overset{OH}{\underset{|}{C}}} = C = O \quad + \; H{\cdot}e \; + \; Energy$

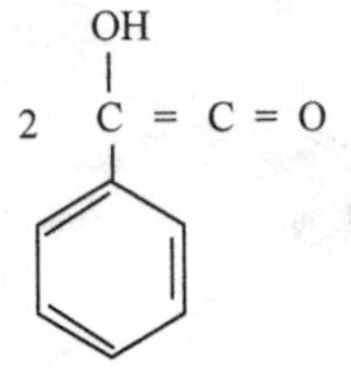

$2H{\cdot}e \; + \; 2HO{\cdot}nn \xrightarrow{\hspace{2cm}} (J) \quad \circlearrowright$ (x)

Overall equation: $2C_6H_5C_2H_5 + 4\underline{O_2} \longrightarrow 2C_6H_5(OH)C = C = O + 4H_2O$

$+ \; Energy(3)$ (xi)

Stage 6:

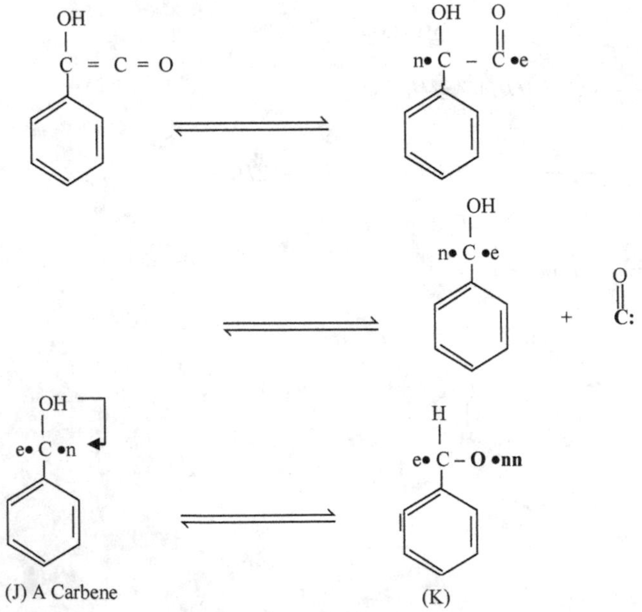

$\underset{C_6H_5}{\overset{OH}{\underset{|}{C}}} = C = O \rightleftharpoons n{\cdot}\underset{C_6H_5}{\overset{OH}{\underset{|}{C}}} - \overset{\overset{O}{\|}}{C}{\cdot}e$

$\rightleftharpoons n{\cdot}\underset{C_6H_5}{\overset{OH}{\underset{|}{C}}}{\cdot}e \quad + \quad \overset{O}{\underset{..}{C:}}$

$e{\cdot}\underset{C_6H_5}{\overset{OH}{\underset{|}{C}}}{\cdot}n \rightleftharpoons e{\cdot}\underset{C_6H_5}{\overset{H}{\underset{|}{C}}} - O{\cdot}nn$

(J) A Carbene   (K)

**MOLECULAR REARRANGEMENT OF THE SECOND KIND
OF THE FIRST TYPE**

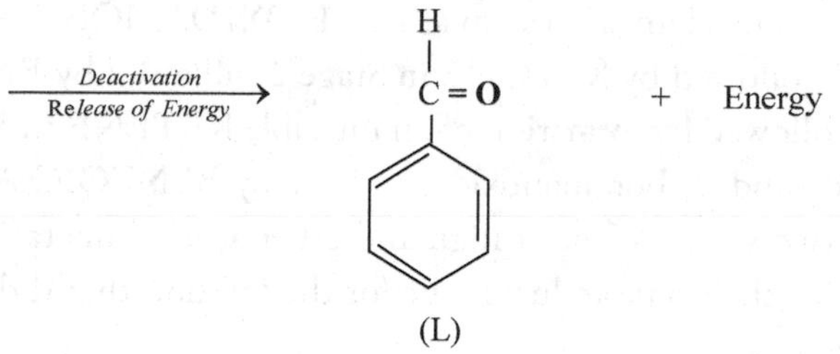

$$\xrightarrow[\text{Release of Energy}]{\text{Deactivation}} \underset{\text{(L)}}{C_6H_5-C\!\!H\!\!=\!\!O} \quad + \quad \text{Energy} \tag{xii}$$

$$\underline{Overall\ equation}: 2C_6H_5C_2H_5\ +\ 4O_2 \longrightarrow 2C_6H_5CHO\ +\ 2CO\ +\ 4$$

$$+\ Energy\,(4) \tag{xiii}$$

Stage 7:

$$O_2 \underset{\text{Oxygen molecule}}{\overset{\text{Equilibruim State Existence of Oxidising}}{\rightleftharpoons}} 2en\cdot\ddot{O}\cdot nn$$

$$2en\cdot O\cdot nn\ +\ \underset{(Stabilized)}{2C=O} \overset{\text{Oxidation}}{\rightleftharpoons} \underset{(B)}{2e\cdot\overset{O}{C}-O\bullet nn}$$

$$(B) \xrightarrow[\text{Release of Energy}]{\text{Deactivation}} 2CO_2\ +\ E_: \tag{xiv}$$

$$\underline{Overall\ Equation}:O_2 + 2CO \longrightarrow 2CO_2\ +\ Energy \tag{xv}$$

Stage 8:

$$O_2 \rightleftharpoons 2nn\bullet O\bullet en$$

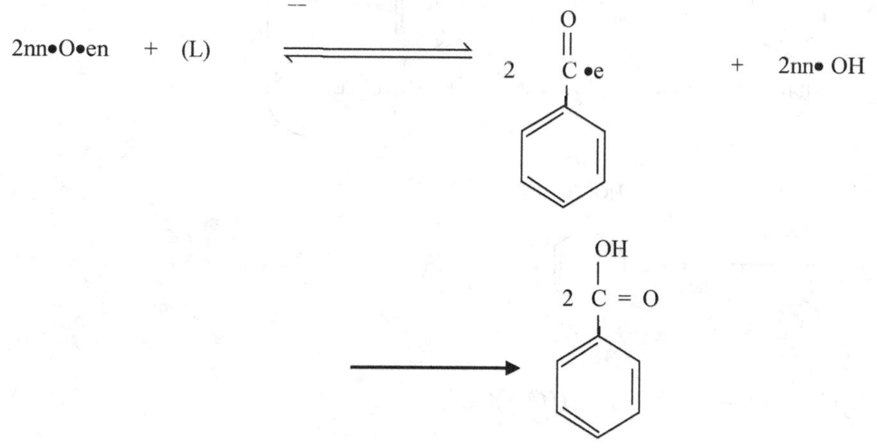

$$2nn\bullet O\bullet en\ +\ (L) \rightleftharpoons 2\,\overset{O}{\underset{}{C}}\bullet e \quad +\ 2nn\bullet OH$$

$$2\,\overset{OH}{\underset{}{C}}=O \tag{xvi}$$

Via Enolization

$$\underline{Overall\ equation}:\ 2C_6H_5C_2H_5\ +\ 6O_2 \longrightarrow 2C_6H_5COOH\ +\ 2CO_2$$

$$+\ 4H_2O\ +\ Energy\,(5) \tag{xvii}$$

Via No Enolization

$$\underline{Overall\ equation}:\ 2C_6H_5C_2H_5\ +\ 6O_2 \longrightarrow 2C_6H_5COOH\ +\ 4H_2O\ +\ 2CO_2$$

$$+\ Energy\,(5) \tag{xviii}$$

an eight stage Equilibrium mechanism system at FULL OXIDATION, beginning with formation of an ALCOHOL in Stage 1, followed by KETONE in Stage 2, followed by ENOL in Stage 3, followed by a DI-ENOL in Stage 4, followed by formation of an unstable KETENE in Stage 5, which decomposed in Stage 6 to Benzaldehyde and carbon monoxide, and finally BENZOIC ACID in the last stage; noting that from Stage 1, the two systems (one at high and other at mild operating conditions) can operate in parallel to the end to give the final products- one for the fraction that did not enolize and the other for the fraction that enolized.

(Laws of Creations for Oxidation of Ethyl benzene)

Rule 1638: This rule of Chemistry for **Oxidation of Propyl Benzene,** states that, when oxidized, the followings are obtained when the operating condition is high-

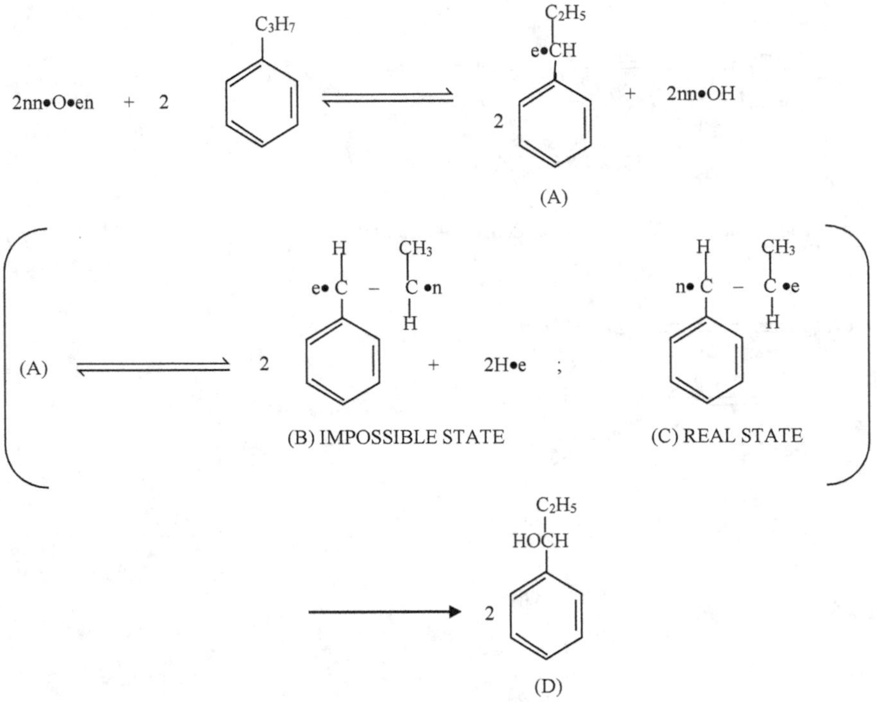

Stage 1:

$$\text{Overall equation}: 2C_6H_5C_3H_7 + O_2 \longrightarrow 2C_6H_5CH(OH)C_2H_5 \qquad (ii)$$

Stage 2:

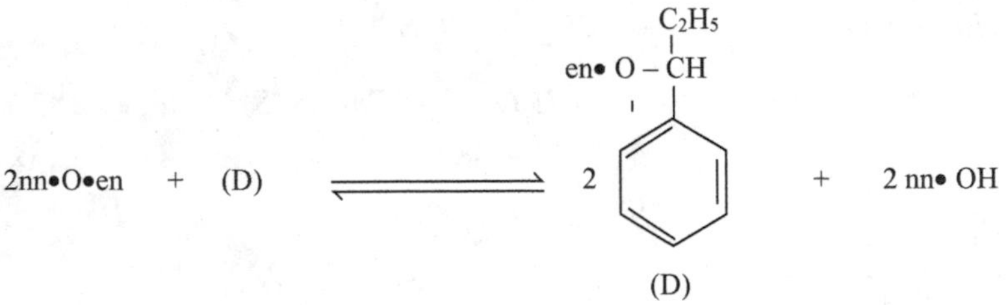

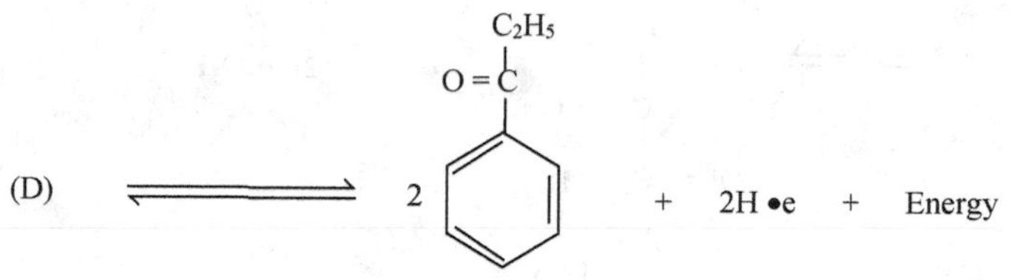

$$(D) \rightleftharpoons 2 \; [C_6H_5-C(=O)-C_2H_5] \; (E) \; + \; 2H\bullet e \; + \; Energy$$

$$2H\bullet e \; + \; 2HO\bullet nn \longrightarrow 2H_2O \tag{iii}$$

Overall equation: $2C_6H_5C_3H_7 + 2\underset{--}{O_2} \longrightarrow 2C_6H_5COC_2H_5 + 2H_2O + Energy$ (iv)

Stage 3:

$$\underset{--}{O_2} \rightleftharpoons 2nn\bullet O\bullet en$$

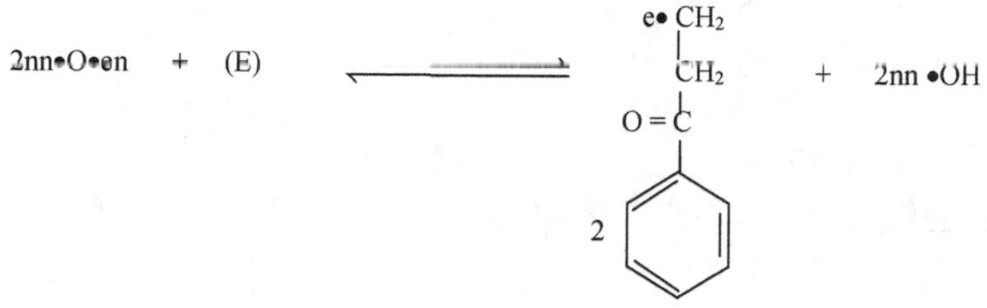

$$2nn\bullet O\bullet en \; + \; (E) \rightleftharpoons (F) \; + \; 2nn\bullet OH$$

where (F) is $e\bullet CH_2-CH_2-C(=O)-C_6H_5$

$$(F) \rightleftharpoons 2\; (G) \; - \; 2H\bullet e$$

where (G) is the radical $n\bullet C(H)-C(H)(\bullet e)(H)$ with $O=C-C_6H_5$

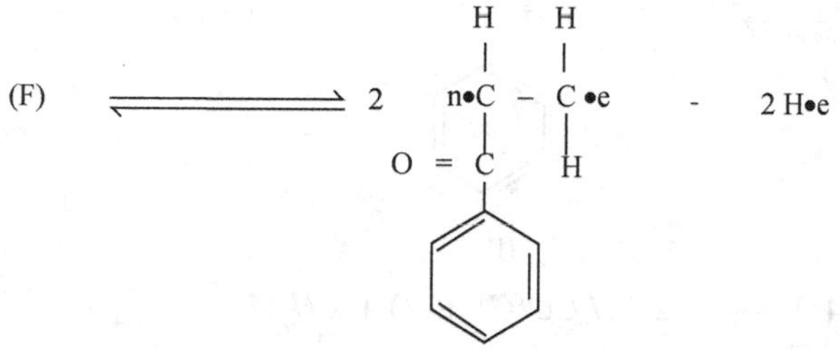

$$2H\bullet e \; + \; 2HO\bullet nn \rightleftharpoons 2H_2O$$

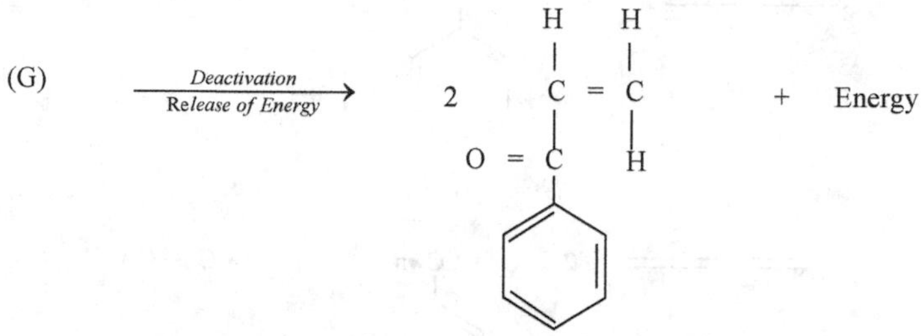

$$(G) \xrightarrow[\text{Release of Energy}]{\text{Deactivation}} 2\; (H) \; + \; Energy$$

where (H) is $HC=CH_2$ with $O=C-C_6H_5$

(v)

Overall equation: $2C_6H_5C_3H_7 + 3\underset{--}{O_2} \longrightarrow 2C_6H_5(CO)HC=CH_2 + 4H_2O$

$+ \; Energy \,(2)$ (vi)

Stage 4:

$$\underset{--}{O_2} \rightleftharpoons 2nn\bullet O\bullet en$$

$2nn \bullet O \bullet en$ + (H) $\rightleftharpoons$

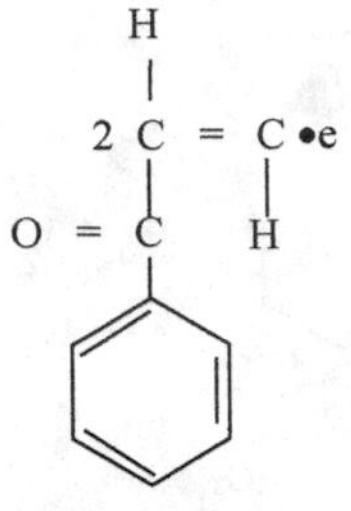

$$2 \underset{\underset{\displaystyle (I^1)}{\overset{\displaystyle O = \underset{\displaystyle C6H5}{C} - H}{|}}}{\overset{\displaystyle H}{\underset{}{C}} = C \bullet e} + 2nn \bullet OH$$

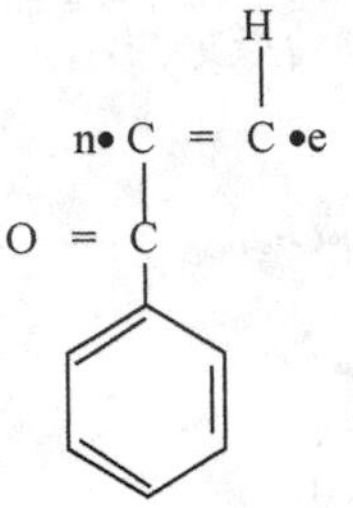

(I^1) $\rightleftharpoons$

$$2 \underset{\underset{\displaystyle (I)}{\overset{\displaystyle O = \underset{\displaystyle C6H5}{C} }{|}}}{n \bullet \overset{\displaystyle H}{\underset{}{C}} = C \bullet e} + 2H \bullet e$$

$2H \bullet e$ + $2HO \bullet nn$ $\rightleftharpoons$ $2H_2O$

(I) $\xrightarrow[\text{Release of Energy}]{\text{Deactivation}}$

$$2 \underset{\underset{\displaystyle (II)}{\overset{\displaystyle O = \underset{\displaystyle C6H5}{C} }{|}}}{\overset{\displaystyle H}{\underset{}{C}} \equiv C} + \text{Energy} \qquad \text{(vii)}$$

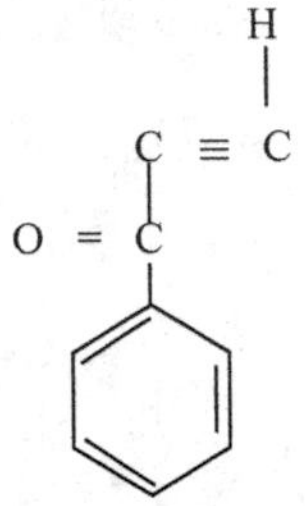

Overall equation: $2C_6H_5C_3H_7 + 4O_2 \longrightarrow 2C_6H_5CO(C \equiv CH) + 6H_2O$

$\qquad\qquad\qquad\qquad\qquad + Energy\,(3)$ \qquad (viii)

Stage 5:

(II) $\rightleftharpoons$

$$2 \underset{\underset{\displaystyle (J)}{\overset{\displaystyle O = \underset{\displaystyle C6H5}{C} }{|}}}{C \equiv C \bullet n} + 2H \bullet e$$

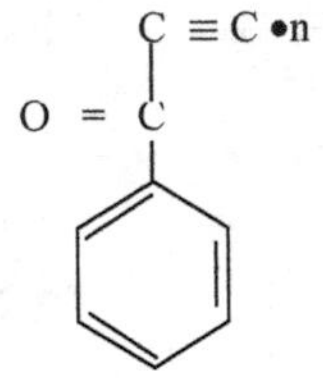

(J) $\rightleftharpoons$

$$2 \underset{\underset{\displaystyle (K)}{\overset{\displaystyle O = \underset{\displaystyle C6H5}{C} \bullet n }{|}}}{} + 2 e \bullet C \equiv C \bullet n$$

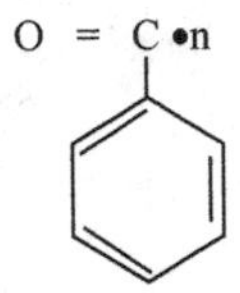

$2e \bullet C \equiv C \bullet n$ $\rightleftharpoons$ $4 \; e \bullet \overset{\displaystyle \bullet n}{\underset{\displaystyle \bullet e}{C}} \bullet n$

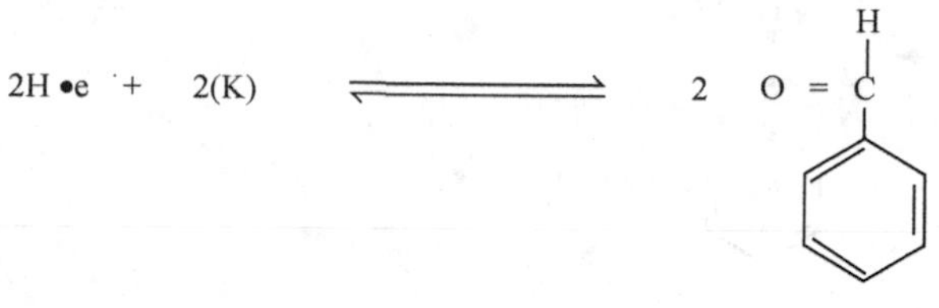

$$2H \bullet e \ + \ 2(K) \quad \rightleftharpoons \quad 2 \ \underset{\text{(L)}}{\left[O = \overset{H}{\underset{|}{C}} - C_6H_5 \right]}$$

$$4 \ e \cdot \overset{\bullet n}{\underset{\bullet e}{C}} \cdot n \quad \xrightarrow[\text{Release of Heat}]{\text{Deactivation}} \quad 4 \ e \bullet \overset{\bullet \bullet}{C} \bullet n \quad + \quad \text{Energy}$$

Activated Carbon $\qquad\qquad$ Carbon Black

(ix)

Overall Equation: $2C_6H_5C_3H_7 \ + \ 4O_2 \longrightarrow 2C_6H_5CHO \ + \ 4C \,(Carbon\ Black)$

$$+ \ 6H_2O \ + \ Energy\,(4) \qquad\qquad (x)$$

Stage 6 &7:

$$O_2 \xrightleftharpoons[\text{Oxygen molecule}]{\text{Equilibrium State Existence of Oxidising}} 2en \cdot \overset{\cdot \cdot}{O}$$

$$2en \cdot O \cdot nn + \ 2 : \overset{\bullet n}{\underset{\bullet e}{C}} \quad \xrightarrow{\text{Oxidation}} \quad 2nn \bullet O - \overset{\bullet e}{\underset{\bullet n}{C}} \bullet e$$

$$\underset{\text{(Stabilized)}}{} \qquad\qquad \underset{(F)}{}$$

$$\xrightarrow[\text{Release of Energy}]{\text{Deactivation}} \quad 2 : C = O \ + \ Energy \qquad\qquad (xi)$$

Overall Equation: $O_2 + 2\overset{\bullet n}{\underset{\bullet e}{C}} : \longrightarrow 2 : C = O \ + \ Energy$ $\qquad\qquad (xi)$

Stage 8&9:

$$O_2 \xrightleftharpoons[\text{Oxygen molecule}]{\text{Equilibrium State Existence of Oxidising}} 2en \cdot \overset{\cdot \cdot}{O} \cdot nn$$

$$2en \cdot O \cdot nn + \ 2C = O \quad \xrightarrow{\text{Oxidation}} \quad 2e \bullet \overset{\overset{O}{\|}}{C} - O \bullet nn$$

$$\underset{\text{(Stabilized)}}{} \qquad\qquad\qquad \underset{(B)}{}$$

$$(B) \quad \xrightarrow[\text{Release of Energy}]{\text{Deactivation}} \quad 2CO_2 \ + \ Energy \qquad\qquad (xiii)$$

Overall Equation : $O_2 + 2CO \longrightarrow 2CO_2 \ + \ Energy$ $\qquad\qquad (xiv)$

Overall equation: $2C_6H_5C_3H_7 \ + \ 8O_2 \longrightarrow 2C_6H_5CHO \ + \ 4CO_2 \ + \ 6H_2O$

$$+ \ Energy\,(6) \qquad\qquad (xv)$$

Stage 10:

$$O_2 \rightleftharpoons 2nn \bullet O \bullet en$$

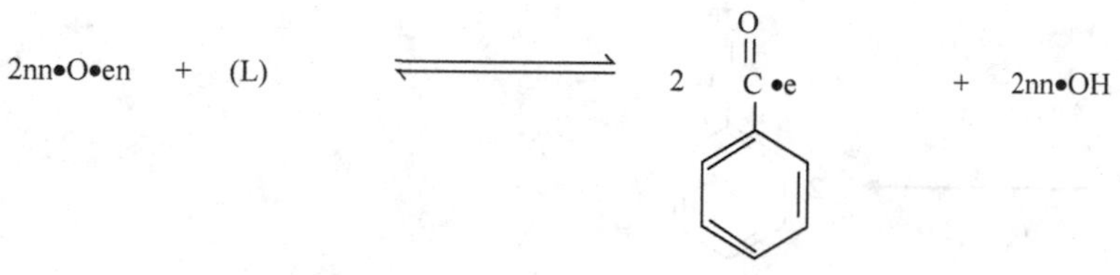

$$2nn \bullet O \bullet en \ + \ (L) \quad \rightleftharpoons \quad 2 \ \underset{(M)}{\left[\overset{\overset{O}{\|}}{C} \bullet e - C_6H_5 \right]} \ + \ 2nn \bullet OH$$

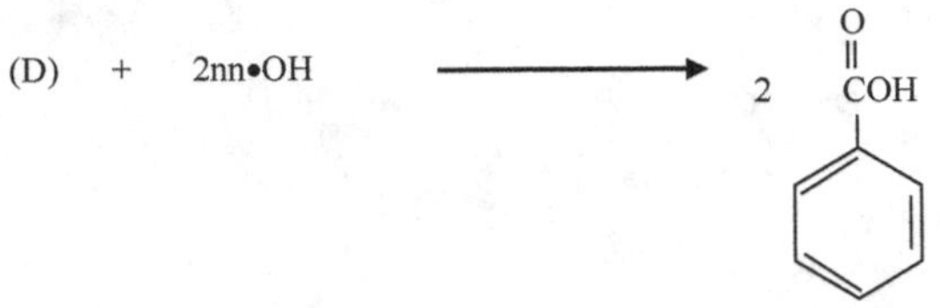

(D) + 2nn•OH $\longrightarrow$

$$2 \quad \underset{\text{(N)}}{\text{C}_6\text{H}_5\text{COOH}}$$

(xvi)

Overall equation: $2C_6H_5C_3H_7 + 9O_2 \longrightarrow 2C_6H_5COOH + 6H_2O + 4CO_2$

$+ \quad Energy(8)$

(xvii)

an eight stage Equilibrium mechanism system at FULL OXIDATION in which the sixth and seventh stages are two in one (i.e., in parallel), beginning with formation of an ALCOHOL in Stage 1, followed by KETONES in Stages 2 and 3 followed by formation of an unstable KETONE in Stage 4, which decomposed in Stage 5 to Benzaldehyde and carbon black, and finally BENZOIC ACID in the last stage; noting that no enol was formed.

(Laws of Creations for Oxidation of Propyl Benzene)

Rule 1639: This rule of Chemistry for **Oxidation of Propyl Benzene,** states that, when oxidized, the followings are obtained when the operating condition is mild-

Stage 1:

$$O_2 \rightleftharpoons 2nn•O•en$$

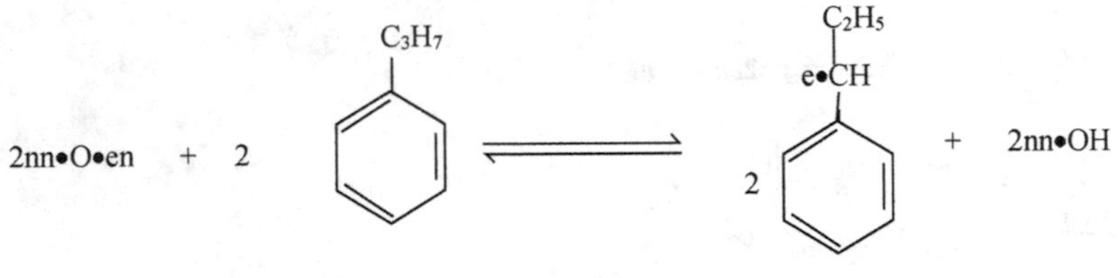

2nn•O•en + 2 [C₃H₇—benzene] $\rightleftharpoons$ 2 [e•CH(C₂H₅)—benzene] + 2nn•OH

(A)

$$(A) \rightleftharpoons 2 \text{ [e•C(H)–C(CH}_3)•n \text{ benzene]} + 2H•e \quad ;$$

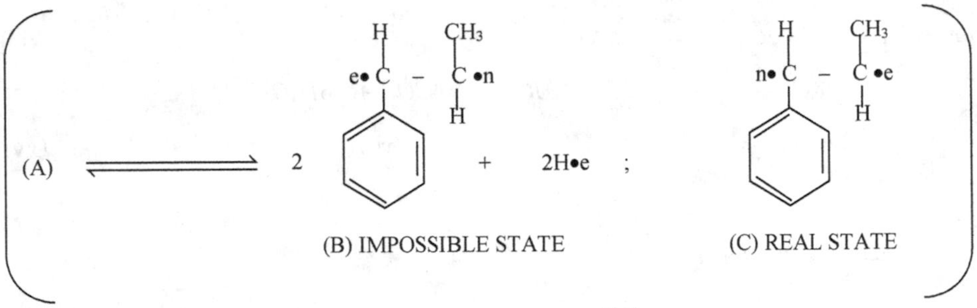

(B) IMPOSSIBLE STATE (C) REAL STATE

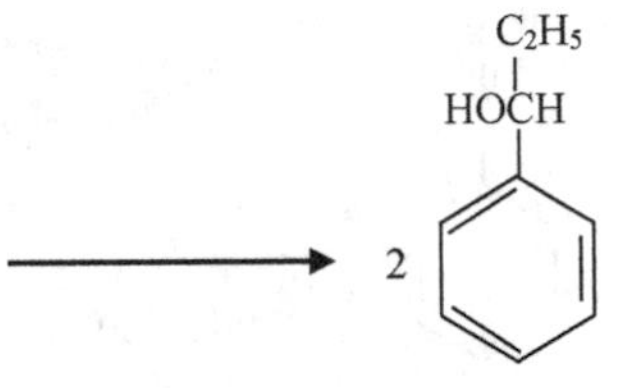

$\longrightarrow$ 2 [HOCH(C₂H₅)—benzene]

(D)

(i)

Overall equation: $2C_6H_5C_3H_7 + O_2 \longrightarrow 2C_6H_5CH(OH)C_2H_5$

(ii)

Stage 2:

$$O_2 \rightleftharpoons 2nn{\cdot}O{\cdot}en$$

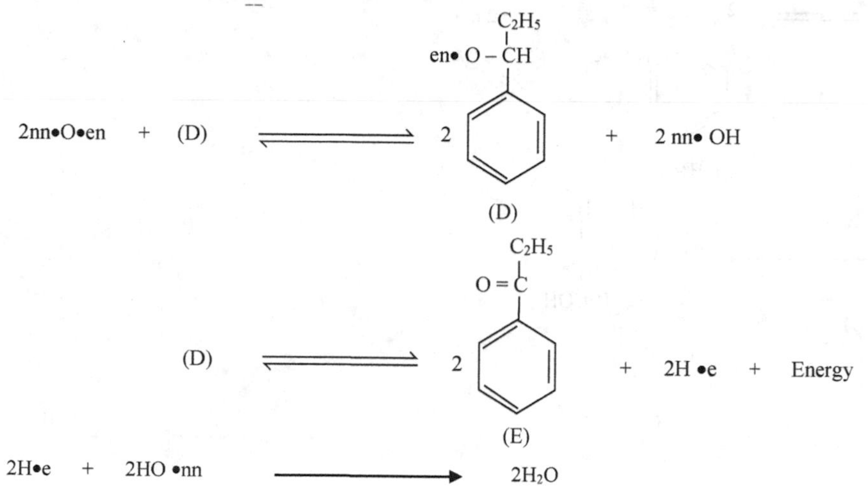

$$2nn{\cdot}O{\cdot}en \quad + \quad (D) \quad \rightleftharpoons \quad 2 \quad \underset{(D)}{\overset{\overset{\displaystyle C_2H_5}{\underset{\displaystyle |}{en{\cdot}\,O-CH}}}{\bigcirc}} \quad + \quad 2\,nn{\cdot}\,OH$$

$$(D) \quad \rightleftharpoons \quad 2 \quad \underset{(E)}{\overset{\overset{\displaystyle C_2H_5}{\underset{\displaystyle |}{O=C}}}{\bigcirc}} \quad + \quad 2H{\cdot}e \quad + \quad Energy$$

$$2H{\cdot}e \quad + \quad 2HO{\cdot}nn \quad \longrightarrow \quad 2H_2O \qquad\qquad \text{(iii)}$$

Overall equation: $2C_6H_5C_3H_7 \;+\; 2O_2 \longrightarrow 2C_6H_5COC_2H_5 + 2H_2O + Energy$ (iv)

Stage 3:

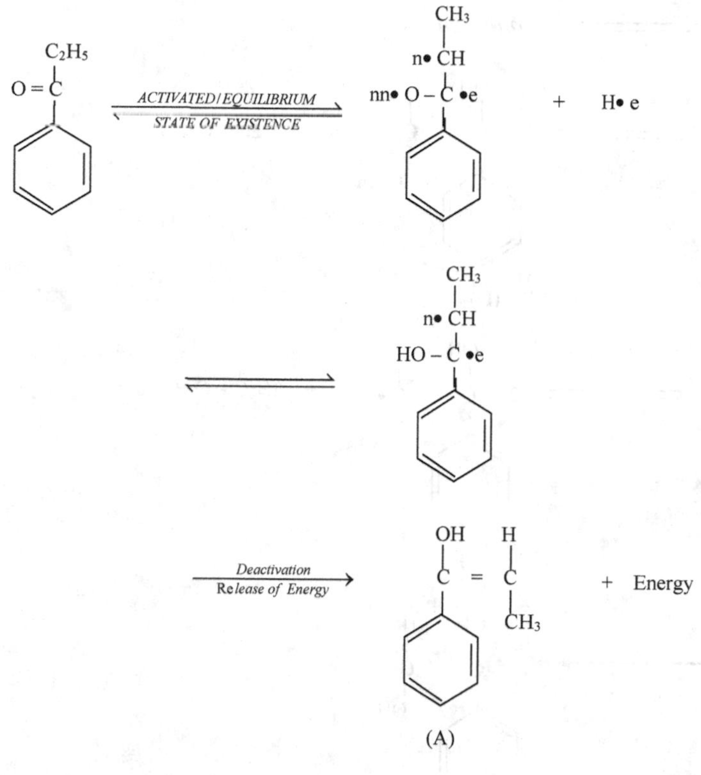

$$\underset{}{\overset{\overset{\displaystyle C_2H_5}{\underset{\displaystyle |}{O=C}}}{\bigcirc}} \quad \underset{\substack{\textit{ACTIVATED / EQUILIBRIUM} \\ \textit{STATE OF EXISTENCE}}}{\longleftrightarrow} \quad \underset{}{\overset{\overset{\displaystyle CH_3}{\underset{\displaystyle |}{n{\cdot}\,CH}}}{\underset{\displaystyle |}{nn{\cdot}\,O-C{\cdot}e}}}\bigcirc \quad + \quad H{\cdot}e$$

$$\rightleftharpoons \quad \underset{}{\overset{\overset{\displaystyle CH_3}{\underset{\displaystyle |}{n{\cdot}\,CH}}}{\underset{\displaystyle |}{HO-C{\cdot}e}}}\bigcirc$$

$$\underset{\substack{\textit{Deactivation} \\ \textit{Release of Energy}}}{\longrightarrow} \quad \underset{(A)}{\overset{\overset{\displaystyle OH \quad H}{\underset{\displaystyle |\;\;\;\;\;|}{C=C}}}{\bigcirc}}_{\underset{\displaystyle CH_3}{|}} \quad + \quad Energy$$

(v)

Overall equation: $2C_6H_5C_3H_7 \;+\; 2O_2 \longrightarrow 2C_6H_5(OH)C=CHCH_3 + 2H_2O$

$+\ Energy\,(2)$ (vi)

Stage 4:

$$O_2 \rightleftharpoons 2nn{\cdot}O{\cdot}en$$

$2nn \bullet O \bullet en$ + (A) $\rightleftharpoons$ 2 [structure: phenyl ring with OH, C=C, H, H₂C•e] (B) + 2 nn• OH

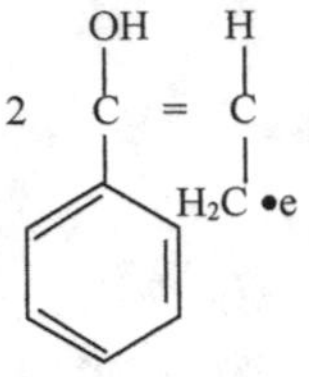

$\longrightarrow$ 2 [structure: phenyl ring with OH, C=C, H, H₂COH] (C)

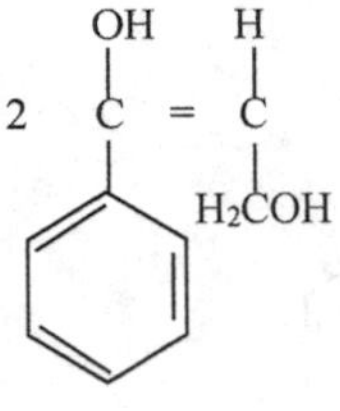

(vii)

$$\underline{Overall \ equation}: 2C_6H_5C_3H_7 \ + \ 3O_2 \longrightarrow 2C_6H_5(OH)C = CH(CH_2OH) + 2H_2O$$

$$+ \quad Energy(2)$$

(viii)

Stage 5: $\qquad O_2 \rightleftharpoons 2nn \bullet O \bullet en$

$2nn \bullet O \bullet en$ + (C) $\rightleftharpoons$ 2 [structure: phenyl ring with OH, C=C, H, H₂CO•en] (D) + 2nn• OH

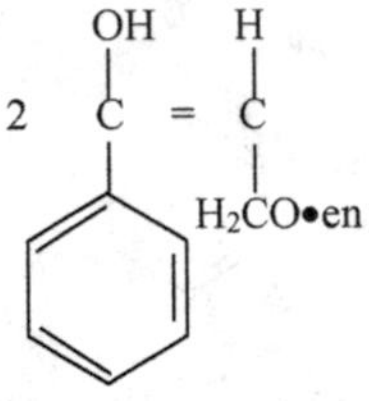

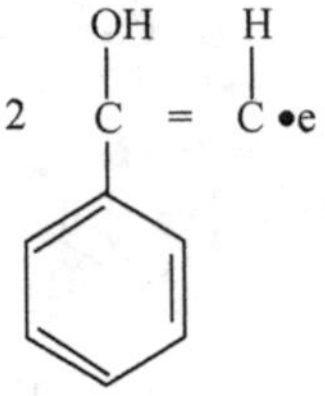

(D) $\rightleftharpoons$ 2 [structure: phenyl ring with OH, C=C, H, •e] (E) + 2H₂C = O + Energy

(E) + 2nn• OH $\longrightarrow$ 2 [structure: phenyl ring with OH, C=C, H, OH] (F)

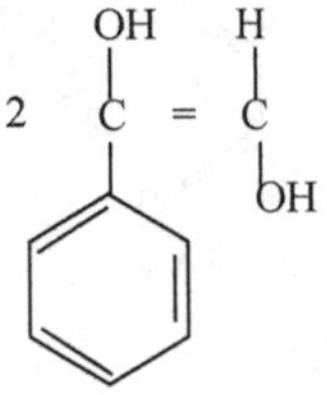

(ix)

$$\underline{Overall \ equation}: 2C_6H_5C_2H_5 \ + \ 4O_2 \longrightarrow 2C_6H_5(OH)C = CH(OH) + 2H_2O$$

$$+ \ 2H_2C = O + \quad Energy(3)$$

(x)

710

Stage 6:

$$O_2 \rightleftharpoons 2nn\cdot O\cdot en$$

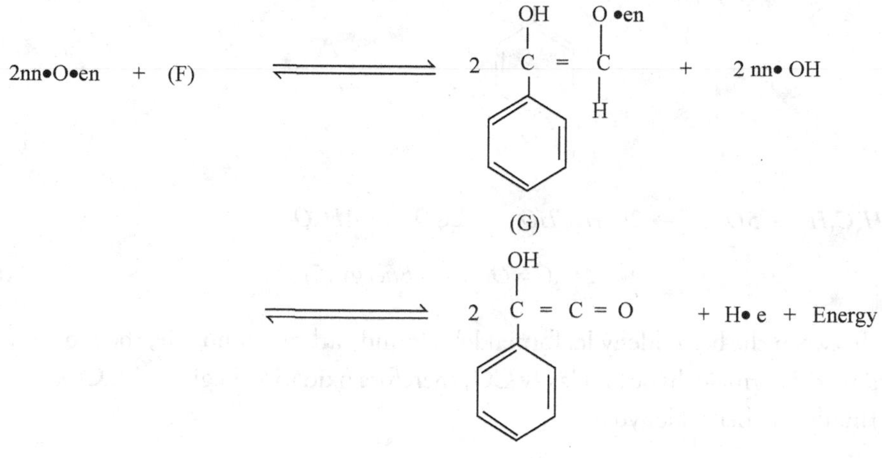

$2nn\cdot O\cdot en \ + \ (F) \ \rightleftharpoons \ 2$

(G)

$+ \ 2\ nn\cdot OH$

$\rightleftharpoons \ 2$

(H)

$+ \ H\cdot e \ + \ Energy$

$2H\cdot e \ + \ 2HO\cdot nn \ \xrightarrow{\hspace{3cm}} \ 2H_2O$ $\hspace{2cm}$ (xi)

Overall equation : $2C_6H_5C_3H_7 \ + \ 5O_2 \ \longrightarrow \ 2C_6H_5(OH)C=C=O \ + \ 4H_2O$

$+ \ 2H_2C=O \ + \ Energy\,(4)$ $\hspace{2cm}$ (xii)

Stage 7:

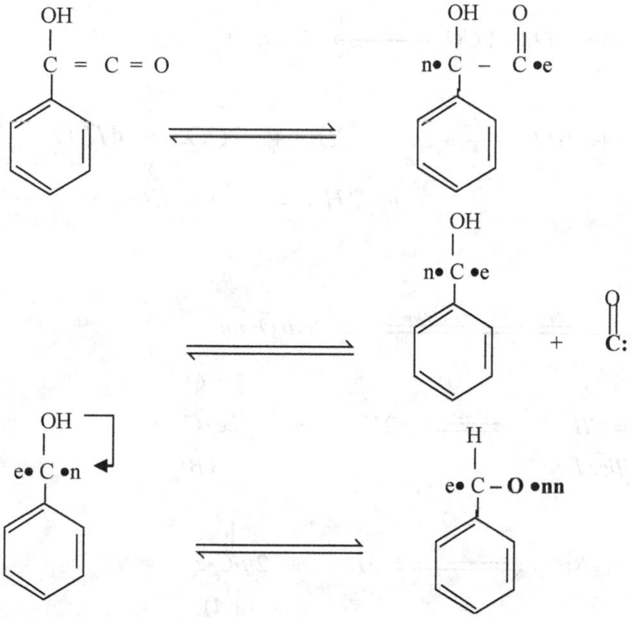

(I) A Carbene $\hspace{3cm}$ (J)

**MOLECULAR REARRANGEMENT OF THE SECOND KIND
OF THE FIRST TYPE**

711

$$\xrightarrow[\text{Release of Energy}]{\text{Deactivation}}$$

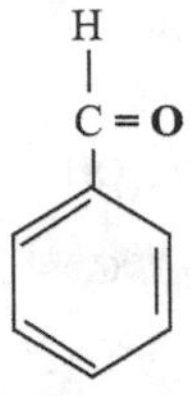

$$\underset{(K)}{\overset{\overset{\displaystyle H}{\underset{\displaystyle |}{C=O}}}{\bigcirc}} \quad + \quad \text{Energy} \qquad \text{(xiii)}$$

Overall equation: $2C_6H_5C_3H_7 + 5\underset{--}{O_2} \longrightarrow 2C_6H_5CHO + 2CO + 4H_2O$

$$+ \quad 2H_2C=O + \quad Energy\,(5) \qquad \text{(xiv)}$$

important to note that between the benzaldehyde, formaldehyde and carbon monoxide, the most stable is benzaldehyde, followed by the formaldehyde and lastly CO; therefore oxidation begins with CO, followed by formaldehyde and finally the benzaldehyde-

Stage 8:

$$\underset{--}{O_2} \underset{\text{Oxygen molecule}}{\overset{\text{Equilibruim State Existence of Oxidi}\sin g}{\rightleftharpoons}} 2en\cdot\ddot{\overset{..}{O}}\cdot nn$$

$$2en\cdot O\cdot nn + \underset{(Stabilized)}{2C=O} \xrightarrow{\text{Oxidation}} \underset{(B)}{2\,e\bullet \overset{\overset{\displaystyle O}{\|}}{C}-O\bullet nn}$$

$$(B) \xrightarrow[\text{Release of Energy}]{\text{Deactivation}} 2CO_2 + Energy \qquad \text{(xv)}$$

Overall Equation : $\underset{--}{O_2} + 2CO \longrightarrow 2CO_2 + Energy \qquad \text{(xvi)}$

Overall equation: $2C_6H_5C_3H_7 + 6\underset{--}{O_2} \longrightarrow 2C_6H_5CHO + 2CO_2 + 4H_2O$

$$+ \quad 2H_2C=O + Energy\,(6) \qquad \text{(xvii)}$$

Stage 9:

$$\underset{--}{O_2} \underset{\text{Oxygen molecule}}{\overset{\text{Equilibruim State Existence of Oxidi}\sin g}{\rightleftharpoons}} 2en\cdot\ddot{\overset{..}{O}}\cdot nn$$

$$2en\cdot O\cdot nn + \underset{(Stabilized)}{2O=CH_2} \xrightarrow{\text{Oxidation}} 2H-O\cdot nn + 2e\cdot\overset{\overset{\displaystyle O}{\|}}{C}-H$$
$$(B)$$

$$(B) \rightleftharpoons 2H\bullet e + \underset{|\,(A)}{2n\bullet\overset{\overset{\displaystyle O}{\|}}{C}\bullet e}$$

$$2H\bullet e + 2nn\bullet OH \rightleftharpoons 2H_2O$$
$$(A) \xrightarrow[\text{release of Energy}]{\text{Deactivation}} 2:C=O + Energy \qquad \text{(xviii)}$$

Overall Equation: $\underset{--}{O_2} + 2O=CH_2 \longrightarrow 2H_2O + 2CO + Energy \qquad \text{(xix)}$

Stage 10:

$$O_2 \underset{Oxygen\ molecule}{\overset{Equilibrium\ State\ Existence\ of\ Oxidising}{\rightleftharpoons}} 2en \cdot \ddot{O} \cdot nn$$

$$2en \cdot O \cdot nn + 2C = O \xrightarrow{Oxidation} 2e \bullet \overset{\overset{O}{\|}}{C} - O \bullet nn$$

$$\text{(Stabilized)} \qquad\qquad (B)$$

$$(B) \xrightarrow[Release\ of\ Energy]{Deactivation} 2CO_2 + Energy \qquad\qquad (xx)$$

$$\underline{Overall\ Equation}: O_2 + 2CO \longrightarrow 2CO_2 + Energy \qquad\qquad (xxi)$$

$$Overall\ equation: 2C_6H_5C_3H_7 + 8O_2 \longrightarrow 2C_6H_5CHO + 4CO_2 + 6H_2O$$

$$+ \ Energy(8) \qquad\qquad (xxii)$$

Stage 11:

$$O_2 \rightleftharpoons 2nn \bullet O \bullet en$$

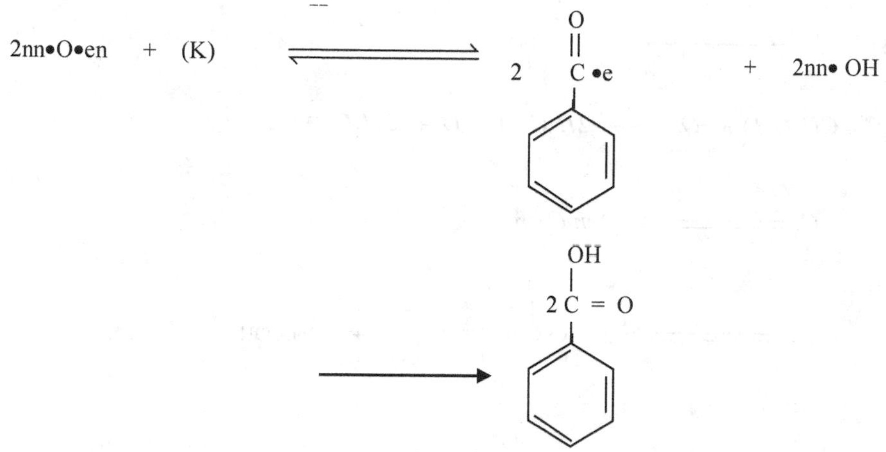

$$2nn \bullet O \bullet en + (K) \rightleftharpoons$$

$$+ 2nn \bullet OH$$

$$\longrightarrow$$

$$(xxiii)$$

Via Enolization

$$\underline{Overall\ equation}: 2C_6H_5C_3H_7 + 9O_2 \longrightarrow 2C_6H_5COOH + 4CO_2$$

$$+ \ 6H_2O + Energy(8)$$

$$(xxiv)$$

Via Non-Enolization

$$Overall\ equation: 2C_6H_5C_3H_7 + 9O_2 \longrightarrow 2C_6H_5COOH + 6H_2O + 4CO_2$$

$$+ \ Energy(8) \qquad\qquad (xxv)$$

an eleven stage Equilibrium mechanism system at FULL OXIDATION, beginning with formation of an ALCOHOL in Stage 1, followed by KETONE in Stage 2, followed by ENOL in Stage 3, followed by a DI-ENOLS in Stages 4 and 5, followed by formation of an unstable KETENE in Stage 6, which decomposed in Stage 6 to Benzaldehyde, formaldehyde and carbon monoxide, and finally BENZOIC ACID in the last stage; noting that from Stage 2, the two systems (one at high and other at mild operating conditions) can operate in parallel to the end to give the final products- one for the fraction that did not enolize and the other for the fraction that enolized.

(Laws of Creations for Oxidation of Propyl Benzene)

713

Rule 1640: This rule of Chemistry for **Vinyl alcohol an unstable monomer,** states that, when allowed to undergo oxidation, the followings are obtained-

Stage 1:

$$O_2 \rightleftharpoons 2nn \cdot O \cdot en$$

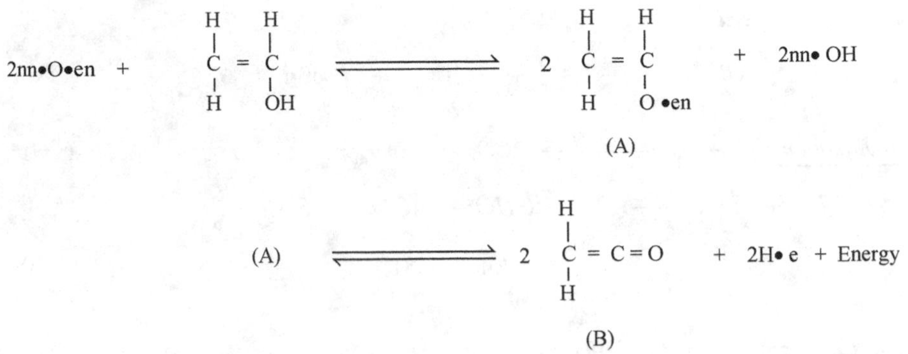

2nn•O•en + [H₂C=C(H)(OH)] $\rightleftharpoons$ 2 [H(H)C=C(H)(O•en)] + 2nn• OH

(A)

(A) $\rightleftharpoons$ 2 [H(H)C=C=O] + 2H• e + Energy

(B)

2H•e + 2HO •nn $\longrightarrow$ 2H₂O (i)

Overall equation: $2H_2C=CH(OH) + O_2 \longrightarrow 2H_2C=C=O + 2H_2O + Energy$ (ii)

Stage 2:

$$O_2 \rightleftharpoons 2nn \cdot O \cdot en$$

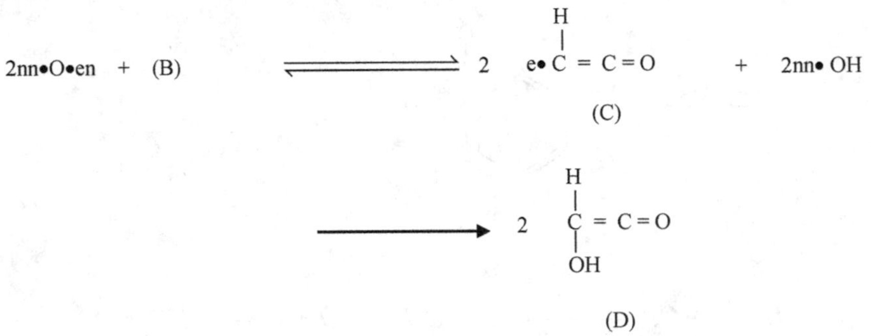

2nn•O•en + (B) $\rightleftharpoons$ 2 [e• C(H)=C=O] + 2nn• OH

(C)

$\longrightarrow$ 2 [C(H)(OH)=C=O]

(D) (iii)

Overall equation: $2H_2C=CH(OH) + 2O_2 \longrightarrow H(OH)C=C=O + 2H_2O + Energy$ (iv)

Stage 3:

$$O_2 \rightleftharpoons 2nn \cdot O \cdot en$$

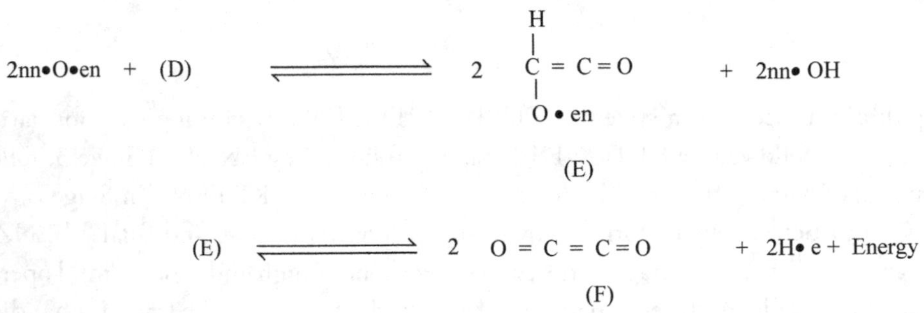

2nn•O•en + (D) $\rightleftharpoons$ 2 [C(H)(O•en)=C=O] + 2nn• OH

(E)

(E) $\rightleftharpoons$ 2 O=C=C=O + 2H• e + Energy

(F)

2H• e + 2nn• OH $\longrightarrow$ 2H₂O (v)

Overall equation: $2H_2C = CH(OH) + 3O_2 \longrightarrow 2O = C = C = O + 4H_2O + Energy$

Stage 4:

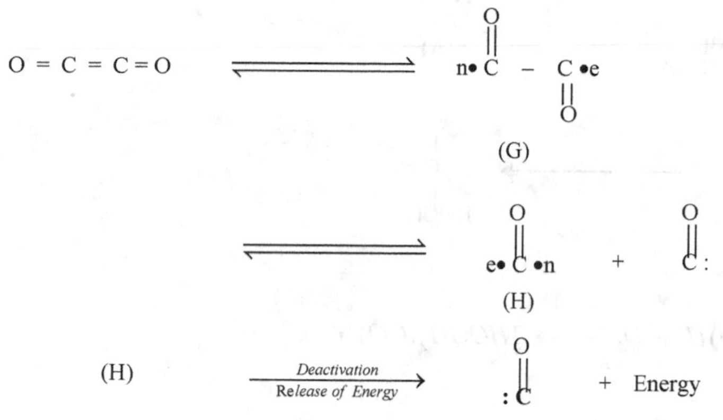

$$O = C = C = O \rightleftharpoons n\bullet \overset{\overset{O}{\|}}{C} - C\bullet e$$

(G)

$$\rightleftharpoons e\bullet \overset{\overset{O}{\|}}{C}\bullet n + \overset{\overset{O}{\|}}{C}:$$

(H)

$$(H) \xrightarrow[\text{Release of Energy}]{\text{Deactivation}} :\overset{\overset{O}{\|}}{C} + Energy$$

Overall equation: $2H_2C = CH(OH) + 3O_2 \longrightarrow 4CO + 4H_2O + Energy$

Stage 5:

$$O_2 \xrightleftharpoons[\text{Oxygen molecule}]{\text{Equilibruim State Existence of Oxidising}} 2en\cdot\ddot{O}\cdot nn$$

$$2en\cdot O\cdot nn + 2C = O \xrightarrow{\text{Oxidation}} 2e\bullet\overset{\overset{O}{\|}}{C} - O\bullet nn$$
$$\text{(Stabilized)} \qquad\qquad (B)$$

$$(B) \xrightarrow[\text{Release of Energy}]{\text{Deactivation}} 2CO_2 + Energy \qquad\qquad (ix)$$

$$\text{Overall Equation}: O_2 + 2CO \longrightarrow 2CO_2 + Energy$$

Stage 6: Same as Stage 5 above for the second two moles of CO.

Overall equation: $2H_2C = CH(OH) + 5O_2 \longrightarrow 4CO_2 + 4H_2O + Energy$ (5)

a five stage explosive Equilibrium mechanism system at FULL OXIDATION wherein the last stage is two in one (i.e., in parallel), explosive in the sense that, energy is released in every stage almost like the case of Benzoic acid; noting that (B) formed in Stage 1, a Ketene is an ELECTROPHILE and so also is (F) in Stage 3, a diketone which is very unstable and since it cannot be oxidized, it decomposes in the fourth stage based on the operating conditions, to give two moles of carbon monoxide with release of energy and this finally oxidized to CO_2.
(Laws of Creations for Oxidation of Vinyl alcohol)

Rule 1641: This rule of Chemistry for **Acetaldehyde an isomer of Vinyl alcohol,** states that, when oxidized, the followings are obtained-

Stage 1: $O_2 \rightleftharpoons 2nn\bullet O\bullet en$

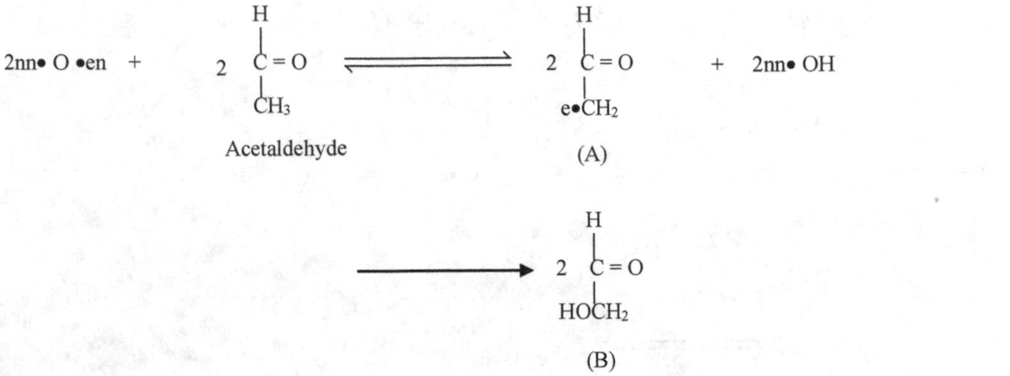

$$2nn\bullet O \bullet en \;+\; 2 \underset{CH_3}{\overset{H}{\underset{|}{\overset{|}{C}}}}=O \;\rightleftharpoons\; 2 \underset{e\bullet CH_2}{\overset{H}{\underset{|}{\overset{|}{C}}}}=O \;+\; 2nn\bullet OH \qquad (vi)$$

Acetaldehyde   (A)

$$\longrightarrow\; 2 \underset{HOCH_2}{\overset{H}{\underset{|}{\overset{|}{C}}}}=O$$

(B)   (i)

Overall equation: $2H_3C(CO)H + \underset{--}{O_2} \longrightarrow 2HOCH_2(CO)H$ (ii)

       (vii)

Stage 2:

       (viii)

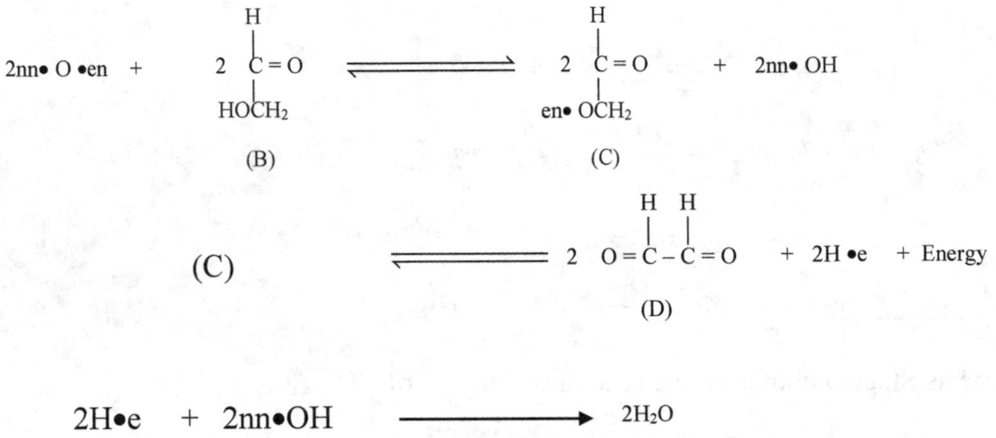

$$2nn\bullet O \bullet en \;+\; 2 \underset{HOCH_2}{\overset{H}{\underset{|}{\overset{|}{C}}}}=O \;\rightleftharpoons\; 2 \underset{en\bullet OCH_2}{\overset{H}{\underset{|}{\overset{|}{C}}}}=O \;+\; 2nn\bullet OH$$

(B)   (C)

$$(C) \;\rightleftharpoons\; 2\; \underset{}{O}=\overset{H}{\underset{|}{C}}-\overset{H}{\underset{|}{C}}=O \;+\; 2H\bullet e \;+\; Energy$$

(D)   (x)

$$2H\bullet e \;+\; 2nn\bullet OH \;\longrightarrow\; 2H_2O$$

      (iii)
      (xi)

Overall equation: $2H_3C(CO)H \;+\; 2\underset{--}{O_2} \longrightarrow 2HCO(HCO) + 2H_2O + Energy$

      (iv)

Stage 3:

$$\underset{--}{O_2} \rightleftharpoons 2nn\bullet O\bullet en$$

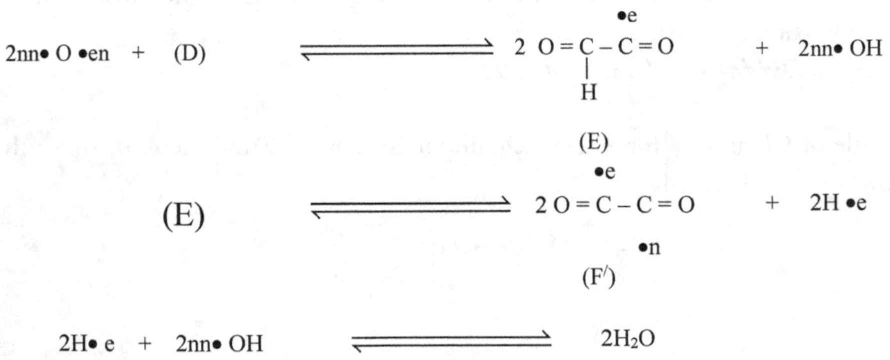

$$2nn\bullet O \bullet en \;+\; (D) \;\rightleftharpoons\; 2\; O=\overset{\bullet e}{\underset{\underset{H}{|}}{C}}-C=O \;+\; 2nn\bullet OH$$

(E)

$$(E) \;\rightleftharpoons\; 2\; O=\overset{\bullet e}{C}-\underset{\bullet n}{C}=O \;+\; 2H\bullet e$$

(F′)

$$2H\bullet e \;+\; 2nn\bullet OH \;\rightleftharpoons\; 2H_2O$$

(F) $\xrightarrow[\text{Release of Energy}]{\text{Deactivation}}$ $2\ O=C=C=O$ $\quad + \quad$ Energy

(G)

$\quad$ (v)

Overall equation: $2H_3C(CO)H + 3O_2 \longrightarrow 2O=C=C=O + 4H_2O + Energy$

$\quad$ (vi) $\quad\quad$ (vi)

Stage 4:

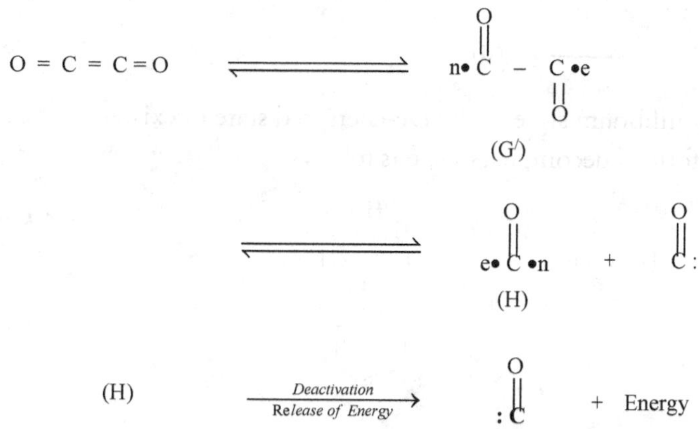

$$O=C=C=O \quad \rightleftharpoons \quad n\bullet \overset{\overset{\displaystyle O}{\|}}{C} - \overset{\underset{\displaystyle \|}{\underset{\displaystyle O}{}}}{C}\bullet e$$

(G')

$$\rightleftharpoons \quad e\bullet \overset{\overset{\displaystyle O}{\|}}{C}\bullet n \quad + \quad \overset{\overset{\displaystyle O}{\|}}{C}:$$

(H)

(H) $\xrightarrow[\text{Release of Energy}]{\text{Deactivation}}$ $:\overset{\overset{\displaystyle O}{\|}}{C}$ $\quad + \quad$ Energy

Overall equation: $2H_3C(CHO) + 3O_2 \longrightarrow 4CO + 4H_2O + Energy$ $\quad$ (viii)

Stage 5:

$$O_2 \xrightleftharpoons[\text{Oxygen molecule}]{\text{Equilibruim State Existence of Oxidising}} 2en\cdot\ddot{O}\cdot nn$$

$$2en\cdot O\cdot nn + 2C=O \xrightleftharpoons{\text{Oxidation}} 2e\bullet\overset{\overset{\displaystyle O}{\|}}{C}-O\bullet nn$$
$$\text{(Stabilized)} \quad\quad\quad\quad\quad\quad (I)$$

(I) $\xrightarrow[\text{Release of Energy}]{\text{Deactivation}}$ $2CO_2$ $\quad + \quad$ Energy $\quad$ (ix)

Overall Equation : $O_2 + 2CO \longrightarrow 2CO_2 + $ Energy $\quad$ (x)

Overall equation: $2H_3C(CO)H + 5O_2 \longrightarrow 4CO_2 + 4H_2O + Energy(5)$ $\quad$ (xi)

Overall equation: $2H_2C=CH(OH) + 5O_2 \longrightarrow 4CO_2 + 4H_2O + Energy(5)$ $\quad$ (xii)

a five stage Equilibrium mechanism system at FULL OXIDATION, wherein heat is released in five stages same as with vinyl alcohol; for which for specific reasons, aldehydes have the added abilities to exist in many states of existences- Stable, Equilibrium, Activated, Activated /Equilibrium, Decomposition and Combination states of existences.

(Laws of Creations for Oxidation of Acetaldehyde)

Rule 1642: This rule of Chemistry for **Ortho-Xylene,** states that, when made to undergo combustion, the followings are obtained-

Stage 1:

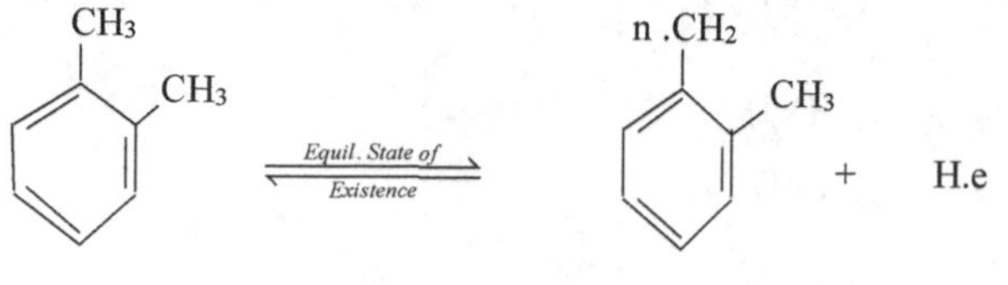

(I)

H.e $+$ O$=$O $\xrightleftharpoons{\text{Activation}}$ H$-$O$-$O.en

H$-$O$-$O.en $+$ (I) $\longrightarrow$ H$-$O$-$O$-$CH$_2$

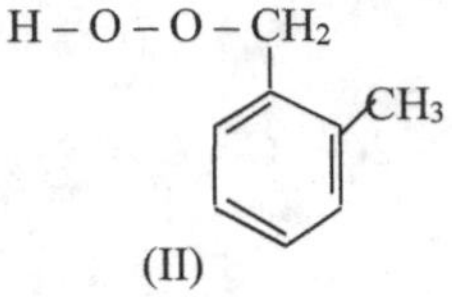

(II) (i)

Overall equation: o-xylene $+$ O$_2$ $\longrightarrow$ (II) (ii)

without o-xylene existing in equilibrium state in the De-energized state of existence, it cannot readily be combusted; the peroxide (II) formed decomposes next as follows-

Stage 2:

(II) $\rightleftharpoons$ H$-$O.nn $+$ en.O$-$C$-$H with CH$_3$

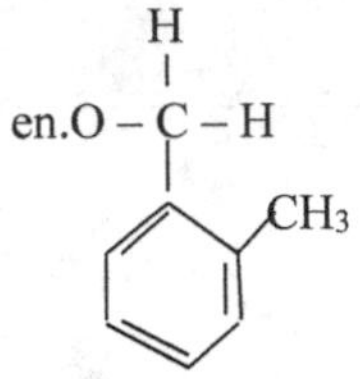

(III)

(III) $\rightleftharpoons$ H.e $+$ O$=$C with CH$_3$ $+$ Energy

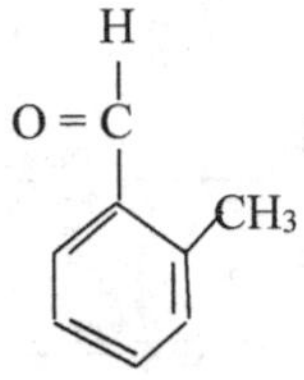

(IV)

H.e $+$ nn.OH $\longrightarrow$ H$_2$O (iii)

Overall equation: (II) $\longrightarrow$ H$_2$O $+$ (IV) $+$ Energy (iv)

Overall overall equation: o-xylene $+$ O$_2$ $\longrightarrow$ H$_2$O $+$ (IV) $+$ Energy (v)

Stage 3:

(IV) $\xrightleftharpoons{\substack{\text{Equil. State of}\\\text{Existence of aldehyde}}}$ O$=$C $\bullet$n with CH$_3$ $+$ H.e

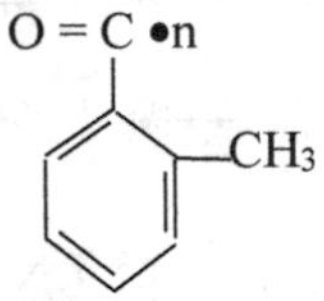

(V)

H.e $+$ O$=$O $\xrightleftharpoons{\text{Activation}}$ H$-$O$-$O.en

H$-$O$-$O.en $+$ (V) $\longrightarrow$ O$=$C with O$-$O$-$H, H, CH

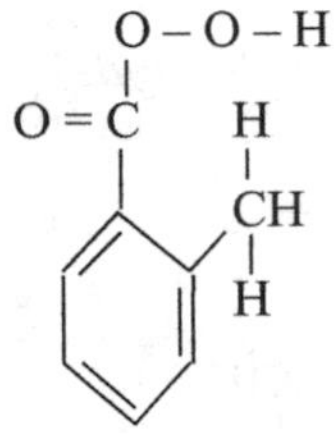

(VI) (vi)

Overall equation: (V) $+$ O$_2$ $\longrightarrow$ (VI) (vii)

718

Stage 4: $\quad$ 2(VI) $\rightleftharpoons$ $\quad$ 2HO •en $\quad$ +

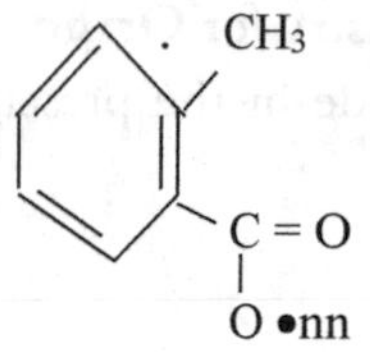

(VII)

2HO •en $\quad$ $\rightleftharpoons$ $\quad$ 2H •e $\quad$ + $\quad$ O_2

(VII) $\quad$ $\rightleftharpoons$

(VIII) $\qquad$ (IX)

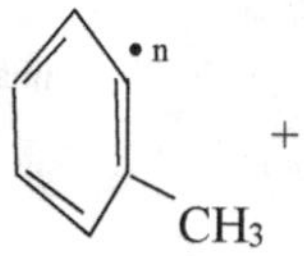

2H •e $\quad$ + $\quad$ 2 (VIII) $\quad$ $\rightleftharpoons$

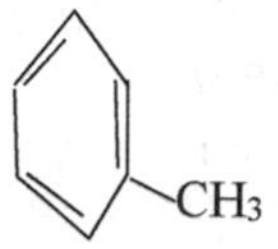

Toluene

(IX) $\qquad\longrightarrow\qquad$ $2CO_2$ $\quad$ + $\quad$ Heat $\hfill$ (viii)

Overall Equation: $\quad$ 2 o-Xylene $\;+\;$ $4O_2$ $\longrightarrow$ $2CO_2$ $\;+\;$ $2H_2O$ $\;+\;$ 2 Toluene

$$+ \quad \text{Energy (2)} \hfill \text{(ix)}$$

Stages 5-11: See Rule 1632 for Combustion of Toluene wherein seven stages exist with the following overall equation-

Final Overall equation: $2C_6H_5CH_3 \;+\; 6O_2 \longrightarrow 2CO_2 \;+\; 4H_2O$

$$\text{(AIR)} \qquad + \;\; 2C_6H_4O_2 \;+\; Energy(5)$$
$$(G) \hfill \text{(x)}$$

Final Overall equation: $2C_6H_4(CH_3)_2 \;+\; 10O_2 \longrightarrow 4CO_2 \;+\; 6H_2O$

$$\text{(AIR)} \qquad + \;\; 2C_6H_4O_2 \;+\; Energy(7)$$
$$(G) \hfill \text{(xi)}$$

Where (G) is Benzoquinone

an eleven stage Equilibrium mechanism system, wherein like other alipharomatic hydro-carbons, the unfriendly aromatic CO_2, benzoquinone from benzene, CO_2 from the aliphatic side, and water are the products, with energy coming from seven stages far less than that from Oxidation.
(Laws of Creations for Combustion of o-Xylene)

719

Rule 1643: This rule of Chemistry for **Ortho-Xylene,** states that, when used as a raw material for the production of Phthalic anhydride in the presence of **vanadium pentoxide,** the overall reaction is as shown below-

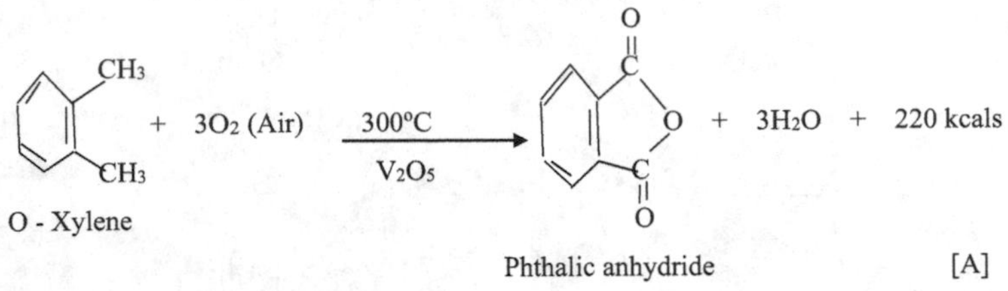

$$O\text{ - Xylene} + 3O_2 \text{ (Air)} \xrightarrow[\text{V}_2\text{O}_5]{300°C} \text{Phthalic anhydride} + 3H_2O + 220 \text{ kcals} \qquad [A]$$

the mechanisms of which are as follows-

Stage 1:

Vanadium pentoxide, (I), (II)

$$(II) \xrightarrow[\text{Energy}]{\text{Release of}} \text{(III)} + nn.O.en + \text{Energy}$$

$$(I) + (III) \xrightleftharpoons[\text{Existence of } V_2O_4]{\text{Equilibrium State of}} O_2V - VO_2$$

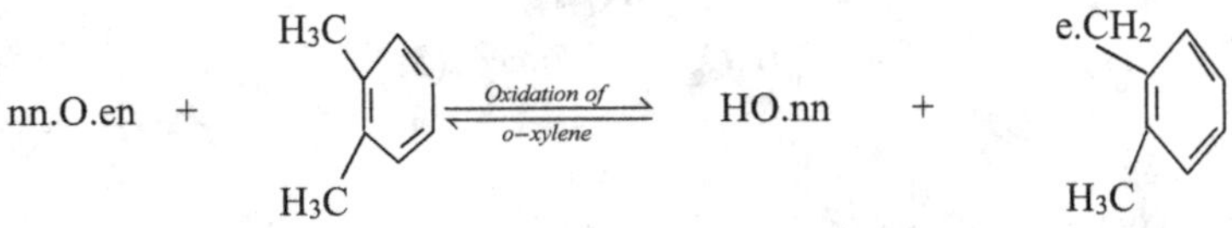

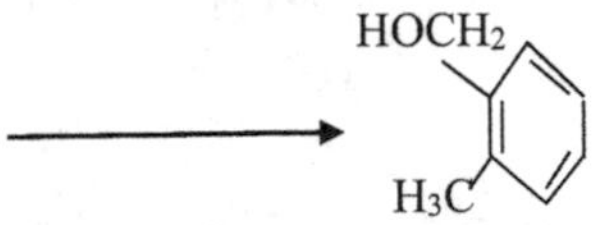

$$nn.O.en + \xrightleftharpoons[o-xylene]{\text{Oxidation of}} HO.nn +$$

(IV) (i)

(ii)

Overall equation: $V_2O_5 + \text{o-xylene} \longrightarrow V_2O_4 + \text{(IV)}$

Stage 2: $O_2V - VO_2 \rightleftharpoons O_2V.e + n.VO_2$

$$O_2V.e + O=O \xrightleftharpoons[O_2]{\text{Activation of}} O_2V - O - O.en$$

$$O_2V - O - O.en + n.VO_2 \longrightarrow O_2V - O - O - VO_2 \qquad \text{(iii)}$$

(iv)

Overall equation: $V_2O_4 + O_2 \longrightarrow O_2V - O - O - VO_2$

Stage 3: $2O_2V - O - O - VO_2 \xrightleftharpoons[\text{of peroxide}]{\text{Equil. State of Exist.}} 2O_2V - O.en + 2nn.O - VO_2$

$2O_2V - O.en \rightleftharpoons 2O_2V.e + 2en.O.nn + 2Energy$

$2en.O.nn \xrightleftharpoons[\text{Existence of Oxidizing } O_2]{\text{Equilibrium State of}} O_2$ (Oxidizing oxygen molecule)

$2O_2V.e + 2nn.O - VO_2 \longrightarrow 2O_2V - O - VO_2$

(v)

Overall Equation: $2O_2V - O - O - VO_2 \longrightarrow O_2 + 2O_2V - O - VO_2 + 2Energy$

(Vanadium pentoxide)

(vi)

Overall overall equation: $2V_2O_5 + 2o\text{-xylene} + 2O_2 \longrightarrow$

$2V_2O_5 + 2(IV) + O_2 + 4Energy$

(vii)

Stage 4: $O_2 \xrightleftharpoons[\text{Existence of } O_2]{\text{Equil. State of}} 2nn.O.en$

$2nn.O.en + 2(IV) \xrightleftharpoons{\text{Oxidation}} 2\,[\text{HOCH}_2\text{---ring---e.CH}_2] + 2nn.OH$

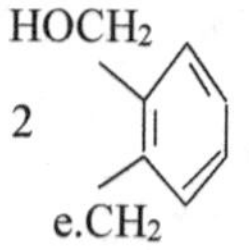

$\longrightarrow 2\,[\text{HOCH}_2\text{---ring---HOCH}_2]$

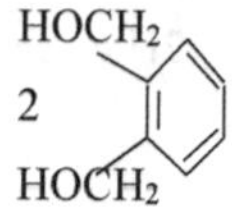

(V)

(viii)

Overall equation: $O_2 + 2(IV) \longrightarrow 2(V)$

(ix)

Stage 5: $2O_2V - O - VO_2 \rightleftharpoons 2O_2V.n + 2en.O - VO_2$

$2O_2V - O.en \rightleftharpoons 2en.O.nn + 2e.VO_2 + 2Energy$

$2nn.O.en + 2(V) \rightleftharpoons 2\,[\text{O.en---CH}_2\text{---ring---HOCH}_2] + 2nn.OH$

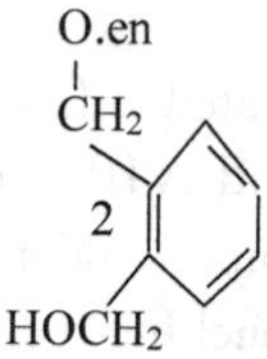

(VI)

$2(VI) \rightleftharpoons 2\,[\overset{O}{\overset{\|}{CH}}\text{---ring---HOCH}_2] + 2H.e + 2x\ Energy$

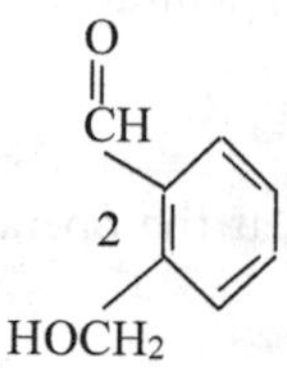

(VII)

$2O_2V.n + 2e.VO_2 \rightleftharpoons 2O_2V - VO_2$

$2H.e + 2nn.OH \longrightarrow 2H_2O$

(x)

Overall equation: $2V_2O_5 + 2(V) \longrightarrow 2H_2O + 2V_2O_4 + 2(VIII)$

$+ 2(1 + x)\ Energy$

(xi)

using another two moles of molecular oxygen from air, V_2O_5 is recovered again as already shown in Stages 2 and 3, i.e.,

Stage 6: The same as Stage 2.

Stage 7: The same as Stage 3.

for these two stages where V_2O_5 is recovered, the overall equation is as follows-

Overall equation: $2V_2O_4 \ + \ 2O_2 \longrightarrow 2V_2O_5 \ + \ \boldsymbol{O_2} \ + \ 2\text{Energy}$ (xii)

and the oxidation of (VII) continues as follows-

Stage 8: $\boldsymbol{O_2} \ \underset{\text{Existence of } O_2}{\overset{\text{Equil. State of}}{\rightleftharpoons}} \ 2\text{nn.O.en}$

$2\text{nn.O.en} \ + \ 2(\text{VII}) \quad \overset{\text{Oxidation}}{\rightleftharpoons} \quad 2 \underset{\overset{\displaystyle CH_2}{\underset{\displaystyle O.en}{|}}}{\overset{\displaystyle O = CH}{\Big|}} \bigcirc \quad + \quad 2\text{nn.OH}$

(VIII)

$2(\text{VIII}) \quad \rightleftharpoons \quad 2 \underset{\displaystyle O = CH}{\overset{\displaystyle O = CH}{\Big|}} \bigcirc \quad + \quad 2\text{H.e} \quad + \quad 2\text{x Energy}$

(IX)

$2\text{H.e} \ + \ 2\text{nn.OH} \longrightarrow 2H_2O$ (xiii)

Overall equation: $\boldsymbol{O_2} \ + \ 2(\text{VII}) \longrightarrow 2H_2O \ + \ 2(\text{IX}) \ + \ 2\text{x Energy}$ (xiv)

Overall overall equation: $2V_2O_5 \ + \ 2\text{o-xylene} \ + \ 4O_2 \text{ (AIR)} \longrightarrow$

$2V_2O_5 \ + \ 2(\text{IX}) \ + \ 4(2+x) \text{ Energy} \ + \ 4H_2O$ (xv)

taking note of the amount of energy generated so far, from two main sources- during release of oxidizing oxygen and the deactivation steps [(VI) and (VIII), noting that energy is required to activate the oxygen molecules; the (IX) formed here is a di-aromatic aldehyde, a very useful product for which one can decide to stop the oxidation at this point, by quenching the reaction-

$2\text{o-xylene} + 4O_2 \xrightarrow{V_2O_5 \text{ Cat. } 360^\circ C} 2(\text{IX}) \text{ [a di-aldehyde]} \ + \ 4(2+x) \text{ Energy} \ + \ 4H_2O$

(xvi)

however, with the presence of V_2O_5 and at the operating conditions, the oxidation of the di-aldehyde continues in the next stage-

Stage 9: $2O_2V - O - VO_2 \ \rightleftharpoons \ 2O_2V.n \ + \ 2\text{en.O} - VO_2$

$2O_2V - O.en \ \rightleftharpoons \ 2\text{en.O.nn} \ + \ 2\text{e}.VO_2 \ + \ 2\text{Energy}$

$2O_2V.e \ + \ 2\text{n}.VO_2 \ \rightleftharpoons \ 2V_2O_4$

722

$$2nn.O.en \quad + \quad 2(IX) \quad \rightleftharpoons \quad 2 \overset{\displaystyle \overset{O}{\underset{\|}{e.C}}}{\underset{O=CH}{\bigcirc}} \quad + \quad 2nn.OH$$

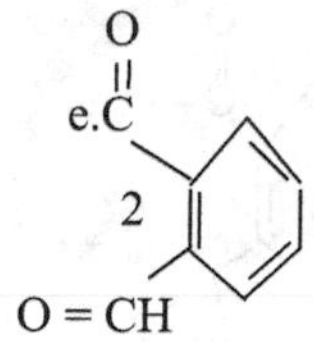

(X)

$$2 \overset{\displaystyle \overset{O}{\underset{\|}{HO-C}}}{\underset{O=CH}{\bigcirc}}$$

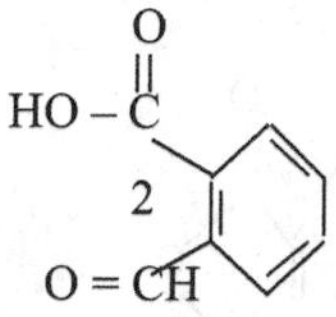

(XI)   (xvii)

Overall equation: $2V_2O_5 \quad + \quad 2(IX) \longrightarrow 2V_2O_4 \quad + \quad 2(XI)$

$+ \; 2 \; Energy$ (xviii)

the vanadium pentoxide is regenerated once again as already shown by Stages 2 and 3 and Equation (xii) and when done in **Stages 10 and 11,** the overall equation is as follows-

Stage 10: Same as Stage 2.

Stage 11: Same as Stage 3.

Overall overall equation: $2V_2O_5 \quad + \quad 2o\text{-xylene} \quad + \quad 6O_2 \; (AIR) \longrightarrow$

$2V_2O_5 \quad + \quad 2(XI) \quad + \quad 2(5+2x) \; Energy \quad + \quad 4H_2O \quad + \quad \boldsymbol{O_2}$

(xix)

Stage 12: $\boldsymbol{O_2}$ $\underset{\text{Existence of } O_2}{\overset{\text{Equil. State of}}{\rightleftharpoons}}$ $2nn.O.en$

$$2nn.O.en \quad + \quad 2(XI) \quad \underset{}{\overset{Oxidation}{\rightleftharpoons}} \quad 2 \overset{\displaystyle \overset{O}{\underset{\|}{HO-C}}}{\underset{\underset{\|}{e.C} \atop O}{\bigcirc}} \quad + \quad 2nn.OH$$

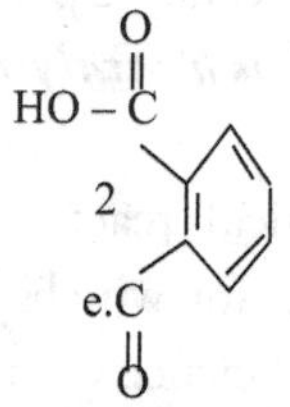

(XII)

$$2 \overset{\displaystyle \overset{O}{\underset{\|}{HO-C}}}{\underset{\underset{\|}{HO-C} \atop O}{\bigcirc}}$$

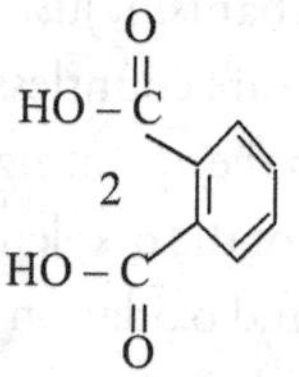

(XIII) Phthalic acid   (xx)

Overall overall equation: $2V_2O_5 \quad + \quad 2o\text{-xylene} \quad + \quad 6O_2 \; (AIR) \longrightarrow$

$2V_2O_5 \quad + \quad 2(XIII) \quad + \quad 2(5+2x) \; Energy \quad + \quad 4H_2O$

(xxi)

Stage 13:

$$2(XIII) \quad \rightleftharpoons \quad 2 \overset{\displaystyle \overset{O}{\underset{\|}{C-O.nn}}}{\underset{\underset{\|}{C-O-H} \atop O}{\bigcirc}} \quad + \quad 2H.e$$

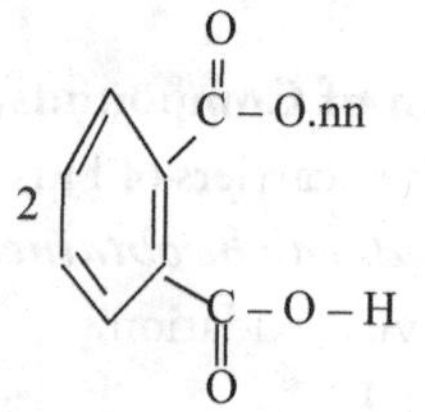

723

(XIV)

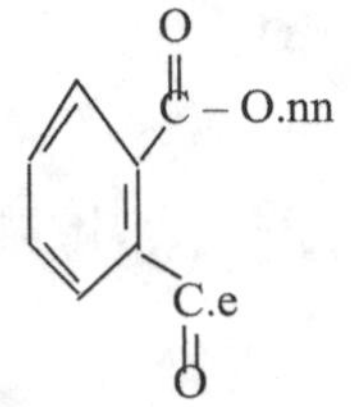

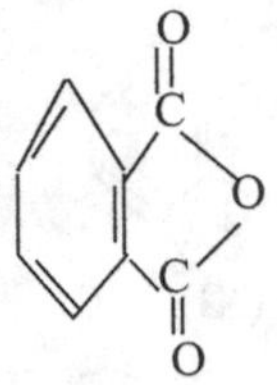

(XV) **Phthalic anhydride**   (xxii)

Overall overall equation: $2V_2O_5$ + 2o-xylene + $6O_2$ (AIR) $\xrightarrow{360^\circ C}$

$2V_2O_5$ + $2(5 + 2x)$ Energy + $6H_2O$ + 2(XV)   (xxiii)

a thirteen stage Equilibrium mechanism system-
the number of moles as indicated from the mechanism must be used in a well-designed reactor where there is perfect mixing in-order to obtain full conversion. *[Don't divide both sides of the equation by two]* Indeed, worthy of note are the followings-

a)   The large amounts of energy released.

b)   That V_2O_5 is indeed a passive and an active catalyst playing *the roles of a stabilizing agent passively, an oxidizing agent actively and as a catalyst actively.* Many oxidiz-ing agents do not operate this way.

c)   The stage-wise operations of some chemical reactions (such as combustion, oxidation, hydrogenation and more or indeed Nature), for which one cannot bypass a particular stage and move to the end, as done in present-day Science with the use of rate determining steps and so on. Every stage is very important and every stage is step-wisely operated in a fixed order. For every reaction, there is one and only one mechanism, just as for every problem, there is one and only one solution, and for every solution there are countless numbers of "unnecessary problems".

d)   Indeed the mechanism of the reaction based on the operating conditions is complete OXIDATION with no combustion taking place, unless part of the o-xylene cannot be stabilized by V_2O_5. If part cannot be stabilized, then both combustion and oxidation will take place side by side in parallel.

e)   While for combustion, six stages are involved, for oxidation thirteen stages are involved with the use of this catalyst and molecular oxygen.

(Laws of Creations for Oxidation of o-Xylene)

Rule 1644: This rule of Chemistry for **Oxidation of Compounds,** states that, based on the manners and nature of Oxygen in the imaginary domain when carriers of H is around its neighborhood, *products coming from different large families of compounds can be obtained if MOLAR ratios of components are properly chosen and controlled,* during and via Oxidation.
(Laws of Creations for Oxidation of Compounds)

Rule 1645: This rule of Chemistry for **Semi-perpetual production of Energy oxidatively using water indirectly from H_2 and Oxygen of the AIR,** states that, to do this the followings are obtained-

Stage 1:

$$H_2 \underset{\text{H_2 in Equilibrium State of Existence}}{\overset{\text{Hyrogen catalyst used to keep}}{\rightleftharpoons}} H \cdot e + n \cdot H$$

$$H \cdot e + O = O \underset{\text{Oxygen(Air)}}{\overset{\text{Activation of}}{\rightleftharpoons}} H - O - O \cdot en$$

$$H - O - O \cdot en + n \cdot H \xrightarrow[\text{Existence of HOOH}]{\text{Combination State of}} H - O - O - H \tag{i}$$

$$\underline{Overall\ Equation}: H_2 + O_2 \xrightarrow{\text{Hydrogen Catalyst}} H - O - O - H \tag{ii}$$

Stage 2:

$$2H - O - O - H \underset{\text{Existence of Hyrogen peroxide}}{\overset{\text{Equilibrium State of}}{\rightleftharpoons}} 2H - O \cdot en + 2nn \cdot O - H$$

$$2H - O \cdot en \underset{\text{Release}}{\overset{\text{Energy}}{\rightleftharpoons}} 2H \cdot e + \underline{O_2} + Energy$$

$$2H \cdot e + 2nn \cdot O - H \xrightarrow[\text{presence of $\underline{O_2}$}]{\text{Suppressed by the}} 2H_2O \tag{iii}$$

$$\underline{Overall\ Equation}: 2HOOH \longrightarrow 2H_2O + \underline{O_2} + Energy \tag{iv}$$

$$\underline{Overall\ equation}: \underset{\text{(Stable)}}{2H_2} + \underset{\text{(AIR)}}{2O_2} \xrightarrow[\text{CATALYST}]{H_2} \underset{\text{(Warm)}}{H_2O} + \underset{\text{(Oxidizing oxygen)}}{\underline{O_2}} + Energy(Heat) \tag{v}$$

Stage 3:

$$\underline{O_2} \underset{\text{Existence of O_2}}{\overset{\text{Aquilibrium State of}}{\rightleftharpoons}} 2en \cdot \ddot{O} \cdot nn$$

$$2nn \cdot \ddot{O} \cdot en + 2H_2O \overset{\text{Abstraction}}{\rightleftharpoons} 2H - O \cdot en + 2nn \cdot O - H$$

$$2H - O \cdot en \underset{\text{Release}}{\overset{\text{Energy}}{\rightleftharpoons}} 2H \cdot e + \underline{O_2} + Energy$$

$$2H \cdot e + 2nn \cdot O - H \xrightarrow[\text{of $\underline{O_2}$}]{\text{Suppressed by presence}} 2H_2O \tag{vi}$$

$$\underline{Overall\ Equation}: \underset{(A)}{\underline{O_2} + 2H_2O} \xrightarrow[\text{Water}]{\text{Oxidation of}} \underset{(B)}{\underline{O_2} + 2H_2O} + Energy \tag{vii}$$

the Two Oxidizing Oxygen Molecules, (A) and (B) above, are very different in terms of the amount of Strain Energy in their rings, for which that of (A) is greater than that of (B); the first law with respect to Energy Conservation in Chemical Engineering must be obeyed; even if energy was added from Stage 1 both rings will still be different in energy content because the reaction is exothermic—

O⬭O	>	O⬭O	>	O⬭O	>	O=O
Most Strained		More Strained		Least Strained		No SE (In Air)

Oxidizing Oxygen Molecules (Rings)

[A]

725

noting that for Stage 3 to exist as a stage in series, just like what exists between NaOH and water where heat is released will require the presence of a passive catalyst such as an ionic metal or ionic compound which cannot be oxidized to suppress the Equilibrium state of existence of water, otherwise the followings take place along with Stage 3 above, since there must be a fraction of water in Stable state of existence-

Stage4:

$$H_2 \underset{CATALYST}{\overset{H_2}{\rightleftharpoons}} H \cdot e + H \cdot n$$

$$H \cdot e + H_2O \rightleftharpoons H_2 + HO \cdot en + Energy$$

$$HO \cdot en \rightleftharpoons H \cdot e + en \cdot O \cdot nn + Energy$$

$$nn \cdot O \cdot en + H_2 \rightleftharpoons HO \cdot nn + H \cdot e$$

$$H \cdot e + n \cdot H \rightleftharpoons H_2$$

$$H \cdot e + nn \cdot OH \longrightarrow H_2O$$

$$\underline{Overall\ equation}: H_2 + H_2O \xrightarrow{H_2}_{CAT.} H_2O + H_2 + Energy \qquad (viii)$$

Stage 5: Same as Stage 4 above in series with continuous presence of fresh water.

Stage 6: Same as Stage 1.

Stage 7: Same as Stage 2.

Stage 8: Same as Stage 3.

Stage 9 Same as Stage 4.

Stage 10.........: Continuation of Stages 1 and 4 in series and in parallel.

and these continue perpetually in the system producing energy in every stage apart from the stage where HOOH is produced with water vapor as part of the products; noting that this large amount of energy can be converted to mechanical energy to drive systems or be converted to electrical energy electrolytically in an ionic media-solution *as a battery and continuous source of energy in the system;* for which if no H_2 gas is available, this can be produced in-situ using chemical methods, wherein the choice of chemical method is important with respect to availability of the raw materials, control of types of bye products and more; and that indeed based on Stage 4, oxygen can be excluded from the system.

(Laws of Creations of Production of Energy from Water/H_2)

Rule 1646: This rule of Chemistry for **the Nucleus of All Atoms,** states that, amongst all atoms, since only Hydrogen (H) and Helium (He) atoms are the only two atoms where when the only radicals carried by them in their lone shell (1s) is removed what is left behind is just the Nucleus, hence all the subatomic particles in the Nucleus are members of the families of H and He with more of H than He, H being the first member of the Family of ATOMS.- for which **as a rule, the sub-atomic particles must be shell-less to reside inside the Nucleus.**
(Laws of Creations for Nucleus of All Atoms).

Rule 1647: This rule of Chemistry for **the Nucleus of all Atoms,** states that, based on the numbers of *sub-atomic particles of Hydrogen and Helium carried by them,* none of them can be the same.
(Laws of Creations for Nucleus of Atoms)

Rule 1648: This rule of Chemistry for **the Nucleus of All Atoms,** states that, there are in general two identical parts inside the Nucleus called *MATTER and ANTI-MATTER,* both mirror images, each

containing four sub-atomic particles- *the proton(s), neutron(s) (both secondary), α-ray(s), β-ray(s) or negatron(s) (both primary)* on the Matter side, *anti-proton(s), anti-neutron(s) (both secondary), Positron(s), anti-α-ray(s) (both primary)* on the Anti-Matter side [The Octet]
(Laws of Creations for Nucleus of an Atom)

Rule 1649: This rule of Chemistry for **the Nucleus of an Atom,** states that, without the nucleus, there is no outer sphere of an atom, for it is the nucleus that determines what can exist in the outer sphere of an atom (and not the other way round) and it is the nucleus that carries the source of *life and living,* through which "THE SPIRIT" all set in MOTION link through "RAYS" and then to the radicals.
(Laws of Creations for Nucleus of an Atom)

Rule 1650: This rule of Chemistry for **the Nucleus of All Atoms,** states that, what happens in the outer sphere of the atom, *does not affect the states of existence of the atomic sub-particles (protons + neutrons + positron + etc.),* just as what goes on, on Earth or Mars or Jupiter etc., does not affect the state of existence of the SUN, the nucleus of our Solar system (a Macro-Atom).
(Laws of Creations for Nucleus of All Atoms)

Rule 1651: This rule of Chemistry for **the Nucleus of All Atoms,** states that, *all chemical and polymeric reactions taking place along the boundary of the outer sphere of an atom do not affect the state of existence of the Nucleus,* but dependent on the type of Nucleus of an atom or Central atom of a Molecule.
(Laws of Creations for Nucleus of All Atoms)

Rule 1652: This rule of Chemistry for **the Nucleus of All Atoms,** states that, since the <u>Atomic Number</u> *of an atom is said to be the number of <u>protons</u> while Atomic weight is the sum of the atomic number and the number of <u>neutrons</u> in which each (neutron) and each (proton) has **one atomic mass unit**,* then the mass of every atom is inside the Nucleus of the atom and not outside the nucleus of the atom, and for every atom, the structure is symbolically represented as shown below, that in which the Nucleus is identified with Atomic weight and Atomic number for every atom top and bottom respectively and the **number of radicals** wrongly called "Electrons" outside the Nucleus are shown in circular orbits around the Nucleus-

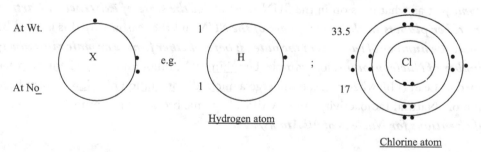

At Wt.   X   e.g.   1   H   ;   33.5   Cl

At No   1   Hydrogen atom   17   Chlorine atom

(Laws of Creations for Nucleus of an Atom)

Rule 1653: This rule of Chemistry for **the Nucleus of All Atoms**, states that, since what exists outside the Nucleus of an Atom are weightless, this clearly indicates that ALL THE PLANETS surrounding the SUN are weightless since the SUN is the Nucleus of our Solar system.
(Laws of Creations for the Nucleus of All Atoms)

Rule 1654: This rule of Chemistry for **the Nucleus of All Atoms,** states that, based on the nomenclatures or symbols for atoms in the world today which are very much in order, a "proton" (i.e. outside the nucleus) is a positive charge carried by the hydrogen center as shown below-

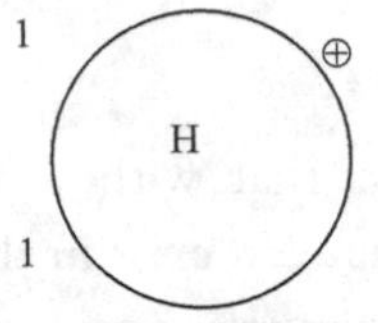

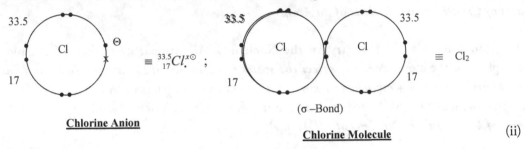

$$\equiv\ {}^{1}_{1}H^{\oplus}\ \equiv\ \text{Pr} oton \tag{i}$$

while for the chlorine anion and chlorine molecule, the followings are obtained-

$$\equiv\ {}^{33.5}_{17}Cl^{x\ominus}_{.}\ ;$$

Chlorine Anion

(σ –Bond)

Chlorine Molecule (ii)

clear indication of the fact that what is inside the Nucleus on the Matter side is not a Proton, but herein called **PROTONUCLEON** distinguished from a Proton as shown below-

$$
{}^{1}_{1}H^{\oplus} \qquad\qquad ; \qquad\qquad {}^{0}_{0}H^{\oplus}\ \equiv\ {}^{1}_{1}H
$$

$$PROTON \qquad\qquad\qquad PROTONUCLEON$$

Outside the Nucleus   *Inside the Nucleus* (iii)

noting that the positive charge carried by the proton is a cation, while that carried by the nucleoproton is not a cation or a charge, but something else.
(Laws of Creations for Nucleus of All Atoms)

Rule 1655: This rule of Chemistry for **the Nucleus of All Atoms,** states that, whatever changes take place in the nucleus of an atom must affect the *state of existence of the radicals in the outer sphere of the atom,* just as what goes on in the SUN must affect *the state of existence of Earth or Mars or Jupiter or other planets in the outer sphere of the SUN* and the reason why this is so, is because first and foremost *the number of radicals in the outer sphere whether female or male must always be equal to the number of Nucleoprotons,* for example, Uranium (U) cannot undergo a chemical reaction and be transformed to Lead (Pb); whereas it can undergo a nuclei disintegration reaction to form Lead, the basic foundation of "Natural radioactivity" and "Artificial Atomic breakdown capture"
(Laws of Creations for Nucleus of All Atoms)

Rule 1656: This rule of Chemistry for **the Nucleus of All Atoms,** states that, *the Proto-nucleon which is secondary and one of the atomic sub-particles of Matter,* is a hydrogen atomic sub-particle with *an atomic weight of 0 and Atomic number of 0 and a Nucleopositive charge **an entity** with a unit mass of 1 as* shown below-

$$
{}^{0}_{0}H^{\oplus} \qquad \equiv \qquad {}^{1}_{1}H
$$

$$PROTONUCLEON$$

with a resultant atomic mass unit of 1 and a resultant positive charge of +1 as shown on the right hand side of the equation above.

(Laws of Creations for Nucleus of All Atoms)

Rule 1657: This rule of Chemistry for **the Nucleus of All Atoms,** states that, *the Neutron which is secondary and one of the atomic sub-particles of Matter,* is a hydrogen atomic sub-particle with *an Atomic weight of 2 and Atomic number of 1 and a Nucleonegative charge __an entity__ with a unit mass of -1* as shown below-

$$ {}_1^2 H^{\odot} \quad \equiv \quad {}_0^1 H $$

$$ NEUTRON $$

with a resultant *Atomic mass unit* of 1 and a resultant charge of zero as shown on the right hand side of the equation above.

(Laws of Creations for Nucleus of All Atoms)

Rule 1658: This rule of Chemistry for **the Nucleus of All Atoms,** states that, *the Anti-Protonucleon which is secondary and one of the atomic sub-particles of Anti-Matter,* is a hydrogen atomic sub-particle with *an atomic weight of 0 and Atomic number of 0 and a Nucleonegative charge __an entity__ with a unit mass of -1* as shown below-

$$ {}_0^0 H^{\odot} \quad \equiv \quad {}_{-1}^{-1} H $$

$$ ANTI - PROTONUCLEON $$

with a resultant atomic mass unit of -1 and a resultant negative charge of -1 as shown on the right hand side of the equation above.

(Laws of Creations for Nucleus of All Atoms)

Rule 1659: This rule of Chemistry for **the Nucleus of All Atoms,** states that, *the Anti-Neutron which is secondary and one of the atomic sub-particles of Anti-Matter,* is a hydrogen atomic sub-particle with *an atomic weight of -2 and Atomic number of -1 and a Nucleopositive charge __an entity__ with a unit mass of 1* as shown below-

$$ {}_{-1}^{-2} H^{\oplus} \quad \equiv \quad {}_0^{-1} H $$

$$ ANTI - NEUTRON $$

with a resultant atomic mass unit of -1 and a resultant charge of zero as shown on the right hand side of the equation above.

(Laws of Creations for Nucleus of All Atoms)

Rule 1660: This rule of Chemistry for **the Nucleus of All Atoms,** states that, *the Negatron or β-ray which is primary and one of the atomic sub-particles of Matter,* is a hydrogen atomic sub-particle with *an atomic weight of 1 and Atomic number of 0 and a Nucleonegative charge __an entity__ with a unit mass of -1* as shown below-

$$ {}_0^1 H^{\odot} \quad \equiv \quad {}_{-1}^{o} H $$

$$ NEGATRON \; OR \; \beta - ray $$

729

with a resultant atomic mass unit of 0 and a resultant negative charge of -1 as shown on the right hand side of the equation above.

(*Laws of Creations for Nucleus of All Atoms*)

Rule 1661: This rule of Chemistry for **the Nucleus of All Atoms,** states that, *the Positron or anti-β-ray which is primary and one of the atomic sub-particles of Anti-Matter,* is a hydrogen atomic sub-particle with *an atomic weight of -1 and Atomic number of 0 and a Nucleopositive charge* **an entity** with a unit mass of 1 as shown below-

$$\,_{0}^{-1}H^{\oplus} \quad \equiv \quad \,_{1}^{0}H$$

POSITRON OR ANTI – β – ray

with a resultant atomic mass unit of 0 and a resultant positive charge of +1 as shown on the right hand side of the equation above.

(*Laws of Creations for Nucleus of All Atoms*)

Rule 1662: This rule of Chemistry for **the Nucleus of All Atoms,** states that, *the α-ray which is primary and one of the atomic sub-particles of Matter,* is a helium atomic sub-particle with *an atomic weight of 2 and Atomic number of 0 and two Nucleopositive charges* **an entity** with a unit mass of 2 as shown below-

$$\,_{0}^{2}He^{2\oplus} \quad \equiv \quad \,_{2}^{4}He$$

α – Ray

with a resultant atomic mass unit of 4 and a resultant positive charge of +2 as shown on the right hand side of the equation above.

(*Laws of Creations for Nucleus of All Atoms*)

Rule 1663: This rule of Chemistry for **the Nucleus of All Atoms,** states that, *the Anti-α-ray which is primary and one of the atomic sub-particles of Anti-Matter,* is a helium atomic sub-particle with *an atomic weight of -2 and Atomic number of 0 and two Nucleonegative charges* **an entity** with a unit mass of -2 as shown below-

$$\,_{0}^{-2}He^{2\odot} \quad \equiv \quad \,_{-2}^{-4}He$$

ANTI – α – ray

with a resultant atomic mass unit of -4 and a resultant negative charge of -2 as shown on the right hand side of the equation above.

(*Laws of Creations for Nucleus of All Atoms*)

Rule 1664: This rule of Chemistry for **the Nucleus of All Atoms,** states that, the use of the word "matter" alone to associate with atoms and all living and so-called non-living systems is very much in place and in order, because while matter is real, cannot leave and can be measured, anti-matter is imaginary, can leave and cannot be measured in our Solar system. [From Matter …..To Matter]

(*Laws of Creations for Nucleus of All Atoms*)

Rule 1665: This rule of Chemistry for **the Nucleus of All Atoms,** states that, based on what the atomic sub-particles are carrying, *none of the sub-atomic particle carry what are called positive or negative charges, since these are imaginary entities with a unit mass of 1 when positive, a unit mass of -1 when negative and unit mass of zero when neutral and these in addition have directions,* that is, they are vector quantities electromagnetic in character and therefore herein called **NUCLEAR CHARGES.** *(Laws of Creations for Nucleus of All Atoms)*

Rule 1666: This rule of Chemistry for **the Nucleus of All Atoms,** states that, *the atomic number carried by the final H or He sub-particles is the number of Nuclear charges carried by the sub-particle,* wherein for H, there are +1 (Protonucleons, Positrons), 0 (Neutrons and Anti-neutrons), -1 (Nucleoelectron or negatron or β-ray, Anti-protons), while for He, there are +2 (α-rays), -2 (Anti-α-rays); from which one can observe that only Neutrons are neutral in character; while the whole nucleus with both matter and anti-matter is also neutral in character when in Stable state of existence. *(Laws of Creations for Nucleus of All Atoms)*

Rule 1667: This rule of Chemistry for **Radioactivity,** states that, an Atom is said to be radioactive if it has the ability to naturally break down or disintegrate spontaneously to release only one **primary atomic sub-particle** one at a time to give a different atom-

$$Radioactive\ atom \xrightarrow[Down]{Breaks} \beta - or\ Anti - \beta - rays \quad + \quad Another\ Atom \tag{i}$$

$$Radioactive\ atom \xrightarrow[Down]{Breaks} \alpha - or\ Anti - \alpha - rays \quad + \quad Another\ Atom \tag{ii}$$

that which cannot be influenced by temperature, chemical combination or any other influence at the command of the Chemist. *(Laws of Creations for Radioactivity)*

Rule 1668: This rule of Chemistry for **Radioactivity,** states that, the spontaneous removal of **primary atomic sub-particles** from the nucleus of an atom depends on the nature of the atom in terms of Atomic numbers, Atomic weight, Stability and most importantly *the mode of transformation.* *(Laws of Creations for Radioactivity)*

Rule 1669: This rule of Chemistry for **Radioactivity,** states that, Mode of transformation during disintegration is that in which a radioactive atom which is *metallic in character cannot disintegrate to give another atom which is non-metallic in character,* for example, consider the followings-

$$\underset{Metal}{^{23}_{11}Na} \quad \not\longrightarrow \quad \underset{Non-metal}{^{19}_{9}F} \quad + \quad ^{4}_{2}He \tag{i}$$

$$\underset{(Isotopic\ sodium)}{^{24}_{11}Na} \quad \not\longrightarrow \quad \underset{Non-metal}{^{20}_{9}F} \quad + \quad ^{4}_{2}He \tag{ii}$$

noting that α-rays are also **primary atomic sub-particles** *(Laws of Creations for Radioactivity)*

Rule 1670: This rule of Chemistry for **Radioactivity,** states that, Stability during disintegra-tion is that in which when β- and anti-β-rays are released, stable isobars must be obtained as shown below, and when a- and anti-a-rays are released, stable atoms isotopic or non-isotopic are obtained-

$$\require{mhchem} {}^{23}_{11}Na \longrightarrow {}^{23}_{12}Mg + {}^{0}_{-1}e$$

(*Stable sodium*)   (*Not an isotope*
<u>*Metal*</u>   *of Mg*) – <u>*Metal*</u>   (i)

$${}^{24}_{11}Na \longrightarrow {}^{24}_{12}Mg + {}^{0}_{-1}e$$

(*Isotopic sodium*)   (*Stable*)   (*β – ray*)   (ii)

$${}^{13}_{7}N \longrightarrow {}^{13}_{6}C + {}^{0}_{1}e$$

(*Isotopic nitrogen*)   (*Isotopic carbon*)   (*Anti – β – ray*)
                                    Positron   (iii)

$${}^{127}_{53}I \quad ; \quad {}^{128}_{53}I \longrightarrow {}^{0}_{-1}e + {}^{128}_{54}Xe$$

Stable Iodine   *Unstable*   *Xenon*   (iv)

for which *invariably what this implies, is that all atoms* <u>*but one*</u>*, are radioactive, their ability to release instantaneously nuclei "particles", depending on the type of atom, in- stability and application of natural laws* and that one is the H atom.
(Laws of Creations for Radioactivity)

Rule 1671: This rule of Chemistry for **Radioactivity,** states that, an atom is stable, when it cannot disintegrate and therefore largely in abundance.
(Laws of Creations for Radioactivity)

Rule 1672: This rule of Chemistry for **the Nucleus of All Atoms,** states that, since Neutrons, Anti-neutrons, Protons and Anti-protons cannot be emitted spontaneously on its own, they are said to be *Secondary primary sub-particles.*
(Laws of Creations for Nucleus of All Atoms)

Rule 1673: This rule of Chemistry for **Nuclear Bombardment,** states that, when a Matter or an Atom is hit by an atomic sub-particle, the phenomenon is said to be **ARTIFICIAL ATOMIC BREAKBOWN,** that in which the Atom moves either from *Stable (ORDER) to Stable atom (ORDER)* or *Unstable (DISORDER) to an Unstable atom (DISORDER)* or from *Stable (ORDER) to an Unstable atom (DISORDER)* and never from <u>Unstable (DISORDER) to Stable atom (ORDER).</u>
(Laws of Creations for Nuclear Bombardment)

Rule 1674: This rule of Chemistry for **Atomic sub-particles,** states that, while those that carry *negative nuclear charge(s) are said to be Females,* those that carry *positive nuclear charge(s) are said to be Males,* for example α-rays, positrons, protonucleons are Males while β-rays, anti-protonucleons and anti-α-rays are Females [And these can be used as Initiators for specific monomers] and those that carry *no charge are neither Males nor Females but Neutral,* for example neutrons and anti-neutrons.
(Laws of Creations for Atomic sub-particles).

Rule 1675: This rule of Chemistry for **Nuclear Bombardment,** states that, when an Atom is hit by a *Female sub-particle,* only Neutral type of atomic sub-particle can be released leaving behind another atom as shown below *using β-rays-*

$$\beta - rays \xrightarrow[Matter]{Hits} \gamma - rays \ are\ emitted$$

$$[Electromagnetic\ waves - 10^{-12} metre\ wavelength] \tag{i}$$

when γ-ray a Neutral one is given by-

$$\gamma - ray \quad \equiv \quad {}_{1}^{1}H^{\odot} \quad \equiv \quad {}_{0}^{0}H$$
$$NEUTRAL \tag{ii}$$

and this looks like the so-called electron a secondary particle, since it is released by force from inside the Nucleus by Nuclear reactions and does not exist inside a Nucleus.

$$\begin{array}{cccccc} {}_{-1}^{0}H & + & {}_{11}^{23}Na & \xrightarrow[Reaction]{Nuclear} & {}_{10}^{23}Ne & + & {}_{0}^{0}H \\ \beta-ray & & (Stable) & & (Unstable) & & \gamma-ray \end{array} \tag{iii}$$

$$\begin{array}{cccccc} {}_{-1}^{0}H & + & {}_{96}^{241}Cm & \xrightarrow[Capture]{\beta-ray} & \gamma-rays & + & {}_{95}^{241}Am \\ \beta-ray & & \underline{Curium} & & & & \underline{Americum} \\ & & (Unstable) & & & & (unstable) \end{array} \tag{iv}$$

(Laws of Creations for Nuclear Bombardment)

Rule 1676: This rule of Chemistry for **Nuclear Bombardment,** states that, when an Atom is hit by a *Female sub-particle,* only Neutral type of atomic sub-particle can be released leaving behind another atom as shown below *using Anti-Protonucleons*

$$_{-1}^{-1}H \quad \xrightarrow[Matter]{Hits} \gamma - rays \tag{i}$$

$$\begin{array}{cccccc} {}_{-1}^{-1}H & + & {}_{11}^{23}Na & \xrightarrow[Reaction]{Nuclear} & {}_{10}^{22}Ne & + & {}_{0}^{0}H \\ (Anti-\Pr otonucleon) & & (Stable) & & (Unstable\ Neon) & & \gamma-rays \end{array} \tag{ii}$$

(Laws of Creations for Nuclear Bombardment)

Rule 1677: This rule of Chemistry for **Nuclear Bombardment,** states that, when an Atom is hit by a *Female sub-particle,* one radical (Free or Non-free) must be lost to form a *New Atom* if the particles are Negatrons or Anti-protonucleons and this loss is what is gained in the formation of a *New sub-particle.* *(Laws of Creations for Nuclear Bombardment)*

Rule 1678: This rule of Chemistry for **Nuclear Bombardment,** states that, when an Atom is hit by a *Female sub-particle,* only Neutral type of atomic sub-particle can be released leaving behind another atom as shown below *using Anti-α-rays -*

$$Anti-\alpha-rays \quad \xrightarrow[Matter]{Hits} \quad \gamma-rays \text{ are emitted}$$

$$[\textit{Electromagnetic waves} - 10^{-12} \textit{metre wavelength}] \tag{i}$$

$$\underset{(Anti-\alpha-rays)}{_{-2}^{-4}He} \quad + \quad \underset{(Stable)}{_{6}^{12}C} \quad \xrightarrow[Reaction]{Nuclear} \quad \underset{(Unstable)}{_{4}^{8}Be} \quad + \quad \underset{\gamma-rays}{_{0}^{0}H} \tag{ii}$$

(Laws of Creations for Nuclear Bombardment)

Rule 1679: This rule of Chemistry for **Nuclear Bombardment,** states that, when an Atom is hit by a *Female sub-particle,* two radicals (free or Non-free) must be lost to form a ***New Atom*** if the particles are Anti- α-rays and this loss is what is gained in the formation of a ***New sub-particle;*** clear indication of the existence of different types of for example γ-rays from different types of Female sub-particles.
(Laws of Creations for Nuclear Bombardment)

Rule 1680: This rule of Chemistry for **Nuclear Bombardment,** states that, when an Atom is hit by a *Male sub-particle,* either a Male or Neutral type of sub-particle can be released with or without leaving behind another Atom as shown below ***using Protonucleons-***

$$\underset{(proton)}{_{1}^{1}H} \quad + \quad \underset{\underline{Stable}}{_{3}^{7}Li} \quad \longrightarrow \quad \underset{The\ Atom\ (\gamma-ray)}{2\,_{2}^{4}He + \,_{0}^{0}H} \quad OR \quad \underset{(Unstable)}{_{4}^{8}Be} \quad + \quad \gamma \tag{i}$$

$$\underset{(proton)}{_{1}^{1}H} \quad + \quad \underset{\underline{Unstable}}{_{3}^{6}Li} \quad \longrightarrow \quad \underset{\underline{\alpha-rays}}{_{2}^{4}He} \quad + \quad \underset{(Unstable)}{_{2}^{3}He}$$
$$[\textit{Not Favored}] \tag{ii)a}$$

$$\underset{(proton)}{_{1}^{1}H} \quad + \quad \underset{\underline{Unstable}}{_{3}^{6}Li} \quad \longrightarrow \quad \underset{(Unstable)}{2\,_{2}^{4}He} \quad + \quad \underset{(Anti-Neutron)}{_{0}^{-1}H}$$
$$\textit{FAVORED} \tag{ii)b}$$

$$\underset{(proton)}{_{1}^{1}H} \quad + \quad \underset{\underline{Unstable}}{_{5}^{11}B} \quad \longrightarrow \quad \underset{(Unstable)}{3\,_{2}^{4}He} \quad + \quad \underset{\underline{\gamma-rays}}{_{0}^{0}H} \tag{iii}$$

noting that with protonucleons, another Atom must be produced and for the second reaction above to be favored, anti-neutron along with two different He Atoms (Isotopes) must be the products.
(Laws of Creations for Nuclear Bombardment)

Rule 1681: This rule of Chemistry for **Nuclear Bombardment,** states that, when an Atom is hit by a *Male sub-particle,* one Free-radical must be gained to form a ***New Atom or Atoms*** if the particles are Protonucleons or positrons and this gain is what is lost in the formation of a ***New sub-particle.***
(Laws of Creations for Nuclear Bombardment)

Rule 1682: This rule of Chemistry for **Nuclear Bombardment,** states that, when an Atom is hit by a *Male sub-particle,* either a Male or Neutral type of sub-particle can be released with or without leaving behind another Atom as shown below ***using α-rays-***

$$_2^4He \quad + \quad _7^{14}N \quad \longrightarrow \quad _1^1H \quad + \quad _8^{17}O$$

$$\text{(Stable)} \qquad\qquad \text{(proton)} \qquad \text{(Unstable)} \tag{i}$$

$$_2^4He \quad + \quad _4^9Be \quad \longrightarrow \quad _0^1H \quad + \quad _6^{12}C$$

$$\text{(Stable)} \qquad\qquad \text{(neutron)} \qquad \text{(Stable)} \tag{ii}$$

noting that with α-rays, another Atom must be produced.
(Laws of Creations for Nuclear Bombardment)

Rule 1683: This rule of Chemistry for **Nuclear Bombardment,** states that, when an Atom is hit by a *Male sub-particle,* one or two radicals (Free or Non-free) must be gained to form a ***New Atom*** if the particles are α-rays and this gain is what is lost in the formation of a ***New sub-particle.***
(Laws of Creations for Nuclear Bombardment)

Rule 1684: This rule of Chemistry for **Nuclear Bombardment,** states that, when an Atom is hit by a *Male sub-particle,* either a Male or Neutral type of sub-particle can be released with or without leaving behind another Atom as shown below ***using positrons-***

$$_1^0H \quad + \quad _6^{13}C \quad \xrightarrow[\text{Reaction}]{Nuclear} \quad _7^{13}N \quad + \quad _0^0H$$

$$Positron \qquad \text{(Unstable)} \qquad\qquad \text{(Unstable)} \qquad \gamma-rays$$

(Laws of Creations for Nuclear Bombardment)

Rule 1685: This rule of Chemistry for **Nuclear Bombardment,** states that, when an Atom is hit by a *Male sub-particle,* one radical (Free or Non-free) must be gained to form a ***New Atom*** if the particles are Positrons and this gain is what is lost in the formation of a ***New sub-particle.***
(Laws of Creations for Nuclear Bombardment)

Rule 1686: This rule of Chemistry for **Nuclear Bombardment,** states that, when an Atom is hit by a *Neutral sub-particle,* only Neutral type of sub-particles can be released which could be the neutron itself or another neutral sub-particle or the neutron itself could be captured as shown below, leaving behind an Isotope of the Atom hit-

$$_0^1H \quad + \quad _{53}^{127}I \quad \longrightarrow \quad \gamma-ray \quad + \quad _{53}^{128}I$$

$$\underline{neutron} \quad \text{(Stable)}-Iodine \qquad\qquad\qquad \text{(Unstable)}-Iodine \tag{i}$$

$$(_0^1H) \quad + \quad _{92}^{238}U \quad \longrightarrow \quad _{92}^{239}U \quad + \quad _0^0H$$

$$\text{(slow)} \qquad \text{(Stable)} \qquad \text{(Unstable)} \quad (\gamma-rays)$$

$$\text{(Neutron capture)} \tag{ii}$$

$$(_0^1H) \quad + \quad _{92}^{238}U \quad \longrightarrow \quad _{92}^{237}U \quad + \quad 2\,(_0^1H)$$

$$\text{(fast)} \qquad\quad \text{(Stable)} \qquad \text{(Unstable)} \qquad \text{(Neutrons)}$$

$$(Non-capture\ disintegration) \tag{iii}$$

noting that, never is there a time when male and female sub-particles are produced.

(Laws of Creations for Nuclear Bombardment)

Rule 1687: This rule of Chemistry for **Nuclear Bombardment,** states that, when an Atom is hit by a *Neutral sub-particle,* no radical is gained or lost to form a ***New Atom or Atoms*** and nothing is gained or lost in the formation of a ***New sub-particle.***
(Laws of Creations for Nuclear Bombardment)

Rule 1688: This rule of Chemistry for **Nuclear Bombardment,** states that, the types of products obtained during Nuclear bombardment, depends on the following operating conditions-

(i)   *The type of sub-particle used for the bombardment in terms of Male, Female and Neutral characters,*
(ii)   *The energy carried by the bombarding particle,*
(iii) *The speed of the particle, that which determines the energy carried,*
(iv)   *The type of atom or matter being bombarded,*
(v)   *The type of atomic sub-particle to be produced since this must be one of the products and*
(vi)   *The environment in which the bombardment is being done.*

(Laws of Creations for Nuclear Bombardment)

Rule 1689: This rule of Chemistry for **Cathode rays,** states that, these are atomic sub-particles which like β-rays are ELECTRONS, since they are one of the products of atomic breakdown of an Atom in a Discharge tube, via Nuclear reactions, but unlike β-rays are secondary in character since they are products from a latent force in the discharge tube and therefore of different capacity from β-rays and since they do not exist inside the Nucleus of an Atom-

$$Atoms \xrightarrow[Discharge-Tube]{Break\ Down\ in} Cathode-rays \quad + \quad Positive-rays$$
$$(High-speed\ electron) \qquad (Re\,sidue\ of\ atoms) \qquad \text{(i)}$$

which when it as well as β-rays are made to hit Matter, emit χ-rays and γ-rays respectively, both of which are like Electrons, but neutral in character.

$$Cathode-rays \xrightarrow[Matter]{Hits} \chi-rays\ are\ emitted$$
$$[Electromagnetic\ waves-10^{-10}metre\ wavelength]$$

$$\beta-rays \xrightarrow[Matter]{Hits} \gamma-rays\ are\ emitted$$
$$[Electromagnetic\ waves-10^{-12}metre\ wavelength] \qquad \text{(ii)}$$

$$Cathode\ rays \equiv {}_{-1}^{0}H \quad ; \quad \beta-rays \equiv {}_{-1}^{0}H$$
$$ELECTRONS$$
$$\gamma-rays \equiv {}_{0}^{0}H \quad ; \quad \chi-rays \equiv {}_{0}^{0}H$$
$$NEUTRAL \qquad \text{(iii)}$$

(Laws of Creations for Cathode rays)

Rule 1690: This rule of Chemistry for **Deuterium and Deuteron,** states that, just as Protonucleon ${}_{1}^{1}\mathbf{H}$ is the atomic secondary sub-particle related to ${}_{1}^{1}\mathbf{H}$•e atom that which has one electro-free-radical in the last shell, so also Deuteron ${}_{1}^{2}\mathbf{D}$ is atomic secondary sub-particle related to ${}_{1}^{2}\mathbf{D}$•e **(Deuterium)** atom that which is said to be an isotope of H and equally has one electro-radical in its last shell.
(Laws of Creations for Deuterium and Deuteron)

Rule 1691: This rule of Chemistry for **Nuclear Bombardment,** states that, when an Atom is hit by a *Male sub-particle,* either a Male or Neutral type of sub-particle can be released with or without leaving behind another Atom as shown below *using Deuteron-*

$$^2_1D \quad + \quad ^2_1D \quad \longrightarrow \quad ^1_1H \quad + \quad ^3_1T$$

(*Deuterium*)   <u>*Deuteron*</u>   (Pr*otonucleon*)   <u>*Tritium*</u>

<u>*Unstable*</u> <u>*Atom*</u>   <u>*Unstable*</u>   (i)

OR

$$^2_1D \quad + \quad ^2_1D \quad \longrightarrow \quad ^3_2He \quad + \quad ^1_0H$$

(*Deuterium*)   *Deuteron*   *Another ray*   (*Neutron*)   (ii)

<u>*Unstable*</u>   *a particle* (*Fast Speed*)

$$^2_1D \quad + \quad ^{23}_{11}Na \quad \longrightarrow \quad ^1_1H \quad + \quad ^{24}_{11}Na$$

<u>*Deuteron*</u>   (*Stable*)   (Pr*otonucleon*)   (*Unstable*)   (iii)

(Laws of Creations for Nuclear Bombardment)

Rule 1692: This rule of Chemistry for **Nuclear Bombardment,** states that, when an Atom is hit by a Unique non-Nucleus *Male sub-particle such as Deuteron,* either no radical (Free or Non-free) is gained or lost to form a *New Atom(s)* or a radical is lost when no New Atom is formed.
(Laws of Creations for Nuclear Bombardment)

Rule 1693: This rule of Chemistry for **Nuclear Bombardment,** states that, when an atom is hit by **any type of atomic sub-particle,** apart from formation of another Atom or no Atom, *another or same type of sub-atomic particle carrying energy with it must be released if the Laws of Conservation of Energy must be obeyed.*
(Laws of Creations for Nuclear bombardment)

Rule 1694: This rule of Chemistry for **Nuclear Bombardment/Radioactivity,** states that, when an atom is hit by a **Male atomic sub-particle,** depending on the type of Atom, after Nuclear reactions via Bombardment has taken place, this could be followed by spontaneous release of a sub-particle (Radioactivity) as shown below *for α-rays-*

$$^4_2He \quad + \quad ^{10}_5B \quad \xrightarrow[\text{BOMBARDMENT}]{\text{NUCLEAR}} \quad ^{14}_7N$$

(α − *rays*)   (*Stable*)   (*Stable*)

Cannot take place unless ^0_0H *is released*   (i)

$$^4_2He \quad + \quad ^{10}_5B \quad \xrightarrow[\text{BOMBARDMENT}]{\text{NUCLEAR}} \quad ^{13}_7N \quad + \quad ^1_0H$$

(α − *rays*)   (*Stable*)   (*Unstable*)   (*Neutron*)   (ii)

$$^{13}_7N \quad \xrightarrow{\text{RADIOACTIVITY}} \quad ^{13}_6C \quad + \quad ^0_1H$$

(*Unstable*)   (*Unstable*)   (*Positron*)   (iii)

noting that the first equation was not possible because no atomic sub-particle carrying energy with it was released; while the second reaction is Nuclear bombardment, the third reaction is Radioactivity wherein a **primary sub-particle** was released.
(Laws of Creations for Nuclear Bombardment/Radioactivity)

Rule 1695: This rule of Chemistry for **Nuclear Bombardment/Radioactivity,** states that, when an atom is hit by a **Male atomic sub-particle,** depending on the type of Atom, after Nuclear reactions via Bombardment has taken place, this could be followed by spontaneous release of a sub-particle (Radioactivity) as shown below *for Deuteron-*

$$\underset{Deuteron}{{}^{2}_{1}D} \quad + \quad \underset{(Stable)}{{}^{23}_{11}Na} \quad \xrightarrow{\substack{Nuclear \\ Bombardment}} \quad \underset{(Pr\,otonucleon)}{{}^{1}_{1}H} \quad + \quad \underset{(Unstable)}{{}^{24}_{11}Na} \qquad (i)$$

$$\underset{(Unstable)}{{}^{24}_{11}Na} \quad \xrightarrow{RADIOACTIVITY} \quad \underset{(Stable)}{{}^{24}_{12}Mg} \quad + \quad \underset{(\beta-rays)}{{}^{0}_{-1}H} \qquad (ii)$$

noting that only **primary sub-particle** can be released in the second reaction.
(Laws of Creations for Nuclear Bombardment/Radioactivity)

Rule 1696: This rule of Chemistry for **Nuclear Bombardment/Radioactivity,** states that, when an atom is hit by a **Male or Neutral atomic sub-particle,** depending on the type of Atom, after Nuclear reactions via Bombardment has taken place, this could be followed by spontaneous release of a sub-particle (Radioactivity) as shown below *for Neutrons-*

$$\underset{neutron}{{}^{1}_{0}H} \quad + \quad \underset{(Stable)-Iodine}{{}^{127}_{53}I} \quad \xrightarrow{\substack{NUCLEAR \\ BOMBARDMENT}} \quad \gamma-ray \quad + \quad \underset{(Unstable)-Iodine}{{}^{128}_{53}I} \qquad (i)$$

$$\underset{(Unstable)}{{}^{128}_{53}I} \quad \xrightarrow{RADIOACTIVITY} \quad \underset{(Stable)}{{}^{128}_{54}Xe} \quad + \quad \underset{(\beta-rays)}{{}^{0}_{-1}H} \qquad (ii)$$

noting that only primary sub-particles can be released in the second reaction.
(Laws of Creations for Nuclear Bombardment/Radioactivity)

Rule 1697: This rule of Chemistry for **Catalytic Nuclear Bombardment,** states that, when Deuterium is hit by an **α-ray a Male atomic sub-particle,** apart from the α-ray acting as a carrier of energy, it does not appear to take part in the reaction as shown below-

$$\underset{(\alpha-rays)}{{}^{4}_{2}He} \quad + \quad \underset{(Deuterium)}{{}^{2}_{1}D} \quad \xrightarrow{\substack{NUCLEAR \\ BOMBARDMENT}} \quad \underset{(\alpha-rays)}{{}^{4}_{2}He} \quad + \quad \underset{(Pr\,otonucleon)}{{}^{1}_{1}H} \quad + \quad \underset{(Neutron)}{{}^{1}_{0}H} \qquad (i)$$

$$i.e., \quad \underset{(Deuterium)}{{}^{2}_{1}D} \quad \xrightarrow{\substack{RADIOACTIVITY \\ \alpha-ray\,as\,CATALYST}} \quad \underset{(Pr\,otonucleon)}{{}^{1}_{1}H} \quad + \quad \underset{(Neutron)}{{}^{1}_{0}H} \qquad (ii)$$

for which in the absence of the α-ray acting as a passive catalyst (Energy carrier), the Deuterium cannot spontaneously disintegrate radioactively to give no Atoms, but two Atomic sub-particles.
(Laws of Creations for Catalytic Nuclear Bombardment)

Rule 1698: This rule of Chemistry for **Nuclear Bombardment,** states that, Nuclear reactions to produce other particle that exist in the Nucleus is limited to use of only positively nuclearly magnetically charged rays- *"protons", "deuterons", α-rays, etc.;* β-rays and "neutrons" cannot do these, but instead produce Non-Nucleus particles or are "captured" to produce another atom for which in general, like Chemical reactions, *only males commence reactions when both are present.*
(Laws of Creations for Nuclear Bombardment)

Rule 1699: This rule of Chemistry for **H atom,** states that, the H atom is the only atom that does not carry a "neutron", an α-ray, and therefore a β-ray and their corresponding antis; it carries only one **Protonucleon and Anti-Protonucleon.**
(Laws of Creations for H Atom)

Rule 1700: This rule of Chemistry for **Deuterium and Tritium isotopes of hydrogen,** states that, these do not carry a-rays and therefore β-rays, but carry one "proton" and one and two neutrons respectively and they both look like combination between one H atom and one and two neutrons respectively (Neutron Capture) as shown below-.

$$\underset{(Hydrogen\ Atom)}{^1_1H} \quad + \quad \underset{(Neutron)}{^1_0H} \quad \xrightarrow[BOMBARDMENT]{NUCLEAR} \quad \underset{(Deuterium)}{^2_1D} \quad + \quad \underset{(\gamma-rays)}{^0_0H} \tag{i}$$

$$\underset{(Deuterium)}{^2_1D} \quad + \quad \underset{(Neutron)}{^1_0H} \quad \xrightarrow[BOMBARDMENT]{NUCLEAR} \quad \underset{(Tritium)}{^3_1T} \quad + \quad \underset{(\gamma-rays)}{^0_0H} \tag{ii}$$

$$\underline{Overall\ Equation}: \quad ^1_1H + 2\,^1_0H \longrightarrow \,^3_1T + 2\,^0_0H \tag{iii}$$

for which one can observe what can be called **Nuclear Combination mechanism.**
(Laws of Creations for Deuterium and Tritium)

Rule 1701: This rule of Chemistry for **Nuclear Combination Mechanisms,** states that, when γ-rays are emitted when β-rays "hit" matter (electron capture) and are also emitted when protons or neutrons "hit" matter as shown below-

$$\underset{\beta-ray}{^0_{-1}H} \quad + \quad \underset{\substack{Curium\\(Unstable)}}{^{241}_{96}Cm} \quad \xrightarrow[Capture]{\beta-ray} \quad \gamma-rays \quad + \quad \underset{\substack{Americum\\(unstable)}}{^{241}_{95}Am} \tag{i}$$

$$\gamma-rays \quad \equiv \quad ^1_1H^{\odot} = \,^0_0H \tag{ii}$$

are indeed not only Nuclear Bombardment reactions, because the word "hit" above is not only bombardment, but also a COMBINATION mechanism system, that largely involving the Capture of a sub-particle; hence in general when γ- or χ- rays are involved as products in the reaction, the mechanism is Nuclear Bombardment Combination mechanism.
(Laws of Creations for Nuclear Bombardment Combination mechanism)

Rule 1702: This rule of Chemistry for **Nuclear Bombardment reactions,** states that, when they take place, it is only *via Decomposition mechanism,* between only *one Atomic sub-particle and one Atom*

or *Central Atom of a Molecule on a One to One basis, one at a time for Nuclear Bombardment reactions,* noting that while the number of sub-particles on the LHS does not have to be equal to the number of sub-particles on the RHS *unlike* Com*bination mechanism where they must be equal,* the fact is that for the one Atom which appeared on the LHS, there must be one Atom on the RHS which could be the same or different.
(Laws of Creations for Nuclear reactions)

Rule 1703: This rule of Chemistry for **Balancing of Nuclear Equations,** states that, the balancing of Nuclear equations has nothing to do with the magnetic charges carried by the particles, since the particles are no atoms or molecules, and has nothing to do with balancing of atoms, since Chemical reactions wherein only radicals along the boundary are used are not involved, *but to do with balancing of Atomic weights and Atomic numbers only on the Left and Right hand sides of the equation;* for which mass and energy cannot be used interchangeably by application of Albert Einstein theory of relativity as shown below-

$$Mass_1 + Energy_1 \neq Mass_2 + Energy_2$$
$$[Mass - energy \ equation]$$

where it looks as if mass and energy are of the same units! (i)

(Laws of Creations for Balancing of Nuclear Equations)

Rule 1704: This rule of Chemistry for **The Electromagnetic spectrum,** states that, this is a spectrum which contains only secondary (some massless) atomic sub-particles of *Radio waves, Microwaves, Infrared, Visible, Ultraviolet, x-ray and Gamma rays* in that decreasing frequency order all electromagnetic radiation waves *which can transport Energy from one location to another* **in Quantum** *at different but fixed frequencies at the speed of Light;* atomic sub-particles completely different from those inside the Nucleus of Atoms some of which are massless and some with negative masses of -1 and -4 and some with positive masses of 1 and 4, the 4 shocking known to be not a contributing factor to the atomic weight of an atom for specific reasons.
(Laws of Creations for The Electromagnetic spectrum)

Rule 1705: This rule of Chemistry for **The Electromagnetic spectrum,** states that, just as some atoms emit spontaneously β-rays, positron rays and other primary sub-particles, so also the SUN emits some other rays such as γ (Gamma)-rays, χ-rays, Ultraviolet rays, Visible light ray, Infrared rays, Microwave, Radio waves (Electromagnetic spectrum) and more.
(Laws of Creations for Electromagnetic Spectrum)

Rule 1706: This rule of Chemistry for **Energy in Atomic sub-particles,** states that, this Energy in general *is manifested only when the particle is in motion (i.e., moving particle) and not with the matter that is being bombarded, noting that each atomic sub-particle inside a nucleus of an atom or matter being bombarded carries the largest form of energy one can ever envisage in humanity, just like that of the SUN in our Solar system* for which one can indeed see **why Energy can never be created or destroyed, for it is already there inside and outside the nucleus of every atom.**
(Laws of Creations for Energy in Atomic sub-particles)

Rule 1707: This rule of Chemistry for **Energy,** states that, there are in general four main kinds or forms of Energy and these are-

(i)   Chemical form of Energy (Mass related)
(ii)   Mechanical form of Energy (Mass related)
(iii)   Electrical form of Energy (Not mass related)
(iv)   Electromagnetic form of Energy (Not mass but Speed related)

in which the last with respect to the nucleus of Atoms that involves the sub-atomic particles in and outside the atom is the largest form of Energy in our world all based on the hydrogen and Helium atoms, noting that conversion from one form to the other are readily possible limitedly based on the operating conditions, with each of the four kinds above having different types.
(Laws of Creations for Energy)

Rule 1708: This rule of Chemistry for **Conservation of Energy during Radioactivity,** states that, when an atomic sub-particle is released instantaneously from an Isotopic radioactive atom, the followings shown below diagrammatically takes place-

ENERGY CONSERVATION DURING RADIOACTIVITY

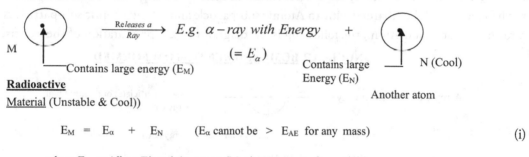

$$E_M = E_\alpha + E_N \qquad (E_\alpha \text{ cannot be } > E_{AE} \text{ for any mass}) \qquad \text{(i)}$$

where E_{AE} = Albert Einstein's energy (Maximum energy for particle)

noting that E_M and E_N are STORED ENERGIES which when at rest are ZERO relatively and when a particle is released from M, the energy is reduced to E_N over a period ranging from fractions of seconds to millions of years, the difference carried by the particle which begins to manifest itself in the particle based on the speed at point of release reaching its peak at the speed of light the point at which N is formed.
(Laws of Creations for Energy Conservation during Radioactivity)

Rule 1709: This rule of Chemistry for **Conservation of Energy during Nuclear Bombard-ment,** states that, when an Atom is bombarded with an Atomic sub-particle, and a New atomic sub-particle is released without the formation of a New Atom, then the following is the diagrammatic representation of the exercise-

NUCLEAR BOMBARDMENT/NO ATOM FORMED

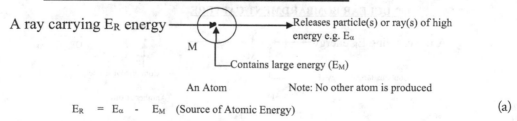

$$E_R = E_\alpha - E_M \quad \text{(Source of Atomic Energy)} \qquad \text{(a)}$$

and this is the origin of the Hydrogen Atomic bomb as shown below-

$$[A]$$

$$^2_1D \quad + \quad ^2_1D \quad \longrightarrow \quad ^3_2He \quad + \quad ^1_0H$$

(*Deuterium*)   *Deuteron*   *Another ray*   (*Neutron*)

Unstable   *a particle* (*High Speed*)   (i)

$$[B]$$

$$^4_2He \quad + \quad ^2_1D \quad \xrightarrow[\text{BOMBARDMENT}]{\text{NUCLEAR}} \quad ^4_2He \quad + \quad ^1_1H \quad + \quad ^1_0H$$

(*α − rays*)   (*Deuterium*)   (*α − rays*) (*Nucleoproton*) (*Neutron*)   (ii)a

$$i.e., \qquad ^2_1D \quad \xrightarrow[\text{α−ray as CATALYST}]{\text{RADIOACTIVITY}} \quad ^1_1H \quad + \quad ^1_0H$$

(*Deuterium*)   (*Nucleoproton*)   (*Neutron*)   (ii)b

for this can only be done with the isotopes of Hydrogen Atom wherein no New Atom can be formed, noting that the energy from (i) is far greater than that from (ii)b.
(Laws of Creations for Hydrogen Atomic Bomb)

<u>**Rule 1710:**</u> This rule of Chemistry for **Conservation of Energy during Nuclear Bombard-ment,** states that, when an Atom is bombarded with an Atomic sub-particle and a New Atomic sub-particle is released with formation of a New Atom, the following is the diagrammatic representation of the exercise-

<u>**NUCLEAR BOMBARDMENT /ATOM FORMED**</u>

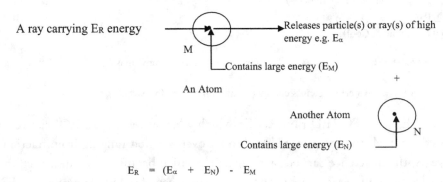

$$E_R = (E_\alpha + E_N) - E_M$$

for which, because E_M has been reduced to E_N, E_R has been increased to E_α manifestation of the increase in Energy, based on the speed of Light, the point at which N is formed.
(Laws of Creations for Energy Conservation during Nuclear Bombardment)

<u>**Rule 1711:**</u> This rule of Chemistry for **Conservation of Energy during Nuclear Bombard-ment,** states that, when an Atom is bombarded with a Neutron, either the neutron is captured with release of γ-rays or another neutron is abstracted to produce additional neutrons as shown below diagrammatically,

<u>**NUCLEAR BOMBARDMENT/CAPTURE**</u>

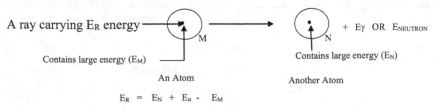

$$E_R = E_N + E_\alpha - E_M$$

for which if this is allowed to continue in chains, only some heavy atoms can be used, in which another form of "Atomic bomb" is created.

(Laws of Creations for Energy Conservation during Nuclear Bombardment)

Rule 1712: This rule of Chemistry for **Energy Content in an Atomic sub-particle,** states that, the Energy content of a particle is proportional to the square of the speed of the particle (p) as shown below-

$$E = Kp^2 \quad ; \quad E_{MAX} = Kc^2 \equiv Energy\ content\ in\ particle\ at\ rest$$
$$[c\ is\ the\ speed\ of\ light]$$

(where K is a unique constant in units of quanta which can be seen as an imaginary mass and has nothing to do with the mass of the particle)

wherein the Energy content of the particle at rest (E_{MAX}) which seems to be zero is manifested at the speed of light, *__is the essence of Albert Einstein's theory of relativity,__* otherwise, why should a particle such as β-ray, anti-β-ray, and all particles with zero or negative mass carry energy with them when this energy is not Mechanical or Chemical, but an Electromagnetic form of energy almost similar to the Electrical form (Based of electro-radicals); for which when the speed is zero, the Energy which is Electromagnetic is there, but not there, noting that the Constant K is a function of the type of sub-atomic particle in terms of an imaginary mass and a universal constant.

(Laws of Creations for Energy Content in an Atomic Sub-particle)

Rule 1713: This rule of Chemistry for **The Cosmic rays,** states that, Cosmic rays based on universal data, are like *Enzymes* in Living systems which carry the Blue-prints of what they want to do, but unlike Enzymes in Living systems which are specific, *these carry the Blue prints of about ninety percent Protonucleons, about nine percent α-particles and remaining one percent for all the NUCLEI of all the Natural elements in the Periodic Tables in the same proportion as they occur in our Solar system;* travelling with varying speed carrying along corresponding non-destructive High energies and as a Catalyst (Passive and Active) from them all kinds of *Matter in our Solar system were created or manifested; they were also* already in existence before creation of **our Universe,** coming from a very far away distant point of SINGULARITY released via Explosion.

(Laws of Creations for Cosmic Rays)

Rule 1714: This rule of Chemistry for **Nuclear Fission,** states that, this is the breakdown of an Atom via Decomposition mechanisms beginning with Nuclear bombardment, followed by Radioactivity in a sequential manner both in series and in parallel all taking place in STAGES down to the smallest possible Atom obeying the Laws of Nuclear reactions, for which many types of bombarding particles must be involved such as the use of Cosmic rays if complete breakdown is desired as shown below for one of the sub-particles- the Neutrons from a Nuclei.

*Stage*1 :

$$^{238}_{92}U \quad + \quad ^{1}_{0}H \quad \xrightarrow[\text{BOMBARDMENT}]{\text{NUCLEAR}} \quad ^{237}_{92}U \quad + \quad 2^{1}_{0}H$$
$$\qquad\quad (Fast) \qquad\qquad\qquad\qquad\qquad (Slow) \qquad\qquad\qquad\qquad (i)$$

Stage2 :

$$\underset{(xFraction)}{^{237}_{92}U} \xrightarrow[\beta-RAYS]{RADIOACTIVITY} \quad ^{237}_{93}Np \quad + \quad \underset{(\beta-rays)}{^{0}_{-1}H}$$

$$\underset{(yFraction)}{^{237}_{92}U} \xrightarrow[Positrons]{RADIOACTIVITY} \quad ^{237}_{91}(Pa) \quad + \quad \underset{(Positrons)}{^{0}_{1}H}$$

$$\underset{(zfraction)}{^{237}_{92}U} \xrightarrow[\alpha-RAYS]{RADIOACTIVITY} \quad ^{233}_{90}Th \quad + \quad \underset{(\alpha-rays)}{^{4}_{2}He}$$

$$\underset{(Remaining\ fraction)}{^{237}_{92}U} \xrightarrow[Anti-\alpha-RAYS]{RADIOACTIVITY} \quad ^{241}_{94}Pu \quad + \quad \underset{(Anti-\alpha-rays)}{^{-4}_{-2}He} \qquad \text{(ii)}$$

Stage3 :

$$\underset{(Slow)}{^{238}_{92}U} \quad + \quad ^{1}_{0}H \xrightarrow[BOMBARDMENT]{NUCLEAR} \quad ^{239}_{92}U \quad + \quad \underset{(\gamma-rays)}{^{0}_{0}H} \qquad \text{(iii)}$$

Stage4 :

$$\underset{(xFraction)}{^{239}_{92}U} \xrightarrow[\beta-RAYS]{RADIOACTIVITY} \quad ^{239}_{93}Np \quad + \quad \underset{(\beta-rays)}{^{0}_{-1}H}$$

$$\underset{(yFraction)}{^{239}_{92}U} \xrightarrow[Positron]{RADIOACTIVITY} \quad ^{239}_{91}Pa \quad + \quad \underset{(Positrons)}{^{0}_{1}H}$$

$$\underset{(zFraction)}{^{239}_{92}U} \xrightarrow[\alpha-RAYS]{RADIOACTIVITY} \quad ^{235}_{90}Th \quad + \quad \underset{(\alpha-rays)}{^{4}_{2}He}$$

$$\underset{}{^{239}_{92}U} \xrightarrow[Anti-\alpha-RAYS]{RADIOACTIVITY} \quad ^{243}_{94}Pu \quad + \quad ^{-4}_{-2}He \qquad \text{(iv)}$$

$$E.T.C.$$

for which one can observe why such reactions are limited to Heavy metals, wherein exists large numbers of Isotopic Elements as can be seen above in the Stages when neutrons are still present in the system; noting that neutrons were used above in the absence of positively nuclearly charged sub-particles in the system, otherwise the sub-particle will be the first to commence the Nuclear bombardment, and in addition the existence of some of the Radioactivity stages will depend on the half-life period of disintegration of an Atom.

(*Laws of Creations for Nuclear Fission*)

Rule 1715: This rule of Chemistry for **β-rays and α-rays two primary sub-particles on Matter side,** states that, the β-rays and α-rays borrow or give "neutrons" and "protons" two secondary sub-particles from the atomic weights and atomic numbers of the original atom for their existence, for which in an atom, they remain as such without contributing to the Atomic weight and Atomic number of an Atom, otherwise why should the atomic weight be limited only to the "neutrons" and "protons" alone when α-particles have weights almost twice as much as the others put together?

(*Laws of Creations for β- and α-rays*)

Rule 1716: This rule of Chemistry for **The Nucleus of All Atoms,** states that, since the a-rays are doubly "positively nuclearly charged" and β-rays are singly "negatively nuclearly charged", two β-rays are required to neutralize one a-ray for which indeed, in the nucleus, the α- and β- rays are paired as shown below-

ELECTROSTATICALLY NUCLEARLY NEGATIVELY
CHARGED-PAIRED BONDS

$$^{2}_{0}He^{\oplus}_{\oplus} \cdots\cdots\cdots ^{\odot}X^{1}_{0} \quad \equiv \quad \alpha^{\oplus}_{\oplus}\cdots\cdots ^{\odot}\beta$$

$$^{1}_{0}X^{\odot} \qquad\qquad\qquad \beta^{\odot} \qquad\qquad\qquad\qquad (A)$$

noting that the charges above as already said are not ionic or covalent or electrostatic or polar, but herein called ***Nuclear Magnetic Electrostatic Charges (Real and Imaginary);*** the nuclearly negatively charged center being real, while the nuclearly positively charged center is imaginary analogous to the types used in conducting electric current in fluids in the outer sphere of the atom; and also noting that though the number of β-rays equals the number of nucleoprotons, after neutralization of α- and β-rays, the presence of nucleoprotons keeps the nucleus still **nuclearly positively charged on the MATTER side;** for in view of the complex character displayed by the two rays above, hence Nuleoprotons and Neutrons can be the only ones that can contribute to the Atomic weight and Atomic number of an atom.
(Laws of Creations for Nucleus of All Atoms)

Rule 1717: This rule of Chemistry for **The Nucleus of All Atoms**, states that, since the nucleus cannot be "charged", when the outer sphere is not charged, and since positrons and anti-protons have been identified, ***then the resultant "positive charge" on the Matter side is neutralized by an equal "negative charge" on the other side of the nucleus, for which therefore, the nucleus is a combination of matter and anti-matter, hence while the matter has positive mass, the anti-matter has negative mass in all our different gravitation fields in our Solar system (Earth, Moon, Jupiter, Mars, etc.)***
(Laws of Creations for Nucleus of All Atoms)

Rule 1718: This rule of Chemistry for **The H Atom/Particles**, states that, since the "particles" in the nucleus can be observed to have their origin from H and He atoms, these are indeed atomic sub-particles, in which the He particle and Neutron can be made to undergo further disintegration as shown below under very harsh operating conditions-

$$^{4}_{2}He\,(^{2}_{0}He^{2\oplus}) \longrightarrow ^{3}_{2}He\,(^{1}_{0}He^{2\oplus}) + ^{1}_{0}H$$
$$(\alpha-rays) \qquad\qquad (Helium-ray) \quad (Neutron) \qquad\qquad (i)$$

$$^{1}_{0}H \longrightarrow ^{0}_{-1}H + ^{1}_{1}H$$
$$(Neutron) \qquad (\beta-rays) \quad (Pr\,otonucleon) \qquad\qquad (ii)$$

and since ***all the sub-particles of H are indeed of the same size*** no matter what it is carrying, hence, the smallest indivisible particle is H and it is also the smallest atom and the only thing PERMANENT (Not CHANGE which is a function of TIME).
(Laws of Creations for the H atom)

Rule 1719: This rule of Chemistry for **The Nucleus of All Atoms,** states that, since the Anti- α-rays are doubly "negatively nuclearly charged" and Anti-β-rays (Positrons) are singly "positively nuclearly charged", two Anti-β-rays are required to neutralize one Anti-α-ray for which indeed, in the nucleus, the Anti-α- and Anti-β- rays are paired as shown below and compared with that of Matter side-

ELECTROSTATICALLY NUCLEARLY POSITIVELY CHARGED

PAIRED-BONDS

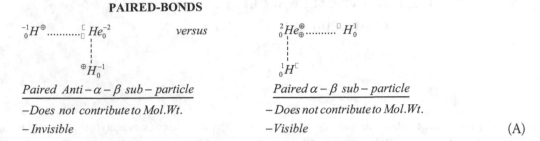

Paired Anti − α − β sub − particle	*Paired α − β sub − particle*
− *Does not contribute to Mol.Wt.*	− *Does not contribute to Mol.Wt.*
− *Invisible*	− *Visible*

(A)

noting that the charges above as already said are not ionic or covalent or electrostatic or polar, but herein called ***Nuclear Magnetic Electrostatic Charges (Real and Imaginary);*** the nuclearly positively charged center being real, while the nuclearly negatively charged center is imaginary analogous to the types used in Chemical and Polymeric reactions in the outer sphere of the atom; and also noting that though the number of Anti-β-rays equals the number of Anti-Protonucleons, after neutralization of Anti-α- and Anti-β-rays, the presence of Anti-nucleoprotons keeps the nucleus still **nuclearly negatively charged on the ANTI-MATTER side;** for in view of the complex character displayed by the two rays above and the fact that the Antis- have negative masses which cannot be measured, hence only Nuleoprotons (or Protonucleons) and Neutrons can be the only ones that can contribute to the Atomic weight and Atomic number of an atom and when the charged character above is combined with the Matter side, one can see that The Nucleus is a NEUTRAL BODY.

(Laws of Creations for Nucleus of All Atoms).

Rule 1720: This rule of Chemistry for **the Nucleus of All Atoms,** states that, the presence of ***Electrostatically Nuclearly Charged Positively and Negatively Paired Bonds,*** the POSITIVE on the Anti-Matter side and the NEGATIVE on the Matter side is the origin of countless numbers of phenomena in our Solar system too countless to list, with respect to for example- Thunder and Lightning, Magnetism, The North and South Poles, and so on in MOTHER NATURE'S world.

(Laws of Creations for Humanity)

Rule 1721: This rule of Chemistry for Radicals in the outer sphere of an atom, states that, Radicals are massless sub-atomic particles wherein there are two kinds- Free and Non-free, each with their own identity of male and female character, and just like the atomic sub-particles in the Nucleus, are carrying hidden energies inside them far less than those in the Nucleus; for which one can indeed see <u>why Energy can never be created or destroyed, for it is already there also outside the nucleus of every atom.</u>

(Laws of Creations for RADICALS)

In order to bridge the gap between the past developments from "The beginning of a New Dawn for Humanity" to the developments so far, three hundred and sixty three rules have been proposed. In the "The beginning of a New Dawn for Humanity", Volume (I), new concepts developed on Oxidation and Combustion, Nitrogen containing compounds and more, the morphology of polymers, dissolution of polymers and engineering aspects of polymerizations processes, had to be stated into rules here in order to provide better understanding of the concepts and complete Volume (III) with the need to ask questions. These are to provide easier understanding of other Volumes ahead. After all the develop-ments so far, stating some of the important rules here, still remain in place. Laws with respect to the Nucleus of Atoms have also begun to be stated herein as this was introduced right from the "The beginning of a New Dawn for Humanity".

As exists in our planet today with all species, there are different types of families of compounds/monomers, all with different types of rules guiding their existences and co-existences with their members and other members. In dealing with the aspect of atoms, compounds, molecules, ions, radicals, charged species, one has inevitably tried as much as possible to deal with more complex compounds. From all the considerations so far, it is important to note that there are so many chemical terminologies which have been used in the past and even presently, but are meaningless. Some of the terminologies do not indeed exist, while so many had to be adequately redefined.

One has deliberately used some of the older or present names in order to provide in a systematic order, a better understanding for the current developments. ***This however marks the beginning of a new dawn for humanity.*** The very unique character of natural laws is the fact that, in all the developments so far, none of the rules have been broken or have had an exception. It is indeed difficult to say categorically which law is important or not important. The more you go into each of the laws, the more you begin to see how NATURE operates. One will begin to see that all disciplines including Religion will be affected. Indeed, we have only just started on the surface. Of the three disciplines of Natural Science, the most unique is CHEMISTRY. For it is inside CHEMISTRY you see MATHEMATICS which one has started to reveal, you see PHYSICS, which one has started to reveal. It is also in the same CHEMISTRY, you see ENGINEERING, which one has started to reveal, you see SOCIAL SCIENCES between families of compounds which one has started to reveal, you see MEDICAL SCIENCES, which one has started to reveal, you will see the ARTS, which one has started to reveal. In that same CHEMISTRY, you see in totality RELIGION, which one has started to reveal. CHEMISTRY is indeed the greatest discipline in Humanity if not the ONLY discipline. Hence, it has been identified as the MOTHER OF ALL DISCIPLINES. Everything in our world revolves round the ATOMS, in particular CARBON which carries the number 12. It is not a coincidence neither is it an accident. It is one of the greatest central origins of our LIVES in the planet EARTH which is like ***A SINGLE RADICAL*** in our solar system IN ITS OUTER SPHERE along its boundary, as laid by the ALMIGHTY INFINITE GOD.

This brings one to the end of Volume (III) dealing particularly with Initiation of chemical and polymeric compounds. Initiation will continue in Vol. (IV), where copolymeri-zation system will largely be considered, in order to fully identify the driving forces favoring the existence of different types of copolymers.

References

1. G. B. Butter and F. L. Ingley. J. Am. Chem. Soc., 73: 894 (1951).
2. G. B. Butler and R. J. Angelo, J. Am. Chem. Soc., 79: 3128 (1957).
3. W. E. Gibbs and J. M. Barton, The Mechanism of Cyclopolymerization of Nonconjugated Diolefins, in G. E. Ham (ed.), "Vinyl Polymerization," Vol. 1, part 1, Chap. 2, Marcel Dekker, Inc.; New York, (1967).
4. T. Holt and W. Simpson, Proc. Roy. Soc. (London), A238 : 154 (1956).
5. M. Gordon and R. J. Roe, J. Polymer Sci., 21: 57, 75 (1956).
6. G. B. Butler, Pure and Appl. Chem., 4 : 299 (1962) ; Polymer Preprints, 8 (1) : 35 (1967).
7. C. S. Marvel and R. D. Vest, J. Am. Chem. Soc., 79 : 5771 (1957); 81 : 984 (1959).
8. C. S. Marvel and J. K. Stille, J. Am. Chem. Soc., 80 : 1740 (1958).
9. W. Cooper, Polyenes, in P. H. Plesch (ed.), "The Chemistry of Cationic Polymeriza-tion," Chap. 8, Pergamon Press Ltd., Oxford, (1963).
10. N. G. Gaylord et al., J. Polymer Sci., A - 1 (4): 2493 (1966); A - 1 (6) ; (1963).
11. N. G. Gaylord and M. Svestka, J. Polymer Sci., B7: 55 (1969).
12. E. M. Fetles (ed.), "Chemical Reactions of Polymers," Inter - science Publishers, John Wiley & Sons, Inc., New York, (1964).
13. C. R. Noller, "Textbook of Organic Chemistry," W. B. Saunders Company, (1966), pg. 656.
14. G. Odian, "Principles of Polymer Systems," McGraw - Hill Book Company, (1970), pgs. 630 - 632.

Problems

15.1. Distinguish between monomers that add Inter - molecularly from those that add Intra - molecularly.

15.2 What are the factors that determine the ability of a monomer to favor ring formations along the chain? Use examples to illustrate the factors.

15.3 The products obtained in the polymerization of 4-methyl- 1,6- heptadiene and 1- methyl- 1,6- heptadiene contains no residual unsaturation.

   (a) What are the character of the monomer?
   (b) What are the chemical structures of the polymers?
   (c) Identify the transfer species in the systems.

15.4 Distinguish between monomers that add Intra - molecularly from those that add Inter -/ Intra – molecularly

15.5 The products obtained in the polymerization of diallyl phthalate, diethylene glycol bisallyl carbonate and triallyl cyanurate are highly cross-linked thermosetting products, with rings in them.

   (a) Identify the route(s) favored for the polymerization of the monomers.
   (b) What are the sizes of rings favored by the monomers?
   (c) Why and how are cross-links favored in these systems and why are they thermo-setting in character?

15.6 (a) Why does p-divinyl benzene not favor cyclization reactions, but o-divinyl benzene and m-divinyl benzene do? Explain.
   (b) How does p-divinyl benzene favor being used as a cross-linking agent between two chains, but the o- and m- divinyl benzene do not?
   (c) Between o- and m-divinyl benzene, which is better for producing better polymeric network system? Explain.

15.7 (a) When 1,4- dienes are involved as monomers, why is it that two monomer units may be required to form bicyclic rings? Can one monomer unit be used to form rings continuously along the chain?

   (b) Why can't 1,3- dienes be used in the same manner 1,4- dienes are used for cyclization?

   (c) How can 1,3- dienes be made to favor cyclization reaction?

15.8 (a) While positively charged polymerization of polymerizable unconjugated dienes is possible, it is not possible with polymeric monomers with internal double bonds. Why?

   (b) Vinyl acetate monomer is a Nucleophile. This monomer is very unique, because its isomer methyl acrylate is an Electrophile whose natural route is nucleo-free-radical. The natural route of vinyl acetate is electro-free-radical which it does not favor. It favors only the nucleo-free-radical route as shown below, something which is not to be expected. It does, because there is no transfer species to abstract.

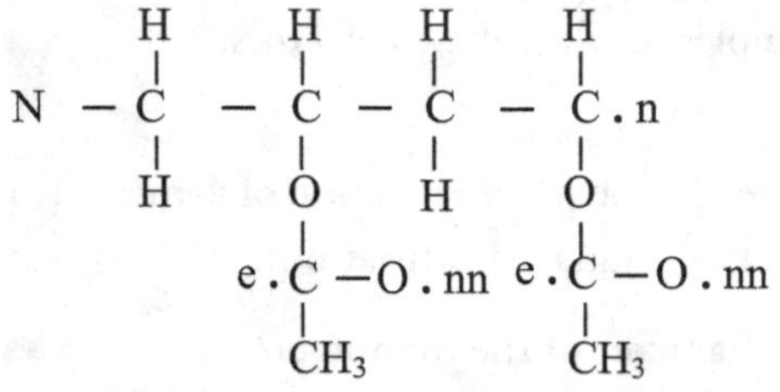

$$
\begin{array}{ccccccc}
 & H & & H & & H & & H \\
 & | & & | & & | & & | \\
N & - C & - & C & - & C & - & C.n \\
 & | & & | & & | & & | \\
 & H & & O & & H & & O \\
\end{array}
$$

$$
e.\overset{\displaystyle O}{\underset{\displaystyle CH_3}{C}}-O.nn \quad e.\overset{\displaystyle O}{\underset{\displaystyle CH_3}{C}}-O.nn
$$

Answer the following questions.

   (i) Unlike some Nucleophiles which favor both routes, this favors only the route not natural to it. Why? What is the meaning of this with respect to humans?

   (ii) How can the growing polymer chain kill itself or how can it be killed?

   (iii) Can the monomer be cyclized during polymerization? Can its dead polymer be used as a polymeric monomer for cyclization?

   (iv) What is unique about its dead polymer?

15.9 (a) Shown below is a reaction of cyclization of polymeric acrylamide as monomer.

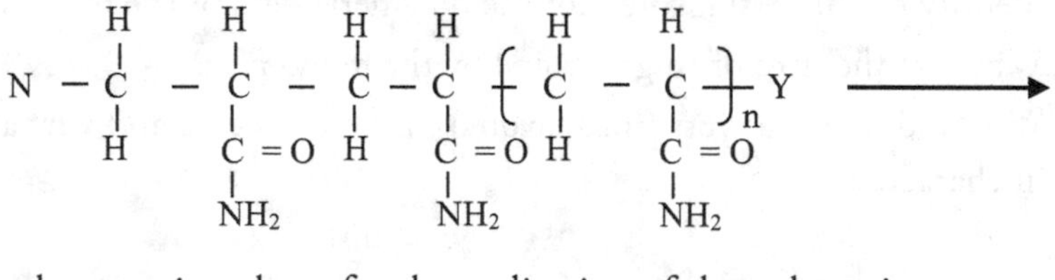

Complete the equation above for the cyclization of the polymeric monomer, identifying the mechanisms of polymerization.

   (b) Identify the problems to be faced if any, in using the compound shown below as a poly-monomer for cyclization.

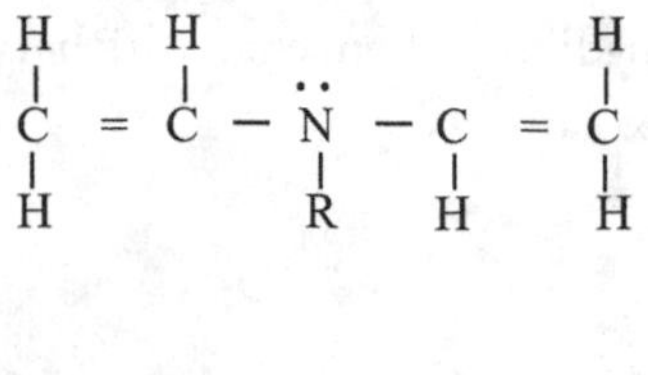

15.10 (a) Shown below are two substituted 1,5- dienes.

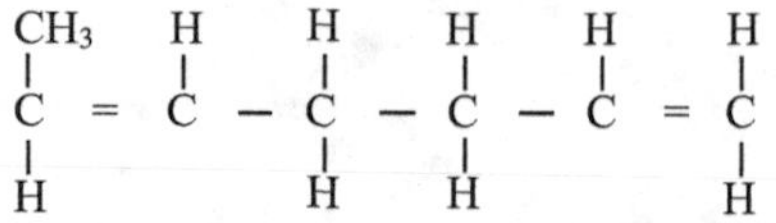

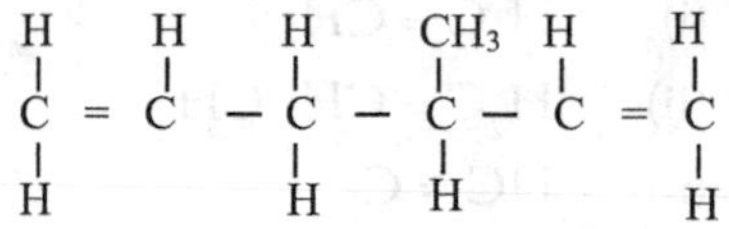

$$\underset{\underset{H}{|}}{\overset{\overset{CH_3}{|}}{C}} = \underset{\underset{H}{|}}{\overset{\overset{H}{|}}{C}} - \underset{\underset{H}{|}}{\overset{\overset{H}{|}}{C}} - \underset{\underset{H}{|}}{\overset{\overset{H}{|}}{C}} - \overset{\overset{H}{|}}{C} = \overset{\overset{H}{|}}{\underset{\underset{H}{|}}{C}} \quad ; \quad \underset{\underset{H}{|}}{\overset{\overset{H}{|}}{C}} = \underset{\underset{H}{|}}{\overset{\overset{H}{|}}{C}} - \underset{\underset{H}{|}}{\overset{\overset{H}{|}}{C}} - \underset{\underset{H}{|}}{\overset{\overset{CH_3}{|}}{C}} - \overset{\overset{H}{|}}{C} = \overset{}{\underset{\underset{H}{|}}{C}}$$

(i) Identify the route(s) favored for the cyclization of the monomers.

(ii) Show the type of rings formed where a route is favored.

(iii) What type of polymers are produced?

(b) Provide the new classification for Unconjugated Dienes.

15.11. Show below is a compound.

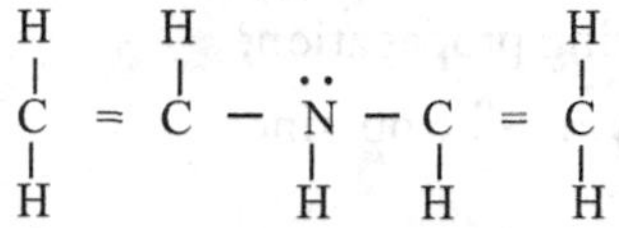

$$\underset{\underset{H}{|}}{\overset{\overset{H}{|}}{C}} = \underset{\underset{H}{|}}{\overset{\overset{H}{|}}{C}} - \overset{\overset{..}{}}{\underset{\underset{H}{|}}{N}} - \underset{\underset{H}{|}}{\overset{}{C}} = \overset{\overset{H}{|}}{\underset{\underset{H}{|}}{C}}$$

(i) Why is it that one of the vinyl groups is not loosely bonded to the nitrogen center?

(ii) Distinguish between this compound and that of Q. 15.9(b).

(iii) What is the difference between the compound and ammonia?

15.12 a) Shown below are two generating sources of electro-free-radical initiators.

NaCN   HOSO_3H/H_2O

(Strong)   (strong)

Show the initiators generated from them and which can be used to polymerize the following monomers, showing the products obtained from them.

(I) 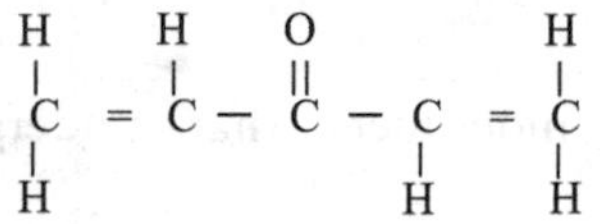
$$\underset{\underset{H}{|}}{\overset{\overset{H}{|}}{C}} = \underset{\underset{H}{|}}{\overset{\overset{H}{|}}{C}} - \overset{\overset{O}{\|}}{C} - \underset{\underset{H}{|}}{\overset{}{C}} = \overset{\overset{H}{|}}{\underset{\underset{H}{|}}{C}}$$

(II) 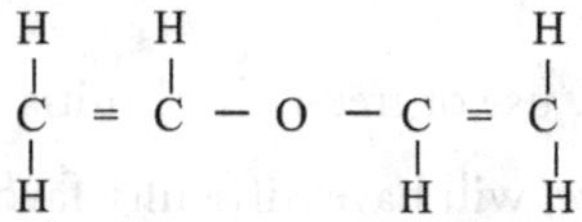
$$\underset{\underset{H}{|}}{\overset{\overset{H}{|}}{C}} = \underset{\underset{H}{|}}{\overset{\overset{H}{|}}{C}} - O - \underset{\underset{H}{|}}{\overset{}{C}} = \overset{\overset{H}{|}}{\underset{\underset{H}{|}}{C}}$$

(III) 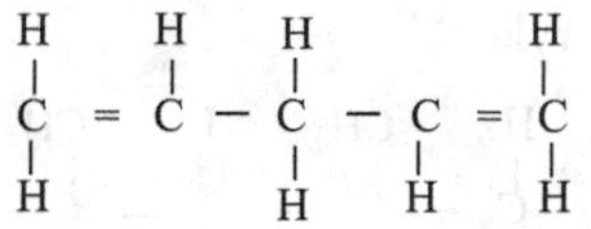
$$\underset{\underset{H}{|}}{\overset{\overset{H}{|}}{C}} = \underset{\underset{H}{|}}{\overset{\overset{H}{|}}{C}} - \underset{\underset{H}{|}}{\overset{\overset{H}{|}}{C}} - \underset{\underset{H}{|}}{\overset{}{C}} = \overset{\overset{H}{|}}{\underset{\underset{H}{|}}{C}}$$

b) What family of Unconjugated Dienes do the monomers belong to?

15.13 (a) Distinguish between the use of 1,4- diyne and 1,4- diene as poly-monomers for cyclization reactions.

(b) Show the use of 1,6- heptadiyne as a poly-monomer and do the same for 1,5-hexadiyne.

15.14 Show how these groups can be re-classified

    (i) $H_2C = CH-$

    (ii) $H_2C = CH- CH_2 -$

    (iii) $HC \equiv C-$

    (iv) $HC \equiv C - CH_2 -$

    (v)   Provide the new classifications for Alkyl groups.

15.15 (a) What are Di- Inter- molecular cyclo homo- and co-polymerizations?

    (b) Distinguish the unique features between this type of cyclo-polymerization from all the other types.

    (c) Which types of 1,3- dienes are known to undergo this type of cyclopolymerization?

    (d) Why is 1,2- addition not favored during propagation?

    (e) How can high-molecular weight polymers be obtained?

15.16 (a) Why is it that monomers such as acrolein, methyl acrylate, acrylamide, acrylonitrile etc. do not favor being used for Di-Inter-molecular cyclopolymerization reactions?

    (b) Can divinyl benzene favor being used as such? Explain.

    (c) Why is it that this type of molecular cyclopolymerization is not favored chargedly?

    (d) Can one use a Ziegler-Natta type of initiator for Di- Inter molecular cyclo homo or copolymerization actively? Then if the answer is No, what is the real initiator that triggers activation of the poly-monomer? Explain.

15.17 Distinguish between the different types of Inter-/Intra- molecular cyclohomo and copolymerizations in terms of the types of functional groups carried by the polymonomers.

15.18 (a) Identify the types of monomers that undergo Intra- molecular alternating cyclocopolymerization.

    (b) Explain the mechanism of the reaction for cyclization under -

        (i) Pyrolytic conditions.

        (ii) Inter- conditions (i.e. use of free-radical Initiators).

    (c) Which of (i) and (ii) above, will have difficulty favoring full cyclization?

15.19 Shown below is a poly(methyl acrylate).

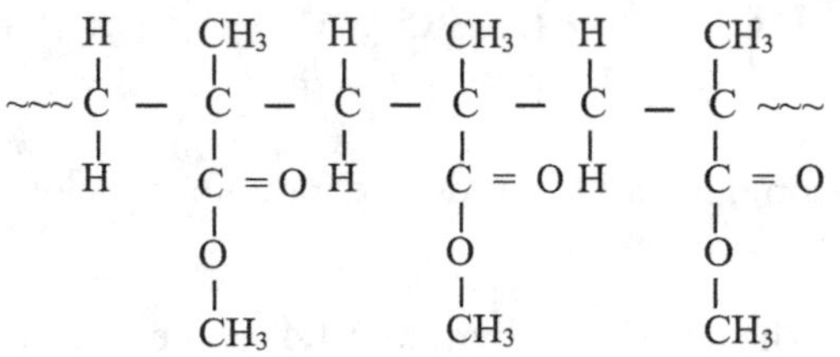

$$\sim\sim\sim \overset{\displaystyle H}{\underset{\displaystyle H}{C}} - \overset{\displaystyle CH_3}{\underset{\displaystyle \underset{O}{\underset{|}{CH_3}}}{\underset{|}{\underset{|}{C=O}}C}} - \overset{\displaystyle H}{\underset{\displaystyle H}{C}} - \overset{\displaystyle CH_3}{C} - \overset{\displaystyle H}{\underset{\displaystyle H}{C}} - \overset{\displaystyle CH_3}{C} \sim\sim$$

On the basis of factors favoring polymer reactions reactivity, the rate enhancement by a neighboring group, usually referred to as Anchimeric assistance, is said to occur primarily when the cyclic intermediate favored is 5- or 6- membered ring.

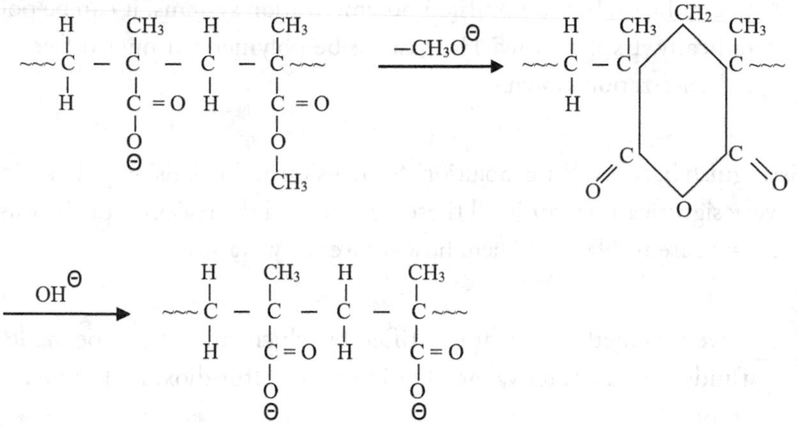

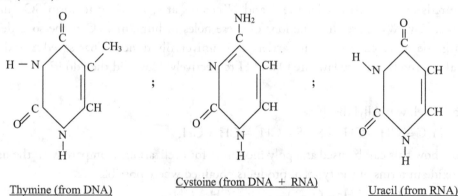

(i) Can the reaction above be classified as Intra- molecular alternating cyclo- copolymerization?

(ii) Does the equation above make sense?

(iii) Under what conditions can the rings be obtained? Complete the equation above, based on the current developments and ascertain if ever it can take place in the absence of a second component, such as water.

15.20 Shown below are three ringed compounds from deoxyribonucleic acid (DNA) and ribonucleic acid (RNA).

Thymine (from DNA) ; Cystoine (from DNA + RNA) ; Uracil (from RNA)

Structural formulas for pyrimidine bases

Based on the New Frontiers, describe the chemistry of the three rings.

15.21 (a) Under what conditions do Inter-, Intra-, Inter/Intra-, and Inter/Intra-/Intra-.... additions take place in polymeric systems?

(b) Can 1-, 3-, 5- triene undergo Di- inter molecular cyclo homopolymerization? Explain showing the types of ring sizes to expect.

15.22 (a) In Di- inter cyclopolymerization, can the initiating ring grow from both sides during polymerization?

(b) Propylene can be polymerized in all systems using electro-free-radicals and not nucleo-free-radicals, but in Emulsion polymerization systems, it can be polymerized nucleo-free-radically. Explain why? How can it be polymerized nucleo-free - radically in Emulsion polymerization systems?

15.23 Distinguish between Bulk, Solution, Suspension and Emulsion polymerization systems. Mixing is a very significant factor in all these systems and the reasons are obvious. What are they? This is not the case in Nature. Then, how do we copy Nature?

15.24. a) Give two methods so far identified by which current can be made to flow through some fluids such as water vapors, liquid water, sulfur dioxide, Iodine heptafluoride (IF7), and so on.

b) Distinguish between the HOLES in following centers-

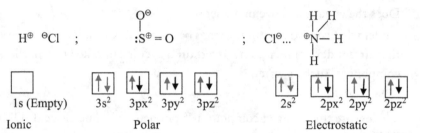

The reds are radicals left for the S and N centers after bonding to form SO_2 and NH_4Cl respectively. What are the functions of these holes in humanity? Can the so-called positive charge be removed from their carriers as is universally done? What saved S and N in not breaking the Boundary laws are O and H respectively. How did they do it?

15.25 Shown below is Allyl disulfide.

$$H_2C = CH - CH_2 - S - S - CH_2 - CH = CH_2$$

Show how this can be used as a poly-monomer for cyclization. Compare with the use of Allyl peroxide in terms of the types of products obtained where possible.

$$H_2C = CH - CH_2 - O - O - CH_2 - CH = CH_2.$$

If not possible, explain why it is so.

15.26 H with the number 1and atomic number 1, is the first Atom in the Periodic Table of Identified Atoms in our Planet Earth. C with the number 6 and atomic number 12, is the sixth Atom in the Periodic Table of Identified Atoms in our Planet Earth. These two Atoms have Elements. The two form the Hydrocarbon family tree alone the largest of all families formed from just two Atoms. Identify what are exceptionally unique with H and C based on the Hydrocarbon family tree.

15.27　How many kinds of Resonance Stabilization phenomena exist and for each kind, how many types exist? Give examples of the kinds and types.

15.28　Divinyl benzenes are unique in the sense that they are both Unconjugated dienes and Conjugated dienes. Show how these behaviors are manifested and applied. Distinquish between p-divinyl benzene with 2-Ethylkene 1,3-diene or 2-Vinyl 1,3-diene [$H_2C = CH – C(CH = CH_2) = CH_2$].

15.29　Under what conditions can a compound with two activation centers be activated both at the same time chargedly and or free-radically. Give examples to illustrate your answers.

15.30.　a)　Shown below is the following reaction.

$$RCH_2CH = CHCH_3 + (CH_3CO)_2NBr \longrightarrow RCHBrCH = CHCH_3 + (CH_3CO)_2NH$$

Answer the following questions.

i)　Why was the H from RCH_2 be the one replaced, instead of the one from CH_3 group, where R is an alkylane group?

ii)　Is one of the products allylic?

iii)　Provide the mechanism for the reaction.

b)　What is the Equilibrium State of Existence of the following compounds?

i) $HC \equiv CH$; ii) $H_2C = CH_2$; iii) $H_3CCH = CH_2$; iv) $H_9C_4CH = CH_2$;
v) $H_9C_4CH = CHCH_3$; vi) $H_9C_4CH = CHC_2H_5$; vii) C_4H_{10} ; viii) ClC_4H_9 ;
ix) $ClCH = CHCH_3$; x) $ClCH_2C = CHCH_3$; xi) $H_2C = C = CH – CH_3$;
xii) $H_2C = C = CH – CH_2Br$

15.31　a)　What is Trans annular migration?

b)　What is Trans annular addition?

c)　Give examples of compounds that undergo the (b) phenomenon to give stable and activated monomers.

d)　How can Trans annular addition be prevented during Di-Inter polymerization?

15.32　Shown below are some ringed monomers-

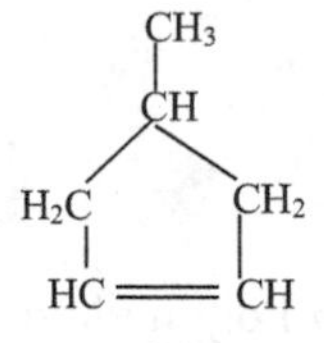

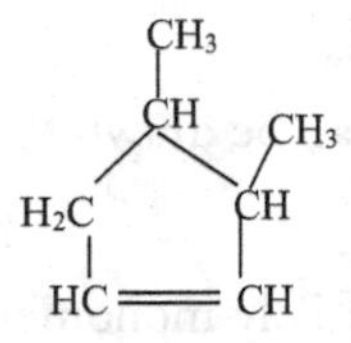

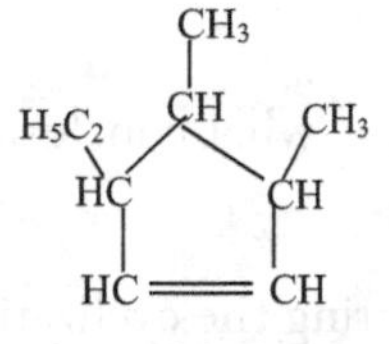

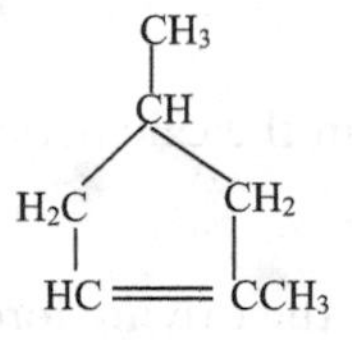

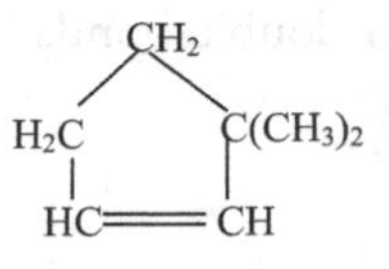

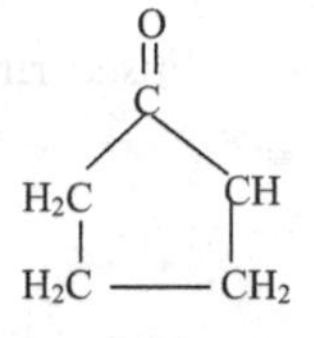

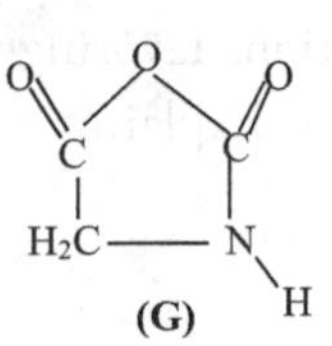

Show the activated states of existence of these monomers. Identify their transfer species and show how they can be polymerized. Name the compounds.

15.33 a) What are Ozone and Ozone layer? Show what happens when ozone and its layer are activated. Can current be made to flow through the layer? Explain.

b) When cyclo butane is activated chargedly and radically, the followings are obtained.

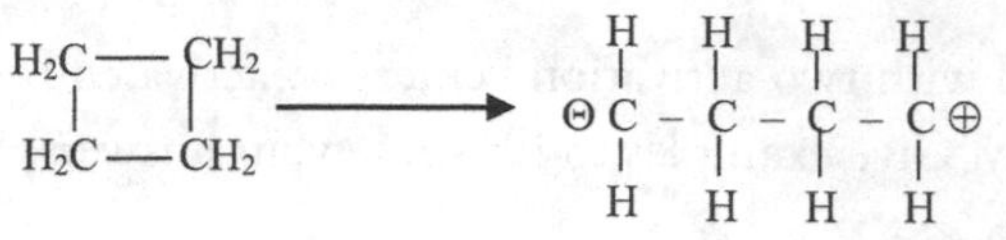

Activated chargedly **(A)** 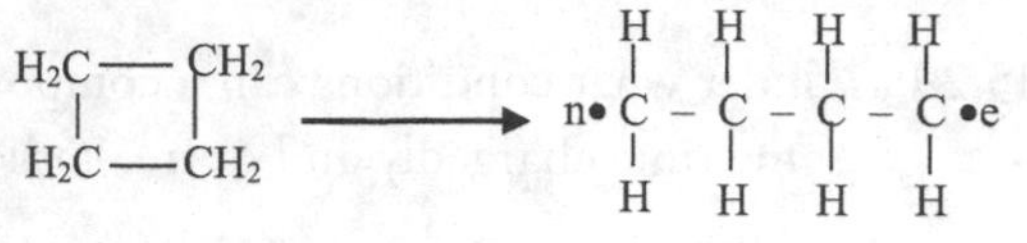   **Activated free-radically** **(B)**

Can (A) be transformed to (B) or (B) to (A)? Explain

15.34 a) What are Poly-monomers?

b) Show the new classifications of Poly-monomers

c) Identify the routes undergone by each member of the family for cyclization.

15.35 a) Based on the New Frontiers, define SE, MRSE and MaxRSE.

b) Determine the SEs in the following rings. whether the ring exists or not.

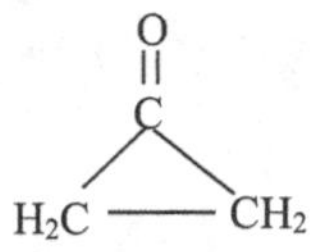

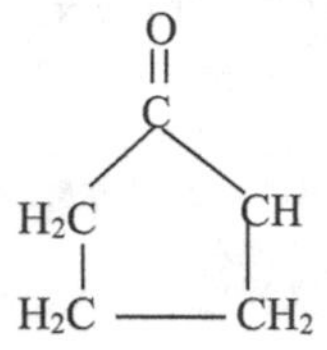

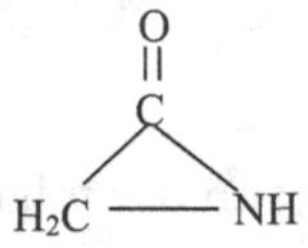

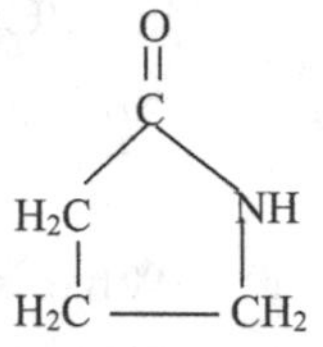

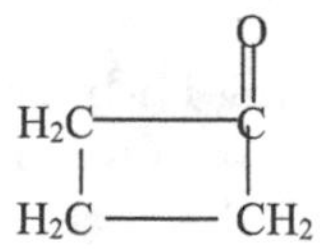

From the calculations above, what conclusions can be drawn.

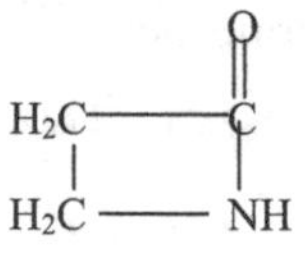

15.36 List the driving forces favoring the cyclization of Poly-monomers and Polymeric monomers.

15.37 List the conditions favoring the existence of Fused rings. Can double bonds be cumulatively placed in a ting? Explain.

15.38   Shown below is a compound (A), and a dead polymer (B).

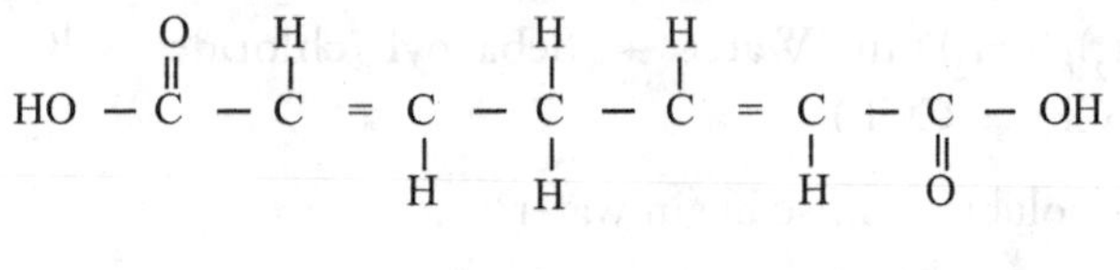

$$HO - \overset{\overset{O}{\|}}{C} - \overset{\overset{H}{|}}{\underset{\underset{H}{|}}{C}} = C - \overset{\overset{H}{|}}{\underset{\underset{H}{|}}{C}} - \overset{\overset{H}{|}}{\underset{\underset{H}{|}}{C}} = C - \overset{\overset{H}{|}}{\underset{\underset{O}{\|}}{C}} - OH$$

(A) A Compound

$$N \left\{ \overset{\overset{H}{|}}{\underset{\underset{H}{|}}{C}} - \overset{\overset{H}{|}}{C} = C - \overset{\overset{H}{|}}{\underset{\underset{C=O}{|}}{C}} \right\}_x \overset{\overset{H}{|}}{\underset{\underset{H}{|}}{C}} - \overset{\overset{H}{|}}{C} = C - \overset{\overset{H}{|}}{\underset{\underset{C=O}{|}}{C}} \left\{ \overset{\overset{H}{|}}{\underset{\underset{H}{|}}{C}} - \overset{\overset{H}{|}}{C} = C - \overset{\overset{H}{|}}{\underset{\underset{C=O}{|}}{C}} \right\}_y Y$$

$$CH_3 \qquad\qquad CH_3 \qquad\qquad CH_3$$

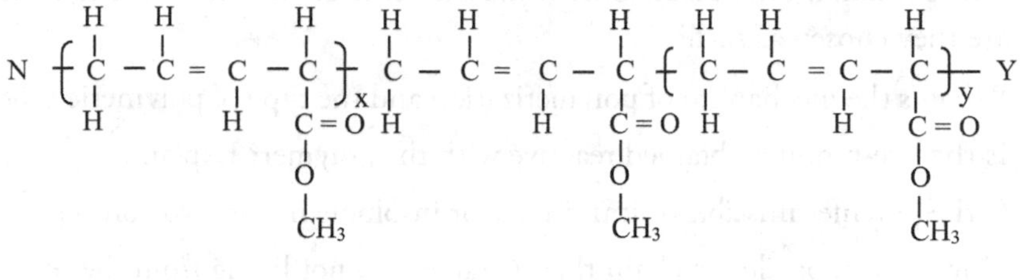

(B) A POLYMERIC CHAIN

i)   Identify if (A) and (B) are allylic.

ii)   Show that (A) cannot be used as a poly-monomer.

iii) Show how (B) can be used as a polymeric monomer.

iv)   If the CH_3 groups in (B) are replaced with H, show what type of polymeric chain that will be obtained.

15.39. Provide the new classifications for Industrial Polymerization processes for water miscible and organic miscible polymers.

15.40   Universally, up to the present moment, Miscibility and Solubility are used inter-changeably as if they are the same, because the process is non-productive. However, based on the New Frontiers, they are not the same, and because of their very extensive applications in for example Unit Separation processes in Chemical Engineering, there was need to distinguish between the two of them as well as Immiscibility and Insolubility.

Answer the following questions.

i)   Distinguish between Dissolution and Miscibility.

ii)   Distinguish between Miscibility and Solubility.

iii)   Distinguish between Solubility and Insolubility.

iv)   What is a Solvent?

v)   What is a Non-solvent?

vi)   What is Immiscibility?

vii)   Is Methanol a Non-solvent for Polyacrylamide as is generally believed to be the case? Explain.

viii) Is acrylamide soluble in water or is polyacrylamide soluble in water?

15.41   Describe the use of the following combinations for Step polymerization.

Diamine ($H_2N(CH_2)_6NH_2$) in Water + Sebacoyl chloride ($ClCO(CH_2)_8COCl$) in Tetrachloroethylene ($Cl_2C = CCl_2$)

i)   Is the Diamine soluble or miscible in water?

ii)   Is the Diacyl chloride soluble or miscible in the tetrachloroethylene?

iii)   One combination is Polar/Ionic while the other combination is Polar/Non-ionic. Why are they chosen as such?

iv)   What is the mechanism of polymerization and the type of polymeric product obtained?

v)   Is the bye-product obtained reactive with the polymer? Explain.

vi)   Is the polymer miscible or immiscible or insoluble in the two combinations? Explain.

vii)   How do you produce a chain that is stable, i.e., not living from the terminals.

viii)   Can cross-links be formed between two or more chains? Explain.

ix)   Why is the polymer produced suspended at the interface of the two combinations?

x)   Does the polymer have to be continuously removed if polymerization is to continue? Describe the entire process.

THIS MARKS THE END OF THE FIRST STAGE OF A LONG JOURNEY FOR HUMANITY. AND SO BE IT, FOR IT IS THE BEGINNING OF A NEW DAWN.

Index

Printed in the United States
By Bookmasters